ENCYCLOPEDIA OF
PHYSICAL SCIENCE
AND TECHNOLOGY
Volume 10   Org–Poll

EXECUTIVE ADVISORY BOARD

ENCYCLOPEDIA OF
PHYSICAL SCIENCE AND TECHNOLOGY

An Encyclopedic Reference Complete in Fifteen
Volumes, with Volume 15 the Index Volume

ROBERT A. MEYERS, EDITOR
TRW, INC.

ACADEMIC PRESS, INC.
Harcourt Brace Jovanovich, Publishers
Orlando   San Diego   New York   Austin
Boston   London   Sydney   Tokyo   Toronto

ACADEMIC PRESS, INC.
Orlando, Florida 32887

United Kingdom Edition published by
ACADEMIC PRESS INC. (LONDON) LTD.
24–28 Oval Road, London NW1 7DX

Library of Congress Cataloging in Publication Data

Encyclopedia of physical science and technology.

Includes index.
1. Science—Dictionaries. 2. Engineering—
Dictionaries. 3. Technology—Dictionaries. I. Meyers,
Robert A. (Robert Allen), Date
Q123.E497 1987   503'.21   86-1118
ISBN 0—12—226910—1 (v. 10 : alk. paper)

PRINTED IN THE UNITED STATES OF AMERICA

87 88 89 90   9 8 7 6 5 4 3 2 1

EDITORIAL ADVISORY BOARD

CONTENTS

GUIDE TO USING THE ENCYCLOPEDIA

Articles in the *Encyclopedia of Physical Science and Technology* are arranged alphabetically by subject. A table of contents appears in each volume. The volumes may be consulted directly by checking the contents of the appropriate volume. Thus, a reader who needs information on Air Pollution Control will find an entire article on that subject in Volume 1 (A–Arc). An article on X-Ray Analysis appears in Volume 14. The final volume in the set, Volume 15, contains a combined subject index for all 15 volumes.

The reader may check the subject index in Volume 15 to find the volume (in boldface) and page numbers (lightface) for the information sought. The reader may also refer to the relational index found in Volume 15 for groupings of related articles. For example, under the heading Structural Engineering, the following articles are listed:

Concrete, Reinforced
Masonry
Mechanics of Structure
Solar Energy in Buildings

Alphabetization within the Encyclopedia follows the word-by-word method. For example,

Volume 1 contains articles in the following order:

Air Pollution (Meteorology)
Air Pollution Control
Aircraft Performance and Design

Volume 4 contains articles in the following order:

Electron Transfer Reactions, General
Electron Transfer, Transition Metal Complexes
Electronic Displays
Electrophoresis

Cross references within articles lead the reader to related articles for background information. Additional cross references lead the reader to the appropriate article. For example:

Acid Rain–see Air Pollution (Meteorology)
Acoustics, Atmospheric—see Atmospheric Acoustics

Each article is designed to present the subject in a standard format. A table of contents precedes each article, and a defining paragraph appears at the beginning of each article.

ORGANIC CHEMICAL SYSTEMS, THEORY

Josef Michl   *University of Utah*

GLOSSARY

Electron affinity: Energy released when an electron is brought from infinity and added to a molecule (it may be negative if energy actually has to be provided).

Electronegativity: Measure of the tendency of an atom or an orbital to attract and accommodate electron density.

Gradient of a surface: The gradient at a point is a vector directed up the steepest slope of the surface at that point; its length is a measure of the slope.

Improper rotational axis of symmetry (of order *n*): A molecule is said to possess such an axis located in a particular direction if rotation by an angle $2\pi/n$ about that direction, followed by mirroring in a plane perpendicular to the direction, converts the molecule back to itself.

Inner-shell electrons: Electrons in orbitals of energy lower than that of the valence shell (closer to the nucleus).

Interaction matrix element: The interaction matrix element of the Hamiltonian H between wave functions ϕ_1 and ϕ_2 is given by $\int \phi_1 H \phi_2 \, d\tau$, where integration is over all space.

Ionization potential: Minimum energy that must be provided to a molecule in order to remove one of its electrons to infinity.

Ortho, meta, para: Designation of relative positions of two substituents on a benzene ring: located on adjacent ring carbons (ortho), on next-nearest-neighbor ring carbons (meta), and on carbons across the ring from one another (para).

Pauli principle: A wave function describing the state of a system containing two or more electrons is antisymmetric with respect to the exchange of all coordinates of any two electrons (i.e., is converted to minus itself on such an exchange). One of the consequences is that two electrons of the same spin cannot reside in the same orbital.

The theory of organic chemistry deals with the fundamental concepts that underlie and unify the experimental observations made by chemists working with organic molecules. It is believed that the behavior of molecules can be understood, in principle, in terms of a few basic laws of physics and that it is only the mathematical complexity of the resulting equations that limits the accuracy with which the behavior of organic molecules can be predicted *a priori*. In spite of the largely approximate nature of the theoretical treatments applicable to large molecules, the theory has made substantial contributions, primarily by providing the language through which the various observed phenomena can be interrelated. It permits the rationalization of trends and at times of individual observations concerning the reactivity and properties of organic molecules, and in some instances it has provided useful predictions. [*See* ORGANIC CHEMISTRY, SYNTHESIS.]

I. Classical Bonding Theory

Although the theory of organic chemistry has now been cast in terms of quantum theory, most of the older qualitative concepts of the classical bonding theory remain useful. The classical theory thus represents a suitable introductory level for the subject at hand.

ENCYCLOPEDIA OF PHYSICAL SCIENCE
AND TECHNOLOGY, VOL. 10

A. STRUCTURAL FORMULAS

In the classical description an organic molecule is represented by a structural formula. This is a collection of atomic symbols (C, H, O, N, etc.) representing atoms (including their inner-shell electrons but excluding their valence-shell electrons; the d electrons of a transition metal atom are included, although they may participate in bonding). At least one, but usually many, of these atoms must be carbon in order for the molecule to qualify as organic. The atomic symbols are connected by a network of single, double, and triple lines, which stand for single, double, and triple covalent bonds, respectively. These are formed by the sharing of electron pairs between atoms, with one bond representing one pair. The number of nearest neighbors to which an atom is attached is called its coordination number.

In addition, short lines (see **2a**) or pairs of dots can be used to represent unshared (lone) electron pairs on an atom, but these are frequently omitted. Single dots represent odd (unpaired) electrons, if such are present (see **2a–c**).

The symbols C and H for the carbon atom and the hydrogens attached to it, respectively, are also frequently omitted. Commonly occurring groups of atoms of well-known internal structure are often indicated by giving the kind and number of atoms involved (e.g., C_2H_5 for ethyl) or by an abbreviation (in this case, Et). A few examples are given in **1a** through **2c**.

The number of bonds formed by an atom (its "covalency") is dictated by the rules of valence. These state that in order for an organic molecule to have reasonable stability under ordinary conditions rather than to appear only as a transient reaction intermediate, if at all, the valence shells of all atoms in the structure have to contain a certain number of electrons: 2 for hydrogen, 8 for other main-group elements, and 18 for transition metal elements. The group of 8 electrons in

the valence shell of an atom is often referred to as a valence octet. In order to determine the number of electrons in the valence shell of an atom, one counts all the unpaired electrons or electrons present in lone pairs on that atom, plus two electrons for each single bond in which the atom is participating (four for a double bond, six for a triple bond). In structures **1a** through **c** all atoms satisfy the rules of valence; in structures **2a** through **c** the terminal oxygen atom does not.

Each type of bond is associated with a contribution to the total energy of the molecule, and these contributions are approximately additive. Typical bond strengths are presented in Table I. These are to be taken only as a rough guide since the immediate environment of the bond, steric strain (Section I,B), and resonance (see Section I,C) can have significant effects.

Atoms with valence shells that contain fewer electrons than demanded by the valence rules are said to be coordinatively unsaturated and usually are carriers of high chemical reactivity (terminal oxygen in **2**, the central carbon in **3**). Atoms with valence shells that contain a larger number of electrons than dictated by the rules are said to be hypervalent. This situation is rare for atoms of the elements of the second row of the periodic table (presumably due to their small size and the resulting steric crowding) but fairly common for those of the third and lower rows, where the number of valence-shell electrons can be 10, 12, or even higher. Molecules containing hypervalent atoms are often stable, particularly if the hypervalent atom is of lower electronegativity than its neighbors (e.g., the tin atom in **4**; the electronegativity of an element increases as

TABLE I. Typical Bond Energies in Organic Molecules[a]

X	X—H	X—C	X—N	X—O	X=C	X≡C
C	100	81	69	84	148	194
N	93	69	38	43	148	213
O	110	84	43	33	172	
F	135	105	65	50		
Si	72	69		103		
P	77	63				
S	83	65			128	
Cl	103	79	48	50		
Br	88	67		53		
I	71	57		57		
N—N	38	N=N	100	N≡N	226	
O—O	33	N=O	145			
S—S	54	O=O	96			

[a] In kcal/mol; 1 kcal = 4.184 kJ.

one moves up and to the right in the periodic table).

Those molecules that contain an odd number of electrons cannot satisfy the rules for all of their atoms and are known as free radicals (e.g., **2**).

In addition to atomic symbols and symbols for bonds, lone pairs, and unpaired electrons, the classical structural formulas of organic chemistry also indicate atomic charges (e.g., **3–5**). The way to determine the charge on an atom is to count the valence-electron ownership of an atom in the molecule and compare it with the number of valence electrons on a neutral isolated atom of the same element. If the two agree, the formal charge is zero. If there is one more electron on the atom in the molecule than on an isolated atom, the formal charge is minus one and so on.

In order to determine the valence-electron ownership of an atom in a molecular structure, one counts all electrons indicated as lone pairs as well as unpaired electrons on the atom, plus one for each bond in which the atom participates. Thus, one assumes that the two electrons of a bond are shared equally between the two atoms that it joins. To indicate that this is unrealistic when the two atoms differ in their electronegativity, the charges are referred to as formal. For molecules containing transition metal elements, the individual formal charges are frequently not indicated at all.

The sum total of formal charges on atoms in a molecule is equal to its net charge, and this is always indicated. Negatively charged molecules are called anions, positively charged ones cations. The electrostatic force of attraction between two oppositely charged ions is sometimes referred to as an ionic bond.

Typical bonding situations in which atoms of elements that are most commonly found in organic molecules find themselves in molecular structures are listed in Table II. Analogous bonding situations are found throughout each column of the periodic table, except that atoms of second-row elements resist hypervalency. [*See* PERIODIC TABLE (CHEMISTRY).]

The hydrogen atom does not suffer from steric constraints in its ordinary univalent state, in which it makes only one bond. It can enter into a special kind of weak hypervalent interaction known as the hydrogen bond, which attaches it to a lone-pair-carrying second atom. The hydrogen bond is indicated by a dotted line. As usual for hypervalent interactions, hydrogen bonding is particularly important if the neighbors of the hydrogen atom are highly electronegative. [*See* HYDROGEN BONDS.]

B. MOLECULAR GEOMETRIES

Classical structural formulas imply molecular geometries. These are determined by bond lengths, valence angles, and dihedral angles and describe the average nuclear positions when the molecule is at equilibrium.

In real molecules, at least some vibrational and internal rotational motion is always present. In many organic molecules, this can be neglected in the first approximation, and the molecules can be considered rigid. Some, particularly those lacking rings and multiple bonds, are definitely floppy at room temperature due to nearly free rotation around single bonds but can be viewed as rigid at sufficiently low temperatures.

1. Bond Lengths

Bond lengths are generally determined by the nature of the two atoms bonded, with minor variations depending on the environment. Each kind of atom can be associated with the value of its "covalent radius." A bond length is approximately equal to the sum of the covalent radii of the participating atoms. Typical lengths of the most common bonds in organic molecules are listed in Table III.

2. Valence Angles

Valence angles are the angles between two bonds on the same atom, generally dictated by the coordination number of the atom. However, if lone pairs are present on an atom, each of these counts for yet another neighbor. On the other hand, the presence of a single unpaired electron on an atom usually has only a minor influence on its valence angles.

For atoms forming two single or multiple bonds and carrying no lone pairs, the valence angles normally are in the vicinity of 180°, for atoms forming three single or multiple bonds and carrying no lone pairs they are ~120°; and for atoms forming two such bonds and carrying one lone pair, they are usually a little smaller.

TABLE II.   Common Building Blocks of Organic Molecules

Building blocks	Number of electrons in valence shell	Coordination number[a]
In agreement with the rules of valence		
—H	2	1
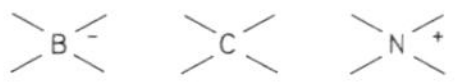 >B< >C< >N<⁺	8	4
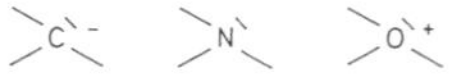 >C:⁻ >N: >O:⁺	8	3
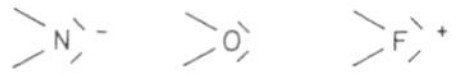 >N:⁻ >O: >F:⁺	8	2
—Ö:⁻ —F̈:	8	1
Coordinatively unsaturated		
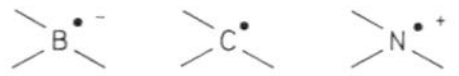 >B·⁻ >C· >N·⁺	7	3
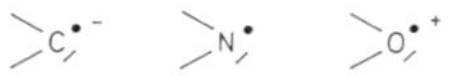 >C·⁻ >N· >O·⁺	7	2
 —N·⁻ —Ö· —F̈·⁺	7	1
 >B— >C·⁻	6	3
>C: >N:⁺	6	2
—N̈ —Ö⁺	6	1
Examples of hypervalent atoms		
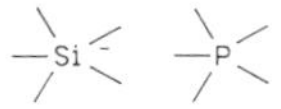 —Si< —P<	10	5
 —S⁄	10	4
 —I⁄	10	3
—H---	(4)	(2)

[a] Parentheses are for hydrogen atom in a hydrogen bond, not always considered hypervalent.

For atoms forming four bonds and carrying no lone pairs they are ~109°; and for atoms carrying a lone pair plus three bonds or two lone pairs plus two bonds they are a little smaller still.

The angles of 120° correspond to the center of an equilateral triangle being connected to its vertices, so that all three bonds are coplanar. The angles of 109° correspond to the center of a regular tetrahedron being connected to its vertices. The steric arrangement around an atom carrying a total of five bonds and lone pairs usually corresponds to a trigonal bipyramid and that around an atom carrying a total of six bonds and lone pairs to a regular octahedron, with the atom in the center in each case.

The valence angles given provide only a rough guide. Their exact values in any real molecule depend on the environment and, in particular, on the number of lone pairs on the atom.

The bending motions are generally relatively easy at 180° valence angles, so that large excursions from the normally preferred angle are possible at room temperature. Similarly, out-of-plane vibrations around atoms characterized by

TABLE III. Typical Bond Lengths in Organic Molecules[a]

X	X—H	X—C	X=C	X≡C
B	1.21	1.56		
C	1.09	1.54	1.34	1.20
N	1.00	1.47	1.30	1.16
O	0.96	1.43	1.22	
F	0.92	1.38		
Si	1.48	1.84		
P	1.42	1.87		
S	1.34	1.81	1.56	
Cl	1.27	1.76		
Br	1.41	1.94		
I	1.61	2.14		
N—O	1.36			
N=O	1.21			
N—F	1.36			
N—Cl	1.75			

[a] In Å; 1 Å = 100 pm.

120° valence angles are also relatively easy, as is the interchange of positions of a lone pair and a single bond. Tetrahedral bond arrangements around an atom are relatively rigid. However, a flipping motion ("umbrella inversion") of a lone pair from one to the other side of the three bonds present is very facile on atoms of the second row (not those of lower rows). The interchange of ligand positions is easy on pentacoordinate atoms, difficult on hexacoordinate ones.

The presence of small rings in the molecule may introduce very large deviations from the usual valence angle values. For example, in cyclopropane the carbon atoms form an equilateral triangle so that the CCC angle is 60°, not much more than half of the normally expected value. Such deviations from the normally preferred angles are energetically unfavorable, and the molecule is said to exhibit angular strain.

3. Dihedral Angles

A dihedral angle is defined as the angle between two planes, both of which pass through the same bond. One of the planes also contains one of the additional bonds formed by one of the bond termini, and the other plane contains one of the additional bonds formed by the other terminus.

The preferred dihedral angles around a double bond are 0° and 180°, once again counting a lone pair on a terminus as another nearest neighbor. Thus, the usual geometries around C=C and N=N double bonds are planar (6–9). However, small twisting distortions from planarity are relatively easy at room temperature.

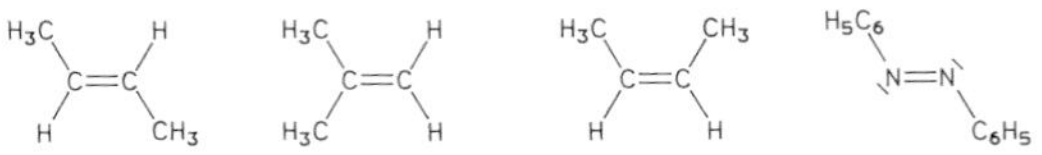

There is a much weaker preference for particular values of the dihedral angle around single bonds, and rotation around such bonds is nearly free. Usually, the value of 0° ("eclipsed") is avoided, and values of around 60° ("staggered") to 90° are somewhat preferred, depending on the number of lone pairs on the termini.

The general rules just stated for bond lengths and angles permit the construction of mechanical molecular models either from balls and sticks or on a computer screen. The size of the balls that represent the volume of individual atoms is given by their van der Waals radii. The sum of the van der Waals radii of two atoms represents the distance of most favorable approach of these two atoms if they are not mutually bonded (e.g., atoms on neighboring molecules in a crystal). Values for these quantities are compiled in Table IV. Molecules in which two or more atoms that are not bonded to one another are located at distances shorter than the sum of the van der Waals radii are strained by steric crowding and

TABLE IV. Atomic van der Waals Radii[a]

Atom	Radius (Å)	Atom	Radius (Å)	Atom	Radius (Å)
H	1.2	O	1.4	F	1.4
N	1.5	S	1.9	Cl	1.8
P	1.9	Se	2.0	Br	2.0
As	2.0	Te	2.2	I	2.2
Sb	2.2				

[a] 1 Å = 100 pm.

are less stable than otherwise expected. Often, it is possible to avoid some of this unfavorable interaction by a distortion of the valence angles.

4. Molecular Mechanics

It is possible to augment the set of bond energies by a set of energy increments for deviations from optimum bond lengths, valence angles, dihedral angles, and van der Waals distances as a function of the magnitude of each and to compute the energy of a molecule as a function of its geometry within the framework of such a "springs and balls" model. Equilibrium geometry can then be found by energy optimization. This approach is known as molecular mechanics and provides a good approximation, particularly for hydrocarbons, for which extensive and carefully optimized parameter sets are available. It runs into difficulties with molecules in which more than one classical valence structure is important.

C. RESONANCE (MESOMERISM)

An important concept in classical structural theory is resonance (mesomerism). This is related to the fact that more than one classical structure can normally be written for a molecule. A double-headed arrow is usually placed between the structures to indicate that all of them contribute to the description of bonding in the molecule. It is important to note that all of the structures refer to the same molecular geometry.

The situation is most easily exemplified in the case of multiple bonds (e.g., **10–12**). The existence of two or more valence structures can also be indicated by curved arrows, as in **10c** to **12c**.

The relative importance of several contributing structures is dictated primarily by their energies (i.e., the energies of hypothetical molecules in which only the structure in question would contribute). The energies can be estimated qualitatively. Low energy is favored by the presence of a large number of bonds in the structure (a highly twisted double bond does not yield much stabilization—steric inhibition of resonance), by the absence of unfavorable charge separations, and by the presence of negative charges on electronegative atoms and positive charges on electropositive atoms.

Two structures contribute equally when they are equivalent by symmetry, as in the allyl radical **13**. A bond that is single in some and absent in other contributing resonance structures is called a partial bond and is often drawn with a dashed line.

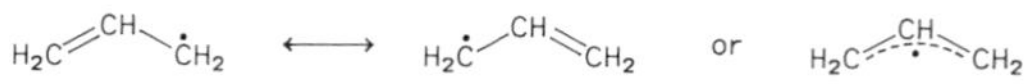

The existence of several contributing structures is associated with a more or less significant effect on the thermodynamic stability of the molecule relative to that expected from the simple rules for any one of the contributing structures. Usually, it results in a stabilization, referred to as resonance energy.

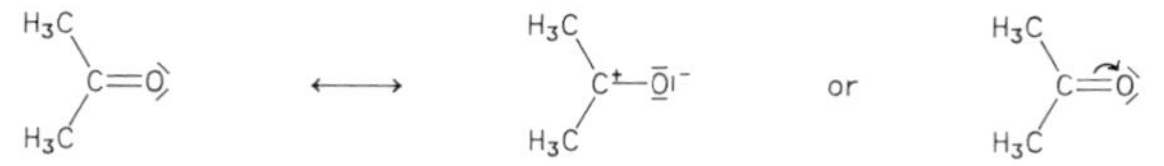

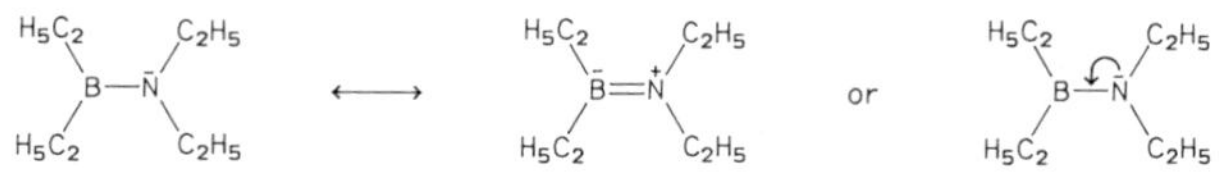

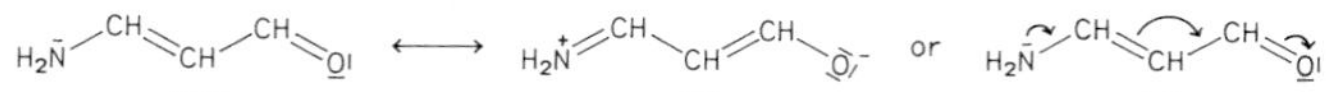

Resonance energy plays a particularly important role in cyclic systems. Those cyclic systems with two equivalent doubly bonded structures that contain $4N + 2$ electrons in the perimeter (N is an integer) exhibit a large stabilization and are known as aromatic. The stabilization is known as the aromatic resonance energy. The archetypical example is benzene (**14**). Those containing $4N$ electrons in the perimeter are actually destabilized (antiaromatic); a good example is cyclobutadiene (**15**).

14

15

In the case of polycyclic ring systems of conjugated double bonds it is more difficult to specify the degree of aromatic stabilization or antiaromatic destabilization. A good rule of thumb is to count the number of possible structures, ignoring those with an even number of double bonds located inside any single ring. The higher the number, the larger the stabilization. These kinds of problems are far more efficiently handled by the more advanced quantum mechanical theories of molecular structure.

In order to simplify notation and to avoid writing a large number of contributing structures, it is customary to draw a circle inside an aromatic ring, as shown for **14**.

Ambiguities in the writing of molecular structures exist even in compounds containing only single bonds. In principle, it is possible to write the structures H^+H^-, H^-H^+, and H—H for molecular hydrogen. The charge-separated structures are ordinarily not written, and their existence is tacitly understood when H—H is written; this is true of all other bonds as well. In some cases the need to write more than one equivalent classical structure for a molecule containing only single bonds is not so easily avoided, and such structures are often called nonclassical (e.g., structures **16a–16d** for the CH_5^+ cation and structures **17a–17d** for the norbornyl cation). Just as in the case of resonance involving double bonds, such systems with several equivalent structures are frequently written with auxiliary symbols such as circles, often with dashed lines, particularly to indicate the delocalization of charge.

16a   16b   16c   or   16d

17a   17b   17c   or   17d

D. ISOMERISM

Molecules that do not differ in overall molecular formula (specified by the number of atoms of each kind and the net charge) but only in internal structure are called isomers.

1. Constitutional Isomers

Constitutional isomers are compounds that differ in connectivity, that is, in the way in which the constituent atoms are connected to one another. Graph theory is an important tool for their enumeration.

An important class of constitutional isomers are positional isomers, in which the functional groups are the same but differ in their location within the molecule (e.g., **6** and **7**, or the ortho, meta, and para isomers **18–20**).

18   19   20

2. Stereoisomers

Stereoisomers have the same connectivity but differ in the way in which the constituent atoms are oriented in space. They can be divided into configurational stereoisomers and conformational stereoisomers. The precise specification of the spatial arrangement of the groups in a configurational isomer is called its configuration, and in a conformational isomer, its conformation.

a. Configurational Stereoisomers. These stereoisomers cannot be made superimposable by any rotations about single bonds. In order to make them superimposable, rotation about a double bond or a dissociation of one or more single bonds, or both, is necessary (e.g., **6** and **8**). Since these processes normally require considerable energy, they usually do not occur at a measurable rate at room temperature. Configurational stereoisomers can normally be isolated from one another and stored essentially indefinitely at room temperature.

b. Conformational Stereoisomers. Often called conformers for short, these stereoisomers can be made superimposable by rotations about single bonds. Examples are the axial (**21**) and equatorial (**22**) conformers of a monosubstituted

cyclohexane. Since rotations about single bonds are normally very facile, it is usually impossible to separate conformers from one another and to handle them separately at room temperature.

c. Chirality. Another important classification of stereoisomers into two groups is related to optical activity, manifested by the rotation of a plane of polarized light on passage through a sample.

A pair of stereoisomers that are related to one another in the same way as an object and its mirror image are called enantiomers (e.g., **23** and **24**). Any pair of stereoisomers that are not related in this way are called diastereomers (e.g., **6** and **8** or **21** and **22**).

A molecule that is not identical with its mirror image is called chiral and occurs as a pair of enantiomers. The necessary and sufficient condition for chirality is the absence of an improper rotational axis of symmetry in the molecule (this includes a center of inversion and mirror plane symmetry elements).

A mixture of equal amounts of two enantiomers is known as a racemic modification and is optically inactive. A pure enantiomer or an unbalanced mixture of two enantiomers is optically active; the two enantiomers have opposite handedness and cause the plane of polarization to rotate in opposite directions.

II. Qualitative Molecular Orbital Model of Electronic Structure

Although the classical structural theory as outlined so far accounts for much of organic chemistry, it leaves quite a few observations unexplained. In order to proceed and relate the chemical behavior of molecules to fundamental physical principles it is necessary to use quantum theory.

This can be done at a more or less rigorous quantitative level. A fairly direct translation of the structural concepts just outlined into mathematical terms then leads to the quantum mechanical valence–bond theory of electronic structure. Although conceptually appealing, computationally this method is quite unwieldy and has not seen much use. An approach that is almost universally adopted nowadays is the quantum mechanical molecular orbital (MO) theory of electronic structure. This is computationally manageable and is described in some detail later.

Even the mathematically simpler MO approach to electronic structure requires the use of large computers for any quantitative applications. Although large-scale computations have been extremely valuable in enhancing the understanding of organic molecules, they are in themselves not appropriate for the day-to-day thinking of bench chemists. However, they have had a great influence on the much simpler qualitative models of electronic structure adopted for daily use by organic chemists and even some effect on that ill-defined body of knowledge generally referred to as the organic chemist's "intuition." The current qualitative model of molecular electronic structure based on the qualitative notions of MO theory represents a significant advance over the structural theory outlined in the preceding sections, but is not easy to describe unequivocally since its form varies among individuals.

Current qualitative thinking about the electronic structure of organic molecules is based on the independent-particle model, in which it is assumed that the motion of each electron is dictated by the field of stationary nuclei and the time-averaged field of all the other electrons. In this model, any correlation of the instantaneous positions of the many electrons present is neglected.

Only in several well-recognized and more or less exceptional situations are correlation effects explicitly introduced. This is particularly true in the treatment of biradicals, which represent an important class of reaction intermediates but which are not discussed here, in the treatment of several other unusual bonding situations, and in the treatment of photochemical processes (e.g., in the consideration of differences in the reactivity of excited singlet and triplet states).

A. Atomic Orbitals

The independent-particle model is well known from the quantum mechanical description of atomic structure. Each electron in an atom is assumed to reside in an atomic orbital (AO) with a maximum of two electrons (of opposite spin) in any one orbital (Pauli principle). The AO is a function of the coordinates of one electron. The square of its magnitude at any point in space gives the probability density for finding the electron at that point. The magnitude itself can be positive or negative, with zero values at the boundaries of the positive and negative regions. The boundaries are referred to as nodal surfaces (planes, spheres, etc.). The energy of an electron residing in an orbital increases with the increasing number of nodal surfaces in the orbital.

Atomic orbitals are characterized by a principal quantum number n (there are $n - 1$ radial nodes in an AO) and a letter indicating their shape as dictated by the angular nodes: An ns orbital has no angular nodes and is of spherical symmetry; each of the three equienergetic np orbitals has one angular mode (usually taken to be a plane through the nucleus, with respect to which the orbital is antisymmetric); each of the five nd orbitals has two angular nodes; and so on. Shorthand abbreviations for orbital shapes are shown in Fig. 1A. These are meant to indicate the regions of space in which the numerical value of the AO is the largest, also known as the lobes of an AO, as well as their signs.

In the ground state of an atom, the AOs are assumed to be occupied by the available electrons in the order of increasing energy of the subshells: $1s$, $2s$, $2p$, $3s$ .... The last at least partially occupied subshells and the more stable ones of the same principal quantum number represent the valence shell [in transition metals this contains the nd, $(n + 1)s$, and $(n + 1)p$ subshells]. Only valence-shell AOs are normally considered important for bonding: the $1s$ orbital in hydrogen and the $2s$ in lithium and beryllium ($2p$ are very close in energy and could be included as well), $2s$ and $2p$ in boron through

neon, $3s$ in sodium and magnesium ($3p$ could be included), $3s$ and $3p$ in aluminum through argon, and so on.

The qualitative theory of bonding also makes use of combinations of valence AOs known as hybrid orbitals. Combining an s with a p orbital produces two equivalent sp hybrids pointing in opposite directions. Combining an s with two p orbitals produces three equivalent sp^2 hybrids pointing to the corners of an equilateral triangle. And combining an s with three p orbitals produces four equivalent sp^3 hybrids pointing to the corners of a regular tetrahedron (Fig. 1B). Inequivalent hybrids can be used for other desired valence angles.

B. MOLECULAR ORBITALS

The independent-particle model can be extended to molecules by assuming that each electron again resides in an orbital of well-defined energy. Now, however, except for the inner-shell AOs, the orbital is spread over the space spanned by the molecular framework and is referred to as a molecular orbital. To a very good approximation, the inner-shell electrons behave in exactly the same way in the molecule as they would in an isolated atom, and we will not be concerned with them further. Instead, we shall concentrate on the valence MOs.

The delocalized form of MOs that we have just described is known as their canonical form. It can be shown that the wave function describing the ground electronic configuration of a molecule, containing a certain number of doubly occupied orbitals, does not change at all when these occupied MOs are mixed with one another in an arbitrary way. This degree of freedom in the wave function permits the construction of MOs that have been mixed in such a way that each one is localized in the smallest amount of space possible, using one of several possible criteria. The price one pays is that these localized MOs no longer have well-defined individual energies, although the total energy of the system is just as well defined as before.

It turns out that each of the localized MOs tends to be fairly well contained within the region of one bond or one lone pair; only in molecules with several important classical structures is this localization poor. On close inspection, the localization is actually never perfectly complete, and each localized orbital possesses weak "tails" extending to other parts of the molecule. Except for the exact nature of these tails, such a

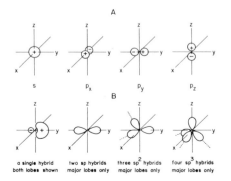

A

s p_x p_y p_z

B

a single hybrid   two sp hybrids   three sp^2 hybrids   four sp^3 hybrids
both lobes shown   major lobes only   major lobes only   major lobes only

FIG. 1.   (A) Atomic orbitals. (B) Hybrid orbitals.

bond orbital often looks very much the same in all molecules that contain that particular bond, say, C—C. It is along these lines that one can begin to understand bond additivity properties and their failure in the case of molecules in which several classical resonance structures play an important role.

The standard qualitative model of molecular electronic structure requires the construction of a number of valence MOs sufficient to hold all the valence electrons. This is normally done separately for the electrons responsible for those bonds that are present in all important classical structures of the molecule ("localized" bonds, hence "localized" electrons, although strictly speaking, they are not really localized) and separately for the electrons responsible for partial bonds that are present in some and absent in other important classical structures ("delocalized" bonds, hence "delocalized" electrons).

The usual approach is based on the recognition of the fact that the formation of bonds in molecules represents only a small perturbation of electronic structure of the constituent atoms, bonding energies being of the order of 1% of total atomic energies. This is because the electric field in the vicinity of any one atom, where an electron spends most of its time, is dominated by the nucleus of this atom, since the force of attraction of an electron by a positive charge grows with the second inverse power of the distance from that positive charge.

This situation is acknowledged by assuming that each MO can be built by mixing AOs or hybrid orbitals.

C. Orbital Interactions

A brief consideration of the ways in which orbitals interact will be useful not only for a description of the procedure in which AOs are mixed to produce MOs, but also for subsequent reference to the mutual mixing of MOs, for instance, those of two molecules reacting with one another.

In the mixing of two orbitals, the important factors are their energies and the interaction matrix element (this is frequently referred to as the resonance integral β between the two orbitals). As a result of the mixing of n orbitals, n new orbitals result.

It is by far easiest to consider the case of only two mutually interacting orbitals ϕ_1 and ϕ_2 (Fig. 2). Let the energy of ϕ_1 be below that of ϕ_2.

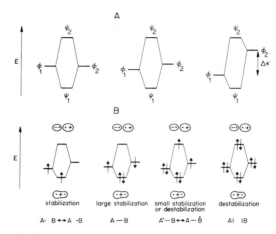

FIG. 2.   Orbital energies. (A) Interaction of orbitals ϕ_1 and ϕ_2 to produce linear combinations ψ_1 and ψ_2. (B) Interaction of atomic orbitals (outside levels) to produce a bonding and an antibonding combination (inside levels, orbital shape indicated). The effect on the net energy of the system is indicated for occupation with one to four electrons. Arrows represent electrons with spin up and spin down. At the bottom, the classical structural representation is shown for comparison.

After the interaction, the energy of the more stable of the two new resulting orbitals ψ_1 will lie below that of ϕ_1, and that of the less stable one ψ_2 will lie above that of ϕ_2, as if the two original energies repelled one another. The amount by which each orbital energy is shifted will depend on the strength of the interaction element β and on the difference $\Delta\varepsilon$ in the original energies. If $\Delta\varepsilon \gg |\beta|$, the shifts are approximately inversely proportional to $\Delta\varepsilon$ (this approximation is known as first-order perturbation theory). When the two orbitals are originally degenerate ($\Delta\varepsilon = 0$), the shifts are the largest.

An important characteristic of the interaction between two orbitals ϕ_1 and ϕ_2 is their overlap integral $S_{12} = \int \phi_1\phi_2 \, d\tau$, where the integration is over all space. For normalized orbitals ($S_{11} = S_{22} = 1$) the value of S_{12} can vary between -1 and $+1$. Usually if S_{12} is positive, the interaction element β is negative, and if S_{12} is negative, it is positive. If $S_{12} = 0$, orbitals ϕ_1 and ϕ_2 are said to be orthogonal; they can still have a nonvanishing resonance integral. The energy shifts caused by the interaction of two orthogonal orbitals are opposite in direction but equal in magnitude. When two nonorthogonal orbitals interact, the less stable new orbital ψ_2 is destabilized more than the more stable new orbital ψ_1 is stabilized.

The more stable of the two new orbitals is referred to as bonding, the less stable as antibonding, with respect to the interaction considered. An orbital with the same energy as before the interaction is called nonbonding; such orbitals often result when more than two AOs are mixed. If two electrons are available to fill each bonding orbital, maximum stabilization will result relative to the situation before orbital interaction. If too many electrons are available and some must be placed into antibonding orbitals, some or all of the stabilization is lost and net destabilization may result. For this reason, the interaction of a doubly occupied with an unoccupied orbital leads to a net stabilization, the interaction of two doubly occupied orbitals to a net destabilization.

These general concepts can now be specialized to the case of mixing of AOs or hybrid orbitals to produce bond orbitals. Two valence-shell AOs or properly constructed hybrid orbitals located at the same atom are always orthogonal. For two orbitals located at the same center, the interaction element β is zero if one or both are pure AOs. However, two hybrid orbitals at the same center have a nonvanishing mutual interaction element. Its magnitude is related to the promotion energy (i.e., the energy difference between the AOs from which the hybrids were constructed).

Two AOs or hybrid orbitals located at different atoms can be, but do not need to be, orthogonal. A nonvanishing overlap integral between two p AOs can be produced in two geometrically distinct ways. When the axes of the two orbitals are aimed at one another, the overlap is said to be of the σ type; when they are parallel, it is said to be of the π type (Fig. 3). Intermediate situations are also possible. The interaction matrix element β between two AOs or hybrid orbitals located at different atoms is approximately proportional to the negative of their overlap integral S_{12}.

The two new orbitals ψ_1 and ψ_2 that result from the interaction are linear combinations of the two old orbitals (Fig. 2). In the bonding orbital ψ_1, the overlapping portions of the entering old orbitals ψ_1 and ψ_2 have the same sign. In the antibonding orbital ψ_2, the overlapping lobes of the original partners are of opposite signs, and its sign changes as one goes across the interaction region from the center of ϕ_1 to the center of ϕ_2. The surface where the sign of ψ_2 changes is called a nodal surface (plane).

In summary, the bonding combination ψ_1

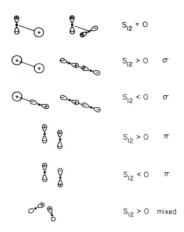

FIG. 3.   Overlap between orbitals.

lacks a nodal plane and has an increased electron density between the atoms (in-phase mixing if $S_{12} > 0$), while the antibonding combination ψ_2 has a nodal plane and a reduced electron density between the atoms (out-of-phase mixing if $S_{12} < 0$). The bonding orbital ψ_1 will accommodate two electrons and thus produce a net stabilization (a chemical bond). This is the origin of the classical "shared electron pair." For maximum bonding stabilization, it is desirable to maximize the absolute value of the overlap integral S_{12}.

A third electron would have to enter the antibonding orbital ϕ_2, causing a loss of much and possibly all of the overall stabilization. A fourth electron would also have to enter ϕ_2, and now interaction would definitely lead to a net destabilization. For this reason, filled AOs (lone pairs) repel and avoid one another ("lone-pair repulsion"). For minimum destabilization, it is desirable to minimize $|S_{12}|$. For instance, if the lone pairs are of the p type and are on adjacent atoms, a dihedral angle of close to 90° will be preferred (e.g., in H_2O_2).

If the original orbitals ϕ_1 and ϕ_2 are degenerate, $\Delta\varepsilon = 0$, they will enter the new orbitals ψ_1 and ψ_2 with equal weights. If the initial energies are unequal, the more stable initial orbital ϕ_1 will enter with a numerically larger coefficient into the more stable resulting orbital ψ_1 and with a smaller coefficient into the less stable resulting orbital ψ_2. The opposite will be true for the less stable initial orbital ϕ_2. This result is quite logical: In the more stable of the two orbitals, the electrons spend more time on the more electronegative atom.

D. CONSTRUCTION OF MOLECULAR ORBITALS

1. "Localized" Bonds

In the first step, AOs on the participating atoms are mixed to produce hybrids pointing in the directions of the desired bonds. These are then combined pairwise, each pair producing a bonding and an antibonding orbital strictly localized at the atoms to be bound. Occupancy of the bonding orbital by two electrons then produces a localized bond, while the antibonding orbital remains unused. If electrons are left over after all bonding orbitals have been occupied twice, they are placed into remaining AOs or hybrid orbitals that have not been used in the aforementioned procedure (nonbonding lone pairs or unpaired electrons). So far, the picture resembles the classical description, without resonance.

At a more advanced level, it is recognized that the several bonding and antibonding orbitals located at a single atom have nonvanishing interaction elements with one another. If they are allowed to mix, they will produce the fully delocalized canonical MOs. This usually has a predictable effect on the total energy, which can be absorbed in bond additivity schemes. Exceptions are the case of cyclic delocalization (e.g., in cyclopropane) and certain cases of stabilization of radicals, carbenes, and biradicals. A well-known exception in which such delocalization must be considered involves the interaction of a lone pair on an atom such as oxygen with the orbitals corresponding to bonds leading to its neighbors, which dictates certain stereochemical preferences known as the anomeric effect.

2. Delocalized Bonds

Although delocalized bonds could be obtained in the same manner in principle, it is more efficient to proceed more directly. All n AOs (and possibly hybrid orbitals) that participate in the "delocalized" system are mixed at once, producing n canonical delocalized MOs directly. Typically, these are sets of orbitals characterized by mutual π overlap and referred to as the π-electron system. A good example are the six p orbitals of the sp^2 hybridized carbons of benzene that are left over after the localized bond framework has been constructed (Fig. 4).

On the one hand, the process of simultaneous mixing of more than two orbitals is more difficult to visualize, but on the other hand, there is usually only one participating orbital on an atom, simplifying matters somewhat. Usually, about

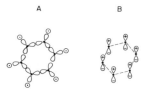

FIG. 4. (A) Hydrogen $1s$ atomic orbitals and carbon sp^2 hybrid orbitals (major lobes only) needed for the description of "localized" σ bonds in benzene. (B) Carbon $2p_z$ atomic orbitals needed for the description of "delocalized" π bonds in benzene.

half of the resulting MOs are stabilized (bonding), perhaps one or a few are nonbonding, and about a half are destabilized and antibonding. Their approximate forms can be guessed from the requirement that a nodal surface is added each time that one goes to the energetically next higher MO. Mnemonic devices for the resulting energy patterns are shown in Fig. 5 for linear and cyclic polyenes. In the latter case, the origin of the special stability ("aromaticity") of $4N + 2$ electron systems is apparent: This is the electron count needed to produce a closed shell. In general, however, at least a back-of-the-envelope numerical calculation is necessary. Here, Dewar's perturbational MO (PMO) method is particularly useful.

The interaction of fully or partly localized orbitals is frequently referred to as conjugation. Interactions between several AOs that belong to a π system can be referred to as π conjugation, or conjugation for short. Interactions of an orbital of the π type with one or more suitably

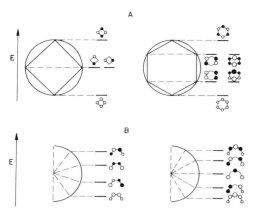

FIG. 5. Conjugated π systems; orbital energies and shapes (top view, white represents positive, black represents negative sign). (A) Cyclic conjugation (annulenes). (B) Linear conjugation (polyenes).

aligned bond orbitals of the σ type are referred to as hyperconjugation. Interactions between adjacent bond orbitals of the σ type or one such orbital and a lone-pair orbital are referred to as σ conjugation. As noted above, they occur through the interaction between two hybrid orbitals located on the same atom.

3. Hypervalent Bonding

The procedures described so far will not work for hypervalent atoms that do not have enough valence AOs to provide a sufficient number of localized bond orbitals. Without going into detail, we note that in this case three-center orbitals can be constructed instead, and a satisfactory simple description results.

The expansion of the electron count in the valence shell of an atom of a main-group element beyond eight is a reflection of the inadequacy of the way in which the classical rules of valence assign electrons into a valence shell. In reality, the true total electron density in the region of space corresponding to an atomic valence shell does not reach these high formally assigned numbers. Fundamental limitations on the latter are given by the Pauli principle, which demands that any electron density in excess of an octet has to occupy orbitals of the next higher principal quantum number (Rydberg AOs). It is a gross oversimplification, however, to pretend that the two electrons of a bond contribute full occupancy of two to the valence shell of each of the atoms connected by the bond, particularly when the two atoms differ greatly in electronegativity. The degree to which the valence shell of the atom of the more electropositive element is actually filled is overestimated by the simple rules, and this accounts for the ease with which it enters into hypervalency.

III. Quantitative Aspects of Molecular Structure

A. POTENTIAL ENERGY SURFACES

1. Construction of Potential Energy Surfaces

Before attempting a quantitative quantum mechanical treatment of molecules it is customary to separate the motions of nuclei from those of electrons. This separation is known as the Born–Oppenheimer approximation, and it underlies all of the current thinking about molecular structure. It is justified by the much larger mass of nuclei compared with that of electrons, which causes the nuclei to move much more slowly than electrons. Thus, electrons adjust their motions essentially instantaneously to any change in the location of the nuclei, as if they had no inertia. [*See* POTENTIAL ENERGY SURFACES.]

From the viewpoint of quantum theory a molecule is a quantum mechanical system composed of atomic nuclei and electrons. It has an infinite number of stationary states, that is, states whose measurable properties do not change with time. Each of these states is characterized by an energy and a wave function. A wave function is a prescription for assigning a numerical value ("amplitude") to every possible choice of coordinates for all particles in the system. A square of the number assigned to any choice of these coordinates represents the probability density that a measurement will find the system at that particular collection of coordinates.

In order to find the stationary states of a quantum mechanical system, their energies, and wave functions, one must solve the Schrödinger equation $\hat{H}\psi = E\psi$, where $\hat{H}$ is the Hamiltonian operator of the system, ψ the wave function, and E the energy of the stationary state.

In order to obtain a quantum mechanical description of a molecule within the Born–Oppenheimer approximation, at least in principle, one proceeds as follows. Fixed molecular geometry is assumed. Mathematically, this corresponds to choosing a point in the nuclear configuration space. As soon as this is done, the Hamiltonian operator for electronic motion is fully defined so that the corresponding Schrödinger equation can be solved for ψ and E. An infinite number of solutions exist, differing in their energies E and wave functions ψ.

One of the characteristics of an electronic wave function is the number of unpaired electrons it contains. Those wave functions in which this number is zero describe singlet states. Those that contain two unpaired electrons describe triplet states, and so on. Wave functions of radicals can have one unpaired electron (a doublet state), three (a quartet state), and so on.

Of the infinite number of stationary wave functions that are solutions to the Schrödinger equation for a chosen nuclear geometry, one is of lowest energy. Almost always this is a singlet wave function if the organic molecule has an even number of electrons and a doublet wave function if it has an odd number of electrons. In the following, we assume an even number of electrons. We label the lowest energy singlet wave function $\psi(S_0)$ and its energy $E(S_0)$. The next higher energy singlet wave function is then identified and labeled $\psi(S_1)$, and its energy is identified and labeled $E(S_1)$. This is the wave function of the first excited singlet state. Similarly, the wave functions of the second and higher electronic excited singlet states can be identified. Among the triplet wave functions the one with the lowest energy is called $\psi(T_1)$ and its energy $E(T_1)$. Similarly, higher triplet state wave functions and their energies are identified.

A different molecular geometry is then chosen and the process repeated. While this is difficult to do in practice, one can at least imagine performing this kind of operation for all possible molecular geometries. In a plot of $E(S_0)$ against the values of the geometrical parameters that describe the molecular structure, a surface will then result. This is the potential energy surface for this particular electronic state of the molecule (the potential energy of the molecule is its total energy minus the energy of overall translational, rotational, and vibrational motion).

The resulting surface is easy to visualize if only one or two geometrical variables are used to describe the molecular structure. As shown in Fig. 6, in the former case the set of points $E(S_0)$ represents a line; in the latter case it represents a two-dimensional surface, often displayed in the form of a contour diagram. For all organic molecules of real interest, the number of independent geometrical variables necessary for the description of the internal geometry is large ($3N - 6$, where N is the number of atoms). The resulting surfaces are multidimensional and difficult to en-

visage. Frequently, they are referred to as hypersurfaces.

What can still be visualized readily are one-dimensional or two-dimensional cross sections through these hypersurfaces, which correspond to only a limited variation of molecular geometries, particularly to specific kinds of intramolecular motion, such as rotation around a bond.

2. Motions on the Surfaces

The gradient of the potential energy surface defines the forces acting on the nuclei. The resulting changes of molecular shape can be represented by a point that moves through the nuclear configuration space. It is possible to visualize the vibrational motions of the molecule as well as its internal rotations as the motions of a marble rolling on the potential energy surface.

If the molecule is isolated, its total energy will remain constant, and the marble will perform endless frictionless motion on the surface, trading the potential energy against the kinetic energy of nuclear motion and vice versa. If the molecule exchanges energy with its environment, it will tend to lose any excess energy it may have and settle in one of the valleys or minima on the surface.

A proper description of the motion that corresponds to vibrations and internal rotations again must be quantum mechanical since even the relatively heavy nuclei really obey quantum rather than classical mechanics. Once again, one can find the stationary states of the vibrational motions and their wave functions and energies by setting up the appropriate Schrödinger equation and solving it. The Hamiltonian operator that enters into this equation now contains the information on the potential energy embodied in the shape of the potential energy surface. An infinite set of possible solutions again exists. A finite number of solutions have energies corresponding to bound states, that is, those with energies below the dissociation limit for the molecule (energy required to break the weakest bond and separate one of the atoms to infinity). The wave function of lowest energy represents the vibrational ground state of the molecule. In this state the kinetic energy of the nuclear motion is not zero since this would violate the uncertainty principle. It is referred to as the zero-point energy.

Most molecules have a well-defined equilibrium geometry that corresponds to a minimum in the $E(S_0)$ surface. The wave function of the

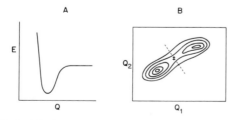

FIG. 6.   Display of a one-dimensional (A) and a two-dimensional (B) cut through a potential energy surface. In (A), energy is plotted against a geometrical variable Q. One minimum is present. In (B), contour lines connect points of equal energy in the Q_1,Q_2 geometrical space. Two minima are present. The separation into two catchment basins is shown by a dashed line; the transition state structure is indicated by a double dagger.

lowest vibrational stationary state is heavily localized near this minimum. If the shape of the potential energy hypersurface in this vicinity can be approximated by a paraboloid, the vibrational motion in the lower vibrational states is harmonic and can be described as a product of $3N - 6$ normal mode motions ($3N - 5$ for a linear molecule), each characterized by a frequency ν. The zero-point energy is obtained by summing the contribution $\frac{1}{2}h\nu$ from each of the normal modes. The value $h\nu$ is equal to the energy separation from the lowest to the next higher energy level in each mode. The transitions between these individual levels lie in the region from several hundred to several thousand wave numbers and are commonly studied by infrared and Raman spectroscopy. [See INFRARED SPECTROSCOPY; RAMAN SPECTROSCOPY.]

In most cases more than one local minimum is found on the $E(S_0)$ hypersurface (Fig. 6B). This means that a given collection of nuclei and electrons has more than one possible equilibrium geometry. Usually, this means that several isomers of the molecule exist, but some of the minima may also correspond to dissociation products, the sets of two or more smaller molecules formed from the same collection of nuclei and electrons. Some of the minima may have equal energies (e.g., a pair of enantiomers). In general, one minimum is of lower energy than all the others and corresponds to the most stable isomer of the molecule.

Each minimum is surrounded by its catchment basin (Fig. 6B). Within a given basin, the potential energy surface slopes toward the minimum in question. Each catchment basin defines the range of geometries that correspond to a given chemical species.

3. Chemical Reactions

Under ordinary conditions molecules are frequently not in stationary vibrational states. Vibrations and internal rotations are affected by collisions with neighboring molecules, which add or subtract small amounts of vibrational energy more or less randomly. At thermal equilibrium the vibrational energy content is $\frac{1}{2}kT$ for each degree of freedom. It is this supply of random thermal energy that permits molecules to escape from one catchment basin to another in a thermal reaction. [See KINETICS (CHEMISTRY).]

In order to move from one minimum to an-

other it is necessary to overcome ridges that separate them (Fig. 6B). This is done most easily by travel through saddle points, which correspond to transition structures and are usually called transition states. Some of these lie only a little higher in energy than a starting minimum. Travel over these saddles is easy even at low temperatures, and a chemical species corresponding to a shallow catchment basin may not be isolable at room temperatures or even below. Conformational isomers are a good example of such chemical species. Other catchment basins may be deeper and the surrounding saddle points difficult to reach. They often correspond to configurational or constitutional isomers.

An important attribute of a saddle point connecting two catchment basins is the width of the saddle. Very narrow saddles are difficult for the rolling marble to find and decrease the probability of travel from one catchment basin to another. Wide saddles accommodate a much larger flux under otherwise identical conditions and lead to fast motion.

For a reaction whose rate is not limited by the rate of diffusion, the reaction rate is well described by Eyring's transition state theory. For the rate constant k, we have

$$k = \kappa \frac{kT}{h} \exp\left(\frac{\Delta S^{\ddagger}}{R} - \frac{\Delta H^{\ddagger}}{RT}\right)$$

where κ is the transmission coefficient, k the Boltzmann constant, T absolute temperature, h Planck's constant, R the gas constant, $\Delta H^{\ddagger}$ the activation enthalpy (i.e., the difference from the lowest vibrational level in the original catchment basin to the lowest level available over the saddle point), and the activation entropy $\Delta S^{\ddagger}$ is a function of the width of the saddle relative to that of the starting minimum.

Isotopic substitution does not have any effect on the potential energy surfaces as usually defined. However, it does affect the dynamics of the vibrational process and the dynamics of the motion over saddle points by changing the magnitude of the energy of zero-point motion and the spacing of the vibrational stationary states and thus the density with which these states are packed. Isotopic substitution thus leads to changes in vibrational spectra and also to changes in reaction rate constants.

Some vibrational stationary states have energies that lie above some of the lower saddle points. In such states the molecule is free to travel from one catchment basin to another and in that sense has no fixed chemical structure.

This sort of situation occurs, for instance, for rotations around single bonds at elevated temperatures.

The perfect quantum mechanical analogy to the rolling-marble description given earlier is described in terms of the motion of a wave packet, that is, of a nonstationary wave function initially more or less strongly localized in a particular region of nuclear geometries. Since the vibrational levels are spaced quite closely together, however, the classical description in terms of the rolling marble is often quite adequate.

In the Born–Oppenheimer approximation, the overall rotation of a molecule can also be uncoupled from other kinds of motion. For an isolated molecule, the solution of the Schrödinger equation for rotational motion leads to a set of very closely spaced stationary levels. Transitions between these levels occur in the microwave region of the electromagnetic spectrum. In solution, the quantization of rotational motion is usually destroyed by intermolecular interactions, so that it has little importance in theoretical organic chemistry.

So far we have concentrated on the lowest singlet hypersurface $E(S_0)$. However, motion can be studied similarly on other hypersurfaces if they can be calculated to start with. Such electronically excited states are usually produced either by absorption of light in the ultraviolet and visible regions, as studied by electronic spectroscopy, or by energy transfer from another electronically excited molecule. The latter is particularly useful in the case of excitation into a triplet state, since excitation from a ground singlet to a triplet state by direct absorption of light has a very low probability.

The study of the processes involving the higher potential energy surfaces is the domain of photochemistry and photophysics. These are very difficult or impossible to understand in terms of the classical picture described in Section I, and it is thus easy to see why photochemists are more likely than any other organic chemists to discuss chemistry in terms of potential energy surfaces. Further discussion of photochemically induced reactions can be found below. [See PHOTOCHEMISTRY, ORGANIC.]

4. Reaction Coordinate Diagrams

For many purposes it is useful to condense the multidimensional complexity of molecular motion from one catchment basin to the next into a one-dimensional reaction coordinate diagram in which the degree of progress from the first minimum over the transition state to the second minimum is plotted horizontally as the so-called reaction coordinate. The quantity plotted vertically can be the potential energy E, which we have been discussing so far, but then all information that has to do with the properties of the surface along dimensions other than the reaction coordinate is lost. It is more common to plot instead the Gibbs free energy $\Delta G = \Delta H - T \Delta S$ corresponding to all degrees of freedom other than the reaction coordinate (ΔH is enthalpy, ΔS is entropy). In this fashion information on the entropic constraints dictated by the shape of the potential energy surface in directions perpendicular to the reaction path is preserved.

The effects of structural perturbations on such reaction diagrams can be relatively easily envisaged. For instance, in many cases a structural factor that will stabilize the product relative to the starting material will also tend to stabilize the transition state, albeit to a smaller degree, as is indicated schematically in Fig. 7. Reactions of this kind are said to follow the Bell–Evans–Polanyi principle. Such reactions tend to obey linear free energy relationships, which tie thermodynamic quantities such as equilibrium constants to rate quantities such as kinetic rate constants: $\Delta G_1^{\ddagger} - \Delta G_2^{\ddagger} = \alpha(\Delta G_1^0 - \Delta G_2^0)$, where $\Delta G^{\ddagger}$ is the free energy of activation and ΔG^0 the free energy of reaction. Perhaps the best known example is the Brønsted law, which relates equilibrium and kinetic acidity. In general, linear free energy relationships interrelate changes in free energies or free energies of activation for a series of reactants, usually differing by substitution. They are useful in mechanistic studies.

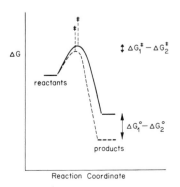

FIG. 7.   Reaction coordinate diagram for two similar reactions differing in ΔG. Location of the transition states is indicated by double daggers.

Cases are also known in which structural perturbations act quite differently on the transition state and the product. Then, reactions with larger equilibrium constants do not necessarily proceed faster than those with smaller ones.

Another consequence of a parallel stabilization of a product and of the corresponding transition state is the displacement of the transition structure toward the starting materials along the reaction coordinate (Fig. 7). Those reactions that obey the Bell–Evans–Polanyi principle thus will have earlier transition states relative to other, similar reactions if they are more favorable thermodynamically and a later transition state if they are less favorable. This statement is known as the Hammond postulate.

At times it is useful to separate motion not only along the reaction coordinate, but also along one other direction selected from all the others and to plot the free energy of the reacting system against two geometrical variables. Diagrams of this kind are particularly popular in the study of substitution and elimination reactions and are known as More O'Ferrall diagrams. The geometries corresponding to the starting material and the product lie at diagonally opposed corners of a square (Fig. 8) and are connected by an energy surface that rises up to a saddle point and then descends to the other corner. The effects of a change in the relative stability of the starting materials and products on the position of the transition state are given by the Hammond postulate: The transition structure moves away from the corner that has been stabilized. The shape of the surface along the other diagonal is just the opposite: The reaction path is of

low energy relative to the two corners. For this reason a change in the relative stability of the geometries represented by the two corners has an effect opposite to that discussed in the Hammond postulate. A stabilization of one of the corners will move the transition point closer to that corner.

While it can be generally assumed that the molecules in the initial catchment basin are almost exactly at thermal equilibrium, which is being perturbed quite insignificantly by the escape of the most energetic molecules over a saddle point or saddle points, this is not always true. In hot ground state reactions a molecular assembly is initially generated with a vibrational energy content much higher than would correspond to the temperature of the surrounding medium, perhaps as a result of being born in a photochemical process or in a very highly exothermic thermal process. Such an assembly can then react at much higher rates.

A study of the reaction rates of molecules starting in individual vibronic levels provides much more detailed information than the study of molecules that are initially at thermal equilibrium but has so far remained a domain of small-molecule chemical physicists and has had little impact on the theory of organic reactions.

An important point to consider is the possibility that a molecule has a choice of more than one saddle over which it can escape from an original catchment basin. The nature of products isolated from such competing reactions can depend on the choice of reaction conditions. Under kinetic control the molecules of the two products are not given an opportunity to travel back over their respective saddles into the initial catchment basin. The preferred product will be that which is being formed faster. On the other hand, if the molecules are provided with an opportunity to return to the initial catchment basin so that an equilibrium is eventually approached, the reaction is said to be run under thermodynamic control and the thermodynamically more stable product will be isolated.

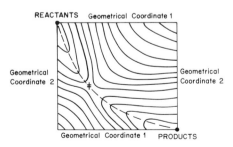

FIG. 8.   More O'Ferrall contour diagram for a reaction requiring two types of nuclear motion that can proceed more or less synchronously. The reaction coordinate is shown as a dashed line, and the structure of the transition state is indicated by a double dagger. The two motions involved could be, for instance, the approach of a nucleophile and the departure of the leaving group in a nucleophilic substitution reaction.

B. ELECTRONIC WAVE FUNCTIONS

Once a wave function of a quantum mechanical state is available, it is possible to calculate its observable properties in a straightforward manner. Unfortunately, the Schrödinger equation is very difficult to solve, and this has not been achieved exactly for any but the simplest molecular systems. The fundamental mathematical problem can be traced to the fact that even for fixed nuclear positions the molecule still represents a many-body system if it contains more than one electron. Nevertheless, it is possible to obtain useful approximate solutions. These become gradually less accurate as the molecule in question increases in size. [See QUANTUM MECHANICS.]

Essentially all of the approximate methods for the calculation of electronic wave functions for an organic molecule start with a basis set of AOs already mentioned in the qualitative discussion. In a so-called minimum basis set description, the AOs assigned to an atom in a molecule are those that would be occupied in a neutral isolated atom of that element plus the remainder of its valence shell. The AOs can be used in two ways for the construction of molecular electronic wave functions.

1. The Valence–Bond Method

In the conceptually simple but computationally very difficult valence–bond (VB) method one first assigns all available electrons to individual AOs in a way that satisfies the Pauli principle, that is, placing no more than two electrons in any one AO and assigning opposite spins to electrons that share the same AO. The method of doing this will not be described in detail here, but it is possible to ensure that the resulting many-electron wave function is properly spin-adapted and represents a singlet, a doublet, a triplet, and so on, as appropriate. An example is the H^+H^- VB structure for the H_2 molecule in which the two available electrons are both assigned to the $1s$ orbital of the hydrogen on the right. Since they are then necessarily spin-paired this represents a singlet structure.

It is generally possible to write a very large number of such VB structures for any given molecule. For instance, for the H_2 molecule we could equally well have written H^-H^+ or four additional structures in which one electron was assigned to each of the available $1s$ AOs and which differed only in the assignment of electron spin. In order to obtain the desired final electronic wave function ψ it is necessary to allow these VB structures to mix.

Mathematically, this corresponds to taking a linear combination of all structures of the same spin multiplicity. In order to determine the coefficients with which each of the structures enters into the final mixture, one uses the variational theorem, which states that the best approximation to the ground wave function $\psi(S_0)$ is given

by those coefficients that lead to a wave function of lowest energy. Determining the coefficients requires the diagonalization of a matrix the dimension of which is equal to the number of VB structures of a given multiplicity possible for the molecule. This increases very rapidly with the size of the molecule, making a full calculation impractical even for quite small organic molecules. Another important difficulty in the calculation is the complicated nature of the computation of each of the elements of the matrix, due ultimately to the fact that the AOs are not orthogonal. Because of these mathematical difficulties the VB method has not had much numerical use, although it remains important as a conceptual tool.

2. The Molecular Orbital Method

The way in which almost all computations of molecular electronic wave functions are performed nowadays is the MO method already mentioned. In this method it is recognized from the outset that in a molecule electrons are delocalized over the whole region of space spanned by the nuclear framework. Accordingly, AOs are first combined into MOs. Unlike the VB structures mentioned earlier, these are still one-electron wave functions.

Mathematically, the MOs are written as linear combinations of atomic orbitals. From n AOs it is possible to construct up to n linearly independent MOs. This is more than is usually needed since in the next step, in which electron occupancies are assigned to these molecular orbitals, each can be occupied by up to two electrons of opposite spins. The unused MOs are referred to as unoccupied (or virtual). Such an assignment of electron occupancies to MOs results in a many-electron wave function known as a configuration. The number of configurations possible is usually very large and is in fact equal to the number of VB structures possible for the same spin multiplicity (singlet, doublet, etc.). In order to obtain the final electronic wave function ψ it is now necessary to mix all possible MO configurations in a fashion analogous to that used to mix the VB structures before. This configuration mixing is referred to as configuration interaction (CI). Indeed, the final wave function ψ obtained by the VB procedure in which all VB structures are included and that obtained by the MO procedure in which all configurations are included will be one and the same. It is referred to as the full configuration interaction (FCI) wave function. It cannot be obtained in practice for any but the smallest organic molecules, because the number of possible configurations increases astronomically with molecular size.

The MO path to the FCI wave function ψ involves one more step than the VB path, and this appears to be more complicated at first sight. However, the availability of the additional step endows the MO method with more flexibility than the VB method, which permits a mathematically much simpler formulation. This flexibility arises in the step of combining AOs in MOs. First, it is easy to ensure that the MOs are mutually orthogonal, and this leads to an immense simplification in the calculation of matrix elements. Second, it is also possible to optimize the choice of the MOs in such a way as to speed up the convergence of the final summation in which configurations are combined to obtain the state wave function.

Indeed, in the most commonly used form of the MO procedure this final step is omitted altogether, and a single configuration built from optimized molecular orbitals is used as an acceptable, if poor, approximation of the FCI electronic wave function ψ. These optimum MOs are known as the self-consistent field (SCF) or Hartree–Fock (HF) MOs. They are obtained using the variational principle and demanding that the energy of the one configuration under consideration be as low as possible. The energy difference between the HF description and the FCI description is referred to as the correlation energy. It is important to note that weak intermolecular interactions (van der Waals interactions), important in processes such as molecular recognition and complexation, cannot be calculated at the SCF level. In calculations of these effects, inclusion of correlation effects is essential.

One of the advantages of the MO method is that it is relatively easy to improve the SCF solution partially without having to go all the way to the unreachable limit of FCI. Several methods for making such improvement are available, such as (1) limited configuration interaction in which the CI expansion is truncated in some systematic fashion well before the FCI limit is reached and (2) methods in which correlation effects are viewed as a small perturbation of the SCF solution and are treated by perturbation theory. The use of these methods is particularly important (1) when the electronic wave function ψ is being calculated for a geometry far removed from the molecular equilibrium geometry, (2)

when the molecule has very low lying excited electronic states, (3) when the molecule is a biradical, (4) if the calculation is performed for an excited electronic state, or (5) if intermolecular forces are to be calculated.

So far we have discussed only the minimum basis set approximation. This might be quite adequate if the AOs were chosen in a truly optimal manner, but it would require a nonlinear optimization and this is normally not done. Even so, in order to obtain a truly accurate solution for the electronic wave function ψ within the Born–Oppenheimer approximation it would be necessary to increase the number of basis set AOs used in the calculation (in principle, to infinity).

The form of the AOs usually adopted in numerical work is normally chosen for computational convenience (Gaussian-type orbitals) in a way that makes them quite different from the optimum orbitals of an isolated atom. Thus, in practice most computations are not performed with a minimum basis set but with extended basis sets, and several such standard sets are in common use.

The electronic wave functions resulting from such large-scale calculations are usually not easy to visualize. Frequently, it helps to draw the resulting electron densities in the form of contour maps. Alternatively, electrostatic potential contour maps can be constructed, indicating the nature of the electric fields to be expected in the vicinity of the molecule.

3. *Ab Initio* and Semiempirical Methods

The approach described so far is of the so-called *ab initio* type, in which the whole computation is done from first principles, taking from experiment only the values of fundamental constants such as electron charge. As already indicated, accurate results cannot be obtained by these methods for most molecules of interest in organic chemistry, but the approximate solutions obtained at the SCF or improved SCF level provide much useful information. For certain properties, such as molecular geometries, dipole moments, and the relative energies of conformers, the agreement with experiment is excellent.

An alternative approach to the problem of molecular electronic structure is provided by semiempirical models. In these no attempt is made to derive the properties of atoms from first principles. Rather, they are taken as described by a set of parameters obtained by fitting experimental data, and an attempt is made to find a model

Hamiltonian that will provide a good description of interatomic interactions. The form of the model Hamiltonian is patterned after the *ab initio* analysis. In the most common semiempirical methods it is still a fairly complicated many-electron Hamiltonian, so that its exact stationary wave functions cannot be found for molecules of interest, and only approximate solutions are obtained. One almost invariably starts with a minimum basis set of AOs and proceeds to an SCF type of wave function, possibly followed by a limited amount of improvement toward the FCI wave function. The parameters that enter the Hamiltonian are optimized so as to bring about close agreement between the molecular properties computed from the approximate wave function (usually SCF) and those observed experimentally. Most commonly, the properties fitted are heats of formation, molecular equilibrium geometries, or suitable spectroscopic properties. In this way one attempts to incorporate intraatomic correlation energies and a large part of interatomic electron correlation energies into the model through parameter choice, although one works only at the SCF level or at least not much beyond it.

The agreement of the calculated properties with experiment is roughly comparable to the agreement obtained by extended basis set *ab initio* methods at the level of the SCF approximation or slightly better, at least for those classes of molecules for which the semiempirical parameters were originally optimized. However, even for molecules quite different from those on which the original optimization was performed, the agreement is frequently striking considering that orders of magnitude less computer time is needed for the semiempirical computations. This great reduction in computational effort is to a large degree due to the almost universal use of the so-called zero differential overlap approximation, which greatly reduces the number of electron repulsion integrals needed in the computations.

Some of the best known examples of semiempirical methods are those developed for the treatment of electronic ground states by Dewar and collaborators (MNDO, AM1, MINDO/3) and the methods developed by Jaffé (INDO/S), Zerner (INDO/S), and their respective collaborators for calculations involving electronically excited states. A very simple procedure is the extended Hückel method popularized by Hoffmann. Others are the older methods developed for the treatment of π electrons only: the PPP method of Pariser, Parr, and Pople and the extremely crude but also extremely simple HMO method of Hückel.

These semiempirical models should not be confused with approximate models that are designed to mimic the results of *ab initio* calculations in a simpler manner rather than to mimic the results of experiments. The best known approximate MO methods are the CNDO and INDO methods developed by Pople and collaborators.

C. MOLECULAR PROPERTIES

Although we have indicated how the electronic wave function of an electronic state and its energy are calculated, we have said very little about the calculation of other molecular properties once the wave function is known.

The calculation of molecular equilibrium geometry in a given electronic state, usually S_0, is performed by varying the assumed nuclear geometry and repeating the calculation of the energy by one of the methods referred to earlier until a minimum is found. This search is normally performed by computer routines that compute surface gradients in order to speed up convergence toward a local minimum in the $(3N - 6)$-dimensional nuclear configuration space. From the computed curvatures of the surface at the minimum one obtains the force constants for molecular vibrations and the form of the normal modes of vibration. For a true local minimum, all the force constants must be positive. A similar type of procedure, minimization of the norm of the gradient, can be used for finding transition states. After a transition point is found, it is essential to convince oneself that a normal mode analysis produces only one vibration with a negative force constant. This mode corresponds to the path from one catchment basin to the other, and the corresponding vibration has an imaginary frequency (the restoring force is negative). Other modes of vibration are ordinary and permit the evaluation of the entropy of the transition state.

Most other molecular properties are normally evaluated only at the equilibrium geometry, although strictly speaking they should be calculated at a large number of geometries and averaged over the vibrational wave function of the state in question. Generally, they are obtained by representing the observable by a quantum mechanical operator and computing the expectation value of this operator over the wave function.

At the HF level, *ab initio* or semiempirical, some of these properties can be obtained in an approximate manner more simply. Thus, the lowest ionization potential of the molecule is approximately equal to the negative of the energy of the highest occupied molecular orbital (HOMO). The electron affinity of the molecule is similarly approximated by the negative of the energy of the lowest unoccupied molecular orbital (LUMO). The energies of electronic excitation are usually more complicated to obtain in that the introduction of CI may be quite necessary.

Those electron excitations that can be well described as a promotion of an electron from one single occupied MO to one unoccupied MO are rare but exist in some molecules. An example is the so-called L_a band in the absorption spectra of aromatic hydrocarbons and the first intense band of polyenes. In the SCF approximation the electronic excitation energy is equal to the energy difference between the orbital out of which the promotion occurs and the orbital into which it occurs, minus the repulsion energy of two electron densities, each provided by an electron in one of these two MOs. This approximates the energy of the triplet excited state. The energy of the singlet excited state is higher by twice the exchange integral between the two MOs involved (the self-repulsion energy of a charge density produced by taking a product of the two orbitals).

Due to the approximate nature of the SCF treatment and to the additional approximations involved in the statements just made, the results are usually more useful for an interpretation of trends within a group of compounds rather than the absolute value for any one compound. The semiempirical methods in particular have had much use in this kind of application.

A property related to this is the formation of intermolecular complexes characterized by a charge-transfer transition, which frequently occurs in the visible region and in which an electron is transferred from one molecule to another. The energy of this transition is related to the ionization potential of the donor and the electron affinity of the acceptor moiety and once again can be correlated with the computed MO energies.

Not only the energies but also the coefficients of the MOs computed in the SCF picture are approximately related to observable properties. Thus, for a given MO, the square of the coefficient on a particular AO is related to the proba-

bility that an electron in that MO will be found in that particular AO. This relation is particularly simple in those semiempirical methods that use the zero differential overlap approximation. Then, the AOs are mutually orthogonal and the relation between the square of the coefficient and the probability is a simple proportionality. In methods that use nonorthogonal AOs, such as the *ab initio* ones, it is more difficult to define electron populations for AOs and atoms. The procedure usually used is known as the Mulliken population analysis. This permits a calculation of electron densities in AOs and of total electron densities on atoms. These in turn can be related to molecular dipole moments, infrared spectral intensities, and, much more approximately, to nuclear magnetic resonance (NMR) shielding constants.

The distributions of unpaired spin obtained in an analogous fashion from the squares of coefficients of a singly occupied orbital are related to the hyperfine coupling constants in electron spin resonance spectroscopy.

There is another class of molecular properties that are not related in a simple way to the expectation value of an operator: the so-called second-order properties. Some of the most important of these are molecular polarizability, Raman intensities, and chemical shielding in NMR spectroscopy. They can be computed by introducing an outside perturbation such as an electric or magnetic field explicitly into the calculation of the molecular wave function. These calculations are more difficult and less reliable, particularly with respect to magnetic properties, where the incompleteness of the basis sets used tends to make the results dependent on the choice of origin of coordinates. Good progress has been made in the calculation of NMR chemical shielding constants and their anisotropies, while the calculation of intensities in Raman spectra still leaves much to be desired.

IV. Reaction Paths

A. THERMAL REACTIONS

A reaction involving motion through a single transition state is referred to as an elementary reaction step. Most reactions of organic compounds involve a sequence of such reaction steps in which the reacting system passes through a series of intermediates and transition states that separate them. The reaction interme-

product, and one of the antibonding and unoccupied orbitals of the reactant becomes bonding and occupied in the ground state of the product. In the region of transition state geometries halfway through the reaction path, both orbitals are approximately nonbonding. Between them, they contain two electrons, and these two electrons do not contribute to bonding in the molecule at the transition state geometry. In effect, the molecule is a biradical and contains one less bond than its number of valence electrons would in principle allow it to have. The transition state is unfavorable and is of the antiaromatic type, containing four electrons in an array of four AOs with all positive overlaps (isoelectronic with cyclobutadiene).

Although it is already apparent which of the two reactions chosen as examples is allowed and which is forbidden, it is useful to consider the construction of the configuration correlation diagram as well (Fig. 10). Here low-energy configurations are constructed by considering all suitable occupancies of the MOs of the reactants on the left-hand side and of the products on the right-hand side. The symmetry of each is again identified as antisymmetric or symmetric with respect to each of the above symmetry elements, using the rule $S \times S = A \times A = S$ and $A \times S = A$. Correlation lines are now drawn from left to right by keeping the occupancy of each MO in each configuration constant. This often produces crossings of configurations of the same symmetry. According to the noncrossing rule these must ultimately be avoided.

This is accomplished by introducing configuration interaction, which converts the configuration correlation diagram to the desired state correlation diagram, as indicated in Fig. 10. Clearly, the crossing of correlating MOs in the case of the ethylene + ethylene cycloaddition causes a similar crossing of lines in the configuration correlation diagram. Since the effects of configuration mixing, which produces the final state diagram, are generally relatively small, a memory of the crossing at the geometry of the transition state survives and results in a large barrier in the energy of the ground state in the middle of the correlation diagram. It is then concluded that the transition state is unfavorable relative to the case of the ethylene + butadiene process, in which no such barrier is imposed by the correlation.

B. PHOTOCHEMICAL REACTIONS

In photochemical reactions, initial electronic excitation is introduced by the absorption of a photon or by an energy transfer from another molecule. It is normally followed by a very rapid, radiationless conversion to the lowest excited singlet or the lowest triplet energy surface, depending on the multiplicity of the initial excited state. Also, any vibrational energy in excess of that dictated by the temperature of the surrounding medium, whether generated by the initial excitation or by the radiationless process, is rapidly lost to the solvent, unless one works in a gas phase at low pressure. Thus, in a matter of a few picoseconds or less the molecule ends up in one or another of the local minima in the S_1 or T_1 surface. Further motion on the surface may follow, depending on the temperature and the height of the barriers surrounding the local minimum. Also, radiationless conversion from the S_1 to the T_1 state, known as intersystem crossing, can occur. This often happens on a nanosecond time scale. Sooner or later a radiationless return to the S_0 state ensues. The final fate of the molecule is further loss of excess vibrational energy and thermal equilibration at the bottom of one or another catchment basin in the S_0 surface, depending on where on the S_0 surface the molecule landed. If this is the same minimum from which the initial excitation occurred, the process is viewed as photophysical. If it is not, a net chemical reaction has occurred and the process is labeled photochemical.

At times the excited S_1 surface may touch or nearly touch the S_0 surface, in which case the

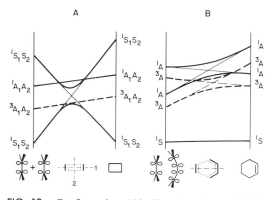

FIG. 10. Configuration (thin lines) and state (thick lines) correlation diagrams for the concerted face-to-face cycloaddition of two ethylene molecules (A) and for the concerted Diels–Alder cycloaddition of ethylene to butadiene (B). Full lines, singlets; dashed lines, triplets. See caption to Fig. 9.

return to S_0 is very fast. Such areas in S_1 are often referred to as funnels since they very effectively return molecules to the ground state.

In order to understand photochemical reaction paths it is thus important to have an understanding of the location of barriers as well as minima and funnels in the S_1 and T_1 surfaces, plus a sufficient understanding of the S_0 surface to allow a prediction or rationalization of the fate of a molecule that lands in a known region of this surface. Correlation diagrams are often useful for this purpose. For instance, the diagram for the face-to-face cycloaddition of two ethylenes shown in Fig. 10 shows the presence of a minimum in the S_1 surface in the general area of geometries at which the pericyclic transition state occurred in the ground state. While the latter was energetically unfavorable in the S_0 state, making the reaction highly unlikely since the molecules will probably find other reaction paths, the minimum in the S_1 state provides an efficient driving force for the photochemical cycloaddition to proceed efficiently. Thus, reactions that fail to occur in the ground state are often smooth when performed photochemically, and vice versa.

In general, by virtue of molecules landing at otherwise improbable and highly energetic areas on the S_0 surface, photochemical processes are capable of producing very highly energetic ground state products. Yet, frequently the same perturbations, such as substituent effects, that increase the stability of a molecule in the ground state also facilitate its photochemical reactions by lowering the barriers encountered along the way. The interplay of these two aspects of the excited-state surfaces—minima and barriers—make the consideration of photochemical processes far more complex than the study of thermal reactions.

BIBLIOGRAPHY

Burkert, Ulrich, and Allinger, Norman L. (1982). "Molecular Mechanics." American Chemical Society Monograph. Washington, D.C.

Čársky, Petr, and Urban, Miroslav (1980). "*Ab Initio* Calculations." Springer–Verlag, Berlin, Germany.

Dewar, Michael J. S., and Dougherty, Ralph C. (1975). The PMO Theory of Organic Chemistry. Plenum. New York, N.Y.

Dewar, Michael J. S. (1984). *J. Am. Chem. Soc.* **106**, 669.

Hehre, Warren J., Radom, Leo, Schleyer, Paul V. R., and Pople, J. A. (1986). "*Ab Initio* Molecular Orbital Theory." Wiley, New York.

Lowry, Thomas H., and Richardson, Kathleen S. (1981). "Mechanism and Theory in Organic Chemistry," 2nd ed. Harper and Row. New York, N.Y.

Salem, Lionel. (1982) "Electrons in Chemical Reactions." Wiley, New York, N.Y.

Schaefer, Henry F. III, and Segal, Gerald A. (eds.) (1977). *Modern Theoretical Chemistry,* Vols. **3**, **4**, **7**, and **8**, Plenum. New York, N.Y.

Simons, Jack (1983). "Energetic Principles of Chemical Reactions." Jones and Bartlett. Boston, Massachusetts.

ORGANIC CHEMICALS, INDUSTRIAL PRODUCTION

H. Harry Szmant *Chemical Consultant, Sanibel, Florida*

GLOSSARY

Chemical feedstocks, organic: Primary raw materials from which organic chemicals are assembled: petroleum, natural gas, coal, and renewable, that is, biomass, resources.

Chemical process industries: Various manufacturing segments of the economy that are based on the production of chemicals and involve chemical transformation processes.

Commodity chemicals: High-volume chemicals of low unit cost that are the fundamental building blocks of the chemical industry.

Fine chemicals: Well-recognized chemicals of exacting quality that require skillful, multistep production and/or isolation processes.

Pseudo-commodity chemicals: Chemicals similar to those of the commodity category except for a smaller demand and somewhat higher unit cost because their production requires some standard material transformations.

Specialty chemicals: High-unit-value, usually proprietary materials designed for problem-solving performance under problematic circumstances.

Of the over 4 million known compounds of carbon, about 60,000 are utilized by industry and about 18,000 are of major industrial interest and thus are listed in comprehensive buyers' guides. Exact production/consumption statistics for most industrial organic chemicals relevant to the United States are difficult to obtain because of the restrictive policies designed to protect confidential industrial information. Hence, the United States International Trade Commission (USITC) releases production data only for some 6000 materials of synthetic origin, and this information represents the lower range of production. Approximate production/demand and unit cost figures as well as major uses of representative industrial organic chemicals are listed in Table I.

I. Nature and Scope

Based on the demand in the marketplace (of the United States) and unit costs, it is convenient to differentiate between several categories of industrial organic chemicals, as shown in Table II.

Categories I and II encompass the basic industrial building blocks, and the intermediates derived from them through relatively simple chemical transformations, as well as common polymers, that is, plastics, elastomers or rubbers, textile fibers, molding and coating compositions, and so on. Category III represents well-known chemicals obtained through multistep synthetic procedures or by way of relatively simple isolation from renewable (biomass) sources without subjecting the natural products to significant chemical transformations. The unit costs of the latter materials depend on the scarcity of sources and the concentrations of the desired components in the raw materials. Examples of such natural products are the essential oils employed as fragrances and flavoring agents, gums, certain carbohydrates and proteins, and alkaloids. Category IV includes sophisticated products developed in response to a

TABLE I. Approximate Production/Demand, Unit Costs, Origins, and Major Uses of Representative Industrial Organic Materials (United States, 1983–1986)

Material[a,f]	Production/demand[b] (B lb/yr)	Unit cost[c] ($/lb)	Source(s)[d,f] and precursor(s)	Major uses[e,f]
Acetaldehyde	0.9	0.33	Ethylene Ethanol	Pyridines (50), peracetic ac, penta-erythritol, glyoxal, 1,3-butylene glycol
Acetaminophen N-Acetyl-p-amino-phenol (APAP)	0.024	3–4	p-Aminophenol/acetic anh	Med
Acetic ac	2.9	0.23	Methanol/sng Acetald/O_2 n-Butane/O_2	Vinyl acetate (45), cellulose acetates (17), terephthalic ac/diMe est (12), acetic ac est (10), acetic anh (4), chloroacetic ac (2)
Acetic anh	1.6	0.42	Acetic ac Methanol/sng	Cellulose acetates (80), aspirin, acet-aminophen, acetic ac est
Acetoacetanilide	0.01	1.3	Acetic ac/aniline	Dyes, syn int
Acetone	1.8	0.22	Cumene/O_2 Isopropyl alc	Sol and Me isobutyl ket (50), Me meth-acrylate (32), bisphenol A (8)
Acetylene	0.28	~0.45[g]	Calcium carbide Nat gas	Vinyl chloride (50), 1,4-butanediol, butyrolactone, N-Me pyrrolidone, syn int
Acrylamide	0.14	0.75	Acrylonitrile	Poly(acrylamide)
Acrylic ac	0.8	0.66	Propylene/O_2	Me, Et, Bu, and 2-Et hexyl est and pol (75), exp (25)
Acrylonitrile (AN)	2.4	0.45	Propylene/NH_3/O_2	Tex fib (28) and pla/rbr (18), adiponitrile (10), acrylamide (4), exp (40)
Acrylonitrile–buta-diene–styrene (ABS)	1.0	1.0	Monomers	Aut/appliances cmp (90), exp (10)
Alkylbenzenesulfonic ac salts	0.2	0.5–1	LAB/H_2SO_4	Detergents
Adipic ac	1.4	0.58	Cyclohexane/O_2	Nylon 66 (82), hexamethylenediamine (8), P urethanes (4), pla est (3)
Alkyd res (also see P est res)	0.4	0.7	Phthalic anh/glycerine, isophthalic ac, etc.	Coatings
Alpha-olefins (AO)			Ethylene	Saa, lub oils
C_6–C_{10}	0.5	0.3	Paraffins	
C_{11} plus	0.6	0.3	Fatty alc	
Anethole	0.0012	4.5	Anisole (80) Turpentine (20)	Frg, flv, anisaldehyde
Aniline	0.8	0.4	Nitrobenzene Phenol/NH_3 Chlorobenzene/NH_3	MDI (60), rbr chm (18), dye int (10), hydroquinone (3)
Aramid pol	0.045	4–45	Terephthalic ac/m- or p-phenylenediamine	Hgh prf fib
Ascorbic ac (vitamin C)	0.035	4	Sorbitol	Vitamin supplement, food add, antioxi-dant
Aspartame	0.002	85	Aspartic ac/phenyl-alanine	Food add
Aspirin	0.031	2	Salicylic ac/acetic anh	Med
Atrazine (2-chloro-4-EtNH-6-i-PrNH-s-triazine)	0.06	2	Cyanuric chloride/amines	Herbicide
Benzene	12	0.14	Ptr rfg, reforming hydrodealkylation (96) Coal tar (4)	Ethylbenzene/styrene (55), cumene (20), cyclohexane (15), aniline (5), alkylben-zenes (3)
Benzoic ac	0.16	0.5–1.7	Toluene	Phenol (50), est pla (15), benzoyl chloride (12), Na salt (8 imp)
Benzoyl peroxide	0.09	2.4	Benzoyl chloride/Na_2O_2	Csm, pol cat
Benzyl chloride	0.017	0.5	Toluene	Bu benzyl phthalate pla (60), benzyl alc and est
Butylamines (mostly di)	0.03	0.8	Bu alc/NH_3	Syn int, quat
Bisphenol A (BPA)	0.8	0.67–0.71	Acetone/phenol	Epoxy, P carbonate and P sulfone res, P etherimides, P arylates

(continued)

TABLE I.   (*Continued*)

Material[a,f]	Production/demand[b] (B lb/yr)	Unit cost[c] ($/lb)	Source(s)[d,f] and precursor(s)	Major uses[e,f]
1,3-Butadiene	2.5 1.2 imp	0.27	Ptr crk byp Butane crk onp	Styrene–butadiene rbr (37), P butadiene rbr (22), hexamethylenediamine (10), styrene–butadiene latex (9), ABS res (6), neoprene rbr (7), nitrile rbr (3)
n-Butane	2.4	0.11	Ptr and LPG	Crk to butenes and butadiene
1,4-Butanediol	0.1	0.8	Acetylene/CH_2O Maleic anh/H_2	Tetrahydrofuran, syn int
n-Butene-1	0.34	0.26	Ethylene Butane crk	LLDPE (42), P butene pol(12), exp (40)
n-Butyl alc	0.9	0.36	Propylene, oxo	Bu acrylate and methacrylate (30), ethers (22), Bu acetate (12)
n-Butyl acetate	0.11	0.52	Bu alc/acetic anh	Sol: nitrocellulose-based lacquers (62), inks (5), adh (2); exp (25)
n-Butyl acrylate	0.2	0.7	Bu alc/acrylic ac	P acrylates
n-Butyl lithium	0.001	15	*n*-Bu chloride/Li	Pol cat, syn rgt
sec-Butyl lithium	0.001	16.5	*sec*-Bu chloride/Li	Syn rgt
t-Butyl lithium	0.001	30	*t*-Bu chloride/Li	Syn rgt
n-Butyl methacrylate	0.01	0.9	Bu alc/methacrylic ac	P methacrylate coatings
t-Butyl peroxide	0.003	1.4	Isobutylene/H_2O_2	Pol cat
Butyl rbr	0.3	0.8	Isobutylene and lower olefins	Tire cmp (80): inner tubes, lining; chlorinated and brominated rbr, exp (14), sealants (6)
n-Butyraldehyde	0.8	0.3	Propylene/oxo	P vinylbutyral, 2-Et hexanol
Caffeine	0.01	4.3–4.8	Coffee Syn	Food add, med
Caprolactam	1.1	0.7	Cyclohexane Phenol	Nylon 6 (92), exp (5)
Carbon black	2.6	0.3–0.4	Ptr and ntg	Rbr reinforcing agt (90), inks (8)
Carbon dioxide (liquid/ solid only)	9	0.04	Ptr and frm	Rfg (30), beverages (25), chm ind (10): salicylic ac, org carbonates
Carbon disulfide	0.3	0.2	Coke/sulfur	Regenerated cellulose (40): rayon, cellophane, sponges; rbr chem (10), org xanthates
Carbon tetrachloride	0.7	0.35	Ntg/Cl_2 CS_2/Cl_2	Fluorocarbons (90), exp (6), sol
N,N'-Carbonyl diimidazole	0.0001	68	Imidazole/phosgene	Syn rgt
Carboxymethylcellulose (CMC, cellulose gum)	0.06	1.6	Chloroacetic acid/ cellulose	Dispersing and thickening agt (40): food, med, paints; drilling muds (30), ppr and tex size (12)
Castor oil Ricinoleic ac K salt	0.12 imp 0.004	0.3–0.75 0.8 1.6	Castor bean Castor oil	Saa, syn int Syn int Saa
Cellulose acetate	0.7	1.3	Cellulose, acetic ac/anh	Cigarette filters (50), fbr and pol (32), exp (10)
Chloroacetic ac	0.1	0.6	Acetic ac/Cl_2	CMC and other carboxymethyl eth (75) thioglycolic ac, glycine
Chlorobenzene (mono- or MCB)	0.25	0.4	Benzene/Cl_2	Sol (42), nitro drv (32), diphenyl oxide, phenylphenols
Chloroform	0.4	0.35	Ntg/Cl_2 CCl_4/Fe, HCl	Fluorocarbon rfg (65), fluorinated pol (28), sol
Chloroparaffins	0.095	0.5	Ptr paraffins/Cl_2 (20–70%)	Lub and metal-working oils (60), flame rtd, pla
Chloroprene	0.26	0.4	Butadiene/Cl_2	Neoprene rbr, P chloroprene latex for fabric coatings, adh
Citric ac	0.25	0.9–1.2	Glucose frm	Food add, phr, stack gas desulfurization, met cleaners
Citronellol	0.0015	3.6	Citronellal/H_2	Frg, syn int: acetate, formate
Choline chloride	0.09	0.3–2.3	TriMe amine/EO/HCl	Feed and food add
Cinnamic ac	0.003	2	Benzaldehyde, acetic anh	Syn int frg, food add

TABLE I. (*Continued*)

Material[a,f]	Production/demand[b] (B lb/yr)	Unit cost[c] ($/lb)	Source(s)[d,f] and precursor(s)	Major uses[e,f]
Coumarin	0.0004 imp	6	Salicylaldehyde, acetic anh	Frg
Coal tar pitch	1.5	0.12	Coal tar	Binder for carbon electrodes, refractories
Cobalt naphthenate	0.002	2	Naphthenic ac, Co salt	Paint drying agt
Codeine and salts	0.09	300–400	Opium Morphine, MeI	Med
o-Cresol	0.05	0.8	Phenol, Me alc	Herbicides (50), epoxy res (25), novalak res
m-Cresol	0.01	1.7	Coal tar Ptr	Syn int (vitamin E)
p-Cresol	0.01	1.2	Coal tar Ptr	Butylated hydroxytoluene (BHT), res, syn int
m,p-Cresol	0.02	0.9	Coal tar	Tricresyl phosphate pla (TCP)
Cresylic ac	0.14	0.6	Coal tar	Antioxidants (20), TCP (15), res (15), wire enamel sol, disinfectant, flt agt
Cumene	3.9	0.15	Benzene/propylene	Phenol/acetone prd (98), α-methylstyrene
Cumene hydroperoxide	0.002	0.8	Cumene/O_2	Pol cat, epoxidation rgt
Cyclohexane	2.1	0.15	Benzene/H_2 Naphtha	Adipic ac (60), caprolactam (30)
Cyclohexylamine	0.009	1.0	Aniline/H_2	Boiler chm (60), rbr chm (2), syn int
Diacetone alc	0.04	0.6	Acetone	Sol, hydraulic fluid
2,6-Di-t-butyl-p-cresol (BHT)	0.01	1.4	p-Cresol/isobutylene	Antioxidant
o-Dichlorobenzene	0.04	0.5	Benzene/Cl_2	Sol (TDI), syn int, 3,4-dichloroaniline
p-Dichlorobenzene	0.06	0.45	Benzene/Cl_2	Deodorant/moth control (40), PPS (24), syn int
Dicyclopentadiene (DCP)	0.1	0.35	Ptr crk byp	Pol (50), UPR (25), D in EPDM (15), syn int
Diethanolamine	0.13	0.5	EO/NH_3	Syn int, morpholine, saa
Diethylenetriamine (DETA)	0.03	1.6	Ethylene chloride/NH_3	Epoxy crg agt; add to ppr, lub, fuel; ssa, cor inh, pentaacetic ac drv (DTPA) chl agt
Diethylene glycol (DEG)	0.45	0.3	EO/H_2O	PU and UPR (30), triethylene glycol (15), antifreeze (10), dioxane, sol
Dimethylaminopropyl-amine	0.005	1.1	AN/diMe amine	Syn int
2,4- and 2,6-Dinitro-toluene (4:1)	0.7	0.3	Toluene/HNO_3	Diamine for TDI prd
Dioctyl adipate	0.03	0.7	2-EH/adipic ac	Pla
Dioctyl phthalate (DOP)	0.3	0.45	2-EH/phthalic anh	Pla
Diphenyl carbonate	0.001	2	Phosgene/phenol	Syn int
Dipropylene glycol (DPG)	0.05	0.45	PO/H_2O	UPR (55), pla (30), PU (10)
Dimethyl terephthalate (DMT)/Purified terephthalic ac (PTA)	7	0.35	p-Xylene/O_2, Me alc	PET (90), PBT (5)
Dithiocarbamic ac drv	0.025	2–4	Amines/CS_2	Fungicides (Zn), rbr vulcanization accelerators
Dodecenylsuccinic anh	0.005	0.9	MA/C_{12} AO	Saa
Dodecyl alc ethoxylated sulfate salts	0.03	0.5–0.8	Dodecyl alc/EO/H_2SO_4	Saa
Dodecyl sulfate salts	0.1	0.7	Dodecyl alc/H_2SO_4	Saa
p-Dodecylphenol	0.2	0.5	Phenol/C_{12} AO	Saa int
Durene	0.002	1	Ptr and MTG byp	Pyromellitic dianh
Enzymes, industrial proteases (50): rennin, bacterial carbohydra-tases (37): amylases, glucoseisomerase	—[h]	—[h]	Rnr	Cat: fod prd (58), saa (15), txt trt, ppr, ltr prd
Epichlorohydrin	0.1	0.86	Allyl chloride, H_2O_2 or $Cl_2/H_2O/CaO$	BPA and other glycidyl eth, syn int

(*continued*)

TABLE I.   (*Continued*)

Material[a,f]	Production/demand[b] (B lb/yr)	Unit cost[c] ($/lb)	Source(s)[d,f] and precursor(s)	Major uses[e,f]
Epoxidized soya oil	0.1	0.5	Soya oil, hydroperoxides	Pla, PVS stabilizer
Epoxy res	0.42	1.3	BPA, epichlorohydrin	Coatings (45), electronic prd (30), cmp (25)
Erythrosine (FD&C no. 3)	0.0003	24	Phthalic anh/phenol/I_2	Food, Drug and Cosmetic certified coloring agt
Other FD&C Dyes	0.006	9 avg	Miscellaneous	
Ethane	6.2	0.1	Ntg	Ethylene
Ethanol	6.7	0.27	Frm (70)	Frm: fuel (80)
	0.1 imp		Ethylene (30)	Syn: sol (55), Et drv (45), vinegar (4.5), est (4.5), eth (4.5), am
Ethanolamines, mono-, di-, and tri-	0.5	0.4	EO/NH_3	Saa (35), ac gas spn (15), syn int
Ethylamines, di-, tri-, mono-	0.04	1	Ethanol/NH_3	Syn int
Ethylbenzene	8	0.22	Ethylene/benzene	Styrene (99)
Ethyl chloride	0.25	0.25	Ethylene/HCl Ethanol/HCl	TetraEt Pb (90), Et cel (8), eth
Ethylene	31	0.15–0.18	Ptr rfg byp Ntg and LPG crk	LDPE and LLDPE (26), HDPE (22), EO/EG (16), ethylene dichloride/VCM (12), ethylbenzene/styrene (8), linear alc (5), VAM (2), ethanol (2)
Ethylene dibromide (EDB)	0.17	0.3	Ethylene/Br_2	Agr agt, fuel add, sol
Ethylene dichloride	13	0.19	Ethylene/Cl_2	VCM (85), Cl drv sol (80), vinylidene chloride (3)
Ethylene glycol (EG)	4.8	0.2	EO/H_2O	PET (50), antifreeze (40), PEG, eth
	0.3 imp			
Ethylene glycol ethers	0.32			Sol, fuel add, ctg, phthalate pla, diMe crown eth
Methyl (2-ME)		0.34		
Methyl acetate (2-MEA)		0.43		
Ethyl (2-EE)		0.51		
Ethyl acetate (2-EEA)		0.55		
Ethylene oxide (EO)	5.8	0.35	Ethylene/O_2	EG (60), ethoxylated drv (10), EG eth (5), ethanolamines (5)
Ethylene-propylene pol (EPM and EDPM rbr)	0.47	1.0	Ethylene/propylene and diene for EPDM	Pla and rbr prd (50), impact add (18), oil add (12)
Ethylenediamine, mono- (EDA), di-, and higher	0.08	1.3	Ethylene dichloride/NH_3	Chl agt (30), pol drv (15), EDTA, ethoxylated drv, agr, PUR
Ethylenediamine tetraacetic ac Na salt (EDTA)	0.02	0.37	EDA/chloroacetic ac	Chl agt
2-Ethylhexyl alc (2-EH), octyl alc	0.5	0.4	Butyrald/H_2	Pla (60), acrylate est (12), diesel fuel add (5)
Eugenol (4-allyl-2-methoxyphenol)	0.0002	3.5	Cloves	Fla, frg
Fluorocarbons (FC)	1.0			
FC-11, CCl_3F		0.57	CCl_4/HF	PU, foam blow agt
FC-12, CCl_2F_2		0.68	CCl_4/HF	Rfg
FC-14, CF_4		0.7	CCl_4/HF; also C/F_2	Prd microelectronic dev
FC-22, $CHClF_2$		1.05	CCl_4/HF	Rfg, $CF_2{=}CF_2$ (TFE), $CF_2{=}CFCF_3$ for Teflon pla and Vitron rbr
FC-113, CCl_2FCClF_2		0.89		Sol, $CClF{=}CF_2$ for pla
FC-114, $CClF_2CF_2Cl$		1.02		Sol
Formaldehyde	5.6 (37% b. wt.)	0.1	Me alc/O_2	UF (25), PF (20), P acetal res (5), 1,4-butanediol (5), pentaerythritol, hexamethylenetetramine, MF, MDI and PMDI

TABLE I.   (*Continued*)

Material[a,f]	Production/demand[b] (B lb/yr)	Unit cost[c] ($/lb)	Source(s)[d,f] and precursor(s)	Major uses[e,f]
Formic ac	0.05	0.4	C_4H_{10}/O_2 Pentaerythritol prd byp CO/NaOH or Me alc	Tex and ltr ind (40), med (20), rbr prd (15)
Fructose				
Syrup (HFCS)	10	0.2	Corn starch/enz	Food add
Crystalline	0.01	0.9		
Fumaric ac	0.03	0.7	Frm MA	Ppr ind (40), food add (20), pol (20)
Furfuraldehyde	0.12	0.75	Bagasse, cereal byp	Furfuryl alc, THF, tetrahydrofurfuryl alc, syn int
Furfuryl res	0.02	0.8	Furfuryl alc	Foundry mold binder
Glucose (dextrose)	0.8	0.41–0.46	Corn starch/enz	Food add, sorbitol, gluconic ac, P glucose, Me glucoside, frm, syn int
Gluconic ac	0.02	0.5	Glucose	Salts: food and saa add, met fin
Glycerine	0.3 0.03 imp	0.9	Fats/oils (67) Allyl chloride/Cl_2/H_2O	Phr and csm prd (25), humectant (18), pol (35), food (15), exp (2)
Glycine	0.004	1.9–2.1	Chloroacetic ac/NH_3	Food add
Glyoxal (40% b. wt.)	0.06	0.45	Acetaldehyde/HNO_3	Txt prd, syn int, ppr prd, pol
Hexamethylenediamine (HMDA)	1	0.8	Adipic ac Butadiene/HCN AN/H_2	Am cmp of nylons
Hexamethylenetetramine (hexamine, hexa)	0.05	0.6	CH_2O/NH_3	Res: PF, UF, MF (40), chl agt (24), cyclonite (8), rbr rgt (8)
1,6-Hexanediol	0.01	1	Butadiene/CO/H_2	PUR, syn ltr
1-Hexene	0.02	0.3	Ethylene	LLDPE film
Hydroxyethylcellulose (HEC)	0.035	2.4	Alkali cel/EO or ethylene chlorohydrin	Vis add (45), saa (35), cos and phr (8)
Hydrogen cyanide	1.0	0.5	AN byp (30) Ntg/NH_3 (70)	Adiponitrile (40), Me methacrylate (35), cyanuric chloride (10), NaCN (5), chl agt, methionine
Inositol	0.0001	8	Corn steep liquor	Vitamin B complex
Ionones	0.0001	13–18	Citral (oil of lemon grass)	Frg, fla, vitamin A syn
Isoamylene	0.04	0.5	Ptr crk	Pla, Me *t*-amyl eth (MTAE)
Isobutane	1	0.14	LPG	Isobutylene
Isobutylene	0.76	0.28	Isobutane, ptr crk	MTBE, butyl rbr
Isobutyraldehyde	0.28	0.4	Propylene/oxo	Neopentyl glycol
Isoprene	0.13	0.23	Ptr ckg Isobutylene/CH_2O	P isoprene rbr
Isopropyl alc	1.3	0.25	Propylene	Sol (48), acetone (25), est, am
Lactose	0.1	0.2–0.6	Whey	Food add, phm
Lidocaine	0.0001	17	2,6-DiMe aniline/diEt-aminoacetic ac	Local anesthetic
Lignin sulfonates	0.85	0.2	Ppr ind byp	Saa, concrete add
Linear alkylbenzene (LAB)	0.55	0.45	AO/benzene Chloroparaffins/benzene	Sulfonate (LAS) saa (90)
rac-Lysine	0.008 0.025 imp	1.1	Cyclohexene/ONCl	Feed add
L-Lysine	0.03 0.009 imp	5.6	Frm	Food add
Maleic anh (MA)	0.37	0.55	Butane/O_2 Benzene/O_2	UPR (50), lub add (12), styrene copol (SMA) (6), agr, fumaric ac, malic ac, dodecenylsuccinic ac
Malic ac	0.015	0.8	MA	Food add (with aspartame), acidulant
Mannitol	0.03	3	HFCS/H_2	Phr, food add
Melamine	0.10	0.5	Urea	MF res
Melamine—formaldehyde res (MF)	0.2	0.6	Melamine/CH_2O	Cmp, mld res
2-Mercaptobenzothiazole	0.016	1.3	Aniline/CS_2	Rbr vulcanization accelerators

(*continued*)

TABLE I.   (*Continued*)

Material[a,f]	Production/demand[b] (B lb/yr)	Unit cost[c] ($/lb)	Source(s)[d,f] and precursor(s)	Major uses[e,f]
Methane (as chem. feedstock is 37% of total)	42	0.05	Ntg	NH_3, MeOH and carbon black (98); ethylene, acetylene, HCN, chlorinated drv
Methanol	8	0.08	Sng	CH_2O (30), MTBE fuel add (15), acetic ac/anh (12), sol (10), Me est, MeCl
Methionine (race-methionine)	0.1	0.87	Acrolein/MeSH/HCN	Feed and food add
Methylamines (di-, mono-, tri-)	0.2	0.55	Me alc/NH_3	Mono-: carbamate pst Di-: dimethylformamide (DMF), syn int Tri-: choline, quat
Methyl bromide	0.045	0.57	Me alc/HBr	Fum (80), syn int (10)
Methyl *t*-butyl ether (MTBE)	3.5	0.2	Me alc/isobutylene	Fuel add
Methyl chloride	0.4	0.26	Me alc/HCl	Silicone pol (72), MC (6), quat (5), methylating agt
Methyl ethyl ket (MEK)	0.55	0.36	*sec*-Butyl alc	Sol (80), lub oil ref
Methyl ethyl ket hydroperoxide	0.012	1.7	*sec*-Butyl alc/O_2	UPR cat
Methyl iodide	0.1	2.6	Me sulfate/NaI	Methylating agt
Methyl isobutyl ket (MIBK)	0.16	0.5	Acetone cnd prd/H_2	Sol
Methyl methacrylate (MMA)	0.85	0.52	Me alc/acetone/HCN	Pol (PMM)
Methylene chloride	0.6	0.35	CH_4/Cl_2 Me alc/Cl_2	Sol (40), aerosols (20), blg agt
4,4′-Methylenedianiline (MDA)	0.4	Capt use, 2	Aniline/CH_2O	MDI (96), epoxy res, dye int epoxy crg agt, dye int, eng pol
4,4′-Methylene diphenyl diisocyanate (MDI) and polymeric MDI (PMDI)	0.4	0.9	MDA/$COCl_2$ MDA/CH_2O/$COCl_2$ Nitrobenzene/CO/ $COCl_2$	PU
α-Methylstyrene	0.1	0.3	Cumene	Pol (ABS, UPR)
p-Methylstyrene	0.03	0.3	Ethylbenzene, Me alc	Pol
Monoethanolamine	0.13	0.4	EO/NH_3	Syn int, saa
Naphthalene	0.3	0.2–0.3	Coal tar (70) Ptr ref (30)	Phthalic anh (55), tetralin, 1-naphthol, Me 1-naphthylcarbamate, syn int
Naphthenic ac	0.035	0.2	Ptr rfg	Oil paint dryers, agr, cat, lub add, cor add, machine oils
1-Naphthol	0.02	1.8	Nph-SO_3H	Dyes, Me carbamate
2-Naphthol	0.01	1.1	Nph-SO_3H	Dyes
Neopentyl glycol	0.1	0.55	Isobutyrald/CH_2O	Alkyd res
Neoprene	0.02	1.3	Chloroprene	Rbr (70), exp (30)
Nitrile rbr	0.12	0.9–1	Butadiene, styrene, AN	Rbr prd (80), latex (20)
Nitrobenzene	1	0.3	Benzene/HNO_3	Aniline (97), syn int
Nitroparaffins	0.1		Propane/HNO_3	Syn int (CH_2O/H_2 prd)
Nitromethane		2.4		Fuel add
Nitroethane		2.5		
2-Nitropropane		0.55		
p-Nitrophenol	0.035	1.0	Phenol/HNO_3	Me and Et thiophosphates pst (85), APAP, syn int
Nonylphenol	0.15	0.5	Phenol/propylene trimer	Saa (65), antioxidant (25), lub add (10)
Nonylphenol, ethoxylated	0.24	0.5	Nonylphenol/EO	Saa
Nonylphenyl phosphite	0.011	0.8	Nonylphenol/PCl_3	Rbr vulcanization accelerator
Nylon 66 ⎫ Nylon 6 ⎬	2.4	1.1–2.0	Adipic ac/HMDA Caprolactam	Carpet prd (60), tex (20), tire cord, eng res
Octyl alc, see 2-ethyl-hexyl alc				
Oxalic ac	0.025 imp	0.45	EG/O_2 Sawdust/NaOH HCO_2Na/O_2	Tex trm, met cleaning

TABLE I. *(Continued)*

Material[a,f]	Production/demand[b] (B lb/yr)	Unit cost[c] ($/lb)	Source(s)[d,f] and precursor(s)	Major uses[e,f]
n-Paraffins			Ptr ref	LAB (70), linear alc (20), chlorinated paraffins, sol
C_6–C_{16}	2.1	0.2		
C_{10} plus	0.7	0.35		
Pentachlorophenol	0.05	0.55	Phenol/Cl_2	Wood prv
Pentaerythritol	0.12	0.7	Acetald/CH_2O	Alkyd res (60), syn lub est (10), tall oil est adh (8), tetranitrate explosive (4), acrylate est pol
Perchloroethylene	0.6	0.30	Ethylene/Cl_2	Dry cleaning and met trt (65), FC-113 syn (28)
Phenol	2.6	0.25	Cumene/O_2	PF (40), bisphenol A (22), caprolactam (18), 2,6-xylenol (6), alkylphenols (4), aniline, *o*- and *p*-cresol, salicylic and *p*-hydroxybenzoic ac
Phenol–formald res (PF)	2.6	0.3	Phenol/CH_2O/Hexa	Adh for construction boards (60), insulation prd (15), mld res (6), foundry core binders (5)
m-Phenylenediamine	0.01	2	*m*-Dinitrobenzene/H_2 ⎫	Aramid res with benzenedicarboxylic ac (*m*- or *p*-), P imide res
p-Phenylenediamine	0.01	4	*p*-Nitroacetanilide ⎬	
Phosgene (mostly capt use)	1.6	0.6	CO/Cl_2	TDI (50), MDI/PMDI (36), P carbonate (8), MeNCO (IC), org carbonates, chloroformates, syn int
Phthalic anh	0.9	0.3	*o*-Xylene Naphthalene	Pla (52), UPR (23), alkyd res (19), halogenated drv fire ret, phenolphthalein, syn int for dyes (2-chloroanthraquinone, quinizarin, rhodamines), anthraquinone, phthalonitrile, phthalimide
β-Pinene	0.04	0.9–1.9	Turpentine *α*-Pinene	Syn int
P acetal res	0.15	1		
P vinyl formal			PVA/CH_2O	Safety glass
P vinyl butyral			PVA/butyrald	Safety glass
P oxymethylene (POM)			CH_2O	Eng res
P acrylates	1.3	0.8–1.4	Me, Et, Bu est of acrylic/methacrylic ac Acrylates of di-, tri-, and tetraalc	Coatings and mld prd
P acrylic ac salts	0.02	0.7	Acrylamide/acrylic ac	Waste water trt
P acrylonitrile (PAN, acrylic fib)	0.65	1.1	AN	Tex for clothing (65), home prd (30)
P amide res				
Nylons	2.7	1.7	Adipic ac/HMDA; caprolactam	Tex fib (90)
Other	0.08	1.0	Rosin ac/NH_3	Hot-melt adh
(also see aramid pol)				
P butadiene rbr	0.7	0.7	Butadiene	Tires (75), high impact pol add (PS and others) (15)
P butenes	0.55	0.3	Isobutylene	Lub oil add (75), sealant (10), adh (4)
P carbonate	0.35	1.8	bisphenol A/$COCl_2$	Mfg of appliances, machines, auto cmp (45), sheets (30), exp
P ester pol				
Saturated				
Alkyd	0.7	0.6–1	Phthalic anh, isophthalic ac, glycerine, pentaerythritol, neopentyl and other glycols	Coatings
Terephthalates				
P ethylene (PET)	3.3	0.8	PTA/DMT/EG	Tex fib (60), film (20), bottles (15), exp (5)

(continued)

TABLE I.   (*Continued*)

Material[a,f]	Production/demand[b] (B lb/yr)	Unit cost[c] ($/lb)	Source(s)[d,f] and precursor(s)	Major uses[e,f]
P butylene (PBT)	0.1	1	PTA/DMT, 1,4-butanediol	Eng pol: mld ind prd
Unsaturated (UPR)	1.2	0.4–0.5	Alkyd/MA/styrene/ styrene drv/fatty acids	Boat, cpt (60), coatings
P ethylene	14			
High density (HDPE)	6	0.4–0.5	Ethylene	Blw (40), ext (30), inj (25) mld prd; chlorinated PE (CPE)
Low density (LDPE) (*d* < 0.94)	8	0.35–0.4		
LDPE/LLDPE (60:40)			Ethylene/1-butene or 1-hexene	Pck film (60)
P ethylene/propylene glycols	1.5	0.6	EO/PO	PU cmp, mono-, diesters saa, phr, csm
P ethylene/propylene (see ethylene/propylene pol)				
P ethylene/vinyl acetate (VAM) and ethylene/ vinyl alc (EVOH)	0.1	0.5	Ethylene/VAM	Adh, barrier film pck
P methyl methacrylate (PMM); (see also P acrylates)	0.4	1	Me methacrylate (MMA)	Sheets, ext prd
P phenylene sulfide (PPS)	0.01	2	*p*-Dichlorobenzene/ Na$_2$S	Eng res, cmp
P propylene (PP)				
Isotactic	5	0.45	Propylene	Inj mld prd
Atactic, amorphous (APP)	0.2	0.3	PP byp	Roofing, pck, adh
P styrene (PS)	4.2			
Straight	2	0.4	Styrene	Mfg ind and dom prd
High impact (PS/rbr)	1.2	0.4	Styrene/rbr blend	Mfg ind and dom prd
Expandable (EPS)	0.6	0.7	Styrene beads	Pck
AN modified (SAN) (also see SBR, ABS)	0.4	0.7	Styrene/AN	Mfg ind and dom prd
P terpene res	0.04	0.7	Pinenes	Adh
P tetramethylene glycol (PTMG)		1.4	THF	PU cmp
P vinyl alc (PVA)	0.18	1.0	PVAc	Tex trt (36), P vinyl acetals (28), adh (18), saa (9)
P vinyl acetate (PVAc)	0.5	0.6	VAM	Ctg
P vinyl chloride (PVC)	7	0.4–0.5	VCM/pla	Ext prd (65), calendered sheet/film (10), mld prd (5), chlorinated prd (CPVC)
P vinylidene chloride (PVDC)	0.03	0.8	Vinylidene chloride	Film co-pol, co-ext pck prd
P urethane (PU)				
Plastics	2	1	TDI/MDI/etc./polyols	Flexible foams (65): bedding, furniture Rigid foams (35): insulation, pck, mfg, coatings
Elastomers	0.2	2	TDI/MDI/polyols block pol	Spandex fbr
Propane	8.5	0.12	LPG	Fuel, ethylene, propylene, nitroparaffins
Propionic ac	0.11	0.3	Ethylene/CO/H$_2$O	Agr (30), Na, Ca salts feed/food prv (20), cel est (20)
Propiophenone	0.01	2	Benzene/propionic ac	Syn int
Propyl *p*-hydroxybenzoate (propylparaben)	0.001	5	Propyl alc/*p*-hydroxybenzoic ac	Antioxidant

TABLE I.   (*Continued*)

Material[a,f]	Production/demand[b] (B lb/yr)	Unit cost[c] ($/lb)	Source(s)[d,f] and precursor(s)	Major uses[e,f]
Propylene	15	0.16	LPG ckg	PP (34), AN (17), PO (12), cumene (8), oxo prd (7), IPA (6), allyl chloride
Propylene glycol (PG)	0.5	0.4	PO	UPR (45), PU polyols, pet food, phr prd, tobacco and cellophane add
Propylene oxide (PO)	2.1	0.47	Propylene/hydroperoxides or Cl_2/H_2)	Polyether polyols for PU (60), PG and oligomers (24), amino alc
Rayon	0.4	0.9	Chm cellulose	Txt fbr
Ricinoleic ac (see castor oil)				
Rosin res (also see P amide res)	0.4	0.4–0.6	Tall oil distillation residue/polyols	Coatings, adh
Salicylic ac	0.045	1.4	Phenol/CO_2	Aspirin (45), est and salts (18), foundry res (10), syn int
Silicones (P siloxanes)	0.1	2–5	Me and phenyl chlorosilanes	Fluids, lub oils (90), res and rbr (10)
Sodium methoxide or methylate	0.01	0.6	Me alc/Na	Syn rgt
Sorbitol	0.27	0.3–0.7	Glucose/H_2	Toothpaste (35), food/phr add (28), ascorbic ac (20), saa
Stearic ac salts other than Na, K	0.1	0.6–1.2	Ac and inorganic salts	Lub add, high pressure greases, csm
Styrene	7.5	0.2	Ethylbenzene	PS (52), exp (16), SBR (8), UPR (6), ABS (10)
Styrene–butadiene				
Rbr (SBR)	1.8	0.35	Styrene, butadiene, carbon black	Tires (72), adh, sealants, footwear
Latex	0.3	0.3	Styrene/butadiene, H_2O	Coatings
Sucrose	5	0.3 (U.S.) 0.05 (World)	Sugar cane and beets	Food (99), syn int
Octaacetate		5	Sucrose/acetic anh	Denaturant for ethanol
Acetate isobutyrate		0.6	Sucrose, acetic anh and butyric ac	Coatings, adh
P ether polyols		0.8	Sucrose/EO/PO/butylene oxide	PU polyols
Tall oil, crude (CTO) nonfuel use	1.7	0.09	Ppr ind byp	Fatty ac (28), rosin (27)
Refined		0.3	CTO fractionation	
fatty ac (TOFA)	0.48	0.3		Phosphate beneficiation, alkyds, dimer ac, saa for rbr prd, cor inhibitor, epoxy and asphalt add
rosin	0.45	0.4		Ppr size, saa
Tetraethyl/methyl lead	0.1	1	Et or Me chloride/PbNa	Octane number enhancement (69), exp
Tetrahydrofuran (THF)	0.12	1	1,4-Butanediol Furfuraldehyde	PTMG
Theophylline	0.00001	5.5	N,N-DiMe urea/Et cyanoacetate	Smooth-muscle relaxant
Thioglycolic ac	0.05	2	Chloroacetic ac/NaSH	Permanent hair wave prd (40), glycerine monoester
Toluene	5	0.2	Ptr rfg	Fuel add (50), hydrodealkylation (HDA) to benzene (25), sol (15), TDI, TNT, benzyl chloride, syn int
Toluene or tolylene diisocyanate (TDI)	0.64	1.0	2,4- and 2,6-Diaminotoluene/$COCl_2$	PU (99): flexible foams (88), coatings (5), rbr (4)
1,1,1-Trichloroethane, methyl chloroform (TCEA)	0.6	0.4	VCM/HCl, then Cl_2	Cleaning agt (70), adh sol (10), syn int
Triethanolamine	0.1	0.5	EO/NH_3	Saa, spn of ac gases

(*continued*)

TABLE I. (*Continued*)

Material[a,f]	Production/demand[b] (B lb/yr)	Unit cost[c] ($/lb)	Source(s)[d,f] and precursor(s)	Major uses[e,f]
Trichloroethylene (TCE)	0.15	0.38	1,1,2,2-Tetrachloroe-thane/CaO	Dry cleaning agt (80), syn int
Triethylene glycol	0.12	0.5	EG/EO	Ntg drying (55), sol (10), P ester res, PU, pla
Urea	14	0.1	NH_3/CO_2	Frt (60), exp (15), feed (10), UF (6), melamine (1), syn int
Urea–formaldehyde res (UF)	1.2	0.2	Urea/CH_2O	Mld prd, insulating foam
Vanillin	0.006	6	Vanilla beans Lignin	Flv, frg add
Vetiveryl acetate	0.0007	27–63	Vetiver oil imp	Frg
Vinyl acetate monomer (VAM)	2	0.4	Ethylene/acetic ac/O_2	PVAc (40), PVA (15), co-pol with ethylene (EVAC) or VCM (12), P vinyl acetals (5)
Vinyl chloride monomer (VCM)	7	0.3	Ethylene/HCl/O_2	PVC (85), co-pol with VAM, VDC; CPVC, mfg of pipe, plumbing cmp
Vitamins, total	0.04	4–11,000	Nat sources or syn	Food add
Vitamin E (α-tocopherol) [see ascorbic ac (vitamin C)]	0.009	17	Pseudoionone	
Xylenes, mixed	5	0.13	Ptr rfg	Sol, fuel, isomer spn
o-Xylene	0.8	0.13	Xylene mix: spn/ isomerization	Phthalic ac (99)
m-Xylene	0.2	0.36		Isophthalic ac
p-Xylene	4	0.2		DMT/PTA (95), aramid pol, sol, exp
2,6-Xylenol	0.0025	1	Phenol/Me alc	P phenylene oxide pol

[a] Carbon dioxide, carbon disulfide, carbon black, and hydrogen cyanide are included in this table, but salts of carbonic acid are not. Acronyms/abbreviations are shown in parentheses and not explained elsewhere.

[b] Demand on merchant market is shown, rather than production capacity, and may include some exports. Only significant imports are shown, and captive use is usually excluded.

[c] Unit prices are based on list or spot prices published during the early part of 1986 and do not take into consideration discounts or "temporary, voluntary allowances" (TVAs). Price ranges depend on degree of purification. The drop in the pricing of crude petroleum and natural gas experienced during the early part of 1986 affects mostly the prices of commodity chemicals.

[d] Commas separate alternative sources or precursors while slashes (/) separate reactants in a single process.

[e] Percentages of given uses are shown in parentheses. Only significant exports are listed.

[f] The following abbreviations are employed in this table: ac = acid(s), acidic; add = additive; adh = adhesive, glue, cement, mucilage, tackifier; agr = agrichemical(s); agt = agent(s); alc = alcohol(s); ald = aldehyde(s); am = amine(s); anh = anhydride(s); apl = application; aut = automotive; avg = average; blw = blow, blowing; Bu = butyl; b. wt. = by weight; byp = byproduct(s); capt = captive; cat = catalyst(s), catalyzed; cel = cellulose; chl = chelating; chm = chemical(s); cmp = component(s); cpt = composite(s); cnd = condensation; cor = corrosion; crg = curing; CRG = carcinogen, proven or suspected; crk = cracking; csm = cosmetic(s); ctg = coatings; dev = device(s); dom = domestic; drv = derivative(s); elt = electric; eln = electronic; eng = engineering; enz = enzyme(s), enzymatic; est = ester(s); Et = ethyl; eth = ether(s); exp = exports; ext = extrusion; FD&C = food, drug, cosmetic dye(s) approved by the FDA; fib = fiber(s); fin = finishing; flt = flotation; flv = flavor(ing); frg = fragrance; frm = fermentation(s); fum = fumigant; HDA = hydrodealkylation; hgh = high; imp = imports; ind = industry(ies), industrial; inj = injection; int = intermediate(s); ket = ketone(s); LPG = liquefiable (liquefied) petroleum gases; ltr = leather; lub = lubricant(s), lubricating; Me = methyl; med = medicinal; met = metal(s); mfg = manufacturing, manufacture; mix = mixture(s); mld = molding, molded; nat = natural; nph = naphthalene; ntg = natural gas; onp = on purpose (production); org = organic; oxo = hydroformylation (process, product); P = poly-; pck = packaging; phr = pharmaceutical; pla = plastic(s); pol = polymer(s), polymerization; ppr = paper; Pr = propyl; prd = product(s), production; prf = performance; prv = preservative; pst = pesticide; ptr = petroleum; quat = quaternary ammonium compound(s); rbr = rubber, elastomer; rct = reaction(s); res = resin(s); ret = retardant; rfd = refined; rfg = refrigerant; rgt = reagent, reactant; rnr = renewable resource(s); saa = surface active agent, emulsifer, syndet, detergent; sng = synthesis gas (CO/H_2); sol = solvent(s); spn = separation; syn = synthesis, synthetic; tex = textile; trt = treatment; vis = viscosity.

[g] Contractural arrangements may result in lower unit cost.

[h] Demand is equivalent to $0.185 B, and unit cost is reported in terms of enzymatic activity units.

TABLE II.   Categories of Industrial Chemicals

Category	Demand[a] approximate range (lb/yr)	Unit cost[b] approximate range ($/lb)
I. Commodity chemicals	10^9–10^{10}	0.1–0.75
II. Pseudo-commodity chemicals	10^7–10^8	0.75–2
III. Fine chemicals	10^5–10^7	2–10^2 plus[c]
IV. High-technology, proprietory, specialty materials	10^5–75×10^6	5–10^3
V. Traditional specialty products	up to 10×10^6	1–100

[a] Demand figures apply to U.S. merchant market.
[b] Unit cost data refer to early 1986. The unit costs of commodity and pseudo-commodity polymers are only somewhat higher than those of the monomers unless complex polymerization processes are involved.
[c] Unit costs are a function of the nature of multistep synthetic processes, costs of raw materials and reagents, and the patent position of producers.

demand by industry, government, or the consumer market for problem-solving or "enabling" materials. Examples of such materials are high-performance adhesives, sealants, lubricants, dispersing, absorption, electronic, and many other types of materials that can be referred to as high-technology specialties. These include high-temperature-resistant, high-strength, electricity-conducting, and other special polymers. Finally, category V refers to traditional, often complex formulations of chemicals produced industrially or isolated from natural sources in order that they may function in a desired fashion in countless applications. Examples of these materials are industrial and domestic cleansing and sanitizing products, personal-care products, traditional glues, inks, paints, and other functional materials such as metal-working, brake, compressor, and hydraulic fluids, and so on. It is noteworthy that the recent trend exhibited by the chemical industry of the United States, Western Europe, and Japan to focus on the production of "specialty products" refers to materials of category IV, as shown in Table III. [*See* PHARMACEUTICALS.]

These categories of industrial organic materials differ in capital investment, scientific and technological requirements, profit margins, production characteristics, marketing approaches, and so on, but a detailed analysis of these differences is beyond the scope of this article.

TABLE III.   Categories of High-Technology Specialty Products[a]

Adhesives and sealants
Agrichemicals
Antioxidants and antiozonants
Antistatics
Biocides
Bioengineering therapeutics:
    insulin, human growth hormone, interferon, and so on
Business machine materials
Catalysts, including enzymes
Cleaners, industrial and institutional
Coal and fuel additives:
    gasoline, diesel, and jet fuels
Coatings
Composites, fiber-reinforced[b]
Conducting materials
Corrosion inhibitors
Cosmetic ingredients
Defoamers
Diagnostic chemicals, including immunoassay reagents[b]
Dispersants
Dyes
Elastomers, high-performance
Electronic materials[b]:
    semiconductors, printed circuit materials, ultra-pure reagents
Fire retardants
Flavors and fragrances
Flocculants
Flotation agents

(continued)

TABLE III. (*Continued*)

Food additives
Foundry chemicals
Herbicides
Laboratory chemicals
Laundry additives:
 bleaches, softeners, antistatics, spot removers
Leather treatment chemicals
Lubricant additives
Lubricants, synthetic[b]
Magnetic materials
Membranes, synthetic
Metal-plating and -finishing chemicals
Microencapsulation systems
Mineral surface-treatment agents
Mining chemicals
Oil-field chemicals:
 drilling-mud additives, oil-recovery agents
Paint additives:
 leveling, suspending, thixotropic agents
Paper additives
Pesticides
Pharmaceuticals
Phase transfer agents
Photographic chemicals
Photovoltaic chemicals[b]
Pigments, organic
Plasticizers
Plastic stabilizers and reinforcing materials
Polymers, high-performance[b]
Printing materials
Radiator and heat-transfer fluids
Refinery and pipeline chemicals
Rubber processing chemicals
Surfactants, emulsifiers
Textile chemicals
Thickeners
Ultraviolet light absorbers
Waste-water treatment materials
Water-boiler chemicals
Water-management chemicals
Wood preservatives

[a] Sales in the United States during 1985 are estimated at $45 B.
[b] Categories of highest annual growth.

II. Origins

In excess of 80% of industrial organic chemicals are currently derived from petroleum and natural gas.

Crude petroleum is subjected to a variety of physical and chemical processes that include fractionation and thermal and catalytic decomposition reactions (cracking, hydrocracking, reforming, and so on) that degrade high-molecular-weight components to molecules suitable for use as gasoline. Ethylene, propylene, butylenes, and also the three most important aromatic hydrocarbons—benzene, toluene, and the xylenes (the BTX compounds)—are formed in the course of the refining operations. These low-molecular-weight hydrocarbons constitute the largest source of industrial organic chemicals, even though some are reassembled to give a high-quality alkylate gasoline, or, as is the case with BTX, they contribute directly to high-octane gasoline.

Methane is the principal hydrocarbon component of natural gas, and it is separated to become the main constituent of natural gas employed as a fuel. The propane and butane components of natural gas are readily isolated as liquefiable petroleum gases (LPG), and these hydrocarbons, as well as ethane, are also chemical feedstocks apart of their use as fuels.

Thus, as is evident from the demand/consumption figures mentioned in Table I, methane, ethylene, propylene, and the BTX compounds, together with some additional cracking products such as butylenes and 1,3-butadiene, represent about 118 billion (B) lb of major starting materials for the production of industrial organic compounds derived from petroleum and natural gas. The USITC reports a minimum of about 215 B lb of synthetic organic chemicals derived in 1983 from these basic building blocks, and records about 173 B lb of inter-company sales valued at about $60 B. Large as these figures are, they are dwarfed by the fuel requirements of the United States.

For some 50 years the major chemical uses of methane were the production of its chlorinated derivatives, carbon black, and its transformation [Eq. (1)] to a mixture of carbon monoxide and hydrogen (synthesis gas or syngas).

$$CH_4 + H_2O \rightarrow CO + 3H_2 \qquad (1)$$

(In this and subsequent equations, the catalyst, temperature, and pressure requirements of the processes are ignored). The availability of synthesis gas stimulated the industrial production [Eq. (2)] of methyl alcohol in the United States. The conditions of the methanol process have been improved over the years, and currently many of the natural-gas-rich countries have embarked on large-scale methyl alcohol ventures.

$$CO + 2H_2 \rightarrow CH_3OH \qquad (2)$$

Consequently, methanol has recently assumed a more important role as a chemical intermediate than its traditional use as a precursor of formaldehyde and methyl esters and ethers. The reaction of methanol with carbon monoxide (carbonylation) competes with older synthetic routes to acetic acid [Eq. (3)], and the reaction of carbon monoxide with methyl acetate (formed *in situ* from acetic acid and methanol) gave rise in 1983 to a 500-MM-lb facility that produces acetic anhydride [Eq. (4)].

$$CO + CH_3OH \rightarrow CH_3CO_2H \quad (3)$$

$$CO + CH_3COOCH_3 \rightarrow (CH_3CO)_2O \quad (4)$$

It is of interest to note that the last-mentioned process employs gasification of coal [represented simplistically by Eq. (5)] as the source of carbon monoxide.

$$C + H_2O \rightarrow CO + H_2 \quad (5)$$

The current world methanol capacity of about 40 B lb begs its uses in the fuel market. On an experimental basis, methanol is utilized directly as a gasoline substitute or additive, it is being converted in New Zealand to a synthetic gasoline [Eq. (6)], and in the United States the production of the octane-enhancing methyl *t*-butyl ether (MTBE) is on the rise [Eq. (7)].

$$n(CH_3OH) \rightarrow C_nH_{2n} + nH_2O \quad (6)$$

where C_nH_{2n} represents a hydrocarbon mixture suitable for direct use as a high-performance gasoline:

$$CH_3OH + CH{=}C(CH_3)_2 \rightarrow CH_3OC(CH_3)_3 \quad (7)$$

Synthesis gas with an appropriately adjusted ratio of carbon monoxide and hydrogen is used in the United States to give approximately 2 B lb of hydroformylation (or "oxo") products from various olefins in which the carbon–carbon double bond is located in a terminal position. This transformation is illustrated by means of propylene, which produces *n*-butyraldehyde and isobutyraldehyde in a ratio that is a function of the catalyst [Eq. (8)]. The presence of additional hydrogen leads to the formation of the corresponding alcohols [Eq. (9)].

$$CO + H_2 + CH_3CH{=}CH_2 \rightarrow$$
$$CH_3CH_2CH_2CHO + (CH_3)CHCHO \quad (8)$$

$$CO + 2H_2 + CH_3CH{=}CH_2 \rightarrow$$
$$CH_3CH_2CH_2CH_2OH + (CH_3)CHCH_2OH \quad (9)$$

The carbon monoxide component of synthesis gas is a very versatile reactant. In addition to the acetic acid and acetic anhydride processes already mentioned, other synthetic applications of industrial importance are the processes that give phosgene [Eq. (10)], propionic acid [Eq. (11)], and formic acid [Eq. (12)].

$$CO + Cl_2 \rightarrow ClCOCl \quad (10)$$

$$CO + CH_2{=}CH_2 + H_2O \rightarrow CH_3CH_2CO_2H \quad (11)$$

$$CO + H_2O \rightarrow HCO_2H \quad (12)$$

The last-mentioned process actually employs sodium hydroxide or methyl alcohol in place of water.

Hydrogen makes an important contribution to the production of industrial organic chemicals by virtue of various hydrogenation processes (benzene to cyclohexane, phenol to cyclohexanol, acetone condensation product to methyl isobutyl ketone, esters of fatty acids to fatty alcohols, olefins to the corresponding saturated hydrocarbons, and so on), and by being the key stepping stone to the production of ammonia. In order to increase the yield of hydrogen from synthesis gas, the latter can be subjected to the water-gas-shift reaction, in which carbon monoxide and steam give carbon dioxide and hydrogen.

Ammonia, in turn, is responsible for the production of urea [Eq. (13)], acrylonitrile [Eq. (14)], triethanolamine [Eq. (15)] and other alkanolamines, aliphatic amines such as trimethylamine [Eq. (16)], and many other nitrogen-containing industrial organic chemicals. Even aniline is currently obtained by the ammination of phenol [Eq. (17)].

$$2NH_3 + CO_2 \rightarrow (H_2N)_2CO \quad (13)$$

$$CH_2{=}CHCH_3 + NH_3 + O_2 \rightarrow CH_2{=}CHCN \quad (14)$$

$$3(CH_2CH_2)O + NH_3 \rightarrow (HOCH_2CH_2)_3N \quad (15)$$

$$3CH_3OH + NH_3 \rightarrow (CH_3)_3N \quad (16)$$

$$(17)$$

Since nitric acid is obtained by the oxidation of ammonia, the origin of such an important chemical as TDI (see Table I), and other aromatic nitro compounds and their derivatives, can be traced back to ammonia and hydrogen.

While most of the large production of urea is destined for use as fertilizer and cattle feed additive, significant amounts serve to produce melamine (2,4,6-triamino-1,3,5-triazine) and its formaldehyde-derived resins (see Table I).

About 3.6 B lb of coal tar—the by-product of the conversion of coal to coke required in the production of steel—is distilled to yield a relatively small fraction of the BTX demand. On the other hand, coal tar is the major source of cresols (cresylic acid) and naphthalene. The gasification of coal mentioned above and other coal conversion processes will surely regain the high-priority attention they achieved in the United States during the decade following the 1973 oil embargo, once the interlude of currently low crude petroleum prices fades into oblivion.

Biomass sources contribute about 12 B lb of chemical feedstocks and intermediates to the production of industrial organic chemicals. This quantity includes the products of biological processes in which the chemical transformations are carried out by microorganisms and higher organisms, or by enzymes isolated from them. The estimate is based on the nonfood and non-fuel consumption of natural rubber; cellulose and its derivatives; fatty acids and glycerine derived from fats and oils; fatty acids and rosin acids obtained by distillation of tall oil; epoxidized soya oil; imported castor oil; starch and its derivatives (obtained primarily from corn) such as glucose, sorbitol, gluconic acid; lactose, furfuraldehyde, lignin and its derivatives, terpenes, polyterpenes, fermentation products such as ethanol and citric acid, and numerous natural products such as essential oils, gums, alkaloids, and so on. It is difficult to estimate precisely the quantitative input of biomass to the total production of industrial organic chemicals because, in the final analysis, industry employs indiscriminately raw materials and intermediates from whatever origin as long as their use is technically convenient and economically prudent. Thus, for example, the production of a cellulose-derived gum like carboxymethyl cellulose (CMC) (see Table I) draws on renewable sources for cellulose (cotton linters or forestry product), and fossil and inorganic sources for the sodium chloroacetate intermediate. The same is especially true in the case of hundreds of commercially significant soaps, detergents, emulsifiers, and other surface-active agents that may be assembled, for example, from a fatty alcohol (obtained from a fatty-acid ester and hydrogen), ethylene oxide derived from petroleum or natural gas and oxygen, and sulfuric acid.

Finally, in addition to the major organic feedstocks already mentioned, we cannot overlook the contribution of inorganic reactants that become incorporated in the structures of industrial organic chemicals, and that function as catalysts, acids or bases. Hydrogen, the halogens (especially chlorine and fluorine, and to a lesser extent bromine and iodine), and ammonia have already been referred to, but the foremost contributor is oxygen used, whenever convenient as air, or in the form of water. It is noteworthy that the most important building blocks derived from petroleum and natural gas are devoid of oxygen, and the oxygen functional groups are introduced most economically by means of molecular oxygen to give such large-volume chemicals as ethylene oxide, phthalic anhydride, terephthalic acid, acetone and phenol (from cumene), acrylic acid, and others. On the other hand, oxygen-containing functional groups introduced by means of carbon monoxide obtained from natural gas are derived from water.

III. Contribution to the Economy

As mentioned in Section I of this article, industrial organic chemicals play an enabling role that affects most manufacturing industries and the consumer market place. It is important to realize that their production exerts a ripple effect that is felt throughout the whole economy. For this reason it is difficult to determine precisely their quantitative contribution, and the results of an analysis depend on the classifications used to distinguish different economic sectors. If we accept the arbitrary definition of the chemical industry to include not only the production of industrial chemicals and synthetic materials, but also pharmaceuticals; detergents; sanitizing, cleansing, and polishing preparations; toiletries and cosmetics; paints and coatings; fertilizers and pesticides; printing inks and carbon black; and adhesives, additives and catalysts, then we arrive at a value of about $214 B for shipments of these products during 1985. This amount represents about 6.7% of the gross national product (GNP). A broader picture of the economic impact of chemicals on the economy is given by the so-called chemical process industries (CPIs), which, again arbitrarily, embrace not only the preceding chemicals and allied products, but also the output of pulp, paper, and paperboard; the processing of petroleum and natural gas; the production of rubber and plastics, as well as of stone, glass, and clay products, and primary nonferrous metals; sugar refining, wet corn milling, and the processing of foods and beverages; textile dyeing and finishing; leather tanning; the manufacture of dry cells, storage batteries and semiconductor materials, carbon and graphite products, and hard-surface floor coverings. Now we arrive at 1985 value of shipments of about $649 B, or 20.5% of the GNP. The justification of attributing the dominant role to industrial organic materials in the output of the CPIs is found in Table IV, which identifies the nature of the different categories of carbonaceous materials and also illustrates the economic benefits that result from the chemical transformations of relatively cheap feedstocks to higher value-added products. It is evident that most of the categories of the CPIs are constituted by organic chemicals or that the latter are instrumental in the processing of the inorganics (for example, as flotation agents, in reflective glass coatings, in safety-glass composites, and as organic pigments in glass or ceramic products).

TABLE IV. Some Sectors of the CPIs

Petroleum refining and natural-gas fractionation
Petrochemical production
Chemicals
Polymers (plastics and rubber)
Coal products
Carbon and graphite products
Carbon black
Inks and other printing materials
Pulp and paper products
Fats and oils
Soap and other surface-active agents
Fertilizers and agrichemicals
Food and beverages
Milk products
Wet corn milling and refining
Sugar refining
Processed food
Glue and gelatin
Leather tanning and finishing
Synthetic leather, oilcloth, and other coated fabrics
Textile fibers: natural and synthetic
Textile dying and finishing
Medicinals
Toiletries and cosmetics
Paints and other coatings
Explosives and ammunitions
Stone, clay, lime, and glass
Cement, ceramic products, tiles, refractories, and
  abrasives
Metallurgical products
  Mining chemicals
  Nonferrous metals
  Ferrous products
  Electroplating, plating, anodizing, polishing of
    metallic products

While the employment in the chemicals and allied products hovers somewhat above the 1 million mark, employment in the CPIs reaches about 4 million people, or about 20% of employment in all manufacturing activities of the United States. Above and beyond all of this, we must recognize that most all other industrial sectors that are not included in the CPIs are affected directly by the availability and performance of chemicals. Thus, for example, progress in reaction injection molding (RIM) of appropriately designed and formulated polymers and the use of fiber-reinforced composites based on unsaturated polyester and epoxy resins (see Table I) are revolutionizing the automotive and other assembly-type manufacturing operations; high-performance plastics enable the construction of lighter and thus more efficient airplanes; new diagnostic, medicinal, and biomedical engineering materials are affecting profoundly the health-care industry; the exponential growth of the microelectronic industry depends to a great extent on the development of superior "electronic chemical specialties"; and the manufacture of progressively superior copying and photographic devices is a function of new chemicals and chemical processes. Again, organic industrial chemicals play the leading role in all of these interactions with the economy, except in such areas as high-performance ceramics, glass fiber optics, germanium–arsenic semiconductors, and metal-based catalysts.

Another way to gauge the economic impact of the chemicals industry is to examine the capital spending for construction of new and modernization of older production facilities, and the expenditures for research and development (R&D) of improved products and processes. During 1985, out of the total capital spending of about $153 B by all manufacturing sectors, the chemicals and allied products industry alone contributed about $16.5 B (or nearly 11%), while all the CPIs accounted for about $43.5 B (or about 28.5%). Similarly, out of the total of $77 B spent on R&D by all manufacturing sectors, the CPIs accounted for $15 B (or 19.5%) and the chemicals industry alone for about $9 B (or 11.5%) of the total. Only the aerospace and machinery manufacturing sectors exceeded the R&D expenditures of the chemicals industry.

Finally, the chemicals industry is one of the few industrial sectors of the economy of the United States that shows a positive trade balance of about $8 B per year. However, the apparently continuous rise in imports of chemicals (currently amounting to about \$14 B) threatens this situation. While organic chemicals constitute over half of the total chemical exports, it is disturbing to note, for example, that the imports of over 3000 chemicals of the benzenoid family increased between 1982 and 1983 from about 681 MM lb to about 2.1 B lb. The fact that the average unit price of these imports decreased from about \$1.76/lb to \$0.58/lb reflects an increase in the imports of commodity chemicals from petroleum-rich countries like Venezuela, but is is obvious that the great majority of the 3000 chemicals belongs to the category of fine chemicals that could be produced domestically.

BIBLIOGRAPHY

Callanan, N. (1980). Who keeps score for the chemical industry. *Chem. Business* March: 31–38.

Meyers, R. A. (1986). "Handbook of Chemical Production Processing." McGraw-Hill, New York.

Szmant, H. H. (1984). Chemistry plus. *CHEMTECH* October:598.

Szmant, H. H. (1986). "Industrial Utilization of Renewable Resources." Technomic Publishing, Lancaster, Pa.

Szmant, H. H. (1987). "Organic Building Blocks of the Chemical Industry." Wiley, New York, in press.

United States International Trade Commission, U.S. Government Printing Office, Washington, D.C.:

"Synthetic Organic Chemicals: United States Production and Sales, 1983," USITC publication 1588, 1984.

Summary of Trade and Tariff Information Reports:

Soap, Detergents, and Surface-Active Agents, March 1983.

Plasticizers, Explosives, and Certain Inorganic Pigments and Pigment-Like Materials, June 1983.

Drugs and Related Products, May 1983.

Certain Fatty Substances; Certain Natural Chemicals and Chemical Products; and Radioactive Elements, Compounds, and Isotopes, and Nonradioactive Isotopes and their Compounds, July 1984.

Certain Organic Chemical Crudes, July 1984.

Miscellaneous Chemicals and Chemical Products, August 1984.

Flavors, Odoriferous Compounds, Perfumery, Cosmetics, and Toilet Preparations, August 1984.

Benzenoid Cyclic Intermediates—Commodity Chemicals, September 1984.

ORGANIC CHEMISTRY: COMPOUND DETECTION

Raphael Ikan
Bernard Crammer   *Hebrew University of Jerusalem*

GLOSSARY

Chromatography: Method of separating two or more substances by distribution between two phases, one fixed (the stationary phase) and the other moving (the mobile phase).

Chromatography of isomers: Chromatographic separation of geometric (cis-trans) and optical (*R* and *S*) enantiomers on adsorbents that have been impregnated with compounds having the ability to complex preferentially or interact with specific functional groups causing the required separation.

Coupled chromatographic and spectroscopic techniques: Method involving a chromatographic system such as GLC or HPLC connected to a mass spectrometer or FTIR spectrometer in order to analyze individual components from a mixture of organic compounds in minute (submilligram) quantities.

Deuterium exchange: Replacement of hydrogen atoms by deuterium atoms usually by means of active compounds containing deuterium such as D_2O and $NaBD_4$. The percentage of deuterium exchange may be determined by spectroscopic analysis.

Droplet countercurrent chromatography (DCCC): Separation technique based on liquid–liquid partition chromatography.

Flash chromatography: Also known as rapid column chromatography.

Fourier transform: Technique in which a short powerful radio-frequency pulse (microseconds) excites either all the ^1H nuclei or ^{13}C nuclei simultaneously. Each nucleus shows a free induction decay (FID) which is an exponentially decaying sine wave with a frequency equal to the difference between the applied frequency and the resonance frequency for that nucleus.

Gas–liquid chromatography (GLC): Technique in which the organic sample is carried through a column by a carrier gas (mobile phase) and the separation of the organic compounds occurs in the stationary phase (the column packing). The compounds are estimated by means of a detector.

Gas–solid chromatography: Technique in which the moving phase is a mixture of gases and the stationary phase is a solid phase. The carrier gas such as nitrogen or helium replaces the solvent in column chromatography. The solid may be finely powdered Celite or kieselguhr. The technique is suitable for organic substances that are volatile without decomposition up to about 300°C.

Gel permeation chromatography (GPC): Technique that separates substances according to their molecular size and shape. Three classes of stationary phases are used: aerogels (porous glass), xerogels (crosslinked dextran), and xerogel-aerogels (polystyrene).

High performance thin-layer chromatography (HPTLC): Technique enabling the separation of very complex mixtures of organic compounds. The plates are prepared from optimized thin adsorbent layers. HPTLC offers greater separation efficiency through smaller plate heights than the conventional TLC plates, shorter analysis time, and detection limits in the nanogram and picogram range.

High pressure liquid chromatography (HPLC): Technique consisting of a stationary phase (a solid surface, a liquid, an ion-exchange resin, or a porous polymer), held in a glass or metal column with the liquid mobile phase being forced through under pressure.

Infrared spectroscopy: Technique in which many functional groups and atoms are characterized by their vibrations and deformations in the 4000–200 cm^{-1} range.

Ion-exchange chromatography (IEC): Technique in which the stationary phase consists of a rigid matrix (polymer), the surface of which carries a net positive (cationic) or negative (anionic) charge to give an ion exchange site R$^+$ or Y$^-$, respectively, which will attract and hold the counterions. Ion exchangers are divided into *anion* and *cation* exchangers.

Liquid–solid (absorption) chromatography (LSC): Separation of compounds (from a mixture) by a liquid mobile phase and a solid stationary phase which reversibly absorbs the solute molecules.

Mass spectrometry: Technique in which a vaporized sample of a substance is bombarded with a beam of electrons, and the relative abundance of the resulting positively charged molecular fragments is determined. The relative abundance versus mass-to-charge ratio that is produced from the substance by the mass spectrometer is called the mass spectrum of the substance.

Nuclear magnetic resonance spectroscopy: Measures the absorption of light energy in the radio-frequency portion of the electromagnetic spectrum. ^{1}H NMR spectroscopy furnishes indirect information about the carbon skeleton of organic molecules. In ^{13}C NMR peaks corresponding to all carbon atoms are recorded.

Paper chromatography (PC): Technique in which the fixed phase is a sheet of filter paper. The sample is placed near the edge of the paper as a small spot. The edge is then dipped in the developing solvent (mobile phase). The solvent rises up the paper by capillary action taking the substance along. The positions of the spots are observed by visible or UV light or by spraying with a chromogen.

Reaction GLC: Certain chemical reactions of organic compounds (such as reduction, oxidation, dehydration) that take place in a gas–liquid chromatographic column. The products are detected by means of the conventional detectors.

Spectroscopy: Instrumental method of assigning structural features and functional groups to organic or inorganic molecules; such features are displayed by intensities and patterns of spectroscopic signals.

Supercritical fluid chromatography (SFC): Technique in which the mobile phase (fluid) is maintained at temperatures somewhat above its critical point. The mobile phases used in SFC are gases such as freon, ethylene, or carbon dioxide. It has superior solution properties and enhances the chromatographic separation of higher molecular weight compounds. The column packings used in SFC are the same as those used in HPLC.

Thermal chromatography: Volatilization of organic compounds at high temperatures and their separation by chromatographic techniques such as TLC.

Thin-layer chromatography (TLC): Chromatographic technique in which the mobile phase is a liquid and the stationary phase is a thin-layer (usually 0.25 mm thick) of an adsorbent (silica gel, alumina, cellulose) spread homogeneously on a flat plate (usually a glass plate) of various dimensions.

X-ray crystallography: Method for determining the molecular structure of crystalline compounds which provides information on the positions of the individual atoms of a molecule, their interatomic distances, bond angles, and other features of molecular geometry.

Compound detection in organic chemistry refers to the methods of separation and identification of organic compounds. In modern technology this involves the use of chromatography (paper, thin-layer, gas–liquid, high-pressure liquid); spectroscopy (infrared, ultraviolet and visible, nuclear magnetic resonance); mass spectrometry; and reaction chromatography (chemical reactions on thin-layer plates or gas chromatographic columns which can be carried out prior to, during, or immediately after the chromatographic separation). Pyrolysis and X-ray crystallography of organic compounds furnish important structural information on the partial structures or on the whole molecule, respectively. The combination (and computerization) of chemical, chromatographic, and spectroscopic techniques has become a more efficient tool for the detection and identification of organic compounds than any of these techniques individually.

I. Introduction

It was only about 40 years ago that chemists had the tedious task of identifying and characterizing unknown organic compounds especially in the area of natural products. This may involve degradation of the molecule followed by synthesis involving many steps. For example Woodward elucidated the structure of strychnine in 1947 and seven-years later successfully synthesized this compound.

The advent of computers and fourier transform completely revolutionized the detection and identification of organic compounds. Modern automated instruments allow very small samples in the nanogram (10^{-9} g) range to be characterized in a very short time. The application of Fourier transform nuclear magnetic resonance (FTNMR) and Fourier transfer infrared (FTIR) allows recovery of the sample in contrast to mass spectrometric (MS) determination which is a destructive but quite often necessary technique.

Modern methods especially in the separation of complex organic mixtures utilizing gas–liquid chromatography (GLC), high-pressure liquid chromatography (HPLC), and droplet countercurrent (DCC) chromatography can separate samples rapidly and efficiently in the picogram range which until fairly recently has been impossible. Coupling the chromatographic instruments to spectrometers enables a partially automated analysis in even less time. The following coupling of chromatographic instruments has been performed: GC–MS, GC–FTIR, GC–UV–VIS, HPLC–MS, HPLC–FTIR, HPLC–FTNMR and MS–MS. (Fig. 1).

These semi-automated systems of analyzing and characterizing small samples are vital to the natural product organic chemist and biochemist for detection of highly active substances in extremely low concentration in living organisms. A typical example is in the field of pheromones which includes insect sex attractants which differ quite markedly in many insects. The concentration has often been found in the 10^{-9}–10^{-12} g range.

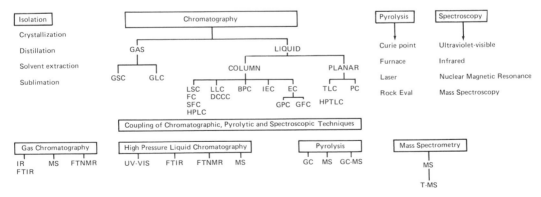

FIG. 1.   Chromatographic and spectroscopic techniques for detection and identification of organic compounds. GC, gas chromatography; GLC, gas–liquid chromatography; GSC, gas–solid chromatography; TLC, thin layer chromatography; HPTLC, high-performance thin layer chromatography; PC, paper chromatography; LSC, liquid–solid chromatography; FC, flash chromatography; SFC, supercritical fluid chromatography; LLC, liquid–liquid chromatography; DCCC, droplet counter current chromatography; PBC, bonded phase chromatography; HPLC, high pressure liquid chromatography; IEC, ion exchange chromatography; EC, exclusion chromatography; GPC, gel permeation chromatography; GFC, gel filtration chromatography; IR, infrared; UV, ultraviolet; NMR, nuclear magnetic resonance; MS, mass spectroscopy; FT, fourier transform; T-MS, Tandem mass spectroscopy.

II. Chromatographic Methods

There is an old Dutch word for chemistry, *Scheikunde*, which literally means the art of separation. Indeed, separation methods form the basis of chemistry, and the definition of a pure chemical substance ultimately depends on separative operations. [*See* CHROMATOGRAPHY.]

Chromatographic methods occupy a rather unique position among modern methods in the field of detection, separation, and identification of organic compounds. The latest chromatographic methods provide simple techniques for separation, purification, and structure determination of organic compounds including the most complicated natural and synthetic macromolecules in biological and geological systems and their metabolic or breakdown products.

Chromatography permits the separation and partial description of substances whose presence is unknown or unsuspected. As an exploratory method, chromatography is indispensable in all sciences dealing with chemical substances and their reactions. Chromatography is, in fact, a physical method of separation in which mixtures are separated by distribution between two phases: a mobile phase and a stationary phase. The mobile phase can be a gas (as in gas chromatography) or a liquid (as in liquid chromatography). [*See* GAS CHROMATOGRAPHY; LIQUID CHROMATOGRAPHY.]

Chromatographic methods can be classified according to the nature of the stationary phase. Gas chromatography is divided into GLC and and gas–solid chromatography (GSC). Liquid chromatography is divided into two main types: column and planar chromatography methods, namely, thin-layer chromatography (TLC) and paper chromatography (PC). Column chromatography is subdivided further into five major column types which constitute HPLC. They are (1) liquid–solid chromatography (LSC), (2) liquid–liquid or partition chromatography (LLC), (3) bonded-phase chromatography (BPC), (4) ion exchange chromatography (IEC), and (5) exclusion chromatography (EC). The latter form includes gel permeation chromatography (GPC) and gel filtration chromatography (GFC).

GLC and HPLC are the most widely used techniques for separation of organic compounds. However, they are distinctly different techniques. For GC analysis the sample must be volatile and stable when the column is heated. Samples for liquid chromatography must be soluble in mobile phase and differentially retarded by the stationary phase. An active mobile phase increases the versatility of the liquid chromatographic technique. By contrast the mobile phase in GC is an inert gas which does not influence the separation. A variety of stationary phases (both polar and nonpolar) can be used. The versatility of liquid chromatography depends on the liquid phase and solid absorbent. Gases, liquids, and solids can be analyzed by GC. The normal range of molecular weights is from 2 to 500, although derivatives of carbohydrates ranging up to 1800 have been analyzed. Liquid chromatography is applicable to liquids and solids that are either ionic or covalently bonded. Molecular weights of 32 to 1,000,000 have been analyzed by LC. The flame ionization detector in GC can readily detect 10^{-11} g and the electron capture detector can detect 10^{-10}–10^{-12} g of many organic compounds. The refractive index detector in liquid chromatography is limited to about 10^{-6} g; the ultraviolet (UV) detector can detect 10^{-8} g for many highly conjugated compounds. The resolution efficiency of a column is expressed in "theoretical plates." Thus, packed gas chromatographic columns of reasonable lengths can generate 2000–10,000 plates. Open tabular column (capillary size columns) can easily generate 50,000–100,000 plates. Total plates available in LC is greatly affected by pressure and analysis time available. Assuming 5,000 psi and flow rates of 0.5 ml/min, a reasonable upper limit is 10,000 plates.

III. Flash Chromatography

The traditional method for preparative separations of simple mixtures of organic compounds by means of column chromatography is giving way to flash chromatography (FC) or low-pressure liquid chromatography (LPLC) which was initially carried out by Still in 1978. FC is a simple, cheaper, and faster technique for routine purification of mixtures, allowing preparative separations from 10 mg to more than 25 g. The resolution associated with FC on a standard 5 in. high column of 40 μm silica gel is as great as 200 theoretical plates. This amount of separating power effectively means that compounds having R_f values on analytical TLC as close as 0.1 may be reliably separated with sample recovery of at least 90%. FC is basically an air-pressure-driven hybrid of medium-pressure and short-column chromatography which has been optimized for rapid separations. Elution of the components is extremely rapid, usually taking about 5 to 10 min. As with other forms of chromatography, sample size is one of the most important variables. The amount of sample that can be separated on a given column is proportional to its cross-sectional area and the degree of separation of the components as indicated by TLC. Still *et al.* successfully separated a 1-g mixture of epimeric alcohols **I** and **II** with only a 65 mg mixed fraction in 7 min on a 40-mm diameter column using 500 mL of 5% ethyl acetate/petroleum ether.

Reversed-phase systems with nonpolar bonded C_8 and C_{18} silica are used in the separation of sugars. A mixture of 200 mg of fructose and sucrose were completely separated in the 3/1 acetonitrile/water mobile phase incorporating a silylamine bonded phase with silica in less than 1 h.

This method has been used for the separation of various products of organic synthesis and small biomolecules.

IV. Droplet Countercurrent Chromatography

The technique of countercurrent chromatography (CCC) has seen a rapid expansion following the introduction of new methods such as droplet countercurrent chromatography (DCCC), rotation locular countercurrent chromatography (RLCC), and coil planet centrifugation. These methods have the advantage of being more rapid and less solvent consuming than traditional CCC. Furthermore, the advent of commercially available, compact apparatus has led to a widespread acceptance of these new liquid–liquid techniques as standard laboratory procedures for the separation of natural products. The detection of compounds that are eluted from a DCCC can be performed by three methods: (1) UV detection for suitable UV-active substances, (2) monitoring of the fractions that are collected by TLC, and (3) weighing of fractions after evaporation of solvent. The majority of DCCC separations involve polar compounds, especially glycosides, which are often difficult to purify. Chloroform–methanol–water systems of varying compositions remain the most widely used, in view of the good formation of droplets

and the convenient viscosity of this combination. The most notable developments in the application of DCCC have occurred in the field of polyphenols, in particular in the separation of tannins. DCCC has also been applied in the separation of natural products such as alkaloids, triterpene glycosides, steroid glycosides, basic steroid saponins, and glycosides of flavonoids. Rotation locular countercurrent chromatography (RLCC) relies on the percolation of one layer of a two-phase solvent system through compartments (loculi) that contain the second layer. During passage of the mobile phase, the loculi (connected into tubes) are constantly rotated, to increase contact between the two phases. Basically, RLCC has the same advantages as DCCC. As in DCCC, the apparatus can be run in either ascending or descending solvent modes but the formation of droplets is not a necessary condition of RLCC. Consequently, a broader range of solvent system is possible, and a system containing ethyl acetate (often incompatible with DCCC) has been used, for example, in the separation of flavonoids—an important application of this method has been the separation of enantiomers of (±)-norephedrine on an instrument consisting of 16 columns and each column containing 37 loculi. The stationary phase was sodium hexafluorophosphate solution at pH 4, and the mobile phase was (*R, R*)-di-nor-5-yl tartrate in 1,2-dichloroethane. Presumably, the enantiomers of (*t*)-norephedrine form different diastereotopic complexes with the tartrate ester, and these complexes are then partitioned differently between the two solvent phases. Separations by RLCC of a range of natural products, including flavones, xanthone glycosides, and antitumor antibiotics have been reported. RLCC provides a useful complementary method to DCCC in instances in which suitable solvent systems are not available.

A. ROTATING COIL METHOD

The introduction of these potentially useful techniques is largely due to Ito in 1981. These methods involve CCD and eliminate the need for solid supports. One of the principal advantages is the speed of operation—the separation time (and consumption of solvent) approximating the level of HPLC in one instance involving the isolation of plant hormones by toroidal coil planet centrifugation.

No direct comparisons with DCCC or RLCC have been reported, and it will be of interest to see whether planet and toroidal coil centrifuges have the possibility of complementing or supplementing DCC or RLCC. Separation of amino acids and peptides have recently been reported utilizing the rotating coil method.

V. High-Pressure Liquid Chromatography

Although the number of compounds detected by GLC could be increased by derivatization of polar functional groups, it has been suggested that only some 15% of all chemicals are capable of existing in the vapor phase.

HPLC has emerged as an instrumental technique offering rapid separations with simultaneous sensitive monitoring of the course of the analysis. HPLC in its most modern form is able to achieve separations in a matter of a few minutes which by previous techniques may have taken hours or days or may not have been possible.

Of particular importance in HPLC development has been the availability of specialized chromatographic column packings and sensitive on-line detection systems for continuous monitoring of the separations being carried out. These developments have led to systems which in favorable instances can on the one hand detect parts per billion (1 in 10^9) levels of organic compounds, and on the other hand be used for collecting gram quantities of pure chemicals by preparative HPLC. The lack of truly universal detectors has resulted in the development of several selective detectors as described in Table I.

Although LC detectors are sufficiently sensitive for trace analysis, they are nevertheless limited in their ability to provide adequate information for unequivocal identification. HPLC is particularly suitable in the separation of high-molecular-weight (up to 6 million) substances

and thermally unstable biologically active products that cannot be volatilized without decomposition; aqueous and nonaqueous samples can be analyzed. HPLC has been applied to the analysis of natural and synthetic products such as amino acids, antibiotics, antioxidants, flavonoids, carotenoids, lipids, flavoring and aroma compounds, herbicides, pesticides, hormones, steroids, mycotoxins, polyaromatics, proteins, sugars, purines, dyes, vitamins, and water and air polutants. An example of HPLC applications includes the detection of the highly toxic aflatoxins sometimes found in peanuts, wheat, corn, and other grain crops. They are produced by fungus on the grain and are considered to be dangerous to human health when their concentration exceeds 20 ppb.

HPLC is an ideal system in pharmaceutical analysis, examination of drug formulation, degradation products of drugs that might be toxic to humans, and detection of drugs in human fluids and tissues. HPLC is useful for polar and ionic compounds of medium to high molecular weight such as drugs and their metabolites.

A. Reversed-Phase HPLC (RP/HPLC)

This technique utilizes a nonpolar stationary phase, usually a fully porous microparticulate chemically bonded alkylsilica and a polar mobile phase are now recognized as the technique par excellence for the separation of polar, ionogenic solutes. For the separation of free amino acids, their derivatives, peptides, and proteins, the octyl (C_8) and octadecyl (C_{18}) hydrophobic phases bonded into 5- and 10-μm fully porous silicas with pore diameters in the range of 60–100 Å and 300–500 Å are being used.

B. Detection Systems

Photometric detectors with variable wavelength capability are most widely used currently

TABLE I.   Comparison of Various LC Detectors

	UV	Refractive index	Fluorescent	Electrochemical	MS Fullscan	SIM[a]
Range of application	Selective	Universal	Very selective	Very selective	Universal	Very selective
Minimum detectable quantity, g.	10^{-9}	10^{-6}	10^{-12}	10^{-12}	10^{-9}	10^{-11}

[a] SIM, Selective ion monitoring.

in LC. As an example, the carcinogenic aflatoxins in cereal products absorb light strongly at both 254 and 365 nm. Many other compounds also absorb light in the former wavelength. At 365 nm, however, most of the sample co-extractives are transparent and no longer interfere with the detection of aflatoxins. Although much work is performed with detectors that operate in the UV region of the spectrum a good deal is also practiced in the visible region.

C. FLUORESCENCE DETECTION

Fluorimetry is well known for its very high selectivity and sensitivity to very small quantities of some samples occurring in biological fluids while being completely insensitive to many other materials such as drugs, vitamins, and steroids. GC detectors such as electrical conductivity and electron capture are also used in LC.

D. RADIOACTIVITY DETECTORS

There are many applications in the studies of the metabolism of drugs, pesticides, etc., in which radioactive samples are employed to enable the compounds of interest to be detected at very low concentrations. The important current trends in HPLC are the use of super critical fluids as eluants and the coupling of HPLC with NMR, MS, or FTIR.

An equally important trend is a reduction in column size and a concomitant increase in analytical speed. Short columns can reduce analysis time, and efficiency may be as high as 5000 theoretical plates since the particle size of the packing material is 3 μm. The analysis time is about 1 minute. A standard HPLC column has a diameter of 4.6 mm as compared to 2, 1, or even 0.5 mm for the new columns. The small size of these columns allows the use of more expensive solvents; deuterated solvents might be used when it is necessary to analyze the collected samples by NMR. The use of ultramicrobore (50-μm diameter) columns allows feeding of the eluted peaks directly into a mass spectrometer or FTIR instrument.

E. PREPARATIVE LIQUID CHROMATOGRAPHY

Most preparative HPLC columns are capable of separating as much as 10 g of sample at a time. Industry often utilizes columns 13 to 22 cm in diameter and 53 cm long having flow rates of 3 to 20 L/min. They can separate from 1 to 10 kg of sample per hour.

F. ENANTIOMERIC ANALYSIS OF AMINO ACIDS BY HPLC

Most of the protein amino acids, except glycine, have at least one asymmetric (chiral) carbon atom and can exist as two isomers, designated D and L enantiomers. When the D and L enantiomers each react with a chiral (optically active) molecule of, for example, the L' configuration, they form the diastereoisomeric compounds DL' and LL'.

Amino acids of L configuration are more abundant in nature; the D-enantiomers are usually found in bacterial cell walls, antibiotic compounds, and rare biological molecules. The DL-form (the racemate) are found in geological specimens such as fossil shells and bones, or they are formed by abiotic synthesis.

Methods for resolving amino acids into their respective enantiomers are of importance in the

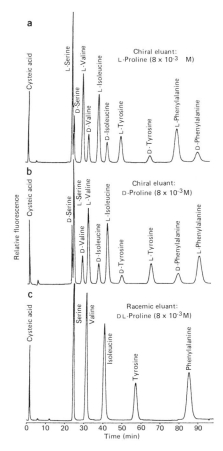

FIG. 2.   Effect of chiral eluant on the separation of D- and L-amino acid enantiomers by ligand-exchange chromatography.

preparation of peptides, drugs, and food additives as well as the extreme complex amino acid mixtures in meteorites in which the ready determination of optical purity is essential.

Two main approaches using HPLC are (1) covalent bonding of chiral ligands [which can complex copper (II) ions] to solid supports (such as polystyrene and polyacrylamide) and resolution of amino acids by eluting with a mobile phase containing copper (II) ions; (2) introduction of chirality into the mobile phase. Metal ions such as Cu(II), Zn(II), Co(II), and Mg(II), in conjunction with chiral ligands are added to the mobile phase. Thus, a Cu(II)–L proline complex as the chiral additive can be operated in conventional cation-exchange resin.

A good enantiomeric resolution of α-amino acids was recently achieved by using chiral complexes of copper (II) with N,N-di-n-propyl-L-alanine (DPA) as the additive in the mobile phase. Actually, the mixture of amino acids is separated into four groups by conventional ion-exchange chromatography and then resolved by means of the chiral DPA reagent (Fig. 2).

VI. Gas Chromotography

A. DERIVATIZATION GAS CHROMATOGRAPHY

Conversion of sample compounds into volatile derivatives make it possible to separate and analyze by GC, groups of compounds for which GC analysis would otherwise be impossible, e.g., amino acids, sugars, prostaglandins, and related compounds. The presence of different polar groups in the molecules of such parent compounds is the most significant source of difficulty associated with their GC analysis.

Carboxyl, hydroxyl, carbonyl, and amino groups, because of their polarity and tendency to form hydrogen bonds, are responsible for the low volatility of the compound and for other phenomena (such as strong absorption on the stationary phase) that make direct GC either difficult or impossible. The separation of closely related compounds is easier after their conversion into suitable derivatives. For example, the sensitive and labile prostaglandins such as $PGF_{2\alpha}$ are converted to their trimethylsilyl derivatives which are stable for GC analysis.

The combination of GC with MS (GC–MS) for which special derivatives are being developed, gives characteristic fragments that make identification and quantitative evaluation easier. Derivatives commonly used to protect these groups

are usually less polar than the original groups. Efforts aimed at developing a single procedure in which several or all functional groups in the molecule could be converted into a suitable derivative in one reaction step led to the development of methods for the preparation of silyl, acyl, isopropyl, and other derivatives.

Esters are common derivatives of carboxyl groups. Methyl esters are the most often used as they have a sufficient volatility even for the chromatography of higher fatty acids contained in fats. A number of methods have been developed for their preparation, often exclusively for the purpose of GC determination.

Elegant methods for esterification with diazomethane and methanolic solutions of BF_3 or BCl_3 are fairly widespread. Reactions with methanol can also be catalyzed with HCl or H_2SO_4. Methyl esters can also be prepared by pyrolysis of tetramethylammonium salts in the inject port, and by esterification on an ion exchanger:

$$R—COOH + CH_2{=}\overset{+}{N}{=}\overset{-}{N}$$
$$\rightarrow R—COOCH_3 + N_2 \uparrow$$
$$R—COOH + (CH_3)_4\overset{+}{N}OH^- \rightarrow RCO\overset{-}{O}\overset{+}{N}(CH_3)_4$$
$$\overset{\Delta}{\rightarrow} RCOOCH_3 + (CH_3)_3N \uparrow$$

Ethers are useful for protecting hydroxyl groups. Hydroxyl groups of high molecular weight compounds, such as sugars and sterols, are converted into ethers by reaction with methyl iodide in the presence of silver oxide.

$$2ROH + 2CH_3I + Ag_2O \rightarrow 2R—OCH_3$$
$$+ 2AgO + H_2O$$

Aryl derivatives are common for hydroxy, amino and thiol groups.

$$\begin{array}{ll}
R—OH & R—O—COR' \\
R—NH \xrightarrow[R' = aryl]{(R'CO)_2O} & R—NH—COR' \\
R—SH & R—S—COR'
\end{array}$$

Silyl derivatives are probably the most commonly used for protecting functional groups of volatile substances in gas chromatography. Trimethylsilyl (TMS) derivatives can be prepared by the reaction of trimethylsilylating agents such as trimethylsilyl chloride with groups containing active hydrogen.

$$\begin{array}{ll}
—OH & —O—Si(CH_3)_3 \\
—COOH & —COO—Si(CH_3)_3 \\
—SH \xrightarrow{Me_3SiCl} & —S—Si(CH_3)_3 \\
—NH_2 & —NH—Si(CH_3)_3 \\
{=}NH & {=}N—Si(CH_3)_3
\end{array}$$

1. Cyclic Derivatives

If two or more functional groups which should be protected occur in the molecule of the substrate, blocking can be accomplished with a bifunctional reagent, thus producing a cyclic product. Cyclic boronates, for example, can be used for the GC of compounds containing *cis*-diol groups in the 1,2- and 1,3-positions.

2. Separation of Enantiomers

Two basic approaches to the separation of enantiomers of amino acids have been applied: (1) amino acid derivatives are chromatographed on optically active stationary phases such as *N*-acyl alkyl esters, ureides or N-acyl alkyl esters of dipeptides and (2) GC separation is performed on conventional stationary phases and the derivatives of amino acids are prepared by reaction with optically active reagents. The bifunctionality of amino acids offers the use of either optically active esters or acyl derivatives.

3. Sugars and Related Compounds

The low volatility of sugars and related compounds such as amino sugars, polyalcohols, and aldonic and uronic acids, caused by the presence of several functional groups in one molecule as well as their high molecular weight and thermal lability are the reasons that they cannot be analyzed by GC methods. Another complication occurs in the formation of α- and β-anomers and pyranose and furanose rings, either during the preparation of the derivative or during the analysis itself.

The hydroxyl groups of saccharides and related sugars are silylated relatively easily; TMS derivatives have been widely applied to this group of substances.

4. Insecticides and Pesticides

This group of substances includes chemically very different compounds, such as ureas and carbamates, organic phosphorus and sulfur compounds, chlorinated hydrocarbons, and heterocyclic compounds. Depending on the type of functional groups in the molecule, common derivatives of these groups are applied. For the analysis of antibiotics TMS derivatives are generally prepared. Vitamins include compounds that differ considerably in their chemistry and therefore the range of possible derivatives is fairly wide.

5. Separation and Identification of Geometrical and Optical Isomers

Lipids are readily separated by chromatographic techniques into various components such as, for example, alcohols, sterols, fatty acids, and esters. However, these are not mostly single compounds but groups of compounds differing in chain length, degree of unsaturation, position of functional groups, and stereochemistry. The separation of some of these groups is accomplished by subjecting them to chromatography on adsorbents that have been impregnated with compounds having the ability to complex preferentially or interact with specific functional groups.

Unsaturated compounds such as fatty acids (oleic, linoleic, etc.) form π complexes: $AgNO_3$ TLC is used extensively to fractionate methyl esters of fatty acids according to their degree of unsaturation. Aromatic hydrocarbons as donors of π electrons are capable of forming donor–acceptor complexes with substances having electron-accepting properties, such as caffeine, tetracyanoethylene, and polynitro substances (e.g. 1,3,5-trinitrobenzene and 2,4,7-trinitrofluorenone).

Chromatographic resolution of optically active isomers requires the introduction of an asymmetric environment either intramolecularly by conversion to diastereomers or intermolecularly by the use of chiral stationary or mobile phases. In GC excellent resolution of derivatized amino acids as been achieved with diastereomers, as well as with chiral stationary phases.

6. Identification Methods Based on Comparison of Retention Data

The most frequently used method for establishing the identity of an eluting component is comparison of the retention data of an "unknown" peak with the retention of a similar injection made under identical operating conditions of a reference substance, which, based on other considerations, e.g., a known synthesis precursor, could probably occur in the sample. Where possible the injection should be repeated with several chromatographic phase systems which exhibit different types of selectivity, i.e., a normal partition, a reversed-phase system, and a liquid–solid (adsorption) system. A vast amount of information exists relating to retention characteristics of samples in GC systems with chemical structure.

7. Identification Methods Using On-Line Selective Detectors

The most common application of this approach is the tentative identification of compounds that contain characteristic UV or visible absorption spectra. The use of more than one detector, linked in series or parallel after the chromatographic solumn, can provide comparative information which reduces the possibility of incorrent assignment of the identity of a component. A simple example is the use of a UV absorbance detector in line with a differential refractive index (RI) detector. The latter will respond to most substances, whereas the former detector is quite selective in its response.

8. Monitoring of Column Effluents by MS

The combination of GC–MS and computerized data handling systems has proved to be one of the most powerful analytical methods for identifying minute (10^{-12} g) components which may be present in chemical samples. The greatest success has been in its application in the fields of forensic science, pollution, and biochemistry.

B. Reaction Gas Chromatography

Reaction GC is a variation of GC in which chemical reaction is coupled with the chromatographic separation. Chemical transformations in analytical reaction gas chromatography always take place in an integral chromatographic system, in a reaction syringe, a precolumn reactor, or the column itself. The combination of the chemical and the chromatographic methods is a more efficient tool for the identification of organic compounds than either of the two individual methods alone.

1. Esterification in Situ

Since organic acids are polar, they are converted into esters prior to analysis. The esterification is carried out either in front of the gas chromatographic column or directly on the column.

2. In Situ Hydrolysis

Hydrolysis is widely used in reaction GC for the identification of unstable and reactive com-

pounds. A consecutive chromatographic separation, saponification, and chromatographic analysis of the resulting products was used for the identification of a mixture of high-boiling esters.

3. Dehydration and Decarboxylation

Monobasic organic acids are decarboxylated to hydrocarbons having one atom of carbon less than the acid. Alcohols are dehydrated and the olefins obtained are hydrogenated to the corresponding hydrocarbons. These transformations are carried out in a reactor at 250–300°C.

4. Carbon Skeleton Determination of Organic Compounds

The technique of carbon-skeleton chromatography is based on removing the functional groups from a compound and reducing the double and triple bonds. A hot tube containing a catalyst is introduced into the GC pathway and hydrogen is used as the carrier gas. As the injected compound passes over the hot catalyst, the compound is chemically degraded to its carbon skeleton. The hydrocarbon products pass into the GC and are identified by their retention times. In this technique hydrogenation, dehydrogenation (hydrogen abstraction), and hydrogenolysis (cleavage of functional groups or heteroatoms) may occur.

Hydrogenation, the saturation of multiple bonds, greatly reduces the number of possibilities in determining the carbon skeleton. Dehydrogenation, the abstraction of hydrogen, takes place with cyclohexane derivatives and forms aromatic compounds at elevated temperatures of about 300–350°C. Hydrogenolysis involves the cleavage of functional groups from a molecule and the addition of a hydrogen atom to each of the cleaved ends.

In typical reactions (catalyst temperature 300°C, H_2 flow) the parent hydrocarbon is obtained from halides, alcohols, and heterocyclic compounds containing sulfur, oxygen, or nitrogen:

$$CH_3CH_2CH_2CH_2CH_2Cl \xrightarrow[\Delta]{H_2} CH_2CH_2CH_2CH_2CH_3$$

$$CH_3CH_2CH\underset{|}{-}CH_3 \xrightarrow[\Delta]{H_2} CH_3CH_2CH_2CH_3$$
$$OH$$

$$\xrightarrow[\Delta]{H_2} CH_3CH_2CH_2CH_3$$

When an oxygen or nitrogen functional group is on the terminal carbon atom (aldehyde, primary alcohol, ester, ether, amine, amide, or carboxylic acid), the next lower homologue of the parent hydrocarbon is obtained, although the parent compound may be produced concurrently:

$$CH_3(CH_2)_{10}CHO \rightarrow CH_3(CH_2)_9CH_3$$

$$CH_3(CH_2)_{15}CH_2OH \rightarrow CH_3(CH_2)_{14}CH_3$$

Carbon-skeleton chromatography is not advanced as a quantitative procedure. Thus, carbon-skeleton chromatography structure was used to elucidate the structure of brevicomin,

brevicomin

an ingredient of the sex attractant of the western pine beetle. A large number of possible structures based on spectral evidence were narrowed down to a few based on the production of nonane in carbon-skeleton chromatography.

Hydrogenation is widely used to determine the structure of unsaturated compounds and to determine the olefin content of a wide variety of mixtures. The simplest means of hydrogenating microgram amounts of a sample "instantaneously" in the chromatographic pathway involves a hydrogenation catalyst and a hydrogen carrier gas. This technique was used for fatty acid methyl esters and a variety of unsaturated compounds (such as alcohols, amides, amines, ketones, esters, ethers, and nitriles) at the microgram level.

5. Locating Double Bond Position

Microozonolysis is used to determine the position of the double bonds. The aldehydes formed by passing ozone through a solution containing olefins is cleaved by triphenylphosphine to aldehydes or ketones (depending on whether the olefin is substituted):

$$RCH=CHR' \rightarrow RCHO + R'CHO$$

$$R'RC=CHR'' \rightarrow R-\overset{\displaystyle O}{\underset{\displaystyle \|}{C}}-R' + R''CHO$$

Thus, the structure of the following terpene alcohol, one of the components of the boll weevil sex attractant, was provided in part by microozonolysis when it yielded 3,3-dimethylcyclohexanone.

3,3-dimethylcyclohexanone

6. Selective Removal of Compounds from Mixtures

Compounds containing certain functional groups may be "subtracted" by specific chemical reagents in the gas chromatographic pathway. These effects are recognized by comparing chromatograms made with and without exposure to the chemical. Thus, primary and secondary alcohols are removed by boric acid, aldehydes and ketones by hydroxylamine, and carboxylic acids by zinc oxide. Subtractions are usually accomplished by including within the system a stainless steel loop consisting of the reactive chemical coated on an inert gas chromatographic support. The loops are most useful for determination of ozonolysis products. This inclusion of a loop containing 5% o-dianisidine on a gas-chromatographic support provides the information by subtracting aldehydes and allowing ketones to pass. Furthermore, $AgNO_3$ is used to hinder the passage of olefins.

A mixture of hydrocarbons can be separated by three selective adsorbents: molecular sieves which retain the n-paraffins, mercuric perchlorate which adsorbs the unsaturated compounds, and a stationary liquid phase which retains the aromatic hydrocarbons. Boric acid on a stationary phase such as Chromosorb P is used for the removal of alcohols from a mixture of organic compounds, the products are nonvolatile esters. A similar technique was used for removal of terpene alcohols from a mixture of terpenoids. Acids can be adsorbed on potassium hydroxide deposited on quartz powder. This technique was found to be suitable for the analysis of compounds having active hydrogens such as fluorene, indene, carbazole, indole, and pyrrole as well as steroids such as estrogens and ketosteroids. o-Dianisidine quantitatively subtracts aldehydes, ketones, and epoxides, and phosphoric acid subtracts epoxides.

The identification of the chemical type of compound can be carried out by dividing the gas chromatographic effluents into many different streams, each of which passes through a suitable color reagent as indicated in Table II.

The GC method is also being applied to the elementary analysis of carbon, hydrogen, oxygen, nitrogen, sulfur, and halogen organic com-

TABLE II.   Color Reagents Used for Detecting Gas Chromatographic Effluents

Class of compound	Reagent	Color
Alcohols	Potassium dichromate/nitric acid	Blue
Aldehydes	2,4-Dinitrophenylhydrazine	Yellow or red precipitate
Ketones	Schiff's reagent	Pink
	2,4-Dinitrophenylhydrazine	Yellow or red precipitate
Esters	Ferrous hydroxamate	Red
Mercaptans	Sodium nitroprusside	Red
	Isatin	Green
	Lead acetate	Yellow precipitate
Sulfides	Sodium nitroprusside	Red (primary)
	Sodium nitroprusside	Blue (secondary)
Nitriles	Ferrous hydroxamate/propylene glycol	Red
Aromatics	Formaldehyde/sulfuric acid	Red
Alkyl halides	Alcoholic silver nitrate	White precipitate

pounds. This permits shorter analysis time, reduction of sample weight, and also increases the accuracy of determination.

7. Combination of TLC and Vapor-Phase Chromatography

The combination of TLC and GLC enables a more complete separation and identification of complex mixtures of organic compounds. In most cases the mixture is subjected to preliminary separation by TLC followed by a complete separation of the preseparated compounds of the mixture by GC. The TLC–GLC technique has been applied in the analysis of citrus and other essential oils and for determination of steroids in urine as well as unsaturated hydroxy acid and glycerides which are separated on a thin layer of silica impregnated with AgNO$_3$ and then subjected to GC. The fractions from the chromatoplates can also be altered prior to running a vapor-phase chromatographic separation. Thus, acids are converted to methyl esters, alcohols and sterols to acetates, etc., prior to GLC. An application of the combination of TLC with GLC is the direct application of the compounds to the thin-layer plate as they emerge from the exit tube of the GC. This technique was used for the analysis of fatty acids, steroids, essential oils found in coffee essence, and many alkaloids. The association of TLC techniques with GC–MS was described for various separations and identifications of organic compounds.

C. PYROLYSIS GAS CHROMATOGRAPHY (Py/GC)

Prolysis is the thermal fission of naturally occurring and synthetic polymers producing a range of smaller molecules. Analysis of these products enables a profile of the original compound to be reconstructed. The pyrolytic technique is usually integrated with a gas chromatograph, a mass spectrometer, or both. This arrangement enables pyrolysis products to be analyzed immediately, so that transfer losses and secondary degradations are minimized. The resulting chromatogram or mass spectrum may provide qualitative information concerning the composition or identity of the sample, quantitative data on its constitution, or it may enable mechanistic and kinetic studies of thermal fragmentation processes.

The combined Py/GC/MS technique requires a minute amount of sample, and the analysis time is very short. Furthermore, such systems are now totally automated and computerized.

Analytical pyrolysis has been used successfully in many disciplines such as polymer chemistry, organic geochemistry, soil chemistry, forensic sciences, food science, environmental studies, microbiology, and extraterrestrial studies involving meteorites and lunar samples. A large number of organic substances found in nature are unsuitable for direct analysis by modern techniques such as column chromatography and mass spectrometry. This may be due to their complex structure and polar and nonvolatile character.

A significant step in the evolution of analytical pyrolysis was the combination of pyrolysis with a sophisticated physiochemical technique for the efficient separation and/or identification of the fragments. In 1959, a combined Py/GC system of polymers was introduced. Improvements in the analysis conditions were obtained by using the high-resolution capabilities of capillary columns. In 1970, a continued interest in the analysis of extraterrestrial samples such as lunar rocks and the Allende meteorite (found in Mexico) was reported, using the modified Py/GC/MS technique. Applying a pyrolysis/mass spectrometry (Py/MS) technique such as the Curie point system offers considerable advantages: rapid analysis, automation of the system, and direct chemical information. Laser microprobe mass analysis (LAMMA) on milligram amount of sample has been reported. An alternative approach is the detection of very small (picogram) amounts of organic matter by direct ionization due to laser irradiation. Thus, digitonin (MW 1228) and sucrose (MW 342) were readily detected.

New techniques such as Py/GC/MS/DS (pyrolysis/gas chromatography/mass spectrometry–data system) and Py/MS/DS have been applied for detection and structural elucidation of complex organic compounds.

1. Pyrolysis of Synthetic Polymers

The thermal degradation of synthetic polymers has proven to be an extremely important analytical technique for revealing composition, structure, and stability profiles. It actually causes unzipping of the polymer chain to yield sequential monomer units. Thus, thermal depolymerization of rubber (at 700°C and atmospheric pressure) yielded isoprene and dipentene, whereas polystyrene yielded a series of monomeric and dimeric hydrocarbons.

2. Pyrolysis of Biological Molecules

The rapid extension of analytical pyrolysis into diverse fields such as taxonomy and soil chemistry has been largely due to progress in the pyrolysis of biological molecules. Such work has shown that the various classes of molecules with the possible exception of nucleic acids give highly characteristic pyrograms and has enabled the detection of the origin of fragments of complex samples such as bacteria and soil constituents. Interest in the pyrolysis of biological molecules is also due to the commercial importance of thermal degradation processes, such as flame retardant, use of biomass such as algae for production of oil and gas (energy), and thermal transformations of organic constituents of food during cooking. Thus, the sensitivity and specificity of Py/GC/MS has enabled octapeptides in infusion fluids and two closely related hormones, lypressin and felypressin to be detected in nanogram amounts in aqueous solutions.

3. Pyrolysis of Drugs and Natural Products

Py/GC, Py/GC/MS, and HPLC/MS are routinely used for the analysis of pure and formulated drugs or for the detection of active principles and metabolites in body fluids. The Py/GC method is used for detection of sulfonamides, barbiturates, and alkaloids such as morphine; heroine, phenacetin and caffeine (often used in forensic science).

The thermal reactions that accompany the smoking of tobacco and cannabis furnished important structural information on tobacco alkaloids. The pyrolysis of tobacco alkaloids yielded products such as quinoline and isoquinoline and nicotinonitrile which are probably derived from nornicotine and mysomine. Pyrolysis of natural polyenes such as β-carotene yielded ionene and small amounts of toluene, m-xylene, and 2,6-dimethylnaphthalene.

Although the identification and quantitation of drugs, e.g., narcotics and alcohol blood levels, are of forensic interest, the major impact of analytical pyrolysis in forensic science is in the identification of complex natural and synthetic substances such as blood, skin, hair, wood, soil, fibers, plastics, and waxes, which might be associated with a suspect and found in the scene of the crime.

4. Pyrolysis of Organic Geopolymers

Pyrolysis methods are used for detection and identification of natural organic polymeric products in the geosphere (such as humic substances and kerogens). The impetus for much of this work has been provided by the US space exploration program, when the efforts in taxonomy, biological molecules, and organic geochemistry which have resulted in Py/GC/MS studies, were undertaken on the surface of the planet Mars. The pyrolysis technique is also being used in the study of environmental pollutants (industrial processes, power stations, and fires). The use of the Py/GC/MS technique furnished important information on the structure of humic substances, melanoidins and coals.

The Rock–Eval method uses a special pyrolysis device (coupled with GC) in which a small sample of sedimentary rock (containing organic matter) is progressively heated to 550°C. During the assay the hydrocarbons already present in the rock (S_1) are first volatilized; pyrolysis of kerogen then results in the generation of hydro-carbons, hydrocarbon-like compounds (S_2), and oxygen containing volatiles, i.e., CO_2 (S_3) and water. Thus, $S_1 + S_2$, expressed in kilograms of hydrocarbons per ton of rock, is an evaluation of genetic potential (abundance and type of organic matter) of the rock.

5. Extraterrestrial Matter

Coupled pyrolysis techniques have been used for the analysis of extraterrestrial organic matter to provide evidence on molecular evolution within the solar system. Samples analyzed so far included meteorites collected on the surface of the earth, lunar rocks transported to earth and certain soils analyzed on the surface of that planet. The fully automated Py/GC/MS/DS analysis of Martian soils revealed only the presence of water and CO.

VII. Supercritical Fluid Chromatography

In supercritical fluid chromatography (SFC), the mobile phase is maintained at a temperature somewhat above its critical point. Since the physical properties of a substance in the super-critical state near the critical point are interme-diate between those of liquids and gases at ambi-ent conditions, it is designated as a fluid. For chromatographic purposes such a fluid has more desirable transport properties than a liquid. SFC is superior to LC in separating efficiency and speed. In comparison to a gas, a fluid shows about a 1000-fold increase in solution capabili-ties. The resultant enhancement in the migration rate of solutes is especially valuable in the analy-sis of higher-molecular-weight compounds. Fur-thermore, some ionic solutes are soluble in a supercritical fluid. This suggests that SFC may be applicable to the analysis of compounds such as the phospholipids, which cannot be volata-lized for GC without decomposition.

The number of compounds that can be ana-lyzed by SFC is potentially enormous. Out of the 10^6 known compounds which are currently more or less well characterized, only about 15% can be volatalized without decomposition. Com-pounds such as proteins, synthetic and natural polymers, lipids, carbohydrates, vitamins, syn-thetic drugs, and metal organic compounds may well be analyzed by SFC.

When a liquid and its vapor in equilibrium with each other are heated in a confined space, the intensive properties of the two coexisting phases become increasingly similar until, at the critical temperature, the two phases coalesce into a *fluid* and acquire the same properties. When this substance is heated beyond the criti-cal temperature, a supercritical phase is ob-tained, the substance is then called supercritical fluid.

For comparative purposes some physical properties of a gas, a liquid, and a supercritical fluid are shown in Table III. The data in Table III show that the viscosity of a supercritical fluid is comparable to that of a gas and its diffusibility is between that of a gas and a liquid.

Any compound which is thermally stable to somewhat beyond its critical point can theoreti-cally be used as the mobile phase. The mobile phases used in SFC include freons, ethylene, pentane, hexane, isopropanol, and carbon diox-ide. It has been observed that the solubility of various solutes increases in supercritical phases. The column packings that are used in SFC are essentially the same as those used in HPLC.

The instrumentation in SFC is quite similar to that of modern high-resolution liquid chroma-tography. A number of modifications are made for LC to be suitable for operation with a super-critical fluid. The separation column can be cou-pled to online detectors other than UV, includ-ing MS, FTIR, FID, and other GC detectors.

Carbon dioxide offers many advantages; it is inexpensive, available in high purity, and innoc-uous. Its near-ambient critical temperature

TABLE III.

Property	Units	Gas	Liquid	Supercritical fluid
Density	g/ml	10^{-3}	1	0.3
Diffusibility	cm³/sec	10^{-1}	$5 \cdot 10^{-6}$	10^{-3}
Dynamic viscosity	poise (g/cm sec)	10^{-4}	10^{-2}	10^{-4}

makes it attractive for use with thermally labile compounds. The UV absorbance of CO_2 is minimal, thus allowing spectrometric detection down to 190 nm.

The advantages of SFC are high resolution per unit time, orthogonal column selectivity compared to GC and HPLC, ease of fraction collection, and analysis of thermally labile molecules. It is likely that 20–40% of the solutes presently separated by HPLC are amenable to SFC separation with supercritical CO_2. It is possible that most separations carried out by normal-phase HPLC may be handled with good advantage by SFC.

VIII. Thin-Layer Chromatography

Thin-layer chromatography (TLC) is a subdivision of liquid chromatography (LC) in which the mobile phase is a liquid and the stationary phase is situated as a thin layer on the surface of a flat plate. TLC is sometimes grouped with paper chromatography under the term *planar liquid chromatography* because of the flat geometry of the paper or layer stationary phases. TLC is a simple, rapid, versatile, sensitive, inexpensive analytical technique for the separation of organic substances. Since numerous stationary phases (sorbents) are available, such as silica, cellulose, alumina, polyamides and ion-exchangers, considerable versatility is available in the type of substances that can be separated.

TLC is a microanalytical procedure and provides for separations and at least tentative identification of substances in the milligram microgram, nonogram, and even picogram (pg) range. Adsorption TLC is very sensitive to differences in configuration that affect the free energy of adsorption onto the layer surface and is, therefore, well suited to the separation of structural isomers. Quantitative estimation of the separated compounds is carried out *in situ* by densitometric estimation of the TLC plates.

A. High-Performance Thin-Layer Chromatography

High-performance TLC enables us to carry out the most complicated separations. The HPTLC plates are prepared from optimized (e.g., particle size and particle size distribution) adsorbent layers and extremely even surfaces. The HPTLC plates offer greater separating efficiency (plate number and resolution) through smaller plate heights than the conventional TLC

plates. Shorter analysis time, detection limits in the nanogram range with UV adsorption detection and in the picogram range with fluorometric detection are additional advantages.

The HPTLC plates may typically deliver some 4000 theoretical plates over a distance of 3 cm in 10 min. This compares to typical values for conventional TLC plates of some 2000 theoretical plates over 12 cm in 25 min.

B. Impregnated Layers

The range of applications of both TLC and HPTLC is considerably expanded by means of different impregnation agents such as acids, bases, or salts added to layers in various concentrations. Stable hydrophilic stationary phases are formed by treatment with agents such as formamide, DMF, ethylene glycol, and various buffers. Lipophilic stationary phases for reversed-phase TLC are obtained by impregnation with liquid paraffin, undecane, and mineral and silicone oils. Impregnation with specific reagents aids the separation of certain types of compounds such as $AgNO_3$ for compounds with double bonds, boric acid or sodium arsenite for vicinal dihydroxy isomers; sodium bisulfite for carbonyl compounds, and trinitrobenzene or picric acid for polynuclear aromatic compounds.

C. Detection and Visualization

Following development, chromatograms are removed from the chamber and are air- or oven-dried to remove the mobile phase, zones are detected by various means. Colored substances may be viewed in daylight without any treatment. Detection of colorless substances is simplest if compounds show self-absorption in the short-wave ultraviolet (UV) region (254 nm) or if they can be excited to produce fluorescence by short-wave and/or by long-wave (365 nm) UV radiation. Otherwise, detection can be achieved by means of chromogenic reagents (producing colored zones), fluorogenic reagents (producing fluorescent zones), or by biological enzymatic methods.

Enzymatic reactions can be monitored on the plate, and the end products can be detected. Biological test procedures are used in the specific detection of biologically active compounds. Thus, detection of hemolyzing compounds such as saponins is achieved by casting a blood–gelatin suspension on the layer and observing hemolytic zones that are transparent and nearly colorless on the turbid red gelatin layer background. Another means of detection is the use of Geiger or flow counters or other specialized means to locate radioactive solutes.

Detection reagents may be impregnated into the layer prior to sample application and development. Chromogenic reagents are of two types: (1) general reagents that react with a wide variety of different compound types and can totally characterize an unknown sample, and (2) specific reagents that indicate the presence of a particular compound or functional group. The universal detection reagent iodine can be used as a 1% alcoholic solution spray, but more frequently, the plate is simply placed in a closed container containing a few iodine crystals. The iodine vapor forms weak charge-transfer complexes with most organic compounds which show up as brown spots on a pale yellow background within a few minutes. Sensitivities in the 0.1–0.5-μg range are often obtained with iodine.

Charring reagents (H_2SO_4) are suitable for glass-backed layers with inorganic (e.g., gypsum) binders only. Many charring reagents produce colored zones when heating is carried out at relatively low temperature; they form black zones at higher temperatures.

Spraying of a chromatogram with a 5% solution of phosphomolybdic acid followed by a brief heating at 110°C gives dark blue spots against a yellow background with a large variety of organic compounds.

A solution of Rhodamine B produces violet spots on a pink background. Antimony trichloride or pentachloride solution in carbon tetrachloride produce spots of different characteristic colors with many organic compounds.

Over 300 spray reagents are known to react more or less specifically with different functional groups to reveal natural products and organic or biochemicals as colored or fluorescent zones. Table IV contains a selection of specific detection reagents. Methods for the quantitation of thin-layer chromatograms can be divided into two categories. In the first, solutes are assayed directly on the layer, either by visual comparison, area measurement or densitometry. In the second, solutes are eluted from the sorbent before being examined further.

D. RADIOCHEMICAL TECHNIQUES

Radioactive isotopes are widely used as tracers or labels for substances separated by TLC for following the causes of chemical and bio-

chemical reactions, determining the distribution of substances in a reaction mixture, elucidating metabolic pathways of drugs, pesticides, pollutants, and natural substances in human, animal and plant tissues, and assessing the purity of isotopes. The most widely used labelled substances in TLRC contain ^3H (tritium) and ^{14}C. [See RADIOACTIVITY.]

The detection of radioactive substances on TLC plates is carried out by liquid scintillation, film registration or autoradiography, and by direct scanning. A thermomicro procedure for rapid extraction and direct application in TLC is the thermomicro application of substances (TAS) method.

The TAS method is a procedure for the isolation and separation of many substances from solid materials and their direct transfer to the starting line on a TLC plate. This is performed by application of heat, either by distillation or sublimation. The emerging vapors are deposited as a spot on the TLC-plate which is then chromatographed in the usual way. The spot can also be scraped off, extracted and analyzed by GLC.

This technique is useful in the fields of drugs, phytochemistry, food additives, and other natural and synthetic organic compounds. Many organic constituents of drug-containing plants were detected by this technique. Examples are constituents of essential oils, purines (e.g., caffeine), narcotics from drugs or tobacco, marijuana constituents, mescaline from Mexican narcotic fungi, organic contituents of fossil fuels, and oil-shales.

E. COUPLED TLC TECHNIQUES

1. TLC–GC

There are a number of ways in which TLC can be combined to advantage with GLC. The spots obtained from TLC may be eluted, concentrated, and then subjected to GLC analysis. This method has been used in the analyses of lipids, steroids, alcohols, fatty acids, esters, glycerides, hydrocarbons, essential oils, and many other natural and synthetic organic compounds. Methyl esters of fatty acids are first separated on silver-nitrate-impregnated silica gel layers according to the degree of unsaturation. The separated fractions are eluted and the products subjected to GLC separation. Furthermore, the resolved methyl esters may be treated to reductive ozonolysis and the resulting fragments analyzed by GC, thus furnishing important structural information.

TABLE IV.   Detection Reagents for Different Functional Groups

Compound class	Reagent	Color
Alcohols	Ceric ammonium sulfate	Brown spots on yellow background
Aldehydes	2,4-Dinitrophenylhydrazine	Yellow to red spots on pale orange background
Ketones	Dragendorff–Munier modification reagent	Orange spots
Alkaloids	Iodoplatinate	Basic drugs yield blue or blue-violet spots
Amides	Hydroxylamine–ferric nitrate	Various colors on white background
Amines and amino acids	Ninhydrin	Yellow-pink-red or violet spots on white background
Carbohydrates	p-Anisaldehyde	Blue-green and violet spots
Carboxylic acids	Cresol green	Yellow color on blue background
Bases	Cresol green	Blue spots on green background
Chlorinated hydrocarbons and chlorine containing particles	Silver nitrate	Gray spots on colorless background
Ethanolamines	Benzoquinone	Red spots on pale background
Heterocyclic oxygen compounds	Aluminum chloride	Flavonoids produce yellow fluorescent spots
Hydrocarbons	Tetracyanoethylene	Aromatic hydrocarbons yield various colors
Hydroxamic acids	Ferric chloride	Red spots on colored background
Indoles	Ehrlich reagent	Indoles : purple. Hydroxy-indoles blue
Feroxides	Ferrous thiocyanate	Red-brown spots on pale background
Phenols	4-Amino antipyrine	Red, orange, or pink spots on pale background
Polynuclear aromatic hydrocarbons	Formaldehyde–sulfuric acid	Various colors on white background
Steroids	p-Toluenesulfonic acid	Fluorescence
Steroid glycosides	Trichloroacetic acid–chloramine T	Digitalis glycosides; blue spots
Terpenes	Diphenyl-phenyl-hydrazyl	Yellow spots on purple background
Vitamins	Iodine-starch	Ascorbic acid; white on blue background

2. GC–TLC

A recent application of the combination of TLC with GLC has been the direct application of the compounds to the thin-layer plate as they emerge from the exit tube of the gas chromatographic apparatus.

It should be pointed out that GLC separates according to the relative volatility, and TLC separates according to the functional groups present. A coupling device was constructed in which the thin layer plate is moved logarithmically with time, while the gas chromatogram operates under isothermal conditions.

3. In Situ Chemical Reactions on TL Plates

In order to detect and identify minute quantities of organic compounds the sample is spotted on a plate which is then covered with a reagent. After a very short time, (mostly a few seconds or minutes) the plate is developed in a suitable solvent system whereupon the reaction products are separated and identified by color or by spectroscopic methods. A great deal of information can be gained from the in situ reaction technique with the expenditure of very small amounts of material. The following in situ reactions on TL plates have been carried out so far: acetylation, dehydration, formation of derivatives: (acetates, dinitrobenzoates, DNPS, methyl esters, phenylisocyanates, and semicarbazones), diazotization, esterification, halogenation, catalytic hydrogenation, acid and alkaline hydrolysis, isomerization, nitration, oxidation, photochemical reactions, reduction, and Diels–Alder reactions.

IX. Spectroscopic Methods

There has been a profound impact on the application of spectroscopic techniques in the detection of organic compounds especially in the areas of natural products and polymers. The commercially available Fourier transform signal handling techniques are rapidly revolutionizing nuclear magnetic resonance (FTNMR) and infrared spectroscopy (FTIR) as well as prototypes of mass spectrometry (FTMS) which are not yet available commercially. Previously organic chemists usually required all the spectroscopic data (UV, IR, NMR, and MS) to supplement microanalysis and general physical characteristics such as refractive index, density, and melting and boiling points in order to elucidate the structure of the organic compound. Furthermore, it was found that at least 50 mg of the material was needed in order to determine the structure. It is now possible with the modern available techniques to determine structures with less than a milligram of material.

A. INFRARED SPECTROSCOPY

Infrared (IR) spectroscopy is probably the quickest and cheapest of the spectroscopic techniques in determining the functional groups of the sample. The samples can be solids, liquids, or gases and can be measured in solution or as neat liquids mulled with KBr or mineral oil. Comparison of IR spectra of substances of known structure has led to many correlations between wavelength (or frequency) of IR absorption and features of molecular structure. Certain structural features can easily be established. For example, in an organic compound that contains only C, H and O, the oxygen can only be present as $C=O$, $O-H$, or $C-O-C$ or a combination of these, such as the ester or carboxylic acid group. [*See* INFRARED SPECTROSCOPY.]

The presence or absence of absorption in the carbonyl region (1730–1670 cm^{-1}) or hydroxy region (3700–3300 cm^{-1}) can serve to eliminate or establish some of these possibilities. One simple application of IR is to determine whether two samples are identical. If the samples are the same, their IR spectra (obtained under identical conditions) must be the same. If the two samples are both pure substances very similar in structure, the differences in the spectra may be so small that it would not be easy to detect them; it may even be beyond the power of the instrument to detect them. The absorption peaks of the spectrum of an impure sample is usually less intense than those of a pure sample and the spectrum will show additional peaks. The IR spectra of enantiomers are identical but those of diastereoisomers are different.

Simple mixtures can be determined from their IR spectra. For example mixtures of cycloalkanones show characteristic differences of the carbonyl group: cyclobutanone (1788 cm^{-1}), cyclopentanone (1746 cm^{-1}), and cyclohexanone (1718 cm^{-1}). IR spectroscopy has been applied to deuteration of organic samples. It is found that the C—H stretch bands disappear from the 3000 cm^{-1} and the C—D stretch bands appear in the 2200 cm^{-1} region. It is possible by calibrating pure compounds to determine the isotopic purity in routine analysis.

Modern IR analysis utilizing high-performance detectors can generate spectra (4000–200 cm^{-1}) in 60 sec. Sample amounts of 10 μg were found to be sufficient for such spectra.

B. FOURIER TRANSFORM INFRARED SPECTROSCOPY

Fourier Transform IR spectroscopy is widely used because of its rapidity of providing high-resolution spectra with samples in the nanogram range. A complete spectrum for a sample of 10^{-9} g can be obtained in less than 1 sec. FTIR permits rapid quantitative characterization of solids, liquids, and gases.

Kinetic processes can be monitored by a technique known as time-resolved spectroscopy which involves FTIR. This method has been applied to analysis of complex materials such as polymer film stretching which can be carried out in milliseconds and chemical transformations involving, for example, coal pyrolysis; it also permits on line analysis of products subject to chromatographic separation methods such as GC and LC. During the past five years GC–IR and GC–FTIR involved separating of mixtures and analysis of the individual compounds by IR spectroscopy. The sensitivity limitation of IR detectors with respect to GC and the time difference between the elution of a GC peak (measured in seconds) and the time scan were two of the problems encountered. GC–FTIR allows an IR spectrum taken from a 5-μg GC peak of isobutylmethacrylate by repeatedly scanning with spectral accumulation and enhancement (Fig. 3). FTIR measurements may be carried out by one of the following techniques: (a) KBr pellets, (b) photoacoustic, and (c) diffuse reflectance methods.

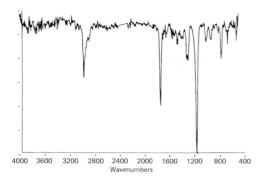

FIG. 3. IR spectrum of a small GC peak (5 μg) of isobutylmethacrylate using spectral accumulation (FTIR/GC cell).

The rapid-scanning property of FTIR spectrometers is having its greatest impact in the field of GC-FTIR. FTIR is now rivaling mass spectrometry for the identification of unknowns eluting from gas chromatographs and in one study was shown to identify more peaks than MS. It is clear that GC–FTIR will be of importance for the analysis of environmental samples. The interface between FTIR and HPLC is still at a premature stage, and most of the HPLC-FTIR results reported to date have involved the use of size-exclusion chromatography with chlorinated solvents, since these have good transmission over much of the infrared. The maximum concentration of most peaks eluting from either normal-phase or reverse-phase HPLC columns rarely exceeds 10 ppm, so that on-line detection in a flow cell is very difficult. For normal-phase and reverse-phase HPLC, continuous elimination of the solvent is usually required before identifiable IR spectra can be measured; no commercial instruments are yet available. For at least one of these systems, a detection limit of 100 ng has been reported for reverse-phase separations with a water–methanol mobile phase.

In conventional HPLC–FTIR, the interferograms are continuously recorded and stored during the analysis. Next, the absorption bands due to the solvent are subtracted from the solution spectra. In order to facilitate these measurements, the volume of the flow cell must be of the order of 1 μL, implying that even with the packings of 3–5-μm diameter, the evolution volumes will be significantly higher than the cell volume. Thus, only a small fraction (less than 1%) of each separated component will be in the cell when the measurements are made. Of the sev-

eral interfaces proposed for solvent elimination, the most promising one involves an initial concentration step in the concentrator tube using N_2 gas and above ambient temperature, followed by the deposition of the concentrated solution on the KCl powder. Further elimination of the solvent is achieved with a stream of air. Finally, the diffuse reflectance infrared Fourier transform (DRIFT) spectra, rather than conventional spectra, are recorded. The DRIFT technique is considered as being multipassing, and thus submicrogram quantities of nonvolatile compounds can be detected. However, DRIFT still has many problems to be solved.

C. MASS SPECTROMETRY (MS)

Mass spectrometry provides a means for studying samples at the molecular level. Although MS is basically a structure-identifying tool, it is not as specific for the detection and determination of functional groups (IR) or relative positions protons or ^{13}C nuclei (NMR) spectroscopy. Mass spectrometry, like UV, IR, and NMR spectroscopy, works best with pure samples. Structural elucidation of mixtures is not possible without some means of separating each constituent in the mixture as in GC–MS.

Besides being a useful structure-elucidating tool, MS can be applied to detect very low levels of specific compounds and elements. Accurate determination of masses can also be determined. An important advantage of the MS technique is its high sensitivity and accuracy. MS is able to provide more specific information per given amount of material than any other analytical technique. Furthermore, even with minute amounts of material (10^{-6}–10^{-9} g), this information can be provided in a reproducible and accurate manner. In the 1940s MS was an important analytical technique for characterizing complex fuel mixtures. About 25 years ago interest was centered on a systematic study of ionic fragmentation mechanisms in order to provide a set of rules that could be applied to elucidate structures of organic compounds. In the late 1960s mass spectrometers were coupled to gas chromatographs. This provided such dramatic improvements in selectivity and sensitivity that MS has since become one of the most generally useful analytical techniques for identification and quantitation of organic substances at ultratrace levels. Chemical ionization mass spectrometry (CIMS) uses reagent ions rather than electrons to ionize a sample. Field desorption and rapid heating techniques are extending mass spectrometry to high-molecular-weight, low-volatility compounds. During the past decade mass spectrometry has undergone a number of significant changes that have far-reaching contributions especially in biology and medicine. The mass range of mass spectrometers has been extended by approximately an order of magnitude in the past decade. Certain types of mass analyzers have been used to reach higher masses (150,000 in some quadrupole experiments). Commercial instruments are now available with mass ranges of 7500 compared with 1000 about ten years ago. Another improvement is desorption ionization [including fast atom bombardment (FAB), secondary ion mass spectrometry (SIMS) and fission fragment methods] which allows ionic, nonvolatile compounds to be examined by MS. A further development is the integration of separation and analysis techniques represented by tandem mass spectrometry (MS–MS) and the much improved liquid chromatography–mass spectrometry (LC–MS) interfaces. The latter capability is already proving its effect with respect to GC–MS but without the limitation to volatile, low-mass compounds. MS–MS is a two-dimensional form of spectrometry which often improves signal-to-noise ratios as well as providing entirely new capabilities, such as that of scanning a mixture for all constituents having particular structural subunits.

The GC–MS combination has had considerable impact on biological and environmental research, providing a specific means of characterizing constituents of mixtures and having sufficient sensitivity and quantitative accuracy to trace constituents. The recent discovery in 1977 of the neural excitotoxin, quinolinic acid, in the mammalian brain at the level of nanomoles per gram was achieved by GC–MS of the volatile hexafluoroisopropanol diester derivative with electron impact (EI) ionization. Quantitation was based on standard addition and single-ion monitoring. The GC–MS technique was applied in the discovery of 19-hydroxylated E prostaglandins by first protecting the unstable β-ketol system by oximation. This discovery, in 1974, implied that the previously identified prostaglandins could be artifacts.

Challenging structural problems, including protein structure determinations, can often be solved by a combination of techniques. Two mass spectrometric methods were used to deduce the structure of the 112-amino-acid antitumor protein macromycin derived from streptomyces culture. Partial acid digestion gave a

mixture of di- to hexapeptides, which was derivatized and analyzed by GC–MS with EI.

Tandem mass spectrometry has been applied in the discovery and confirmation of the structure of the metabolites of the drugs primidone, cinromide, and phenytoin in plasma and urine extracts; analyses were completed in less than an hour, using concentrations of 1 to 50 μg/mL. The study is predicted on the speed and flexibility of MS–MS scans made with a triple quadrupole instrument, and on the realization that metabolites often retain a large portion of the parent drug structure.

Analysis of complex mixtures has often involved time-consuming procedures such as extraction, centrifugation, and chromatography. MS–MS and LC–MS are recent developments that reduce such time-consuming methods. By linking two mass spectrometers in tandem it is possible to employ the first as a separator and the second as an analyzer and, hence, to perform direct analysis of mixtures. The two principal advantages of this system can be illustrated by considering a complex coal liquid mixture. The signal due to a dioxin spike is lost in the chemical noise from the other constituents and single-stage mass spectrometry is not capable of analyzing for it. MS–MS filters against chemical noise and allows a high-quality spectrum of the dioxin to be recorded. In addition to improving detection limits in this way, tandem mass spectrometry provides alternative scan modes which can be employed to search the data bank for particular information. For example, chlorinated dioxins are characterized by the loss of COCl, so a scan for the reaction reveals all dioxins present in the mixture. The GC–MS–MS system can decrease GC–MS detection limits by an order of magnitude. It has been reported that the drug, isosorbide-5-mononitrate, a coronary vasodilator, is metabolised to the glucuronide, which can be determined in urine by a simple MS–MS procedure to 0.1 ng/mL. The improved detection limit in the MS–MS experiment is the direct result of minimizing interferences. Both the sensitivity and the speed of analysis with MS–MS can be illustrated with the administration of the drug tetrahydrocannabinol in doses of 0.1 mg/kg for eight days down to 10^{-11} g/ml by using a combination of GC–MS with simple MS–MS to avoid extensive sample cleanup. Sensitivities in parts per trillion have been reported in MS–MS studies on animal tissue. High-resolution mass spectrometry and MS–MS have been used to achieve absolute detection limits of less than 10^{-12} g for tetrachlorodibenzodioxin and a GC–MS–MS combination has produced spectra with 250 fg samples (<20 parts per trillion) at the rate of 30 samples per day. In terms of sample throughput, the determination of trichlorophenol in serum at concentrations as low as 1 ppb and a rate of 90 samples per hour was reported. Characterization of the foodstuff contaminant (and chemical warfare analog) vomitoxin at 25 pg in wheat at a rate of 10 minutes per sample was also reported.

A different approach to the characterization of mixtures of nonvolatile compounds is LC–MS. The first practical LC–MS interface was based on complete removal of the solvent and temporary storage of the solute during transport by a moving belt or wire into the ion source. In the source the sample is either thermally desorbed and ionized by electron or chemical ionization or the belt is bombarded by an energetic beam to create secondary ions. The large pumping capacities of chemical and atmospheric ionization sources make it possible to work at flow rates consistent with normal column operation, for example, 2 mL of aqueous mobile phase per minute. The solvent itself acts as the reagent gas in these experiments. An alternative, the thermospray procedure does not use any external ionization technique. An aerosol generated in the interface is evaporated, and separation of charges present in the nominally neutral solution allows positive- and negative-ion mass spectra to be recorded. The performance of LC/MS for β-hydroethyltheophylline is 10 pg (selected ion monitoring) or 1 ng (full spectrum) with respect to its detection limits. The method involving direct liquid introduction and CI gives comparable data; for example, 50 ng of vitamin B_{12} gives a negative-ion spectrum of high quality.

Fourier transform mass spectrometry (FTMS) illustrates the speed with which instrumental developments are transforming mass spectrometry. The high resolution of FTMS is probably its most important aspect. Impressive performance data have been reported, such as a resolution of 1.4×10^6 for $m/z = 166$ from tetrachloroethane and 10^8 for $m/z = 18$ from water. FTMS instruments are capable of performing MS—MS experiments. Unlike a conventional MS–MS experiment, in which the different stages of analysis are separated in space, the separation here is achieved in time. This allows the extension of the experiment to three (MS–MS–MS) or more stages. Because FTMS instruments require very low pressures for optimum performance, interfacing with chromatography is a problem, although a GC–FTMS has been reported by Wilkins in 1982.

One aspect of LC–MS which is rapidly being developed is the HPLC–MS and has great promise for the analysis of many compounds. The primary obstacle to easy coupling of HPLC to MS arises from the fact that the flow for conventional HPLC columns is approximately an order of magnitude greater than can be accommodated by the commonly used ion sources and pumping systems. The requirement for a transfer of a maximal quantity of sample and a minimal amount of solvent into the ion source of the mass spectrometer has necessitated the development of different interfaces. The need for introduction of smaller amounts of HPLC effluents into a mass spectrometer has stimulated the development of narrow bore columns. The application of HPLC–MS with a direct liquid introduction has been applied to the identification of marine sterol peroxides by Djerassi and Sugnaux in 1982. An ultrasphere ODS column of i.d. 5 μm and methanol–water (99:1, v/v) solvent system were used.

At present it is difficult to predict if the HPLC–MS technique with still many difficulties will become a routine tool for detection and identification of HPLC solutes. It is clear, however, that this technique remains the only coupled MS method by providing unambiguous identification of solutes in HPLC effluents at room temperature or even lower than room temperature. Many natural products that have been identified by GC–MS techniques can be verified by HPLC–MS.

D. NUCLEAR MAGNETIC RESONANCE SPECTROSCOPY

It was over 35 years ago that NMR spectroscopy became attractive to organic chemists when Knight reported in 1949 that the precise frequency of energy absorption depends on the chemical environment of hydrogens. The frequency of radiation required for NMR absorption depends on the isotope and its chemical environment: the number of absorption peaks for magnetic nuclei and the intensity of the absorption peaks is proportional to the number of nuclei. Before the advent of Fourier transform, NMR spectra were measured by continuously varying the field-frequency ratio seen by the analytical sample and is referred to as the continuous-wave or slow passage method. During the past 15 years FTNMR spectrometers have appeared which have the advantage over continuous-wave NMR in that an increased sensitivity for an equal amount of instrument time is obtained, because the entire spectrum is observed with each pulse. [*See* NUCLEAR MAGNETIC RESONANCE.]

NMR is a valuable technique in molecular elucidation and verification. In some instances, the NMR spectrum is sufficient for identifying relatively simple unknown compounds while in other applications structural information from NMR spectra compliments that of other chemical and spectroscopic methods. FTNMR is very useful for studying reaction kinetics and chemical equilibria as well as quantitatively analyzing mixtures. For example, it was found that in the reduction of 5,5-dimethylcyclohexen-2-one with LiAlH₄ two products were obtained, 5,5-dimethylcyclohexen-2-ol and 15% of 5,5-dimethylcyclohexanol. This impurity was not previously detected by other methods.

It was only from FTNMR that chemical kinetics and structural analyses of the acid-catalyzed rearrangement of *trans*-methyl chrysanthemate to methyl lavandulyl esters could be accurately analyzed.

Simplification of NMR spectra of more complex molecules such as natural products and polymers can be done in three ways:

1. Selective decoupling spin–spin interactions by the double resonance technique. The sample is simultaneously irradiated with two radio-frequency fields; one frequency gives the resonance pattern of interest, and the other frequency is the resonance frequency of the coupled nuclei. When the second proton is irradiated with sufficient intensity, the spin–spin coupling collapses giving simple spectra.

2. Deuterium exchange of protons. Substitution of deuterium for hydrogen often gives spec-

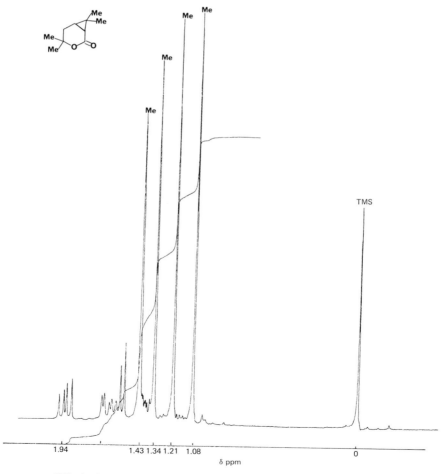

FIG. 4. ¹H NMR spectrum (300 MHz) of (±) *cis*-δ-lactone in CDCl₃.

tra having no observable coupling to deuterium because of its small magnetogyric ratio. Protons attached to heteroatoms are easily exchanged by deuterium such as OH in alcohols, SH in mercaptans, and NH in amines. The use of reagents containing deuterium such as $LiAlD_4$, D_2SO_4, $NaBD_4$, CD_3COCl, D_2O and $CD_3ND_2 \cdot DCl$ are just a number of reagents that can be used in synthesizing complex structures in which the NMR spectra would be much simpler. It should be realized that selective decoupling can then be applied to deuterated compounds, thus producing even simpler spectra. Simple proteins can be examined in this way.

3. Complexation of the organic molecule by certain paramagnetic metal ions can result in useful changes in the NMR spectrum. It has been shown that certain transition elements such as europium and praesodymium induce large shifts in the proton NMR spectra of organic compounds that coordinate to the metal, for example, molecules that have certain functional groups that contain oxygen or nitrogen. The chloroform soluble *tris*(dipivalomethanato) europium (III) complex, $Eu(DPM)_3$, has been used as a shift reagent to simplify the proton NMR spectra of compounds, including alcohols, steroids and terpenoids such as the racemic mixture of the *cis*-δ-lactone of cis chrysanthemic acid (Fig. 4). In this instance the singlets of the methyl moieties are clearly shifted in the presence of the enantiomer (−)-2,2,2-trifluoro-1-(9-anthryl)ethanol or Pirkle's reagent (Fig. 5).

A routine sample on a 60-MHz instrument

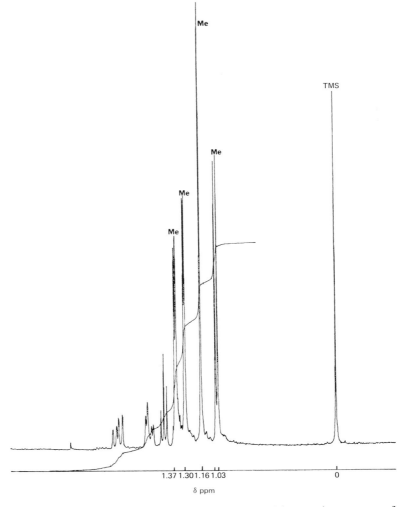

FIG. 5. ¹H NMR spectrum (300 MHz) of (±) *cis*-δ-lactone in presence of (−)-2,2,2-trifluoro-1-(9-anthryl)ethanol (Pirkle's reagent).

would require about 5–50 mg of the sample in about 0.4 mL solvent. A microtube consisting of a thick-wall capillary allows spectra to be obtained on less than 1 mg of sample. The use of FTNMR makes it possible to obtain spectra on amounts as small as 5 μg. The ideal solvent should contain no protons in its structure, be low-boiling point, nonpolar, and inert. Carbon tetrachloride is often used in non-FTNMR spectroscopy, depending if the sample is sufficiently soluble in it. The most widely used solvent is deuterated chloroform ($CDCl_3$).

1. ^{13}C-NMR Spectroscopy

The reason that ^{13}C NMR was developed much later than the proton NMR was due to the low natural abundance of the ^{13}C nucleus (1.108%) as compared to ^{1}H NMR (100%) and the low gyromagnetic ratio. These two problems were overcome during the past 15 years due to the arrival of Fourier transform where the accumulation of scans over a period of time overcomes the problem of the low abundance of ^{13}C. The advantages of ^{13}C NMR compared to ^{1}H NMR is that better resolution is obtained since the ^{13}C absorptions for most of the organic molecules are spread over 200 ppm instead of 10 ppm. Secondly carbons bearing no protons are revealed and finally a count of the number of

protons attached to each carbon results from comparison of the broad-band decoupled ^{13}C NMR spectrum with the off resonance ^{13}C spectrum. Therefore, the number of methyl (quartet), methylene (triplet), methinyl (doublet), and quarternary (singlet) carbons in a fairly complex molecule such as in the natural sweetener, stevioside, $C_{38}H_{60}O_{18}$ (Fig. 6) is easily determined by ^{13}C rather than by ^{1}H NMR.

On the other hand there are a number of disadvantages in ^{13}C NMR. A larger sample size (up to 100 mg) and longer sampling time (up to several days for small samples) is required. If 100 mg of a sample is available and that sample is of high solubility, the time required may be only a few minutes. A good spectrum can be obtained on as little as a 1 mg sample when several days are available for scanning the sample provided the sample is completely stable in the deuterated solvent. Another disadvantage is that due to variations in relaxation times and nuclear Overhauser effects (NOE), the areas of absorption differ for individual carbons (up to a factor of about 10). Thus, it is not as easy to tell relative numbers of carbons from ^{13}C as it is protons from ^{1}H NMR. Finally protons attached to heteroatoms are not visible. For the time being ^{13}C NMR spectroscopy is a method that compliments other spectroscopic techniques necessary to elucidate structures of organic molecules.

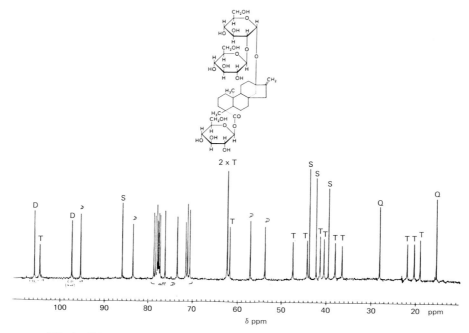

FIG. 6. ^{13}C NMR spectrum of the natural sweetener stevioside ($C_{38}H_{60}O_{18}$).

One commercial application of ^{13}C NMR is the determination of the purity of certain types of organic compounds. For example, whether a reduction, oxidation, or esterification processes have gone to completion can easily and also quite rapidly (if sufficient material is available) be determined.

2. ^{13}C-CP/MAS NMR

It has been observed that line broadening of NMR spectra of solid powders arises jointly from ^{13}C–^{1}H static dipolar interactions and from chemical shift anisotropy. These effects are eliminated or greatly reduced by dipolar decoupling (dephasing) of protons and by spinning the sample at an angle of 54.7° (magic angle) to the applied field. For solids a greater rate of spectral accumulation is required, this method is called cross polarization (CP).

High resolution ^{13}C-NMR which utilizes both cross polarization with dipolar dephasing and magic angle spinning (MAS) is called ^{13}C-CP/MAS NMR. This technique directly measures the organic carbon distribution in terms of aliphatic and aromatic carbon structures in oil shales, coals, humic substances, melanoidins, and other polymeric substances.

3. Applications of FTNMR

Because of the known utility of NMR as a fingerprinting technique, there is recent interest in the combination of this method with HPLC. Following the initial work on NMR spectroscopy of static HPLC fractions, considerable efforts have been focused on the development of analyses of flowing liquid chromatographic fractions in the presence of hydrogen containing solvents. The coupling of these two techniques may offer some advantages compared with the HPLC–MS systems since there are no volatility (or molecular weight) requirements, and the HPLC–FTNMR technique is more informative and nondestructive than MS (especially for studies of molecular stereochemistry). Furthermore, both HPLC and NMR measurements are carried out in solution and, thus, no phase transformations are required. Spectra can be obtained at low temperatures for nonstable compounds such as the 19-hydroxy prostaglandins that were previously mentioned. The development of suitable interfaces has been hampered by several problems. Primarily, the intensity of the NMR signal is dependent on the flow rate; the extent of line broadening and, consequently, the spectral res-

olution, depend on solvent viscosity and detector volume. Because of the relatively low sensitivity of NMR a compromise must be found between the optional flow rate and the detector volume. Moreover, the HPLC—FTNMR operation is limited to certain types of solvents. Since most commonly used HPLC solvents contain hydrogen, the memory of the minicomputer used for data acquisition will be saturated with the solvent signal, thus overshadowing the weak sample signal. This has necessitated the use of special suppression techniques for solvent signals, such as filtering, selective presaturation, selective excitation, and spin-echo techniques, or increasing the dynamic range of the minicomputer. It is possible to use deuterated solvents as the HPLC solvents but this involves considerable expense. During the past decade FTNMR has been applied to solid states especially to elucidate microstructures of polymers. The term microstructure pertains to the level of isomerism of a polymer chain. For example oxidized sites in polymers (carboxylic acids, ketones or alcohols) can be detected and quantified at levels as low as two sites per thousand residues from the ^{13}C NMR spectrum. Although carbon is ubiquitous in synthetic polymers, other nuclei have been used to advantage for polymer structure determination. For example, ^{15}N NMR for nylon polymers, ^{31}P NMR for poly(phosphazenes), and ^{19}F NMR for fluoropolymers. For some time ^{1}H FTNMR has been regarded as less informative for synthetic polymers than the carbon resonance; it is now the focus of renewed interest. Two dimensional nuclear Overhauser effects (NOE), coupling constant-resolved (*J*-resolved), and coupling constant-correlated (*J*-correlated) two-dimensional spectra all show promise for the elucidation of microstructures in polymers.

Perhaps one of the recent exciting and fruitful areas of FTNMR research is in biochemistry. It is only been a decade since Moon and Richards at California Institute of Technology and Hoult at Oxford demonstrated that FT–^{31}P NMR spectroscopy is an effective probe of the *in vivo* generation and utilization of phosphate-bond energy. FTNMR spectroscopy has since been applied to the study of many diverse systems, progressing from intact cells to perfused organs to animals and man. *In vivo* FTNMR is noninvasive, and thus event-induced changes in the concentration of metabolites can be followed over a time period. ^{31}P is by far the most widely used nucleus for biochemical FTNMR determinations, as the key metabolite, ATP, adenosine diphosphate (ADP), creatine phosphate, and sugar phosphates, for example, all contain phosphorus. The NMR sensitivity of phosphorus is high (100%) and offers selectivity not obtained with nuclei such as ^{13}C or ^1H because it is not ubiquitous, but rather occurs primarily in cellular metabolites. Metabolic pathways are now being followed by ^{31}P FTNMR. Techniques using whole animals (or humans) require that the NMR measurements be localized to a particular organ or to a certain portion of the body. More recently is the application of FTNMR in the study of the metabolism of ^{13}C-labeled drugs. The fate of the labeled carbons can be followed as a function of time using ^{13}C NMR spectroscopy. Also new pulse sequences have now been developed to observe only those protons that are attached to the labeled carbons.

X. X-Ray Crystallography

The purpose of X-ray analysis of crystal structures is to provide information on the positions of the individual atoms of a molecule, their interatomic distances, bond angles, and other features of molecular geometry such as the planarity of a particular group of atoms, the angles between the planes, and torsion angles around the bonds. The resulting three-dimensional representation of the atomic contents, rapidly determined by modern computerized techniques, of the crystal establishes the complete conformational structure and geometrical details hitherto unknown. [*See* X-RAY ANALYSIS.]

This information is of primary interest to chemists and biochemists who are concerned with the relation of structural features to chemical properties. It was only X-ray analysis that finally determined the double helix of two DNA molecules held together by hydrogen bonds that verified Watson and Crick's proposed model for DNA. Thus, X-ray analysis of organic compounds provides an unambiguous complete three-dimensional picture of the molecule, whereas other chemical and physical methods of structure analysis involved different aspects which collectively one can deduce the number and nature of the atoms bonded to each atom present (the topology of the molecule) or, for relatively simple molecules, provide some quantitative information from which geometrical details can be derived. There are instances whereby only X-ray analysis will decide absolutely the spatial arrangement of groups of at-

oms in simple organic molecules. It was X-ray analysis that confirmed that the cyclic sulphite (Fig. 7) had a boat–chair conformation with the S=O group trans to the cyclopropane ring and in an equatorial position.

The apparent disadvantages of X-ray analysis is that it is time consuming even with the use of automated computerized techniques. Furthermore, X-ray analysis can only be carried out on pure crystalline materials. The crystals must be a certain size and regularity in order that a good X-ray structure of the compound is obtained.

XI. Future Prospects

The combination and computerization of spectroscopic and chromatographic methods is a more efficient goal for the detection and identification of organic compounds than either of the two methods individually.

In his book, "Philosophy of Sciences" published in 1836, Andre Marie Ampere made use of the strange term *Cybernetique*, which in its modern concept denotes the multiscient robot of self-organizing machine. It is reasonable to assume that computer-controlled *cybermachines*, such as the coupled instruments, will soon become essential tools in every modern analytical laboratory.

It seems that the rapid development of the new techniques for the detection and identification of organic compounds is in accordance with

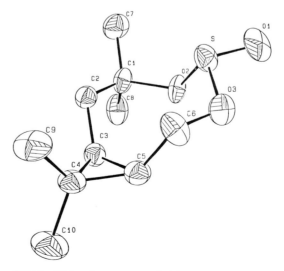

FIG. 7. Molecular structure of eight-membered cyclic sulfite as revealed by X-ray analysis.

the idea of the ancient Roman historian Tacitus who said: *Piscus crastinus papylo qui sapientam hodiernam continet,* meaning "Tomorrow's fish is wrapped in the paper which contains today's wisdom."

BIBLIOGRAPHY

Cooper, J. W. (1980). "Spectroscopic Techniques for Organic Chemists." Wiley, New York.

Grob, R. L., and Kaiser, M. A. (1982). "Environmental Problem Solving Using Gas and Liquid Chromatography," Journal of Chromatography Library, Vol. 21. Elsevier.

Gudzinowicz, B. J., and Gudzinowicz, M. J. (1979). "Analysis of Drugs and Metabolites by Gas-Chromatography–Mass Spectrometry," Chromatographic Science Series, Vol. 6, Marcel Dekker, New York.

Heftman, E. (1983). "Chromatography," Journal of Chromatography Library, Vol. 22B, Elsevier.

Ikan, R. (1982). "Chromatography in Organic Microanalysis." Academic Press, New York.

Irwin, W. J. (1982). Analytical Pyrolysis. "Chromatographic Science Series," Vol. 22. Marcel Dekker. New York.

Levy, G. C., Lichter, R. L., and Nelsen, G. L. (1980). "Carbon-13 Nuclear Magnetic Resonance Spectroscopy." Wiley, New York.

Pryde, A., and Gilbert, M. T. (1979). "Applications of High Performance Liquid Chromatography." Chapman and Hall, London.

ORGANIC CHEMISTRY—*SEE* PHYSICAL ORGANIC CHEMISTRY

ORGANIC CHEMISTRY, SYNTHESIS

John Welch   *State University of New York at Albany*

GLOSSARY

Asymmetric synthesis: Stereoselective preparation of a single enantiomer.

Biomimetic synthesis: Construction of molecules by mimicking biosynthetic processes.

Diastereoselectivity: Tendency of a reagent or reactant to form a single diastereomer in a reaction.

Electrophilicity: Tendency of a reactant that is electron deficient to satisfy this deficiency in a reaction.

Enantioselectivity: Tendency of a reagent to form a single enantiomer in a reaction.

Functional group: Structural feature of a larger molecule which imparts to that molecule a particular chemical reactivity. Functional groups often contain heteroatoms but may also result from changes in hybridization or bonding.

Natural products: Molecules formed by biological organisms. Natural products are frequently chiral and often complex.

Nucleophilicity: Tendency of a reactant that has an electron pair available for bonding to contribute this electron pair in a reaction.

Regioselectivity: Ability of a reactant to discriminate and principally form a single regioisomer.

Retrosynthetic analysis: Systematic bond disconnection of a synthetic target for purposes of planning a synthesis.

Organic synthesis, the science of the preparation of organic molecules, is a crucial element of organic chemistry. It is synthesis that makes possible the preparation of materials with novel structures and facilitates the study of naturally occurring substances otherwise available only after tedious and exacting isolation. The synthetic chemist is often called upon to test the latest advances in theory by preparing new molecules; yet the observations of the synthetic chemist frequently contribute to the development of new theories. [*See* ORGANIC CHEMICAL SYSTEMS, THEORY.]

Organic chemistry, as the chemistry of carbon compounds, requires the functionalization of these compounds to prepare new substances. Once functionalized, construction of new carbon molecular frameworks is possible. Historically, the preparation of natural products has been the most important application of synthetic methods. Contemporary chemists not only seek to prepare the chiral natural compounds efficiently by asymmetric synthesis but, by utilizing biomimetic strategies, to copy the biosynthetic pathways.

Limitations of space restrict this article to highlights from this diverse field. Throughout the discussion, name reactions and procedures are cited in parentheses.

I. Functional Group Manipulation

As discussed in the introductory section, a key element of directed organic synthesis is the selective introduction and modification of functional groups. Manipulation of a functional unit may provide discrimination in chemical reactivity. Functional groups typically contain heteroatoms such as halides, oxygen as in alcohols or carbonyls, or nitrogen as in amines. Variation of the hybridization and bonding of carbon as in alkenes, alkynes, and arenes also constitutes functionality. Recently, carbon bonds to most of the elements of the periodic chart have been exploited by organic chemists; however, the chemistry of most of these more esoteric molecules is illustrated by those groups discussed earlier. [*See* HETEROCYCLIC CHEMISTRY.]

A. HALOGENATION

Halogenation is one of the most effective means to introduce functionality in saturated or unsaturated hydrocarbons. The relatively low reactivity of alkanes and alkenes requires reactive reagents for substitution. A large number of well-established methods for the interconversion of organic halides to other functionalities increases the utility of halogenated hydrocarbons as intermediates in the introduction of other functional groups. Conversely selective preparation of halides from alcohols carbonyls, or amines is also an important process. [*See* HALOGEN CHEMISTRY.]

1. Reactions of Alkanes and Alkenes

Reaction of an alkane with a halogen usually procedes via a free radical pathway. The reactions of chlorine and bromine are more useful than those of fluorine or iodine. Direct fluorination of hydrocarbons with molecular fluorine is a very exothermic process resulting in complex product mixtures and extensive decomposition. Recently, it has been possible to improve the selectivity of the direct reaction of fluorine by dilution with an inert gas and by using very efficient cooling in a special apparatus. [*See* FLUORINE CHEMISTRY.]

Reactions with bromine or chlorine must be initiated either photochemically or by the use of a free radical initiator:

$$X_2 \xrightarrow{h\nu} 2X\cdot$$
$$RH + X\cdot \rightarrow R\cdot + HX$$
$$R\cdot + X_2 \rightarrow RX + X\cdot \tag{1}$$
$$R\cdot + X\cdot \rightarrow RX$$

As with fluorination, a mixture of products is formed. Photochemically induced iodination suffers from side reactions resulting from the hydrogen iodide formed. [*See* PHOTOCHEMISTRY, ORGANIC.]

Although arenes will undergo radical addition reactions with loss of aromaticity, the Lewis acid promoted electrophilic halogenation of arenes is a substitution reaction where aromaticity is not lost.

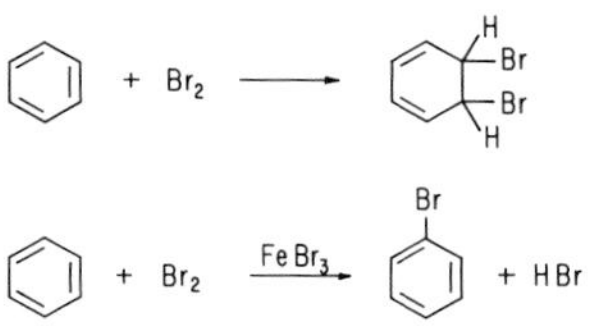

$$(2)$$

As an electrophilic substitution reaction, halogenation is governed by directing effects and can be regioselective.

In contrast to alkanes, alkenes readily add bromine and chlorine in electrophilic addition reactions:

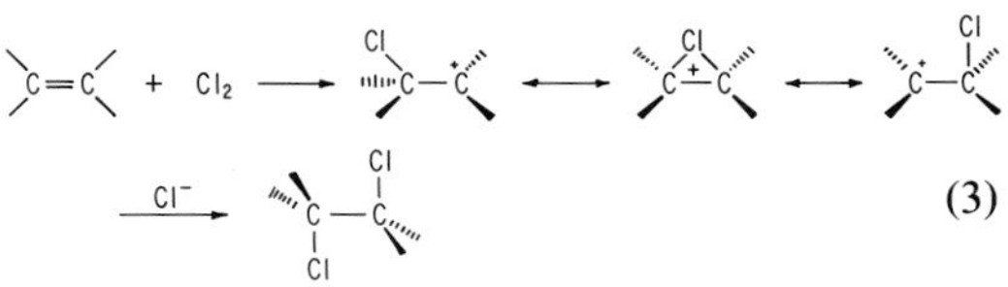

$$(3)$$

Hydrogen halides also add to alkenes following Markovnikov's rule, where the halide attaches to the more highly substituted carbon:

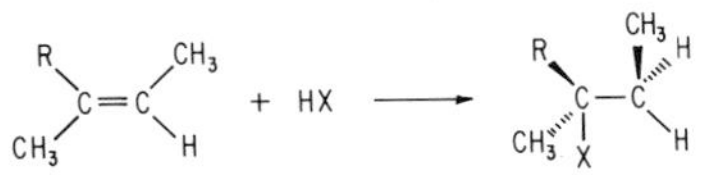

$$(4)$$

Hydrogen halide addition is occasionally accompanied by side reactions resulting from cationic rearrangements. Additions of hydrogen halides to alkynes are generally slow, forming, after addition of 2 moles of hydrogen halide, the geminal dihalides.

Alkenes may be treated with reagents such as *N*-bromosuccinimide in the presence of an initiator such as peroxide or light:

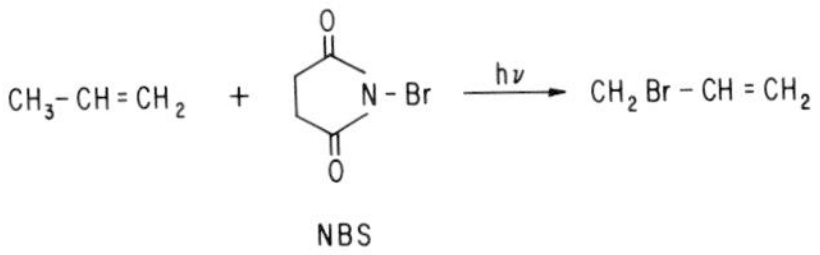

$$(5)$$

Under these conditions, where products of free radical attack are favored, the reaction can be very specific for allylic halogenation.

2. Reactions of Alcohols

Alcohols are easily converted to halides by displacement reactions. It is necessary to improve the leaving group ability of the hydroxyl by protonation or by conversion to a sulfonate or phosphate ester for successful displacement:

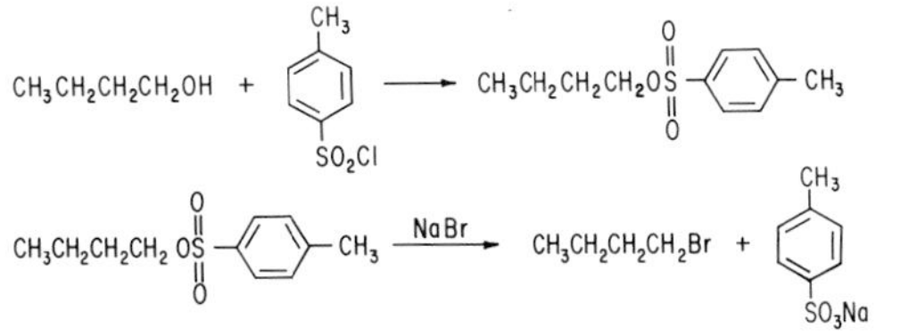

$$(6)$$

In reactions of primary and secondary alcohols with hydrogen fluoride, hydrogen chloride, or hydrogen bromide, competing rearrangements of the cationic intermediate may lessen the selectivity of the reaction:

$$
\underset{CH_3}{\overset{CH_3}{>}}CH-CH\underset{CH_3}{\overset{OH}{<}} + HBr \longrightarrow \underset{CH_3}{\overset{CH_3}{>}}CBr-CH_2CH_3 \quad (7)
$$

Treatment of an alcohol with thionyl chloride, sulfur tetrafluoride, phosphorous tribromide, or phosphorous pentachloride can yield the unrearranged halide regioselectively.

3. Haloalkylation Reactions

Halogenation can also be effected simultaneously with carbon–carbon bond forming reactions employing halogenated reactants. The chloromethylation reaction of aromatics with chloromethyl methyl ether and a Lewis acid is probably the best known example of such a reaction. It is possible to avoid the use of the toxic chloromethyl methyl ether by *in situ* formation of the reagent with formaldehyde and hydrogen chloride:

$$
\text{C}_6\text{H}_6 + HCHO \xrightarrow{HCl/ZnCl_2} \text{C}_6\text{H}_5CH_2Cl \quad (8)
$$

Alpha haloketones may be prepared by the reaction of carboxylic acid halides with diazomethane. Direct treatment of a diazoketone with a hydrogen halide, in a related reaction also yields alpha haloketones or, under modified conditions, dihaloketones substituted with different halogens:

$$
\text{Ar-CO-CHN}_2 \xrightarrow{HF} \text{Ar-CO-CH}_2F + N_2 \quad (9)
$$

Dihalomethylation may also be effected by reaction of dihalocarbanions with carbonyl compounds or by the alkylation of a carbanion with chloroform:

$$
\underset{R}{\overset{O}{R-C-R}} + LiCHCl_2 \longrightarrow \underset{R}{\overset{OH}{R-C-CHCl_2}} \quad (10)
$$

Halomethylenation is possible by reaction of an appropriately halogenated Wittig reagent with an aldehyde or ketone.

B. Hydroxylation

The selective direct hydroxylation of alkanes is not a practical general laboratory transformation although enzymatic hydroxylation of unactivated carbons is a major metabolic reaction in animals. Preparatively hydroxylation is more easily accomplished by hydrolysis of a halide.

The ease of hydrolysis, iodide > bromide > chloride, parallels the synthetic utility of the reaction:

$$
CH_3CH_2CH_2CH_2I \xrightarrow{H_2O} CH_3CH_2CH_2CH_2OH \quad (11)
$$

Hydrolysis of fluorides is of little consequence. In order to minimize side reactions, such as elimination to an alkene or rearrangement, S_N2 reaction conditions are desirable. However, under some conditions elimination reactions can be useful or desired. The synthesis of phenol from treatment of chlorobenzene with hydroxide ion first procedes by elimination to form benzyne followed in a second step by the addition of water to form phenol.

1. Addition of Water to Alkenes

The direct hydroxylation of alkenes with aqueous acid results in the Markovnikov addition of water. Under these conditions, rearrangements of the intermediate cation are possible with a corresponding loss of regiospecificity.

$$
\underset{CH_3}{\overset{CH_3}{>}}C=CH_2 \xrightarrow{H^+/H_2O} \underset{CH_3}{\overset{CH_3}{>}}\overset{+}{C}-CH_3 \longrightarrow (CH_3)_3COH \quad (12)
$$

Regiospecific addition of water in an anti-Markovnikov sense is readily possible via hydroboration of an olefin. The addition of borane or alkylborohydrides can be highly regioselective. Oxidation of the boron–carbon bond with hydrogen peroxide generates the desired alcohol:

$$
3\,CH_3CH=CH_2 + BH_3 \longrightarrow (CH_3CH_2C_2)_3B \xrightarrow{H_2O_2/OH^-} 3\,CH_3CH_2CH_2OH \quad (13)
$$

When addition in a Markovnikov sense is required, oxymercuration can be very effective and is not accompanied by rearrangements. Addition of an electrophilic mercury species to an olefin, followed by trapping of the intermediate cation, leads to the formation of an oxygenated organomercurial. The carbon–mercury bond may be cleaved reductively with sodium borohydride:

$$
CH_3CH_2CH=CH_2 + Hg(OAc)_2 \longrightarrow CH_3CH_2\overset{OAc}{\underset{|}{CH}}-CH_2HgOAc
$$
$$
\xrightarrow{NaBH_4} CH_3CH_2\overset{OH}{\underset{|}{CH}}-CH_3 + Hg^{\circ} \quad (14)
$$

2. Reduction of Carbonyl Compounds

Aldehydes, ketones, carboxylic acids, and carboxylic acid derivatives can be reduced to alcohols. Historically, catalytic reduction with hydrogen and a catalyst such as platinum oxide or Raney nickel was employed to form the hydroxylic product in high yields. More recently, the more convenient metal hydride reducing agents such as lithium aluminum hydride or sodium borohydride and their derivatives have been employed to form alcohols at ambient or subambient temperatures:

$$R \overset{O}{\underset{}{C}} R + LiAlH_4 \longrightarrow \xrightarrow{NH_4Cl/H_2O} R \overset{OH}{\underset{R}{C}} H \quad (15)$$

A variety of additional methods exist for the reduction of carbonyl compounds to alcohols. A typical example is the Meerwein–Pondorf–Verley reduction of carbonyls with aluminum alkoxides:

$$\text{(structure)} =O + Al(OCH(CH_3)_2)_3 \xrightarrow{H^+} \text{(structure)} \overset{OH}{\underset{}{C}} CH_3 \quad (16)$$

3. Electrophilic Oxygenation

Contemporary work has focused on the development of methods for the electrophilic oxygenation of alkanes, alkenes, and arenes:

$$O_3 \xrightarrow{H^+} [HO_3^+] \xrightarrow{\text{(benzene)}} \text{(phenol)} \overset{OH}{\underset{}{}} + H^+ + O_2 \quad (17)$$

Treatment of ozone or hydrogen peroxide with very strong acids, such as hydrogen fluoride–antimony pentafluoride or fluorosulfuric acid–antimony pentafluoride, is proposed to lead to the formation of protonated ozone or hydrogen peroxide, which are highly electrophilic reagents.

4. Hydroxymethylation

By simple analogy with haloalkylation chemistry, alcohols can also be introduced by carbon–carbon bond forming reactions. The addition of formaldehyde to a nucleophilic carbon such as the alpha carbon of an enol or organometallic reagent leads to hydroxymethylation:

$$\text{(Ar)}{-}MgBr + HCHO \xrightarrow{H^+/H_2O} \text{(Ar)}{-}CH_2OH \quad (18)$$

In a different approach, treatment of an alkyl boron compound with carbon monoxide leads to carbon monoxide insertion in the alkyl carbon boron bond which can be reduced to form the hydroxymethyl group:

$$3 \overset{R}{\underset{H}{C}}=\overset{H}{\underset{H}{C}} \xrightarrow{BH_3} (RCH_2-CH_2)_3 B \xrightarrow{CO} (RCH_2CH_2)_2 \overset{O}{\underset{}{B}}C CH_2CH_2R$$

$$\xrightarrow{LiBH_4} \xrightarrow{H_2O} RCH_2CH_2CH_2OH \quad (19)$$

The Williamson ether synthesis, in which treatment of an alkyl halide with the alkali metal salt of an alcohol yields an ether, appends an alkoxy chain to an alkyl halide.

$$CH_3I + CH_3CH_2CH_2CH_2ONa \longrightarrow CH_3CH_2CH_2CH_2OCH_3 + NaBr$$

$$(20)$$

Alkoxymethylation is possible by reaction of substituted ethers with an appropriate nucleophile.

B. AMINATION

Selective formation of amines is extremely important for the synthesis of many natural products and biologically active compounds. The introduction of amines in the presence of other functionality is a particularly challenging synthetic problem. The selectivity required for practical utility is not generally associated with the more rigorous conditions required for direct amination; therefore, functional group transformations are employed.

1. Displacement of Halides

The reaction of an alkyl halide with ammonia often leads to a mixture of primary, secondary, and tertiary amines where more than one alkyl halide molecule has reacted with a single ammonia. Selective formation of the primary amine may be possible if the alkyl halide is soluble in the presence of a large excess of ammonia. On a laboratory scale, the Gabriel synthesis or one of its modifications may be a more practical approach. The alkyl halide is treated with phthalimide or an alkali metal salt of phthalimide. The product alkylated phthalimide is hydrolyzed to selectively form the primary amine:

$$\text{(phthalimide)}N-Na + CH_3CH_2CH_2Br \longrightarrow \text{(phthalimide)}N-CH_2CH_2CH_3 + NaBr$$

$$\xrightarrow{NaOH} \text{(structure)}\overset{ONa}{\underset{ONa}{}} + H_2NCH_2CH_2CH_3 \quad (21)$$

Selective formation of secondary or tertiary amines by reaction of an alkyl or dialkyl amine is generally ineffective for the same reason that the reaction of ammonia is not selective. However, direct reaction of an amine with an excess of an alkyl halide can be a useful method to the preparation of quaternary ammonium salts.

2. Derivatization of Carbonyl Compounds

An especially useful way to prepare secondary and tertiary amines is via an intermediate imine or iminium ion. For example, reaction of formaldehyde with a secondary amine in the presence of formic acid leads directly to a new tertiary amine:

$$\begin{array}{c} R \\ \diagdown \\ R' \end{array} NH + HCHO \longrightarrow \begin{array}{c} R \\ \diagdown + \\ R' \end{array} N = CH_2 + HCO_2H \longrightarrow \begin{array}{c} R \\ \diagdown \\ R' \end{array} N - CH_3 + CO_2 + H^+ \tag{22}$$

Dimethyl amines are conveniently prepared by reaction of formaldehyde with a primary amine in the presence of formic acid:

$$CH_3CH_2CH_2NH_2 + HCHO + HCO_2H \longrightarrow CH_3CH_2CH_2N(CH_3)_2 \tag{23}$$

Acids may be converted to amides through reaction of the acid chloride with an amine:

$$R COCl + R'CH_2NH_2 \longrightarrow R \overset{O}{\underset{\|}{C}} NH CH_2R'$$

$$R \overset{O}{\underset{\|}{C}} NH CH_2R' \xrightarrow{LiAlH_4} R - CH_2NH CH_2R' \tag{24}$$

Reduction of the acid amide to a new amine is possible by catalytic hydrogenation using a metal catalyst (such as platinum oxide, palladium oxide, palladium on barium sulfate, or Raney nickel). Metal hydrides such as lithium aluminum hydride, aluminum trihydride(alane), or diborane will also conveniently reduce amides to amines.

The direct conversion of acids to amines via an acyl nitrene intermediate is also possible by the Hofmann, Schmidt, or Curtius rearrangement:

$$R \overset{O}{\underset{\|}{C}} N_3 \longrightarrow RN = C = O + N_2 \xrightarrow{H_2O} RNH_2 + CO_2 \tag{25}$$

3. Allylic Amination

Allylic amination of alkenes via sulfur and selenium reagents is also possible. In the latter case, the reagent is prepared by treatment of selenium with chloramine T, in the former, by reaction of thionyl chloride with *p*-toluenesulfonamide:

$$2 CH_3 - \bigcirc - SO_2NH_2 + SOCl_2 \longrightarrow CH_3 - \bigcirc - SO_2N = S = NSO_2 - \bigcirc - CH_2$$

$$\xrightarrow{RCH_2 - CH = CH_2} \begin{array}{c} R - CH - CH = CH_2 \\ | \\ NH \\ | \\ SO_2 - \bigcirc - CH_3 \end{array} \tag{26}$$

4. Mannich Reaction

Condensation of formaldehyde with an amine forms an iminum ion which can act as an electrophile in further synthetic transformations. One result is the addition of a carbon bearing an amine to the nucleophile. Commonly the reaction is effected with enolizable carbonyl compounds; however, the reaction will work with any compound with an activated hydrogen:

$$HCHO + \begin{array}{c} CH_3 \\ \diagdown \\ CH_3 \end{array} NH \longrightarrow \begin{array}{c} CH_3 \\ \diagdown + \\ CH_3 \end{array} N = CH_2 \xrightarrow{\begin{array}{c} OH \\ | \\ CH_2 \diagup \diagdown CH_3 \end{array}} \begin{array}{c} CH_3 \\ \diagdown \\ CH_3 \end{array} N - CH_2 - CH_2 \overset{O}{\underset{\|}{C}} CH_3 \tag{27}$$

The intermediate iminium ion is even electrophilic enough to react with arenes.

C. CARBONYLATION AND CARBOXYLATION

One of the most useful functional groups in organic synthesis is the carbonyl group. The carbonyl function reacts at the carbonyl carbon as an electrophile or may enolize and react at the alpha carbon as a nucleophile. Commensurate with the importance of carbonyl compounds to synthetic chemistry, a large number of methods have been developed for the preparation of these compounds.

1. Oxidation of Alcohols

A tremendous variety of oxidants are known for the conversion of alcohols to carbonyl containing compounds. The most difficult of these transformations is the oxidation of a primary alcohol to an aldehyde, the aldehyde often being more susceptible to oxidation, than the primary alcohol. Typical of the reagents employed to effect these oxidations are chromium and manganese oxides, dimethyl sulfoxide and dicyclohexylcarbodiimide (Pfitzner–Moffat), and dimethyl sulfoxide and oxalyl chloride (Swern):

$$RCH_2OH \xrightarrow{[Ox]} RCHO \tag{28}$$

Oxidation of secondary alcohols to ketones is simpler if only because ketone products are more stable to further oxidation. Typically chromium reagents (CrO_3) or permanganate salts ($KMnO_4$) are used:

$$R'R'CH-OH \xrightarrow{[Ox]} R'R'C=O \qquad (29)$$

The conversion of primary alcohols or aldehydes to carboxylic acids is normally a facile reaction using $KMnO_4$. Aldehydes can be easily oxidized by air to acids, a result of the lability of the aldehydic hydrogen.

2. Carbonylation

The direct addition of a carbonyl group to olefins is promoted by metal catalysts such as rhodium, iridium, and cobalt. The oxo reaction, the commercial route to the preparation of aldehydes which are reduced to alcohols important as plasticizers, relies on such a carbonylation. [See CATALYSIS, HOMOGENEOUS; ORGANIC CHEMICALS, INDUSTRIAL PRODUCTION.]

$$CH_2=CH_2 + CO + H_2 \xrightarrow{HCo(CO)_4} CH_3CH_2CHO \qquad (30)$$

The formylation of arenes by carbon monoxide and Lewis acids (Gatterman–Koch) is an important well-established method for the synthesis of arene aldehydes:

$$\text{(31)}$$

The use of carbon monoxide can be avoided by reaction of dimethylformamide in the presence of phosphorous oxychloride (Vilsmeier–Haack). The electrophilic intermediate chloroiminium ion adds to activated arenes. The substitution product is readily hydrolyzed to an aldehyde:

$$\text{(32)}$$

3. Nucleophilic Reagents

Carbonyl compounds may also be prepared by the reaction of a nucleophilic reagent such as an organolithium or Grignard reagent with formamide (Bouveault) or with an orthoester. In these examples, the initial product of reaction is a hemiaminal or acetal, respectively, which is unreactive to additional equivalents of the nucleophilic reagent but is readily hydrolyzed under acid conditions to reveal an aldehyde:

$$CH(OR')_3 + R\,MgBr \longrightarrow R-CH(OR')_2 + ROMgBr$$
$$RCH(OR)_2 + H^+/H_2O \longrightarrow RCHO \qquad (33)$$

The reactions are not limited to Grignard or organolithium reagents; any activated hydrogen compound will give similar results. In related reactions an organometallic compound will react directly with carbon dioxide to form on hydrolysis a carboxylic acid.

D. INTRODUCTION OF SULFUR

The importance of sulfur in natural products and in products of commerce should be mentioned. Sulfur-containing compounds also have significant utility in directing further synthetic transformations.

Sulfonation of arenes is one of the oldest and best-known reactions for the introduction of sulfur into organic molecules. Treatment with sulfuric acid and oleum or chlorosulfonic acid leads to the preparation of sulfonated aromatics that are useful as surfactants, dye constituents, or chemical intermediates:

$$\xrightarrow{H_2SO_4 / SO_3} \qquad SO_3H \qquad (34)$$

Sulfonation under acidic conditions is of little utility in the preparation of aliphatic sulfonates. Alkyl sulfonates are selectively prepared by reaction of alkanes with sulfur dioxide and chlorine. When the reaction is conducted in the presence of ultraviolet light, side reactions such as chlorination are effectively suppressed:

$$RCH_3 + SO_2 + Cl_2 \xrightarrow{h\nu} RCH_2SO_2Cl \qquad (35)$$

The initial product of these reactions, as well as the product of the reaction of arenes with chlorosulfuric acid, are sulfonyl chlorides which may be hydrolyzed to the desired acids.

1. Sulfide Formation

Sulfides may be prepared by the reaction of alkali metal salts such as sodium sulfide with alkyl halides. Dialkyl sulfides are prepared by the reaction of the alkali metal salt of an alkyl sulfide with a second molecule of an alkyl halide:

$$RSNa + R'CH_2X \longrightarrow R'CH_2SR + NaX \qquad (36)$$

Alternatively alkyl halides may be reacted with thiourea to form sulfides upon hydrolysis. Sulfides are also formed by the Markovnikov addition of hydrogen sulfide to alkenes. Alkyl sulfides will add in a conjugate manner to

α,β-unsaturated carbonyl compounds, giving β-ketodialkyl sulfides.

II. Carbon–Carbon Bond Forming Reactions

The preparation of the functional groups most important in organic synthesis has been described, but in the design of a synthesis, the construction of the carbon skeleton is often the greatest challenge. Frequently the desired functionality is either protected to prevent undesired side reactions or is carried through a synthetic sequence in a masked form, to be liberated after other transformations have been accomplished. These protected or masked functional groups have been described as synthons, structural units which can be formed or assembled by known or conceivable synthetic transforms. The retrosynthetic analysis of the carbon skeleton determines which carbon–carbon bonds should be formed and in what order they should be assembled. After analyzing the molecule and determining convenient fragments, the emphasis now focuses on determining which carbon–carbon bond forming reactions can be employed.

Carbon–carbon bond forming reactions can be described as being ionic processes, in which one fragment is electron deficient (electrophilic) and the other fragment is electron rich (nucleophilic), or as radical processes in which each fragment contributes a single electron. Traditionally, ionic processes have been the better understood and more effectively manipulated. However on occasion, reactions with radical intermediates have been described and effectively employed as well. Ionic processes will be described as resulting from nucleophilic or electrophilic components. Such a distinction is somewhat artificial, but parallels traditional descriptions for ionic carbon–carbon bond forming reactions in which one component must be electrophilic and the other nucleophilic. Typical electrophilic reactions such as alkylation and acylation and typical nucleophilic reagents such as organometallic compounds, enamines, and deprotonated imines will be discussed. [*See* ELECTRON TRANSFER REACTIONS, GENERAL.]

A. Alkylation

As mentioned earlier alkylations are described as electrophilic from the perspective of the alkylating agent. The electrophilic alkylation of alkanes or alkenes is known but is only of limited utility because of the difficulty of achieving selectivity. Nonetheless, the electrophilic alkylation of arenes is a reaction of substantial significance.

1. Alkyl Halides

Reports of the reaction of alkyl halides with arenes in the presence of Lewis acids appeared in the literature as early as 1877. Although alkyl fluorides are the most reactive of the alkyl halides, alkyl chlorides and bromides are the most widely used. Alkyl iodides are less commonly used as a result of accompanying side reactions and decomposition. Tertiary and benzylic halides are the most reactive, with secondary halides less reactive but more reactive than primary halides. Dialkyl halonium ions prepared by treatment of excess alkyl halide with antimony pentafluoride are especially reactive alkylating reagents. However, the most reactive alkylating agents are the methyl fluoride–antimony pentafluoride and ethyl fluoride–antimony pentafluoride complexes:

$$RCH_2Cl \ + \ \bigcirc \ \xrightarrow{AlCl_3} \ \bigcirc^{CH_2R} \qquad (37)$$

$$2\,CH_3I \ + \ SbF_5 \ \longrightarrow \ CH_3\overset{+}{I}CH_3$$

$$CH_3\overset{+}{I}CH_3 \ + \ \bigcirc \ \longrightarrow \ \bigcirc^{CH_3} \ + \ CH_3I \qquad (38)$$

$$CH_3CH_2{}^{\text{···}}FSbF_5 \ + \ \bigcirc \ \longrightarrow \ \bigcirc^{CH_2CH_3} \qquad (39)$$

These powerful electrophiles will even react with poor nucleophiles such as alkanes.

2. Alkenes, Alcohols, Esters, and Ethers

In addition to alkyl halides, alkenes will act as electrophiles in the presence of Lewis or Brønsted acids. Alcohols, esters, and ethers also will act as carbon electrophiles under acidic conditions.

B. Acylation

Acylation is another important electrophilic reaction for the substitution of arenes. The electrophilic reagent, an acyl cation, is stabilized by resonance. Frequently used as acylating reagents in the presence of a Lewis acid are acyl chlorides, anhydrides, acids, and esters. Generally the reaction is performed by the addition of the Lewis acid catalyst to a mixture of the reactant arene and acyl halide:

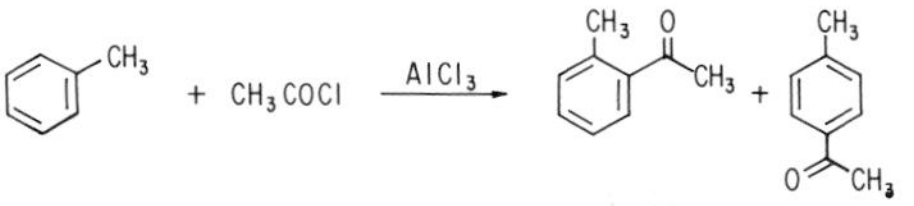

$$\text{(40)}$$

$$\text{(41)}$$

Similar to the reactions of dialkyl halonium ions or methyl fluoride–antimony pentafluoride complexes, stable acylium ions, prepared by reaction of an acid fluoride with antimony pentafluoride, are extremely reactive acylating agents.

C. ALKYLATION OF ORGANOMETALLIC REAGENTS

Organometallic reagents often react as carbanions and as such are nucleophilic. The electrophilic component of these reactions is generally aliphatic, but reactions of aromatic compounds are known. Although the alkylation of anionic carbon is well known, the mechanism of the reaction is still not clearly understood.

1. Alkyl Halides

The reactions of organometallic reagents with alkyl halides are coupling reactions. The reactions of alkyl chlorides, bromides, or iodides procede well when the nucleophilic partner is a cuprate or a Grignard reagent (organomagnesium halide) in the presence of a copper, iron or nickel catalyst:

$$(CH_3)_2 \, CuLi \; + \; RI \; \longrightarrow \; R-CH_3 \qquad (42)$$

$$CH_3MgBr \; + \; RI \; \xrightarrow{\;Cu^+\;} \; RCH_3 \qquad (43)$$

Allylic and propargylic halides also have been employed to alkylate organometallic reagents.

2. Alkyl Sulfonates

Often the alkylation of organometallic reagents is more effectively accomplished with an alkyl sulfonate. Sulfonates will undergo displacement reactions by both Grignard reagents and cuprates, frequently more cleanly and in better yield than the corresponding halides.

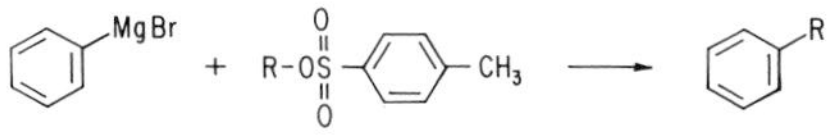

$$\text{(44)}$$

3. Epoxides

Epoxides may serve as a source of electrophilic carbon for reactions with organometallic reagents. The epoxide is opened, alkylating the

organometallic while simultaneously forming a new alcohol.

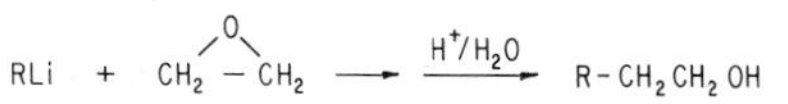

$$RLi \; + \; CH_2 \!-\! CH_2 \; \xrightarrow{\;H^+/H_2O\;} \; R-CH_2CH_2OH \qquad (45)$$

4. α,β Unsaturated Carbonyl Compounds

Nucleophilic organometallic reagents, particularly cuprates, will also add in a conjugate manner to α,β unsaturated carbonyl compounds to form an enolate, stabilized by resonance. Bulky alkyl groups (R) enhance this reaction relation to carbonyl addition. Similarly, α,β-unsaturated sulfones will add organometallics to form a stabilized anionic product:

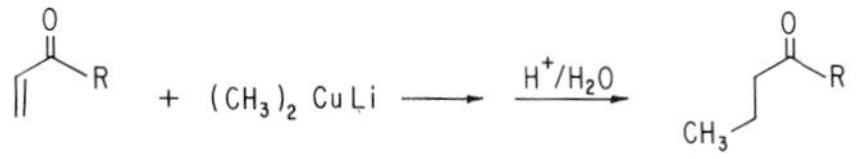

$$\text{(46)}$$

5. Alkylation of Active Hydrogen Compounds

Active hydrogen compounds are those in which the substituents present are capable of stabilizing the conjugate base formed on deprotonation of the starting material. Stabilization of the resultant anion significantly enhances the acidity of the proton, hence "active hydrogen" compounds. Common substituents which in this way stabilize negative charge are esters, aldehydes, ketones, acids, nitriles, nitro groups, sulfoxides, sulfones, sulfonamides, and sulfonates. Even further increased acidity is found on disubstitution, as in malonic esters. Deprotonation by even a weak base is sufficient to form the anion which may then be trapped by an alkyl halide. The reaction is a general one and applies to the other groups described as well:

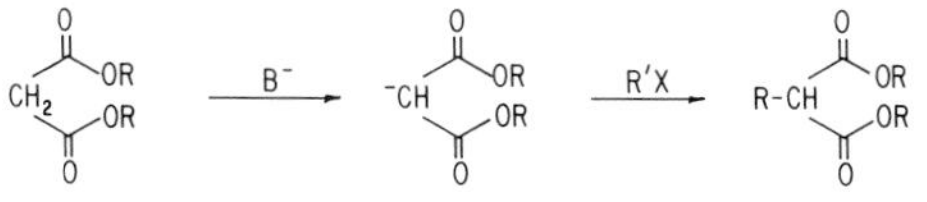

$$\text{(47)}$$

Imines, prepared by condensation of a primary amine with an aldehyde or ketone, may be deprotonated with a strong base to form the nitrogen equivalent of an enolate. The ready alkylation of these anions has found utility in enantioselective processes to be discussed later:

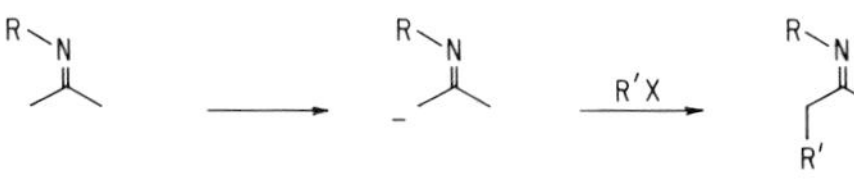

$$\text{(48)}$$

In addition to the oxidized forms of sulfur, dithioacetals and ketals, such as 1,3-dithianes, also have activated hydrogens. Dithiane is readily deprotonated and easily alkylated. The aldehyde or ketone functionality masked by the dithioacetal or ketal is easily released under a variety of conditions:

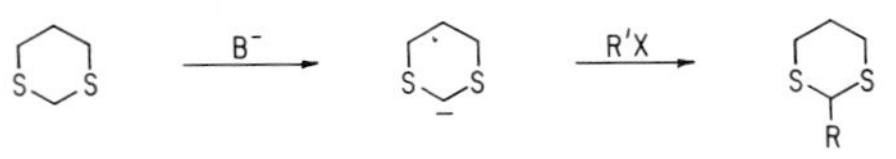

$$\text{(49)}$$

6. Alkylation of Organoboron Reagents

Trialkyl boron reagents react readily with alpha halocarbonyl compounds, displacing the halide with an alkyl group. The reaction is of broad, general utility. Related reactions occur with diazoalkanes, esters and ketones:

$$R_3B \;+\; BrCH_2\overset{O}{\underset{\|}{C}}R' \longrightarrow RCH_2\overset{O}{\underset{\|}{C}}R' \qquad (50)$$

D. REACTION OF ORGANOMETALLIC REAGENTS WITH CARBONYL COMPOUNDS

1. Grignard and Related Reagents

The addition of Grignard reagents to carbonyl compounds to form alcohols is an extraordinarily useful and general reaction. The reaction is facile and relatively simple. Ketones, aldehydes, and esters which require two equivalents of the Grignard reagent) react well:

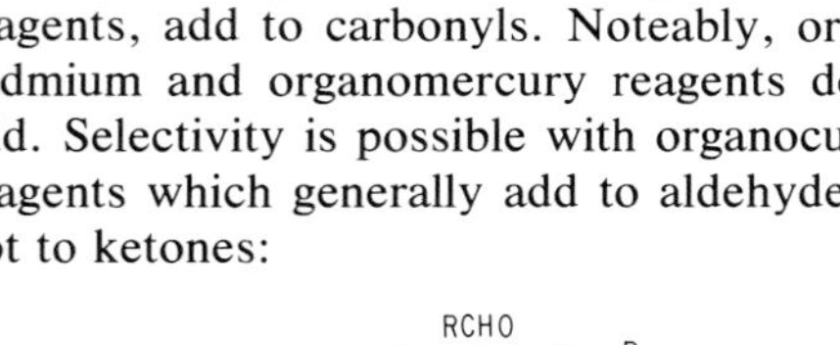

$$\text{(51)}$$

Other organometallic compounds, such as organolithium organozinc, and organoaluminum reagents, add to carbonyls. Noteably, organocadmium and organomercury reagents do not add. Selectivity is possible with organocuprate reagents which generally add to aldehydes but not to ketones:

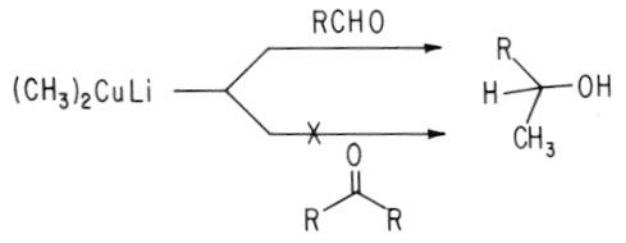

$$\text{(52)}$$

As mentioned earlier, addition of Grignard reagents to α,β-unsaturated carbonyl compounds often results in 1,4- rather than 1,2-addition reactions. In closely related chemistry, α bromo

esters, acids, amides, and nitriles will add to carbonyl compounds in the presence of zinc dust. Presumably an organozinc reagent (Reformatsky), formed *in situ*, reacts like a Grignard reagent and adds to the carbonyl compound:

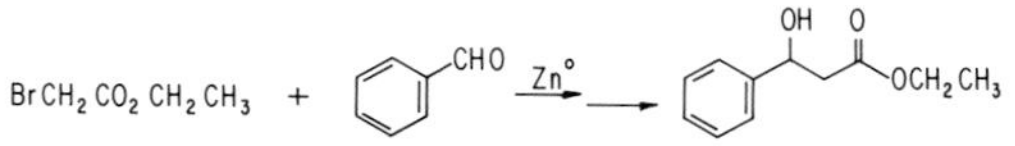

$$\text{(53)}$$

Organolithium reagents will add to the lithium salt of carboxylic acids to yield, on acidification, ketones:

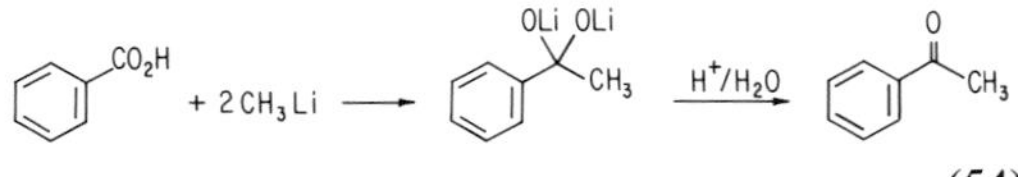

$$\text{(54)}$$

2. Addition of Active Hydrogen Compounds to Carbonyls

On deprotonation, active hydrogen compounds such as those substituted with carbonyl groups (as in esters, aldehydes, ketones, or acids) as well as nitriles, nitro groups, sulfoxides, sulfones, sulfonamides, and sulfonates will add to carbonyl compounds. The best known of these condensations is the aldol reaction. An enolate adds readily to ketones or aldehydes to form a beta hydroxy carbonyl compound:

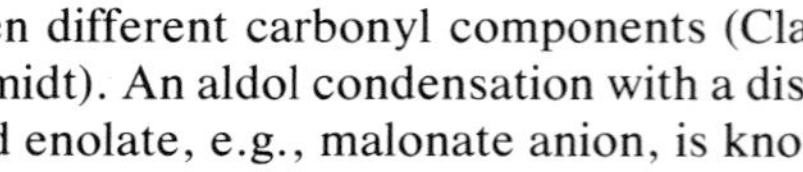

$$\text{(55)}$$

Under more vigorous reaction conditions this product is readily dehydrated to give the alpha, beta unsaturated carbonyl compound. The reaction may involve self condensation or may be a directed (crossed) aldol condensation between different carbonyl components (Claisen–Schmidt). An aldol condensation with a disubstituted enolate, e.g., malonate anion, is known as the Knoevenagel reaction:

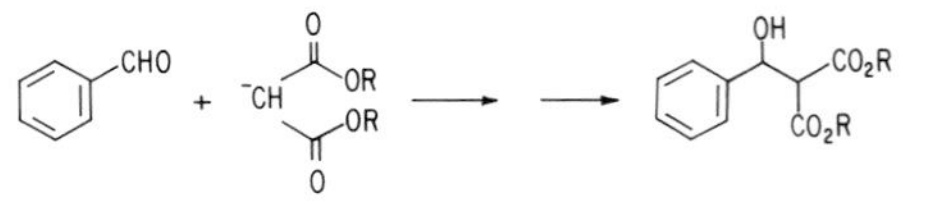

$$\text{(56)}$$

If substituted with an alpha carboxylic acid, the aldol product is especially susceptible to decarboxylation.

3. Wittig Reactions

One of the most useful reactions of a stabilized anion with a carbonyl compound is the condensation of a phosphorous ylide with an aldehyde or ketone to form an olefin. The ylide is usually prepared from an alkyl triphenylphosphonium salt by deprotonation with a strong base. Addition of the ylide to the carbonyl compound initially forms a betaine intermediate which collapses to the olefin and triphenylphosphine oxide. The phosphonium salts are commonly prepared with triphenylphosphine and an alkyl halide; however, other less common phosphines have been employed as well:

$$Ph_3P + RCH_2X \longrightarrow Ph_3\overset{+}{P}CH_2R\ X^- \qquad (57)$$

$$Ph_3\overset{+}{P}CH_2R + B^- \longrightarrow Ph_3\overset{+}{P}-\underset{}{\overset{-}{C}}H-R \qquad (58)$$

$$Ph_3\overset{+}{P}\underset{}{\overset{-}{C}}HR + R'\overset{O}{\underset{}{\|}}R' \longrightarrow \underset{Ph_3}{\overset{\overset{\overset{R'}{|}}{\overline{O}-C-R'}}{\overset{|}{P}-CHR}} \longrightarrow$$

betaine

$$\qquad (59)$$

$$\underset{H}{\overset{\overset{R'}{|}}{Ph_3\ \overset{}{P}{\underset{}{\overset{/|}{-}}}\overset{O}{\underset{}{-}}R}} \longrightarrow Ph_3P=O\ + \underset{R'}{\overset{R'}{\underset{}{}}}\!\!\overset{H}{\underset{R}{\diagup\diagdown}}$$

Ylides may also be prepared form phosphonates to yield a slightly more reactive ylide, in a variation known as the Horner–Emmons reaction:

$$(RO)_2\overset{O}{\underset{}{\|}}{P}\ CH_2R' \longrightarrow (RO)_2\overset{O}{\underset{}{\|}}{P}\ \overline{C}HR' \xrightarrow{PhCHO} PhCH=CHR'\ + \underset{(RO)_2\overset{}{\overset{O}{\|}}PO^-}{}$$

$$\qquad (60)$$

The olefin product stereochemistry may be controlled by choice of reaction conditions to form selectively either the cis or trans product.

It is also possible to form reactive ylides with sulfur compounds. Typically, dimethyl sulfonium methylide will add to carbonyl compounds to form epoxides rather than alkenes:

$$\underset{+}{\overset{CH_2}{\underset{}{|}}}{CH_3\overset{}{S}-CH_3} \xrightarrow{B^-} \underset{CH_3}{\overset{CH_3}{\underset{}{|}}}{S^+-\overline{C}H_2} \xrightarrow{PhCHO} PhCH\overset{O}{\overset{}{\diagup\diagdown}}CH_2 + CH_3SCH_3$$

$$\qquad (61)$$

III. Natural Product Synthesis

The application of synthetic methods to the preparation of naturally occurring compounds has always been an important part of organic chemistry. From the early days when separation and structural elucidation was the principal object, organic chemists have been interested in natural products. Rapidly, however, the synthesis of these materials came to be of importance,

in part to verify proposed structural assignments and, of increasing importance today, to prepare quantities of natural materials with interesting pharmacological properties otherwise available only in very limited amounts. The preparation of natural compounds has also had a significant impact on the theory of organic chemistry, for example, the principle of the conservation of orbital symmetry may have resulted in part from insights developed during the synthesis of vitamin B-12. In this section, several examples have been taken from the synthesis of steroids and prostaglandins to illustrate the development of synthetic strategy. [See PHARMACEUTICALS.]

A. STEROIDS

The early work of Robinson on the synthesis of the steroid nucleus may be contrasted with the elegant Woodward approach to the total synthesis of steroids. Robinson's approach requires the masterful manipulation of functionality so that through the course of the synthesis, relay compounds may be reached. Relay compounds are materials reached by alternate pathways, frequently by the degradation of a readily available natural product. The Robinson synthesis of androsterone, which spanned several years, required four stages, each stage comprised of numerous steps. The first stage was the transformation of 2,5-dihydroxynapthalene to the Reich diketone. The second stage was the transformation of the Reich diketone to the Koester and Logemann (KL) ketone. The third stage was the transformation of the KL ketone to dimethyl aetioallobilianate benzoate. The fourth required the conversion of dimethyl aetioallobilianate benzoate to androsterone. Both the KL ketone and aetioallobilianic acid were available from natural materials (Fig. 1).

In contrast to the lengthy manipulation of functionality to facilitate relay synthesis, Woodward's synthesis of methyl dl-3-keto-Δ4,9(11).16-etiocholatrienoate is considerably shorter and illustrates an advance in synthetic strategy. This efficient approach relies strongly on an understanding of stereochemical relationships and the efficient choice of functionality (Fig. 2).

B. PROSTAGLANDINS

The prostaglandins are a closely related family of compounds discovered as early as 1930 but whose structure was not determined until the

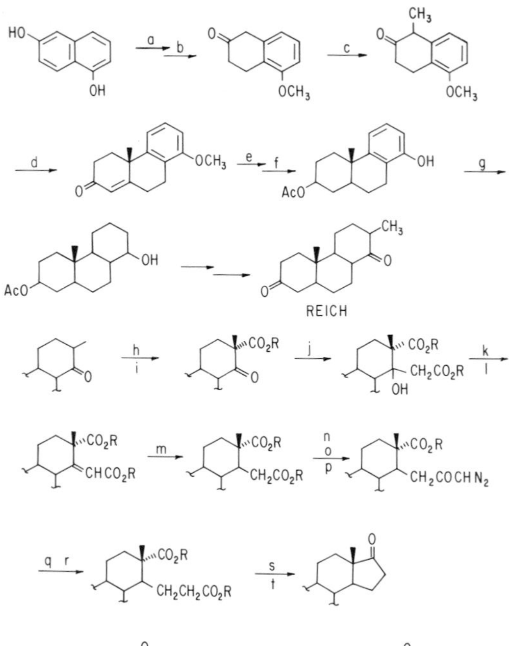

FIG. 1. Robinson synthesis of androsterone. (a) Sodium methoxide, dimethyl sulfate; (b) sodium, ethanol; (c) sodium methoxide, methyl iodide; (d) diethylaminobutanone, methyl iodide, potassium; (e) hydrogen iodide; (f) platinum oxide, hydrogen; (g) palladium–strontium carbonate, hydrogen; (h) sodium triphenyl methide, carbon dioxide; (i) diazomethane; (j) ethyl bromoacetate, zinc; (k) platinum oxide, hydrogen; (l) phosphorous oxychloride; (m) platinum oxide, hydrogen; (n) potassium hydroxide, methanol; (o) oxalyl chloride; (p) diazomethane; (q) silver nitrate; (r) potassium hydroxide, methanol; (s) acetic anhydride; (t) heat.

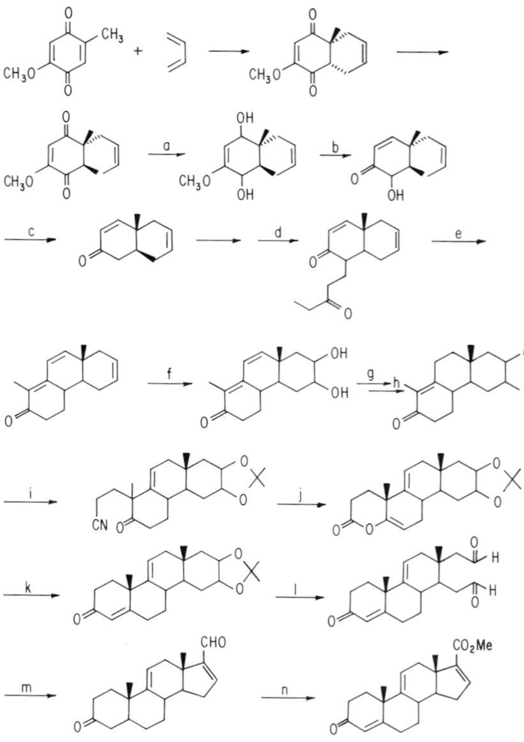

FIG. 2. Woodward synthesis of methyl d,l-3-keto-$\Delta^{4,9(11),16}$-etiocholatrienoate. (a) Lithium aluminum hydride; (b) acid, dioxane; (c) acetic anhydride, zinc; (d) ethyl vinyl ketone, potassium t-butoxide; (e) base, dioxane; (f) osmium tetroxide; (g) acid, acetone; (h) palladium–strontium carbonate, hydrogen; (i) acrylonitrile, Triton-B; (j) base; (k) methyl magnesium bromide, base; (l) periodic acid; (m) aqueous dioxane; (n) potassium dichromate, diazomethane.

1960s. These compounds had a variety of physiological effects, but were only available in very minute quantities. As such they were ideal targets for synthesis. Biosynthetic analysis has shown that fatty acids, in particular arachidonic acid, are the precursors of the prostaglandins. A common feature of the prostaglandins was the cyclopentane ring, with as many as four adjacent stereocenters (Fig. 3). It was demonstrated by Corey and widely adopted by others that a common intermediate, "Corey's lactone," could be used to synthesize a number of the prostaglandins (Fig. 4). Prostaglandin synthesis from this common intermediate can therefore be convergent, the side chains being prepared by separate synthetic procedures and coupled intact to the cyclopentanoid system. This is in sharp contrast to both the Robinson and Wood-

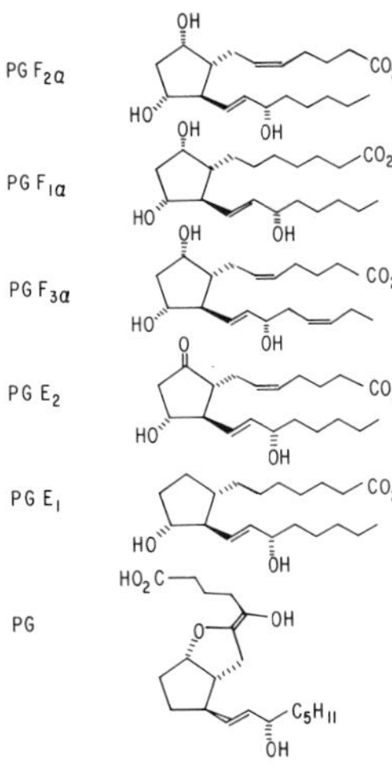

FIG. 3. Prostaglandins.

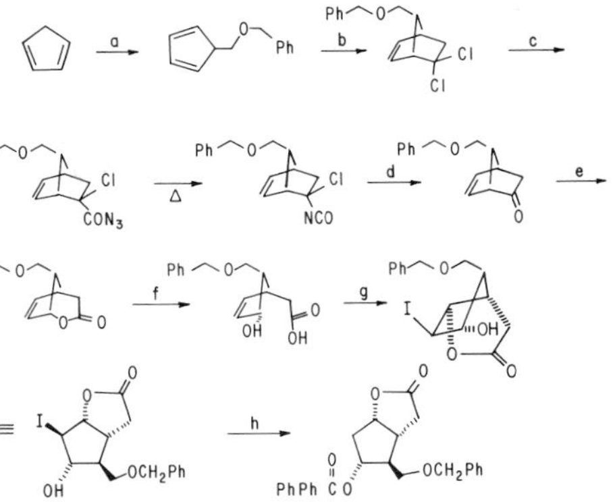

FIG. 5. Synthesis of protected Corey lactone by cycloaddition to a functionalized cyclopentadiene. (a) Thallium sulfate, benzyl chloromethyl ether; (b) 2-chloroacryloyl chloride; (c) sodium azide; (d) aqueous acetic acid; (e) m-chloroperbenzoic acid; (f) base, carbon dioxide; (g) potassium iodide–iodine; (h) p-phenylbenzoyl chloride, tri-butyl tin hydride.

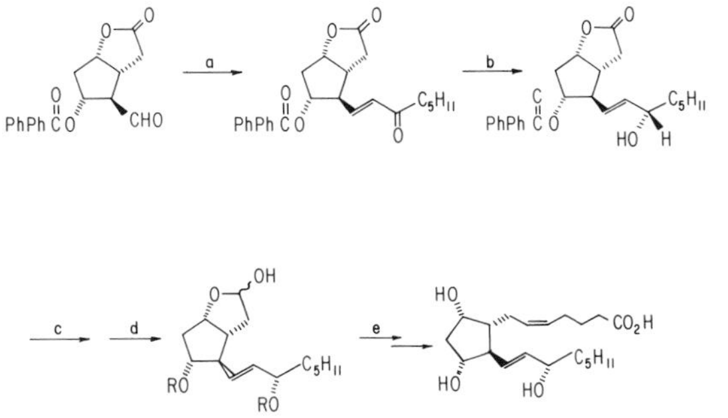

FIG. 6. Elaboration of the Corey lactone to PGF$_{2\alpha}$. (a) $(CH_3O)_2POCHCOC_5H_{11}$; (b) lithium triethyl borohydride; (c) potassium carbonate, dimethoxyethane; (d) diisobutylaluminum hydride; (e) Ph$_3$P= CH(CH$_2$)$_3$CO$_2$Na.

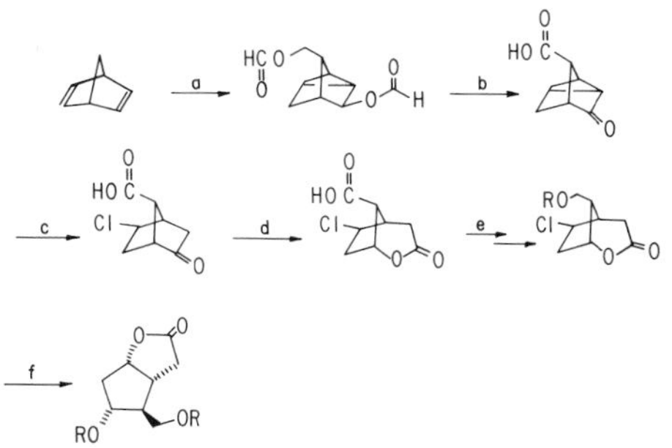

ward syntheses, which were highly linear, and demonstrates another advance in synthetic strategy.

Because of its utility, a number of syntheses for the Corey lactone have been developed. The variety of approaches which may be employed to reach this key intermediate demonstrates the power of convergent synthetic strategies. A few of the many routes to this compound are discussed in the following illustrations.

A functionalized cyclopentadiene was allowed to react in a Diels–Alder reaction to yield a bicyclo-2.2.1-heptane molecule, which on further elaboration contained the correct relative stereochemistry and functionality (Fig. 5). The lactone was carried on to PGF$_{2\alpha}$ by Horner–Emmons reaction, by Wittig reaction, reduction, then deprotection (Fig. 6).

Bicyclo-2.2.1-heptadiene may also serve as a starting material. Under acidic conditions the functionalized nortricyclic intermediate was prepared. After a few additional manipulations the prescribed lactone was revealed (Fig. 7).

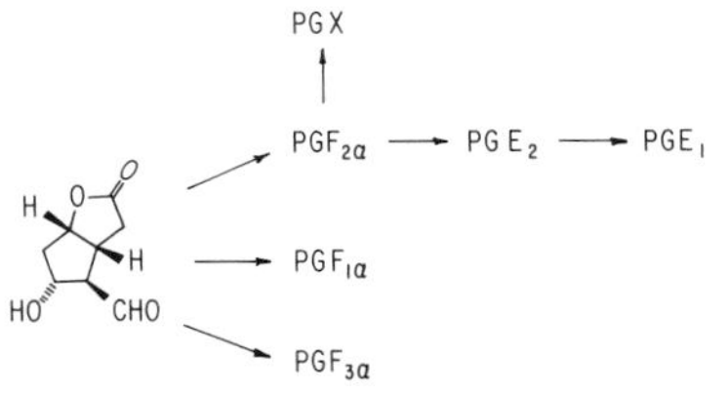

FIG. 7. Bicyclo-2.2.1-heptadiene route to the Corey lactone. (a) Formaldehyde, formic acid; (b) chromic acid; (c) hydrochloric acid; (d) m-chloroperbenzoic acid; (e) ethyl chloroformate, zinc borohydride, dihydropyran; (f) base, hydrogen peroxide.

FIG. 4. Synthetic tree demonstrating the utility of the Corey lactone in preparing prostaglandins.

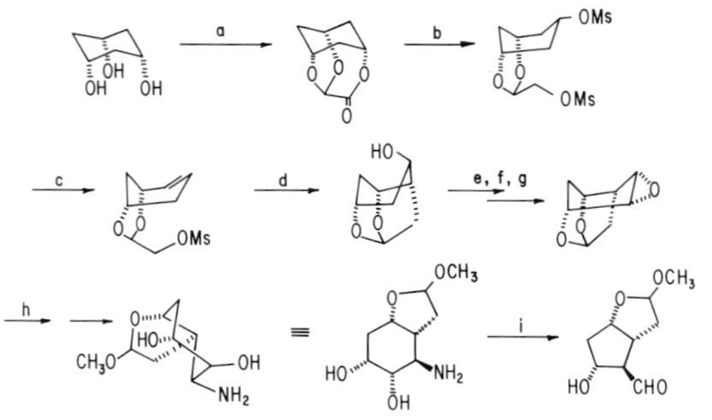

FIG. 8. Woodward synthesis from 1,3,5-cyclohexanetriol. (a) Glyoxylic acid; (b) sodium borohydride, methanesulfonyl chloride; (c) potassium hydroxide; (d) potassium carbonate, dimethoxyethane; (e) methanesulfonyl chloride; (f) potassium hydroxide; (g) hydrogen peroxide; (h) aqueous ammonia, methanolic hydrogen chloride; (i) sodium nitrite, acetic acid.

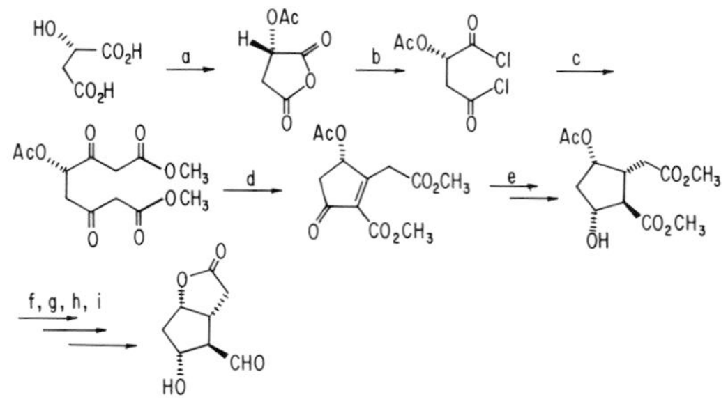

FIG. 10. Chiral synthesis of the Corey lactone from malic acid. (a) Acetyl chloride; (b) dichloromethyl methyl ether, zinc chloride; (c) $HOOCCH_2CO_2CH_3$, base; (d) hydroxide; (e) hydrogenation; (f) potassium hydroxide, methanol; (g) acetic anhydride; (h) dichloromethyl methyl ether, zinc chloride; (i) sodium borohydride.

The lactol formed on reduction of the Corey lactone was also the target of several synthetic approaches. 1,3,5-Cyclohexanetriol was converted into the lactol via a ring contraction sequence. This approach could be made enantiospecific via resolution of an intermediate alcohol (Fig. 8).

Ring contraction was also the key step in the transformation of 1,3-cyclohexadiene to the lactol. An ene reaction was employed in this synthetic scheme to introduce the appropriate substitution pattern (Fig. 9).

An asymmetric synthesis of the lactol from S malic acid required a homologation sequence

followed by an intramolecular aldol condensation (Fig. 10).

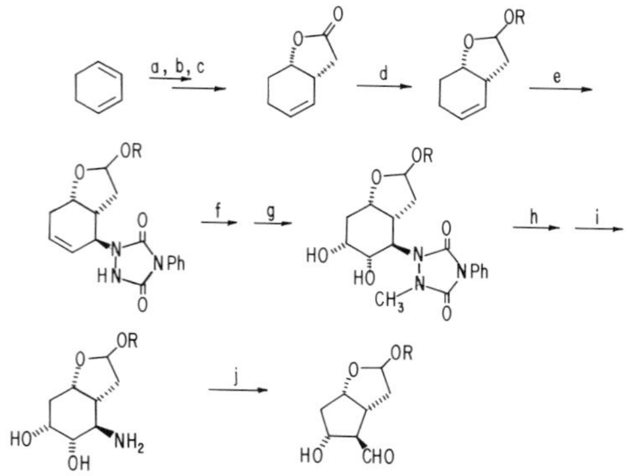

IV. Asymmetric Synthesis

As was clear from the preceeding discussions of steroids and prostaglandins, natural products are often optically active. Their biological effects may be dependent on a specific configuration. Traditionally enantiomers were separated by resolution, i.e., the separation of the pair of diastereomers formed by reaction of the racemic product and an optically pure auxiliary. The use of resolution to isolate the desired enantiomer can be feasible, but with the drawback that one-half of the material separated will be the undesired enantiomer as well as requiring an element of luck in the crystallization of the desired material. [See ORGANIC CHEMISTRY, COMPOUND DETECTION.]

Synthethc chemists have made remarkable strides in enantioselective reactions, where the desired enantiomer is the principle product. Asymmetric syntheses have successfully employed a variety of techniques, but we will limit discussion to enantioselective reducing agents, alkylations, and directed aldol reactions, which are typical of many other reactions and reagents.

For asymmetric synthesis it is necessary to employ a chiral component to direct the further transformations of the substrate. It is most desirable that this component be readily available and inexpensive, therefore common natural products such as terpenes or amino acids are frequently employed. The synthetic chemist

FIG. 9. Preparation of prostaglandin precursor via an ene reaction. (a) Dichloroacetyl chloride; (b) zinc, acetic acid; (c) hydrogen peroxide; (d) diisobutyl-aluminum hydride; (e) N-phenyltriazolinedione; (f) sodium hydroxide, methyl iodide; (g) osmium tetroxide; (h) potassium hydroxide, methanol; (i) platinum oxide, hydrogen; (j) sodium nitrite, acetic acid.

may opt to recover the chiral auxiliary, in the case of reductions, or may chose to incorporate the chiral fragment in the carbon skeleton, as in directed aldol reactions.

A. ENANTIOSELECTIVE REDUCTIONS

Among the most efficient and well-developed enantioselective transformations are asymmetric reductions of alkenes and carbonyl compounds.

1. Hydroboration

Selective hydroborating reagents have been developed from readily available terpenes such as α-pinene and longifolene. The most useful reagents would react in high chemical and optical yields, yet permit recycling of the chiral auxiliary. One successful reagent is diisopinocampheyl borane, prepared by reaction of two equivalents of α-pinene with borane:

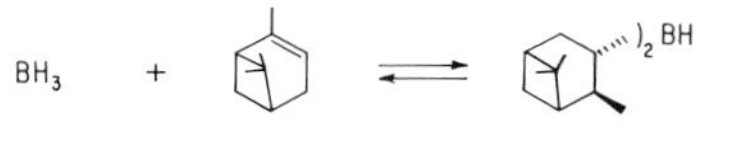

$$(62)$$

Reaction of the diisopinocampheyl borane with 2-methyl-1-alkenes, leads on oxidation to a disappointing 21% enantiomeric excess (e.e., the excess of one enantiomer relative to the other):

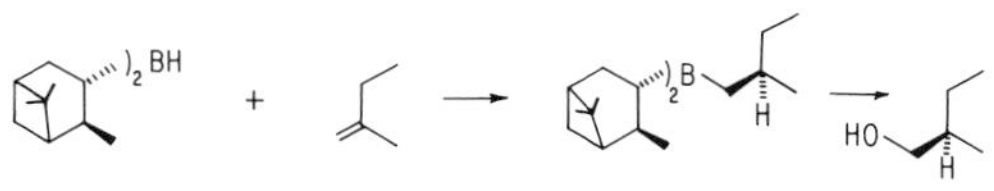

$$(63)$$

However treatment of *cis*-2-butene leads to e.e.'s as high as 98%. The general reaction of cis olefins appears to be limited only by the optical purity of the pinene starting material:

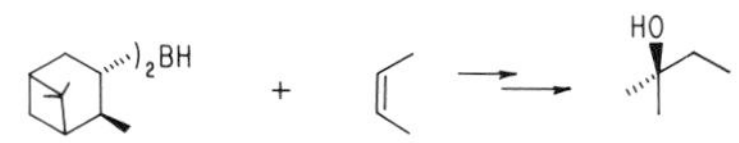

$$(64)$$

Dilongifolylborane, prepared by partial hydroboration of the longifolene, the most abundant sesquiterpene in the world, leads to only poor asymmetric induction in the reaction of 2-methyl-1-butene:

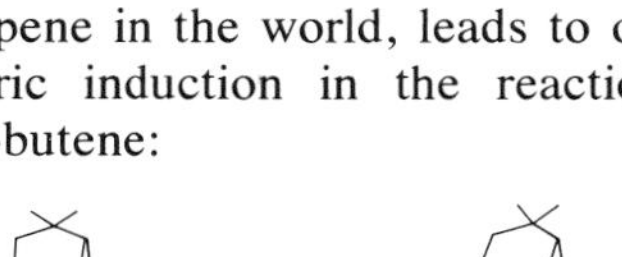

$$(65)$$

However reaction of the cis alkenes is again more selective, e.e.'s as high as 78% are possible.

Limonylborane, a boraheterocycle with non-equivalent alkyl groups bound to boron, is pre-

pared from limonene and has led only to disappointing 60% e.e.'s:

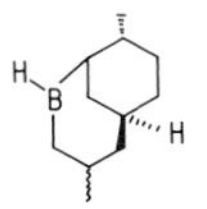

$$(66)$$

The most reactive of the hydroborating agents, monoisopinocampheyl borane reacts with cis olefins with 70% e.e. Also, the high reactivity of the reagent has led to its use with less reactive alkenes, with high e.e.'s having been reported:

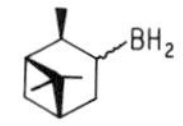

$$(67)$$

2. Chiral Borohydrides

The enantiospecific reduction of carbonyl compounds to alcohols is an extremely useful reaction. Aldehydes have been successfully reduced with chiral trialkylboranes to give products with very high optical purity. The hydroboration of α-pinene with 9-borabicyclononane (9-BBN) forms a chiral trialkylborane which reduces aldehydes with e.e.'s from 85 to 99%:

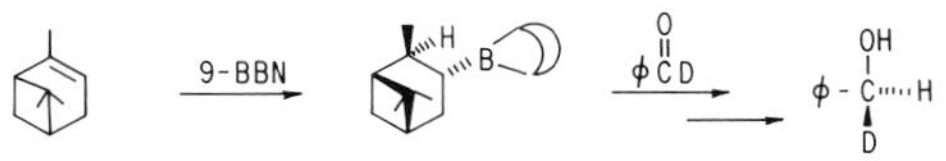

$$(68)$$

The reduction of normal ketones does not proceed well. However, the less sterically demanding acetylenic ketones can be reduced with 72–98% e.e.'s:

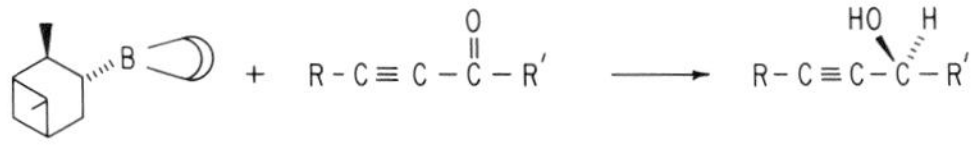

$$(69)$$

Complexed borohydrides have not yet lived up to their potential as enantiospecific reducing reagents. When the hydride reagent, lithium B-3-pinananyl-9-BBN-hydride, prepared by treatment of the α-pinene-9-BBN reagent with *t*-butyl lithium, was allowed to react with aldehydes the asymmetric induction was a disappointing 17–36%:

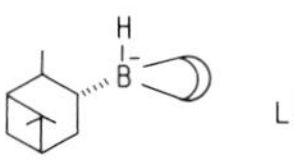

$$(70)$$

However, a related reagent prepared from nopol benzyl ether and 9-BBN gave e.e.'s as high as 70%. The presence of the benzyl ether side chain was predicted to improve asymmetric induction by improving coordination of the cation:

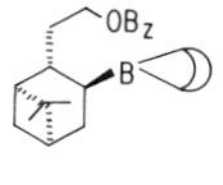

$$(71)$$

3. Complexed Lithium Aluminum Hydrides

Enantioselective reduction of aldehydes and ketones has also been possible with lithium aluminum hydride reagents complexed in a chiral environment. The lability of the ligands around lithium aluminum hydride reagents has so far limited the effectiveness of these reductions. The most effective ligands have been those containing at least one nitrogen. N-substituted amino methyl pyrrolidines, prepared from proline, have been used in the reduction of ketones to give optical yields as high as 96%:

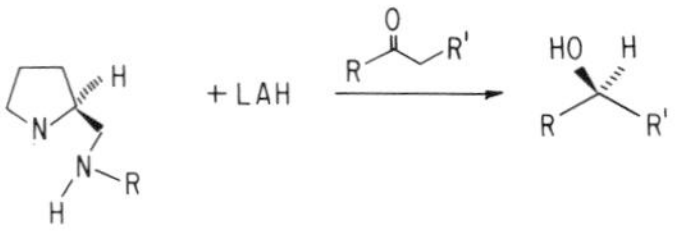

$$(72)$$

1,2-Amino-alcohols, such as ephedrine, also have been successfully used to reduce ketones with e.e.'s from 88 to 90%:

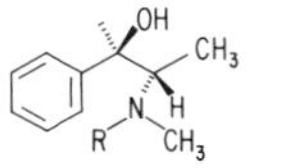

$$(73)$$

1,3-Amino-alcohols, such as Darvon alcohol, have been successfully employed in the enantioselective reduction of acetylenic ketones:

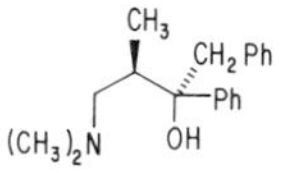

$$(74)$$

Although asymmetric reductions may be very useful procedures, it is important to recognize that the reductions are generally carried out at very low temperatures with excess reagents in order to maximize both the optical and chemical yields. These conditions often make these reagents prohibitively expensive for larger-scale operations.

B. ASYMMETRIC SYNTHESES WITH ENOLATES

The carbonyl group is one of the most useful functional groups because of its ability to act as an electrophile or, in the derived enolate, as a nucleophile. As described earlier, the enolate can react with alkylating agents or with carbonyl compounds in two very useful synthetic reactions. The stereoselectivity of an enolate is directly related to the stereochemistry of the enolate which is in turn affected by the stereochemistry of the parent molecule.

1. Alkylation Reactions

Stereoselective formation of the enolate is essential for stereoselective alkylation. An enolate may be either E or Z using the convention that the highest priority is always assigned to the OM group:

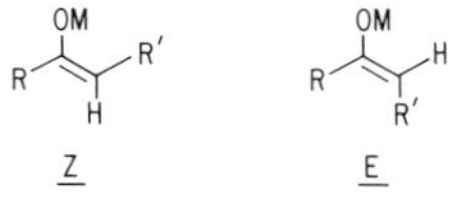

$$(75)$$

Enolate geometry may be controlled by careful choice of the deprotonation conditions. Use of sterically demanding lithium 2,2,6,6-tetramethylpiperidide (LiTMP) favors formation of the E enolate. Addition of hexamethylphosphorous triamide (HMPT) to the reaction will reverse this selectivity to favor formation of the Z enolate. Use of the bulky base lithium hexamethyldisilazide (LiHMDS) will also favor formation of the Z enolate.

With control of the enolate geometry, in cyclic systems alkylation tends to follow the pathways which minimize steric interactions:

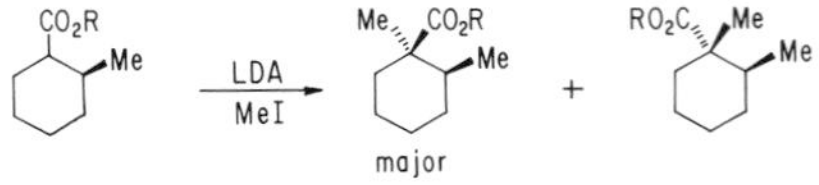

$$(76)$$

Employment of these interactions, with optically active functional groups, leads to control of the stereochemistry of alkylation.

In acyclic systems, even with control of enolate geometry, control of the alkylation reaction requires additional interactions. It is possible to create steric interactions comparable to those seen in cyclic systems by chelation effects with the enolate. If the chelating functionality is asymmetric, then stereochemical control will be possible. The result of such chelation is obstruction of one face of the enolate during the alkylation reactions:

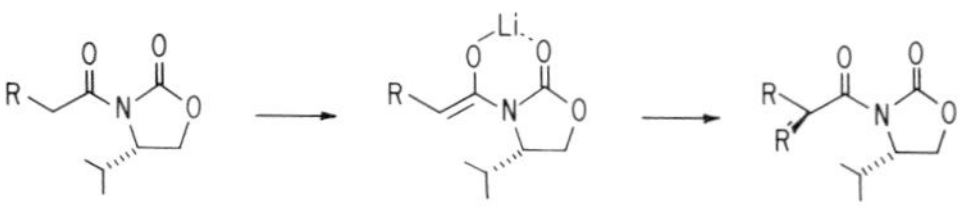

$$(77)$$

A number of chiral auxiliaries derived from available chiral substrates such as valine, norephedrine, or proline have been successfully employed.

2. Directed Aldol Reactions

The stereochemical selectivity of directed aldol reactions is also dependent on enolate geometry. Diastereoselectivity has been postulated to result from steric interactions in a six-membered cyclic transition state (Fig. 11). Diastereofaceselectivity results from addition to a chiral aldehyde, where the two new asymmetric centers may be controlled by a third center present in the starting aldehyde.

As illustrated in Fig. 11, as a general rule Z enolates tend to form syn aldols and E enolates tend to form anti aldols. The rule holds better for the reactions of Z enolates than E enolates. Selectivity increases with the steric demand of the enolate substituents. Diastereoselectivity can be poor even with good control of the enolate geometry if the steric interactions are not significant. Selectivity is also effected by the enolate counterion. The tighter the transition state, the more effective the steric repulsions will be in effecting diastereoselectivity. Metals with shorter bonds to oxygen, therefore, will be more effective. Boron enolates are generally more selective for a given carbonyl compound than other metal enolates.

Diastereofaceselectivity describes the preference of an enolate to react with one face of a chiral aldehyde over the other. This problem has been analyzed many ways; however, Cram's rule is adequate in many cases. Cram's rule states that a nucleophile will react with a carbonyl from the same face as the smallest substituent when the carbonyl group is flanked by the largest substituent. Cram's rule is illustrated by a Newman projection below.

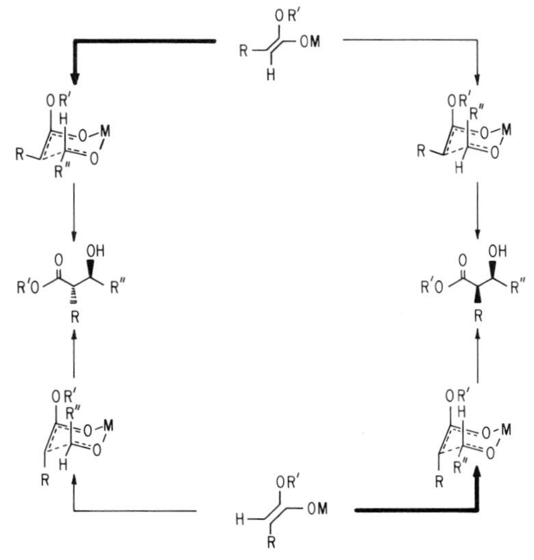

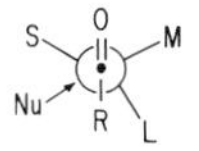

$$\tag{78}$$

More complete analyses, taking into consideration the nature of the flanking substituents, which employ orbital overlap arguments, dipole effects, and chelation, have been advanced to explain results which can not be understood by the simple Cram model. As might be deduced from this discussion, diastereofaceselectivity is remarkably dependent on the nature of the reactants.

Asymmetric induction is also possible in the reaction of chiral enolates with achiral aldehydes and ketones. This approach has been successfully employed with chiral imides, amides, and alpha hydroxyketones:

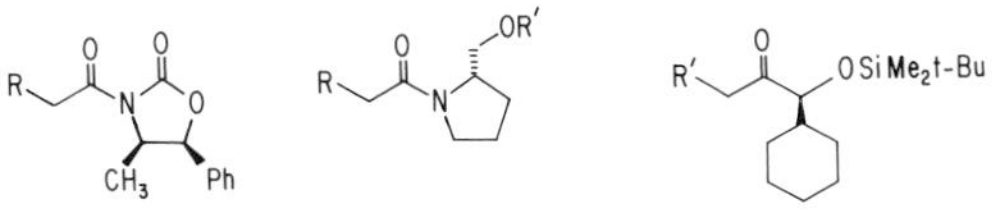

$$\tag{79}$$

The asymmetric inductions realized with these chiral reagents may be amplified remarkably by their reaction with chiral aldehydes which have a complimentary asymmetry that tends to induce formation of the same enantiomer:

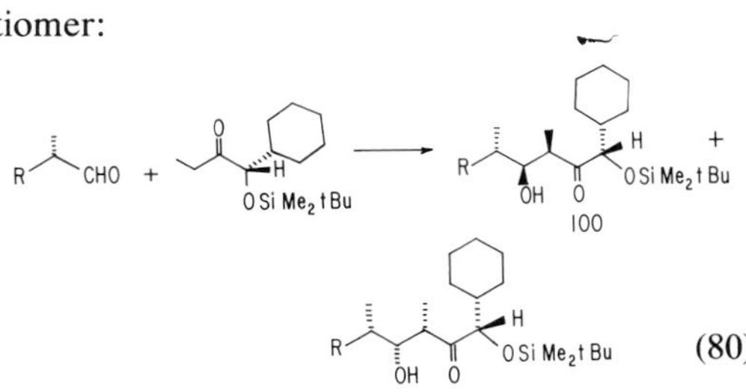

$$\tag{80}$$

FIG. 11.   Zimmerman–Traxler transition state hypothesis for the directed aldol reaction.

V. Biomimetic Synthesis

As discussed earlier natural products have stimulated the development of synthetic chemistry by providing challenging targets since the nineteenth century. In attempting to prepare known natural substances, it is only recently that chemists have turned to mimicking the synthetic approaches employed by nature. This discussion is limited to two different types of biomimetic synthesis, one based on the polyene cyclization reaction and the other the use of biomimetic reagents, functionalized cyclodextrins, as guest–host enzyme models.

A. BIOMIMETIC POLYENE CYCLIZATION

The independent proposal of groups at Columbia University and the ETH in Zurich that the stereoselectivity observed on the cyclization of polyenes is a consequence of the olefin stereochemistry constitutes the basis of the Stork–Eschenmoser hypothesis. The epoxide-initiated cyclization of squalene results in the stereospecific formation of dammaradienol. The all-trans double bond geometry of squalene results in the trans–anti stereochemistry of the ring junctions in dammaradienol:

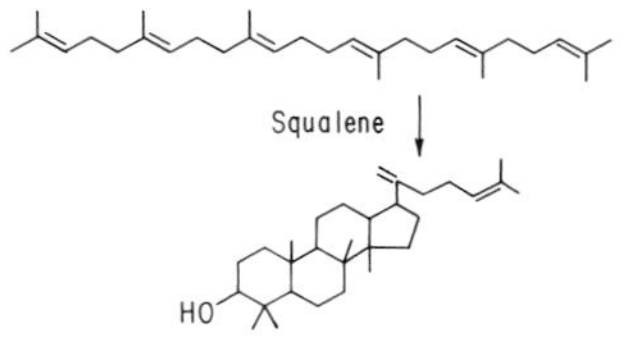

Squalene

(81)

This postulate led W. S. Johnson to study the preparative utility of polyene cyclizations. The Johnson group studied numerous functional groups to both initiate and terminate cyclization of trans polyenes. The treatment of allylic alcohols under acidic conditions to form an allylic cation proved to be one of the more useful initiating functional reactions. Termination of cyclization by an alkyne, to form an intermediate vinyl cation which was then trapped, proved to be exceptionally efficient:

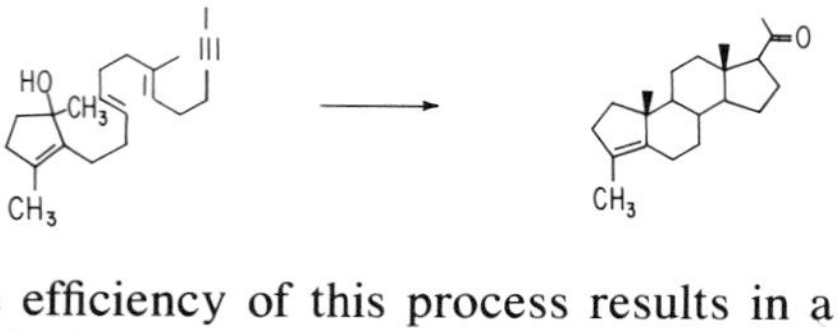

(82)

The efficiency of this process results in a 78% yield of tetracyclic material with very good control of the ring junction stereochemistry. This can be contrasted to the work of Robinson and Woodward described earlier. Particularly exciting was the observation that asymmetric cyclization was possible even when the asymmetric center was not involved in the bond forming process. Cyclization of a pro-C-11 hydroxy polyene with one of the best terminators, a vinyl fluoride, resulted in a 79.5% yield of the compound shown.

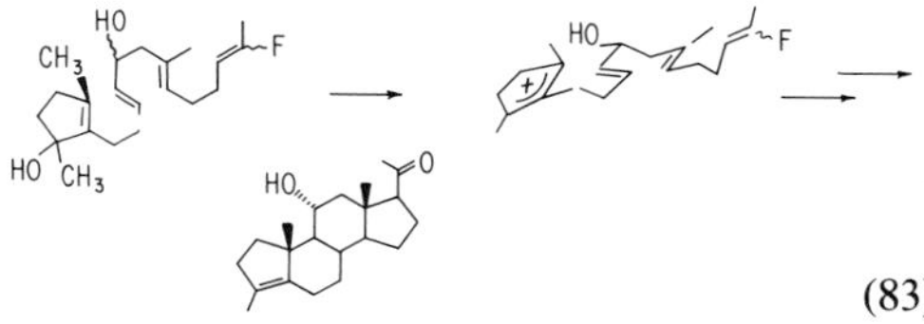

(83)

This material, with the correct configuration at C-11, may be efficiently converted to 11-hydroxy-progesterone, an intermediate in a commercial synthesis of hydrocortisone acetate. Johnson has shown that excellent asymmetric induction in polyene cyclizations is possible by treating acetals derived from (2S,4S)-pentanediol with Lewis acids:

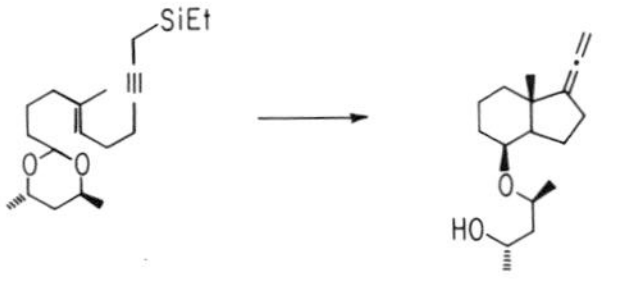

(84)

B. GUEST–HOST INCLUSION ENZYME MODELS

Enzymes have unique catalytic, regulatory, and transport properties. The development of reagents that mimic these properties would be a significant synthetic advance. In early work it was possible to prepare model compounds that could duplicate simplistically some of the transport properties but with little substrate selectivity. Currently enzyme models are being prepared which not only have the ability to recognize the substrate but also can effect site specific reactions. [See ORGANIC MACROCYCLES.]

To promote substrate selectivity several types of inclusion host molecules have been developed. Remarkable successes in chiral recognition have been achieved with cyclic polyether compounds known as crown ethers:

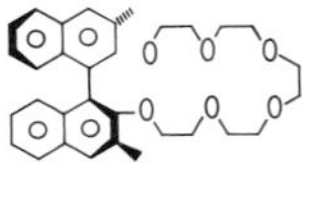

(85)

Cyclodextrins, oligosaccharides forming a distorted torus whose cavity may be occupied by other molecules, have also been explored as selective host molecules:

(86)

Cyclodextrins can have very regular cavity sizes. The cavity will incorporate aromatics to a uniform depth, often tilting the substrate away from the axis of the truncated conical cyclodextrin, most probably to maximize Van der Waals contact. The limited size of the cavity results in substrate selectivity. It has also been possible to selectively functionalize cyclodextrins as illustrated by the preparation of a cyclodextrin substituted with imidazole units:

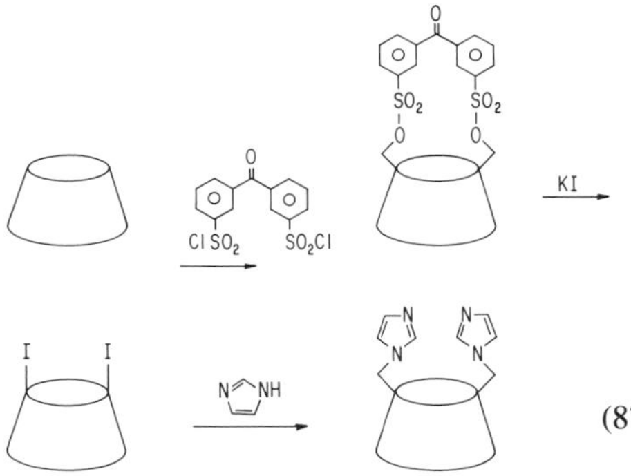

(87)

After the development of methods to prepare these molecules, the preparation of compounds more closely resembling enzymes was possible. Typical of these are cyclodextrins bearing a histamine-zinc complex and found to function as anhydrase models or polyamine metal complexes which act as carboxylic hydrolases:

(88)

There are numerous other examples of biomimetic synthesis. This narrow selection has been chosen to illustrate the progress and potential of this area.

BIBLIOGRAPHY

ApSimon, J. (ed.) (1973). "The Total Synthesis of Natural Products." Wiley, New York.
"Comprehensive Organic Chemistry," Pergamon, Oxford, 1979.
"Fieser and Fieser's Reagents for Organic Synthesis." Wiley, New York.
March, J. (1985). "Advanced Organic Chemistry," 3rd ed. Wiley, New York.
Morrison, J. D. (ed.) (1984). "Asymmetric Synthesis." Academic Press, Orlando.
Mueller, E. (1952). "Houben-Weyl's *Methoden der organischen Chemie*" Georg Thieme Verlag, Stuttgart.
"Organic Syntheses." Wiley, New York. Annual publication of tested procedures since 1921. Summarized thru 1970 in five Collected volumes.
Streitweiser, A., and Heathcock, C. H. (1985). "Introduction to Organic Chemistry," 3rd ed. Macmillan, New York.
"Theilheimer's Synthetic Methods of Organic Chemistry." Karger, Basel.
"Compendium of Organic Synthetic Methods." Wiley-Interscience, New York.

ORGANIC GEOCHEMISTRY—*SEE* GEOCHEMISTRY, ORGANIC

ORGANIC MACROCYCLES

John D. Lamb, Reed M. Izatt, Jerald S. Bradshaw, and
James J. Christensen   *Brigham Young University*

GLOSSARY

Binding constant: Equilibrium constant associated with a reaction in which a ligand binds to a substrate.

Biomimetic: That which mimics biological systems.

Calixarene: Class of compounds consisting of a ring of phenol moities connected by methylene bridges.

Crown ether: Cyclic organic compound consisting of a number of connecting ethylene oxide units.

Cryptand: Organic compound consisting of two or more bridges connecting nitrogen atoms to give multiple rings.

Cyclic polyether: See crown ether.

Cyclodextrin: Class of cyclic polysaccharide compounds.

Guest: Chemical species that can be trapped by a host compound.

Host: Compound that can trap a guest species.

Macrobicyclic effect: Extra stability of complexes of macrobicyclic (cryptands) over those of crown ethers.

Macrocycle: Cyclic ligand large enough to accommodate a substrate in the central cavity.

Macrocyclic effect: Extra stability of complexes of cyclic ligands over those of analogous acyclic ligands.

Receiving phase: Phase in a membrane system to which chemical species move.

Source phase: Phase in a membrane system from which chemical species move.

Spherand: Type of macrocycle (see compound 11, Fig. 1).

The term *macrocycle* has been applied to a large group of heterocyclic organic compounds that can bind cationic, anionic, or neutral substrates by entrapment within the cavity created by the macrocyclic structure. The selectivity of the macrocycle for certain chemical species is a function of many parameters, an important one being the match between substrate size and macrocycle cavity size. When the substrate is surrounded by the macrocycle structure, it is partly or completely isolated from the solvent. By this means, it is possible to solubilize bound substrates into solvents or membranes in which the unbound substrate is not soluble. Furthermore, the change in chemical reactivity of the bound substrate may be exploited to yield catalytic and biomimetic substrate transformation.

I. Macrocyclic Structure and Metal Cation Complexation

Macrocycles, also called macrocyclic compounds, macrocyclic ligands, macropolycycles, and so forth, include a variety of basic structures. The major headings below group currently known macrocycles into broad, general classes. However, the term macrocycle is not confined to the limited number of representative compounds presented in this article.

A. CROWN ETHERS

The name *crown ether* was applied to cyclic polyether molecules such as compounds 1–3 in

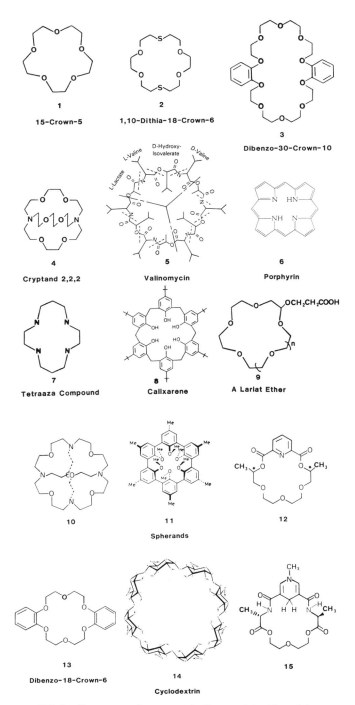

FIG. 1.   Structures of compounds discussed in this article.

Fig. 1, by Pederson, who first reported their preparation in 1967. A trivial nonrigorous nomenclature is commonly used to streamline naming of these complex molecules. Names are structured as follows: (1) principal ring substituents, (2) heteroatoms substituted for oxygen, (3) number of atoms in the principal ring, (4) the name *crown,* and (5) the number of heteroatoms in the principal ring. Thus, compound 1 is named 15-crown-5, compound 2 is 1,10-dithia-18-crown-6, and compound 3 is dibenzo-30-crown-10.

Crown ethers are particularly interesting ligands for two reasons: They measurably bind alkali metal cations in water solution, and they demonstrate size-based selectivity of metal ions. These features are illustrated for the ligands 15-crown-5, 18-crown-6, and 21-crown-7 in Fig. 2, where the thermodynamic equilibrium constant K for the reaction in methanol

$$M^{n+} + \text{18-crown-6} = (M-\text{18-crown-6})^{n+}$$

is plotted versus cation radius. Of the monovalent metal cations, K^+ is bound most strongly by 18-crown-6. X-ray crystallographic determination of the structure of the K^+–18-crown-6 complex shows that the K^+ ion sits at the center of the ligand cavity surrounded by the six ligand oxygen atoms as shown schematically in Fig. 3. The K^+ ion is nearly the correct size to fill the ligand cavity and is bound most strongly. The Na^+ ion is smaller than the 18-crown-6 ligand cavity, so the ligand must fold slightly to permit all six oxygens to associate with the cation. Both Rb^+ and Cs^+ are too large to fit into the ligand cavity. Thus, the relative sizes of cation and ligand cavity explain the selectivity of 18-crown-6 for K^+. Likewise, among alkaline earth cations, the Ba^{2+} ion both fits best in the 18-crown-6 ligand cavity and is bound most strongly. Table I, which lists the ionic radii of a number of metal

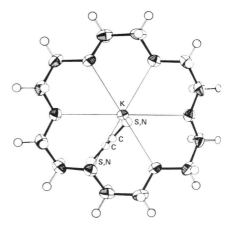

FIG. 3.   Structure of the potassium thiocyanate-18-crown-6 complex based on an X-ray crystallographic determination.

cations, shows that K^+ and Ba^{2+} are of almost equal size.

Figure 2 shows that 21-crown-7, like 18-crown-6, binds the monovalent cation whose size matches that of the ligand cavity, that is, Cs^+. However, the selectivity of 15-crown-5 is not easily explained on the basis of relative size. While the cation Na^+ best matches the cavity in size, K^+ is bound slightly more strongly. This case illustrates that size is not the only factor and is often not the determining factor that con-

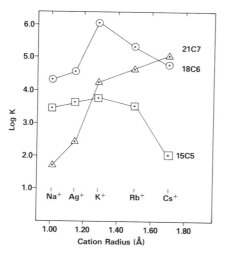

FIG. 2.   Log K for the reaction in CH_3OH at 25°C of several univalent cations with 15-crown-5, 18-crown-6, and 21-crown-6 versus cation radius.

TABLE I.   Radii of Cations and of Crown Ether Cavities

Cation	Radius (Å)
Li^+	0.74
Na^+	1.02
K^+	1.38
Rb^+	1.49
Cs^+	1.70
Mg^{2+}	0.72
Ca^{2+}	1.00
Sr^{2+}	1.16
Ba^{2+}	1.36
Ag^+	1.15
Tl^+	1.50
Pb^{2+}	1.18
Hg^{2+}	1.02
Cd^{2+}	0.95

Macrocycle	Radius (Å)
15-crown-5	0.85
18-crown-6	1.3
21-crown-7	1.7

trols cation selectivity. The problem arises from the fact that Na^+ ion is slightly too large to fit into the 15-crown-5 cavity. For such a case, cation solvation energies dominate in the free energy cycle

$$\begin{array}{ccc} M_g + L_g & \longrightarrow & ML_g \\ \uparrow \quad \uparrow & & \downarrow \\ M_s + L_s & \longrightarrow & ML_s \end{array}$$

for complex formation. (The subscripts g and s indicate species in the gas and solvent phases, respectively.) Compared to Na^+, the larger K^+ is less strongly solvated because of its lower charge-to-radius ratio, so less energy is expended in removing solvent molecules in the complexation process. Because the range of macrocycle sizes is much larger than the range of cation sizes, it is relatively rare that selectivity is governed by size predominantly. It is more often the case that solvation, ligand flexibility, and the effective charge on the binding sites play the dominant roles.

Crown ethers have affinity for metal ions besides those of the alkali and alkaline earth series. Figure 4 shows the binding constants of 18-crown-6 with the series of trivalent lanthanide cations, which decrease in size across the series. Table II shows the binding constants of several

TABLE II. Stability Constants (log K) of Crown Ethers with Heavy Metal Ions in Water at 25°C

Ligand	Stability constant			
	Pb^{2+}	Hg^{2+}	Tl^+	Ag^+
15-Crown-6	1.85	1.68	1.23	0.94
18-Crown-6	4.27	2.42	2.27	1.50
Dibenzo-18-crown-6	1.89	—	1.50	1.41
Dicyclohexano-18-crown-6 (cis-anti-cis)	4.43	2.60	1.83	1.59
1,10-Dithia-18-crown-6	3.13	>5	0.93	4.34
1,10-Diaza-18-crown-6	6.90	17.85	—	7.8

simple crown ethers with Pb^{2+}, Ag^+, Tl^+, and Hg^{2+}.

When the oxygen heteroatoms of crown ethers are replaced by nitrogen or sulfur, the selectivity of the ligands changes markedly. For example, sulfur-containing analogs of 18-crown-6 have lower affinity for alkali and alkaline earth cations and greater affinity for more polarizable cations such as Tl^+ and Hg^{2+}. When nitrogen is substituted, the affinity for alkali and alkaline earth cations also drops, while that for Pb^{2+} and Ag^{2+} increases.

Substitution of aliphatic or aromatic groups onto the heterocyclic backbone of crown ethers has a destabilizing effect on complex stability. Table III shows that the stabilities of complexes of dicyclohexano-18-crown-6 are much more like those of 18-crown-6 than are those of dibenzo-18-crown-6. In the latter case, the benzene rings withdraw electron density from the oxygen atoms, lowering the energy of the ion-dipole interaction in the complex. This explanation is borne out in the observation that if electron-withdrawing substituents are added to the benzene rings, cation complex stability constants drop even further.

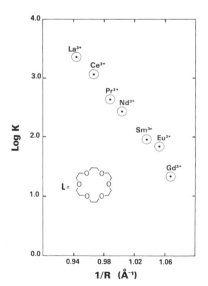

FIG. 4.   Log K for the reaction $M^{3+} + L = ML^{3+}$ (L = 18-crown-6) in methanol at 25°C versus reciprocal of cation radius. No reaction was observed with Tb^{3+}, Dy^{3+}, Ho^{3+}, Er^{3+}, Tm^{3+}, Yb^{3+}, or Lu^{3+}.

TABLE III.   Effect of Substituent Groups on Cation Complex Stability (log K) with 18-Crown-6 and Its Analogs in Methanol at 25°C

Crown ether	Stability constant		
	Na^+	K^+	Ba^{2+}
18-Crown-6	4.36	6.06	7.04
Cyclohexano-18-crown-6	4.09	5.89	—
Dicyclohexano-18-crown-6 (cis-anti-cis)	3.68	5.38	—
Benzo-18-crown-6	4.35	5.05	5.35
Dibenzo-18-crown-6	4.36	5.00	4.28

Bound metal ions show markedly different redox properties from unbound ions. It is possible, using macrocycles such as crown ethers, to stabilize oxidation states of metal ions such as Eu^{2+}.

B. CRYPTANDS

Cryptands, or macrobicyclic ligands, are similar in structure to crown ethers, differing in the addition of a bridge that reaches across the ring to give football-shaped structures such as compound 4 in Fig. 1. The trivial nomenclature used for these ligands is given by the number of ethylene oxide units in each of the three bridges connecting the two nitrogen heteroatoms. Thus compound 4 with two oxygen atoms in each bridge is designated 2.2.2. The additional bridge facilitates more effective encapsulation of metal ions. Consequently, the complexes of these ligands are in general more stable than those of crown ethers. Figure 5 shows the binding constants of a series of cryptands for alkali metal ions. Comparison of Fig. 5 with Fig. 2 shows the

higher stability of the cryptand complexes. It also shows that the cryptands have a high degree of selectivity and that there is a cryptand of the correct size to be selective for each of the ions in the alkali metal sequence. The selectivity in all these cases is largely a result of the match between cation and cavity sizes. The effects of substituting other heteroatoms for oxygen and of organic substitution on the ring backbone are similar to those for crown ethers.

C. NATURALLY OCCURRING MACROCYCLES

Before crown ethers or cryptands were synthesized, macrocyclic ligands of various types had been isolated as natural products from microbial species. Compounds such as valinomycin and enniatin B have been studied extensively because of their ability selectively to bind alkali and alkaline earth metal ions and to transport such ions through biological membranes. Valinomycin is used in K^+ ion selective electrodes because of its high (10,000:1) $K^+ : Na^+$ selectivity.

Valinomycin (compound 5, Fig. 1) is similar to a cyclic protein in structure. In the unfolded configuration, it is too large to accommodate metal ions. However, intramolecular hydrogen bonds cause the ring to tighten in on itself, providing a nearly octahedral arrangement of the oxygen atoms of the correct size to accommodate a K^+ ion. The exterior of the ligand is hydrophobic, making it soluble in lipid membranes. Careful kinetic studies demonstrate that the selectivity of the ligand for K^+ is a function of the rate of cation release from the complex, there being little difference in the rate of cation uptake into the ligand.

D. TETRAAZA MACROCYCLES

Macrocyclic ligands containing four nitrogen heteroatoms separated by various organic bridges have been studied for many decades. Examples are porphyrin (compound 6, Fig. 1) and its analogs, which are the metal-binding sites in many metalloenzymes, hemoglobin, and other naturally occurring compounds. There are also many synthetic tetraaza macrocycles that have affinity for divalent (and other) transition metal ions. Their binding constants with these cations are generally much higher than those typical of crown ethers and cryptands, so the metal ion is held virtually irreversibly. For example, the binding constant (log K) of com-

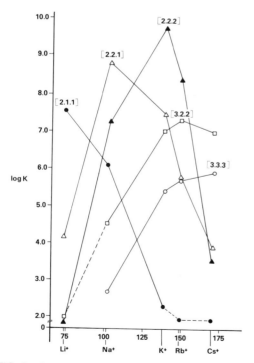

FIG. 5. Variation of equilibrium constant K in 95 volume percent methanol for the reaction of several cryptands (see compound 4, Fig. 1) with the alkali metal ions Li^+–Cs^+ (plotted according to increasing metal ion radius).

pound 7, Fig. 1, with Hg^{2+} is 23 and with Ni^{2+} is 23.5. The importance of these complexes lies in the binding of additional ligands at the axial sites of the bound ion, which serves as the enzyme's active site. The degree and type of aromaticity in the tetraaza ligand structure has a profound influence on the electronic properties of the bound metal ion, which in turn affects the strength and nature of binding to additional ligands.

E. Other Macrocycles

A wide variety of macrocycle types has been reported in addition to the general categories discussed above. The calixarene ligands (compound 8, Fig. 1), which are water insoluble, have a strong, selective affinity for Cs^+. They form neutral complexes through loss of a proton. The lariat ethers (compound 9, Fig. 1) form neutral complexes by the same mechanism, resembling a crown ether with an arm that can reach around to provide ligation at the axial position. Macrotricyclic cryptands (compound 10, Fig. 1) provide essentially spherical or cylindrical neutral traps for metal ions. Spherands (compound 11, Fig. 1) likewise offer elegant binding geometries in which metal ions are bound. The list of macrocycles is far greater than can be presented in this limited space. [See INCLUSION (CLATHRATE) COMPOUNDS.]

II. Complexation of Organic Cations

Ammonium and organosubstituted ammonium cations bind to crown ethers and other macrocycles by the formation of hydrogen bonds to the ligand heteroatoms. An example is the complex of an alkylammonium cation with 18-crown-6 shown in Fig. 6a. The stability of such

TABLE IV. Stability Constants (log K) for Reaction of 18-Crown-6 and with Several Organic Ammonium Cations in Methanol at 25°C

Cation	Log K
RNH_3^+ cations	
NH_4^+	4.27 ± 0.02
$HONH_3^+$	3.99 ± 0.03
$NH_2NH_3^+$	4.21 ± 0.02
$CH_3NHNH_3^+$	3.41 ± 0.02
$CH_3NH_3^+$	4.25 ± 0.04
$CH_3CH_2NH_3^+$	3.99 ± 0.03
$CH_3CH_2OC(O)CH_2NH_3^+$	3.84 ± 0.04
$CH_3(CH_2)_2NH_3^+$	3.97 ± 0.07
$CH_3(CH_2)_2NH_3^+$	3.90 ± 0.04
$CH_2CHCH_2NH_3^+$	4.02 ± 0.03
$CHCCH_2NH_3^+$	4.13 ± 0.02
$(CH_3)_2CHNH_3^+$	3.56 ± 0.03
$CH_3CH_2OC(O)CH(CH_3)NH_3^+$	3.28 ± 0.02
$(CH_3)_3CNH_3^+$	2.90 ± 0.03
$PhCH(CH_3)NH_3^+$	3.84 ± 0.01
$PhNH_3^+$	3.80 ± 0.03
$2\text{-}CH_3C_6H_4NH_3^+$	2.86 ± 0.03
$4\text{-}CH_3C_6H_4NH_3^+$	3.82 ± 0.04
$2,6\text{-}(CH_3)_2C_6H_3NH_3^+$	2.00 ± 0.05
$3,5\text{-}(CH_3)_2C_6H_3NH_3^+$	3.74 ± 0.02
$R_2NH_2^+$	
$NH_2C(NH_2)NH_2^+$	1.7 ± 0.2
$(CH_3)_2NH_2^+$	1.76 ± 0.02
$(CH_3CH_2)_2NH_2^+$	
R_3NH^+ cations	
$(CH_3)_3NH^+$	No complex
R_4N^+ cations	
$(CH_3)_4N^+$	No complex

complexes is influenced by the number of hydrogen bonds that can form and by the degree of steric hindrance for approach of the substrate to the ligand. Table IV lists the binding constants for a number of ammonium cations with 18-crown-6. The stability drops dramatically as the number of available hydrogen bonds is reduced from 3 to 2 to 1. Furthermore, anilinium ions, which contain ortho substituents, form weak or no complexes because the substituents sterically hinder approach of the $-NH_3^+$ group to the ligand.

Unlike ammonium cations, diazonium cations complex to crown ethers by insertion of the positive moiety into the cavity, as in Fig. 6b. Table V shows that complex stability deteriorates markedly with ortho substitution in benzenediazonium cation because of steric hindrance. Figure 7 shows that complex stability is a regu-

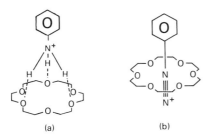

(a)   (b)

FIG. 6. Diagrammatic representation of mode of binding of (a) anilinium and (b) benzenediazonium cations to 18-crown-6.

TABLE V. Stability Constants (log K) for Reaction in Methanol at 25°C of Arenediazonium and Anilinium Cations with 18-Crown-6

Cation	Log K
$PhNH_3^+$	3.80
$2\text{-}CH_3C_6H_4NH_3^+$	2.86
$2,6\text{-}(CH_3)_2C_6H_3NH_3^+$	2.00
$PhNN^+$	2.50
$2\text{-}CH_3C_6H_4NN^+$	a
$2,6\text{-}(CH_3)_2C_6H_3NN^+$	a

a No measurable reaction.

lar function of the electron density in the N_2^+ moiety. As electron density increases, stability declines.

The binding of organoammonium-type cations to macrocycles is the subject of intense interest due to the ability of such systems to mimic enzymes. Specifically, it is possible to add functionalities to the macrocyclic structure to permit chiral recognition in substrates. Chiral macrocycle 12 in Fig. 1, in the (S,S)-form, for example, binds one enantiomer of α-(1-naphthyl)ethylammonium perchlorate more strongly than the other isomer [log $K = 2.47 \pm 0.01$ for the (R)-ammonium salt and 2.06 ± 0.01 for the (S)-salt]. The difference in binding constants results from a greater steric hindrance to the approach of the substrate for one isomer due to the presence of the bulky functionalities.

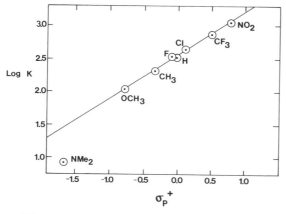

FIG. 7. Plot of log K for formation in methanol at 25°C of the 18-crown-6 complex of $p\text{-}RC_6H_4NN^+$ versus Hammett σ_p^+ values of R.

III. Complexation of Anions and Neutral Molecules

Considerably less attention has been given to the binding of anions to macrocycles than to that of cations. Basically two types of anion binding are known. The first involves macrocyclic structures containing basic sites that provide positively charged binding sites when protonated. Examples of these are compounds 10 (in protonated form) and 14 in Fig. 1, and binding constants with typical anions are found in Table VI. Structure 14 is based on cyclodextrin, which is a large cyclic polysaccharide.

Cyclodextrins are able to accommodate neutral molecules as well as anions. The large cavity contains numerous hydrogen-bonding sites if needed. In general, the cavity simply provides a comfortable microenvironment for many neutral species, especially when the solvent environment is less than ideal.

IV. Applications of Macrocyclic Ligands

A. EXTRACTANTS

One of the first identified uses for crown ether and cryptand ligands was as phase transfer agents in catalyzing synthetic organic reactions. The macrocycle can be used to solubilize salts having oxidizing or reducing anions into hydrophobic solvents. For example, $KMnO_4$ can be solubilized into benzene by 18-crown-6. The "naked" MnO_4^- ion that accompanies the K^+–18-crown-6 complex in solution is a very powerful oxidizing agent in this medium. Use of naked anions of this type has provided a method to enhance the efficiency of many synthetic reactions.

Macrocycles have also been proposed as metal ion extractants in separation processes. It has been shown by J. McDowell and his co-workers at Oak Ridge National Laboratory that

TABLE VI. Stability Constants (log K) for Reaction of Anion with Macrocyclic Ligand 10

Anion	Log K, reaction with ML 10 (protonated)
Cl^- (in water)	>4.0
Br^- (in water)	<1.0
Br^- (in 90% methanol)	1.75

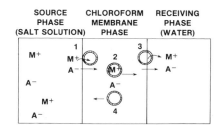

FIG. 8. Facilitated transport of cation M$^+$ through a liquid membrane: (1) Cation M$^+$ forms complex with carrier at first interface; (2) complex plus co-anion A$^-$ diffuse across membrane; (3) M$^+$ and A$^-$ are released into receiving phase; (4) carrier diffuses back across membrane to repeat cycle.

synergistic effects occur when crown ethers are used as coextractants with traditional extractants like diethyl hexylphosphoric acid (HDEHP). Specifically, the degree of metal ion extraction is greater when both crown and HDEHP are used together than the sum of extraction efficiencies when each is used separately. Furthermore, by using the crown, the selectivity of extraction processes can be altered in this manner.

B. MEMBRANE CARRIERS

The naturally occurring macrocycles like valinomycin attracted attention initially because they make biological membranes permeable to metal cations. It was originally thought that such

macrocycles might be responsible for much of the cation permeation of cell membranes observed in nature. More recently, it has been shown that the formation of tubular cation channels through membranes is the most common mechanism. [*See* MEMBRANES, SYNTHETIC (CHEMISTRY).]

Macrocyclic ligands serve as cation carriers in liquid membranes by shuttling the cation from one water interface to the other (Fig. 8). It is the ability of the ligand to solubilize the cation in the hydrophobic membrane that permits this application. If the macrocycle is selective for one cation, it facilitates the transport predominantly of the selected cation and thereby effects a separation of cations.

When neutral macrocycles are used as cation carriers, an anion must accompany the cation through the membrane to provide electrical neutrality (Fig. 8). On the other hand, if the macrocycle contains one or more acidic moieties, neutral complexes may be formed with cations when the macrocycle deprotonates. This phenomenon can be exploited to drive cations through membranes against their concentration gradient (Fig. 9). For protons to transfer from the acidic receiving phase to the basic source phase, metal ions must be transported in the opposite direction. If the proton gradient is higher than that of metal ions, the transported metal ions can be concentrated in toto in the receiving phase. This process is termed coupled transport and is widely exploited in membrane separations.

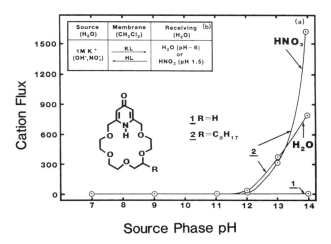

FIG. 9. (a) Plot of cation flux [mol × 10^8/(5 m^2)] versus source phase pH using K$^+$ and the indicated macrocycles. (b) Scheme of coupled K$^+$–H$^+$ transport.

TABLE VII. Rate of Transport of Nitrate Salts through a Stirred Chloroform Membrane Containing 1 mM Carrier

Carrier	Transport rate (mol × 10⁷/24 hr)						
	Na^+	K^+	Rb^+	Cs^+	Ca^{2+}	Sr^{2+}	Ba^{2+}
15-Crown-5	28	3.4	1.0	0	0	0	3.3
18-Crown-6	11.3	280	210	34	26	320	18
21-Crown-7	4.8	67	72	48	4.2	36	41
Dibenzo-18-crown-6	4.2	101	12	1.1	0	0	1.6
Dicyclohexano-18-crown-6	23	340	240	32	160	490	300
Dithia-18-crown-6	1.5	0	0	0	0	0	0.70
2.2.1	130	290	550	28	4.7	3.0	7.5
2.2.2	240	88	205	5	3.1	13	2.8

Table VII demonstrates the selectivity of nonionizable macrocycle-containing membranes among cations. The membrane selectivity is related to, but not necessarily identical to, the macrocycle selectivity in homogeneous solution. Differences in selectivity occur when a cation is bound too strongly to be released from the membrane once it is taken up into the membrane. The optimal range in binding of the cation to the ligand carrier corresponds to a binding constant of 10^6 in methanol.

C. Ion Exchange Sites

Macrocycles have been bound covalently to polymers for use as ion exchange sites for chromatographic separation of cations. In this mode, the selectivity mimics the thermodynamic, selectivity of the simple monomer. Separation of alkali and alkaline earth cations has been achieved using resin-bound crown ethers. Certain macrocyclic compounds have been bonded to polymers and the polymers used for the separation of biogenic ammonium cations as well as the enantiomers of some amino acid salts.

V. Synthesis of Macrocyclic Compounds

The synthetic organic macrocyclic ligands come in all shapes and varieties. Early work was with the polynitrogen macrocyclic compounds such as the cyclams (compound 7, Fig. 1). More recent innovations have been with the macrocyclic polyether (crown) compounds. Charles J. Pedersen of duPont first reported these compounds in 1967. Pedersen was preparing the *bis*-phenol substituted polyether shown in Eq. (1). He isolated a good yield of his intended product but persisted in purifying the by-product to obtain dibenzo-18-crown-6 (compound 13, Fig. 1) which proved to be a remarkable complexing agent for cations. Compound 13 was produced from the catechol impurity in the starting phenol shown in Eq. (1). When the same reaction was carried out with catechol, a good yield of compound 13 was isolated [Eq. (2)]. This reaction is a Williamsen ether synthesis.

The synthesis of the crown compounds has been accomplished by a number of different cycloaddition methods. The basic 2 unit plus 2 unit addition as shown in Eq. (2) has been most used for the simple crowns. Often part of the macrocycle is first synthesized and then a second part added on, as in Eq. (3). This is particularly useful when preparing unsymmetrical crowns such as dibenzo-21-crown-7. The last part of the synthesis shown in Eq. (3) is a simple 1 unit to 1 unit cycloaddition. This process has been used to prepare other types of macrocyclic compounds such as pyridino diester-18-crown-6 [Eq. (4)]. This is a transesterification reaction and gives excellent yields. Indeed, most of these cycloaddition reactions exhibit a template effect in that greater yields are realized when a cation, or in some cases, a neutral molecules, that fits into the macrocyclic cavity is used in the reaction.

There is one other reaction of note for the formation of simple crown compounds. A polyethylene glycol, when treated with one mole of *p*-toluene-sulfonyl chloride, cyclized into a crown compound [Eq. (5)]. The sulfonyl chloride reacts with one of the alcohol units to form the mono tosylate ($RO_3SC_6H_4CH_3$) which is a good leaving group for the internal Williamsen ether synthesis reaction. The potassium cation acts as a template for the cyclization, giving a 75% yield of 18-crown-6, as in Eq. (5).

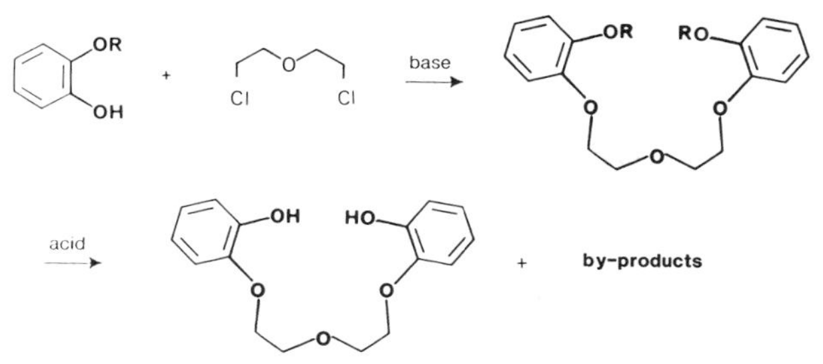

(1)

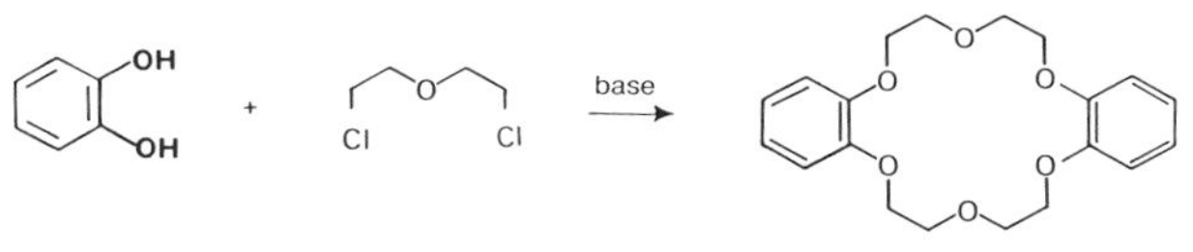

(2)

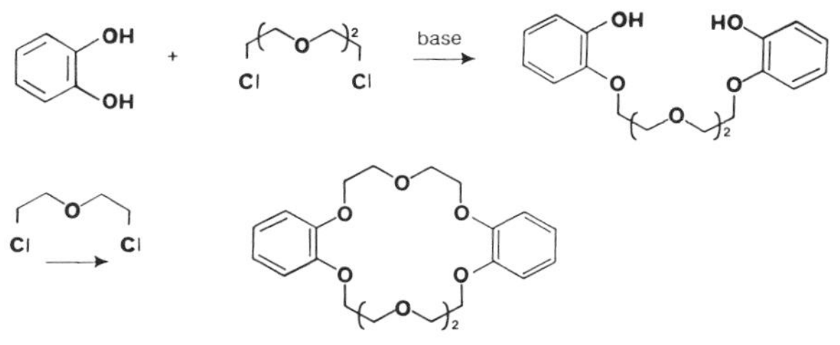

Dibenzo-21-Crown-7

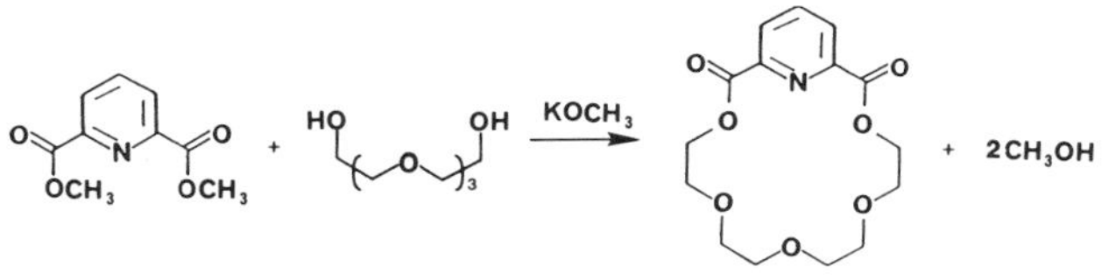

(3)

Pyridino diester-
18-Crown-6

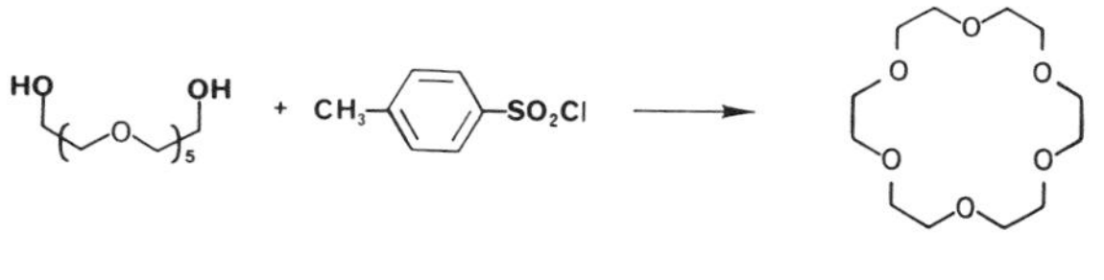

(4)

18-Crown-6

(5)

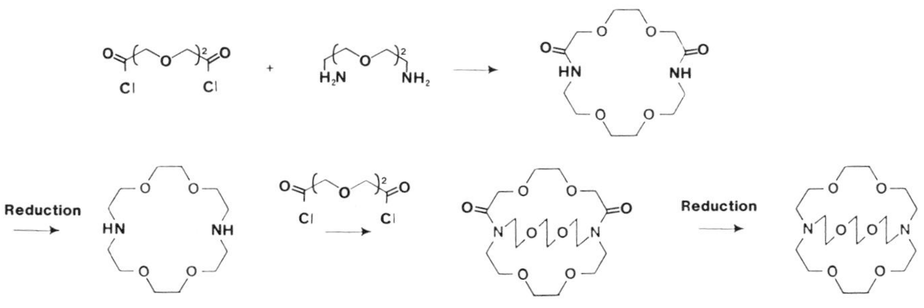

(6)

Some interesting macrobicyclic multidentate compounds have been prepared in the past ten years. These compounds, containing bridgehead nitrogens, require a multistep synthesis [Eq. (6)]. Cryptands with one or more branches containing one to three heteroatoms have also been prepared.

The synthesis of chiral macrocyclic compounds for the study of enantiomeric recognition is an important new development in crown chemistry. Compound 12 (Fig. 1) is just one example of these important new compounds. These compounds have shown chiral recognition for the enantiomers of various organic ammonium salts. Chiral crown 15 has been used to effect an NADH type of asymmetric reduction of certain aromatic ketones.

BIBLIOGRAPHY

Bradshaw, J. S., and Stott, P. E. (1980). *Tetrahedron* **36,** 461–510.

Izatt, R. M., Bradshaw, J. S., Nielsen, S. A., Lamb, J. D., and Christensen, J. J. (1985). *Chemical Reviews* **85,** 271–339.

Izatt, R. M., and Christensen, J. J., eds. (1979, 1981, in press). "Progress in Macrocyclic Chemistry," Vol. 1 (1979), Vol. 2 (1981), and Vol. 3 (in press). Wiley Interscience, New York.

Melson, G. A., ed. (1979). "Coordination Chemistry of Macrocyclic Compounds." Plenum, New York.

P

PALYNOLOGY

Rosemary A. Askin   *Colorado School of Mines*
Stephen R. Jacobson   *Chevron U.S.A.*

GLOSSARY

Angiosperms: Flowering seed plants, including monocotyledons (e.g., grasses, lilies, orchids, palms) and dicotyledons (e.g., oaks, maples, daisies, ragweed).

Biostratigraphy: Study of the distribution of fossils in rock strata.

Gametophyte: Sexual generation that alternates with the asexual sporophyte generation in the life cycle of all plants.

Gymnosperms: Plants that bear naked seeds, including conifers, cycads, ginkgos, extinct seed-ferns, and others.

Hystrichospheres: Outdated term for organic-walled fossil microphytoplankton cysts with spines or projections. This heterogeneous group includes acritarchs, which are microfossils of varied and uncertain affinities, and dinoflagellate cysts.

Kerogen: Organic matter insoluble in either aqueous alkaline or organic solvents. It includes palynomorphs, and other organic components such as woody fragments, cuticle, and amorphous matter.

Microfossils: Fossils so small that a microscope is required for their study. Palynomorphs are included in this group.

Microphytoplankton: Small, usually single-celled plants that float or drift in the water column. This group includes palynomorphs such as dinoflagellate cysts and acritarchs,

as well as other non-palynomorphs like diatoms and coccolithophoroids with siliceous or calcareous skeletons.

Pollen analysis: Analysis of Quaternary pollen assemblages using relative abundances to determine vegetation changes and climatic history.

Range: Horizontal and vertical geological occurrence of a fossil taxon.

Sporopollenin: High molecular weight polymer of carbon, hydrogen, and oxygen that is destroyed by oxidation. It is the durable substance of variable composition that forms the protective covering in spores, pollen, dinoflagellate cysts, and acritarchs.

Taxon: Term coined in the 1940s by botanists as a substitute for "taxonomic group." It includes species, genus, family, order, and so on. (Plural, taxa)

235144.

Palynology is the study of acid-resistant, organic-walled microscopic entities derived from both plants and animals. These include modern and fossil pollen, spores, dinoflagellate cysts, acritarchs, colonial algae, fungi, scolecodonts, and others, as well as the extinct chitinozoa. Interpretations based on palynology increase our understanding of geologic history, and have applications to botany, anthropology, medicine, criminology, and law.

I. Introduction

Palynology is a term coined in 1944 by H. A. Hyde and D. A. Williams to replace the cumbersome and restrictive term "pollen analysis." The neologism is derived from the Greek *paluno* (to strew or sprinkle) and expands the concept far beyond recent pollen and spores. The word "palynomorph" was introduced in 1961 by R. H. Tschudy of the U.S. Geological Survey, who attributed the term to colleague R. A. Scott. It was defined as an "inclusive appellative encompassing the types of microfossils found in palynological preparations." Palynology has come to involve studies of modern and fossil entities, primarily pollen, spores, dinoflagellate cysts, acritarchs, and chitinozoa. Palynologists now may investigate all the diverse organic components concentrated by special preparation techniques. The other recognizable components encountered in palynological preparations include fragmental plant matter such as cuticle and woody tissue, fungal remains, and the zoological residues of polychaete worms, microforaminiferal linings, graptolites, crusta-

ceans, eurypterids, and insects, as well as the less understood melanosclerites.

Palynomorphs include tiny remnants of the oldest sexually reproducing organisms, acid-resistant organic-walled cells from the Proterozoic Era (1.4 billion years old), as well as the modern hayfever-inducing ragweed pollen (*Ambrosia* spp.). Their study requires various chemical and mechanical preparation techniques and the use of either optical, fluorescence, transmission electron, or scanning electron microscopy. Palynology addresses the questions of morphology, function, and distribution of palynomorphs in both space and time, their paleoecological and paleoenvironmental implications, and the geothermal history of their entombing sediments.

A fortuitous combination of characteristics

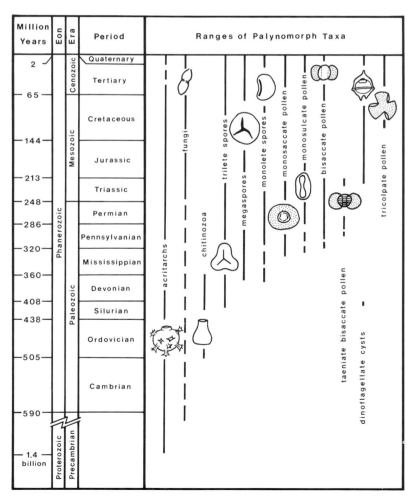

FIG. 1.   Geologic ranges of major palynomorph taxa.

make palynomorphs, as a group, invaluable fossils for geology. These characteristics include their small size (usually less than 150 μm, but ranging from 2 to several thousand μm), the extreme durability of the walls (perhaps the most resistant naturally occurring organic substances known), abundance in small samples, great morphologic variability and rapid evolution, and preservation in a wide variety of marine and non-marine sedimentary rocks. Modern and recently fossilized pollen and spores are also of use to several other scientific disciplines.

The most frequent geological applications of palynology are for dating and correlating the rapidly evolving assemblages of palynomorphs found in sedimentary rocks. Appearances (and disappearances) of major groups through geologic time are indicated in Fig. 1. [*See* GEOLOGIC TIME.]

Palynomorphs have also answered questions posed by anthropologists regarding origins of agriculture, cave dwellers' diets, and environments of early humans; by allergists concerned with seasonal dispersal of ragweed and other pollen responsible for hayfever; by paleoclimatologists reconstructing local and regional evolution of swamps, grasslands, and forests; by plant taxonomists seeking to compare various plant taxa; and by criminologists, proving the origin of mud on suspects' boots.

II. Discussion of Palynomorph Types

A. SPORES

Spores are microscopic propagative bodies, with a single nucleus, whose primary function is plant dispersal and reproduction. Spores are produced by "lower" plants, which include mosses, liverworts, clubmosses (lycopods), horsetails, and ferns.

Spores, produced in a sporangium (spore case), are initially attached to each other in groups of four, called tetrads. These occasionally remain together as obligate tetrads, but usually separate, leaving scars or lines of attachment, vestiges reflecting the spore arrangement in the tetrad. Trilete (Y-shaped) scars result from tetrahedral tetrads, and monolete (− shaped) scars result from tetragonal tetrads (Fig. 2). This trilete or monolete scar functions as an area of weakness which ruptures, allowing emergence of the developing gametophyte. The trilete or monolete scars may be simple, barely

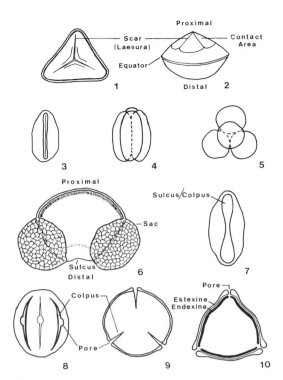

FIG. 2. Diagrammatic sketches illustrating morphologic types and descriptive terms for spores and pollen. (1) trilete spore, proximal view; (2) trilete spore, equatorial/lateral view; (3) monolete spore, proximal view; (4) tetragonal tetrad, derivation of monolete spores; (5) tetrahedral tetrad, derivation of trilete spores; (6) bisaccate pollen, lateral view; (7) monosulcate pollen, distal view; (8) tricolporate pollen, equatorial view; (9) tricolporate pollen, polar view; (10) triporate pollen, polar view.

visible traces, or they may be accompanied by a variety of structures including thickened and raised lips, or, in extreme cases, by high membranous extensions (Fig. 3.13). Some spores, called alete, may show no trace of a scar.

Trilete spores may be circular to triangular, and monolete spores oval to kidney shaped. The spore wall, or exine, may be a single thin or thick layer, or it may be differentiated into two or more layers. These layers may be separated to varying degrees and each layer may be of different thickness and/or ornamentation.

Spore wall ornamentation is highly diverse, reflecting interspecific variation. The spore wall may be smooth or ornamented with an almost infinite assortment of pits, canals, granules, warts, spines, rods; anchor shaped, drumstick shaped, or complex projections; flanges; ridges

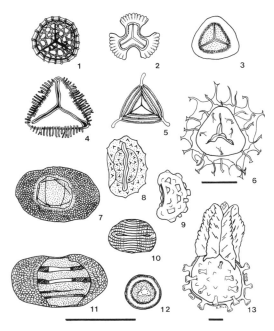

FIG. 3. Diagrammatic sketches of selected spores and gymnosperm pollen. The bar is equivalent to 50 μm [note that (6) and (13) are at smaller magnifications]. (1) *Lycopodiumsporites austroclavatidites* (Cookson) Potonié, Cretaceous-Tertiary; (2) *Tripartites vetustus* Schemel, Mississippian; (3) *Lycospora pellucida* (Wicher) Schopf, Wilson & Bentall, Mississippian-Pennsylvanian; (4) *Reinschospora triangularis* Kosanke, Pennsylvanian; (5) *Appendicisporites unicus* (Markova) Singh, Early Cretaceous; (6) *Ancyrospora ancyrea* (Eisenack) Richardson, Middle Devonian; (7) *Florinites antiquus* Schopf, Pennsylvanian; (8) *Aratrisporites parvispinosus* Leschik, Triassic; (9) *Polypodiisporites* sp., Cenozoic; (10) *Weylandites lucifer* (Bharadwaj & Salujha) Foster, Permian; (11) *Lunatisporites pellucidus* (Goubin) Balme, Early Triassic; (12) *Classopollis classoides* Pflug, Jurassic-Cretaceous; (13) megaspore *Arcellites disciformis* Miner, Cretaceous.

arranged in many different patterns (striate, annulate, reticulate); or in combinations of two or more sculptural types. Ornamentation may vary from one site to another on individual spores. The spore surface bearing the scar, called the proximal face, commonly has reduced sculpture, reflecting the contact area within a tetrad, compared to the opposite (distal) face (Fig. 2). Thickenings (or thinnings) may be restricted to certain areas, for example, at the ends of scars or commonly as a zone around the equator (Fig. 3). It is this ornamentation, along with shape, type of scar, and other structures that make

spores diagnostic and useful for documenting evolving plant assemblages through time.

Some lower plants may be heterosporous; that is, they produce two kinds of spores, small and large, as in some lycopods. The small spores are true microspores that develop into male gametophytes, and the large spores are megaspores that develop into female gametophytes. Other lower plants are homosporous (= isosporous); they produce usually small, nearly equal-sized (within a species) isospores whose gametophyte performs both male and female functions, as in most ferns. In practice, true microspores, isospores, and small megaspores cannot readily be distinguished from each other.

Some spore-like obligate tetrads of uncertain affinities first appeared in the Ordovician Period. Simple trilete spores associated with land-plants first occurred in the fossil record near the end of the Silurian Period. Morphologic diversity increased from these initial simple smooth forms. Many distinctive, increasingly more complex spore types arose during the Devonian Period, including some with distinctive anchor-shaped projections (Fig. 3.6).

Development in plants of the ability to produce two kinds of spores is called heterospory. This occurred during the middle Late Devonian and was a major evolutionary step. (Megaspores are considered the precursors of seeds.) Later, monolete spores became established early in the Pennsylvanian Period, although a few are known from the Late Devonian (e.g., *Archaeoperisaccus* spp.). Spores are particularly valuable for biostratigraphy in Upper Paleozoic and Mesozoic rocks.

B. POLLEN

A pollen grain is a male gametophyte with several nuclei, whose primary function is fertilization of the female ovum. Pollen are produced by higher plants: gymnosperms (including seed ferns, cycads, conifers, and their allies) and angiosperms (flowering plants).

1. Gymnosperm Pollen

The germinal aperture (for emergence of the pollen tube during fertilization) is on the distal face of most gymnosperm pollen and is usually in the form of a single furrow or thinned area of the pollen wall (Fig. 2.6). Some species retain a vestigial proximal trilete or monolete scar.

Some gymnosperm pollen may have a simple elliptical shape with a single furrow (sulcus or colpus), about which the wall may flex to prevent dessication. These "monosulcate" pollen with a single furrow (Fig. 2.7) are common to several orders, for example cycads, ginkgos, and the extinct Bennettitales.

In most gymnosperm pollen, the wall layers are widely separated, with a thin outer layer forming a sac or air bladder (saccate). The most common saccate pollen have a central body with two internally reticulate and usually distally inclined sacs (Fig. 2.6). These bisaccate pollen are produced by most conifers and by the extinct seed-ferns. The proximal surface on the central body of a saccate pollen may bear relatively simple ornamentation, the most striking being the striations (Fig. 3.11) that characterize a variety of seed-fern pollen of the Permian and Triassic Periods. Other saccate pollen have one large radially or bilaterally symmetrical sac (Fig. 3.7), called monosaccate, or three or more sacs called multisaccate.

Other gymnosperm pollen morphologies range from simple spheres as in the extant larch (*Larix* sp.), to the more complex pollen *Classopollis* (Fig. 3.12) and related forms, produced by an extinct group of conifers common during the Jurassic and Cretaceous Periods.

First occurrences of the different morphologic types of gymnosperm pollen are geologically useful. Monosaccate pollen appeared in the Late Mississippian (although superficially similar pseudosaccate spores appeared during the Devonian); bisaccate pollen during the Middle Pennsylvanian (the first rare forms in Early Pennsylvanian); and striate bisaccate pollen in the late Middle Pennsylvanian. Monosulcate pollen became well established at the beginning of the Permian, although early forms (*Schopfipollenites*) appeared at the end of the Mississippian. Monosulcate and most saccate pollen types range to the present; however, striate bisaccate pollen with their parent plants, the seed-ferns, became extinct at the end of the Triassic.

2. Angiosperm Pollen

Wall structure is one of the distinguishing features of angiosperm pollen. It is the outermost main layer, the exine, of pollen and spore walls that is preserved in fossils, the inner easily destroyed cellulosic wall or intine being lost. Two or more layers within the exine itself may be discernible, usually an endexine and outer ektexine. In angiosperm pollen additional differentiation of the ektexine occurs with a charac-

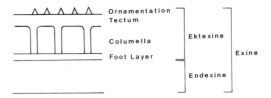

FIG. 4. Diagrammatic cross-section showing exine or outer wall structure of angiosperm pollen.

teristic layer of columns (columella) supporting the outermost tectum (Fig. 4).

The first unequivocal angiosperm pollen (*Clavatipollenites*, Fig. 5.8) appeared during the Early Cretaceous. Angiosperm-like pollen are reported from the Triassic, but these grains do not possess all the angiosperm wall characteristics listed by J. A. Doyle (with N. F. Hughes, one of the principal workers on early angiosperm evolution) as reticulate sculpture, columellar infratectal structure, and lack of a laminated endexine. The aperture morphology of angiosperm pollen rapidly diversified from the primitive single sulcus (as in *Clavatipollenites*) to more complex types and arrangements. The basic aperture types are the colpus (colpate) and pore (porate) or a combination of the two (colporate) (Fig. 2). Colpate apertures (colpi) may occur singly, especially in some monocotyledons such as lilies, but typically three or sometimes more per grain are arranged around the equator. Porate grains may possess a single pore (as in grasses, Fig. 5.7) or two, three, or more, arranged *in loci* or scattered randomly, and the wall structure around the pores may be highly modified. As with spores, the outer wall layer or ektexine of angiosperm pollen may be equipped with spines, reticulate sculpture, or numerous other structures and ornament.

The first appearance of monosulcate angiosperm pollen during the Early Cretaceous was followed by simple tricolpate forms by the late Early Cretaceous, after which rapid diversification continued until the late Tertiary by which time modern plant families were established. Porate apertures developed, first the colporate types then polyporates and triporates. A distinctive group known as "Normapolles," triporate pollen with highly modified pore structure (Fig. 5.3), characterize Late Cretaceous–Early Tertiary assemblages in parts of the Northern Hemisphere.

Various nomenclatural systems are used for fossil angiosperm pollen. For Quaternary pol-

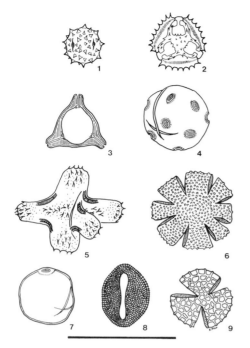

FIG. 5. Diagrammatic sketches of selected angiosperm pollen. The bar is equivalent to 50 μm. (1) *Ambrosia* sp., ragweed, Late Tertiary-Quaternary; (2) *Taraxacum officinale* Weber, dandelion, Tertiary-Quaternary; (3) *Nudopollis* sp., latest Cretaceous-Early Tertiary; (4) *Juglans* sp., walnut, Tertiary-Quaternary; (5) *Aquilapollenites attenuatus* Funkhouser, Late Cretaceous; (6) *Nothofagidites asperus* (Cookson) Stover & Evans, middle Tertiary; (7) Graminae, grass pollen, middle Tertiary-Quaternary; (8) *Clavatipollenites hughesi* Couper, Early Cretaceous; (9) *Tricolpites* sp., Cretaceous.

len, genus (and sometimes species) names of the living parent plants are used. For pre-Quaternary assemblages, however, there is controversy over whether such names can properly be applied. Many pollen and spores appear identical with, or at least closely resemble, their presumed living counterparts and some workers therefore give them the "natural" names of living plants. Others use "half-natural" names that indicate similarity to the living form, adding a suffix such as "*-pollenites*," "*-sporites*," or "*-idites*" to indicate their fossil status. For example, *Alnipollenites* is used for pre-Quaternary pollen resembling those produced by *Alnus*, the modern alder. Fossil analogues may or may not be related to the modern pollen-bearing plants. Pollen, or spores, represent but a single plant organ. A plant species may be defined botanically by a variety of organ characters, in the

flower, leaves, and stems. Many different plant species may produce virtually indistinguishable pollen, a good example being the simple monosulcate gymnosperm pollen; or as has been demonstrated from both living and fossil plants, a single species may produce several morphologic variations.

The justification for natural or half-natural names lessens the farther back in the Tertiary Period that the pollen-bearing plants grew. For this reason, some workers prefer a wholly artificial nomenclatural system of "form-genera" and "form-species," which was first used to name spores from Paleozoic coals. A form name may record the morphology; the rock formation, location, or region it came from; or it may honor a noted palynologist or other person. An example of a form-genera is *Tricolporites*, and a form-species *microreticulatus*. Form-genera are used to name fossil pollen for which there are no known modern analogues. Nomenclatural procedures should follow the rules of the International Code of Botanical Nomenclature (ICBN).

C. ACRITARCHS

Acritarchs are single-celled, organic-walled microphytoplankton of uncertain affinity. Acritarchs were recognized as fossils long before they were named by W. R. Evitt in 1963 to accommodate those hystrichospheres not referable to dinoflagellate cysts. This neologism (Greek *acritos*, uncertain; and *arche*, origin) has come to include a variety of algal cysts made of acid-resistant sporopollenin known from the fossil record in rocks from at least 1.4 billion years old to the Cenozoic. Microfossils now known as acritarchs were first illustrated by microscopist M. C. White in 1862 from specimens seen in thin sections of Devonian and Ordovician rocks of central and western New York.

Acritarchs are remains of hundreds of extinct species predominantly preserved in Paleozoic marine rocks. The central body or vesicle can have one or more wall layers and may be spherical, ellipsoidal, polyhedral, fusiform or flattened triangular, circular, or bacilliform. Surfaces can be smooth or ornamented. The acritarch may have numerous, few, or no processes. The processes can have pointed, rounded, or forked tips, with fins or spines, and so on. Openings that allowed release of cellular material, called excystment apparatuses, include flaps, splits, and tube-like projections, as well as circular (e.g., the circular pylome in Fig. 6) or polygonal

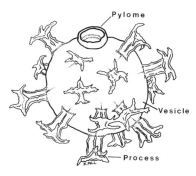

FIG. 6. Diagrammatic sketch of an Ordovician acritarch, *Peteinosphaeridium intermedium* (Eisenack).

openings. This great variety reflects the probable inclusion of many different algal taxa within the category "acritarchs."

Acritarchs described by T. V. Jankauskas from rocks about 1.4 billion years old may be the oldest proof of eukaryotic (nucleated) life. It is the development of the nucleus of the eukaryotic cell that first illustrated the capacity for true sexual reproduction; although some have claimed to have found fossilized nuclei, it is the increased vesicle size that presumes a nucleus. These oldest sexual organisms, albeit single-celled plants, had thin acid-resistant cyst walls that were unornamented and spherical. By the Late Precambrian, processes and ornament had appeared. Rapid evolution of morphology continued through the Devonian, at the end of which many of the organic-walled microphytoplankton disappeared. This is viewed by most workers as a major extinction event, although some acritarchs persisted into the Mesozoic and Cenozoic. Acritarchs are useful as age, paleoenvironmental, and paleogeographic indicators.

There has been an attempt by systematists (e.g., C. Downie) to group the species of acritarchs by various geometric features and to legitimize a classification scheme. Most workers prefer the informal designations based on shape: sphaeromorphs for spherical forms, acanthomorphs for forms with processes, and so on, and thus utilize the morphologic group terms for description without implying botanical affinity. Taxonomic descriptions of species are commonly listed alphabetically by genus.

D. DINOFLAGELLATE CYSTS

Sporopollenin dinoflagellate cysts are important because they are the most abundant and widespread marine palynomorphs preserved in

Upper Triassic to Recent marine rocks. They are also found in rocks recording estuarine and fresh-water paleoenvironments. Only a relatively small percentage of living dinoflagellates, however, produce cysts; some estimates suggest 10 percent. It is reasonable to assume that the approximately 2500 fossil species assigned to about 500 genera represent only a small part of the dinoflagellate lineage.

Living forms anatomically and physiologically seem to demonstrate primitive eukaryotic characteristics suggesting that dinoflagellates, probably noncyst-forming species, could have developed in the Precambrian. Some workers, notably W. R. Evitt and W. A. S. Sarjeant, feel that a single pre-Triassic form, assigned in 1964 to *Arpylorus antiquus* Calandra and found in Upper Silurian rocks of North Africa, represents a link to older dinoflagellates.

Much is known about living dinoflagellates, yet a life cycle demonstrating the cyst forming during a resting period was not observed until D. Wall and B. Dale cultured dinoflagellates in the 1960s. Dinoflagellates are single-celled algae or phytoplankton that live in both marine and nonmarine waters. Many have a "theca" made of cellulosic armor plates, although other substances are known. This cellulose theca is easily dissolved and unknown at present in the fossil record. Some species, however, secrete an acid-resistant, internal cyst made of sporopollenin capable of being fossilized. This internal cyst reflects some of the thecal structures, including the cingulum and sometimes the tabulation of the plates. Because the cysts themselves record only reflected structure, the terms "paracingulum" and "paratabulation" are used for cysts (Fig. 7).

In 1836, C. Ehrenberg, while studying Cretaceous rock chips, made the first observation and illustrations of fossil sporopollenin dinoflagellate cysts. The important understanding of sporopollenin cyst morphology has been worked out through many years of dedication by W. R. Evitt. Palynologists now accept a classification of these cysts based on

1. general features, including orientation criteria, shape, and size;

2. wall structure and surface features;

3. indications of paratabulation, including position of the paracingulum; and

4. the archeopyle.

In most cases, these features reflect the pre-cyst thecate stage of a dinoflagellate. In cysts, sur-

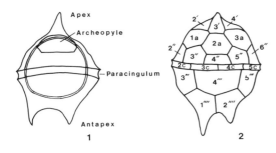

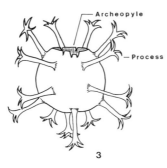

FIG. 7. Diagrammatic sketches of dinoflagellate cysts to illustrate morphology and descriptive terms. (1) Peridinioid proximate cyst, dorsal view, *Deflandrea* sp., Cenozoic; (2) hypothetical peridinioid cyst showing Kofoid system of paratabulation; (3) Chorate cyst, *Oligosphaeridium* sp., Cretaceous-Tertiary.

face features and particularly indications of paratabulation have allowed reconstruction of a thecal plate formula. Two main formula systems have been proposed: the older Kofoid system (Fig. 7.2), and the recently proposed Taylor–Evitt system that has evolutionary implications. The formulas, coupled with shape and position of the excystment apparatus (the archeopyle) and other morphologic features have collectively allowed palynologists to construct a taxonomic framework for dinoflagellate cysts.

There are two basic types of cysts: proximate, which closely resemble the form of the theca and include the important peridinioid cysts (Fig. 7.1); and chorate, where the cyst is widely separated from the theca by processes and include the dinoflagellate hystrichospheres of earlier workers (Fig. 7.3). These morphologically variable, and sometimes highly complex, solid or hollow processes may be scattered randomly over the cyst surface or arranged according to the paratabulation, along plate boundaries, in the center of a plate, and so on. For chorate cysts, these processes are useful for indicating the paratabulation.

Dinoflagellate cysts are valuable in oil and gas exploration because they are the dominant palynomorphs preserved in marine Mesozoic and Cenozoic rocks, frequently the source rocks for hydrocarbons.

E. TASMANITIDS AND COLONIAL ALGAE

Tasmanitids are a distinctive group of fossil unicellular green algae including *Tasmanites* (Fig. 8) and related forms. On the basis of cell wall structure, these algal cysts are thought to be produced by prasinophycean algae. The thin radial canals and associated openings on the wall surface characterize this well-known component of several important petroleum source rocks. Other prasinophycean cysts have a surrounding wing-like structure (ala).

Fossil colonial algae are predominantly green algae and include such forms as *Botryococcus* and *Pediastrum*. They are often useful as paleoenvironmental indicators, and are sometimes age-diagnostic.

F. FUNGAL SPORES AND FRUITING BODIES

Fungal spores and fruiting bodies are frequently found preserved in rocks of early Mesozoic age to the present. Other kinds of fungal remnants are preserved in rocks as old as the Proterozoic. The fungal affinities of material in palynomorph assemblages has been recognized for some time, although it was not until the mid-1970s that systematic descriptive and stratigraphic studies of fungi were undertaken, mainly by W. C. Elsik. Some fungal spores are useful in biostratigraphy.

Fungal spores may lack a pore, or have one or more pores. They may be divided into two or more interconnecting cells by septa (partitions, Fig. 9.1). The spore wall may be smooth or ornamented with granules, spines, ridges (striate, re-

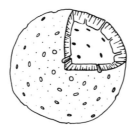

FIG. 8. Diagrammatic sketch of a prasinophycean alga, *Tasmanites* sp., with section removed to show wall structure. This type Paleozoic to present.

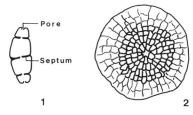

FIG. 9. Diagrammatic sketch of fungi. (1) fungal spore, *Diporicellaesporites* sp., Tertiary; (2) microthyriaceous fruiting body, *Asterothyrites* sp., Tertiary.

ticulate), and so on. They range from as small as 2 μm to greater than 600 μm across. Maximum morphologic diversity seems to have been reached in the early Cenozoic.

Shield-shaped fruiting bodies with radiating cell arrangement (Fig. 9.2) of leaf fungi range from the Early Cretaceous to the present.

Morphologic terms used for fossil fungal remains follow, where possible, terms for living fungi; and their nomenclature is primarily artificial, using form-genera.

G. Chitinozoa

Chitinozoa are flask-shaped, organic-walled, extinct microfossils of uncertain affinity but probably of animal origin (Fig. 10). They were originally reported in 1930 and named the following year by A. Eisenack.

Chitinozoa are known from the earliest Ordovician to the latest Devonian. Some workers have postulated fungal, protozoan, or metazoan affinities for them. On the basis of similar pseudochitin acid-resistant walls and nearly identical stratigraphic ranges, W. A. M. Jenkins suggested an association with the extinct colonial zooplankton known as graptolites.

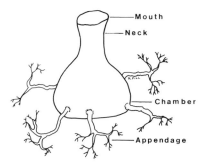

FIG. 10. Diagrammatic sketch of an Ordovician chitinozoa.

Chitinozoans are thought by many to be egg cases, because of the frequent occurrence of serially connected chains of the bottle-shaped tests. Apparently "unhatched eggs," however, have not been shown to contain any diagnostic organic relics.

Chitinozoa occur in the same rocks as graptolites and acritarchs. They evolved rapidly and are effective biostratigraphic markers in marine rocks throughout their range. Their large size (60–2000 μm) allows them to be concentrated, after rock dissolution, with nondestructive techniques like sieving.

H. Scolecodonts

Scolecodonts are the chitinous acid-resistant elements of jaw apparatuses of marine polychaete worms.

They were recognized by G. J. Hinde in 1879, and have more recently been found by Z. Kielan-Jaworowska and D. R. Edgar, in multielement apparatuses (Fig. 11), much like the enigmatic conodonts (phosphatic dental elements of extinct animals). Unlike conodonts, however, which are made of mineral fluorapatite and are

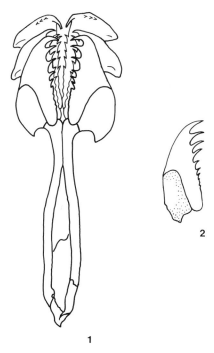

FIG. 11. (1) Diagrammatic sketch of a Silurian scolecodont jaw apparatus, showing pairs of elements; (2) dispersed scolecodont element as found in palynological preparations.

dense, the scolecodont element is of relatively lighter weight organic chitin. This density difference affects the sedimentological distribution of disaggregated scolecodont elements, making their statistical treatment more complicated than conodonts. Palynologists have usually ignored scolecodonts as biostratigraphic markers, although their common occurrence from the Ordovician through the Mesozoic with rare occurrences in the Tertiary has some potential.

III. Characteristics of Palynomorphs; Their Preservation and Recovery

A. COMPOSITION OF SPOROPOLLENIN

The applied science of palynology depends entirely on the durability of sporopollenin, and rarely pseudochitin. Sporopollenin is a term first coined by F. Zetzche in 1932 as "a collective appellation for the resistant wall materials found in spores of pteridophytes and pollen of gymnosperms and angiosperms."

This family of organic chemical compounds was initially examined by observing its reactions to various substances including acids and stains. Beginning in the 1970s, analytical instrumentation was utilized for studying organic compounds by pyrolysis, mass spectrometry, chromatography, and other methods. With these methods sporopollenin has been found to include oxidative polymers of carotenoids and/or carotenoid esters. More thorough recognition of the exact composition and variability of sporopollenin is an area of active study in the 1980s.

The composition of sporopollenin varies for different taxa. Gymnosperm pollen, from a "higher plant" for example, produces a pyrolyzate chemically different from that produced by lycopod spores (from a "lower plant"). Although not yet studied systematically, it seems that suprageneric differences in composition of sporopollenin are the rule rather than the exception. Variation in chemistry of sporopollenin is found between algae (like the acritarchs, tasmanitids, and dinoflagellate cysts), lower plant spores, and pollen of higher plants.

Sporopollenin provides a protective coating for preservation of protoplasmic ingredients critical to the reproduction of virtually all plants. Properties and structures of sporopollenin functionally protect against desiccation, ward off microbial attack, and maintain a physical separation from hostile external conditions. Sporopollenin is elastic and impermeable, consequently allowing spores, pollen, and cysts to respond to microclimatic changes in temperature, humidity, and pressure. By protecting plant propagules, the evolution of sporopollenin was a crucial event allowing the spread of plant life in the early oceans, onto land, and over the land surface.

In the same way that sporopollenin endows palynomorphs with resistance to physical, chemical, biological, and thermal damage, it permits them to survive the vigorous chemical treatments required to liberate palynomorphs from their entombing sedimentary rocks.

B. PRODUCTIVITY

The almost ubiquitous occurrence of palynomorphs in ancient and modern sedimentary basins partly accounts for their impressive utility. This is the result of extremely high productivity of many encysting algae, and the multitudes of spores and pollen dispersed by both lower and higher plants.

Some plants rely on wind for dispersal of their pollen. They produce extraordinarily high numbers of pollen grains, thus ensuring successful pollination. A single pine tree, for example, can produce billions of its bisaccate pollen grains from small male cones each season. Yellow clouds of pollen, which settle to form dense flotsam, block drains, or cause medical problems, attest to the high productivity of wind-pollinated plants. Insect-pollinated plants, however, have evolved complex structures for their flowers to attract insects and require far fewer numbers of pollen to ensure their reproductive success. Insect-pollinating grains display complex ornaments facilitating transport on the legs, antennae, or hair-like projections of insects. In the fossil record, there is a corresponding numerical over-representation of wind-dispersed gymnosperm and angiosperm pollen (for example, pine or *Pinus;* southern beech or *Nothofagus*); whereas pollen of insect-pollinated plants are comparatively rare. Lower plants that rely on their spores to disseminate their progeny may also release large numbers of spores.

Cyst-forming marine microphytoplankton (dinoflagellates and acritarchs) also exist in ample numbers and exhibit population explosions or "blooms," as demonstrated by the modern "red tides" often including billions of dinoflagellates.

C. Factors Affecting Transport and Deposition

Palynomorphs are small (most are 10–150 μm, and range from 2 μm for small fungal spores to several thousand μm for some megaspores), and they have low density and high buoyancy. They are therefore frequently transported away from their site of origin, except for some spores of plants growing in peat bogs and swamps. Some pollen and spores are transported by wind, others by water currents, sometimes for great distances. High global wind currents are the reason contemporary arboreal pollen have been recovered from ice cores taken from the Antarctic ice sheet, a continent where no trees grow today.

Most palynomorphs are eventually deposited in relatively low energy environments. Palynomorphs behave as sedimentary particles in the fine silt and finer range, and therefore are deposited with fine-grained sediments. Coarse-grained sediments, deposited in high energy environments, generally lack palynomorphs.

Pollen and spores from land plants may be carried by rivers and streams into nonmarine depositional environments such as lakes or river meanders, or they may be carried farther to the river mouth or out to sea. Transported fossil pollen and spore assemblages thus may combine representatives of several plant communities from varying habitats, like high altitude forests and coastal lowlands.

The majority of dinoflagellates and acritarchs were probably planktonic. Some may have formed cysts close to the water surface or others near the seafloor. They may have settled and been deposited beneath their aqueous area of life, whereas others were probably subjected to the vagaries of currents and tides, and may have been transported great distances.

D. Preservation and Durability

Palynomorphs are preserved in unoxidized, fine-grained, primarily dark-colored (grey to black) sediments, including mudstones, siltstones, coals, micritic limestones, and cherts. Cannel coal is a type of coal made almost wholly of palynomorphs, and the Australian "white coal" contains abundant *Tasmanites*. Some rich palynomorph-bearing rocks have been shown to contain millions of specimens per gram.

The small size and low specific gravity of palynomorphs affords them protection against abrasion during transport. They do undergo compression during burial, which is sometimes destructive to larger forms. Mild to severe mechanical damage may occur during lithification of the sediments when crystals, mainly pyrite, grow into the palynomorph wall. This may leave the damaged sporopollenin skeletons barely recognizable as palynomorphs. The palynomorph wall has evolved protecting its contents from the hazards of dessication, heat and cold, and microbial attack. Despite this, the cell contents are eventually destroyed by dessication, but the extremely durable walls survive intact in anoxic environments like swamps, some lake and ocean layers, or through an animal's alimentary canal. They can withstand prolonged transport and burial. Palynomorphs are, however, ultimately susceptible to some biological degradation and to oxidation. Different palynomorphs and different wall layers of a single palynomorph have different tolerances to the various destructive agents. Most palynomorphs are destroyed before fossilization and some of the preserved specimens have lost one or more outer wall layers. Once buried, some species of palynomorphs are more resistant to postburial ravages of heat and corrosion than are other palynomorph species in the same rock.

Biological degradation by bacteria and fungi may occur during initial deposition and burial, or after subsequent surface exposure of the rock.

Oxidation can occur during transport and deposition. Pollen and spores may be rare or poorly preserved when exposed to oxidative conditions in soils or in zones with a fluctuating water table. They are almost never found in red beds that owe their color to oxidation (rusting of iron compounds). Effects of oxidation and devolatization by heating after burial are discussed in Sec. V,A,3. Oxidation also occurs during weathering of exposed rock.

E. Preparation Techniques

For palynomorphs recovered from living plants and from unoxidized soils, a technique called acetolysis is used for clearing the internal organic substances. Extracting palynomorphs trapped in rocks, however, requires additional procedures for mineral dissolution and palynomorph concentration for efficient microscopic examination. The growth of palynology stems in part from the development of such preparation techniques. Early workers in the nineteenth century had to observe palynomorphs microscopically in rock chips or in thin rock slices with reflected or transmitted light.

Treatment by various acids selectively dissolves minerals (e.g., hydrochloric acid for carbonates; hydrofluoric acid for silicates), while concentrating the insoluble residue. Subsequent removal of acid-insoluble minerals utilizes the differences in specific gravity between minerals and insoluble organic matter. Panning and, more recently, separation by heavy liquids are two such methods. The acid-insoluble residue (kerogen) may be examined at the pre-oxidation stage to estimate amount of carbonization of the palynomorphs and thus determine thermal maturation, and to observe the unaltered kerogen assemblage, including palynomorphs, for interpretations of organic paleoenvironments and hydrocarbon generative potential.

Slightly oxidative treatments with additional acids and bases concentrate the palynomorphs. This is the primary procedure in coal maceration. This oxidation technique selectively destroys and disperses both organic and inorganic components while preserving palynomorphs. Sieving or heavy liquid procedures permit separation and removal of specific size fractions (very fine, very coarse, or very heavy matter obscures palynomorphs). Staining palynomorphs with dyes improves resolution of their morphology in otherwise transparent grains. The final concentrated residue is then mounted on a glass microslide or an electron microscopy stub. Numerous methods of palynomorph preparation have been described, and two of the most recent published accounts are by I. Doher of the U. S. Geological Survey and D. Phipps and G. Playford from Australia.

IV. Developmental History of Palynology as a Science

A. EARLY DESCRIPTIVE WORK

The first recognized palynomorphs were pollen from living plants, illustrated and described independently in 1682 by Nehemiah Grew and in 1687 by Marcello Malpighi as a direct consequence of the development of Hooke's microscope in 1665. Interpretation of palynomorphs' function and their value as problem solvers developed slowly while their description and nomenclature appeared sporadically in botanical

and geological literature until the late nineteenth century.

In 1836, H. R. Goeppert reported fossil pollen in an upper Tertiary German lignite. Some of C. G. Ehrenberg's "xanthidia" and other fossil remains presented in 1836 from Cretaceous flint chips of Germany were dinoflagellate cysts. Ehrenberg also recognized and identified pollen grains in these rocks. In the same year H. Potonié reported megaspores in Paleozoic rocks. In 1884, P. F. Reinsch published remarkably detailed illustrations and descriptions of many Carboniferous spores.

Most notable among early descriptive work on modern pollen grains is that of the northern European botanists J. E. Purkinje, C. J. Fritzsche, H. von Mohl, and H. Fischer in the mid-1800s to early 1900s.

In 1935, R. P. Wodehouse, one of the founders of the modern science of palynology, published as part of his textbook on "Pollen Grains" a comprehensive discussion of early descriptive work on pollen morphology, theories of sexuality in plants, and the like, citing ancient observations made by the Assyrians, to those made by his own contemporaries.

B. DEVELOPMENT OF TECHNIQUES

F. Schulze, in the mid-1850s was investigating methods for breaking down coal. These and other techniques led to the recovery of palynomorphs from rocks and were developed by the 1870s, when E. T. Newton separated the algal palynomorph *Tasmanites punctatus* from Australian white coal. He treated it "in a finely divided condition, with hydrochloric and hydrofluoric acids, and separating a small proportion of whitish sand by decanting." Additional palynomorph extraction techniques were developed by Scandinavian geologists after 1900 in the study of Quaternary glacial and post-glacial sediments. More complicated procedures to liberate palynomorphs from rocks have been developed subsequently.

Pollen analysis has its roots in Sweden. N. G. Lagerheim and L. von Post calculated relative abundances for different pollen species, tabulated in "pollen diagrams," to document Quaternary vegetation changes at specific locations. Results of studies were published in 1916 by von Post who realized the geological potential and refined the method. In Germany, thesis research by U. Steusloff in 1905 (subsequently described by G. Erdtman) on pollen and other fossils from

Quaternary lake beds was ahead of its time in several areas, including discussion of factors affecting pollen relative abundances.

C. FOUNDATIONS OF PALYNOLOGY

In 1920, R. Thiessen recognized the variation in morphology of "spore-exines" recovered from Paleozoic coals of the eastern United States and foresaw their value for characterizing and correlating coal beds. This idea led students of coal to devise taxonomic criteria and rules of nomenclature for spores and pollen culminating in the 1930s, 1940s, and early 1950s in the works of J. M. Schopf, L. R. Wilson, R. M. Kosanke, G. K. Guennel, A. T. Cross, and others in the United States; E. Knox and A. Raistrick in England; A. C. Ibrahim, A. A. Luber, and S. N. Naumova in the Soviet Union; J. Zerndt in Poland; and F. Loose and R. Potonié in Germany, to name a few. Form-genera and form-species were proposed. Schopf, Wilson, and Bentall's benchmark work in 1944 represented the first attempt to provide a general index of fossil spores. R. Potonié compiled his "Synopsis of *Sporae dispersae*," a series beginning in the 1950s. Potonié, with G. O. W. Kremp, also proposed a suprageneric classification system, one of several artificial taxonomic systems suggested by various workers. None of these systems is universally accepted, many palynologists preferring to simply group fossil spores and pollen in a more informal fashion when the living affinities are unknown or tenuous.

In the 1930s, descriptive work on fossil dinoflagellate cysts was initiated by O. Wetzel, M. Lejeune-Carpentier, A. Eisenack, and G. Deflandre. A series of independent publications by the latter two and their co-workers formed the basis for understanding dinoflagellate cyst morphology.

G. Erdtman in Sweden and R. P. Wodehouse in the United States produced pollen catalogues that systematically described and illustrated modern pollen. The foundations of modern palynology were laid for classification and nomenclature, as was interpretation of Quaternary pollen analyses. Textbooks by Erdtman, Wodehouse, and by Faegri and Iversen remain the principal references for modern and Quaternary palynology today. Also essential for modern, Quaternary, and late Tertiary studies are the reference collections of modern pollen species held by institutions and individuals.

Subsequent, invaluable publications produced since the 1950s include *The Catalogue of Fossil Spores and Pollen,* begun at Pennsylvania State University in 1957 by G. O. W. Kremp, W. Spackman, and H. T. Ames; the "Genera File of Fossil Spores and Pollen," a comprehensive card file including English translations begun in 1976 by J. Jansonius and L. V. Hills at the University of Calgary; G. O. W. Kremp's computer file "Palynodata;" and for microphytoplankton, G. and M. Deflandre's microfiche file of dinoflagellates and acritarchs; and A. Eisenack's catalogues of dinoflagellates and acritarchs. Various other indexes and catalogues on dinoflagellate cysts include those by W. A. S. Sarjeant and C. Downie; L. E. Stover and W. R. Evitt; G. J. Wilson and C. D. Clowes; and a series by J. K. Lentin and G. L. Williams.

D. DEVELOPMENT OF PALYNOLOGY IN THE PETROLEUM INDUSTRY

In the 1950s, the petroleum industry recognized the value of palynology for determining ages of rocks encountered in their drilling operations. The ability to determine age, paleoenvironment, and more recently the thermal maturation of rocks sequentially encountered in drillholes had economic advantage for the industry, and palynology addresses these problems.

Samples obtained from drilling oil wells are in the form of rock chips or cuttings, produced by the rotary action of the drilling bit. These cuttings are brought to the surface by drilling mud circulated down the borehole through the drill pipe and up through the annulus between the pipe and the wall of the borehole. The drilling mud has properties of viscosity and density that allow the cuttings to be removed as the drilling procedure continues. The mud is passed over a shale shaker that acts as a filter, sieving the cuttings from the viscous mud. The sequence of cuttings collected from the shale shaker provides a record, in order, of the rocks encountered by the drill bit. Dating this sequence of cuttings samples with palynomorphs, along with other methods, allows petroleum geologists to reconstruct the stratigraphic history of the rocks encountered, and to correlate this subsurface rock sequence to other sequences from additional wells or nearby surface exposures. The low volumes of cuttings samples produced and the small size of the rock chips, often 3–5 mm or less, make tiny, ubiquitous, and durable fossils such as palynomorphs ideal tools for biostratigraphy.

In major oil company research laboratories and in their operating companies, palynology has become a routine technique for these stratigraphic determinations. Research has been conducted on specific palynomorph types by studying the fossil record and living analogues when possible. Progress in understanding the marine microphytoplankton and chitinozoa, as well as spores and pollen, is the result of industry and university studies and by collaboration with government agencies throughout the world during the 1950s and 1960s. New industry-oriented professional societies, publications, and thesis research sprang up, often with industry sponsorship or support.

In the 1960s and early 1970s, the petroleum industry recognized that thermal maturation of rocks containing the oil-forming organic matter was an important factor in understanding petroleum formation, migration, accumulation, and retention in reservoirs. In order to measure the heat to which the rocks were exposed, indicators previously developed by the coal industry like coal rank, BTU content, and vitrinite reflectance were adopted for coal-bearing cuttings and for microscopic coaly fragments preserved in the rocks. Eventually, variation in spore coloration and fluorescence provided an additional method whereby both age and thermal maturation could be calibrated.

In modeling the burial history of a sedimentary basin, as suggested by the Russian geologist N. V. Lopatin in 1971, a function of time and temperature was derived to recognize the accumulated heat to which petroleum source rocks have been exposed. This method has led to routine theoretical calculations of thermal maturities related to heat flow, thermal conductivity, and time of burial for specific geologic strata at any specific location. The theoretical calculations require calibration to independently obtainable time–temperature measurements. Palynology provides both time and temperature information for this rapidly growing modeling method used in basin analysis.

V. Applications

A. GEOLOGY

1. Biostratigraphy

The primary geological application for fossil palynomorphs is in age interpretation of strata (rock layers) and in the correlation of strata from one place to another. Palynomorphs are ideal fossils for correlating and dating the rocks in which they are preserved because they have varied, yet distinct morphologies and have evolved rapidly. Assemblages of palynomorph taxa and their ranges characterize rock units representing relatively short periods of geologic time. Some of the major evolutionary steps have been noted in Sec. II. Geologic ranges of broad taxonomic and morphologic groups are shown in Fig. 1.

The transport and incorporation of land-plant-derived pollen and spores into marine rocks makes these fossils especially well suited for correlation between marine and continental sediments. The biostratigraphic zonations of land-derived palynomorphs can serve as links for correlating zonations of many fossil groups restricted to either marine or nonmarine realms.

Palynomorphs are effective tools for correlating subsurface rocks during drilling operations for oil exploration. Rich assemblages of palynomorphs may be recovered from tiny rock chips, where larger fossils are often absent or broken and unidentifiable.

Their small size and great durability make palynomorphs excellent candidates for reworking (or recycling). This is the erosion, transport (as discrete grains or embedded within sedimentary particles), and redeposition of older palynomorphs into younger sediments. During deposition sediments may incorporate palynomorphs from plants or plankton growing at that time, as well as recycled grains. When large amounts of geologic time and of thermal maturation separate the two or more episodes of deposition, the recycled specimens may be easily distinguished from the younger "in place" assemblage by their distinctive morphologies, darker color (see Sec. V,A,3) or lesser susceptibility of the palynomorph wall to staining procedures. Recycled palynomorph assemblages may be useful for interpreting ancient transport patterns, or together with other geologic clues on the source of the sediments themselves, they may yield information about rocks that are no longer exposed or no longer exist. Palynomorphs have, for example, been particularly useful around the coast of Antarctica where recycled assemblages from young seafloor sediments enable interpretation of geologic history for rocks now inaccessible on that largely ice-covered continent.

Conversely, reworking may be a disadvantage by complicating biostratigraphic interpretations.

When the recycled palynomorphs closely resemble the "in-place" assemblage in age, morphologies, color, fluorescence, and reaction to stains, they may be essentially impossible to distinguish. This has the effect of extending species ranges and blurring geologic correlations.

One of the more interesting geologic applications of palynology is in unravelling the order of tectonic events involving folding, faulting, uplift, and so on. Recognizing when such events occurred is crucial for understanding the nature and timing of these continuous or intermittent episodes. The dating of such events is routinely used in modeling plate tectonics and mountain building.

The method is simple. Fragments of pre-existing rocks are transported and redeposited during certain tectonic events such as compressional, reverse (thrust) faulting or tensional, normal (gravity) faulting. The resulting clastic sediments include boulders, cobbles, or smaller fragments of the older rocks held together by a matrix or cement. Age of the older rock fragments may be determined from the contained reworked fossils as previously described. Within the matrix or cement, the rocks may also incorporate palynomorphs either of the same age as, or younger than, the tectonic event. By determining the age of the youngest palynomorph assemblage in the rock, a constraint on the timing of the tectonic episode such as faulting can be made. Thus, tectonic events and the sediments thereby produced may be dated.

For valid palynological analysis, carefully collected samples are required. Rock samples of less than five grams may provide adequate palynomorphs for determination of age, paleoenvironment, and thermal maturation. However, because of palynomorph size, ubiquity, and incorporation in other substances, there is a risk that samples can be contaminated by palynomorphs from another source. In drilling operations, contamination is common from components added to the drilling mud. Mud additives such as bentonite and lignite can be rich in palynomorphs. Fragments of uphole palynomorph-bearing rocks can fall downhole during drilling operations, thus contaminating lower rocks. Rocks collected from outcropping surface exposures can be contaminated by modern pollen and spores. Contamination with identical pollen from living plants is potentially a major problem for Quaternary studies. Contamination may occur during sample storage and laboratory preparation. Therefore, care in collection and preparation may be the critical factor in obtaining reliable palynologic results.

2. Paleoenvironment

Palynomorphs can offer much information on paleoenvironment, including paleoclimatic conditions, depositional environments, and water depth. These interpretations depend on the principle of uniformitarianism, the doctrine suggesting that the present is the key to the past, and therefore that ancient environments may have modern analogues. For example, it may be assumed that a palm tree growing 40 million years ago preferred similar conditions to one growing today.

Palynological analyses by Lagerheim and von Post in the early 1900s exemplified this theory. Their techniques for studying Quaternary palynology are followed today, and are explained fully by K. Faegri and J. Iversen in their textbook on pollen analysis. Quaternary pollen analysis involves the detailed sequential sampling of sediments such as lake muds or peats from bogs. Results of percentage counts of all the represented pollen and spores (usually grouped by genera or families) are tabulated in "pollen diagrams" with the vertical series of samples plotted against relative abundances for each taxa. Major changes in abundances of trees and shrubs versus herbaceous plants (such as small plants and grasses) are also recorded. The temporal reconstruction of vegetational change for each site may then be interpreted and local climatic inferences made. Regional patterns may be seen for particular times by drawing "vegetation maps" of vast areas, such as arctic North America, Scandinavia, and northwestern Russia. Much global information, mainly for high latitudes, has been gleaned this way for the succession of Quaternary Glacial and Interglacial Stages, and Postglacial times of the past 10,000 years.

Attributing known Quaternary climatic requirements (such as temperature and rainfall) to similar Tertiary assemblages is a subject of much debate. As noted for angiosperm pollen (Sec. II,B,2), some workers do not use the natural or even half-natural names for fossil pollen and spores, although the palynomorphs are similar or identical to extant species, and many believe inferring similar habitats is even less reliable. Often several independent lines of evidence (e.g., palynomorph data, leaf occurrence and morphology, fruits, width of tree rings, invertebrate and vertebrate fossils, and sedimentary structures) lead to definitive results about paleoenvironments. This multifaceted approach, which includes palynology, has indicated that land masses occupying mid-latitudes supported warm temperate to subtropical vegetation during much of the early Tertiary Period, while cool temperate floras flourished in areas (e.g., Alaska) that now have tundra vegetation.

Making paleoclimatic interpretations from palynomorph assemblages of Mesozoic and Paleozoic age is difficult. These assemblages may include extinct groups, or pollen and spore species for which living analogs are not known with certainty. Again, integration with other fossil evidence is required. Corroboration from meteorological models used in reconstructing rainfall patterns, wind directions, or temperatures for various paleogeographic settings can be helpful. As organs of plants, pollen or spores may themselves have evolved particular functional adaptations; for example, it is suspected that at least some of the striate pollen like the Permian *Vittatina–Weylandites* types (Fig. 3.10), and later striate pollen, may have developed striae as an adaptation for arid conditions. Striations permit expansion and contraction of the pollen body during large-scale diurnal temperature or moisture changes. Other sedimentological and paleobotanical data from these areas suggest at least a seasonally arid habitat for the parent plants. It is probable that many, if not most, structural and/or sculptural features of a pollen or spore can be related to some palynomorph function.

In the fossil record, pollen and spores are most frequently preserved as isolated entities. Consequently there is no obvious link between most palynomorphs and their parent plants. In fortuitous circumstances, however, paleobotanists have found fossilized pollen sacs and sporangia still containing extractable pollen and spores, connected to other fossilized plant parts. In this way it has been possible to relate various Mesozoic and Paleozoic pollen and spores to their parent plants. Knowledge of a palynomorph's affinity, in even a broad sense, allows paleoenvironmental interpretations by analogy.

Interpreting depositional environment from fossil palynomorphs is less ambiguous. Chitinozoa, scolecodonts, most dinoflagellate cysts and acritarchs indicate marine conditions, based on their consistent association with other marine fossils, along with lithologic evidence, and in some cases, comparison with modern forms. Some dinoflagellate cysts and acritarchs record proximity to shoreline, salinity, water depths, and other factors. Swarms of small (10–20 μm) spiny acritarchs are believed to mark marine transgressive (rise in relative sealevel) phases. Spores of water ferns and some colonial algae are good indicators of freshwater paleoenvironments.

A simple and effective measure of depositional environment is the marine : nonmarine ratio. Relative abundance counts in samples from one area can trace the sequence of marine transgressions and regressions for that area. Such counts can be useful in exploration for potential oil reservoirs by indicating channels, estuaries, and shorelines, and for predicting the occurrence of hydrocarbon source rocks.

In paleoenvironmental studies, all insoluble organic components of a palynological preparation (collectively termed "kerogen") can be employed in so-called "palynofacies analysis." Important factors in such an analysis are the relative abundances of palynomorphs, plant cuticle, woody material, and amorphous material (which may include marine or freshwater algal matter, or biologically degraded woody fragments). Palynomorph diversity and abundance of certain palynomorph types, marine : nonments (large, brittle pieces of cuticle cannot travel far), and degree of sorting (ultra-fine material may be winnowed out) are other pertinent considerations in paleoenvironmental reconstructions.

3. Thermal Maturity

In the 1960s, the oil industry began utilizing palynomorphs to measure the heat to which deeply buried rocks had been exposed. Noting that palynomorphs showed a uniform color change when exposed to increasing heat, a group of palynologists led by C. C. M. Gutjahr and F. Staplin calibrated the change. This measurement is called Thermal Alteration Index (TAI) and is applied by assigning numerical values to the color changes pollen and spore walls undergo, from clear greenish-yellow in uneffected grains, to yellow, orange, brown, to opaque black. This technique is invaluable for determining if organic matter in the rocks has been heated enough (indicated by orange-brown palynomorphs) to generate oil molecules, heated further (indicated by dark brown palynomorphs) to "crack" these into gas molecules, or overheated (indicated by opaque black palynomorph remnants) thus destroying the hydrocarbons.

B. Botany

Botanists were the first to recognize pollen and have continued to describe pollen of living plant species. Plant evolution can be traced through pollen morphology, and taxonomic relations among extant plant groups may be revealed. Pollen and spores display functional morphology for specific reproductive strategies. Recent research has shown, for example, that pollen morphology frequently has aerodynamic properties allowing certain flower architecture to select its own species of wind-blown pollen.

Size, shape, and number of palynomorphs produced by a particular species frequently form the basis for survival of that species. Pollen morphology affects how a grain may be transported by wind, insects, birds, or water. A pollen grain may also display morphological adaptations that effect protection, dispersal, or release of genetic matter under particular climate conditions of wet, dry, windy, or cool seasons.

Understanding the palynology of modern plants allows reconstruction, by analogy, of ancient environmental conditions, useful to geologists, archeologists, and occasionally lawyers. Data valuable for such extrapolations include the numbers of palynomorphs produced by a single plant or by particular species, the distance of dispersal from parent plants, and the number of grains likely to be preserved, all adding up to the palynomorphs' relationship to modern plant geography.

C. Anthropology and Intervention by Man

Knowledge of pollen and spores of modern plants allows reconstruction of events of the recent past, pertinent to the development of man and his environment. Distribution of forests and their harvest, evolution of agriculture, and destruction or alteration of habitat may be recognized. Diet of early humans, as well as contemporaneous animals, may be discovered by examining the palynology of their excrement. Historical events, including the cultivation of corn, production of mead, and the geographic implications of pollen on the Shroud of Turin, can also be interpreted.

D. Crime and Law

Examination of pollen from mud on a suspect's shoe or car tires may be useful in the apprehension of a criminal by placing that person at a specific site. Also, identification and geographic origin of illegal drug plants may be supported by recognition of pollen species. Palynology has also been used in legal proceedings for documenting the former presence of wetlands beneath landfills by identifying the diagnostic saltmarsh components.

E. Medicine

No discussion of modern pollen would be complete without reference to summer weather reports of the pollen count and Wodehouse's discussion of this count. The pollen count is reported to inform sufferers of hayfever of the number of pollen in the air and the potential severity to allergic persons. The primary cause of early spring hayfever is pollen produced by the trees that spread enormous numbers of pollen grains during their short period of flowering. In the northern hemisphere these plants include elm, oak, poplar, birch, maple, and willow. Early summer hayfever may be related to the flowering and concomitant pollen dispersal of the grasses. The effect grass pollen produces is usually more intense than that of the early spring trees. Late summer hayfever is related primarily to pollen of ragweed, although cockleburs and sagebrush can also be major offenders. Hayfever is by no means caused only by these few groups. Sensitivity to specific pollen types may be determined by a skin test administered by a physician using either pollen or pollen extracts.

VI. Present Research

Nearly 900 individuals in more than 50 countries and more than 100 institutions from 27 countries were members in 1985 of the American Association of Stratigraphic Palynologists, the largest palynological society in North America. Of these, nearly half are employed directly or as consultants to the energy industry, with most of the others working at universities, government agencies, and museums. Of those at universities, more than half are in geology departments, with most of the others in botany and anthropology departments. Although Soviet palynologists are not represented in these numbers, it is estimated that more than 1000 are devoted to palynology in a variety of Soviet institutions.

Palynological research is published in four principal journals, *Palynology, Review of Palaeobotany and Palynology, Pollen et Spores,* and *Grana,* as well as a host of other paleontological, geological, botanical, and anthropological journals.

BIBLIOGRAPHY

Doher, L. I. (1980). Palynomorph preparation procedures currently used in the paleontology and stratigraphy laboratories, U.S. Geological Survey, *U.S. Geological Survey Circular* 830, 1–29.

Evitt, W. R. (1985). "Sporopollenin Dinoflagellate Cysts, Their Morphology and Interpretation." Hart Graphics, Austin, Texas.

Kremp, G. O. W. (1965). "Morphologic Encyclopedia of Palynology." Univ. of Arizona Press, Tucson, Arizona.

Muller, J. (1981). Fossil records of extant angiosperms. *The Botanical Review,* **47**(1), 1–142.

Phipps, D., and Playford, G. (1984). Laboratory techniques for extraction of palynomorphs from sediments. *Papers,* Dept. of Geology, Univ. of Queensland, **11**(1), 1–23.

Sarjeant, W. A. S. (1974). "Fossil and Living Dinoflagellates." Academic Press, London and New York.

Tappan, H. (1980). "The Paleobiology of Plant Protists." Freeman, San Francisco, California.

Tschudy, R. H., and Scott, R. A. (eds.). (1969). "Aspects of Palynology." Wiley (Interscience), New York.

PAPER—*SEE* PULP AND PAPER

PARTIAL DIFFERENTIAL EQUATIONS

Martin Schechter *University of California, Irvine*

GLOSSARY

Boundary: Set of points in the closure of a region not contained in its interiors.

Bounded region: Region that is contained in a sphere of finite radius.

Eigenvalue: Scalar λ for which the equation $Au = \lambda u$ has a nonzero solution u.

Euclidean n dimensional space $\mathbb{R}^n$: Set of vectors $x = (x_1, \dots, x_n)$ where each component x_j is a real number.

Partial derivative: Derivative of a function of more than one variable with respect to one of the variables keeping the other variables fixed.

A partial differential equation is an equation in which a partial derivative of an unknown function appears. The order of the equation is the highest order of the partial derivatives (of an unknown function) appearing in the equation. If there is only one unknown function $u(x_1, \dots, x_n)$, then a partial differential equation for u is of the form

$$F(x_1, \dots, x_n, u, \frac{\partial u}{\partial x_1}, \dots, \frac{\partial u}{\partial x_n},$$

$$\frac{\partial^2 u}{\partial x_1^2}, \dots, \frac{\partial^k u}{\partial x_1^k}, \dots) = 0$$

One can have more than one unknown function and more than one equation involving some or all of the unknown functions. One then has a system of j partial differential equations in k un-

known functions. The number of equations may be more or less than the number of unknown functions. Usually it is the same.

I. Importance

One finds partial differential equations in practically every branch of physics, chemistry, and engineering. They are also found in other branches of the physical sciences and in the social sciences, economics, business, etc. Many parts of theoretical physics are formulated in terms of partial differential equations. In some cases the axioms require that the states of physical systems be given by solutions of partial differential equations. In other cases partial differential equations arise when one applies the axioms to specific situations. [*See* ALGEBRAIC EQUATIONS; DIFFERENTIAL EQUATIONS, ORDINARY.]

II. How They Arise

Partial differential equations arise in several branches of mathematics. For instance the Cauchy–Riemann equations

$$\frac{\partial u(x, y)}{\partial x} = \frac{\partial v(x, y)}{\partial y}, \qquad \frac{\partial u(x, y)}{\partial y} = -\frac{\partial v(x, y)}{\partial x}$$

must be satisfied if

$$f(z) = u(x, y) + iv(x, y)$$

is to be an analytic function of the complex variable $z = x + iy$. Thus, the rich and beautiful branch of mathematics known as analytic function theory is merely the study of solutions of a particular system of partial differential equations.

As a simple example of a partial differential equation arising in the physical sciences, we consider the case of a vibrating string. We assume that the string is a long, very slender body of elastic material that is flexible because of its extreme thinness and is tightly stretched between the points $x = 0$ and $x = L$ on the x axis of the x, y plane. Let x be any point on the string, and let $y(x, t)$ be the displacement of that point from the x axis at time t. We assume that the displacements of the string occur in the x, y plane. Consider the part of the string between two close points x_1 and x_2. The tension T in the string acts in the direction of the tangent to the curve formed by the string. The net force on the segment $[x_1, x_2]$ in the y direction is

$$T \sin \varphi_2 - T \sin \varphi_1$$

where φ_i is the angle between the tangent to the curve and the x axis at x_i. According to Newton's second law, this force must equal mass times acceleration. This is

$$\int_{x_1}^{x_2} \rho \, \partial^2 y / \partial t^2 \, dx$$

where ρ is the density (mass per unit length) of the string. Thus in the limit

$$T \frac{\partial}{\partial x} \sin \varphi = \rho \frac{\partial^2 y}{\partial t^2}$$

We note that $\tan \varphi = \partial y / \partial x$. If we make the simplifying assumption (justified or otherwise) that

$$\cos \varphi \approx 1, \qquad \frac{\partial}{\partial x} \cos \varphi \approx 0$$

we finally obtain

$$T \, \partial^2 y / \partial x^2 = \rho \, \partial^2 y / \partial t^2$$

which is the well-known equation of the vibrating string.

The derivation of partial differential equations from physical laws usually brings about simplifying assumptions that are difficult to justify completely. Most of the time they are merely plausibility arguments. For this reason some branches of science have accepted partial differential equations as axioms. The success of these axioms is judged by how well their conclusions describe past observations and predict new ones.

III. Some Well-Known Equations

Now we list several equations that arise in various branches of science. Interestingly, the same equation can arise in diverse and unrelated areas.

A. Laplace's Equation

In n dimensions this equation is given by

$$\Delta u = 0$$

where

$$\Delta = \frac{\partial^2}{\partial x_1^2} + \cdots + \frac{\partial^2}{\partial x_n^2}$$

It arises in the study of electromagnetic phenomena (e.g., electrostatics, dielectrics, steady currents, magnetostatics), hydrodynamics (e.g., irrotational flow of a perfect fluid, surface waves), heat flow, gravitation, and many other branches of science. Solutions of Laplace's equation are called harmonic functions.

B. Poisson's Equation

$$\Delta u \equiv f(x), \qquad x = (x_1, \ldots, x_n)$$

Here the function $f(x)$ is given. This equation is found in many of the situations in which Laplace's equation appears, since the latter is a special case.

C. Helmholtz's Equation

$$\Delta u \pm \alpha^2 u = 0$$

This equation appears in the study of elastic waves, vibrating strings, bars and membranes, sound and acoustics, electromagnetic waves and the operation of nuclear reactors.

D. The Heat (Diffusion) Equation

This equation is of the form

$$u_t = a^2 \, \Delta u$$

where $u(x_1, \ldots, x_n, t)$ depends on the variable t(time) as well. It describes heat conduction or diffusion processes.

E. The Wave Equation

$$\Box u \equiv (1/c^2) u_{tt} - \Delta u = 0$$

This describes the propagation of a wave with velocity c. This equation governs most cases of wave propagation.

F. The Telegraph Equation

$$\Box u + \sigma u_t = 0$$

This applies to some types of wave propagation.

G. The Scalar Potential Equation

$$\Box u = f(x, t)$$

H. The Klein–Gordon Equation

$$\Box u + \mu^2 u = 0$$

I. Maxwell's Equations

$$\nabla \times \mathbf{H} = \sigma\mathbf{E} + \varepsilon\, \partial\mathbf{E}/\partial t$$

$$\nabla \times \mathbf{E} = -\mu\, \partial\mathbf{H}/\partial t$$

Here $\mathbf{E}$ and $\mathbf{H}$ are three-dimensional vector functions of position and time representing the electric and magnetic fields, respectively. This system of equations is used in electrodynamics.

J. The Cauchy–Riemann Equations

$$\frac{\partial u}{\partial x} = \frac{\partial v}{\partial y}, \qquad \frac{\partial u}{\partial y} = -\frac{\partial v}{\partial x}$$

These equations describe the real and imaginary parts of an analytic function of a complex variable.

K. The Schrödinger Equation

$$-\frac{\hbar^2}{2m}\Delta\psi + V(x)\psi = i\hbar\frac{\partial\psi}{\partial t}$$

This equation describes the motion of a quantum mechanical particle as it moves through a potential field. The function $V(x)$ represents the potential energy, while the unknown function $\psi(x)$ is allowed to have complex values.

L. Minimal Surfaces

In three dimensions a surface $z = u(x, y)$ having the least area for a given contour satisfies the equation

$$(1 + u_y^2)u_{xx} - 2u_x u_y u_{xy} + (1 + u_x^2)u_{yy} = 0$$

where $u_x = \partial u/\partial x$, etc.

M. The Navier–Stokes Equations

$$\frac{\partial u_j}{\partial t} + \sum_k \frac{\partial u_j}{\partial x_k} u_k + \frac{1}{\rho}\frac{\partial p}{\partial x_j} = \gamma\Delta u_j$$

$$\sum_k \frac{\partial u_k}{\partial x_k} = 0$$

This system describes viscous flow of an incompressible liquid with velocity components u_k and pressure p.

N. The Korteweg–deVries Equation

$$u_t + cuu_x + u_{xxx} = 0$$

This equation is used in the study of water waves.

IV. Types of Equations

In describing partial differential equations, the following notations are helpful. Let $\mathbb{R}^n$ denote Euclidean n dimensional space, and let $\mathbf{x} = (x_1, \ldots, x_n)$ denote a point in $\mathbb{R}^n$. One can consider partial differential equations on various types of manifolds, but we shall restrict ourselves to $\mathbb{R}^n$. For a real- or complex-valued function $u(x)$ we shall use the following notation:

$$D_k u = \partial u/i\partial x_k, \qquad 1 \leqq k \leqq n$$

If $\mu = (\mu_1, \ldots, \mu_n)$ is a multi-index of nonnegative integers, we write

$$x^\mu = x_1^{\mu_1} \cdots x_n^{\mu_n}, \qquad |\mu| = \mu_1 + \cdots + \mu_n$$

$$D^\mu = D_1^{\mu_1} \cdots D_n^{\mu_n}$$

Thus D^μ is a partial derivative of order $|\mu|$.

A. Linear Equations

The most general linear partial differential equation of order m is

$$Au \equiv \sum_{|\mu| \leqq m} a_\mu(x)D^\mu u = f(x). \qquad (1)$$

It is called linear because the operator A is linear, that is, satisfies

$$A(\alpha u + \beta v) = \alpha Au + \beta Av$$

for all functions u, v and all constant scalars α, β. If the equation cannot be put in this form, it is nonlinear. In Section III, the examples in Sections A–K are linear.

B. Nonlinear Equations

In general, nonlinear partial differential equations are more difficult to solve than linear equations. There may be no solutions possible, as is the case for the equation

$$|\partial u/\partial x| + 1 = 0.$$

There is no general method of attack, and only special types of equations have been solved.

1. Quasilinear Equation

A partial differential equation is called quasilinear if it is linear with respect to its derivatives of highest order. This means that if one replaces the unknown function $u(x)$ and all its derivatives of order lower than the highest by known functions, the equation becomes linear. Thus a quasilinear equation of order m is of the form

$$\sum_{|\mu|=m} a_\mu(x, u, Du, ..., D^{m-1}u)D^\mu u$$

$$= f(x, u, Du, ..., D^{m-1}u) \qquad (2)$$

where the coefficients a_μ depend only on x, u and derivatives of u up to order $m - 1$. Quasilinear equations are important in applications. In Section III, the examples in Sections L–N are quasilinear.

2. Semilinear Equation

A quasilinear equation is called semilinear if the coefficients of the highest-order derivatives depend only on x. Thus, a semilinear equation of order m is of the form

$$Au \equiv \sum_{|\mu|=m} a_\mu(x)D^\mu u = f(x, u, Du, ..., D^{m-1}u)$$

$$(3)$$

where A is linear. Semilinear equations arise frequently in practice.

C. ELLIPTIC EQUATIONS

The quasilinear equation (2) is called elliptic in a region $\Omega \subset \mathbb{R}^n$ if for every function $v(x)$ the only real vector $\xi = (\xi_1, ..., \xi_n) \varepsilon \mathbb{R}^n$ that satisfies

$$\sum_{|\mu|=m} a_\mu(x, v, Dv, ..., D^{m-1}v)\xi^\mu = 0$$

is $\xi = 0$. It is called uniformly elliptic in Ω if there is a constant $c_0 > 0$ independent of x and v such that

$$c_0|\xi|^m \leq \left| \sum_{|\mu|=m} a_\mu(x, v, Dv, ..., D^{m-1}v)\xi^\mu \right|$$

where $|\xi|^2 = \xi_1^2 + \cdots \xi_n^2$. The equations in Sections A–C and J–L are elliptic equations or systems.

D. PARABOLIC EQUATIONS

When $m = 2$, the quasilinear equation (2) becomes

$$\sum_{j,k} a_{jk}(x, u, Du) \frac{\partial^2 u}{\partial x_j \partial x_k} = f(x, u, Du). \qquad (4)$$

We shall say that equation (4) is parabolic in a region $\Omega \subset \mathbb{R}^n$ if for every choice of the function $v(x)$, the matrix $\Lambda = (a_{jk}(x, v, Dv))$ has a vanishing determinant for each $x \in \Omega$. The equation in Section III,D is a parabolic equation.

E. HYPERBOLIC EQUATIONS

Equation (4) will be called ultrahyperbolic in Ω if for each $v(x)$ the matrix $\Lambda = (a_{jk}(x, v, Dv))$ has some positive, some negative, and no vanishing eigenvalues for each $x \in \Omega$. It will be called hyperbolic if all but one of the eigenvalues of Λ have the same sign and none of them vanish in Ω. The equations in Sections III,E–H are hyperbolic equations.

The only time equation (4) is neither parabolic nor ultra hyperbolic is when all the eigenvalues of Λ have the same sign with none vanishing. In this case (4) is elliptic as described earlier.

F. EQUATIONS OF MIXED TYPE

If the coefficients of equation (1) are variable, it is possible that the equation will be of one type in one region and of another type in a different region. A simple example is

$$\partial^2 u/\partial x^2 - x \, \partial^2 u/\partial y^2 = 0$$

in two dimensions. In the region $x > 0$ it is hyperbolic, while in the region $x < 0$ it is elliptic. (It becomes parabolic on the line $x = 0$.)

G. OTHER TYPES

The type of an equation is very important in determining what problems can be solved for it and what kind of solutions it will have. As we saw, some equations can be of different types in different regions. One can define all three types for higher-order equations, but most higher-order equations will not fall into any of the three categories.

V. Problems Associated with Partial Differential Equations

In practice one is rarely able to determine the most general solution of a partial differential equation. Usually one looks for a solution satisfying additional conditions. One may wish to prescribe the unknown function and/or some of its derivatives on part or all of the boundary of the region in question. We call this a boundary value problem. If the equation involves the variable t (time) and the additional conditions are prescribed at some time $t = t_0$, we usually refer to it as an initial value problem.

A problem associated with a partial differential equation is called *well posed* if

(a)   a solution exists for all possible values of the given data,

(b)   the solution is unique, and
(c)   the solution depends in a continuous way on the given data.

The reason for the last requirement is that in most cases the given data come from various measurements. It is important that a small error in measurement of the given data should not produce a large error in the solution. The method of measuring the size of an error in the given data and in the solution is a basic question for each problem. There is no standard method; it varies from problem to problem.

The kinds of problems that are well posed for an equation depend on the type of the equation. Problems that are suitable for elliptic equations are not suitable for hyperbolic or parabolic equations. The same holds true for each of the types. We illustrate this with several examples.

A. DIRICHLET'S PROBLEM

For a region $\Omega \subset \mathbb{R}^n$, Dirichlet's problem for equation (2) is to prescribe u and all its derivatives up to order $\frac{1}{2}m - 1$ on the boundary $\partial\Omega$ of Ω. This problem is well posed only for elliptic equations. If $m = 2$, only the function u is prescribed on the boundary. If Ω is unbounded, one may have to add a condition at infinity.

It is possible that the Dirichlet problem is not well posed for a linear elliptic equation of the form (1) because 0 is an eigenvalue of the operator A. This means that there is a function $w(x) \neq 0$ that vanishes together with all derivatives up to order $\frac{1}{2}m - 1$ on $\partial\Omega$ and satisfies $Aw = 0$ in Ω. Thus any solution of the Dirichlet problem for (1) is not unique, for we can always add a multiple of w to it to obtain another solution. Moreover, it is easily checked that one can solve the Dirichlet problem for (1) only if

$$\int_\Omega f(x)w(x)\,dx$$

is a constant depending on w and the given data. Thus we cannot solve the Dirichlet problem for all values of the given data. When Ω is bounded, one can usually remedy the situation by considering the equation

$$Au + \varepsilon u = f \qquad (5)$$

in place of (1) for ε sufficiently small.

B. THE NEUMANN PROBLEM

As in the case of the Dirichlet problem, the Neumann problem is well posed only for elliptic operators. The Neumann problem for equation

(2) in a region Ω is to prescribe on $\partial\Omega$ the normal derivatives of u from order $\frac{1}{2}m$ to $m - 1$. [In both the Dirichlet and Neumann problems exactly $\frac{1}{2}m$ normal derivatives are prescribed on $\partial\Omega$. In the Dirichlet problem the first $\frac{1}{2}m$ are prescribed (starting from the zeroth-order derivative—u itself). In the Neumann problem the next $\frac{1}{2}m$ normal derivatives are prescribed.]

As in the case of the Dirichlet problem, the Neumann problem for a linear equation of the form (1) can fail to be well posed because 0 is an eigenvalue of A. Again this can usually be corrected by considering equation (5) in place of (1) for ε sufficiently small.

When $m = 2$, the Neumann problem consists of prescribing the normal derivative $\partial u/\partial n$ of u on $\partial\Omega$. In the case of Laplace's or Poisson's equation, it is easily seen that 0 is indeed an eigenvalue if Ω is bounded. For then any constant function is a solution of

$$\Delta w = 0 \quad \text{in} \quad \Omega, \qquad \partial w/\partial n = 0 \quad \text{on} \quad \partial\Omega \qquad (6)$$

Thus, adding any constant to a solution gives another solution. Moreover, we can solve the Neumann problem

$$\Delta u = f \quad \text{in} \quad \Omega, \qquad \frac{\partial u}{\partial n} = g \quad \text{on} \quad \partial\Omega \qquad (7)$$

only if

$$\int_\Omega f(x)\,dx = \int_{\partial\Omega} g\,ds$$

Thus we cannot solve (7) for all f and g. If Ω is unbounded, one usually requires that the solution of the Neumann problem vanish at infinity. This removes 0 as an eigenvalue, and the problem is well posed.

C. THE ROBIN PROBLEM

When $m = 2$, the Robin problem for equation (2) consists of prescribing

$$Bu = \alpha\, \partial u/\partial n + \beta u \qquad (8)$$

on $\partial\Omega$, where $\alpha(x)$, $\beta(x)$ are functions and $\alpha(x) \neq 0$ on $\partial\Omega$. If $\beta(x) \equiv 0$, this reduces to the Neumann problem. Again this problem is well prosed only for elliptic equations.

D. MIXED PROBLEMS

Let B be defined by (8), and assume that $\alpha(x)^2 + \beta(x)^2 \neq 0$ on $\partial\Omega$. Consider the boundary value problem consisting of finding a solution of (4) in Ω and prescribing Bu on $\partial\Omega$. On those parts of $\partial\Omega$ where $\alpha(x) = 0$ we are prescribing Dirichlet data. On those parts of $\partial\Omega$ where $\beta(x) = 0$ we are prescribing Neumann data. On the remaining sections of $\partial\Omega$ we are prescribing Robin data. This is an example of a mixed boundary value problem in which one prescribes different types of data on different parts of the boundary. Other examples are provided by parabolic and hyperbolic equations to be discussed later.

E. GENERAL BOUNDARY VALUE PROBLEMS

For an elliptic equation of the form (2) one can consider general boundary conditions of the form

$$B_j u \equiv \sum_{|\mu| \leq m_j} b_{j\mu} D^\mu u$$

$$= g_j \quad \text{on} \quad \partial\Omega, \qquad 1 \leq j \leq m/2 \quad (9)$$

Such boundary value problems can be well posed provided the operators B_j are independent in a suitable sense and do not "contradict" each other or the equation (2).

F. THE CAUCHY PROBLEM

For equation (2) the Cauchy problem consists of prescribing all derivatives of u up to order $m - 1$ on a smooth surface S and solving (2) for u in a neighborhood of S. An important requirement for the Cauchy problem to have a solution is that the boundary conditions not "contradict" the equation on S. This means that the coefficient of $\partial^m u/\partial n^m$ in (2) should not vanish on S. Otherwise the equation (2) and the Cauchy boundary conditions

$$\partial^k u/\partial n^k = g_k \quad \text{on} \quad S, \, k = 0, \ldots, n - 1 \quad (10)$$

involve only the function g_k on S. This is sure to cause a contradiction unless f is severely restricted. When this happens we say that the surface S is *characteristic* for equation (2). Thus for the Cauchy problem to have a solution without restricting f, it is necessary that the surface S be noncharacteristic. In the quasilinear case, the coefficient of $\partial^m u/\partial n^m$ in (2) depends on the g_k. Thus the Cauchy data (10) will play a role in determining whether or not the surface S is characteristic for equation (2). The Cauchy–Kowalewski theorem states that for a noncharacteristic analytic surface S, real analytic Cauchy data g_k and real analytic coefficients a_μ, f in (2), the Cauchy problem (2) (10) has a unique real analytic solution in the neighborhood of S. This is true irrespective of the type of the equation. However, this does not mean that the Cauchy

problem is well posed for all types of equations. In fact the hypotheses of the Cauchy–Kowalewski theorem are satisfied for the Cauchy problem

$$u_{xx} + u_{yy} = 0, \qquad y > 0$$

$$u(x, 0) = 0, \qquad u_y(x, 0) = n^{-1} \sin nx$$

and indeed it has a unique analytic solution

$$u(x, y) = n^{-2} \sinh ny \sin nx$$

The function $n^{-1} \sin nx$ tends uniformly to 0 as $n \to \infty$, but the solution does not become small as $n \to \infty$ for $y \neq 0$. It can be shown that the Cauchy problem is well posed only for hyperbolic equations.

VI. Methods of Solution

There is no general approach for finding solutions of partial differential equations. Indeed there exist linear partial differential equations with smooth coefficients having no solutions in the neighborhood of a point. For instance the equation

$$\frac{\partial u}{\partial x_1} + i\frac{\partial u}{\partial x_2} + 2i(x_1 + ix_2)\frac{\partial u}{\partial x_3} = f(x_3) \quad (11)$$

has no solution in the neighborhood of the origin unless $f(x_3)$ is a real analytic function of x_3. Thus if f is infinitely differentiable but not analytic, (11) has no solution in the neighborhood of the origin. Even when partial differential equations have solutions, we cannot find the "general" solution. We must content ourselves with solving a particular problem for the equation in question. Even then we are rarely able to write down the solution in closed form. We are lucky if we can derive a formula that will enable us to calculate the solution in some way, such as a convergent series or iteration scheme. In many cases even this is unattainable. Then one must be satisfied with an abstract theorem stating that the problem is well posed. Some times the existence theorem does provide a method of calculating the solution; more often it does not.

Now we describe some of the methods that can be used to obtain solutions in specific situations.

A. SEPARATION OF VARIABLES

Consider a vibrating string stretched along the x axis from $x = 0$ and $x = \pi$ and fixed at its end points. We can assign the initial displacement and velocity. Thus we are interested in solving the mixed initial and boundary value problem for the displacement $u(x, t)$

$$\Box u = 0, \qquad 0 < x < \pi, \quad t > 0 \quad (12)$$

$$u(x, 0) = f(x),$$

$$u_t(x, 0) = g(x), \quad 0 \leq x \leq \pi \quad (13)$$

$$u(0, t) = u(\pi, t) = 0, \qquad t \geq 0 \quad (14)$$

We begin by looking for a solution of (12) of the form

$$u(x, t) = X(x)T(t)$$

Such a function will be a solution of (12) only if

$$\frac{T''}{T} = c^2 \frac{X''}{X} \quad (15)$$

Since the left-hand side of (15) is a function of t only and the right hand side is a function of x only, both sides are constant. Thus there is a constant K such that $X''(x) = KX(x)$. If $K = \lambda^2 > 0$, X must be of the form

$$X = Ae^{-\lambda x} + Be^{\lambda x}$$

In order that (14) be satisfied, we must have

$$X(0) = X(\pi) = 0 \quad (15')$$

This can happen only if $A = B = 0$. If $K = 0$, then X must be of the form

$$X = A + Bx$$

Again, this can happen only if $A = B = 0$. If $K = -\lambda^2 < 0$, then X is of the form

$$X = A \cos \lambda x + B \sin \lambda x$$

This can satisfy (15') only if $A = 0$ and $B \sin \lambda \pi = 0$. Thus the only way that X should not vanish identically is if λ is an integer n. Moreover, T satisfies

$$T'' + n^2 c^2 T = 0$$

The general solution for this is

$$T = A \cos nct + B \sin nct$$

Thus

$$u(x, t) = \sin nx (A_n \cos nct + B_n \sin nct)$$

is a solution of (12), (14) for each integer n. However, it will not satisfy (13) unless $f(x)$, $g(x)$ are of a special form. The linearity of the operator $\Box$ allows one to add solutions of (12), (14). Thus

$$u(x, t) = \sum_{n=1}^{\infty} \sin nx (A_n \cos nct + B_n \sin nct)$$

$$(16)$$

will be a solution provided the series converges. Moreover, it will satisfy (13) if

$$f(x) = \sum_{n=1}^{\infty} A_n \sin nx,$$

$$g(x) = \sum_{n=1}^{\infty} ncB_n \sin nx$$

This will be true if $f(x)$, $g(x)$ are expandable in a Fourier sine series. If they are, then the coefficients A_n, B_n are given by

$$A_n = \frac{2}{\pi} \int_0^\pi f(x) \sin nx \, dx,$$

$$B_n = \frac{2}{nc\pi} \int_0^\pi g(x) \sin nx \, dx$$

With these values, the series (16) converges and gives a solution of (12)–(14).

B. FOURIER TRANSFORMS

If we desire to determine the temperature $u(x, t)$ of a system in $\mathbb{R}^n$ with no heat added or removed and initial temperature given, we must solve

$$u_t = a^2 \Delta u, \qquad x \varepsilon \mathbb{R}^n, \quad t > 0 \quad (17)$$

$$u(x, 0) = \varphi(x), \qquad x \varepsilon \mathbb{R}^n \quad (18)$$

If we apply the Fourier transform

$$\hat{f}(\xi) = (2\pi)^{-n/2} \int e^{-i\xi x} f(x) \, dx \quad (19)$$

where $\xi x = \xi_1 x_1 + \cdots + \xi_n x_n$, we obtain

$$\hat{u}_t(\xi, t) + a^2 |\xi|^2 \hat{u}(\xi, t) = 0$$

The solution satisfying (18) is

$$\hat{u}(\xi, t) = e^{-a^2 |\xi|^2 t} \hat{\varphi}(\xi)$$

If we now make use of the inverse Fourier transform

$$f(x) = (2\pi)^{-n/2} \int e^{i\xi x} \hat{f}(\xi) \, d\xi$$

we have

$$u(x, t) = \int K(x - y, t) \varphi(y) \, dy$$

where

$$K(x, t) = (2\pi)^{-n} \int e^{i x \xi - a^2 |\xi|^2 t} \, d\xi$$

If we introduce the new variable

$$\eta = a t^{1/2} \xi - \tfrac{1}{2} i a^{-1} t^{-1/2} x$$

this becomes

$$(2\pi)^{-n} a^{-n} t^{-n/2} e^{-|x|^2/4a^2 t} \int e^{-|\eta|^2} \, d\eta$$

$$= (4\pi a^2 t)^{-n/2} e^{-|x|^2/4a^2 t}$$

This suggests that a solution of (17), (18) is given by

$$u(x, t) = (4\pi a^2 t)^{-n/2} \int e^{-|x-y|^2/4a^2 t} \varphi(y)\, dy \quad (20)$$

It is easily checked that this is indeed the case if φ is continuous and bounded. However, the solution is not unique unless one places more restriction on the solution.

C. FUNDAMENTAL SOLUTIONS, GREEN'S FUNCTION

Let

$$K(x, y) = \frac{|x - y|^{2-n}}{(2 - n)\omega_n} + h(x), \qquad n > 2,$$

$$K(x, y) = \frac{\log 4}{2\pi} + h(x), \quad n = 2$$

where $\omega_n = 2\pi^{n/2}/\Gamma(\tfrac{1}{2}n)$ is the surface area of the unit sphere in $\mathbb{R}^n$ and $h(x)$ is a harmonic function in a bounded domain $\Omega \subset \mathbb{R}^n$ (i.e., $h(x)$ is a solution of $\Delta h = 0$ in Ω). If the boundary $\partial\Omega$ of Ω is sufficiently regular and $h \in C^2(\bar{\Omega})$, then Green's theorem implies for $y \in \Omega$

$$u(y) = \int_\Omega K(x, y)\, \Delta u(x)\, dx$$

$$+ \int_{\partial\Omega} \left(u(x)\, \frac{\partial K(x, y)}{\partial n} - K(x, y)\, \frac{\partial u}{\partial n} \right) dS_x$$

$$\tag{21}$$

for all $u \in C^2(\bar{\Omega})$. The function $K(x, y)$ is called a *fundamenal solution* of the operator Δ. If, in addition, $K(x, y)$ vanishes for $x \in \partial\Omega$, it is called a *Green's function*, and we denote it by $G(x, y)$. In this case

$$u(y) = \int_{\partial\Omega} u(x)\, \frac{\partial G(x, y)}{\partial n}\, dS_x, \qquad y \in \Omega \quad (22)$$

for all $u \in C^2(\bar{\Omega})$ that are harmonic in Ω. Conversely, this formula can be used to solve the Dirichlet problem for Laplace's equation if we know the Green's function for Ω, since the right-hand side of (22) is harmonic in Ω and involves only the values of $u(x)$ on $\partial\Omega$. It can be shown that if the prescribed boundary values are continuous, then indeed (22) does give a solution to the Dirichlet problem for Laplace's equation.

It is usually very difficult to find the Green's function for an arbitrary domain. It can be computed for geometrically symmetric regions. In the case of a ball of radius R and center 0 it is given by

$$G(x, y) = K(x, y) - (|y|/R)^{2-n} K(x, R^2|y|^{-2}y)$$

D. HILBERT SPACE METHODS

Let

$$P(D) = \sum_{|\mu|=m} a_\mu D^\mu$$

be a positive, real, constant coefficient, homogeneous, elliptic partial differential operator of order $m = 2r$. This means that $P(D)$ has only terms of order m and

$$c_0|\xi|^m \le P(\xi) \le C_0|\xi|^m, \qquad \xi \in \mathbb{R}^n \quad (23)$$

holds for positive constants c_0, C_0. We introduce the norm

$$|v|_r = (\int |\xi|^m |\hat{v}(\xi)|^2 d\xi)^{1/2}$$

for function v in $C_0^\infty(\Omega)$, the set of infinitely differentiable functions that vanish outside Ω. Here Ω is a bounded domain in $\mathbb{R}^n$ with smooth boundary, and $\hat{v}(\xi)$ denotes the Fourier transform given by (19). By (23) we see that $(P(D)v, v)$ is equivalent to $|v|_r^2$ on $C_0^\infty(\Omega)$, where

$$(u, v) = \int_\Omega u(x)\overline{v(x)}\, dx$$

Let

$$a(u, v) = (u, P(D)v), \qquad u, v \in C_0^\infty(\Omega) \quad (24)$$

If $u \in C^m(\bar{\Omega})$ is a solution of the Dirichlet problem

$$P(D)u = f \quad \text{in} \quad \Omega \tag{25}$$

$$D^\mu u = 0 \quad \text{on} \quad \partial\Omega, \qquad |\mu| < r \tag{26}$$

then it satisfies

$$a(u, v) = (f, v), \qquad v \in C_0^\infty(\Omega) \tag{27}$$

Conversely, if $\partial\Omega$ is sufficiently smooth and $u \in C^m(\bar{\Omega})$ satisfies (27), then it is a solution of the Dirichlet problem (25), (26). This is readily shown by integration by parts. Thus one can solve (25), (26) by finding a function $u \in C^m(\bar{\Omega})$ satisfying (27). Since $a(u, v)$ is a scalar product, it would be helpful if we had a theorem stating that the expression (f, v) can be represented by the expression $a(u, v)$ for some u. Such a theorem exists (the Riesz representation theorem) provided $a(u, v)$ is the scalar product of a Hilbert space and

$$|(f, v)| \le Ca(v, v)^{1/2} \tag{28}$$

We can fit our situation to the theorem by completing $C_0^\infty(\Omega)$ with respect to the $|v|_r$ norm and making use of the fact that $a(v, v)^{1/2}$ and $|v|_r$ are equivalent on $C_0^\infty(\Omega)$ and consequently on the completion $H_0^r(\Omega)$. Moreover, inequality (28) follows from the Poincare inequality

$$\|v\| \leq M^r |v|_r, \qquad v \in C_0^\infty(\Omega) \qquad (29)$$

which holds if Ω is contained in a cube of side length M. Thus by Schwarz's inequality

$$|(f, v)| \leq \|f\| \, \|v\| \leq \|f\| M^r |v|_r \leq ca(v, v)^{1/2}$$

The Riesz representation theorem now tells us that there is a $u \in H_0^r(\Omega)$ such that (27) holds. If we can show that u is in $C^m(\bar\Omega)$, it will follow that u is indeed a solution of the Dirichlet problem (25), (26). As it stands now, u is only a *weak solution* of (25), (26). However, it can be shown that if $\partial\Omega$ and f are sufficiently smooth, then u will be in $C^m(\bar\Omega)$ and will be a solution of (25), (26).

The proof of the Poincare inequality (29) can be given as follows. It suffices to prove it for $r = 1$ and Ω contained in the slab $0 < x_1 < M$. Since $v \in C_0^\infty(\Omega)$,

$$v(x_1, \ldots, x_n)^2 = \left(\int_0^{x_1} v_{x_1}(t, x_2, \ldots, x_n) \, dt \right)^2$$

$$\leq x_1 \int_0^{x_1} v_{x_1}(t, x_2, \ldots, x_n)^2 \, dt$$

$$\leq M \int_0^M v_{x_1}(t, x_2, \ldots, x_n)^2 \, dt$$

Thus

$$\int_0^M v(x_1, \ldots, x_n)^2 \, dx_1 \leq$$

$$M^2 \int_0^M v_{x_1}(t, x_2, \ldots, x_n)^2 \, dt$$

If we now integrate over $x_2, \ldots, x_n$, we obtain

$$\|v\| \leq M \|v_{x_1}\|$$

But by Parseval's identity

$$\int |v_{x_1}|^2 \, dx = \int |\hat v_{x_1}|^2 \, d\xi$$

$$= \int \xi_1^2 |\hat v|^2 \, d\xi \leq \int |\xi|^2 \, |\hat v|^2 \, d\xi = |v|_1^2$$

E. ITERATIONS

An important method of solving both linear and non-linear problems is that of *successive approximations*. We illustrate this method for the following Dirichlet problem

$$\Delta u = f(x, u), x \in \Omega \qquad (30)$$

$$u = 0 \quad \text{on} \quad \partial\Omega \qquad (31)$$

We assume that the boundary of Ω is smooth and that $f(x, t) = f(x_1, \ldots, x_n, t)$ is differentiable with respect to all arguments. Also we assume that

$$|f(x, t)| \leq N, \qquad x \in \Omega, \quad -\infty < t < \infty \quad (32)$$

$$|\partial f(x, t)/\partial t| \leq \psi(t), \qquad x \in \Omega \qquad (33)$$

where $\psi(t)$ is a continuous function.

First we note that for every compact subset G of Ω there is a constant C such that

$$\max_G |v| \leq C(\sup_\Omega |\Delta v| + \sup_\Omega |v|) \qquad (34)$$

for all $v \in C^2(\Omega)$. Assume this for the moment, and let $w(x)$ be the solution of the Dirichlet problem

$$\Delta w = -N \quad \text{in} \quad \Omega, \qquad w = 0 \quad \text{on} \quad \partial\Omega \quad (35)$$

It is clear that $w(x) \geq 0$ in Ω. For otherwise it would have a negative interior minimum in Ω. At such a point one has $\partial^2 w/\partial x_k^2 \geq 0$ for each k, and consequently $\Delta w \geq 0$, contradicting (35). Since $w \in C(\bar\Omega)$, there is a constant C_1 such that

$$0 \leq w(x) \leq C_1, \qquad x \in \Omega$$

Let

$$K = \max_{|t| \leq C_1} \psi(t)$$

Then by (33)

$$|\partial f(x, t)/\partial t| \leq K, \qquad |t| \leq C_1 \qquad (36)$$

Consequently

$$f(x, t) - f(x, s) \leq K(t - x) \qquad (37)$$

when $-C_1 \leq s \leq t \leq C_1$. We define a sequence $\{u_k\}$ of functions as follows. We taken $u_0 = w$ and once u_{k-1} has been defined, we let u_k be the solution of the Dirichlet problem

$$Lu_k \equiv \Delta u_k - Ku_k = f(x, u_{k-1}) - Ku_{k-1} \quad (38)$$

in Ω with $u_k = 0$ on $\partial\Omega$. The solution exists by the theory of linear elliptic equations. We show by induction that

$$-w \leq u_k \leq u_{k-1} \leq w \qquad (39)$$

To see this for $k = 1$ note that

$$L(u_1 - w) = f(x, w) - Kw - \Delta w + Kw \geq 0$$

From this we see that $u_1 \leq w$ in Ω. For if $u_1 - w$ had an interior positive maximum in Ω, we would have

$$L(u_1 - w) = \Delta(u_1 - w) - k(u_1 - w) < 0$$

at such a point. Thus $u_1 \leq w$ in Ω. Also we note

$$\Delta(u_1 + w) = f(x, w) + K(u_1 - w)$$

$$+ \Delta w \leq K(u_1 - w)$$

This shows that $u_1 + w$ cannot have a negative minimum inside Ω. Hence $u_1 + w \geqq 0$ in Ω, and (39) is verified for $k = 1$. Once we know it is verified for k, we note that

$$L[u_{k+1} - u_k] = f(x, u_k) - f(x, u_{k-1})$$
$$- K(u_k - u_{k-1}) \geqq 0$$

by (37). Thus $u_{k+1} \leqq u_k$ in Ω. Hence

$$\Delta(u_{k+1} + w) = f(x, u_k) - Ku_k + Ku_{k+1}$$
$$+ \Delta w \leqq K(u_{k+1} - u_k)$$

Again we deduce from this that $u_{k+1} + w \geqq 0$ in Ω. Hence (39) holds for $k + 1$ and consequently for all k. In particular we see that the u_k are uniformly bounded in Ω, and by (38) the same is true of the functions Δu_k. Hence by (34), the first derivatives of the u_k are uniformly bounded on compact subsets of Ω. If we differentiate (38), we see that the sequence $\Delta(\partial u_k/\partial x_j)$ is uniformly bounded on compact subsets of Ω (here we make use of the continuous differentiability of f). If we now make use of (34) again, we see that the second derivatives of the u_k are uniformly bounded on compact subsets of Ω. Hence by the Ascoli–Arzela theorem, there is a subsequence that converges together with its first derivatives uniformly on compact subsets of Ω. Since the sequence u_k is monotone, the whole sequence must converge to a continuous function u that satisfies $|u(x)| \leqq w(x)$. Hence u vanishes on $\partial\Omega$. By (38), the functions Δu_k must converge uniformly on compact subsets, and by (34), the same must be true of the first derivatives of the u_k. From the differentiated (38) we see that the $\Delta(\partial u_k/\partial x_j)$ converge uniformly on bounded subsets and consequently the same is true of the second derivatives of the u_k by (34). Since the u_k converge uniformly to u in Ω and their second derivatives converge uniformly on bounded subsets, we see that $u \in C^2(\Omega)$ and $\Delta u_k \to \Delta u$. Hence

$$\Delta u = \lim \Delta u_k$$
$$= \lim[f(x, u_{k-1}) + K(u_k - u_{k-1})] = f(x, u)$$

and u is the desired solution.

It remains to prove (34). For this purpose we let $\varphi(x)$ be a function in $C_0^\infty(\Omega)$ which equals one on G. Then we have by (21)

$$\varphi(y)v(y) = \int_\Omega K(x, y) \, \Delta(\varphi(x)v(x)) \, dx$$

(the boundary integrals vanish because φ is 0 near $\partial\Omega$). Thus if $y \in G$

$$v(y) = \int_\Omega K\{\varphi \, \Delta v + 2 \, \nabla\varphi \cdot \nabla v + v \, \Delta\varphi\} \, dx$$

$$= \int_\Omega \{\varphi K \Delta v - 2v \, \nabla K \cdot \nabla\varphi - vK \, \Delta\varphi\} \, dx$$

by integration by parts. We note that $\nabla\varphi$ vanishes near the singularity of K. Thus we may differentiate under integral sign to obtain

$$\frac{\partial v(y)}{\partial y_j} = \int_\Omega \left\{ \varphi \, \frac{\partial K}{\partial y_j} \, \Delta v - 2v \right.$$
$$\left. \frac{\partial \, \nabla K}{\partial y_j} \cdot \nabla\varphi - v \, \frac{\partial K}{\partial y_j} \, \Delta\varphi \right\} \, dx$$

Consequently

$$\left| \frac{\partial v}{\partial y_j} \right| \leqq \sup_\Omega |\Delta v| \int_\Omega \left| \frac{\partial K}{\partial y_j} \right| dx$$

$$+ \sup_\Omega |v| \int_\Omega \left\{ |\nabla\varphi| \left| \frac{\partial \, \nabla K}{\partial y_j} \right| + |\Delta\varphi| \left| \frac{\partial K}{\partial y_i} \right| K \right\} dx$$

and all of the integrals are finite. This gives (34), and the proof is complete.

F. Variational Methods

In many situations methods of the calculus of variations are useful in solving problems for partial differential equations, both linear and nonlinear. We illustrate this with a simple example. Suppose we wish to solve the problem

$$-\sum \frac{\partial}{\partial x_k} \left[p_k(x) \, \frac{\partial u(x)}{\partial x_k} \right] + q(x)u(x) = 0 \quad \text{in} \quad \Omega$$
$$(40)$$

$$u(x) = g(x) \quad \text{on} \quad \partial\Omega \qquad (41)$$

Assume that $p_k(x) \geqq c_0$, $q(x) \geqq c_0$, $c_0 > 0$ for $x \in \Omega$, Ω bounded, $\partial\Omega$ smooth, and that g is in $C^1(\partial\Omega)$. We consider the expression

$$a(u, v) = \int_\Omega \left\{ \frac{1}{2} \sum p_k(x) \, \frac{\partial u(x)}{\partial x_k} \, \frac{\partial v(x)}{\partial x_k} \right.$$
$$\left. + q(x)u(x)v(x) \right\} dx$$

and put $a(u) = a(u, u)$. If $u \in C^2(\Omega) \cap C^0(\bar\Omega)$, satisfies (41) and

$$a(u) \leqq a(w) \qquad (42)$$

for all w satisfying (41), then it is readily seen that u is a solution of (40). For let v be a smooth function which vanishes on $\partial\Omega$. Then for any scalar β

$$a(u) \leqq a(u + \beta v) = a(u)$$
$$+ 2\beta a(u, v) + \beta^2 a(v)$$

and consequently

$$2\beta a(u, v) + \beta^2 a(v)^2 \geqq 0$$

for all β. This implies $a(u, v) = 0$. Integration by parts now yields (40). The problem now is to find a u satisfying (42). We do this as follows. Let H denote the Hilbert space obtained by completing $C^2(\bar{\Omega})$ with respect to the norm $a(w)$, and let H_0 be the subspace of those functions in H that vanish on $\partial\Omega$. Under the hypotheses given it can be shown that H_0 is a closed subspace of H. Let w be an element of H satisfying (41), and take a sequence $\{v_k\}$ of functions in H_0 such that

$$a(w - v_k) \to d = \inf_{v \in H_0} a(w - v).$$

The parallelogram law for Hilbert space tells us that

$$a(2w - v_j - v_k) + a(v_j - v_k)$$
$$= 2a(w - v_j) + 2a(w - v_k).$$

But

$$a(2w - v_j - v_k) = 4a(w - \tfrac{1}{2}(v_j + u_k)) \geqq 4d.$$

Hence

$$4d + a(v_j - v_k) \leqq 2a(w - v_j)$$
$$+ 2a(w - v_k) \to 4d.$$

Thus

$$a(v_j - v_k) \to 0$$

and $\{v_k\}$ is a Cauchy sequence in H_0. Since H_0 is complete, there is a $v \in H_0$ such that $a(v_k - v) \to$

0. Put $u = w - v$. Then clearly u satisfies (41) and (42), and hence $a(u, v) = 0$ for all $v \in H_0$. If u is in $C^2(\Omega)$, then it will satisfy (40) as well. The rub is that functions in H need not be in $C^2(\Omega)$. However, the day is saved if the $p_k(x)$ and $q(x)$ are sufficiently smooth. For then one can show that functions $u \in H$ which satisfy $a(u, v) = 0$ for all $v \in H_0$ are indeed in $C^2(\Omega)$.

BIBLIOGRAPHY

Bers, Lipman, John, Fritz, and Schechter, Martin (1979). "Partial Differential Equations," American Mathematical Society, Providence, Rhode Island.

Courant, Richard, and Hilbert, David (1953, 1962). "Methods of Mathematical Physics, I, II," Wiley Interscience, New York.

Gilbarg, David, and Trudinger, N. S. (1983). "Elliptic Partial Differential Equations of Second Order," Springer, Berlin.

John, Fritz (1978). "Partial Differential Equations," Springer-Verlag, New York and Berlin.

Lions, J. L., and Magenes, Enrico (1972). "Non-Homogeneous Boundary Value Problems and Applications," Springer-Verlag, Berlin and New York.

Schechter, Martin (1977). "Modern Methods in Partial Differential Equations, An Introduction," McGraw-Hill, New York.

Schechter, Martin (1986). "Spectra of Partial Differential Operators," North–Holland, Amsterdam.

Treves, Francois (1975). "Basic Linear Partial Differential Equations," Academic Press, New York.

PARTIALLY COHERENT PROCESSING

Francis T. S. Yu *Pennsylvania State University*

GLOSSARY

Achromatic lens: Lens without longitudinal chromatic aberration.

Coherent noise: Noise generated due to highly coherent light.

Coherent processor: Signal processor that utilizes coherent light.

Correlation: Mathematical operation involving integration of two functions that are mutual translated with respect to each other.

Diffraction grating: Periodic structure transparency to diffract light.

Impulse response: Response from a linear system with a very short duration of excitation.

Image sampling: An image transparency is on–off sampling by a linear periodic grating.

Mutual intensity functions: Mutual coherence function defined at zero delay time (i.e., $\tau = 0$).

Narrow spectral band filter: Spatial filter with a small finite-spectral bandwidth.

Partially coherent processor: Signal processor that utilizes partially coherent light.

Pseudo-color image: False color image.

Source encoding: Spatial source sampling.

Spatial coherence: Transversal correlation that describes the mutual coherence between any two points on a transversal space (e.g., a plane).

Speech spectrogram: Plot of frequency versus time to describe a spectral content of a speech.

Temporal coherence: Longitudinal correlation that describes the mutual coherence between any two points along the longitudinal direction of wave propagation.

White-light processor: Optical signal processor that utilizes a white-light source.

Partially coherent processing is an optical signal processing technique that utilizes the partially coherent properties of light instead of a strict coherent light. The light may be either spatially or temporally partially coherent.

I. Introduction

Although coherent optical processors can perform a myriad of complicated signal processings, coherent processing systems are usually plagued with coherent artifact noise. These difficulties have prompted us to look at optical processing from a new standpoint: to consider whether it is necessary for all optical processing operations to be carried out by pure coherent sources. We have found that many optical processings can be carried out by partially coherent light or white light. The basic advantages of partially coherent processing are (1) the suppressing of coherent artifact noise; (2) that the partially coherent sources are usually inexpensive; (3) that the constraints on the processing environment are generally very relaxed; (4) that the partially coherent system is relatively easy and economical to operate; and (5) that the partially coherent processor is particularly suitable for color image processing.

ENCYCLOPEDIA OF PHYSICAL SCIENCE
AND TECHNOLOGY, VOL. 10

II. Optical Processing under Partially Coherent Regime

Optical systems under the partially coherent regime have been studied since 1967. These studies have shown that there are difficulties in applying the linear system concept to the evaluation of the performance at high spatial frequencies. These difficulties are primarily due to the inapplicability of the linear system theory under a partial coherence regime. However, under the strictly coherent regime (i.e., spatially and temporally coherent), the optical system operates in a complex wave field. The output complex light field can be described by a complex amplitude convolution integral, such as

$$g(\alpha, \beta) = \int\!\!\int_{-\infty}^{\infty} f(x, y)h(\alpha - x, \beta - y) \, dx \, dy \quad (1)$$

which yields a spatially invariant property of the complex amplitude transformation, where $f(x, y)$ is a coherent complex wave field at the input end of an optical system, and $h(x, y)$ describes the spatial impulse response of the optical system. Equation (1) can be written in the Fourier transform form

$$G(p, q) = F(p, q)H(p, q) \quad (2)$$

where $G(p, q)$, $H(p, q)$, and $F(p, q)$ are the Fourier transforms of $g(x, y)$, $h(x, y)$, and $f(x, y)$, respectively. Thus Eq. (2) describes the optical system operation in the spatial frequency domain, where $H(p, q)$ can be referred to as the *coherent* or *complex amplitude transfer function* of the optical system.

On the other hand, if the optical system is under a strictly incoherent regime, the system operates in intensity, such that the output irradiance is described by the intensity convolution integral

$$I(\alpha, \beta) = \int\!\!\int_{-\infty}^{\infty} f_i(x, y)h_i(\alpha - x, \beta - y) \, dx \, dy \quad (3)$$

which yields a spatially invariant form of the intensity operation, where $f_i(x, y) = |f(x, y)|^2$ is the input irradiance and $h_i(x, y) = |h(x, y)|^2$ is the intensity spatial impulse response of the optical system. Similarly, Eq. (3) can be written in Fourier transform form, that is,

$$I(p, q) = F_i(p, q)H_i(p, q) \quad (4)$$

where $I(p, q)$, $F_i(p, q)$, and $H_i(p, q)$ are the Fourier transforms of $I(x, y)$, $f_i(x, y)$, and $h_i(x, y)$, respectively, and $H_i(p, q)$ can be referred to as the *incoherent* or *intensity transfer function* of the optical system.

We shall now define the following normalized quantities:

$$\bar{I}_i(p, q) = \frac{\int\!\!\int_{-\infty}^{\infty} I(\alpha, \beta)\exp[-i(p\alpha + q\beta)] \, d\alpha \, d\beta}{\int\!\!\int_{-\infty}^{\infty} I(\alpha, \beta) \, d\alpha \, d\beta}$$

$$(5)$$

$$\bar{F}_i(p, q) = \frac{\int\!\!\int_{-\infty}^{\infty} f_i(x, y)\exp[-i(px + qy)] \, dx \, dy}{\int\!\!\int_{-\infty}^{\infty} f_i(x, y) \, dx \, dy}$$

$$(6)$$

$$\bar{H}_i(p, q) = \frac{\int\!\!\int_{-\infty}^{\infty} h_i(x, y)\exp[-i(px + qy)] \, dx \, dy}{\int\!\!\int_{-\infty}^{\infty} h_i(x, y) \, dx \, dy}$$

$$(7)$$

These normalized quantities would give a set of convenient mathematical forms and also provide a concept of image contrast interpretation. Since the quality of a visual image depends to a large extent on the contrast (or the relative irradiance) of the image, a normalized intensity function would certainly enhance the information bearing capacity. With the application of the Fourier convolution theorem to Eq. (3), the following relationship can be written:

$$\bar{I}_i(p, q) = \bar{F}_i(p, q)\bar{H}_i(p, q) \quad (8)$$

where $\bar{H}_i(p, q)$ is commonly referred to as the *optical transfer function* (OTF) of the optical system, and the modulus $|\bar{H}(p, q)|$ is known as the *modulation transfer function* (MTF) of the optical system.

We further note that Eq. (7) can also be written as

$$\bar{H}_i(p, q)$$

$$= \frac{\int\!\!\int_{-\infty}^{\infty} H(p', q')H^*(p' - p, q' - q) \, dp' \, dq'}{\int\!\!\int_{-\infty}^{\infty} |H(p', q')|^2 \, dp' \, dq'}$$

$$(9)$$

By changing the variables $p'' = p' - p/2$, $q'' = q' - q/2$, Eq. (9) results in the following symmetrical form:

$$\bar{H}_i(p, q) = \frac{\int\limits_{-\infty}^{\infty}\!\!\int H(p'' + p/2, q'' + q/2) \times H^*(p'' - p/2, q'' - q/2)\, dp''\, dq''}{\int\limits_{-\infty}^{\infty}\!\!\int |H(p'', q'')|^2\, dp''\, dq''}$$

(10)

We note that the definition of OTF is valid for any linearly spatial invariant optical system, regardless of whether the system is with or without aberrations. Furthermore, Eq. (10) serves as the primary link between the strictly coherent and strictly incoherent systems. There are, however, threefold limitations associated with the strictly coherent processing. First, coherent processing requires the dynamic range of a spatial light modulator to be about 100,000 : 1 for the input wave intensity, or a photographic density range of 4.0. Such a dynamic range is often quite difficult to achieve in practice. Second, coherent processing systems are susceptible to coherent artifact noise, which frequently degrades the image quality. Third, in coherent processing, the signal being processed is carried out by a complex wave field. However, what is actually measured at the output plane of the optical processor is the output wave intensity. The loss of the output phase distribution may seriously limit the applicability of the optical system in some applications.

The drawbacks of coherent processing, can be overcome by reducing either the temporal coherence or the spatial coherence, and there are several techniques available that use those approaches. In the following, we briefly mention four frequently used concepts. The first method is based on a spatially partially coherent source, and the other three methods utilize a broad spectral band light source (e.g., a white-light source) to perform complex wave field operations.

A. SPATIALLY PARTIALLY COHERENT PROCESSING

Techniques of utilizing a spatially partially coherent light to perform complex data processing have been proposed. These techniques share a basic concept: the optical system is characterized by the use of a point spread function (PSF), which is not constrained to the class of nonnegative real functions used in conventional incoherent processing. The output intensity distribution can be adjusted by changing the PSF, that is,

$$I(x', y') = \iint O(x, y)h(x', y'; x, y)\, dx\, dy \quad (11)$$

where $h(x, y; x', y')$ is the PSF, and $I(x', y')$ and $O(x, y)$ are the image and object intensity distributions, respectively. If the pupil function of an optical system is $P(\alpha, \beta)$, the PSF is equal to the square of the Fourier spectrum of the pupil function. Therefore, the output intensity distribution can be adjusted by selecting an adequate pupil function. A typical processing system, as proposed by Rhodes, is shown in Fig. 1. This optical processing system, which is characterized by an extended pupil region, has two input pupil functions, $P_1(\alpha, \beta)$ and $P_2(\alpha, \beta)$. With the optical path lengths of the two arms of the system being equal, the overall system pupil function is given by the sum of $P_1(\alpha, \beta)$ and $P_2(\alpha, \beta)$. However, if the path length in one arm is changed slightly, for example, by moving mirror M_2 a small distance, a phase factor is introduced in one of the component pupil functions, with the result

$$P(\alpha, \beta) = P_1(\alpha, \beta) + P_2(\alpha, \beta)\exp[i\phi(\alpha, \beta)] \quad (12)$$

It is evident that if the pupil transparencies $P_1(\alpha, \beta)$ and $P_2(\alpha, \beta)$ are recorded holographically, arbitrary PSFs can then be synthesized using the $0°-180°$ phase-switching operation. Thus we see that this optical system is capable of performing complex data processing with spatially partially coherent light.

B. ACHROMATIC OPTICAL PROCESSING

An achromatic optical signal processing technique with broad-band source (e.g., white light) is shown in Fig. 2. Lens L_0 collimates the white-light point source S illuminating hologram R.

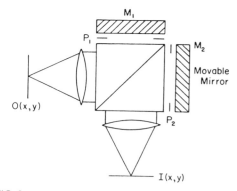

FIG. 1.   A two-pupil incoherent processing system.

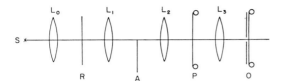

FIG. 2.   An achromatic optical processing system.

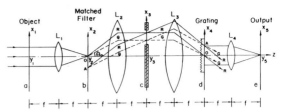

FIG. 3.   Band-limited partially coherent processing system.

The diffraction wavefront is spatially filtered by the aperture A, which removes the zero-order term and the lower sideband. The upper sideband is imaged onto the signal plane P; lens L_3 Fourier-transforms the resulting wavefront; and observation in the transform plane is confined to the optical axis by a slit in the output plane O. The convolution of the demodulated input signal with the desired reference function is recorded by synchronously translating the signal and output films. Such a system, termed an achromatic system, thus has the flexibility of a coherent optical signal processor, along with the potential for the noise immunity of an incoherent system.

C. BAND-LIMITED PARTIALLY COHERENT PROCESSING

A technique of a matched filter for operation with band-limited illumination is shown in Fig. 3. The matched filter consists of a frequency-plane holographic filter, an achromatic-fringe interferometer, a color-compensating grating, and an achromatic doublet. The matched filter, a Fourier hologram, is made by recording the object spectrum and a collimated reference beam with an exposure wavelength λ_0. The reference-beam angle is at θ_0 with respect to the object beam axis. In the correlation operation, the object is illuminated using a broad-spectral source. The various spectral components are dispersed in angle due to the grating-like structure of the matched filter. A lateral dispersed correlation signal is generated in plane c, which is the conventional correlation plane. These spectral com-

ponents are recombined by imaging the matched filter into plane d, where a compensation grating produces a color dispersion that is equal but opposite to that introduced at plane b. The color-corrected correlation signal is observed in the output plane. However, the compensated signal is still wavelength-dependent, since the illumination wavelength changes the scale of the object spectrum at plane b. Insertion of a slit in plane c provides a convenient way to bandlimit the correlation signal to $\Delta\lambda$. Bandlimiting improves the signal-to-noise ratio (SNR) of the compensated correlation output. Broadband illumination can be used for automatic scale search and object size determination. Since spatially incoherent light as well as coherent light can be used to perform matched filtering, an extended white-light source is used together with a slit at plane c to provide a bandlimited partially coherent processing.

D. ACHROMATIC PARTIALLY COHERENT PROCESSING

An achromatic partially coherent optical processing technique with a white-light source was introduced by Yu in 1978. Figure 4 illustrates the processing system, where all the transform lenses are assumed achromatic. A high-diffraction-efficiency phase grating with an angular spatial frequency P_0 is used at the input plane P_1 to disperse the input object spectrum into rain-

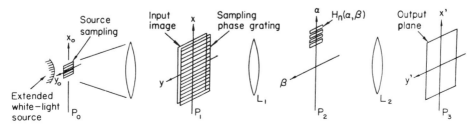

FIG. 4.   An achromatic partially coherent optical processing system.

bow color in the Fourier plane P_2. Thus it is permitting a stripwise design of a complex spatial filter for each narrow spectral band in the Fourier plane. The achromatic output image irradiance can therefore be observed at the output plane P_3. The advantages of this technique are that each channel (e.g., each spectral band filter) behaves as a partially coherent channel, while the overall output noise performance is primarily due to the incoherent addition of each channel, which behaves as if under incoherent illumination. Thus, this partially coherent system has the capability for coherent noise suppression.

III. White-Light Optical Processing

We shall now describe a partially coherent processor that can be carried out by a white-light source, as shown in Fig. 4. The partially coherent processing system is similar to a coherent processing system, except that it uses an extended white-light source, a source encoding mask, a signal sampling grating, multispectral band filters, and achromatic transform lenses. For example, if we place an input object transparency $s(x, y)$ in contact with a sampling phase grating, the complex wave field, for every wavelength λ, at the Fourier plane P_2 would be (assuming a white-light point source)

$$E(p, q; \lambda) = \iint s(x, y)\exp(ip_0 x)$$

$$\exp[-i(px + qy)]\, dx\, dy = S(p - p_0, q) \quad (13)$$

where the integral is over the spatial domain of the input plane P_1; (p, q) denotes the angular spatial frequency coordinate system; p_0 is the angular spatial frequency of the sampling phase grating; and $S(p, q)$ is the Fourier spectrum of $s(x, y)$. If we write Eq. (13) in the form of a spatial coordinate system (α, β), we have

$$E(\alpha, \beta; \lambda) = S\left(\alpha - \frac{\lambda f}{2\pi} p_0, \beta\right) \quad (14)$$

where $p = (2\pi/\lambda f)\alpha$, $q = (2\pi/\lambda f)\beta$, and f is the focal length of the achromatic transform lens. Thus we see that the Fourier spectra would disperse into rainbow colors along the α axis, and each Fourier spectrum for a given wavelength λ is centered at $\alpha = (\lambda f/2\pi)p_0$.

In complex spatial filtering, we assume that a set of narrow-spectral-band complex spatial filters is available. In practice, all the input objects are spatial-frequency-limited. The spatial band-

width of each spectral band filter $H(p_n, q_n)$ is therefore

$$H(p_n, q_n) = \begin{cases} H(p_n, q_n) & \alpha_1 < \alpha < \alpha_2 \\ 0 & \text{otherwise} \end{cases} \quad (15)$$

where $p_n = (2\pi/\lambda_n f)\alpha$, $q_n = (2\pi/\lambda_n f)\beta$, λ_n is the main wavelength of the filter, $\alpha_1 = (\lambda_n f/2\pi)(p_0 + \Delta p)$ and $\alpha_2 = (\lambda_n f/2\pi)(p_0 - \Delta p)$ are the upper and lower spatial limits of $H(p_n, q_n)$, and Δp is the spatial bandwidth of the input image $s(x, y)$.

Since the limiting wavelengths of each $H(p_n, q_n)$ are

$$\lambda_l = \lambda_n \frac{p_0 + \Delta p}{p_0 - \Delta p} \quad \text{and}$$

$$\lambda_h = \lambda_n \frac{p_0 - \Delta p}{p_0 + \Delta p} \quad (16)$$

its spectral bandwidth can be approximated by

$$\Delta\lambda_n = \lambda_n \frac{4p_0(\Delta p)}{p^2 - (\Delta p)^2} \approx \frac{4(\Delta p)}{p_0}\lambda_n \quad (17)$$

If we place this set of spectral band filters side by side and position them properly over the smeared Fourier spectra, the intensity distribution of the output light field can be shown as

$$I(x, y) \approx \sum_{n=1}^{N} \Delta\lambda_n |s(x, y; \lambda_n) * h(x, y; \lambda_n)|^2 \quad (18)$$

where $h(x, y; \lambda_n)$ is the spatial impulse response of $H(p_n, q_n)$ and $*$ denotes the convolution operation. Thus, the proposed partially coherent processor is capable of processing the signal in a complex wave field. Since the output intensity is the sum of the mutually incoherent narrow-band spectral irradiances, the annoying coherent artifact noise can be suppressed.

IV. Propagation of Mutual Intensity Function

In investigating the behavior of a partially coherent processor, we would establish a transformational relationship of the mutual intensity function, which would determine the coherence requirement of a partially coherent processor. We shall use Wolf's theory (from 1955) to develop a transformational formula for the white-light optical system.

One of the most remarkable and useful properties of a converging lens is its inherent ability to perform a two-dimensional Fourier transformation of a complex wave field. Under partially coherent illumination, a thin lens would take a four-dimensional Fourier transformation of the mutual intensity function. Evidently, when a monochromatic plane wave passes through a thin lens, phase-delay transformation takes place, such as

$$T(\xi, \eta) = \exp(ik\eta\Delta_0)\exp[-ik(\xi^2 + \eta^2)/2f] \quad (19)$$

where η is the refractive index of the thin lens, f is the focal length, Δ_0 is the maximum thickness of the lens, $k = 2\pi/\lambda$, λ is the wavelength, and (ξ, η) is the spatial coordinate system of the thin lens.

Let us now consider an object transparency, inserted at a distance d_0 in front of the lens and illuminated by a spatially partially coherent light. If the mutual intensity function at the object plane is $J_1(x_1, y_1; x_2, y_2)$, as depicted in Fig. 5, then the mutual intensity function at the output (α, β) plane can be shown as

$$J_4(\alpha_1, \beta_1; \alpha_2, \beta_2)$$

$$= C \exp\left(i \frac{k}{2d_1}\right.$$

$$\times \left\{1 - \frac{1}{\alpha_1[(1/d_1) + (1/d_0) - (1/f)]}\right\}$$

$$\times [(\alpha_1^2 + \beta_1^2) - (\alpha_2^2 + \beta_2^2)]\bigg)$$

$$\times \int\!\!\int\!\!\int\!\!\int_{-\infty}^{\infty} J_1(x_1, y_1; x_2, y_2)\exp\left(i \frac{k}{2d_0}\right.$$

$$\times \left\{1 - \frac{1}{d_0[(1/d_1) + (1/d_0) - (1/f)]}\right\}$$

$$\times [(x_1^2 + y_1^2) - (x_2^2 + y_2^2)]\bigg)$$

$$\times \exp\left\{-ik \frac{(\xi_1 x_1 + \eta_1 y_1) - (\xi_2 x_2 + \eta_2 y_2)}{d_0 d_1[(1/d_1) + (1/d_0) - (1/f)]}\right\}$$

$$\times dx_1\, dy_1\, dx_2\, dy_2 \quad (20)$$

where C is an appropriate constant. This equation shows that the output mutual intensity function can be obtained by a four-dimensional integral equation.

Up to this point we have disregarded the finite extent of the lens aperture. Such an approxima-

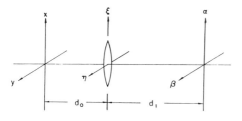

FIG. 5.   A partially coherent optical system.

tion is an accurate one, if the distance d_0 is sufficiently small to place the input transparency deep within the region of Fresnel diffraction with respect to the lens aperture. This condition is satisfied in the vast majority of problems of interest, particularly for optical processing.

There are, however, two special cases worth mentioning:

1. For $d_1 = f$, the output plane is located at the back focal plane. Equation (20) reduces to

$$J_4(\alpha_1, \beta_1; \alpha_2, \beta_2)$$

$$= C \exp\left\{i \frac{k}{2f}\left(1 - \frac{d_0}{f}\right)\right.$$

$$\times [(\alpha_1^2 + \beta_1^2) - (\alpha_2^2 + \beta_2^2)]\bigg\}$$

$$\times \int\!\!\int\!\!\int\!\!\int_{-\infty}^{\infty} J_1(x_1, y_1; x_2, y_2)\exp\left\{-i \frac{k}{2f}\right.$$

$$\times [(\alpha_1 x_1 + \beta_1 y_1)$$

$$- (\alpha_2 x_2 + \beta_2 y_2)]\bigg\}\, dx_1\, dy_1\, dx_2\, dy_2 \quad (21)$$

Thus, except for a quadratic phase factor, the output mutual intensity function is essentially the Fourier transform of the input mutual intensity function.

2. For $d_0 = d_1 = f$, the input and output planes are located at the front and back focal length of the lens. The quadratic phase factor vanishes and Eq. (21) reduces to

$$J_4(\alpha_1, \beta_1; \alpha_2, \beta_2)$$

$$= C \int\!\!\int\!\!\int\!\!\int_{-\infty}^{\infty} J_1(x_1, y_1; x_2, y_2)$$

$$\times \exp\left\{-i \frac{k}{2f}[(\alpha_1 x_1 + \beta_1 y_1)\right.$$

$$- (\alpha_2 x_2 + \beta_2 y_2)]\bigg\}\, dx_1\, dy_1\, dx_2\, dy_2 \quad (22)$$

which is exactly a four-dimensional Fourier transformation, between the input and output mutual intensity functions.

In partially coherent processing, an extended incoherent source is usually used at the front focal plane of a collimating lens to illuminate the input object transparency. Thus the mutual intensity function at the source plane (x_0, y_0) can be written as

$$J(x_0', y_0'; x_0'', y_0'')$$

$$= \begin{cases} \gamma(x_0, y_0) & \text{for } x_0' = x_0'' = x_0, \\ & \quad y_0' = y_0'' = y_0 \quad (23) \\ 0 & \text{otherwise} \end{cases}$$

where $\gamma(x_0, y_0)$ is the intensity distribution of the light source, and (x_0, y_0) is the spatial coordinate system of the source plane. If we assume that the input object transparency is located at the back focal plane of the collimator, then the mutual intensity function at the input plane reduces to

$$J_1(x_1 - x_2; y_1 - y_2)$$

$$= \iint_\infty^\infty \gamma(x_0, y_0)\exp\left\{-i\frac{k}{2f}\right.$$

$$\left. \times [(x_1 - x_2)x_0 + (y_1 - y_2)y_0]\right\} dx_0\, dy_0$$

$$(24)$$

which is essentially the Van Cittert–Zernike theorem. It is evident that the mutual intensity function at the input plane is a spatially invariant function.

A. GENERAL FORMULATION

With reference to the partially coherent processor of Fig. 4, the mutual intensity function at the input plane P_1 due to the source irradiance $\gamma(x_0, y_0; \lambda)$ is

$$J(x_1, y_1; x_2, y_2; \lambda)$$

$$= \iint \gamma(x_0, y_0; \lambda)\exp\left\{-i\frac{2\pi}{\lambda f}\right.$$

$$\left. \times [(x_1 - x_2)x_0 + (y_1 - y_2)y_0]\right\} dx_0\, dy_0 \quad (25)$$

where the integration is over the source plane P_0. The mutual intensity function immediately behind the sampling phase grating can be written as

$$J'(x_1y_1; x_2, y_2; \lambda)$$

$$= J(x_1, y_1; x_2, y_2; \lambda)s(x_1, y_1)$$

$$\times s^*(x_2, y_2)\exp(i2\pi\nu_0 x_1)\exp(-i2\pi\nu_0 x_2)$$

$$(26)$$

where the superscript * denotes the complex conjugate, and ν_0 is the spatial frequency of the sampling grating. Similarly, the mutual intensity function at the Fourier plane P_2 can be written as

$$J(\alpha_1, \beta_1; \alpha_2, \beta_2; \lambda)$$

$$= \iint \gamma(x_0, y_0; \lambda)S(x_0 + \alpha_1 - \lambda f\nu_0, y_0 + \beta_1)$$

$$\times S^*(x_0 + \alpha_2 + \lambda f\nu_0, y_0 + \beta_2)\, dx_0\, dy_0$$

$$(27)$$

where the integration is over the source plane P_0, and $S(\alpha, \beta)$ is the Fourier spectrum of $s(x, y)$. If we assume that a set of narrow-spectral-band spatial filters $H_n(\alpha, \beta)$ are inserted at the Fourier plane, the output mutual intensity function would be

$$J(x_1', y_1'; x_2', y_2'; \lambda)$$

$$= \iiiint J'(x_1, y_1; x_2, y_2; \lambda)$$

$$\times \exp\left[-i\frac{2\pi}{\lambda f}(x_1'\alpha_1 + y_1'\beta_1\right.$$

$$\left. - x_2'\alpha_2 - y_2'\beta_2)\right] d\alpha_1\, d\beta_1\, d\alpha_2\, d\beta_2 \quad (28)$$

Thus the corresponding output intensity distribution is

$$I_n(x', y'; \lambda)$$

$$= \iint \gamma(x_0, y_0; \lambda)$$

$$\times \left|\iint S(x_0 + \alpha - \lambda f\nu_0, y_0 + \beta)H_n(\alpha, \beta)\right.$$

$$\times \exp\left[-i\frac{2\pi}{\lambda f}\right.$$

$$\left. \times (\alpha x' + \beta y')\right] d\alpha\, d\beta\Bigg|^2 dx_0\, dy_0$$

$$\text{for} \quad \lambda_{ln} \leq \lambda \leq \lambda_{hn} \quad (29)$$

where λ_{ln} and λ_{hn} are the lower and upper wavelength limits of $H_n(\alpha, \beta)$. Let us denote $S(\lambda)$ and $C(\lambda)$ as the relative spectral intensity of the light source and the relative spectral response of the output detector. The output irradiance resulting from each narrow spectral band filter would then be

$$I_n(x', y')$$

$$= \int_{\lambda_n - \Delta\lambda_n/2}^{\lambda_n + \Delta\lambda_n/2} \iint \gamma(x_0, y_0; \lambda)S(\lambda)C(\lambda)$$

$$\times \left|\iint S(x_0 + \alpha - \lambda f\nu_0, y_0 + \beta)\right.$$

$$\times H_n(\alpha, \beta)\exp\left[-i\frac{2\pi}{\lambda f}\right.$$

$$\left. \times (x'\alpha + y'\beta)\right] d\alpha\, d\beta\Bigg|^2 dx_0\, dy_0\, d\lambda$$

$$\text{for} \quad n = 1, 2, \ldots, N \quad (30)$$

where λ_n and $\Delta\lambda_n$ are the center wavelength and the bandwidth of the nth narrow-spectral-band filter $H_n(\alpha, \beta)$, respectively. It is therefore apparent that the overall output irradiance would be the incoherent addition of $I_n(x', y')$, that is,

$$I(x', y') = \sum_{n=1}^{N} I_n(x', y') \qquad (31)$$

where N is the total number of the spectral-band filters. Furthermore, if the image processing is a one-dimensional operation, then a fan-shaped spatial filter can be utilized; thus overall output irradiance is reduced to the form

$$
\begin{aligned}
I(x', y') \\
= \iiint \gamma(x_0, y_0; \lambda)S(\lambda)C(\lambda) \\
\times \left| \iint S(x_0 + \alpha - \lambda f\nu_0, y_0 + \beta)H(\alpha, \beta) \right. \\
\times \exp\left[-i \frac{2\pi}{\lambda f} \right. \\
\left. \left. \times (x'\alpha + y'\beta) \right] d\alpha\, d\beta \right|^2 dx_0\, dy_0\, d\lambda \qquad (32)
\end{aligned}
$$

It is evident that the coherence requirement for partially coherent processing can be determined with either Eq. (31) or Eq. (32).

V. Source Encoding and Image Sampling

We shall now discuss a linear transform relationship between the spatial coherence (i.e., mutual intensity function) and the source encoding. Since the spatial coherence depends on the image processing operation, a more relaxed coherence requirement may be used for specific image-processing operations. The object of source encoding is to alleviate the stringent coherence requirement so that an extended source can be used. In other words, the source encoding is capable of generating an appropriate spatial coherence for a specific optical signal processing, such that the available light power from the source may be efficiently utilized.

A. Source Encoding

We begin our discussion with Young's experiment under an extended source illumination, as shown in Fig. 6. First, we assume that a narrow slit is placed in the source plane P_0 behind an extended monochromatic source. To maintain a high degree of coherence between the slits Q_1

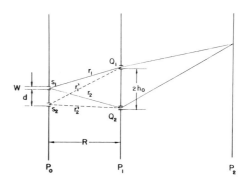

FIG. 6.   Concept of source encoding.

and Q_2 at plane P_2, the source size should be very narrow. If the separation between Q_1 and Q_2 is large, then a narrower slit size S_1 is required. Thus the slit width should be

$$w \leqq \frac{\lambda R}{2h_0} \qquad (33)$$

where R is the distance between planes P_0 and P_1, and $2h_0$ is the separation between Q_1 and Q_2. Let us now consider two narrow slits S_1 and S_2 located in source plane P_0. We assume that the separation between S_1 and S_2 satisfies the path-length relation

$$r_1' - r_2' = (r_1 - r_2) + m\lambda \qquad (34)$$

where the r's are the respective distances from S_1 and S_2 to Q_1 and Q_2, m is an arbitrary integer, and λ is the wavelength of the extended source. Then the interference fringes due to each of the two source slits S_1 and S_2 should be in phase, and a brighter fringe pattern can be seen at plane P_2. To further increase the intensity of the fringes, one would simply increase the number of slits in appropriate locations in plane P_0 such that the separation between slits satisfies the fringe condition of Eq. (34). If the separation R is large, that is, if $R \gg d$ and $R \gg 2h_0$, then the spacing d becomes

$$d = m \frac{\lambda R}{2h_0} \qquad (35)$$

Thus by properly encoding an extended source, it is possible to maintain a high degree of coherence between Q_1 and Q_2 and, at the same time, to increase the intensity of the fringes.

To encode an extended source, we would first search for a coherence function for a specific image processing operation. With reference to the partially coherent processor of Fig. 4, the mutual intensity function at the input plane P_1 can be written as

$$J(\mathbf{x}_1, \mathbf{x}_1') = \iint \gamma(\mathbf{x}_0)K(\mathbf{x}_0, \mathbf{x}_1)K^*(\mathbf{x}_0, \mathbf{x}_1')\, d\mathbf{x}_0$$

(36)

where the integration is over the source plane P_0, superscript $*$ denotes the complex conjugation, $\gamma(\mathbf{x}_0)$ is the intensity distribution of the encoding mask, and $K(\mathbf{x}_0, \mathbf{x}_1)$ is the transmittance function between the source plane P_0 and the input plane P_1, which can be written

$$K(\mathbf{x}_0, \mathbf{x}_1) \approx \exp\left(i2\pi \frac{\mathbf{x}_0\mathbf{x}_1}{\lambda f}\right) \tag{37}$$

By substituting $K(\mathbf{x}_0, \mathbf{x}_1)$ into Eq. (36), we have

$$J(\mathbf{x}_1 - \mathbf{x}_1') = \iint \gamma(\mathbf{x}_0)\exp\left[i2\pi \frac{\mathbf{x}_0}{\lambda f}(\mathbf{x}_1 - \mathbf{x}_1')\right] d\mathbf{x}_0$$

(38)

From the above equation, we see that the spatial coherence and source encoding intensity form a Fourier transform pair, that is,

$$\gamma(\mathbf{x}_0) = f[J(\mathbf{x}_1 - \mathbf{x}_1')] \tag{39}$$

and

$$J(\mathbf{x}_1 - \mathbf{x}_1') = f^{-1}[\delta(\mathbf{x}_0)] \tag{40}$$

where f denotes the Fourier transformation operation. It is evident that the relationship of Eqs. (39) and (40) is the well-known Van Cittert–Zernike theorem. In other words, if a required spatial coherence is given, then a source encoding transmittance can be obtained through the Fourier transformation. In practice, however, the source encoding transmittance should be a positive real quantity that satisfies the physically realizable condition:

$$0 \leq \gamma(\mathbf{x}_0) \leq 1 \tag{41}$$

B. IMAGE SAMPLING

There is, however, a temporal coherence requirement for partially coherent processing. If we restrict the Fourier spectra, due to wavelength spread, within a small fraction of the fringe spacing d of a narrow spectral band filter $H_n(\alpha, \beta)$, then we have

$$\frac{P_m f \Delta\lambda_n}{2\pi} \ll d \tag{42}$$

where $1/d$ is the highest spatial frequency of the

filter, P_m is the angular spatial frequency limit of the input image transparency, f is the focal length of the achromatic transform lens, and $\Delta\lambda_n$ is the spectral bandwidth of $H_n(\alpha, \beta)$. The temporal coherence requirement of the spatial filter is, therefore,

$$\frac{\Delta\lambda_n}{\lambda_n} \ll \frac{\pi}{h_0 P_m} \tag{43}$$

where λ_n is the central wavelength of the nth narrow-spectral-band filter, and $2h_0 = \lambda_n f/d$ is the size of the input object transparency.

VI. Noise Performance

White-light optical processors are known to perform better under noisy conditions than do their coherent counterparts. We shall now show some effects of the noise performance under white-light illumination.

An experimental setup for the study of noise performance of a partially coherent processor is shown in Fig. 7. We shall first investigate the noise performance under the spatially coherent illumination, that is, the effects due to source size (ΔS). For convenience, we assume that the source irradiance is uniform over a square aperture at source plane P_0, which can be written as

$$\gamma(x_0, y_0) = \text{rect}\left(\frac{x_0}{a}\right) \text{rect}\left(\frac{y_0}{a}\right) \tag{44}$$

where

$$\text{rect}\left(\frac{x_0}{a}\right) \equiv \begin{cases} 1, & |x_0| \leq \dfrac{a}{2} \\[2mm] 0, & |x_0| > \dfrac{a}{2} \end{cases}$$

For simplicity, we assume that the input signal is a one-dimensional object independent of the y axis. The Fourier spectrum would also be one-dimensional in the β direction, but smeared into rainbow colors along the α direction. Let the width of the nth narrow-spectral-band filter be $\Delta\alpha_n$. If the filter is placed in the smeared Fourier spectra, the spectral bandwidth of the filter can be written as

$$\Delta\lambda_n = \frac{\Delta\alpha_n}{\nu_0 f} \tag{45}$$

The total number of filter channels can therefore be determined as

$$N \equiv \frac{\Delta\lambda}{\Delta\lambda_n} \approx \frac{\Delta\lambda \nu_0 f}{\Delta\alpha_n} \tag{46}$$

where $\Delta\lambda$ is the spectral bandwidth of the white-light source. Thus, we see that the degree of temporal coherence at the Fourier plane increases as the spatial frequency of the sampling grating ν_0 increases.

For the noise performance in the temporally coherent regime, we would use a variable slit representing a broad spectral filter in the Fourier plane. The output noise fluctuation can be traced out with a linearly scanning photometer, as illustrated in Fig. 7. It is apparent that the output noise fluctuation due to the spectral bandwidth of the slit filter and due to the source size can be separately determined. We shall use the following definition of output signal-to-noise ratio (SNR) to evaluate the noise performance of the partially coherent processor:

$$\mathrm{SNR}_n(y') \equiv E[I_n(y')]/\sigma_n(y') \qquad (47)$$

where $I_n(y')$ is the output irradiance due to nth channel, $E[\ ]$ denotes the ensemble average, and $\sigma_n^2(y')$ is the variance of the output noise fluctuation, that is,

$$\sigma_n^2(y') \equiv E[I_n^2(y')] - \{E[I_n(y')]\}^2 \qquad (48)$$

Evidently, the output intensity fluctuation $I_n(y')$ can be traced out by a linearly scanning photometer. The dc component of the output traces is obviously the output signal irradiance (i.e., $E[I_n]$), and the mean square fluctuation of the traces is the variance of the output noise (i.e., σ_n^2). Thus, we see that the effect of the output SNR due to spectral bandwidth (i.e., temporal coherence) and source size (i.e., spatial coherence) can be readily obtained.

We shall first demonstrate the noise performance due to perturbation at the input plane.

For the amplitude noise at the input plane, the experiments have shown that there is no apparent improvement in noise performance under the partially coherent illumination. The result is quite consistent with the 1978 prediction by Chavel and Lowenthal. However, for the phase noise at the input plane, the noise performance of the system is largely improved with a partially coherent illumination. We shall first utilize a weak phase model as an input noise. We consider the situation of an input object transparency superimposed with a thin phase noise at the input plane. The effect on the noise performance of the optical system can then be obtained by varying the source size and the spectral bandwidth of the slit filter. Figure 8 shows a set of output images with photometer traces illustrating the output noise due to the spectral bandwidth of the slit filter. From these pictures, we see that the SNR improves as the spectral bandwidth of the slit filter increases.

A quantitative measurement of the noise performance due to phase noise at the input plane is plotted in Figs. 9 and 10. From these figures, we see that the output SNR increases monotonically as the spectral bandwidth $\Delta\lambda_n$ of the slit filter increases, and is linearly increasing as the source size enlarges. Thus, the noise performance for a partially coherent processor improves as the degree of coherence (i.e., temporal and spatial coherence) relaxes. In other words, to improve the output SNR of the partially coherent processor, one can relax either the spatial coherence, the temporal coherence, or both.

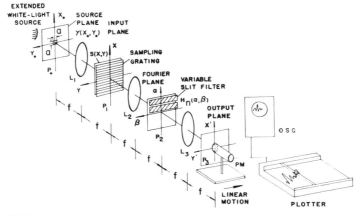

FIG. 7. Noise measurement of a partially coherent processor: $\gamma(x_0, y_0)$, source intensity distribution; P_0, source plane; P_1, input plane; P_2, input plane; P_3, output plane; $s(x, y)$, object transparency; $H_n(\alpha, \beta)$, slit filters; PM, photometer; OSC, oscilloscope.

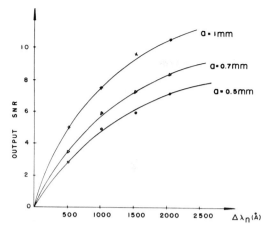

FIG. 9.   Output signal-to-noise ratio for amplitude noise at input plane for various source signs.

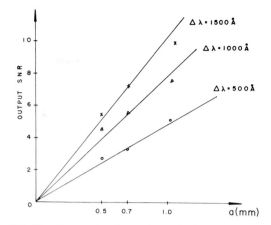

FIG. 10.   Output signal-to-noise ratio for phase noise at input plane for various spectral bandwidths.

Let us now demonstrate the effect of the noise performance due to strong phase noise. A conventional shower glass was used as an input phase noise. Figure 11 shows a set of results obtained under various spectral bandwidth illuminations. Figure 11a shows an output result obtained under entire broad-band white-light illumination. Although this image is somewhat aberrated due to the thick phase perturbation, the image is relatively immune to random noise fluctuation. Comparing the results obtained from Figs. 11a to 11d, we see that the output SNR decreases rather rapidly as the spectral bandwidth of the slit filter decreases. Figure 11d shows the result obtained with a HeNe laser. Aside from the poor noise performance, we have noted that the output image is severely corrupted by coherent artifact noise.

We shall now demonstrate the noise performance due to noise at the Fourier plane. The effects of amplitude noise at the Fourier plane are plotted in Fig. 12. In contrast with the amplitude noise at the input plane, we see that the output SNR increases monotonically as the spectral bandwidth of the slit filter increases. The output SNR also improves as the source size enlarges. Thus, for amplitude noise at the Fourier plane, the noise performance of a partially coherent processor improves as the degree of temporal and spatial coherence decreases.

Figure 13 shows the noise performance due to phase noise at the Fourier plane. From this fig-

ure, once again we see that the output SNR is a monotonically increasing function of the spectral bandwidth $\Delta\lambda_n$. The SNR also increases as the source size increases. However, as compared with the case of phase noise at the input plane of Fig. 10, the improvement of the noise performance is somewhat less effective.

We shall now provide the result of noise performance due to thin phase noise along the optical axis (i.e., Z axis) of the optical system. Figure 14 shows the variation of output SNR due to phase noise inserted in various planes of the optical system. From this figure, we see that the output SNR improves drastically for phase noise inserted at the input and at the output plane under temporally and spatially partially coherent illumination. The noise performance is somewhat less effective for the phase noise at the Fourier plane, even under partially coherent illumination. Nonetheless, the phase noise at the Fourier plane can, in principle, be totally eliminated under very-broad-band illumination, if each of the noise channels is uncorrelated. We have also noted that the output SNR is somewhat lower for higher spatial coherent illumination. In other words, the output SNR can also be improved with extended source illumination.

The noise performance due to amplitude noise along the Z axis is plotted in Fig. 15. From this figure, we notice that the output SNR improves as the noise perturbation is moved away from the input and the output plane, and that the opti-

FIG. 8.   Effect on output image (with a section of photometer traces) due to phase noise at input plane for different spectral bandwidths. (a) For $\lambda_n = 1500$ Å. (b) For $\lambda_n = 1000$ Å. (c) For $\lambda_n = 500$ Å.

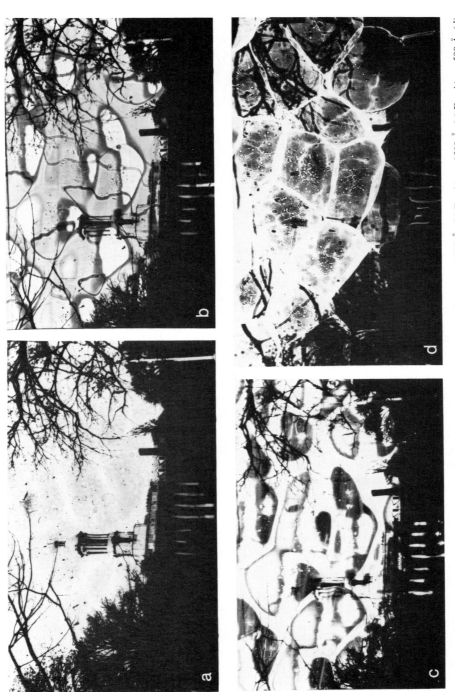

FIG. 11.   Effect on output image due to strong phase perturbation at input plane. (a) For $\Delta\lambda_n = 3000$ Å. (b) For $\Delta\lambda_n = 1500$ Å. (c) For $\Delta\lambda_n = 500$ Å. (d) Obtained with an HeNe laser.

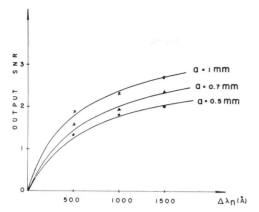

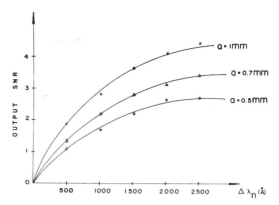

FIG. 12. Output signal-to-noise ratio for amplitude noise at Fourier plane for various source sizes.

FIG. 13. Output signal-to-noise ratio for phase noise at Fourier plane for various source sizes.

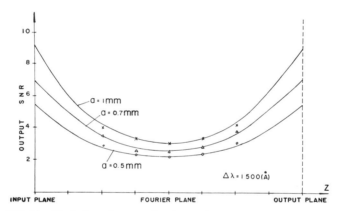

FIG. 14. Variation of output signal-to-noise ratio due to thin phase noise as a function of the Z direction for various source sizes.

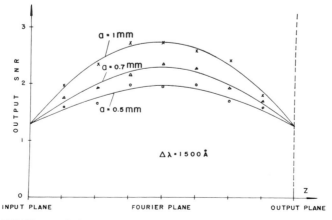

FIG. 15. Variation of output signal-to-noise ratio due to thin amplitude noise as a function of the Z direction for various source sizes.

mum SNR occurs at the Fourier plane. Again, we see that the output SNR is somewhat higher for a larger source size (i.e., lower degree of spatial coherence). It is apparent that if the amplitude noise is placed either at the input or at the output plane, the noise performance cannot be improved, under either partially coherent or incoherent illumination.

VII. Applications

It would occupy innumerable pages to describe all the applications of partially coherent processing. In this section, we restrict ourselves to a few applications that are of interest to our readers. As we have noted in the previous discussion, the partially coherent processor is very suitable for color-image processing. Therefore, in the following examples we confine ourselves to color-image processing.

A. Restoration of Blurred Images

One of the interesting applications of partially coherent processing is the restoration of blurred color photographic images. Since a linearly smeared image is a one-dimensional restoration problem and the deblurring operation is a point-by-point filtering process, a *fan-shape* type deblurring filter is used to compensate for the scale variation of the smeared Fourier spectra due to wavelength.

Let us now describe a color-image deblurring technique utilizing a fan-shape deblurring filter. We assume a blurred image due to linear motion described by

$$\hat{s}(x, y) = s(x, y) * \text{rect}\left(\frac{y}{W}\right), \qquad (49)$$

where $\hat{s}(x, y)$ and $s(x, y)$ denote the blurred and unblurred images, respectively, and

$$\text{rect}\left(\frac{y}{W}\right) \equiv \begin{cases} 1 & y \leq \dfrac{W}{2} \\ 0 & \text{otherwise} \end{cases}$$

where W is the smeared length.

If we insert this blurred image transparency into the input plane P_1 of the partially coherent processor of Fig. 4, the complex light distribution for every wavelength λ at the back focal length of the achromatic transform lens L_2 would be

$$E(\alpha, \beta; \lambda) = S\left[\alpha - \left(\frac{\lambda f}{2\pi}\right) p_0, \beta\right] \text{sinc}\left(\frac{\pi W}{\lambda f} \beta\right) \qquad (50)$$

where p_0 is the angular spatial frequency of the phase grating and $S[\cdot]$ represents the smeared Fourier spectrum of $s(x, y)$. Equation (50) shows that the image spectrum is smeared into a rainbow color spectrum.

We use a fan-shaped deblurring filter as described in the following equation:

$$H(\alpha, \beta; \lambda) = \delta\left(\alpha - \frac{\lambda f}{2\pi} p_0, \beta\right)\left\{\int \left[\text{rect}\left(\frac{y}{W}\right)\right.\right.$$

$$\left.\left. \exp\left(-i \frac{2\pi}{\lambda f} \beta y\right)\right] dy\right\}^{-1}$$

$$= \delta\left(\alpha - \frac{\lambda f}{2\pi} p_0, \beta\right)$$

$$\times \left[\text{sinc}\left(\frac{\pi W}{\lambda f} \beta\right)\right]^{-1} \qquad (51)$$

If this deblurring filter is inserted in the spatial frequency plane, the complex light distribution for every λ at the output image plane would be

$$g(x, y; \lambda) = \mathscr{F}^{-1}\left[S\left(\alpha - \frac{\lambda f}{2\pi} p_0, \beta\right) H(\alpha, \beta; \lambda)\right]$$

$$= s(x, y)\exp(ip_0 x) \qquad (52)$$

where $\mathscr{F}^{-1}$ denotes the inverse Fourier transform. The resultant output image intensity is therefore

$$I(x, y) = \int_{\Delta\lambda} |g(x, y; \lambda)|^2 \, d\lambda \approx \Delta\lambda |s(x, y)|^2 \qquad (53)$$

which is proportional to the spectral bandwidth $\Delta\lambda$ of the white-light source. Thus, we see that the partially coherent processor is capable of deblurring color images.

Figure 16a shows a black-and-white picture of a deblurred color image obtained with the white-light processing technique. Compared with the original blurred image of Fig. 16b, we see that the "stop" sign, the posts, the cars, and the trees are far more distinguishable in the deblurred one. The color reproduction of the deblurred image is rather faithful, and the coherent artifact noise is virtually nonexistent.

B. Image Subtraction

Another interesting application of white-light optical processing is color-image subtraction. Two color-image transparencies are inserted in the input plane of the white-light optical processor of Fig. 4. At the spatial frequency plane, the complex light distribution for each wavelength λ

FIG. 16.   Restoration of blurred images. (a) A black-and-white deblurred color picture due to linear motion. (b) A black-and-white blurred color image.

of the light source may be described as

$$E(\alpha, \beta; \lambda)$$

$$= S_1 \left(\alpha - \frac{\lambda f}{2\pi} p_0, \beta\right) \exp\left(-i \frac{2\pi}{\lambda f} h_0 \beta\right)$$

$$+ S_2 \left(\alpha - \frac{\lambda f}{2\pi} p_0, \beta\right) \exp\left(i \frac{2\pi}{\lambda f} h_0 \beta\right) \tag{54}$$

where $S_1(\alpha, \beta)$ and $S_2(\alpha, \beta)$ are the Fourier spectra of the input color images $s_1(x, y)$ and $s_2(x, y)$, $2h_0$ is main separation of the two color images, and p_0 is the angular spatial frequency of the sampling phase grating. Again we see that the two input image spectra disperse into rainbow colors along the α axis of the spatial frequency plane.

For image subtraction, we should insert a sinusoidal grating in the spatial frequency plane. Since the scales of the Fourier spectra vary with respect to the wavelength of the light source, we must utilize a fan-shaped grating to compensate for the scale variation. Let us assume the transmittance of the fan-shaped grating is

$$H(\alpha, \beta; \lambda) = \left[1 + \sin\left(\frac{2\pi}{\lambda f} h_0 \beta\right)\right] \quad \text{for all } \alpha \tag{55}$$

Then the output image irradiance is

$$I(x, y) \approx \Delta\lambda[|s_1(x, y - h_0)|^2 + |s_2(x, y + h_0)|^2$$

$$+ \tfrac{1}{2}|s_1(x, y) - s_2(x, y)|^2$$

$$+ |s_1(x, y - 2h_0)|^2 + |s_2(x, y + 2h_0)|^2] \tag{56}$$

where $\Delta\lambda$ is the spectral bandwidth of the white-light source. Thus, the subtracted color image can be seen at the optical axis of the output plane. In practice, it is difficult to obtain a true white-light point source. However, this shortcoming can be overcome with the source encoding technique described in Section V,A.

To ensure a physically realizable source encoding function, we let a spatial coherence function with an appropriate point-pair coherence requirement be

$$\Gamma(|y - y'|) = \frac{\sin[(N\pi/h_0)|y - y'|]}{N \sin[(\pi/h_0)|y - y'|]}$$

$$\times \text{sinc}\left(\frac{\pi w}{h_0 d} |y - y'|\right) \tag{57}$$

where N is a positive integer $\gg 1$, and $w \ll d$. Equation (57) represents a sequence of narrow pulses that occur at every $|y - y'| = nh_0$, where n is a positive integer, and their peak values are weighted by a broad sinc factor. Thus, a high degree of spatial coherence can be achieved at every point pair between the two input color-image transparencies. By applying the Van Cittert–Zernike Theorem of Eq. (39), the corresponding source encoding function is obtained:

$$\gamma(|y|) = \sum_{n=1} \text{rect} \frac{|y - nd|}{w} \tag{58}$$

where w is the slit width, $d = (\lambda f/h_0)$ is the separation between the slits, f is the focal length of the achromatic collimated lens, and N is the total number of slits. Alternatively, Eq. (58) can be written in the form

$$\gamma(|y|) = \sum_{n=1}^{N} \text{rect} \frac{|y - n\lambda f/h_0|}{W} \tag{59}$$

for which we see that the source encoding mask is essentially a fan-shape grating. To obtain lines of rainbow-color spectral light sources for the subtraction operation, we would use a linear extended white-light source with a dispersive phase grating, as illustrated in Fig. 17. For broad-spectral-band image subtraction operation, a fan-shape sinusoidal grating is utilized in the Fourier plane, such as

$$G = \tfrac{1}{2}[1 + \sin(2\pi\alpha h_0/\lambda f)] \tag{60}$$

The output image irradiance around the optical axis can be shown as

$$I(x, y) = K|s_1(x, y) - s_2(x, y)|^2 \tag{61}$$

Thus, we assert that a color-subtracted image can readily be obtained.

Figures 18a and 18b show a set of black-and-white pictures of input color transparencies. The first picture shows an F-16 fighter plane taking off over a mountainous coastline area, and the second picture shows the same scenic view but without the fighter plane. Figure 18c shows the corresponding subtracted image at the output plane. From this figure, we see that the shape of the F-16 fighter plane can be recognized. Although the resolution of the result is still below the generally acceptable standard, this drawback may be overcome by utilizing higher-quality optics.

C. Complex Signal Detection

One of the most important applications of optical signal processing is complex signal detection. We shall now describe a white-light optical correlator that can exploit the spectral content of the object under observation. In other words, the correlator is capable of recognizing multiple color objects of different shapes. The need for color-pattern recognition is very diverse and offers many applications. There are, however, two large categories where color is extremely important: natural color variations, and objects deliberately colored for identification, for example, in robotic vision applications.

We assume that a broad-band matched filter, as described in the following equation, is provided:

$$
\begin{aligned}
H(\alpha, \beta) \approx \sum_{\substack{n=-N \\ n\neq 0}}^{N} &\Big\{ K_1|S(\alpha, \beta + f\nu_0\lambda_n)|^2 \\
&+ K_2|S(\alpha, \beta + f\nu_0\lambda_n)| \\
&\times \cos\Big[\frac{2\pi h}{\lambda_n f}\alpha + \phi(\alpha, \beta + f\nu_0\lambda_n)\Big]\Big\}
\end{aligned}
\tag{62}
$$

where $S(\alpha, \beta; \lambda)$ is the input object spectrum, ν_0 is the sampling grating frequency, K_s, C_s, and h are arbitrary constants, and λ_n is the main wavelength of the nth spectral-band filter. If this filter is inserted in the Fourier plane of a white-light processor, as depicted in Fig. 4, then the output complex light filter can be shown to be

$$
\begin{aligned}
g(x, y; \lambda) = &\sum_{n=1}^{N} C_1 s(x, y; \lambda_n) * s(x, y; \lambda_n) \\
&* s^*(-x, -y; \lambda_n) \\
&+ \sum_{n=1}^{N} C_2 s(x, y; \lambda_n) \\
&* [s(x + h, y; \lambda_n) \\
&+ s^*(-x + h, -y; \lambda_n)]
\end{aligned}
\tag{63}
$$

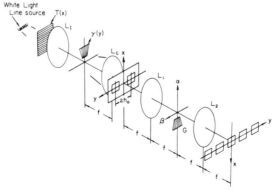

FIG. 17.   A white-light image subtraction system: T(x), phase grating; L_1, imaging lens; L_C, collimated lens; L_1 and L_2, achromatic transform lenses; $\gamma(y)$, fan-shaped source encoding mask; G, fan-shaped diffraction grating.

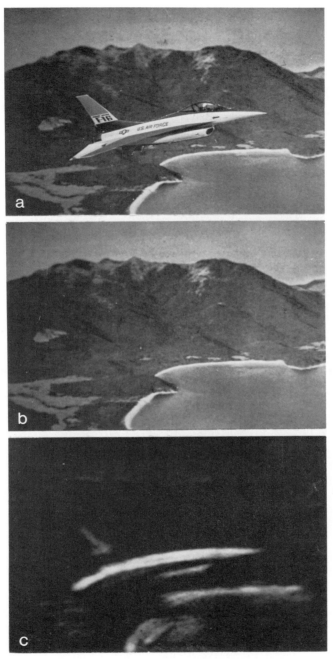

FIG. 18. Image subtraction. (a) and (b) Black-and-white pictures of input color objects. (c) A black-and-white picture of the output subtracted color image.

FIG. 19.   Complex signal detection. (a) A black-and-white picture of an input color object. (b) A black-and-white picture of the output correlation spot.

where $s(x, y; \lambda_n)$ is the color input object, $*$ denotes the convolution operation, and the superscript $*$ represents the complex conjugate. The corresponding output irradiance is

$$I(x, y) = \int_{\Delta\lambda_n} |g(x, y; \lambda)|^2 \, d\lambda$$

$$\approx \sum_{n=1}^{N} \Delta\lambda_n \{K_1 | s(x, y; \lambda_n) * s(x, y; \lambda_n)$$

$$\circledast \, s^*(x, y; \lambda_n)|^2$$

$$+ K_2 | s(x, y; \lambda_n) * s(x + h, y; \lambda_n)$$

$$+ s(x, y; \lambda_n) \circledast s^*(x - h, y; \lambda_n)|^2 \}$$

$$(64)$$

where $*$ denotes the correlation operation, and K_s are the proportionality constants. Thus, we see that an autocorrelation of the input image is diffracted at $x = h$ at the output plane. Since we utilized a broad-spectral-band white-light source, the spectral content of the image can be exploited.

Figure 19a shows a black-and-white photograph of a color aerial photograph with a (red) guided missile in flight. Figure 19b shows a visible (red) correlation spot obtained at the output plane of the correlator. Thus the white-light correlator has a unique feature for color pattern recognition based on both the spectral content and the shape of an input object.

D. COLOR-IMAGE RETRIEVAL

We shall now describe a spatial encoding technique for color-image retrieval. A color transparency is used as an object to be encoded by sequentially exposing with the primary colors of light, onto a black-and-white film. The encoding takes place by spatially sampling the primary color images of the color transparency with a specific sampling frequency and a predescribed direction onto a monochrome film. In order to avoid the moire fringe pattern, the primary color images are sampled in orthogonal directions. Thus the intensity transmittance of the encoded films can be shown as

$$T(x, y) = K\{T_r(x, y)[1 + \text{sgn}(\cos \omega_r y)]$$

$$+ T_b(x, y)[1 + \text{sgn}(\cos \omega_b x)]$$

$$+ T_g(x, y)[1 + \text{sgn}(\cos \omega_g x)]\}^{-\gamma} \quad (65)$$

where K is an appropriate proportionality constant; T_r, T_b, and T_g are the red, blue, and green color-image exposures; ω_r, ω_b, and ω_g are the respective carrier spatial frequencies; (x, y) is

the spatial coordinate system of the encoded film; γ is the film gamma; and

$$\text{sgn}(\cos x) \equiv \begin{cases} 1 & \cos x \geq 0 \\ -1 & \cos x \leq 0 \end{cases}$$

To improve the diffraction efficiency of the encoded film, it is advisable to bleach the encoded film to convert it into a phase-type transparency. The amplitude transmittance of the bleached transparency can be written as

$$t(x, y) = \exp[i\phi(x, y)]$$

where

$$\phi(x, y) = M\{T_r(x, y)[1 + \text{sgn}(\cos \omega_r y)]$$

$$+ T_b(x, y)[1 + \text{sgn}(\cos \omega_b x)]$$

$$+ T_g(x, y)[1 + \text{sgn}(\cos \omega_g x)]\} \quad (66)$$

and M is an appropriate proportionality constant.

If we place this bleached encoded film in the input plane of the white-light optical processor of Fig. 20, then the first-order complex light distribution for every λ at the spatial frequency plane P_2 can be shown as

$$S(\alpha, \beta; \lambda) \approx \hat{T}_r\left(\alpha, \beta \pm \frac{\lambda f}{2\pi} \omega_r\right)$$

$$+ \hat{T}_b\left(\alpha \pm \frac{\lambda f}{2\pi} \omega_b, \beta\right)$$

$$+ \hat{T}_g\left(\alpha \pm \frac{\lambda f}{2\pi} \omega_g, \beta\right)$$

$$+ \hat{T}_r\left(\alpha, \beta \pm \frac{\lambda f}{2\pi} \omega_r\right)$$

$$* \hat{T}_b\left(\alpha \pm \frac{\lambda f}{2\pi} \omega_b, \beta\right)$$

$$+ \hat{T}_r\left(\alpha, \beta \pm \frac{\lambda f}{2\pi} \omega_r\right)$$

$$* \hat{T}_g\left(\alpha \pm \frac{\lambda f}{2\pi} \omega_g, \beta\right)$$

$$+ \hat{T}_b\left(\alpha \pm \frac{\lambda f}{2\pi} \omega_b, \beta\right)$$

$$* \hat{T}_g\left(\alpha \pm \frac{\lambda f}{2\pi} \omega_g, \beta\right) \quad (67)$$

where $\hat{T}_r$, $\hat{T}_b$ and $\hat{T}_g$ are the Fourier transforms of T_r, T_b, and T_g, respectively, $*$ denotes the convolution operation, and the proportionality constants have been neglected for simplicity. We note that the last cross-product term would introduce a moire fringe pattern, which can be easily masked out at the Fourier plane. It is apparent that, by proper color-filtering of the

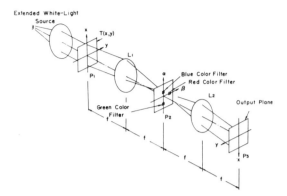

FIG. 20.   A white-light processor for color-image retrieval.

smeared Fourier spectra, a true color image can be retrieved at the output image plane. Thus, the output image irradiance would be

$$I(x, y) = T_r^2(x, y) + T_b^2(x, y) + T_g^2(x, y) \quad (68)$$

which is a superposition of three primary encoded color images.

Let us now provide a result of the retrieved color image obtained with this technique, as shown in Fig. 21. As compared with the original color transparency, we can see that the retrieved color is spectacularly faithful and has virtually no color cross-talk. The resolution and contrast are still below the acceptance for widespread applications; however, these drawbacks may be overcome by utilizing a more suitable film.

E. DENSITY PSEUDO-COLOR ENCODING

Most of the optical images obtained in various scientific applications are gray-level density images—for example, the scanning electron micro-

FIG. 21.   A retrieved color image.

graphs, multispectral-band aerial photographic images, X-ray transparencies, infrared scanning images, and many others. However, humans can perceive details in color better than in gray levels. In other words, a color-coded image can provide better visual discrimination.

We shall now describe a density pseudo-color encoding technique for monochrome images. We assume that a gray-level transparency (called T_1) is available for pseudo-coloring. By a contact printing process, a negative, and a product (called T_2 and T_3, respectively) image transparency can be made. It is now clear that the spatial encoding onto a monochrome film can take place using the procedure described for color image retrieval. The intensity transmittance of the encoded film can be written as

$$T(x, y) = K\{T_1(x, y)[1 + \text{sgn}(\cos \omega_1 y)]$$
$$+ T_2(x, y)[1 + \text{sgn}(\cos \omega_2 x)]$$
$$+ T_3(x, y)[1 + \text{sgn}(\cos \omega_3 x)]\}^{-\gamma} \quad (69)$$

where K is an appropriate proportionality constant. To improve the diffraction efficiency of the encoded film, again we utilize a bleaching process such that the phase encoded transparency becomes

$$t(x, y) = \exp[i\phi(x, y)]$$

where

$$\phi(x, y) = M\{T_1(x, y)[1 + \text{sgn}(\cos \omega_1 y)]$$
$$+ T_2(x, y)[1 + \text{sgn}(\cos \omega_2 x)]$$
$$+ T_3(x, y)[1 + \text{sgn}(\cos \omega_3 x)]\} \quad (70)$$

where M is an appropriate proportionality constant. If we insert this encoded phase transparency at the input plane of the white-light processor of Fig. 20, the complex light distribution due to $t(x, y)$ for every λ at the spatial frequency plane is

$$S(\alpha, \beta, \lambda)$$

$$\approx \hat{T}_1\left(\alpha, \beta \pm \frac{\lambda f}{2\pi} \omega_1\right) + \hat{T}_2\left(\alpha \pm \frac{\lambda f}{2\pi} \omega_2, \beta\right)$$

$$+ \hat{T}_3\left(\alpha \pm \frac{\lambda f}{2\pi} \omega_3, \beta\right)$$

$$+ \hat{T}_1\left(\alpha, \beta \pm \frac{\lambda f}{2\pi} \omega_1\right)$$

$$* \hat{T}_2\left(\alpha \pm \frac{\lambda f}{2\pi} \omega_2, \beta\right)$$

$$+ \hat{T}_1\left(\alpha, \beta \pm \frac{\lambda f}{2\pi} \omega_1\right)$$

$$* \hat{T}_3 \left(\alpha \pm \frac{\lambda f}{2\pi} \omega_3, \beta \right)$$

$$+ \hat{T}_2 \left(\alpha \pm \frac{\lambda f}{2\pi} \omega_2, \beta \right)$$

$$* \hat{T}_3 \left(\alpha \pm \frac{\lambda f}{2\pi} \omega_3, \beta \right) \qquad (71)$$

where $\hat{T}_1$, $\hat{T}_2$, and $\hat{T}_3$ are the smeared Fourier spectra of the positive, negative, and product images respectively, $*$ denotes the convolution operation, and the proportionality constants have been neglected for simplicity. Again, we see that the last cross product (i.e., the moire fringe pattern) can be avoided by spatial filtering. Needless to say, by proper color filtering of the first-order smeared Fourier spectra, a moire-free pseudo-color encoded image can be obtained at the output plane. The corresponding pseudo-color image irradiance is therefore

$$I(x, y) = T_{1r}^2(x, y) + T_{2b}^2(x, y) + T_{3g}^2(x, y) \quad (72)$$

where T_{1r}^2, T_{2b}^2, and T_{3g}^2 are the red, blue, and green intensity distributions of the three spatially encoded images.

Figure 22 shows a black-and-white picture of a color-coded X-ray image of a human skull. The positive image was encoded in blue, the negative image in red, and the product image in green. By comparing the color-coded image with the original X-ray transparency, we have noted that the

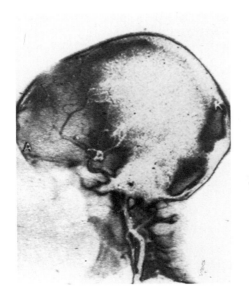

FIG. 22. A black-and-white picture of a density pseudo-color encoded X-ray image.

soft-tissue objects in the color-coded image can be more easily differentiated by the human eye.

F. GENERATION OF SPEECH SPECTROGRAMS

We shall now describe a technique of generating a multicolor speech spectrogram with a white-light optical signal processor. This technique utilizes a cathode-ray-tube (CRT) scanner to convert a temporal signal to a spatial signal recorded onto a moving photographic film. If this encoded film is transporting over the input optical window of a white-light spectrum analyzer, as shown in Fig. 23, then a speech spectrogram can be generated at the output plane.

We stress that for time to spatial signal conversion, the speed of the film motion is required to satisfy the inequality

$$v \geq \nu/R, \qquad (73)$$

where R is the spatial resolution of the film and ν is the highest-frequency content of the time signal.

If the encoded signal format is transporting over the input plane of the white-light processor of Fig. 23, a slanted set of nonoverlapping smeared color spectra in the spatial frequency plane can be observed. It is now clear that if a slanted narrow slit is utilized at the Fourier plane, a frequency color-coded spectrogram can be recorded at the output plane. We further note that, by simply varying the width w of the input optical window, one would obtain the so-called *wide-band* and *narrow-band* spectrograms. In other words, if a broader optical window is utilized, then a narrow-band speech spectrogram, which corresponds to higher spectral resolution, can be obtained. On the other hand, if the optical window is narrower, then a wide-band spectrogram will be obtained. The loss of the spectral resolution due to a narrower optical window would, however, improve the time resolution of the spectrogram. This time–bandwidth relationship is, in fact, the consequence of the well-known Heisenberg's uncertainty relation in quantum mechanics.

Figure 24 shows a black-and-white picture of a frequency color-coded speech spectrogram obtained with the proposed white-light technique. The frequency content is encoded with red for high frequency, green for intermediate frequency, and blue for low frequency. This speech spectrogram represents a sequence of English words spoken by a male voice. These words are "testing, one, two, three, four."

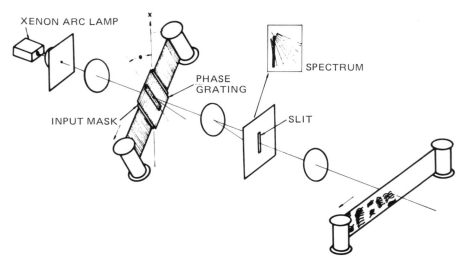

XENON ARC LAMP

PHASE GRATING

SPECTRUM

INPUT MASK

SLIT

FIG. 23.   A white-light optical sound spectrograph.

From this speech spectrogram, the format variation can readily be identified. As compared with the electronic and digital counterparts, the optical technique simplified the processing procedure, and the system is rather versatile to operate.

VIII. Concluding Remarks

We have presented the basic concept of optical processing with partially coherent light. We have reviewed several partially coherent optical processing techniques. A mathematical formulation for a partially coherent processor with a white-light source was obtained. The perfor-mance of a partially coherent processor can be, in principle, evaluated with this generalized formulation. The principle of source encoding and image sampling as applied to the coherence requirement of a partially coherent processor has also been discussed. The objective of utilizing a source encoding technique for a partially coherent processor is to alleviate the stringent spatial coherence requirement so that a practical extended source can be used. The image sampling is to improve the degree of temporal coherence in Fourier plane, so that the image can be processed in complex light field for a broadband light source. Since the partially coherent processing can be carried out by a white-light .

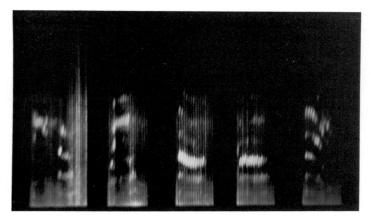

FIG. 24.   A black-and-white picture of a frequency color-coded speech spectrogram.

source, the technique is very suitable for color-image processing. We have also, in this chapter, demonstrated several applications. We have shown that the partially coherent processing system is capable of performing image deblurring, image subtraction, complex signal detection, color image retrieval, pseudo-color encoding, and speech spectrogram generation. We have noted that image processing with white-light is generally very economical and that the technique is rather easy to implement, in contrast with its coherent counterparts.

While the electronic digital signal processor offers flexibility, the partially coherent technique offers the advantages of higher resolution, wavelength diversity, simplicity, and reduced cost. With the current advances of electrooptical spatial light modulators, and optical-digital interface devices, we would expect a gradual merging of the optical and digital techniques; namely, hybrid or microcomputer-based optical signal processing. By utilizing the strengths and merits of both processing techniques, we would expect that many fruitful results will be reported. Furthermore, we believe that partially coherent processing is at the threshold of widespread application. We hope that this chapter will serve as an introductory entry toward the vast application of partially coherent processing.

BIBLIOGRAPHY

Chang, B. J., and Winick, K. (1980). Silver-halide gelatin holograms. *SPIE* **215**, 1972.

Chao, T. H., Zhuang, S. L., Mao, S. Z., and Yu, F. T. S. (1983). Broad spectral band color image deblurring. *Appl. Opt.* **22**, 1439.

Chavel, P., and Lowenthal, S. (1978). Noise and coherence in optical image processing. II. Noise fluctuations. *J. Opt. Soc. Am.* **68**, 721.

Leith, E. N. (1980). Image deblurring using diffraction grating. *Opt. Lett.* **5**, 70.

Leith, E. N., and Roth, J. (1977). White-light optical processing and holography. *Appl. Opt.* **16**, 2565.

Morris, G. M., and George, N. (1980). Matched filtering using band-limited illumination. *Opt. Lett.* **5**, 202.

Yu, F. T. S. (1976). "Optics and Information Theory." Wiley-Interscience, New York.

Yu, F. T. S. (1985). "White Light Optical Signal Processing." Wiley-Interscience, New York.

Yu, F. T. S., and Hsu, F. K. (1985). Complex signal detection with broad band white-light processing. *Appl. Opt.* **24**, 2135.

Yu, F. T. S., and Wu, S. T. (1982). Color image subtraction with extended incoherent sources. *J. Opt.* **13**, 183.

Yu, F. T. S., Zhuang, S. L., and Wu, S. T. (1982). Source encoding for partially coherent optical processing. *Appl. Phys.* **B27**, 99.

Yu, F. T. S., Chen, X. X., and Chao, T. H. (1984). Density pseudocolor encoding with three primary colors. *J. Opt.* **15**, 55.

Yu, F. T. S., Lin, T. N., and Xu, K. B. (1985). White-light optical speech spectrogram generation. *Appl. Opt.* **26**, 836.

Yu, F. T. S., Chen, X. X., and Zhuang, S. L. (1985). Progress report on archival storage of color films utilizing a white-light processing technique. *J. Opt.* **16**, 59.

Yu, F. T. S., Zheng, L. N., and Hsu, F. K. (1985). Noise measurement of a white-light optical signal processor. *Appl. Opt.* **24**, 173.

PARTICLE ACCELERATORS

F. T. Cole   *Fermi National Accelerator Laboratory*
M. Tigner   *Cornell University*

GLOSSARY

Betatron: Circular induction accelerator for electrons. The magnetic guide field rises during acceleration to keep particles on a circle of constant radius.

Circular accelerator: Cyclic accelerator in which particles are bent by magnetic fields around closed paths, passing many times through the same accelerating system.

Colliding beams: System in which the fixed target is replaced by a second beam of accelerated particles moving in the opposite direction. The collisions of moving particles produce very high-energy phenomena.

Cyclic accelerator: Particle accelerator in which each particle passes many times through a small potential difference to be accelerated to high energy.

Cyclotron: Circular accelerator in which protons or heavy ions spiral outward from the center as they are accelerated by a radio-frequency voltage.

Electron volt (eV): Unit used to describe the energy of particles in an accelerator.

Fixed target: Target fixed in location that the beam strikes after acceleration to produce physical changes of interest.

Focusing system: System for maintaining divergent particles in a beam close to the ideal orbit during the course of acceleration and storage.

High-voltage accelerator: Particle accelerator in which each particle passes once through a high potential. The main types of high-

voltage accelerators are Cockcroft–Walton accelerators, Van de Graaff generators, and Marx generators.

Induction accelerator: Particle accelerator in which particles are accelerated by electric fields that are generated from a changing magnetic field by Faraday's law of induction.

Linear accelerator: Cyclic accelerator in which acceleration takes place along a straight line as particles pass sequentially through repeated accelerating units in synchronism with an electromagnetic wave.

Microtron: Circular accelerator in which electrons move in circles that are all tangent at one point where a radio-frequency voltage accelerates them.

Storage ring: Ring like a synchrotron with magnetic field fixed in time so that one or more beams of particles can circulate continuously to produce colliding beams.

Strong-focusing system: System of alternating focusing and defocusing lenses which produce a strong net focusing effect.

Synchrotron: Circular accelerator in which particles are kept in a circle of constant radius by a magnetic guide field that rises in time as they are accelerated by a radiofrequency voltage.

A particle accelerator is a machine that increases the kinetic energy of atomic particles. These particles can be electrons, protons, or heavier ions. They are always electrically charged. The particles are accelerated by electromagnetic forces and they all move in the same direction. They form a beam, somewhat like the beam of a flashlight.

Particles are accelerated in order to be used in

scientific research, in medicine, or in industry. After the particles are accelerated to the desired energy, the beam is steered to strike a target. The accelerated particles cause physical changes in the target. These changes and the study of them are the desired objective of a particle accelerator.

I. Introduction

A. Units

The final energy of the accelerated particles is the most important parameter of a particle accelerator. The particles have charge equal to the electron charge e or a multiple of it, and they are accelerated by potentials measured in volts (V), so a natural unit of energy is the electron volt (eV), the energy acquired by one electron charge in passing through a potential difference of 1 V.

The electron volt is a very small unit of energy ($1\ eV = 1.6 \times 10^{-19}$ J), more directly applicable to energy levels in atoms than to accelerators. There are therefore multiples of the electron-volt that are used to describe accelerators. For example,

$$1\ keV = 10^3\ eV$$
$$1\ MeV = 10^6\ eV = 10^3\ keV$$
$$1\ GeV = 10^9\ eV = 10^3\ MeV$$
$$1\ TeV = 10^{12}\ eV = 10^3\ GeV$$

The energies of particles accelerators now being operated range from a few hundred keV to 1 TeV. The sizes of particle accelerators range from small devices that fit on a table to devices stretching over several miles. One of the largest, the Fermi National Accelerator Laboratory near Chicago, is shown in an aerial view in Fig. 1.

A second important parameter used to describe an accelerator is its intensity, usually the number of particles accelerated per second.

B. Uses of Accelerators

1. Science

Particle accelerators are essential tools of high-energy, or elementary-particle, physics, as it is also called. They are used to accelerate particles to very high energy so that the particles can probe the innermost structure of matter and the forces that govern its behavior. In the interaction between a projectile particle and the target particle it strikes, new kinds of particles can be produced that provide clues to the nature of matter. These particles are usually short lived and decay radioactively in a time much less than a microsecond. Such reactions were produced copiously in the first moments of the "big bang" from which our universe is believed to have evolved, but are now produced only infrequently in nature by cosmic rays. Systematic study of these particles and their interactions requires controlled, copious production using accelerators. [*See* COSMIC INFLATION; COSMIC RADIATION; NUCLEAR PHYSICS.]

Several different kinds of things can happen when a beam particle strikes a particle in the target:

1. The beam particle can be scattered. The distribution of scattered particles that is observed is used to determine the size and shape of the scatterer, as Rutherford did in the classic experiments that demonstrated the existence of the atomic nucleus. [*See* ATOMIC PHYSICS.]

2. The target nuclei or atoms can be changed into a different state by the collision. The nucleus can be excited to a higher energy level and the radiation observed when it drops back to its ground state. In other cases, the target nucleus can be broken up into small fragments, or the beam particle can be fused into the target nucleus. The product nuclei are then observed.

3. New kinds of particles can be created in the collision. For example, positrons, antiprotions, mesons, and hyperons can be created and their properties observed. In some cases, the particles created are formed into a beam and this secondary beam is used for experiments.

When an accelerated particle strikes a stationary particle in a target, a large fraction of the energy so laboriously put into the particle goes to move all the products forward in the direction of motion of the beam, because momentum is conserved in all collisions. Thus a 400-GeV proton accelerator has available only approximately 27 GeV for making new particles. When the proton energy is increased to 1000 GeV, the available energy increases to only 43 GeV. [*See* COLLISION CROSS SECTIONS (ATOMIC PHYSICS).]

A way to make all the energy useful is to utilize a second accelerated beam as a target. If the two beams are moving in opposite directions, the total momentum of the system is zero and none of either beam's energy need be used in moving products downstream. This colliding-beams method is a more economical method of

FIG. 1.   Aerial view of the world's highest energy accelerator, the Fermi National Accelerator Laboratory. The accelerator is in an underground tunnel; a road and a cooling-water pond outline the tunnel on the surface. The accelerated beam is extracted for use in experiments in the areas stretching toward the top left of the photograph. (Photo courtesy of Fermilab.)

creating new, higher-energy interactions and new particles, as depicted schematically in Fig. 2.

The collision rate in colliding beams is much lower than in a fixed target, because the particle density is much less than in a solid target, and there are therefore many fewer particles to interact with in a beam. This difficulty can be partially overcome by storing two circulating beams of particles in a storage ring. There they pass many times through one another to increase the effective collision rate.

In all these kinds of experiments, a physical

change is produced by the beam particles, which are used as a probe to induce these changes. Higher energies are needed to probe to smaller distances within the atomic nucleus, and there has been a constant drive toward higher energy. The highest energies now used for physics experiments are approaching 1 TeV in fixed-target research and two beams each of 1 TeV energy in colliding-beam research. At this time, a collider with beams up 20 TeV each is being designed. These accelerators are very large in size. They are also very complex. Large accelerators are among the most complex devices built by human

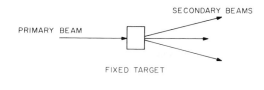

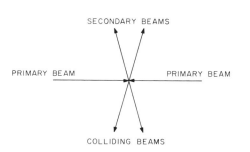

FIG. 2. Schematic diagram of fixed-target and colliding-beam experimental methods.

beings and large numbers of people are needed to operate them. As an example, the accelerator and tunnel of Fermilab are shown in Fig. 3.

An analogy can be made between a light source–microscope combination and an accelerator–detector combination. In a microscope, light passes through a target under study and scatters from it. The transmitted or scattered light is gathered by the microscope and directed

FIG. 3. The Fermilab accelerator tunnel. The upper ring of magnets is the Main Ring, completed in 1972, that accelerates protons to 400 GeV. It is now used to inject 150-GeV protons into the lower ring, the superconducting 1-TeV Tevatron. Above the rings are cooling-water lines, bus tubes for magnet current, and trays for signal wires from diagnostic equipment. The worker is riding an electric golf cart, the preferred means of getting around the 6.3-km tunnel. (Photo courtesy of Fermilab.)

to the eye or to a photographic film where an image is focused that can be analysed. In an accelerator–detector combination, a target is illuminated by the beam from the accelerator. Scattered or transmitted particles or new particles created in the high-energy interaction are then registered in a detector for analysis. The detector can be a simple photographic film or a complex electronic device containing tens of thousands of individual elements and weighing hundreds of tons. The fineness of detail that can be studied with the microscope depends upon the wavelength of the light used; the shorter the wavelength, the finer the detail. Similarly, with illumination by accelerated particles; the higher the energy or the shorter the deBroglie wavelength, the smaller the scale of matter that can be studied.

The analogy is not complete, because the transmitted or scattered light in a microscope makes only small changes in the object under study and the purpose is to study the unchanged object. In an accelerator, major changes can be made in the target. For example, new particles can be produced and studied.

Another scientific application of accelerators that has wide use is in the production of synchrotron radiation in the form of intense ultraviolet light or X rays. This intense source has opened up many new areas of atomic physics, biology, and chemistry for study.

2. Medicine

Smaller accelerators are used in both great arms of medicine, diagnosis and treatment. Their use is very much like that of the familiar X-ray machines. In fact, an X-ray machine is a particle accelerator. In it, electrons are accelerated and made to strike a target to produce X-rays. These X-rays are used to examine tissue, bones, and organs inside the body as an aid in diagnosis. In CAT scanning, computers are used to reconstruct images. [*See* COMPUTERIZED TOMOGRAPHY.]

X rays are also used in treatment to kill cancer tumors by irradiating them. Here higher-energy electrons are used to produce more penetrating X rays. The highest energies are approximately 0.5–10 MeV. In addition, accelerators of heavier particles are used to produce protons, neutrons, or heavy ions for use in cancer therapy. In many cases, these heavier particles offer a better method of radiation therapy than electrons or X rays, because heavy particles deposit their energy over a much smaller depth of tissue than electrons or X rays, and the energy deposition can be much better confined to the tumor site. Heavy particles of energies from 10 MeV to several GeV are used in this therapy.

Particle accelerators are also used in medicine to produce radioactive isotopes, which are then used to trace the movement of chemicals through the human system as an aid in diagnosis. [*See* RADIOACTIVITY.]

3. Industry

The uses of accelerators in industry have some similarity to the uses in medicine. The particle energies are usually low, hundreds of kilovolts to 10 MeV in most cases. A major industrial use is in diagnosis and testing. Pressure vessels, boilers, and other large metal castings are routinely X-rayed to search for internal flaws and cracks, which would lower the strength of the casting. Particle energies of 20 MeV or more are often used for greater penetration of thick castings. [*See* X-RAY ANALYSIS.]

Particle accelerators are also used in materials treatment. Precise concentrations of impurity ions are implanted in metal surfaces for solid-state electronics (semiconductor) manufacture. Particle beams are used to etch microchips in the production of integrated circuits. [*See* RADIATION EFFECTS IN ELECTRONIC MATERIALS; RADIATION PHYSICS.]

Many manufactured objects are sterilized by accelerators. Such sterilization is the preferred method for bandages and surgical instruments, because it damages them less than heat sterilization. The accelerator energies used for sterilization are low enough that no radioactivity is induced in the object being sterilized.

Materials are also changed chemically by accelerator radiation. A notable application is in the polymerization of plastics. Transparent shrink wrapping is treated by accelerators to produce the desired shrinkability with the application of heat.

Development is being carried out to test the applicability of accelerators to controlled-fusion nuclear reactions to produce energy. In the systems envisaged, beams of high-energy particles would be used to bombard a very small deuterium–tritium pellet and heat it by implosion to the point at which energy-producing deuterium–tritium reactions would take place. A very large system would be needed to produce energy economically, and the development is expected to require many years of work. [*See* FUSION DEVICES EXPLOSIVE.]

Food preservation by accelerator radiation is also being carried out, mostly on a trial basis at this time, but with some large-scale application by the military services.

A recent accelerator application is to the destruction of harmful bacteria in sewage by accelerator beams, in order that the treated sewage can then be used as fertilizer.

Some military uses have been suggested, but none have as yet been put into actual practice.

C. TYPES OF ACCELERATORS

Accelerators can be divided into two classes, those in which acceleration is carried out by use of a high dc voltage and those in which acceleration is carried out by a lower voltage through which particles pass many times, which are called cyclic accelerators.

High-voltage accelerators charge a terminal to high voltage with respect to ground. In some high-voltage accelerators, such as the Cockcroft–Walton set and Marx generator, a voltage-multiplying circuit is used and charge is carried to the terminal electronically. In others, such as the Van de Graaff accelerator, known also as the electrostatic generator, charge is carried mechanically to the terminal by a moving belt system. The Van de Graaff can be expanded to the Tandem Van de Graaff, in which negative ions are accelerated to high voltage, then stripped of electrons so that they become positive ions and are accelerated back to ground, thus receiving twice the kinetic energy corresponding to the terminal potential. An electrostatic generator installation is shown in Fig. 4.

High-voltage accelerators are limited in peak energy by practical problems of holding high voltage against sparking or discharges to ground. Cockcroft–Walton and Marx generators can reach approximately 1 MV. Van de Graaff generators have reached 25 MV and tandems have accelerated particles to 50 MeV.

Cyclic accelerators always have accelerating fields that vary in time. The particle's motion is synchronized with this variation. (If the fields were constant, they would be high-voltage accelerators, with the energy limitations of that class.) In linear accelerators, each particle passes once through a sequence of accelerating structures. In circular accelerators, a magnetic field is used to bend the particles around a closed path to pass repeatedly through the same accelerating structure.

Linear accelerators for electrons or other par-

FIG. 4.   Column of 14-MeV tandem Pelletron electrostatic generator. In operation, it is enclosed in a steel tank 5.5 m in diameter and 22 m tall containing sulfur hexafluoride gas at a pressure of 7 atm. (Photo courtesy of National Electrostatics Corp.)

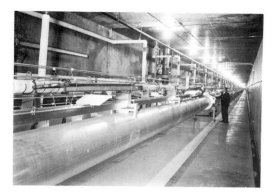

FIG. 5.   Tunnel of the Stanford linear accelerator. The accelerator waveguide is the white horizontal tube at the man's eye level. It sits on a strong girder system for alignment and stability. Radiofrequency power is fed to the waveguide from klystron amplifiers in separate buildings through the rectangular tubes. The white tubes supply cooling water. (Photo courtesy of SLAC.)

ticles moving at speeds close to the speed of light make use of a traveling electromagnetic wave in a special waveguide. Electrons ride the crest of the wave like a surfer on an ocean wave. The largest electron linear accelerator is the 50-GeV Stanford accelerator, which is 2 miles long. It is shown in Fig. 5.

For particles moving much slower than the speed of light, a different structure is needed because the wave velocity cannot be slowed down enough for the particles to keep in step. In this case, drift tubes are used to shield particles from decelerating fields. The largest proton linear accelerator is the 800-MeV accelerator at Los Alamos National Laboratory.

The cyclotron was the first circular accelerator. Particles are injected at the center of a circle and spiral outward in the constant magnetic field as they gain energy. Acceleration takes place as the particles cross the gap between two D-shaped drift tubes. The particles must stay in step with the radiofrequency accelerating voltage. For higher energy, this synchronism cannot be maintained and the accelerating voltage is frequency modulated in the synchrocyclotron. More recent cyclotrons utilize azimuthally varying fields (AVFs) to maintain synchronism. A modern cyclotron is shown in Fig. 6. Cyclotrons have been built to operate up to 800–900 MeV. For electrons, the cyclotron is not a useful device. Instead, the microtron, in which particles travel longer paths as they gain energy, is appropriate.

FIG. 6.   Superconducting AVF cyclotron at Michigan State University. The magnet coils are in a cryostat by which the scientist is standing. Radiofrequency power is brought in through the tubes at the top. (Photo courtesy of Michigan State University.)

The circular accelerator that is used to go beyond 1 GeV is the synchrotron, in which the magnetic field increases in time to keep particles circulating at a constant radius. The frequency of the accelerating voltage is varied to maintain synchronism with the particles as they are accelerated. The largest synchrotron, which utilizes superconducting magnets, is the Tevatron at Fermilab; it accelerates protons to almost 1000 GeV. It is the lower ring of magnets in Fig. 3.

Storage rings are similar to synchrotrons, but the magnetic field is held constant so that particles can continue to circulate. Storage rings with particles circulating in both directions are utilized for colliding-beam experiments.

Particles can be accelerated by the electric fields induced by changing magnetic fields (Faraday's Law), and there are smaller accelerators, the linear induction accelerator and the circular betatron, which use this principle to accelerate electrons to energies of the order of 100 MeV.

Although it is possible in principle to build a synchrotron for an arbitrarily high energy, it eventually becomes impossible economically. There are accelerator scientists looking beyond to new ways of accelerating particles, perhaps using the intense electromagnetic fields possible in lasers or in plasmas for acceleration and guidance of particles. This work is still in the research phase, with many interesting ideas and configurations of these collective accelerators.

D. Physical Principles

Particle accelerators operate according to well-established laws of nature, the laws of electromagnetic fields and the laws of electrodynamics governing the motion of charged particles in these fields. Research and development of accelerators is an active field of science, but it is a field that utilizes known laws of nature rather than seeking new laws. Accelerators are tools used in seeking new laws. [*See* ELECTROMAGNETICS; RADIATION PHYSICS.]

A charged particle moving in an electric field experiences a force in the direction of the field (or the opposite direction if it is a negative charge) and is accelerated in that direction, as shown in Fig. 7a. Thus, its velocity and kinetic energy will increase. The force $F = qE$, where q is the particle charge and E the electric field. If q is measured in coulombs and E in volts per meter, then F is given in newtons.

This is the basic mechanism of operation of a particle accelerator. The electric field can be ei-

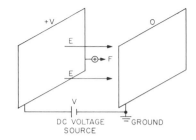

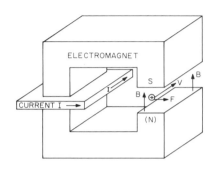

(b) MAGNETIC FIELD and FORCE

FIG. 7.   (a) Electric and (b) magnetic forces.

ther constant in time, in which case particles are accelerated continuously and come out in a steady stream, or varying in time, in which case particles are accelerated only when the electric field is in the right direction and come out in bunches.

A charged particle in a static magnetic field experiences a force in the direction perpendicular to the plane formed by two vectors, the magnetic-field vector and the particle's velocity vector, as shown in Fig. 7b. A particle moving in the direction of the magnetic field experiences no magnetic force. Because the force on a particle in a static magnetic field is perpendicular to its velocity, no work is done on the particle and its energy does not change. The energy increase in a circular accelerator comes from the accelerating system, not from the magnetic field.

The force on a particle moving at velocity v perpendicular to a magnetic field B is

$$F = qvB \qquad (1)$$

If B is measured in Tesla (1 T $= 10^4$ gauss) and v in meters per second, the force F is in newtons. When the magnetic field provides the centripetal force that bends the particle in a circle of radius ρ, then

$$mv^2/\rho = qvB$$

and there is a relation between the momentum p and the product of radius and field

$$p = mv = q\rho B \qquad (2)$$

Thus, as the momentum of a particle is increased during acceleration, either the radius of curvature must increase, as in a cyclotron, or the magnetic field must increase, as in a synchrotron. This relation $p = q\rho B$ is very basic for circular accelerators. It holds for slow and fast particles, including effects of special relativity. For practical calculations, if B is in Tesla, ρ in meters, and p in MeV/c (where c is the speed of light), then

$$p = 300B\rho \qquad (3)$$

There is a sense in which magnetic fields can increase the energy of particles. The linear induction accelerator and betatron discussed above make use of a magnetic field changing in time to induce an electric field that increases the energy of the particles.

In circular accelerators, the paths of particles during the course of acceleration are very long, sometimes many thousands of kilometers. A particle injected at an angle with respect to the ideal path will stray farther and farther during this long distance and will leave the confines of the magnetic field and be lost, unless some means is provided to focus the particles back toward the ideal path. This focusing is accomplished by building in carefully-designed gradients (variations in space) of the magnetic field. Focusing is one of the most important elements of accelerator design.

We emphasize here the difference between a particle accelerator and a nuclear reactor. In the accelerator, the particles are focused and form a beam, all moving in the same direction with the same energy, whereas in a reactor, the particles are heated and move with the random directions and wide energy distribution of a hot gas. In addition, the reactor accelerates particles by nuclear forces, whereas the accelerator makes use of electromagnetic forces. These electromagnetic forces cease when the electric power supplied to the accelerator is interrupted. Thus, the radiation from an accelerator stops, except for small residual effects, when the accelerator is turned off. [*See* NUCLEAR REACTOR THEORY.]

E. History

During the nineteenth century, physicists experimented with Crookes tubes, evacuated glass systems containing internal electrodes. A current of electrons will flow when a large enough voltage is applied between these electrodes. J. J. Thomson used a Crookes tube in his discovery of the electron in 1890. Roentgen discovered X rays using a Crookes tube in 1896. The X-ray tube was later made into a practical device for use in medicine by Coolidge.

Rutherford spurred the development of particle accelerators for use in nuclear physics research in a famous lecture in 1920. He pointed out the need for higher-energy particles for the further development of understanding of the atomic nucleus. During the next decade, both high-voltage and cyclic accelerators were invented. Many different methods of producing high voltage were demonstrated, but there was always great difficulty in providing an accelerating tube that would not spark at high voltage. Cockcroft and Walton, in Rutherford's Cavendish Laboratory, developed a successful accelerating tube. They used an existing voltage-multiplying circuit and built a 300-keV proton accelerator, which they used in 1932 to do the first nuclear physics experiment with accelerated particles.

Ising was the first to conceive (in 1925) a cyclic accelerator, a drift-tube linear accelerator. Wideroe expanded the idea and built a working accelerator in 1928. It accelerated mercury ions to twice the impressed radio-frequency voltage.

Perhaps the most important consequence of Wideroe's work was to stimulate Lawrence to conceive the cyclotron. Lawrence and Livingston built the first operating cyclotron in 1932, and a succession of cyclotrons was built in Lawrence's laboratory through the 1930s.

Lawrence's cyclotrons and electrostatic generators, which had been conceived and demonstrated by Van de Graaff in 1931, were used for nuclear physics research throughout the 1930s. Both were limited to energies of 15 MeV or less and reaching energies beyond this limit was a major topic of research in the 1930s. Beams conceived the traveling-wave linear accelerator and demonstrated it, but radiofrequency power sources were not adequate at that time to build useful accelerators. Thomas proposed the AVF cyclotron in 1938, but his work was not understood until much later. Kerst built the first successful betatron in 1941 and built a second 20-MeV machine before World War II intervened. This model was built in large quantities for use in X-ray testing of large castings, particularly for the armor of military tanks.

The next great step in energy began in 1944 and 1945 when Veksler in the USSR and McMillan in the United States independently conceived the principle of phase stability, permitting frequency modulation of the accelerating voltage in a cyclotron to overcome effects of relativity and making the synchrotron possible. In 1946, the first synchrocyclotron was operated and a number of 300-MeV electron synchrotrons came into operation in the next few years. The research done with these accelerators was important in learning the properties of the pi meson.

World War II radar work had stimulated the development of high-frequency power sources and these were used in linear accelerators. Traveling-wave electron linear accelerators, at frequencies of approximately 3 GHz, were extensively developed by Hansen, Ginzton, Panofsky, and their collaborators, leading over many years to the 50-GeV Stanford Linear Center. Alvarez extended the concept of the drift-tube accelerator for heavier particles and built the first of many drift-tube accelerators used in physics and chemistry research and as injectors for synchrotrons.

Work also began in the late 1940s to build proton synchrotrons. The first proton synchrotron, the 3-GeV Cosmotron at Brookhaven, New York, came into operation in 1952 and the 6-GeV Bevatron at Berkeley, California, came into operation in 1954. An interesting precursor, a 1-GeV proton synchrotron, conceived in 1943 independently of the principle of phase stability, was built in Birmingham, United Kingdom; shortages of money slowed its construction, and it did not operate until after the Cosmotron. The proton synchrotrons were used for research in heavier mesons, the "strange" particles that had been observed in cosmic-ray experiments, and in antiprotons.

Speculative discussions aimed toward increasing the highest accelerator energy led to the conception of the strong-focusing principle by Courant, Livingston, and Snyder in 1952. It was later found that Christofilos had developed the principle independently in 1950. Strong or alternating-gradient focusing keeps the oscillations of particles about the ideal orbit much smaller and makes possible much more compact and more economical magnets.

The discovery of strong focusing led to an explosion of ideas, knowledge, and techniques in particle accelerators. In 1953, Kitigaki and White independently conceived the separated-function strong-focusing synchrotron, which makes possible higher guide fields and more economical designs. In 1958, Collins conceived the long straight section, which makes possible economical configurations with space for injection, acceleration, extraction, and detection equipment. As well as making it possible to go to higher energy with proton synchrotrons, the strong-focusing principle gave impetus to new thinking in many other directions. Linear accelerators were greatly improved in performance by the addition of strong focusing along the orbit, first conceived by Blewett. The AVF principle was rediscovered by a number of people, among them Kolomensky, Ohkawa, Snyder, and Symon. It was extended to spiral-sector focusing by Kerst in 1954. Kerst then proposed that successively accelerated beams could be "stacked" in circulating orbits in a fixed-field alternating gradient (FFAG) accelerator, a variant of the AVF configuration. The intense stacked beam can then be used in colliding-beam experiments. The concept of colliding beams had been known for many years (it was patented in 1943 by Wideroe), but beam stacking is essential to achieve useful rates of collisions. Shortly afterward, a number of people (Newton, Lichtenberg and Ross and, independently, O'Neill) proposed the concept of a storage ring separate from the accelerating device. The storage ring is a better colliding-beam system than the FFAG accelerator because it is less costly and because it makes it possible to provide much more free space for detectors.

Experimental confirmation of these ideas did not lag far behind. The first strong-focusing electron synchrotron was operated by Wilson and his collaborators in 1954, followed by several other electron synchrotrons. The FFAG principle and beam stacking were demonstrated by Kerst and his collaborators in the 1950s. Traveling-wave electron linear accelerators reached the 1-GeV energy range in this same era and a series of important experiments on electron–proton scattering was done that elucidated the structure of the proton. The first electron storage rings were built and operated in the early 1960s. Two large proton synchrotrons of 28 and 33-GeV energy were built by CERN, a new international laboratory in Europe and by the Brookhaven Laboratory. These became the foundation of major advances in high-energy particle physics, with discovery of many new particles and the beginning of a conceptual ordering among them and understanding of them.

The electron synchrotrons and the Stanford linear accelerator, which reached 20 GeV in 1966, added to this understanding.

In the late 1960s, the first major proton storage ring, the ISR, was built at CERN. It stored proton beams of 28 GeV each, equivalent to a fixed-target accelerator of over 1500 GeV energy.

The second generation of strong-focusing synchrotrons also began to be built in the late 1960s. These incorporated the more efficient separated-function magnet system and long straight sections. A proton synchrotron that reached 400 GeV was completed at the New Fermilab in Illinois in 1972. A similar synchrotron was later built at CERN. These workhorses incorporated new beam-sharing methods and each could provide beams simultaneously to several targets

and a dozen major experiments. An important feature that has made this multiple use possible is the very high degree of precision in beam handling and manipulation. The data from these and ISR experiments led to the development of quantum chromodynamics and electroweak theory, large advances in our understanding of the basic building blocks of nature.

Important experimental evidence also came from the second generation of electron storage rings, now always with positrons as the second beam, the first of which reached 3 GeV in each beam at Stanford in 1972. Electron–positron storage rings have now reached 30 GeV per beam. As in the case of proton synchrotrons, electron–positron storage rings have been developed to a high art. Radiofrequency systems to replace the energy lost in synchrotron radia-

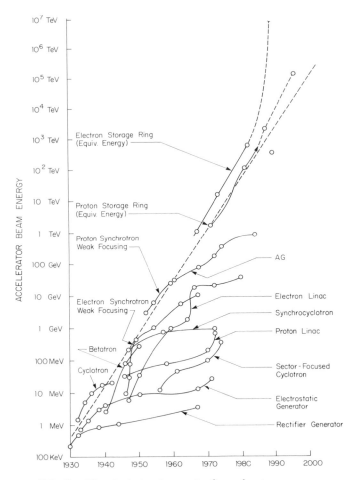

FIG. 8. Historical development of accelerator energy.

tion are a major factor in the design and cost of electron storage rings, but at the same time the synchrotron radiation can make the beam very small in size, making precise manipulations possible and increasing the colliding-beam interaction rate. In fact, synchrotron radiation has become a valuable experimental tool in its own right for use in atomic physics and materials science research, and a number of single-beam electron storage rings have been built for this purpose.

The two large proton synchrotrons were developed further in quite different directions. At Fermilab, superconducting magnets underwent long, arduous development, and a superconducting magnet ring was built and installed in the tunnel of the 400-GeV accelerator. It reached 800 GeV in 1983. The CERN synchrotron was converted to a proton–antiproton storage ring by the addition of a small ring to accumulate antiprotons, making use of the new technique of beam cooling invented by van der Meer. Colliding-beam experiments have been carried out there, culminating in the discovery of the W and Z particles in 1983. These particles have rest energies close to the 100 GeV and cannot be produced even by a 1-TeV fixed-target proton accelerator.

The spectacular successes of these accelerators and storage rings have led to a number of new initiatives. An electron-positron ring (LEP) to initially reach energies of 50 GeV in each beam, with 100 GeV per beam as the ultimate goal, is being built at CERN. An electron–proton ring (HERA) is being built in Germany. A 3-TeV proton synchrotron, convertible to a proton–antiproton storage ring, is being built in the USSR. A single-pass colliding-beam system, SLC, is being built at Stanford. Here the two beams collide only once in a linear system, not a circulating configuration. It is believed that a useable event rate will be achieved by very small beam sizes (thus increasing the density) and very high repetition rates. Design work is in progress in the United States on a proton–proton collider, SSC, with 20 TeV in each beam. Research work has been carried out for many years on methods of collective acceleration, with the goal of achieving much larger accelerating fields. The historical development of the energy of particle accelerators, with a look to the future, is plotted in Fig. 8. One can see from the chart that the development of each new type of accelerator gave an energy increase. Each accelerator type is eventually replaced by another.

II. Types of Accelerators

A. High-Voltage Accelerators

In a high-voltage accelerator, a terminal or electrode is charged to high voltage and particles are accelerated from it to ground potential. If the terminal is charged to a voltage $+V$, a singly charged positive ion will gain energy eV in the accelerator. Thus, the terminal is at potential V with respect to ground, and the maximum possible voltage is limited by problems of holding voltage. Above some voltage, the terminal will spark to ground. The voltage is also limited by less dramatic corona discharge. With ample space to ground and scrupulous attention to detail in design, terminals have been built that hold 25 MV.

The simplest way to produce high voltage is with an ac electrical step-up transformer system. X-ray machines produce voltages up to 1 MV by this method. The beam is only accelerated on one-half the ac cycle and varies in energy throughout the pulse.

In order to avoid scattering of beam particles by residual gas, there must be an accelerating tube that is evacuated to low pressure. The voltage drop must be distributed somewhat uniformly along this tube to avoid sparking. For voltages above approximately 1 MV, the accelerating tube is almost always insulated outside the vacuum by a pressurized gas of high dielectric strength, often sulfur hexafluoride.

To produce a dc voltage, electric charge is brought to the terminal to produce high voltage. Charge may be brought electronically by voltage-multiplying circuits or mechanically by moving devices.

1. Voltage-Multiplying Devices

The first successful high-voltage accelerator, the Cockcroft–Walton accelerator, made use of a voltage-doubling circuit, the Greinacher circuit, shown in Fig. 9. Two rectifiers act on opposite sides of the ac sine wave to charge a capacitance to twice the voltage. The principle can be extended to many stages. Cockcroft and Walton used the Greinacher circuit and developed the first accelerating tube. They accelerated protons to 300 keV and in 1932 demonstrated the first nuclear reaction with artificially accelerated particles, bombarding lithium with protons to produce two alpha particles (helium nuclei). Modern Cockcroft–Walton generators are available commercially up to voltages of approximately

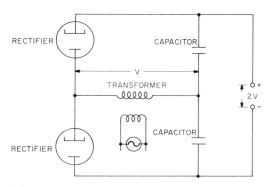

FIG. 9. Schematic diagram of a voltage-multiplying circuit.

1 MV. Special pressurized systems have been built to 3 MV. Cockcroft–Walton generators are widely used as the first stage of higher-energy accelerator systems because they produce beams with very good energy regulation.

The Marx generator is similar in principle. The rectifier system is external to the capacitor stack. In essence, the capacitors are charged in parallel, then discharged in series through spark gaps. Marx generators were originally used in the 1920s to produce surges of high voltage to test electrical generating and transmitting equipment. They are used in modern accelerator work to produce very intense (1000–10,000-A) short (10–50-nsec) pulses of 1–10-MeV particles. The particle energy is very poorly regulated during the pulse.

2. Charge-Carrying Devices

The most prominent of the accelerators in which charge is carried mechanically to the terminal is the electrostatic generator or Van de Graaff accelerator. In this device, depicted schematically in Fig. 10, charge is carried by a moving belt to the high-voltage terminal. In most modern accelerators, the belt is made of a series of insulated metal links, looking a little like a bicycle chain. Van de Graaff accelerators played an important role in the development of knowledge in nuclear physics during the 1930s and are still in use in that field. They are now frequently used in industrial applications.

An electrostatic generator has produced a terminal voltage of 25 MV. It is possible to double the effective voltage by accelerating negative ions (ions with extra electrons) from ground to a high positive voltage, then stripping electrons off by passing the beam through a metal foil and accelerating the positive ions back to ground po-

tential. This tandem Van de Graaff produces a higher energy, but with lower intensity because the process of stripping cannot be made perfectly efficient. The dc beam current in a single electrostatic generator can be as large as 10 to 20 μA, but the dc beam current in a tandem is of the order of 1 μA.

B. Cyclic Accelerators

There are two kinds of accelerators in which a low voltage is used repeatedly to produce high-energy particles: linear accelerators, in which each particle passes once through a sequence of accelerating structures, and circular accelerators, in which each particle traverses a closed path (not necessarily exactly a circle) and passes repeatedly through the same accelerating structure. A magnetic field is used to bend the particles around the closed path. In a cyclic accelerator, the accelerating forces must vary with time, in contrast with the dc forces in high-voltage accelerators. If the forces did not vary with time, they could not be used over and over again to produce high energy.

1. Linear Accelerators

In the linear accelerator, or linac, the particles being accelerated follow paths that are straight to a very close approximation. These particles are accelerated in the desired direction by the action of electric fields, as shown in Fig. 7. In the transverse directions, the particles are confined or focused into a beam by the action of lenses employing static electric or magnetic fields, or in some cases by time-varying har-

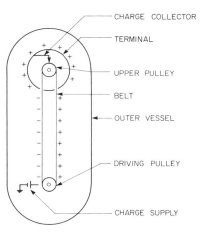

FIG. 10. Schematic diagram of an electrostatic generator or Van de Graaff accelerator.

monic fields, as in the radiofrequency quadru-pole (RFQ) focusing system.

In the accelerators of today, the electric field that provides the accelerating force is shaped by conductors or dielectric media near the beam. The kinetic energy increase of the beam parti-cles as they pass through the accelerator is gained at the expense of energy stored in the accelerating electric field. Because of the con-servative nature of the static electric field, the high-voltage dc accelerators discussed in section II,A, the most elementary form of linac, are lim-ited in maximum particle kinetic energy gain by the maximum constant electric potential that can be supported by an array of conductors. In practical cases this is a few million volts. If the accelerating electric field is time varying, con-tinuous acceleration can be achieved, and there is no physical limit to the maximum particle ki-netic energy that can be achieved. In the time-varying field linac, a substantial fraction of the accelerating field energy is contained in an ac-celerating wave.[1] Particles traveling with the crest of the wave gain energy from it much as a surfer gains energy from a traveling water wave near a beach. The accelerating force will be pro-portional to the strength of the electric field along the axis of the accelerator. That is, $F = qE$. The traveling accelerating wave field will vary with time t and with distance z along the accelerator as

$$E = E_0 \cos \left[\omega \left(t - \frac{z}{v_w} \right) + \phi_i \right] \quad (4)$$

where ω is the frequency of the wave, v_w the phase velocity of the wave and ϕ_i a constant that measures the initial value of the wave strength at the beginning of acceleration, i.e., $t = 0$, $z = 0$. If the particles being accelerated have a velocity v_p along the accelerator axis their position is given by

$$z = v_p \cdot t \quad (5)$$

As a consequence, if the linac is designed so that $v_w = v_p$, the particles are accelerated by a con-stant force in the z direction

$$F = qE_0 \cos \phi_i$$

As a consequence of the wave nature of the acceleration, only particles near the wave

crests, i.e., $\phi_i = 0, 2\pi, 4\pi, \ldots$ receive useful acceleration. Thus, the beams from linacs em-ploying time-varying fields are bunched, the sep-aration of the bunches being $\lambda \cdot v_\omega / c$, where λ is the free-space wavelength of the accelerating harmonic field. As the particles are accelerated, their velocity increases according to the relation

$$\frac{v_p}{c} = \left\{ 1 - \left[\frac{1}{1 + (T/M_0 c^2)} \right]^2 \right\}^{1/2}$$

where T is the kinetic energy of the particle be-ing accelerated and $M_0 c^2$ its rest energy, c being the velocity of light. For kinetic energies much less than the rest energy, $T \ll M_0 c^2$, this expres-sion simplifies to

$$v_p/c \sim (2T/M_0 c^2)^{1/2}$$

the classical relation between velocity and ki-netic energy. For T much greater than the rest energy, $T \gg M_0 c^2$,

$$v_p/c \sim 1$$

Preaccelerators for linacs are often high-voltage dc or pulsed-dc accelerators, operating at from a few hundred kilovolts to a few million volts. Protons and heavier ions have rest energies (masses) of one to many giga electron volts so that they emerge from the preaccelerator with velocities only a small fraction of c. Electrons have a mass of 511 keV so that even a preaccel-erator of only 80 keV boosts their velocity to $0.5c$. The arrangements needed for efficient gen-eration of accelerating waves depend markedly on the desired wave velocity and the designs of linacs for proton or heavier ions and for elec-trons differ markedly.

a. Proton and Heavy-Ion Linacs. The trav-eling-wave system does not work for particle speeds very much less than c, because the wave cannot be efficiently slowed down enough to match the particle speed. If the wave velocity is greater than the particle velocity, the wave passes by each particle. As it passes, a particle will experience decelerating forces during the negative-field portion of the sinusoidal variation of the wave. Instead of a traveling wave, a standing wave is used, and conducting drift tubes are placed around the particle trajectories in the regions of negative fields in order to shield particles from these fields. These drift tubes are shown schematically in Fig. 11. Acceleration then takes place in the gaps between drift tubes and the particles drift through the tubes unaf-fected by negative fields. In most standing-wave

---

[1] A large part of good accelerator design lies in config-uring the field-shaping conductors or dielectrics to maximize the fraction of the total field energy in the accelerating wave component of the field.

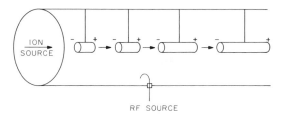

FIG. 11. Schematic diagram of a standing-wave drift-tube linear accelerator.

linear accelerators, the drift tubes contain focusing devices to contain the transverse motion. Drift-tube linear accelerators are used for acceleration of protons or heavy ions. They are capable of producing high intensities of accelerated beam.

The presence of drift tubes strongly affects the choice of frequency in a standing-wave linac. The accelerating frequency cannot be so high that the drift tubes are too small to contain focusing elements. Proton linear accelerators usually utilize frequencies of approximately 200 MHz. Heavy-ion linear accelerators, which inject at even lower speed, often have frequencies in the 60–80-MHz range.

The linac cavities that form the outer envelope of the accelerator and contain the electromagnetic field are built in sections for convenience in manufacture. A 200-MeV proton linac is approximately 500 ft long. The pulse length is short enough that the average rf power is only a few kilowatts. The separate amplifiers for each cavity are synchronized by a master oscillator. Peak currents of 200 mA of 200-MeV protons are achieved in injectors for proton synchrotrons. Linear accelerators with longer pulse length are built for applications in which very high intensity is desired. For higher energy, a standing-wave linac can be used to inject into a traveling-wave linac when the particle speed is comparable to the speed of light.

The copper lining and drift-tubes of a standing-wave linac can be replaced by superconducting metals such as a niobium or lead, and the entire system can be cooled down to superconducting temperature (2–4 K). A superconducting heavy-ion linac, ATLAS, is in operation. The use of superconductivity is economically justifiable for long-pulse or continuous wave (CW) linacs.

Particles can also be accelerated by the electric fields induced by time-varying magnetic fields according to Faraday's Law. In a linear induction accelerator, magnetic fields are pulsed sequentially in synchronism with particle motion through the accelerator. Linear induction accelerators are particularly useful for producing short (10–50-nsec) pulses of very high peak intensity (1000 A) at energies of 10–50 MeV.

b. The Electron Linac. In this kind of linac, the wave velocity is made constant at just less than the velocity of light in free space over almost the entire length. An efficient conductor arrangement that supports the needed longitudinal electromagnetic accelerating wave is shown in Fig. 12. This waveguide is a cylindrical pipe periodically loaded with diaphragms spaced between one-fourth and one-half of the free-space wavelength of the driving harmonic field. The wave velocity is controlled by the diameter of the pipe, about equal to the wavelength, while the rate at which power flows down the waveguide is controlled by the size of the hole in the diaphragm. The operating wavelength of such linacs is set by the simultaneous need for efficient acceleration and for efficient generation of the microwave power carried by the accelerat-

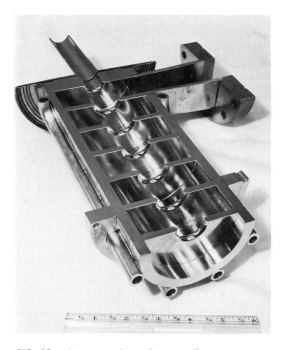

FIG. 12. Cutaway view of a traveling-wave waveguide. The radiofrequency wavelength used is approximately 10 cm. The beam enters from the upper left and passes through the small hole. Radiofrequency power is brought in at the upper right, and the wave travels with the beam. (Photo courtesy of SLAC.)

ing wave. Operating wave lengths between 30 and 3 cm have been used, with 10 cm being the most common today.

Rather tight tolerances must be maintained in construction of the waveguides. At 10-cm operating wavelength, tolerances of about 0.02 mm must be held. At shorter wavelengths, the tolerances are correspondingly smaller. Maintenance of the correct wave velocity also requires regulation of the waveguide temperature to a fraction of a degree.

With proper synchronization by a master oscillator, a large number of units such as those of Fig. 12 can be strung together end to end to produce as high a beam energy as needed. Copper accelerating waveguides available today can maintain effective accelerating fields of 15–20 MV per meter of length, for a power expenditure of 3–5.4 MW per meter of accelerating waveguide. Thus, the output energy of the 3000-m Stanford Linear Accelerator will be about 50 GeV with each of its 240 power amplifiers pulsing at 50 MW.

While magnetron tubes are sometimes used as power sources for linacs of a few million electron volt output energy, klystron amplifiers are the usual choice at microwave frequencies. Today, accelerator klystrons capable of 50 MW peak power with pulse lengths of a few microseconds are being made. Tubes capable of up to 1 GW for a fraction of a microsecond are being contemplated.

As with proton or heavy-ion linacs, it is possible to replace the normal copper conductor with a superconductor, such as niobium or lead, operating at liquid helium temperatures (2–4.5 K). The amount of microwave power needed to establish the accelerating fields is then reduced by a factor of 10^5–10^6. This benefit is somewhat offset by the need to provide powerful refrigerators to maintain the linac at close to absolute zero temperature. The technology has proven difficult but is beginning to come into limited use for accelerators requiring continuous operation, as contrasted with the short-pulse operation mode of conventional linacs.

2. Circular Accelerators

Like linear accelerators, circular accelerators utilize time-varying fields to accelerate particles. In addition, there must be magnetic fields to bend particles around a closed path that returns them to the accelerating structure. The configuration of the accelerating structures and magnetic fields can take many different forms in different circular accelerators.

a. The Cyclotron. The earliest circular accelerator was the cyclotron, invented by Lawrence and first operated by Lawrence and Livingston in 1932. In the cyclotron, particles are injected at the center of the cyclotron and spiral outward as they are accelerated. A uniform, time-independent magnetic field provides the bending to make these spirals. The drift tubes in a cyclotron are in the form of hollow boxes called "dees," after their shape. Acceleration takes place as a particle crosses the gap between dees. Figure 13 is a sketch of cyclotron orbits through the dees.

A particle of higher energy moves faster along the orbit, but has a longer distance to go between gap crossings. The frequency of gap crossings, and therefore accelerations, is constant in the cyclotron, and it produces a steady stream of bunches of accelerated particles. Quantitatively, from Eq. (2) $mv = e\rho B$, but $v = \omega\rho$, so that $m\omega = eB$, and ω is independent of the radius ρ and the energy.

According to the laws of motion when a particle's velocity approaches the velocity of light, i.e., special relativity, the mass of a particle increases as its energy is increased. As a consequence, for the same increase in energy, the velocity of a particle of higher energy does not increase as much. The gap-crossing frequency therefore decreases, and particles fall out of step. This effect makes a noticeable difference at 15 MeV for protons, which limits the peak energy of the Lawrence cyclotron. Several systems to circumvent this problem have been in-

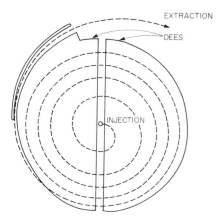

FIG. 13.   Schematic diagram of cyclotron orbits.

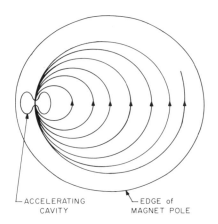

FIG. 14.   Schematic diagram of microtron orbits.

b. The Microtron. The relativistic effects that limit cyclotrons set in at much lower energies for electrons, and a different configuration, the microtron, is appropriate. In the simplest microtron, electron orbits are a series of circles tangent at the position of the accelerating cavity. The periods of revolution on each of these circles differ by an integral number of periods of the accelerating voltage and the electron bunches therefore stay in phase with the accelerating voltage. These simple orbits are sketched in Fig. 14. Microtrons can also be built in the shape of racetracks and have been used to accelerate electrons to energies of more than 100 MeV.

c. The Synchrotron. Energies greater than 1 GeV require a different configuration, the synchrotron. The magnetic guide fields in all the circular accelerators discussed previously are constant in time. In the synchrotron, the magnetic field is increased in time as the particle energy is increased by a radio-frequency accelerating voltage. A graph of a synchrotron cycle is shown in Fig. 15. The radius of the particle orbit is held constant. Synchrotron magnetic fields need to extend over a relatively small aperture rather than over the entire circle, as in constant-field configurations. The synchrotron is therefore a far more economical design for energies in the giga electron volt range. The largest synchrotron, four miles (6 km) in circumference, accelerates protons to an energy close to 1 TeV (Fig. 3).

Electron synchrotrons, although they share the principle and magnetic-field configuration

vented. One of these is the synchrocyclotron, in which the frequency of the accelerating voltage is reduced to keep in step with a group of bunches of particles as they are accelerated. The synchrocyclotron produces bursts of a series of bunches of accelerated particles. Synchrocyclotrons have been built to produce protons of 750-MeV energy. They have been largely superseded by azimuthally varying field (AVF) cyclotrons, which are also called sector-focused cyclotrons. Here the magnetic field varies periodically around the azimuth of the cyclotron in such a way as to keep the frequency of gap crossings constant. Like cyclotrons, AVF cyclotrons produce a steady stream of bunches. The intensity possible is much larger than in a synchrocyclotron, and there are many being used for science research.

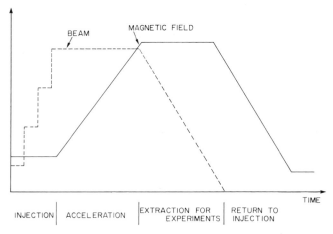

FIG. 15.   Operation cycle of a synchrotron.

with proton synchrotrons, have a separate feature. Charged particles radiate electromagnetic waves when they are accelerated, as they are in following a curved path. In accelerators, this is called synchrotron radiation. For light particles like electrons, the energy radiated away is large enough that it must be replaced if the electrons are not to spiral inward and strike the walls. The energy loss per revolution ΔT by an electron of kinetic energy T following a circle of radius ρ is

$$\Delta T = 88.5 T^4/\rho$$

where ΔT is in keV if T is in GeV and ρ is in meters. Thus, the energy lost in the form of synchrotron radiation increases very rapidly as T is increased. The accelerating system must make up this energy loss, as well as provide voltage for acceleration. Radiofrequency systems to carry out these functions become so large that electron synchrotrons become uneconomical compared with linear accelerators at very high energies.

d. The Storage Ring. Storage rings are very similar in general configuration to synchrotrons. The magnetic guide fields are constant in time,

and a beam of particles circulates continuously. In some designs, two storage rings are intertwined with one another and beams of particles circulate in opposite directions, colliding at the intersection points. The intersection point can be clearly seen in the photograph of the CERN Intersecting Storage Rings in Fig. 16. In other designs, beams of particles and antiparticles (electrons and positrons or protons and antiprotons) circulate in opposite directions on the same path in the same magnetic field, often with small auxiliary fields to keep them from colliding except at the designated collision points. Like electron synchrotrons, electron storage rings are limited in energy by synchrotron radiation. Present-day technology utilizing superconducting microwave cavities could be used in a storage ring of a few hundred GeV.

e. The Betatron. As with linear accelerators, it is possible to accelerate particles by induction in a circular configuration. Circular induction accelerators are called betatrons. They are used to accelerate electrons; many betatrons are used to provide 20–30-MeV electrons for medical or industrial work. The largest betatron produced 300-MeV electrons.

FIG. 16.   The CERN Intersecting Storage Rings. The large blocks are gradient magnets. The intersection of the two vacuum chambers can be clearly seen. (Photo courtesy of CERN.)

III. Particle Motion

The motion of particles in an accelerator is one of the most important aspects of accelerator science. The importance of particle dynamics arises in part from the fact that in many kinds of accelerators particles travel very large distances in the course of acceleration or storage (sometimes many millions of miles or kilometers). Stability against small perturbations and errors is vital if the particles are to stay in the accelerator for such distances.

In addition, in most accelerators it is desirable to accelerate as high an intensity of particles as possible. At high intensity, the mutual electromagnetic forces among the charged particles, space charge, can be important, thus disturbing the stability that each particle would have by itself. To produce a stable beam in an accelerator, particles will execute only small oscillations about the ideal path during acceleration. These oscillations will occur in both of the two directions transverse to the ideal path and in the longitudinal direction along that path. Although there can be situations in which the transverse and longitudinal motions are coupled, in most accelerator configurations the coupling is very small, and the two kinds of motion can be discussed separately, as we shall do.

A. TRANSVERSE MOTION

1. Betatron Oscillations

Transverse oscillations are called betatron oscillations because Kerst and Serber gave the first clear discussion of them in connection with Kerst's development of the first successful betatron. The equations of motion of transverse oscillations had, in fact, already been given by Walton and in a form more useful for cyclotrons by Thomas.

The overall objective is to make the transverse motion dynamically stable, so that particles injected in the vicinity of the ideal orbit will remain in that vicinity. In some short accelerating systems, this can be achieved by focusing the beam at the particle source, but in longer accelerators, restoring forces along the orbit are required in order to achieve this stability. These restoring forces are supplied by external electric and magnetic fields.

We can neglect the effects of the acceleration process for the discussion of transverse motion. We also neglect here the space–charge forces between particles and thus discuss the motion of a single particle in an external field that is constant in time. This field is not necessarily constant in space and these variations with distance are important to focusing.

2. Circular Accelerators

Let us consider the case of the transverse magnetic field in a circular accelerator, the field that bends the particle around a closed path. Here r is the radial direction in the plane of the closed-orbit path, the median plane, and z is the dimension perpendicular to the orbit plane. The distance along the orbit is s. If the vertical magnetic field B_z varies as a function of r, that is, if it has a gradient, then there is a radial field B_r at positions off the median plane, as follows from the Maxwell equation $\nabla \times \mathbf{B} = 0$. A particle moving in the s direction experiences a vertical force whenever it is away from the median plane. If the vertical field decreases with radius ($dB_z/dr < 0$), the force deflects the particle back toward $z = 0$ for z either positive or negative. This force is depicted in Fig. 17. On the other hand, the force of the particle is always in the direction away from the median plane if $dB_z/dr > 0$.

Thus, if the guide field decreases with radius, motion off the median plane is stable in the sense that a particle starting off the median plane will not move to ever-larger z. This vertical focusing was found experimentally in the earliest cyclotrons and understood qualitatively at that time. It was made quantitive by Kerst.

In a decreasing field, the radial force on a particle decreases with radius. However, the centripetal force mv^2/r needed to keep the particle of mass m and speed v in a circle radius r decreases as $1/r$. Thus, if the field decreases less rapidly than $1/r$, a particle at larger radius feels a larger force focusing it back toward the ideal

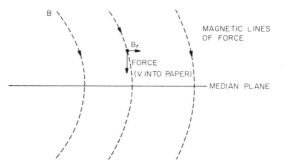

FIG. 17. Radial field and resulting vertical force in a gradient magnet.

orbit and the particle has horizontal or radial focusing. Kerst expressed these results in terms of the relative derivative or field index

$$n = -\frac{r}{B}\frac{\partial B_z}{\partial r} \tag{6}$$

and the condition for focusing in both transverse directions is

$$0 < n < 1 \tag{7}$$

In this weak focusing, horizontal and vertical focusing are complementary to one another in the sense that each decreases as the other increases, and a balance must be struck in design between vertical and horizontal aperture. For example, the fields in cyclotrons decrease very little with radius, in order to keep as close as possible to isochronous motion, and the vertical focusing is very weak.

Many synchrotrons were built with weak focusing and operated well. But the amplitudes of oscillations about the ideal orbit are larger in a larger accelerator (approximately proportional to the radius), reaching 10–20 cm in the Cosmotron (3 GeV) and Bevatron (6 GeV) proton accelerators. It would be extremely costly to make use of weak focusing in a much higher-energy accelerator, when oscillation amplitudes could be as large as several meters.

Strong focusing overcomes this difficulty by using an alternating series of gradients, thus alternating-gradient focusing, to focus both horizontally and vertically. An alternating series of gradients focuses a particle in a manner similar to an alternating series of optical lenses, as we can see in Fig. 18. A ray is farther from the axis in the converging (focusing) lenses than in the diverging (defocusing) lenses and so is bent more sharply, so that the net result is focusing.

A gradient that is focusing for horizontal motion is defocusing for vertical motion, but the alternation produces focusing in both. The complementarity that limits weak focusing is avoided, and the focusing can be much stronger.

The gradients vary periodically around the circumference of the accelerator, with a fixed number of periods per revolution. Oscillation amplitudes in a large synchrotron are a few centimeters or less and the vacuum chambers and magnet apertures are correspondingly small.

A particle that is not injected on the ideal orbit will execute betatron oscillations about that orbit. These oscillations are usually characterized by ν, the number of complete oscillations per revolution around the accelerator. There are separate values for horizontal (ν_r) and vertical (ν_z) oscillations. In a weak-focusing accelerator, the oscillations are sinusoidal, and both ν values are less than unity. In a strong-focusing accelerator, these oscillations are sinusoidal on the average, with periodic excursions around the average sine wave, shown in Fig. 19, and ν_r and ν_z are usually considerably larger than unity. Thus, in a weak-focusing accelerator, the oscillations have the form

$$r = r_0 + A_r \cos(\nu_r s/R + \theta_r)$$
$$z = A_z \cos(\nu_z s/R + \theta_s) \tag{8}$$

where the amplitudes A_r and A_z and the phases θ_r and θ_z are determined by the initial conditions at injection.

In a strong-focusing accelerator, the oscillations have the form

$$r = r_0 + A_r \sqrt{\beta_r(s)} \cos[\phi_r(s) + \theta_r]$$
$$z = A_z \sqrt{\beta_z} \cos[\phi_z(s) + \theta_s] \tag{9}$$

$$\phi_r(s) = \int_0^s \frac{ds}{\beta_r(s)} \tag{10}$$

and

$$\phi_z(s) = \int_0^s \frac{ds}{\beta_z(s)}$$

The periodic functions β_r and β_s are the betatron amplitude functions. By Eq. (9), the amplitude

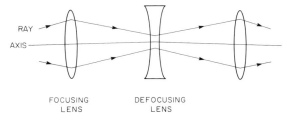

RAY

AXIS

FOCUSING
LENS

DEFOCUSING
LENS

FIG. 18.   Focusing of rays in a series of lenses.

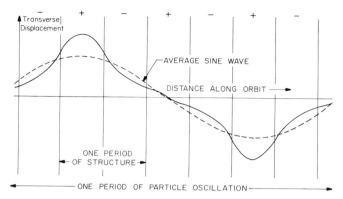

FIG. 19.   Particle oscillation in an alternating-gradient structure: +, focusing; −, defocusing.

varies periodically with $\sqrt{\beta}$. By Eq. (10), the phase advances as $1/\beta$, so that β is the instantaneous wavelength of the oscillation. The amplitude function and the ν value are related because the total phase advance per revolution is

$$\phi(2\pi R) = 2\pi\nu = \int_0^{2\pi R} \frac{ds}{\beta(s)} \qquad (11)$$

for either r or z.

A group of particles is injected with angles and positions distributed around the ideal orbit. It is instructive to plot the motion of the group of particles in a space whose axes are position and angle at a given point s, as in Fig. 20. This is called a phase space. As the group moves around the accelerator and s varies, the envelope containing the group will vary in shape, but its area will remain constant. If instead of the angle the product of angle and total momentum is taken as a phase–space coordinate, the phase–space area remains constant during ac-

celeration. This is an example of a general dynamical rule called Liouville's Theorem. We shall return to this theorem in the discussions of beam stacking and cooling below.

The horizontal and vertical motions are independent in an ideal accelerator and each has a separate two-dimensional phase space. In a real accelerator, nonlinear restoring forces, magnetic-field imperfections or magnet misalignments can introduce horizontal–vertical coupling, and the two motions can affect each other. The phase spaces are then not independent, but the four-dimensional volume of the combined phase spaces occupied by particles remains constant.

Magnetic-field errors and magnet misalignments can also distort the beam path by inducing forced oscillations in the beam. The central orbit then moves in a periodic forced oscillation, the closed orbit, and all particles oscillate about this closed orbit. The occupied region of phase space then moves in this forced oscillation. If the ν value is close to an integer, the closed-orbit oscillation becomes very large, and the beam can rapidly leave the accelerator. This integral resonance can be kept under control by careful construction and alignment of the magnets and by careful control of the ν value to avoid integers.

In a strong-focusing accelerator, errors in magnetic-field gradients can make the oscillations about the closed orbit unstable if the ν value is a half-integer. In this case, the occupied region of phase space becomes elongated, even though area is still preserved, and particles reach very large oscillation amplitudes. Half-integral resonances are not as serious as integral resonances, but care must be taken in construction and alignment to avoid them.

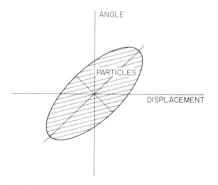

FIG. 20.   Transverse phase space. The shaded elliptical area is a group of particles. This ellipse will oscillate as the group goes around the accelerator.

Even in an ideal accelerator, not all particles will have exactly the same momentum. Each particle's momentum will also vary relative to the ideal momentum during longitudinal oscillations. A particle whose momentum is different from the central momentum will undergo forced transverse oscillations about the closed orbit. These will appear in phase space as overlapping groups of particles. This phenomenon is called dispersion in analogy to optics.

To be focused toward one another, two particles on different orbits must encounter different magnetic fields, as in a gradient magnet or quadrupole, or go through different lengths of field. Different lengths can be achieved by building magnets whose edges are not perpendicular to particle orbit as shown in Fig. 21. This edge focusing is used in AVF cyclotrons. If an edge is slanted so that the path length increases with radius, the edge is horizontally defocusing and vertically focusing.

In a radial sector AVF cyclotron, both upstream and downstream magnet edges are vertically focusing. Radial focusing is provided by the increase of guide field with radius. This system is called Thomas focusing. In a spiral-sector AVF cyclotron, one edge is vertically focusing and the other is vertically defocusing, giving alternating-edge focusing, analogous to alternating-gradient focusing.

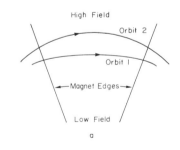

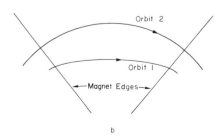

FIG. 21.   (a) Gradient and (b) edge focusing.

3. Linear Accelerators

The paths of particles in high-voltage accelerators are short enough that adequate focusing for the entire accelerator can be achieved at injection. Small radiofrequency electron linear accelerators also need no external focusing. Longer electron linear accelerator structures are interrupted periodically for focusing magnets.

Proton linear accelerator beams need focusing, partly because the transverse and longitudinal motions are more closely coupled than those in synchrotrons and energy from the longitudinal motion can drive transverse oscillations. There is no centripetal force in a linear accelerator, so there is no analog of weak focusing. Before strong focusing was developed, many proton linear accelerators had wire grids installed in the drift-tube bore opening to change the variation of the electric field with longitudinal distance and radius to make it focus particles rather than defocus them. But the grids intercepted many beam particles and were unsuitable for high-intensity beams because they were heated and burned by beam.

After strong focusing was developed, quadrupole focusing magnets were built into the drift-tube interiors, and linear accelerators became high-intensity accelerators. More recently, methods of shaping the radiofrequency field to produce quadrupole focusing (RFQ) have been developed and provide even higher intensities and smaller beam losses.

Beams extracted from an accelerator and secondary beams produced in a target can also be steered and focused by sequences of bending magnets and strong-focusing lenses. These beam lines are used to bring beams to the point of use in an optimally focused configuration.

4. Synchrotron Lattices

The lattice of a synchrotron is the periodic arrangement of bending and focusing magnets around the circumference. The focusing, amplitude functions, and dispersion all depend on the lattice. Two developments make it possible to vary these functions and to achieve optimal desired orbit properties.

a. Separated-Function Magnets. Particles are bent around the accelerator by dipole fields that are independent of radius. They are focused to stay close to the central orbit by quadrupole fields that vary linearly with distance from the center. In the original conception of strong focusing, these two functions of bending and focusing were combined in one gradient magnet, in a combined-function lattice. A separated-function lattice carries out these two functions in separate magnets. This lattice is more efficient because the bending field is the same throughout the magnet and is not limited by the maximum field attainable at the high-field side of the magnet. The bending field in a separated-function conventional iron dipole can easily be 1.8–2 T, while in a gradient magnet it is difficult to achieve more than 1.3 T without significant field

distortion. The difference is even more striking in superconducting magnets, where it is more difficult to design a gradient magnet. The focusing is also more efficient in a separated-function lattice, because focusing magnets are concentrated at locations where the amplitude function β is large and defocusing magnets are concentrated at locations where β is small. Combined-function and separated-function magnets are shown in cross section in Fig. 22. Separated-function lattices are now almost always used in synchrotrons and storage rings.

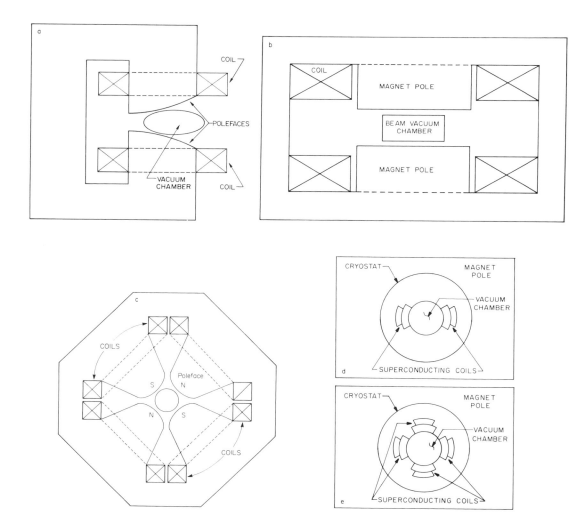

FIG. 22. Alternating gradient synchrotron magnet cross sections. (a) Gradient (combined-function) magnet, (b) dipole (bending) magnet, (c) quadrupole (focusing) magnet, (d) superconducting dipole magnet, and (e) superconducting quadrupole magnet.

b. Long Straight Sections. The usual lattice has straight sections, field-free spaces between magnets whose maximum length is of the order of the magnet length. For the introduction of necessary auxiliary apparatus, such as accelerating cavities, injection equipment and particle detectors for experiments in storage rings, the long straight section has proven most useful. If the normal bending arcs of the accelerator are simply interrupted by field-free regions of size necessary to emplace necessary auxiliary equipment, the natural divergence of the beam will result in excessive aperture requirements. This divergence can be avoided by use of a few separate focusing lenses (quadrupoles) to maintain the focusing properties of the lattice without significantly encumbering the needed space. Periodic arrays, or sub lattices, of concentrated lenses permit design of straight, almost field-free regions of arbitrary length.

5. Experimental Measurement of Transverse Motion

It is possible to "fly blind" and operate a particle accelerator without measurements of the beam. Indeed, early accelerators operated this way, with the only indication of beam coming from the final accelerated beam striking a target and producing X-rays or radioactivity. However, an accelerator can be operated much more easily and at much higher intensity if the beam position and size are known during acceleration or storage.

The first methods of beam measurement were movable probes to stop the beam at an adjustable radius in a cyclotron. In early synchrotrons, probes were replaced by fluorescent screens which were observed through transparent windows. These rudimentary devices are still sometimes used in the early stages of searching for circulating beams, although nowadays the energy is high enough that fluorescent screens are viewed remotely using television cameras.

The center of mass of a circulating beam can be measured continuously by detecting the electric fields of the beam bunch with pickup electrodes or the magnetic fields with pickup coils, in each case surrounding the beam. By using these methods, it is possible to measure the transverse position at given locations and the phase of the beam bunches relative to the accelerating voltage. The transverse position as a function of azimuth is just the closed orbit discussed above. The beam is seen as a tube of charge centered on the closed orbit. The closed-orbit information can be used to set currents in correction magnets to reduce its distortion and to analyze magnet misalignments. The closed-orbit and beam-phase information are used together as input to the rf feedback systems discussed in Section III,B. A measured closed orbit is shown in Fig. 23a.

Another important practical aspect of particle bending and focusing is scattering by the residual gas in the accelerator. Neglecting special space–charge effects, single charged particles or low-intensity beams suffer significant scattering and diffusion in even a few feet of air at normal pressure. Accordingly, the beam must pass through an evacuated space. Usually a vacuum-tight metal or ceramic tube surrounds the beam and is evacuated by pumps. In most accelerators, a residual pressure of 10^{-8} atm is acceptable. In storage rings, where the effective path length may be several billion kilometers, a substantially better vacuum (10^{-11}–10^{-12} atm) is required.

It is also possible to measure the beam shape making use of ionization of a dilute gas by the beam or by analysis of the frequency spectrum of the beam electric field, with knowledge of the dispersion function at the location of the pickup. In places where each beam particle passes only once, grids of wires can be used in connection with ionization or a single wire can be moved rapidly through the beam. Two measurements arriving at the acceleration system at different locations can be combined to give the distribution in phase space. An example of such a measurement is shown in Fig. 23b.

B. Longitudinal Motion

1. Introduction

In dc high-voltage accelerators or in induction accelerators (betatrons), particles that are accelerated at different times experience the same accelerating field. But in accelerators that utilize radiofrequency fields to accelerate particles (linear accelerators, microtrons, cyclotrons, and synchrotrons), particles at different times will experience different accelerating fields and will consequently have different motions.

2. Longitudinal Stability and Acceleration

How the longitudinal motion evolves will depend on how the frequency of revolution de-

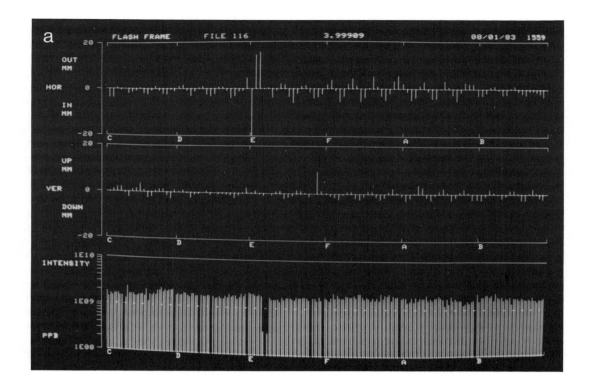

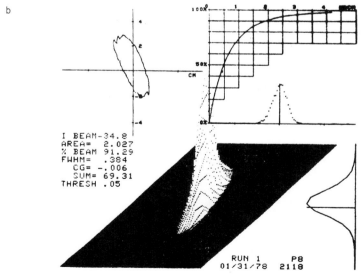

FIG. 23.   Measured orbit properties. (a) Closed-orbit displacement around a circular accelerator. (b) Emittances. The horizontal axis of the three-dimensional plot is displacement; the diagonal axis is angle. The vertical is beam intensity. At the lower right is the distribution in angle. At the upper left is a phase–space plot of the contour containing a stated fraction of the beam. At the upper right is a plot of the fraction of beam contained within the emittance of the horizontal axis. (Photos courtesy of Fermilab.)

pends on particle energy. Let us illustrate this by a simple physical example. In Fig. 24, the ordinate is the accelerating voltage and the abscissa is time. Thus, on this plot the accelerating voltage is a sine wave. Consider now two particles that cross the accelerating gap at the same time on one revolution. The voltage is rising as they cross. Particle 1 is in step with the accelerating voltage and crosses the gap the second time at the same phase. Particle 2 has higher energy. Let us consider two alternatives:

1. Revolution frequencies increases with energy ($df/dE > 0$). The revolution period is smaller for particle 2, and it will arrive earlier, lower on the sine wave and will gain less energy than particle 1. The energy difference between the two particles will be decreased after the second pass. Similarly, a particle of lower energy would arrive later, higher on the wave, and would gain more energy, again decreasing the energy difference. Thus, the energy difference across the entire group of particles will not increase and they will be accelerated together. The accelerating longitudinal motion is stable.

2. Revolution frequency decreases with energy ($df/dE < 0$). Now particle 2 arrives later and gains more energy than particle 1. It continues to gain more energy at every pass and the energy difference increases continuously. Thus, the longitudinal motion is unstable, and the group of particles will break up and not be usefully accelerated.

Luckily, there is a saving grace. There is another side to the voltage wave, where the voltage is still in the right direction to accelerate, but is falling. Now the arguments of less or more voltage are just reversed and the acceleration longitudinal motion is stable in an accelerator with $df/dE < 0$.

What affects df/dE? There are two factors: a particle with higher energy goes faster, which always increases its frequency; but in some accelerators, it goes on a different orbit between gap crossings, which in almost all cases decreases its frequency of revolution. These two factors are in opposite directions; how they balance depends on the kind of accelerator.

In linear accelerators, all particles travel the same path and there is no difference in path length, only difference in speed; thus df/dE is always positive. On the other hand, in weak-focusing synchrotrons, orbits corresponding to different energies are relatively widely separated because the guide-field variation with radius is small. Here the path-length difference always overbalances the speed difference and df/dE is always negative. The same is true in a microtron. In strong-focusing synchrotrons, the speed difference is larger than the path-length difference at low energy in proton accelerators, so $df/dE > 0$, but the path-length difference is constant and the speed difference decreases as particle speeds approach the speed of light, so that $df/dE < 0$ at high energy. There is thus a transition energy at which df/dE is zero. At this energy, the radiofrequency accelerating voltage must be turned off, then turned back on within a few milliseconds at a different phase relative to the beam particles. Acceleration then continues on the back side of the wave. This has not been difficult in practice in proton synchrotrons. It is not necessary at all in strong-focusing electron synchrotrons because electrons move at close to the speed of light at much lower energy and the transition energy in an electron synchrotron is therefore lower than the injection energy and is never crossed.

In this discussion, cyclotrons are an anomaly. In principle, speed differences and path-length

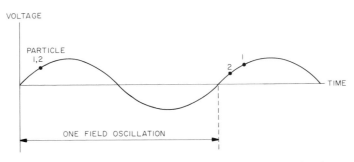

FIG. 24. Particles riding on a voltage wave during acceleration.

differences just balance in cyclotrons, and they are always exactly at transition energy. In practice, there are small effects beyond our present discussion that give enough marginal stability that particles are accelerated.

3. Phase Oscillations

The quantitative discussion of longitudinal stability can be made quantitative. The results of quantitative study can be described graphically in a plot like that of Fig. 25, giving angular momentum of the particle (almost the same thing as energy) against phase of a particle relative to the radiofrequency accelerating voltage. This phase can vary between 0 and 2π (360°). If the accelerating voltage is turned off and a group of particles is simply coasting around the ring, this group is represented by a band stretching horizontally across the entire range of 2π in phase and stretching vertically across a range in angular momentum (and energy) that corresponds to the energy spread of the group.

When there is an accelerating voltage, there is a region of closed curves representing stable oscillations. These curves surround an equilibrium phase, which appears on the plot as a point at the center. A particle that starts at this phase and angular momentum will remain at it as the whole plot rises vertically during acceleration. Particles that start within the stable region will move on a closed curve around the equilibrium phase, oscillating in momentum and phase. These oscillations are called phase oscillations or synchrotron oscillations. In almost all cases, the frequency of these oscillations is very much smaller than the frequency of revolution, so that many revolutions are needed to complete one circuit around the diagram. Particles of different energy have different orbits in a circular accelerator and there is a radial oscillation corresponding to the energy oscillation of phase oscillations.

The stable region is called a bucket. The edge of a bucket is called the separatrix. Particles starting beyond the separatrix will slip in phase relative to the accelerating voltage and will not be accelerated continuously.

During acceleration, the particles form a bunch within the bucket. Even if they fill the bucket to the separatrix, they do not occupy the full 2π in phase; a bucket that accelerates occupies less than 2π, because the voltage is decelerating for one-half the range in phase.

It is also possible to have a stationary bucket, where the equilibrium particle crosses the accelerating gap at the moment the voltage is zero. In a stationary bucket, particles can occupy the entire 2π in phase. It is also possible to accelerate a bunch in a stationary bucket by starting it near the bottom and continuing through one-half a phase oscillation, where it is near the top. This is how acceleration takes place in an electron linear accelerator.

In circular accelerators, the accelerating frequency is an integral multiple of the revolution frequency. Use of a higher frequency will make it possible to use smaller accelerating cavities and power amplifiers. It may also make it possible to use radio-frequency components developed for other applications. The multiple h is the harmonic number. There are now h buckets stretched across the range of 2π in phase. Each of these buckets has the properties discussed above.

The operation of a synchrotron accelerating system can be improved considerably by a beam feedback system. Pickup electrodes of the kind described in Section II,B are used to measure the phase and radius of the beam bunches. This information is fed back electronically to correct the phase and voltage of the accelerating system to keep the beam centered in the vacuum chamber and the beam and radio-frequency accelerating system in phase with each other. External signals can be used at particular times in the

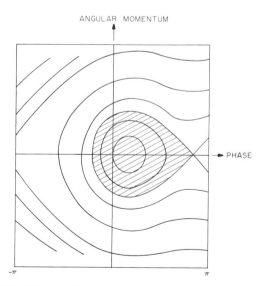

ANGULAR MOMENTUM

PHASE

$-\pi$

π

FIG. 25.   Accelerating bucket. Particles in the shaded area will move on closed curves as the bucket moves up in energy during acceleration. Particles outside the bucket will slip in phase and be lost.

cycle to move the beam either laterally or longitudinally for purposes of extraction, targeting, or stacking.

In this plot of angular momentum versus phase, the area occupied by a group of particles remains constant during acceleration. This combination is a longitudinal phase space, analogous to the phase spaces discussed in connection with transverse motion.

Even though phase-space area is preserved, in many cases the bunch will filament into many small threads that wind around the bucket. The empty areas between filaments are carried along with bunches in acceleration and the effective area of the bunch can increase, thus "mixing in empty space." It is possible to avoid this decrease of density by turning accelerating voltages on and off very slowly or by overfilling buckets at the start, so they stay full through acceleration, while some particles are thrown away.

4. Beam Stacking

A batch of accelerated particles can be left to circulate in a storage ring. Then another batch can be injected, accelerated, and put next to the first bunch in energy. This stacking process can be repeated many times. If care is taken to avoid mixing in empty space, the total phase-space area occupied is the sum of the areas of the individual batches. The particle density in physical space can be increased greatly by beam stacking and this makes colliding beams feasible.

Synchrotron radiation in electron or positron storage rings helps, because the emission of radiation damps motion in a properly designed ring and increases the density in phase space and the particle density in physical space, which increases the interaction rate.

C. Many-Particle Effects

1. Introduction

In almost every use of accelerators, higher intensity is desirable. However, higher intensities bring with them new phenomena arising from the electromagnetic forces between particles. These forces can affect the focusing of particles and can also introduce new kinds of instabilities.

2. Effects on Focusing

The electrostatic repulsion between particles in a beam decreases the restoring forces that focus the beam. This decreases the transverse oscillation frequency of the focusing. As more particles are added, the frequency continues to decrease until it reaches a resonance. Then additional particles will be driven to large amplitude by the resonance and lost. The beam is space-charge limited.

Two charges moving on parallel paths repel each other electrostatically, but they also form two parallel currents, which are attracted to each other by magnetic forces. The magnetic forces reduce the electrostatic repulsion and increase the space-charge limit. The repulsive force is always greater, but the magnetic force increases as the particles are speeded up, so space-charge forces are very small at high energies. Both the electrostatic and magnetic forces are modified at high energy by the presence of the conducting vacuum chamber surrounding the beam walls.

It is possible to cancel space-charge forces by injecting charges of the opposite sign, electrons in a positive-ion beam or positive ions in an electron beam. These attract beam particles and the combination of this charge neutralization and magnetic attraction can completely cancel space-charge effects, making it possible for the beam to "pinch" down to a thread. But this pinched beam is like the free end of a spurting water hose—it kinks and lashes around like a snake, so that it is highly unstable and not useful for accelerating particles.

3. Instabilities

The kink instability is an example of the phenomena that arise from the electromagnetic fields of the beam. These fields can be reflected back from the metal vacuum chamber to react on the beam. They can drive the beam either longitudinally or transversely. These phenomena are called coherent instabilities because all the beam particles feel the same forces and move coherently together. Because it is coherent, the motion can be detected by a beam pickup and an opposing electromagnetic force applied to electrodes inside the chamber. The bad effects of coherent instabilities can be overcome to some extent by such feedback systems. In some cases, however, the frequencies involved are too high to make feedback practical, and the instabilities must be controlled by design changes.

D. Beam Cooling

We have emphasized in our discussion the constancy of phase-space area, both longitudinal and transverse. There are methods to reduce the phase-space area and thus to increase the density of particle beams, which is advantageous for colliding beams. These methods are called beam cooling.

In all kinds of beam cooling, there is an interaction with another physical system. The phase-space area of the total physical system is constant, but the area of a smaller part can be decreased by transferring phase-space area to the rest of the system.

1. Synchrotron Radiation

All charged particles radiate electromagnetic energy when accelerated. The only effects usually significant for particle accelerators occur for the lightest charged particles, electrons and positrons, although at the multiterra electron volt energies now being discussed, synchrotron radiation is also important for protons. As they are bent around the curved path by centripetal acceleration in a circular accelerator or storage ring, electrons and positrons emit a narrow cone of radiation, tangent to their instantaneous orbits. The radiated energy is mostly in the ultraviolet and X-ray regions. The energy radiated increases rapidly as the particle energy is increased. Most of the acceleration voltage in a multigiga electron volt electron synchrotron is needed to make up the energy radiated away.

This synchrotron radiation decreases both longitudinal and transverse particle oscillations about the equilibrium orbit. Longitudinal or energy oscillations are reduced because particles with energy exceeding the equilibrium value will radiate at a greater rate than equilibrium particles and thereby be damped toward the equilibrium energy. Particles with energy less than the equilibrium value radiate at a smaller rate than equilibrium particles and are restored toward the equilibrium energy by the accelerating system. Transverse oscillations are damped because the synchrotron radiation is emitted along the direction of motion, reducing both longitudinal and transverse momentum components. The accelerating system restores only the longitudinal momentum component. Neither longitudinal nor transverse oscillations are reduced all the way to zero because the sudden random emission of the photons that make up the synchrotron radiation itself excites small longitudinal and transverse oscillations. The equilibrium beam size, typically about a millimeter, results from the balance between the average damping effect and this stochastic quantum excitation.

In synchrotron radiation, the phase-space area of the beam is decreased, while the phase-space area of the electomagnetic field is increased by the addition of the radiated photons.

2. Electron Cooling

A gas of protons or heavy ions and a gas of electrons will interact with one another by Rutherford scattering. If the proton gas has more thermal energy, it will give this energy to the electrons through the scattering, thus decreasing the phase-space area of the proton gas while increasing that of the electron gas.

This exchange will also occur in a frame of reference in which both gases are moving together in the same direction, that is, if they are proton and electron beams. This electron cooling can reduce both the longitudinal and transverse energy spreads of the beam. Electron cooling is done in practice by arranging an electron beam to move at the same speed as the proton beam through a straight section of a ring. At the same speed, the electron beam has much less momentum and is easily bent in and out of the proton beam at the ends of the straight section. The proton beam is repeatedly cooled by multiple traversals of the straight section as it circulates.

Electron cooling is more effective at lower energy, although higher-energy cooling has been discussed. It can be made much more effective by the addition of an external longitudinal magnetic field.

3. Stochastic Cooling

Decrease of the amplitude of a coherent oscillation by means of electronic feedback was discussed in Section III,B. The root mean square amplitude of a beam can also be reduced by a feedback system. This is stochastic cooling. The basic plan of stochastic cooling is a beam pickup that measures the average position of the beam, an amplifier system, and a kicker that transmits the amplified signal to the beam.

Consider a ring full of N circulating particles and break the circulating beam up into N parcels, so that each parcel around the ring contains only one particle. If the electronic system has enough frequency bandwidth to respond to this small a parcel, it can give a signal to the kicker to correct each individual particle. Thus, particles give up their phase-space area to the electronic system.

If, as in practice, the electronic system has smaller bandwidth, each parcel has more than one particle. Each of the other particles in a parcel gives electronic noise interfering with a given particle's correction signal and the system corrects more slowly. Eventually, as more particles are added, the noise masks the signal completely and cooling ceases.

Stochastic cooling has been used as the basis for a spectacularly successful effort to cool antiproton beams and to use them in a colliding-beam system with countercirculating particles. Important new results in physics have been achieved with this system.

IV. Accelerators of the Future

As discussed in the historical section of the introduction, remarkable improvements in accelerator capabilities have been achieved. In applications to science, medicine, and industry, these improvements have increased maximum energy capabilities by a factor of more than 10 million in 50 years, increased beam current capacity by an even larger factor, increased the variety of atomic particles that can be accelerated, and lowered the cost of particle acceleration dramatically. These improvements have been brought about by a combination of means: new acceleration methods have been devised and technological improvements to existing methods have continually come about.

Developments in accelerators for basic scientific research have emphasized energy increase at lowered cost per unit beam energy. Improvements in accelerators for medicine have focused on increasing the variety of particles that can be accelerated, the precision of control over the beams and the compactness and cost effectiveness of the devices. Industrial applications have continuously broadened through reduction in accelerator costs, increases in beam current and variety of accelerated particle and through miniaturation and increased portability.

Even though accelerators have become very sophisticated, the rate of improvement has remained steady. While this progress has occurred across the whole range of accelerator applications, the most dramatic developments have usually occurred in accelerators to be used for basic science. It is the expected developments in this area that are emphasized here. We may classify these developments as improvements to existing accelerator types and as new and improved accelerator methods.

A. Improvements to Existing Research Accelerator Types

Developments of existing methods come through deepening understanding of the physics of the method and through application of new and improved materials and techniques both in the design and in the manufacture of the accelerator. We can foresee significant improvements to both circular accelerators and microwave linacs.

1. Circular Accelerators

Basic particle physics research with colliding proton beams up to 100 TeV or more per beam can probably be carried out with circular accelerators based on the synchrotron principle. Continuing improvement in our understanding of nonlinear particle dynamics and of collective effects in dense beams will permit use of smaller and therefore more economical magnet and beam-channel cross sections while accelerating denser beams. Superconducting materials that are available now could be used to construct a 20-TeV proton storage ring of 90-km circumference, having a peak magnetic field of 6 T. Materials now under study in the laboratory may make possible accelerator magnets that operate at up to 15 T. With such magnets, an accelerator of 100 TeV with 190-km circumference would be possible. Expected improvements in cryogenic technology and electronic controls will make it possible to operate such an accelerator with a few people and at power demands about the same as those of present research accelerator complexes.

Circular accelerators producing synchrotron radiation for basic physical, biological, and chemical research will also be markedly improved through basic understanding of the physics of these accelerators and of the mechanisms of coherent radiation production. The next generation of such machines is likely to be about 1 km in circumference and be outfitted with special devices called wigglers and undulators for enhancing the emitted radiation for particular research purposes. The wavelength of the radiation will be adjustable and concentrated in a narrow band and its intensity will be increased, permitting a broader range of use.

B. Microwave Linacs

Because of the intense synchrotron radiation emitted by electrons confined in circular accelerators, basic particle physics research with electrons at energies much above 100 GeV will probably have to be carried out with colliding beams produced by linacs. We can foresee the possibility of extending current microwave linac technology up to perhaps 1 Tev per beam. Such a collider might employ two 10-km linacs. Required for economic viability will be improved understanding of beam collective effects, permitting acceleration of denser beams, as well as accelerating waveguides capable of better energy transfer to the beam while supporting effective accelerating fields of about 100 MV per meter. Such gradients have been produced in the laboratory. An additional ingredient must be microwave power sources capable of several hundred megawatts peak pulse power with high efficiency. A microwave generator based on a relativistic electron beam formed by photoemission from a pulsed laser-irradiated photocathode, the lasertron, gives some initial promise of meeting these goals. Other generators under development may also turn out to be useful. Although less far along in development, the superconducting microwave linac may play a future role in research applications requiring long pulses or continuous beams. Such machines for low-energy nuclear physics research are now in operation and an intermediate-energy machine is under construction.

C. New and Improved Acceleration Methods

Although circular accelerators for special purposes are likely to be used for many years to come, the dissipative synchrotron radiation to which all charged particles following curved orbits are subject will limit the viability of any type of circular accelerator at some energy. We may thus expect a growing emphasis on linear accelerators. In all such devices the particles being accelerated gain their energy at the expense of an electromagnetic wave of some type. To be useful, the wave must travel at a velocity very close to that of the particle and have a significant electric field component in the desired direction of motion of the particle. Different existing or potential accelerator types are distinguished by how the accelerating electromagnetic wave is created and how its velocity is controlled. In principle, these necessary conditions for acceleration can be arranged in free space near far from any material body, in a space near a specially designed array of conductors or dielectories, or in some material medium.

In free space, only plane electromagnetic waves exist. In these waves electric and magnetic fields are transverse to the wave velocity. A single-frequency plane wave is incapable of continuous acceleration of a charged particle. If two waves of different frequencies are used, however, continuous acceleration can be achieved, one wave serves to give the particle a slightly sinuous orbit so that its velocity has a component parallel to the accelerating field of the second wave. This is the inverse free-electron laser. Because the orbit is slightly curved, it also is limited in its maximum energy capability by synchrotron radiation. Although no such accelerator has yet been operated, it is estimated that electron energies of up to 300 GeV might be achieved. As currently conceived, it would not be a useful accelerator for protons or heavier particles.

Various arrangements of conductors have been devised to support longitudinal electromagnetic waves at wave velocities close to that of light. The classical microwave linac is one such example. By operating such a device at shorter wavelengths, it is believed that higher accelerating fields might be supported. At wavelengths of 1 cm or less, a free-electron laser might be used as the source of the driving electromagnetic wave and is believed to have the potential for high conversion efficiency. The low-energy, high-current beam of the laser might be parallel and close to the low-current, high-energy beam being accelerated, thereby minimizing coupling losses. This idea has been called the two-beam accelerator. A free-electron laser of almost 100 MW peak power at 1 cm has been operated. The maximum accelerating field capability of all accelerators using conductor arrays to control the wave type and velocity is limited by damage to the conducting material. A possible way to avoid this fundamental limit is to use a different array for each pulse so that damage is not relevant. The possibility of forming a suitable periodic array of liquid droplets made conducting by creating a plasma on their surfaces has been suggested. To take advantage of the enormous peak power available from lasers as an electromagnetic wave source, the droplets would be of microscopic size. No such accelerator has been constructed and many practical details remain to be worked out. Accelerating fields of several hundred million volts per meter might be possible.

The accelerators discussed above are driven by harmonic power sources. A wide frequency band conductor-controlled accelerator, the wake field transformer accelerator, has been proposed. Similar in some respects to the two-beam accelerator, it employs as an energy source a high-current, low-energy beam consisting of a series of rings propagating along their axis of symmetry at essentially the velocity of light. The energy carried by these rings is deposited as a pulse of energy near the periphery of a conducting cylinder. This pulse of energy propagates toward the center of the cylinder being compressed thereby. The resulting high fields at the center are then used to accelerate the low-current, high-energy beam. Acceleration fields of as much as 200 MV/m are expected. Prototypes are under construction.

In a medium such as a gas, plasma, or charged-particle beam, electromagnetic waves propagate more slowly than in free space. One of these can be used to achieve continuous acceleration in a variety of ways. One of the most interesting is the beat-wave accelerator. In this device, two superposed laser beams of slightly different frequency travel coaxially or almost coaxially with and just ahead of the particle beam to be accelerated. If the difference of the two laser frequencies is just equal to the plasma frequency, the laser pulses resonantly organizes the plasma into dense clumps with ratified spaces between, like beads on a string along the path of the laser pulses. The electric fields resulting from these charge separations are calculated to be enormous, as much as several gigavolts per meter. The existence of beat waves has been established experimentally as has their ability to accelerate particles. Suitable lasers and plasma parameters needed to engender the enormous fields predicted have not yet been achieved.

Improvements in accelerator development will result from all of these studies. Which, if any, of these approaches will provide the front-line accelerators of the future remains to be seen.

BIBLIOGRAPHY

Lawson, J. D., and Tigner, M. (1984). The physics of particle accelerators. *Ann. Rev. Nuclear Part. Sci.,* **34,** 99.

Livingston, M. S., and Blewett, J. F. (1962). "Particle Accelerators." McGraw Hill, New York.

Scharf, W. (1986). "Particle Accelerators and Their Uses." Harwood Academic Publishers, New York.

"Physics of High Energy Particle Accelerators." AIP Conference Proceedings No. 87, (R. A. Carrigan, F. R. Huson, and M. Month, eds.) American Institute of Physics, New York, 1982.

"Physics of High Energy Particle Accelerators." AIP Conference Proceedings No. 105, (M. Month, ed.) American Institute of Physics, New York, 1983.

"Physics of High Energy Particle Accelerators." AIP Conference Proceedings No. 127, (M. Month, P. F. Dahl, and M. Dienes, eds.) American Institute of Physics, New York, 1985.

Wilson, R. R. (1980). *Sci. Amer.,* **242,** 26.

PARTICLE BEAMS—*SEE* LASER AND PARTICLE BEAMS

PASCAL AND PASCAL-SC

L. B. Rall *University of Wisconsin–Madison*

GLOSSARY

Compiler: Program which translates programs written in a given language into actual machine instructions, a process called compilation.

Implementation: Compiler and libraries used with a specific computer and operating system to translate and execute programs written in some programming language.

Program: Set of instructions to perform a given computation.

Program library: Collection of programs, usually task oriented.

Programming language: Notational system for writing programs.

Pascal is a high-level, structured computer programming language. High-level languages allow writing computer programs in a notation which is similar to the one customarily employed in descriptions of the problem to be solved. Structured programming languages have a high degree of logical organization which provides a framework for the development of a program from the statement of the problem to its solution. Pascal has a relatively simple structure, which makes it easy to learn and use, and enough features to make it useful for many applications. Pascal is described as a top-down language, which means that quantities and subroutines employed in the program are defined before they are used. This logical organization assists the understanding and modification of programs already written in Pascal, and makes it possible to produce simple and efficient Pascal compilers. Pascal-SC (Pascal for Scientific Computation) is an extension of Pascal with additional features specifically directed to the needs of computation in physical science and technology. These include standards for accuracy of arithmetic operations, the availability of directed rounding of results of operations, more general function and operator definitions, dynamic arrays, string manipulation, and the possibility of developing programs in modular form. [*See* PROGRAMMING LANGUAGES.]

I. Pascal

A. GENERAL DESCRIPTION

The Pascal programming language was developed by Niklaus Wirth of the Swiss Federal Institute of Technology and introduced about 1970. Pascal is named for the French mathematician and philosopher Blaise Pascal (1623–1662). Hence, Pascal is a proper noun, not an acronym. The basic characteristics of Pascal are a well-organized logical structure and a fairly simple set of fundamental rules and concepts. Pascal was deliberately designed to be easy to teach and learn, and also to enable simple and efficient translation of programs written in this language into actual machine instructions. At the same time, Pascal includes sufficient capabilities to solve problems of interest in a number of application areas.

The use of Pascal quickly became widespread. At the present time, Pascal is one of the most widely used programming languages for instruction and actual applications. It is available for most computer systems from small personal microcomputers to the largest central installations. In 1982, the International Organization for Standardization (ISO) issued a definition of Standard Pascal (ISO-7185-1982), to provide a firm, agreed basis for description and implementation of the language.

The logical structure of Pascal not only makes programs easy to write in this language but also simplifies the understanding and modification of programs already written in Pascal. This facilitates revision and correction of Pascal programs. This process of program maintenance is often of considerable importance, particularly in connection with large programs used for commercial or other applications.

The construction of a computer program written in Pascal follows the same logical design as the solution of most problems in mathematics, physical science, and technology. Starting with well-defined concepts and quantities, known relationships and transformations are applied to obtain the final conclusions or results giving the solution of the problem. The rules of the Pascal language enforce this orderly progression of program development. Experience with Pascal has shown it to be a convenient and efficient language to use to write useful computer programs.

B. Basic Structure of Pascal Programs

All Pascal programs are divided into three parts:

1. program heading,
2. definition and declaration part, and
3. statement part.

The program heading gives the name of the program and establishes its means for communication with the outside world for the input of data and output of results.

In the definition and declaration part, constants and data types to be used in the program are defined, and labels, variables, functions, and procedures are declared. This is done in the following order:

1. label declaration,
2. constant definition,
3. type definition,
4. variable declaration,
5. procedure and function declaration.

In general terms, constants, data types, and variables refer to quantities that will be processed by the program. Labels are used as statement numbers for control purposes. Functions and procedures are subprograms, or subroutines, which perform special computations or other chores, such as reading data or writing results. The use of subroutines allows the main task of the program to be broken up into subtasks in an efficient way and gives a large program a clearer and more compact structure. The declaration and definition part of a Pascal program performs much the same function as the glossary of a scientific paper. In this part, the notation to be used in the rest of the program is introduced. With only two minor exceptions, the general rule of Pascal programming is that an object must be defined or declared before it can be used. This provides a logical, consistent basis for programming computations.

Finally, the statement part of the program specifies exactly the actions to be performed in terms of quantities and subroutines which have already been declared or defined. These include a small number of predeclared or predefined quantities and subroutines which are considered to be part of the Pascal language.

Together, the definition and declaration part and the statement part comprise what is called the *block* of the Pascal program. The definition and declaration part can be empty for the case in which the statement part uses only predeclared and predefined objects.

C. Example of a Pascal Program

The general remarks in Section I,B will be illustrated by an actual Pascal program for solving the quadratic equation

$$ax^2 + bx + c = 0$$

with real coefficients a, b, and c. Although simple, this example, Program 1, shows a number of the important features of Pascal programming.

Even though the notation of this program is peculiar to Pascal, it is not difficult to see, even without the comments, that it automates the use of the quadratic formula

$$x = \frac{-b \pm \sqrt{b^2 - 4ac}}{2a}$$

along the same lines that a person would ordinarily follow. In the statement part of the program, the coefficients are requested and read, following which the discriminant $d = b^2 - 4ac$ is computed. Finally, d is examined to see if it is positive, zero, or negative, according to which the corresponding solutions of the quadratic equation are computed and written out.

D. Notation

In most areas of science and technology, much of the terminology used involves words based on Greek or Latin. Computer science,

PROGRAM 1.

```
program solvequad(input,output);    {Program Heading}
            {Definition and Declaration Part}
        var a,b,c : real;    {coefficients}
            d : real;    {discriminant}
            r1,r2 : real;    {solutions}
    begin    {Statement Part}
        writeln('Enter a,b,c:');
        read(a,b,c);    {enter coefficients}
        d := b*b − 4*a*c; {calculate discriminant}
        if d > 0 then
            begin    {roots are real and unequal}
                r1 := (−b − sqrt(d))/(2*a);    {smaller root}
                r2 := (−b + sqrt(d))/(2*a);    {larger root}
                writeln('x1 = ',r1);    {output x1}
                writeln('x2 = ',r2)    {output x2}
            end    {roots are real and unequal}
        else if d = 0 then
            begin    {roots are real and equal}
                r1 := −b/(2*a);    {root}
                writeln('x1 = x2 = ',r1)    {output root}
            end    {roots are real and equal}
        else begin    {roots are complex conjugate}
                r1 := −b/(2*a);    {real part}
                r2 := sqrt(abs(d))/(2*a);    {imaginary part}
                writeln('x1 = ',r1,' + ',r2,'i');    {output x1}
                writeln('x2 = ',r1,' − ',r2,'i');    {output x2}
            end    {roots are complex conjugate}
    end.    {program solvequad}
```

however, takes its terminology directly from English. Thus, care should be taken to distinguish between the technical and ordinary use of a word. In what follows, a word being used in its technical sense will be shown in **boldface** to avoid confusion or an awkward construction.

Pascal programs are written using the English alphabet, the Arabic numerals 0, 1, ..., 9, separators, and a small set of symbols. Separators are blank spaces, ends of lines, and the symbols { and } or alternatively (* and *), which are used to enclose comments. Comments are explanatory statements entered in the program text for clarity and are ignored during the process of compilation of the program into machine instructions.

The symbols used in a Pascal program are classified as word symbols or special symbols. The 35 word symbols are

and	array	begin
case	const	div
do	downto	else
end	file	for
function	goto	if

in	label	mod
nil	not	of
or	packed	procedure
program	record	repeat
set	then	to
type	until	var
while	with	

Each word symbol has a specific purpose and cannot be used in any other way in a program.

The 21 special symbols are:

$$+ \; - \; * \; / \; = \; < \; >$$
$$. \; , \; : \; ; \; (\;) \; [\;] \; \uparrow$$
$$<= \; >= \; <> \; := \; ..$$

The last five special symbols are formed from two of the previous symbols, but are regarded as distinct, individual symbols.

As in the case of the separators { and }, alternatives are provided for some of the special symbols which may not be available on certain keyboards, namely (. and .) for [and], and @ or ˆ for ↑, respectively. In addition to the special symbols, the apostrophe ' is used to enclose character strings (for example, 'Enter a,b,c:' in the example program above). In order to include an apostrophe itself in a character string, it is written as '', as in 'Newton''s method', for example.

In addition to the special symbols and word symbols provided by Pascal, programs will ordinarily contain names of variables and other entities, as well as numerical values. In the example program above, the coefficients of the quadratic equation were denoted by a, b, and c, and the constants 2 and 4 appeared in the statement part of the program. Letters, digits and perhaps other characters are used to construct these identifiers and constants, following certain simple rules. An identifier used to name a variable or function must begin with a letter, which can be followed by other letters and digits. Thus, the programmer is free to use notation in a Pascal program which follows the terminology customarily used in the problem area. In addition, Pascal provides a set of predefined identifiers for useful entities, such as the function sqrt used to evaluate the square root in the example program above.

E. Syntax and Semantics

Strict rules of formation, or syntax, must be followed in writing a Pascal program. This enables the compiler to determine during the process of mechanical translation if the program and its statements conform to the Pascal language. This analysis is known as parsing the program. For example, an entire Pascal program must consist of a program heading, followed by a semicolon, followed by a block, followed by a period. This particular syntax rule, or production, is symbolized in extended Backus–Naur form (EBNF) by

Program = ProgramHeading ";" Block ".".

The double quotes surrounding the semicolon and period indicate that these are terminal symbols, that is, symbols actually used in writing the program. The program heading and block have their own syntactic specifications, which can be unraveled in turn until finally we arrive at the terminal symbols with which the program is actually written. For example, the production for the formation of an identifier to name a Pascal object is

Identifier = Letter {Letter | Digit}.

This means that an identifier starts with a letter, followed by 0 or more occurrences of a letter or

a digit. The braces {...} denote 0 or more occurrences of their contents, while the vertical bar | denotes or. The end of the production is denoted by a period. This contains no terminal symbols, but the productions for letter and digit list all possible corresponding terminal symbols

Letter = "a" | "b" | "c" | ⋯ | "y" | "z".

Digit = "0" | "1" | "2" | ⋯ | "8" | "9".

Thus, according to these rules, e, ei, ei0, eiei0 are legal identifiers, but 0, 0i, 0i0 are not. The syntax rule for identifier does not specify any upper bound for the length of an identifier, which will be limited in each actual implementation. In addition, most implementations allow the use of uppercase as well as lowercase letters.

The ISO Pascal Standard 7185-1982 gives a complete set of syntax rules for Standard Pascal. These are essential in order to write a Pascal compiler, which is a program that translates programs written in Pascal into actual machine instructions. In the following, essential syntax rules will be cited and illustrated by examples.

The writing, or synthesis, of a Pascal program proceeds in the opposite direction. We start with the terminal symbols and use the syntax rules as a guide to the construction of a correctly formed program.

In addition to syntactic correctness, there is also the question of the meaning, or semantics, of the program. The situation is the same in a natural language, such as English, in which gramatically correct sentences can be meaningless or self-contradictory. An example of a semantic error in a Pascal program would be an attempt to assign the value of 1.5 to a variable which has been defined to have only integers as values. The semantics of Pascal is defined by verbal descriptions.

F. The Program Heading

The program heading of the example program given is

program solvequad(input,output)

The word **program** is a word symbol of Pascal which is used exclusively to designate the point at which a Pascal program starts. It is a signal to the Pascal compiler that translation is to begin at this point. From the standpoint of syntax, word symbols are terminal symbols, i.e., they can actually appear in programs. (Some must appear, for example, **program, begin,** and **end.**)

In this example, **solvequad** is the name given to the program by the programmer. Usually, names of programs are chosen to suggest the function of the program, but not necessarily. The name is arbitrary, except that it must follow the Pascal syntax rules for an identifier. Formally, the syntax of a program heading is

ProgramHeading = "program"
Identifier[ProgramParameterList].

Square brackets indicate 0 or 1 occurrences of their contents. Furthermore,

ProgramParameterList
   = "("IdentifierList")".
IdentifierList = Identifier {"," Identifier}.

Thus, the program identifier list consists of a sequence of identifiers separated by commas and enclosed in parentheses. These identifiers are the ones used internally in the program. The ones given in the example, **input** and **output,** are predefined in Pascal. The program will expect to read data in ordinary text form from **input,** and write results in the same format to **output.** When the program is executed, names of input/output devices or information files will be specified as the actual values of the parameters in the program parameter list. For example, if a cathode ray terminal (CRT) is used as an input/output device, then **input** will be referred to its keyboard, and **output** to the screen. The actual name of the terminal, for example, could be con:, in which case the program solvequad could be executed by entering an instruction of the form

execute solvequad con: con:

to indicate the use of the terminal for input and output. Actual names of devices, files, and formats for execution instructions are completely implementation dependent.

G. Label Declaration

Labels can be used to distinguish certain program statements so that control may be transferred to them from other parts of the program, thus disrupting the normal flow of execution of statements in sequential order. A label is a digit sequence, that is, a digit possibly followed by one or more digits. The length of a label is usually limited to a maximum of four digits. A label (say, 10) is declared as follows:

label 10;

Several labels, separated by commas, are declared by a single statement:

label 10,15,1000,1500;

A statement in the statement part of the program can be prefixed by a label followed by a colon; for example,

10: writeln('Enter a,b,c:');

A statement so labeled can serve as the target of a goto (unconditional jump) statement, such as

goto 10;

This causes the statement with label 10 to be executed next and thus can interrupt the normal sequential execution of statements. Properly formulated Pascal programs rarely require the use of labels and goto statements.

H. Constant Definition

It is convenient to be able to define certain constants at the beginning of the program in order to abbreviate notation in the following program code and to make it easy to change programs to suit other applications (for example, to process different amounts of data specified by the constant). Constant definitions are given in the form

const dim = 10;

For several constants, the form is

const dim = 10;
      maxreal = 9.99999999999E+99;
      E = 'Enter a,b,c:';
      minreal = $-$maxreal;

The alignment given is just for readability. The syntax rules are

ConstantDefinitionPart = ["const"
   ConstantDefinition ";" {ConstantDefinition
      ";"}].
ConstantDefinition = Identifier "="
   Constant.
Constant = [Sign](UnsignedNumber |
   ConstantIdentifier) | CharacterString.

The notation (A | B) means either A or B. Thus, a constant can be an integer, a real (i.e., a floating-point number), a character string (a sequence of characters enclosed by single quotes), or a previously defined constant identifier, perhaps preceded by a sign. For example, following the constant definitions, the statement

writeln(E);

would have the same effect as

writeln('Enter a,b,c:');

Constant identifiers can be used as convenient abbreviations for quantities or phrases which are used throughout the program and can also serve the purpose of making a program easy to modify. For example, if the dimensions of vectors and matrices processed by a given program are keyed to the constant dim, then adaptation of the program to handle vectors and matrices of different sizes can be accomplished simply by changing the value of dim in the constant definition part.

In Pascal, the following are standard, predefined constants:

true, false, maxint

The first two are the corresponding logical values of type Boolean, and the last denotes the largest integer available in the given implementation.

I. DATA STRUCTURES AND TYPE DEFINITION

One of the most powerful tools Pascal provides to the programmer is the ability to organize data into coherent, identifiable, and meaningful structures, called data types or simply types. This is done in the type definition part of the program, which has the syntax

TypeDefinitionPart = ["type"
TypeDefinition ";" {TypeDefinition ";"}].
TypeDefinition = Identifier "=" Type.
Type = SimpleType | StructuredType |
PointerType.

and so on. The meanings of simple, structured, and pointer types will be explained later.

For example, the type definition part of a program could consist of

type dimtype = 1..dim;
vector = array[dimtype]of real;

where it is assumed that the integer constant dim has been defined previously. The first definition organizes the subrange 1, 2, ..., dim of the integers into a specific structure called **dimtype,** while the second uses the elements of dimtype to index an array of real (floating-point) numbers called **vector.**

As a more elaborate example, information about the position, velocity, temperature, and pressure in a fluid at a given time can be organized in the following way:

type coordinate = (x,y,z);
fluid = record
position : array
[coordinate]of real;
velocity : array
[coordinate]of real;
temperature : real;
pressure : real;
time : real
end;

The first type, coordinate, is what is called an enumerated data type, and consists of the identifiers of three coordinates directions. Type **fluid** is a record type, in which identifiers and types are assigned to the various quantities of interest. If **gas** is declared to be a variable of type fluid, then gas.position[z] would be the z-coordinate of a point in the fluid at which the velocity in the y direction is gas.velocity[y], the temperature is gas.temperature, and the pressure is gas.pressure, all at time gas.time. The use of ordinary terminology of the problem area for quantities, rather than cryptic abbreviations, is a great convenience for other users of the program, and increases its readability.

Starting with a small number of data types which are predefined in Pascal, additional types can be constructed in various ways. In the definitions given above, for example, type dimtype is constructed using the predefined type integer, while type vector uses type dimtype and the predefined type real.

1. Predefined Data Types

The predefined, or standard, data types offered by Pascal are the simple data type real, the ordinal data types Boolean, char, and integer, and the file type text. The example program solvequad given in Section I,C used only these standard types.

a. Type Real.   Type real designates the floating-point numbers available in the given implementation of Pascal, which are a finite subset of the real number system of mathematics. Pascal provides two formats for the representation of floating-point numbers but otherwise leaves the specific details to the implementation. The first format includes a decimal point and a fractional part, for example,

$$123.45, \quad -0.0979, \quad 1.0e + 45,$$
$$-78.456e - 01, \quad 3.333e + 01.$$

The notation e + 45 indicates multiplication of the previous decimal number by the scale factor 10^{45}. When the sign is omitted, the number is interpreted as positive. At least one digit must precede and follow the decimal point. The second format for type real omits the fractional part:

$$1e + 00, \quad -4354e + 21, \quad -87695e - 03.$$

The inclusion of the letter e and the following scale factor distinguishes these numbers from integers: 1 is interpreted to be of type integer, while 1e + 00 is the floating-point number of type real with equal numerical value.

What are called **simple** data types consist of type real and the ordinal data types, with elements which can be put into correspondence with a subset of the integers. There are three predefined ordinal types in Pascal, type Boolean, type char, and type integer.

b. Type Boolean. Type Boolean consists of only two elements, represented by the predefined constants **false** and **true.** The ordering of these elements is false < true. This type appears in conditional statements in which decisions are made between alternative courses of action.

c. Type Char. This type consists of a finite and ordered set of characters, which may vary from one implementation to another. It includes the letters, digits, and symbols with which Pascal programs are written. Elements of type char are denoted in programs by enclosure in apostrophes, or single quotes, for example, 'a', 'A', etc. The apostrophe itself is doubled as a character or in a character string (sequence of characters), as noted previously.

For implementation, the most commonly used character set is the American Standard Code for Information Interchange (ASCII) set. Each character in the set corresponds to a two hexadecimal (hex) digit integer in base sixteen, with the first digit in the range 0–7 and the second in the range 0–F. There are thus 128 characters in the set, which is given in Table I. Usually, ASCII characters are represented internally in computers by one byte of information, which consists of eight binary or two hexadecimal digits.

Thus, the order of the characters in the ASCII set is from the top to bottom of each column in

TABLE I. The ASCII Character Set

Second hex digit	First hex digit							
	0	1	2	3	4	5	6	7
0	NUL	DLE		0	@	P	`	p
1	SOH	DC1	!	1	A	Q	a	q
2	STX	DC2	"	2	B	R	b	r
3	ETX	DC3	#	3	C	S	c	s
4	EOT	DC4	$	4	D	T	d	t
5	ENQ	NAK	%	5	E	U	e	u
6	ACK	SYN	&	6	F	V	f	v
7	BEL	ETB	'	7	G	W	g	w
8	BS	CAN	(	8	H	X	h	x
9	HT	EM	)	9	I	Y	i	y
A	LF	SUB	*	:	J	Z	j	z
B	VT	ESC	+	;	K	[	k	{
C	FF	FS	,	<	L	\	l		
D	CR	GS	–	=	M	]	m	}
E	SO	RS	.	>	N	^	n	~
F	SI	US	/	?	O	–	o	DEL

Table I, continuing to the top of the next column to the right. For example, the character + corresponds to the hexidecimal integer B2, which has the decimal value 43, and so on. The 32 characters listed in the first two columns and DEL are called control characters. The remaining 95, each of which corresponds to a single recognizable symbol, are called printable characters.

d. Type Integer. The elements of type integer are an implementation-dependent subset of the integers. The ordering of this type is the customary ordering of integers, and the maximum integer available in the given implementation is denoted by the predefined constant **maxint.** The set of integers will include all the integers with absolute value less than or equal to maxint, that is, −maxint, ..., −1, 0, 1, ..., maxint. Negative integers are denoted by a minus sign followed by a digit sequence. An integer is considered to be positive if the sign is omitted.

e. Type Text. Type text is a file type, which is one of the structured types in Pascal. Basically, a file is a sequence of unspecified length of objects of the same type. A file of type text, or textfile, is a file of lines of characters, that is, elements of type char. In other words, a textfile is similar in structure to a document produced on an ordinary typewriter and is essentially a legible form for transmission of information. However, the lines in a textfile can be of arbitrary length, at least up to limits imposed by the implementation, so textfiles include special markers for ends of lines and for the end of the file in addition to the elements of type char. The end of file marker will follow an end of line marker directly. The predefined variables **input** and **output** are of type text.

2. Ordinal Data Types

An ordinal type has elements that can be put into one-to-one correspondence with a finite subset of the integers and thus ordered in an unambiguous way. In addition to the predefined ordinal types Boolean, char, and integer, the user can define new ordinal types in two ways, by enumeration or as a subrange of a previously defined ordinal type.

a. Enumerated Types. The type coordinate given in the second example above is an enumerated type, constructed by simply listing the identifiers of its elements, separated by commas and enclosed in parentheses. Another example is

type size = (small,medium,large);

Identifiers of elements of various enumerated types must be distinct. The order of elements is the same as in which they are enumerated (in this example, small < medium < large).

b. Subrange Types. A new ordinal data type can be constructed as the subrange of an existing (that is, previously defined) ordinal type, called the host type, by listing the smallest and largest elements of the host type to be included, separated by two periods. These lower and upper bounding elements, as well as all elements of the host type in between them, will be elements of the new type. Subrange types which have type integer as their host type are used extensively for indexing arrays such as vectors, matrices, terms of series, and so on. For example, with the definitions

const dim = 10;
    type dimtype = 1..dim;

type dimtype will consist of the subrange 1,2,3,...,9,10 of the predefined ordinal type integer. Given the definition above of the ordinal type size, subrange types lowsize and highsize can be defined by

type lowsize = small..medium;
    highsize = medium..large;

Type lowsize will consist of the elements small, medium of type size, while highsize will consist of medium, large.

3. Structured Data Types

In addition to the simple data types (type real and the ordinal types), Pascal provides several structured types for more elaborate organization of information. These are the array, file, record, and set types.

a. Array and String Types. An array consists of a specified number of elements of the same type which are called components of the array. Examples of arrays which occur most frequently in physical science and technology are vectors, matrices, and terms of sequences or series. There are no predefined array types in Pascal.

The components of the array are indexed for access by means of one or more ordinal or subrange types, called index types. Examples of array definitions are:

type   vector = array[dimtype]of real;
        matrix = array[dimtype]of vector;
        series = array[0 . . 100]of integer;
    truthtable = array[1 . . 5,1 . . 10]of
                Boolean;

It is assumed that the subrange type dimtype has already been defined. The form of an array definition is thus the identifier, followed by an equal sign, the word symbol **array,** one or more index types enclosed in square brackets and separated by commas, the word symbol **of,** and finally the identifier specifying the type of the components. The component type can be a previously defined structured type as well as a simple type, as in the definition of type matrix above.

Access to components of a variable of array type is gained by specifying values of the corresponding index type or types, which are enclosed in square brackets following the variable identifier. For example, if v is a variable of type vector, then its fifth component is v[5], which denotes a variable of type real. Similarly, if t is of type truthtable, then t[3,8] denotes the Boolean variable located in its third row and eighth column. The access to an element of a variable m of type matrix can be done by specifying the eighth component of the row vector m[3] as m[3][8], or more simply as m[3,8]. Multidimensional arrays are actually accessed row-wise, that is,

    matrix = array[dimtype]of vector;
truthtable = array[1 . . 5,1 . . 10]of
            Boolean;

are abbreviations for

> matrix = array[dimtype]of array[dimtype]
>   of real;
> truthtable = array[1..5]of array[1..10]of
>   Boolean;

Thus, t[3] refers to the third row of the truthtable t, which is an array[1..10]of Boolean.

In order to conserve storage, an array can be defined to be a packed array, for example,

> truthtable = packed array[1..5,1..10]of
>   Boolean;
> name = packed array[1..12]of char;

Some computers use 32 or more binary digits (bits) to store a specific piece of information, while representation of elements of type Boolean requires only one bit (0 or 1), and only 7 or 8 bits are needed for elements of type char. This saving in storage may entail an increase in the time required to access components of the array.

The second type defined above, name, is an example of a special array type which is called a string type, that is, a one-dimensional packed array of elements of type char.

In Pascal, the definition of arrays is static, that is, the sizes of arrays are set once and for all in the type definition part of the program. For example, variables of type vector as defined above will have exactly dim components. This makes it awkward to write programs to handle vectors of different sizes, even though the same algorithm is used in each case, because each change in the definition of the constant dim requires recompilation of the program. One solution is to define dim to be large, and pad smaller vectors out to length dim with zeros. In what the standard refers to as a level 1 implementation of Pascal as distinguished from level 0 we can use a **conformant array schema,** for example,

> v : array[low..high: integer]of real;

in parameter lists in procedure and function headings. This allows a certain amount of dynamics in dealing with arrays of different sizes. In Pascal-SC, arrays of unspecified sizes can be defined in order to allow a single program to process arrays with sizes that are not known in advance but are specified by the data of the problem being solved.

b. File Types. Files consist of sequences of an unspecified number of elements which are all of the same type. This corresponds to the organization of information utilized by the mass storage and input/output devices of a given computer system. In particular, the predefined file type text for textfiles described above has a structure which corresponds to the input of information from the keyboard of a computer terminal, namely, a sequence of lines of variable length of characters. This encyclopedia article was first prepared as a textfile in this way. The end of a file is indicated by a special end of file marker.

In general, file types are defined as illustrated by the following examples:

> type primes = file of integer;
>   roster = file of name;
>   gasdata = file of fluid;

It is assumed that the data types name and fluid have been defined previously.

The access to components of a file is sequential, that is, one component at a time. This is in contrast to the random access to individual components of an array. Access to a file component is by means of a buffer variable. If, for example, lawyer denotes a file of type roster, then the buffer variable is denoted by lawyer ↑ (or alternatively by lawyer^ or lawyer@). The variable lawyer ↑ can have a single element of type name as its value. In order to inspect (read) a file, predeclared Pascal procedures are available to load the buffer variable with elements of the file, or to append the value of the buffer variable to the end of a file which is being generated (written).

The unspecified extent of files and the limitation to sequential access to them creates certain difficulties in handling them in the same way as other data types. Consequently, it is not allowed to assign a file, or a structured type which has a file as a component, as the value of a variable in the statement part of the program. For example, if A and B are variables of type **roster,** then it is illegal to try to assign the value of B to A by the assignment statement A := B. All other data types are said to be assignable types.

c. Record Types. Record types are similar to arrays in that they consist of a specified number of components but differ from arrays and files in that their components can be of different types. The component types, of course, must have been previously defined. There are no predefined record types in Pascal. Type fluid, defined above, is a record type, as is

type complex = record re,im: real end;

which can be used to represent complex numbers. Thus, in the definition of a record type, the word symbol **record** is followed by a list of components and their types, separated by semicolons if necessary, and the definition is terminated by the word symbol **end.** Components of a record are accessed by their identifiers, preceded by a period. Thus, if z is a variable of type complex, its real and imaginary parts can be accessed as z.re and z.im, respectively. The identifiers re,im of the record defining type complex are called fields of the record. The fields of this record are called fixed fields, and type complex is said to be defined by a fixed record.

In certain applications, it is useful to define a record type with fields that may differ in various circumstances. Such records are said to have a variant part, and the corresponding fields are called variants. For example, such a variant record can be defined to express the solutions of a quadratic equation as two real numbers, or as the complex conjugate pair in case the discriminant is negative. For example,

type roots = record
case dneg: boolean of
true : (c : array[1..2]of
complex);
false : (r : array[1..2]of
real)
end;

The field dneg is called the tag field of the variant part of this record, and its value is the case selector. If x is a variable of type roots with x.dneg = true, then x.c[1] and x.c[2] will represent the complex conjugate solutions of the quadratic equation, while if x.dneg = false, then x.r[1] and x.r[2] are the corresponding real solutions.

A variant record can have fixed parts, which must precede the variant part. The variant part of a record can include fields which are themselves variant records. Names given to various fields of a record must be distinct, including names of fields in the variant part.

d. Set Types. A set containing N objects has 2^N subsets, including the empty set and the set itself. In order to have a convenient way to keep track of elements and subsets of a given set and to perform certain set operations, Pascal provides set types. A set type in Pascal consists of all subsets of the set of all elements of a given

ordinal type, for example,

type sizeset = set of size;

defines a set called sizeset which consists of all eight subsets of the set size = (small,medium,large). The ordinal type size is called the base type of the set type sizeset. The cardinality (number of elements) of the base type, or set, is usually limited by the implementation, often to 256. It is possible to define a set type as packed and to give the base type explicitly as a subrange or enumerated type, for example,

type indexset = packed set of −4..100;
caseset = set of (I,II);

There are no predefined set types in Pascal.

For purposes of illustration, a set type over a base type with N elements can be considered to be of type array[1..N] of Boolean. A variable s of this type will have s[i] = true if it contains the ith element of the base type, otherwise, s[i] = false. This provides a way to perform set operations, such as union and intersection, and to check for inclusion of elements or subsets without explicitly listing the contents of the subsets.

A value can be assigned to a variable s of set type by means of a set constructor, which lists the elements of the base set to be included in square brackets, separated by commas. For example, [medium,large] could be assigned as a value to a variable s of type sizeset. The result of this in the Boolean array analogy would be s[small] = false, s[medium] = true, s[large] = true. To actually determine the content of s, it would be necessary to test for inclusion of each element of the base type in turn.

4. Pointer Types

Pointer types differ from other types in that their elements (variables of pointer type) do not represent data directly, but rather indicate the location of data of a given type. Thus, the elements of a pointer type function in the same way as addresses of storage locations in a computer. Pointer types can be used to add and delete items to lists dynamically, that is, the length of the lists do not require prior specification. In order to define a pointer type **pref** with elements which refer, or point, to elements of a target type **ref**, the form is

type pref = ↑ ref;

for example,

type vector = array[dimtype]of real;
pvec = ↑ vector;

Here, a variable p of type pvec points to an element of type vector, denoted specifically by p ↑, with real components p ↑ [1],...,p ↑ [dim].

The target type ref need not be previously defined, which is one of the two deviations of this kind allowed in Pascal. At this point in the translation of the program, it is simply noted that ref is a type identifer. Of course, a definition of ref is expected at some place in the type definition part of the program. This flexibility allows the construction of data types which include pointers to other elements of the same type. In turn, this permits information to be ordered into lists. This information can be referred to by pointers, rather than predeclared identifiers, so that lists can be reordered and items entered or deleted dynamically in the course of the execution of a program. Thus, for example, the definitions

```
type pref =  ↑ ref;
     ref = record
              vec : vector;
              next : pref
           end;
```

allow one variable r of type ref to point to another, namely r.next ↑. The definition of the record type ref is legal, because pref is a previously defined type. The definition of type ref can be regarded as completing the definition of the pointer type pref, which is said to be bound to type ref.

Pascal provides the word symbol **nil** as the value of a pointer variable which points to no element of its target type. There are no predefined pointer types in Pascal.

5. Compatibility of Types

In order for various operations to be performable in Pascal, the types of the operands must be compatible. Types are compatible if they are the same, if one is a subrange of the other or both are subranges of the same host type, if both are either packed or unpacked set types over compatible base types, or if both are string types with the same number of components. Thus, types of operands are checked for compatibility as part of the semantic analysis of a Pascal program during the translation process. This helps to ensure the correctness of the program.

J. VARIABLE DECLARATION

The variable declaration part of a Pascal program serves much the same purpose as the glossary of a scientific paper; it identifies the symbols to be used and the information structure of each, either explicitly or by reference to a previously defined type. The syntax of the variable declaration part is

```
VariableDeclarationPart = ["var"
VariableDeclaration ";"
{VariableDeclaration ";"}].
VariableDeclaration = IdentifierList ":"
Type.
IdentifierList = Identifier {"," Identifier}.
```

All variables representing data and results must be declared before they can be used in the statement part of the program. Pascal provides two predeclared variables, input and output, which are of type text and identify textfiles used for the indicated purposes. The identifiers and types of all other variables to be used in the program must be explicitly declared. For example,

```
var air : fluid;
    a,b,c,d,r1,r2 : real;
    v : vector;
```

In addition to reference to a previously defined type, the structure of a variable can be stated explicitly, as in

```
var u : array[dimtype]of real;
```

Here, the variable u and the variable v declared previously have the same structure, but u is not of type vector. The values of one of these variables cannot be assigned to the other directly, only componentwise. Information with the same structure should generally be organized into a single data type so that it can be processed more simply.

The names of all variables must be distinct from each other, regardless of type, and from names of functions, procedures, and predefined quantities used in the program.

K. PROCEDURE AND FUNCTION DECLARATIONS

It is usually convenient to organize the main task of a program into a number of subtasks to be performed by subprograms. For example, reading or at least writing a textfile occurs in almost every program. Pascal offers two forms of subprograms, called procedures and functions, respectively. These will be referred to collectively as subroutines. A number of predeclared subroutines are available in Pascal to perform routine computational chores. Most Pascal programs of any size will also include subroutines constructed and declared by the programmer.

The basic distinction between a function and a procedure is that a function yields a single result, identified by the function identifier. Furthermore, the result type of a function is limited to simple and pointer types in Pascal. The number and types of results produced by a procedure are specified in the parameter list of the procedure heading and are essentially arbitrary.

1. Predeclared Procedures and Functions

The predeclared procedures and functions of Pascal will be listed according to the types of their arguments.

a. Type real. There are no predefined procedures for type real. The predefined functions that take a real argument are round, trunc, abs, arctan, cos, exp, ln, sin, sqr, and sqrt. The functions round and trunc yield an integer result, that is, round(r) and trunc(r) are of type integer, where r denotes a variable of type real. The remaining functions yield real results:

round(r)   The closest integer to r. If r is exactly halfway between two integers, then round(r) has the value of the one further from 0.

trunc(r)   The fractional part, if any, of r is discarded to obtain round(r), which is the integer closest to r in the direction of 0.

abs(r)   The absolute value $|r|$ of r.

arctan(r)   The principal value of the arctangent of r, sometimes denoted by $\tan^{-1} r$.

cos(r)   The cosine of r.

exp(r)   The exponential e^r.

ln(r)   The natural logarithm of r.

sin(r)   The sine of r.

sqr(r)   The square r^2 of r.

sqrt(r)   The square root $\sqrt{r}$ of r.

b. Ordinal Types. There are no predeclared procedures for ordinal types. The three predeclared functions for ordinal types are ord, pred, and succ. The function ord yields an integer result, while pred, and succ, if defined, yield results of the same ordinal type as their argument. Let o denote a variable of some ordinal type, then

ord(o)   The integer giving the ordinal number of o, for example, ord('+') = 43 for the element + of the ASCII character set.

pred(o)   The predecessor of o in the ordering of the ordinal type, or an error if o is the first element of its type. For example, pred(2) = 1 in type integer,

succ(o)   The successor of o in the ordering of its ordinal type, for example, succ('+') = ',' in the ASCII character set, or an error if o is the last element of its type.

c. Type integer. The predeclared ordinal functions ord, pred, and succ apply to type integer. In addition, there are the predefined functions chr and odd with results of types char and Boolean, respectively:

chr(i)   The character, if it exists, with ordinal number i, for example, chr(43) = '+'. Otherwise, an error.

odd(i)   The Boolean value true if i is an odd integer, otherwise, false.

There are no predeclared procedures for type integer.

d. Array Types. The predeclared procedures pack and unpack for arrays are provided to transfer information between packed and unpacked arrays. Generally speaking, access to components of an unpacked array is faster, while packed arrays occupy less storage.

pack(u,i,p)   Packs the ith and following components of the unpacked array u into the packed array p.

unpack(p,u,i)   Transfers the components of the packed array p into the unpacked array u, starting at the ith component of u.

There are no predeclared functions for array types.

e. File Types. A number of predeclared functions and procedures are provided for various chores involving the transfer of information to and from files. Some of these subroutines apply to files in general, while others are formulated only for textfiles, which are the kind most commonly used for input and output. The predeclared procedures and functions allow the file identifier to be omitted in case of reference to the appropriate standard file input or output.

i. GENERAL FILES. The function provided for general files is eof, which tests for the end of file marker and gives a Boolean value as result.

eof(f)   The Boolean value true if the file f is in generation (write) mode or if the end of a file in inspection (read) mode has been reached, otherwise, false. If f is omitted, then input is assumed.

Procedures provided for general files are get, put, read, reset, rewrite, and write. The procedures get and put are used for sequential access to the components of a file for inspection and generation, respectively. They advance by one component at a time, if possible. The procedures read and write are somewhat more versatile for input and output of information. The mode of a file (inspection or generation) is set by the corresponding procedure reset or rewrite:

get(f) Assigns the value of the next component of the file to the buffer variable $f\uparrow$ unless the end of the file marker is next, in which case $f\uparrow$ is undefined and eof(f) is set equal to true. Call of get(f) with eof(f) = true or f not in inspection mode is an error. If f is omitted, then input is assumed.

put(f) Appends the value of $f\uparrow$ to the end of the file f, following which $f\uparrow$ becomes undefined. Call of put(f) with $f\uparrow$ undefined or f not in generation mode is an error. If f is omitted, then output is assumed.

read(f,v) Bypasses the use of the buffer variable $f\uparrow$ by assigning the current component of f directly to the variable v, followed by advancing to the next component of f. The effect is the same as the assignment $v := f\uparrow$ followed by get(f). Several components of f can be read sequentially and their values assigned to the variables v1,...,vn by use of the format read(f,v1,...,vn). If f is omitted, then input is assumed.

reset(f) Puts the file f in inspection mode. The value of the first component of the file is assigned to $f\uparrow$ and eof = false, unless the file is empty, in which case $f\uparrow$ is undefined and eof = true.

rewrite(f) Puts the file f in generation mode. The file will be empty (cleared), and eof = true.

write(f,v) Appends the quantity identified by v to the end of the file f, abbreviating the assignment of the value of v to $f\uparrow$ followed by put(f). The format write(f,v1,...,vn) appends the values identified by v1,...,vn to f in order. If f is omitted, then output is assumed.

The component types of given files determine the types of data which can be read from or written to them. Many implementations apply reset and rewrite automatically to input and output, respectively, but details may differ. However, it is usual to consider that input is always in inspection mode, and output is always in generation mode, so that it is unnecessary to apply reset or rewrite to the corresponding standard textfile explicitly.

ii. TEXTFILES. Additional predeclared subroutines are provided for textfiles, which are the usual means for communication of legible information. Some of these are useful in connection with the organization of textfiles into lines. The procedure read(f,v1,..,vn) for a textfile f requires that v1,...,vn represent either quantities of type real, integer, char, or subrange types of char, integer. The same restrictions apply to write(f,v1,...,vn), except that string types are also allowed as possible values for v1,...,vn.

The predeclared function is

eoln(f) This gives the Boolean value true if the end of line marker has been reached in a textfile in inspection mode, otherwise, false. Call of eoln(f) is an error if eof(f) = true. If f is omitted, then input is assumed.

The procedures page, readln, and writeln apply only to textfiles. The procedures page and writeln are used to format the file for subsequent inspection, display, or printing. The procedure readln provides a way to skip (or absorb) end of line markers when reading a textfile:

page(f) This is used to direct subsequent lines of f to a new page and will vary considerably from one implementation to another.

readln(f,v) Performs read(f,v), and then moves the point at which f is being inspected to the beginning of the next line, if possible. The procedure readln(f,v1,...,vn) modifies read(f,v1,...vn) in a similar way. If f is omitted, then input is assumed.

writeln(f,v) Performs write(f,v), after which an end of line marker is appended to the file f. Thus, writeln(f) will produce a blank line in f. The procedure writeln(f,v1,...,vn) modifies write(f,v1,...,vn) in a similar way. If f is omitted, then output is assumed.

f. Set Types. There are no predeclared procedures or functions which have variables of set type as arguments.

g. Pointer Types. The predeclared procedures **new** and **dispose** are provided for the respective purposes of creating and eliminating storage dynamically for target variables of a pointer type. For example, if the pointer variable p points to a record, then new(p) allocates storage for a record of the same type, and p is assigned the corresponding address to refer to that storage area. Similarly, dispose(p) deallocates the storage previously allocated to the record to which it points, and the value of p becomes undefined. A variable of target type will be inaccessible if no variable of pointer type points to it, but it will continue to occupy storage space. Furthermore, it is impossible to dispose of an inaccessible target variable:

new(p)   Creates the pointer variable p, and allocates storage for the target variable p↑ to which p points. If the target variable is a record with variant parts, new(p,f1,...,fn) can be used to create a pointer p to a record which has only the variant fields f1,...,fn, and allocate the corresponding storage as well.

dispose(p)   Deallocates storage assigned to p↑, and results in an undefined value of p, an error if p is nil or undefined. The pointer variable p must have been created by new(p). If p was created by new(p,f1,...,fn), then it can only be destroyed by dispose(p,f1,...,fn), with exactly the same fields f1,...,fn. If p is nil or undefined, then calling either form of dispose is an error.

If the target type of p is a variant record, then new(p) will allocate sufficient storage for all variants of the resulting record p↑. There are no predeclared functions for variables of pointer type.

2. Subroutine Declarations

The structure of a subroutine (procedure or function) declaration in Pascal is essentially the same as the structure of the Pascal program itself. There is a subroutine heading, followed by a semicolon, followed by a block, followed by another semicolon. The block again consists of a declaration and definition part, followed by a statement part. In the declaration and definition part, labels, constants, types, variables, and more functions and procedures can be declared or defined, just as in the corresponding part of a program.

The functions of the various parts of a subroutine declaration are also similar to the corresponding parts of a program. The subroutine heading gives the name of the subroutine, which is the identifier which will be used to call (that is, use) the subroutine in the statement part of the program. The identifier of the subroutine can then be followed by a parameter list, enclosed in parentheses, that lists various identifiers, called formal parameters, which identify data and results of the subroutine. This formal parameter list sets up communication with the main program or other subroutines in much the same way that the program parameter list of the main program sets up communication with the outside world. The declaration and definition part will specify any labels, constants, types, variables, functions, and procedures that pertain to the subroutine itself. Finally, the statement part will list the actions that perform the work of the subroutine.

For example, suppose that many quadratic equations have to be solved in a program for some scientific or engineering application. It would then be efficient to organize this task as a procedure, such as Program 2.

The procedure heading in this case gives the identifiers and types of the variables a, b, c, r1, r2 which will be used, as before, as data and results in the subroutine itself. In addition, a Boolean variable dneg has been introduced to enable the real and complex cases to be distinguished, because in the real case, the values of r1 and r2 will be solutions of the quadratic equation, while in the complex case, these values give the real and imaginary parts of one of the pair of complex conjugate solutions of the equation.

3. Formal and Actual Parameter Lists

The identifiers given in the parameter list in the subroutine heading are called formal parameters, because these are the names of the variables (including possibly functions and procedures) which will be used by the subroutine itself. Thus, these quantities can have different identifiers in the program or other subroutine in which the given subroutine is used. For example, the coefficients of the quadratic equations to

PROGRAM 2.

```
                    {heading}
procedure solvequad(a,b,c: real; var r1,r2: real; var dneg: boolean);
                    {definitions and declarations}
        var d : real;      {discriminant}
                    {statements}
    begin        {procedure solvequad}
        d := b*b − 4*a*c;      {calculate descriminant}
        if d >= 0 then
            begin    {real case}
                r1 := (−b − sqrt(d))/(2*a);       {first root}
                r2 := (−b + sqrt(d))/(2*a);       {second root}
                dneg := false                      {roots are real}
            end      {real case}
        else
            begin    {complex case}
                r1 := −b/(2*a);                    {real part}
                r2 := sqrt(abs(d))/(2*a);          {imaginary part}
                dneg := true                       {roots are complex}
            end      {complex case}
    end;             {procedure solvequad}
```

be solved in the main program could be denoted by p0, p1, and p2, and the desired solutions by x1 and x2. In addition, the Boolean variable cc could be used, where cc = true would indicate complex conjugate roots. Then, the procedure could be called by the statement

solvequad(p0,p1,p2,x1,x2,cc);

Here, p0,p1,p2,x1,x2,cc is called an actual parameter list, and its entries are called actual parameters. In the same way, the list of files given in the heading of a program is a formal parameter list, and the sequence of names of files or devices to be used when the program is executed is an actual parameter list. Since the formal parameter list gives identifiers and types of quantities used in the subroutine, it performs some of the functions of the declaration and definition part of the subroutine heading.

A function heading differs slightly from a procedure heading in that the result type is stated, and the function identifier serves as the identifier of its result. For example,

```
function discriminant(a,b,c: real): real;
    begin       {discriminant}
        discriminant := b*b − 4*a*c
    end;          {discriminant}
```

Here, the expression discriminant(p0,p1,p2) denotes a real variable for p0, p1, and p2 real. In particular, discriminant(4.0,2.0,1.0) has the real value 8.0. We could also write this as discriminant(4,2,1), because it is possible to assign integer actual parameters to real formal parameters.

4. Calls by Reference and Value

In the case of a function, the formal parameter list in the heading usually denotes data, while the result is designated by the function identifier. Thus, the data of function discriminant are denoted by the formal parameters a, b, and c.

In the case of a procedure, data and results can be distinguished in the formal parameter list by using the word symbol **var** before the latter, as was done for the variables r1,r2,dneg in the heading of procedure solvequad. The addresses of these quantities, which must be single variables, are furnished to the subroutine, which can then alter their values. Such variables are said to be called by reference in the formal parameter list. Only the values of other entries in the formal parameter list are passed to the subroutine, for example, a, b, and c in solvequad or discriminant are called by value in this way. Such value parameters can be expressions, or formulas, instead of single variables. These are evaluated when the subroutine is called, and the resulting values are passed to the subroutine. For example, suppose the coefficients in the quadratic equations to be solved are in arithmetic progression, such as a, a+2, and a+4. Then, solvequad could be called by

solvequad(a,a+2,a+4,x1,x2,cc);

Thus, quantities called by value are transferred directly or else computed on the spot. Consequently, they can only serve as data.

5. Block Structure of Programs

The organization of a Pascal program is basically as a main program, in which various subroutines may be declared. These, in turn, may contain declarations of further subroutines, and so on. This establishes a heirarchy, at which the main program can be considered to be at level 1, subroutines declared in it at level 2, and so on, with subroutines declared at level n being at level n + 1. A subroutine at level n is said to be within the *scope* of the subroutine at level n − 1 in which it is declared.

The heirarchy of subroutines in a Pascal program establishes what is called a block structure for the program, corresponding to a geometric visualization of the hierarchy. For example, if the main program M includes subroutines A, B, and C, then the blocks of these subroutines can be considered to be enclosed by the block of M (in the scope of M). Similarly, if A includes AA and AB, then the blocks of AA and AB are enclosed by the block of A and the block of M.

The basic method of transfer of information between a subroutine and the block which encloses it is by means of the parameter list in the subroutine heading. On the one hand, a subroutine should operate as independently of its enclosing blocks as possible. This will enable it to be incorporated into other programs, if desired. On the other hand, the subroutine should have access to information in enclosing blocks such as type definitions, declarations of other subroutines, and so on, to permit the subroutine to be written concisely, and limit the transfer of information to only essential quantities. In Pascal, these somewhat conflicting requirements have been resolved by making all unconcealed quantities in enclosing blocks available, or visible, to the subroutine. A quantity in an outer block can be concealed by declaring or defining a quantity with the same identifier in an inner block. Action on this quantity in the subroutine has no effect on the quantity with the same identifier in the outer block. The predeclared and predefined quantities of Pascal can be considered to have been introduced in a block at level 0 which encloses the main program M.

Ideally, the inner workings of a subroutine should be invisible to the program or subroutine in an enclosing block that uses it. The calling subroutine in the outer block is mainly interested in giving the inner one it calls some data and getting back the corresponding results. This is analogous to the way in which work is parceled out to assistants or the way in which an already written computer program is used. In Pascal, quantities declared or defined in a subroutine are not accessible, or visible, from blocks that enclose the given subroutine. The scope of a declaration or definition of an identifier consists of the block in which it occurs and all blocks enclosed by that block.

Basically, when a subroutine is called, or activated, in the course of the execution of a program or another subroutine, storage space is assigned to the quantities used by the subroutine. When its task is finished and the results sent to appropriate locations, this working space is released. Thus, from the standpoint of the calling routine, the quantities used internally, or locally, by the called subroutine are nonexistent, not merely inaccessible.

According to the scope rules of Pascal, it is possible for a subroutine to alter a quantity in an enclosing block directly by reference to its identifier, instead of by way of the identifier list in the subroutine heading. This kind of side effect of a subroutine should be strictly avoided in programming, since it can produce errors that are difficult to correct and limits the usability of the subroutine to a particular setting.

6. Recursion

In Pascal, a subroutine may call itself. This is called recursion, and the calling of a subroutine by itself is termed a recursive call. For example, consider the following function for computing a^k, where a is a real number and k is an integer:

```
function power(a: real; k: integer): real;
    begin
        if k <= 0 then
        begin
            a := 1/a;
            k := abs(k)
        end;
        if k = 0 then power := 1
            else power := a*power(k,k−1)
    end;
```

In this example, calling power(a,4) would result in the calls of power(a,3), power(a,2), power(a,1), and finally power(a,0) in turn to obtain the result a^4. What happens essentially in the case of a recursive call is that a sufficient number of copies of the subroutine are made to accomplish its purpose. After power(a,0) has been evaluated and its result returned to power(a,1), then the storage space allocated to it is released, and so on until power(a,4) returns its result to the calling routine and is deactivated. The number of times a recursive call can be made can be limited by the implementation or simply the amount of storage space available on a given computer.

7. Forward Declaration of Subroutines

The basic philosophy of the Pascal language is that everything must be declared or defined before it can be used. One exception has been noted previously in connection with pointer types, which can be defined in advance of their target types. The other exception is the forward declaration of subroutines. The need for this deviation from the standard order arises if two (or more) subroutines call each other. For simplicity, suppose that A calls B and B calls A. To solve the dilemma of which to declare first, the heading of A, for example, is given, followed by the predefined Pascal directive forward; then, B can be declared in its entirety. In the translation of B, A will be treated as previously declared. Following the declaration of B, the declaration of A will consist of its identifier only followed by its block. References to B in the block of A will then be to a previously declared subroutine. Supposing that A is a function called alphonse and B a procedure called gaston, the form of such a forward declaration is

```
function alphonse(x,y: integer): real;
    forward;
procedure gaston(a,b: integer; var r: real);
    begin
        if a <> b then r := alphonse(a,b)
        else r := 1
    end;
function alphonse;
    var q: real;
    begin
        if x < y then x := x + 1
        else x:= y;
        gaston(x,y,q);
        alphonse := q
    end;
```

Eventually, this interdependent pair of subroutines will produce 1 as the value of both the function alphonse and the result r of the procedure gaston.

L. STATEMENTS

The statement part of a Pascal program specifies the actions to be taken by the program in order to perform its task. This part of the program is of primary interest to the programmer, who must decide how to use the tools available in the language to achieve the desired purpose. The previous parts of the program serve to define and organize information, which will then be processed by means of execution of the instructions given in the statement part. Pascal provides specific forms, called statements, for the codification of these instructions. The kinds of statements available in Pascal are thus the basic operational tools available to the programmer in this language. The statement part of the program consists of a sequence of such statements, enclosed by the word symbols **begin** and **end**, where end is followed by a period to denote termination of the text of the program. Statements can be classified as simple, conditional, repetitive, or compound. In addition, there is a special statement called the **with** statement, which is useful in processing variables of record type. It is also convenient to consider the empty statement, which performs no action.

Simple statements denote a single action which is performed unconditionally. The simple statements are the **assignment**, **procedure**, and **goto** statements. Conditional statements are used to select among alternative actions depending on parameter values. The conditional statements are the **if** statement and the **case** statement. Repetitive statements, as their name implies, are used to repeat an action. The repetitive statements are the **for**, **while**, and **until** statements. The number of repetitions of the action is specified in a **for** statement. In the **while** and **until** statements, the number of repetitions is controlled by a Boolean parameter, instead of being specified in advance. Compound statements consist of a sequence of statements, enclosed by the word symbols **begin** and **end**, and separated by semicolons. A compound statement thus denotes a series of actions. The **with** statement permits access to components of a stated record by their field names alone, which can simplify manipulation of a complicated record.

1. Assignment Statements

The syntax of an assignment statement is

AssignmentStatement = (Variable | FunctionIdentifier) ":=" Expression.

For example,

    d := b*b − 4*a*c;
    discriminant := d;

In the first, a variable identifier stands before the assignment symbol ":=", while a function identifier occupies the same position in the second. Following the assignment symbol is an expression, or formula, indicating a computation to be performed. The meaning of the first assignment statement is "Evaluate b*b − 4*a*c and assign the result to d as its value." The identifiers in an expression refer to values before its evaluation; for example,

    n := n + 1;

means that one is to be added to the current value of n, and the result assigned as the new value of n.

In order for an assignment statement to be semantically valid, the type of the value obtained for the expression must be assignment compatible with the type of the variable or function result. This will be true only if result and expression are both of the same assignable type, or the type of result is real and expression is integer, or result and expression have compatible ordinal or set types, with the value of expression an element of the type of result, or result and expression are compatible string types (of the same length).

a. Expressions. Expressions are fundamental elements of computation in a Pascal program. Expressions can consist of a single variable identifier or a sequence of operands and operators. Operators are denoted by the corresponding operator or word symbols provided in Pascal. Operands can be constant or variable identifiers or function calls (function identifiers with actual parameter lists), such as in the assignment statement

    r1 := (−b + sqrt(d))/(2*a);

The idea in the formation of expressions is to follow ordinary mathematical notation as closely as possible but in a linear format. Another deviation from common usage is that the operation of multiplication is denoted explicitly by "*", rather than by juxtaposition. This is necessary to allow identifiers to consist of more than one letter, thus, m*n denotes m times n, while mn identifies a variable different from m and n.

b. Operators. Pascal provides operator and word symbols for the formation of expressions involving a few of the basic data types. The order in which various operations will be performed in the evaluation of an expression is determined to a certain extent by a precedence assigned to operators of various classes. The Boolean negation operator **not** has the highest precedence, followed by multiplication operators, addition operators, and relational operators in turn. Operators of equal precedence are evaluated from left to right. Parentheses are used, if necessary, to specify the order of evaluation of an expression in an unambiguous way. Thus, in

    d := b*b − 4*a*c;

the multiplications are performed before the subtraction, yielding the correct result. However, if the parentheses are omitted in the denominator in the expression for one of the roots of the quadratic equation, then (−b + sqrt(d))/ 2*a will be evaluated as ((−b + sqrt(d))/2)*a, which is not the value specified by the quadratic formula.

A summary of Pascal operators is given below, in order of precedence.

i. BOOLEAN NEGATION.

not   Applied to a Boolean operand with value true or false, yields the other value.

ii. MULTIPLICATION OPERATORS.

*   Multiplication. If both operands are integers, then the resulting product is of type integer. If at least one operand is real, then the product is of type real.

*   Set intersection. Yields the intersection of two operands of the same set type.

/   Division. The operands can be either integer or real; the resulting quotient is always of type real. Division by 0 is a error.

div   Quotient. Yields the integer quotient in the division of its integer operands. Division by 0 is an error.

mod   Modulus. Yields the remainder in division of its first integer operand by the second, which must be positive. The result is positive if both operands have the same sign, otherwise, it is negative.

and   Logical and. Yields the value true if both of its Boolean operands are true, otherwise, false.

iii. ADDITION OPERATORS.

+   Identity. Does not alter the value of a following single real or integer operand.

+   Addition. Applied to two real or integer operands, yields their sum as type integer if both operands are integer, or of type real if at least one operand is real.

+   Set union. Yields the union of two operands of the same set type.

−   Sign inversion. Inverts the sign of a following single real or integer operand.

−   Subtraction. Applied to two real or integer operands, yields their difference as type integer if both operands are integer, or as type real if at least one of the operands is real.

−   Set difference. Yields the difference of two operands of the same set type. That is, $a - b$ denotes the set of elements of a which are not elements of b.

or   Logical or. Yields the value true if at least one of its Boolean operands is true, or false if both are false.

iv. RELATIONAL OPERATORS.

The result type of all relational operators is Boolean.

=   Equality. The result for operands of the same simple, string, set, or pointer type is true if they are equal, otherwise, false.

<>   Inequality. The result for operands of the same simple, string, set, or pointer type is true if they are unequal, otherwise, false.

<=   Not greater. Operands of the same simple or string type are compared, with the result true if the first operand is less than or equal to the second, otherwise, false.

<=   Subset. Tests two operands of the same set type for inclusion of the first in the second. True if the first is a subset of the second, otherwise, false.

>=   Not less. Operands of the same simple or string type are compared, with the result true if the first operand is greater than or equal to the second, otherwise, false.

>=   Superset. Tests two operands of the same set type for inclusion of the second in the first. True if the second is a subset of the first, otherwise, false.

in   Element inclusion. The second operand is of set type, and the first of the ordinal base type of the set type. If the first operand is an element of the second, then the result is true, otherwise, false.

2. Procedure Statements

A procedure statement consists of the procedure identifier and its actual parameter list (or actual write list in the case of the standard procedures write and writeln), for example,

```
gaston(-10,a,r);
writeln('r = ',r);
```

A procedure statement activates the corresponding procedure with the given parameters, that is, its actions will be carried out before the next statement of the program is executed.

3. Goto Statements

A goto statement such as

```
goto 10;
```

will interrupt the normal sequential execution of statements by causing the statement with label 10 to be executed next. The labeled statement must be in the same statement sequence as the goto statement or in an enclosing block. Normally, well-organized Pascal programs do not require the use of labels or goto statements.

4. If Statements

The syntax of **if** statements is

IfStatement = "if" BooleanExpression "then" Statement ["else" Statement].

In the if...then form, Statement will be executed only if BooleanExpression has the value true. If BooleanExpression is false, then no action occurs, and execution of the program proceeds to the next statement. The **if** statement in this case is equivalent to an empty statement. In the if ...then...else form, the first Statement will be executed if BooleanExpression is true, otherwise, the second Statement will be executed. This form of **if** statement was used in the program solvequad to distinguish between the various cases for solutions, depending on the value of the discriminant. **If** statements thus allow the choice between two alternative actions, either Statement or no action in the first form, or one Statement or the other in the second form.

5. Case Statements

Case statements allow selection between several alternative actions. This is done by means of an expression called the case selector, with result of ordinal type, called the case index type.

Constants of this type are used to label the corresponding statements. The statement with the label equal to the result of the case selector is executed. If the value of the case selector does not occur as a label, then the case statement is equivalent to the empty statement, that is, no action is performed and execution proceeds with the following statement. For example, suppose that i is a variable of type integer. Then, we can use

```
case i of
    1:  write('First');
    2:  write('Second');
    3:  write('Third');
    4:  write('Fourth')
end;
```

to output the names of the corresponding ordinal values. Empty statements can also be included in the list of cases, if desired. The second form of the **if** statement could be expressed by the case statement

```
case BooleanExpression of
    true:  statement1;
    false: statement2
end;
```

6. For Statements

A **for** statement governs the repetition of an action by specification of the initial and final values of a control variable. For example, suppose that a, b, and c are of type vector = array[dimtype]of real as defined previously. Then, the sum of a and b can be assigned to c by means of the **for** statement

```
for i := 1 to dim do c[i] := a[i] + b[i];
```

The control variable i must be declared to be of type dimtype in the block which contains the **for** statement. Furthermore, the value of the control variable is to be protected from attempted change of its value by statements in the scope of the block in which it is declared. The values of the control variable proceed successively from the given initial value to the given final value. In the example above, the control variable is incremented by 1 until the governed statement has been executed for i = dim, after which execution of the program proceeds with the next statement. The order can be reversed, for example,

```
for i := dim downto 1 do c[i] := a[i] + b[i];
```

adds the components of the vectors a,b to obtain the corresponding components of c starting at the other end of each vector. In the first form of

the for statement, the value i of the control variable is followed by succ(i), while it is followed by pred(i) in the second form. Thus, in the case

```
for control := initial to final do statement;
for control := final downto initial do
    statement;
```

the number of times that statement is to be executed is specified in advance by the number of elements of the type of the control variable in the subrange initial..final.

In general, **for** statements are particularly suitable for processing arrays using control variables belonging to the index type.

7. While Statements

An action governed by a **while** statement is repeated as long as a Boolean expression returns the value true. The syntax of a **while** statement is

```
WhileStatement = "while"
    BooleanExpression "do" Statement.
```

The governed statement would ordinarily be a compound statement which can effect the change in value of the Boolean expression to false to terminate the repetition after the desired result has been obtained. If the value of BooleanExpression is false when the **while** statement is encountered, it will then be equivalent to the empty statement, and the governed statement will not be executed.

While statements are useful to govern iterative computations, where the number of iterations required to reach a certain termination condition is not known in advance. For example, suppose that functions f(x) and fprime(x) for the value of a real function and its derivative have been declared. Also, suppose that the real variable x is a satisfactory approximate solution to the equation f(x) = 0 if abs(f(x)) < epsilon, a given real number. Then, Newton's method for solving f(x) = 0 to a satisfactory degree of approximation can be formulated as the **while** statement

```
while abs(f(x)) < epsilon
    do x := x − f(x)/fprime(x);
```

8. Repeat Statements

Ordinarily, for a sequence of statements to be controlled by a conditional statement, they have to be stated as a single compound statement, that is, enclosed between the word symbols begin and end, and separated by semicolons. In the case of a **repeat** statement, the word symbols **repeat** and **until** enclose the sequence of statements to be repeated. The general form of a **repeat** statement is

repeat
    $statement_1$;
    $statement_2$;
    ...
    $statement_n$
until BooleanExpression;

The repetition of the sequence of statements enclosed by repeat...until will continue until BooleanExpression = true. The sequence of statements governed by a **repeat** statement will always be executed at least once, because BooleanExpression is not evaluated until after the sequence is executed. This is in contrast to a while statement, in which the controlling expression is evaluated before the governed statement is executed.

For example, the statement

repeat
    writeln('Do you want
    to continue (y/n)?');
    read(c)
until c in ['y','Y','n','N'];

where c is a variable of type char, will send the question to output and read the file input until one of the characters listed in the set constructor is encountered. Any other characters read from input would be ignored.

As in the case of **while** statements, **repeat** statements are useful for the control of iterative processes, where the number of iterations necessary to reach a termination condition is not known in advance. Suppose that it is desired to terminate the Newton iteration of the previous section when the correction $-f(x)/fprime(x)$ is small in absolute value, say $<$ delta, where delta is a given real variable. Then, one can write

repeat
    cx := $-f(x)/fprime(x)$;
    x := x + cx
until abs(cx) $<$ delta;

to accomplish this purpose. Of course, cx would also need to have been declared to be a variable of type real in the variable declaration part of the program.

9. With Statements

With statements provide a way to abbreviate the names of record fields. For example, if

carbondioxide has been declared to be a variable of type fluid, the position of one of its particles could be set by the sequence of statements

carbondioxide.position[x] := 1.5;
carbondioxide.position[y] := 2.3;
carbondioxide.position[z] := 4.2;

Alternatively, a **with** statement could be used for the same purpose:

with carbondioxide do
    begin
        position[x] := 1.5;
        position[y] := 2.3;
        position[z] := 4.2
    end;

The syntax of a **with** statement is

WithStatement = "with"
  RecordVariableList "do" Statement.

A record variable list is simply a sequence of record identifiers, separated by commas. This allows the fields of several records to be referenced only by their field identifiers in Statement. Of course, the field identifiers must be distinct, and different from identifiers for other quantities declared in the scope which includes the with statement.

II. Pascal-SC

A. General Description

Pascal-SC (Pascal for Scientific Computation) is an extension of Pascal with additional features directed especially toward scientific and technological computation. Pascal-SC was developed during the 1970s by a group lead by Ulrich W. Kulisch at the University of Karlsruhe in Germany. An initial implementation for microcomputers appeared in 1980 and has been followed by others. Implementations of Pascal-SC at present are based on reports pending future standardization.

The principal motivation for the development of Pascal-SC was the realization that no existing programming language provided adequate requirements or capabilities for floating-point arithmetic to enable the assessment of error or validation of numerical results in a convenient way which could also be portable from one implementation to another. The theory of floating-point arithmetic showed that operators with directed rounding as well as rounding to the closest floating-point number are necessary. In addition, the requirements for accuracy and validation of results in vector and matrix arithmetic showed that the scalar product of vectors must be calculated to the closest floating-point number and with directed rounding.

The introduction of operators for directed rounding in Pascal-SC is based on a general operator concept, which allows use of operator symbols and notation for arbitrary numerical data types. Thus, for example, the symbols +, −, *, and / can be used for the corresponding operations between complex numbers or for interval arithmetic and are not limited to type real. Similarly, vector and matrix operations can be written in ordinary notation. The concept of function is also extended in Pascal-SC to allow the result type of a function to be an arbitrary assignable type, rather than being limited to simple types as in Pascal. Thus, in Pascal-SC, expressions involving structured numerical data types can be written in a form which follows ordinary mathematical notation. In Pascal, the same information can only be processed by a sequence of procedure calls, which usually results in programs that are more tedious to write and difficult to understand.

Higher-level implementation of Pascal-SC provide dynamic array types with unspecified numbers of components, general access to arrays (for example, to columns as well as rows of matrices), a separate string type, and a module concept. The dynamic array concept of Pascal-SC removes the limitation of the level 0 implementations of Pascal to static arrays in a more general way than the conformant array schemata of level 1 Pascal. The ability to handle arrays with sizes specified by input data makes it possible to write general programs for a number of scientific and technological applications in a convenient way. The string type of Pascal-SC supplements the string type of Pascal and has features more suitable for text processing and symbolic computation. Finally, the module concept allows programs to be developed in several parts rather than as a single entity. Furthermore, modules that have already been programmed and compiled can be incorporated into new programs in such a way that declarations and definitions in the module are available to the program. The program which uses the module can then be checked completely for syntactic and semantic correctness.

As an extension of Pascal, Pascal-SC makes use of the many good features of Pascal, and adds some additional strengths, particularly for scientific and technological computation. A programmer familiar with Pascal does not need to learn a completely new language to use Pascal-SC and take advantage of its additional features. Since Pascal is a sublanguage of Pascal-SC, programs already written in Pascal are still useful in a Pascal-SC environment.

B. FLOATING-POINT ARITHMETIC

In Pascal-SC, arithmetic for type real must conform to strict requirements of accuracy based on the theory of floating-point arithmetic. First of all, the basic operations +, −, *, and / must yield the closest floating-point number to the exact value of their results. In case of a tie, the rounding is antisymmetric, either toward zero or away from zero. Directed rounding is available, and eight additional operator symbols are introduced for this purpose,

$$+> \quad -> \quad *> \quad />$$

for upward rounding to the closest floating-point number greater than the exact result, and

$$+< \quad -< \quad *< \quad /<$$

for downward rounding to the closest floating-point number less than the exact result. With these operations, it is possible to implement correctly rounded interval arithmetic, which provides a means for error estimation and validation of computed results.

Input and output of floating-point numbers can also be accompanied by the same roundings as for arithmetic operations. Current implementations of Pascal-SC use decimal arithmetic internally, to minimize input/output conversion errors. Directed rounding allows numbers that cannot be converted exactly to be represented as intervals with the downward and upward rounded results as endpoints.

The standard functions abs, arctan, cos, exp, ln, sin, sqr, sqrt for type real in Pascal-SC are required to be computed accurately enough so that there is no floating-point number in the interval between the exact and the computed result. In most cases, the closest floating-point number to the exact result is obtained.

C. SCALAR PRODUCTS

Computations with vectors and matrices are important not only in physical science and technology, but also in statistics, economics, and other fields. The scalar product of vectors $\mathbf{a}$ =

$(a_1, a_2, ..., a_{dim})$ and $\mathbf{b} = (b_1, b_2, ..., b_{dim})$,

$$\mathbf{a} \cdot \mathbf{b} = \sum_{i=1}^{dim} a_i * b_i$$

is fundamental to many calculations. In Pascal and other languages, the scalar product is approximated by a sum of products, for example,

```
function scalarprod(a,b : vector): real;
    var  s : real;
         i : integer;
    begin
        s := 0;
        for i := 1 to dim do s := s +
        a[i]*b[i];
        scalarprod := s
    end;
```

Because of roundoff error, the accuracy of this kind of simulation of the scalar product varies widely and is generally unknown. In order to extend to matrix and vector arithmetic the possibilities for error estimation and validation of results available in real arithmetic, Pascal-SC provides the standard function scalp, which is called by scalp(a,b,r). The arguments a and b of scalp must be of the same vector type, that is, one-dimensional arrays of the same length. The parameter r is an integer that controls the rounding of the result. The value r = 0 gives the closest floating-point number to the exact scalar product, while r = −1 rounds the result downward and r = 1 upward.

The accurate scalar product with directed rounding is used in a number of Pascal-SC utility procedures for numerical linear algebra and other applications which provide answers with sharp, guaranteed error bounds. These include procedures for the solution of linear systems of equations, matrix inversion, calculation of eigenvalues and eigenvectors of matrices, and the evaluation of polynomials and their roots.

The function scalp is implemented by using an accumulator in storage which is long enough to hold the sums of all the double-length products a[i]*b[i]. The actual length of this accumulator depends in a simple way on the length of the floating-point numbers used and the range of their exponents. For example, for 12 decimal-digit floating-point numbers with exponents from −99 to +99, about 440 digits are sufficient.

The accurate scalar product function scalp has applications in many settings. For example, to calculate the sum of the components a[1],...,a[dim] of a vector a, and the sum of squares of its components, both to the closest floating-point numbers, the statements

```
for i := 1 to dim do ones[i] := 1;
sum := scalp(a,ones,0);
sumsquares := scalp(a,a,0);
```

can be used. Such sums are useful in statistics and are usually not calculated with the accuracy available in Pascal-SC. Of course, it is assumed above that a and ones have been declared to be of type array[dimtype]of real, and that i is of type dimtype.

It is possible to add a new scalar product to a previously calculated one by increasing the value of the rounding parameter r by 4. For example, if it is desired to compute

$$s = \sum_{i=1}^{dim} [a_i + a_i^2]$$

to the closest floating-point number, then this can be done by the statements

```
for i := 1 to dim do ones[i] := 1;
sum := scalp(a,ones,0);
s := scalp(a,a,4);
```

In advanced implementations of Pascal-SC, type dotprecision is provided. The variables of this type each require the storage space allotted to the calculation of a scalar product by scalp. In terms of this type, the above calculation would be written as

```
s := #*(a*ones + a*a);
```

Expressions of type dotprecision are rounded to the closest floating-point number by #*, as indicated above, downward by #<, upward by #>, and to the smallest interval containing the exact value by ##. Type interval has the standard definition

```
type interval = record inf,sup : real end;
```

If d is a variable of type dotprecision, then an assignment statement has the form

```
d := #(a*ones + a*a);
```

D. OPERATOR DECLARATION

Two advantages of Pascal-SC over Pascal are that functions can be declared to have arbitrary assignable result types and that operators can be declared to manipulate arbitrary data types using ordinary notation. Functions are declared in Pascal-SC in exactly the same way as in Pascal. Operator declarations, which occur in the declaration and definition part of the Pascal-SC program, specify an operator symbol or name, one or two operands, and the result. For example, to subtract complex numbers, the operator − can be declared as follows:

```
type complex = record re,im : real end;
operator − (a,b : complex) diff : complex;
   begin
         diff.re := a.re − b.re;
         diff.im := a.im − b.im
   end;
```

In addition to function and procedure, **operator** is a word symbol in Pascal-SC.

Following the declaration of this operator, all that is required in the statement part of the program to assign the difference of the complex variables x and y to the complex variable z is

$$z := x - y;$$

This would have to be done by a procedure statement in Pascal, because type complex is a structured type, not a simple type.

The parameter list for an operator consists of two operands for a binary operator or one operand for a unary operator. For example, to invert the sign of a complex number, the operator

```
operator − (a : complex) neg : complex;
   begin
         neg.re := − a.re;
         neg.im := − a.im
   end;
```

could be used. The value of the result of an operator is assigned to a result identifier, which is given along with its type following the operator parameter list.

In Pascal-SC implementations, operators, utility functions, and procedures are collected into libraries or modules for the numerical data types most commonly used in scientific and technological computation. These provide complex, interval, and complex interval arithmetic, as well as vector and matrix arithmetic for component types real, complex, interval, and complex interval. The advantage of using ordinary operator notation when computing with such quantities is that programs are much easier to write and understand than if an equivalent sequence of procedure calls is used.

There are two ways to identify operators in Pascal-SC. The first is by use of an operator symbol, such as − in the example above. This is sometimes referred to as "overloading" the symbol. The second is to use an ordinary identi-

fier to denote the operator. These operators are called "named" operators. In Pascal-SC, the order of declarations and definitions is not as rigid as in Pascal; however, every object must be declared or defined before it can be used, with the exception of pointer types and forward declarations as in Pascal.

1. Overloading of Operator Symbols

In addition to the 21 special symbols of Pascal and the 8 Pascal-SC symbols for arithmetic operations with directed rounding, the word symbols

and div in mod or not

can be overloaded, and Pascal-SC also provides the additional symbols

**   +*   ><

for the convenience of the programmer. In Pascal-SC, the unary operator **not** and the signs + and − have the highest priority, followed by multiplication operators, which include all that begin with * or /, then the addition operators, including all which begin with + or −, and finally the relational operators involving the =, <, and > symbols. For example, *> and ** have the priority of multiplication operators, while −< and +* have the priority of addition operators.

The operator given above for complex subtraction is typical of an overloaded operator. The compiler distinguishes among these operators by the sequence and types of their operands when the program is translated into machine instructions, so that defining − for type complex has no effect on the use of the same symbol for subtraction of integers, reals, intervals, vectors, matrices, etc. Functions and procedures can also be overloaded in Pascal-SC; once again, subroutines with the same identifier are distinguished by the sequence and types of their operands. However, the result types of operators and functions are not used to distinguish these subroutines, so that it would be an error, for example, to introduce declarations of functions with the same formal parameter lists but different result types.

2. Named Operators

If an identifier is used instead of a symbol to denote an operator, then the operator declaration must be preceded by a declaration of its priority, which has the syntax

OperatorPriorityDeclaration = "priority"
Identifer "=" (↑ | * | + | =) ";".

The order of priority, from highest to lowest, is symbolized by ↑,*,+,=. For example, to declare the operator xor for "exclusive or" with the priority of addition, the form would be

```
priority xor = +;
operator xor (a,b : boolean) res : boolean;
  begin
        res := a <> b
  end;
```

To use xor in the program, it would be written between its operands in the same way as the standard operators and, or are used:

if rain xor night then light
:= blue else light := red;

This is referred to as "infix" notation.

E. DYNAMIC ARRAYS

The restriction on the level 0 implementations of Pascal to static arrays is a handicap in scientific and technological computation, where it is desireable, for example, to have a single program that will process vectors and matrices of various sizes. The conformant array schemata of level 1 implementations of Pascal provide the possibility of dealing with this problem, but in a less general way than in Pascal-SC, which also provides general function and operator concepts for dynamic arrays.

Another advantage of Pascal-SC is easy access to subarrays. If, for example,

```
type truthtable  =  array[1..5,1..10]
  of Boolean;
var t: truthtable;
```

then Pascal can only access *t* by rows, that is, t[2] will be the second row of t, which is an array[1..10]of Boolean. In Pascal-SC, the third column of t is accessed by t[*,3], and is an array[1..5]of Boolean. Subarrays of higher dimensional arrays are accessed in a similar way.

To illustrate the introduction of dynamic arrays in Pascal-SC, consider the type definitions

```
type vector  =  dynamic array[*]of real;
     matrix  =  dynamic array[*,*]of real;
```

Variables can be declared by

var v : vector[1..10];

in the main program, or with limit expressions such as

var b : matrix[1..m,1..n];

in subroutines. In the latter case, the values of m,n must be available at the time of the subroutine call, because the limits are evaluated when the variable declaration is processed. Standard functions lbound(a,k) and ubound(a,k) are provided to evaluate the lower and upper bounds of the kth index range of the array a. If k is omitted, then k = 1 is assumed. All index ranges are assumed to be subranges of type integer.

In terms of dynamic arrays, an operator for matrix by vector multiplication can be formulated as follows:

```
operator * (var a: matrix; var b: vector)
  res: vector[lbound(a)..ubound(a)];
    var i  : integer;
    begin
        if ubound(a,2) − lbound(a,2) <>
        ubound(b) − lbound(b) then
        error(2);
        for i := lbound(a) to ubound(a)
        do res[i] := scalp(a[i],b,0)
    end;
```

It is assumed that a procedure called **error** has been defined to print a suitable error message and terminate the computation in case of incompatible ranges for these operations. Pascal-SC provides the standard procedure svr(0) to abort the execution of the program and return to the operating system.

F. STRINGS

Strings, or sequences of characters, are introduced only in a very limited way in Pascal, where a string variable s is declared by

var s : packed array[1..length]of char;

for example, where the integer constant length has been defined previously. The only possibilities for use of a string variable in Pascal are assignment to a string variable of equal length, or for output as a parameter in a write parameter list.

In addition to the strings available in Pascal, which will be called a-strings (for array strings), Pascal-SC allows definition of string types, for example, by

type shoestring = string[24];

which specifies the maximum length of string variable of type shoestring to be 24 characters. Variables of a defined string type, for example,

var a,b,c : shoestring;

will be called s-strings. Value assignments can be made between s-string variables of arbitrary lengths; s-strings can be input, and s-strings can be manipulated in certain ways. For example, the noncommutative operator + is provided to append one s-string to another, and some standard functions are available for s-string manipulation. The function image will form the result of an integer or real expression into an s-string, while the functions ival and rval will give the integer or real values of s-strings which are syntactically integer or real constants, respectively. The function length will determine the actual length of an s-string, while maxlength gives the maximum number of characters permitted for the string. The function pos will indicate the position of the first occurrence of a given substring in an s-string, while substring will extract a substring with given position and length from a given s-string.

The string type and the utilities provided for string manipulation give Pascal-SC additional capabilities for symbolic computation and text processing as compared to Pascal.

G. Modules

In Pascal, a program is a single entity that must be developed in its entirety to be checked for syntactic and semantic correctness. Most implementations of Pascal and Pascal-SC make use of external libraries of pretranslated code for certain purposes, such as complex arithmetic, to save the labor of writing and translating simple subroutines. However, the inclusion of such external subroutines in a program is usually done without the possibility of checking for syntactic and semantic correctness. Furthermore, the programming cannot proceed unless the external library is complete. Modules, on the other hand, provide a way to develop programs in independent parts, with syntactic and semantic checking across module boundaries. Furthermore, various modules that have been completed do not have to be retranslated due to changes in the main program or when other modules are completed. Collections of modules can be assembled to extend the capabilities of the language in a way that can be shared directly among users. Modules can also be distributed in

translated form, in case the source code is regarded as proprietary.

The syntax for a module is similar to that for a program, but without a statement part:

Module = ModuleHeading ";"
[DeclarationAndDefinitionPart] "end"
".".
ModuleHeading = "module" Identifier.

The declaration and definition part of a module is essentially the same as for an ordinary Pascal-SC program, and specifies labels, constants, types, variables, functions, procedures, operator priorities, and operators. However, quantities to be exported from the module, that is, employed in a program which uses the module, have to be prefixed by the reserved word **global**. For example, the module

```
module complexarithmetic;
    global type complex = record re,im
    : real end;
    global operator + (a,b: complex)
    sum: complex;
    begin
            sum.re := a.re + b.re;
            sum.im := a.im + b.im;
        end;
end.
```

can be used in the program main:

```
program main(input,output);
    use complexarithmetic;
    var x,y,z : complex;
    begin
            ...
        z := x + y;
            ...
    end.
```

to introduce type complex and the operator + for complex addition. The use clause for modules must follow the program heading directly.

Quantities not labeled as global in the module belong to the module itself and are invisible in a program that uses the module. This is similar to the way in which quantities introduced in a subroutine are not visible outside the scope of that subroutine.

As an example of independent development of programs and modules, suppose that the program main uses all the operations of complex arithmetic, but the module for complex arithmetic has not yet been completed. However, if all global identifiers are specified in the incomplete module, then it can be translated, and the program that uses the module can be checked completely for syntactic and semantic correctness. For example, the incompletely programmed module above could be used for the development of a main program involving rational complex expressions while the module for complex arithmetic is being completed.

```
module complexarithmetic;
    global type complex = record re,im :
    real end;
    global operator + (a,b: complex) sum:
    complex;
        begin
            sum.re := a.re + b.re;
            sum.im := a.im + b.im;
        end;
    global operator − (a,b: complex) diff:
    complex;
        begin end;
    global operator * (a,b: complex) prod:
    complex;
        begin end;
    global operator / (a,b: complex) quo:
    complex;
        begin end;
end.
```

As in the case of the other extensions of Pascal in Pascal-SC, the module concept is designed to follow the ideas of Pascal programming closely, rather than being as general as possible.

BIBLIOGRAPHY

Alefeld, G. and Herzberger, J. (1983). "Introduction to Interval Computations." Academic Press, New York.

Bohlender, G., Rall, L. B., Ullrich, C., and Wolff von Gudenberg, J. (1987). "Pascal-SC: A Microcomputer Language for Scientific Computation." Academic Press, Orlando.

Jensen, K. and Wirth, N. (1985). "Pascal User Manual and Report," 3rd ed, (revised by A. B. Mickel and J. F. Miner). Springer Verlag, Berlin and New York.

Kulish, U. W. and Miranker, W. L. (1981). "Computer Arithmetic in Theory and Practice." Academic Press, New York.

Kulisch, U. W. and Miranker, W. L. (eds.) (1983). "A New Approach to Scientific Computation." Academic Press, New York.

Moore, R. E. (1979). "Methods and Applications of Interval Analysis." Society for Industrial and Applied Mathematics, Philadelphia, Pennslyvania.

PATTERN RECOGNITION

Theo Pavlidis *SUNY at Stony Brook*

GLOSSARY

Clustering: Partition of a set of samples into groups such that each group corresponds to a different pattern.

Design set: Samples used during the design of a classifier. Also called learning or training set.

Discriminant function: One of a set of functions defined over the input objects and the patterns, and having the property that the function that achieves the maximum value corresponds to the correct pattern.

Feature: Number obtained from a measurement performed on the physical input with the hope that a set of such numbers will be more useful to a classifier than the raw data.

Grammatical pattern recognition: See Syntactical Pattern Recognition.

Graph grammars: Formalism for producing graphs, analogous to using natural language grammars to produce sentences of words.

Learning (supervised): Estimation of parameters for a pattern classifier from samples that have been labeled with the names of the pattern they represent.

Learning (unsupervised): Estimation of parameters for a pattern classifier from unlabeled samples.

Sequential pattern recognition: Method where features are computed as needed until reliable classification is achieved.

Statistical pattern recognition: Pattern recognition for which the input objects are mapped into arrays of numbers and subsequent pro-

cessing is based on the statistical distributions of such arrays.

Structural pattern recognition: Pattern recognition emphasizing object descriptions in terms of simpler parts. The relation between such parts is also emphasized.

Syntactical pattern recognition: Pattern recognition for which the input objects are mapped into strings of symbols. Subsequent processing is done by techniques taken from the theory of formal languages and grammars.

Test set: Object samples used to confirm the performance of a classifier.

Template matching: Recognition by direct superposition and comparison, often pixel by pixel.

In common use the word "pattern" means something which is offered as a perfect example to be imitated. Therefore, pattern recognition (by a person or a machine) is the identification of the ideal which a given object was made after. In technical practice the term is used to describe the replication by machine of the human function of identifying something seen or heard. Such terms as speech recognition, character recognition, etc., emphasize the applications where pattern recognition is used.

I. Illustration of the Problem

While humans and animals are very good at recognizing patterns, progress in mechanical pattern recognition has been very slow. Most people familiar with the Roman alphabet recognize the lines of Fig. 1 as being examples of the phrase SAMPLE OF LETTERS even though the form of the individual letters varies a lot. If we were to compare the black areas of the paper for all the E's, we would find that they have very

SAMPLE OF LETTERS

SAMPLE OF LETTERS

SAMPLE OF LETTERS

SAMPLE OF LETTERS

SAMPLE OF LETTERS

SAMPLE OF LETTERS

SAMPLE OF LETTERS

SAMPLE OF LETTERS

SAMPLE OF LETTERS

SAMPLE OF LETTERS

SAMPLE OF LETTERS

SAMPLE OF
LETTERS

FIG. 1.   Samples of Text. The first ten lines have been produced from fonts available in the phototypesetter that printed this document (Roman, Constant Width, Italic, Helvetica Heavy, Futura, Century Bold Italic, Souvenir, Palatino, etc.), while the rest have been drawn using a bitmap graphics editor.

little in common. People have a concept of the ideal form of each letter and recognize specific printed characters as being instances of that particular letter. The ideal is not even defined explicitly, and its perception may differ from person to person. An example of an even more challenging problem is designing a machine that will recognize pictures of dogs from pictures of other animals.

Early attempts at pattern recognition focused on a holistic (Gestalt) approach: simple transformations were performed on the input data and then statistical techniques were used to devise pattern matchers. It soon became apparent that this was not always a fruitful approach, and efforts have been made to devise transformations of the input that have some relation to human perception. Typically, a complex object is described in terms of simple components or characteristics. For example, alphanumeric characters are described as aggregates of simple strokes. Some writers reserve the term "pattern recognition" for the holistic approach and use the terms *pattern analysis* or *structural pattern recognition* for the descriptive approaches. At the same time that the importance of analysis was realized, it became apparent that recognition must be considered in context, something which is certainly true for human recognition.

For example, in some of the lines of Fig. 1 the two T's are joined, and the same shape would have been called a capital Greek pi (Π) if it had been seen within a mathematical equation or Greek text. Thus, text understanding cannot be separated entirely from character recognition. The interaction between analysis, synthesis, and recognition has caused some confusion in the use of terms. More often than not such terms as pattern analysis, pattern synthesis, image analysis, image understanding, text understanding, computer vision, machine vision, robot vision and speech understanding, reflect the methodology used (or even the source of research funding) rather than the problem being solved.

A meaningful distinction between analysis (e.g., image analysis) and recognition is that recognition is like taxonomy while analysis is using the taxonomy to understand what is happening. However, the analysis and recognition process alternate in any nontrivial task. If a given taxonomy results in something that does not make sense, another taxonomy may be attempted.

EXAMPLE 1.   Given the raw data of an image, one must search for areas corresponding to objects of interest, typically by segmenting the image into regions that are uniform by some definition. In the case of a reading machine we must look for dark versus white regions. In the case of a natural scene we must look for such classes as sky, clouds, foliage, etc. In the first example the recognition is rather obvious and it may be an overkill to call it pattern recognition. In the second case the decision about the type of an image area is nontrivial, and it may make sense to look for features that describe a sky area versus a clouds area. The shape of an area may be important and therefore pattern recognition may be called on again to distinguish one from the other. Finally, the aggregate of areas produces a description that may lead to further pattern recognition. For example, we may wish to identify the subject of a piece of text, or the city where a picture was taken.

EXAMPLE 2.   We wish to design a robot that can recognize oak trees from other trees. What is the ideal oak tree? How can we teach the ideal to the robot? Clearly, we need at least two methods: one when the robot is close enough to identify the shape of individual leaves and another for identifying (tentatively) from a distance. We need image analysis techniques for picking up object outlines and/or individual leaves so that their shape can be examined.

The concept of the ideal can be formed deductively or inductively. In the latter case the observer abstracts the concept by observing a series of imperfect examples. The term *learning,* or *inference,* are used to describe that process. The examples may be labeled or unlabeled. (The latter case is also called *unsupervised learning* or *clustering*). In the technical literature the term pattern recognition is often used to describe the formation of a concept as well as the identification of an object with a concept. Since the latter step involves the selection from amongst a finite number of alternatives, it is often called *classification.*

The close relationship between recognition and the other steps of signal processing has caused a certain dichotomy in the field. In particular, different methodologies are used for visual and auditory signals, and this article will focus primarily on visual patterns. [*See* ACOUSTIC SIGNAL PROCESSING; SIGNAL PROCESSING, DIGITAL.]

II. Introduction to Pattern Recognition by Machine

The ease with which people perform pattern recognition encouraged early research on the subject because of the many military and industrial applications that automatic pattern recognition would have. However, practical progress has been very slow. It seems that nature has endowed the nervous system with mechanisms for processing vital information that are far more complex and powerful than the most advanced computers. If we have any hope of replicating human pattern recognition by machine, we must investigate carefully the various steps of the process.

During either learning or classification, physical objects must be represented in a form that can be manipulated by a computer. For example, pictures are converted into matrices of integers. But this is not enough, as Fig. 1 illustrates. While the silhouettes of all instances of each character differ significantly in the area of the paper they occupy, human observers still perceive them as examples of the same letter. Therefore, we must apply further transformations such that objects are represented in a mathematical form that admits a *similarity measure* so that objects belonging to the same class will be near each other according to that mea-

sure. The following are two of the most common approaches.

In *statistical pattern recognition* objects are mapped into a set of numbers, the features, which are interpreted as coordinates of points in a high dimensional Euclidean space. The ideal object is assumed to be the mean of a statistical distribution generating the objects. Learning then amounts to the inference of the parameters of the distribution, and classification is usually done according to a maximum likelihood criterion.

In *syntactical pattern recognition* objects are mapped into a string of symbols from a finite alphabet, and it is assumed that there exist grammars that generate the strings of all objects that are forms of the same ideal and no others. Learning amounts to the inference of grammars, and classification is done by parsing.

In addition, the term *structural pattern recognition* is used for approaches that emphasize the expression of objects of complex form as aggregates of simpler parts, and perform the classification in two steps: classifying the simpler parts first, and then deducing the class of the aggregate from the parts and their interrelationships. Structural pattern recognition could use either statistical or syntactic classifiers, or a mixture of both.

The experience of the last thirty years suggests that the transformation of physical input to a mathematical entity (*feature generation*) is the most critical step for the success of pattern recognition by computer. Research in various applications has identified some transformations as more promising than others, but there is no general theory and it does not seem likely that there will ever be one. Indeed, the features depend so much on the problem that the field of pattern recognition has lost some of the unity that it had when the emphasis was on classification. Pictorial pattern recognition is closely related to image analysis, while speech recognition is closely related to speech processing and so forth.

Some sets of features are information preserving in the sense that one could reconstruct the object (or a close approximation of it) from the features. For example, the Fourier transform of a waveform provides such a set of features. This is not necessarily a desirable property. Quite often we would like to have economical descriptions, where each feature takes a significant value for only one class. The following example illustrates this. [*See* SIGNAL PROCESSING, GENERAL.]

EXAMPLE 3. Consider the shapes of Fig. 2. We examine two problems. One where line *AB* divides the two classes (Problem I) and another where line *CD* divides the two classes (Problem II). In the first case we must distinguish smooth outlines from rough ones and in the second circular outlines from oblong outlines. The moment invariants of a plane shape may be used to solve the second problem. If $[x(s), y(s)]$ is a parametric representation of the contour with the length s as a parameter, then we can compute moments M_{ij} as

$$M_{ij} = \int x(s)^i y(s)^j \, ds \qquad (1)$$

Clearly, M_{00} is the length of the contour and M_{10}/M_{00} and M_{01}/M_{00} the coordinates of its center of gravity. We may compute higher order moments with respect to that center rather than the origin. (Actually, there is a simple transformation between the two.) Circular contours will have moments M_{20} and M_{02} nearly equal, so the ratio M_{20}/M_{02} could be used as a single feature to distinguish circles from ovals. One may try to use higher order moments for solving Problem I as well, but it is more natural to look for a measure of contour roughness. If we plot the angle ϕ that a tangent to the contour makes with the horizontal, as a function of the parameter s, then the function $\phi(s)$ will be monotonic for smooth contours, but not for rough contours. (See Fig. 3.) A good feature for recognizing roughness will be the number of times the first derivative of that function becomes zero, or the number of zero crossings of $\phi'(t)$. A combination of the two features would allow us to recognize four kinds of patterns: smooth oblong, smooth circular, rough oblong, and rough circular.

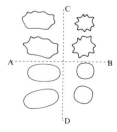

FIG. 2. Samples of contours that can be classified according to two criteria: their roughness and their circularity.

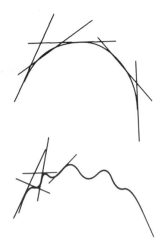

FIG. 3. The angle with the horizontal of the tangent to a contour is a steadily increasing function for smooth contours, but it fluctuates for rough contours.

In order to have a working system we still need to specify a number such that when the ratio M_{20}/M_{02} is below it, then we shall infer circularity. Also, we may wish to allow some zero crossings of $\phi'(t)$ before calling the contour rough. We set the thresholds by obtaining samples from each class of contours, and then using them to estimate these numbers. Such samples constitute the design or training set. Then we test our solution on a second set of samples, the testing set. This process of inferring the parameters used in the solution from samples lies at the heart of mechanical pattern recognition and it has many outward similarities to human learning. Since such processes must be implemented on a machine (the digital computer) they must be specified mathematically in great detail. One cannot escape from implementing a mathematical parameter estimation procedure. In contrast, human learning has no such limitations and this explains the relative poor performance of mechanical pattern recognition compared to human pattern recognition.

Problems that may be encountered during the training/testing process include the following.

1. It may not be possible to separate the classes. For example, we may find that contours labeled as smooth have between 0 and 6 zero crossings while rough contours have 4 to 30. We have an overlap of classes in the feature space. This cannot be fixed without modifying the feature definition.

FIG. 4. Partial occlusion and overlap make the contour recognition problem far more difficult for machines than for humans.

2. Even if we separate the classes during training, we have no guarantee that they could be separated later.

Before going much further we should ask ourselves whether we have a practically useful pattern recognition system. The answer is no because we have assumed perfect definitions for the contours. In practice such contours are extracted as outlines of objects that may touch or even overlap each other, may be obscured by other objects, etc. A human observer has little trouble with the recognition of patterns in Fig. 4, but the problem is much tougher for a machine.

III. Feature Generation for Pictorial Patterns

Classification of pictorial patterns falls into two broad categories: recognition of the shape of the silhouette or recognition of the interior. There are two kinds of shape descriptors: integral descriptors and structural descriptors. The latter attempt to imitate empirically human perception and they comprise two broad classes: contour features and skeletal features. Objects that are compact can be expressed by the dents and protrusions of their contour. Objects that have an elongated shape can be described by their skeletons.

The term "preprocessing" is used loosely to include all operations that are performed on the data before the application of the pattern recognition algorithm. Such operations are assumed to be of a general nature and independent of the recognition problem. Typically they include *digitization* (sampling and quantization), *filtering* (noise elimination), and *segmentation* (separation of objects from the background and from each other). In many practical applications the right kind of preprocessing can facilitate significantly the solution of the pattern recognition problem. [*See* COMPUTER PROCESSING OF PICTURES.]

EXAMPLE 4. Pages of printed text are supposed to be black and white images. However, digitizers tend to average the brightness over an area, so that the digitized image has intermediate values. A common method for converting this image back to black/white form is thresholding: pixels above a given value are called white and the others are called black. Because of the previous averaging, this process may not reproduce the original image and, in particular, it may create broken or touching characters. Thus, the quality of the printing presented to the computer program is poorer than that of the original. The digitizing process has added noise (subtracted detail)!

A. INTEGRAL DESCRIPTORS FOR PLANAR SHAPES

These are computed over the area or the contour of a silhouette. Typical amongst them are moments, Fourier shape descriptors, etc. If $b(x, y)$ is the brightness function of the input and $[x(t), y(t)]$ a parametric description of the contour, then integral shape descriptors are usually in one of the following two forms:

$$f_i = \int_{\text{AREA}} b(x, y)g_i(x, y)\, dx\, dy$$

or

$$f_i = \int_{\text{LENGTH}} F[x(t), y(t)]\, dt$$

For an example of the function $F[\,]$ see Eq. (1) above. The kernel $g_i(x, y)$ can be thought of as a mask. Such integral forms capture the overall shape of the object, but they tend to miss highly localized characteristics. In the example of Fig. 5 an integral descriptor is more likely to partition the pictures in groups (A, D) and (B, C) rather than (A, B) and (C, D). Of course, the application determines which is the most desirable grouping.

B. CONTOUR DESCRIPTORS FOR PLANAR SHAPES

The curvature computed along a contour conveys important shape information and extrema

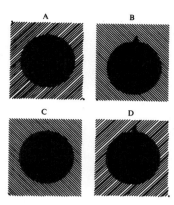

FIG. 5.   The above four pictures can be paired in two ways: (a) according to the form of the background (lines at 45 or 135 degrees); (b) according to the shape of the foreground (circle with or without protrusion).

of curvature can be used as features. The major problem with curvature is that it uses derivatives and therefore its computation is sensitive to noise. We would like to find only "essential" curvature maxima and ignore "wiggles" as shown in Fig. 6. One solution to the noise sensitivity of curvature is to prefilter the data. If $f(s)$ denotes the curvature as a function of arc length and $g(s)$ is the impulse response of a filter, then the filtered output will be

$$\int_{-\infty}^{\infty} f(s)g(t - s)\, dt$$

If $g(s)$ is a broad based function (low pass filter), then quick changes in curvature will be averaged out. Asada and Brady have used a Gaussian fil-

ter with

$$g(t) = (1/\sqrt{2\pi}\sigma)\, \exp(-t^2/2\sigma^2)$$

(where σ is the bandwidth of the filter) followed by differentiation to detect feature points.

For some applications concavities along the contour are sufficient shape descriptors. Because the raw data are noisy, it is common to do some curve fitting first before attempting to search for concavities. A polygonal approximation is often sufficient, but higher order spline approximations are also used. When polygonal approximation is used in the context of pattern recognition, the number as well as the locations of the vertices must be determined and this results in a nonlinear optimization problem. However, suboptimal solutions are often good enough. Fig. 7 shows a contour shape descriptor used by S. Mori and K. Yamamoto for a handwritten numeral recognition system. First, a polygonal approximation of the contour is found and the convex hull of that polygon is determined. Features are computed from the differences between the convex hull and the polygon. Such differences correspond to concavities in the contour. The following features were used for shape description: (a) the length of the polygon along the concavity, L_1; (b) the length of the chord across the concavity, L_2; (c) the distance ρ between the chord and the center of gravity CG of the sides of polygon that are part of the concavity; (d) the direction of the normal to the chord θ; etc.

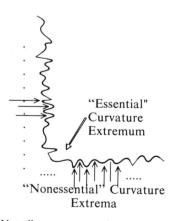

FIG. 6.   Not all curvature maxima are equally important for shape description. Human observers may ignore the "wiggles" and perceive the above curve as two nearly straight lines that form a corner.

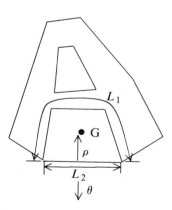

FIG. 7.   Features generated from the concavity of contour used for handwritten character recognition. (Adapted from Yamamoto, K. and S. Mori, "Recognition of Handprinted Characters by Outermost Point Method," *Proc. Fourth Intern. Joint Conf. Pattern Recognition,* Kyoto, Japan, Nov. 7–10, 1978, pp. 794–796.)

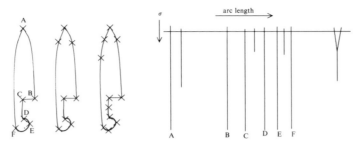

FIG. 8.   Critical points overlayed on a contour (left) and their locus on the σ–arc length plane. (Adapted from Figure 23 of H. Asada and M. Brady "The Curvature Primal Sketch," *IEEE Trans. Pattern Analysis and Machine Intelligence,* **PAMI-8** (1986), pp. 2–14.)

In order to find an approximation, one must choose a parameter for the closeness of fit. Such a parameter determines which wiggles will be called noise and smoothed out. The corresponding problem in filtering is the bandwidth of the filter. (The parameter σ in the Gaussian filter described above.) Multiscaling has been proposed as an answer to this difficulty. Instead of a single parameter one uses a sequence of filters and the features are selected not from amongst a set of points, but from a set of lines as shown in the right side of Fig. 8. In this figure some feature points are present over a wide range of the parameter σ, while others are not.

C. SKELETAL DESCRIPTORS FOR PLANAR SHAPES

There are many mathematical techniques for converting a planar region (a blob or silhouette) into a set of thin lines. (Not necessarily straight lines.) The set of such lines is called the skeleton of the blob and the transformation is called thinning. Figure 9 shows two examples. While in theory such transformations may be used for any shapes, they are more appropriate for objects that were meant to be thin to start with,

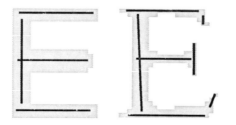

FIG. 9.   Stroke vectors overlayed on digitized printed characters. (Adapted from Fig. 3 of T. Pavlidis "A Vectorizer and Feature Extractor for Document Recognition.")

such as alphanumeric characters. (These are assumed to be written with pen or pencil strokes.) The details of the thinning algorithms are beyond the scope of this article and the conversion of blobs into line drawings is only the first step in feature generation. The collection of the lines resulting from thinning must be processed further in order to produce useful features such as endpoint or junction locations.

D. FEATURES FOR TEXTURED AREAS

Texture expresses the roughness of an area and it is usually described by second order statistics: sets of numbers that describe how much the brightness changes from pixel to pixel. This means that texture must be computed over an area rather than over a single pixel. If we select too small an area, then we have an unreliable estimate of the statistics. If the area is too large, then it is likely to contain more than one texture, and the estimate is unreliable again. The strategy for area selection is beyond the scope of this article, and we will discuss only examples of features.

The simplest texture statistic is the cooccurrence matrix. For a picture with G gray levels, it is a $G \times G$ matrix whose ijth element equals the number of pairs of adjacent pixels the first of which has intensity i and the second has intensity j. If a picture has uniform brightness, then all the elements of the cooccurrence matrix will be zero except along the diagonal. In general, smooth areas yield matrices with diagonal dominance. A 256×256 picture with 256 gray levels will yield a cooccurrence matrix of the same size as the picture itself, so we have not gained anything. Therefore, it is necessary to use some numbers computed over the matrix as features. One possibility is to merge gray levels and create a smaller matrix, say 8×8, and use its elements as features. Or we may use the fraction of the elements in a diagonal zone as a single feature.

The power spectrum (i.e., the absolute values of the Fourier transform) of an area provides texture information because rough areas produce high frequencies. Of course further data reduction is needed. One possibility is to use a model for the generation of the texture and then to estimate its parameters from the spectrum. The fractal dimension of an area provides a measure of its roughness, and fractal models have been used in computer graphics to generate natural looking images. If the fractal hypothesis is correct, then the power spectrum $P(f)$ should be proportional $f^{-(2H+1)}$ for some parameter H. By taking logarithms we find that

$$\log(P(f)) = -(2H + 1) \log(f) + k$$

If we compute $P(f)$ over an area, then we can estimate H from the slope of the best fitting straight line for the $\log(P(f))$ versus $\log(f)$ plot. Thus H is a texture feature. This method was proposed by A. Pentland and used to recognize areas of distinct texture.

E. SOLID OBJECTS

Solid objects can have different orientation with respect to the viewer so their projections on the image plane may differ from instance to instance. In some applications the recognition does not stop with the object identification, but continues until certain parts of it are identified. For example, a robot not only must find out which object is, say, a hammer, but it also must identify the handle of the hammer. One may debate whether object description is pattern recognition, and from a philosophical viewpoint the debate makes sense. However, structural pattern recognition identifies the patterns from their descriptions (rather than global measurements) so the distinction does not make much practical sense.

There are two basic approaches to solid object recognition. One is to divide the image into regions of uniform brightness and then attempt to recognize the object from these parts. (This method can be used with other criteria: for example, uniform texture or constant gradient of the brightness function.) The other is to attempt a direct detection of the surfaces. Figure 10a shows an example where each uniform area corresponds to a plane face of a cube. Figure 10b

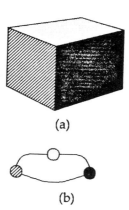

(a)

(b)

FIG. 10.   The lower part of the figure shows a graph whose nodes correspond to regions of the top part. Such a graph can be used to generate features for the recognition of solids.

shows a graph whose nodes correspond to regions and whose branches link adjacent regions. Binary features can then be defined on the basis of labeled subgraphs by setting a feature value to one, if the corresponding subgraph is found in an image. Nonbinary features are also possible if one is willing to define partial graph matching.

Direct detection of surfaces can be made on the basis of illumination models, or directly by fitting range data. However, there are limits on the complexity of the surfaces to be detected, so that even if one uses higher order surfaces, complex objects must be recognized as labeled graphs, similar to that of Fig. 10b. The recognition of the shape of a surface from the illumination values is based on the observation that for diffuse illumination the brightness of a surface is proportional to the cosine of the angle between the normal to the surface and the direction of light. Let n_x, n_y, and n_z denote the components of the normal to the surface that we seek to determine at a point $(x, y, z(x, y))$. (The relation between z and x and y is established from the definition of the surface.) If the measured brightness at that point is $f(x, y)$, the light intensity I, the surface reflectance ρ_s, and the illumination is parallel along a direction u_x, u_y, and u_z, then the following equation must be valid.

$$f(x, y) = I\rho_s(n_x u_x + n_y u_y + n_z u_z)$$

In general only f is known, so this equation has too many unknowns. Therefore, one must use more than one point (x, y), which may lead to a system of overdetermined equations. In such a case we look not for an exact solution but for a

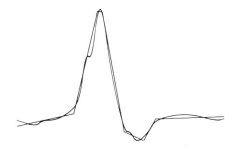

FIG. 11. A waveform with its polygonal approximation.

solution that minimizes the error between the left hand and right hand sides of the equations. We should also point out that if two surfaces have the same gradient function they cannot be distinguished by this method.

F. Waveforms

Waveforms are often, but not always, functions of time. In traditional signal processing waveforms are processed by spectral techniques, but in pattern recognition we are often interested in features that are of short duration, which may be difficult to detect from the spectrum. In some cases the spectrum itself is treated as a waveform.

Probably the most important type of waveforms are those encountered in speech processing applications. Other types include electrocardiograms, well logs from geophysical exploration, etc. For the most part, pattern recognition of such waveforms relies heavily on specialized knowledge of the application. This is particularly true for speech. Still there are some problems that are application independent, such as the detection of major peaks. Because of the noise, they cannot be found as extrema where the first derivative becomes zero. Therefore, much of the research has focused on alternative techniques. A piecewise linear approximation of the waveform will yield a sequence of slopes that can be used as input to a syntactic classifier. Figure 11 shows an example of a waveform and its polygonal approximation.

IV. Inference (Learning) and Classification

Early efforts in pattern recognition had focused on learning and classification and it was not until the late sixties that the importance of preprocessing and feature generation was realized. While we know now that classification techniques cannot provide by themselves the solution to the pattern recognition problem, they are still a necessary part of the overall process and they should not be overlooked.

Let $\mathcal{X}$ be the space where the objects have been mapped after the preprocessing and feature generation. The space $\mathcal{X}$ may be a (Euclidean) space of n—tuples of numbers, or a set of strings from particular alphabet, etc. We also have sets of labeled samples $S_1, S_2, ..., S_N$ with instances of each pattern. Inference amounts to finding partitions of $\mathcal{X}$ that are well-defined mathematically and have the property that all the members of S_i are in the ith partition. Figure 12 shows an example where $\mathcal{X}$ is the plane and there are four patterns (classes). Note that we can solve the problem of pattern recognition without having explicit prototypes: the partition is enough. Classification is equivalent to finding the partition where a new object falls. On the other hand, if we select prototypes and a distance measure, then pattern recognition is equivalent to finding the nearest prototype to a new object. Of course, this implies a partition as Fig. 13 shows.

If we assume that $\mathcal{X}$ is given, then there is a tradeoff between the number of patterns we can handle and the variation from the ideal we allow. If we allow only a little variation, then the samples will fall on top of each other and we have space for many partitions. If we allow large variations, then we need a large area around each prototype and we can fit only a few partitions within $\mathcal{X}$. This trade off is well known in practice. In speech recognition we have systems with small vocabulary (few patterns) that can understand many speakers (large variations from the ideal) or systems with large vocabulary

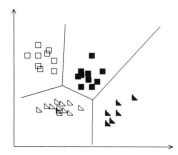

FIG. 12. Sample points from four different classes of patterns displayed on the feature space (a plane in this case).

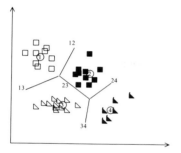

FIG. 13.   The numbers in circles denote prototypes, while the labels next to the line segments refer to the two prototypes that are equidistant from that segment.

(many patterns) from few speakers (small variations from the ideal). Similarly we have reading machines that recognize handwritten numerals only (few patterns) written by the general population (large variations from the ideal) or machines that recognize all alphanumeric characters and punctuation marks (many patterns) from a single typewriter font only (the only variations from the ideal are due to noise). A simultaneous increase in both number of patterns and allowed variability implies an enlargement of the space $\mathcal{X}$, which amounts to either an increase in computational costs (possibly more features) or an advance in the state of the art.

A. TEMPLATE MATCHING

It may be possible to construct templates (e.g., pixel matrices) for each pattern and directly match objects with these templates. Template matching is a prime example of the holistic approach where patterns are matched without regard to their detailed structure. Let $f_i(x, y)$ be a set of templates (ideal patterns) defined over a region R and $g(x, y)$ be an object that must be recognized. Let $d(g, f_i)$ be a distance measure defined as

$$d(g, f_i) = \int_R [g(x, y) - f_i(x, y)]^2 \, dx \, dy \quad (2)$$

Then we can recognize g as the f_j pattern if

$$d(g, f_j) = \min_i [d(g, f_i)] \quad (3)$$

Equation (2) provides a rather naive distance measure but it will serve to illustrate some problems. One is the registration problem, namely the lining up of the input with each template. The other problem is distorting transformations, such as scaling, rotation, etc. If T_k is a set of such transformation, then Eq. (2) can be re-

placed by

$$d(g, f_i)$$

$$= \min_k \int_R [T_k(g(x, y)) - f_i(x, y)]^2 \, dx \, dy \quad (4)$$

The transformation family may depend on more than one parameter, so the process of finding the best matching transformation can be rather expensive. In particular, one may include coordinate transformations as well as value transformations.

B. STATISTICAL CLASSIFICATION

Statistical pattern recognition assumes that each object can be represented by a vector $\mathbf{x}$ whose components are the features and one of the following two conditions holds:

1.   The vectors $\mathbf{x}$ obey statistical distributions that have different parameters when conditioned for different patterns.
2.   There exists a distance measure defined over the vectors so that $d(\mathbf{x}, \mathbf{y})$ is small when $\mathbf{x}$ and $\mathbf{y}$ are instances of the same pattern and large when they are from different patterns.

Let $p(A \mid \mathbf{x})$ be the probability that we are observing pattern A given the feature vector $\mathbf{x}$. Then we recognize patterns by finding the maximum of $p(A \mid \mathbf{x})$ over all possible patterns. If we have a set of samples from each class, then it is easy (at least in principle) to compute the statistics of $\mathbf{x}$ for each pattern $p(\mathbf{x} \mid A)$ using the design set. Then we can find $p(A \mid \mathbf{x})$ from Bayes' formula.

$$p(A \mid \mathbf{x}) = \frac{p(\mathbf{x} \mid A)p(A)}{\sum_B p(\mathbf{x} \mid B)p(B)} \quad (5)$$

This is a generally valid formula, but there are difficulties with its application because experimental data for $p(\mathbf{x} \mid A)$ may be difficult to organize. We examine two cases. (a) $\mathbf{x}$ is a binary vector (all its elements are either zero or one). In practice $\mathbf{x}$ has a large number of components, often over 100. But even with 20 components we have $2^{20} = 1,048,576$ possible vectors. It is practically impossible to make a table of how often each of these patterns occurs for each class. (b) $\mathbf{x}$ takes real values. While such vectors usually have fewer components than binary vectors, the designer is faced with the impossibility of covering all values, since each component now takes infinite values. While the mathematical theory of sieves offers some tools for dealing with this case, the problem is practically intractable without any simplifying assumptions.

The inapplicability of tabular methods leads into making assumptions about the functional forms of the probabilities involved. Thus we sacrifice generality for practical feasibility. A common assumption is that $p(\mathbf{x} \mid A)$ is a Gaussian distribution

$$p(\mathbf{x} \mid A) = k \, [\det(\mathbf{Q}_A)]^{-1/2}$$
$$\times \exp[-\tfrac{1}{2}(\mathbf{x} - \mathbf{m}_A)'\mathbf{Q}_A^{-1}(\mathbf{x} - \mathbf{m}_A)] \tag{6}$$

where k is a constant, $\mathbf{m}_A$ the mean and $\mathbf{Q}_A$ the autocovariance matrix for class A. The prime denotes transpose. Then during the training step we must compute $\mathbf{m}_A$ and $\mathbf{Q}_A$ for each pattern.

If the features are binary and we assume that they are statistically independent, then we can write

$$p(\mathbf{x} \mid A) = \prod_i p_{Ai}^{x_i}(1 - p_{Ai})^{1-x_i} \tag{7}$$

where p_{Ai} is the probability that the ith binary feature is one, given that the true pattern is A. Then, training reduces to the estimation of the parameters in the above equations.

It is possible to simplify things further by estimating and comparing logarithms of probabilities rather than the probabilities themselves and also by removing from the comparison, constants that do not depend on the pattern. The resulting functions are called discriminant functions. In many cases they turn out to be linear functions of the form

$$g_A(\mathbf{x}) = \mathbf{w}_A'\mathbf{x} + w_A^0 \tag{8}$$

which can be simplified into

$$g_A(\mathbf{x}) = \mathbf{w}_A'\mathbf{x} \tag{9}$$

by extending the vector $\mathbf{x}$ with a component equal to one and then including w_A^0 inside $\mathbf{w}_A$. Classification is performed by computing $g_A(\mathbf{x})$ for each A and deciding that $\mathbf{x}$ belongs to class B if $g_B(\mathbf{x})$ is greater than all other $g_A(\mathbf{x})$'s for $A \neq B$. There are methods for estimating $\mathbf{w}_A$ and w_0 without necessarily knowing anything about the underlying distribution. In particular, if there exist vectors $\mathbf{w}_A$ such that we can correct decisions over all the samples of a design set X, then we can find them as follows:

1. Start with an arbitrary set of $\mathbf{w}_A$'s.
2. Examine all samples. If a given sample $\mathbf{x}_k$ that belongs to class A is placed in class B, correct the vectors by adding $\mathbf{x}_k$ (possibly scaled by a constant) to $\mathbf{w}_A$ and subtracting it from $\mathbf{w}_B$.

3. If any changes were made in the $\mathbf{w}_A$'s, repeat step 2.

It can be shown mathematically that this method always finds a solution if one exists. This simple algorithm has often been described in pseudo-psychological terms: rewarding the classifier for correct decisions (adding to $\mathbf{w}_A$) and punishing for wrong (subtracting $\mathbf{w}_B$).

Distance versus Probability. If we take the logarithm of both sides of Eq. (6) we find that

$$\log[p(\mathbf{x} \mid A)] = \log k - \tfrac{1}{2}\log[\det(\mathbf{Q}_A)]$$
$$-\tfrac{1}{2}(\mathbf{x} - \mathbf{m}_A)'Q_A^{-1}(\mathbf{x} - \mathbf{m}_A) \tag{10}$$

Then, a maximum likelihood decision rule is equivalent to classification by minimum distance, if the distance is defined as the quadratic expression in the right hand side of the equation. If the covariance matrices $\mathbf{Q}_A$ do not depend on the class, then it is possible to transform the data in such way that we can use the Euclidean distance as a similarity measure. Furthermore, one can show that in this case it is possible to devise a linear discriminant function like Eq. (9).

Good and Bad Models. In many problems we have no exact knowledge of the distributions and people use either the Euclidean distance or linear discriminant functions for classification. Figure 14 shows two sets of curves (thin and thick lines) that delineate regions of the plane where points from each pattern would lie with a certain probability. For example, if a point is from pattern a, it will lie with probability p_1^a to the left of the leftmost thin curve, while if it is from pattern b, it will lie with probability p_1^b to the right of the rightmost thick curve, etc. The use of a linear discriminant function or the Euclidean distance is equivalent to drawing a straight line and deciding that points to its left represent pattern a and points to its right represent pattern b. A statistically optimal procedure

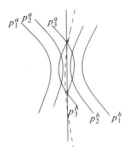

FIG. 14. Parts of contours of equal probability from two distributions.

might produce another curve, such as that shown dashed in Fig. 14. If the values of $p_1^{[ab]}$ are large, then it will be unlikely that any point will fall near the critical decision region and the difference between the two decision rules will be insignificant. On the other hand if $p_1^{[ab]}$ is small, then there will be many such points and the error will depend significantly on the rule. So the larger the error, the more important the choice of the rule is! This mathematical observation explains the practical experience that simple classifiers are good enough when one has good features and elaborate classifiers do not help if the features are not properly chosen.

Nearest neighbor rules attempt classification by finding the closest point from the design set to an unknown sample. They do not assume any previous efforts for finding prototypes or distributions. Of course, the use of a distance measure implies a distribution as we just saw. Let X be a set of n feature vectors $\mathbf{x}_i$ each with a label c_i denoting its correct class. Then a new sample $\mathbf{x}$ is placed in class j if

$$d(\mathbf{x}, \mathbf{x}_j) = \min_i[d(\mathbf{x}, \mathbf{x}_i)]$$

If we assume an underlying probability distribution for the patterns and n to be very large, then it can be shown that the nearest neighbor rule provides a decision that is wrong no more than twice as often as an optimal statistical decision.

C. SYNTACTICAL CLASSIFICATION

Syntactical pattern recognition assumes that each object can be represented by a string s and that there exist grammars G_i such that each grammar generates all the strings corresponding to a particular pattern and only those strings. For example, a contour can be expressed as a sequence of small, approximately circular arcs. We may decide to map each one of them to one of three symbols: c if it is concave, v if it is convex, and f if it is flat. Then, any contour could be expressed as a string using the three symbols c, v, and f. The smooth contours in Fig. 2 will be represented in the form v^n while the rough contours will be concatenations of groups of symbols of the form $(c^{k_i}f^{l_i}v^{m_i}f^{n_i})$ where l_i and n_i could be zero. This simple encoding makes the discrimination between rough and smooth contours very easy, but it is not very helpful in separating ovals from circles. One could use a larger alphabet so that not all convex arcs are mapped into the same symbol, but even then the syntactic approach would not be very useful.

Syntactic methods are well suited for detecting local patterns or where there is recursion, but they become awkward for nonrecursive global characterizations. The following example shows a good application of syntactic techniques.

We want to distinguish triangular from square pulses as shown in Fig. 15. Let p denote a small segment with positive slope, n a small segment with negative slope, and z a small segment with nearly zero slope. Then both the triangular and the square pulses will be of the form $p^k z^m n^s$. However, m will be small (below two for example) for triangular pulses and large (greater than three for example) for square pulses. We have now two grammars generating each class of pulses

Triangular pulses	Square pulses
$I \rightarrow PZN$	$I \rightarrow PZN$
$P \rightarrow pP/p$	$P \rightarrow pP/p$
$N \rightarrow nN/n$	$N \rightarrow nN/n$
$Z \rightarrow$ null$/z$	$Z \rightarrow zZ/z^4$

String grammars can handle only one-dimensional situations and therefore tree and graph grammars have also been used.

A major obstacle in the application of syntactic techniques is the difficulty of grammatical inference: given two sets of strings find a grammar that accepts all strings of the first set and none of the strings of the second set. This is the counterpart of the problem of estimating the discriminant functions which we discussed above. The first difficulty of grammatical inference is that one can construct rather easily a finite automaton that will accept exactly the strings of the first set, therefore appearing to solve the problem. However, such an automaton is useless unless the sample of strings included all possible strings of the first class, something that is unlikely in practice. Therefore, one must place constraints on the structure of the automaton or grammar so that it accepts other strings besides those in the sample. But what strings? A person may answer, "strings similar to those in the sample." Unfortunately, we have no measures

FIG. 15.   A sequence of pulses of two kinds of shape: triangular and square.

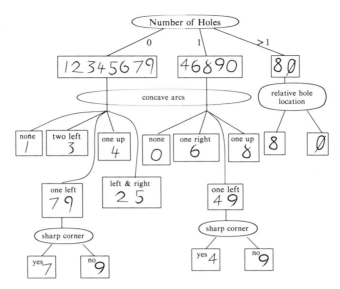

FIG. 16.   Decision tree for the recognition of handwritten numerals. Features are listed inside ovals and results of a decision in squares. The tree is incomplete because it does not contain a discriminator between "2" and "5." We leave it to the reader to think of one.

of string similarity that reflect the similarity seen by a human observer. If the number of classes is small, then it is possible to have interactive inference, and grammars are useful in such cases. For example, the literature contains examples of syntactic pattern recognition for numerals (ten classes) but not for the complete set of alphanumerics and punctuation marks.

D. HIERARCHICAL CLASSIFICATION

Hierarchical pattern recognition does not classify objects immediately, but it uses a sequence of preclassifications using a decision tree. First, we classify objects into a small number of classes, then each class is subdivided into more, and so forth. Some problems lend themselves naturally to such a hierarchy (for example, medical diagnosis) but a hierarchical process is sometimes desirable for efficiency. If we have N classes and we attempt to place an object in one of them, the computational effort will be proportional to N. On the other hand if we use a sequence of bisections (i.e., a binary decision tree), then the computational effort will be proportional only to the logarithm base 2 of N. In addition, we may save on the computation of the features. We need not compute all possible features for each sample, but only those needed

to reach a decision. In this respect, we have sequential pattern recognition. Figure 16 shows an example of hierarchical classification. Note that while there are only ten true classes, the hierarchical classifier yields fifteen but the average number of features computed is less than three. The reader should be warned that a practical numeral recognition system requires a much bigger decision tree because of noise and style variability.

V. Applications

Tasks which are repetitive and uninteresting for humans or which are performed in dangerous or unhealthy environments are good candidates for mechanization and offer the challenge to replace human operators by computers. For example, spray painting in car assembly lines was one of the first tasks where human workers were replaced by robots. Balancing the desirability for automation is the feasibility of building a system that can match the human performance. Tasks which involve human visual and auditory pattern recognition seem particularly challenging. It is unlikely that one could build a pattern recognition system in the foreseeable future that would outperform a highly motivated, interested, well trained, and comfortable human worker. On the other hand, the experience of the last ten years indicates that mechanical pattern recognition can compete successfully with poorly motivated, bored, poorly trained, and/or uncomfortable human workers.

Regardless of the desirability, a critical factor in determining feasibility is the degree to which we have control of the environment. Thus, industrial applications where illumination is under complete control tend to be easier than military applications where not only the environment is beyond our control, but we may also be facing an adversary situation (camouflage). The list of applications given below is based primarily on the published literature. (It is possible that other successful applications of pattern recognition have been kept secret for commercial or security reasons.)

There is a big step in going from a method that appears theoretically attractive, to one that is practically useful. Most practical applications impose very severe performance requirements on pattern recognition systems. It is usually desirable that there should be no false recognition (substitution) at all, and any errors should be cases where the automatic system flags things it cannot classify (rejection) in order to have them corrected later by a human observer. In industrial inspection or medical diagnosis problems, it is important that no defective objects or pathological cases be missed, and any errors should be false alarms that can be analyzed either by a more sophisticated system or by a human specialist.

Clearly, zero error rates are impossible, so specifications define some finite value that a system must meet before it can be accepted. That value depends on the specific application. A substitution error in an automatic postal sorter means a piece of mail is sent first to the wrong town, thus increasing the eventual delivery time by a factor of two or three; an unpleasant but not catastrophic consequence. On the other hand, letting a defective piece past an inspection stage may result in serious human injury or damage to other expensive equipment. In such cases the substitution error rate should be 10^{-6} or less. Since most published methods report far higher rates (and some do not even make the distinction between the two types of errors), one is justified in being skeptical about the true significance of the published solutions. (Note, also, the difficulty of certifying such low error rates.)

The practically acceptable rejection error rates are also lower than what is usually reported in the literature. There the penalty is mostly financial and may eliminate the cost effectiveness of the automatic system. For example, the literature often contains descriptions of character recognition systems with a 5% error rate. It requires significant ingenuity to develop such a system, and its creator may have good reason to be proud of it. However, it is practically useless: a typical page of typewritten text contains about 2000 characters so that error rate translates into 100 errors per page. It will take a good typist longer to correct these errors than to retype the whole page! A more desirable practical solution is to need one correction every few, say five, pages. That translates into a 0.01% error rate (which is what most commercially available devices claim).

The following is a summary of some of the most common applications in the public domain. They are listed roughly in order of decreasing control over the environment (and therefore increasing difficulty).

Text Recognition. This is one of the earliest applications and there are reports that Alan Turing himself tried his hand on this problem as early as 1945. It is usually referred to as Optical Character Recognition (OCR) for historical reasons. The word "optical" is used to distinguish it from magnetic ink readers and the word "character" means alphanumeric character. This is an application where sufficient progress has been made so that there are many commercially available machines doing this task, at least within certain constraints. For example, at the time of this writing some department stores used OCR to read price labels that have been printed with a special font. Also, the Japanese and United States post offices, amongst others, were using OCR for reading postal codes. (However, at this time there is no machine available that would handle successfully the example of Fig. 1.) The first application of OCR was the reading of handwritten programs into the machine. Time sharing and video terminals made that application obsolete long before any substantial progress was made, but other applications appeared in the meantime. These include postal automation (reading of postal codes or complete address for mail sorting), office automation (OCR is used as a link between the paper and the electronic world), reading machines for the blind, etc.

Recognition of Graphics. One obstacle to the introduction of Computer Aided Design

(CAD) is the existence of paper drawings that are still useful. (For example, plans of the telephone network.) Therefore, there is a need to convert the paper drawing into computer readable form. This requires the recognition not only of text but also of graphics and symbols. For example, all the objects of Fig. 17 should be recognized as diodes.

Industrial Inspection. Specific applications include shape analysis for quality control, inspection of wired boards, inspection of VLSI masks, surface inspection for smoothness, etc. Most such applications belong more to image processing than to pattern recognition because they emphasize detection of anomalies. A special challenge is the need for very high throughput. This has led to suggestions of a multistep process. A simple but fast algorithm with a high false alarm rate is used to identify a potentially troublesome area, which then can be processed by a slower but more powerful method.

Robotics. While the first three classes of applications dealt with plain shapes, robotics vision includes the recognition of solid objects. Some may be objects that the robot must pick up and others are part of its environment, such as obstacles it must avoid. At this time very few industrial robots use vision (less than ten percent according to some industry sources).

Geophysical. Pattern recognition has been used, at least in laboratories, for exploration of petroleum deposits using two kinds of primary measurements. One kind is well logs that are records of physical variables as functions of depth. Another kind is seismic data that are records of propagation and reflection of seismic waves produced from explosive charges. The well log analysis is reduced to a waveform analysis, and multiscaling was first developed for that purpose. The immediate goal is to segment the waveform into regions that can be recognized as different kinds of rock, sand, etc.

Biomedical. Because of the high cost of professional or even technical medical labor, automation has been offered as a means of pre-

screening. (Few patients would trust a completely automatic diagnosis even if it were possible.) The following are applications reported in the technical literature:

1. Chromosome classification: Early methods focused on the analysis of the shape of contours using syntactic techniques. The development of banding techniques has caused the introduction of gray scale imaging techniques into the feature generation process.

2. Cell counting: This requires the recognition of different kinds of blood cells and reports the number of each type in a given blood sample.

3. Analysis of Electrocardiograms: The shape of the waveform is supposed to contain information that allows the diagnosis of certain pathologies. The literature contains examples of successful detection of severe pathologies, but the challenge is to detect subtle abnormalities if the test is going to have any value as a broad screening tool.

4. Radiography: The major applications seems to be screening of chest X-rays.

5. Medical Diagnosis Systems: Such systems fall more properly under the category of "expert systems."

Fingerprint Recognition. This requires the matching of a fingerprint to one in a labeled data base in order to identify a suspect. While the fingerprints in the data base are usually of good quality, the unlabeled fingerprints are often of poor quality. As of this writing there are claims of successful automated systems in Japan, at least for prescreening: the computer extracts a small set of fingerprints matching a given sample and a human operator does the final matching.

Military. Target recognition is a frequently cited application, especially from low resolution infrared image. Other applications include general image analysis for automatic vehicle navigation and detection of abnormalities in signals that are the output of monitoring devices (signifying, for example, the presence of an intruder).

BIBLIOGRAPHY

Comput. Vision, Graphics, Image Processing (periodical).
Duda, R. O., and Hart, P. E. (1973). "Pattern Classification and Scene Analysis." Wiley, New York.
Grenander, U. (1976). "Pattern Synthesis." Springer-Verlag, Berlin and New York.
Fu, K. S., ed. (1977). "Syntactic Pattern Recognition, Applications." Springer-Verlag, Berlin and New York.

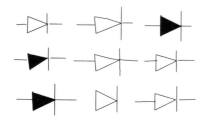

FIG. 17.   Examples of symbols of diodes.

IEEE Trans. Pattern Analysis Machine Intelligence (periodical).

Pavlidis, T. (1977). "Structural Pattern Recognition." Springer-Verlag, Berlin and New York.

Gonzales, R. C., and Thomason, M. G. (1978). "Syntactic Pattern Recognition." Addison–Wesley, Reading, Massachusetts.

Fu, K. S. (1982). "Syntactic Pattern Recognition and Applications." Prentice Hall, Englewood Cliffs, New Jersey.

Grenander, U. (1981). "Abstract Inference." Chap. 8. Wiley, New York.

Kittler, J., Fu, K. S. and Pau, L. F. eds. (1982). "Pattern Recognition Theory and Applications." NATO Advanced Study Institute Series. Reidel, Boston.

Pattern Recognition (periodical).

Pavlidis, T. (1982). "Algorithms for Graphics and Image Processing." Computer Science Press, Rockville, Maryland.

PERCOLATION

A. Aharony *Tel Aviv University*
D. Stauffer *University of Cologne*

GLOSSARY

Animals: Clusters in which every cluster containing the same number of sites has the same importance in averages over all clusters; correspond to percolation clusters with concentration $p \to 0$.

Bond: Connection between two neighboring sites.

Cluster: Group of neighboring occupied sites.

Concentration: Probability p that a site is occupied.

Conductivity: Ratio of electric current density to electric field.

Diffusion: Motion on a lattice where at every step the new direction is selected randomly, that is, independent of the previously selected direction.

Fractal: Object the mass of which increases with $(\text{length})^D$, where D differs from the Euclidean dimension d of the lattice into which the fractal is embedded.

Incipient infinite network: Infinite network at the percolation threshold.

Infinite network: In an infinite lattice, an infinitely large cluster; in finite systems, usually the largest cluster.

Site: Single element of a lattice.

Threshold: That concentration p of occupied sites at which for the first time an infinite network appears in an infinite system if p increases; denoted as p_c and often called the percolation threshold.

The theory of percolation aims to obtain quantitative estimates for the properties of disordered systems. In its simplest mathematical version, one considers a periodic lattice of sites, each of which is randomly occupied (with probability p) or empty (with probability $1 - p$). Such a lattice contains clusters of neighboring occupied sites. A simple example is shown in Fig. 1. As the concentration increases from zero, larger and larger clusters appear. The mean size of these clusters grows with p and diverges at a well-defined threshold concentration p_c. For $p > p_c$ there exists an "infinite" cluster, which connects the two sides of an arbitrarily large sample.

I. Introduction

Although the percolation problem is easily defined by one simple rule, it cannot be solved exactly. As we show in Section II, the model has very interesting properties, which exhibit universal features (independent, e.g., of details of the lattice). These features are closely related to the special geometric structure of the infinite cluster at p_c, which exhibits self-similarity, that is, invariance under the change of length scale. Graphs that have such properties, called fractals, can be described as having noninteger (fractal) dimensionality. [*See* FRACTALS.]

In this article we restrict ourselves to the simplest meaning of the complex concept of self-similarity: A structure is self-similar if subsections of the structure have a mass M proportional to some power of their linear dimension L, at least for large L:

$$M \propto L^D$$

ENCYCLOPEDIA OF PHYSICAL SCIENCE
AND TECHNOLOGY, VOL. 10

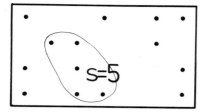

FIG. 1.   Example of site percolation and clusters.

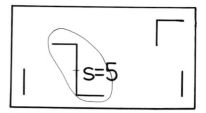

FIG. 2.   Example of bond percolation and clusters.

In special cases, like the Sierpinski structures to be mentioned in Section III, the concept of self-similarity attains a more direct geometric meaning, which for percolation clusters is valid only in the average sense.

The absence of a basic length scale inherent in self-similar structures is shared by many systems that undergo continuous phase transitions. Power laws like $M(L) \propto L^D$ appear in these structures for similar and other quantities. Scaling theory describes the interrelation of the exponents of the power laws describing different quantities. The study of the changes in the properties of systems when the length scale is changed was the basis for 1982 Nobel Laureate K. Wilson's renormalization group theory explaining scaling and universality.

Although the percolation problem is defined in geometric and statistical terms, it has had many applications in the physical sciences. Historically, the first percolation theory came from chemistry in connection with polymer gelation during World War II. Usually, polymers are linear structures, but sometimes branches occur. Macromolecules with many branches may form an infinite network that is no longer a liquid solution. The formation of pudding is an example of this so-called sol-to-gel transition or gelation; the pudding is a jelly. Another example is the boiling of an egg, where the initially more liquid egg becomes more solid after it is heated for some time. In these cases, the molecules are always there, but the chemical bonds between them are either formed or not formed. For the primitive case of our lattice this means that we need here a modified form of percolation: All sites are occupied, but bonds between neighboring sites are formed randomly with probability p and remain absent with probability $1 - p$. A cluster is a group of neighboring sites connected by bonds formed between them.

We call the second type of problem bond percolation (Fig. 2), whereas the first case defined earlier is called site percolation (Fig. 1). P.

Flory, and by a different method W. Stockmayer, solved the gelation problem with approximations that make it equivalent to bond percolation on the so-called Bethe lattice of Fig. 3, which contains no cyclic links and which ignores the constraint that two different parts of the macromolecule have to keep a certain minimum distance from one another and cannot occupy the same volume. This Flory–Stockmayer, or classical, theory plays a role in gelation similar to that of the van der Waals equation for the gas-to-liquid transition or the molecular field approximation for the ferromagnetic Curie point. In particular, all three examples predict a sharp phase transition; that is, a qualitative change takes place in the system at a certain value of a continuous variable. With increasing temperature T, the difference between a liquid and its vapor vanishes at the critical point, and the spontaneous magnetization of a ferromagnet vanishes at the Curie temperature. Similarly, with increasing time the heated polymer solution gels at the gel point.

What is the nature of this phase transition? Figures 1 and 2 make it plausible that the polymer solution becomes more solidlike once an infinite network of connected molecules is formed, that is, once an infinite cluster is formed in an infinitely large sample at the percolation threshold p_c. In the Flory–Stockmayer theory that threshold is

$$p_c = 1/(f - 1) \tag{1}$$

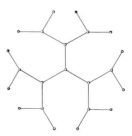

FIG. 3.   Central part of a Bethe lattice.

where the functionality f is the maximum number of bonds a molecule can form with its neighbors. Equation (1) holds for both site and bond percolation on the Bethe lattice ($f = 3$ in the example of Fig. 3). The name percolation and its application to regular lattices, as in Figs. 1 and 2, were introduced in 1957.

In spite of its history, the gelation phenomenon is not that application on which modern percolation theory is usually tested. The main interest of physicists in percolation centered on the critical phenomena in the immediate neighborhood of the percolation threshold p_c, described by critical exponents to be defined later. Not many experiments are performed with real polymers to determine reliably the critical exponents of gelation. The most precise information exists about the behavior of the viscosity, which diverges as one approaches the sol-to-gel transition from the sol. However, here the connection between the viscosity and the geometric cluster properties of percolation theory is not entirely clear. Clearer are the predictions of percolation theory for the fraction of mass contained in the infinite network or for the average mass and diameter of clusters as measured by light scattering. However, few precise measurements of critical exponents exist for these quantities. Thus, where theory is reliable, experiment is not, and vice versa. Therefore, even after several decades of research we cannot give a clear answer to the question whether percolation theory correctly describes the gelation process of polymers.

More successful was the application of percolation to thin films of materials sputtered onto a surface. Small droplets form, which grow to larger sizes if more material is added. Here a one-to-one correspondence between laboratory experiment and computer simulation of percolation was established, for example, for the fractal nature of the largest cluster, as reflected by the dependence of its "mass" on the system size.

II. Simple Static Properties of Percolation

Perhaps the most fundamental question of percolation theory is, what is the value p_c of the percolation threshold; that is, what is the probability p_c for which an infinite network is first formed in an infinite lattice? This problem can be investigated by exact mathematics, by expansion of suitable quantities into power series in p or $1 - p$, and by computer simulation. The results for common two- to four-dimensional lattices are given in Table I. Here the results for triangular site, square bond, triangular bond, and honeycomb bond percolation are exact; the others are numerical estimates, usually accurate to the last decimal given here. We see that Eq. (1) is not fulfilled but describes the general trend: The more neighbors a site has, the smaller is the percolation threshold.

It is rather easy to produce by computer a percolation picture, as in Fig. 4. One simply lets the computer go through a large two-dimensional array and, at every array element, decides with the help of a random number whether that site is full or empty. If the random number is distributed evenly between zero and unity, the site is taken as occupied for random numbers smaller than p, and empty otherwise.

It is not so easy to count clusters in a sample like Fig. 4 or to determine whether there is a cluster connecting the top and the bottom of the sample. That aim can be achieved efficiently with the Hoshen–Kopelman algorithm. There is a fortran program for counting clusters in bond percolation on simple cubic lattices. Only one line of a two-dimensional lattice, or one plane in three dimensions, has to be stored at any one time in the computer. Each cluster that is started new when we go through the lattice is given an index, and an index of index. If later two initially separated clusters merge into one cluster, this fact is recorded by changing the index of the involved index. The index itself remains unchanged; we do not have to go back in the lattice to relabel previously investigated sites. Fast computers can handle each site within a few microseconds; as of January 1987 the largest sys-

TABLE I. Percolation Thresholds for Common Lattices

Lattice	Site	Bond
Honeycomb	0.696	0.653
Square	0.593	0.500
Triangular	0.500	0.347
Simple cubic	0.312	0.249
bcc[a]	0.245	0.178
fcc[b]	0.198	0.119
$d = 4$[c]	0.197	0.160

[a] bcc, Body centered cubic.
[b] fcc, Face centered cubic.
[c] $d = 4$, Four-dimensional hypercubic lattice.

FIG. 4. Clusters of sites connected by bonds formed in random site percolation on a square lattice (a) below ($p = .543$), (b) at ($p = .593$), and (c) above ($p = .643$) the percolation threshold. The largest cluster is denoted by stars, the smaller clusters containing >10 sites by dots.

tem simulated to our knowledge was a 160,000 × 160,000 square lattice.

One may also investigate which fraction P_∞ of sites belongs to the infinite network in an infinite system or to the largest cluster in a finite system (in the limit of system size → ∞). Obviously for probability p below the threshold p_c this fraction, often called the percolation probability P_∞, is zero. One may instead look at the mean cluster size S, which is found if one selects randomly an occupied lattice site, counts how large the cluster is to which it belongs, and averages over all lattice sites selected in this way. It is plausible that this mean cluster size S diverges if the percolation threshold p_c is approached from below since, above this threshold, an infinite network is present. Finally, one may define a correlation length ξ as the root mean square average distance between two randomly selected occupied sites within the same cluster. In short, ξ is a typical cluster radius; and this length must also diverge if $p \to p_c$ from below. For p above p_c one can define a finite mean cluster size S and a finite correlation length ξ by restricting the initially selected sites to be on some finite cluster, not on the infinite network.

How do the quantities P_∞, S, and ξ vanish or diverge at the threshold? Figure 5 shows the variation of P_∞ and S on the square and simple cubic lattices (bond percolation). For p very close to p_c the curves may be fit by power laws:

$$P_\infty \propto (p - p_c)^\beta, \qquad p > p_c \qquad (2a)$$

$$S \propto (p_c - p)^{-\gamma}, \qquad p < p_c \qquad (2b)$$

$$\xi \propto (p_c - p)^{-\nu}, \qquad p < p_c \qquad (2c)$$

These proportionalities are supposed to be valid only asymptotically for $p \to p_c$, just as $\sin(x) = x$ is valid only for $x \to 0$. With the above restriction to finite clusters, Eqs. (2b) and (2c) also apply to the case $p > p_c$, with $p_c - p$ replaced by $p - p_c$ and with a different proportionality constant. The quantities β, γ, and ν and similar exponents governing other power laws are called critical exponents. A large part of modern percolation research has centered on these critical exponents. Similar exponents have been thoroughly investigated for critical phenomena near many other phase transitions. These critical phenomena refer to the asymptotic region $p \to p_c$, length $\to \infty$, time $\to \infty$, . . . , on which we also concentrate for percolation.

Values of the exponents β, γ, and ν are listed in Table II. Unlike Table I, this table contains only one entry for all the two-dimensional (and one for all the three-dimensional) lattices, all of which have the same exponents. This universality is one of the main reasons that critical exponents for percolation theory have attracted so much attention. Our understanding of universal-

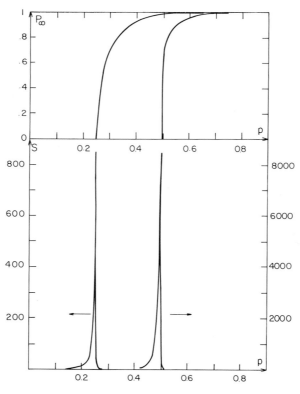

FIG. 5. Variation of percolation probability P_∞ and mean cluster size S for bond percolation on the square ($p_c = .50$) and simple cubic ($p_c = .249$) lattice.

TABLE II. Values of Percolation Exponents in Two and Three Dimensions and for High Dimensions[a]

Exponent	$d = 2$	$d = 3$	$d \to \infty$
β	$\frac{5}{36}$	0.4	1
γ	$\frac{43}{18}$	1.8	1
ν	$\frac{4}{3}$	0.9	$\frac{1}{2}$
σ	$\frac{36}{91}$	0.45	$\frac{1}{2}$
τ	$\frac{187}{91}$	2.2	$\frac{5}{2}$
$D(p = p_c)$	$\frac{91}{48}$	2.5	4
$D(p < p_c)$	1.56	2	4
$D(p > p_c)$	2	3	4
μ	1.3	2.0	3

[a] Rational numbers presumably are exact; the others are numerical estimates [μ is defined in Eq. (15)].

ity is based in part on self-similarity, as explained in Section III.

III. Fractals, Scaling, and Renormalization

One of the powerful methods of confirming Eq. (2) uses extrapolation of data obtained in numerical simulations. These are done on finite lattices of size L^d ($L \times L$ in two dimensions, $L \times L \times L$ for three dimensions). Above p_c, if the correlation length ξ is small compared with L, the largest cluster fills the sample uniformly, with a density equal to P_∞. Thus, the number of sites in the sample that belong to this cluster is

$$M(L) = L^d P_\infty \propto L^d (p - p_c)^\beta \propto L^d \xi^{-\beta/\nu} \quad (3)$$

This situation is exhibited in Fig. 4c. As p decreases toward p_c, the correlation length ξ gradually increases. Since ξ represents a typical size of a finite cluster, it also represents a typical size of the "empty" areas in the infinite cluster. As can be seen from Fig. 4c, at $p = p_c + 0.05$ there appear empty areas of *all* sizes up to ~ 7 lattice units, which is a rough estimate of ξ. The appearance of all these sizes reflects the property of self-similarity at distances smaller than ξ.

If one continues to decrease p all the way to p_c (Fig. 4b), ξ diverges to infinity and the picture in Fig. 4b contains empty islands (i.e., lattice sites not belonging to the largest cluster) at *all* length scales. A direct counting of the number M of sites on the largest cluster within the sample now yields

$$M(L) \propto L^D \quad (4)$$

with D having the noninteger values listed in

Table II. Since D replaces d in the power of L, as compared with Eq. (3), it is natural to consider D as a generalized "fractal" dimensionality.

A power law dependence like Eq. (4) characterizes fractal graphs, an example of which is shown in Fig. 6. In this example, called the Sierpinski carpet, the number of filled squares changes by a factor of 8 when the length scale changes by a factor of 3, so that $8 = 3^D$, or $D = \log 8/\log 3 = 1.893$. Indeed, fractals like the Sierpinski carpet have been widely used to investigate some of the properties of the infinite cluster at p_c.

Looking at Fig. 4c, one may divide it into unit squares of sizes $\xi \times \xi$. Within each square the cluster is self-similar, obeying Eq. (4), whereas the squares form a uniform distribution. Since the number of squares is $(L/\xi)^d$, and their "mass" (number of occupied sites on the largest cluster) scales as ξ^D, we find that the total mass scales as $\xi^D (L/\xi)^d \propto L^d \xi^{D-d}$. Comparison with Eq. (3) now yields

$$D = d - \beta/\nu \quad (5)$$

The power laws in Eq. (2) are direct consequences of the self-similarity.

More generally, Eqs. (3) and (4) can be com-

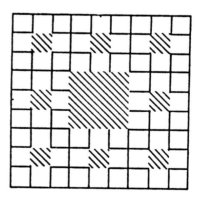

FIG. 6. Example of a Sierpinski carpet as a model for large clusters at the percolation threshold. Empty squares are shadowed; all other squares are occupied and will be divided in the next iteration into nine squares, of which the center square will be empty. The length L of this structure is the number of elementary squares on each side; it increases by a factor of 3 at each iteration. The "mass" M of the cluster is the number of occupied elementary squares in it; it increases by a factor of 8 at each iteration. Thus, $M = L^D$, and the fractal dimension $D = \log 8/\log 3 = 1.893$ is close to that of the largest percolation cluster at the threshold, $D = \frac{91}{48} = 1.896$.

bined into a scaling form (with + for $p > p_c$ and − for $p < p_c$):

$$M(L, \xi) = L^D m_\pm(L/\xi) = L^D \tilde{m}((p - p_c)L^{1/\nu}) \quad (6)$$

We see that ξ is the only relevant length scale in the problem, so that the dependence of the ratio M/L^D on L should be only through the ratio L/ξ (i.e., L is measured in units of ξ). The scaling function $m_+(z)$ has the limiting behavior $m_+ \to$ const for $z \ll 1$ and $m_+(z) \propto z^{\beta/\nu}$ for $z \gg 1$. Equation (6) can be interpreted as describing a crossover from the fractal dimension D at $L \ll \xi$ to the Euclidean dimension d at $L \gg \xi$. For $L \gg \xi$ one may thus say that $D(p > p_c) = d$, at least for $d < 6$ (Table II).

Measurements of $M(L, \xi, p)$ yield the exponents D, β, and ν and indeed confirm the scaling form [Eq. (6)]. A direct indication of universality arises from the fact that Eq. (5) has been shown to apply not only to simple model computer experiments, but also to the electron microscope picture of real sputtered films (Fig. 7). Note that our theories refer to percolation on lattices, whereas these experiments are done on a continuum. However, computer simulations of continuum percolation seem to give the same exponents as lattice percolation.

Scaling forms like Eq. (6) can be written for other quantities, for example, for the mean cluster size,

$$S = L^{\gamma/\nu} s_\pm(L/\xi) = L^{\gamma/\nu} \tilde{s}((p - p_c)L^{1/\nu}) \quad (7)$$

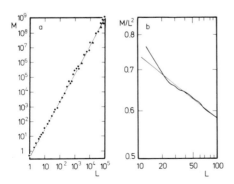

FIG. 7. (a) Log–log plot of the number of sites, $M(L)$, in the largest cluster of an $L \times L$ triangular lattice at its site percolation threshold. The slope of the straight line is the fractal dimension $D = \frac{91}{48}$. (b) Log–log plot of the fraction of sites, $M(L)/L^d$, measured experimentally on sputtered metal films of size $L \times L$ at their two-dimensional percolation threshold. The slope of the straight line is the prediction $D - d = -\beta/\nu = -\frac{5}{48}$ of percolation theory. The experimental data are summarized by the wavy line.

They are called finite size scaling and are crucial in the analysis of data from small samples. In many computer simulations, exponent ratios like γ/ν can be estimated more accurately through Eq. (7) and its analogs than the exponents γ and ν separately through Eq. (2).

As can be seen qualitatively from Fig. 4, finite clusters also exhibit self-similar features. Indeed, similar to Eq. (6), the average linear size R_s of a finite cluster with s sites is expected to behave as

$$R_s = s^{1/D} r_\pm(s/\xi^D) = s^{1/D} \tilde{r}((p - p_c)s^\sigma) \quad (8)$$

with

$$\sigma = 1/(D\nu) \quad (9)$$

Assumption (9) shows that the finite clusters have the same fractal dimension as the largest cluster as long as $R_s \ll \xi$, or $s \ll \xi^D$. The probability of finding much larger finite clusters of size $R_s \gg \xi$ is exponentially small. However, the function $r_\pm$ or $\tilde{r}$ yields a different behavior in that limit. For $p > p_c$ these rare large clusters are very compact, and $R_s \propto s^{1/d}$, that is, $D(p > p_c) = d$. In contrast, the rare large clusters found for $p < p_c$ are very loosely structured. These clusters turn out to have a lower fractal dimension $D(p < p_c)$, called that of lattice animals and listed in Table II. These lattice animals (percolation clusters for $p \to 0$) may be a model for branched polymers in a very dilute solution, where interactions between different macromolecules are negligible. In contrast, percolation clusters near p_c may be a model for gelation, the formation of an infinite network out of many intermediate-size macromolecules.

Scaling relations like Eq. (6) or (8) were introduced into percolation theory as phenomenological assumptions, in analogy to critical phenomena. Renormalization group arguments give further theoretical support for their validity. In a typical example of the renormalization group analysis one transforms a given dilute lattice into a coarse-grained version of itself, in which the length scale is enlarged by a factor $b(b = \sqrt{3}'$ in Fig. 8). Basic cells, of b^d sites (three sites in the example), are replaced by single new sites, which are considered to be occupied if two opposite edges are connected by a cluster of previous sites, and empty otherwise. The result is a renormalized concentration p' of new sites; p' is a function of the concentration p of previous sites, $p' = R(p)$. In the simple example of Fig. 8,

$$p' = p^3 + 3p^2(1 - p)$$

since a triangle is defined as occupied if all three, or at least two, of its sites are occupied, thus connecting at least two of the three corners.

One may coarse-grain the new sites again, that is, replace b^d of the new sites by one super-site with occupation probability $R(p')$, and so on. The relation $p' = R(p)$ applies to each of these "renormalizations." After many iterations, the renormalized concentration approaches unity if one started with p near 1, and approaches zero if initially p was near 0. There exists one invariant fixed point $R(p^*) = p^*$ ($0 < p^* < 1$) of the transformation; in our example this fixed point is at $p^* = p_c = \frac{1}{2}$. If one starts very close to p^*, one can linearize $R(p) = p^* + (p - p^*)\Lambda + \cdots$ and set $\Lambda = b^y$:

$$p' - p^* = (p - p^*)b^y \qquad (10)$$

Since the renormalization changed the length scale by a factor b, one has $\xi' = \xi/b$. Using $\xi \propto (p - p^*)^{-\nu}$ [Eq. (2c)], one identifies $y = 1/\nu$, at least for large b.

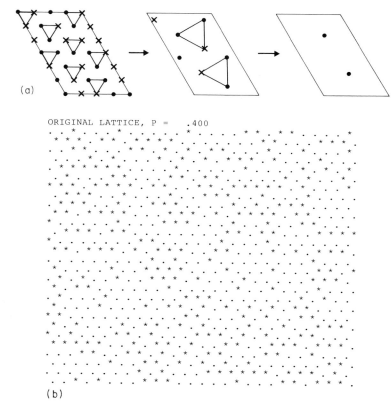

(a)

ORIGINAL LATTICE, P =   .400

(b)

FIG. 8. (a) Renormalization of small cells on the triangular lattice. Every lattice site (dot for occupied, cross for empty sites) belongs to one triangle; each triangle is renormalized (coarse-grained, averaged) into a new site. The new sites form again a triangular lattice, with a lattice constant $b = \sqrt{3}$ times larger than the old distance between nearest neighbors. A new site is occupied if in the small cell that it represents two opposite edges are connected. (b)–(p) Parts of a large computer-generated sample (stars for occupied and dots for empty sites). The large lattices are renormalized at an initial concentration of .4 (which is approaching zero as indicated through subsequent renormalizations), of .5 (which stays at this fixed point $\frac{1}{2}$ throughout all further iterations), and finally of .6 (which is approaching unity as indicated through subsequent renormalizations). Only for the threshold $p = .5$ is there self-similarity in the sense that the renormalized lattice has on the average the same properties as the unrenormalized lattice, apart from a change in lengths. This self-similarity is the basis of fractal properties and of scaling laws.

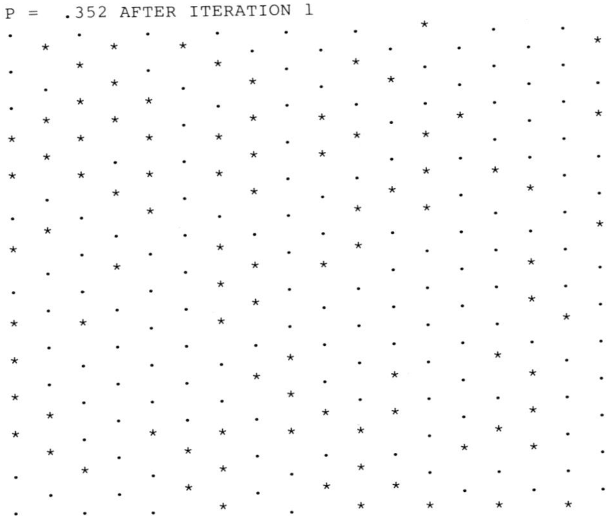

P = .352 AFTER ITERATION 1

(c)

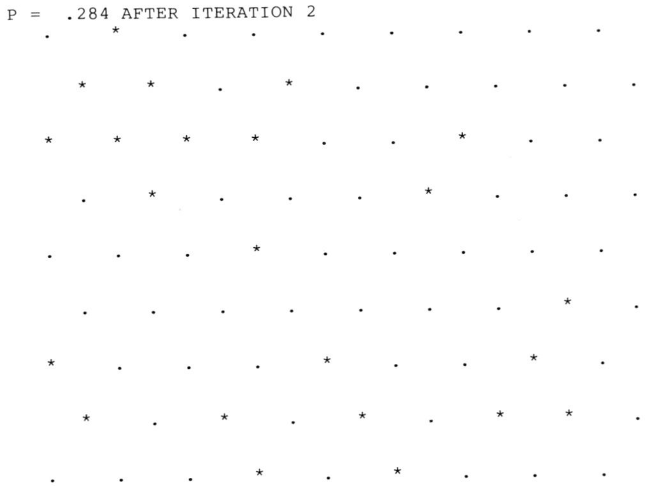

P = .284 AFTER ITERATION 2

(d)

FIG. 8. (*Continued*)

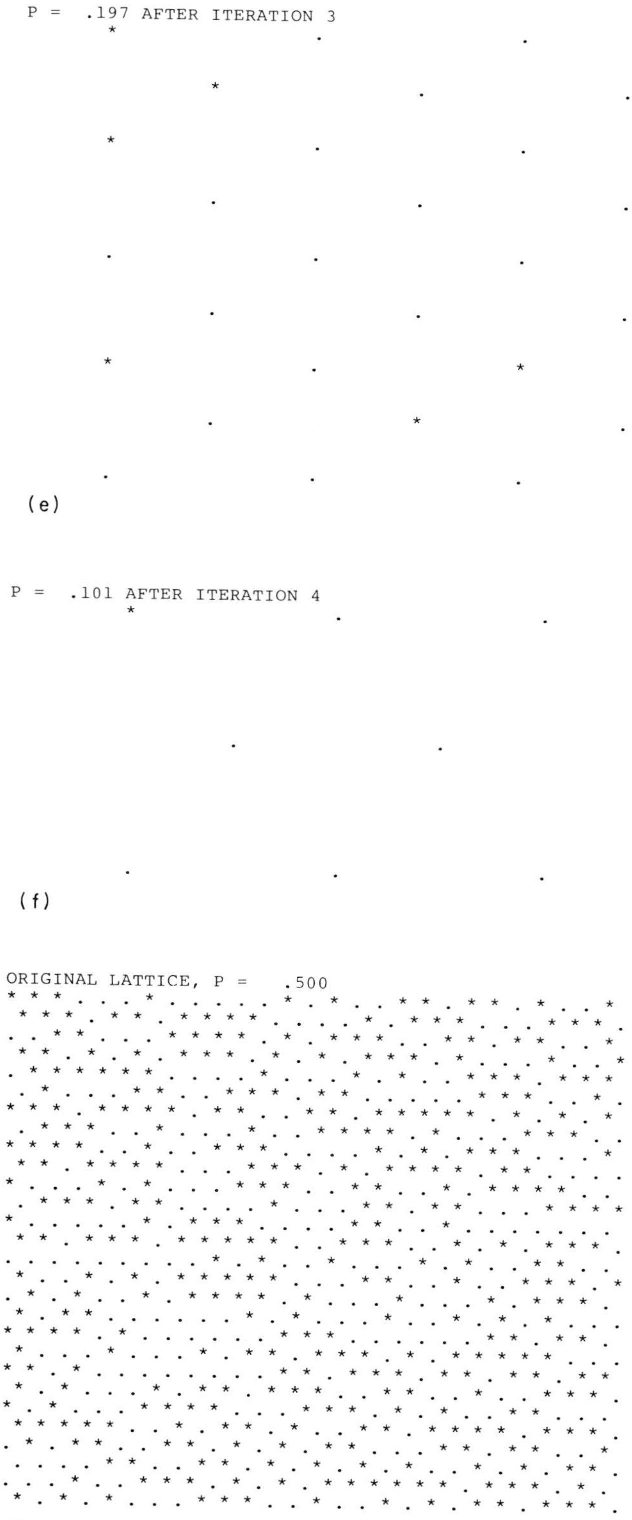

P =   .197 AFTER ITERATION 3

(e)

P =   .101 AFTER ITERATION 4

(f)

ORIGINAL LATTICE, P =   .500

(g)

FIG. 8.   (*Continued*)

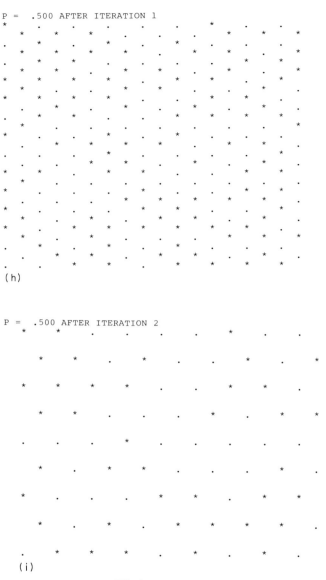

(h)

(i)

FIG. 8. (*Continued*)

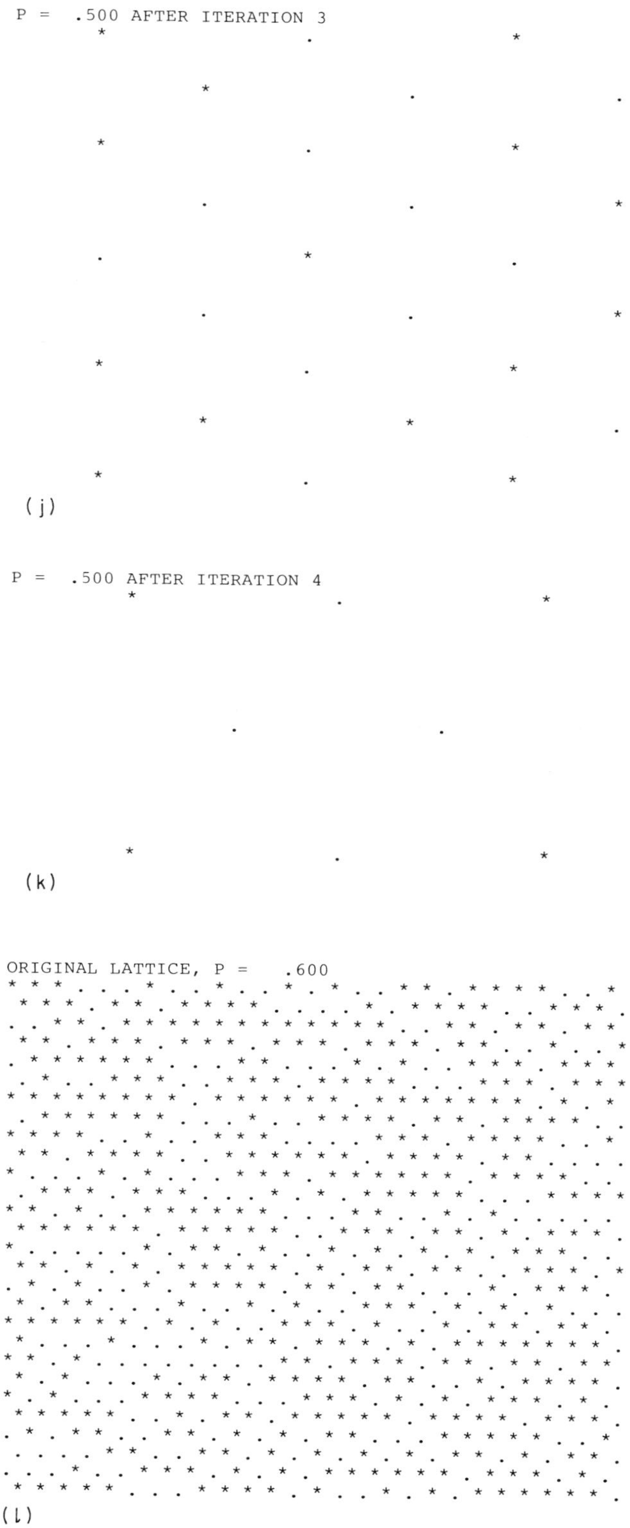

P = .500 AFTER ITERATION 3

(j)

P = .500 AFTER ITERATION 4

(k)

ORIGINAL LATTICE, P = .600

(l)

FIG. 8. (Continued)

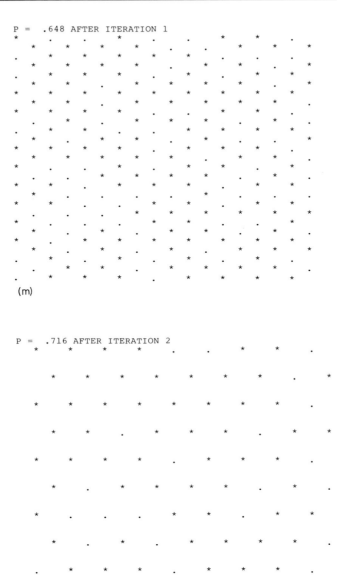

(m)

(n)

FIG. 8. (*Continued*)

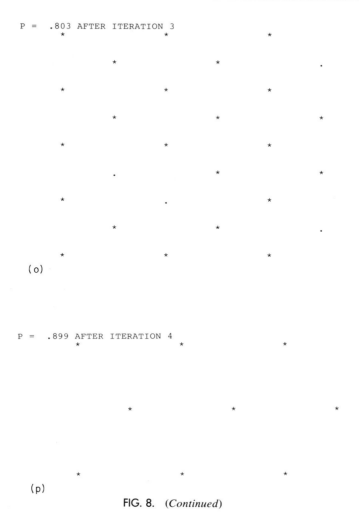

FIG. 8. (*Continued*)

In our example, $b = \sqrt{3}$, $b^y = \frac{3}{2}$ from expanding $R(p)$ about $p^* = \frac{1}{2}$, and thus $y = 1/\nu = \log\frac{3}{2}/\log\sqrt{3} = 0.738$, close to the correct result $y = \frac{3}{4}$ (Table II). This renormalization method for finite b is only approximate, because all collective phenomena are simulated by a small cell with only three sites. The approximation is better for larger values of b; large cells can be studied by computer simulation, whereby one checks with what probability $R(p)$ opposite edges are connected.

If one had a different (e.g., square) lattice, details of the renormalization group transformation $p \rightarrow p' = R(p)$ would change. However, the properties near the fixed point, like the values of the critical exponents, reflect the behavior at very large length scales. On these scales, local short-range differences become irrelevant, and universality results.

Equations like (10) imply power law behavior at $p = p_c$ for the quantities of interest and yield scaling forms like Eq. (6) or (8). Another quantity of interest is the average number n_s of clusters containing s connected sites each. (We normalize these cluster numbers by dividing them by the total number of lattice sites.) This quantity is related to the finite clusters only: Since there seems to be at most one infinite network present in percolation for $d < 6$, the number of infinite networks divided by the lattice size goes to zero for infinite lattices. As in the above examples [Eqs. (6) and (8)], the cluster numbers n_s obey the scaling form

$$n_s = s^{-\tau}N_\pm(s/\xi^D) = s^{-\tau}\tilde{N}((p - p_c)s^\sigma) \quad (11)$$

with the new critical exponent τ. One can now relate the mean cluster size S and the percolation probability P_∞ to the cluster numbers via

$$S \propto \sum_s s^2 n_s \tag{12a}$$

$$P_\infty = p - \sum_s s n_s \tag{12b}$$

at least close to the threshold. Here the sums run over all finite cluster sizes, with $s = 1$ corresponding to isolated sites. [In Eq. (12b), p has to be replaced by unity for bond percolation.] These equations lead to the scaling relations

$$\tau = 2 + \beta/(\beta + \gamma); \qquad \sigma = 1/(\beta + \gamma) \tag{13}$$

Generally (also for other phase transitions), one describes relations between critical exponents as scaling laws. They are derived from similarity assumptions like Eqs. (6), (7), (8), and (11). Equation (11) states that the cluster size distributions near sizes $s = s_1$ and $s = s_2$ are the same apart from a factor, provided that we scale the distance $p - p_c$ proportional to $s^{-\sigma}$ in order to keep the argument of the scaling function f in Eq. (11) the same.

The number n_s of small clusters can be evaluated exactly. For example, on a square lattice each pair of neighboring occupied sites has six empty neighbor sites (also called the perimeter) and can be oriented either horizontally or vertically. Thus, $n_s = 2p^2(1 - p)^6$ for $s = 2$. For $s = 3$ we may have either a horizontal or a vertical straight line of three occupied sites [contribution $2p^3(1 - p)^8$] or one of four corner configurations [contribution $4p^3(1 - p)^7$]. Thus,

$$n_3/p^3 = 2(1 - p)^8 + 4(1 - p)^7$$

Similar, but more complicated formulas have been derived for s up to 10 or 20. They are used to express quantities like the mean cluster size S as a power series in p or $1 - p$, an important technique for analyzing critical behavior. We see that for $p \to 1$ only the most compact configurations with the smallest perimeter survive. Thus, in this limit (and actually for all very large clusters above p_c) the clusters have a rather compact structure, with the radius R_s increasing as little as possible with increasing mass s, that is, $R_s \propto s^{1/d}$, as mentioned in relation to Eq. (9). In the opposite limit, $p \to 0$, for fixed size s, all cluster configurations occur equally often and one ends up with the lattice animal limit.

All these scaling forms [Eqs. (6), (8), and (11)] are valid only asymptotically for large clusters at or very close to the percolation threshold. Only here are the various critical exponents defined. Through the scaling laws [Eqs. (5), (9), and (13)] relating these exponents to one another, we

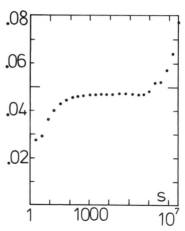

FIG. 9. Semilogarithmic plot, versus s, of the number of clusters containing at least s sites, multiplied by $s^{\tau-1} = s^{96/91}$, at the site percolation threshold $p = .5927$ of a $160,000 \times 160,000$ square lattice with periodic boundary conditions. Each site took on average ~ 3 μsec on an IBM 3081 computer. According to the scaling assumption [Eq. (11)] $n_s \propto s^{-\tau}$, the quantity plotted here should approach a plateau for large s. The deviations for the largest cluster sizes come from the boundaries or inaccuracies in p_c (or both).

can predict all these exponents once we know two of them. Again, this property is shared by many other phase transitions and their static critical phenomena.

These scaling theories of percolation seem quite well confirmed by computer or laboratory experiments. For example, at the percolation threshold the cluster numbers n_s follow the simple power law $s^{-\tau}$ predicted by Eq. (11), as Fig. 9 indicates. In two dimensions these are good arguments, though no mathematically rigorous proof, that the critical exponents of percolation are rational numbers like $\nu = \frac{4}{3}$. Similar rational exponents also arise in many models for magnets and fluids where the two-dimensional exponents are known exactly. In three dimensions, no exact exponent is known, either for percolation or for magnets and fluids.

IV. Kinetic Aspects of Percolation

In this section we deal with percolation processes rather than percolation structures. Thus, we consider time-dependent phenomena connected with percolation.

A certain numerical method of building a percolation cluster by computer simulation can be interpreted as a time-dependent process. One starts with a given isolated site. Then one looks at the shell of neighbors surrounding that site and determines for each neighbor whether that site is occupied or empty; this process completes the first time step. In the second time step, one looks at the neighbors of the occupied sites that were not visited before, and so on. Once a site is identified as empty or occupied, it will always have this status during the growth process. In this way one cluster of s sites is created with a probability proportional to $n_s s$. In addition, one can count how many time steps were needed to build this cluster. For example, to build a large two-dimensional cluster of s sites, one needs a time

$$t_s \propto s^{0.6} \propto R_s^{1.1} \tag{14}$$

at the percolation threshold. In this time-dependent interpretation, percolation can be used as a model for the spread of epidemics or forest fires. At any given time, however, the cluster structure is that of random percolation, as described in the previous section. We restrict ourselves in this section to those kinetic percolation models that lead to the same static behavior as random percolation. The time t can also be interpreted as the "chemical" or "topological" distance in the sense of the shortest connection of a point with the origin, with all steps in between taken within the cluster [See KINETICS (CHEMISTRY).]

A much more extensively studied transport property of percolation is the conductivity of random resistor networks. Imagine every occupied site (or bond) in the site (or bond) percolation problem to be a conductor, and every empty site or bond to be an insulator. Electric currents can flow only between neighboring conductors. The purpose of such a network is to model the electric properties of real alloys of metals with insulators, the flow of fluids through porous rocks, and so on. This random resistor network has a conductivity Σ, which is zero for $p < p_c$ and nonzero for $p > p_c$, since only in the presence of an infinite network of neighboring conductors can electric charges flow from one end of the sample to the other. For p approaching the threshold p_c from above we define a new critical exponent μ through

$$\Sigma \propto (p - p_c)^\mu \tag{15}$$

Since at present the relation between μ and the static critical exponents like β and γ is unclear, we listed this exponent μ in Table II as if it were independent of the others. At the threshold $p =$

p_c, in a finite system of linear dimension L the conductivity varies as $L^{-\mu/\nu}$, entirely analogous to Eq. (6) for the percolation probability: $P_\infty = M(L)/L^d \propto L^{-\beta/\nu}$.

One can also replace the conductors in the above random resistor network by superconductors (zero resistivity) and the insulators by resistors. Then the conductivity is infinite for $p > p_c$ due to the infinite superconducting cluster. The conductivity therefore diverges if p approaches p_c from below. The exponent for this divergence, often called s in the literature, equals μ in two dimensions and is ~ 0.75 in three.

The dynamic aspect of the conductivity becomes clearer if one looks at diffusion processes. Let us put some particle, or "walker," on an occupied site in the site percolation problem. At every time step, the walker selects randomly one of its nearest neighbors. If that neighbor is occupied, then the walker moves there; otherwise, it stays where it is. Then, in the next time step, the walker again selects a random neighbor of its current site and moves there if it is occupied. Repeated again and again, this process is like an ant in a labyrinth, as P. G. de Gennes pointed out. Many different "ant" investigations have been undertaken since then.

If all sites are occupied ($p = 1$), the above process is the usual diffusion process of a random walker on a regular lattice; the averaged squared distance traveled in time t is $R^2 = t$. This rms distance R (as the crow flies) must be distinguished from the cluster radius R_s or the renormalization function $R(p)$ of the preceding section and is also, of course, different from the chemical distance, the shortest path connecting the present site with the origin of the walk. The relation between the time t (number of steps) and the averaged distance R can be generally written as

$$t \propto R^{d_w} \tag{16}$$

with d_w representing the fractal dimensionality of the random walk. For usual diffusion (at $p = 1$) one has simply $d_w = 2$. In other words, we identify here the "mass" M of Eq. (4) with time t, a similarity that is particularly plausible if the walk is taken as a model for polymers.

If nearly all sites are empty ($p \to 0$), most walkers will start on an isolated occupied site where they cannot move at all: $R = 0$. For all $p < p_c$, only finite clusters exist, and R thus cannot increase to infinity for time $t \to \infty$. More precisely, the ant species defined above visits all sites of a finite cluster equally often, if given enough time. Thus, ants starting on clusters with s sites will walk a distance R of the order of the cluster radius R_s. An ant starts on such an s cluster with probability $n_s s$ since for the origin of a walk a large cluster is selected more often than a small cluster. Thus, R^2 approaches for long times below p_c a limit proportional to $\sum_s R_s^2 n_s s$. The scaling assumptions of the previous section predict this sum to diverge for $p \rightarrow p_c$:

$$R^2 \propto (p_c - p)^{-m}, \qquad m = 2\nu - \beta \quad (17)$$

This static limit was confirmed by accurate computer simulations after initial difficulties were overcome.

For p above p_c but below unity, the many holes in the infinite network slow down the diffusion of the walking ant: $R^2 < t$. Moreover, if the walk starts on a finite cluster, which it may still do even above p_c, the distance will again be limited by the cluster radius. Thus, the ant still diffuses, $R^2 \propto t$, but the diffusivity R^2/t is diminished compared with the pure case, $p = 1$. If p approaches p_c from above, this diffusivity goes to zero as $(p - p_c)^\mu$, since Einstein predicted diffusivities to vary as the conductivity [Eq. (15)]. (The current density is proportional to the velocity and the density of charge carriers. The velocity is the product of mobility and electric field. The ratio of current density to electric field is the conductivity. According to Einstein the mobility varies as the diffusivity.) Thus, we have for very long times near the threshold

$$R^2 \propto (p - p_c)^\mu t \quad (18a)$$

If we restrict the starting point to be on the infinite network only, and no longer on any occupied site, then the finite clusters no longer enter the average and the distance increases by a factor $1/P_\infty \propto (p - p_c)^{-\beta}$:

$$R^2 \propto (p - p_c)^{\mu-\beta} t \quad (18b)$$

Both types of averages, diffusion only on the infinite network or anywhere on the lattice, have been investigated by computer simulations and were found to agree with the above predictions.

Similar to the scaling assumptions [Eqs. (3), (6), and (8)] we may postulate

$$R^2 = t^{2k} f_\pm(t/t_\xi) = t^{2k} \bar{f}((p - p_c)t^x) \quad (19a)$$

with a characteristic time

$$t_\xi \propto (p - p_c)^{-1/x} \quad (19b)$$

The scaling function $\bar{f}(y)$ varies asymptotically in such a way that Eq. (18a) is recovered for large positive argument y and Eq. (17) for large

negative y. This requirement leads to

$$2k = mx; \quad x = 1/(m + \mu); \quad m = 2\nu - \beta$$

$$(20a)$$

with the asymptotic behavior $\bar{f} \propto y^\mu$ and $\propto (-y)^{-m}$ for large positive and negative arguments y, respectively, provided that the random walk starts on any occupied site. If instead it starts on the infinite cluster only, we have to match Eq. (18b) instead of Eq. (18a) in the diffusion regime. Denoting the corresponding exponents and scaling functions by primes, one finds

$$2k' = 2\nu x \quad (20b)$$

with the same $x' = x$ but a different scaling function $\bar{f}'(y)$ varying as $y^{\mu-\beta}$ for $y \rightarrow \infty$.

Right at the percolation threshold $p = p_c$, the scaling functions $\bar{f}$ and $\bar{f}'$ approach some constants for zero argument, and one has

$$R \propto t^k \propto t^{(\nu-\beta/2)/(2\nu+\mu-\beta)} \quad (21a)$$

for walks anywhere at the percolation threshold, and

$$R \propto t^{k'} \propto t^{\nu/(2\nu+\mu-\beta)} \quad (21b)$$

for walks on the largest cluster at $p = p_c$. Thus, the squared distance R^2 does not increase linearly with t nor does it approach a constant limit for large times. Instead, it increases with some smaller power of the time, an effect also called anomalous diffusion.

From Eqs. (19b) and (20b) we see that the characteristic time varies as $t_\xi \propto \xi^{1/k'}$; it is the time to travel a distance ξ on the infinite cluster [Eq. (21b)]. The exponent $1/k'$, which is equal to the anomalous fractal dimension of the random walk on the largest cluster, is analogous to the dynamic critical exponent z in second-order phase transitions. Combined with Eq. (4), the volume spanned by the sites visited within time t is of the order $R^D \propto t^{Dk'}$. In comparison with the same quantity for random walks on the full ($p = 1$) lattice, $t^{d/2}$, the combination $d_s = 2Dk'$ has been called the fracton or spectral dimension. Numerically, d_s seems to be very close (although not exactly equal) to $\frac{4}{3}$ for percolation clusters in $d > 1$ dimensions. Were this Alexander–Orbach relation, $d_s = \frac{4}{3}$, exact, it would imply an explicit dependence of μ on the static exponents β and ν. However, because this seems to be only an approximation and no alternative relations are at present generally accepted, the critical exponent μ (and therefore k, k', or d_s) is regarded here as an independent new exponent. Possible relations between the dynamic and static exponents were the subject of much research.

Computer simulations have confirmed the above scaling theory within their numerical errors. For too short times, the exponents "measured" for k or k' are too high; a million time steps on lattices containing millions of sites might be necessary to get the three-dimensional exponent $k = 0.20$ or $k' = 0.27$ with good accuracy (Fig. 10). In two dimensions these systematic deviations for short times are smaller, and $k = 0.33$ or $k' = 0.35$ can be determined more accurately. The measurement of the diffusivity slightly above p_c [Eq. (18a)] is particularly difficult, since Eq. (19) tells us that the normal diffusion behavior can be observed only for times much longer than the characteristic time $\propto (p - p_c)^{\beta - \mu - 2\nu}$. For times much shorter than this characteristic time, one observes the same anomalous diffusion as for $p = p_c$, even if p is slightly different from p_c.

As in biology, one can study more than one ant species. It does not matter much if the diffusing ant, having tried to move in a prohibited direction, immediately tries another one before time goes by. (In the earlier definition time goes by even after an unsuccessful attempt.) For this species the diffusion process is faster, but the critical exponents seem to remain the same (same dynamic universality class). Very complicated phenomena seem to occur if a bias is introduced, due to which one direction is chosen more often than the others. This bias corresponds to an electric field, if we identify the random walker with an electron in a disordered material; however, in that case interactions

between electrons may be important and must be studied by computer simulations. A magnetic field incorporated into the "ant" model of electron diffusion does not seem to change the universality class. "Butterflies" can jump to more distant sites and not only to nearest neighbors. Finally, "termites" are diffusing random walkers on a superconductor–resistor network, and "parasites" are ants diffusing on lattice animals, as defined earlier.

V. Modifications and Generalizations

The butterfly example suggests that in percolation one need not look only at clusters defined via nearest-neighbor connections. In the square site problem, when one allows sites to be connected to a cluster if they are nearest or next nearest neighbors, then the percolation threshold is lowered to $1 - p_c(nn)$, where $p_c(nn) = 0.593$ is the nearest-neighbor threshold discussed so far. For other lattices no such simple relation is known. Of particular interest is the limit of long-range connections allowed between cluster sites, when $p_c \to 0$.

In all previous examples the sites (or bonds) were occupied randomly. There may also be correlations between the sites. For example, neighboring occupied sites might be bound together with a certain energy $-J$, which favors such a configuration with a probability $\propto \exp(-J/k_B T)$, where k_B is Boltzmann's constant and T the absolute temperature. Then we have the clustering problem of the lattice gas or Ising model. It seems that its critical exponents are the same as those for random percolation except when phase separation occurs due to the site–site interaction J.

For normal bond percolation, a bond once formed connects its two ends. One may instead define a bond as an arrow in which one end is connected with the other, like a one-way street, but not backward. If the preferred directions are the same throughout the lattice we get the different behavior of "directed percolation."

Among the kinetic generalizations of percolation we mention a polymerization model in which diffusion of "initiators" forms macromolecules: Chemical bonds are formed along the path of these catalysts, but not more than f bonds can emanate from one site. This process may belong to a universality class different from random percolation.

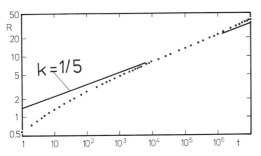

FIG. 10. Distance R versus time t for "ant" diffusion at the percolation threshold of a 256^3 simple cubic lattice. Each step took ~ 0.2 μsec per "ant" on a Cyber 205 vector computer. The scaling assumption [Eq. (19a)] predicts for large times $R \propto t^{0.20}$, as symbolized by the straight line in this log–log plot. The deviations for very large t are due to the finite lattice size.

More drastically different from usual percolation are diffusion-limited aggregates in which single particles diffuse from far away to a cluster, stick to it, and thus let the cluster grow. In cluster–cluster aggregation models, whole clusters diffuse toward one another and thus form larger clusters. Some aspects of this approach are well described by the Smoluchowski equation, formulated at the beginning of this century, in which the coagulation probability for two different cluster sizes s and s' is simply the product $n_s n_{s'}$, multiplied by a coefficient $K(s, s')$; by suitable assumptions for $K(s, s')$ one can reproduce at least one of the percolation exponents.

VI. Conclusion

We have seen that percolation, though very simple to define, has a rather rich behavior. This article has centered on its critical exponents, where two exponents (e.g., ν and D) determine the static behavior and a third (e.g., μ) seems necessary (at least presently) for the dynamic aspects. Thus, the fractal dimension D of the large clusters at the percolation does not completely determine the critical exponents, but it is certainly an important parameter for studying similarities and differences between percolation and, say, the aggregates of the preceding section.

Percolation is a good way to enter modern research without too much background in physics or mathematics. It can be used, for example, to teach scaling laws, as well as renormalization group and critical phenomena.

BIBLIOGRAPHY

Aharony, A. (1986). *In* "Directions in Condensed Metaphysics" (G. Grinstein and G. Mazenko, eds.), p. 1. World Scientific, Singapore.

Aharony, A. (1986). *In* "Scaling Phenomena in Disordered Systems" (R. Pymn and A. T. Skjeltorp, eds.), p. 289. Plenum, New York.

Alexander, S., and Orbath, R. (1982). *J. Phys. (Paris)* **43**, L625.

Alexandrowitz, Z. (1980). *Phys. Lett.* **80A**, 284.

Balberg I., and Binenbaum, N. (1985). *Phys. Rev.* **B32**, 527. But see B. I. Halperin, S. Feng, and P. S. Sen, *Phys. Rev. Lett.* **54**, 2391 (1985).

Brodbent, S. R., and Hammersley, J. M. (1957). *Proc. Cambridge Philos. Soc.* **53**, 629.

Burkhardt, T. W., and van Leuwen, J. M. J., eds. (1982). "Real-Space Renormalization," Topics in Current Physics, Vol. 30. Springer-Verlag, New York.

Coniglio, A. (1986). *Physica* **140A**, 51.

de Gennes, P. G. (1976). *La Recherche* **7**, 919.

Deutscher, G., Zallen, R., and Adler, J., eds. (1983). "Percolation Structures and Processes." Adam Hilger, Bristol. In particular, Chap. 10 by G. Deutscher, A. Kapitulnik, and M. Rapaport.

Domb, C., Green, M. S., and Lebowitz, J. L., eds. (1972–1984). "Phase Transitions and Critical Phenomena," Vols. 1–9. Academic Press, Orlando, Florida.

Family, F., and Landau, D. P., eds. (1984). "Kinetics of Aggregation and Gelation" North Holland, Amsterdam.

Gefen, Y., Aharony A., and Alexander, S. (1983). *Phys. Rev. Lett.* **53**, 77.

Gouker, M., and Family, F. (1983). *Phys. Rev.* **B28**, 1449.

Grassberger, P. (1983). *Math. Biosciences* **62**, 157.

Grassberger, P. (1985). *J. Phys.* **A18**, L 215.

Havlin, S., and Nossal, R. (1984). *J. Phys.* **A17**, L 427.

Herrmann, H. J. (1986). *Phys. Rep.* **136**, 153.

Hoshen, J., and Kopelman, R. (1976). *Phys. Rev.* **B13**, 3428.

Kapitulnik, A., Aharony, A., Deutscher, G., and Stauffer, D. (1983). *J. Phys.* **A16**, L 269.

Kesten, H. (1982). "Percolation Theory for Mathematicians." Birkhauser, Boston.

Mandelbrot, B. B. (1982). "The Fractal Geometry of Nature." Freeman, San Francisco.

Nienhuis, B. (1982). *J. Phys.* **A14**, 199.

Rapaport, D. C. (1985). *J. Phys.* **A18**, L 175.

Rapaport, D. C. (1986). *J. Phys.* **A19**, 291.

Stauffer, D. (1985). "Introduction to Percolation Theory." Taylor and Francis, London.

Stauffer, D., Coniglio, A., and Adam, M. (1982). *Adv. Polym. Sci.* **44**, 103.

PERIODIC TABLE (CHEMISTRY)

A. Truman Schwartz   *Macalester College*

GLOSSARY

Atom: Smallest characteristic constituent particle of an element.

Atomic mass (atomic weight):   Relative mass of an atom. By current international agreement, the standard for all atomic masses is the isotope carbon-12 which is arbitrarily assigned an atomic mass of exactly 12. Atomic mass was the original numerical basis for the periodic table.

Atomic number: Number of protons in the nucleus of an atom. It is also the nuclear charge (in conventional electronic units) and the number of electrons in a neutral atom. The atomic number is the current numerical basis for the periodic classification of the elements.

Electron: Fundamental subatomic particle with a charge of 1.602×10^{-19} C and a rest mass of 9.110×10^{-31} kg. The number and arrangement of electrons in atoms are largely responsible for the chemical properties of elements and many of their physical properties.

Electron configuration: Designation conveying the arrangement of electrons in the various energy levels, sublevels, and orbitals of an atom.

Electron shells (levels):   Principal energy levels occupied by the electrons in an atom, specified by the principal quantum number n.

Element: One of approximately 100 pure, simple substances that cannot be decomposed into even simpler ones by chemical means. Each individual element is made up of atoms of identical atomic number.

Family (group):   Vertical column of the periodic table containing elements with similar chemical and physical properties and similar electron configurations, especially for the outer electrons. There are 18 such columns in the modern periodic table, although earlier versions sometimes grouped the members of certain adjacent columns into one family.

Isotopes: Atoms of the same element which differ in atomic mass, that is, in number of neutrons.

Metalloid: Element having properties intermediate between those of a metal and a nonmetal or exhibiting some metallic and some nonmetallic properties.

Orbital: Electron cloud, generated from a wave function, that represents the probability of locating an electron as a function of three-dimensional spatial coordinates.

Period: Horizontal row of the periodic table, over which the chemical and physical properties of elements usually change gradually with increasing atomic number. There are seven periods, ranging in length from 2 to 32 elements.

Quantum number: Number, in most cases an integer, specifying the fact that the associated energy, momentum, or other property of a system is restricted to certain values.

Series: Sequence of similar elements that make up part of a period. Typically, the elements in a series are characterized by a progressive increase in the number of electrons in a particular sublevel. For example, in the first transition series, elements 21–30 differ in the number of $3d$ electrons present in each atom. Other series include the second and third transition series, the lanthanides, and the actinides.

The periodic table of the elements summarizes the taxonomy of chemistry. It organizes the elements—those hundred-odd simple substances that cannot be broken down into still simpler ones. However, elements can be combined in a vast number of ways to form essentially everything material that we know. Consequently, to study the elements and their characteristics is a fundamental step in comprehending the universe. Such study has led to the periodic system of the elements. Observation and experiment reveal the certain elements exhibit similar physical and chemical properties. When the elements are arranged in order of increasing atomic mass (or, more precisely, increasing atomic number), elements with similar properties appear at regular, periodic intervals. The periodic table reveals this regularity. Elements with common characteristics fall in vertical columns called families or groups. These properties generally change quite gradually along each horizontal row or period. The reasons for the similarities and differences have now been illuminated through experimental and theoretical understanding of the arrangement of electrons in atoms. The chemical and physical properties of the elements are, in large measure, determined by the number of electrons in each characteristic atom and their energetic and spatial distributions. The periodic system is, in effect, a classification based on atomic structure. The modern table thus discloses the connectedness of bulk properties and atomic structure, experiment and theory, the macroscopic and the microscopic. It not only represents and presents the stuff of chemistry, it also epitomizes the science.

I. Development of the Periodic System

A. ELEMENTARY AND ATOMIC CONCEPTS

The periodic table, the great classificatory scheme of chemistry, is based on two of the most fundamental concepts in physical science—elements and atoms. In their rudimentary forms, both of these ideas were inventions of the pre-Socratic Greek philosophers. It was probably a quest for permanence and order beneath the change and diversity of observed phenomena that led Thales of Miletus (ca. 624–545 B.C.) to suggest that all matter is ultimately derived from water. Other philosophers also subscribed

to the idea of one elementary substance, but offered other candidates for the central role. Thus, Anaximenes proposed air, Heraclitus favored fire, and Anaximander invented *apeiron,* an eternal and unlimited element that did not correspond exactly to any earthly substance.

Some time around 450 B.C., Empedocles (ca. 490–430 B.C.) appears to have synthesized some of these ideas in his argument that all matter is constituted of various mixtures of four primordial substances—earth, air, fire, and water. The two greatest philosophers of classical antiquity, Plato (428–347 B.C.) and Aristotle (384–322 B.C.), adopted Empedocles' four elements, though Aristotle added a fifth, the *quita essentia* which made up the crystalline spheres of the heavens. Aristotle's four earthly elements figured prominently in his natural philosophy, and the persistence of that philosophy meant that earth, air, fire, and water were accepted by at least some scholars as late as the seventeenth century. There was even empirical evidence that seemed to support the validity of this hypothesis. For example, when a green stick burns, fire is obviously present, hot gases escape, water droplets form along the stick, and an earthy ash remains. However, it is important to realize that these four elements were more philosophical constructs than specific entities with chemical consequences.

Similarly, atoms were proposed by the ancient Greeks not so much to explain specific natural phenomena as to account for permanence amid change. Atomists like Democritus of Abdera (ca. 460–370 B.C.) argued that a body could not be infinitely subdivided. The process ultimately had to stop at the level of an "uncuttable" particle, literally, ατομος. The Roman poet, Lucretius (ca 100–55 B.C.), claimed that atoms and the void constituted all things and elaborated that idea to explain meteorology and geology, sensation and sex, cosmology and sociology, and even life and the mind. In spite of such ambitious comprehensiveness, atoms were not as widely accepted as were the four elements, largely because Aristotle rejected them. (His chief argument was not with the atoms, but with the nothingness between them. Aristotle appears to have abhorred a vacuum even more than nature does.) Nevertheless the concept of ultimate particles of matter were occasionally invoked as a working hypothesis by scientists such as Gassendi, Galileo, Boyle, and Newton. Indeed, Newton went so far as to express this opinion: "It seems probable to me, that God in the Beginning form'd Matter in solid, massy, hard, impenetrable, movable Particles.''

Thus, by 1700, the idea of atoms was gaining scientific currency while the progress made in chemistry was calling the idea of the four elements into serious question. The century which followed did much to further establish the former and to overthrow the latter. Antoine Laurant Lavoisier (1743–1794), the Frenchman often credited with being the "Father of Modern Chemistry" ennunciated a working definition of "simple substances" in his 1789 text, *Traité Élémentaire de Chimie*. The first edition included a table (Fig. 1) of 33 elementary substances that could not be decomposed by chemical operations. Even here we see an attempt at classification. Most of the metals listed by Lavoisier (antimony, silver, copper, iron, and so on) had been known and used for centuries. So too with certain of the nonmetals such as carbon and sulfur. However, Lavoisier's list also included hydro-gen, oxygen, and nitrogen, three gases which had recently been isolated and, thanks to his theoretical system, correctly identified as specific, elementary substances. Essentially all of these substances still appear in a modern periodic table. Lavoisier's guess that the five "earths" he includes—lime, magnesia, baryta, alumnia, and silica—may in fact be complex substances has been substantiated. We now know them to be oxides of calcium, magnesium, barium, aluminum, and silicon. Two of his entries, light and heat, are not material at all. Nevertheless, Lavoisier's concept of simple substances and his preliminary list proved to be of inestimable importance for the development of chemistry and the periodic classification of the elements.

For example, a clearly defined working concept of an element was essential for the elaboration of the atomic theory by John Dalton (1766–1844). In 1808, this English schoolmaster

TABLEAU DES SUBSTANCES SIMPLES.

	NOMS NOUVEAUX.	NOMS ANCIENS CORRESPONDANTS.
Substances simples qui appartiennent aux trois règnes, et qu'on peut regarder comme les éléments des corps.	Lumière...........	Lumière.
	Calorique..........	Chaleur. Principe de la chaleur. Fluide igné. Feu. Matière du feu et de la chaleur.
	Oxygène..........	Air déphlogistiqué. Air empiréal. Air vital. Base de l'air vital.
	Azote............	Gaz phlogistiqué. Mofette. Base de la mofette.
	Hydrogène........	Gaz inflammable. Base du gaz inflammable.
Substances simples, non métalliques, oxydables et acidifiables.	Soufre...........	Soufre.
	Phosphore........	Phosphore.
	Carbone..........	Charbon pur.
	Radical muriatique....	Inconnu.
	Radical fluorique....	Inconnu.
	Radical boracique.....	Inconnu.
Substances simples, métalliques, oxydables et acidifiables.	Antimoine........	Antimoine.
	Argent	Argent.
	Arsenic...........	Arsenic.
	Bismuth..........	Bismuth.
	Cobalt...........	Cobalt.
	Cuivre...........	Cuivre.
	Étain............	Étain.
	Fer.............	Fer.
	Manganèse........	Manganèse.
	Mercure..........	Mercure.
	Molybdène........	Molybdène.
	Nickel...........	Nickel.
	Or..............	Or.
	Platine..........	Platine.
	Plomb...........	Plomb.
	Tungstène........	Tungstène.
	Zinc............	Zinc.
Substances simples, salifiables, terreuses.	Chaux...........	Terre calcaire, chaux.
	Magnésie.........	Magnésie, base de sel d'Epsom.
	Baryte...........	Barote, terre pesante.
	Alumine..........	Argile, terre de l'alun, base de l'alun.
	Silice...........	Terre siliceuse, terre vitrifiable.

FIG. 1. Lavoisier's table of simple substances or elements. [Reprinted from "Traité Élémentaire de Chemie," Vol. 1 of "Oeuvres de Lavoisier." Imprimerie Impériale, Paris, 1864, Johnson Reprint, New York, 1965, p. 135.]

published *A New System of Chemical Philosophy* in which he put forth his postulates about the structure of matter. Each element, Dalton argued, is constituted of identical, immutable, and uniquely characteristic atoms. When elements combine to form compounds, their atoms unite in a fixed ratio that is characteristic of the compound. For example, water is made up of molecules containing specific numbers of hydrogen and oxygen atoms. Because these numbers are fixed, the elementary composition of the compound, by mass, is also constant. This latter property could be quite accurately determined, even in Dalton's time. However, Dalton had no direct way of ascertaining the correct atomic ratio of the elements in a water molecule or in any other compound particle. As a consequence, he could not calculate, with confidence, the relative masses of the atoms of the various elements—their atomic masses or atomic weights. The atomic masses of the elements and the atomic ratios characteristic of specific compounds are manifested in the elementary mass composition of the compounds. Once the correct atomic ratio is known, atomic masses can be readily calculated from mass composition. Alternatively, knowledge of atomic masses permits conversion of composition by mass to composition by atomic ratio. However, Dalton faced the dilemma of an equation with two unknowns.

That dilemma continued to plague chemistry for 50 years and hence limited the utility of the atomic theory. Because of the importance of mass relationships in chemical reactions, there was a concensus that the characteristic atomic mass of an element was a significant property. However, there was little agreement on what the correct values were. Hydrogen was generally recognized to be the "lightest" element, and hence was typically assigned an atomic mass of 1. Relative to this standard, the atomic mass of oxygen was approximately 8, according to some scientists, or 16, according to others. The situation had reached such a sorry impass that the progress of chemistry was being impeded. Therefore, a special international conference was convened in Karlsruhe, Germany in 1860, with the expressed aim of resolving the confusion over atomic masses. Significantly, the two scientists most clearly associated with the development of the periodic table, the Russian Dmitri Ivanovich Mendeleev, and the German, Julius Lothar Meyer, were among those in attendance. Once a set of reliable atomic masses was accepted by the scientific community [thanks in large measure to the application of Amadeo Avogadro's Hypothesis by Stanislao Cannizzaro (1826–1910)], the time was ripe for the formulation of the periodic law and the table based on it.

B. EARLY DISCOVERIES OF SIMILARITIES IN ELEMENTARY PROPERTIES

The fact that certain elements and compounds exhibit similar physical and chemical properties was well known long before the adoption of an accurate set of atomic masses. Indeed, some elements were first prepared in what appeared to be families. Thus, Humphry Davy (1778–1829) succeeded in isolating six hitherto undiscovered metals from some very common compounds. Early in the nineteenth century, he used an electric current to decompose potash and soda to yield potassium and sodium, respectively. These soft metals proved to be highly reactive toward oxygen, water, and many other chemicals. Exposed to air, they almost immediately lose their metallic luster; placed in water they react violently to form hydrogen gas and alkaline solutions. Hence, they have come to be called alkali metals. This family includes four other members—lithium, rubidium, cesium, and francium. Not only do these elements share many common chemical and physical properties, they form many compounds with similar properties. For example, lithium chloride, potassium chloride, rubidium chloride, and cesium chloride have much in common with sodium chloride, ordinary table salt.

Davy also prepared magnesium, calcium, strontium, and barium by the electrolytic decomposition of their compounds. These metals, known as the alkaline earths, are also quite soft and reactive, but not as reactive as their chemical cousins, the alkali metals. Hence, they appear to constitute a different family. With improved analytical methods, other elementary families were found, among them the halogens, a group of reactive and widely distributed nonmetals including fluorine, chlorine, bromine, and iodine. Clearly, this was evidence for order in the great diversity of matter. [*See* HALOGEN CHEMISTRY.]

C. DÖBEREINER'S TRIADS

Classification has always been an important enterprise in the study of natural phenomena and the systematization of knowledge, especially in the early stages of a discipline. The nineteenth century saw Linneus' taxonomy of plants and the great classification of animals by Cuvier. It is therefore not surprising that it was also in this century that Johann Wolfgang Döbereiner (1780–1849) proposed one of the first systems to organize the mineral world. In 1817, this German scientist pointed out that elements could be grouped in triads of related substances. His first trio was actually composed of the oxides of calcium, strontium, and barium, but he later turned his attention to the elements themselves. He found that when he arranged the elements in order of increasing atomic mass, imperfect as those values were, they fell in a series in which the intermediate atomic mass was close to the arithmatic mean of the atomic masses of the other two. Moreover, other properties such as density, melting point, and boiling point behave similarly. The tabulation below illustrates this pattern with modern data for calcium, strontium, and barium. Döbereiner constructed similar tables of triads containing lithium, sodium, and potassium on the one hand, and chlorine, bromine, and iodine on the other. Between 1827 and 1858, various chemists, including Josiah Parsons Cooke (1827–1894), added other elements to Döbereiner's triads. For example, magnesium was included in the alkaline earth series, and fluorine was added to the halogen trio. In addition, several other families were characterized and, as always, atomic mass was one of the properties tabulated.

D. The "Telluric Helix" of de Chancourtois

In his definitive history of the periodic table, the Dutch chemist, J. W. van Spronsen, credits six individuals with major and more or less independent contributions to the creation of the periodic system of the elements. They are Alexandre Emile Béguyer de Chancourtois (1820–1886), John Alexander Reina Newlands (1827–1898), William Odling (1829–1921), Gustavus Detlef Henrichs (1836–1923), Julius Lothar Meyer (1830–1895), and Dmitri Ivanovich Mendeleev (1834–1907). De Chancourtois, professor of geology at the Ecole des Mines in Paris

and a great systematizer of minerals and geological specimens, subscribed to the principle that the properties of the elements are the properties of numbers. To demonstrate this, he arranged the elements in a three-dimensional spiral in order of increasing atomic mass. When thus ordered, elements with similar properties fell in vertical columns. Because this helix had the element tellurium at its center, de Chancourtois gave it the peculiar name, *Vis tellurique*. Figure 2, which is based on his system of 1862, strongly suggests the cyclic or periodic nature of elementary properties, though de Chancourtois did not exploit the idea.

E. Newlands' "Law of Octaves"

The elementary system of J. A. R. Newlands was, in a sense, ahead of its time. In 1863 or 1864, he introduced what he called "ordinal numbers" for the elements. He simply assigned each of them a whole number, in order of increasing atomic mass. Thus, hydrogen was 1, lithium was 2 (helium was as yet undiscovered), beryllium was 3, and so on. He then arranged the elements in tabular form, in the sequence of their ordinal numbers. The first seven, hydrogen through oxygen, were placed in a row. Next came fluorine, an element with some similarity to hydrogen. In recognition of that fact, Newlands placed fluorine below hydrogen. Sodium followed, below its fellow alkali metal, lithium. The table continued in this fashion, through element number 14, sulfur, placed below its chemical kinsman, oxygen. Other series followed, with family resemblances being generally observed.

Newlands was well aware that "the eighth element starting from a given one is kind of a repetition of the first. This particular relationship," he went on to write, "I propose to provisionally term the Law of Octaves." The term proved an unfortunate one. Most likely Newlands meant the reference to the musical scale as only an analogy, but to some of his contemporaries it must have smacked of numerology or purported evidence of cellestial harmonies. (As we shall

Property	Calcium	Strontium		Barium
		Mean	*Measured*	
Atomic mass	40.0	88.7	87.6	137.3
Density (g/cm³)	1.6	2.6	2.6	3.5
Melting point (°C)	850	777	770	704
Boiling point (°C)	1490	1564	1370	1638

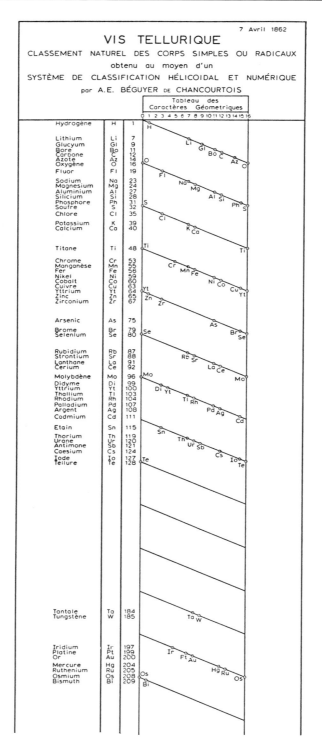

FIG. 2.   The Telluric Helix of de Chancourtois. [Redrawn by
J. W. van Spronsen from de Chancourtois' in *Compt. Rend.*
55, 600 (1862). Reprinted from van Spronsen, J. W. (1969).
"The Periodic System of Chemical Elements: A History of
the First Hundred Years." Elsevier, Amsterdam.]

see, there were, in fact, a number of rather out-landish attempts to quantize the elements ac-cording to some sort of Pathagorean or Platonic system.) When Newlands read his paper before the Chemical Society of London in March, 1866, one of his auditors even went so far as to wag-gishly inquire if he had considered placing the elements in alphabetical order and looking for similar properties. Newlands' paper was not ac-cepted for publication because it was judged too speculative. In addition, there were a number of obvious errors in Newlands' table. In an early version he had left vacant spaces for undiscov-ered elements, and even ventured to make pre-dictions of the properties of these missing sub-stances. Unfortunately, most of the predictions were wrong. Subsequently, he eliminated many of the vacancies, with the result that he was left with some rather strange elementary bedfel-lows—very different elements that occupied the same column.

For all its failings, Newlands' system was an important and independent approximation to the periodic classification of Mendeleev and Meyer, which ultimately prevailed. Unfortunately, Newlands had to wait for recognition, even in his own country. Mendeleev and Meyer re-ceived the Davy Medal of the Royal Society in 1882 for their contributions to the organization of the elements, but Newlands was not similarly honored until five years later. Newlands final vindication came in 1913 when another English-man, H. G. J. Moseley, established the physical reality behind the ordinal numbers of the ele-ments.

F. The Periodic Table of Mendeleev

Dmitri Ivanovich Mendeleev is one of the most colorful characters in the history of chem-istry. An account of his childhood reads like a plot by Dostoevsky. Of Russian and Mongolian ancestry, Mendeleev was born in Siberia in 1834, the youngest of 17 or 18 children. While Dmitri was still a child, his father, director of a secondary school, went blind. In order to sup-port her family, Maria Kornileva Mendeleeva managed a glass factory. However, after some years, the factory burned down and the elder Mendeleev died of tuberculosis. Dmitri had just completed his secondary education and his obvi-ous intelligence clearly called for further train-ing. Therefore, his mother took her 16-year old son and one of her daughters on a 1000-mile journey, by horse-drawn vehicle, to Moscow.

Her efforts to obtain a place for Dmitri at the university there failed, but ultimately he was en-rolled in the Central Pedagogic Institute at St. Petersburg.

Following his studies in St. Petersburg and a brief period as a school teacher, Mendeleev was allowed to continue his education and research in Paris and Heidelberg. Shortly after his return to Russia, he was appointed professor of chem-istry at the Technological Institute of St. Peters-burg. He held that position until 1890 when he resigned in a policy dispute with the administra-tion. Subsequently, he assumed the post of Di-rector of the Bureau of Weights and Measures. A strong individualist, Mendeleev appears to have been bold and outspoken in his educa-tional, social, and political views. He was simi-larly courageous in putting forth his scientific ideas, a characteristic which is very evident in his approach to the classification of the ele-ments. Indeed, it is probably Mendeleev's bold-ness in adhering to his classificatory scheme in the face of apparent contradictions and in mak-ing predictions based upon that scheme that has led to his identification with the periodic table.

Mendeleev also had another attribute essen-tial for the task at hand—an encyclopedic knowledge of the chemical properties of the ele-ments and thousands of their compounds. In 1869, he was summarizing much of that knowl-edge in a textbook, *Principles of Chemistry*. As part of this project, he was searching for a way to organize the great diversity of information into a pedagogically and scientifically sound sys-tem. According to the historical account, on February 17, 1869, Mendeleev sat at his desk, arranging a stack of cards. Each carried the name, symbol, and atomic mass of a different element. In what has been likened to a game of patience or solitaire, he laid out the cards in rows, in order of increasing atomic mass. Again he saw what others had observed: when ar-ranged in this fashion, elements naturally fall into families exhibiting similar chemical proper-ties which periodically repeat themselves. In fact, the properties of the elements appear to be a function of the atomic mass, or, in Mende-leev's words, "the size of the atomic weight de-termines the nature of the elements." That con-clusion appeared in a paper on "The Relation of the Properties to the Atomic Weights of the Ele-ments," read within a month of the February discovery.

Later that year, the first version of Mende-leev's periodic table appeared in a Russian sci-

Reihen	Gruppe I. — R²O	Gruppe II. — RO	Gruppe III. — R²O³	Gruppe IV. RH⁴ RO²	Gruppe V. RH³ R²O⁵	Gruppe VI. RH² RO³	Gruppe VII. RH R²O⁷	Gruppe VIII. — RO⁴
1	H=1							
2	Li=7	Be=9,4	B=11	C=12	N=14	O=16	F=19	
3	Na=23	Mg=24	Al=27,3	Si=28	P=31	S=32	Cl=35,5	
4	K=39	Ca=40	—=44	Ti=48	V=51	Cr=52	Mn=55	Fe=56, Co=59, Ni=59, Cu=63.
5	(Cu=63)	Zn=65	—=68	—=72	As=75	Se=78	Br=80	
6	Rb=85	Sr=87	?Yt=88	Zr=90	Nb=94	Mo=96	—=100	Ru=104, Rh=104, Pd=106, Ag=108.
7	(Ag=108)	Cd=112	In=113	Sn=118	Sb=122	Te=125	J=127	
8	Cs=133	Ba=137	?Di=138	?Ce=140	—	—	—	— — —
9	(—)	—	—	—	—	—	—	
10	—	—	?Er=178	?La=180	Ta=182	W=184	—	Os=195, Ir=197, Pt=198, Au=199.
11	(Au=199)	Hg=200	Tl=204	Pb=207	Bi=208	—	—	
12	—	—	—	Th=231	—	U=240	—	— — —

FIG. 3. Mendeleev's periodic table of 1871–1872. Originally published in *Annalen der Chemie*, Supplemetal Vol. 8, 1872. [Reprinted from Ihde, A. (1964). "The Development of Modern Chemistry," p. 245. Harper and Row, New York.]

entific journal. The version shown in Fig. 3 dates from 1872 and was published in supplemental volume 8 of a German journal, *Annalen der Chemie*. It lists 63 known elements, and leaves spaces for 31 more. The elements are arranged in eight major groups, seven of which contain two subgroups each to further refine the classification. The formulas heading each column represent characteristic formulas of the specified compounds of the elements below. Thus, the formula R^2O for Group I means that these elements form oxides that correspond to this atomic ratio, in modern symbolism, H_2O, Li_2O, etc.

Note in particular the spaces left in Group III for elements with atomic masses 44 and 68 and in Group IV for an element with atomic mass 72. If Mendeleev had not left these spaces, he could not have continued to group similar elements in the same vertical columns. They soon would be scattered all over the table. Titanium would have appeared under boron, and the rest of row 4 would have been shifted one space to the left. Two additional leftward shifts would have occurred in row 5, and so on.

Characteristically, Mendeleev was not content to merely predict the existence of undiscovered elements; he also had the temerity to predict the properties of these elements and some of their compounds. His most successful predictions were for what he called ekaboron (the element below boron in his 1872 tabulation), ekaaluminium, and ekasilicon. When scandium, gallium, and germanium were ultimately discovered, their measured properties showed an amazing agreement with Mendeleev's calcula-

tions. Table I illustrates the accuracy with which he predicted the properties of germanium. Such predictive triumphs did much to establish the periodic table and assure its international acceptance.

G. PERIODIC SYSTEMS OF LOTHAR MEYER AND OTHERS

Although Mendeleev is the major figure in the development of the periodic system of the elements, Julius Lothar Meyer proposed a very similar arrangement at about the same time. Trained as a physician, Meyer ultimately became professor of chemistry and rector of the University of Tübingen in southern Germany. Like Mendeelev, his motivation for organizing the elements was associated with writing a textbook. Some of Meyer's preliminary attempts at

TABLE I. Predicted Properties of Ekasilicon and Observed Properties of Germanium

	Predicted	Observed
Atomic mass	72	72.6
Color of element	gray	gray
Density of element (g/cm³)	5.5	5.36
Formula of oxide	XO_2	GeO_2
Density of oxide (g/cm³)	4.7	4.703
Formula of chloride	XCl_4	$GeCl_4$
Density of chloride (g/cm³)	1.9	1.887
Boiling point of chloride (°C)	<100	86

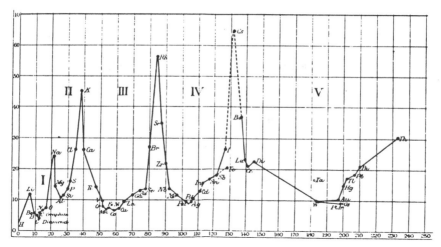

FIG. 4.   Meyer's plot of atomic volume versus atomic mass. Redrawn by T. Bayley for *Philosophical Magazine,* 1882, from Meyer's graph in *Annalen der Chemie,* Supplementary Vol. 7, 1870. [Reprinted from Ihde, A. (1964). "The Development of Modern Chemistry," p. 251. Harper and Row, New York.]

a periodic classification date from 1864, but his first real publication on the subject was in 1869. Meyer stressed the importance of elementary combining powers or valence in his tabulation and emphasized the periodicity of certain physical properties. An example of the latter is Meyer's plot of atomic volume (the volume occupied by one gram atomic mass of an element under specified conditions of temperature and pressure) versus atomic mass. This plot, shown in Fig. 4, illustrates the cyclic or periodic nature of the property. The maxima of the curve are occupied by lithium, sodium, potassium, and the other alkali metals that Mendeleev placed in his Group I.

Although Meyer and Mendeleev appear to have made their discoveries independently, they became involved in an unfortunate priority dispute. In retrospect, it seems clear that both deserve credit for their contributions to the periodic table, as do several of their contemporaries. Newlands and de Chancourtois have already been mentioned. Other, less well known contributors include William Odling and Gustavus Detlef Henrichs. Odling's system was based, in large measure, on the properties of compounds and the recognition of trends in the formulas of analogous compounds. Thus, he noted the pattern formed by the nonmetal hydrides: CH_4, NH_3, H_2O, and HF. Odling was also willing to deviate from the sequence of increasing atomic mass in order to place tellurium and iodine in their appropriate families.

Henrichs was a European-born chemist who spent much of his career as a professor of physical science at the University of Iowa. He appears to have been a fascinating character with a strong commitment to the importance of number in nature. For example, he attempted to make comparisons between the distances between the planets and the intervals between atomic masses. In another, equally quixotic enterprise, he tried to relate the wavelengths of spectral lines to the sizes of atoms. He even went so far as to resurrect the ancient idea that the shapes of atoms are the regular Platonic solids. However, for all of this rather dubious numerology, Henrichs did come up with an elementary classification which suggested a periodic variation of chemical and physical properties as a function of atomic mass.

II. Features of the Modern Periodic Table

A. THE SHORT FORM TABLE

Since the time of Mendeleev, the periodic table has repeatedly proved its utility as an invaluable scheme for summarizing information, predicting properties, and explaining chemical phenomena. Today, the basis for elementary periodicity is known to be atomic structure. This knowledge, which is discussed in Section III, has led to a number of revisions in the form of

Periods	Group I	Group II	Group III	Group IV	Group V	Group VI	Group VII	Group VIII	Group 0
I	H 1 1.0080								He 2 4.003
II	Li 3 6.940	Be 4 9.02	B 5 10.82	C 6 12.010	N 7 14.008	O 8 16.000	F 9 19.00		Ne 10 20.183
III	Na 11 22.997	Mg 12 24.32	Al 13 26.97	Si 14 28.06	P 15 30.98	S 16 32.06	Cl 17 35.457		A 18 39.944
	A B	A B	A B	A B	A B	A B	B A		
IV Even Series	K 19 39.096	Ca 20 40.08	Sc 21 45.10	Ti 22 47.90	V 23 50.95	Cr 24 52.01	Mn 25 54.93	Fe 26 55.85, Co 27 58.94, Ni 28 58.69	
IV Odd Series	Cu 29 63.57	Zn 30 65.38	Ga 31 69.72	Ge 32 72.60	As 33 74.91	Se 34 78.96	Br 35 79.916		Kr 36 83.7
V Even Series	Rb 37 85.48	Sr 38 87.63	Y 39 88.92	Zr 40 91.22	Cb 41 92.91	Mo 42 95.95	Ma 43	Ru 44 101.7, Rh 45 102.91, Pd 46 106.7	
V Odd Series	Ag 47 107.880	Cd 48 112.41	In 49 114.76	Sn 50 118.70	Sb 51 121.76	Te 52 127.61	I 53 126.92		Xe 54 131.3
VI	Cs 55 132.91	Ba 56 137.36	La 57 138.92 Also rare-earth metals 59 to 71*	Ce 58 140.13					
VI	Au 79 197.2	Hg 80 200.61	Tl 81 204.39	Hf 72 178.6 Pb 82 207.21	Ta 73 180.88 Bi 83 209.00	W 74 183.92 Po 84	Re 75 186.31 —85	Os 76 190.2, Ir 77 193.1, Pt 78 195.23	Rn 86 222.
VII	—87	Ra 88 226.05	Ac 89	Th 90 232.12	Pa 91 231	U 92 238.07			
Formulas of Oxides	R₂O	RO	R₂O₃	RO₂	R₂O₅	RO₃	R₂O₇	RO₄	
Formulas of Hydrides	RH	RH₂	RH₃	RH₄	RH₃	RH₂	RH		

*Pr 59 140.92 | Nd 60 144.27 | Il 61 | Sa 62 150.43 | Eu 63 152.0 | Gd 64 157.26 | Tb 65 159.2 | Dy 66 162.46 | Ho 67 164.94 | Er 68 167.64 | Tm 69 169.4 | Yb 70 173.6 | Lu 71 175.0

Short Periods — Long Periods

FIG. 5. Short form of the periodic table. [Reprinted from Sneed, M. C., and Maynard, J. L. (1944). "General College Chemistry," 7th printing, p. 53. Van Nostrand-Reinhold, New York.]

the table. The fundamental features, however, have remained unchanged for over a century. Figure 5 is a periodic table from an American general chemistry text published in 1944. It follows the pattern established by Mendeleev. The elements are arranged in repeating horizontal rows or periods, in order of increasing atomic number, which very nearly corresponds to the sequence of increasing atomic mass. Elements with similar properties fall in vertical columns which represent groups or families. Following Mendeleev, the first seven of these columns are divided into subgroups, here designated A and B. Group VIII contains three elements (originally termed the "transition elements") in each period, but the group is not subdivided. In addition, the 1944 table contains a ninth column, designed Group 0. These are the rare, noble, or inert gases, discovered in the late nineteenth century. The ease with which the periodic table accommodated these generally unreactive elements was another triumph of the system. Conversely, the existence of the periodic table made it evident, after William Ramsay (1852–1916) first isolated argon in August of 1894, that there must be at least four more unknown elements to complete this new family of "lazy, hidden strangers." In fact, a total of six rare gas elements (three of them bearing names with those meanings) have been found, characterized, and placed in the appropriate column.

As more data about elementary properties and more information about atomic structure were amassed, it became apparent that the "short-form" of the periodic table compromised correctness for compactness. The subfamily arrangement had always been awkward. In many cases, the members of the A and B subgroups had rather little in common. For example, it is difficult to justify putting a reactive gas like oxygen in the same family with chromium, a shiny bright metal. Rather than trying to keep such ill-matched elementary households together, they were split in the "long-form" of the periodic table, which has become the standard representation.

B. The Modern Long Form Table

Figure 6 is a modern version of a long-form periodic table. The particular one reproduced here is very similar to that approved in 1983 by the American Chemical Society Committee on Nomenclature. It is expected that a similar version will be approved by the International Union of Pure and Applied Chemistry. The reason for revision is international inconsistency in identifying the various columns or families. As already noted, Mendeelev divided what he called Groups I through VII into A and B subgroups. Over the years, however, American and European chemists applied the A and B designations in opposite senses. Thus, on most periodic tables in the United States, boron (B), aluminum (Al), and gallium (Ga) are members of Group IIIA. According to European usage, these elements fall into Group IIIB. The proposed compromise does away with the A and B notation and gives each column of the table its own number, ranging from 1 through 18. This new notation will generally be used in this article, though there will be occasional references to the still-prevalent traditional American system.

The table shows that the length of a period is by no means constant. The first is only two elements long—hydrogen (H) and helium (He). Next come two periods with eight elements each including some of the most abundant and important substances known. Nitrogen (N) and oxygen (O) are the two chief components in the atmosphere and are, of course, essential for life. Carbon (C) is so ubiquitous that an entire subdiscipline, organic chemistry, has been created to study its compounds. Silicon (Si) and aluminum (Al) are among the most plentiful elements in the earth's crust. These elements are members of a fairly large collection known as main group or representative elements. This category includes Groups 1, 2, (sometimes 12), 13, 14, 15, 16, and 17. In the traditional American usage, these are essentially the A group elements, with the exclusion of Group VIIIA. They are called representative because they exhibit a wide range of properties and serve as prototypes for other elements. [See MAIN GROUP ELEMENTS.]

Because of the uncoupling of the subgroups, the fourth and fifth periods have 18 elements each. The ten metals from scandium (Sc) through zinc (Zn) head columns labeled 3 through 12 in the new system. They constitute the first transition series (though some authors exclude the elements of Group 12 from this classification and assign them to the representative elements). Members of the transition series exhibit certain similarities in chemical and physical properties. Changes from one element to the next are less pronounced than in the main group elements. Included among the transition elements are the metals that are most important in construction, alloying, and plating—vanadium

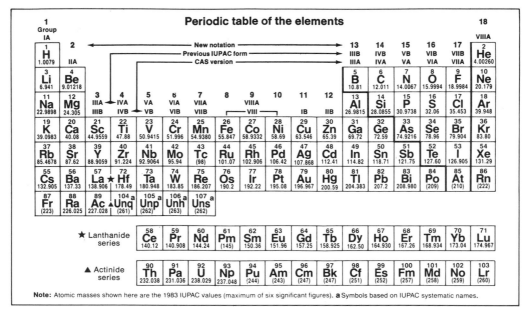

FIG. 6. Modern long form of the periodic table illustrating various numbering conventions for elementary groups. [Reprinted from *Chem. Eng. News* p. 27, February 4, 1985.]

(V), chromium (Cr), manganese (Mn), iron (Fe), nickel (Ni), copper (Cu), and zinc (Zn). After the transition elements comes gallium (Ga)—the ekaaluminum of Mendeleev, right where he predicted it. Five more elements, culminating in krypton (Kr) follow. The same pattern is repeated in row 5: an alkali metal, an alkaline earth, the ten metals of the second transition series, and six more elements in Groups 13 through 18.

Row 6 demonstrates that the long-form periodic table is not long enough. Element number 58, cerium (Ce), does not have sufficient chemical or physical similarity to titanium (Ti) or zirconium (Zr) to merit inclusion in Group 4. Instead, it is much more like 13 other elements called the lanthanide, rare earth, or inner transition series. These elements are similar and exhibit an even more gradual change in properties than do the transition metals. For this reason, and for reasons related to atomic structure, the lanthanides are grouped together in a row along the bottom of the table, with the understanding that they really fit between lanthanum (La) and hafnium (Hf). (In some tables, lanthanum is also included in the rare earth series.) The third transition series follows and the row is completed with main group elements. [*See* RARE EARTH ELEMENTS AND MATERIALS.]

The seventh period begins, as do the others, with an alkali metal and an alkaline earth. Actinium (Ac) is followed by a series of 14 actinide elements corresponding to the lanthanides. All of these elements are radioactive, and those beyond uranium (U) do not occur in nature—at least not in appreciable quantities. Rather, they are all artificially made via nuclear reactions. They are among the products of modern atomic research, and their names refer to the places (Berkeley and California) and the people (Fermi and Lawrence) involved. [*See* ACTINIDE ELEMENTS.]

The latest and heaviest elements start a fourth transition series, but they do not have the fascinating, informative, and often romantic names that characterize the other elements. Romance has been eliminated for the sake of international harmony. It seems that the nations currently making new elements, sometimes only a few atoms at a time, have found it difficult to agree on what to call their short-lived progeny. Therefore, the International Union of Pure and Applied Chemistry has decreed that after lawrencium, all new elements will be given names based on their Latinized atomic numbers. Thus, element 104 is officially known as unnilquadrium and its symbol is Unq. Kurchatovium or rutherfordium are more interesting, if less quantitative. In any case, it is fortunate that the new nomenclature rules were not in force when element 101 was named. The first element of the second hundred is, appropriately, mendeleevium.

C. OTHER REPRESENTATIONS OF THE PERIODIC SYSTEM

The long form of the periodic table, described above and illustrated in Fig. 6, is familiar from textbooks and wall charts. However, it and the short form discussed earlier, are far from the only representations of the periodic system of the elements. Over the years, a great variety of forms have been proposed, and they have exhibited various virtues. Only a few can be mentioned here.

One class of two-dimensional graphical representations is based on a triangle or planar pyramid. The apex of the triangle contains only two elements—hydrogen and helium. Periods 2 and 3, with eight elements each, form the next two rows. Below them are Periods 4 and 5, each containing 18 elements. The figure rests on two rows of 32 elements (the bottom one incomplete), corresponding to Periods 6 and 7. Members of the same family are connected by diagonal lines. This arrangement is not without merit, especially in its relationship to electronic levels and sublevels, but family relationships are somewhat obscured.

On the other hand, the periodic similarities between and among members of the same groups are strongly emphasized in the helical representation of elementary order. This arrangement, originally proposed by de Chancourtois, has many variants. Although planar, two-dimensional spirals have been published, the three-dimensional forms are more effective in conveying periodicity. For example, the helix of elements of increasing atomic number has been wound around regular cylinders and cones. The aim of most of these arrangements has been to keep similar elements in vertical alignment. However even here, some strange juxtapositions occur, so there have been efforts to more precisely group the elements on the bases of their properties. Such attempts have led to some highly complex figures.

One favorite form has been based on the lemniscate, a two-looped "figure 8." Constructed in three dimensions, such a system looks rather like a complicated parking ramp or a very elaborate pretzel. The loops of the lemniscate can be equal or unequal in size, and some versions have featured three or more loops. In some cases, the geometry is ingenious but tends to obscure rather than clarify the chemistry.

Given the importance of the periodic table and the human itch to find order—even when none exists—it is not surprising that a number of questionable curiosities have been proposed. Some schemes have been built on numerology; several organized the elements in the form of a family tree; and others claimed to divine relationships among atomic masses, spectral lines, and the tones of the musical scale. Today they seem as fanciful as Kepler's attempts to explain the orbits of the planets in terms of nested Platonic solids.

III. Atomic Structure and Elementary Periodicity

One of the great achievements of modern atomic theory is the way in which it accounts for the periodic properties of the elements—the similarities in chemical and physical properties within a family and the gradual changes in properties as one moves down or across the periodic table. The fundamental correctness of the scheme developed by Mendeleev and his contemporaries has been established through physical measurements and the computations of quantum mechanics. Today, the reason for the periodicity of the elements is well established; the explanation lies in the structure of the atom. Therefore, a discussion of tabular trends will be deferred until after this brief consideration of atomic structure and, in particular, the arrangement of electrons within atoms. [*See* ATOMIC PHYSICS.]

A. RADIOACTIVITY, ISOTOPES, AND ATOMIC MASS

The contemporary conception of atomic structure is a product of the twentieth century. Shortly before the turn of the century, a number of important experimental discoveries were made, chiefly in Europe. In late 1895, Wilhelm Röntgen (1845–1923) discovered X-rays in his Würzburg laboratory. Several months later, while investigating what he believed to be the same phenomenon, the French physicist, Henri Becquerel (1852–1908) first observed natural radioactivity. In 1897, J. J. Thompson (1856–1923) of Cambridge University detected, in the rays emitted from the cathode of a partially evacuated gas discharge tube, negatively charged particles, later called electrons, which proved to be much less massive than hydrogen atoms. Further investigation of these messages from inner space made it clear that the atom could not be "a nice hard fellow, red or gray in colour, according to taste." The phrase is from Ernest Rutherford (1871–1937) describing his early education in New Zealand. More than anyone else, Rutherford, through his brilliant experiments, changed all of that and disclosed the unseen structure within the atom.

One of Rutherford's early contributions was to demonstrate that radioactive emmanations consisted of three major types: alpha particles with a charge of $+2$ and a mass of about 4 on the standard atomic mass scale, beta particles which are identical to cathode ray particles or electrons, and gamma rays, massless manifestations of electromagnetic radiation. In collaboration with the chemist, Frederick Soddy (1877–1956), Rutherford went on to show that when an element emits alpha particles, it is spontaneously transmuted into the element lying two places to the left in the periodic table. Thus, uranium (U) becomes thorium (Th) after giving off alpha radiation. On the other hand, loss of beta particles transforms the emitting element (thorium, for example) into the element situated immediately to its right (in this case, protactinium or Pa). "Ye Gods, Soddy," Rutherford observed, "they'll have us out for alchemists!" Yet the evidence was undisputable: the ancient goal of the alchemist, to change one metal into another, was occurring naturally. Elements, and hence their atoms, are not immutable. [See RADIOACTIVITY.]

To further complicate, but ultimately clarify the issue, atomic mass now lost its uniquely characteristic quality. In the above example, loss of a beta particle by an atom of thorium does not significantly alter the mass of the atom. Nevertheless, the identity of the atom is changed. Atomic mass must not be a singularly identifying property of an element. Indeed, as Soddy demonstrated, not all of the atoms of any given element are identical. Most elements naturally exist in a mixture of atoms with differing atomic masses. These different forms Soddy called isotopes, literally emphasizing that all isotopes of the same element belong in the "same place" in the periodic table. The atomic masses determined for the elements from combining masses, gas densities, or other traditional chemical measurements are in fact weighted averages of the isotopic masses in the naturally occurring mixture. Thus, hydrogen, the lightest of all elements consists almost exclusively of atoms of mass 1. However, one atom in 6500 has a relative mass of 2, an isotope called deuterium. In addition, there is a third isotope of hydrogen, a radioactive form called tritium which has a mass of 3. The atomic mass for hydrogen, 1.0079, includes small contributions due to deuterium and tritium. For elements where the natural mixture of isotopes is less strongly biased towards a single species, the average atomic masses deviate considerably from whole numbers. Chlorine (atomic mass 35.453) and copper (atomic mass 65.546) are cases in point.

B. Nuclear Charge and Atomic Numbers

The existence of isotopes clearly indicates that, the periodic law notwithstanding, the properties of the elements cannot be a simple function of atomic mass. Atomic mass is not a sufficiently fundamental property. However, if atomic mass does not determine the identity and properties of an element, what does? The answer again came from Rutherford's laboratory. In 1909 he performed one of his most famous experiments, firing a stream of alpha particles at a thin gold foil and observing their trajectories. Most of the positively charged projectiles passed through the foil with very little change in path. Occasionally, however, the alpha particles would be deflected through large angles, in some cases almost completely reversing direction. "It was quite the most incredible event that has ever happened to me in my life," Rutherford observed, as an old and honored scientist looking back at this crucial experiment. "It was almost as incredible as if you fired a 15-inch shell at a piece of tissue paper and it came back and hit you." From this bizarre behavior, Rutherford calculated that an intense positive charge and most of the mass of the atom must be concentrated in a very small region of space. He assumed this extremely dense region would be at the center or nucleus of the atom. Thus was born the familiar and somewhat inaccurate picture of the atom as a miniature solar system, with electrons orbiting the positive and massive nucleus.

Four years later, two young men working with Rutherford made giant theoretical and experimental strides towards elaborating that model. The theoretician was the Dane, Niels Bohr (1885–1962), who borrowed a revolutionary idea from Max Plack (1858–1947) and Albert Einstein (1879–1955). The two German physicists had postulated that energy is grainy: electromagnetic radiation travels in packets called photons or quanta, and the energy associated with each of these packets is proportional to the frequency of the radiation. Bohr assumed that the energy of the single electron in a hydrogen atom is also quantized, that is, it can only have certain values, each corresponding to a circular orbit of specifically defined radius. Moreover, he derived mathematical expressions for both the energy and the radius. When an electron falls from a larger orbit with higher energy to a smaller orbit with lower energy, a photon of light, corresponding to the energy difference, is emitted. This light appears as a spectral line of distinctly characteristic frequency. Bohr's calculations reproduced with amazing accuracy the frequencies which had been observed and measured for the spectral lines of hydrogen. Here, at last, was an atomic explanation for a phenomenon which had been used to identify the rare gases and many other elements. [See QUANTUM THEORY; ATOMIC SPECTROSCOPY.]

The experimentalist, Henry G. J. Moseley (1887–1915), was measuring the frequency of spectral lines—not in the visible region of the spectrum, but in the high-frequency X-ray re-

gion. He produced the X-rays by bombarding a number of metallic targets with electrons. He found that the frequencies of the most intense lines fell into a pattern. When the square roots of the frequencies are plotted against the atomic masses of the elements, as in Fig. 7, a reasonably straight line is obtained. However, the line is not straight enough to satisfy a scientist's passion for order. So instead, Mosely plotted the same data against the number obtained by assigning 1 to hydrogen, 2 to helium, and so on, continuing sequentially throughout the periodic table. These results are also included in the figure. The points exhibit almost perfect linearity. Such agreement simply cannot be coincidental, it must be causal. The conclusion is clear: the order number or atomic number, as Moseley called it, must represent a real physical property. "This quantity," he wrote, "can only be the charge on the central positive nucleus." The atomic number, not the atomic mass, determines the elementary identity of an atom.

Thanks to Moseley's work, and that of his successors, we now know that the atomic number of an element is equal to the positive charge on the nucleus, determined by the number of positive particles or protons it contains. This number also equals the number of electrons in

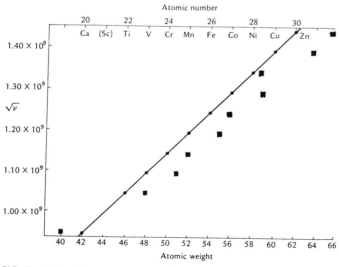

FIG. 7. Plot of the square root of the frequency of the major X-ray spectral line versus atomic mass (atomic weight) (■) and atomic number (●) of the emitting element. Plotted by A. T. Schwartz from H. G. J. Moseley's original data in *Philosophical Magazine* **26**, 1031 (1913). [Reprinted from Schwartz, A. T. (1973). "Chemistry: Imagination and Implication," p. 158. Academic Press, New York.

the electrically neutral atom. It increases integrally from hydrogen, atomic number 1, through uranium, atomic number 92, and beyond. All the isotopes of any given element have the same atomic number, that is, the same number of electrons and protons per atom. Thus, all atoms of carbon (atomic number 6) are made up of six electrons surrounding a nucleus containing six protons. However, carbon atoms differ in mass because of differences in the number of neutrons, the neutral particles also found in atomic nuclei. The most common isotope of carbon has a mass of 12—exactly 12 since it is defined as such and forms the basis of the internationally accepted atomic mass scale. Each atom of carbon-12 contains, six neutrons in addition to its six electrons and six protons. An atom of carbon-13 is identical, except that its nucleus includes seven neutrons. Similarly, carbon-14, the radioactive isotope used in dating artifacts, consists of atoms with eight neutrons in each nucleus.

The fact that the atomic number of an element is a more fundamental property than its atomic mass means that the periodic law must be modified: the properties of the elements are periodic functions of their atomic numbers, not their atomic masses. This dependence also explains the several instances in the periodic table where the correct elementary placement results in a deviation from the normal trend of increasing atomic mass. Argon (Ar) has an atomic mass of 39.948, while potassium (K) has an atomic mass of 39.0983. Yet no one who knows anything about the properties of these elements would think of putting the highly reactive metal in Group 18 along with the rare gases, or classifying the unreactive gas as one of the alkali metals. Although argon does have a higher average atomic mass than potassium, its atomic number is 18 and that of potassium is 19. Similar atomic mass inversions occur for tellurium (Te, atomic number 52, atomic mass 127.60) and iodine (I, atomic number 53, atomic mass 126.905) and for cobalt (Co, atomic number 27, atomic mass 58.9332) and nickel (Ni, atomic number 28, atomic mass 58.69).

C. Quantum Numbers, Electron Shells, and Atomic Orbitals

The atomic number determines the identity of an element because the chemical properties of an element are almost exclusively due to its electrons and its outer or valence electrons in

particular. The quantum mechanical model, which was first proposed in the 1920s, treats matter as if it had wavelike characteristics. Solution of the Schrödinger wave equation for an atom yields a set of mathematical wave functions that can be related to the probabilities of locating the electrons both spatially and energetically. This function, when plotted in three-dimensional space, generates a probability cloud. We cannot be absolutely certain where the electron will be at any instant, but we do know the region of space that is most probably occupied over time. [See QUANTUM MECHANICS; QUANTUM CHEMISTRY.]

The wave functions include four quantum numbers that emerge from the calculation. These numbers in effect provide a unique energetic and spatial "address" for each electron in an atom. Such information, in turn, helps inform our understanding of chemical properties and periodicity. The principal quantum number, symbolized n, is the chief indicator of the energy of the electron and the size of the electron cloud—in other words, how far, on the average, the electron is from the nucleus. This number can take on positive integral values: 1, 2, 3, etc. The fact that n can only have whole number values means that the energy of the atom is restricted to certain values, just as Bohr assumed in his model for the hydrogen atom. Electrons with the same value for the principal quantum number are said to be in the same shell or level.

The angular momentum of the electron is also quantized, and the azimuthal quantum number, l, specifies the permissible values. The numerical values that can be assumed by l are determined by the value of n. Thus, if n equals 1, l can only equal 0. If n equals 2, l can assume two values, 0 or 1. The general relationship is $l = 0$, 1, 2, ..., $(n - 1)$, and consequently there are n different values of l for any particular value of n. Each value of l represents a subshell or sublevel of the principal shell. These subshells are usually identified by an alphabetical code based on old spectroscopic terms. Thus, $l = 0$ is designated an s sublevel, $l = 1$ is called a p state, $l = 2$ is symbolized as d, $l = 3$ corresponds to an f subshell, and so on, in alphabetical order. The electronic distributions associated with various values of the azimuthal quantum number are also identified by this code. The designation is important because l reflects the shape of these electron clouds.

A third number which emerges from the wave equation is the magnetic quantum number, m_l. Its permissible integral values are determined by *l*. When *l* is 0, m_l can only be 0; if *l* equals 1, m_l can have values of +1, 0, or −1. In general, m_l can assume $2l + 1$ values for each value of *l*: $m_l = 0, \pm 1, \pm 2, ..., \pm l$. Physically, the various values of the magnetic quantum number indicate the possible values of the quantized *z* component of the angular momentum of the electron. This means that m_l determines the spatial orientation of the corresponding electron cloud. Each distinct orientation is termed an atomic orbital or simply, an orbital. An *s* orbital, for which *l* = 0 and $m_l = 0$, is spherically symmetric about the

nucleus—there is no angular dependence of the distribution and hence no detectable difference in orientation. For *l* = 1, however, there are three *p* orbitals corresponding to the three permissible values of m_l: +1, 0, and −1. The probability distribution given by the wave function is bilobed, much like a dumbell, and the three orbitals are oriented along the three rectilinear coordinates, as illustrated in Fig. 8. Similarly, there are five different *d* orbitals, reflecting the five values of m_l consistent with *l* = 2.

The spin quantum number, m_s, completes the

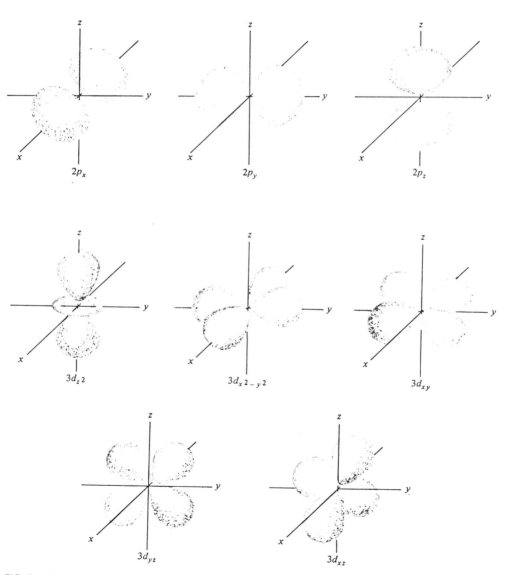

FIG. 8.   Representation of the 2*p* and 3*d* electron orbitals. [Reprinted from Bailar, J. C., *et al.* (1984). "Chemistry," 2nd ed., pp. 217, 218. Academic Press, Orlando.

set of four. It can assume only two possible values, $+\frac{1}{2}$ and $-\frac{1}{2}$. This differentiation takes into account the fact that an electron can be regarded as spinning either clockwise or counterclockwise.

D. BUILDING UP THE ELEMENTS: ELECTRON CONFIGURATIONS

The rules governing the relationships among the four quantum numbers are used in the "build-up" of the elements. We start with the simplest atom, hydrogen, and successively add electrons to generate new atoms and elements. It should be emphasized that this is a pencil-and-paper exercise which may not mimic the processes by which the various elements actually were and are formed. However, the *aufbau* process, to use its common German name, works well as a heuristic device. It is guided by the Pauli Exclusion Principle which stipulates that no two electrons in the same atom can have identical sets of all four quantum numbers. Therefore, as electrons are added, each is assigned a unique set of n, l, m_l, and m_s. The order of assignment is dictated by increasing energy. For each successive atom, the most stable electronic arrangement (or ground state) is that with the lowest energy. In general, the energy of an electron follows the sequence on increasing values for n and l, but there are exceptions that manifest themselves in the periodic table.

The lowest energy level available to the single electron in a hydrogen atom is characterized by $n = 1$, $l = 0$, $m_l = 0$, and $m_s = +\frac{1}{2}$ or $-\frac{1}{2}$ (the value of the spin quantum number is of no energetic consequence in this case). The electronic configuration is designated $1s$, 1 for $n = 1$ and s for $l = 0$. The electron can occupy other orbitals, corresponding to other values of n, l, and m_l, but these represent "excited" states of higher energy. Next comes helium, with atomic number 2. Its two electrons share the same values for the first three quantum numbers: $n = 1$, $l = 0$, $m_l = 0$. However, in conformity with the Pauli Principle, they must differ in m_s. Thus, $m_s = +\frac{1}{2}$ for one of the electrons and $-\frac{1}{2}$ for the other. It is meaningless to ask which is which. What is important is the fact that the maximum capacity of any orbital is two electrons of opposite spin. Since both electrons in a helium atom are in the $n = 1$, $l = 0$ orbital, the ground state for the element is written $1s^2$, the superscript indicating double occupancy. Note that these two sets of quantum numbers are the only sets

for which $n = 1$. Given the rules, there are no more combinations without duplication. Therefore, helium represents a completed $n = 1$ electron shell, as well as a fully filled $1s$ orbital. This, then, is an electronic explanation for why hydrogen and helium are the only members of the first row of the periodic table.

The second row contains eight members. It begins with lithium (atomic number 3). The two electrons with lowest energy occupy the $1s$ orbitals as in helium. The third electron must enter a more energetic orbital, further removed from the nucleus. It is reasonable to expect that this third electron will occupy a state for which $n = 2$, and experiment and calculation show this to be the case. Recall, however, that $n = 2$ is compatible with two values for l: 0 and 1. The energy corresponding to the former value is the lower, so the outermost electron in lithium is characterized by the quantum numbers $n = 2$, $l = 0$, $m_l = 0$, $m_s = +\frac{1}{2}$ or $-\frac{1}{2}$. The ground state electronic configuration of the element is $1s^2 2s$. Beryllium (atomic number 4) follows the pattern established by helium and has a $1s^2 2s^2$ electronic arrangement. This fills the $2s$ orbital, but not the entire $n = 2$ shell. The second shell also contains p orbitals for which $l = 1$. Moreover, there are three $2p$ orbitals in different orientations, corresponding to $m_l = +1$, 0, and -1. Each of these three orbitals has a potential population of two electrons, one with $m_s = +\frac{1}{2}$ and the other with $m_s = -\frac{1}{2}$. Thus, the maximum capacity of the $2p$ orbitals is six electrons. The limit is explained by the fact that six and only six unique sets of the four quantum numbers can be made from $n = 2$, $l = 1$, $m_l = +1$, 0, -1, and $m_s = +\frac{1}{2}$, $-\frac{1}{2}$. This means six different electron position and energy "addresses" and the possibility of six new elements—boron (atomic number 5) through neon (atomic number 10). Their electronic configurations appear in Table II. Note that the values of m_l and m_s are not specified; indeed, for some of these atoms, their values cannot be unambiguously assigned.

With neon (atomic number 10), the $n = 2$ electron shell becomes completely filled. Element number 11, sodium, must therefore make use of the $n = 3$ shell. Building on the neon structure, the 11th (and 12th) electrons enter the $3s$ orbital. After magnesium, the $3p$ orbitals are added to until the electronic arrangement of argon is attained and the third row of the periodic table is completed. The electronic configurations of the elements in this row are also included in Table II. The similarities of the electron arrangements

TABLE II.   Electronic Configurations of the Elements in Periods 2 and 3

	Period 2				Period 3		
Name	Symbol	Atomic number	Electronic configuration	Name	Symbol	Atomic number	Electronic configuration
Lithium	Li	3	$1s^22s$	Sodium	Na	11	$1s^22s^22p^63s$
Beryllium	Be	4	$1s^22s^2$	Magnesium	Mg	12	$1s^22s^22p^63s^2$
Boron	B	5	$1s^22s^22p$	Aluminum	Al	13	$1s^22s^22p^63s^23p$
Carbon	C	6	$1s^22s^22p^2$	Silicon	Si	14	$1s^22s^22p^63s^23p^2$
Nitrogen	N	7	$1s^22s^22p^3$	Phosphorus	P	15	$1s^22s^22p^63s^23p^3$
Oxygen	O	8	$1s^22s^22p^4$	Sulfur	S	16	$1s^22s^22p^63s^23p^4$
Fluorine	F	9	$1s^22s^22p^5$	Chlroine	Cl	17	$1s^22s^22p^63s^23p^5$
Neon	Ne	10	$1s^22s^22p^6$	Argon	Ar	18	$1s^22s^22p^63s^23p^6$

for members of the same family are apparent. Of particular importance is the configuration of the valence electrons. Atoms with the same number of outer electrons in similar orbitals tend to gain, lose, or share electrons in similar ways. This means that they exhibit similar chemical properties, including valence or "combining power."

Potassium, element 19, is a good test of this argument. The point is that the $n = 3$ orbitals are not fully filled in argon. The $l = 0$ $(3s)$ and $l = 1$ $(3p)$ sublevels are, but the $l = 2$ $(3d)$ orbitals remain empty. If the nineteenth electron occupied a $3d$ orbital, there would be little reason to expect a high degree of chemical and physical similarity between potassium and sodium. But, as we have noted, these two elements clearly belong in the same family. The inference is that the outer electronic configurations of the two substances must be similar. Since sodium has a single $3s$ electron in its valence shell, we write, by analogy, $1s^22s^22p^63s^23p^64s$ for the electronic configuration of potassium. Experimental results, especially spectral lines and ionization energies, and quantum mechanical calculations confirm the correctness of this assignment. In potassium, the $4s$ energy level is lower than the $3d$. Similarly, the outer electrons in calcium are in the $4s^2$ configuration, in conformity with the pattern established for the alkaline earths.

It is significant that the properties of scandium (Sc), the next element in order of increasing atomic number, do not replicate those of boron and aluminum. Therefore, scandium is not placed in column 13 (or IIIA as it was previously numbered). Rather, it heads a new column now labeled 3. Electronic arrangement again provides an explanation: scandium adds a $3d$ electron to the configuration of calcium. Scandium thus launches a new series of elements—the

transition metals—in which the $3d$ orbitals are progressively filled. The fact that there are ten elements in this first transition series reflects the total electron capacity of these orbitals. In terms of quantum numbers, ten and only ten unique sets of n, l, m_l, and m_s can be created from $n = 3$, $l = 2$, $m_l = +2, +1, 0, -1, -2$, and $m_s = +\frac{1}{2}$, $-\frac{1}{2}$. One would expect the $3d$ orbitals to become fully filled in the ground state of zinc, element number 30. In fact, this occurs in copper (atomic number 29). Because the completely occupied d orbitals are particularly stable, the configuration $[Ar]3d^{10}4s$ is energetically favored over $[Ar]3d^94s^2$. After zinc, $[Ar]3d^{10}4s^2$, electrons are added to the $4p$ orbitals and the next six elements have the outer electronic arrangements characteristic of the main group elements in columns 13 through 18.

The fifth row of the periodic table duplicates in properties and electronic configuration the fourth row. Building on the structure of krypton (Kr), electrons are first added to the $5s$ orbitals, then to the $4d$, and finally to the $5p$. The row concludes with xenon (Xe).

Row 6 begins with an alkali metal, cesium (Cs), followed by an alkaline earth, barium (Ba), suggesting electronic configurations of $[Xe]6s$ and $[Xe]6s^2$, respectively. A $5d$ electron is added in lanthanum (La), which is placed in Group 3, immediately below yttrium (Y). But the next element, cerium (Ce), is not located in column 4. It appears at the bottom of the table, the first member of the lanthanides. The fact that there are fourteen of these rare earths is evidence of the fourteen sets of quantum numbers that include $n = 4$ and $l = 3$. This means that the $4f$ orbitals are filled in this series. Because these orbitals lie well within the composite electron cloud, they do not significantly influence the chemical properties of the lanthanides. The similar chemistry of these elements is attributable to similar configurations of $6s$ and $5d$ electrons which are furthest from the nucleus and least tightly bound. Thus, all the lanthanides tend to lose three electrons in forming chemical bonds. After this series is completed with element 71, lutetium, subsequent electronic addition is to the $5d$ orbitals. The third transition series results. Mercury (Hg) is the final member of this collection of rather rare and valuable metals. It is followed by six elements in which the outer electrons are found in the $6p$ orbitals.

The seventh and last period mirrors the sixth. Like all the others, row 7 begins with elements in Groups 1 and 2. Actinium (atomic number 89) is placed in Group 3, but it is followed by the fourteen radioactive members of the actinide series. These elements, which involve electron addition to the $5f$ orbitals, are homologs of the lanthanides and are placed directly below them. The remaining new synthetic elements, with atomic numbers of 104 or greater, apparently have electrons in the $6d$ orbitals and thus should be similar to the transition metals.

One way to keep track of where the electrons are going is to identify blocks of elements by the electron orbitals added to during the build-up process. Because the outer electron configuration of the alkali metals and the alkaline earths are always ns and ns^2, the elements of columns 1 and 2 (Groups IA and IIA) are known as members of the s block. Groups 3 through 12 (IIIB–VIII, IB, and IIB) constitute the d block, because in these transition elements electrons are introduced into the d orbitals. The p block contains the elements in Groups 13 through 18 (IIIA–VIIIA). Finally, the lanthanides and actinides are known as f-block elements.

The assignment of quantum numbers indicates that orbital energy levels do not always follow the sequence of increasing n and l values. For example, the $4s$ orbitals are filled before the $3d$, and the $5s$, $5p$, and $6s$ are filled before the $4f$. One can, however, write the orbitals in the general order of increasing energy and, hence, filling: $1s$ $2s$ $2p$ $3s$ $3p$ $4s$ $3d$ $4p$ $5s$ $4d$ $5p$ $6s$ $4f$ $5d$ $6p$ $7s$ $5f$ $6d$. This sequence, plus the rules governing the relationships and assignments of quantum numbers, make it theoretically possible to write the electronic configuration of any element. In practice, minor adjustments are sometimes required because the higher orbitals are very close in energy. Factors such as the stability conferred by fully filled and half-filled subshells can cause electron shifts in the ground state. When such differences occur, they are manifested in the fine detail of chemical and physical properties.

IV. Tabular Trends

A. ATOMIC RADIUS

Chemical and physical properties of the elements and their compounds constituted the evidence for the periodic system in the last half of the nineteenth century. Modern understanding of the arrangement of electrons within the atoms of the various elements has provided an explanation of these properties and their periodic nature. One property that is quite obviously associated with atomic structure is atomic size. For metallic elements, atomic radii can be determined by subjecting crystals to X-ray diffraction. The angles at which the X-rays bounce off the atoms are the data from which the calculations are made. Another experimentally based technique involves computing the distance between chemically bonded atoms from spectroscopic measurements. Quantum mechanical calculations also yield atomic radii, including those plotted in Fig. 9. [See CRYSTALLOGRAPHY; X-RAY ANALYSIS.]

Figure 9 is a graph of atomic radius against atomic number. A periodic pattern is visible: the maxima are occupied by the alkali metals and the minima by the rare gases. Within any column or family, the radius increases with increasing atomic number. This is to be expected, since the atomic number equals the number of electrons in the atom. Each new shell increases the size of the atom. However, within any row or period, the atomic radius decreases with increasing atomic number. Here the added electrons make the atom smaller. This latter trend is attributable to the fact that the positive nuclear charge (the number of protons) also increases along a row. The added outer electrons do not effectively screen the attractive force due to the protons, and hence these forces increase with atomic number. This means the electrons are pulled closer to the nucleus and the atomic radius decreases.

B. METALLIC CHARACTER AND IONIZATION ENERGY

One readily observable periodic property is the variation in the metallic character of the ele-

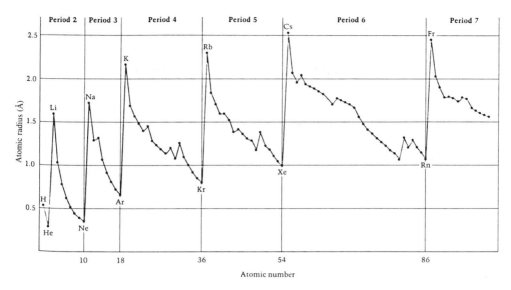

FIG. 9. Plot of atomic radius versus atomic number. Data from Waber, J. T., and Cromer, D. T. (1965). *J. Chem. Phys.* **42,** 4116. [Reprinted from Ebbing, D. D. (1984). "General Chemistry," p. 181 Houghton Mifflin, Boston.

ments. The members of the first two columns of the table are highly reactive metals—so reactive that the alkali metals and alkaline earths are never found naturally in their uncombined forms. In the process of chemical combination, each alkali metal atom loses one electron. The positive particle that results has one more proton than electron. Hence, it is an ion with a charge equal to that of a proton (+1 in conventional units). This electron loss is termed oxidation, and the amount of energy required to remove an electron from a gaseous atom is called the ionization energy or ionization potential.

The electrons removed in the ionization process are those least strongly attracted to the positive nucleus. In the case of the Group 1 elements, the single electron that is normally lost during a chemical reaction comes from the outermost s orbital: $2s$ for lithium, $3s$ for sodium, $4s$ for potassium, and so on. This oxidation process yields ions with essentially the same electronic arrangement as the adjacent rare gases. Thus, Na^+ has the same electronic configuration as neon: $1s^2 2s^2 2p^6$. The two species are said to be isoelectronic. The fact that the rare gases are very unreactive (indeed, they were considered and called "inert" until 1962, when the first compounds of xenon were made), indicates that they have very little tendency to lose, gain, or share electrons. In other words, their electronic

configurations, which involve fully filled subshells, are especially stable. Once that configuration is attained, as in Na^+, it is very difficult to remove a second electron. Sodium, in essentially all of its compounds, exists in this ionic form, and the other alkali metals behave similarly.

In the ground state of the alkaline earth atoms (Group 2), two s electrons occupy the outermost orbital. Both of these electrons can be quite easily removed to form ions of charge +2. Once again, these ions are isoelectronic with the rare gases.

This tendency to lose electrons and form positive ions is characteristic of metals. Moreover, the relative mobility of electrons within a bulk sample of these elements gives rise to electrical and thermal conductivity, malleability and ductility, and the bright, shiny appearance associated with metals. About 75% of the elements can be classified as metals. The degree of metallic character increases moving down any column of the periodic table, but it decreases moving from left to right along a row. In Period 3, for example, aluminum is clearly metallic, silicon has some metallic properties, phosphorous is more nonmetallic, and sulfur, chlorine, and argon are pretty obviously nonmetals. The stair step line on Fig. 6 is meant to represent the boundary between the metals and the nonmetals, but like many differentiations, it is not absolute or unambiguous. The elements that flank the boundary are sometimes called metalloids because they are intermediate in properties. Indeed, it is not uncommon for metalloids to exist in two crystalline forms or allotropes, one more metallic and the other more nonmetallic. [See ELECTRICAL RESISTIVITY OF METALS.]

Not surprisingly, the trends in metallic character parallel ionization energies, but in different directions. The most aggressively reactive metals are those that are most easily ionized. Thus, metallic character increases as ionization energy decreases. Cesium, for example, has a lower ionization energy and greater metallic reactivity than sodium. The $6s$ electron in the larger atom is further from the nucleus than the $3s$ electron in sodium, and hence it is easier to remove. Within a row or period, ionization energy increases with atomic number, as the increasing unshielded nuclear charge enhances the electrostatic attraction experienced by the electrons.

These trends are represented graphically in Fig. 10, a plot of the ionization energy required to remove one electron versus atomic number. The rare gases occupy the maxima on the curve, indicating how difficult it is to form positive ions from their atoms. The small zig-zags in the graph are important because they can be related to the details of electronic configuration. For example, the experimental values for ionization energy give evidence of energy differences between s and p orbitals in the same electron shell and the stability associated with half-filled and fully filled subshells. In fact, some of the best evidence for the correctness of the quantum mechanical model of the atom comes from such data.

C. Electron Affinity

A good deal of energy is required to form positive ions from the elements of Group 17. Instead, fluorine and the other halogens tend to gain one electron per atom in chemical reactions—a process called reduction. The negative ions that result are, once more, isoelectronic with the adjacent rare gas atoms. Thus, a chlorine atom, with a $1s^2 2s^2 2p^6 3s^2 3p^5$ electron configuration, becomes converted to a chloride ion, Cl^-, with the stable configuration of an argon atom: $1s^2 2s^2 2p^6 3s^2 3p^6$. The stability of the rare gas electron arrangement also helps explain why the atoms of oxygen and the other Group 16 elements gain two electrons during reaction to form ions such as O^{2-}. The formation of negative ions is one characteristic of reactive nonmetals.

A measure of the energy involved when a neutral atom in the gaseous state acquires one electron is called the electron affinity. When the negative ion thus formed is stable, energy is released and the electron affinity has a negative sign. When the ion is unstable, the electron af-

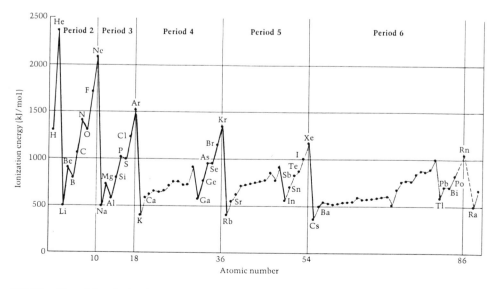

FIG. 10.   Plot of first ionization energy versus atomic number. [Reprinted from Ebbing, D. D. (1984). "General Chemistry," p. 184. Houghton Mifflin, Boston.]

finity is positive. Thus, chlorine should (and does) have a large negative electron affinity. Figure 11 reveals that a plot of electron affinities as a function of atomic number is somewhat more complicated than a graph of ionization energies, but the results are again interpretable in terms of electron configurations.

D. CHEMICAL REACTIVITY AND COMBINING RATIOS

Ionization energies, electron affinities, and electron configurations are important predictors of chemical reactivity. With such knowledge, and the guidance of the periodic table, chemists can make educated guesses about the compounds most likely to be formed from reacting elements. For example, a metal with a low ionization energy, such as sodium, and a nonmetal with a high electron affinity, like chlorine, will react vigorously. In the process, one electron is transferred from each participating sodium atom to each reacting chlorine atom, forming stable Na^+ and Cl^- ions. The resulting compound, sodium chloride, is a crystalline solid composed of equal numbers of the two ions, occupying alternating sites in a regular three-dimensional array. The one-to-one atomic ratio is characteristic of compounds formed between alkali metals and halogens. Similarly, the number of electrons readily lost and gained by the atoms of other reacting elements determine the formulas of the ionic compounds formed. [See ELECTRON TRANSFER REACTIONS, GENERAL.]

Even when electrons are not transferred and ions are not formed, electron configuration remains a key to the chemistry of the elements. In covalent compounds, the atoms are linked by shared electron pairs called covalent bonds. Here again, the number of electrons shared and, hence, the atomic ratio of the elements in the compound are largely determined by the electron configurations of the atoms involved. The formulas CH_4, NH_3, H_2O, and HF reflect the fact that the number of valence electrons increases from four in carbon to seven in fluorine. In order to achieve the high stability associated with eight outer electrons (as in the rare gases) each carbon atom has a tendency to effectively acquire four more electrons through sharing with four hydrogen atoms. Similarly, a nitrogen atom shares a pair of electrons with each of three hydrogen atoms, oxygen shares with two hydrogen atoms, and fluorine shares with one hydrogen. Moreover, once the pattern is established, there is good reason to expect it to be followed in the subsequent rows, for example, in SiH_4, PH_3, H_2S, and HCl.

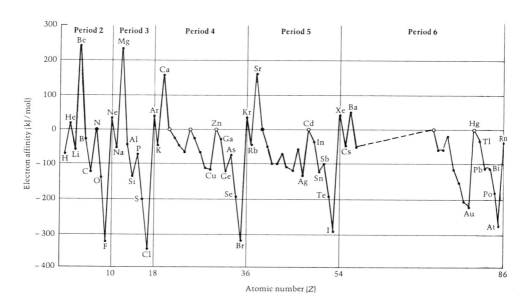

FIG. 11. Plot of electron affinity versus atomic number. Elements indicated by open circles at zero form unstable negative ions. Their electron affinities are probably negative, but their exact values are unknown. [Reprinted from Ebbing, D. D. (1984). "General Chemistry," p. 186. Houghton Mifflin, Boston.]

There are, of course, exceptions to these and other generalizations, but they only serve to keep chemistry interesting. The elegant order of the periodic table reflects an equally elegant order in nature—at both the phenomenological and structural levels. Indeed, there have been few if any scientific classification schemes that have proved more potent in prediction, more effective in explanation, and more adaptable to new knowledge.

BIBLIOGRAPHY

Cotton, F. A., and Wilkinson, G. (1980). "Advanced Inorganic Chemistry," 4th ed. Wiley, New York.

Greenwood, N. N., and Earnshaw, A. (1984). "Chemistry of the Elements." Pergamon Press, Oxford.

Laidler, K. J., and Ford-Smith, M. H. (1970). "The Chemical Elements." Bogden and Quigley, Tarrytown-on-Hudson, New York.

Mazurs, E. G. (1974). "Graphic Respresentations of the Periodic System During One Hundred Years." University of Alabama Press, University, Alabama.

Powell, P., and Timms, P. L. (1974). "The Chemistry of the Non-Metals." Chapman and Hall, London.

Puddephatt, R. J. (1972). "The Periodic Table of the Elements." Clarendon Press, Oxford.

van Spronsen, J. W. (1969). "The Periodic System of Chemical Elements: A History of the First Hundred Years." Elsevier, Amsterdam.

PERSONAL COMPUTER—*SEE* MICROCOMPUTER DESIGN

PETROLEUM GEOLOGY

Richard C. Selley *Imperial College, University of London*

GLOSSARY

Anticline: Upfold of strata, a common form of petroleum trap.

Cap rock: Layer of impermeable strata sealing petroleum within a reservoir.

Evaporites: Group of rock-forming minerals resulting from the evaporation of sea water.

Kerogen: Solid hydrocarbon, insoluble in petroleum solvents, that generates petroleum when heated.

Petroleum: Fluid hydrocarbons, both liquid (crude oil) and gaseous.

Reservoir: Porous permeable rock, generally a sediment, capable of storing petroleum.

Seal: Layer of impermeable strata that caps a petroleum reservoir.

Sedimentary basin: Part of the earth's crust containing a thick sequence of sedimentary rocks.

Source rock: Fine-grained organic-rich sediment capable of generating petroleum.

Trap: Configuration of strata, such as an anticline, capable of trapping petroleum.

Unconformity: Break in deposition, often marked by an angular discordance between two sets of strata.

Petroleum geology is concerned with the way in which petroleum forms, migrates, and may become entrapped in the crust of the earth. It studies the generation of organic matter on the surface of the earth, the way in which this may be buried and preserved in sediments, and the physical and chemical processes that lead to the formation of petroleum and its expulsion from sedimentary source beds. Petroleum geology also embraces the dynamics of fluid flow in porous permeable rocks. It is concerned with the depositional environments and postdepositional history of the sandstones and limestones that commonly serve as petroleum reservoirs. Petroleum geology also studies the ways in which tectonic forces may deform strata into configurations that may trap migrating petroleum.

I. Field of Petroleum Geology

Petroleum is the name given to fluid hydrocarbons, both the gases and liquid crude oil. It is commonly noted that petroleum occurs almost exclusively within sedimentary rocks (sandstones, limestones, and claystones). Petroleum is seldom found in igneous or metamorphic rocks. Thus petroleum geology is very largely concerned with the study of sedimentary rocks. It deals with processes operating on the surface of the earth today to discover how organic detritus may be buried and preserved in fine-grained sediments. It considers where and how porous sands become deposited, because these may subsequently serve as reservoirs to store petroleum. It is also important to know how porosity and permeability are distributed in sedimentary rocks. This is essential to understand the way in which petroleum emigrates from the source beds, how it may subsequently be distributed in reservoir beds, and how it may flow once that reservoir is tapped. Petroleum geology is also concerned with the structural configuration of the crust of the earth. This interest includes both large and small scale crustal features. On the broadest scale petroleum geology is concerned with the formation of sedimentary basins. This is because some basins are highly favorable for petroleum generation, while others are less suit-

ENCYCLOPEDIA OF PHYSICAL SCIENCE
AND TECHNOLOGY, VOL. 10

able. The deposition of petroleum source beds varies with basin type, as does the heat flow that is responsible for generating oil and gas from the source rock. [*See* SEDIMENTARY PETROLOGY.]

On a smaller scale petroleum geology is concerned with the way in which strata may be deformed into folds and how they may be ruptured and displaced by faults. Such structural deformation may form features capable of trapping migrating petroleum. [*See* STRATIGRAPHY.]

II. Properties of Petroleum

Petroleum, as previously defined, is the name given to the fluid hydrocarbons. A closer examination of the physical and chemical properties of petroleum is perhaps necessary as a prerequisite to an analysis of the geology of petroleum. The term kerogen is applied to solid hydrocarbons that are insoluble in the normal petroleum solvents, such as carbon tetrachloride, yet which have the ability to generate fluid petroleum when heated. One of the best known examples of kerogen is coal and its associates peat, lignite, and anthracite. Kerogen is also found disseminated in many fine-grained sediments and sedimentary rocks (the clays, claystones, and shales). This is seldom of direct economic value itself, but is of great interest to petroleum geologists because it is the source material from which commercial quantities of petroleum are derived. [*See* COAL GEOLOGY.]

Liquid petroleum is generally termed crude oil, oil, or simply "crude." In appearance oils vary from colorless, to yellow-brown or black. The former are generally of low viscosity and low density, while the black oils tend to be heavier and more viscous. The gravity of petroleum is commonly expressed by reference to a scale proposed by the American Petroleum Institute. The API gravity is calculated as follows:

API gravity = (141.5/specific gravity at 60°F)

− 131.5

Chemically petroleum consists principally of hydrogen and carbon, but also contains small percentages of oxygen, nitrogen, sulfur, and traces of metals, such as vanadium, cobalt, and nickel. The common organic compounds include alkanes (paraffins), naphthenes, aromatics, and heterocompounds. The specific gravity of oil is variable. Heavy oils are defined as those with a specific gravity greater than 1.0 (10°API). These are rich in heterocompounds and aromatics. The heavy oils grade with increasing density and viscosity into the plastic hydrocarbons (generally termed pitch, tar, or asphalt). The light oils are rich in alkanes and naphthenes. They grade into "condensate." This is petroleum that is gaseous in the earth's crust, but which condenses to a liquid at the surface. [*See* PETROLEUM REFINING.]

Petroleum gas is generally referred to as "natural gas," though this ignores the non-hydrocarbon natural gases of the atmosphere. Natural gas is generally composed of methane and varying amounts of ethane, propane, and butane. Petroleum gas composed almost exclusively of methane is termed "dry gas." Petroleum gas with substantial amounts of the other gaseous alkanes is termed "wet gas." Petroleum gases are also associated with traces of other gases, notably hydrogen sulfide, carbon dioxide, and nitrogen.

Gas hydrates are compounds of frozen water that contain gas molecules. Naturally occurring gas hydrates contain substantial amounts of petroleum gas, principally methane. These are only stable in certain critical pressure/temperature conditions. They occur today in the surface sediments of deep ocean basins, and in permafrost. At present petroleum in gas hydrates cannot be extracted commercially.

III. Origin of Petroleum

The origin of petroleum has long been a matter for debate. From the days of Mendele'ev some chemists and astronomers have argued that petroleum is of inorganic abiogenic origin. Petroleum geologists have seldom doubted that it is of organic origin.

It is known that hydrocarbons can form in space. Jupiter and Saturn are believed to be composed of substantial amounts of methane. A particular type of meteorite (the stony meteorites, or chondrites) contain complex hydrocarbons up to the level of amino acids and isoprenoids. Petroleum is found in some igneous rocks (rocks formed from the cooling of molten magma).

In most of these occurrences the petroleum infills fractures and has obviously migrated into the rock long after it cooled. Rarely, however, it occurs in gas bubbles within the rock itself. In some cases it can be demonstrated that the magma has intruded organic-rich sedimentary rocks from which the petroleum may have been acquired, but this is not universally true. Methane occurs in many rift valley floors, both on land (as in the East African rifts) and in the mid-ocean ridge rift systems. Carbon isotope analysis is able to show that this is not always of shallow biogenic origin, but that it is often of deep thermal origin. These observations have caused some people to argue that petroleum is of inorganic abiogenic origin; that petroleum occurs in the mantle of the earth from which it seeps to the surface, being intermittently expelled up fault systems aided by earthquakes. The fundamental reaction is believed to be between iron carbide and water (analogous to the reaction whereby acetylene is produced by calcium carbide and water):

$$CaC_2 + 2H_2O = C_2H_2 + Ca(OH)_2$$

These ideas are popular with some astronomers, but, except in the USSR are seldom held by geologists. Most geologists, and certainly those engaged in the commercial quest for petroleum, believe that it is of organic origin.

There are many lines of evidence to support the thesis that commercial petroleum accumulations are of organic origin. Some of the evidence is chemical and some geological. Oil exhibits levorotation, the ability to rotate polarized light; this property is peculiar to biogenically synthesized organic compounds. Oil contains traces of complex organic molecules that are common in living plant and animal tissues. Examples include porphyrins, steroids, and derivatives of chlorophyll. In many instances oil is found in porous permeable reservoirs totally enclosed by impermeable shale. It is hard to see how petroleum could have migrated into the reservoir, either from the earth's mantle or from outer space. In such instances the enclosing shales generally contain kerogen. Gas chromatography demonstrates the similarity of petroleum generated from the shale kerogen with that found in the reservoir. Finally there is the ubiquitous observation that petroleum is commonly found in sedimentary rocks and not in igneous or metamorphic ones.

These observations lead to the conclusion that although minor traces of petroleum may form inorganically in space and in the earth's mantle, commercial accumulations of petroleum are formed from the thermal maturation of biologically synthesized organic compounds.

The foregoing analysis of the origin of petroleum points us toward the study of sedimentary rocks and their contained fluids. It is now gener-

ally recognized that the four basic requirements for a commercial petroleum accumulation are (1) a thermally mature source rock, (2) a porous permeable reservoir to retain the petroleum, (3) an impermeable cap rock, or seal, to retain the petroleum within the reservoir, and (4) a configuration of rock, such as an anticline, to trap the petroleum within the reservoir. These items will now be considered in turn.

IV. Generation and Migration of Petroleum

A. THE PRODUCTION AND PRESERVATION OF ORGANIC MATTER

Photosynthesis is the process on which all organic matter is based. In this reaction carbon dioxide and water are turned into sugar and oxygen in the presence of sunlight.

$$6CO_2 + 12H_2O = C_6H_{12}O_6 + 6H_2O + 6O_2$$

The sugar that is produced by photosynthesis is the starting point for the construction of the more complex organic molecules found in plants and animals. In the ordinary way when life forms die they are eaten or broken down by the normal processes of bacterial decay. Thus organic matter is seldom preserved in sedimentary rocks. There are, however, certain depositional environments that favor the preservation of organic matter. In lakes a thermal stratification of water sometimes develops. Sunlight warms the shallow water and encourages blooms of algae. As these give off oxygen they provide the base for animal food chains. In the deeper waters sunlight does not penetrate. The water is thus cooler and denser. Lacking photosynthesizing algae this zone may become anoxic. The density stratification inhibits the mixing of oxic and anoxic waters. Organic detritus falling from above may be preserved in the stagnant sediments of the lake floor. Thus ancient lacustrine deposits are often petroleum source beds.

A similar stratification, with an upper oxygenated and a lower anoxic layer of water, sometimes develops in arid climate lagoons. In this environment the stratification is due not to thermal layering, but to evaporation. Dense anoxic brines concentrate on the lagoon floor and favor the pickling of organic detritus in the underlying muds.

Organic sediments may also be deposited in ocean basins. In modern oceans the shallow waters are oxygenated by plankton photosynthesis. The deep ocean basins are seldom anoxic. But an oxygen-deficient zone commonly occurs between 200 and 1500 m on the eastern sides of the major ocean basins. The upper limit of this zone commonly coincides with the edge of the continental shelf. Thus organic matter may be preserved in sediments deposited on the continental slope and rise. Global rises of sea level cause marine transgressions and favor the deposition of blanket petroleum source beds across continental shelves. Today the flow of cold dense oxygenated polar water into the ocean basins of lower latitudes prevents their floors from becoming stagnant. It has been argued that when the earth has had a uniform equable climate this mixing would not occur. Global anoxic events may then favor the ubiquitous preservation of organic matter on the ocean basin floors. These are the four types of sedimentary environment that favor the preservation of organic detritus in fine-grained sediments.

B. GENERATION OF PETROLEUM FROM KEROGEN

As muds are buried they compact, lose porosity, and are lithified into shales or mudstones. The organic matter that they contain undergoes many changes. These can be conveniently grouped into three stages: diagenesis, katagenesis, and metagenesis. Diagenesis occurs in the shallow subsurface at near normal temperatures and pressures. Methane, carbon dioxide, and water are driven off from the organic matter to leave a complex hydrocarbon termed kerogen (as previously defined this is a hydrocarbon that is insoluble in normal petroleum solvents, but expels petroleum when heated). There are three main types of kerogen. Type I, sapropelic kerogen, is formed from algal matter. It has a high potential for generating oil. Type II, liptinic kerogen, can generate both oil and gas. Type III, humic kerogen, is predominantly gas prone. This last type of kerogen forms as peat on the surface of the earth. As this is buried and matured it evolves through lignite, bituminous coal, and anthracite to graphite (pure carbon).

With increasing burial the source bed is subjected to rising temperature and pressure. Diagenesis merges into categenesis. This is the principal phase of petroleum generation. Initially oil is given off once temperatures are above about 60°C. The peak of oil generation is at about 120°C. Above this temperature oil generation declines, to be replaced by gas genera-

tion. Methane gas emission peaks at about 150°C and ceases at about 200°C. This temperature marks the transition between catagenesis and metamorphism. At this stage the kerogen has yielded up all its petroleum. The residue is pure carbon, graphite.

C. PETROLEUM MIGRATION

The migration of petroleum has long been a matter for debate. It is normal to differentiate primary from secondary migration. Primary migration refers to the emigration of petroleum from the source rock into a permeable carrier bed. Secondary migration refers to the migration of petroleum from the carrier bed into the reservoir, and to subsequent flow during petroleum production. Secondary migration is relatively simple to understand. Oil and gas move in response to buoyancy and pressure differentials within large pore systems. Primary migration presents a different problem. Shales are impermeable and normally provide excellent seals that inhibit upward petroleum movement (see later). How then can petroleum emigrate from an impermeable source bed into a permeable carrier formation? It is not enough to invoke simple compaction of clay during burial. When clays compact most of the porosity is lost in the first kilometer or so of burial. Temperatures are still too low to permit kerogen maturation. At the depths at which petroleum generally forms (some 3 to 4 km) little porosity loss remains to take place. Many different mechanisms have been proposed to explain primary migration. These invoke solution of oil and gas in water, improbably high temperatures, abnormally high pressures, and the use of soapy micelles as petroleum solvents which then vanish once they have done their task. Many mechanisms are now current to explain primary petroleum migration. There is no single theory that has received universal acceptance. Perhaps an understanding of the mechanism is not too important. The significant fact is that gas chromatography can correlate kerogen in source beds with the trapped petroleum that once emigrated from them.

Thus the first requirement for a commercial accumulation of petroleum is the existence of a thick sequence of organic-rich shale, with an excess of 1.5% organic carbon. Whether the kerogen yields oil or gas, or both, is dependent on the type of kerogen and its level of thermal maturation. There parameters are commonly dis-

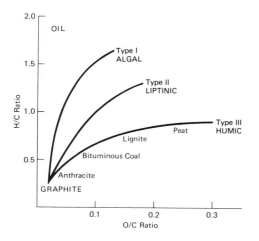

FIG. 1. Graph of hydrogen/carbon ratio plotted against oxygen/carbon ratio. This shows the composition of the different types of kerogen. Type I kerogen generates principally oil with minor gas. Type II kerogen generates both oil and gas. Type III kerogen generates only gas. From Selley, R. (1983). "Petroleum Geology for Geophysicists and Engineers," by courtesy of International Human Resources Development Corporation.

played on graphs that plot the ratio of hydrogen : carbon against oxygen : carbon (Fig. 1).

V. Petroleum Reservoirs and Cap Rocks

A. POROSITY AND PERMEABILITY

The second requirement for a commercial accumulation of petroleum is the existence of a reservoir capable of storing the oil or gas. There are two basic requirements for a reservoir: porosity and permeability. Porosity is the storage capacity of the rock. This is sometimes expressed as the void ratio, or more commonly as a percentage:

Porosity = (void volume/rock volume) × 100%

Permeability is the rate of flow of fluid through the rock. It is expressed by Darcy's Law. This states that the rate of flow of a homogeneous fluid in a porous medium is proportional to the pressure gradient, and inversely proportional to the fluid viscosity. This is generally expressed mathematically thus:

$$Q = [K(P_1 - P_2)A]/\mu L$$

Where Q is the rate of flow, K a constant, ($P_1 -$ P_2) the pressure drop, A the cross-sectional area of the sample, L the length of the sample, and μ the viscosity of the fluid.

Permeability is measured in Darcies (D). One Darcy is the permeability that allows a fluid of 1 cP viscosity to flow at a velocity of 1 cm/sec for a pressure drop of 1 atm/cm. Since most petroleum reservoirs commonly have permeabilities of less than 1 D the millidarcy (1/1000 D) is usually used.

Theoretically any rock can act as a petroleum reservoir so long as it has these two essential properties of porosity and permeability. Igneous and metamorphic rocks are generally composed of a tight interlocking mosaic of mineral crystals. They thus seldom have porosity and permeability unless they have been fractured or weathered beneath an unconformity (an old land surface). Only about 10% of the world's oil reserves have been found in igneous and metamorphic basement. Most petroleum reservoirs are found in the sedimentary rocks (sandstones, limestones, and shales) because, by their very nature, they form from the deposition of granular particles with microscopic pores between the grains. Shales (compacted mud) are commonly porous. But, because of the small diameter of their pores, shales are generally impermeable and seldom act as reservoirs. About 45% of the world's petroleum reserves have been found in sandstones and about 45% in carbonates (limestones and dolomites).

B. PORE SYSTEMS

Table I lists the various types of pore systems encountered in petroleum reservoirs. There are two main types of pore: primary (or syndepositional) and secondary (or postdepositional). Primary pores are those that form when a sediment

TABLE I.   A Classification of Reservoir Porosity

Primary or syndepositional porosity
    Intergranular or interparticle
        (characteristic of sandstones)
    Intragranular or intraparticle
Secondary or postdepositional porosity
    Solution porosity (common in limestones)
        Moldic (fabric selective)
        Vuggy—cavernous (cross-cuts rock fabric)
    Intercrystalline
        (characteristic of dolomites)
    Fractures
        (develop in any brittle rock)

is deposited. They may be either intergranular, occurring between grains, or intragranular, occurring within grains. The former is common in terrigenous sediments and is therefore the major type of pore found in sandstone reservoirs (Fig. 2). Intraparticle pores occur within shells in skeletal lime sands. During burial, however, these pores are often destroyed by compaction or by cementation.

There are several different types of secondary porosity. Solution due to migrating fluids may generate moldic or vuggy porosity. Moldic porosity is fabric selective. That is to say, only one element of the rock is leached out, such as fossil shell fragments. Vuggy pores cross-cut the fabric of the rock, grains, matrix, and cement. Vugs are thus commonly larger than moldic pores. They grade up with increasing size into cavernous porosity. Solution porosity is characteristic of limestone reservoirs. This is because calcium carbonate is less stable in the subsurface than the quartz (silica) of which sandstones are largely composed. Unlike primary intergranular porosity solution pores are often isolated from one another. Thus, such rocks may have considerable porosity but negligible permeability. The other main type of porosity is that which is caused by fracturing.

Fractures can develop in any brittle rock, including igneous or metamorphic basement, limestone, or cemented sandstone. Fracturing is generally caused by tectonic processes, such as folding and faulting. Once initiated, fractures may be enlarged by solution or infilled by cementation. Fracturing may not increase the porosity of a rock very much, but it may cause a dramatic increase in permeability. Thus if a well

has drilled into a petroleum reservoir of low permeability it is common to fracture the rock adjacent to the borehole by explosive charges. The last main type of porosity to consider is the intercrystalline variety. This is the type of pore system that is typical of dolomite reservoirs. Dolomite is a mixed calcium magnesium carbonate ($CaMg \cdot [CO_3]_2$). Some dolomite forms at the surface of the earth. This is generally fine grained, and, though porous, generally tight. In the subsurface, however, dolomite may form by the replacement of limestone. The reaction is a reversible one:

$$2CaCO_3 + Mg^2 \rightleftharpoons CaMg(CO_3)_2 + Ca^2$$

When calcite is replaced by dolomite there is an overall volume shrinkage of as much as 13%. This results therefore in a considerable increase in porosity if rock bulk volume remains unchanged. The resultant dolomite rock commonly has a friable saccharoidal appearance. These secondary dolomites may thus serve as very effective petroleum reservoirs because they exhibit both porosity and permeability.

Figure 3 shows the relationship between porosity and permeability for some of the different pore systems that have just been described. This shows the very complex relationship that exists between the texture of a rock and its reservoir potential. A major part of petroleum geology is concerned with understanding the depositional environments of sediments, so as to predict their distribution in the subsurface. Similar interest is shown in the postdepositional geochemical reactions that take place between sediments and migrating pore fluids. These cause sediments to be cemented and turned into solid rock, thus destroying primary porosity. But selective leaching can subsequently give rise to secondary solution pore systems of erratic distribution.

C. CAP ROCKS OR SEALS

The third essential requirement for a commercial accumulation of petroleum is a cap rock or seal. This is a sedimentary stratum that immediately overlies the reservoir and inhibits further upward movement. A cap rock need have only one property. It must be impermeable. It can have porosity, and may indeed even contain petroleum, but it must not permit fluid to move through it. Theoretically any impermeable rock may serve as a seal. In practice it is the shales and evaporites that provide most examples. Shales are probably the commonest, but evapo-

FIG. 2. Photomicrograph of a thin section through a sandstone petroleum reservoir showing primary intergranular porosity between the quartz framework grains. Width of photograph is 3 mm.

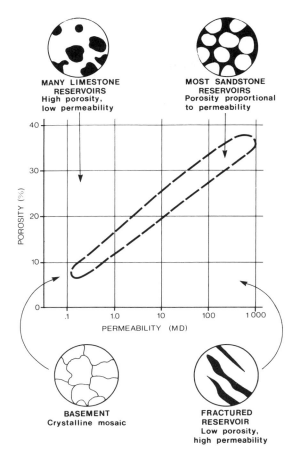

FIG. 3. Graph of porosity plotted against permeability showing the reservoir characteristics of the different types of pore systems. Courtesy of R C Selley & Co. Ltd.

rites are the more effective. We saw earlier how mud is compacted during burial into mudstone or shale. These rocks are commonly porous, but because of the narrow diameter of the pore throats, they have negligible permeability. Thus shales generally make excellent seals to stop petroleum migration. When strata are folded or faulted, however, brittle shales may fracture. As described earlier fractures enhance permeability most dramatically. In such instances petroleum may leak from an underlying reservoir and ultimately escape to the surface of the earth.

The evaporites are a group of sedimentary rocks that were originally thought to have formed from the evaporation of sea water. They are composed of crystalline layers of halite, or rock salt (NaCl), anhydrite ($CaSO_4$), carnallite (KCl), and other minerals. The evaporites exhibit a number of anomalous physical properties. Unlike most sedimentary rocks, when they are subjected to stress they do not respond by fracturing, but by plastic flow. Thus, a petroleum saturated reservoir may undergo all sorts of structural deformation and may even be fractured, but an overlying evaporite may still provide an effective unfractured impermeable seal. [See EVAPORITES.]

VI. Petroleum Traps

A. TERMINOLOGY AND CLASSIFICATION

The last major requirement for a commercial accumulation of petroleum is the existence of a trap. Source rock, reservoir, and seal must be arranged in such a geometry that migrating petroleum is trapped in some configuration of strata and cannot escape to the surface of the earth. The simplest type of trap is an upfold of strata termed an anticline. Figure 4 illustrates such a structure and some of the terms applied to traps. A trap may contain oil, gas, or both. This will be a function of the pressure/temperature conditions in the reservoir, but will also be related to the type of kerogen that exists in the source bed, and in its level of thermal maturation. Gas is lighter than oil and oil lighter than water. There is thus a gravity segregation of the reservoir fluids. The top may consist of a gas zone, or cap, separated by the gas : oil contact from the oil zone. The oil zone is separated from the water zone by the oil–water contact. The reservoir beneath the oil zone is termed the bottom water, and that adjacent to the field area is termed the edge water. The vertical interval from the crest of the reservoir to the petroleum : water contact is termed the pay zone. Not all of this interval may be productive. It may also contain impermeable strata. Thus it is usual to differentiate between the gross pay and the net effective pay. The vertical interval from the crest of a reservoir to the lowest closing contour on a trap is termed the "closure." The lowest closing contour is termed the spill plane. The nadir of the spill plane is termed the "spill point." Depending on the amount of petroleum available a trap may or may not be full to the spill point. The term "field" is applied to a petroleum-productive area. An oil field may contain several separate oil "pools." A pool is a petroleum accumulation with a single petroleum : water contact.

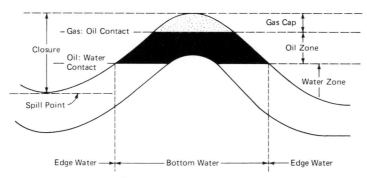

FIG. 4.   Cross-section through a simple anticlinal petroleum trap to illustrate terminology. From Selley, R. 1983. "Petroleum Geology for Geophysicists and Engineers" (Fig. 28), by courtesy of International Human Resources Development Corporation.

There are many different types of petroleum trap. They are commonly classified into five main categories and associated subcategories as shown in Table II. These will now be defined and then described in turn. Structural traps are those caused by tectonic forces in the earths' crust. They thus include anticlines and fault traps. Diapiric traps are due to density contrasts in sedimentary rocks, normally evaporites and overpressured clays. Stratigraphic traps are due to depositional, erosional, or diagenetic processes. Hydrodynamic traps are caused by water flow. Combination traps are due to a combination of any two or more of the above processes.

Statistical studies have shown that about 75% of the worlds' petroleum comes from structural traps, principally anticlines, about 5% from diapirs, and about 6% from stratigraphic traps. The rest are of combined origin. It must be remembered, however, that these figures pertain to the petroleum that has been discovered to date. No one knows what the actual proportions

TABLE II.   A Classification of Petroleum Traps

Structural traps
  Anticlines
  Fault traps
Diapiric traps
  Salt diapirs
  Mud diapirs
Stratigraphic
  Unconformity related (truncation and onlap)
  Unassociated with unconformities (channels, bars, and reefs)
  Diagenetic (due to solution or cementation)
Hydrodynamic traps—due to water flow
Combination

are. These figures thus reflect our ability to find oil, and probably indicate our obsession with drilling anticlinal prospects.

B. STRUCTURAL TRAPS

Structural traps are those that are caused by earth movements such as folding or faulting. Folds may be caused by compression or tension of the earths' crust. Compression causes a shortening of the crust, deforming it into a series of upfolds, or anticlines, and downfolds or synclines. Compressive anticlines are the main types of trap for the oil that comes from the foothills of the Zagros Mountains of Iran. When the earth's crust is subjected to tension it begins to subside developing into a basin several hundred kilometers or more in diameter. This basin may be subsequently infilled by sediments. The floor of such a basin commonly splits into a mosaic of fault blocks. The uplifted blocks are termed "horsts" and the downdropped blocks "grabens." As the basin floor is buried by sediments the strata are draped over the horsts. Subsequent compaction enhances the vertical closure. As the topography is mantled by successive layers of sediment the closure of the anticlines diminishes upward. These anticlines are therefore different in origin from those due to compression. They are commonly termed drape or compactional anticlines. The Forties field of the North Sea is a classic example of a compactional anticlinal trap.

Fault traps are of several varieties, but in all cases it is essential for the fault to be impervious and sealed. This is by no means always the case and many faults are open conduits to fluid movement. It is not possible to establish whether a fault is open or closed without drilling to test. Tension of the crust generates "normal" faults across which part of the sedimentary section is absent (Fig. 5B). The Fahud field of Oman is a good example of a normal fault petroleum trap. When a normal fault moves, the strata on the downthrown side often flop down into the space created by the fault movement. These are called "rollover" anticlines. It is often noted that strata thicken when traced from the upthrown to the downthrown side of the fault and that the throw of the fault increases when measured downward from sediment increment to increment (Fig. 5C). This demonstrates that the fault is removed repeatedly through time. Such faults are thus termed growth faults. Petroleum is often trapped when associated with growth faults, both where reservoir beds are truncated and sealed by the fault, and in adjacent rollover anticlines. Examples are common in the Tertiary deltas of Nigeria and the Gulf Coast of the United States.

Compression causes reverse faults that give rise to the vertical repetition of strata (Fig. 5A). Thrust faults, as these are often called, are commonly associated with compressive anticlines. Such traps are common along the edges of mountain fronts. The Turner Valley and Winterton fields of the Rocky Mountains are examples of such traps. Transverse movements of the crust give rise to wrench faults. These commonly consist of a single fault at a depth that bifurcates upward into several branches. These commonly generate an anticline coincident with the wrench fault. These are commonly termed "flower" structures. Oil fields in such traps are common in the Los Angeles basin of California.

C. DIAPIRIC TRAPS

Diapiric traps are caused by variations in the density of sediments. In the ordinary way as sediments are buried they become compacted due to the overburden pressure. They lose porosity and their density increases. There are two exceptions to this general rule in which low-density sediments may be overlain by denser sediment. This situation is inherently unstable. The deeper less dense material is displaced upward as the overburden bears down. The upward movement is localized into domes or diapirs only a few kilometers across. These structures can be foci for petroleum entrapment. The two rocks that can generate diapirs are the evaporites and overpressured clay. The evaporites commonly have a density of about 2.03 g/cm^3. This is greater than that of freshly deposited sand and clay. But as normal sediments compact their density exceeds that of evaporites at about 800 m of burial. Below this depth therefore diapiric deformation may be expected to commence. Movement may be initiated by structural forces, but can also apparently occur spontaneously. The second type of diapir is produced by overpressured clays. In some environments, especially deltas, sedimentation is so rapid that some sediments are unable to dewater before they are buried and sealed by younger detritus. With increasing burial the situation can occur in which dense compacted sediment overlies less dense undercompacted clay. This may be squeezed upward in cylindrical diapirs analogous to those produced by salt movement. There are many ways in which petroleum may become trapped over and adjacent to a diapir (Fig. 6). This can range from simple domal traps over the crest to cylindrical fault traps in which the dome has pierced through the surrounding strata. Salt dome oil fields that occur include the Ekofisk and associated fields of the North Sea. Clay diapirs trap petroleum in the Beaufort sea of Arctic Canada. Both salt and mud diapirs generate petroleum traps in the Gulf of Mexico province of Texas and Louisiana. [See STRATEGIC PETROLEUM RESERVE.]

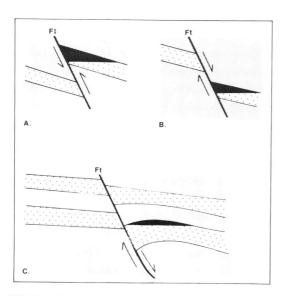

FIG. 5. Cross-sections to illustrate the different types of fault trap. (A) Reverse fault caused by compression. (B) Normal fault caused by tension. (C) Growth fault with petroleum trapped in adjacent rollover anticline. Courtesy of R C Selley & Co. Ltd.

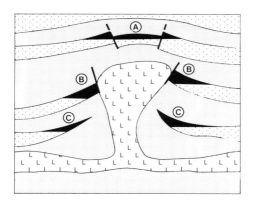

FIG. 6. Cross-section through a salt dome to show the different types of associated petroleum trap. (A) Crestal dome. (B) Flank fault traps. (C) Flank pinchout traps. Courtesy of R C Selley & Co. Ltd.

D. Stratigraphic Traps

As previously defined, stratigraphic traps are due to depositional, erosional, or diagenetic processes (Table II). Though the beds involved may be tilted from the horizontal, folding and faulting are absent in a pure stratigraphic trap (Fig. 7). Depositional, erosional, and diagenetic stratigraphic traps will now be described in turn. The three main types of depositional stratigraphic traps are reefs, bars, and channels.

Reefs are a major type of depositional stratigraphic trap. Modern coral reefs are highly porous and permeable. They may, in the fullness of time, be buried beneath organic-rich marine muds, thus becoming potential petroleum traps (Fig. 7A). There are many ancient petroliferous reefs. Today reefs thrive only in warm clear shallow sea water. The evidence of ancient reefs suggests a similar ecology, though corals have only become important reef-building organisms in recent geological time. Calcareous algae have always been important reef builders, as have bryozoa, bivalves, and several extinct groups of lime-secreting colonial organisms. Recent reefs are highly porous and permeable. But they are composed of calcium carbonate which is highly unstable in the subsurface. Acid rain water may leach lime from the upper part of a reef and reprecipitate it as a cement in the primary pore spaces deeper down. Subsequent uplift and weathering may generate secondary solution and fracture porosity. Thus ancient reefs are not always porous and permeable. Even where they possess these properties their distribution may be unrelated to the initial arrangement when the reef first formed. Despite these problems there are many ancient reefs that serve as stratigraphic traps for petroleum. Notable examples occur in the Devonian rocks of Alberta, and the Cretaceous rocks of Mexico and the Arabian Gulf.

Beaches, barrier islands, and offshore bars are all sedimentary environments that deposit clean well-sorted sands of high porosity and permeability. Where these are enveloped in organic-rich marine shales they may become stratigraphic traps (Fig. 7B). Notable examples occur in the Cretaceous basins of the Rocky Mountain foothills from Alberta in the north to New Mexico in the south.

There are several sedimentary environments

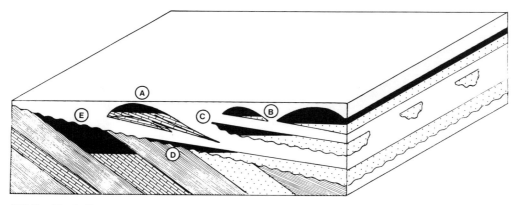

FIG. 7. Block diagram to illustrate the different types of stratigraphic traps. (A) Fossil reef limestone. (B) Coastal barrier-bar sand. (C) Sandstone infilled channel. (D) Onlap of sand above unconformity. (E) Truncation of limestone beneath unconformity. Courtesy of R C Selley & Co. Ltd.

in which channels become infilled with sand that may serve as a petroleum reservoir (Fig. 7C). These range from alluvial flood plains, through deltas, to deep water submarine channels at the foot of the continental slope. In all of these situations sand channels may have cut into and been overlain by impermeable muds. These muds may act as source beds and seals, generating petroleum that is then trapped within the channel sands. There are many examples of alluvial channel sand traps in the Cretaceous basins of Colorado and Wyoming and western Canada. Petroliferous deltaic distributary channel sand traps are common in the Pennsylvanian sediments of Illinois and Oklahoma.

Channel and bar stratigraphic traps both have linear geometries. They are thus colloquially referred to as "shoestring" sands. It is important to note, however, that channels will tend to trend down the old depositional slope into the center of a sedimentary basin. Barrier bar and beach sands, however, will tend to be elongated at right angles to channels, being aligned parallel to the paleoshoreline.

The second main group of stratigraphic traps to consider are those that are associated with unconformities. An unconformity is a surface that marks a major break in the depositional history of an area. The underlying rocks may be of any type, igneous, metamorphic, or sedimentary. Where the latter is the case the strata may have been folded or tilted before the deposition of the overlying sediments. There is often evidence of weathering and fracturing in the rocks beneath the unconformity, though subaerial exposure has not always occurred. Unconformities play an important role in petroleum entrapment for two main reasons. Weathering beneath the unconformity may generate secondary porosity and permeability in all sorts of rocks, both sedimentary and basement. These weathered zones may serve both as reservoirs and as conduits to permit petroleum migration. Second, unconformities permit the juxtaposition of source beds and reservoirs. Sometimes source overlies reservoir and sometimes the reverse occurs. Thus there are two types of stratigraphic traps associated with unconformities: onlap or pinchout traps and subcrop or truncation traps.

At its simplest, an onlap trap may be a blanket sand that pinches out up dip (Fig. 7D). It is sealed by impermeable rocks beneath and by an onlapping shale (generally the source rock, as well as the cap). Many unconformities are old land surfaces, however, and sands may be deposited in old topographic lows. Alternating hard and soft sediments may have been weathered and eroded to form scarps, dip slopes, and strike vallies. Fluvial or shallow marine sands may have been deposited along the old vallies and sealed by marine muds. Stratigraphic traps of this type are known as the Mississippian : Pennsylvanian unconformity of Oklahoma. Alternatively, the unconformity may have been a planar land surface that was locally incised by alluvial vallies. These may have been sand filled and drowned by an advance of the sea that deposited impermeable source beds above them. Such alluvial valley sands may thus act as stratigraphic petroleum traps.

Subcrop truncation traps can be developed in both sandstones or limestones (Fig. 7E). Some sort of closure, either depositional or structural, is required to seal the reservoir off along the strike of the truncated reservoir. Some of the world's largest fields occur in truncation traps. The East Texas field and the Messla field of Libya are two such examples.

The third type of stratigraphic trap is due to diagenesis. This term encompasses the physical and chemical changes that take place in a sediment after it has been deposited. Diagenesis generally leads to a gradual destruction of porosity as the sediment is compacted due to the weight of the overlying detritus, and as primary porosity is destroyed by the precipitation of minerals from percolating ground water. Sometimes, however, diagenesis leads to the generation of secondary porosity. This is either by leaching, giving rise to molds, vugs, and caverns, or by mineral replacement, such as dolomitization, giving rise to intercrystalline porosity. There are few petroleum traps that can be solely attributed to diagenesis. There are, however, many traps in which diagenesis has played a major role in the distribution of porosity and permeability. This is more true of carbonate than sandstone reservoirs, because calcium carbonate is chemically less stable than silica in the subsurface environment. We have already observed the effects of diagenesis on the porosity evolution of reefs and subunconformity traps.

E. HYDRODYNAMIC AND COMBINATION TRAPS

The last two types of trap to consider are those due to hydrodynamic flow, and those caused by a combination of factors. In some sedimentary basins ground water flows down toward the basin center. In so doing it may en-

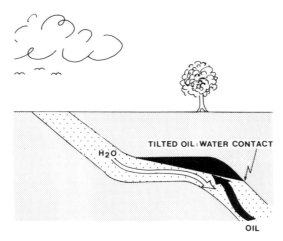

FIG. 8.   Cross-section through a hydrodynamic trap. Petroleum moving up through permeable strata is trapped in a minor flexure by downward flowing water. Note the tilted oil:water contact. Courtesy of R C Selley & Co. Ltd.

counter oil migrating up toward the surface. Oil is lighter than water so within a given permeable bed oil will tend to occupy the upper part and water the lower part. Strata sometime contain flexures which lack vertical closure. That is to say a regional dip may have local horizontal wrinkles to it. In the ordinary way oil could not be trapped beneath such benches. But where there is hydrodynamic flow entrapment can take place in such situations. The oil:water contact will, of course, be tilted, and the petroleum will remain trapped in the flexure only as long as the downward flow of water continues (Fig. 8). Very few pure hydrodynamic traps have been discovered to date. There are, however, many oil fields around the world in which the oil:water contact is not a horizontal plane, but is in fact tilted. The lateral flow of water beneath such traps is one of several causes of tilted oil:water contacts.

This leads logically to a discussion of combination petroleum traps. These were previously defined as those due to a combination of two or more trap-forming processes: structural,

diapiric, stratigraphic, or hydrodynamic. There is a multiplicity of ways in which these several processes may combine to form traps. The interplay of hydrodynamic and diagenetic processes on the more common trap types has been already noted. Probably the commonest type of combination trap is where a structural high has been growing during sedimentation. The uplift itself may have been initiated by a fold, fault, or diapir. Such highs are likely to be subject to erosion that can cause unconformity-related traps. These will include crestal subunconformity truncations as well as depositional pinchouts around the edges. Carbonate reefs may also be located on syndepositional structural highs. Large fields that are formed by a combination of structural and stratigraphic processes include the Brent and associated fields of the North Sea, and the Prudhoe bay field on the North Slope of Alaska. The permutations of ways in which combination traps may form are infinite.

BIBLIOGRAPHY

Chapman, R. E. (1983). "Petroleum Geology." Elsevier, Amsterdam.

Dickey, P. A. (1981). "Petroleum Development Geology," 2nd ed. Pennwell, Tulsa.

Hobson, D. G., and Tiratsoo, E. N. (1981). "Introduction to Petroleum Geology," 2nd ed. Applied Science Publishers, Barking.

Hunt, J. M. (1979). "Petroleum Geochemistry and Geology." Freeman, New York.

Kinghorn, R. R. F. (1983). "An Introduction to the Physics and Chemistry of Petroleum." John Wiley, New York.

Levorsen, A. I. (1967). "Geology of Petroleum." Freeman, New York.

Link, P. K. (1982). "Basic Petroleum Geology." Pennwell, Tulsa.

Selley, R. C. (1983). "Petroleum Geology for Geophysicists and Engineers." International Human Resources Development Corporation, Boston.

Selley, R. C. (1985). "Elements of Petroleum Geology." Freeman, New York.

Tissot, B. P., and Welte, D. H. (1978). "Petroleum Formation and Occurrence," 2nd ed. (1984). Springer-Verlag, Berlin.

PETROLEUM REFINING

James H. Gary   *Colorado School of Mines*

GLOSSARY

Aromatic: Type of hydrocarbon compound containing at least one benzene ring.

Barrel: Volumetric measure of refinery feedstocks and products equal to 42 U.S. gallons.

Conradson carbon residue: Measure of carbon-forming potential of a petroleum (amount of coke formed) fraction expressed in weight percent as determined by a standard laboratory test procedure.

Motor octane number (MON): Measure of uniformity of burning (resistance to knocking) of gasoline under laboratory conditions which simulate highway driving conditions.

Naphtha: Product stream which boils in the gasoline boiling range.

Naphthene: Cycloparaffin compound, which is a paraffin with a ring structure.

Octane number: A comparative number used to express the resistance to knocking of a gasoline. Refers to the volume percent of isooctane in an isooctane, normal-heptane mixture which knocks in a standard test engine under the same conditions as the sample tested.

Olefin: Unsaturated nonring hydrocarbon compound which has a double-bond between two of the carbon atoms in the molecule.

Paraffin: Saturated nonring hydrocarbon compound in which all carbon atoms are connected by single bonds.

Ramsbottom carbon residue: Measure of the carbon-forming potential (amount of coke formed) of a petroleum fraction expressed in weight percent of the fraction. This is determined by a standard laboratory test procedure.

Research octane number (RON): Measure of uniformity of burning (resistance to knocking) of gasoline under laboratory conditions which simulate city driving conditions.

Resid, residuum: Undistilled portion of a crude oil. Usually the atmospheric or vacuum tower bottom streams.

Crude petroleum oils, as found in nature, require processing to improve quality and to change boiling ranges of the components in order to meet the product quantities and specifications necessary to supply the transportation fuels and heating oils needed. As each crude oil has its own specific characteristics, diverse refining processes are required and each processing plant, familiar to us as a refinery, is different from other refineries that process various crudes. Refining processes are very complex and require sophisticated equipment and instrumentation. As a result, the petroleum refining industry has a very high investment cost per worker. Even with its complexity the industry is very efficient and refining costs per gallon of product are low because of the high quantities of crude oils processed. In the United States about 12,000,000 barrels of crude oil are charged to refineries each day. This amounts to approximately 600 millions tons per year and dwarfs the output of most of the "heavy industries."

Historically, crude oils have been classified as paraffin, intermediate, or naphthene base because of the characteristics of products produced upon distillation and of the residuum remaining after distillation. This was especially true when U.S. refineries processed U.S. crudes exclusively because it gave indications of the amount of processing required and the characteristics of the products. Today, because a significant amount of the crude oil processed in U.S. refineries is imported, this general classification, although helpful, does not adequately indicate the cost of processing or the quality of the products and more complete specifications are required.

The more important characteristics of crude oils with respect to processing requirements and costs are the density, sulfur and nitrogen content, distillation range, carbon residue, metals content, and the amount of salt and water in the crude oil. Transportation requirements are indicated by the viscosity and pour point of the oil. The density of the oil is expressed as °API or grams/milliliter and the sulfur and nitrogen contents in weight percent, if greater than 0.1%, or parts per million by weight (ppm), if less than 0.1%. One percent is equal to 10,000 ppm. Distillation range can be expressed according to either ASTM or true boiling point [TBP or 15/5 (distillation in a column containing 15 theoretical stages at a 5/1 reflux ratio)] distillations. Carbon residue is given in weight percent as determined by either the Conradson or Ramsbottom standard test procedures. Metal contents (nickel, vanadium, and iron) are given in parts per million by weight. Traditionally, the salt content is expressed as sodium chloride equivalents in pounds per 1000 barrels (PTB) but more recently as parts per million by weight. Conversion between these units varies with crude oil gravity but 1 PTB is approximately equivalent to 3 ppm.

The capacities of U.S. refineries are expressed in barrels per day (BPD or B/D). In truth, today there is no such item as a barrel but the measurement is equivalent to 42 U.S. gallons at 60°F. In other parts of the world, refinery capacities are usually given in metric tons per year or per day. For an average density crude oil there are approximately 7.3 barrels per metric ton.

Crude oils are complex mixtures of hydrocarbons and organic compounds. The components can range from very light (low-boiling point) molecules, such as methane with a molecular weight (MW) of 16, to very heavy (high-boiling) molecules having molecular weights in the thousands and boiling points above 700°C (1300°F). The general classes of compounds include paraffins, cycloparaffins (naphthenes), and aromatics. Olefins do not occur naturally in crude oils but are created in processing. There are so many different hydrocarbons in crude oils that only

those with molecular weights of less than 100 are separated into individual compounds and the higher molecular weight components are separated into fractions by boiling point ranges (i.e., 90–180°C). Many of the compounds containing seven or more carbon atoms consist of aromatic rings with paraffinic side chains and behave as either paraffins or aromatics depending upon the relative amounts of the aromatic and paraffinic parts and the situation.

As taken from the ground, less than half of the average barrel of crude can be used for transportation fuels (gasoline, jet fuel, and diesel). The remainder of the barrel consists of compounds which have boiling points too high or too low to be included. Since the large-volume profitable products are transportation fuels, it is necessary to convert as much of the barrel to transportation fuels as is economically possible. Refineries which have the necessary conversion equipment to produce a full slate of products from the crude are called integrated refineries while those without conversion processes are called topping or hydroskimming refineries. The topping refinery usually has an atmospheric distillation unit to separate the transportation fuel boiling range fractions and a catalytic reformer to improve the gasoline octane. In addition to these units, the integrated refinery contains conversion processes that change the molecular weights of the higher and lower molecular weight compounds to those that have boiling points in the transportation fuel boiling range. If the crude oils contain unacceptable levels of contaminants (usually sulfur and nitrogen), integrated refineries will also have processing equipment to remove them. This additional equipment adds to both the capital and operating cost of the refinery and each processing unit must be justified economically. A simplified block-flow diagram for an integrated refinery is shown in Fig. 1.

One significant aspect of refinery operation that is frequently overlooked is the rather limited flexibility with respect to pricing and product distribution. In other industries products are priced frequently on the basis of raw materials costs plus the additional processing and overhead costs necessary to convert them into the finished products. There are problems in using this technique in refinery operations. A refinery must operate with a limited amount of storage and as a result must dispose of all products at essentially the same rate as crude oil is charged to the refinery. This means that if the market for the heavy, high-boiling point components (those

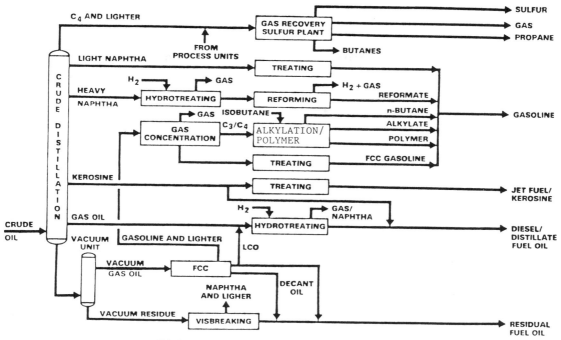

FIG. 1.   Refinery flow diagram. (Courtesy of UOP.)

too heavy to be included in the transportation fuels) is saturated, the refinery must either convert these into salable products, shut the refinery down, or drop the price low enough that they can be sold. Usually, this means this material is sold for less than the raw material cost and these losses must be absorbed by increasing the selling prices of the transportation fuels. This is one of the reasons for today's refineries being so capital intensive. A successful refinery operation is based on economic comparisons of alternative processing costs and product values for all available crude oil feedstocks.

There can be no description of an average refinery operation because all refineries operate in a different manner. A specific refinery's operation is a function of the types of processing equipment it has, the properties of the crude oils charged to the refinery, and the product slate it has to make to meet market demand. Starting with a heavy, high-sulfur California crude oil, the type of processing units in a refinery determines the proportion of the crude that can be converted into transportation fuel. Figure 2 illustrates the wide range of product slates that can be produced from this type of crude as a function of the kind of residuum processing equipment used. The greater the amount con-

verted to transportation fuels, the higher the capital and operating costs. Each case has to be independently evaluated and, because of the number of choices available, the operating cost of each refinery is different from that of other refineries even though comparable products are produced.

I. Refinery Operation

There are several ways to classify specific refinery operations and for here we will classify them as preparation, conversion, quality improvement, and auxiliary. Some operations can fall into more than one class, for example, alkylation which not only converts low-boiling gaseous hydrocarbons into higher boiling gasoline blending stocks but also produces a very high octane material. The processes discussed are as follows:

1. Preparation
   Desalter
   Crude oil distillation
   Gas plant
2. Conversion
   Coker
   Fluid catalytic cracker

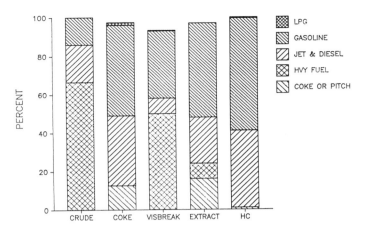

FIG. 2. Effects of residua processing on refinery yields.

Hydrocracker
Alkylation
Polymerization
3. Quality Improvement
Hydrotreating
Catalytic reforming
Solvent extraction
Visbreaking

The processing units will be discussed in the general order shown for a fuels-type integrated refinery processing high-sulfur crude oils. If low-sulfur crude oils only are processed, the hydrotreating units are not needed and processing costs are substantially lower.

II. Crude Distillation

The first major processing unit in the refinery is the crude oil distillation unit (Fig. 3) consisting of an atmospheric distillation followed by a vacuum distillation unit. These units separate the crude oil into fractions according to boiling point. Two units are necessary because hydrocarbons thermally decompose (crack or pyrolyze) at temperatures above 370°C (700°F) and, by reducing the pressure at which distillation takes place, separation of hydrocarbons with boiling points of up to 566°C (1050°F) at atmospheric pressure can be achieved at distillation temperatures low enough that little cracking takes place. The products from the crude units are used as feed stocks to other processing units or blended into products.

The crude oil entering the refinery has water-containing dissolved salts emulsified with it. When heated to temperatures used in refinery processing, these salts dissociate to form corrosive compounds and act to poison catalysts which increase operating costs significantly. These salts are removed in a desalting unit by washing the crude oil with hot water and separating the water by using chemical demulsifiers or electrostatic precipitators.

FIG. 3. Atmospheric and vacuum crude distillation columns. (Courtesy of Foster Wheeler, Inc.)

The crude oil is heated in a furnace to approximately 400°C (750°F) and charged to the flash zone of the atmospheric distillation column. In this column the crude oil is separated into a bottoms product (atmospheric reduced crude) having an initial boiling point of about 345°C (650°F), a gaseous overhead product boiling at temperatures less than 0°C (32°F), and several liquid side-stream products withdrawn at intervals down the tower (see Fig. 4). The desired properties of each side-stream product are controlled by the initial and final boiling points (cut points) of the stream.

The bottoms stream from the atmospheric distillation column is sent to the vacuum distillation unit. Here it is heated in a furnace to over 425°C (800°F) and charged into the flash zone of the vacuum tower. Cracking is minimized by injecting steam into the oil at the furnace inlet to give high velocities and short residence times in the furnace tubes. The distillation tower is operated at absolute pressures in the range of 7 to 20 kPa (50 to 150 mm Hg). Steam injected into the bottom of the vacuum tower reduces the effective hydrocarbon partial pressure to 1.3–5.3 kPa (10–40 mm Hg) and thereby decreases the boiling points of the hydrocarbon compounds by as much as 150°C (300°F). This permits producing a bottoms product (vacuum reduced crude) with an initial boiling point of 565°C (1050°F). The

vaporized components are condensed and removed from the column as vacuum gas oils (see Fig. 5).

The low pressures for the vacuum column are created external to the tower by use of vacuum pumps or barometric condensers and steam ejectors. Because the most significant pressure is that at the flash zone, vacuum towers are designed with internals having low pressure drops to make the pressure at the flash zone as close to the pressure at the top of the column as practical. This usually means the vacuum distillation has a much larger diameter than the atmospheric tower even though the atmospheric tower has a higher feed rate.

The number of side-streams removed from the vacuum distillation column can range from one to four depending upon the products to be made. If only used as feed stocks to conversion units, usually only light and heavy vacuum gas oils are removed. If used as feed stocks to make blending stocks for manufacture of lubricants, as many as four side streams having narrower boiling ranges may be removed.

For a given crude oil the relative amounts of the side-streams can be varied by adjusting the cut point temperatures within a limited range. During the summer when gasoline demand is at its highest, the quantity of the heavy straight-run naphtha stream can be increased by raising its

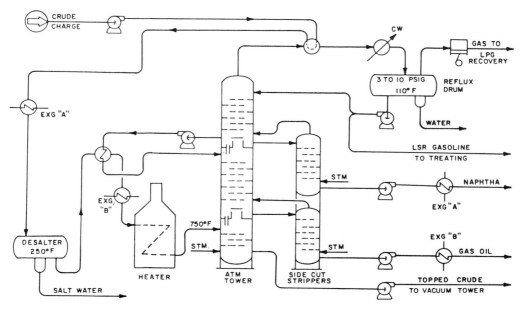

FIG. 4.   Atmospheric distillation unit. [From Gary, J. H., and Hardwerk, G. E. (1984). "Petroleum Refining, Technology and Economics," 2nd ed. Marcel Dekker, New York.]

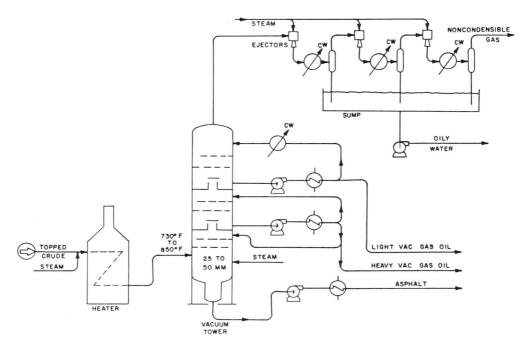

FIG. 5. Vacuum distillation unit. [From Gary, J. H., and Hardwerk, G. E. (1984). "Petroleum Refining, Technology and Economics," 2nd ed. Marcel Dekker, New York.]

end-point to 195°C (380°F). During the winter when there is a smaller demand for gasoline and a greater need for home heating oil, the end-point for gasoline can be decreased to 177°C (350°F), thus increasing the kerosine yield. Typical boiling ranges for the crude unit products are shown in Table I.

III. Gas Plant

A refinery contains one or more gas-processing units to separate the low-boiling components (butanes and lighter) contained in the crude oil

and produced during processing into individual components, or groups of components as needed for products or as feedstocks to other units. Usually, gas streams from distillation columns are sent to a saturated gas plant (from crude unit, reformer, hydrocracker, etc.) if they contain only small amounts of olefins. Those gas streams from processes producing significant concentrations of olefins (FCC, coker, and visbreaker) are sent to a cracked gas plant (Fig. 6).

In most refineries, the methane and ethane components of the gas streams are separated

TABLE I. Crude Distillation Unit Products

Fraction	Typical boiling ranges		Disposition
	°C	°F	
Gases	−160 to 0	−260 to 32	To gas plant
LSR naphtha	25 to 85	80 to 180	To isomerizer or gasoline blending
HSR naphtha	90 to 180	200 to 350	To reformer
Kerosine	160 to 260	320 to 500	To jet, diesel, or #1 or #2 HO blending
Atmospheric gas oil	200 to 350	400 to 650	To diesel or #2 HO blending
Light vacuum gas oil	310 to 400	600 to 750	To FCC or HC feed
Heavy vacuum gas oil	370 to 560	700 to 1050	To FCC or HC feed
Residuum	565+	1050+	To asphalt, #6 FO, or coker feed

FIG. 6.   Gas plant. (Courtesy of Phillips Petroleum.)

into a single fuel-gas stream because there is refinery use for methane or ethane alone. Propane, propylene, *n*-butane, isobutane, and the butylenes are separated because they are either used as feedstocks for other processing units or used as components in LPG (liquefied petroleum gas).

The fuel gas (methane and ethane) is separated from the propane and higher boiling components by absorption into a heavy naphtha lean oil stream. The propane and higher boiling components are then separated into individual components by a series of distillation columns. Impurities, such as hydrogen sulfide and ammonia, are removed from the gas streams and concentrated by amine oil absorption units. The high concentration hydrogen sulfide stream is converted into elemental sulfur by use of the Claus process.

In the Claus sulfur unit the hydrogen sulfide is converted into elemental sulfur and water by burning one-third of the hydrogen sulfide with air to produce sulfur dioxide, then reacting the sulfur dioxide with the remaining hydrogen sulfide by using a Bauxite catalyst. The sulfur is sold as one of the refinery products.

IV. Conversion Processes

A. COKING PROCESSES

Crude oil components that compose the vacuum tower bottoms are difficult to process because the components of the crude having poor processing characteristics have been concentrated in the residua. The high coke-forming potential (high Conradson and Ramsbottom carbon residues), high metals contents (especially

nickel and molybdenum), and high sulfur and nitrogen compositions cause rapid catalyst deactivation. As a result, noncatalytic processing methods are usually preferred although in some cases hydroprocessing or solvent extraction is used. The most common processes used are the noncatalytic thermal cracking processes, coking, and visbreaking.

As the molecular weight of hydrocarbon compounds increases the hydrogen-to-carbon ratio decreases. To make acceptable lower boiling point (also lower molecular weight) materials the hydrogen-to-carbon ratio must be increased. This can be done in two ways: by adding hydrogen (usually requiring catalysts) and by removing carbon (no catalysts required). Thermal cracking processes remove carbon to improve the hydrogen-to-carbon ratio of the lower molecular weight liquid and gaseous products from the reactions. In the case of coking, the carbon is removed in the form of a solid product called coke.

The oldest and most widely used coking process is delayed coking (Table II and Fig. 7). Coking is a time- and temperature-dependent process and a combination of a furnace and cok-

FIG. 7.   Delayed coking unit. (Courtesy of Foster Wheeler, Inc.)

TABLE II. Delayed Coker Yields[a]

Products	Light resid		Medium resid		Heavy resid	
°API	15.4		8.1		3.8	
K_w	11.71		11.6		11.39	
CCR (wt%)	8.54		17.1		21.19	
	Usual	Special	Usual	Special	Usual	Special
C_4 and Ltr	7.46	6.39	9.21	7.89	10.90	9.51
C_5–169°C	13.33	11.36	12.90	10.48	12.66	10.10
169–266°C	15.03	13.42	15.52	13.54	15.69	13.60
266–343°C	12.55	11.18	12.96	11.28	13.10	11.34
343°C+	33.73	42.18	19.21	30.00	10.25	22.54
Coke	17.90	15.47	30.20	26.81	37.40	32.91

[a] Courtesy of Continental Oil Company.

ing drum is used to control the temperature and reaction time so the coke is deposited in the coke drum for later removal. At least two coke drums are required with one drum on-stream while coke is being removed from the other drum. The coke drums are from 6 to 9 m in diameter and up to 30 m tall. Typical on-stream times for a drum are 16–24 hr with a corresponding time needed to remove the coke from the drum and get it ready to be put back on-stream. A simplified process flow diagram for a delayed coking unit is shown in Fig. 8.

Usually the vacuum tower bottoms (VRC) is sent directly to the coking unit. The hot feed is charged to the coker fractionation tower several trays above the hot vapor inlet from the coking drum. This gives control over the amount of recycle and the end-point of the coker gas oil withdrawn as a side-stream above the feed inlet. The feed is withdrawn from the bottom of the tower, heated in a furnace to 480 to 525°C (900 to 975°F), and charged into the bottom of the coke drum. The pyrolysis reactions are endothermic and the temperature in the drum decreases as the cracking reactions proceed. Temperatures of the cracked product vapors leaving the drum range from 435 to 460°C (815 to 860°F) and the drums are operated from 70 to 690 kPa (15 to 100 psig) pressure.

The vapors leaving the coke drum are fed into the coker fraction tower near the bottom of the column. This fractionator is operated in a similar

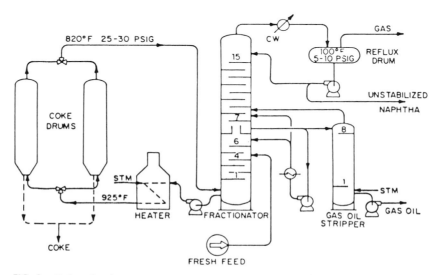

FIG. 8. Delayed coker. [From Gary, J. H., and Hardwerk, G. E. (1984). "Petroleum Refining, Technology and Economics," 2nd ed. Marcel Dekker, New York.]

manner to the atmospheric crude distillation tower and a gaseous overhead stream consisting of butanes and lighter components is taken off the top with liquid side-streams removed in the naphtha and gas oil boiling ranges. All of the liquid and gaseous streams are rich in olefinic hydrocarbons produced by the cracking reactions. The butane and lighter gases are sent to the gas plant for separation into components. The propylenes and butylenes in this stream are used as feed stocks to alkylation or polymerization units to be converted into high octane gasoline blending stocks. After hydrogenation, the naphtha stream is sent to the catalytic reformer to be upgraded into a high-octane gasoline blending stock and the gas oil boiling-range streams are used as FCC or hydrocracker feeds or hydrotreated for blending into diesel fuel or home heating oils.

After a coke drum is filled with coke, it is taken off-stream and replaced with the other drum. The coke is cut from the drum with high-pressure hydraulic drills using water at pressures from 17,250 to 31,000 kPa (2500 to 4500 psig). The two major types of coke produced are sponge coke and needle coke. Most of the sponge coke is sold for fuel and the major market for the remainder is for the manufacture of anodes for the production of aluminum. Approximately 1 ton of anode coke is required for production of 2 tons of aluminum. High-quality needle coke is sold for the production of electrodes for electric steel furnaces.

A fluid-bed coking process was developed in the early 1950s but very few units have been built. In the 1970s, a modification of this process called flexicoking was introduced and several of these units have been constructed. The major advantage of this process is that 90% of the coke produced in the coking operation is converted into a low Btu fuel gas (373 to 450 kJ/m³ or 100 to 120 Btu/scf) which can be used as a low-cost energy source for use within the refinery. Energy is the major operating cost for a refinery and use of this low-cost fuel can significantly reduce refinery operating costs. However, the capital and operating costs for a flexicoking unit are up to 50% greater than for a delayed coker and this unit must be justified on an economic basis.

B. FLUID CATALYTIC CRACKING (FCC)

Catalytic cracking is the most widely used process for converting high boiling point hydro-carbons into molecular compounds boiling in the transportation fuel boiling range. A typical barrel of crude oil charged to U.S. refineries contains about 20% gasoline boiling range material. By charging distillate and bottoms streams having boiling points too high to be blended into diesel fuel, the gasoline yield can be increased to 50% by volume of crude oil charged to the refinery. With a very few exceptions, the fluid catalytic cracking (FCC) unit (Fig. 9) is the only type of catalytic cracking unit in use in the United States today. These units are named for their use of a very small particle size (average about 50 μm) microspherical catalyst which is fluidized by passing gases through the catalyst bed at a velocity high enough to expand the bed to the point the particles move about in random motion and the resulting mixture of solid particles and gases behaves similar to a liquid.

Typical feedstocks to the FCC unit include vacuum gas oils, coker gas oils, and some atmospheric reduced crudes. The reduced crudes produce more carbon than distillate feeds and also contain metals that quickly reduce the catalytic activity of the catalyst. Some especially designed FCC units can handle large quantities of

FIG. 9. Fluid catalytic cracking unit. (Courtesy of ARCO.)

some atmospheric reduced crudes but the average FCC unit feed contains only about 5% atmospheric reduced crude oil.

Coke formed by the cracking reaction is deposited on the catalyst and rapidly lowers its activity. The catalytic activity is restored by burning the coke from the catalyst with air after the catalyst is separated from the hydrocarbon stream. The oil is mixed with the catalyst and the reaction takes place in the riser and the carbon is burned from the catalyst in the regenerator. The cracking reactions are endothermic and the heat necessary to heat the hydrocarbon to the reaction temperature and to supply the heat of reaction is provided by the heat of combustion from burning the coke from the catalyst.

Reaction temperatures range from 480 to 525°C (900 to 975°F) and pressures in the riser reactor from 203 to 274 kPa (15 to 25 psig). Regenerator temperatures range from 620 to 815°C (1150 to 1500°F) and pressures from 203 to 274 kPa (15 to 25 psig). A simplified process flow diagram for a fluidized-bed reactor unit is shown in Fig. 10a and for the reactor-regenerator section of a riser-type FCC unit in Fig. 10b. The feed and recovery section of the riser unit is similar to that shown in Fig. 10a.

The feed to the FCC unit is preheated by heat exchange with hot product streams to about 260°C (500°F) and then brought to reaction temperature when mixed with hot regenerated catalyst at the bottom of the riser. The feed is cracked and vaporized while traveling together up the riser and then separated quickly to prevent overcracking and a loss in gasoline production. After separation, the catalyst is stripped with steam to remove adsorbed hydrocarbons before being sent to the regenerator. The hydrocarbon gases pass through a cyclone separator to remove catalyst particles larger than 7 μm and are then sent to a distillation column for separation into fractions. The distillation column is operated similar to the crude unit atmospheric distillation column and products with comparable distillation ranges are produced.

The flue gas resulting from burning the carbon from the catalyst is passed through several stages of cyclone separators and an electrostatic precipitator to remove as much of the entrained catalyst as possible. The hot gases (from 620 to 815°C) are used to generate steam, compress combustion air, and generate electricity by expansion through turbines.

The catalyst used in the FCC unit is a mixture of amorphous and crystalline (zeolitic) silica-alumina structures. The amorphous part of the catalyst gives it a high mechanical strength that resists abrasion and the zeolites give it a high cracking activity. The high cracking activity permits riser cracking with reaction times of less than 6 sec and very high selectivities. Whole families of these catalysts are made that offer a range of properties such as higher octanes of the gasoline produced, greater resistance to poisoning by sulfur and metals, and higher gasoline yields.

FCC yield structures are functions of catalyst type and activity, feedstock characteristics, and operating conditions. The primary operating variables are riser temperature, catalyst-to-oil ratio (C/O), and catalyst activity. A representative product distribution at an 80% conversion level is approximately 65% by volume of C_5-216°C (C_5-420°F) gasoline (93 RON, 82 MON), 18% butane-butylenes, 11% propane-propylene, and 4% by weight of ethane and lighter. The remaining 20% by volume of the product is cycle gas oil. All of the liquid and gaseous products contain substantial amounts of olefins. As a result the gases make good feed stocks for alkylation and polymerization units.

Several licensors and designers of FCC units have built units to process bottoms streams having Conradson or Ramsbottom carbon values up to 10%. The major differences between these units and the distillate feed FCC units are a larger regenerator (sometimes two regenerator stages) and air supply in order to burn more carbon per barrel of feed and some means of removing the extra heat of combustion over that needed to heat the feed to reaction temperature and to supply the heat of reaction. This can take the form of steam coils in the regenerator or of an external catalyst cooler through which the catalyst is circulated. The cooler may generate steam or transfer the heat to a hot fluid system which supplies heat to other processes. Such units, however, are frequently limited by the cost of the make-up catalyst needed to maintain the catalyst activity or the metals loading on the catalyst at the desired level.

C. HYDROCRACKING

In part, hydrocracking (Fig. 11) is a combination of fluid catalytic cracking and hydrotreating where the products are paraffinic and aromatic rather than olefinic and aromatic. However, hydrocracking has another characteristic which makes it much more important to refinery opera-

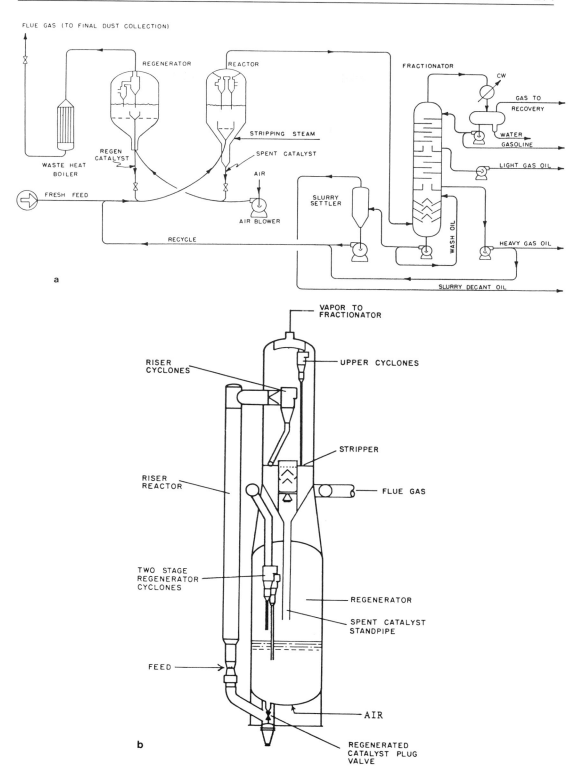

FIG. 10.   (a) FCC unit, Model III. (Courtesy Exxon Research and Engineering.) (b) Riser FCC and regenerator unit. (Courtesy of The M. W. Kellogg Company.)

FIG. 11.   Hydrocracking unit.

tions than the combination of the other two processes. This factor is the ability to crack aromatic ring compounds selectively into gasoline, diesel, and jet fuel. When catalytic cracking units are operated at conditions severe enough to crack aromatic rings the products are mostly light gases and coke. In hydrocracking, some light gases and coke are made but the major portions of the products are gasoline and middle distillate transportation fuels. The FCC unit and the hydrocracker complement each other in refinery operations where the FCC unit operates best with virgin or hydrotreated gas oils as feedstocks and the hydrocracker takes cycle gas oils and coker distillates as feedstocks.

Specially designed hydrocrackers can use atmospheric or vacuum-reduced crudes as feedstocks but the majority of hydrocrackers are designed for distillate feeds. The hydrocrackers designed for distillate feedstocks are all fixed-bed catalytic units while the hydrocrackers feeding bottoms streams can be either fixed-bed or ebullated-bed units. Ebullated-bed (also called expanded-bed) units are, in principle, similar to fluidized-bed units except that the catalyst particles are expanded using a mixture of liquids and dense gases rather than gases alone.

A simplified group of class reactions that take place in hydrocracking operations is shown in Fig. 12. The cracking reactions are endothermic and the hydrogenation reactions are highly exothermic with the result that heat is generated and temperatures rise in the fixed-bed reactors. In multibed reactors, cold hydrogen quench streams are introduced between beds, and also between reactors, to control temperatures. High temperatures increase the rates of reactions but the thermodynamics of exothermic reactions are favored by low temperatures. The result is a

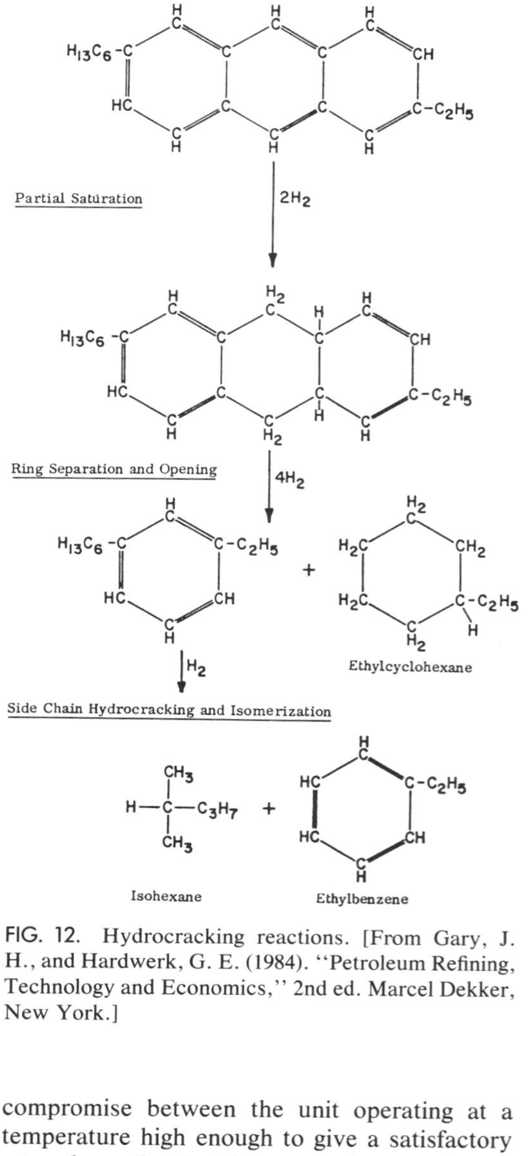

Partial Saturation $2H_2$

Ring Separation and Opening $4H_2$

Ethylcyclohexane

Side Chain Hydrocracking and Isomerization

H_2

Isohexane   Ethylbenzene

FIG. 12.   Hydrocracking reactions. [From Gary, J. H., and Hardwerk, G. E. (1984). "Petroleum Refining, Technology and Economics," 2nd ed. Marcel Dekker, New York.]

compromise between the unit operating at a temperature high enough to give a satisfactory rate of reaction but low enough to produce a more selective effect to get higher yields of the desired products. Typically, hydrocrackers operate at temperatures between 290 and 425°C (550 and 800°F) and pressures between 8275 and 20,700 kPa (1200 and 3000 psig). High hydrogen recycle rates are used to maintain catalyst activity by reducing coke laydown on the catalyst and thereby obtaining long run times between shut-downs for catalyst regeneration.

For fixed-bed hydrocracking units proper feedstock preparation is necessary to minimize catalyst poisoning and give high yields and a long catalyst life. Many hydrocracking units use a hydrotreating reactor before the hydrocracking reactors to convert organic sulfur and nitrogen compounds to hydrogen sulfide and ammonia to prevent reduction of hydrocracking catalyst activity and deterioration of yield structure. During a run, the reactor temperatures are gradually increased to maintain a constant severity of operation (constant degree of sulfur or nitrogen removal or constant conversion level). Increasing the reaction temperature increases the cracking rates faster than the hydrogenation rates and more of the feed is cracked to gases and coke. This results in lower yields of desirable products and less profitable operations.

A simplified process flow diagram for a single-stage distillate hydrocracking unit is shown in Fig. 13. The feed is mixed with the hydrogen recycle stream, heated to about 330°C (625°F), and fed into the top of the first reactor. The first reactor usually contains a cobalt–molybdate catalyst to reduce the sulfur and nitrogen in the oil stream and to provide a mild hydrocracking reaction. The effluent from the first reactor is mixed with the recycle stream from the bottom of the distillation stream, quenched with cold hydrogen to regulate the temperature, and charged into the hydrocracking reactor. The hydrocracking reactor (one or more) contains catalysts selected for its hydrocracking properties, which usually consist of a precious or transition metal (palladium, platinum, etc.) on a zeolitic-type silica–alumina base. The effluent from the last hydrocracking reactor is cooled and sent to a high-pressure separator where the recycle hydrogen is separated from the liquid. The hydrogen gas is mixed with the make-up hydrogen stream and sent to the recycle compressor. The liquid from the high-pressure separator goes through a pressure reduction valve into a low-pressure separator where low-boiling gaseous hydrocarbons and some hydrogen are separated and sent to a gas plant for separation. The liquid from the low-pressure separator is heated and fed into an atmospheric distillation column for separation into products having the desired boiling range. The bottoms stream from the fractionator is mixed with the feed to the unit and recycled. By regulating operating conditions, the hydrocracker can make all gasoline and lighter products, or can maximize jet fuel or diesel fuel products.

Liquid product yields from the hydrocracker

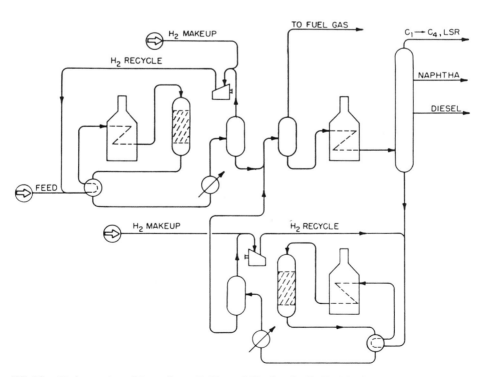

FIG. 13. Hydrocracker. [From Gary, J. H., and Hardwerk, G. E. (1984). "Petroleum Refining, Technology and Economics," 2nd ed. Marcel Dekker, New York.]

can range from 115 to 130 vol% of hydrocracker feed and hydrogen consumptions range from 90 to 355 std m^3/m^3 feed (500 to 2000 scf/bbl feed). The liquid and gaseous products compare with virgin products of the same boiling range with respect to quality and degree of saturation. Gasoline octanes are in the 80s and the heavy hydrocrackate [boiling range of 82–195°C (180–380°F)] frequently is sent to the reformer to improve its octane. The hydrocracker is a major source of isobutane for feed to alkylation units.

Theoretically hydrocracking units have the ability to handle any type of feedstock. In practice, however, the rate of catalyst deactivation has a major effect on the endpoints of the feeds that are used as well as the effects of impurities. The higher the endpoints of the feeds the greater the laydown of carbon on the catalyst and the greater the risk of having metal compounds in the feed. As a result, most of the hydrocrackers are designed to charge distillate feeds only. New units have been developed with special provisions for handling feedstocks containing impuri-

ties and high-boiling components. The majority of these are fixed-bed units that have process flows similar to that shown in Fig. 14 but operate with 5 to 10 times as much catalyst per unit of feed as the distillate feed units and at much higher pressures [18,960 vs 10,340 kPa (2750 vs 1500 psig)] to slow the build-up of coke on the catalyst. There are two processes that use ebullating-bed reactors that permit the maintenance of catalyst activity by the withdrawal of spent catalyst and the addition of fresh catalyst while the unit is operating. These are the H-Oil and LC-fining processes. A simplified process flow diagram of this type of process is shown in Fig. 14.

The ebullated-bed type process is similar to a fluidized-bed reactor process in that the catalyst is maintained in random motion by the passage of hydrogen and hydrocarbons upward through the reactor and catalyst can be withdrawn and added while the unit is on-stream. The heavy feed (usually atmospheric reduced crude) is mixed with recycle hydrocarbons and hot hy-

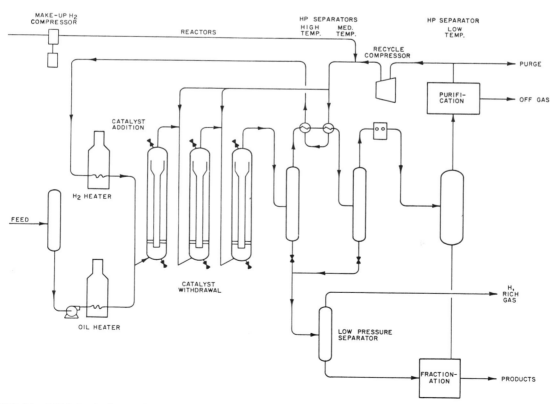

FIG. 14.   LC-fining hydrocracker. [From Gary, J. H., and Hardwerk, G. E. (1984). "Petroleum Refining, Technology and Economics," 2nd ed. Marcel Dekker, New York.]

drogen and introduced into the bottom of the first reactor. Here it mixes with the recirculating hydrocarbon stream and the catalyst in the reactor. The recirculating hydrocarbon stream within the reactor is necessary to maintain a high enough velocity in the reactor to keep the catalyst in an ebullated state. A combination of desulfurization, denitrogenation, demetalizing, and hydrocracking reactions takes place in the reactors. The H-Oil and LC-fining units have three reactors in series in each train and conversions up to 65% on a once-through basis can be attained based on a 525°C+ (975°F+) feed. After the last reactor the flow is similar to that of the fixed-bed units with hydrogen sulfide and ammonia absorption units on the hydrogen recycle stream to maintain the hydrogen quality required. The catalysts used contain cobalt, molybdenum, nickel, or tungsten supported on Bauxite, alumina, or silica–alumina bases as needed.

The products from hydrocracking units are separated in an atmospheric distillation column and used as feedstocks to other refinery processing units or, if the quality permits, sent to product blending.

D. ALKYLATION

Alkylation, as used in the petroleum refining industry, refers to the process for reacting olefins with an isoparaffin to produce a higher molecular weight paraffin. Specifically the process reacts propylenes and butylenes with isobutane to produce branched chain paraffins in the gasoline boiling range. The product is called alkylate and is a premium gasoline blending stock because it has high octane numbers and a low sensitivity.

There are two basic processes, one using concentrated hydrofluoric acid (Fig. 15) as the catalyst and the other using concentrated sulfuric acid. Although the product yields and quality are essentially the same, there are significant differences in process operation because of differences in catalyst characteristics. The process using hydrofluoric acid as catalyst is performed at temperatures between 30 and 43°C (85 and 110°F) while the process using sulfuric acid operates from 4 to 13°C (40 to 55°F). The reactions are highly exothermic and the heat created by the reactions must be removed from the reactor in order to keep the reactions under control and to achieve good yields and high alkylate quality. For the hydrofluoric acid process the reaction

FIG. 15. Hydrofluoric acid alkylation unit. (Courtesy of Phillips Petroleum.)

temperature is sufficiently high that the temperature can be controlled by removing the heat of reaction with cooling water. For the sulfuric acid process it is necessary to use refrigeration to remove the reaction heat and control the temperature. The need for refrigeration increases the operating cost significantly over that of the hydrofluoric acid process.

If a stream of mixed olefins (propylene and butylene) is reacted with isobutane, the product stream leaving the reactors consists of

1. C_5+ alkylate
2. Propane
3. *n*-butane
4. Alkylate tar.

The alkylate tar is a viscous brown oil consisting of polymerization products of high molecular weight. Only a small amount is made, usually less than 0.1% of the olefin feed.

For the mixed olefin feed the C_5+ alkylate product has a research octane number (RON) of 94 and a motor octane number (MON) of 91. For each volume of olefin feed, 1.5 volumes of alkylate are produced. These numbers are approximate because the propylene/butylene ratio in the feed and operating conditions affect the product quality and quantity.

A simplified process flow diagram for the hydrofluoric acid process is shown in Fig. 16. The olefin and isobutane feed streams are dehydrated by passing them through a solid-bed desiccant unit. Good dehydration is necessary to prevent equipment corrosion and to prevent dilution of the acid catalyst. The dehydrated streams are mixed with each other and with the hydrofluoric acid catalyst at a pressure suffi-

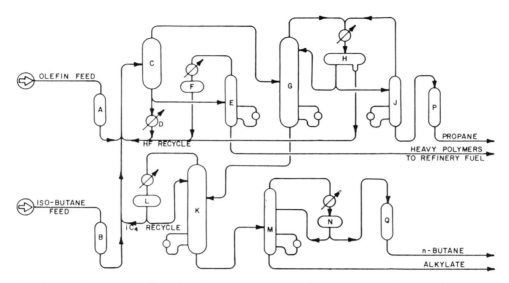

FIG. 16. HF alkylation. A, Olefin feed drier; B, isobutane feed drier; C, acid settler; D, acid cooler; E, acid rerun column; F, acid accumulator; G, depropanizer; H, depropanizer accumulator; J, acid stripper; K, deisobutanizer; K, deisobutanizer accumulator; M, debutanizer; N, debutanizer accumulator; P, propane caustic treater; Q, butane caustic treater. [From Gary, J. H., and Hardwerk, G. E. (1984). "Petroleum Refining, Technology and Economics," 2nd ed. Marcel Dekker, New York.]

ciently high to prevent vaporization of the components. The mixture is passed into a separating vessel and separated into two layers. The top layer of reacted hydrocarbons and excess isobutane is sent through a series of distillation columns to separate the propane, isobutane, *n*-butane, and C_5+ alkylate into streams. The propane and *n*-butane are sent to storage, the isobutane is recycled, and the alkylate is sent to product blending.

The acid stream settling to the bottom of the separating vessel is recycled but a slip stream is withdrawn to separate the alkylate tar from the hydrofluoric acid in a distillation column. The pure acid is put into the acid recycle stream and the alkylate tar is sent to an accumulator.

Isobutane is recycled to give a high ratio of isobutane to olefins in the reactor. Typically recycle ratios of isobutane to olefins in the feed are from 3 to 10.

The heat of reaction can be removed by passing the reactants through a water-cooled heat exchanger before entering the separating vessel or by passing the recycled acid through a cooler.

The process flow for a sulfuric acid system is similar to that of the hydrofluoric acid unit with two major exceptions. First, sulfuric acid is more viscous than hydrofluoric acid and it is necessary to use mixers to get good contact between the hydrocarbons and the acid catalyst.

This introduces a mechanical heat of mixing that must be removed in addition to the heat of reaction in order to maintain the reaction temperature. Second, a refrigeration system with a compressor and condenser is required to remove the heat of reaction and mixing by vaporizing the propane and some of the isobutane in the system, compressing the gases produced, and condensing them for reintroduction into the system as a liquid. The distillation system to separate the components in the product stream is similar to that for the hydrofluoric acid process.

The capital costs for the two systems are similar because the acid recovery system and safety requirements for the hydrofluoric system balance the refrigeration system cost of the sulfuric acid unit. Both units are constructed of carbon steel but the hydrofluoric acid unit requires that pump impellers, valve trim, and portions of the system where water–acid mixtures are a possibility be constructed of Monel metal or Monel clad steel.

E. Polymerization

Propylene and butylenes can be reacted to form polymers in the gasoline boiling range. The product is olefinic with a RON = 97 and a MON = 83. It is a good gasoline blending stock but has a high sensitivity (sensitivity = RON − MON) and does not perform as well in engines as does alkylate. Approximately 0.7 volumes of polymer gasoline is produced per volume of olefin feed because of the higher molecular weight and liquid density of the product.

Even though polymer gasoline is not as valuable as alkylate, the capital and operating costs for a polymerization unit are substantially lower than for an alkylation unit and payout times are usually much shorter. As a result, the polymerization process is a viable process for producing high octane blending stocks for unleaded gasoline production.

The most widely used catalyst for the polymerization process is phosphoric acid on an inert support such as kieselguhr, carbon, or quartz. Sulfur and oxygen in the feed adversely affect the catalyst and basic materials neutralize the acid. Some water in the feed is necessary to ionize the acid to make it active but an excess dilutes the acid. As a result the catalyst must be replaced every 2 to 4 months depending upon the ratio of the volume of feed per volume of catalyst. Typical reaction conditions are 205°C (400°F) and 3550 kPa (500 psig). A simplified process flow diagram for a polymerization unit is shown in Fig. 17.

The polymerization reaction is highly exothermic and propane and butanes are included in the feed stream to act as heat sinks to give better control of the reaction temperature. In addition, cold propane quenches are injected into the reactor to provide adequate temperature control. Prior to entering the reactor, the feed is caustic washed to remove mercaptans, contacted with an amine solution to remove hydrogen sulfide, washed with water to remove any entrained caustic or amines, dried by passing through a desiccant bed, and injected with a controlled amount of water (350–400 ppm) to activate the catalyst. The reaction products leaving the reactor are sent through a series of distillation columns to separate the propane and butanes from the polymer gasoline product.

V. Quality Improvement Processes

A. HYDROTREATING

Hydrotreating processing units (Fig. 18) have been installed in increasing numbers in refineries because the decreasing availability of low-sulfur crude oils has required that more high-sulfur

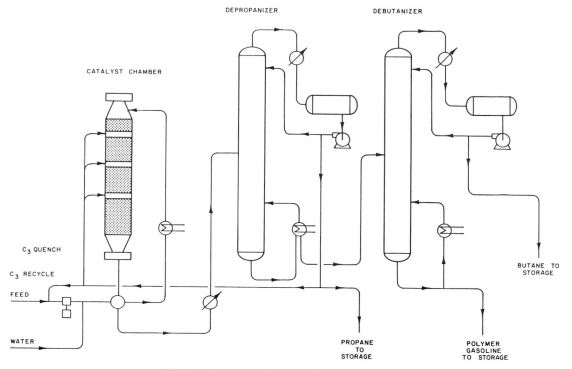

FIG. 17.   Polymerization unit. (Courtesy of UOP.)

FIG. 18. Hydrotreating unit.

crudes be processed. Hydrotreating improves the quality of an oil by removing sulfur and nitrogen and increasing the hydrogen content by saturating olefins and some of the aromatics. The process is performed by using catalysts to react hydrogen at high pressures with the feedstocks. The processes can be called hydrotreating, hydrodesulfurization (HDS), or hydrodenitrogenation (HDN).

A simplified process flow diagram for a hydrotreating unit is shown in Fig. 19. The oil feed to the unit is mixed with the recycle hydrogen stream, passed through a furnace to heat it to the reaction temperature of 260–430°C (500–800°F), and then fed into the top of the first reactor. The reactor temperature is a function of the density of the feed, the degree of sulfur removal desired, and the activity of the catalyst. As the oil–hydrogen stream passes down-flow through the reactor it contacts the catalyst and the hydrogenation reactions take place. These reactions are highly exothermic and it is necessary to inject cold hydrogen quench streams at various places in the reactor train in order to control the reaction temperature. The hydrogen reacts with sulfur to form hydrogen sulfide and with nitrogen to form ammonia. After leaving the last reactor, the stream is cooled to approximately 38°C

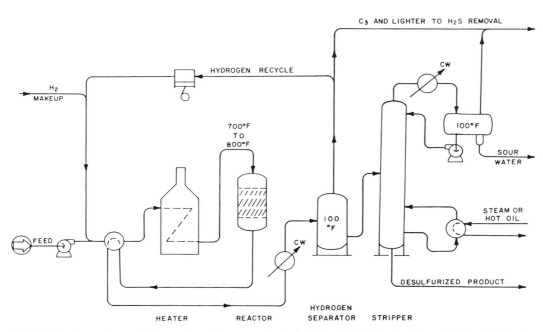

FIG. 19. Hydrotreating unit. [From Gary, J. H., and Hardwerk, G. E. (1984). "Petroleum Refining, Technology and Economics," 2nd ed. Marcel Dekker, New York.]

(100°F) and fed to a separating vessel where the recycle hydrogen is separated from the liquid. The liquid from the separator flows through a pressure-reducing valve, a heater, and then into a distillation column where the gases and lighter compounds produced by secondary reactions are separated from the hydrotreated product.

A sophisticated family of catalysts has been developed which has catalysts specific with respect to the operation desired. Cobalt–molybdenum sulfides on alumina or silica–alumina selectively reduce the sulfur content although nitrogen removal and hydrocarbon saturation take place. Nickel–molybdenum sulfides on alumina or silica–alumina supports have higher hydrogenation activities and are more selective for nitrogen removal and hydrocarbon saturation. Operating at higher temperatures and hydrogen pressures increases the degree of sulfur and nitrogen removal and hydrocarbon saturation but the rates of secondary cracking reactions increase more rapidly with temperature than the hydrogenation reactions and result in significantly lower yields of hydrotreated product.

During the period of a run the catalyst activity decreases because of the build-up of coke on the catalyst. The temperature of the process is increased to counterbalance this but eventually the decrease in product yield caused by excessive cracking requires that the catalyst be replaced or regenerated. Hydrogen is consumed by these operations and the amount used is related to the degree of sulfur and nitrogen removal and the quantity of olefins and aromatics that are saturated. In addition to sulfur and nitrogen removal, oxygen and halides are also removed and result in consumption of hydrogen.

B. CATALYTIC REFORMING

The virgin naphthas from the crude distillation unit have very low octane numbers, usually ranging from the 40s to the low 60s. Today's gasolines have octanes in the high 80s and low 90s and it is necessary to increase the octanes of the virgin naphthas to make them suitable for gasoline blending. As a general rule, the octane number of a hydrocarbon increases as the molecule becomes more compact. For example, n-octane has a RON of −19, a more compact isooctane (3-methylheptane) has a RON of 35, and the most compact isooctane (2,2,4-trimethylpentane) has a RON of 100. An aromatic molecule containing the same number of carbon atoms (1,4-dimethylbenzene) has a RON of 146. The octane numbers of straight-run (virgin) gasolines can be improved by converting paraffins to isoparaffins or aromatics. Two processes are used, one (isomerization) for compounds containing six or fewer carbon atoms [light naphthas with boiling ranges from 0 to 82°C (30 to 180°F)] and the other (reforming) for compounds containing seven or more carbon atoms [heavy naphthas with boiling ranges from 82 to 193°C (180 to 380°F)]. Isomerization increases octane by converting normal paraffins to isoparaffins and reforming increases octane by converting paraffins to aromatics. Compounds with six or less carbons get the greatest octane improvement by isomerization. Many refineries do not require isomerization units but all refineries need reforming units.

The main reactions taking place during the catalytic reforming operation are the dehydrogenation of cycloparaffins to produce aromatics. These reactions are highly endothermic and the temperature decreases as the hydrocarbons proceed through the reactor. To maintain an adequate rate of reaction it is necessary to remove the cool reactants and increase their temperature by passing them through a furnace before returning them to a catalyst bed. Typical reactor temperatures are from 495 to 525°C (925 to 975°F). Other endothermic reactions also take place, such as dehydrocyclization and dehydroisomerization of paraffins, which increases the necessity to reheat the reactants at periodic intervals. The gasoline boiling range product from a reformer is called reformate.

The primary catalyst used for catalytic reforming operations is platinum–rhenium supported on a silica–alumina base. These metals are quickly poisoned by sulfur and nitrogen compounds and the heavy naphtha feed to the unit must first be hydrotreated to reduce the sulfur and nitrogen contents to an acceptable level.

Although high hydrogen partial pressures in the reactor retard the desired reactions, some undesirable secondary reactions result in coke formation on the catalyst and it is necessary to maintain a suitable hydrogen partial pressure to keep the rate of coke deposition (and catalyst deactivation) at a low level. Typical reactor pressures vary from 790 to 1825 kPa (100 to 250 psig) depending upon the type of processing unit and the frequency of catalyst regeneration.

Reforming processes are classified as continuous (Fig. 20), cyclic, or semiregenerative depending upon the frequency and method of catalyst regeneration. The continuous unit is

FIG. 20. Continuous catalytic reforming unit. (Courtesy of The Standard Oil Company.)

designed to permit the steady movement of catalyst through the reactor system to a regenerator and back into the reactors. Continuous regeneration maintains high catalyst activity at low hydrogen pressures in the reactors and thereby gives high conversion levels and high reformate yields. The semiregenerative unit has three or more reactors in series and must be shut down to regenerate the catalyst and restore its activity. Times between catalyst regeneration vary from 6 to 18 months and it is necessary to operate at high hydrogen pressures to obtain long on-stream times at the expense of lower reformate yields as the activity of the catalyst decreases. The cyclic unit is similar to the semiregenerative

unit except that it contains a spare or "swing" reactor to replace one of the regular reactors while it is taken off-line to be regenerated. This permits all of the reactors to be regenerated, one at a time, while the unit is operating.

A simplified process flow diagram for a continuous catalytic reforming unit is shown in Fig. 21. The hydrotreated feed is combined with the hydrogen recycle, heated in the first pass of a furnace to 495°C (925°F), and fed into the first reactor. The primary reaction occurring in the first reactor is the highly endothermic dehydrogenation of cycloparaffins to produce aromatics and the temperature drops as the hydrocarbons pass through the reactor. The effluent from the first reactor is sent through the second furnace pass and is reheated to 495°C or above. From the second furnace pass the hydrocarbons go to the second reactor where the major reactions are the dehydroisomerization and dehydroalkylation of the normal and isoparaffins. These reactions are also endothermic but less so than those occurring in the first reactor. After the second reactor the hydrocarbons are again heated to 495°C or above before being sent to the third reactor. In the third reactor, the major reaction is hydrocracking of normal paraffins to light gasses to remove the unreacted low-octane components from the gasoline boiling range hydrocarbons and thus increasing the octane of the reformate product. This adversely affects the reformate yield but permits the achievement of the

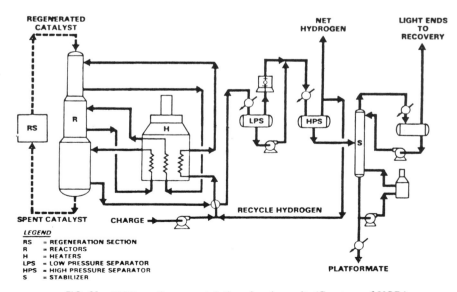

FIG. 21. UOP continuous catalytic reforming unit. (Courtesy of UOP.)

desired octane improvement. From the last reactor the reactants are cooled and go to a separator where the gaseous hydrogen is separated from the liquid products. The liquid products are separated in a distillation column into a C_5+ reformate and butanes and lighter product. The severity of the reforming operation is expressed as the RON of the reformate.

For a given feedstock, the higher the severity the lower the yield of reformate. Typical values are yields of 79% at a 100 RON and 85% at 94 RON. The reformate yield at a given severity is a function of the aromatic and cycloparaffin content of the feedstock. The reformate has a very high sensitivity which frequently ranges from 10 to 12 numbers. For example, a 98 RON reformate may have an 86.5 MON. The reforming reactions are a source of a large amount of hydrogen for refinery use. At a 100 RON severity, 710 to 1600 std M^3 of hydrogen is produced per cubic meter of reformer feed (400 to 900 scf/bbl feed).

C. ISOMERIZATION

Typically the octane numbers of light virgin naphthas [C_5 to 82°C (C_5-180°F)] are in the 60s. As separated from the crude oil this fraction is a mixture of normal and isopentanes and hexanes. By converting all of the normal paraffins to isoparaffins the octane numbers of this fraction can be increased by approximately 20 numbers. The process to convert normal paraffins to isoparaffins is called isomerization and utilizes a catalyst comprised of platinum supported on alumina or silica–alumina.

During a single pass of the naphtha through the reactor the reaction mixture approaches thermodynamic equilibrium which results in the product containing significant amounts of unreacted normal paraffins and an octane improvement of only 13 numbers is attained. If the unreacted n-pentane is separated from the product and recycled through the unit, an octane improvement of approximately 17 numbers is

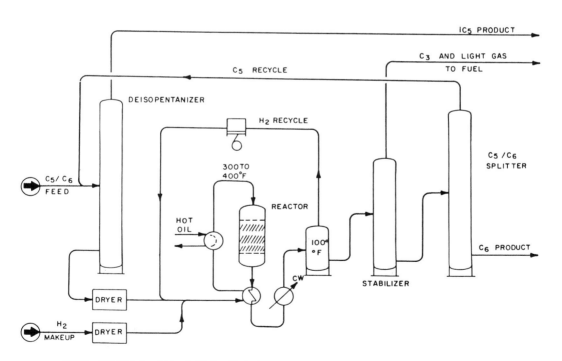

FIG. 22.   Isomerization unit. Reboil and reflux facilities on fractionators are not shown for simplification. [From Gary, J. H., and Hardwerk, G. E. (1984). "Petroleum Refining, Technology and Economics," 2nd ed. Marcel Dekker, New York.]

achieved. To obtain a 20 number octane improvement, both the *n*-pentanes and the *n*-heptanes must be recycled. The normal paraffins can be separated from the product stream and recycled by either distillation or by adsorption on activated carbon and molecular sieves.

The chemical reactions involved are relatively independent of pressure and little hydrogen is produced or consumed by the process. Hydrogen is recycled to minimize catalyst deactivation by coke build-up on the catalyst and the units usually operate at pressures of 1830–2515 kPa (250–350 psig). Reaction temperatures are much lower than for reforming and range from 150 to 205°C (300 to 400°F). Because of the low reactor temperatures very little cracking occurs and product (isomerate) yields of 98% and greater can be expected.

A simplified process flow diagram for a light naphtha isomerization unit is shown in Fig. 22. The process flow is similar to that of a hydrotreating unit except for the distillation columns to separate the normal paraffins from the reactor product so they can be recycled. The process flow shown is for the system that recycles the *n*-pentane but does not separate and recycle the *n*-hexane. For the system shown, an octane improvement of about 17 numbers can be achieved. The isomerate product is blended into gasoline.

D. VISBREAKING

Unless cracked or sold as asphalt, the vacuum reduced crude (VRC) is usually sold as a #6 Fuel Oil or Bunker Fuel Oil for its heating-value-equivalent price. These heavy fuel oils have viscosity and pour point specifications they must meet in order to be sold. It is usually necessary to dilute the VRC with lighter gas oils in order to meet the #6 or Bunker Fuel Oil specifications. These gas oils (commonly called cutting stocks when used for this purpose) are more valuable to the refinery for use as feed stocks for conversion processes and it is desirable to minimize the amount of cutting stocks used. A mild thermal cracking process called visbreaking can reduce the viscosities and pour points of VRC and reduce the amount of cutting stocks required. Typically, the amount of heavy fuel oil production can be decreased by 25–30% with a visbreaker operation. A simplified process flow diagram for a coil-type visbreaker is shown in Fig. 23.

The degree of viscosity and pour point reduction is a function of the characteristics of the residua feed stock. Residua with high wax contents achieve a higher reduction than do residua having higher contents of asphaltenes. All feeds have viscosity and pour point reduction limitations due to the product becoming unstable or actually having increases in these properties due to polymerization of cracked products. The properties of the cutting stocks used to blend with the visbreaker products also have an effect of the severity of the visbreaker operation that produces an acceptable product with highly aromatic cutting stocks, such as FCC cycle gas oils, producing the higher quality heavy fuel oils.

There are two types of visbreaker operations,

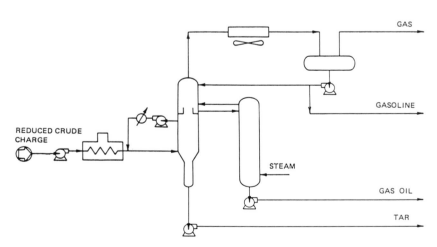

FIG. 23. Coil visbreaker. [From Gary, J. H., and Hardwerk, G. E. (1984). "Petroleum Refining, Technology and Economics," 2nd ed. Marcel Dekker, New York.]

TABLE III. Visbreaking Results: Agha-Jari Short Resid[a]

	Feed	Product
Yields (wt%)		
Butane and lighter	2.4	
C_5-166°C naphtha	4.6	
Gas oil, 349°C EP	14.5	
Tar	78.5	
Product properties		
Naphtha		
°API		—
Sulfur (wt%)		—
RONC		—
Gas oil		
°API		32.2
Sulfur (wt%)		—
Tar or feed		
On crude (%)	—	
°API	8.2	5.5
Sulfur (wt%)	—	—
Viscosity, cS, 50°C	100,000	45,000

[a] From Rhoe, A., and de Blignieres, C. (1979). *Hydrocarbon Proc.* **58**(1), 131–136.

coil and soaker cracking. As in all types of cracking processes, the reactions are time and temperature dependent. The coil visbreaker uses shorter cracking times and higher furnace temperatures while the reverse is true for soaker cracking. The coil operation uses simpler equipment providing shorter turn-around times while the soaker unit permits lower furnace outlet temperatures with lower fuel consumption and longer run times between shutdowns to clean the coke from the furnace tubes. The products from the two operations are essentially equal in yields and properties.

Either atmospheric-tower or vacuum-tower bottoms is used as feed to the units. The coil visbreaker uses furnace outlet temperatures from 470 to 500°C (880 to 930°F) and coil residence times of 1 to 3 min. Soaker visbreaking uses soaker residence times of 10 min or longer and furnace outlet temperatures of 425 to 445°C (800 to 830°F). The products from visbreaking are a butane and lighter gas stream, a gasoline boiling range naphtha stream, a gas oil stream, and a tar or bottoms stream. The characteristics of these products are very much like the similar boiling range products from delayed coking as they contain substantial amounts of olefinic hydrocarbons. Typical results for a visbreaking operation are shown in Table III.

The factors now active that are affecting product quality, demand, and distribution, causing greater attention to health and environmental effects, and requiring the processing of poorer quality crude oils are making major changes in refinery equipment and operations. Many refineries have closed because of the inability to meet these changes. It is a challenge to those remaining to comply with the new requirements and regulations while remaining competitive and profitable.

BIBLIOGRAPHY

Bland, W. F., and Davidson, R. L. (1967). "Petroleum Processing Handbook." McGraw-Hill, New York.

Gary, J. H., and Handwerk, G. E. (1984). "Petroleum Refining, Technology and Economics," 2nd ed. Marcel Dekker, New York.

Meyers, R. A. (1986). "Handbook of Petroleum Refining Processes." McGraw-Hill, New York.

PHARMACEUTICALS

Bryan G. Reuben   *Polytechnic of the South Bank, London*

GLOSSARY

Antibacterial: Synthetic compound that inhibits the growth of microorganisms.

Antibiotic: Chemical produced by microorganisms that inhibits the growth of other microorganisms especially disease-causing bacteria.

Antihistamine: Drug that is a histamine blocker and is used to counter allergy and stomach ulcers.

Antiinflammatory: Drug that blocks inflammatory prostaglandins and is used as an analgesic and to ease rheumatism and arthritis.

Anxiolytic drug: Drug used to counter anxiety; sometimes known as a minor tranquilizer.

Cardiovascular disease: Disease of the heart and blood vessels regarded as a single system.

Chemotherapeutic agent: Chemical that will damage one kind of living matter (e.g., bacteria or cancer cells) without harming any other.

Drug: Material that can alter the structure or function of the living organism. In this article, "drug" and "pharmaceutical" have been used synonymously to refer to the pharmacologically active ingredient in a medicinal formulation.

Hormone: Chemical secreted into the bloodstream which acts on cells elsewhere in the body to produce an effect. Hormones are usually steroids or polypeptides but may be simpler molecules such as epinephrine. They are produced by glands known collectively as the endocrine system.

Medicine: Formulation containing a pharmaceutical together with excipients, emulsions, etc., which make the drug suitable for administration.

Neuroleptics: Major tranquilizers used for schizophrenia and manic depression.

Pharmacodynamic agent: Anesthetic that is nontoxic to all tissues and whose affect is wholly reversible.

Steroid: Group of chemical compounds based on the perhydrocyclopentenophenanthrene ring system.

Sulfonamide: Drug based on p-aminobenzene sulfonamide (sulfanilamide).

Thiazide: Group of chemical compounds based on 3,4-dihydro-$2H$-1,2,4-benzothiadiazine.

A pharmaceutical is a therapeutically active drug, that is, a chemical substance that alters the structure or function of a living organism to cure or alleviate disease. Most pharmaceuticals are purchased by the public in the form of prescription drugs. They are produced by the pharmaceutical industry under strict regulatory procedures by methods closer to those of conventional laboratory chemistry than of the bulk petrochemical sector of the chemical and allied producers industry (CPI).

I. Key Developments

Since 1935 the pharmaceutical industry has become a large and profitable sector of the Chemicals and Allied Products Industry. It is research intensive, and competition is frequently between patent-protected products. Many products were discovered by "molecular roulette," but recently there has been progress toward rational drug design based on an understanding of how drugs work.

In an industry which offers that most prized of all possessions—good health—issues of quality control and testing and safety of new drugs are of paramount importance. About 50% of all prescriptions are for the 50 most widely prescribed drugs, and these in turn are aimed at seven main types of illness. Discussion of the top 50 drugs provides a concise overview of the industry.

II. The Chemotherapeutic Revolution

The modern drug industry can be said to have started in 1935 with the introduction of the sulfonamide antibacterials. Prior to 1935, it was unusual for a physician to be able to prescribe a drug to cure a specific disease. Digitalis had been available since the eighteenth century for heart disease; aspirin, barbital, and the local anesthetic procaine had been synthesized at the turn of the twentieth century. Heparin was used as an anticoagulant and suramin against trypansomiasis since 1916, insulin for diabetes since 1921, pamaquin against malaria since 1926. A number of vaccines and vitamins were available. In 1910, Ehrlich had synthesized salversan—the first true chemotherapeutic agent—for use against syphilis. Meanwhile, in the 1920s, six drugs, singly or in combination made up 60% of all British prescriptions.[1] The 1932 pharmacopoeia contained just 36 synthetic drugs.

The sulfonamide antibacterials were active against the hemolytic *Streptococcus,* a microorganism involved particularly in puerperal fever (childbed fever) and pneumonia, and the death rate from these dropped dramatically.

Penicillin was the next antibacterial and the first antibiotic to be developed. Noted in a chance observation by Fleming in 1928, penicillin was isolated by Florey and Chain in 1939. Large-scale production was achieved in the United States during World War II. By 1950, most of the major groups of antibiotics had been identified. By 1960, infectious disease had ceased to be a major cause of death in developed countries.

If the 1940s were the decade of antibiotics, the 1950s were the decade of psychotropic medicines. Chlorpromazine was discovered in 1950, meprobamate in 1954, and chlordiazepoxide (Librium) in 1960. Thus, drugs were available for treatment of schizophrenia, anxiety, and acute depression.

[1] Statistics given throughout this article are for the United States except where specified as British.

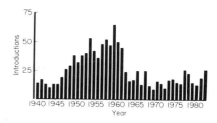

FIG. 1. New single entity drug introductions to the U.S. market, 1940–1982 (Source: New Chemical Entity Data Base/Pharmaceutical Manufacturers' Association).

Heart drugs were developed more slowly. Reserpine was discovered in 1952 and methyldopa in 1960, but the golden age was the late 1960s and early 1970s.

The period from 1935 to the late 1960s has been termed the first chemotherapeutic revolution. Many of the drugs were lifesaving. The regulatory system was none too strict. The drugs were discovered by a sort of molecular roulette involving the screening of millions of more or less arbitrarily chosen compounds, and scientists had little idea how they worked. These pioneering days have now passed. New drugs are being developed within a much stricter regulatory framework. Their function frequently is more to improve the quality of life than to save life. Most significant of all, many of them are sought on the basis of an understanding of the biochemistry of the systems they are intended to influence. This new approach has been termed the second chemotherapeutic revolution.

The pattern of new drug introductions is shown in Fig. 1. The late 1950s were the height of the first chemotherapeutic revolution, but the stringent new regulations in the wake of the thalidomide disaster (1960) reduced the rate of innovation. After a minimum in the mid-1970s, the rate has started to climb again as the second chemotherapeutic revolution gets under way.

III. Characteristics of the Industry

In 1982, the U.S. pharmaceutical industry produced about 313 million pounds of pharmaceutical chemicals compared with 300 billion pounds of chemicals produced by the organic chemicals industry. This tonnage, however, accounted for one-eighth of the sales of the CPI and 18% of the value added. By this measure, only the industrial organic chemicals sector of the industry is larger. [*See* ORGANIC CHEMICALS, INDUSTRIAL PRODUCTION.]

In 1982, expenditure on health care in the United States was second only to food, drink, and tobacco. The average American spent $1305 on health care of which $95 or 7.3% went for pharmaceuticals. In 1984, about $3 billion was spent by hospitals for pharmaceuticals and $14.5 billion by consumers.

The United States manufactures about one-third of the world's pharmaceuticals and of this about one-third is sold abroad, making drugs one of the country's most important exports. The United States, together with Switzerland, West Germany, the United Kingdom, and France, are clear world leaders in pharmaceutical innovation and manufacture. Figure 2 shows the positive trade balances of these countries in pharmaceuticals together with their research and development expenditures and, as a measure of drug consumption, the number of prescriptions per head filled annually. Japan's trade balance is expected to increase rapidly making her a member of the "club," and there are indications that France may drop out. Prescription data indicate that the major consumers of drugs are not necessarily the major producers.

The drug industry makes small quantities of complicated organic chemicals on general purpose equipment. The high value added means that expensive reagents familiar in the laboratory but rare in the bulk chemicals sector may be used. Batch processes, which permit rigorous quality control, are the norm. Hence, capital outlays are low by chemical industry standards, and the productivity of capital is high. Research expenditures are very high amounting in the United States to 36% of research and development spending in the CPI in 1982. Put another way, a typical drug firm spends more than twice the proportion of its sales revenue on research and development than does a chemical company. [*See* BATCH PROCESSING, (CHEMICAL ENGINEERING).]

The reason is not hard to find. In 1970, (the last date for which figures are available) member companies of the Pharmaceutical Manufacturers' Association prepared or isolated 1,260,000 substances and pharmacologically tested 703,900. A total of 1013 compounds reached the stage of clinical trials while only 16 new compounds reached the market and not all of them were a success. By 1982, the cost of developing a single new drug was estimated to be $75 million and innovation and testing took 7–10 years, thus consuming an appreciable fraction of its patent life.

Most of this time and money is spent on expensive pre-clinical testing and even more expensive clinical testing to satisfy the standards of the Food and Drug Administration (FDA) and to ensure that a drug is safe when it comes on the market. No drug can be guaranteed 100% safe and the industry and the FDA have the unenviable task of preventing dangerous drugs from being launched, on the one hand, and making sure that the community benefits from advances in medical treatment on the other.

High research expenditure leads to a high level of innovation. Of drugs available in 1977, 95% originated after 1950. The industry has grown since 1950 typically at 7–15% each year. It is easily the most profitable sector of the CPI averaging a pre-tax profit in 1983 of 19% of sales compared with 9% for the CPI and 6% for all manufacturing industry. Nonetheless, drug prices have lagged behind other health care prices and have dropped from 13.3% of total health expenditure in 1950 to 7.5% in 1981 even though a greater range of more effective drugs is available.

IV. How Drugs Work

The mode of action of drugs is usually explained by receptor theory, an idea which originated with Ehrlich ("The Magic Bullet") and led to the invention of Salversan. The theory states that there are chemically active sites within the body (or on microorganisms) that combine with complementary functional groups in a chemical either produced by the body naturally or introduced into it as a drug. The chemical group can

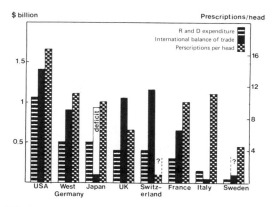

FIG. 2. Research and development expenditure, international balance of trade and numbers of prescriptions per head in some developed countries for the pharmaceutical industry (Sources: Office of Health Economics, UN Commodity Trade Statistics) (Japanese prescriptions/head is an estimate).

be almost any type known to the chemist; it can attach by almost any bonding mechanism but its configuration is crucial. The trans isomer of diethylstilbestrol, for example, behaves as a female sex hormone; the cis isomer has only 7% of its activity. Binding to a receptor may result in the promotion or inhibition of some bodily process such as the contraction of a muscle or the release of a hormone.

Receptors may be active sites on enzymes, nucleic acids, permeases (nonenzyme proteins) or small molecules. Receptor molecules are present in the body in very small amounts (receptor proteins in the brain make up less then one millionth of brain protein), and they cannot as yet be isolated. Binding to them, however, can be shown with the aid of radioactive ligands.

Most drugs fall into one of three classes—agonists (replacement drugs), redesigned or modified agonists, and antagonists or blockers.

Agonists are identical with substances usually used in the body but which it cannot synthesize. The reason may be an illness (e.g., diabetes) or inherent (e.g., vitamins, essential amino and unsaturated fatty acids). Hence, insulin is an agonist for treating diabetes and vitamin C is an agonist, also called a diet supplement, for treating scurvy. They have the same effect on receptors as the natural substances.

Modified agonists are similar to natural agonists, but their action has been modified in some way. They share the affinity of agonists for receptors and to some extent their efficacy. For example, progesterone (1) is a natural agonist which blocks ovulation when a woman is pregnant but which cannot be taken by mouth. Norethindrone (2) mimics the effect of progesterone and can be taken by mouth; hence, it is used in various contraceptive pills.

(1) Progestrone   (2) Norethindrone

Antagonists share the affinity of agonists for receptors but lack their efficacy. Hence, they are able to block access of other chemicals to the receptor. An example is the use of propranolol to relieve angina.

The heart contains so-called β_1 receptors. In response to exercise, fear, or anxiety, the adrenal medulla secretes a hormone called epinephrine. This interacts with the β_1 receptors of the heart. The heart is stimulated. It tries to prepare the body for "fight or flight" and demands more blood and oxygen. In the case of a person with narrowed coronary arteries, this leads to angina and perhaps damage to the heart. The structure of the anti-anginal drug, propranolol (3), is similar to that of epinephrine (4). It attaches to the β_1 receptors and prevents epinephrine from reaching them. Because it lacks the efficacy of epinephrine, it does not stimulate the heart and the anginal attack is avoided. Hence propranolol is called a beta blocker.

(3) Propranolol   (4) Epinephrine

The majority of modern drugs are receptor antagonists. By their nature, they are toxic; but the useful ones are selectively toxic. This is not a new concept. Maimonides, the twelfth century philosopher, and Withering, the eighteenth century British physician, both saw drugs as "poisons that heal."

V. Drug Delivery Systems

An effective drug must interact with the correct receptors and have the appropriate action with no or only minor side effects. In addition, however, it must be formulated in such a way that it is transported to the site where it is to act in correct concentrations and for the appropriate time. This is the so-called bioavailability of the drug. To achieve the correct bioavailability, the rates of absorption, distribution, biotransformation, and excretion of the drug in the body must be carefully balanced. Bioavailability varies from person to person depending on age, sex, body weight, presence of food in the stomach, and many other personal factors. In a pill to be taken by mouth, it also depends on various factors such as the drug's crystal size and form, presence of excipients and other additives and the compressive force used in making the pill. Formulation and dosage forms are thus important aspects of drug administration.

A few drugs act locally. They are applied topically or taken by mouth and then not absorbed from the gut or urine. Most drugs, however, must enter the bloodstream in order to reach receptors. There are four methods of administration:

1.   Enteral. The drug is absorbed from the gastro-intestinal tract and taken (except suppositories) by mouth. Examples are pills, soft and hard gelatin capsules, liquids, powders, solutions, suppositories, enteric coated pills, and time release formulations.

2.   Percutaneous. The drug is absorbed through the skin. Examples are tablets beneath the tongue, films placed on the eye, and pads on the skin near the heart.

3.   Inhalation. The drug is absorbed through the pulmonary epithelium in the lungs. Gases may be inhaled, liquids given via aerosols or atomizers and solids by insufflators.

4.   Parenteral. The drug is given by injection into a nerve or the spinal cord, into a vein (intravenous), into a muscle (intramuscular), or under the skin (subcutaneous).

The method of administration governs the rate of absorption of the drug and may also be dictated by the stability of the drug. For example, many cephalosporin antibiotics are unstable to stomach acid and so cannot be taken by mouth and must be injected.

VI. The Top 50 Ethical Pharmaceuticals

Every drug has at least three names—a chemical name, a generic name and one or more brand names. Thus, structure (5) is described chemically as *p*-acetamidophenol and is a nonnarcotic analgesic. Even for so simple a chemical as this, the chemical name is quite long and the drug is known by the generic name, acetaminophen, and by many brand names (given by the companies who market it) including Tylenol.

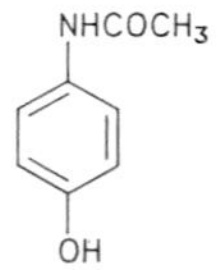

NHCOCH₃

(5) Acetaminophen

Pharmaceuticals generally regarded as safe may be bought by the patient and are known as proprietary, or over-the-counter, drugs. They are backed by advertising to the general public and are often based on traditional recipes. Aspirin, acetaminophen, and certain antihistamines are common ingredients. Vitamin and mineral preparations are sold over the counter, and the antiinflammatory drug, ibuprofen, has recently become the first addition for many years to the drugs that can be sold in this way. Sales of over-the-counter drugs in the United States in 1981 amounted to $6 billion.

Potentially dangerous drugs are known as ethical drugs and must be prescribed by a physician. They are usually pharmacologically more active than over-the-counter drugs and are advertised only to the medical profession. Whenever possible they are patent protected. Where a drug is not patent protected, there is a distinction between a prescription for the drug by its brand name and by its generic name, the latter usually being cheaper. Regulations as to what the pharmacist should do are in a state of flux. Sales in the United States of ethical pharmaceuticals in 1981 amounted to $13.5 billion.

There are over 1000 commonly prescribed ethical pharmaceuticals. In 1983, 1.5 billion prescriptions were dispensed of which about 90% were for brand-name products. Details of the most widely prescribed drugs by brand name are recorded by the National Prescription Audit by IMS America and published annually in the Pharmacy Times. The classification is by brand name, however, and not by chemical entity. Table I shows the 50 most widely prescribed drugs in 1983 in the United States by chemical entity (estimated by the author). They account for about one-half of all prescriptions. Which drugs are the most important depends on the criteria chosen; the ones listed here provide the bulk of the pharmaceutical industry's cash flow, keep it in business, and finance its research programs.

Table II shows the categories of drug in the top 50. Drugs for cardiovascular disease head the list, reflecting the fact that over one-half of U.S. deaths are attributed to such disease and that drugs do much to prolong the life of sufferers and improve its quality. Second come antibiotics, reflecting the conquest of infectious disease in spite of the high incidence of minor infections. Third come drugs affecting the central nervous system. Most of the ones in the top 50 are anxiolytics or sleeping pills. 20 million Americans are said to have mental problems, but many of the consumers of these pills may well be unhappy or lonely rather than suffering from true mental disorder. Diazepam (Valium) was the most widely prescribed drug for many years but has recently slipped to eighth place as physicians have recognized the dangers of tolerance and addiction.

TABLE I. The 50 Most Widely Prescribed Ethical Drugs in the United States in 1983[a]

Rank	Generic name	Typical brand names	Use
1.	Hydrochlorothiazide	Dyazide (combined with #2)	Diuretic
2.	Triamterene	Dyazide (combined with #1)	Diuretic
3.	Codeine[b]	Tylenol/codeine, Empirin/codeine	Analgesic, antipyretic
4.	Amoxicillin	Amoxil, Wymox	Antibiotic
5.	Erythromycin	EES, E-mycin	Antibiotic
6.	Propranolol	Inderal	Beta-blocker
7.	Penicillin-V potassium salt	V-cillin K, Pen-V-K, Beepen	Antibiotic
8.	Diazepam	Valium	Minor tranquilizer, skeletal muscle relaxant
9.	Digoxin	Lanoxin	Cardiac glycoside
10.	Tetracycline	Achromycin V	Antibiotic
11.	Ampicillin	Omnipen, Amcill, Penbritin	Antibiotic
12.	Cimetidine	Tagamet	Antihistamine, H2-blocker for ulcers
13.	Norethindrone with Mestranol	Ortho-Novum	Oral contraceptive
14.	Potassium chloride	Slow-K, K-tab, Klotrix	Electrolyte replenisher
15.	Methyldopa	Aldoril (with #1), Aldomet	Antihypertensive
16.	Furosemide	Lasix	Diuretic
17.	Ibuprofen	Motrin, Brufen	Antiinflammatory
18.	Propoxyphene	Darvocet, Darvon	Analgesic
19.	Sulfamethoxazole and trimethoprim	Bactrim, Septra, co-trimoxazole	Antibacterial
20.	Nitroglycerin	Nitrostat, Nitro-dur	Coronary vasodilator
21.	Cephalexin	Keflex	Antibiotic
22.	Conjugated estrogenic hormone	Premarin Oral	Estrogen hormone therapy
23.	Naproxen	Naprosyn, Anaprox	Antiinflammatory
24.	Metoprolol	Lopressor	Beta-blocker
25.	Ethynylestradiol with Norgestrel	Ovral, Lo-Ovral	Oral contraceptive
26.	Brompheniramine	Dimetapp	Antihistamine, nasal decongestant
27.	Flurazepam	Dalmane	Hypnotic
28.	Amitriptyline	Elavil, Triavil	Antidepressant
29.	L-Thyroxine sodium salt	Synthroid	In thyroid deficiency
30.	Chlordiazepoxide	Librium, Librax (with clidinium bromide)	Minor tranquilizer (anticholinergic)
31.	Diphenhydramine	Benadryl	Antihistamine, nasal decongestant
32.	Isosorbide dinitrate	Isordil	Coronary vasodilator
33.	Theophylline	Theo-Dur	Bronchodilator
34.	Promethazine	Phenergan, Synalgos	Antihistamine
35.	Chlorpropamide	Diabinese	Hypoglycemic
36.	Lorazepam	Activan	Minor tranquilizer
37.	Butalbital	Fiorinal	Hypnotic/sedative
38.	Chlorpheniramine	Ornade	Antihistamine
39.	Phenytoin	Dilantin	Anticonvulsant, antiepileptic
40.	Atenolol	Tenormin	Beta-blocker
41.	Clorazepate	Tranxene	Minor tranquilizer
42.	Prednisone	Deltasone	Antiinflammatory, antiallergic
43.	Piroxicam	Feldene	Antiinflammatory
44.	Indomethacin	Indocin	Antiinflammatory
45.	Insulin	Iletin	Hypoglycemic
46.	Oxycodone[b]	Percodan, Percocet	Analgesic
47.	Miconazole	Monistat	Fungicide
48.	Chlorthalidone	Hygroton	Diuretic, antihypertensive
49.	Doxepin	Sinequan, Adapin	Anti-depressant
50.	Nifedipine	Adalat	Calcium antagonist

[a] Source: Author's estimates based on Pharmacy Times, April 1984.
[b] Codeine and oxycodone occur in formulations with acetaminophen, aspirin, and various other drugs. Acetaminophen and aspirin are over-the-counter drugs and have not been listed separately.

TABLE II.   Categories of Drugs in the Top 50 (U.S. 1983)

Category	Number of drugs	Number of prescriptions (millions)
Cardiovascular drugs	13	316
Antibiotics and antibacterials	8	182
Drugs affecting the central nervous system	9	86
Analgesics (excluding aspirin and acetaminophen)	3	65
Antihistamines	5	52
Steroid drugs	4	45
Nonsteroid antiinflammatory drugs	4	41
Miscellaneous	4	32

The fourth group is the antihistamines used to counter allergic conditions and (in the case of cimetidine) for the treatment of stomach ulcers. The other important groups are analgesics and nonsteroid antiinflammatories used to counter pain (especially arthritic pain), and steroid drugs which are used both in oral contraceptives and as antiinflammatories.

Table III lists the sources of drugs. By value of products, fermentation is the most important since almost all antibiotics are produced by this method. Even the chemically modified antibiotics require a fermentation step to provide their precursors, and modern biotechnology processes may also be classified under this heading.

By tonnage, chemical synthesis is the most important. It provides the great majority of heart drugs, central nervous system drugs, antihistamines and analgesics. In the following discussion, drugs whose source is not specified are made by chemical synthesis.

TABLE III.   Sources of Drugs

Source	Examples
Chemical synthesis	Heart drugs, psychotropic drugs, antihistamines, analgesics. Also chemical modification of materials from other sources.
Fermentation	All antibiotics except chloramphenicol, steroids from diosgenin, vitamins B12 and C. Biotechnology.
Animal extracts	Heparin, thyroid, insulin, steroid hormones.
Vegetable extracts	Alkaloids (e.g., digoxin, papaverine, atropine, codeine, and quinidine), steroids.
Biological sources	Vaccines and serums (e.g., against diphtheria, polio, and whooping cough).

A. CARDIOVASCULAR DRUGS

The thirteen cardiovascular drugs in the top 50 fall into several categories. Digoxin is a heart stimulant and is used in congestive heart failure. It is extracted from the leaves of the white foxglove, *Digitalis lanata* and is easily the oldest drug in widespread use, the crude extract having been used clinically by the British physician, William Withering, in 1785. It is a glycoside made up of one molecule of digoxigenin (6) and three molecules of a sugar, digitoxose.

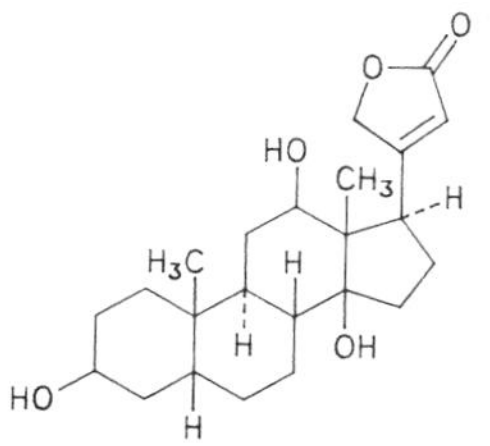

(6) Digoxigenin

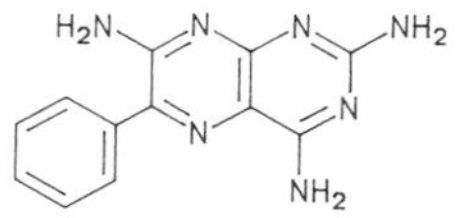

(7) Triamterene

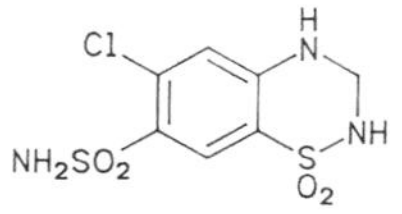

(8) Hydrochlorothiazide

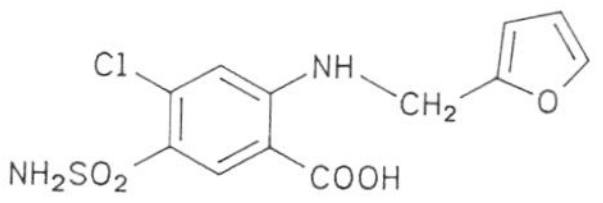

(9) Furosemide

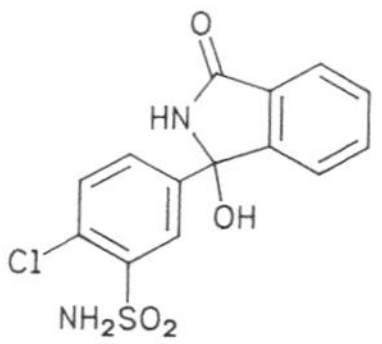

(10) Chlorthalidone

Cases of heart failure are often accompanied by fluid retention (edema), and it is helpful to tackle this by diuretics such as triamterene (7), hydrochlorothiazide (8) furosemide (9), and chlorthalidone (10). Furosemide is called a loop diuretic because it acts on the loop of Henle in the kidney. Hydrochlorothiazide and chlorthalidone are both thiazides. All three cause unacceptable loss of potassium ions along with the fluid and are taken together with a potassium replenisher, notably potassium chloride tablets. Triamterene is a weaker diuretic but does not cause such serious potassium loss.

Methyldopa (11) is an antihypertensive, a drug for the control of high blood pressure. It acts on the nervous system, which has two branches, one being the autonomic nervous system which is responsible for involuntary processes such as heartbeat. The autonomic nervous system is further subdivided into the

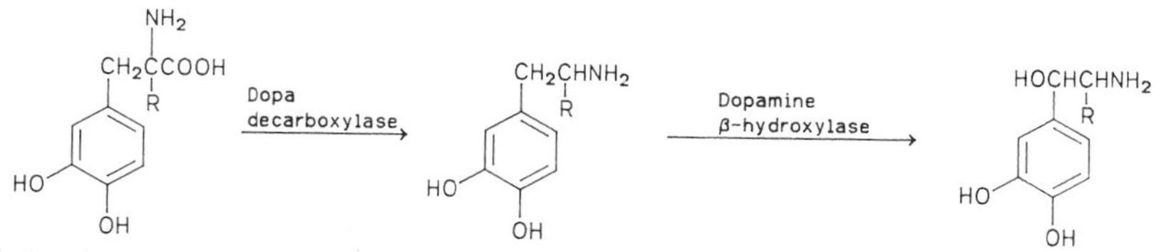

(11) R = CH₃ Methyldopa   (12) R = CH₃ Methyldopamine   (13) R = CH₃ Methylnorephinephrine
(14) R = H   Dopa   (15) R = H   Dopamine   (16) R = H   Norephinephrine

adrenergic or sympathetic nervous system, which is usually concerned with expenditure of energy and in which the nerve synapses release norepinephrine (16), and the cholinergic or parasympathetic nervous system, which usually conserves and restores energy and in which the synapses release acetylcholine. Norepinephrine stimulates the adrenergic nervous system including the heart and is said to be a sympathomimetic amine. As mentioned earlier, stimulation is also caused by the hormone epinephrine (4), which is therefore a norepinephrine agonist.

Norepinephrine is produced in the body by the action of two enzymes on the amine dopa (14). Methyldopa is a modified dopa agonist and differs from it by a methyl group. It competes with dopa for supplies of dopa decarboxylase, so less dopamine (15) and norepinephrine are produced. The methylnorepinephrine (13) produced instead is a weaker stimulant than norepinephrine, hence blood pressure is reduced.

The remainder of the heart drugs listed are used for treatment of coronary heart disease due to narrowing of the coronary arteries by atheroma (fatty deposits). The disease may lead to angina (the characteristic agonizing pain across the chest), myocardial infarction, and sudden death. The drugs used to control angina consist of vasodilators and beta-adrenergic agents (beta-

blockers). The vasodilators relax the involuntary muscles especially those in the walls of the blood vessels near the heart. The blood vessels dilate and the heart's workload is reduced.

Nitroglycerin (17) has been used for over a century for this purpose. Isosorbide dinitrate (18) is a similar drug. Their mode of action is not understood. Nifedipine (19) is a drug about 20 years old which has attracted renewed interest in recent years. Its mode of action appears to be to inhibit the inward flow of calcium ions across the membranes of the heart cells, and this appears to reduce their demand for oxygen. Other "calcium antagonists" have also been developed.

The beta-blockers act, as described earlier, by blocking the β_1 receptors in the heart. Propranolol (20), metoprolol (21), and atenolol (22) are examples. They differ in their effect on the β_2 receptors in the lung, and metoprolol and atenolol are better for asthmatic patients but reduce blood pressure less.

Groups of cardiovascular drugs not represented in the top 50 include antiarrhythmics to steady an irregular heart beat, anticholesteremics to reduce blood cholesterol levels, anticoagulants to reduce blood clotting, and angiotensin-converting enzyme inhibitors which reduce blood pressure by inhibiting formation of angiotensin II in the body.

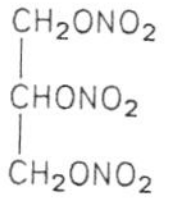

(17) Nitroglycerine

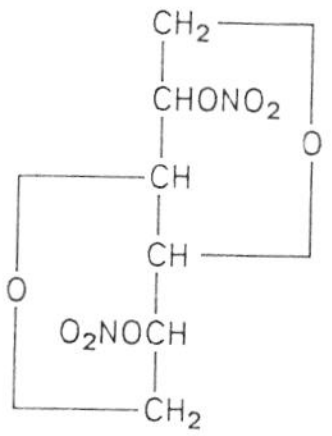

(18) Isosorbide dinitrate

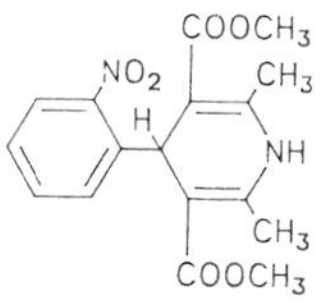

(19) Nifedipine

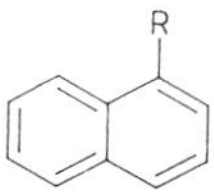

(20) Propranolol

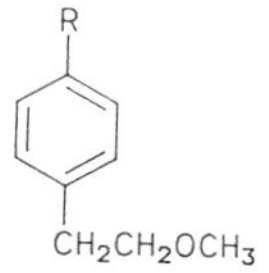

(21) Metoprolol

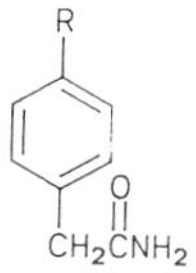

(22) Atenolol

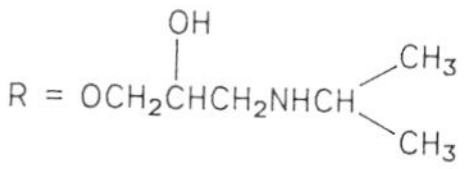

R = OCH₂CHCH₂NHCH(CH₃)₂ ... OH

B. ANTIBIOTICS AND ANTIBACTERIALS

Four antibiotics containing a β-lactam ring appear in the top 50. Three are penicillins—amoxicillin, penicillin V, and ampicillin—and there is a single cephalosporin, cephalexin. Their structures are based on the penicillin (23) and cephalosporin (29) nuclei, and the products are marketed as free acids or sodium or potassium salts. The original penicillin was penicillin G (benzylpenicillin) (24), which is proliferated by the mold *Penicillium chrysogenum*. It had three drawbacks. It is unstable to stomach acid and so must be injected. It has a narrow spectrum of activity, and after it had been in use for some years a strain of bacterium, *Staphylococcus aureus*, appeared which produced the enzyme β-lactamase capable of destroying the penicillin.

Penicillin V (phenoxymethylpenicillin) (25) was a major advance and was obtained by replacement of phenylacetic acid by phenoxyacetic acid in the fermentation substrate. It is stable to acid and can be taken by mouth.

The semisynthetic penicillins followed. It proved possible to cleave the side chain from Penicillins V and G either by chemical means or by immobilized enzymes. The nucleus, 6-aminopenicillanic acid (26), could have other side chains grafted onto it. Ampicillin (27a) was the first major compound of this group and amoxicillin (27b) is a newer one. Both are much broader spectrum antibiotics than penicillin V and can be taken by mouth. Neither is resistant to β-lactamase, and the most widely prescribed compound in this class is the sodium salt of dicloxacillin (28).

Cephalosporins differ from penicillins in having a sulphur-containing ring with six rather than five members. Cephalosporin C (30) is produced by fermentation with *Cephalosporium acrimonium*, which was first found in Sardinian sewage. It is pharmacologically inactive and is cleaved to give 7-aminocephalosporanic acid (31) to which side chains may be attached by chemical means. Cephalexin (32) is the most widely prescribed drug in this class.

Erythromycin (33) is the most widely prescribed non-β-lactam antibiotic and is proliferated by *Streptomyces erythreus*. It is a member of the class of macrolide antibiotics and consists of a macrocyclic lactone ring attached to units of the sugars L-cladinose and D-desosamine. It is particularly useful for patients who are sensitive to penicillin especially children whose teeth might be discolored by tetracyclines.

Tetracyclines are also produced by fermentation. *Streptomyces rimosus* gives chlortetracycline (34) and catalytic hydrogenation converts it to tetracycline (35), the most important mem-

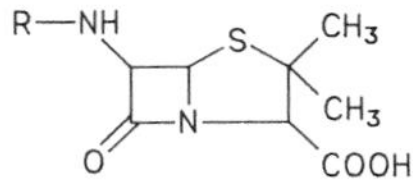

(23) Penicillin nucleus

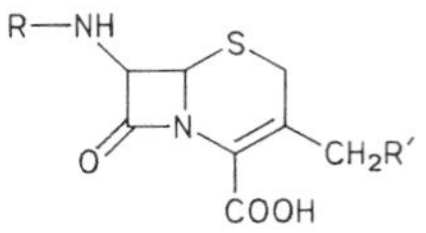

(29) Cephalosporin nucleus

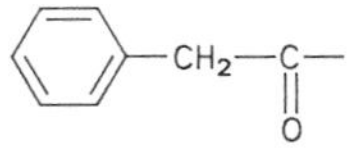

(24) Penicillin G

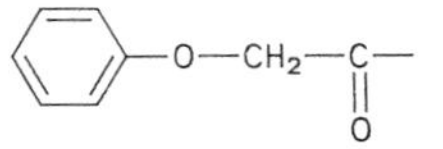

(25) Penicillin V

(26) 6-Aminopenicillanic acid

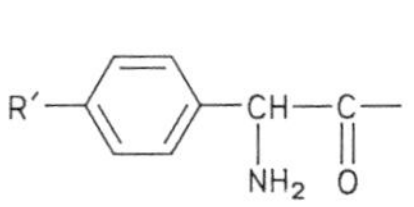

(27a) R' = H Ampicillin
(27b) R' = OH Amoxicillin

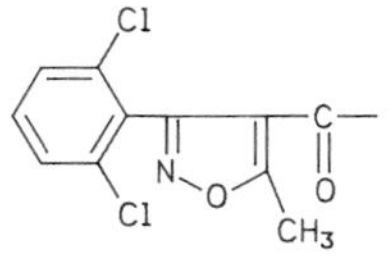

(28) Dicloxacillin

(30) Cephalosporin C

(31) 7-Aminocephalosporanic acid

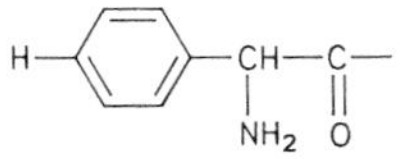

(32) Cephalexin

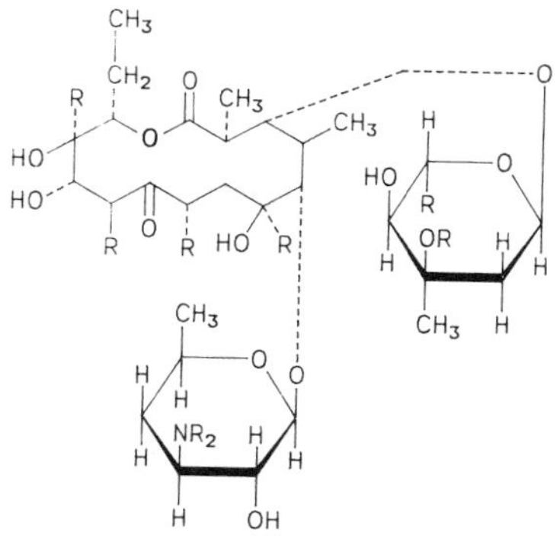

(33)   R = CH₃   Erythromycin

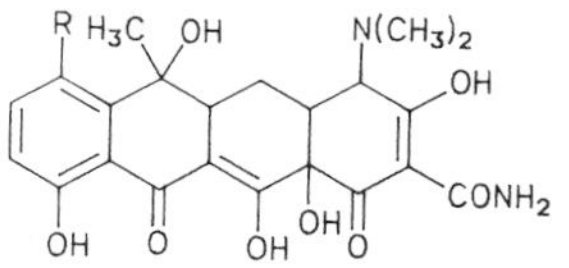

(34)   R = Cl   Chlortetracycline
(35)   R = H   Tetracycline

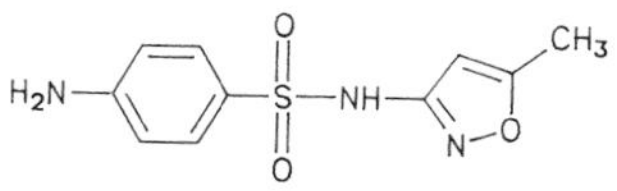

(36)   Sulfamethoxazole

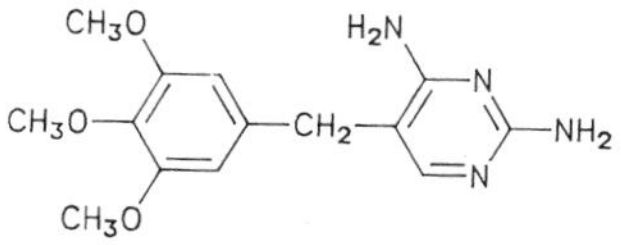

(37)   Trimethoprim

Cotrimoxazole

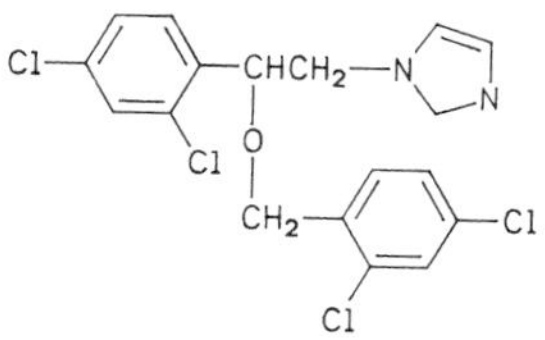

(38)   Miconazole

ber of the group. Tetracyclines have the widest spectrum of antibiotic activity of any group yet discovered and are active against rickettsial infections, a group of rare but usually fatal diseases including Rocky Mountain spotted fever.

Cotrimoxazole is a mixture of a sulfonamide, sulfamethoxazole (36), and the antimalarial, trimethoprim (37). Although these compounds have been known for many years, they were discovered to have synergy only recently, and the combination has risen rapidly up the rank order of drugs. Many sulfonamides are effective but sulfamethoxazole has the closest rate of absorption, distribution, biotransformation, and excretion to trimethoprim. Cotrimoxazole is used primarily in chronic bronchitis and urinary tract infections. Both its constituents are synthesized chemically and, hence, are antibacterials rather than antibiotics.

Miconazole (38) is also chemically synthesized and is used specifically against vaginal fungi. Fungal infections are far less important in the United States than bacterial or viral infec-

tions. It is noteworthy, however, that in the Third World, even though tuberculosis and leprosy are deadly bacterial diseases, the major diseases are due to infection by protozoa (amebiasis, malaria, trypansomiasis, sleeping sickness, and leishmaniasis) and worms (tapeworm, hookworm, schistosomiasis, etc).

C. DRUGS AFFECTING THE CENTRAL NERVOUS SYSTEM

Drugs affecting the central nervous system, otherwise known as psychotropics, fall into six therapeutic categories, apart from the psychotomimetics which are drugs of abuse. The categories represented in the top 50 are the anxiolytics or minor tranquilizers [diazepam (39), chlordiazepoxide (40), lorazepam (41), and clorazepate (42)], the hypnotics/sedatives [flurazepam (43), butalbital (44)], the antidepressants [amitriptyline (45), doxepin (46)], and the anticonvulsants [phenytoin (47)]. Unrepresented are the neuroleptics (major tranquilizers) which have revolutionized the treatment of schizo-phrenia and manic depression and the stimulants and appetite suppressants.

The anxiolytics in the top 50 are benzodiaze-pines (39–42). The hypnotic/sedative, fluraze-pam (43), is also in this class. The mechanism of action of the anxiolytics is uncertain. Indeed, it is difficult to find the biochemical processes in-volved in so diffuse an illness as anxiety. Anx-iolytic treatment should be limited to short peri-ods because tolerance develops within four months of continuous use. Withdrawal can lead initially to acute anxiety and the patient, con-vinced he is still ill, returns to the physician for a repeat prescription and can eventually become dependent on the drug.

Amitriptyline (45) and doxepin (46) are used in severe depressive illness in which the depres-sion is coupled perhaps with suicidal tendencies and behavioral changes such as social with-drawal. Anxiolytics are depressant drugs and will only exacerbate such a condition. The widely prescribed antidepressants are tertiary amines and have a tricyclic structure. They probably inhibit the uptake of sympathomimetic amines such as norepinephrine into nerve syn-apses and hence increase their concentrations in the body.

Flurazepam is the main hypnotic/sedative in the top 50. Barbiturates were the traditional hypnotics/sedatives but have gradually been re-placed by less dangerous compounds. The sur-viving top 50 barbiturate is butalbital (44), the other compounds in this class differing in their alkyl side chains.

Phenytoin (47) is an anticonvulsant used in epilepsy. It has largely replaced the barbiturates which for a long time were drugs of choice both as hypnotics/sedatives and as anticonvulsants. It is sometimes prescribed together with pheno-barbital.

D. ANALGESICS

Pain accompanies most disease and the relief of pain is an important if secondary role of medi-cal science. Pain relievers are divided into two categories—narcotic analgesics such as mor-phine which prevent the appreciation of pain by the nerve centers of the brain and non-narcotic analgesics such as aspirin which block the pain at its site and prevent pain impulses from being sent to the brain.

There are three narcotic pain relievers in our top 50. Propoxyphene (48) is widely used on its own and codeine (49b) and oxycodone (50) are ingredients with acetaminophen (51) and aspirin (52) in many formulations. Propoxyphene is made synthetically, but codeine is either ex-tracted from opium alkaloids from the poppy, *Papaver Somniferum*, or made by methylation of morphine (49a) from the same source. Narcotic alkaloids may produce addiction.

Aspirin and acetaminophen are the most widely consumed analgesics. Without codeine they may be bought over the counter: it is the codeine that makes the formulations ethical. Consequently they have been omitted from the list of the top 50 ethical pharmaceuticals; they are both non-narcotic analgesics. Aspirin, and probably other compounds in this class, act by inhibiting the release of two materials which would otherwise cause inflammation of the tis-sues. They are the inflammatory prostaglandins PGE_2 (53) and $PGE_{2\alpha}$. The drawback of aspirin

Anxiolytics:

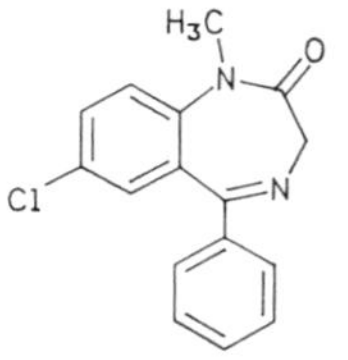

(39) Diazepam

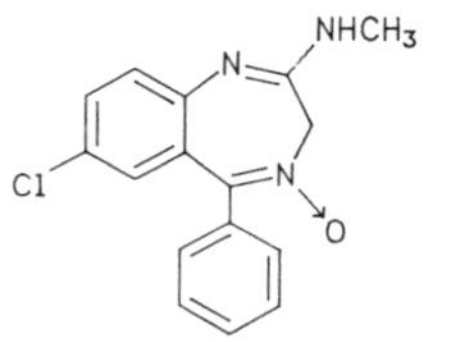

(40) Chlordiazepoxide

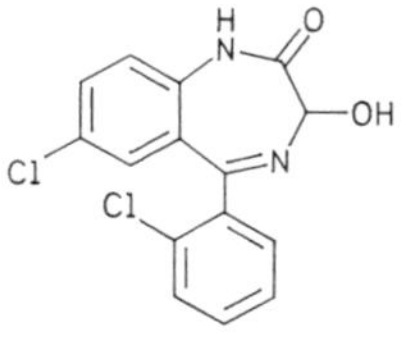

(41) Lorazepam

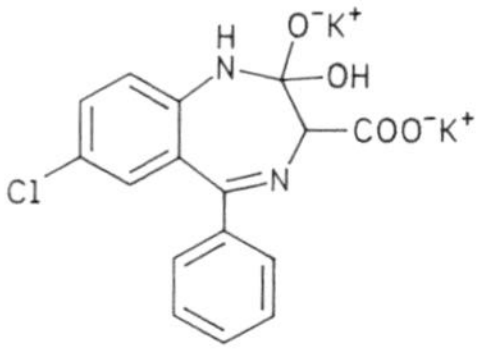

(42) Chlorazepate, dipotassium salt

Hypnotics/Sedatives:

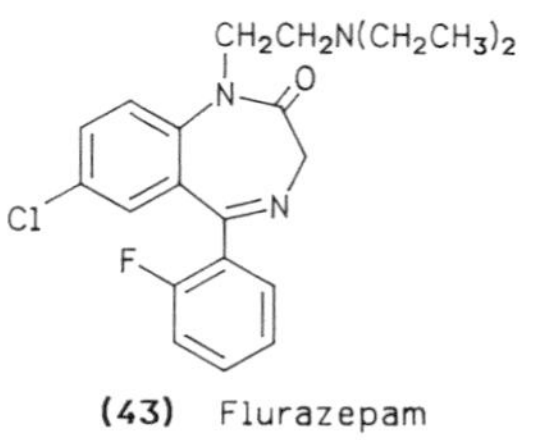

(43) Flurazepam

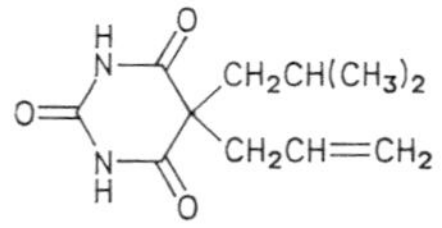

(44) Butalbital

Antidepressants:

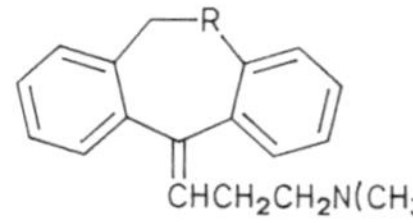

(45) R = CH₂   Amitriptyline
(46) R = O   Doxepin

Anticonvulsant:

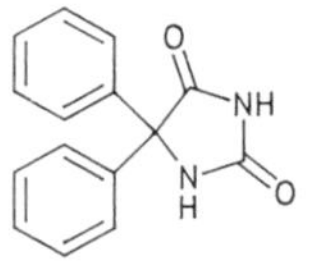

(47) Phenytoin

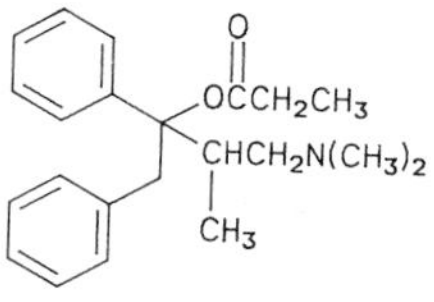

(48) Propoxyphene

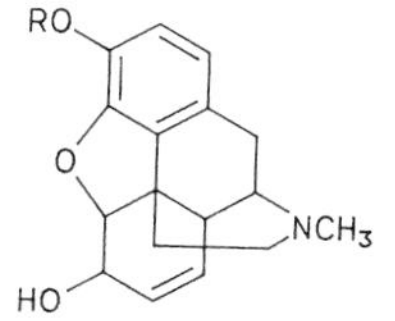

(49a) R = H   Morphine
(49b) R = CH3 Codeine

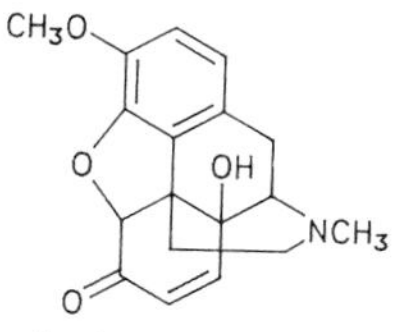

(50) Oxycodone

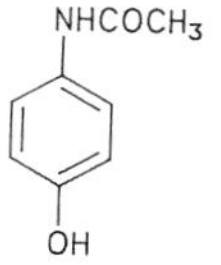

(51) Acetaminophen

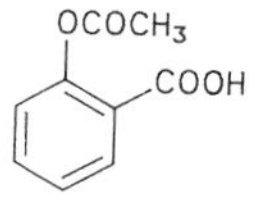

(52) Aspirin

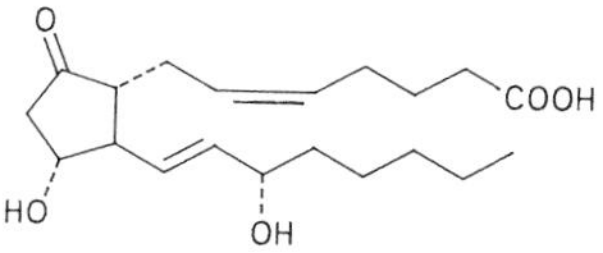

(53) PGE2

is that it causes irritation and bleeding of the stomach lining. Acetaminophen is considerably better in this respect and has made dramatic inroads into the aspirin market in recent years. U.S. production figures for 1982 showed them neck and neck with about 23 million pounds of each produced. Only vitamin C is made on a similar scale and these are the only pharmaceuticals that approach the scale of other industrial chemicals.

E. ANTIHISTAMINES

Antihistamines are divided into two groups, the H1-blockers [brompheniramine (54a), diphenhydramine (55), promethazine (56), and chlorpheniramine (54b)] and the H2-blockers [cimetidine (57)]. H1-blockers are used to counter allergic conditions. Allergy is characterized by the release of histamine (58) into the body from mast cells which are found primarily in connective tissue. It can be triggered by cell damage and by reaction to foreign proteins to which the body has become sensitized. The body reacts by the runny eyes and nose and sore throat of hay fever or by allergic rashes.

H1-blockers act to block the site at which histamine produces these symptoms. They are characterized by the grouping R—X—C—C—N=, where X may be nitrogen, oxygen, or carbon. It is presumably this grouping that bonds to the receptor. Promethazine resembles the phenothiazine neuroleptics related to amitriptyline.

The H1-blockers have been on the market for many years. In addition to their antiallergic action, they produce drowsiness, and this is made use of in a number of over-the-counter sleeping products. Other H1-blockers are used to counter vertigo, nausea, and motion sickness.

Cimetidine is a relatively new antihistamine used to counter gastric and duodenal ulcers. It had been known that excess hydrochloric acid in the stomach was an exacerbating factor in ulcers and that histamine stimulated the secretion of acid. In spite of this, antihistamines had no effect on ulcers. It transpired that there were two kinds of histamine receptors designated H1 and H2. The former were involved in allergic reactions and were blocked by conventional antihistamines. The latter were responsible for hydrochloric acid secretion. Cimetidine was found

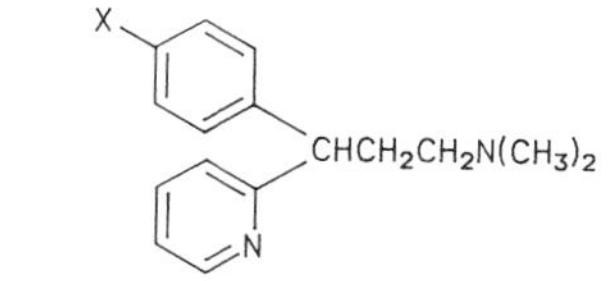

(54a) X = Br   Brompheniramine
(54b) X = Cl   Chlorpheniramine

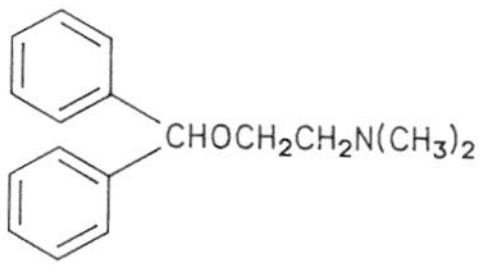

(55) Diphenhydramine

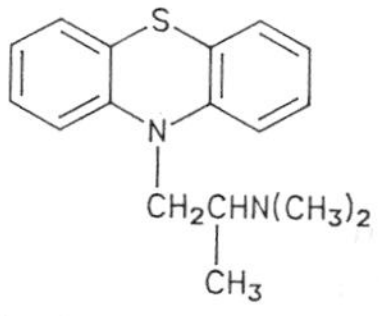

(56) Promethazine

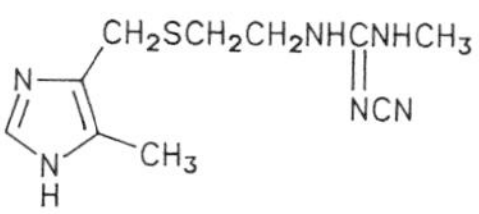

(57) Cimetidine

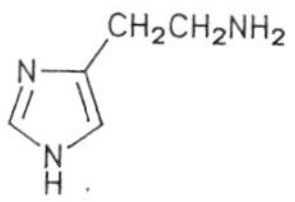

(58) Histamine

to be an H2-blocker and its structure is related to that of histamine. It is said to have been commercially the most successful drug of all time.

F. STEROID DRUGS

Steroid drugs have wide application and are used as oral contraceptives, sex hormones, anabolic agents, and antiinflammatory drugs. In our top 50, norethindrone/mestranol (**59a/61a**) and norgestrel/ethynylestradiol (**59b/61c**) are oral contraceptives. Conjugated estrogenic hormone is used to alleviate the symptoms of menopause. Prednisone (**60b**) is an antiinflammatory used especially in rheumatoid arthritis.

Oral contraceptives consist of a monthly cycle of an estrogen (female sex hormone, 18 carbon atoms) followed by a progestogen (which would normally act to suppress ovulation and maintain pregnancy after insemination). Neither progesterone (**59c**), the body's progestogen, nor the natural estrogens such as estrone (**61b**) can be taken orally, so the contraceptives are modified agonists. Norethindrone and norgestrel are progesterone agonists. Ethynylestradiol and mestranol are estrogen agonists.

Conjugated estrogenic hormone is a mixture of female sex hormones—estrone (**61b**), equilin (**62a**) and dihydroequilin (**62b**)—which occur in the human body and are extracted from the urine of pregnant mares. They compensate for the decreased production of female hormones during menopause.

Prednisone is a cortisone (**60a**) and hydrocortisone agonist which counters inflammation and allergy. When given as drugs, the two natural compounds have unpleasant side effects. Prednisone retains many of their beneficial effects with fewer drawbacks.

Steroids were originally obtained from meat packing house wastes but are now obtained by chemical modification of plant steroids. One precursor is diosgenin found in the barbasco yam and another is stigmasterol, a component of the distillate from the steam deodorization of soybean oil.

G. NONSTEROID ANTIINFLAMMATORY AGENTS

The terms rheumatism and arthritis cover a multitude of problems from a minor pain in a single joint to a serious and disabling illness. These conditions are widespread and, as the sufferers rarely die and rarely get better, there is a huge market for drugs to relieve pain and swelling and reduce inflammation. Aspirin and acetaminophen are widely used. Steroids are saved for the most serious cases because of their side effects. In between is a range of drugs known as nonsteroid antiinflammatory agents and about 70 are currently on the market. The four most popular are ibuprofen (**63**), naproxen (**64**), piroxicam (**65**), and indomethacin (**66**). All except piroxicam are substituted acetic acids.

They have been very successful in short-term treatment of bursitis or cartilage inflammations,

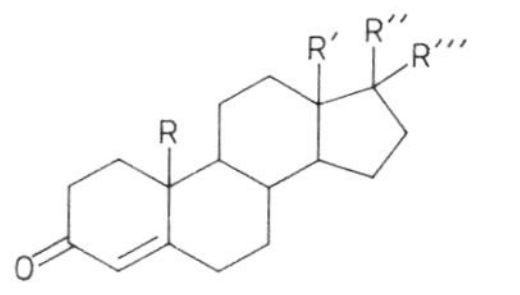

(**59a**) R = H, R′ = CH₃, R″ = OH, R‴ = C≡CH
        Norethindrone
(**59b**) R = H, R′ = CH₂CH₃, R″ = OH, R‴ = C≡CH
        Norgestrel
(**59c**) R = CH₃, R′ = CH₃, R″ = COCH₃, R‴ = H
        Progesterone

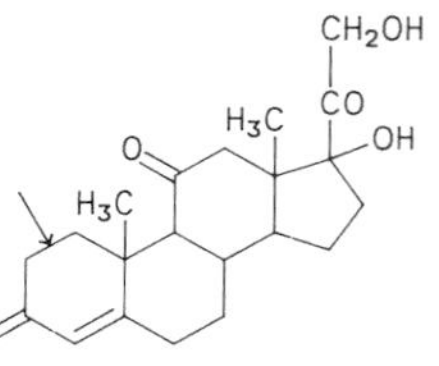

(**60a**) Cortisone
(**60b**) Double bond at arrow: Prednisone

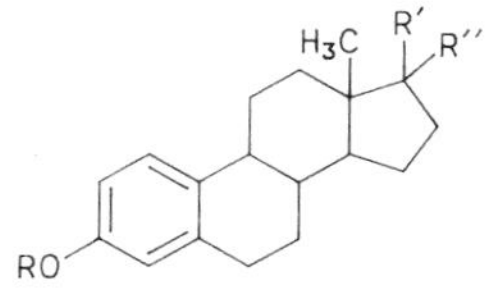

(**61a**) R = CH₃, R′ = OH, R″ = C≡CH
        Mestranol
(**61b**) R = H, R′, R″ = ═O
        Estrone
(**61c**) R = H, R′ = OH, R″ = C≡CH
        Ethynylestradiol

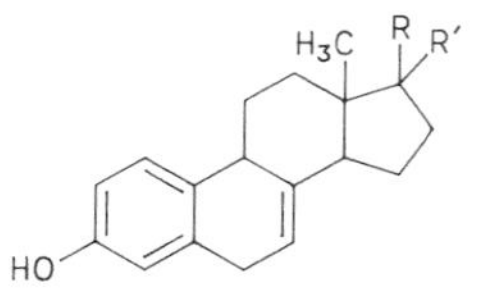

(**62a**) R, R′ = ═O   Equilin
(**62b**) R = OH, R′ = H   Dihydroequilin

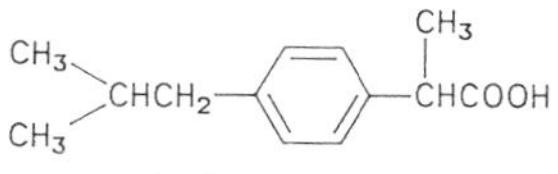

(63) Ibuprofen

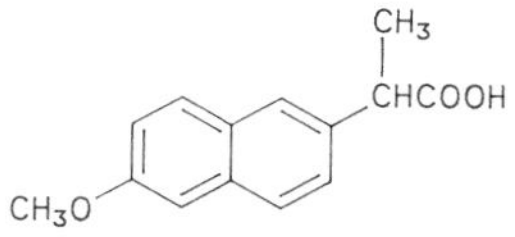

(64) Naproxen

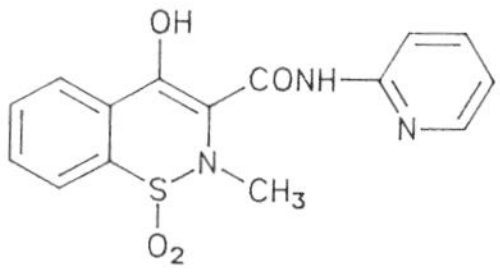

(65) Piroxicam

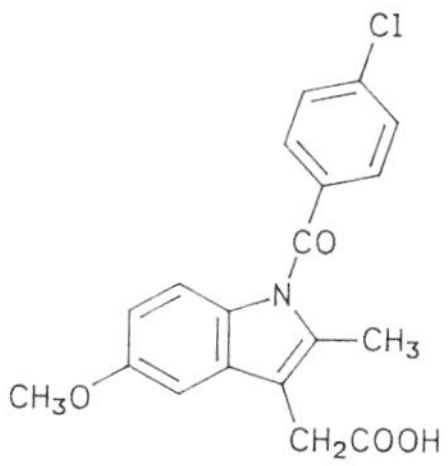

(66) Indomethacin

but in long-term treatment of arthritis, although useful, they have the drawback of aspirin in that they may irritate the stomach.

H. MISCELLANEOUS DRUGS

The four remaining drugs are L-thyroxine, theophylline, chlorpropamide and insulin. L-Thyroxine (67) is normally produced by the thyroid gland. If insufficient amounts are produced (hypothyroidism) patients become mentally slow, lethargic, and cold. L-Thyroxine, obtained from pork or beef thyroid glands or by synthesis, acts as an agonist.

Theophylline (68) is an antiasthma drug which dilates the bronchii. It has gained in popularity in recent years as sustained-release capsules have become available.

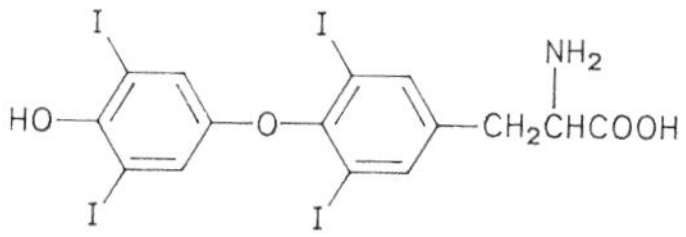

acids joined by peptide linkages with sulphur bridges between chains. Insulin must be injected but chlorpropamide can be taken orally. It is structurally similar to the sulfonamide drugs but has no antibacterial action. Instead, it stimulates the pancreas to produce more insulin.

The preceding discussion indicates that the top 50 drugs are prescribed for a surprisingly restricted range of illnesses. They are nonetheless backed up by equivalent or second-choice drugs that are not in the top 50. In addition, there are many other illnesses which are less common than those in Table III but for which drugs exist, for example gout and glaucoma. There are widespread diseases such as cancer and virus infections for which drugs are only in the development stage. There are rare diseases (e.g., Wilson's disease) for which drugs are

(67) L-Thyroxine

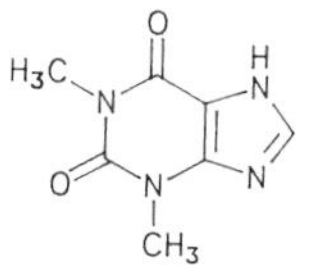

(68) Theophylline

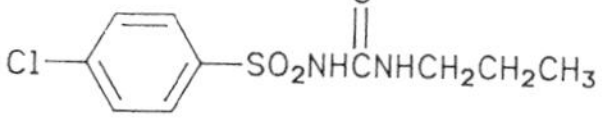

(69) Chlorpropamide

Chlorpropamide (69) and insulin are both drugs for diabetes, a condition characterized by a deficiency or reduced effectiveness of insulin in the body. Insulin, obtained from the pancreas of cows and pigs or by a novel biotechnological process involving recombinant DNA, is an agonist and supplements bodily insulin. It is a large molecule consisting of a sequence of amino

available but for which the market is small. There are also other distressing but rare illnesses like Huntingdon's chorea for which no medication is of much value. Finally, there are also tropical diseases such as leprosy or schistosomiasis from which hundreds of millions of mankind suffer but which are not prevalent in the United States. The aim of the pharmaceutical industry is to make the drugs currently in the top-50 obsolete and to fill some of the above gaps.

BIBLIOGRAPHY

American Chemical Society (1977). "Chemistry in Medicine—The Legacy and the Responsibility." American Chemical Society, Washington D.C.

British Medical Association and the Pharmaceutical Society of Great Britain, The British National Formulary, London, biannually.

Burger, J. (1980). "The Basis of Medicinal Chemistry," 4th ed., (M. E. Wolff, ed.), 3 Vols. Wiley, New York.

Goodman, A. G., Goodman, L. S., and Gilman, A. (1980). "The Pharmacological Basis of Therapeutics," 6th ed. Macmillan, London.

Huff, B. B. (ed.) (1982). "The Physician's Desk Reference," 36th ed. Medical Economics Company, Oradell, New Jersey.

Kleemann, A., and Engel, J. (1978). "Pharmazeutische Wirkstoffe: Synthese, Patente, Anwendungen." Thieme, Stuttgart.

Lednicer, D., and Mitscher, L. A. (1977, 1980, 1984). "The Organic Chemistry of Drug Synthesis," 3 vols. Wiley, New York.

Pharmaceutical Manufacturers' Association (1980). "Pharmaceutical Industry Fact Book." Pharmaceutical Manufacturers' Association, Washington D.C.

Wilson, C. O., Giswold, O., and Doerge, R. F. (1977). "Textbook of Organic Medicinal and Pharmaceutical Chemistry," 7th ed. Lippincott, Philadelphia, Pennsylvania.

Wittcoff, H. A., and Reuben, B. G. (1980). "Industrial Organic Chemicals in Perspective," Vol. 2. Wiley, New York.

Wittcoff, H. A., and Reuben, B. G. (1985). "The Pharmaceutical Industry—Chemistry and Concepts," ACS Audiotape. American Chemical Society, Washington D.C.

PHASE TRANSFORMATIONS, CRYSTALLOGRAPHIC ASPECTS

U. Dahmen   *Lawrence Berkeley Laboratory*

GLOSSARY

Anisotropy: Directional dependence of properties or behavior.

Coincidence site lattice (CSL): Lattice formed by the translations common to two misoriented grains.

Domains: Symmetry-related crystalline regions. As a result of a symmetry-breaking transition (such as ordering) a single crystal transforms to an assembly of domains. Any two domains are related by one of the broken symmetries.

Habit plane: Major plane of contact between two phases.

Homogeneous: When referring to elastic inclusions: having the same elastic constants as the matrix; when referring to nucleation: random nucleation without the aid of defects. When referring to strain: everywhere the same, as in a linear transformation; opposite of heterogeneous.

Invariant plane (line) strain: Form of strain that leaves a plane (line) unstretched and unrotated.

Motif: Group of atoms associated with each lattice point in a crystal lattice; also called basis.

Orientation relationship: Relative orientation of two crystals, specified by pairs of parallel planes and directions, by a rotation tensor, or by the angle and axis of rotation.

Parent/product: Phase before/after transformation. Also referred to as matrix/precipitate, matrix/inclusion, high-temperature phase/low-temperature phase, group/subgroup, austenite/martensite, disordered phase/ordered phase.

Space group: Set of symmetry operations that leaves a crystal structure invariant.

Special point: Point of high symmetry lying at intersection of symmetry elements in a lattice.

Strain accommodation: Process of elastic or plastic deformation around an inclusion allowing a change in shape, orientation, or volume during a transformation.

Variant: One of a set of crystallographically equivalent inclusions; for example [1 0 0], [0 1 0], and [0 0 1] needles are variants of the same type of ⟨1 0 0⟩ needle precipitates. The term is also used to describe the four (usually nonequivalent) solutions of the martensite problem.

A phase is a structurally and chemically homogeneous volume of material. A phase transformation (sometimes called transition) entails a change in either structure or composition, or both. The structural aspects of phase transformations in crystalline solids include the diffusionless martensitic transformation which is a purely structural change and proceeds by the athermal movement of a glissile interface. However, structure is also important in transformations involving short-range diffusion such as order–disorder reactions, polymorphic changes, or recrystallization and grain growth. Even in those transformations that require long-range diffusion such as precipitation, eutectoid, or discontinuous reactions, the structure of the inter-

ENCYCLOPEDIA OF PHYSICAL SCIENCE
AND TECHNOLOGY, VOL. 10

face between parent and product phase has important implications for the morphology or the growth mechanism of the new phase.

Two types of structural change may be distinguished, a purely crystallographic change due, for example, to a rearrangement of atoms within a rigid lattice, and a dimensional change due to a distortion of the lattice. The former type is described by the concepts of pure crystallography and group theory and forms the basis of the theory of group–subgroup transitions. The latter type focuses on the change in lattice dimensions, largely ignoring the atomic arrangement in the lattice; it forms the basis of the theory of elastic inclusions and of martensite theory. Most transformations induce both types of structural change, crystallographic and dimensional, but usually one or the other dominates.

A general homogeneous distortion can be written as the distortion tensor $\mathbf{A}$, which may be separated into a pure distortion $\mathbf{D}$ and a rigid body rotation $\mathbf{R}$:

$$\mathbf{A} = \mathbf{RD}$$

The pure distortion may be decomposed further into a pure shear $\mathbf{S}$ and a pure volume expansion $\mathbf{V}$:

$$\mathbf{A} = \mathbf{RSV}$$

This decomposition is useful because it separates the changes in orientation ($\mathbf{R}$), shape ($\mathbf{S}$), and volume ($\mathbf{V}$). Grain boundaries are described by a change in orientation, martensite transformations are dominated by the shape change, and precipitation reactions often have large orientation, shape, and volume changes, but each component has a different effect on the final morphology.

I. Crystallography Principles

In most solids the atoms are arranged in a periodic network such that their mutual coordination is optimized. The basic repeat unit of this network or crystal lattice is the unit cell. The atomic structure remains invariant under translation through any vector $\mathbf{t}$ that is the sum of integral multiples of three basic vectors $\mathbf{a}_1$, $\mathbf{a}_2$, and $\mathbf{a}_3$. The set of all such translation vectors $\mathbf{t}$ forms the translational group T. When a mathematical point in space is repeated through the vectors of the translation group the Bravais lattice is obtained (Fig. 1a). For certain angles and ratios between the basis vectors additional symmetries such as reflection, inversion, or rotation

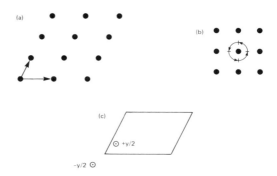

FIG. 1.   Illustration of three basic symmetry elements: translation symmetry leads to the Bravais lattice (a), rotation symmetry exists for special angles (here 90°) between basis vectors of the Bravias lattice (b), and screw axis symmetry results from special distance (here $y = \frac{1}{2}$) between atoms in the motif (c).

arise, for example, the 4-fold rotation axis in Fig. 1b. The set of all such point symmetry elements $\mathbf{R}$ that leave a given Bravais lattice invariant is called its point group G. [See CRYSTALLOGRAPHY.]

If all lattice points of the Bravais lattice are occupied by identical atoms we have a simple (primitive) crystal with all the translation and point symmetry elements of the Bravais lattice. Not all crystals are this simple, however, and most solids contain more than one atom per lattice site. The group of atoms that occupies a lattice point is called the basis or motif. The position of atoms in the basis and the arrangement of identical bases in the Bravais lattice are dictated by bonding requirements. For certain bond angles and distances in the motif two new symmetry elements arise (screw axes and glide mirrors) that are a combination of a point symmetry element $\mathbf{R}$ with a fractional translation τ in the unit cell (e.g., $\frac{1}{2}$ 0 0). An example of a motif giving rise to either a 2-fold axis or a 2-fold screw axis, or no symmetry, depending on the bond distance y is shown in Fig. 1c. A Bravais lattice is projected along its 2-fold b axis. The two identical atoms in the motif are a vertical distance y apart. If y is irrational the crystal has no rotational symmetry, if $y = 0$ it has a 2-fold rotation axis and if $y = \frac{1}{2}$ it has a 2-fold screw axis, that is, an identical position to any point in the cell is reached by rotation through 180° followed by a translation of $\tau = \frac{1}{2}$ along the rotation axis. Applying this operation twice is the same as a simple translation along the axis. The set of all such symmetry elements $(\mathbf{R}|\tau)$ (including simple point symmetry elements for which $\tau = 0$) forms the space group.

Thus translation group T, point group G, and space group S are characterized by their specific symmetry operations t, R, and $(R|\tau)$. Only a relatively small number of different crystallographic translation, point, and space groups exists, the 14 Bravais lattices, the 32 point groups, and the 230 space groups. Any crystal belongs to one of the 230 space groups. Each space group has a characteristic point group and Bravais lattice. Since the translations τ in the space group elements are on the order of unit cell dimensions, they are considered microscopic symmetry elements, in contrast to point symmetry operations that concern directions only and therefore represent the macroscopic symmetry of a crystal. A complete listing of crystallographic symmetry groups is compiled in the International Tables for Crystallography. A brief account of the concepts and notations necessary to represent and understand symmetry groups will be given below.

A. Translation Group

Many of the physical properties of crystals result from their invariance under translations t of the translation group. It is this periodicity that allows investigation of crystal structures and phase transformations by diffraction techniques. Some aspects of phase transformations are best described in terms of Fourier series as the most natural way to express periodic functions. Crystallography makes use of the fact that in order to describe a crystal structure it is sufficient to specify the dimensions and content of a single unit cell. The strain tensor of a transformation from one crystal structure to another is derived from the correspondence between unit cells. These and other aspects of translational symmetry are utilized in the theoretical and experimental investigation of phase transformations.

B. Point Group

For certain dimensions of the unit cell a Bravais lattice is invariant under rotation as well as translation. The set of rotations, reflection, or inversion operations R that leave the lattice points of such unit cells invariant is called the point group. Point symmetry in general characterizes the set of equivalent directions, hence it describes the macroscopic symmetry of an object. For example, many flowers have 5-fold rotational symmetry, a cylinder has infinite rota-

FIG. 2. Construction illustrating the rotation angles α that are consistent with translation symmetry t.

tional symmetry ($R = \infty$) about its axis, and a sphere has infinite rotational symmetry about any axis. However, only 1-, 2-, 3-, 4-, and 6-fold rotational symmetries are compatible with translation symmetry and therefore only those rotations occur in the 32 crystallographic point groups. To illustrate this consider the row of atoms in Fig. 2, which is a section of a larger space lattice with translation period t. If the whole assembly is rotated about a lattice point by $\pm\alpha$ degrees the new lattice points must again conform to translational symmetry, that is, they must be an integral multiple N of horizontal translations t apart. Thus

$$X = Nt = 2t \cos \alpha, \quad \text{or } \cos \alpha = N/2$$

The five possible rotation symmetries and their notation are listed in Table I. Each type of rotation axis (or rotor) is characteristic for a particular crystal system. In addition there are two more crystal systems that arise from combinations of these elements: the orthorhombic systems with two diads at right angles and the cubic system with three tetrads and four triads. Each of these seven crystal system has a characteristic space lattice. The angles and edge lengths of the unit cell of each space lattice must be compatible with the symmetry of the crystal system. The unit cells of the seven space lattices are shown in Fig. 3. It can be seen that each cell has more symmetry elements than just translation and the unique n-fold rotor. For example the monoclinic unit cell which has a diad through each lattice point automatically has other diads going through the face and edge centers [see Fig.

TABLE I. The Five Rotation Angles Compatible with Translation

N	α	n	Notation	Crystal system
2	360°	1	Monad	Triclinic
−2	180°	2	Diad	Monoclinic
−1	120°	3	Triad	Rhombohedral
0	90°	4	Tetrad	Tetragonal
1	60°	6	Hexad	Hexagonal

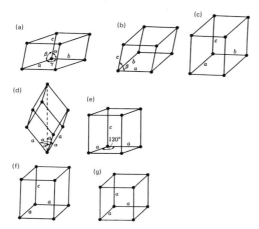

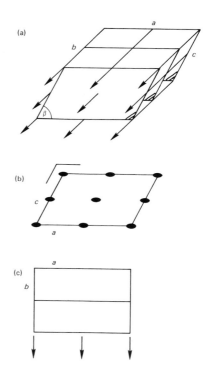

FIG. 3. Unit cells of the seven primitive space lattices: triclinic (a), monoclinic (b), orthorhombic (c), rhombohedral (d), hexagonal (e), tetragonal (f), and cubic (g).

4a). In addition to mirror planes on the faces normal to the unique axis it also has a mirror halfway between them. Any of these symmetry elements, when applied to the monoclinic unit cell, will reproduce its lattice points in identical positions, that is, leave the lattice invariant. These symmetry elements are shown in two projections (Fig. 4b and c) using diad symbols and mirror notation to indicate their locations. A complete list of standard symbols used for rotation axes, mirror planes, centering translations, screw axes, and glide mirrors is given in Table II.

C. GROUP PROPERTIES

The properties of the point group are easily visualized using the point symmetry of the monoclinic crystal system, characterized by a diad. The lowest symmetry monoclinic point group contains only two elements, the identity (1) and the diad (2). This set is written as $G = \{1, 2\}$, in shorthand $G = 2$, and it represents a group since (1) it contains the identity operation (1), (2) it is closed, that is, the product of any two successive operations is equivalent to a single operation which is itself part of the group (e.g., 2 followed by 2 is the identity 1), and (3) each element $\mathbf{R}$ has an inverse $\mathbf{R}^{-1}$ such that $\mathbf{RR}^{-1} = \mathbf{R}^{-1}\mathbf{R} = 1$.

There are two other monoclinic point groups, $G = m$ and $G = 2/m$, again characterized by the diad but containing additional symmetry elements. The holohedral (highest symmetry) monoclinic group contains four elements: $G = \{1, \bar{1}, 2, m\}$. In shorthand this is referred to as

FIG. 4. Monoclinic space lattice with location of diads and mirrors: perspective view (a), projection along b axis (b), and perpendicular to b axis (c) in notation used in International Tables for Crystallography (see Table II).

$G = 2/m$ with the oblique line indicating that the diad and the mirror refer to the same axis. The operation of such a group is best illustrated in a multiplication table (Table III) which lists the products (successive application) of any two elements. It is clear from this table that any two successive operations are equivalent to a single operation which is itself an element of the group. In our example, each element happens to be its own inverse, that is, yields the identity when applied twice in succession. This need not always be the case, for example, a tetrad needs to be applied four times to produce the identity. This is known as the order of the symmetry element. Thus a hexad is of order six and a mirror of order two. The inverse of a $60°$ rotation (6^1) is a $300°$ rotation (6^5).

D. STEREOGRAPHIC REPRESENTATION

A convenient graphic representation of the point group symmetry is the stereographic projection. Consider a general direction, indicated by a pole in the stereogram (Fig. 5a) in a crystal with $2/m$ monoclinic symmetry, that is, with point group $G = \{1, \bar{1}, 2, m\}$. Operation on this

TABLE II.   Crystallographic Symmetry Elements and Their Notation

Type of symmetry element	Written symbol	Graphical symbol	
Center of symmetry	$\bar{1}$	o	
		Perpendicular to paper	In plane of paper
Mirror plane	m	———	⌐ ⌐
Glide planes	$a\ b\ c$	- - - - -	← ↓
		Glide in plane of paper	Arrow shows glide direction
		.... Glide out of plane of paper	
	n	–·–·–·	⬈
Rotation	2		
	3		——————→
	4		
	6		
Screw axes	2_1		——————→
	$3_1, 3_2$		
	$4_1, 4_2, 4_3$		
	$6_1, 6_2, 6_3, 6_4, 6_5$		
Inversion axes	$\bar{3}$		
	$\bar{4}$		
	$\bar{6}$		

crystal by all four symmetry operations will take this pole to the four positions shown in Fig. 5b (open and closed circles represent poles in different hemispheres). The number of crystallographically equivalent poles in a stereogram is equal to the number of symmetry elements in its points group. This number is known as the order of the group. Stereograms characterizing all 32 crystallographic point groups are shown in Fig. 6. Note that they are subdivided into the seven crystal systems and that each crystal system has more than one possible point group. The holohedral group describes the symmetry of the space lattice of that crystal system. Thus the monoclinic space lattice has $2/m$ symmetry.

Only when the lattice points are occupied by atoms or groups of atoms does the space lattice become a crystal. And only when the group of atoms at each lattice site, the motif, has lower symmetry than $2/m$ does the crystal have a lower than holosymmetric point group. An example is shown in Fig. 7 where the motif is made up of two dissimilar atoms aligned with the diad. This motif is not invariant under inversion or reflection perpendicular to the rotation axis. The crystal therefore cannot have this symmetry either. The point symmetry of this crystal is the set of elements common to the lattice $G_0 =$

TABLE III.   Multiplication Table for $2/m$ Monoclinic Point Group

	1	$\bar{1}$	2	m
1	1	$\bar{1}$	2	m
$\bar{1}$	$\bar{1}$	1	m	2
2	2	m	1	1
m	m	2	$\bar{1}$	1

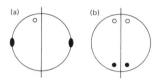

FIG. 5.   Stereographic representation of point symmetry. The general point in (a) is taken to all its equivalent positions by the symmetry elements of the point group $2/m$ (b). Open circles are in upper hemisphere, solid dots in lower hemisphere.

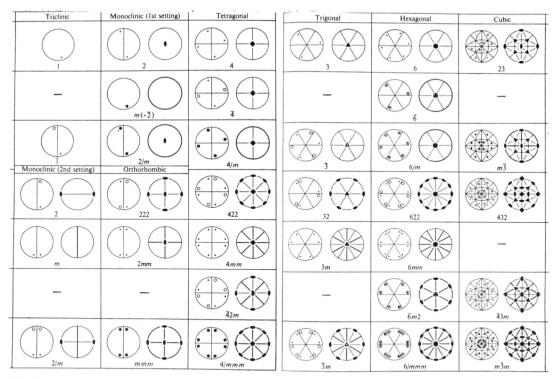

FIG. 6. Stereographic representation of the 32 crystallographic point groups. The order of a group is the multiplicity of a general point.

$\{1, \bar{1}, 2, m\}$ and the motif $G_1 = \{1, \infty, m\}$, also known as the intersection group $H = G_0 \cap G_1 = \{1, 2\}$ which is the lowest symmetry monoclinic group: 2. Note that the mirrors m of the lattice and the motif are not parallel and are therefore not part of the intersection group H. This is an important point: the orientation of the symmetry elements is essential in forming the intersection group. For example two cubic point groups share only the identity $\{1\}$ or inversion $\{1, \bar{1}\}$ if their symmetry axes are not aligned. The point symmetry of a crystal depends therefore on the space lattice, that is, the dimensions and angles of the unit cell, and on the motif, that is, the atomic arrangement about each lattice point.

E. Tensor Representation

The point symmetry elements, $2, m, \bar{1}$ etc. can be represented more explicitly by transformation tensors. Consecutive application of two symmetry operations is equivalent to matrix multiplication of the corresponding tensors. As an example a reflection in a plane perpendicular to the z axis of an orthogonal coordinate system is simply written as the tensor

$$m = \begin{pmatrix} 1 & 0 & 0 \\ 0 & 1 & 0 \\ 0 & 0 & \bar{1} \end{pmatrix}$$

Such symmetry elements **R** represent proper (det **R** $= 1$) or improper (det **R** $= -1$) rotations. Improper rotations change the handedness of a crystal. They are unitary operators, hence $\mathbf{R}^T = \mathbf{R}^{-1}$, (the transpose $\mathbf{R}^T$ is equal to the inverse $\mathbf{R}^{-1}$). In addition, as shown before, only certain

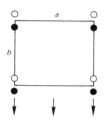

FIG. 7. Projection of monoclinic unit cell as in Fig. 4c illustrating absence of mirror due to two dissimilar atoms in the motif.

rotations are compatible with translational symmetry. Some simple point symmetry elements are listed in Table IV. Under a symmetry operation **R**, vectors **r** and tensors **M** transform as

$$\mathbf{r}' = \mathbf{R}\mathbf{r} \quad \text{and} \quad \mathbf{M}' = \mathbf{R}\mathbf{M}\mathbf{R}^{-1}$$

F. SPECIAL POINTS

The intersections of symmetry elements, for example the line where two mirror planes meet, are special positions in a lattice. When symmetry elements intersect in a point this becomes a special point. As shown later the special points of a lattice are of fundamental importance in the theory of phase transitions. The lattice points at the corners of the monoclinic cell shown in Fig. 4 are special points since they lie at the intersection of a diad with the mirror plane. The coordinates of these special points are characterized by the coordinates of the origin, (0 0 0), since the others can be derived from it by simple lattice translations **t**. However, there are other special points, for example at $(\frac{1}{2}\,0\,0)$, $(\frac{1}{2}\,\frac{1}{2}\,0)$, and $(\frac{1}{2}\,\frac{1}{2}\,\frac{1}{2})$, that do not coincide with lattice points of the primitive Bravais lattice. It is possible to place

lattice points on special points without altering the point symmetry of the lattice. A lattice point at $(\frac{1}{2}\,\frac{1}{2}\,0)$, for example, will change the primitive (symbol P) monoclinic Bravais lattice into a base-centered one (symbol C) without changing the number or location of point symmetry operations. The only change is an additional translation symmetry, the $(\frac{1}{2}\,\frac{1}{2}\,0)$ base centering translation. Centering translations are denoted A, B, C for base, F for face, and I for body centering. If all such centering translations are added to the 7 primitive space lattices, a total of 14 different Bravais lattices is formed.

G. SPACE GROUP

As we have seen, the point group is concerned with macroscopic symmetry, that is, the equivalence of directions in an object. Directions are not changed by translations, although translations do limit the number of crystallographic point groups to 32 through the requirement of compatibility with translational symmetry. The space group is concerned with microscopic symmetry, including translations as well as directions. If a motif of lower symmetry is placed on a lattice of higher symmetry the space group is the set of operations that leaves this pattern (or crystal) invariant. In a unit cell of the monoclinic lattice, outlined in Fig. 8, it is apparent that the motif or group of atoms at the center of the cell cannot be obtained by a centering translation, but by reflection across the dashed mirror line followed by a fractional ($\tau = \frac{1}{2}$) translation parallel to the line. This is the operation of a glide mirror. Alternatively the same arrangement can be obtained by the 2-fold screw axes indicated by half arrows. However, this is no longer the case if the symmetry of the motif is lowered further by differentiating between its two atoms. The crystal structure then either has a screw

TABLE IV.   Tensors **R** for Some Point Symmetry Elements

$1 = \begin{pmatrix} 1 & 0 & 0 \\ 0 & 1 & 0 \\ 0 & 0 & 1 \end{pmatrix}$	Identity
$\bar{1} = \begin{pmatrix} \bar{1} & 0 & 0 \\ 0 & \bar{1} & 0 \\ 0 & 0 & \bar{1} \end{pmatrix}$	Inversion
$m_z = \begin{pmatrix} 1 & 0 & 0 \\ 0 & 1 & 0 \\ 0 & 0 & \bar{1} \end{pmatrix}$	Mirror perpendicular to z axis
$2_z = \begin{pmatrix} \bar{1} & 0 & 0 \\ 0 & \bar{1} & 0 \\ 0 & 0 & 1 \end{pmatrix}$	Diad along z axis
$3 = \begin{pmatrix} 0 & 1 & 0 \\ 0 & 0 & 1 \\ 1 & 0 & 0 \end{pmatrix}$	Triad along cube body diagonal
$4 = \begin{pmatrix} 0 & 1 & 0 \\ \bar{1} & 0 & 0 \\ 0 & 0 & 1 \end{pmatrix}$	Tetrad along z axis

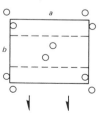

FIG. 8.   Projection of monoclinic unit cell perpendicular to b axis with four-atom motif illustrating a-glide mirror perpendicular to b axis (dashed lines) and screw axis parallel to b axis (half arrows).

axis or a glide mirror but not both. By taking into account these small translations τ important in microscopic symmetry, the 32 point groups are further differentiated into 230 space groups.

In a manner similar to point symmetry elements, space group operations such as 2_1, n, 4_3 etc. (see Table II) may be represented by an operator $(\mathbf{R}/\tau)$ that denotes a point symmetry operation $\mathbf{R}$ followed by a translation τ. For example a 2-fold screw axis along the z direction would be

$$2_1 = \begin{pmatrix} \bar{1} & 0 & 0 & 0 \\ 0 & \bar{1} & 0 & 0 \\ 0 & 0 & 1 & \frac{1}{2} \end{pmatrix}$$

that is, a 2-fold rotation followed by a translation of $\frac{1}{2}$ along the rotation axis. Some transformation and multiplication rules for such tensors are

$$(\mathbf{R}|\tau)\mathbf{r} = \mathbf{R}\mathbf{r} + \tau$$

$$(\mathbf{R}|\tau)(\mathbf{Q}|\varepsilon) = \mathbf{R}\mathbf{Q} + \mathbf{R}\varepsilon + \tau$$

$$(\mathbf{R}|\tau)^{-1} = (\mathbf{R}^{-1}|-\mathbf{R}^{-1}\tau)$$

H. Notation

The International (Hermann–Mauguin) notation used to describe space groups consists of two parts: (1) a letter (A, B, C, F, I, P) indicating the centering type of the unit cell, and (2) a set of characters giving symmetry elements along one, two, or three principal symmetry directions in the crystal. Table II lists the space group symbols along with their graphic representation. For example, the space group symbol $P2_1/c$ represents the primitive monoclinic crystal structure in Fig. 8 with the characters 2_1 (2-fold screw axis) and c (c-glide mirror) referring to the same principal symmetry direction in the crystal. The two symbols are therefore separated by an oblique line. On the other hand, the symbol $Im\bar{3}m$, representing a body-centered cubic structure, has three characters referring to the cube edge, cube diagonal, and face diagonal, respectively. By convention the first character after the letter indicating the centering type describes the characteristic symmetry direction in each crystal system, for example, [0 0 1] in the tetragonal and hexagonal systems and [1 1 1] in the rhombohedral system.

To summarize: crystals are a periodic arrangement of atoms on a space lattice that has translational and rotational symmetries. Only five rotational symmetries are compatible with translation, the 1-, 2-, 3-, 4-, and 6-fold axes. When combined in space, these give rise to 7 crystal systems, triclinic, monoclinic, orthorhombic, tetragonal, rhombohedral, hexagonal, and cubic, each with a characteristic shape of its unit cell as prescribed by symmetry and shown in the 7 primitive Bravais lattices. Without being occupied by atoms, these 7 simple space lattices have a number of point symmetry operations (rotation, inversion, mirror) in addition to the ones required for the crystal system. The set of these operations is called the point group of the system. Each of these seven holosymmetric point groups has a limited number of subgroups with lower symmetry but still compatible with the shape of its characteristic unit cell. This leads to a total of 32 possible point groups.

In addition to the primitive lattice translations, centering translations within the unit cell such as A, B, C, F, I are possible without changing the point group symmetry. When these centering translations are combined with the 7 primitive space lattices, the 14 different Bravais lattices result. Translations within the unit cell, which are either pure or combined with mirror or rotation symmetry, are registered in the space group. Due to these microscopic symmetries the 32 point groups are further differentiated into the 230 space groups.

Translation groups are characterized by symmetry operations $\mathbf{t}$, point groups by $\mathbf{R}$, and space groups by $(\mathbf{R}|\tau)$.

I. Symmetry Principles

The physical properties of crystals depend on their crystal structure and are often anisotropic, that is, different for different directions of the crystal. Any macroscopic physical property of a crystal must have at least the point symmetry of the crystal itself. Stated differently, this is known as Neumann's principle: "The symmetry elements of any physical property of a crystal must include the symmetry elements of its point group." Thus some physical property, for example, thermal expansion, of a crystal can be isotropic while the crystal itself is not. However, the converse is not true: if a physical property is anisotropic, the crystal structure must have the same anisotropy. For example, the spontaneous polarization in ferroelectric crystals has polar symmetry (∞m). According to Neumann's principle, the crystal point group must share the same polar symmetry, that is, it must be a subgroup of (∞m). This allows only point groups with a unique rotor and any mirrors parallel to it. Of the 32 crystallographic point groups only 10 (1, 2, 3, 4, 6, m, $2mm$, $3m$, $4m$, $6mm$) are compatible with ferroelectricity.

As a result of Neumann's principle a proper description of the physical properties of a crystal must be invariant under the operation of its symmetry group. Physical properties can be described by matter tensors that give the response of the crystal to an external stimulus such as temperature or a stress field. The type and rank of the tensor depend on the property it describes. For example the tensor of thermal expansion is of rank two. It describes the strain field of a crystal when the stimulus temperature is applied. Other matter tensors of order two relate a vector stimulus to a vector response, for example, the tensors of electrical or thermal conductivity, diffusivity, permittivity, etc. Others relate a pair of vectors to a third as in the third-order tensor of the Hall constants. The elastic behavior of crystals is described by fourth-order tensors relating two second-order tensors, the stress field, and the strain field. Any of these matter tensors must remain invariant under all the symmetry operations $\mathbf{R}$ of the point group of the crystal. A tensor of rank two transforms as $\mathbf{M}' = \mathbf{R}\mathbf{M}\mathbf{R}^{-1}$. Since a matter tensor $\mathbf{M}$ must be invariant under the symmetry operations $\mathbf{R}$ of the point group, $\mathbf{M}' = \mathbf{M}$ for all symmetry operations $\mathbf{R}$ of the crystal. Using the rules of matrix multiplication, it can be shown, for example, that any second-order matter tensor in a tetragonal system must have the form

$$\mathbf{M} = \begin{pmatrix} M_1 & & \\ & M_1 & \\ & & M_3 \end{pmatrix}$$

and for a cubic system must be isotropic

$$\mathbf{M} = \begin{pmatrix} M_1 & & \\ & M_1 & \\ & & M_1 \end{pmatrix}$$

This means that diffusion, thermal expansion, or any physical property described by a second-order tensor is isotropic in cubic crystals. These results can be derived directly from Neumann's principle. A more general statement of this principle was given earlier by Curie: "When definite causes produce definite effects the elements of symmetry of the causes should be apparent in the effects." In this form the principle is applicable not just to crystallography but to any physical phenomenon. If the cause is the crystal structure and the effect the physical properties we recover Neumann's principle. If, in the case of a phase transformation, the cause is a decrease in temperature and the effect is an arrangement of ordered domains due to the loss of symmetry in the phase transition, then the symmetry of the cause (isotropic) should be apparent in the symmetry of the effect, that is, all possible domains should appear at random. On the other hand, if the cause has a lower symmetry, such as a temperature gradient, uniaxial stress, or magnetic field then this symmetry should be apparent in the distribution of domains formed under the influence of this field. In order to predict and detect such effects we must answer the question of how many different domains can form in a given phase transition.

II. Modulated Structures

A. Characteristics

A large class of phase transformations can be described by a modulation of some physical quantity (e.g., composition, magnetization, displacement) associated with a crystal. In such transformations, the crystal lattice is modified with the periodicity of a static plane wave. For example, a sinusoidal composition modulation with wave length λ can be written as the (static) concentration wave

$$n(\mathbf{r}) = c + Q \sin 2\pi \mathbf{k}\mathbf{r} \qquad (1)$$

where $\mathbf{k}$ is the wave vector ($|\mathbf{k}| = 1/\lambda$), and $n(\mathbf{r})$ the probability that the lattice site at position $\mathbf{r}$ is occupied by an atom whose concentration in the alloy is c. The amplitude Q of the modulation is proportional to the order parameter η which varies between zero for the disordered structure and unity for the fully ordered structure. An example is seen in Fig. 9 which illustrates the disordered and ordered state of a binary alloy with a simple cubic lattice. The ordered structure is tetragonal with the c axis parallel to the wave vector $\mathbf{k}$. When applied to an fcc lattice this describes the type of ordering found in equiatomic Cu–Au alloys. The wavelength $\lambda = na$ is an integer multiple n of the lattice periodicity a (in this particular case $\lambda = a$) and the wave is commensurate with the crystal lattice. The correspond-

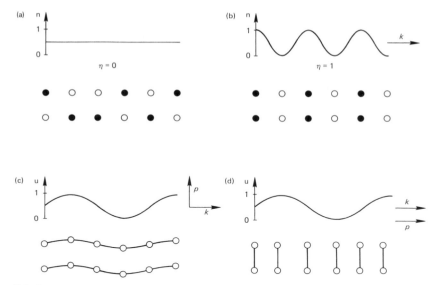

FIG. 9.   Illustration of modulated structures: disordered solid solution (a), composition modulation (b), transverse displacement modulation (c), and longitudinal displacement modulation (d).

ing wave vector $\mathbf{k} = \langle 1\ 0\ 0 \rangle$ is on the Brillouin zone boundary, that is, the wavelength λ is on the order of the lattice parameter, a characteristic of an ordering modulation. A concentration wave with wave vector near the origin ($\lambda \gg a$) describes a clustering modulation. Continuous phase transitions that proceed by the gradual amplification of certain composition fluctuations are called spinodal ordering ($\lambda \simeq a$) and spinodal clustering or spinodal decomposition ($\lambda \gg a$).

When the wavelength is not an integral multiple of a lattice spacing the modulation is incommensurate. In the direction of the wave vector, the crystal then loses its true translational symmetry. Most ordering modulations are commensurate. Although incommensurate structures exist incommensurations are often found to be a mixture of two commensurate phases. If the quantity that is modulated is not a scalar such as concentration but a vector such as displacement or magnetization, the wave description of the position $\mathbf{u}(\mathbf{r})$ of an atom originally at $\mathbf{r}$ becomes

$$\mathbf{u}(\mathbf{r}) = \mathbf{r} + \mathbf{p} \sin 2\pi \mathbf{k} \mathbf{r} \qquad (2)$$

where $\mathbf{p}$ is the polarization vector and $\mathbf{k}$ the wave vector. Since $\mathbf{p}$ is a vector quantity we now have to distinguish between transverse ($\mathbf{p} \perp \mathbf{k}$) and longitudinal ($\mathbf{p} \parallel \mathbf{k}$) modulations, illustrated in Fig. 9c and d. Short wavelength ($\lambda \simeq a$) commensurate displacement waves describe shuffles whereas long wavelength ($\lambda \gg a$)

modulations describe lattice distortions such as the premartensitic "tweed" effect. A simple example of a short wavelength longitudinal displacement wave is the ω-transformation found in Ti and Zr base alloys. It can be described as the local collapse of every other pair of $\{1\ 1\ 1\}$ planes of the bcc lattice, or a wave with $\mathbf{k} = \frac{2}{3} \langle 1\ 1\ 1 \rangle$ and $\mathbf{p} = \frac{1}{6} \langle 1\ 1\ 1 \rangle$.

Due to their periodic nature modulated structures give rise to diffraction peaks. Long wavelength modulations lead to satellite reflections near the main Bragg peaks while their short wavelength counterparts cause extra reflections at rational positions between Bragg peaks of the unmodulated structure. As an example, the ordered structure of Fig. 9b would give rise to extra reflections at the positions of the wave vector ($\mathbf{k} = \langle \frac{1}{2} 0\ 0 \rangle$) halfway between the Bragg peaks of the disordered structure. The wave vector $\mathbf{k}$ can thus be read directly from a diffraction pattern as the position of the extra reflections in the first Brillouin zone (superlattice reflections). This is shown in Fig. 10a for a short wavelength composition modulation (ordering). The equivalent for a long wavelength (clustering) modulation is shown in Fig. 10b where satellites are seen around each Bragg peak as well as the origin.

In a displacement modulation not all satellite reflections are allowed due to the directional nature of the polarization $\mathbf{p}$. Satellite reflections are forbidden near all Bragg peaks $\mathbf{g}$ for which

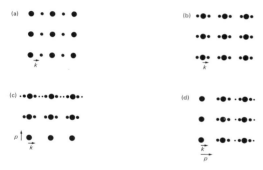

FIG. 10. Schematic diffraction patterns from modulated structures: short wavelength composition modulation (ordering) (a), long wavelength composition modulation (clustering) (b), and long wavelength transverse (c) and longitudinal (d) displacement modulation.

gp = 0. Long wavelength modulated transverse and longitudinal displacements would thus give rise to the diffraction patterns shown in Fig. 10c and d. The short wavelength equivalent can lead to either extra reflections or extinctions depending on the magnitude and direction **p** of the modulation, that is, the nature of the shuffle.

The structures considered so far were simple modulations by a single plane wave. More complex structures are obtained from combinations of several waves. For composition modulations this is expressed as a sum of concentration waves

$$n(\mathbf{r}) = c + \sum_j Q_j \exp(2\pi i \mathbf{k}_j \mathbf{r}) + c.c. \quad (3)$$

where Q_j is the amplitude of the jth wave with wave vector $\mathbf{k}_j$ defined in the first Brillouin zone, $c.c.$ is the complex conjugate, and $n(\mathbf{r})$ is the probability of site $\mathbf{r}$ to be occupied by a particular species whose concentration in the alloy is c.

Due to crystal symmetry some of the wave vectors $\mathbf{k}_j$ are crystallographically equivalent, for example [0 0 1] and [1 0 0] in cubic systems. Such vectors are said to belong to the same star. The star of a wave vector **k** is the set of vectors obtained by applying all the symmetry elements of the point group of the crystal. If the summation is rewritten as the sum over different stars s it becomes

$$n(\mathbf{r}) = c + \sum_s \eta_s \sum_j \gamma_s(j) \exp(2\pi i \mathbf{k}_j \mathbf{r}) + c.c. \quad (4)$$

Here the summation j is carried out over the equivalent vectors in a star and the summation s

runs over the different stars. The $\gamma_s(j)$ are coefficients determining the relative contribution of each of the equivalent waves in a star and the η_s are the long-range order parameters.

B. SPECIAL POINT ORDERING

Whether or not an ordered structure is thermodynamically favorable depends on the specific interaction between the different atoms. Generally the wave vector **k** of a stable ordered phase will change continuously with the form of this interaction. However, there are some special wave vectors, determined by the crystal symmetry alone, whose positions are independent of atomic interactions. It can be shown that at such special points any function with the periodicity of the lattice must have a symmetry-dictated extremum, that is a minimum, saddle point, or maximum. Not surprisingly, ordered structures with special point wave vectors are often found to be the most stable phases over a wide range of composition and temperature. Special point wave vectors are the reciprocal space equivalent of special points in a unit cell (see Section I,F) and are thus easily determined. For each disordered structure it is therefore possible to enumerate all possible special point ordered structures simply by summing over all compatible combinations or special wave vectors. This leads to an elegant derivation of the most commonly found ordered structures. Examples for phases based on the fcc lattice (special points are the stars of $\langle 0\ 0\ 0\rangle$, $\langle 1\ 0\ 0\rangle$, $\langle\frac{1}{2}\frac{1}{2}\frac{1}{2}\rangle$, $\langle 1\ \frac{1}{2}\ 0\rangle$) and the bcc lattice (special points are the stars of $\langle 0\ 0\ 0\rangle$, $\langle 1\ 1\ 1\rangle$, $\langle\frac{1}{2}\frac{1}{2}\frac{1}{2}\rangle$, $\langle\frac{1}{2}\frac{1}{2}\ 0\rangle$) together with some alloys forming such phases and the operating special point wave vectors are listed in Table V. Some of these structures are shown in Fig. 11. Note that the method applies to substitutional and interstitial phases alike since it is based entirely on symmetry and not on the kind of atomic interaction.

C. FIRST- AND SECOND-ORDER TRANSFORMATIONS

First-order transformations are accompanied by drastic changes in macroscopic properties of the material and the coexistence of two phases at the transformation temperature. Typical examples of such transitions are melting, evaporation, precipitation, or polymorphic changes. In second-order transformations the thermodynamic properties of a material change continuously but their second derivatives such as spe-

TABLE V. Wave Vectors of Some Common Special-Point Ordered Structures

k	Substitutional	Interstitial
fcc based		
[0 0 1]	CuAuI	$(Fe,Ni)_2N$
[1 0 0], [0 1 0], [0 0 1]	Cu_3Au	Fe_4N
[1/2 1/2 1/2]	CuPt	
[0 0 0], [1/2 1/2 1/2], [1/2 1/2 $\bar{1}$/2]	$CuPt_3$	
[0 0 1], [1 0 1/2]	Al_3Ti	Ni_4NII
[1 0 0], [0 1 0], [0 0 1],	$CuPt_7$	Fe_8N
[$\bar{1}$/2 1/2 1/2], [1/2 $\bar{1}$/2 1/2]		
[1/2 1/2 $\bar{1}$/2], [1/2 1/2 1/2]		
bcc based		
[1 1 1]	CuZn	
[1/2 1/2 0]		Ta_2O
[1/2 1/2 1/2]	NaTl	
[1 1 1], [1/2 1/2 1/2]	Fe_3Al	
[1 1 1], [1/2 1/2 0], [1/2 $\bar{1}$/2 0]		Ta_4O
[1 1 1], [1/2 1/2 0],		Fe_8N
[0 1/2 $\bar{1}$/2], [1/2 0 1/2],		
[1/2 0 $\bar{1}$/2], [1/2 $\bar{1}$/2 0]		

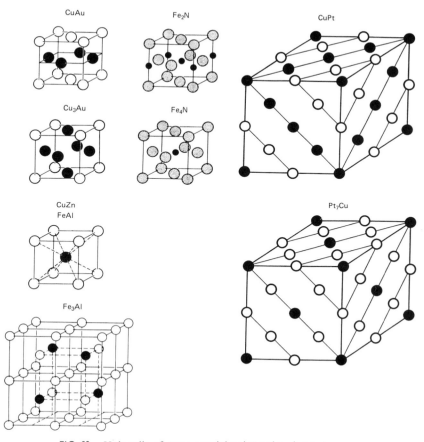

FIG. 11.   Unit cells of some special point ordered structures.

cific heat undergo a discontinuous change. Second-order transformations arise from instability to small fluctuations such as, for example, static concentration waves in order–disorder reactions. The two phases involved in the transition cannot coexist at any temperature. [*See* CHEMICAL THERMODYNAMICS.]

Landau and Lifshitz in a symmetry-based theory have derived criteria necessary for a second-order transition:

1.   The symmetry of the product phase must be a subgroup of that of the parent phase.
2.   The transition must be generated by special point wave vectors.
3.   The sum of any three wave vectors $\mathbf{k}$ of the star generating the transition must not be equal to a reciprocal lattice vector $\mathbf{g}$ of the parent structure, that is

$$\mathbf{k}_1 + \mathbf{k}_2 + \mathbf{k}_3 \neq \mathbf{g}$$

This third criterion is a simplified version of the original Landau criterion.

All group–subgroup transitions fulfill the first condition. This includes, for example, order–disorder, ferroelectric, ferroelastic, or magnetic transitions. CuAu and Cu_3Au for instance (see Fig. 11) are both subgroups of the fcc structure and are generated by the wave vectors of the $\langle 1\ 0\ 0 \rangle$ star. The second criterion predicts this transition to be of first order since $\mathbf{k}_1 + \mathbf{k}_2 + \mathbf{k}_3 = \langle 1\ 1\ 1 \rangle = \mathbf{g}$. On the other hand for structures generated by the $\langle \frac{1}{2}\ \frac{1}{2}\ \frac{1}{2} \rangle$ star, such as CuPt, a second-order transition is allowed by symmetry.

D. Long Period Superlattices

In addition to special point structures many other ordered structures can form and have been observed. These can be enumerated by considering all the possibilities of distributing different atomic species on a given lattice, and their stability can be evaluated theoretically. A large class of such structures arises from a long period modulation of the stacking order in the basic structure, that is, periodic stacking faults. In the basic structure of SiC many periodic arrangements of stacking faults, termed polytypes, have been found. If the change in stacking alters only the modulated, or ordered, structure the stacking faults are called antiphase boundaries and the resulting structure is a long period superlattice. An example of a one-dimensional long period superlattice is that found in Cu–Au alloys. As illustrated in Fig. 12, an antiphase boundary in every fifth unit cell changes the tetragonal

FIG. 12.   CuAu II long period superlattice due to periodic stacking shifts.

structure of CuAu I to the orthorhombic structure of CuAu II. Similar one-dimensional long period superlattices are found in Cu–Al alloys. Au–Mn alloys exhibit both one- and two-dimensional examples of these structures. In semiconductors long period superlattices have been produced artificially to form multiple quantum well structures with unique properties.

III. Domain Structures

A. Group–Subgroup Transitions

In an order–disorder reaction a compositionally disordered crystal transforms to an ordered one on crossing the critical ordering temperature during cooling. The low-temperature product phase has lower symmetry than the high-temperature parent phase. A single crystal of the parent phase can therefore transform to several orientations of the product phase and the resulting regions of different orientation are called domains. The simplest case, that of antiphase domains, is illustrated below using the β-brass order–disorder transition based on the body-centered cubic (bcc) lattice. In a solid solution of Cu and Zn in a ratio of 1:1 each lattice site is occupied at random by Cu and Zn atoms with a probability of $n(\mathbf{r}) = 0.5$ [see Eq. (1)]. Upon crossing the critical ordering temperature the ordered structure shown in Fig. 11 forms. Due to the ordering the corner sites are no longer equivalent to the body-centered sites and thus two different domains are possible depending on whether the Cu atoms occupy one or the other sites. Both possibilities are equally likely and do in fact occur. Where two such domains meet they form an antiphase domain boundary such as the ones shown schematically in Fig. 13a. These can be made visible in electron micrographs and an example is shown in Fig. 13b. The boundaries mark the impingement of domains nucleated out of phase in different regions (thermal boundaries).

The order–disorder reaction in Cu–Zn is a simple example of the more general class of group–subgroup transitions. It illustrates an im-a

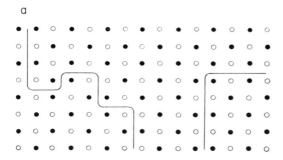

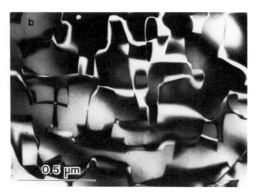

FIG. 13. Antiphase domain boundaries: schematic representation (a) and TEM micrograph of antiphase boundaries in Ni–Al alloy (b). (Courtesy K. H. Westmacott.)

portant point: domains are due to broken symmetries. Generally, every symmetry element lost in a transition will give rise to a possible domain boundary. For CuZn, only one symmetry element is lost, the $\frac{1}{2}\langle 1\,1\,1\rangle$ body-centering translation. Hence only one type of domain boundary can be formed (either thermally or by glide). In general the loss of translation elements gives rise to antiphase domains, the loss of inversion symmetry causes inversion (enantiomorphic) domains, the loss of a mirror symmetry results in twin domains, and the loss of a rotation axis leads to orientation domains.

It was shown earlier that the symmetry lowering in a group–subgroup transition may be described as a commensurate modulation of a physical quantity associated with the lattice, such as concentration, magnetization, or displacement. Extra spots (superlattice reflections) will appear in a diffraction pattern at the positions of the wave vectors $\mathbf{k}$ generating the ordered structure, (such as $\mathbf{k} = \langle 1\,1\,1\rangle$ in CuZn). Domain boundaries can be made visible in electron microscopy when one of these wave vectors fulfills the Bragg condition (see Fig. 13b). Thus a $(1\,1\,1)$ superlattice reflection could be

used to image antiphase domains in CuZn and a $(1\,0\,0)$ reflection in CuAu.

Since each symmetry element lost in a transition gives rise to a domain, the number of possible boundaries is equal to the index n of the space group symmetry G_1 of the product crystal in the group G_0 of the parent crystal, that is, the ratio of the orders of the two groups:

$$n = \text{order of } G_0/\text{order of } G_1 \qquad (5)$$

When changes in the unit cell size are taken into account, this number is multiplied by the ratio of the primitive unit cell volumes: V_1 and V_0:

$$n = (V_1/V_0)(\text{order of } G_0/\text{order of } G_1) \qquad (6)$$

For example in the Cu_3Au ordering transformation (see Fig. 11) the parent phase is $Fm\bar{3}m$ and the product phase $Pm\bar{3}m$, $n = 192/48 = 4$. These are the four possible antiphase domains, separated by the three possible domain boundaries, each of which is characterized by a lost translation of the type $\langle \frac{1}{2}\,\frac{1}{2}\,0\rangle$. Note that for n domains there are only $n - 1$ boundaries.

Each boundary is characterized not only by one but by a whole set of symmetry elements, the coset of the boundary. For example if a symmetry element $(\mathbf{R}|\tau)$ generates a boundary between domains, then $(\mathbf{R}|\tau)g_1$ describes the same boundary if g_1 is an element of the symmetry group G_1 of the product phase. This is true for all elements of G_1, and the set $(\mathbf{R}|\tau)G_1$ contains all the symmetry elements generating this boundary. Each boundary has its own unique coset. Two cosets have either no elements in common or are identical. If two symmetry elements belong to the same coset they generate the same boundary. The parent symmetry group G_0 can be decomposed into a sum of cosets characteristic for the domains formed in the transition to G_1:

$$G_0 = G_1 + \Sigma_i(\mathbf{R}_i/\tau_i)G_1$$

where $(\mathbf{R}_i/\tau_i)$ are the lost symmetry elements. As an example, in Cu_3Au ordering, the coset decomposition becomes

$$Fm\bar{3}m = Pm\bar{3}m + \{(1|\tfrac{1}{2}\,\tfrac{1}{2}\,0)$$
$$+ (1|\tfrac{1}{2}\,0\,\tfrac{1}{2}) + (1|0\,\tfrac{1}{2}\,\tfrac{1}{2})\}Pm\bar{3}m$$

B. The Effect of Strain

The change of symmetry in a transition is usually associated with a distortion of the unit cell. In most group–subgroup transitions this transformation strain is small ($\leq 1\%$) and it is usually neglected in determining the possible domain boundaries. However, larger transformation strains have a significant effect on the possible domain configurations and the orientation of the domains themselves. Even for small strains, domain boundaries are not arbitrary in their orientation but tend to assume positions of minimum strain. The location $\mathbf{u}$ of such boundaries is given by

$$\mathbf{u}(\mathbf{s} - \mathbf{s}')\mathbf{u} = 0 \tag{7}$$

where $\mathbf{s}$ and $\mathbf{s}'$ are the small strain transformation tensors for the two domains meeting at the boundary. Since $\det |\mathbf{s} - \mathbf{s}'| = 0$, the solutions $\mathbf{u}$ describe invariant planes and Eq. (7) is the small strain equivalent of the invariant plane condition in martensite theory, as described later. Consequently domains usually meet along planar interfaces, a condition that becomes more stringent for larger strains.

In the case of the cubic to tetragonal transformation the strain tensors have the form

$$\mathbf{s} = \begin{pmatrix} e & & \\ & e & \\ & & -2e \end{pmatrix} \quad \text{and} \quad \mathbf{s}' = \begin{pmatrix} e & & \\ & -2e & \\ & & e \end{pmatrix}$$

Equation (7) has solutions $\mathbf{u} = \langle h\,k\,k \rangle$, hence the invariant plane interfaces are of the $\{0\,1\,1\}$ type. Figure 14 illustrates that two domains meeting at a boundary undergo a slight rotation toward each other. The rotation angle $\alpha = (3/2)e$ is on the order of one degree if the strain e is on the order of 1%. Since the domains will rotate in opposite directions when they meet along two different $\{0\,1\,1\}$ planes (see Fig. 14a and b) they strictly become four different domains. The total transformation that leads to only 3 orientation domains if the small strains are neglected will form 12 orientation domains if the rotations re-

sulting from the strain are taken into account. The number of domains (also called variants) is thus no longer given by the index n of G_1 in G_0, since due to the small rotation the 4-fold axes of the parent and product phase are no longer exactly parallel.

C. VARIANTS

Even though it does not change the symmetry of the low-temperature phase itself, the rotation due to strain causes the loss of symmetry elements common to parent and product phases. The set of remaining common symmetry elements is given by the intersection H of the parent group G_0 and product group G_1 in its slightly rotated orientation, $H = G_1 \cap G_0$. The number of domains is then the index of H in G_0:

$$n = \text{order of } G_0/\text{order of } H \tag{8}$$

If the product phase coexists with the parent phase as in any first-order transformation, isolated domains or inclusions form in the matrix crystal. Different crystallographically equivalent inclusions are called variants. Their rotation from a symmetrical orientation determines the orientation relationship and their number is given by Eq. (8) with G_0 and G_1 being the point groups rather than space groups since the transformation strain usually destroys all common translations. The generation of variants is again due to those symmetry elements of the parent phase that are not shared by the product phase, that is, the broken symmetries. Figure 15 illustrates this process in a hypothetical cutting and welding operation similar to the ones used to find the strain energy of an inclusion (see Fig. 32) or the dislocation network in a grain boundary (see Fig. 20). A spherical volume of the matrix containing the inclusion is cut out, rotated, inverted, or reflected, and rewelded in the ma-

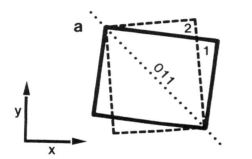
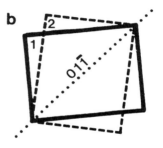

FIG. 14.   Two tetragonal domains 1 and 2 meeting on $(0\,1\,1)$ plane (a) and $(0\,1\,\bar{1})$ plane (b). A slight rotation of the domains toward their common boundary leads to relative changes in orientation.

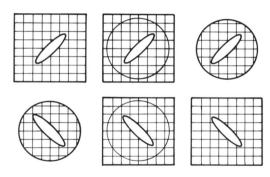

FIG. 15. Hypothetical sequence of operations illustrating the generation of precipitate variants.

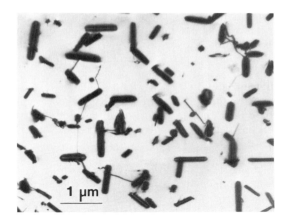

FIG. 16. TEM micrograph with $\langle 0\,0\,1 \rangle$ beam direction showing needle-shaped precipitates along $\langle 6\,5\,1 \rangle$ directions in Cu–Cr alloy.

trix. The matrix crystal is undisturbed by this operation because by definition it is invariant under a symmetry operation. However, the inclusion is in a new orientation, that is, it has become a different variant. Symmetry elements that are common to matrix and inclusion will not generate new variants. These elements form the intersection group $H = G_1 \cap G_0$ and an example of one such element is the 2-fold rotation about the plane normal in Fig. 15.

The electron micrograph in Fig. 16 shows a distribution of many variants of Cr precipitate needles along $\langle 6\,5\,1 \rangle$ directions in a Cu matrix. The intersection group of the matrix $G_0 = m\bar{3}m$ and the precipitate $G_1 = m\bar{3}m$ is only $H = \bar{1}$ of order 2 because of the low-symmetry orientation relationship adopted by the precipitate lattice. The number of variants is therefore $n = 48/2$. The distribution of the 24 variants can be visualized as poles on a stereogram showing a general direction (see Fig. 6) and can be enumerated by permutation of the indices $\langle 6\,5\,1 \rangle$. Notice that both the product and parent lattices are cubic but are misoriented so that they have only an inversion center in common. The orientation relationship is one of low symmetry since it is dictated by the invariant line or invariant plane criterion discussed in Sections V and VI. The symmetry of the orientation relationship is given by the intersection group H.

D. EQUILIBRIUM INCLUSION SHAPE

It is easy to see that H is also the symmetry group of the equilibrium inclusion shape: the structure and energy of a given interface plane are identical for two planes related by a symmetry operation of H and hence all equivalent faces

are expected to be found bounding a particle. The needles shown in Fig. 16 have a low-symmetry equilibrium shape, that of a pinacoid ($H = \bar{1}$). Particles with higher intersection symmetries H are shown in cross section in Fig. 17a–c. In all three cases the germanium particle and the aluminum matrix share a mirror parallel to the image plane and a 2-fold axis perpendicular to it. When there are no other common elements (see Fig. 17a) the intersection group is monoclinic, $H = 2/m$. Higher symmetry intersection groups are seen in Fig. 17b ($H = mmm$, orthorhombic) and

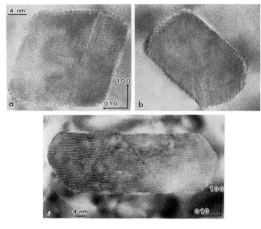

FIG. 17. TEM micrographs of Ge needle precipitates in Al with different common symmetries, seen in cross-section; monoclinic (a), orthorhombic (b), and tetragonal (c). Note corresponding shapes.

Fig. 17c ($H = 4/mmm$, tetragonal). Note that the overall particle shapes conform to the expected symmetries, that is, they are close to their equilibrium shape.

Usually the more symmetrical groups H (higher order) are preferred over those of low symmetry. When H is such that any misorientation would lower its symmetry it is said to be at a symmetry-dictated extremum.

IV. Grain Boundaries

Grain boundaries are interfaces separating two crystals identical in structure but different in orientation. The misorientation between two grains is described by the axis l and angle θ of misorientation (axis/angle pair) or alternatively by a rotation tensor R. If in addition to the misorientation there is also a shift τ between the grains the interface operation is $(R|\tau)$. Most geometrical properties can be derived from this operation. However, a boundary is completely described only if the boundary plane is specified as well. Thus nine parameters are necessary for a complete description of a grain boundary, three for the axis and angle of rotation, three for the direction and magnitude of the translation, and three for the position of the boundary plane.

Existing grain boundary models are based either on geometric or energetic criteria. The latter usually require large computer programs and the results depend strongly on the interatomic potential chosen. The former generally have simpler, analytical solutions but are not directly related to the energy. However, experience has shown boundaries with optimum geometry are usually low-energy boundaries. Some geometrical models of grain boundaries are described below.

A. LOW-ANGLE BOUNDARIES

If the misorientation between two grains is less than $\sim 10°$ their boundary is usually considered a low-angle boundary. The dislocation array shown in Fig. 18a is a simple example of a low-angle boundary and a high-resolution image of such a boundary in molybdenum is shown in Fig. 18b. The misorientation θ is related to the dislocation spacing d as

$$\tan \theta = b/d \qquad (9)$$

where b is the Burgers vector of the dislocation. At $10°$ misorientation the dislocations would thus be spaced $d \cong 6b$ apart, causing consider-

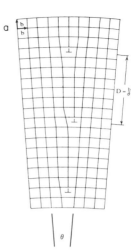

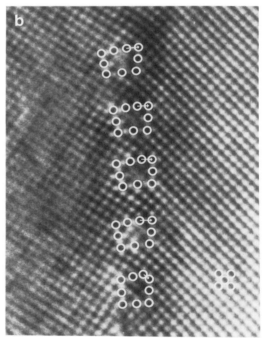

FIG. 18. Low-angle tilt boundaries: schematic of symmetrical boundary (a) and TEM micrograph of nonsymmetrical boundary in Mo (b). (Courtesy of J. M. Pénisson and R. Gronsky.)

able overlap between their strain fields. It would therefore be physically meaningless to describe larger misorientations as arrays of individual dislocations, and other constructions become important. These will be discussed under high-angle boundaries. Some further principles can be illustrated using low-angle boundaries. The example shown in Fig. 18 is a symmetrical tilt boundary; it is located on the plane of symmetry between the two crystals and it contains the axis of misorientation. Because of their simplicity, symmetrical tilt boundaries figure prominently in the modeling of grain boundaries.

Another high-symmetry configuration is the pure twist boundary shown in Fig. 19. Such boundaries are normal to the axis of misorientation and require not a single but at least two sets (a network) of dislocations. As for a symmetrical tilt boundary, the angle of misorientation is related to the spacing of each array in the network by Eq. (9).

For the same axis/angle pair, the boundary can take arbitrary orientations which then of course lack the special symmetry of pure twist or tilt boundaries. One way to visualize all the possible boundary orientations is to treat one grain as an inclusion in the other. The grain boundary is then a closed surface and the dislocation network a set of closed loops on that surface. To obtain their spacing and orientation, consider the misorientation of the enclosed grain to be the result of a phase transformation whose sole effect is a small change in orientation of the lattice without changing the volume, shape, or

crystal structure of the "new phase." In order to accommodate this change in orientation it is necessary to deform the grain by slip, resulting in a dislocation network in the interface. This is best seen in the schematic sequence in Fig. 20. From a single crystal cut a region (a), remove and rotate it (b), deform it plastically such that it will fit into its previous space (c) and (d), and insert and reweld it in the original crystal (e). The slip steps formed in (c) and (d) now become dislocation loops as shown in the perspective drawing (f). By extension it is easy to see from this description how a small-angle grain boundary of any axis/angle pair and on any plane could be constructed. In particular, notice that the same misorientation could be achieved with a different set of dislocations, depending on the slip systems operating in steps (c) and (d). Also note that in the particular example shown the planes of the dislocation loops intersect in a line parallel to the rotation axis. The rotation axis is an invariant line of the "transformation" and must be unaffected by the dislocations. Not every rotation axis conveniently lies at the intersection of two crystallographic slip planes. More than two sets of dislocation loops are then necessary to accommodate a rotation around an irrational axis but again the total set must leave the rotation axis unchanged.

Since the rotations involved are small, the rotation matrix can be written as

$$\mathbf{R} = \mathbf{I} + \boldsymbol{\omega} \qquad (10)$$

where $\mathbf{I}$ is the identity matrix and $\boldsymbol{\omega}$ is an antisymmetrical tensor, that is, $\omega_{ij} = -\omega_{ji}$. This "rotational strain" tensor $\boldsymbol{\omega}$ must be countered by an equal and opposite plastic strain $-\boldsymbol{\omega}$, made up of dislocation arrays. Each array of dislocation loops of given orientation, spacing, and Burgers vector has a characteristic strain tensor. For example, an array of shear loops on

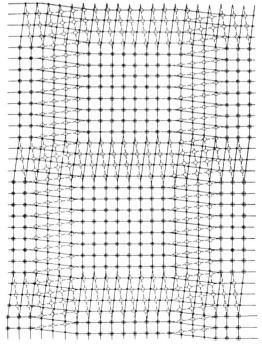

FIG. 19. Low-angle twist boundary seen face-on. [From W. T. Read, Jr., "Dislocations in Crystals." Copyright 1953, with permission of McGraw-Hill.]

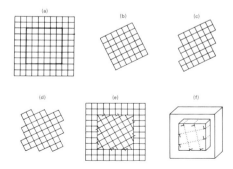

FIG. 20. Sequence of hypothetical operations illustrating the formation of an arbitrary grain boundary.

(0 0 1) planes with Burgers vector **b** in the [1 0 0] direction is written as the simple shear strain:

$$\mathbf{s} = \begin{pmatrix} 0 & 0 & b/d \\ 0 & 0 & 0 \\ 0 & 0 & 0 \end{pmatrix}$$

By adding appropriate arrays to form an antisymmetrical tensor $\boldsymbol{\omega}$ any small angle boundary can be described. This addition applies only for small strains since the superposition principle (strains are additive) holds only when products of strains are negligible.

B. O-Lattice

Another method to determine the Burgers vector content **B** along a vector **p** in a boundary is by Frank's formula:

$$\theta(\mathbf{l} \times \mathbf{p}) = \mathbf{B}$$

where $(\mathbf{l}, \theta)$ is the axis/angle pair. Frank's formula is valid only for small rotation angles θ. The concept of measuring the misfit **B** in an arbitrary direction **p** in the boundary is very useful and is central to the more general theory of surface dislocations. By decomposing the misfit **B** into a sum of lattice translations it becomes possible to devise a corresponding set of dislocation lines that must be traversed by the vector **p**. If **p** is selected so that the misfit along it is equal to a Burgers vector **b** then **p** becomes an O-lattice vector $\mathbf{x}^0$, a basic concept in O-lattice theory. For low-angle boundaries

$$\boldsymbol{\omega}\mathbf{p} = \Sigma_i \mathbf{b}_i \tag{11a}$$

or

$$\boldsymbol{\omega}\mathbf{x}^0 = \mathbf{b} \tag{11b}$$

where $\boldsymbol{\omega}$ is the antisymmetrical tensor describing the "rotational strain." For misorientations greater than ~10°, the small-angle approximation is no longer valid and Eqs. (11a,b) become

$$(\mathbf{I} - \mathbf{R}^{-1})\mathbf{p} = \Sigma_i \mathbf{b}_i \tag{12a}$$

$$(\mathbf{I} - \mathbf{R}^{-1})\mathbf{x}^0 = \mathbf{b} \tag{12b}$$

A dislocation is necessary in the boundary every time the rotational mismatch $(\mathbf{I} - \mathbf{R})$ measured along a vector $\mathbf{x}^0$ equals a Burgers vector **b**. Perhaps this is most apparent for coincidence site lattices (CSL) for which both **b** and $\mathbf{x}^0$ are lattice translation vectors.

C. Coincidence Site Lattice

When two identical interpenetrating lattices are rotated from initial coincidence around a lattice point, there are certain discrete rotation angles for which lattice points other than the origin coincide. An example is shown in Fig. 21a where a black lattice and a white lattice are outlined by a dashed and a solid square. The coincident points (solid dots) form a lattice themselves, termed the CSL, shown in Fig. 21b. The CSL is characterized by the inverse density of coincidence sites, Σ. For the example illustrated in Fig. 21, $\Sigma = 5$. The rotation angle θ is 36.9° ($\tan \theta/2 = 1/3$) due to the coincidence of the [3 1 0] vector in one lattice with the [3 $\bar{1}$ 0] vector in the other lattice. Any rational $\langle h\,k\,l \rangle$ lattice vector can be used to generate a CSL characterized by $\Sigma = h^2 + k^2 + l^2$. If $h^2 + k^2 + l^2$ is even, Σ is its largest odd divisor. For example a $\langle 3\,1\,0 \rangle$ vector $\Sigma = 10$ and a $\langle 2\,1\,0 \rangle$ vector $\Sigma = 5$ generate the same $\Sigma 5$ CSL shown in Fig. 21. In

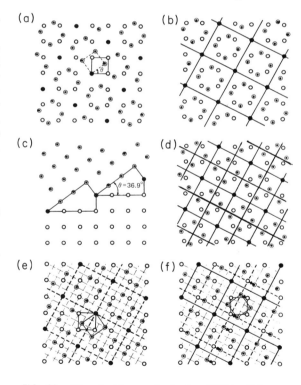

FIG. 21. $\Sigma 5$ coincidence site lattice (CSL); dichromatic pattern of two interpenetrating square lattices of black and white atoms with solid dots showing coincidence points (a), CSL outlined (b), structural units along possible boundary plane (c), O-lattice outlined (d), DSC lattice outlined (e), and effect of DSC shift on CSL (f). [From H. F. Fischmeister, *J. Phys. Colloque* C4-3 (1983).]

Fig. 21c a boundary plane has been chosen and the black (white) lattice discarded from below (above) the plane of the boundary. Structural units typical of this boundary are outlined. The density Γ of coincidence sites in a boundary plane depends on its position. Special grain boundaries have high values of Γ and low values of Σ.

Any lattice vectors of the CSL are solutions $\mathbf{x}^0$ of the O-lattice equation (12b) since at each coincidence site the rotational displacements between the two lattices amount to a lattice translation $\mathbf{b}$. However, not all O-lattice vectors $\mathbf{x}^0$ are CSL vectors. The complete O-lattice for the $\Sigma 5$ boundary is shown in Fig. 21d. When the misorientation between the lattices is not exactly equal to one necessary for a CSL, all the coincidence sites (except the center of rotation) are lost. However the O-lattice is still maintained with lattice vectors $\mathbf{x}^0$ which are now irrational although close to the rational CSL vectors. Thus the O-lattice changes continuously between the discrete CSLs.

In order to find the geometrically necessary network of dislocations in an interface from the solution of Eq. (12b) one constructs cell walls midway between any two O-lattice points. These represent the positions of worst match since the O-lattice points are the positions of

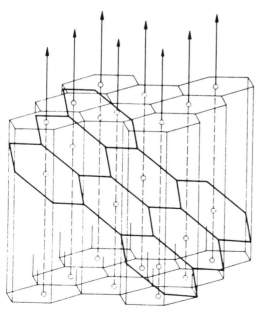

FIG. 22.   Hexagonal O-lattice due to misorientation of two grains around a common 3-fold axis. Heavy outlines show predicted dislocation network where the boundary plane intersects the O-lattice cell walls.

best match. An example is shown in Fig. 22 in which the O-lattice is a hexagonal set of parallel lines and the cell walls are shown as hexagonal prisms. The vectors $\mathbf{x}^0$ corresponding to three coplanar Burgers vectors $\mathbf{b}$ lie in the basal plane. The predicted dislocation network lies at the intersection of the chosen boundary plane with the cell walls (heavy outline). Notice that the boundary could also be curved or even a closed surface as considered earlier. The same dislocation network as shown for example in Fig. 20 would be predicted. However, in this view, the physical origin of the dislocations is not considered since they arise only as a geometrical necessity.

Even though the equations are valid for any angle of misorientation this construction is physically meaningful only for low-angle boundaries since, as shown earlier, for misorientations $>10°$, their cores are too close to each other.

D. HIGH-ANGLE BOUNDARIES

High-angle boundaries are best treated as small deviations from the nearest CSL. They are then similar to low-angle boundaries that are a small deviation from the $\Sigma 1$ CSL, that is, a single crystal. Each CSL, in particular $\Sigma 1$, may be considered a low-energy configuration. Usually the lower Σ is, the lower the energy. Any deviation from a CSL is accommodated by dislocations, lines of high local distortion in favor of relaxed or undistorted regions of the CSL in between. In the case of low-angle boundaries the dislocations are primary dislocations and the boundary between them perfect crystal ($\Sigma 1$). This is shown in Fig. 18. In the case of high-angle boundaries the dislocations are secondary dislocations and the boundary between them perfect CSL. These secondary dislocations have small Burgers vectors given by the smallest difference vectors between lattices 1 and 2 in the exact CSL orientation. A dislocation with such a Burgers vector will translate the complete CSL and is part of the so-called DSC (displacement shift complete) lattice. The example of the $\Sigma 5$ DSC lattice is shown in Fig. 21e and the effect of a DSC lattice translation is illustrated in Fig. 21f. Any misorientation from the perfect $\Sigma 5$ orientation can be accommodated by these dislocations with areas of perfect $\Sigma 5$ boundary in between.

When a boundary is between two adjacent low-energy CSL orientations it could approximate one or the other with an appropriate set of dislocations. In fact, experience has shown that it does both. By mixing well-matched structural units characteristic for each of the neighboring CSLs in the right proportion the boundary energy is minimized. Any degree of misorientation can be accommodated in this manner and a change in misorientation is similar to a change in the structural "composition" of the boundary. Rearrangement of structural units in a boundary, by separation or mixing, may in a sense be considered a phase transformation. This must be accompanied by a local change in boundary orientation. Structural units are a refinement of the purely geometrical CSL concept since local atomic relaxations and relative translations between the two lattices are taken into account. This can be done by computer simulation assuming an interatomic potential, or more simply by geometrical criteria based on rigid sphere packing. The latter model is successful, at least in fcc metals where the hard sphere model is a good approximation.

E. Grain Boundary Crystallography

In the wider context of interphase boundaries the concepts of CSL, DSC, and O-lattice can be derived alternatively as follows. A general interphase boundary described by a transformation $\mathbf{A}$ followed by a translation τ is called a heterophase boundary, characterized by the operation $(\mathbf{A}|\tau)$. If $\mathbf{A}$ leaves the crystal structure invariant $(\mathbf{A} = \mathbf{R})$ it describes a homophase boundary. This can be a domain boundary if $(\mathbf{R}|\tau)$ is a symmetry operation of the parent phase, or a grain boundary if $(\mathbf{R}|\tau)$ is a general rotation/translation. The translation may be irrational, for example, if due to relaxation at a grain boundary, or a rational lattice translation $\mathbf{t}$, for example, at an antiphase domain boundary.

Due to crystal symmetry (crystal space group G) the description of a boundary by a single operator $(\mathbf{R}|\tau)$ is ambiguous because any symmetrically equivalent orientation of either crystal will give a different but equivalent boundary operation. The boundary is completely described by the set of all such operations $(\mathbf{R}|\tau)G$, called the coset of the boundary. Of this set, often the operation with the smallest rotation angle is used to characterize the boundary.

If $(\mathbf{R}|\tau)$ is a symmetry operation of the crystal, that is, an element of the space group G, it leaves the entire lattice invariant. Each lattice point is brought to an equivalent position. For a general boundary, $(\mathbf{R}|\tau)$ is not a symmetry element of the crystal. Thus formation of a boundary is a symmetry-breaking transition and the boundary operation $(\mathbf{R}|\tau)$ brings lattice points to nonequivalent positions. However, there may still be geometrical locations $\mathbf{r}$ in the lattice which remain invariant:

$$(\mathbf{R}|\tau)\mathbf{r} = \mathbf{r} \qquad (13)$$

Such locations $\mathbf{r}$ exist if $(\mathbf{R}|\tau)$ is a reducible operator. For these positions, the effect of the rotation $(\mathbf{R}\mathbf{r})$ equals the effect of the translation, $(\mathbf{r} - \tau)$. This is shown schematically in Fig. 23a.

If τ (but not $\mathbf{r}$) is a lattice translation $\mathbf{t}$, one of the elements of the translation group T_1 of lattice 1 (for example a Burgers vector $\mathbf{b} = -\mathbf{t}$), then $\mathbf{R}\mathbf{r}$ becomes an O-lattice vector $\mathbf{x}^0$ (see Fig. 23b) and Eq. (13) becomes

$$(\mathbf{R}^{-1}|\mathbf{b})\mathbf{x}^0 = \mathbf{x}^0$$

This is identical to Eq. (12b) and the solutions $\mathbf{x}^0$ to these equations describe points, lines, or planes that remain invariant in the transition $(\mathbf{R}|\mathbf{t})$. Finally, if both $\mathbf{r}$ and τ in Eq. (13) are members $\mathbf{t}_1$ and $\mathbf{t}$ of the translation group T, then $\mathbf{r}$ becomes a CSL vector $\mathbf{t}_1$:

$$(\mathbf{R}|\mathbf{t})\mathbf{t}_1 = \mathbf{t}_1$$

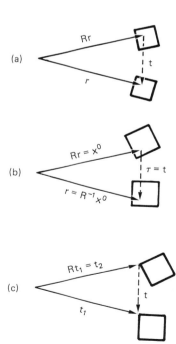

FIG. 23. Schematic illustration of rotation/translation operation: general operation (a), generation of O-lattice (b), and generation of CSL (c).

as shown schematically in Fig. 23c. If this is rewritten as

$$\mathbf{R}\mathbf{t}_1 = \mathbf{t}_1 - \mathbf{t}$$

the left-hand side is a translation in crystal 2 while the right-hand side is a translation in crystal 1. The CSL is thus the group of those vectors $\mathbf{t}_1$ which are translations in both crystals, that is, the intersection of the translation groups

$$\text{CSL} = T_1 \cap \mathbf{R}T_1 \qquad (14)$$

The DSC lattice is simply the corresponding union of the translation groups

$$\text{DSC} = T_1 \cup \mathbf{R}T_1 \qquad (15)$$

It was shown that the operation $(\mathbf{R}|\tau)$, if it is reducible, leaves a set of locations $\mathbf{r}$ invariant, that is $(\mathbf{R}|\tau)$ is a symmetry operation of the bicrystal. The entire set of bicrystal symmetries is the union of the symmetry elements common to both crystals $G_1 \cap G_2$, with those operations $(\mathbf{R}|\tau)G_1$ that transform crystal 1 into crystal 2 while simultaneously transforming crystal 2 into crystal 1: $G_1(\mathbf{R}|\tau)^{-1}$. Thus the bicrystal symmetry is

$$(\mathbf{G}_1 \cap \mathbf{G}_2) \cup [(\mathbf{R}|\tau)G_1 \cap G_1(\mathbf{R}|\tau)^{-1}]$$

Operations that interchange the two crystals are also called colored symmetry operations, referring to the colors, that is, black and white, assigned to the two identical but misoriented crystals shown for example in Fig. 21. The corresponding pattern is called the dichromatic pattern. Its symmetry is described by the Shubnikov groups, and it is interesting to note that it can have symmetries not present in either lattice. If, for example, $(\mathbf{R}|\tau)$ is the twinning operation in an fcc crystal where $\mathbf{R}$ is a $\{1\ 1\ 1\}$ mirror m and $\tau = 0$, the bicrystal symmetry of the resulting $\Sigma 3$ CSL is

$$(\bar{3}m) \cup (m/o) = 6/mmm$$

V. Martensite

A. CHARACTERISTICS

Martensite transformations are characterized by a surface relief that indicates that a shape change is associated with the transformation. The surface relief can be measured from the displacement of scratches placed on a flat polished surface before transformation (see Fig. 24). The interface plane between the martensite inclusion and the matrix (habit plane) remains macroscop-

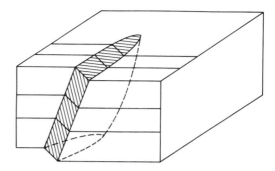

FIG. 24. Schematic of surface relief where a martensite plate intersects the surface.

ically undistorted (invariant plane) and the total deformation is an invariant plane strain (IPS). Due to the constraint from the solid matrix, martensite inclusions take the form of plates, laths, or needles, with an internal structure resulting from slip or twinning. The transformation is diffusionless and fast, proceeding at rates near the speed of sound. This implies that large numbers of atoms transfer rapidly and in an orderly fashion from the high-temperature structure (austenite) to the low-temperature structure (martensite). Nearest neighbors remain nearest neighbors.

The best-known and technologically most important martensite transformation after which the whole class of transformations is named is that in steel. It occurs on rapid cooling below the transformation temperature M_s which depends on carbon concentration. The volume fraction of martensite formed is determined by an equilibrium between the temperature-dependent chemical driving force and the strain energy of accommodating the shape change. In the case of steel the parent austenite is fcc and the product martensite bct. A characteristic orientation relationship between austenite and martensite is usually observed which is generally nonrational and of low symmetry. This leads to a large number of crystallographically equivalent orientation variants and complex morphologies. The semicoherent habit plane, usually also irrational, must be glissile, a condition that imposes restrictions on the interface dislocations.

Not all of these characteristics are typical for martensite transformations alone. Orientation relationships, habit planes, surface relief, and geometrically glissile interfaces are often also observed in diffusion-controlled precipitation reactions. For example the Bainite reaction in steel bears all of these marks but is not martensitic because interstitial carbon diffuses during the reaction, sweeping ahead of the interface and eventually stopping the reaction front. Other transformations that exhibit a surface relief while allowing some diffusion are sometimes classified as bainitic. Diffusionless rapid transformations that do not show surface relief are termed massive transformations.

B. IPS Geometry

The phenomenological theory of martensite transformations is based on the observation of an IPS deformation, and well-defined orientation relationship. Since the habit plane is undistorted and unrotated it can be modeled geometrically rather than on some energetic criterion but it can be shown that an IPS is favored also by strain energy considerations. This is similar to the structure of grain boundaries where optimized geometry usually coincides with minimum energy configurations.

An exceptionally simple example of a martensitic transformation is that from fcc to hcp cobalt. As shown in Fig. 25a the only strain is due to a change in stacking of the close-packed planes. The close packed plane is undistorted and naturally becomes the habit plane. In this case the transformation strain itself is an IPS of the form

$$\mathbf{B} = \begin{pmatrix} 1 & 0 & s \\ 0 & 1 & 0 \\ 0 & 0 & 1 \end{pmatrix}$$

where s is the shear due to the change in stacking sequence. For the two-dimensional case illustrated in Fig. 25a this reduces to

$$\mathbf{B} = \begin{pmatrix} 1 & s \\ 0 & 1 \end{pmatrix}$$

The habit plane and orientation relationship both have simple rational form. However, this is not generally the case, and usually there is no lattice plane that remains unstretched and unrotated in the transformation. If, for example, the close packed plane in Fig. 25a were stretched by a factor a in the transformation it could no longer be an undistorted habit plane, and a slightly rotated, irrational habit plane would

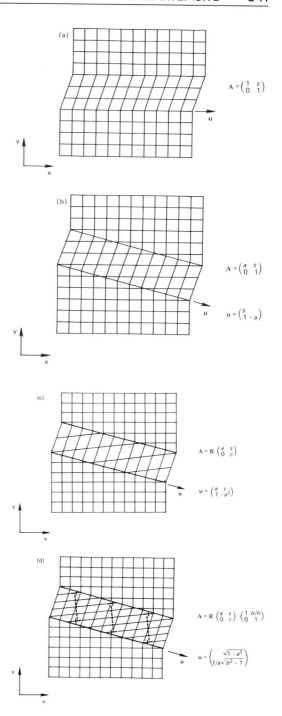

FIG. 25. Two-dimensional illustration of IPS geometry; simple shear on the habit plane (a), additional expansion a in the habit plane (b), additional expansion c normal to the habit plane (special case) (c), and additional lattice-invariant shear (general case) (d).

have to be adopted. The transformation tensor would be

$$\mathbf{B} = \begin{pmatrix} a & s \\ 0 & 1 \end{pmatrix}$$

and the invariant plane (or here in two dimensions the invariant line), given by the condition $\mathbf{Bu} = \mathbf{u}$, hence $\mathbf{u}$ $\binom{s}{1-a}$. This is shown in Fig. 25b where the perfect match along the direction $\mathbf{u}$ is apparent. At an angle

$$\tan \theta = (1 - a)/s \qquad (16)$$

to the x axis, this habit plane is irrational but the orientation relationship is still rational with the close-packed planes exactly parallel.

If we further allow the spacing of the close-packed planes to change by a factor c the transformation

$$\mathbf{B} = \begin{pmatrix} a & s \\ 0 & c \end{pmatrix}$$

no longer has an invariant interface. However, if there were a direction that was undistorted but not unrotated, it could simply be rotated back by a rigid body rotation $\mathbf{R}$ so that $\mathbf{RBu} = \mathbf{u}$. In the present example, this requires that

$$\det|\mathbf{B}^2 - \mathbf{I}| = 0 \qquad (17)$$

that is, the factors a, c, and s must follow the relations $(a^2 - 1)(c^2 - 1) = s^2$. If this relationship happens to be fulfilled by the transformation matrix, the coherent interface along the invariant line $\mathbf{u}$ shown in Fig. 25c is obtained at an angle

$$\tan \theta = (1 - a^2)/as \qquad (18)$$

However, usually this relationship will not hold and must be fulfilled by adding a lattice-invariant deformation. This can be achieved by either twinning or slip in a number of ways. For example, an array of edge dislocation loops with Burgers vector b and spacing h would add a lattice-invariant deformation

$$\mathbf{S} = \begin{pmatrix} 1 + b/h & 0 \\ 0 & 1 \end{pmatrix}$$

or an array of shear loops with the same Burgers vector and spacing h would be

$$\mathbf{S} = \begin{pmatrix} 1 & b/h \\ 0 & 1 \end{pmatrix}$$

Considering the requirement of a glissile interface the array of shear loops is more realistic.

The combined transformation

$$\mathbf{A} = \mathbf{RBS} \qquad (19)$$

now has an invariant interface if the dislocation spacing h in the lattice-invariant shear $\mathbf{S}$ is adjusted so that $\mathbf{RBSu} = \mathbf{u}$. This requires that

$$\det|(\mathbf{BS})^2 - \mathbf{I}| = 0 \qquad (20)$$

or explicitly $(a^2 - 1)(c^2 - 1) = (s + ab/h)^2$. This leads to the semicoherent interface shown schematically in Fig. 25d with the same angle of inclination as for the corresponding coherent interface in Fig. 25c, now given by

$$\tan \theta = \frac{1}{a} \sqrt{\frac{1 - a^2}{1 - c^2}} \qquad (21)$$

Figure 25d is the two-dimensional analog of the crystallography of a general martensite transformation. It shows the shape strain $\mathbf{A}$, the lattice invariant shear $\mathbf{S}$ in the form of a shear dislocation array, and the lattice rotation $\mathbf{R}$.

To illustrate schematically the same situation in three dimensions becomes more difficult. However, the same principles apply and the phenomenological theory of martensite transformations theory provides an algorithm to determine the habit plane and orientation relationship for a given transformation and slip system.

C. THE PHENOMENOLOGICAL THEORY

In order to find the transformation strain $\mathbf{B}$, a plausible correspondence between the austenite and martensite lattices must first be established. In some cases the choice is obvious, especially when the distortions are small and the relation between the two lattices is easily recognized. The fcc to bct transformation in steel is not so obvious because the transformation strains are large. The most widely used lattice correspondence for this case is the Bain correspondence shown in Fig. 26. The bct cell shown in heavy outline within the fcc lattice must be deformed along three orthogonal axes to the correct dimensions of the martensite ($c/a \simeq 1.08$). $\mathbf{B}$ is the matrix of distortions along these principal axes

$$\mathbf{B} = \begin{pmatrix} \eta_1 & & \\ & \eta_1 & \\ & & \eta_3 \end{pmatrix} \qquad (22)$$

In most martensite transformations, the volume change (given by $\det|\mathbf{B}|$) is small compared to the shape change, and $\mathbf{B}$ causes almost a pure shear deformation. The transformation $\mathbf{B}$ alone

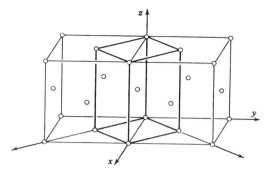

FIG. 26.   Bain correspondence between fcc and bct lattices. The bct lattice can be deformed into a bcc lattice by three orthogonal strains.

leaves no direction or plane invariant although it leaves all directions **u** unextended for which

$$(\mathbf{B}\mathbf{u})^2 = \mathbf{u}^2 \qquad (23)$$

In general,

$$\det|\mathbf{B}^2 - \mathbf{I}| \neq 0$$

and the solutions **u** of this equation lie on a cone of unextended lines. If, as shown above for the two-dimensional case, we add a lattice-invariant shear **S** of appropriate magnitude such that

$$\det|(\mathbf{B}\mathbf{S})^2 - \mathbf{I}| = 0$$

the unextended lines **u** given by the equation

$$(\mathbf{B}\mathbf{S}\,\mathbf{u})^2 = \mathbf{u}^2 \qquad (24)$$

lie in a plane. By adding the appropriate rotation **R**, this undistorted plane becomes an invariant plane, and

$$\mathbf{A} = \mathbf{R}\mathbf{B}\mathbf{S}$$

is an invariant plane strain. The normal to the invariant or habit plane can be found from the eigenvectors **e** of the symmetrical transformation matrix $\mathbf{A}^2 = (\mathbf{B}\mathbf{S})^2$, that is, the solutions of

$$\mathbf{A}^2\mathbf{e} = \lambda^2\mathbf{e} \qquad (25)$$

Since $\mathbf{A}^2$ is a symmetrical matrix it has orthogonal eigenvectors, and since $\det|\mathbf{A}^2 - \mathbf{I}| = 0$ one eigenvalue, say λ_2^2 is equal to one, that is, the corresponding eigenvector is an invariant line. If the other two eigenvalues are $\lambda_1^2 < 1$ and $\lambda_3^2 > 1$, a second unextended line can be found between $\mathbf{e}_1$ and $\mathbf{e}_3$ at an angle θ to the eigenvector $\mathbf{e}_1$, given in a form similar to Eq. (21):

$$\tan\theta = \sqrt{\frac{1 - \lambda_1^2}{\lambda_3^2 - 1}} \qquad (26)$$

An explicit normalized expression for the nor-mal **n** to the habit plane can be written

$$\mathbf{n} = \sqrt{\frac{\lambda_3^2 - 1}{\lambda_3^2 - \lambda_1^2}}\,\mathbf{e}_3 + \sqrt{\frac{1 - \lambda_1^2}{\lambda_3^2 - \lambda_1^2}}\,\mathbf{e}_1 \qquad (27)$$

D. Special Solutions

This equation still requires the solution of the eigenvector equation, Eq. (25), but for two special cases of high symmetry, analytical solutions are available for the fcc → bct transformation in terms of the principal strains η_1 and η_3. For fcc slip on (1 0 1) [1 0 $\bar{1}$], or bcc twinning, the solution is $\mathbf{n} = [h\ k\ l]$ with

$$h = \frac{1}{2\eta_1}(M - N)$$

$$k = \frac{1}{\eta_1}\sqrt{\frac{\eta_1^2 - 1}{1 - \eta_3^2}}$$

$$l = \frac{1}{2\eta_1}(M + N)$$

where

$$M = \sqrt{\frac{\eta_1^2 + \eta_3^2 - 2\eta_1^2\eta_3^2}{1 - \eta_3^2}}$$

and

$$N = \sqrt{\frac{2 - \eta_1^2 - \eta_3^2}{1 - \eta_3^2}}$$

and for simultaneous slip on the two conjugate bcc slip systems (0 1 1) [0 $\bar{1}$ 1] and (0 $\bar{1}$ 1) [0 1 1] the solution is $\mathbf{n} = [h\ k\ l]$ with

$$h = \sqrt{\frac{\eta_1^2(\eta_1^2 - 1)}{2C}}$$

$$k = \sqrt{\frac{\eta_1^2(\eta_1^2 - 1)}{2C}}$$

$$l = \sqrt{\frac{\eta_1^2 - \eta_3^2(2\eta_1 - 1)^2}{C}}$$

where $C = \eta_1^4 - \eta_3^2(2\eta_1 - 1)^2$.

E. Orientation Relationship

The orientation relationship **R** can be described by the rotation axis **w** and rotation angle α:

$$R_{ij} = \delta_{ij}\cos\alpha + w_iw_j(1 - \cos\alpha)$$
$$- \varepsilon_{ijk}w_k\sin\alpha$$

where ε_{ijk} is the permutation symbol. Since **A**, **B**, and **S** are known, **R** may be back-calculated as

$$R = AS^{-1}B^{-1}$$

A simpler, more direct way of finding the orientation relationship is based on the fact that the lattice-variant part of the transformation **RB** and its inverse **B**$^{-1}$**R**^{T} are invariant line strains as is apparent from

$$RB = AS^{-1} \qquad (28)$$

Both **A** and **S**$^{-1}$ are IPS tensors and the two invariant planes must intersect along an invariant line. The invariant line **i** must therefore lie where the slip plane **q** of **S**$^{-1}$ intersects the cone of unextended lines of **B**:

$$(Bi)^2 = (i)^2 \qquad \text{and} \qquad iq = 0 \qquad (29)$$

A conjugated relation may be written for the inverse transformation valid for plane normals. The unextended normal **m** is given by

$$(B^{-1}m)^2 = (m)^2 \qquad \text{and} \qquad mp = 0 \quad (30)$$

where **p** is the slip direction. Using Euler's theorem the rotation axis **w** is then simply the cross-product of the rotational strains

$$w = (B - I)i(B^{-1} - I)m/D$$

where

$$D = (B + I)i(B^{-1} - I)m$$

The rotation angle α is

$$\alpha = 2 \text{ arc tan } |w|$$

This procedure is illustrated stereographically in Fig. 27 where the cone of unextended lines of **B** (initial cone), solution of Eq. (23) is shown as a solid circle. The transformation **B** moves directions **u** from the initial to the final cone and plane normals **h** in the opposite sense. The traces of the slip plane **q** and of the plane perpendicular to the slip direction **p** are shown as great circles. The possible solutions for the invariant lines **i** [Eq. (29)] and invariant normals **m** [Eq. (30)] are marked as solid and open circles. There are four combinations of invariant lines and normals, each giving a different orientation relationship *R*. The four corresponding habit planes and magnitudes of the lattice-invariant shear *s* and the shape shear *g* are also different. The interface energy is related to *s* through the density of dislocations necessary to produce an invariant plane whereas the strain energy is related to *g*, the shape change that remains to be accommodated by the matrix. This is how the geometrical solutions are related to energetic criteria, and

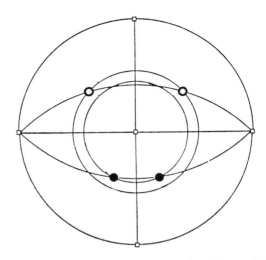

FIG. 27. Stereogram showing invariant lines (solid dots) at intersection of slip plane with initial cone of unextended lines and invariant normals (open circles) at intersection of plane normal to the slip direction with final cone of unextended lines.

the solutions with the smallest values of *g* and *s* are physically most realistic.

Refinements of the basic theory have included a uniform dilatation δ of the habit plane and multiple lattice invariant shear systems

$$A = RBS_1S_2S_3 \cdots$$

These have improved the match with experimental results in some cases but have done so at the sacrifice of the clarity and relative simplicity of the basic theory.

Electron microscope observations of the interface structure can identify the slip plane(s) of the lattice-invariant shear *S*. This is easily seen by considering the lattice-invariant part **RB** of the transformation. As shown above this must be an invariant line strain since **RB** = **AS**$^{-1}$, with the invariant line at the intersection of the habit plane (interface) with the slip plane. All the dislocations in the interface must lie along this direction so that the shear **S** does not interfere with the invariant line. The density of these dislocations can be measured provided they are not too closely spaced, a condition identical to low-angle grain boundaries.

F. TWINNED MARTENSITE

It has been assumed in the theory outlined above the lattice invariant shear *S* is accomplished by slip. Twinning is an alternative deformation mode. A schematic comparison of slip-

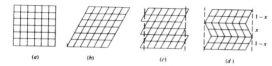

FIG. 28. Lattice-variant deformation (b) of crystal (a) with equal and opposite lattice-invariant deformation by slip (c) and twinning (d).

and twinning-produced lattice-invariant shear is shown in Fig. 28. If two twin-related regions of martensite form simultaneously, their relative volume fraction x determines the net amount of shear produced this way. Algebraically this is written as

$$\mathbf{A} = \mathbf{R}[x\mathbf{B}_1 + (1 - x)\mathbf{R}_2\mathbf{B}_2] \qquad (31)$$

where $\mathbf{B}_1$ and $\mathbf{B}_2$ are different variants of the Bain strain. One eigenvector of $\mathbf{A}^2$ is made an invariant line by adjusting the volume fraction x appropriately. The remainder of the procedure is as before. Slip and twinning are equivalent modes of deformation and the resulting habit planes and orientation relationships are identical.

The line of reasoning applied in the crystallographic theory of martensite transformations is most apparent for small strains. The decomposition of the macroscopic IPS $\mathbf{A} = \mathbf{RBS}$ where $\mathbf{R}$ is a rigid body rotation, $\mathbf{B}$ the Bain strain (approximately a pure shear), and $\mathbf{S}$ a simple shear becomes

$$\mathbf{A} = (\mathbf{I} + \boldsymbol{\omega})(\mathbf{I} + \boldsymbol{\varepsilon})(\mathbf{I} + \mathbf{s})$$

where $\boldsymbol{\omega}$ is the antisymmetrical matrix of a small rotational strain, $\boldsymbol{\varepsilon}$ the symmetrical matrix of a pure shear strain, and $\mathbf{s}$ the strain associated with a simple shear. Neglecting products of small strains we have

$$\mathbf{A} - \mathbf{I} = \boldsymbol{\omega} + \boldsymbol{\varepsilon} + \mathbf{s} \qquad (32)$$

the rotation, strain, and shear are additive and their order is unimportant (superposition principle). The shear strain $\mathbf{s}$ is chosen so as to reduce one principal strain to zero [equivalent to Eq. (20)], and a second invariant line is found between the other two principal strains by adjusting the orientation relationship $\boldsymbol{\omega}$ accordingly [equivalent to Eq. (28)]. If twinning is the deformation mode, Eq. (31) for small strains reduces to

$$\mathbf{A} - \mathbf{I} = \boldsymbol{\omega} + x\boldsymbol{\varepsilon}_1 + (1 - x)\boldsymbol{\varepsilon}_2 \qquad (33)$$

The procedure is as before: (1) choose the volume fraction x to make one invariant line, and (2) use $\boldsymbol{\omega}$ to produce a second invariant line.

An example of a small-strain martensite transformation with twinning as the lattice-invariant shear is the transformation in In–Tl alloys. The lattice correspondence is obvious since the transformation involves only small distortions. The transformation strain tensor is

$$\boldsymbol{\varepsilon}_1 = \mathbf{B}_1 - \mathbf{I} = \begin{pmatrix} -e & & \\ & -e & \\ & & 2e \end{pmatrix}$$

and for the twin

$$\boldsymbol{\varepsilon}_2 = \mathbf{B}_2 - \mathbf{I} = \begin{pmatrix} -e & & \\ & 2e & \\ & & -e \end{pmatrix}$$

and therefore

$$\mathbf{A} - \mathbf{I} = \boldsymbol{\omega} + x\boldsymbol{\varepsilon}_1 + (1 - x)\boldsymbol{\varepsilon}_2$$

$$= \boldsymbol{\omega} + \begin{pmatrix} -e & & \\ & (2 - 3x)e & \\ & & (3x - 1)e \end{pmatrix}$$

With $x = 2/3$, the transformation strain becomes

$$\mathbf{A} - \mathbf{I} = \boldsymbol{\omega} + \begin{pmatrix} -e & & \\ & 0 & \\ & & e \end{pmatrix}$$

It is apparent that another unextended line can be found midway between the remaining two strains at an angle $\tan \Theta = 1$ from the principal axes. This becomes an invariant line by addition of a small rotation

$$\boldsymbol{\omega} = \begin{pmatrix} 0 & 0 & -e \\ 0 & 0 & 0 \\ e & 0 & 0 \end{pmatrix}$$

which turns the pure shear into a simple shear and which determines the orientation relationship. This geometry is illustrated in Fig. 29a. Note that in this case the shape strain $\mathbf{A} - \mathbf{I}$ is a simple shear parallel to the habit plane. Another twinned martensite plate can be found with the same habit plane but opposite shear direction:

$$\mathbf{A}_1 - \mathbf{I} = \begin{pmatrix} -e & 0 & -e \\ 0 & 0 & 0 \\ e & 0 & e \end{pmatrix}$$

and

$$\mathbf{A}_2 - \mathbf{I} = \begin{pmatrix} e & 0 & e \\ 0 & 0 & 0 \\ -e & 0 & -e \end{pmatrix}$$

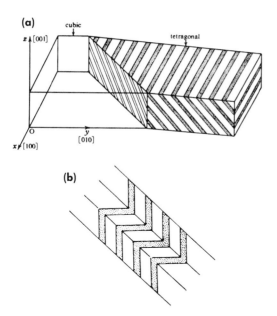

FIG. 30.   TEM micrograph of homogeneous distribution of θ'' precipitates in Al–Cu alloy.

FIG. 29.   Twinned martensite: habit plane of a single twinned plate in contact with matrix showing shape strain (a) and four self-accommodating twinned variants with no net shape strain (b). [From A. Kelly and G. W. Groves, "Crystallography and Crystal Defects." Copyright 1970, with permission of Addison-Wesley.]

Adding equal amounts of both variants leads to a vanishing total shape strain. Thus when formed in groups of four variants in the manner shown in Fig. 29b a self-accommodating configuration is achieved of the kind of that is commonly found in shape–memory alloys.

VI. Precipitation

A. CHARACTERISTICS

Precipitation reactions are first-order phase transformations in which the parent and product phase (matrix and precipitate) coexist at the transformation temperature. Usually the precipitate is different from the matrix in composition, crystal structure, orientation, and atomic volume, causing a barrier to its nucleation and growth. Uniform distributions of fine precipitates are used in precipitation hardening alloys such as the Al–Cu alloy shown in the electron micrograph in Fig. 30. During nucleation and in the early stages of growth precipitates tend to be coherent. They lose coherency during continued growth, a process that is usually accompanied by a loss in alloy strength. The shape and distri-

bution of precipitates depend strongly on such crystallographic factors as the relationship between parent and product lattices, elastic anisotropy, and the mode and mechanisms of accommodation of the new phase in the old. It is difficult to predict exact precipitate shapes with any accuracy and for most purposes it is sufficient to distinguish among three basic shapes: spheres, plates, and needles. The most widely encountered precipitate shape is that of a flat plate such as the particles seen in Fig. 30. The reasons why plates are preferred will become clear in the course of this section. Semicoherent precipitates are derived from coherent ones by introducing dislocations in the interface to relieve the elastic distortions. All three types of precipitates (schematic examples for each type are shown in Fig. 31), coherent, semicoherent, and incoherent, are subject to crystallographic constraints.

In this section we will first examine the factors that determine the optimum shape of a coherent precipitate (also called inclusion), and then describe a simple physical model for the loss of coherency in plates. It is possible to make a qualitative prediction of the shape of an inclusion and a quantitative prediction of its orientation. Examples of applications to real alloys will be given.

B. ELASTIC INCLUSIONS

The theory of coherent elastic inclusions has been worked out in some detail and in this section some of its results are presented along with the underlying physical reasoning.

The strain energy of coherent inclusions in an

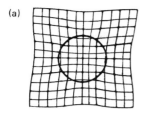

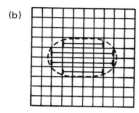

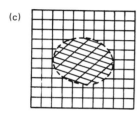

FIG. 31.   Coherent (a), semicoherent (b), and incoherent (c) inclusion.

elastic solid is a function of inclusion shape and orientation. To find the optimum shape and orientation of an inclusion transforming under elastic constraint from a solid matrix is the main objective of the strain energy approach. Without constraint the inclusion would undergo the stress-free transformation strain ε_{ij}. On the other hand, if the matrix were infinitely rigid it would constrain the inclusion completely back to the shape it had before transforming, reversing the transformation strain elastically and setting up a stress σ_{ij}. The work W (per unit volume) done in this process is

$$W = \tfrac{1}{2}\sigma_{ij}\varepsilon_{ij}$$

(Throughout Einstein's summation convention is implied for repeated indices.) Using Hooke's law:

$$\sigma_{ij} = c_{ijkl}\varepsilon_{kl}$$

where c_{ijkl} are the elastic constants of the inclusion, the strain energy becomes

$$W = \tfrac{1}{2}c_{ijkl}\varepsilon_{ij}\varepsilon_{kl}$$

Note that at this stage the matrix is strain free and all the strain energy is contained in the in-

clusion. However, since in real materials the matrix is not rigid it will not constrain the inclusion completely. Instead, part of the transformation strain will be accommodated by elastic deformation of the matrix. As a result the strain energy is no longer contained entirely in the precipitate but is divided between the work done in constraining the precipitate partially and the work necessary to accommodate part of the transformation strain in the matrix. The hypothetical sequence of cutting, straining, and welding operations shown in Fig. 32 illustrates this point.

1.   Make a cut around a volume within the matrix (a), remove it (b), and transform it stress free (c) (ε_{ij} is the transformation strain, $\sigma_{ij} = 0$).

2.   Apply a stress σ_{ij} that results in an elastic strain $-\varepsilon_{ij}$, reversing the transformation strain elastically (d).

3.   Replace and reweld the inclusion (e) and relax the applied stress (f). The first step corresponds to the extreme case of a transformation in an infinitely soft matrix; since $\sigma_{ij} = 0$ it causes no strain energy. The second step corresponds to the opposite extreme of transformation in an infinitely rigid matrix. The strain energy is $W = \tfrac{1}{2}\sigma_{ij}\varepsilon_{ij}$. In the third step the matrix is given more realistic elastic properties resulting in a total strain energy between zero and $\tfrac{1}{2}\sigma_{ij}\varepsilon_{ij}$. If the matrix is elastically identical to the precipitate, the inclusion is called homogeneous. If the elastic constants of matrix and precipitate are different, the inclusion is called heterogeneous.

The degree to which the inclusion is accommodated in the matrix depends on its shape and orientation. This is physically most apparent for an incoherent inclusion that, by definition, exerts only a hydrostatic pressure (no shear stresses), for example, an amorphous precipi-

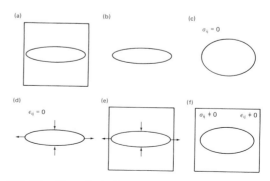

FIG. 32.   Hypothetical sequence of operations in the formation of an elastic inclusion.

tate or a gas bubble. Hence, we will first consider the simplest case of an incoherent inclusion.

C. INCOHERENT PRECIPITATES

For an incoherent inclusion no atomic correspondence is maintained across the interface. The strain energy is thus caused only by the volume difference between the inclusion and the hole it occupies. This energy depends on the shape of the hole. In an elastically isotropic material a spherical hole will yield uniformly in all directions. A disk-shaped hole on the other hand will yield unevenly. The flat faces are displaced easily under the internal pressure set up by the volume difference while the radial displacement is small. When the disk is very thin the entire volume change is accommodated by yielding of the flat faces. As a result a very thin incoherent precipitate is essentially stress free and therefore causes essentially no strain energy. To a good approximation the strain energy per unit volume of an incoherent plate is proportional to its aspect ratio c/a.

Thus of the three main possible shapes (spheres, plates, and needles), plates will always be preferred because they are easily accommodated. The smaller the aspect ratio the smaller is the strain energy and the flattening of a given incoherent precipitate is limited only by the increasing surface energy. A typical example of an incoherent inclusion is amorphous SiO_2 forming in Si.

D. COHERENT PRECIPITATES

Most applications of the strain energy approach deal with coherent rather than incoherent inclusions. Under constraint, incoherent inclusions are always in a state of hydrostatic stress whereas coherent inclusions can support shear stresses. This makes it necessary to distinguish between different types of stress-free transformation strains for coherent precipitates. These are best characterized by their principal strains e_1, e_2, and e_3. For many transformations this is self-evident. A cubic to tetragonal distortion, for example, is completely described by elongations or contractions along the three orthogonal cube axes. The strain tensor is thus automatically given in the diagonal form.

Other transformations are described by a general (nondiagonal) strain tensor e_{ij} which can always be decomposed into an antisymmetrical

part $\omega_{ij} = \frac{1}{2}(e_{ij} - e_{ji})$ and a symmetrical part $\varepsilon_{ij} = \frac{1}{2}(e_{ij} + e_{ji})$. For small strains ω_{ij} represents a rigid body rotation and does not contribute to the strain energy. The symmetrical part ε_{ij} (the pure strain) can always be referred to three orthogonal principal axes (its eigenvectors). In this coordinate frame ε_{ij} is diagonal with the principal strains e_1, e_2, and e_3 (the eigenvalues) as the diagonal elements:

$$\varepsilon_{ij} = \begin{pmatrix} e_1 & & \\ & e_2 & \\ & & e_3 \end{pmatrix}$$

The three principal strains are all different in the general (orthorhombic) strain tensor. If two strains are equal, the strain tensor is cylindrically symmetric (sometimes called tetragonal), and if all three strains are equal the spherically symmetric strain tensor represents a pure dilatation. In addition, the strains can be all of the same sign or of mixed sign. This yields six cases that will be considered individually below.

The solution to the inclusion problem would be simple if the strain energy caused by a coherent inclusion were a simple analytical function of the principal strains and the variables "shape" and "orientation." Unfortunately no such function exists for the general case. However, for the special case of an oblate spheroidal inclusion (a flat ellipsoid of revolution) undergoing a stress-free strain (e_1, e_2, e_3) along its axes a, a, and c, an analytical expression that includes the shape effect explicitly through the aspect ratio c/a has been derived. The three shapes of plate, sphere, and needle can be approximated by $c/a < 1$, $c/a = 1$, and $c/a > 1$, respectively. The inclusion orientation in this case is fixed through the assumption that the axes of the spheroid are aligned with the principal axes of the strain. For this case, the strain energy takes the form $W = f(e_1, e_2) + g(e_1, e_2, e_3)c/a$.

It can be shown that for a pure dilatation ($e_1 = e_2 = e_3$) the total strain energy is independent of the shape: $g(e_1, e_2, e_3) = 0$. For a plate, the contribution of the strain e_3 normal to the plate disappears as the aspect ratio c/a approaches zero. Physically this means that the strain e_3 normal to the plane of the plate is accommodated well, as in the incoherent case, and a very thin plate is essentially free to expand in thickness without setting up a stress. An example of this is an interstitial dislocation loop that may be regarded as an inclusion plate undergoing an expansion e_3 with $e_1 = e_2 = 0$. The extra plane is incorporated with the same plane spacing as the equivalent planes in the matrix, that is, accommodation is essentially complete.

On the other hand, the strains e_1 and e_2 in the plane of the plate are constrained by the matrix, setting up large stresses and as the aspect ratio c/a goes to zero, the shape-independent term of the strain energy $f(e_1, e_2)$ remains finite.

These results may be summarized as follows: a plate-shaped inclusion with very small aspect ratio is stressed in its plane and essentially stress free normal to its plane (a state of plane stress). The strain energy in this situation is caused only by the principal strains e_1 and e_2 in the plane, as indicated by the 2×2 block in the upper left of the strain tensor

$$\varepsilon_{ij} = \begin{pmatrix} e_1 & & \\ & e_2 & \\ & & e_3 \end{pmatrix}$$

This corresponds to the case of the inclusion axes parallel to the principal strains.

Consider the more general situation with the inclusion not parallel to the principal strains. Referred to the inclusion frame the strain then takes the form of a general symmetrical tensor

$$\varepsilon_{ij} = \begin{pmatrix} \varepsilon_{11} & \varepsilon_{12} & \varepsilon_{13} \\ \varepsilon_{21} & \varepsilon_{22} & \varepsilon_{23} \\ \varepsilon_{31} & \varepsilon_{32} & \varepsilon_{33} \end{pmatrix}$$

The same reasoning used before to illustrate easy accommodation of the expansion $\varepsilon_{33} = e_3$ for incoherent and coherent plates implies that the shears ε_{13} and ε_{23} are also easily accommodated by the matrix and thus contribute little to the strain energy of a plate. A shear dislocation loop is perhaps the best example to illustrate that almost all the strain (and hence strain energy) is located in the matrix outside the loop (or inclusion). Therefore the strain components ε_{11}, ε_{22}, and the shear ε_{12} in the plane of the plate (indicated by a block in the tensor ε_{ij}) dominate the strain energy. The approach used in the following sections seeks to minimize only the dominant part of the strain energy.

As seen above the strain energy of a homogeneous inclusion in the case of a pure dilatation is independent of the inclusion shape. For a spherical precipitate ($c/a = 1$) the total strain energy is only approximately one-third of the work that would be necessary to compress the particle elastically as in a rigid matrix. Thus the

fact that the transformation strain in the particle is partially accommodated by elastic distortion of the matrix reduces the total strain energy by approximately two-thirds. Of this total strain energy one-third resides in the spherical particle and two-thirds in the matrix. This distribution of the strain energy between matrix and precipitate is reversed as the inclusion becomes flattened ($c/a \to 0$). Hence, as shown in the previous section, most of the strain energy resides in the inclusion, for a thin plate. Consider now an inclusion with elastic constants different from those of the matrix, that is, a heterogeneous inclusion. Having established that in the homogeneous case a spherical inclusion contains one-third of the strain energy, let one phase become slightly softer than the other. Clearly the energy savings are greater if the matrix softens since it carries two-thirds of the total strain energy. This is reversed for plates in which the strain energy reduction is larger if the particle softens rather than the matrix. Therefore the total strain energy of a heterogeneous inclusion is minimized if its major fraction in the corresponding homogeneous case is located in the softer phase: if the precipitate is elastically softer than the matrix it should be a plate, but if it is harder than the matrix a sphere would be preferred.

Extending this analysis to the case of anisotropic elasticity is also straightforward. As shown above, a thin plate is in a state of plane stress which causes the major fraction of the total strain energy. If the plane of the stress is parallel to an "elastically soft plane" of the crystal the strain energy is smaller than for any other orientation. It can be shown that the strain energy of an elastically anisotropic plate inclusion on a plane $\{h\,k\,l\}$ depends on a term $(1 - A)(h^2k^2 + k^2l^2 + l^2h^2)$, where $A = 2c_{44}/(c_{11} - c_{12})$ is the Zener anisotropy ratio. The strain energy is then minimized for $\{1\,0\,0\}$ plates if A is larger than unity, as, for example, in all fcc metals. Conversely $\{1\,1\,1\}$ plates are preferred if $A < 1$. Since a very thin plate leaves the matrix essentially unstressed the strain energy is independent of the matrix elastic constants.

A heterogeneous anisotropic inclusion should form a plate if it is softer than the matrix. The only difference to the case of isotropic elasticity is that now two elastic shear moduli must be compared with that of the matrix. Thus an anisotropic inclusion with shear moduli $\frac{1}{2}(c_{11} - c_{12})$ and c_{44} in an isotropic matrix with shear modulus μ will form a plate if min $[(\frac{1}{2}c_{11} - c_{12}), c_{44}] < \mu$. Otherwise a sphere is preferred.

In the previous section it was shown how the elastic strain energy of a coherent precipitate undergoing a pure dilatation can be minimized through an inclusion shape and orientation that concentrates the major part of the strain energy in the softer phase. The high symmetry of the dilatational transformation strain allows a degree of freedom in the orientation of the precipitate which is lost in the case of a cylindrical strain ($e_1 = e_2 \neq e_3$). The precipitate will be a plate in the plane containing the two equal strains e_1, if $e_1 < e_3$ and a needle if $e_1 > e_3$.

Consider a cylindral strain with mixed signs ($e_3/e_1 < 0$), for example, an expansion e_3 and a contraction e_1, in the plane normal to it. The strain energy of a thin plate on the plane normal to e_3 is dominated by the uniform strain e_1 in this plane. However, another orientation can be found in which the strain energy is still lower, if the plate is inclined to contain a direction in the interface in which the transformation strain is zero. Such a direction must exist since the strain changes sign between e_1 and e_3. The angle θ of the unextended line with the (0 0 1) plane is given by

$$\tan \theta = \sqrt{-e_1/e_3} \qquad (34)$$

If a thin disk takes this orientation it is elastically distorted in its plane only by a uniaxial extension $-e_1$ constraining the transformation strain e_1, while the direction normal to it is free of transformation strain (unextended line). This is the geometrical condition of an invariant (unextended and unrotated) line in the interface as it is used for instance in martensite theory [Eqs. (21) and (26)].

In an isotropic crystal the uniaxial tension constraining the plate could be in any direction in the (0 0 1) plane containing the uniform strain e_1. In an anisotropic cubic crystal this stress would be expected to lie in a soft direction in the (0 0 1) plane. Thus the elastic distortion of the plate causes minimum strain energy if it occurs in the $\langle 1\ 0\ 0 \rangle$ direction for $A > 1$ and the $\langle 1\ 1\ 0 \rangle$ direction for $A < 1$. The resulting habit plane will contain this "tensile axis" and lie at an angle θ to the (0 0 1) plane. This leads to $\{0\ k\ l\}$ habits for $A > 1$ and $\{h\ h\ l\}$ habits for $A < 1$.

Orthorhombic strains ($e_1 \neq e_2 \neq e_3$) with unmixed strains will lead to precipitates whose dimensions tend to be inverse to the transformation signs. The elastic behavior plays only a secondary role in determining the shape, and the orientation is entirely given by the principal strains.

In the case of orthorhombic strains with mixed signs the same principles as for the cylindrical strain apply. A plate should be inclined to have a stress-free line in the interface and most of the strain energy is caused by the uniaxial stress in the plane of the plate. Thus the smaller one of the two principal strains with equal sign will lie in the plate being elastically constrained and the stress-free line is formed between the remaining two principal strains. This determines the orientation uniquely and elastic anisotropy plays only a secondary role. The angle θ of inclination is given by Eq. (34).

E. LOSS OF COHERENCY

Beyond a critical size it becomes energetically favorable for a coherent particle to lose coherency. This size depends on the difference between the interface energy of a semicoherent inclusion (proportional to the surface area) and the strain energy of a coherent inclusion (proportional to the volume). At the critical size the accommodation of the transformation strains switches from an elastic to a plastic mode. A simple example of this transition is the punching out of prismatic dislocation loops. Interstitial loops are punched into the matrix leaving behind vacancy loops in the interface (or vice versa). A TEM micrograph of rows of punched loops around a He gas bubble in Nb–Zr alloy is shown in Fig. 33a. The corresponding schematic (Fig. 33b) emphasizes the fact that a spherical inclusion will punch out loops in all crystallographically equivalent directions, for example, in the case of Nb–Zr in all $\langle 1\ 1\ 1 \rangle$ directions. Such rows of punched prismatic loops will leave behind an equal number of opposite dislocation loops in the interface and render the particle semicoherent while accommodating the volume component of the transformation strain.

In order to accommodate the shear component of the transformation strain, shear loops are more efficient since they need not be generated in pairs. As a result the shape change is easier to accommodate than the volume change. Martensite plates and many semicoherent precipitates can be modeled successfully on the assumption that an array of shear dislocations accommodates part of the transformation strain. A schematic of a precipitate plate containing shear loops in its interface is shown in Fig. 34. The crystallographic constraints in this case are less apparent than in the case of a spherical inclusion under a pure dilatation (Fig. 33). One such con-

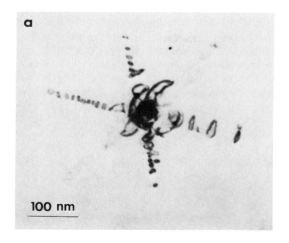

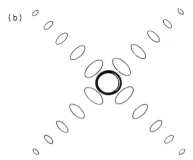

FIG. 33. Punched-out dislocation loops due to volume strain: TEM micrograph (a) and schematic (b) of punched loops around a helium bubble in Nb–Zr alloy. (Micrograph C. Echer.)

straint is that the loop plane must be a slip plane and contain the Burgers vector; but it is not immediately clear which of the crystallographically equivalent loop arrays are preferred for a given plate and what will be its habit plane.

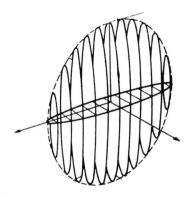

FIG. 34. Schematic of lens-shaped precipitate with array of shear loops in the interface.

An answer to these questions is provided by the theory of martensite transformations which gives an algebraic solution for the habit plane and orientation relationship if the slip plane and direction and of course the transformation strain are known. A more physical description on the loss of coherency and resulting interface structure based of the analogy with slip in a tensile test is summarized below.

Consider a thin coherent plate with mixed principal strains. As outlined above it will be in a state of uniaxial stress if it is oriented for minimum strain energy, that is, such that it contains an invariant line in the interface.

If it deforms plastically it is most likely to do so on the slip system with the highest resolved shear stress. This depends on the relative position of the tensile axis and the slip system; and the resolved shear stress is a maximum if both the slip plane and the slip direction make an angle of 45° with the tensile axis. The problem in the elastic case of finding a direction with a low Young's modulus in the plane of unmixed strains now translates to finding the tensile direction in this plane with the highest Schmid factor $R = \cos \phi \cos \lambda$ for a crystallographic slip system (where ϕ and λ are the angles of the tensile axis with the slip direction and the slip plane normal, respectively). As a first approximation it may be assumed that the Schmid factor is maximized when the tensile axis makes an angle of $\lambda = 45°$ with the slip plane. For slip on the (1 0 1) plane this direction is [1 0 0], as shown in the stereogram in Fig. 35a. The invariant line (open circle) must be located at the intersection of the cone of unextended lines with the slip plane so that it remains unaffected by the plastic deformation, a condition well known from martensite theory. The resulting habit plane (pole marked by star) contains the tensile axis and the invariant line. Applications of these principles to coherent and semicoherent precipitates are given below.

F. Applications

The transformation strain for α'' $Fe_{16}N_2$ (bct) in Fe (bcc) is cylindrically symmetric and

$$\varepsilon_{ij} = \begin{pmatrix} -0.0023 & & \\ & -0.0023 & \\ & & 0.0971 \end{pmatrix}$$

Since the anisotropy ratio A is greater than unity for Fe the habit plane predicted for α'' is (0 k l)

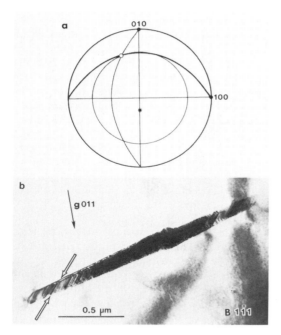

FIG. 35. γ' precipitate in Fe–N alloy: stereogram illustrating predicted habit plane (a) and corresponding TEM image showing inclined γ' plate with interface striations along invariant line direction (b).

with $k/l = \sqrt{0.0023/0.971}$, a plane that is inclined about 8.5° with respect to (0 0 1). This inclination is in agreement with experimental observations.

As for α'' $Fe_{16}N_2$, the transformation strain for γ' Fe_4N (fcc iron sublattice) in Fe (bcc) has cylindrical symmetry; with $a_{bcc} = 0.28678$ nm, $a_{fcc} = 0.3795$ nm, we obtain

$$\varepsilon_{ij} = \begin{pmatrix} -0.0643 & & \\ & -0.0643 & \\ & & 0.3233 \end{pmatrix}$$

These strains are large and the habit plane must be determined by the criterion of maximum Schmid factor. γ' slips on {1 1 1} fcc planes derived from the inclined {1 0 1} bcc slip planes as shown stereographically in Fig. 35a. The predicted $(0\ k\ l)$ habit plane defined by the $[\bar{4}\ 9\ 4]$ invariant line and the [1 0 0] tensile axis is $(0\ \bar{4}\ 9)$, in agreement with the experimental observations. Figure 35b shows a TEM image of a typical γ' plate. The parallel striations are interfacial dislocations along the invariant line direction.

With the lattice parameters $a_{bcc} = 0.3147$ nm, $a_{hcp} = 0.3002$ nm, $c_{hcp} = 0.4724$ nm, the trans-

formation strain tensor for Mo_2C (hcp) in Mo (bcc) is

$$\varepsilon_{ij} = \begin{pmatrix} 0.17 & & \\ & 0.06 & \\ & & -0.05 \end{pmatrix}$$

when referred to [1 1 0], [1 $\bar{1}$ 0], and [0 0 1] axes. These directions transform to [0 0 0 1], [0 1 $\bar{1}$ 0], and [2 $\bar{1}$ $\bar{1}$ 0], respectively. Since all these principal strains are different the cone of unextended lines now has an elliptical cross section. Due to the crystallography of the hexagonal Mo_2C precipitate, (0 0 0 1) is the slip plane. This is derived from the vertical (1 1 0) bcc slip plane. Again, the tensile axis in the (0 0 1) plane (the plane of unmixed strains) at 45° to the (1 1 0) bcc/(0 0 0 1) hcp slip plane is [1 0 0]. The invariant line at the intersection of the cone of unextended lines with the (1 1 0) slip plane is ~[$\bar{1}$ 1 3]. The resulting $(0\ k\ l)$ habit plane containing the [1 0 0] tensile axis and the [$\bar{1}$ 1 3] invariant line is $(0\ \bar{3}\ 1)$, again in agreement with experimental observations.

The main results of this section can be summarized as follows:

1. For principal strains of unmixed sign the precipitate dimensions tend to be inverse to the strains.

2. For principal strains of mixed sign the interface is oriented so as to include an invariant line.

3. For coherent precipitates remaining degrees of freedom for shape and orientation are used to concentrate the major part of the equivalent homogeneous isotropic strain energy in the softer phase or orientation.

4. For plate-shaped inclusions the tensile axis must be a soft direction in the plane of unmixed strain [in our examples the (0 0 1) plane]. For coherent precipitates the tensile axis must be elastically soft, that is, have a low elastic modulus, and for semicoherent precipitates it must be plastically soft, that is, be oriented for the maximum Schmid factor on the slip system.

G. MECHANISMS

The mechanism of a phase transformation depends upon thermodynamic, kinetic, and structural parameters. If at a given temperature the parent phase is unstable with respect to small fluctuations of a physical quantity such as composition or atomic displacement then no barrier to nucleation exists and the transformation can proceed by the gradual amplification of some

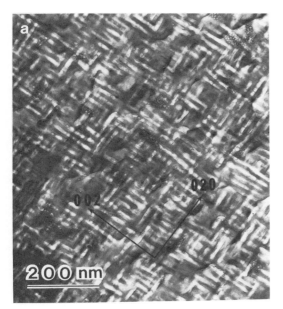

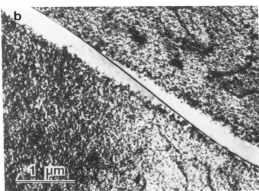

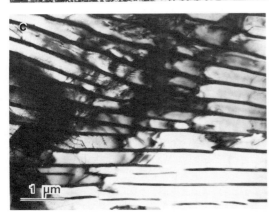

FIG. 36. TEM micrographs showing microstructures typical for different mechanisms of decomposition: spinodal decomposition in Cu–Ni–Fe alloy (a) (micrograph K. Kubarych), nucleation and growth in Al–Si alloy (b), and eutectoid reaction in Fe–C alloy (c).

fluctuations to form a modulated structure. Because of their barrier-free nucleation such structures are homogeneous and independent of preexisting microstructures and defects. This is illustrated in Fig. 36a with a Cu–Ni–Fe alloy which has undergone spinodal decomposition. On the other hand, when a barrier to nucleation exists the resulting two-phase microstructure depends on the type and distribution of defects available to help overcome the barrier. This situation is more commonly encountered than that of modulated structures and usually leads to less homogeneous microstructures such as the Al–Si alloy shown in Fig. 36b. Vacancies are necessary in this alloy to overcome the nucleation barrier due to the large volume increase on precipitation. As a result precipitate-free zones develop in regions depleted of vacancies such as grain boundaries, surfaces, or dislocations. Defects other than vacancies can also lead to preferred nucleation. Dislocations, stacking faults, grain boundaries, and other defects have all been found to provide heterogeneous nucleation sites. If the alloy decomposes only along an advancing interface, usually starting from a grain boundary, the reaction is called discontinuous, as opposed to the continuous reactions described above. Discontinuous reactions lead to characteristic cellular morphologies. Similar cellular morphologies are also observed in eutectoid decomposition (see Fig. 36c) where both reaction products are different in structure from the parent phase. If no suitable defects are available to allow precipitation of the equilibrium phase it is often found that intermediate metastable phases form with a structure and composition approximating that of the stable equilibrium phase. In fact the properties of most precipitation hardening alloys are based on such metastable phases. Structurally, the mechanism of a phase transformation depends, therefore, not only on preexisting microstructure and defects but also on available intermediate phases.

Acknowledgments

This work is supported by the Director, Office of Energy Research, Office of Basic Energy Sciences, Materials Sciences Division of the U.S. Department of Energy under Contract No. De-AC03-76SF00098.

BIBLIOGRAPHY

Aaronson, H. I., Laughlin, D. E., Sekerka, R. F., and Wayman, C. M. (eds.) (1982). Proc. Conf. on "Solid-Solid Phase Transformations." AIME.

Boccara, N. (ed.) (1981). "Symmetries and Broken Symmetries in Condensed Matter Physics." IDSET, Paris.

Christian, J. W. (1975). "The Theory of Transformations in Metals and Alloys," Part I, 2nd ed. Pergamon Press, Oxford.

Hahn, T. (ed.) (1983). "International Tables for Crystallography," Vol. A., D. Reidel, Boston.

Kelly, A., and Groves, G. W. (1970). "Crystallography and Crystal Defects." Addison-Wesley, London.

Khachaturyan, A. G. (1983). "Theory of Structural Transformations in Solids." John Wiley, New York.

Nye, J. F. (1957). "Physical Properties of Crystals." Oxford University Press, New York.

Shubnikov, A. V. (1974). "Symmetry in Science and Art." Plenum, New York.

Wayman, C. M. (1964). "Introduction to the Crystallography of Martensitic Transformations." Macmillan, New York.

PHOTOACOUSTIC SPECTROSCOPY

H.-H. Perkampus   *University of Düsseldorf*

GLOSSARY

Absorption coefficient: Based on the Bouguer–Lambert–Beer law, the absorbance A is given by $2.303A = \varepsilon cd = \beta d$. The absorption coefficient β is expressed by $\beta = 2.303A/d$ with the unit $[\beta]$ in reciprocal centimeters.

Composite-piston (CP) effect: Takes into consideration the superposition of both the heat waves described by the RG theory and the thermal expansion of a sample.

Depth-profile: Depth-specific characterization of a sample based on the frequency dependence of the thermal diffusion length.

Modulation frequency: By use of a light beam chopper, the amplitude modulated incident light flux produces a photoacoustic signal of the same frequency.

Normalized photoacoustic (PA) signal: Amplitude of a photoacoustic signal normalized to that of a reference standard under tha same recording conditions.

Optical absorption length l_β: Penetration depth of the incident light flux, for which the intensity is diminished by e^{-1}. The optical absorption length l_β is defined as $l_\beta = \beta^{-1}$ with the unit $[l_\beta]$ in centimeters.

Phase angle of PA signal: Phase-angle shift relative to a given modulation frequency. The phase angle depends on instrumental parameter as well as on thermal and optical properties of the sample.

Radiationless deexcitation: One of the relaxation pathways of an optically excited molecule, along which the excitation energy is converted to heat (internal conversion). This is the basis of the PA effect.

Reference standard: Sample that is photoacoustically saturated in the wavelength region of interest. Its PA signal is directly proportional to the incident light flux.

Rosencwaig–Gersho (RG) theory: Describes the production of a PA signal in condensed media by the use of a one-dimensional model for that case of gas-coupled PA spectroscopy.

Signal saturation: Thermal diffusion length μ_{th} is greater than the optical absorption length l_β. The PA signal becomes independent of the optical absorption coefficient.

Thermal diffusion length μ_{th}: Given by $\mu_{\text{th}} = \sqrt{2\alpha/\omega}$, where α is the thermal diffusivity and ω the chopping frequency of the incident light beam in radians per second. The unit is in centimeters.

The photoacoustic effect is based on the conversion of absorbed light energy into heat by means of radiationless deexcitation processes. On irradiation of a solid or liquid sample by intensity modulated light, a heat wave of the same frequency is generated in the sample. The heat wave is transferred by diffusion to the surface of the sample, where oscillatory thermal effects are generated in the coupled gas. These can be detected as an acoustic signal. This technique is known as gas-coupled photoacoustic spectroscopy.

ENCYCLOPEDIA OF PHYSICAL SCIENCE
AND TECHNOLOGY, VOL. 10

I. History

The photoacoustic effect was discovered in 1880–1881 by Alexander Graham Bell while experimenting with wireless telegraphy. If a solid sample enclosed in a gas-filled cavity is irradiated with light modulated at the frequency $f[\sec^{-1}]$, a signal can be found by an acoustic detector coupled to the system; Bell used an ear trumpet. Inspired by these experiments, Tyndall and Röntgen investigated this phenomenon in gases. Bell found that the optoacoustic (now photoacoustic) effect not only depends on the intensity I_0 of the irradiating light but also on the absorptivity of the irradiated sample. Because both parameters, I_0 of the light source and the absorptivity of the sample characterized by the absorption coefficient $\beta[\mathrm{cm}^{-1}]$ depend on the wavelength, the variation of wavelength λ of the irradiating light directly leads to the photoacoustic spectra. As early as 1881 Bell constructed the first photoacoustic spectrometer, called Spectrophon, for the visible range of light.

Although the principal physical bases of photoacoustic spectroscopy (PAS) were known from the beginning, it was as late as the 1970s that the PA effect developed to an applicable physical method. Decisive contributions were the development of sensitive microphones as detectors, the progress of amplifier techniques, the construction of intense light sources, and the development of a theory of the PA effect for condensed media by Rosencwaig and Gersho in 1975.

II. Basics of the Photoacoustic Effect

For the understanding of the photoacoustic effect, we refer to the simplified term scheme of a free molecule in Fig. 1. After absorption of the light quanta $h\nu_1$ or $h\nu_2$, the molecule is in the electronically excited state S_1 or S_2. From S_2 or vibronic states of S_1 ($v > 0$) the molecule is deactivated to S_1 with the vibration quantum number $v = 0$ by radiationless transfer; this state has an average lifetime of 10^{-8} to 10^{-9} sec.

From this state, two competing deactivation processes are possible: fluorescence with the rate constant k_{FM} and radiationless deactivation with k_{IM}. For the photoacoustic effect, the radiationless processes are of importance, since the exciting energy $h\nu$ is converted into heat produced during the average lifetime of the excited states. A system continuously irradiated also produces heat continuously, which can be mea-

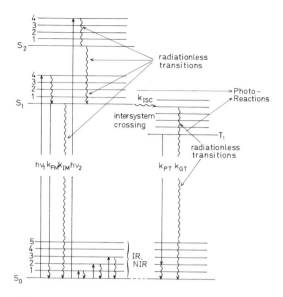

FIG. 1. Jablonski diagram showing the radiative and radiationless transitions of a molecule.

sured by means of a calorimeter; this is the principle of photocalorimetry. By excitation directly to S_1 ($v = 0$) the quantity of heat produced depends only on the relationship of the competing processes k_{FM} and k_{IM}; this is a way to estimate the quantum yield of fluorescence. A system irradiated with light modulated at the frequency $f[\sec^{-1}]$ where this modulation frequency is low compared with the number of deactivation steps per time unit also produces heat at the frequency $f[\sec^{-1}]$. In a gas-filled cavity, a periodic change of pressure results (i.e., a pressure or sound wave) that can be measured by a microphone. The process of photoacoustic (signal generation) spectroscopy consists in changing the excitation energy $h\nu$ into heat and detecting it in the form of mechanical work.

While the generation of a PA signal in gases is relatively simple to understand, matters are much more complicated in condensed systems such as solids and liquids. Although the primary step, the absorption of light and successive radiationless deactivation, is the same, the heat produced periodically in the bulk of the sample has to reach the surface to be transferred to the adjacent thin layer of gas. This layer of gas, whose effective thickness depends upon its heat conductivity, can be considered as a piston that moves with frequency f and transfers a sound wave to the surrounding gas molecules in the closed cavity, which can be measured by a mi-

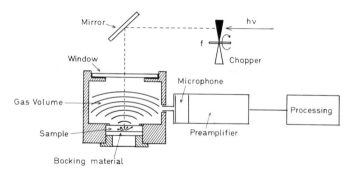

FIG. 2. Schematic representation of a gas-coupled PA cell for the measurement of solid or liquid samples.

crophone (piston model). From this short description it can be seen that heat production in the bulk is followed by heat diffusion processes, and therefore the PA effect in a condensed system is influenced by the thermal properties of the sample and the adjacent gas.

For thin samples on a substrate, the thermal properties of the substrate also have to be taken into account. A gas-coupled PA cavity is depicted in Fig. 2. A PA cavity of this type offers the advantage that unabsorbed light can leave the cavity without being reflected by the substrate into the cell, which is significant for numerous applications (liquids, thin films, sheets, pressings, and glasses). A cavity of this type can also be considered as a Helmholtz resonator where the resonance frequency is lower for longer distances between gas and microphone membrane. At this resonance frequency, the PA cavity has its highest sensitivity. For many purposes, however, it is useful to vary the modulation frequency in a wide range. A cavity of ≤ 2 cm^3 total volume can be used in the frequency range from 2 to 4000 [sec.$^{-1}$]. Figure 3 shows the signal that is characteristic of such a PA cavity in dependance of the modulation frequency; the resonance frequency is about 4000 [sec.$^{-1}$]. It should be mentioned that there is no general method for construction of PA cavities; every problem demands its own optimized PA cavity.

III. Theory

A. GASES

For the theoretical treatment of the PA effect, we refer to the simple term scheme with two levels in Fig. 4. The rate constants k and k' label radiant and nonradiant transfers. The constant k_{01}, which describes the excitation process $E_0 \rightarrow E_1$, is given by

$$k_{01} = \rho_\nu B_{01} + A_{01} \qquad (1)$$

Accordingly, for the deactivation $E_1 \rightarrow E_0$,

$$k_{10} = \rho_\nu B_{10} + A_{10} \qquad (2)$$

where $B_{01} = B_{10}$ is the Einstein B coefficient for stimulated radiation transfers ($0 \rightarrow 1$: absorption; $1 \rightarrow 0$: emission), A_{01} and A_{10} are the Einstein A coefficients for spontaneous radiation transfers with $A_{01} = 0$, and ρ_ν is the radiation density at the transfer energy $E_\nu = h\nu = E_1 - E_0$.

The change of occupation density N_1 in the state E_1 with respect to time is given by the difference between the number of molecules in the unity volume reaching and leaving E_1.

$$dN_1/dt = (k_{01} + k'_{01})N_0 - (k_{10} + k'_{10})N_1$$

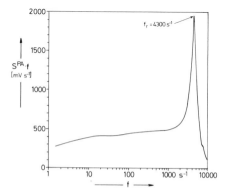

FIG. 3. Frequency dependence of a PA cell used for solid samples (resonance frequency, 3400 sec^{-1}; inner volume of the cell, 0.724 cm^3; microphone, Brüel and Kjaer type 4166 (sensitivity 49 mV/Pa).

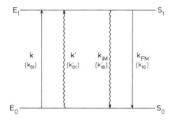

FIG. 4.   Two-level scheme explaining the photoacoustic effect for a gas.

Taking Eqs. (1) and (2) into account gives

$$dN_1/dt = (\rho_\nu B_{01} + A_{01} + k'_{01})N_0$$
$$- (\rho_\nu B_{10} + A_{10} + k'_{10})N_1$$

which for $A_{01} = 0$, $k_{01} = 0$, and $B_{01} = B_{10}$ gives

$$dN_1/dt = \rho_\nu B_{01}(N_0 - N_1) - (A_{10} + k'_{10})N_1$$

Since $A_{10} = k_{FM}$ is the rate constant of spontaneous emission (fluorescence) and $k'_{10} = k_{IM}$ for radiationless deactivation, for $k_M = k_{FM} + k_{IM} = \tau_M^{-1}$,

$$dN_1/dt = \rho_\nu B_{01}(N_0 - N_1) - k_M N_1$$
$$= \rho_\nu B_{01}(N_0 - N_1) - \tau_M^{-1} N_1 \qquad (3)$$

Analog:

$$dN_0/dt = -\rho_\nu B_{01}(N_0 - N_1) + \tau_M^{-1} N_1 \qquad (4)$$

For $N = N_0 + N_1$ and $N_0 = N - N_1$ it follows from Eqs. (3) and (4) for the stationary state $dN_1/dt = dN_0/dt$ that

$$N_1 = \rho_\nu B_{01}/(2\rho_\nu B_{01} + \tau_M^{-1})N \qquad (5)$$

and

$$N_0 = [(\rho_\nu B_{01} + \tau_M^{-1})/(2\rho_\nu B_{01} + \tau_M^{-1})]N \qquad (6)$$

For the approximation $\rho_\nu \simeq I_\nu(h\nu/c)$, where I is the light intensity, c the velocity of light, and $h\nu$ the excitation energy, the definition $B = B_{01}h\nu/c$ follows

$$N_1 = \frac{B I_\nu}{2BI_\nu + \tau_M^{-1}} N \qquad (5a)$$

$$N_0 = \frac{BI_\nu + \tau_M^{-1}}{2BI_\nu + \tau_M^{-1}} N \qquad (6a)$$

By modulating the exciting light with the angular velocity $\omega = 2\pi f$, the intensity I_ν is

$$I_\nu = I_{0\nu}(1 + \delta\, e^{i\omega t}), \qquad 0 \le \delta \le 1$$

It follows from Eq. (5a) that

$$N_1 = \frac{BI_{0\nu}(1 + \delta e^{i\omega t})}{2BI_0(1 + \delta e^{i\omega t}) + \tau_M^{-1}} N \qquad (7)$$

Equation (7) gives the number of molecules elevated from state E_0 to E_1 by the irradiating light. When the molecule returns by radiationless deactivation to the ground state, the equivalent energy difference $\Delta E = E_1 - E_0$ is freed as heat and kinetic energy, which means an increase of internal energy in the adiabatic system. It follows that

$$dU = dE_{kin}$$

Considering the time-dependence of this process,

$$dU/dt = dE_{kin}/dt = k_{IM}N_1 E_1 \qquad (8)$$

The expression $k_{IM}N_1 E_1$ reveals the energy per time unit liberated as heat by radiationless deactivation. According to the first law of thermodynamics,

$$dU = dE_{kin} = \left(\frac{\partial E_{kin}}{\partial T}\right)_V dT + \left(\frac{\partial E_{kin}}{\partial V}\right)_T dV$$

For $V = $ const, it follows that

$$dE_{kin} = \left(\frac{\partial E_{kin}}{\partial T}\right)_V dT = C_V\, dT \qquad (9)$$

Integration of Eq. (9) delivers

$$E_{kin} = C_V T + f(V) \qquad (10)$$

where C_V is specific heat, and $f(V)$ is a function depending upon volume only, not on temperature.

In this consideration, it is also assumed that the kinetic energy comes in the form of translation energy only. Rotations and vibrations are neglected.

From Eq. (10) it immediately follows that

$$T = (E_{kin} - f(V))/C_V \qquad (11)$$

According to the law of ideal gases,

$$p = kNT = (k/C_V)N[E_{kin} - f(V)] \qquad (12)$$

with k the Boltzmann constant and N the number of molecules per volume; the temperature T is expressed by Eq. (11). The change of pressure with time (i.e., the pressure wave) follows from Eq. (12), taking Eq. (8) into account:

$$\frac{dp}{dt} = \frac{k}{C_V} N \frac{dE_{kin}}{dt} = \frac{k}{C_V} N(k_{IM}N_1 E_1)$$

The substitution of N_1 from Eq. (7) gives

$$\frac{dp}{dt} = \frac{k}{C_V} \frac{N^2 E_1}{\tau_{IM}}$$

$$\times \frac{BI_{0\nu}[1 + \exp(i\omega t)]}{2BI_{0\nu}[1 + \delta\exp(i\omega t)] + \tau_M^{-1}} \qquad (13)$$

The simplification of this expression by progression allows integration, which leads to the expression

$$p = \frac{kE_1N^2}{C_V\omega}$$

$$\cdot \frac{2\,\tau_{\text{IM}}^{-2}BI_{0\nu}\delta}{(2BI_{0\nu} + \tau_{\text{M}}^{-1})[(2BI_{0\nu} + \tau_{\text{M}}^{-1}) + \omega^2]^{1/2}}$$

$$\cdot e^{i(\omega t - \gamma - \pi/2)} \qquad (14)$$

In this equation $\gamma = \omega\tau_{\text{M}}$. From Eq. (14) the photoacoustic microphone signal follows, which is $q = -p$.

Equation (14) leads to two border-line cases:

a. $2BI_0 \ll \tau_{\text{M}}^{-1}$, that is, I_0 and B are small. In this case,

$$q \simeq \frac{kE_1N^2}{C_V\omega^2}\left[\frac{\tau_{\text{M}}}{\tau_{\text{IM}}}\right]^2 \frac{2BI_0}{(1 + \omega^2\tau_{\text{M}}^2)^{1/2}}\, e^{i(\omega t - \gamma + \pi/2)}$$

$$(14a)$$

The photoacoustic signal is proportional to N^2 and changes linearly with BI_0. Taking the definition of τ_{M} and τ_{IM} into account, it can be seen that the expression $\tau_{\text{M}}/\tau_{\text{IM}}$ is equivalent to the quantum yield ϕ_{IM} of the radiationless deactivation:

$$\tau_{\text{M}}/_{\text{IM}} = k_{\text{IM}}/(k_{\text{FM}} + k_{\text{IM}}) = \phi_{\text{IM}}$$

This means that the signal is proportional to the square of the quantum yield of the radiationless deactivation. Since k_{IM} increases with temperature, the PA signal also increases.

b. $2BI_0 \gg \tau_{\text{M}}^{-1}$, that is, I_0 and B are very large. It follows for low frequencies ($2BI_0 \gg \omega$) that

$$q \simeq \frac{kE_1N^2}{C_V\omega^2}\,\tau_{\text{IM}}^{-2}\,\frac{1}{2BI_0}\,\delta e^{i(\omega t - \gamma + \pi/2)} \quad (14b)$$

The PA signal is saturated and proportional to I_0^{-1}. Saturation in this case means that as a result of the high pumping energy significantly more molecules reach the state S_1 than undergo deactivation during the same time interval. In Eqs. (14), (14a), and (14b) the correlation with the absorption properties of the sample is implicitly given by the Einstein-B coefficient. The theory of the PA effect in gases sketched here in the form of a two-level scheme explains in a simple manner, the most important facts. A more detailed consideration has to take rotations and vibrations into account, and it also has to be kept in mind that the model given here is only a simple approximation to molecular reality.

B. Solids and Fluids

As mentioned above, the radiationless deactivation of excited gas molecules directly increases the kinetic energy of this adiabatic system. In condensed systems (solids and fluids), however, the heat produced periodically in the bulk of the sample has to reach the surface to be transferred to the coupled gas layer. Processes of heat diffusion (and temperature conductivity) are of great importance for the treatment of the generation of the PA signal.

On the basis of the known differential equations of thermal diffusion, Rosencwaig and Gersho developed a theory (RG theory) for the treatment of the PA effect in condensed systems; the one-dimensional model of their theory is depicted in Fig. 5.

In a cylindrical PA cavity, the optical and thermal homogeneous sample with thickness l_s is deposited on a backing material of thickness l_b. This arrangement is irradiated by sinusoidally modulated light. The energy absorbed by the sample is deactivated completely without radiation. Light penetration and heat diffusion in the sample are strictly perpendicular to the sample surface. For the absorption of the electromagnetic radiation in the sample the Bouguer–Lambert–Beer law is valid.

In Table I all parameters of the RG theory are collected.

The following equations are the most important for the practical use of PAS:

$$\alpha_i = \kappa_i/\rho_i C_i \qquad (15)$$

$$a_i = (\omega/2\alpha_i)^{1/2} \qquad (16)$$

$$\alpha = a_i^{-1} = (2\alpha_i/\omega)^{1/2} \qquad (17)$$

Considering the heat distribution stated by the RG theory, we obtain three one-dimensional heat diffusion equations. Those for the backing material and the gas phase are identical, while for the sample the absorption of modulated light leading via radiationless deactivation to periodic heat production has to be taken into account.

In the framework of the RG theory, the three differential equations for the stationary state are solved with the help of four boundary conditions. These follow from the demand for continuity of heat flow and temperature at the boundary layers of the sample and support and of the sample and gas. Moreover, it is assumed that heat convection in the gas phase can be neglected and that cavity walls are at room temperature. The explicit solution leads to a complex

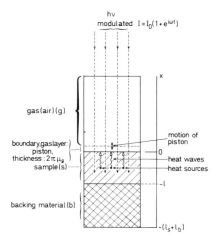

FIG. 5.   Model of the Rosencwaig–Gersho theory.

expression that is rather unpracticable for most purposes. Of physical interest is the explicit solution $\theta(0, t)$ since it describes the amplitude of the periodic temperature change at the sample surface ($x = 0$; compare Fig. 5). The value of $\theta(0, t)$ can be calculated for the parameters of the given experimental arrangement from the explicit solution. In the gas phase the heat wave described by the analogous function $\theta(x, t)$ is damped completely within a layer of $2\pi\mu_g(x > 0$; Fig. 5). According to Eq. (17), μ_g depends upon the modulation frequency ω and the thermal diffusivity of the gas.

The gas layer of $2\pi\mu_g$ thickness acts as an acoustic piston on the rest of the gas (see Fig. 5).

The periodic movement of this gas piston is described in the RG theory by the ideal gas law, taking an averaged temperature change across the spatial extension of the piston into account. Assuming an adiabatic reaction of the gas in the cavity upon the piston pressure, it follows for the photoacoustic signal Q that

$$Q = (\gamma_0 P_0 \mu_g / 2 l_g T_0)\theta(0, t) \tag{18}$$

Herein $\gamma = C_p/C_v$ is the relation of the specific heats of the coupled gas, P_0 the average pressure, and T_0 the average temperature of the sample. For the usual experimental conditions of PA spectroscopy, the increase of temperature in the sample due to the absorption can be neglected; therefore, T_0 is equal to room temperature.

The real physical pressure change $p(t)$ is

$$p(t) = Q_1 \cos(\omega t - \tfrac{1}{4}\pi) - Q_2 \sin(\omega t - \tfrac{1}{4}\pi) \tag{19}$$

where Q_1 and Q_2 stand for the real 2nd imaginary part of the photoacoustic signal Q. The explicit expression for the PA signal is given for completeness:

$$Q = \frac{\beta I_0' \gamma p_0 \mu_g}{2\sqrt{2}\kappa_s l_g(\beta_2 - \sigma^2)T_0}$$
$$\times \frac{\begin{array}{l}(r - 1)(b + 1)\exp(\sigma l_s) - (r + 1) \\ \times (b - 1)\exp(-\sigma l_s) \\ + 2(b - r)\exp(-\beta l_s)\end{array}}{\begin{array}{l}(g + 1)(b + 1)\exp(\sigma l_s) - (g - 1) \\ \times (b - 1)\exp(-\sigma l_s)\end{array}} \tag{20}$$

TABLE I.   Parameters for PAS in Condensed Media (RG Theory)

Symbol	Definition	Dimension
κ_i [a]	Thermal conductivity	J cm^{-1} sec^{-1} K^{-1}
ρ_i	Density	g cm^{-3}
C_i	Specific heat	J g^{-1} K^{-1}
α_i	Thermal diffusivity	cm^2 sec^{-1}
a_i	Thermal diffusion coefficient	cm^{-1}
μ_i	Thermal diffusion length	cm
$\beta = 2.303 \ (A/d)^b$	Absorption coefficient	cm^{-1}
l_β	Optical absorption length	cm

[a] The subscript i stands for b (backing material), s (sample), or g (gas); all of these parameters must be taken into account for all phases contacting each other at the boundary layers.

[b] A = absorbance; d = layer thickness of sample in cm.

The definitions of r, b, g, and σ are

$$r = (1 - i) \frac{\beta}{2} \left(\frac{2\alpha_s}{\omega} \right)^{1/2}, \qquad g = \frac{\kappa_g}{\kappa_s} \left(\frac{\alpha_s}{\alpha_g} \right)^{1/2}$$

$$b = \frac{\kappa_b}{\kappa_s} \left(\frac{\alpha_s}{\alpha_b} \right)^{1/2} \qquad \sigma = (1 + i) \left(\frac{\omega}{2\alpha_s} \right)^{1/2}$$

With the help of this relation, Q can be calculated for the given experimental conditions if all thermal data are known. The complexed result delivers the amplitude q of the PA signal as

$$q = \sqrt{Q_1^2 + Q_2^2} \qquad (21)$$

The phase angle ψ is

$$\psi = \arctan(Q_2/Q_1) \qquad (22)$$

Equations (21) and (22) are very important for the interpretation of PA signals, since these signals are obtained by a phase-sensitive lock-in amplifier technique where either amplitude or phase angle can be chosen.

C. BORDERLINE CASES OF RG THEORY

Although Eq. (20) can be used for the theoretical calculation of the PA signal provided that all optical and thermal data including light intensity are known, the consideration of borderline cases simplifying this complex expression proved to be of practical use. These borderline cases are sufficient approximations for many applications and shall be presented in short form.

For this purpose we define the optical absorption length l_β and the thermal diffusion length μ according to Eq. (17):

$$l_\beta = 1/\beta \quad \text{(cm)} \qquad \text{and} \qquad \mu_s = \sqrt{2\alpha_s/\omega} \quad \text{(cm)}$$

where l_β is the reciprocal of the absorption coefficient. For $l = l_\beta$, the intensity I_0 is according to $I = I_0 e^{-\beta l_\beta}$ reduced to $I = 0.37 I_0$.

The thermal diffusion length μ_s depends upon the thermal diffusivity and hence upon the modulation frequency ω. Considering the thickness l_s of the sample, we can distinguish the following cases, which are collected in Fig. 6:

1. Optically transparent, $l_\beta > l_s$
2. Optically opaque, $l_\beta < l_s$

In each case one can also make a distinction on the basis of

thermally thin samples, $\mu_s > l_s$ (cases 1a, 1b, and 2a); and

thermally thick samples, $\mu_s < l_s$ (cases 1c, 2b, 2c).

The solutions for these border-line cases are given below.

1. Cases 1a and 1b

$$Q = (1 - i) \frac{\mu_g I_0}{2} \cdot \frac{\mu_b}{\kappa_b} (\beta l_s) \cdot B$$

$$B = \frac{\gamma p_0}{2\sqrt{2}\, T_0 V} \qquad (23)$$

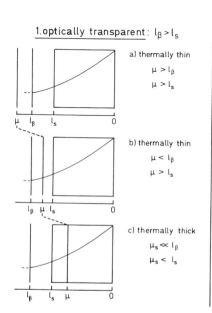
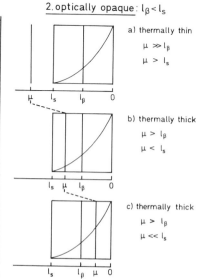

FIG. 6.   Special cases of the RG theory.

The signal is proportional to the absorption coefficient β and depends upon the thermal properties of the backing material. Because of the definition of μ in Eq. (17), the signal Q is proportional to ω^{-1} provided the phase is constant. V introduced instead of l_g is the volume of the cavity.

2. Case 1c

$$Q = -i \left(\frac{\mu_g I_0}{2}\right) \left(\frac{\mu_s}{\kappa_s}\right) (\beta\mu_s)B \qquad (24)$$

The signal depends upon β and the thermal properties of the sample and is proportional to $\omega^{-3/2}$.

3. Case 2a

$$Q = (1 - i) \left(\frac{\mu_g I_0}{2}\right) \left(\frac{\mu_b}{\kappa_b}\right)B \qquad (25)$$

In this case, frequency dependence and phase of the signal are the same as in Eq. (23), but the PA signal does not depend upon the optical properties of the sample; the signal is completely saturated.

4. Case 2b

$$Q = (1 - i) \left(\frac{\mu_g I_0}{2}\right) \left(\frac{\mu_s}{\kappa_s}\right)B \qquad (26)$$

The signal is also saturated as in case 2a and therefore independent of the optical properties of the sample. In contrast to Eq. (25), the thermal properties are to be taken into account. The phase is the same as in Eqs. (23) and (25), and the dependence of the PA signal is also analogous to cases 1a and 2a.

5. Case 2c

$$Q = -i \left(\frac{\mu_g I_0}{2}\right) \left(\frac{\mu_s}{\kappa_s}\right) (\beta\mu_s) \,\text{B} \qquad (27)$$

This case is identical to case 1c. The signal depends upon the optical properties of the sample, and its intensity is proportional to $\omega^{-3/2}$.

Cases 1c, 2b, and 2c are thermally thick samples that can be summarized for random optical densities:

$$Q = \frac{\beta(r - 1)I_0\mu_g}{\kappa_s(g + 1)(\beta^2 - \sigma^2)} B \qquad (28)$$

In Eqs. (23)–(28), the signal Q is given in units of pressure (Pa). For practical purposes Q has to be transferred into voltage (i.e., the sensitivity S^M of the microphone given in mV/Pa has to be known):

$$S^{PA} = Q^{PA}S^M$$

Solutions (23)–(28) contain a real part giving the amplitude of the PA signal (intensity) and an imaginary part informing about the phase angle ψ.

If only the amplitude of the PA signal is to be considered, the following relationship between absorption coefficient and the PA signal, corrected for intensity, is obtained from Eq. (28) and the definitions for g, r and σ:

$$S^{PA} = \frac{\beta[(\beta\mu_s/F) - 1]}{\beta^2 - (2/\mu_s^2)} F \qquad (29)$$

The factor F contains, besides the parameters hidden in g, r and σ, the microphone sensitivity, the constant B [compare Eq. (23)] and the relative radiation intensity. F is constant for the run of one PA spectrum. If the thermal diffusion length μ_s of the sample and the absorption coefficient β for one wavelength are known, this factor can be calculated for the respective PA signal according to Eq. (29).

Hence Eq. (29) can be used to correlate corrected PA signals (in arbitrary units) and absorption coefficients β (cm^{-1}). This follows because of the saturation effect (compare case 2b) in the range of larger absorption coefficients to a non-linear scale of the ordinate. A linear dependence of the PA signal on the absorption coefficient exists only for the condition $\mu_s < 1/\beta = l_\beta$. Because $\beta\mu_s \ll 1$ and $\beta^2 \ll \mu_s^{-2}$, Eq. (29) simplifies to

$$S^{PA} = (\beta\mu_s^2/2)F \qquad (30)$$

This corresponds to the borderline cases 1c and 2c [see also Eq. (27)] if constants mentioned there or fixed parameters are included in F.

An extension of the RG theory is the concept of the composite piston: the heat conductivity acoustic waves that are formed in the sample by thermoelastic effects contribute to the signal.

For the simplified model of the composite piston (CP), the PA signal for thermally thick samples is

$$Q = \left(\frac{-iI_0\gamma P_0}{2\omega\rho_s C_s}\right) \frac{\beta\mu_g}{T_0(1 + i)(g + 1)(r + 1)}$$
$$+ \alpha_T[1 - \exp(-\beta l_s)] \qquad (31)$$

Besides the parameters mentioned in the RG theory, the PA signal also depends upon the linear thermal expansion coefficient α_T of the sample, especially in optic transparent samples of larger thicknesses; and at high modulation frequencies, deviations from the RG theory result.

IV. Saturation Effect

In the borderline cases 2a and 2b we get signal saturation, (i.e., the PA signal is independent of the absorption coefficient β of the investigated sample). Equations (25) and (26) show that the PA signal depends only upon the radiation intensity I_0. Since I_0 of the light source has a spectral intensity distribution, the resulting PA spectrum corresponds to the spectrum of the light source.

A body that completely absorbs all incident light over a wide spectral range can be considered as an ideal black body. In PAS such a body can be approximated by a carbon standard, which can be simply made by sooting a glass or quartz plate. The PA spectrum of this standard would, according to case 2b, correspond to the lamp spectrum, to be exact to the combination of lamp and monochromator. In Fig. 7 such a spectrum is depicted. The saturation effect of a black standard can be used to generate a reference signal in PA measurements to correlate the signal of the sample to the radiation intensity I_0 at the same wavelength. This property is exceptionally important for the construction of a two-beam PA spectrometer. In all other cases this saturation effect is mostly unwanted although frequent in strongly absorbing systems.

The conditions for borderline case 2b show that the terminal diffusion length μ_s is larger than the optical absorption length l_β: $\mu_s > l_\beta$. Very small values of l_β are to be expected if the values of β are large. An absorption coefficient $\beta = 10^4$ cm^{-1} corresponds to an optical absorption length $l_\beta = 1$ μm. The thermal diffusion length, however, is in the range of 100–50 μm for modulation frequencies of 100 sec^{-1} (i.e., the conditions for signal saturation $\mu_s > l_\beta$ (case 2b) are given).

Since absorption coefficients of $10^3 < \beta < 5 \times 10^4$ cm^{-1} are very common with many inorganic and especially, organic compounds, one has to be aware of this effect in all PA investigations of solids.

However, PA spectroscopy itself offers the possibility of eliminating the saturation effect. According to Eq. (17), the thermal diffusion length is inversely proportional to the square root of the modulation frequency $\omega = 2\pi f$. By increasing the modulation frequency ω the thermal diffusion length μ_s decreases. By variation of ω, μ_s can be made smaller than l_β, that is, we move from case 2b to case 2c, where according to Eq. (27) the PA signal is linear to the absorption coefficient β.

These remarks point out that the saturation effect is often responsible if the PA spectrum does not correspond to the familiar absorption spectrum. A similar effect is also known in reflection spectroscopy if intensely colored solids are investigated. In such cases the regular remission prevails compared with the diffused part. The problem can be circumvented using the dilution method, by adding an indifferent nonabsorbing standard (MgO, NaCl, BaSO$_4$, SiO$_2$, etc.) to the material to be investigated. This technique can also be applied in PA spectroscopy.

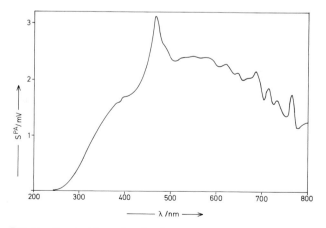

FIG. 7.   Spectral intensity distribution of a xenon high-pressure arc lamp (type Osram XBO/450 W); $S^{PA} = f(\lambda)$ of a carbon black reference; monochromator, Amko type Metrospec (spectral band width 8 nm; blazed at 500 nm); modulation frequency $f = 105$ sec^{-1}).

V. Design of a Photoacoustic Spectrometer

In Fig. 8 the scheme of a two-beam PA spectrometer with computer control is depicted. The optical part with light source and monochromator corresponds to the arrangement in conventional absorption spectroscopy. Beam splitting and test cells are also analogous. Only the photomultiplier is substituted for reference (PA_R) and also for test cell (PA_s) by coupled microphones (e.g., condenser microphones 4166 by Bruel and Kjaer) and their preamplifiers. The microphone signals are fed into two phase-sensitive lock-in amplifiers and via ADC into a computer.

The lock-in amplifiers are locked to the chopper whose frequency can be varied between 2 and about 4000 sec^{-1}. The intensity of the light source can be electronically modulated up to 10,000 sec^{-1}. Since the lamp intensity can be kept constant to ±1% for a longer period of time by a feed-back system, a one-beam arrangement is also possible. In this case the reference spectrum of the carbon standard is retained in the computer so that the successive spectra of the sample can be referred to this standard.

A fixed, partially transmittant mirror serves as a beamsplitter, which passes about 10% of the light intensity to the reference cell and reflects 90% into the sample cell. All results can be supervised during measurements on the monitor of the computer, recorded on a diskette (disk 2), and finally printed or plotted.

The two-beam PA spectrometer depicted in Fig. 8 has a working range from 250 nm to near infrared. To measure within this wide spectral range, the gratings of the monochromator have to be adapted correctly (correct blaze angle). This is also valid for the selection of the light sources. The intensity of the xenon lamp diminishes below 300 nm so that PA spectra for $\lambda <$ 250 nm are usually very noisy. Significantly higher light intensities are obtained with lasers. The advantage of high intensities and hence high PA signals is obscured by the disadvantage that only a few selected wavelengths can be used. By using a tunable dye laser, the wavelength can be varied in a limited range.

Photoacoustic spectrometers with lasers as light sources are individually constructed for special problems. In these cases, quite often electret microphones or piezoelectric detectors are used in gas-coupled cells instead of the mechanically very sensitive condenser microphones. Laser–PA spectrometers are being used for PA investigations in gases in the near IR.

VI. Depth Profiles

Examining the borderline cases of RG theory reveals that the PA signal depends very decisively upon the optical absorption length l_β and the thermal diffusion length μ_s in the sample. Although l_β depends upon the absorptivity of the sample (i.e., upon the wavelength), μ_s is determined by the modulation frequency $\omega = 2\pi f$ according to Eq. (17). The larger ω the smaller μ_s becomes, and the heat-transporting layer shifts towards the sample–gas boundary layer. This behavior can be used to perform depth profiles

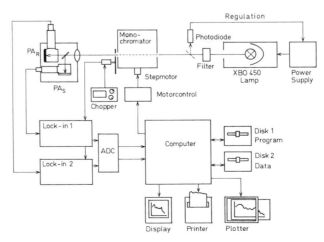

FIG. 8.   Setup of a PA double-beam spectrometer.

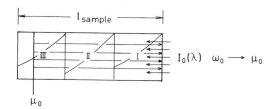

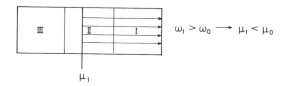

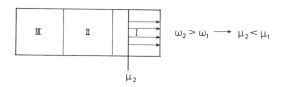

$$\mu = \sqrt{\frac{2\alpha}{\omega}} \quad ; \quad Q_{PA} \sim \frac{\beta \mu_{sample}}{2a_g}\left(\frac{\mu_{sample}}{\varkappa_{sample}}\right) K I_0$$

FIG. 9. Explanation for the measurement of depth profiles.

by PA measurements without destroying the sample. See Fig. 9 for an understanding of this method. If we assume, schematically, three different absorbing layers (e.g., three different dyes) that are arranged on top of each other with relatively pronounced boundaries, dissolved in an evenly distributed matrix, the heat transport is determined by the thermal properties of the matrix, which also depends on the modulation frequency $\omega = 2\pi f$. At low modulation frequencies, μ_s is rather high. The irradiating light is absorbed in all three layers and transformed into heat which generates the PA signal (i.e., the PA spectrum).

In this case the measured spectrum would consist of the superposition of all three PA spectra. Increasing the modulation frequency to ω_1 results in a smaller value for μ_1, and only the heat wave generated in the space between μ_1 and the surface of the sample is effective and delivers a PA spectrum. The proportion of the absorbing layer III in Fig. 9 is not perceived. The PA spectrum is now equivalent to the superposition of the spectra of layers II and I. With a

still higher modulation frequency ω_2, because $\mu_1 < \mu_2$, only that proportion of the PA signal generated in layer I is noticed. The combination of measurements at ω_0, ω_1, and ω_2 in this way delivers a picture of the arrangement of these three layers from inside to outside (i.e., a so-called depth profile).

Examples of this type of use of PAS are the investigation of green leaves in vivo and of petals where the outer layer is usually a protecting wax layer that absorbs in the near UV only. Figure 10 shows the spectrum of a green leaf. At the modulation frequency of 105 sec^{-1}, the PA spectrum shows the chlorophyll- and carotinoid-containing inner layers, whereas at 3600 sec^{-1}, only the outer protecting wax layer can be observed.

Typical three-layer systems are color papers and films. Figure 11 gives an example of the depth profile of an Afga color-reversal paper. Similar PA measurements were also performed on reversal films. This method can also be used to determine the inhomogeneous distribution of an absorbing species in a matrix after diffusion; in such cases a concentration profile from front to back can be found in the sample. The PA signals differ very strongly if measured from the front (diffusion side) or back.

As a special case of a depth profile this can be called lateral specific behavior. In this case, use can also be made of the variation of modulation frequency. Figure 12 shows an example of such

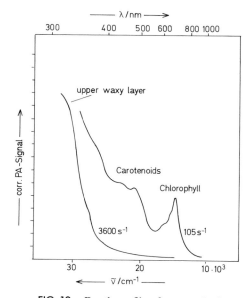

FIG. 10. Depth profile of a green leaf.

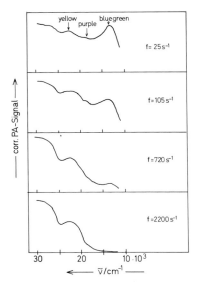

FIG. 11. Depth profile of photographic color paper.

a system: a glass into which colloid silver was diffused from one side, so that from surface to inside there is an inhomogeneous distribution of colloid silver in the glass. The presented PA spectra with modulation frequencies of 15, 105, and 7400 sec^{-1} show a distinctive dependence upon the modulation frequency. The absorption of the glass can be seen in the range $\lambda < 300$ nm across the whole bulk of the sample. The thermal diffusion length μ_s decreases in the three reported measurements with increasing frequency.

The result is that the absorption of the glass in the PA spectra is reduced in relation to the ab-

sorption of the color centers, as can be clearly seen at 7400 sec^{-1}. This allows the conclusion that the color centers are situated directly below the sample surface. In agreement with this conclusion the absorption band of the colloid silver at $\lambda = 400$ nm cannot be detected if the glass is being measured from the back side. The conventional absorption measurement fails completely in such cases, since transmission measurements integrate across the whole bulk.

VII. Determination of Thermal Data

According to Eq. (23), in case 1c (optically transparent and thermally thick) the PA signal is directly proportional to the absorption coefficient β. By continually increasing β, we reach the range of saturation (i.e., the signal becomes independent of β: case 2b, optically opaque and thermally thick [compare Eq. (25)]. The dependence of the PA signal upon the absorption coefficient is given in Fig. 13. For practical evaluations Eqs. (23) and (25) can be combined. With the relations $\alpha_s = \kappa_s/\rho_s C_s$ and $\mu_s = (\omega_s/2\pi f)^{1/2}$ [see Eqs. (15) and (17)], the real part of Eq. (24) for the signal amplitude can be brought into the form

$$S_{PA}^{trans} = \left[K' I_0 \frac{I}{\sqrt{(2\pi f)^3}} \frac{1}{\rho_s C_s} \right] \beta \qquad (32)$$

The slope of the graphic representation in the linear part of Fig. 13 is given by the expression in brackets. Factor K' contains all constant apparatus parameters [compare Eq. (22) and fol-

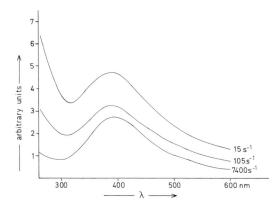

FIG. 12. Colloidal silver that has been diffused into one side of a glass plate. Example for a side-specific photoacoustic effect.

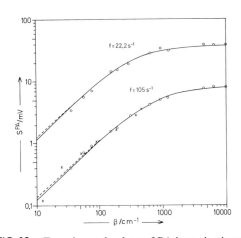

FIG. 13. Experimental values of PA investigations of solid solutions of phenol red (○) and crystal violet (×) in polyvinylpyrolidone (PVP) compared with calculations based on the R–G (———) and CP (– – –) model.

lowing], microphone sensitivity, and the amplification factor for different modulation frequencies. With the knowledge of K', I_0, f, and ρ_s the specific heat of the solid matrix, respectively, of the solvent can be determined from the slope. The light-absorbing and hence the PA signal-generating species are present in the matrix at relatively low concentrations, so the thermal properties of the matrix present in abundance determine the propagation of the PA signals. The dissolved material plays the role of an indicator that does not change the thermal properties of the matrix or solvent. This material, however, must not fluoresce in order to eliminate the competing process of PAS. Within the saturation range in Fig. 13, Eq. (26) is valid and can be written as

$$S_{PA}^{sat} = K'I_0(1/2\pi f)(\sqrt{\alpha_s}/\kappa_s) \qquad (33)$$

Dividing Eq. (32) by (33) and taking Eq. (15) into account gives

$$\alpha_s = [S_{PA}^{trans}(\beta)/S_{PA}^{sat} \cdot \beta]^2 2\pi f \qquad (34)$$

From C_s and α_s, κ_s can be calculated according to Eq. (15). Usable systems for such investigations are, for instance, polyvinyl alcohol (PVA), polyvinylpyrolidon (PVP), polyethylene (PE) as matrix, and phenol red as nonfluorescing indicator compounds. As presented here, thermal data in combination with exact (optical) absorption measurements [compare Eq. (32)] are accessible by means of PA spectroscopy provided the RG theory is valid for the thermal thick case.

VIII. Determination of Fluorescence Quantum Yields

In Fig. 1 it was pointed out that fluorescence is the competing process to the PA effect. If other processes that have to be considered in certain systems are excluded in a first approximation, it is valid for the quantum yield

$$\phi_{IM} = 1 - \phi_{FM} \qquad (35)$$

This equation offers a simple relation between the quantum yield of the radiationless process ϕ_{IM} and that of fluorescence ϕ_{FM}. With the knowledge of ϕ_{IM}, ϕ_{FM} is directly accessible.

The molecule can be excited into the vibrational ground state of the first excited electronic state S_1 by energy of the wavenumber $\bar{\nu}_0$ or into higher vibrational states (see Fig. 1) by $\bar{\nu}_n$, followed by a fast radiationless transition to the vibrational ground state of S_1. From here radiating and radiationless deactivations are compet-

ing. It has to be taken into account that radiating transitions into vibrational excited states of S_0 are also possible; when these vibrational states relax, additional heat is produced (see Fig. 1). Therefore an averaged fluorescence wavenumber $\bar{\nu}_f$ is defined that corresponds to the central point of the corrected fluorescence spectrum. When Eq. (35) is taken into account, the PA signal S^{FM} of a fluorescing sample is

$$S^{FM} = K^{FM}P_{abs}^{FM} \frac{\bar{\nu}_n - \bar{\nu}_0}{\bar{\nu}_n}$$
$$+ \frac{\bar{\nu}_0}{\bar{\nu}_n}\left[1 - \phi_{FM} + \phi_{FM}\frac{\bar{\nu}_0 - \bar{\nu}_f}{\bar{\nu}_0}\right] \qquad (36)$$

The parameter K^{FM} is a constant that includes the thermal properties of the system and apparatus parameters. P_{abs}^{FM} represents the light energy absorbed by the system. For nonfluorescing samples, where all the energy is deactivated radiationless, it is analogous to

$$S^{IM} = K^{IM}P_{abs}^{IM} \qquad (37)$$

Combining Eqs. (36) and (37) delivers for the fluorescence quantum yield

$$\phi_{FM} = \frac{\bar{\nu}_n}{\bar{\nu}_f}\left[1 - \frac{S^{FM}}{S^{IM}}\frac{K^{IM}}{K^{FM}}\frac{P_{abs}^{IM}}{P_{abs}^{FM}}\right] \qquad (38)$$

The determination of fluorescence quantum yields by PA measurements is hence independent of $\bar{\nu}_0$.

Equation (38) can be simplified by standardizing the PA signals S^{IM} and S^{FM} for samples of equal extinction. Under this condition, it is referred to the same absorbed radiation energy in both samples, so that the quotient $P_{abs}^{IM}/P_{abs}^{FM}$ takes the value 1.

According to the theoretical predictions of the RG theory, the constants K^{IM} and K^{FM} of both samples are equal, so that the quotient K^{IM}/K^{FM} can also be assumed to have the value 1. With these suppositions Eq. (38) simplifies to

$$\phi_{FM} = \frac{\bar{\nu}_n}{\bar{\nu}_f}\left[1 - \frac{S^{FM}}{S^{IM}}\right] \qquad (39)$$

In Fig. 14, the PA signals of phenol red (S_{PA}^{IM}) and rhodamine 6G (S_{PA}^{FM}), having similar absorption coefficients but strongly differing fluorescence quantum yields, dissolved in PVA, are depicted as functions of film thickness. Beyond a film thickness of 30 μm, the PA signals are constant in both cases. This means that the backing material does not influence the signal and that the films can be considered thermally thick. These measurements can be evaluated according to Eq. (39). In Fig. 15 the fluorescence quan-

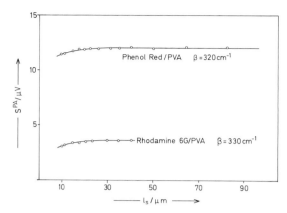

FIG. 14. PA signal of solid solutions of phenol red (nonfluorescent) and rhodamine 6G (fluorescent) in polyvinylalcohol (PVA) of same absorption coefficient but different film thicknesses.

tum yields ϕ_{FM} of rhodamine 6G are depicted with respect to its concentration. For $c \leq 10^{-4}$ mol liter^{-1} in PVA, a value of $\phi_{FM} = 0.85$ is obtained, which agrees very well with values given in the literature. It can also be seen in this figure that at higher concentrations, even in PVA as a solvent, a concentration quenching is observable. Other dyes (rhodamine B, γ-eosin, fluorescein, and acridinorange) dissolved in PVA also show such a concentration dependence.

IX. Phase Angle and Phase-Angle Spectra

The photoacoustic signal can be considered as a complex quantity Q characterized by the real

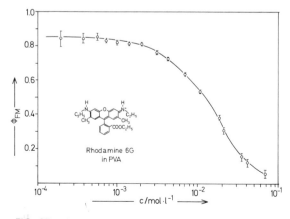

FIG. 15. Quantum efficiencies of rhodamine 6G in PVA at different concentrations.

component Q_1 and the imaginary component Q_2:

$$Q = Q_1 + iQ_2 \qquad (40)$$

From Q_1 and Q_2, the amplitude q and the phase angle ψ of the photoacoustic signal can be derived from Eqs. (21) and (22).

With a phase-sensitive lock-in amplifier for the detection of the PA signal, the values for the real component Q_1 ($A \cos \psi$) and the imaginary component Q_2 ($A \sin \psi$) are obtained directly at the two outputs.

Since the phase of the photoacoustic signal is influenced by the apparatus parameters (such as cell resonance, type of microphone, and microphone amplifier), the detected phase ψ_{exp} corresponds to

$$\psi_{exp} = \psi + \psi_{app} \qquad (41)$$

The value for ψ_{app} is constant for given apparative conditions, so that the change of phase is caused by the photoacoustic signal only.

The phase of the PA signal is independent of the intensity of the irradiating light, the reflection at the sample surface, and the quantum yield, since these factors influence the real and imaginary components in the same way and therefore have no influence on the phase angle given by Eq. (22).

The phase dependence of the PA signal can be traced to three reasons. Phase shifts are caused by

1. the delay in diffusion of the generated heat from the bulk to the surface,
2. thermal expansion of the sample, and
3. the lifetime of the excited state.

The theoretical phase angle of the photoacoustic signal for the six borderline cases given in the accompanying tabulation were derived by Rosencwaig and Gersho [compare Eqs. (23)–(27)].

	Borderline case	Phase angle
Transparent	1a	$-\frac{1}{2}\pi$
	1b	
	1c	$-\frac{3}{4}\pi$
Opaque	2a	$-\frac{1}{2}\pi$
	2b	
	2c	$-\frac{3}{4}\pi$

Based on the RG theory, an expression for thermally thick samples could be derived in which the phase angle is determined by the absorption coefficient β and the thermal diffusion length μ_s

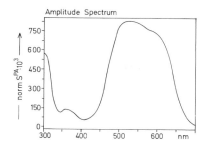

Amplitude Spectrum

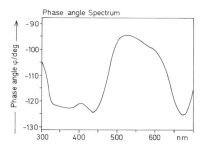

Phase angle Spectrum

FIG. 16. PA amplitude and phase-angle spectra of a saturated aqueous solution of crystal violet (modulation frequency $f = 28$ sec^{-1}, $\lambda_{max} = 526$ nm, $\beta_{max} = 2470$ cm^{-1}, and sample thickness $l_s = 0.65$ cm).

of the sample ($\mu_s = \sqrt{2\alpha/\omega}$):

$$\psi = -\pi + \arctan(1 + \beta\mu_s) \qquad (42)$$

A calibration of the phase angle according to Eq. (42) can be made by using a standard of high optical density (borderline case 2b of the RG theory), whose phase angle is $-\pi/2$ according to this theory. The agreement between experiment and phase angles calculated following Eq. (42) is well established for high values of β. However, the acoustic signal contributed by the thermal expansion of the sample is not considered in the RG theory (CP effect). This effect leads to an inversion of the phase angle with decreasing values of the absorption coefficient β.

The coherence between the phase angle ψ and absorption coefficient β given in Eq. (42) is the basis for the measurement of the phase-angle spectra.

The phase-angle spectrum of a saturated solution of crystal-violet in water is depicted in Fig. 16. The comparison with the amplitude spectrum shows agreement in the range of strong absorption ($450 \leq \lambda \leq 650$ nm). The deviations in the phase-angle spectrum below 450 nm and above 650 nm can be explained in accordance with the basic theory.

X. Photoacoustic and Diffuse Reflectance Spectroscopy

A critical evaluation of PA spectroscopy (PAS) as an analytical method for the investigation and characterization of optical properties of opaque and light-scattering samples has to include the results of diffuse reflectance spectroscopy (DRS) as an established reference method. The preceding theoretical and experimental investigations emphasize the coincidence of both methods by postulating the proportionality of the photoacoustic (S^{PA}) and the photometric ($1 - R_\infty$) signal.

To predict the dependence of the PA spectrum upon the optical properties of the sample, the local intensity distribution function of the heat sources generated by absorption processes including their phase and frequency-dependent contribution to the PA signal has to be calculated as a first step.

The Kubelka–Munk theory (KM theory) associates the diffuse reflectance of scattering materials to phenomenologically defined absorption and scattering coefficients (K, S). Since photometric, as well as photoacoustic, experiments are based upon the same optical processes, the calculation of the heat source distribution with the aid of the K–M theory allows a comparison of the methods. This calculation yields a heat source distribution function that very well represents the overall heat production of an opaque sample irradiated with the light intensity I_0 and measured by calorimetric or photometric methods ($x = -\infty \rightarrow x = 0$):

$$\dot{q} = I_{abs} = I_0(1 - R_\infty)bS \int_{-\infty}^{0} \exp\{bSx\}\, dx$$

$$\dot{q} = I_0(1 - R_\infty) \qquad (43)$$

with $b = (a^2 - 1)^{1/2}$ and $a = (K + S)/S$. Each heat source element $d\dot{q}(x)$ contributes only a fraction of its phase and frequency dependent energy to the volume work of the photoacoustic cell, which functions according to the principle of the Carnot machine; most of the absorbed electromagnetic energy, however, is lost by temperature equalization (increase of entropy). Assuming, in analogy to heat transfer through a material free of heat sources, a damping function of the form exp Ωx with the surface distance x of the source element and a modulation and sample-specific part Ω, we obtain the following expression, proportional to the PA signal:

$$S^{PA} \sim I_0(1 - R_\infty)bS \int_{-\infty}^{0} \exp\{(bS + \Omega)x\}\, dx$$

$$S^{PA} \sim I_0(1 - R_\infty) \frac{bS}{bS + \Omega} \qquad (44)$$

Equation (44) shows that the expressions $(1 - R_\infty)$ and S^{PA} are correlated by a spectral independent factor as well as a term that according to the value of Ω depends upon the spectroscopic properties of the sample. For Ω, a frequency-dependence analog to the material free of heat sources is assumed:

$$\Omega = \sqrt{\omega/2\alpha} \qquad (45)$$

where ω is the angular frequency and α the thermal diffusion coefficient. Equation (44) allows the study of the influence of the modulation frequency upon the difference of DR and PA spectra.

Taking the different optical properties b_1S_1 and b_2S_2 at the wavelengths λ_1 and λ_2 as a substitute for a spectrum, we get the photometric effect

$$\Delta R = (1 - R_{\infty 1}) - (1 - R_{\infty 2}) \qquad (46)$$

The photoacoustic effect, however, is proportional to

$$\Delta S^{PA} \sim (1 - R_{\infty 1})b_1S_1/(b_1S_1 + \Omega)$$
$$- (1 - R_{\infty 2})b_2S_2/(b_2S_2 + \Omega) \qquad (47)$$

From this follows an increasing sensitivity of the PA signal upon changes of optical sample properties with increasing modulation frequency;

this is especially pronounced in samples with low values of R_∞ and S. This influence of the modulation frequency can be seen in Fig. 17, where the PA spectra of a yellow iron oxide pigment (Bayferroxgelb 920) are depicted, by comparing the varying relation of the band maxima and band shoulders respectively at $\lambda = 635$ nm and $\lambda = 460$ nm to each other and relative to the saturation range ($\lambda < 380$ nm). Especially eye catching is the difference between DR and PA spectra in investigations of pigments with semiconductor characteristics; their dyeing properties are determined decisively by position and form of the absorption edge. In such cases, a PA spectrum obtained at high modulation frequencies shows a realistic picture of the relative absorption intensities in the range of the absorption edge while an uncritical KM analysis of the DR spectrum yields a rather distorted equivalent of the absorption spectrum. As an example the DR and PA spectra of a cadmium sulfide pigment (Cadmopur Goldgelb S) are shown in Fig. 18.

In the forementioned consideration only the ideal case of completely diffuse reflected light was treated. However, especially in samples of pure pigments, a partial regular reflectance contributes to the difference between DR and PA spectra, since the PA signal, in contrast to the reflectance, is independent of the angle distribution of the reflected light. This will be demonstrated for two copper phthalocyanine samples (Hostapermblau AR). Figure 19a shows DR and

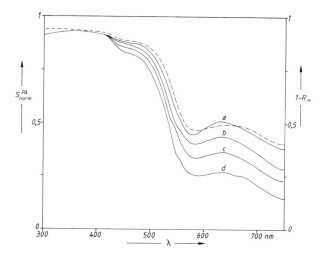

FIG. 17. PA spectra at modulation frequencies of 5 cycles/sec (curve a), 250 cycles/sec (curve b), 1000 cycles/sec (curve c), and 2700 cycles/sec (curve d). Also, diffuse reflectance spectrum (–––) of a yellow iron oxide pigment.

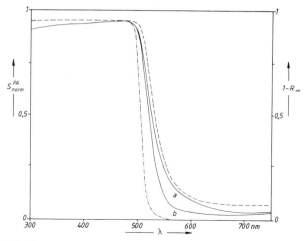

FIG. 18.   PA spectra at modulation frequencies of 5 cycles/sec (curve a) and 1000 cycles/sec (curve b). Also, DR spectrum (– – –) and K–M function (–·–·–) of a cadmium sulfide pigment.

PA spectra of a sample with a dull surface and Fig. 19b the equivalent spectra of a glossy surface made by pressing the sample against a polished plate. At high modulation frequencies, in both cases identical PA spectra are obtained, since the influence of the different scattering ability decreases with decreasing thermal penetration depth.

DR and PA spectra of scattering samples differ in their relative intensities of differently absorbing bands. Because of the signal propagation by a damped thermal wave (which leads to a reduction of the effectively excited layer) the photometric saturation range in the PA spectrum is reduced. On the other hand, a low absorption already causes a strong reduction of reflectance. While the last mentioned effect can be compensated by a calculation of the Kubelka–Munk function $F(R_\infty)$ of a DR spectrum, the photometric estimation of the absorption spectrum of a dark sample is only possible by dilution with a white pigment.

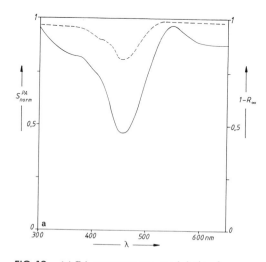

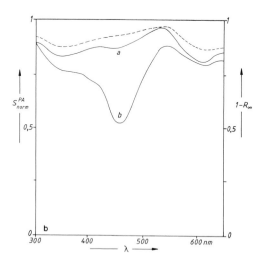

FIG. 19.   (a) PA spectrum at a modulation frequency of 2700 cycles/sec and DR spectrum of a copper phthalocyanine with dull surface. (b) PA spectra at modulation frequencies of 5 cycles/sec (curve a) and 1000 cycles/sec (curve b); and DR spectrum of a copper phthalocyanine with a glossy surface.

In cases in which an alteration of the absorption spectrum is possible by this method (e.g., by adsorption at the white pigment surface or changes of a characteristic crystal structure or size), PAS is the better choice. This is also true for samples with a surface structure or gloss (except for the gloss caused by binding materials) which prevent a photometric or ellipsometric measurement of absorption.

The investigation of fine-structured, small-banded spectra of weakly absorbing compounds and color measurements, however, are the uncontested domains of diffuse reflectance spectroscopy.

XI. Various Applications of Photoacoustic Spectroscopy

A. SOLID SAMPLES

1. Rare Earth Oxides (Ho_2O_3, Er_2O_3). Holmium oxide and erbium oxide have low optical absorption coefficients. Even for low modulation frequencies their PA Spectra are not saturated and can be used for test measurements. By diffuse reflectance spectroscopy, however, a better spectral resolution is achieved because much smaller slit widths than for PA spectroscopy can be used. Thus, $\Delta\lambda_{Rem} > 0.25$ nm and $\Delta\lambda_{PAS} > 2$ nm, impairing the information that can be taken out of the PA spectra.

2. Inorganic Pigments (e.g., Fe_2O_3, Cr_2O_3). The powdery pigments have medium values of the optical absorption coefficient β. At low modulation frequencies the PA signal is saturated. If the pigments are diluted with a white standard (e.g., TiO_2), however, the PA spectra are comparable to the remission spectra. At high modulation frequencies ($f > 1000$ sec^{-1}) even the pure pigments have well-resolved PA spectra. Furthermore, the phase angle spectrum yields information about the absorption properties even if low modulation frequencies are used.

3. Soot. The infrared PA spectra of powdery samples of soot have been recorded. In the visible and ultraviolet range, soot serves as a reference standard. As a practical application, PAS can be used to determine the quantity of dispersed soot on filters for air and exhaust gases.

4. Organic Pigments (e.g., quinacridones metallo–phthalocyanines and azo dyes). The powdery pigments have very high β values. At low

modulation frequencies saturation dominates. Diffuse reflectance spectroscopy does not show a spectral resolution for the pure pigments. However, the dilution technique (e.g., TiO_2) yields spectra that are identical for both spectroscopic methods. At high modulation frequencies ($f > 1000$ sec^{-1}), the PA spectrum of the pure pigment is well resolved in contrast to its remission spectrum. In a dilution series the PA spectra deliver information about interactions of the crystals with the white standard. The determination of different particle sizes in pigmented dyes is possible. The phase-angle spectra contain information about the absorption properties even at low modulation frequencies.

5. ZnS, CdS, CdSe, GaP. The band edge of these semiconductors can be determined by PAS, which is superior to diffuse reflectance spectroscopy.

6. TiO_2. Determination of the band edge and a differentiation between both crystal forms, anatase and rutile, is possible.

7. VO_2(0.8% Al). From the temperature dependence of the PA signal, a phase transition of the polycrystalline material can be observed.

8. TlJ. The purity of a solid TlJ film can be checked by PAS.

9. Potassium Dichromate. The powders of this compound show absorption coefficients in the medium range; their PA spectra are comparable to the solution spectra.

10. Colored Lacquer Layers. These layers contain organic and inorganic pigments with high β values dispersed in lacquer layers (e.g., automobile lacquers). By the use of PAS disturbances of the dispersion, sedimentation and flocculation can be measured.

11. Colored Glasses, Filter Glasses. These glasses have absorption coefficients β in the range of $0 < \beta < 1000$ cm^{-1} and thicknesses between $1 < l_s < 5$ mm. For a glass (e.g., $l_s = 1$ mm and β values above 90 cm^{-1}), the optical transmission T cannot be measured by the use of simple spectroscopic methods because $T < 10^{-4}$ ($A > 4$). The PA spectra, however, are well resolved and allow the determination of spectroscopic data in a transmission range $10^{-4} > T > 10^{-8}$ ($4 < A < 8$). PAS permits nondestructive determination of the dyeing compound (metal cations). Analytical use is made for the nondestructive examination of old glasses (e.g., from church windows).

12. Adsorption at Solid Surfaces. The possibility of a quantitative analysis of adsorbed species by PA methods has been shown. For potassium dichromate and copper sulfate, PAS not only allows a quantitative analysis of adsorbed organic dyestuffs but also investigation of interactions with the substrate or between adsorbed molecules (energy transfer).

13. Sublimated Films. Thin, sublimated films of copper phthalocyanine on quartz plates as backing material can be used as a model system for thermally thin samples. Because of their high absorption coefficients, the PA spectra can be measured easily, even at very small thicknesses ($l_s \approx 20$ nm). Furthermore, the PA spectra of thin films of aromatic hydrocarbons (coronene, hexabenzocoronene, tetracene, and benzo[b]chrysene) can be studied. Since some of these compounds also fluoresce, even in the solid state, photophysical processes in the solid state can be studied by PAS.

14. Polymers. The thermal diffusivity α and the thermal conductivity κ of colored polymers can be determined by PAS (e.g., after mechanical stress); the dyestuff serves only as the producer of the heat, while the heat transfer is determined by the thermal properties of the polymer material. Another application of PAS is the determination of absolute fluorescence quantum yields of a dyestuff dissolved in a solid matrix of PVA or PVP. In the infrared spectral region PA measurements allow a determination of the crystallinity of the polymers.

15. Electrode Surfaces. The optical properties of copper oxide films on Cu electrodes have been measured in situ.

B. SOLUTIONS

1. Phenol Red in Water. Phenol red does not fluoresce and hence allows one to scrutinize the theoretical models of the PA effect (R–G theory and CP model).

2. Methyleneblue, Fe(bipy)$_3$Br$_2$ in Water. Investigations of signal saturation and determination of optical absorption data by evaluation of the phase of the photoacoustic signal have been made.

3. Rhodamine 6G and Other Dyes in Water. Determination of the absolute quantum yield of fluorescence is possible.

4. Eosin in Methanol. Determination has been made of the rate constant k_{TM} for radiationless transition into the triplet state.

5. Rhodamine 6G and Malachite Green in Ethanol. Proof of energy transfer in solution by PAS has been obtained.

DIPHENYLDIAZOMETHANE IN BENZENE (DEGASSED). Determination of the reaction heat on the formation of diphenylcarbene can be obtained.

C. BIOLOGICAL SYSTEMS

1. Cytochrome C in Water. Demonstration of the oxidized and reduced form is possible.

2. Blood. Because of the high absorption coefficient and strong light scattering, absorption measurements of undiluted blood are not feasible. The PA spectrum, however, shows the complete absorption properties.

3. Plant Leaves. At low modulation frequencies the PA spectrum reveals the absorption spectrum of an undisturbed green leaf with the absorption bands of the chlorophyll and the carotinoids. Increasing the modulation frequencies delivers PA spectra in which only the UV protecting wax layer can be recognized (depth profile).

4. Flower Petals. The PA spectra show at low modulation frequencies the absorptions of the flower dyes and proof of the UV wax layer at high modulation frequencies (compare above). Frequency-dependent measurements of PA spectra deliver information about pH depth profiles in flower petals.

5. Marine Algae. PA spectra give information about their absorption bands.

6. Marine Plankton. These are the same as marine algae; however, in this case PA spectra reveal more detailed spectroscopic information than conventional absorption measurements.

7. Hair. PAS studies at different modulation frequencies permit the demonstration of UV protection layers, which can draw by hair treatment.

8. Chloroplasts. PA contributions of photosynthesis in suspensions with and without electron transport inhibitor have been made

9. Cell Membranes. The purple membrane of Halobacterium halobrium contains bacteriorhodopsin bound to retinal. PA investigations with simultaneous radiation from the side allow statements about the cyclic photochemical process in the membrane.

BIBLIOGRAPHY

Hess, P. (1983). Resonant photoacoustic spectroscopy. *In* "Topics in Current Chemistry" (F. L. Boschke, ed.), Vol. 111, pp. 1–32. Springer-Verlag, Berlin, Heidelberg, New York.

Kinney, J. B., and Staley, R. H. (1982). Application of photoacoustic spectroscopy. *Ann. Rev. Mater. Sci.* **12** 295–321.

Pao, Y.-H. (1977). "Optoacoustic Spectroscopy and Detection." Academic Press, New York.

Perkampus, H.-H. (1985). Photo-Akustik-Spektroskopie. *In* "Analytiker-Taschenbuch" (W. Fresenius, H. Günzler, W. Huber, I. Lüderwald, G. Tölg, and H. Wisser eds.), Vol. 5, pp. 93–134. Springer-Verlag, Berlin, Heidelberg, New York, Toronto.

Rosencwaig, A. (1980). "Photoacoustics and Photoacoustic-Spectroscopy" [Chemical Analysis Series, Vol. 57 (P. J. Elsing and J. D. Wineforder, eds.).] Wiley, New York.

Rosencwaig, A. (1980). Photoacoustics and deexcitation processes in condensed media. *In* "Radiationless Processes" (B. D. Bartolo and V. Goldberg, eds.), pp. 431–463. Plenum, New York.

Somoano, R. B. (1978). Optoakustische Spektroskopie in kondensierten Phasen. *Angew. Chemie* **90,** 250–258. (See also international edition.)

Tam, A. C. (1983). Photoacoustics: Spectroscopy and other Applications; In "Ultrasensitive Laser Spectroscopy" (D. Kliger, ed.), pp. 1–108. Academic Press, New York.

West, G. A., Barret, J. J., Siebert, D. R., and Reddy, K. V. (1983). Photoacoustic spectroscopy. *Rev. Sci. Instrum.* **54,** 797–817.

PHOTOCHEMISTRY, ORGANIC

David I. Schuster   *New York University*

GLOSSARY

Chemiluminescence: Emission of light as a consequence of a chemical reaction occurring in the dark, that is, as a result of thermal generation of electronic excited states.

Electronic excited states: High-energy states of a molecule arising from promotion of electrons from orbitals occupied in the electronic ground state to higher-energy, previously unoccupied orbitals, usually by absorption of light energy.

Emission of light: Process whereby an electronic excited state is deactivated to a lower energy state (usually the ground state) with the excess energy given off as radiation; the emission is classified as either fluorescence or phosphorescence, depending on whether it derives from singlet or triplet excited states, respectively.

Energy transfer: Process by which electronic excitation is transmitted intermolecularly or intramolecularly by either collisional or long-range mechanisms.

Excimer: Molecular aggregate formed by loose association of an excited state and a ground state of the same compound, where such association does not occur between two ground state molecules.

Exciplex: Association of an excited state and a ground state of different molecules or of two dissimilar groups in the same molecule, where such an association does not take place for the species in their electronic ground states.

Flash photolysis: Process by which electronic excitation is induced by a short pulse of light (often by lasers), allowing the observation of short-lived reaction intermediates and measurement of ultrafast kinetic processes.

Internal conversion: Nonradiative transition from higher to lower energy states with the same spin multiplicity, e.g., radiationless decay from the S_2 to the S_1 state or the T_2 to the T_1 state of a molecule.

Intersystem crossing: Nonradiative transition between two electronic states of different spin multiplicity, most commonly S_1 to T_1 and T_1 to S_0.

Photosensitization: Process in which the sensitizer or donor molecule is photoexcited and then transfers excitation to an acceptor molecule, resulting in a population of specific electronic excited states of the acceptor molecule.

Quantum yield: Efficiency of a given photochemical or photophysical process, which is the ratio of the number of molecules undergoing that process to the number of quanta of light initially absorbed.

Quenching: Process by which a given reaction or photophysical decay mode of an electronic excited state is inhibited in the presence of an added species; this may be due to energy transfer, electron transfer, chemical reaction, and so on.

Radiationless decay: Process whereby an electronically excited state is deactivated to a lower energy state of the system, with conversion of the excess energy into heat.

Sensitized excitation: Process whereby one molecule (the sensitizer) is excited by light

ENCYCLOPEDIA OF PHYSICAL SCIENCE
AND TECHNOLOGY, VOL. 10

to an electronic excited state, which then induces reaction of some other molecule, usually as a result of energy or electron transfer.

Singlet excited state: Electronic excited state in which the spins of the two electrons in different molecular orbitals are paired.

Triplet excited state: Electronic excited state in which the spins of the two electrons in different molecular orbitals are unpaired.

Triplet sensitizer: Compound that, upon electronic excitation, forms triplet states with high quantum efficiency and then interacts with another molecule by transfer of the triplet excitation.

Organic photochemistry is the chemistry of electronic excited states of organic molecules, which are generated by excitation of organic compounds using appropriate wavelengths of light in the ultraviolet or visible regions of the spectrum. Excited states of organic molecules exist with either paired (singlet) or unpaired (triplet) electron spins, and these states have different physical properties (e.g., energies, lifetimes, and molecular geometries) and different chemical reactivities. These reactivities are very different from the reactivity of the compound in its ground state. Chemical reactions of electronic excited states must be extremely fast to compete with rapid deactivation of excited states by inherently fast radiative (fluorescence and phosphorescence) and nonradiative pathways (internal conversion and intersystem crossing). Typical reactions include isomerization around carbon–carbon double bonds, molecular rearrangements and fragmentations, intra- and intermolecular hydrogen abstraction, cycloadditions, and dimerization.

I. Molecular Excitation

Organic photochemistry concerns the physical and chemical processes that occur as a result of the absorption of light by organic molecules. It is well established that these processes result from the excitation of molecules into higher electronic states, due to the promotion of electrons from occupied molecular orbitals into higher energy unoccupied orbitals. The minimum energy required for such excitation involves excitation of electrons in the highest occupied molecular orbital (HOMO) into the lowest unoccupied molecular orbital (LUMO),

as shown in Fig. 1, and corresponds to the lowest energy (longest wavelength) band in the electronic absorption spectrum of the molecule. For molecules containing unsaturation in the form of $C{=}C$, $C{=}O$, or aromatic functionality, such absorption occurs in the ultraviolet or visible regions of the spectrum. Saturated compounds such as aliphatic hydrocarbons and alcohols do not absorb above 200 nm, and therefore special short-wavelength light sources must be used to excite these systems. Conjugated unsaturation shifts the absorption spectrum to longer wavelengths by reducing the HOMO–LUMO energy gap, and the absorption shifts into the visible region of the spectrum (400–700 nm), resulting in color for highly conjugated systems such as carotene (**1**) and polynuclear aromatic systems such as naphthacene (**2**).

The electronic excited states produced have an excitation energy E given as

$$E = h\nu = hc/\lambda = hc\bar{\nu} \tag{1}$$

where λ is the wavelength of the absorbed light and $\bar{\nu}$ the wave number, which is equal to the reciprocal of the wavelength in units of reciprocal centimeters or in conventional energy units such as kcal/mol, Joules/mol, or electron volts, using the following conversion factors:

$$E\left(\frac{kcal}{mol}\right) = \frac{2.86 \times 10^4}{\lambda \ (nm)}$$

$$= 2.86 \times 10^{-3} \ \bar{\nu} \ \ (cm^{-1})$$

$$= 23.06 \ E \ \ (eV) \tag{2}$$

$$E \ (J/mol) = 8.36 \times 10^{-2} \ \bar{\nu} \ \ (cm^{-1})$$

$$= 2.39 \times 10^{-4} \ E \ \ (kcal/mol) \tag{3}$$

$$E \ (eV) = 4.34 \times 10^{-2} \ E \ \ (kcal/mol)$$

$$= 1.24 \times 10^{-4} \ \bar{\nu} \ \ (cm^{-1}) \tag{4}$$

The amount of energy associated with absorp-

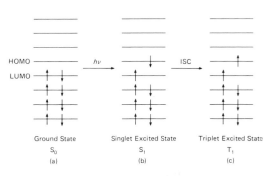

FIG. 1. Orbital description of ground state and singlet and triplet electronic excited states.

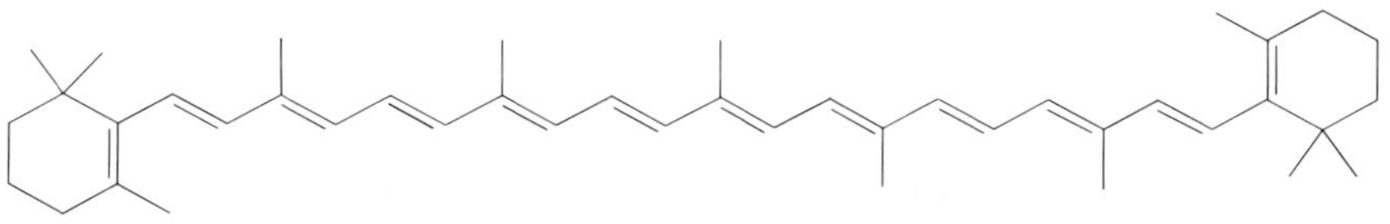

(1)

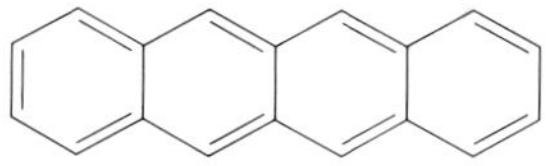

(2)

tion of one mole (6.02×10^{23}) of photons is called an einstein and ranges from 143 kcal/mol at the short wavelength edge of the ultraviolet spectrum (200 nm) to 40.8 kcal/mol at the long wavelength edge of the visible spectrum (700 nm). This light is absorbed in about 10^{-15} sec or one femtosecond, a time that is so short compared to times associated with rotational and vibrational molecular motion (10^{-12} sec or longer) that the excited state must be "born" with exactly the same molecular geometry that it possessed in the ground state at the moment it encountered the photon that induced the excitation.

Because of quantum mechanical restrictions associated with the excitation process that relate to conservation of energy and momentum (discussed in detail in the reference works on this subject), the excited state must be initially produced with the same electronic spin state as the species being excited. In the vast majority of cases, the electrons in the ground state molecule undergoing excitation are all paired; that is, the occupied molecular orbitals each contain two electrons with paired spins, denoted as α and β, corresponding to spin quantum number of $+1/2$ and $-1/2$, respectively. Such an electronic configuration corresponds to a singlet state, so the ground state is designated the S_0 state. In the state produced on direct electronic excitation, where the electrons now reside in different molecular orbitals, as in the HOMO–LUMO situation illustrated in Fig. 1b, the electrons still possess antiparallel spins, and this therefore corresponds to a singlet excited state. The situation in Fig. 1, for example, corresponds to the S_1 state, but there are clearly a large number of higher energy singlet excited states S_n corresponding to excitation of each of the electrons in

the molecule to any one of the empty high-energy molecular orbitals.

Only a finite number of these states (often only one, two, or three) can be populated using conventional ultraviolet or visible light sources, depending on the absorption spectrum of the molecule. In most cases of organic molecules in condensed phases (liquid or solid), the upper singlet states decay to the S_1 state by a process called internal conversion (with energy given off to the surroundings as heat) in a time ($\sim 10^{-12}$ sec) that usually does not allow even for unimolecular chemical reactions to occur from the upper state much less bimolecular reactions that depend upon collisions with other molecules. It is for this reason that wavelength-dependent photochemistry is a relatively rare phenomenon, since usually only the lowest excited states possess lifetimes commensurate with chemical reactivity. However, an increasing number of such cases have been documented in recent years, usually involving unimolecular isomerizations or fragmentations, indicating this rule may require reexamination. [See KINETICS (CHEMISTRY).]

Singlet excited states can spontaneously decay directly to the ground state with emission of a photon by a process known as fluorescence. This radiative process usually occurs in 10^{-7} to 10^{-10} sec. The quantum-mechanical coupling of the S_0 and S_1 states, which involves overlap of the vibrational and the electronic portions of the wave functions of the two states, governs the dynamics associated with light absorption as well as fluorescence. [See LUMINESCENCE.]

The fluorescence spectrum is always redshifted relative to the absorption spectrum of a molecule, and they overlap only (if at all) at the wavelength corresponding to the 0–0 transition, that is, the transition between the lowest vibrational levels in each of the two states. In appropriate cases where the molecular geometry is relatively rigid, as in aromatic hydrocarbons such as anthracene, the two spectra have a mirror-image appearance. Because of the short lifetimes of singlet excited states as a consequence of inherently rapid radiationless decay, there are severe constraints on other processes proceeding from these states. Thus, radiationless decay from S_1 to the ground state is relatively unimportant. However, radiationless decay from S_1 to the triplet state T_1, or in general from S_i to T_j, can be a kinetically important process. This process, known as intersystem crossing, is associated with a flip in the spin of one of the electrons (see Fig. 1c). Since direct excitation of the ground state to triplet states is strongly forbidden because of spin conservation rules, the possibility of populating triplet states directly is vanishingly small (corresponding to molecular extinction coefficients for light absorption of less than 10^{-2}) and is significant only under highly special circumstances. Therefore, the usual route to triplet excited states involves intersystem crossing from excited singlet states, which is also formally a spin-forbidden process, but which occurs in a number of systems because of relaxation of these restrictions due to quantum-mechanical coupling of the spin and orbital angular momenta of the electron. This so-called spin–orbit coupling is particularly favorable for carbonyl compounds (particularly aldehydes and ketones), some heterocyclic compounds, and systems containing a heavy atom (e.g., Br, I, Xe, and metals) which may be present in the molecule or may simply be part of the molecular environment (e.g., the solvent). Thus, the T_1 state of benzophenone is formed with an efficiency of unity on direct excitation of this compound, indicating that intersystem crossing dominates all other possible decay paths of the S_1 state in this case. For simple unsaturated hydrocarbons such as alkenes and dienes, the efficiency of intersystem crossing ϕ_{st} is effectively zero, while this quantity varies widely for aromatic hydrocarbons depending sensitively on molecular structure and the medium.

Since the direct decay from triplet excited states to the ground state S_0 is a spin-forbidden process, it is relatively slow, occurring in milliseconds or seconds, rather than nanoseconds as in the case of singlet excited states. Thus other processes, including even bimolecular chemical reactions, can compete with unimolecular decay

of triplet excited states. Luminescence from triplet states is known as phosphorescence, which is distinguishable from fluorescence mainly by its relatively slow rate, on the order of 10^1–10^4 sec^{-1}. Both types of luminescence are enhanced at low temperatures with samples in rigid matrices such as glasses, where radiationless decay processes including chemical reaction are strongly inhibited.

The various processes that interconvert electronic states of molecules are schematically pictured in Fig. 2, known as a Jablonski diagram.

In principle, chemical reactions can take place from either singlet or triplet excited states, although the restrictions imposed by the inherent excited state lifetime give triplets an advantage as reaction precursors. Since the excited states possess large amounts of energy over the corresponding ground states, a number of reaction pathways exist that have no counterpart in the ground state (i.e., thermally induced) reactions of the same compound. Nonetheless, electronic excitation of organic molecules does not result in random cleavage of covalent bonds despite the fact that the excitation energy typically exceeds the dissociation energies of many bonds in the molecule. That is, of many available paths that are energetically feasible, only a relatively small number are actually observed. This indicates that energy maxima and minima on the potential surfaces of excited states, which depict the changes in the energy of the excited state associated with alterations in molecular geome-

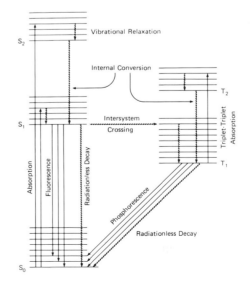

FIG. 2.   Modified Jablonski diagram.

try and with particular reaction pathways, determine the ultimate fate of these species, just as in the ground state. This subject will be discussed later in more detail. [See ORGANIC CHEMICAL SYSTEMS, THEORY; POTENTIAL ENERGY SURFACES.]

II. Methods of Generating Electronic Excited States

Electronic excited states are produced on absorption of light of wavelengths corresponding to promotion of electrons from occupied to unoccupied molecular orbitals. These electronic transitions correspond to the ultraviolet and visible absorption spectra of organic molecules, and excitation using light sources must correspond to those regions of the spectrum where the compound shows such absorption, except in the unusual case of biphotonic absorption using high-energy lasers. There are basically two types of electronic excitation, one involving continuous exposure of the sample to the source and the other involving excitation with light pulses of short duration, known as flash excitation. The first type of experiment is usually associated with light-induced conversion of starting materials to products, which are then isolated and identified, and with certain types of kinetic studies (such as sensitization and quenching, described below). Pulse excitation is most frequently used to study the properties of transient intermediates such as their absorption spectra and the rates of their formation and/or decay upon absorption of a photon by the molecule of interest. Depending on the time resolution of the instrumentation, which has been dramatically improved in recent years, such experiments can be carried out on millisecond, microsecond, nanosecond, picosecond (10^{-12} sec), and even femtosecond (10^{-15} sec) time scales, using a variety of electronic and optical techniques (which are not described here but can be found in the bibliography and current literature). It is now possible to observe directly the singlet and triplet excited states of organic molecules and reaction intermediates derived from them. By carrying out such experiments at very low temperatures (e.g., 8 K) one can detect initial kinetically unstable photoproducts which at higher temperatures are converted into the products that are actually isolated. This long list includes highly reactive systems such as ketenes, cyclobutadiene, and benzyne and highly strained systems such as trans-cyclohexenes

and cyclopropanones. Studies of the formation and decay of transient intermediates from the moment of excitation to relatively long times has allowed specification of the sequence of events involved in many important photochemical processes, such as vision and photosynthesis.

III. Energy Transfer and Electron Transfer

Under appropriate conditions, energy can be transferred intermolecularly from one molecule to another or intramolecularly from one chromophore to another to generate an excited state different from that generated in the initial light absorption step. There are several possible mechanisms for energy transfer. The so-called trivial mechanism involves absorption by an acceptor molecule (A) of the light emitted by the donor excited molecule (D), which is characterized by changes in the donor emission spectrum, no change in the donor lifetime, and no dependence on the viscosity of the medium. A more important mechanism involves Coulombic or dipole–dipole interaction of the donor and acceptor molecules, also known as resonance transfer, the theory of which was worked out by Förster. The rate of energy transfer by this mechanism is inversely proportional to the sixth power of the distance between the donor and the acceptor, the extent of their spectral overlap, and dipole orientation factors. Transfer of singlet excitation,

$$^1D^* + A_0 \rightarrow D_0 + {}^1A^* \tag{5}$$

can take place by the Förster mechanism in favorable cases over long distances, on the order of 50 Å or more. The parameter R_0, which is equal to the separation of donor and acceptor at concentrations of A where the rate of energy transfer is equal to the rate of decay of the excited state of D, shows good agreement with values calculated from Förster theory. Good agreement between theory and experiment is also found in cases of intramolecular energy transfer between donor and acceptor moieties attached to a rigid molecular framework (such as a steroid backbone) at a fixed distance.

Transfer of excitation from a donor triplet to a ground state acceptor to give the donor ground state and an acceptor singlet

$$^3D^* + A_0 \rightarrow D_0 + {}^1A^* \tag{6}$$

by a Coulombic mechanism is predicted by theory to be doubly forbidden and is generally not observed in solution, but has been observed in matrices at low temperature. Triplet–triplet energy transfer,

$$^3D^* + A_0 \rightarrow D_0 + {}^3A^* \qquad (7)$$

does occur in solution but by an exchange interaction (as formulated by Dexter) involving collision of the donor and acceptor molecules. The process is diffusion controlled, and therefore dependent on the viscosity of the medium, in cases where the triplet excitation energy of the donor is greater than that of the acceptor by at least 3 kcal/mol. When this is not the case, the rate of triplet transfer falls off dramatically, and in endothermic processes the deficit in energy appears as an activation energy.

The ideal arrangement for transfer of triplet excitation is shown in Fig. 3, where the singlet excitation energy of the donor is less than that of the acceptor, allowing selective excitation of the donor, while the triplet excitation energy of the donor is greater than that of the acceptor, allowing efficient triplet energy transfer. This situation is rather common, as in the prototypical examples of benzophenone as donor and naphthalene as acceptor, where the triplet transfer rate measured in benzene solution is $5 \times 10^9 \, M^{-1} \, \text{sec}^{-1}$. Analogous intramolecular transfer of triplet excitation is shown by compounds (3) and (4), where excitation of the benzoyl chromophore at 350 nm leads exclusively to phosphorescence of the naphthalene moiety at 460 nm, indicating 100% efficient triplet transfer,

whereas in the steroid (5), triplet transfer is only 35% efficient because of the larger separation of the two chromophores. Of course, it is necessary that the efficiency of intersystem crossing of the donor be very high to produce high yields of acceptor triplets, and this is fortunately the case for a large number of compounds, most notably aromatic ketones such as benzophenone and acetophenone, as well as a number of aromatic hydrocarbons. The relatively long lifetime of the donor triplets helps to insure efficient capture by acceptor molecules in a collisional process.

Since direct excitation of organic molecules can lead to both singlet and triplet excited states, it is not possible a priori to specify which products are derived from each of these states. Using triplet–triplet energy transfer, however, one can generate triplet excited states of acceptors, completely bypassing the corresponding singlet excited states, using appropriate triplet donors known as sensitizers:

$$\text{Sens}_0 + h\nu \rightarrow {}^1\text{Sens}^* \rightarrow {}^3\text{Sens}^*$$
$$^3\text{Sens}^* + A \rightarrow \text{Sens}_0 + {}^3A^* \qquad (8)$$

where Sens is the sensitizer and A the acceptor.

By this technique, the chemistry of the triplet state can be determined without the complication of concomitant singlet reactions. This triplet sensitization technique can also be used to

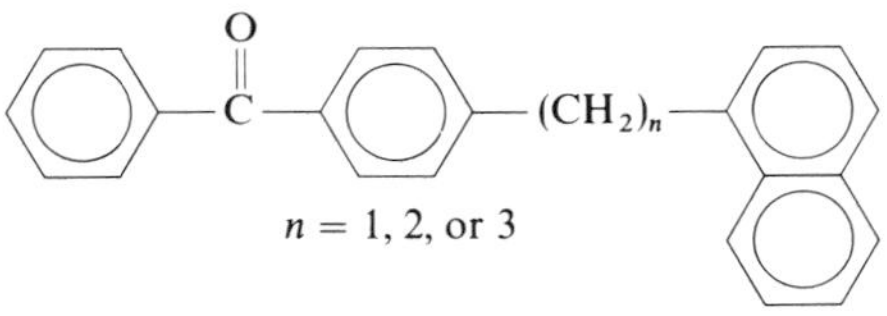

(3)

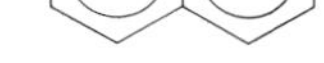

(4)

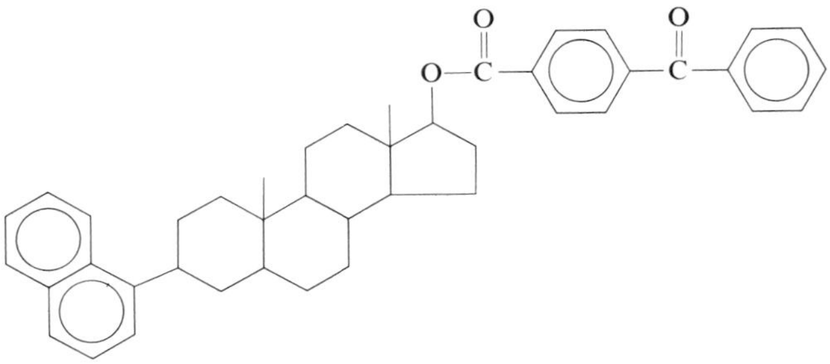

(5)

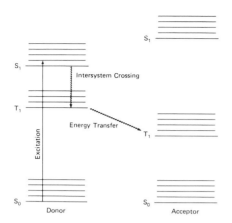

FIG. 3. Schematic description of transfer of triplet excitation.

populate triplet states of molecules, which are formed inefficiently if at all on direct excitation, as in the case of unsaturated aliphatic hydrocarbons, and thus to allow study of the chemistry associated with these states. Conversely, one can use appropriate acceptors as quenchers of the chemistry of triplet donors. For example, photochemical reactions of ketones which proceed via triplet excited states are efficiently quenched in the presence of compounds such as naphthalene or 1,3-dienes as a result of transfer of triplet excitation to these compounds from the ketone triplets.

Sensitizers function not only as energy transfer agents but also in electron transfer processes. The free energy change for a photoinduced electron transfer process is given by the Weller equation,

$$\Delta G = E\left(\frac{D}{D^+}\right) - E\left(\frac{A^-}{A}\right) - E_{0,0} - \frac{e_0^2}{\alpha\varepsilon} \quad (9)$$

where $E(D/D^+)$ is the oxidation potential of the donor, $E(A^-/A)$ the reduction potential of the acceptor, $E_{0,0}$ the excitation energy of the excited species, and $e_0^2/\alpha\varepsilon$ the energy gained by bringing the two radical ions to the encounter distance α in a solvent of dielectric constant ε. The last term is very small in polar solvents (e.g., ~0.06 eV in acetonitrile, which is a common solvent for such studies) but can be significant in nonpolar media.

It is thought that the first step in such processes involves formation of a complex between the excited and ground state donor–acceptor pair. Such exciplexes have been directly detected in several systems by a characteristically broad fluorescence that is red-shifted relative to the fluorescence of the excited sensitizer. If

electron transfer in a donor–acceptor pair is exothermic [i.e., $\Delta G < 0$ in Eq. (9)], radical anions and radical cations may be formed in polar solvents, and reactions are observed that are often quite distinct from those of the same compound in their electronic ground and excited states.

For example, excitation of 9,10-dicyanoanthracene (DCA) in the presence of 1,1-diphenylethylene (DPE) leads to efficient electron transfer, as shown in Eq. (10).

$$DCA + h\nu \rightarrow {}^1DCA^* \xrightarrow{DPE} DCA^{\overline{\cdot}} + DPE^{\overset{+}{\cdot}}$$

$$(10)$$

If the interaction takes place in an alcohol solvent, the result is a net anti-Markovnikov addition to the C=C bond, according to the sequence shown in Eq. (11).

$$[Ph_2C{=}CH_2]^{\overset{+}{\cdot}} + ROH \xrightarrow{-H^+}$$

$$Ph_2\dot{C}{-}CH_2OR \xrightarrow{DCA^{\overline{\cdot}}}$$

$$DCA + Ph_2\overline{C}{-}CH_2OR \xrightarrow{H^+} Ph_2CH{-}CH_2OR$$

$$(11)$$

Another example involves sensitized photooxygenation, in which electron transfer is followed by reaction of the radical anion $A^{\overline{\cdot}}$ with O_2 to give superoxide $O_2^{\overline{\cdot}}$, which reacts with $D^{\overset{+}{\cdot}}$ to give the final product. An example of a sensitizer that can act as an electron donor is 9,10-dimethoxyanthracene (DMA), which with the same alkene leads to electron transfer in the opposite direction, as shown in Eq. (12).

$$DMA + h\nu \rightarrow {}^1DMA^* \xrightarrow{DPE} DMA^{\overset{+}{\cdot}} + DPH^{\overline{\cdot}}$$

$$(12)$$

A full discussion of electron transfer processes is beyond the scope of this article, and the reader is referred to several excellent review articles on this subject.

Excited states can also be generated in special circumstances by ground-state reactions. The phenomenon is well known in chemiluminescent systems, such as that responsible for light production from fireflies, in which high-energy molecules are thermally converted to products in their electronic excited states. Two simple examples are shown in Eqs. (13) and (14).

$$\text{⬡} \xrightarrow{\text{heat}} \text{⬡} + light \quad (13)$$

(6)

$$\underset{(7)}{\overset{\overset{\displaystyle O\text{---}O}{\underset{\displaystyle CH_3\;CH_3}{\underset{\displaystyle |}{CH_3\text{---}\!\!\overset{|}{}\!\!\text{---}\!\!\overset{|}{}\!\!\text{---}CH_3}}}}{}} \;\xrightarrow{\text{heat}}\; \overset{\displaystyle O^*}{\diagup\!\!\diagdown} + \overset{\displaystyle O}{\diagup\!\!\diagdown} \qquad (14)$$

In the first, the highly strained Dewar benzene (6) undergoes thermal ring opening to give an

(6)

excited state of benzene, which can be detected most easily by energy transfer to the highly luminescent 9,10-dibromoanthracene; the unusually large activation energy of the ring-opening reaction, despite its overall exothermicity, is a consequence of the fact that this is a symmetry-forbidden process according to the principles first enunciated by Woodward and Hoffmann. In the second example, the tetramethyldioxetane (7) opens to give a pair of acetone molecules,

$$\underset{(7)}{\underset{\displaystyle CH_3\;CH_3}{H_3C\text{---}\!\!\overset{\displaystyle O\text{---}O}{\overset{|}{}\!\!\text{---}\!\!\overset{|}{}}\!\!\text{---}CH_3}}$$

one of which is electronically excited. Interestingly, only triplet excited acetone is produced. This type of reaction has become a convenient way to generate triplet electronic excited states of ketones in the absence of a light source, and has found utility in a number of studies.

IV. Excitation Sources

There is a large variety of commercial excitation sources, from simple inexpensive hand-held UV lamps to large high-energy lasers. One relatively standard arrangement for continuous irradiation involves high-intensity mercury lamps placed inside cooling jackets surrounded by a sample compartment. Usually there is space for cylindrical glass filters to be placed around the lamp and an arrangement for stirring the solution either magnetically or by a flow of an inert gas through the solution. Aliquots of the solution can be removed for analysis (spectroscopic or chromatographic) either through a stopcock or a sidearm on the main reaction chamber. A variation of this apparatus involves a rotating turntable (merry-go-round) which is placed around the light source in the center with compartments for a number of test tubes, all placed the same distance from the lamp. This allows for irradiation of a number of samples simultaneously, which is useful in exploratory studies (e.g., studies of the effect of different solvents on the photochemistry of a particular system) or mechanistic studies (particularly quenching studies, described below). The entire apparatus can be placed in a cooling bath for temperature control, if desired. A second type of experimental setup places the sample to be irradiated in the center of a chamber containing a battery of lamps around the inside of the apparatus. Lamps with output at different wavelengths are available, and the solution (or several samples) in the center can be stirred magnetically and placed in cooling baths, if desired. This apparatus is available in a large size for preparative-scale photochemical studies. Another type of setup, for smaller sample sizes, utilizes an optical bench, which is a metal rail to which the lamp, lenses, holders for glass or solution filters, and sample holders are attached. This apparatus is particularly convenient for quantum yield measurements (see Section VII) or for studies in which the excitation wavelength needs to be carefully controlled. All of these setups use medium-power lamps, which are sufficient for most organic photochemical research, but very high-power lamps (which usually require special arrangements for efficient cooling) are available for large-scale photochemical studies.

Until recently, flash photolysis systems were mostly homemade, but now a number of microsecond and nanosecond flash systems are available commercially in a variety of designs. With the advent of lasers for excitation in the ultraviolet region of the spectrum, often involving overtones of the basic laser output, a number of organic photochemists have been using laser flash photolysis for studying kinetic processes occurring in the nanosecond time domain. Of course, an appropriate detection system, often controlled by a minicomputer, is necessary for such experiments. Another apparatus in wide use involves measurement of fluorescence lifetimes by single photon counting, also in the nanosecond domain. An electronically integrating actinometer for measurement of quantum yields without the somewhat tedious use of chemical actinometers has recently become available. Finally, a large variety of lamps, lenses, filters, photomultipliers, and monochromators are commercially available for just about any conceivable photochemistry research project.

The above is meant to be illustrative rather than comprehensive. Interested readers can refer to the bibliography and current papers in the literature for more specific information and names of commercial suppliers.

V. Properties of Electronic Excited States

Referring to the Jablonski diagram in Fig. 2, one sees that processes that interconvert states with the same multiplicity (e.g., internal conversion, fluorescence) are generally much faster than those that interconvert states of different spin multiplicity (intersystem crossing, phosphorescence), although there are exceptions such as the very rapid intersystem crossing seen with aromatic ketones. S_1 and T_1 are the most important excited states, at least in condensed phases, since they are most often responsible for the reactions and luminescence seen when the molecules are excited. There are some notable exceptions. For example, the fluorescence of azulene (**8**) is due to decay directly

(**8**)

from the S_2 state to the ground state; in this system, the energy gap between the S_2 and S_1 states is unusually large, which sharply reduces the rate of internal conversion and allows the radiative decay from the upper state to dominate. Another example is anthracene (**9**) and

(**9**)

many substituted anthracenes, which transfer triplet excitation efficiently from their T_2 states in competition with radiationless decay from T_2 to T_1, again due to an unusually large energy gap between these two states. In general, the rate of radiationless decay depends inversely on the energy gap between the two states and on the overlap of the zero point vibrational wave function of the upper electronic state with some high

isoenergetic vibrational wave function of the lower state, since the crossing between the two states (i.e., intersection of potential energy curves) must occur with minimum changes in the position or momentum of the nuclei according to the Franck–Condon principle. In general, molecular rigidity tends to favor radiative over radiationless decay from the S_1 and T_1 states to the S_0 state. Thus, the phosphorescence lifetimes of perdeuterated aromatic hydrocarbons tend to be much higher than their protio analogs at 77 K [e.g., benzene-d_6 (26 sec) versus benzene-h_6 (7 sec); naphthalene-d_8 (22 sec) versus naphthalene-h_8 (2.3 sec)]. While the triplet states of the deutero and protio compounds possess the same excitation energy, this energy corresponds to a higher vibrational level of the ground state in the deutero than in the protio compound, since the vibrational levels are more closely spaced the heavier the molecule. This leads to poorer coupling between the excited state and the ground state and slower (and hence less efficient) radiationless decay for the deuterated compounds. Nearly every deuterated triplet emits, while in the protio compounds only a relatively small fraction of the triplets decay by the radiative pathway, since the radiationless decay pathway predominates.

If we consider the nature of the one-electron excitation that occurs when a simple alkene is irradiated in its longest wavelength absorption band, this corresponds to promotion of an electron from the bonding π orbital (HOMO) to the antibonding π^* orbital (LUMO), as shown in Fig. 4a. This electronic transition is called a $\pi \rightarrow \pi^*$ transition, and the excited state produced is a singlet π, π^* state ($^1\pi, \pi^*$). Such states generally have a low probability of intersystem crossing to the corresponding triplet π, π^* states ($^3\pi, \pi^*$), but these states can be produced using triplet sensitization techniques, as described earlier. The planar configuration of alkenes in their ground states (torsional angle $\theta = 0°$) is due to the overlap of the p-orbitals on the sp^2-hybridized olefinic carbons, which is optimum when these orbitals are parallel, leading to a π-bond. In the singlet and triplet excited states, where one electron is in an antibonding π-orbital (see Fig. 4a) while one electron remains in a bonding π-orbital, to a first approximation there is no longer any net π-bonding, and therefore no compulsion for the molecule to remain planar. In fact, these excited states relax to a more stable geometry, corresponding to $\theta = 90°$, in which the electrons are separated from each other to

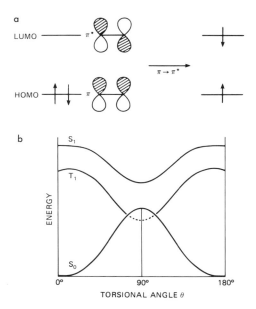

FIG. 4.   Electronic excitation of alkenes.

the maximum extent, reducing Coulombic repulsion. This perpendicular geometry corresponds to the minimum in the potential energy surface of the excited states and to the maximum in the potential energy of the ground state, as shown in Fig. 4b. Thus, the minimum energy separation between the ground and excited state potential surfaces occurs at the perpendicular geometry, which is therefore the most likely geometry (often designated a "hole" or "funnel") for radiationless decay from the excited state back to the ground state. In the case of alkenes, once the molecule finds itself at the maximum point on the ground state surface, it can proceed to either of the more stable planar geometries, with $\theta = 0°$ or 180°, which correspond respectively to cis and trans alkenes when there is appropriate substitution on the double bond. Thus, one can readily understand why direct or sensitized irradiation of alkenes results in facile cis–trans isomerization. Since each of the isomers in a cis–trans mixture can undergo excitation and isomerization, a steady-state equilibrium mixture eventually results, which is called a photostationary state. The composition of this mixture is usually different from that resulting from thermal equilibration of the two isomers.

If we extend this type of analysis to a 1,3-diene, we can see from Fig. 5 that the lowest energy excitation (HOMO to LUMO) represents promotion of an electron from the second to the third π-MO. The relative signs of the p orbitals

in each of the four π-MOs are depicted in Fig. 5. Thus, the lowest $\pi \rightarrow \pi^*$ excitation in this case corresponds to promotion of an electron from an orbital that is bonding between C1–C2 and C3–C4 and antibonding between C2–C3 to one which is bonding between C2–C3 and antibonding between C1–C2 and C3–C4. Thus, the π-bonding between C1–C2 and C3–C4 should be weakened and that between C2–C3 strengthened in the lowest singlet and triplet excited states vis-à-vis the ground state. In the ground state of conjugated dienes, there is restricted rotation (i.e., possibilities of Z and E isomers) around the double bonds and relatively free rotation (giving so-called s-*cis* and s-*trans* conformational isomers) around the central single bond. In the excited state, the situation is expected to be exactly reversed, which indeed corresponds to the experimental observations: isomerization takes place around the terminal C=C bonds (i.e., interconversion of Z and E isomers), whereas rotation around the central bond becomes more restricted. This type of argument can be extended to long-chain conjugated polyenes, such as carotene and retinal, where cis–trans (Z–E) isomerization plays a critical role in the biological activity of these compounds on exposure to light.

The arrangement of orbitals for simple carbonyl compounds (aldehydes and ketones) is slightly more complicated, as in Fig. 6, which shows the energy levels for formaldehyde. In this case, the HOMO is not the carbonyl π-MO but rather the nonbonding p_y orbital on oxygen (designated n) which contains two electrons, us-

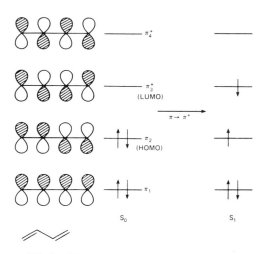

FIG. 5.   Electronic excitation of 1,3-dienes.

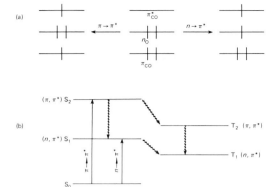

FIG. 6. Orbital and state energy levels of formaldehyde.

ing the coordinate system indicated in the figure. The lowest energy transition represents promotion of an n electron into the carbonyl π^*-MO, but because of the orthogonality of these orbitals this $n \rightarrow \pi^*$ transition is first-order forbidden; although observed in UV spectra of such compounds, it has very low intensity. The higher intensity $\pi \rightarrow \pi^*$ transition occurs at shorter wavelengths in the UV. Thus, in this case the lowest energy singlet excited state S_1 has an n, π^* electronic configuration while the S_2 state has a π, π^* configuration. These two states have different distributions of electron density, different dipole moments, different relaxation modes to minimize their energies, and different chemical reactivities. Corresponding to each of these is a triplet excited state, ^3n, π^* and $^3\pi$, π^*. The singlet–triplet energy gaps are very different for the two configurations, being much larger for π, π^* states than for the n, π^* state. This can be understood qualitatively in terms of differences in electron repulsion in the two states. The electrons in both n, π^* states occupy different regions of space, while there is more to be gained energetically in π, π^* states by requiring the electrons to be separated in the triplet as compared with the singlet excited state. Thus, for simple carbonyl compounds, the two triplet states are much closer in energy than the corresponding singlets, ^3n, π^* being the lowest triplet for formaldehyde, acetone, and benzophenone. For acetophenone, however, the two triplets are so close in energy that solvent polarity governs their ordering: ^3n, π^* is the lowest triplet in nonpolar media and $^3\pi$, π^* the lowest in polar media. Because of differences in the coupling of the two triplets with the ground state, their phos-

phorescence lifetimes at low temperature are quite different, ^3n, π^* states having shorter lifetimes. As a general rule, radiationless transitions between states of different electronic configurations are faster than between states of the same configuration, so intersystem crossing from S_1 (^1n, π^*) to $^3\pi$, π^* occurs in preference to crossing to ^3n, π^* as seen with benzophenone ($k_{st} \sim 10^{11}$ sec^{-1}). In acetone, intersystem crossing from the S_1 state must necessarily take place to ^3n, π^* because $^3\pi$, π^* lies energetically above S_1, and therefore is much slower ($k_{st} \sim 10^9$ sec^{-1}). The S_1 state of acetone is sufficiently longer lived than the corresponding state of benzophenone that prompt fluorescence of acetone but not of benzophenone can be observed, although the efficiency of intersystem crossing is nearly 100% in both cases. Moreover, reactions occurring via acetone singlets (e.g., cycloaddition to alkenes) are known, while no such reactions are seen in the case of benzophenone.

Because of the presence of a lone electron in the nonbonding p_y orbital on oxygen in n, π^* excited states of aldehydes and ketones, such states might be expected to show chemical reactivity analogous to that of alkoxy radicals RO·, and indeed this is the case. Alkoxy radicals abstract hydrogen readily from a variety of donors SH to give alcohols and donor-derived radicals S·, and analogous behavior is shown by prototypical n, π^* triplets such as the benzophenone triplet, as illustrated in Eq. 15. This reaction is the first step in the well-known photoreduction of benzophenone, discovered by Ciamician and Silber over 80 years ago. Further examples will be discussed later.

$$^3Ph_2CO^* + S\text{—}H \rightarrow Ph_2\dot{C}\text{—}OH + S\cdot \quad (15)$$

Other types of unsaturated molecules which possess nonbonding pairs of electrons are also expected to have n, π^* states, such as pyridine (**10**) and aniline (**11**). In the latter case, the orien-

(10)   (11)

tation of the lone pair with respect to the aromatic ring is changeable with rotation around the C—N bond, so a variety of electronic transitions are possible and are indeed observed. When a molecule contains both electron-donating and electron-withdrawing groups attached to the π-system, it is possible to get electronic transitions in which electron density is intramolecularly transferred from the donor to the acceptor group. These charge-transfer transitions are generally of very high intensity, occur at relatively long wavelengths, and are red-shifted in polar solvents, as seen in the case of Michler's ketone (4,4'-dimethylaminobenzophenone).

VI. Excimers and Exciplexes

The fluorescence of polynuclear aromatic compounds such as pyrene (**12**) changes as a

(**12**)

function of its concentration above 10^{-5} M, with a broad emission appearing at longer wavelengths in place of the structured monomer emission. The monomer fluorescence is regenerated if the sample is diluted. The broad fluorescence is attributed to a pyrene excimer, a reversibly formed face-to-face complex of an excited and a ground-state molecule, whose electronic stabilization is considerable (see Fig. 7). With time-resolved fluorescence spectros-

copy, only the monomer emission can be observed 1 nsec after exposure of a 10^{-3} M solution to a brief excitation pulse, while the excimer emission can be detected after 20 nsec and dominates the spectrum after 100 nsec. Excimers can also be formed intramolecularly when the two fluorescent moieties are connected as in compounds of the type (A—(CH$_2$)$_n$—A (**13**). The probability of forming

$$\text{(13)}$$

excimers depends critically on n, which determines the ease of achieving a conformation in which the units A are face to face.

Although triplet excimers are not expected to have the same stabilization energy as their singlet counterparts, several examples have been found, detected as in the singlet analogs by red shifts and broadening of the phosphorescence spectra. For example, compound (**13**), where A is an α-naphthyl group and $n = 3$, forms an intramolecular triplet excimer. Other examples derive from spectroscopic studies of broken dimers, that is, where a molecule derived from face-to-face dimerization of two aromatic systems (e.g., anthracene dimer) is irradiated at low temperatures in a matrix to form the two monomers in close proximity under conditions where they cannot diffuse apart, allowing optimal spectroscopic interaction.

As mentioned earlier, a complex A*D formed between an electronically excited species A* and a ground-state molecule D (or between D* and A) is called an exciplex or, less often, a heteroexcimer. As in the case of excimers, these species can sometimes be detected by their emission spectra. For example, the fluorescence of aromatic hydrocarbons such as anthracene (**9**) is readily quenched by amines such as N,N-diethylaniline (**14**), and a new structureless

(**14**)

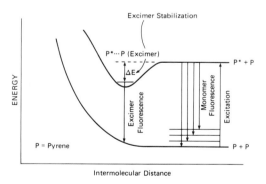

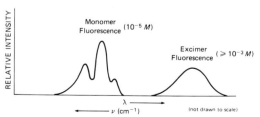

FIG. 7.   Stabilization and fluorescence of excimers.

emission appears to the red of the anthracene fluorescence. There is no concomitant change in the absorption spectrum of the hydrocarbon–amine mixture, indicating no ground-state complexation is occurring. Time-resolved fluorescence indicates that, at very short times after the light pulse, only the fluorescence of the uncomplexed hydrocarbon is observed, while the exciplex emission appears after 100 nsec.

There is very good evidence that exciplexes are stabilized by charge-transfer interaction. Thus, for a series of related donors and acceptors, the fluorescence maxima of the exciplexes are linearly related to the quantity $E(D/D^+)$ − $E(A^-/A)$, as defined earlier [see Eq. (9)]. In the example above, the amine is the donor and the aromatic hydrocarbon the acceptor. Even in cases where the exciplexes could not be seen spectroscopically, the rates of quenching of fluorescence of aromatic hydrocarbons by dienes was shown to depend on the ionization potential of the diene (linear relation observed between $\log k_q$ and IP). From measurements of the dependence of the emission maximum of the exciplex on the polarity of the solvent, it is possible to determine the dipole moment of the exciplex, as illustrated by DCA with 1,3-cyclohexadiene (11 Debye units) and 1,2-dimethoxybenzene (9 Debyes). In some cases where an addition product was formed between A and D, it was possible to show kinetically that the exciplex was an obligatory intermediate along the reaction pathway and not simply an energy-wasting side reaction. The intensity of exciplex emission is usually sharply decreased in solvents of increased polarity, because of rapid dissociation of the exciplex into solvated radical ions D^+ and A^-. The exciplex can decay back to the ground states of D and A in competition with formation of radical ions and/or other products, and the radical ions themselves can revert to ground states of D and A by back electron transfer.

One added wrinkle is that quenching of exciplex fluorescence by a third molecular species (which could be a second molecule of the donor or the acceptor) has been occasionally observed, and it is assumed that this involves formation of a termolecular complex, or terplex. If a new exciplex is formed by substitution of one partner by another, the process has been called S_{ex}, by analogy with nucleophilic (S_N) and electrophilic (S_E) substitution reactions in ground-state chemistry. With time-resolved emission spectroscopy, the conversion of one exciplex D_1A^* to another, D_2A^* has been demonstrated.

VII. Kinetic and Mechanistic Studies

A. QUANTUM YIELDS

The efficiency of a photochemical process, or quantum yield, is defined as the number of molecules undergoing that process divided by the number of quanta of light absorbed in a given period of time:

$$\phi_i = \frac{\text{number of molecules undergoing process } i}{\text{number of quanta absorbed}}$$

(16)

The process may be a radiative transition (fluorescence or phosphorescence), a radiationless transition (internal conversion, intersystem crossing), or a chemical reaction. In the last case, one could measure the quantum efficiency for formation of a given product or series of products and/or disappearance of the reactant. It is frequently important to compare quantum yields for a given reaction on direct excitation vis-à-vis triplet sensitization.

Quantum yields are critical quantities since they indicate the relative importance of the various pathways available in a particular system for dissipation of the energy of the electronically excited molecule initially generated on absorption of light. Thus, a quantum yield for a given reaction of 1.0 would indicate that the pathway leading to the product from the initial electronic excited state is 100% efficient, that is, all the intermediates along the reaction path are produced quantitatively, and side reactions, as well as light emission and radiationless decay back to the ground state, do not take place to a significant extent. Note, however, that this observation alone gives no indication about the actual rate of any of these processes, except that the reactions leading to the product are much faster than the competitive processes available to the system. If the quantum yield for a reaction is much larger than unity, a chain reaction mechanism initiated by light absorption is indicated.

If a given product D is produced by a series of steps, that is,

$$A + h\nu \rightarrow A^* \rightarrow B \rightarrow C \rightarrow D$$

where B and C represent intermediates along the path from A to D, the quantum yield for formation of D is equal to the product of the quantum yields of each of the steps leading from A^*:

$$\phi_D = \phi_{A^*\rightarrow B}\phi_{B\rightarrow C}\phi_{C\rightarrow D} \qquad (17)$$

Thus, if a given product is formed from the triplet state of A, the quantum yield of the reaction ϕ_r is given by the quantum yield of formation of the triplet ϕ_{ST} multiplied by the efficiency with which the triplet goes on to the product:

$$\phi_r^{dir} = \phi_{ST}\phi_r^T \qquad (18)$$

In a triplet-sensitized reaction, the efficiency of the same reaction is given by the efficiency with which triplet energy transfer takes place from the sensitizer to the reactant ϕ_{et} multiplied by the same factor:

$$\phi_r^{sens} = \phi_{et}\phi_r^T \qquad (19)$$

If conditions for completely efficient triplet energy transfer are established, which is easily accomplished, the relationships of Eqs. 20 and 21 hold:

$$\phi_r^{sens} = \phi_r^T \qquad (20)$$

$$\phi_r^{dir} = \phi_{ST}\phi_r^{sens} \qquad (21)$$

Thus, if the quantum yield for formation of product via a triplet state on direct excitation is equal to the quantum yield for the same reaction on triplet-sensitized excitation under conditions where $\phi_{et} = 1.0$, one can safely conclude that the triplet is formed with 100% efficiency on direct excitation.

The lifetime of an excited state is equal to the reciprocal of the sum of the rate expressions for all of the processes available to that state. For a singlet excited state, which can fluoresce with rate constant k_f, form triplets with rate constant k_{ST}, undergo radiationless decay with rate constant k_d, and undergo chemical reaction with a rate constant k_r, the singlet excited state lifetime τ_s is given by

$$\frac{1}{\tau_s} = k_f + k_{ST} + k_d + k_r \qquad (22)$$

The quantum yield of fluorescence in this case is equal to the rate of that process divided by the sum of all the processes for depletion of that state:

$$\phi_f = k_f/(k_f + k_{ST} + k_d + k_r) \qquad (23)$$

or

$$\phi_f = k_f\tau_s$$

In general,

$$\phi_i = k_i\tau_s$$

Thus, the rate constant for a given singlet excited state process can be determined from measurements of the quantum yield of that process and the lifetime of the singlet excited state.

The corresponding expression for a process i occurring via a triplet excited state is

$$\phi_i^T = \phi_{ST}k_i^T\tau_T \qquad (24)$$

If a given photochemical reaction occurs rapidly in the laboratory time frame, the light energy is being used efficiently by the system, which means that the quantum yield for reaction is relatively large. For the example of Eq. 22, this means that k_r is large relative to k_f, k_{ST}, and k_d, but it does not say anything about the absolute magnitude of any of the rate constants. One can easily imagine a situation in which a reaction of a certain compound A is more efficient than that of compound B (meaning that more product is formed per unit time on irradiating A than on irradiating B using a given light source), even though k_r for A is smaller than k_r for B, because the lifetimes of the excited states are different, that is, $k_r^A\tau_{s(t)}^A > k_r^B\tau_{s(t)}^B$. The reactivity of any chemical species, including electronic excited states, is reflected by the rate constants for the various processes by which that species reacts or decays. Thus, discussions of chemical reactivity of excited states should focus on rate constants and not on quantum yields, although the latter are obviously important parameters that are directly related to the chemical yields of the various photoproducts.

There are a number of techniques available for measurement of quantum yields for formation of products or disappearance of starting materials. For a direct measurement, one has to measure the light flux through the sample (usually over a narrow wavelength range, requiring use of a monochromator) and changes in the quantities of the materials of interest. The latter is readily accomplished by standard spectroscopic or chromatographic analysis. For the former, one usually calibrates the optical system using a chemical actinometer that undergoes a readily measurable change (such as UV absorbance) of known quantum efficiency, so that the magnitude of the change can be related to the light intensity (einsteins per unit time). The most commonly used system is potassium ferrioxalate, which undergoes reduction to the Fe^{2+} state (easily measured by the UV absorbance of its phenanthroline complex) with well-established efficiency. More recently, a variety of easily handled reversible photochromic systems have been employed for this purpose. Many other actinometers whose parameters are suitable for the purposes of specific studies have been utilized. The usual procedure employs a focused beam of monochromatic light on an optical bench (see Section IV), with splitting of the beam into two beams at right angles, one proceeding to the sample cell (with a backup cell to measure light that passes through the sample) and the second beam to a cell containing an actinometer. Recently, an apparatus has become available that accomplishes the same purpose without use of chemical actinometry, by measurement of the flux difference in front of and behind the sample cell by the absorption-fluorescence capacity of a Rhodamine B scintillator coupled with semiconductor photoelements.

An alternative procedure involves simultaneous irradiation (usually using a merry-go-round experiment) of the unknown sample and a chemically related system whose quantum efficiency has been previously determined. The success of this approach depends on making sure that both samples have equivalent optical densities at the irradiating wavelengths (assuring equal light absorption) and equivalent exposure to the light source, which typically is a high-intensity broad-spectrum lamp of the type used for preparative-scale irradiations.

For measurements of fluorescence and phosphorescence quantum efficiencies, one uses appropriate light-emitting standards and compares the integrated intensity of the emission over a similar wavelength range, correcting for differences in light absorption (optical density) of the sample and the standard. Full details are given in standard reference works, including several in the bibliography.

B. Lifetimes of Electronically Excited States

The direct measurement of excited state lifetimes is not possible using the standard kinds of apparatus available in most organic photochemistry laboratories. The basic procedure is to produce relatively high concentrations of excited states, which are detected spectroscopically, and to measure the rate of their decay. This obviously requires very short but very intense excitation pulses and appropriate time resolution in the detection system. For many (but not all) triplet states, excitation with a microsecond or even millisecond pulse provides sufficient time resolution, with spectroscopic detection in the ultraviolet–visible region using a second light beam (usually at right angles to the exciting beam). This requires appropriate photomultipliers and display capability so that the decay of the absorption of the transient intermediate can be monitored directly. Of course, one must demonstrate that the transient species being observed is indeed a triplet state and not some other species (such as a free radical) derived from the triplet; such a distinction is usually based on quenching experiments using oxygen and standard triplet quenchers such as naphthalene, ferrocene, and 1,3-dienes, which increase the rate of triplet decay. For analogous studies of singlet excited states, whose lifetimes are inherently much shorter, a nanosecond or even picosecond pulse width is required, as produced from a laser. Often the frequency of the light emitted by the laser must be doubled or tripled (using appropriate crystals) to obtain wavelengths in the UV suitable for excitation of organic compounds. The detection system is necessarily more sophisticated to enable kinetic measurements in the nanosecond time domain. Such experiments often involve collaboration of organic photochemists with physical chemists or the use of equipment in special government-sponsored laboratories, although an increasing number of organic photochemists have either built or purchased their own apparatus for such studies.

Another way to measure lifetimes of singlet excited states is to monitor the rate of decay of their fluorescence. There are various ways to do this, the most common of which is the single-photon counting technique. This involves nanosecond pulse excitation of the sample and measurement of the time delay between excitation of the sample and detection of a single photon of the emitted light (following filtering out of random noise) at a very sensitive photomultiplier. The experiment is repeated many times per second, and the data (in terms of number of photons detected as a function of their delay times) are stored in a multichannel analyzer and displayed, when desired, in the form of a decay curve. This can then be analyzed in terms of the best fit to one, two, or more simultaneous exponential decays. In cases where the fluorescence lifetime is short and becomes comparable to the width of the excitation pulse, deconvolution of the experimental data using standard computer programs can be used to extract the true sample fluorescence decay times.

Phosphorimeters are usually used to measure phosphorescence lifetimes. A standard microsecond flash apparatus can also be adapted for such measurements by monitoring the emission rather than the UV absorption of the transient triplet excited state.

C. Quenching Studies

Flash experiments can be used to measure rates of quenching of excited states directly by monitoring transient lifetimes as a function of the quencher concentration. The quenching simply provides another route for decay of the excited states, and consequently, the overall decay rate increases (i.e., the excited state lifetime decreases) linearly with the quencher concentration:

$$\frac{1}{\tau} = \frac{1}{\tau_0} + k_q[Q] \quad (25)$$

where τ_0 is the lifetime in the absence of quencher Q and k_q the quenching rate constant.

To determine rates of quenching by oxygen, one can measure fluorescence decay rates or fluorescence intensities of samples that are degassed, purged with air, or purged with oxygen, corresponding to three different oxygen concentrations. Oxygen solubilities in various solvents given in standard reference sources can then be used to obtain values of k_q using Eq. (25).

The quantum efficiency of process i, which proceeds from an electronic excited state A, will be reduced in the presence of some species Q that quenches that excited state; the quenching mechanism may involve energy transfer, exciplex formation, or chemical reaction. If the quenching is a bimolecular interaction of the quencher Q and the excited state A* with rate constant k_q, the quenching expression is given in Eq. (26), which is the Stern–Volmer equation:

$$\phi_i^0/\phi_i^Q = 1 + k_q\tau_A[Q] \quad (26)$$

where ϕ_i^0 is the quantum yield for process i in the absence of quencher Q, ϕ_i^Q the quantum yield for the same process in the presence of the quencher at a given concentration [Q], and τ_A the lifetime of the excited state in the absence of the quencher. Thus, a plot of the relative efficiency of process i as a function of [Q] should be a straight line with an intercept of 1.0 and a slope of $k_q\tau_A$. For example, the fluorescence of 9,10-dicyanoanthracene (DCA) in acetonitrile is quenched by 2,2,7,7-tetramethyl-3,5-cycloheptadienone (15), as shown in Fig. 8a. A plot of the

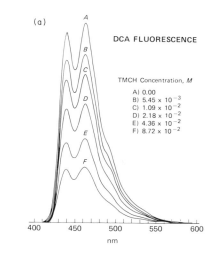

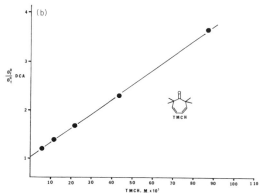

FIG. 8. Quenching of the fluorescence of 9,10-dicyanoanthracene by compound (15): (a) spectra and (b) Stern–Volmer plot.

been shown to lead to radical ions by electron transfer. Using Eq. (26) in the opposite way, one can determine excited state lifetimes if k_q is known or can be estimated. For example, a value of $k_q = 2 \times 10^{10}$ M^{-1} sec^{-1} has been observed for quenching by oxygen of the fluorescence of a large number of compounds; this value can therefore be used in Eq. (26) to obtain singlet lifetimes from data for fluorescence quenching by oxygen.

The most common application of Eq. (26) is in studies of quenching of reactions proceeding from triplet excited states. Direct measurements by flash techniques of rates of quenching of triplet states by the triplet energy transfer mechanism demonstrate that this collisional process occurs at a diffusion-controlled rate as long as the triplet excitation energy of the donor is greater than that of the acceptor by at least 3

(15)

relative intensity of the fluorescence versus the concentration of (15) is a straight line (Fig. 8b). From the slope and the known lifetime of ^{1}DCA* in acetonitrile (15.2 nsec), a value for k_q of 2.0 $\times$ 10^9 M^{-1} sec^{-1} was obtained; this interaction has kcal/mol, in which case values of k_q (for a given donor and acceptor) are found to vary inversely with the viscosity of the medium. For example, k_q for quenching of benzophenone triplets by naphthalene is $5 \times 10^9 \, M^{-1} \, \text{sec}^{-1}$ in benzene and solvents of similar viscosity. Parenthetically, flash experiments conclusively show that this process results in the formation of naphthalene triplets which can be detected by their characteristic UV absorption. The rate of triplet energy transfer falls off markedly when the process is endothermic.

If the triplet energy of the donor and acceptor are known with some degree of certainty (e.g., from phosphorescence spectra or by analogy to other compounds of related structure), one can choose conditions such that quenching by a triplet transfer mechanism will occur at a known (i.e., diffusional) rate. In that case, Stern–Volmer analysis of quenching of photoreactions allows a good estimate of the triplet lifetime of the reacting state. For example, the lifetimes of aromatic and aliphatic ketone triplets estimated from quenching of their photoreactions by conjugated dienes and by aromatics such as naphthalene, assuming diffusional rates, are in excellent agreement with lifetimes subsequently determined by laser flash photolysis. Figure 9 shows an example from our studies of the photochemistry of phenanthrone (**16**) which gives five

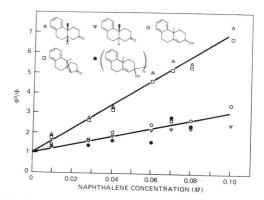

FIG. 9. Quenching by naphthalene of the photochemical transformations of compound (**16**): Stern–Volmer plots.

different products on irradiation in isopropyl alcohol. Triplet quenching studies using naphthalene demonstrate clearly that two of the products arise from one triplet state, postulated to be a $^3\pi, \pi^*$ state, and three from another kinetically distinct triplet, probably the $^3n, \pi^*$ state. The slope $k_q\tau_T$ of the steeper plot agrees exactly with values of k_q and τ_T obtained independently on laser flash excitation of this compound in isopropyl alcohol.

D. MECHANISTIC STUDIES: DIRECT CHARACTERIZATION OF REACTION INTERMEDIATES

Many types of reaction intermediates can be generated from electronic excited states. The nature of the products often is strongly sugges-

tive of the intermediacy of specific reactive species—for example, carbocations, carbenes, and free radicals—by analogy to the chemical behavior of such intermediates when generated thermally. More often than not, the chemistry observed is very similar when the same reactive intermediates are generated in solution either thermally or photochemically, so such comparisons are indeed valid. In addition, techniques are available, as will be described in the following, for the direct observation of reaction intermediates, which can be used to confirm their role in photochemical transformations. However, the sensitivity of these techniques is such that molecular species are sometimes detected that are not in fact intermediates in the reaction of interest but are involved in a side reaction or decay process of the excited state. Thus, a mechanistic connection must be made between an intermediate detected by the techniques described below and the photochemical reactions of the system under investigation.

1. Electron Spin Resonance

Chemical entities with one or more unpaired electrons can be detected by electron spin resonance (ESR), which has been used to study organic free radicals, triplet diradicals and radical pairs and triplet carbenes. [*See* ELECTRON SPIN RESONANCE.]

Each species has its own characteristic spectrum, which can be used for diagnostic purposes. For example, the $(CH_3)_2\dot{C}OH$ radical produced by irradiating acetone in isopropyl alcohol has a characteristic seven-line spectrum resulting from coupling of the nuclear spins of the methyl protons with the electron spin on the carbinol carbon. Since a steady-state free radical concentration of about 10^{-8} M is necessary for observation using ESR, high-intensity light sources are often required for these experiments. One can similarly observe the ESR spectrum of $Ph_2\dot{C}OH$ radicals generated by irradiation of benzophenone in appropriate media [see Eq. (15)]. The same species has been detected by using conventional microsecond flash photolysis. This radical has a characteristic absorption at 500–550 nm, and the rate of its decay by second-order kinetics leading to benzpinacol [Eq. (27)] was directly measured.

$$2\ Ph_2\dot{C}OH \rightarrow Ph_2C(OH)C(OH)Ph_2 \quad (27)$$

The ESR spectra of radical ions resulting from photoinduced electron transfer can be observed. A simple example is the spectrum of $DCA^{\cdot}$ which is seen on excitation of DCA in the presence of a variety of electron donors.

It is usually not possible to observe ESR spectra of triplet carbenes, diradicals, and radical pairs in fluid media, and recourse must be made to studies in matrices at low temperatures. For example, the characteristic spectrum of triplet methylene $:CH_2$ produced on irradiation of diazomethane CH_2N_2 in various matrices at 77 K has been observed. Detailed analysis of the spectrum showed that the H—C—H angle in triplet $:CH_2$ was 136° and not 180° as had been predicted theoretically and initially reported from spectroscopic data. Many other triplet carbenes have been generated and observed using this technique, and it has been concluded that in most cases (the main exception being some halomethylenes) the triplet carbene is of lower energy than the corresponding singlet. Some triplet diradicals that have been generated photochemically from the corresponding azo or diazo compounds in matrices at low temperatures and detected by their ESR spectra are shown below:

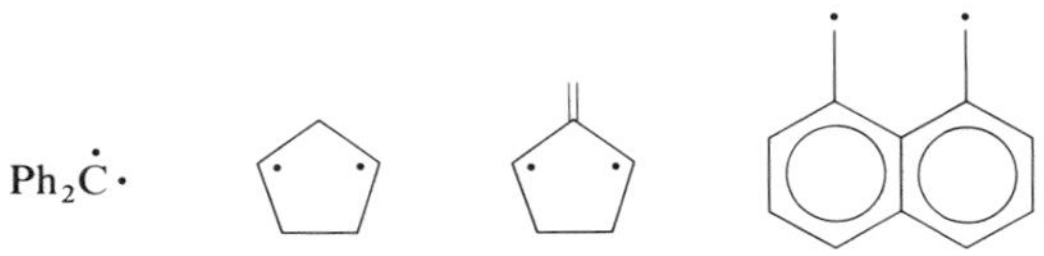

Ph$_2\dot{C}\cdot$

In an analogous manner, triplet radical pairs produced by the photofragmentation of the corresponding azo compound in a matrix at low temperature can be detected by ESR. In the following case, an average separation of the unpaired electrons of 6–7 Å has been determined:

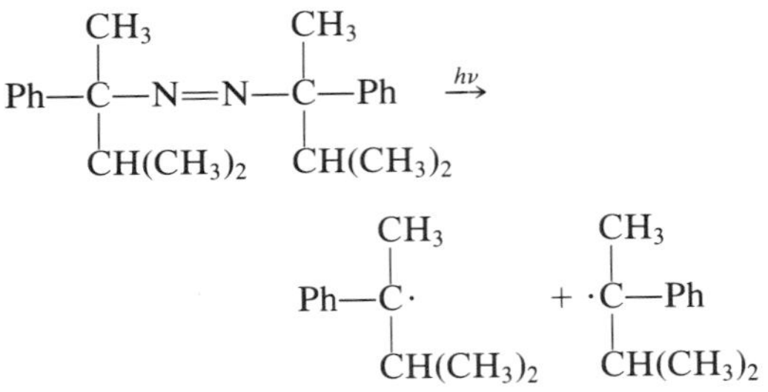

2. Chemically Induced Dynamic Nuclear Polarization

An increasingly common way to detect the intermediacy of radical pairs in photochemical reactions is the use of chemically induced dynamic nuclear polarization (CIDNP), in which enhanced signals (absorption and emission) in the nuclear magnetic resonance (NMR) spectra of photoproducts and/or starting materials are observed. A detailed analysis of polarized spectra, using the radical-pair theory of Kaptein, Closs, and Oosterhoff, allows (1) determination of the nature (singlet or triplet) of the radical-pair precursor to the photoproduct and therefore of the multiplicity of the excited state leading to the radical pair, and (2) differentiation between formation of the product by coupling of radicals in a solvent cage or by reaction of radicals that have escaped from the initial solvent cage. This technique represents an important advance in the use of NMR for mechanistic investigations in photochemistry; the interested reader is referred to Turro's text and to primary references in this field for more specific details.

As an example, it could be shown from CIDNP studies that excitation of the *trans* azo compound (**17**) gives the corresponding *cis* isomer (**18**), as shown in Eq. (28), which then thermally breaks down to give a geminate diazenyl-cumyl singlet radical pair (**19**), which can recombine in the solvent cage to regenerate the starting material [Eq. (29)]. Caged radical pairs are indicated schematically by a solid line above the two radicals. Radical pair (**19**) competitively loses N_2 to give a longer-lived phenyl-cumyl secondary radical pair (**20**), which is the direct precursor of α-methylstyrene, benzene, and 2,2-diphenylpropane [Eq. (30)]. Escape of cumyl radicals from either of the two cages leads to "free" radicals that give rise to bicumyl and cumene [Eq. (31)]. It should be explicitly noted that the only photochemical process is the conversion of (**17**) into (**18**), the remainder being the chemistry of the photogenerated free radicals. The entire scheme is shown below:

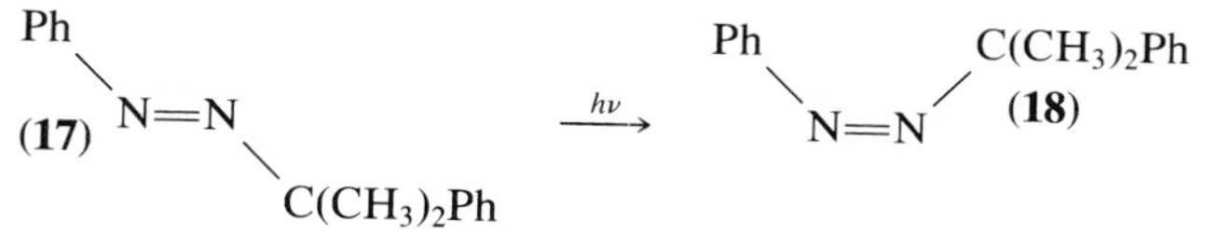

$$(17) \quad \overset{Ph}{\underset{C(CH_3)_2Ph}{N=N}} \quad \xrightarrow{h\nu} \quad \overset{Ph}{\underset{}{N=N}} \overset{C(CH_3)_2Ph}{} \quad (18) \tag{28}$$

$$(18) \to \overline{PhN_2\cdot \quad \cdot C(CH_3)_2Ph} \to (17) \tag{29}$$
$$(19)$$

$$(19) \to \overline{Ph\cdot \quad \cdot C(CH_3)_2Ph} + N_2 \to \tag{30}$$
$$(20)$$

$$PhH + PhC(CH_3){=}CH_2 + CH_3C(Ph)_2CH_3$$

$$(19) \text{ and/or } (20) \to \text{free } Ph\dot{C}(CH_3)_2 \to PhC(CH_3)_2H + PhC(CH_3)_2C(CH_3)_2Ph \tag{31}$$

CIDNP has also been used in studies of photoinitiated radical-ion reactions. Examples include radical cations formed on photoinduced electron transfer from cyclopropanes and dienes to quinones such as chloranil.

3. Matrix Isolation: Infrared Spectroscopic Detection of Reactive Intermediates

It is now generally appreciated that a number of photochemical reactions at ambient temperatures involve initial formation of highly reactive compounds that undergo secondary thermal transformations to give the observed products. By carrying out the irradiations at very low temperatures under the conditions of matrix isolation, one can detect and identify these initial photoproducts. The use of inert matrices, such as argon at 8 K, not only limits the thermal activation of the system but also physically isolates reactive intermediates and prohibits their translational as well as rotational motion. Thus, irradiation of the lactone (21), itself derived photochemically from α-pyrone (22), in an argon matrix at 8 K gave cyclobutadiene (23), identified by its characteristically simple infrared (IR) spectrum [Eq. (32)].

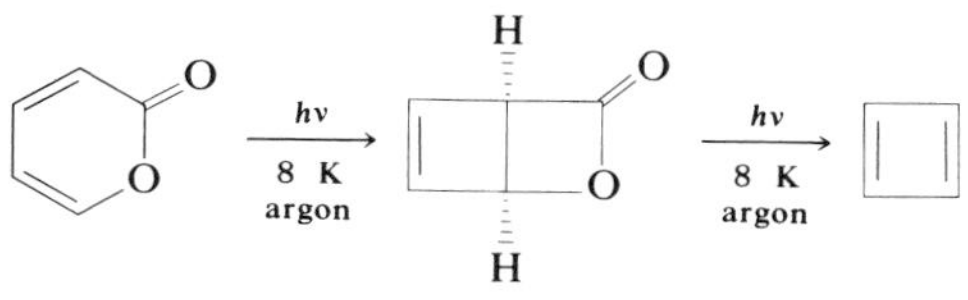

(22)   (21)   (23)

$$(32)$$

This was the first synthesis of (23) under conditions where it was sufficiently stable to be examined spectroscopically. The IR absorption of (23) disappears on over-irradiation or warming of the system above 35 K, and cyclobutadiene

dimers, the products of room-temperature irradiation, are formed. Cyclobutadiene (23) was also formed by irradiation of pyridine in argon at 8 K, presumably via initial formation of the bicyclic isomer shown in Eq. (33).

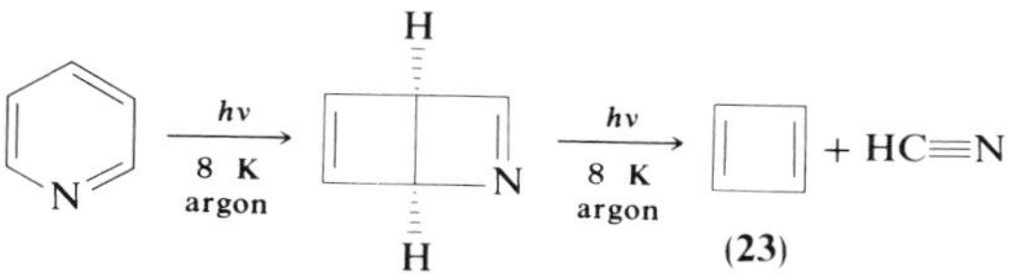

$$(33)$$

Another example is phenyl azide, whose photochemistry is of interest in connection with the use of aryl azides in labeling biologically active sites in macromolecules and commercial use in photoresists. A number of products are formed when phenyl azide is irradiated in solution at ambient temperatures, but at 8 K in argon, the only product is the azacycloheptatetraene (24) [Eq. (34)], which is indeed the precursor of the observed products. It is not clear whether (24) is formed directly from the electronically excited azide or from singlet phenylnitrene.

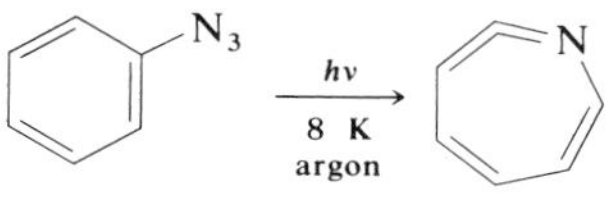

$$(34)$$

(24)

A final example involves the photochemical isomerization of the bicyclic ketone (25a) into (25b), which was initially rationalized in terms of a one-step concerted electron redistribution process. Low-temperature irradiations with IR detection showed that a thermally unstable ketene is actually an intermediate in this reaction, as shown in Eq. (35). The ketene is stable in the dark below −180°C but rapidly disappears to give a mixture of (25a) and (25b) at higher temperatures.

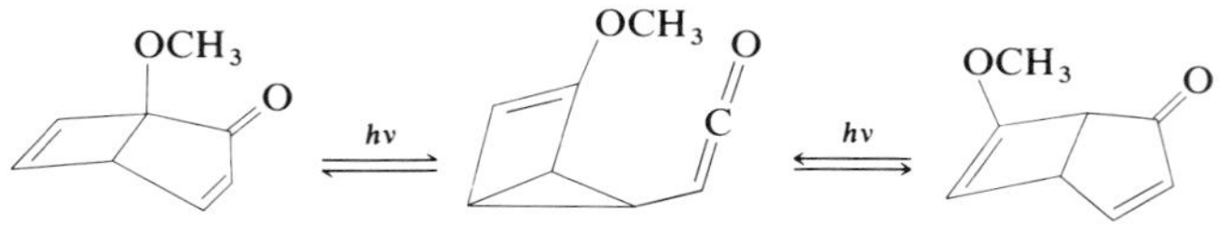

(35)

(25a)   (25b)

These three examples demonstrate the importance of studying photochemical reactions at very low temperatures to elucidate the nature of the primary photoprocesses that occur on photoexcitation of organic molecules, since many steps may intervene on the way to the products observed under ambient conditions.

4. Detection of Reactive Intermediates by Flash Photolysis

It has already been mentioned that flash photolysis has been used to detect free radicals and to measure the kinetics of their formation and decay in light-induced reactions. The classic example involves the ketyl radical $Ph_2\dot{C}OH$ formed on excitation of benzophenone (Ph_2CO) in the presence of benzhydrol Ph_2CHOH in benzene solution. A more recent example from our own work involves formation of the radical cation of DABCO (26) accompanied by solvated

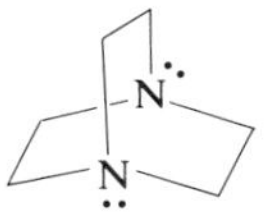

(26)

electrons when this diamine is excited at 254 nm in a polar solvent such as acetonitrile. The radical cation of (26), which has an absorption centered at 460 nm, is relatively long lived under these conditions. The same absorption appears when the cyclic enone (27) and related compounds

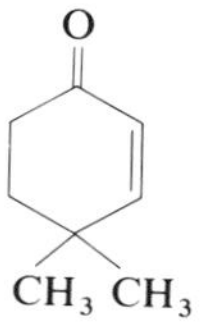

(27)

pounds are excited at 353 nm in the presence of DABCO using a laser pulse, indicating electron transfer is occurring. The rate of growth of the absorption due to the DABCO radical cation can be readily measured as a function of DABCO concentration, giving both the rate constant for the electron transfer step and the lifetime of the

triplet state of (27), which is the electron acceptor under these conditions.

When 1-phenylcyclohexene (28) is excited in

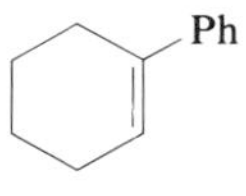

cis-(28)

solution with a laser pulse at 265 nm, a transient absorption is seen in the 300–430 nm range, which is unaffected by oxygen. This species is, however, sensitive to acid. It has been concluded from a detailed kinetic analysis that the transient is the ground state trans-(28), which is

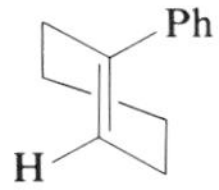

trans-(28)

necessarily highly twisted, and that the reaction with H^+ involves formations of carbocation (29),

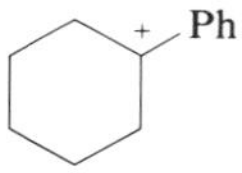

(29)

which in alcohol solutions gives ethers (30). It

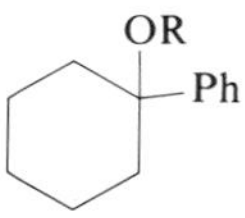

(30)

had been previously proposed that twisted trans-cycloalkenes might be intermediates in the acid-catalyzed photoaddition of alcohols to cycloalkenes, but this was the first firm evidence to substantiate such a mechanism. The formation of dimer (31) on direct or triplet-sensitized excitation of (28) in neutral methanol at −75°C is the result of a Diels–Alder reaction between the highly reactive ground state trans-(28) and the starting cis-(28), thus providing strong chemical evidence for the structure assignment made in the flash study.

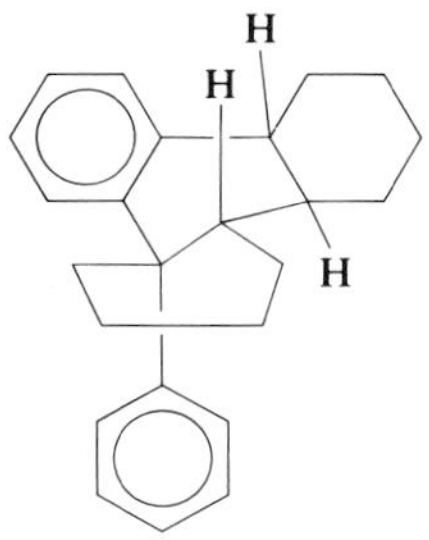

(31) 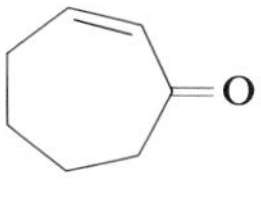   (32) 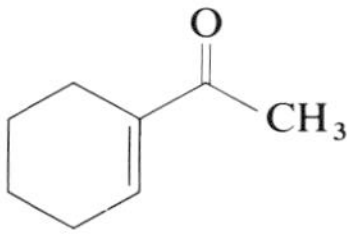   (33)

Another type of system in which highly reactive *trans*-cycloalkenes are generated photochemically is illustrated by cycloheptenone (32) and acetylcyclohexene (33). In both cases, long-lived ground-state twisted *trans*-isomers could be detected by laser flash photolysis, as were short-lived triplet state precursors. The latter were concluded to be "relaxed" triplets that were stabilized energetically by twisting around the C=C bond, analogous to the stabilization of simple alkenes discussed earlier. Flash excitation of cyclohexenones including (16) and (27) has provided evidence for formation of analogous twisted triplet states, but ground-state *trans*-isomers of the starting material have not yet been detected, presumably because the strain imposed by having three sp^2 carbons in a six-membered ring is energetically prohibitive. A combination of flash and steady-state kinetic studies indicated that twisted triplet states of these cyclohexenones are intermediates on the pathway to photoisomers ((34) and (35), respectively) and products involving cycloaddition to alkenes (see later discussion).

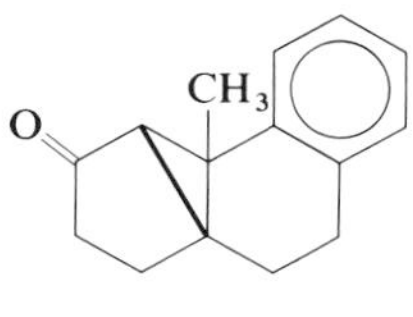

(34) 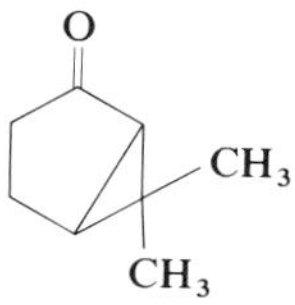   (35)

A final example illustrating the utility of laser flash photolysis in mechanistic investigations concerns the well-known Norrish Type II reaction of ketones. The basic reaction scheme in the case of γ-methylvalerophenone (36) is shown in Scheme I. Previous work (mainly quenching studies) established unequivocally that this reaction occurs exclusively from triplet excited states of the ketones, formed with 100%

efficiency in the case of aryl ketones. By analogy with the hydrogen-abstraction reactions of simple ketones, it was proposed that the mechanism involves intramolecular H-abstraction to give a triplet 1,4-biradical, which then either cyclizes to cylobutanols or fragments to an alkene and an enol, which then tautomerizes to acetophenone. Reverse H-transfer from oxygen to carbon regenerates the starting material. It was also assumed that a spin flip to give a singlet biradical must precede the cyclization and fragmentation reactions.

Triplet quenching studies as well as direct laser flash measurements demonstrated that the triplet lifetimes in these systems are very short, for instance, that of (36) is only 4.7 nsec. This lifetime is determined entirely by the rate constant for intramolecular γ-H abstraction. Laser flash excitation of (36) leads to transient absorption between 380 and 510 nm attributed to the triplet biradical, with a lifetime of 97 nsec in methanol. Lifetimes in analogous cases in a variety of solvents at room temperature run from 25 to 2200 nsec. There is a clear enhancement in biradical lifetimes in polar versus nonpolar solvents that is attributed to an increase in the average distance between the two radical sites in media which can stabilize the system by H-bonding, thus reducing the rate of intersystem crossing to singlet biradicals. Typical triplet quenchers do not affect the lifetime of this transient but do reduce its yield, clearly implicating a triplet excited state precursor. The transient is quenched by addition of typical radical scavengers, such as thiols and tri-*n*-butyl tin hydride, which are known to reduce the yields of reaction products. This reaction, shown in Scheme I, involves H-transfer to the alkyl radical site to give a ketyl monoradical that actually has a longer lifetime than the original biradical. Quenching of the biradical by a variety of electron acceptors was also observed, which is presumed to in-

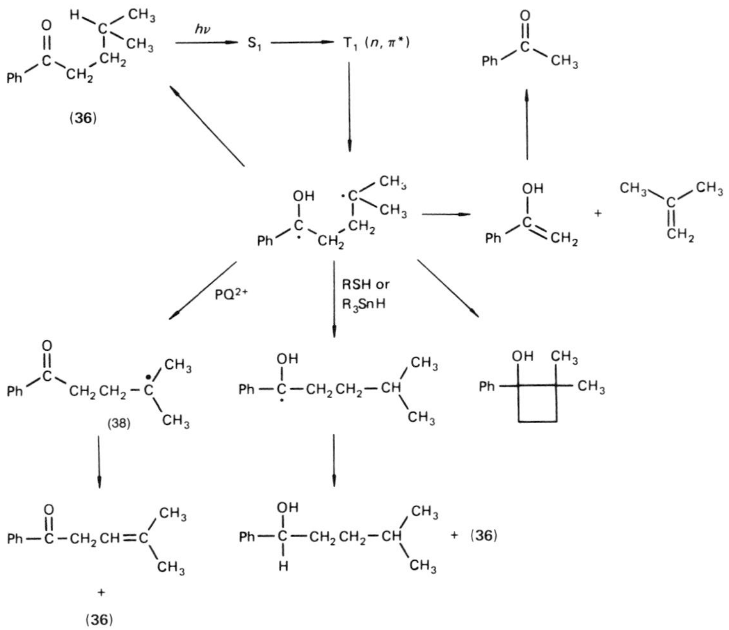

SCHEME I.

E. Mechanistic Studies: Indirect Characterization of Reaction Intermediates

volve electron transfer from the ketyl radical site, as shown in Scheme I for the case of methyl viologen (paraquat) dications PQ^{2+} (**37**). By fol-

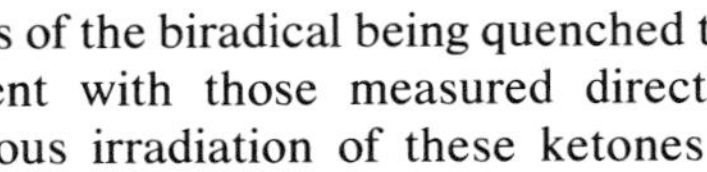

(37)

lowing the buildup of PQ^{+}, which has a strong and characteristic UV spectrum, one can obtain rate constants for the electron transfer step and lifetimes of the biradical being quenched that are consistent with those measured directly. On continuous irradiation of these ketones in the presence of electron acceptors, the normal Norrish Type II reaction products are not observed, but rather the reaction course is altered as shown in Scheme I to give products derived from radical (**38**).

Thus, these experiments using flash techniques allow direct observation of the previously proposed 1,4-biradicals and, together with earlier studies on the ketone triplet states, provide a detailed picture of the dynamics of the Norrish Type II reaction of aromatic ketones. At the present time, this is one of the best understood photochemical reaction systems.

In many instances where reaction intermediates are not directly detectable by physical methods, or at least where they have escaped detection, their involvement can be inferred from the nature of the products and/or kinetic data. Two examples will be presented to illustrate this point.

Xylene-sensitized excitation of 1,2-dimethylcyclohexene (**39**) in ether that contains a trace of sulfuric acid leads to the two isomeric alkenes shown in Eq. (36a).

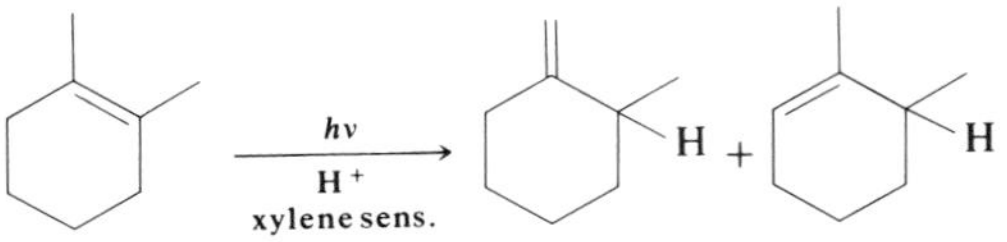

(36a)

The most reasonable mechanism involves triplet-sensitized *cis–trans* isomerization to the highly strained *trans*-cyclohexene, which on protonation gives a tertiary carbocation that leads to the observed products (see Scheme II). Indeed, similar reactions in CH_3OH give good yields of the tertiary methyl ether, while use of

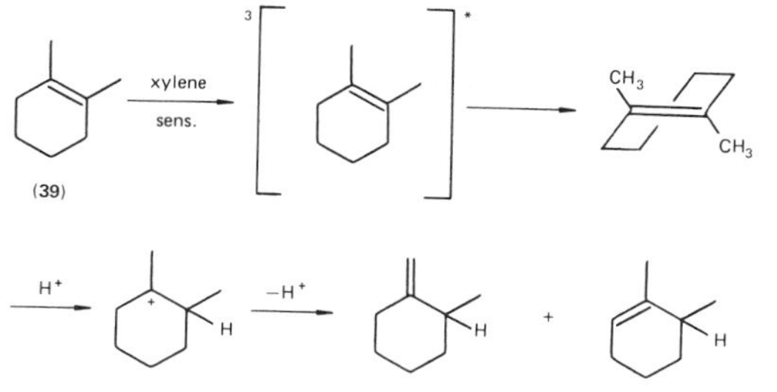

SCHEME II.

CH$_3$OD results in extensive incorporation of deuterium in the expected positions. It should be noted that triplet-sensitized excitation of the exocyclic alkene does not result in isomerization back to the endocyclic isomer, since rotation around the C=C bond in the exocyclic alkene simply interchanges the hydrogens on the double bond. Direct irradiation of (39) in pentane containing a trace of sulfuric acid affords the same two alkenes as in Eq. (36a) in addition to the three new products shown in Eq. (36b).

proposed by Kropp that carbenes are generated from the Rydberg π, R(3s) state of the alkene, which is generated by excitation of one of the π-electrons into a C—H antibonding orbital that can be represented as a 3s heliumlike orbital. The main point is that in a Rydberg state the π-electron is in a molecular orbital in which the average distance from the core of the molecule is relatively large (a kind of "holding pattern"), leaving the core of the molecule relatively electron deficient. It has been proposed that such states are accessible from highly substituted alkenes using low-wavelength UV light and are responsible not only for generation of carbenes but also for certain nucleophilic trapping reactions, as illustrated in the case of tetramethyl-ethylene in Scheme IV.

A completely different type of reaction system is that of the cross-conjugated cyclohexadienones, which undergo a complex series of photoisomerizations illustrated by the conversion of dienone (40) into the so-called lumiketone (41), which undergoes secondary photoiso-

(36b)

The formation of these products is most readily rationalized in terms of the intermediacy of a carbene, as shown in Scheme III. Thermal decomposition of the corresponding tosyl hydrazone, which should give the proposed carbene, does indeed afford the same three products. Rearrangement via carbene intermediates seems to be a general pathway for alkenes on direct but not on triplet-sensitized excitation. It has been

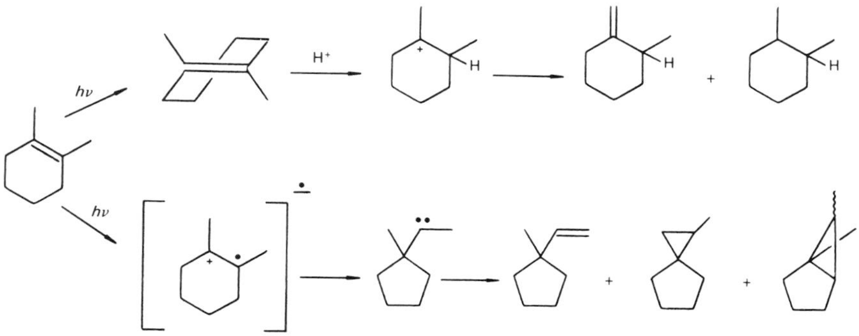

merization into phenols (42) and (43) and the ketene (44); the latter in aqueous media is converted into carboxylic acid (44a) [see Eq. (37)]. Similar deep-seated molecular rearrangements are well known on irradiation of steroid dienones such as dehydrotestosterone (45) and the sesquiterpene α-santonin (46).

In each case the reaction sequence is initiated by the conversion of the dienone into the corre-

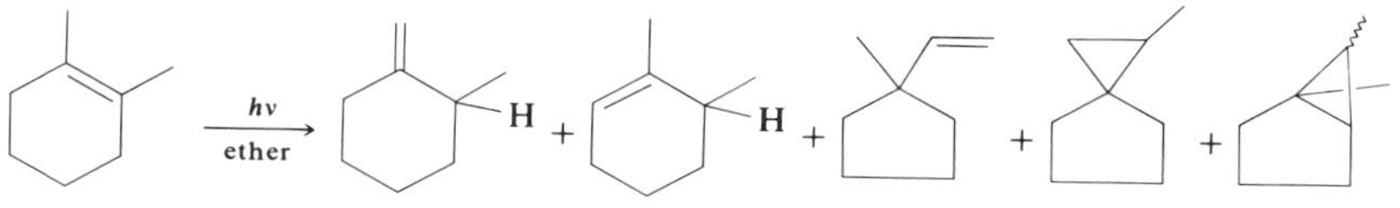

SCHEME III.

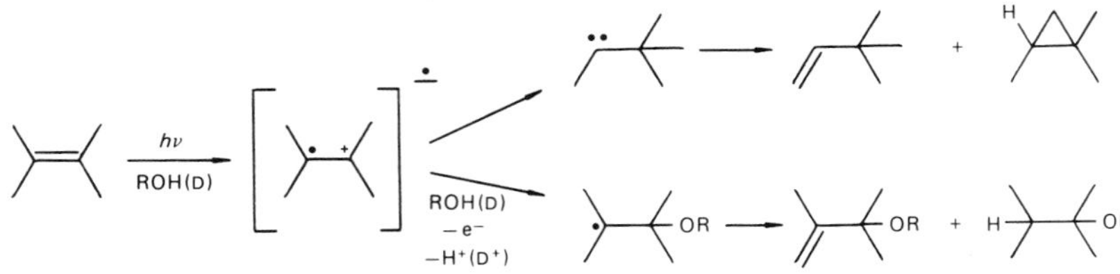

SCHEME IV.

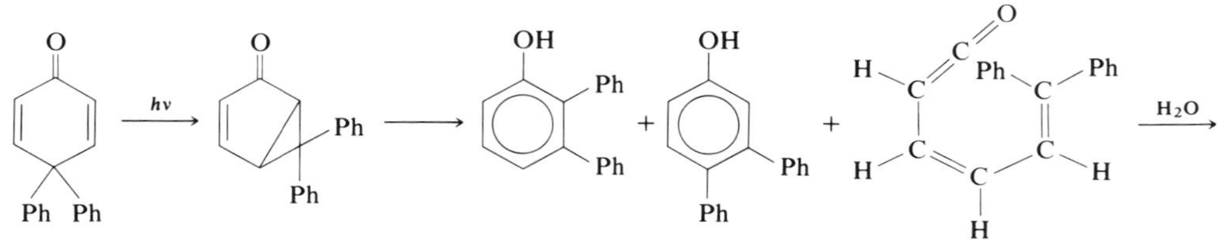

(40)   (41)   (42)   (43)   (44)

$$HOOC-CH_2CH=CH-CH=C\begin{smallmatrix}Ph\\\\Ph\end{smallmatrix}$$ (37)

(44a)

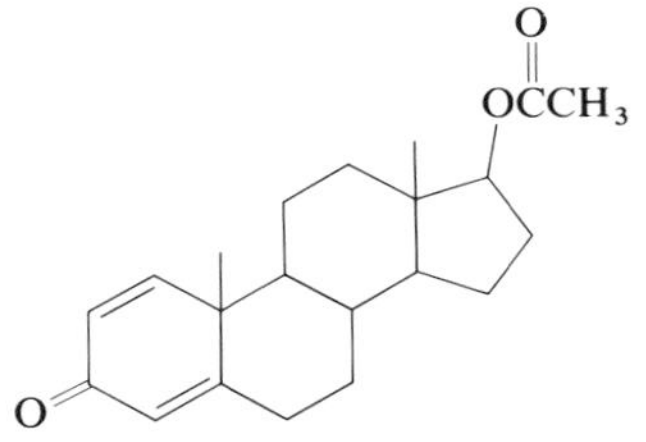

(45) 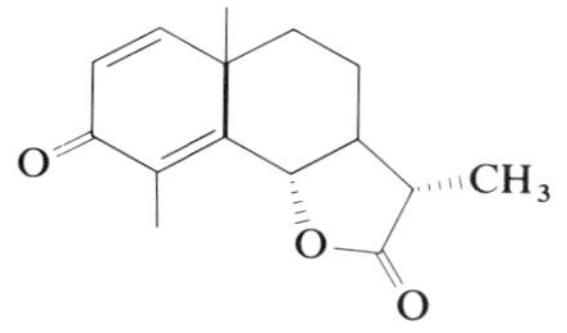   (46)

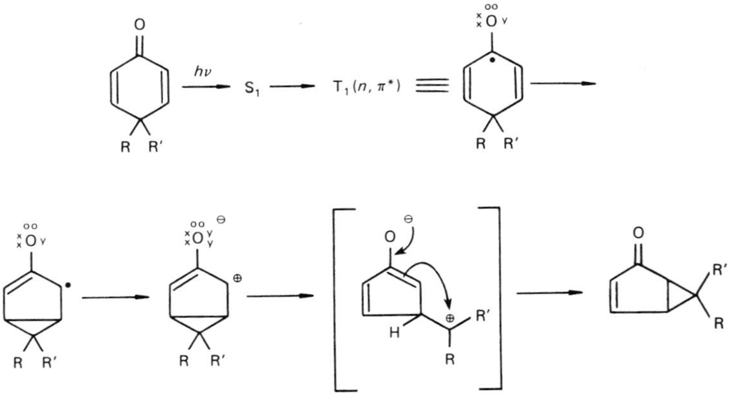

SCHEME V.

sponding lumiketone. These transformations defied mechanistic interpretation until the proposal by Zimmerman (in 1961) that the reaction proceeded via an intermediate ground-state zwitterion formed from the triplet n, π^* state of the dienone, initiated by bonding between C3 and C5 and decay to the ground-state potential surface, followed by demotion of an electron from the LUMO of the excited state to a nonbonding orbital on oxygen. Formation of the lumiketone from the zwitterion then occurs by a ground-state molecular rearrangement that has analogy with transformations of related types of cations. These steps are outlined in Scheme V. The subsequent phototransformations of the lumiketones and of other compounds further along the reaction pathway could also be readily rationalized in terms of zwitterion intermediates.

Such zwitterions have not been directly detected by flash techniques although their intermediacy in these reaction systems is supported by two types of studies. One approach was to generate the same zwitterion by a ground-state reaction to see whether the same rearrangement products would be observed. This is illustrated in Eq. (38), where treatment of the bromoketone (47) with base was anticipated to lead to a bridged zwitterion by a Favorskii-type reaction; indeed, rearrangement to the lumiketone was observed.

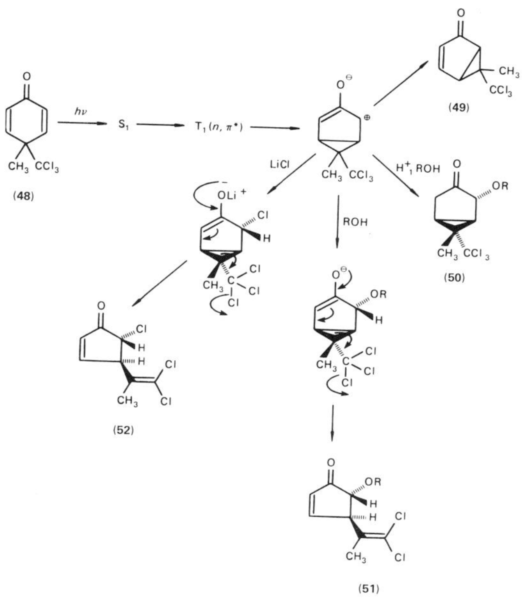

SCHEME VI.

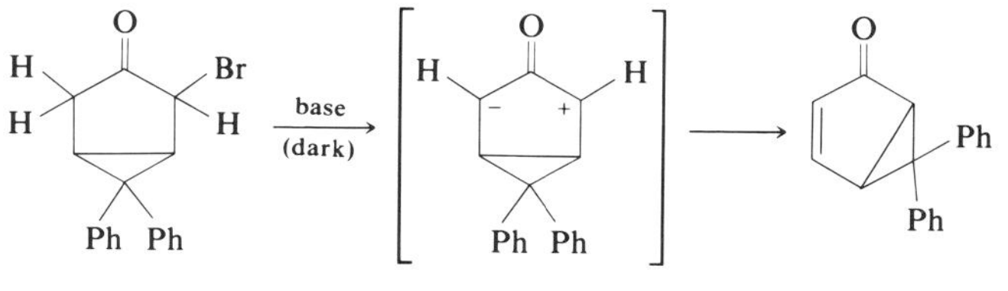

terms of nucleophilic interception of an intermediate zwitterion, as shown in Scheme VI. Indeed, formation of these products would be difficult to explain by a fundamentally different mechanism. When (48) was irradiated in a nonnucleophilic medium, the formation of (49) was

Although this result lends support to the proposal that zwitterions are intermediates in the photorearrangement of the dienone, it in itself does not require such a conclusion, since there may be significant differences in the intermediates generated in the two systems; that is, different solvents are used and counterions are present in the ground-state reaction of the bromoketone.

A different tack was taken by us in studies involving the trichloromethyl-substituted dienone (48). Irradiation of this compound in a variety of solvents gave the corresponding lumiketone (49), while irradiation in nucleophilic media such as alcohols gave products (50) and (51), whose formation could be readily rationalized in

quenched by added LiCl concomitant with the formation of the corresponding chloroadduct (52); kinetic studies showed that formation of (49) and (52) occurred via a common intermediate. The nature of the chemistry strongly suggests that this intermediate is a zwitterion. The possibility that LiCl is quenching an earlier reaction intermediate, such as a triplet state, is excluded by the observation that LiCl does not affect other triplet photoreactions of (48) not involving zwitterion intermediates.

Thus, these studies provide good circumstantial evidence for the involvement of zwitterions in the reactions of cross-conjugated dienones, although direct observation of such species has yet to be achieved.

VIII. Typical Organic Photochemical Reactions

In this section, photochemical reactions are categorized and a number of new reactions are introduced. Reference is also made to reactions already presented.

A. Proton Transfer Reactions

Photoinduced proton transfer is a process that plays an important role in many photobiological reaction systems. In general, the acidity and basicity of organic compounds in the excited state are quite different from the same compounds in the ground state. Thus, phenols are considerably stronger as acids in the excited state than in the ground state by perhaps five orders of magnitude, as seen by their conversion to the corresponding phenolate anions at much lower pH than in the ground state. Since fluorescence of the phenolate is observed, proton transfer must be taking place, to give the product (phenolate) in its excited state. This is a rare example of an adiabatic photoreaction, since decay to the ground-state potential surface typically occurs prior to formation of the final product in most organic photoreactions.

Another type of photoinduced proton transfer involves proton transfer not to an electronic excited state but to a reaction intermediate at a later stage. Thus, when ketone triplets interact with amines, electron transfer to the ketone radical anion occurs. This is followed by proton transfer either from nitrogen (in the case of primary and secondary amines) or from the carbon next to nitrogen to generate a radical pair, as shown in Scheme VII below. These radicals can go on to products or can undergo a reverse hydrogen-atom transfer to regenerate starting materials, as demonstrated clearly in flash studies. It has been suggested that a number of photoreactions involving net transfer of hydrogen atoms may proceed by two steps involving initial electron transfer followed by proton transfer, but this is still a subject of debate. Other types of reactions involve proton transfer to a reactive ground-state intermediate that is generated photochemically, as in the triplet-sensitized photoaddition of alcohols to cyclic alkenes [see (28)] and the sensitized anti-Markovnikov addition of alcohols to alkenes illustrated in Eqs. (9) and (10).

B. Cis–Trans Isomerization of Alkenes and Azo Compounds

As discussed earlier (see Section V), direct and triplet-sensitized excitation of alkenes causes isomerization around the C=C bond via singlet and triplet excited states, respectively, resulting in photostationary mixtures of alkene isomers. In the latter case, the ratio of cis to trans (Z to E) isomers is a function of the triplet excitation energy of the sensitizer, remaining constant as long as that energy is at least 3 kcal/mol greater than the triplet excitation energy of either isomer. As the sensitizer energy is lowered, and the rate of triplet transfer to the olefins begins to differ, the composition of the stationary state can be drastically altered, as shown in Fig. 10. In some systems, it is possible to promote isomerization preferentially in one direction or the other with an appropriate choice of the triplet sensitizer.

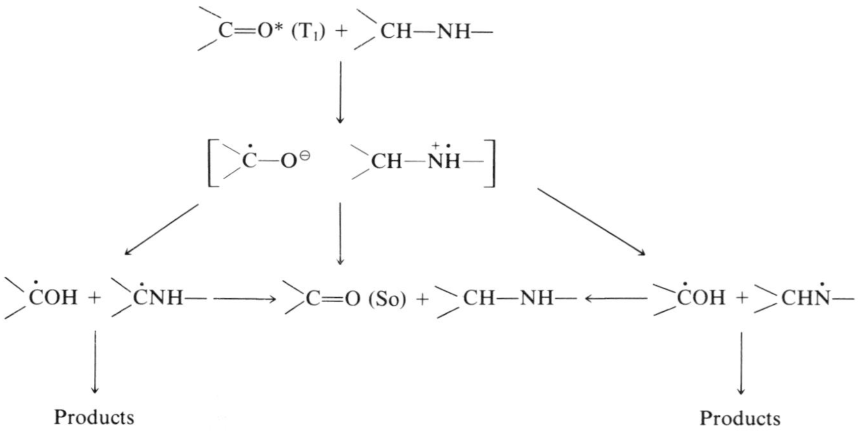

SCHEME VII.

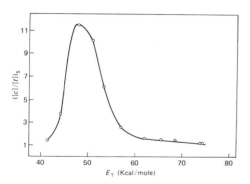

FIG. 10. Ratio of photostationary state concentrations of *cis*- and *trans*-stilbene versus triplet excitation energies of photosensitizers.

The photoisomerization of 11-*cis*-retinal to all-*trans*-retinal [Eq. (39)] is the key reaction in the visual process.

$$(39)$$

The visual pigment rhodopsin consists of 11-*cis*-retinal bound to the protein opsin. The *trans*-retinal dissociates from the protein, which initiates a series of biochemical events culminating in transmission of appropriate signals from the eye to the brain. Picosecond and nanosecond flash studies at low temperatures show that a series of intermediates are formed sequentially, called prelumirhodopsin, lumirhodopsin, meta-rhodopsin I, and metarhodopsin II, ultimately leading to *trans*-retinal. It has been shown that the cis–trans isomerization occurs very early, within a few picoseconds of the initial excitation of rhodopsin; a ''hula-twist'' mechanism for the isomerization has been proposed. Details can be found in the primary literature and monographs.

Isomerization around N=N bonds in azoben-

zene (Ph—N=N—Ph) and related compounds is also induced by direct or triplet-sensitized excitation, as illustrated by the reaction in Eq. (28). The UV–visible absorption spectra of the two isomers are sufficiently different that the course of the reaction can be followed by conventional UV spectroscopy. In some cases the reaction system is photochromic because of the difference in color of the two isomers. For example, the photoisomerization of *trans*- to *cis*-azobenzene can be induced by excitation at wavelengths between 280 and 370 nm and reversed by irradiation at 254 nm without significant side reactions. This system has found use as a reversible UV actinometer.

C. PHOTOCYCLOADDITIONS AND PHOTODIMERIZATIONS

1. Intermolecular Photodimerization of Alkenes

Direct excitation of concentrated solutions of alkenes frequently results in dimerization to give cyclobutanes, corresponding to a $[\pi^2 + \pi^2]$ cycloaddition. This reaction is illustrated for *cis*-2-butene in Eq. (40),

$$(40)$$

where two cyclobutanes are formed with preservation of the cis stereochemical relationship in the starting material. Under these conditions, the reaction generally occurs by attack of an alkene singlet excited state on a ground-state alkene and involves suprafacial attack (i.e., attack at both carbons of the C=C bond on the same face of the molecule) on each component. This reaction competes with cis–trans isomerization and therefore has an appreciable quantum efficiency only when the alkene is present in high concentration, suggesting that singlet excimers may be intermediates. Photodimerization of alkenes can also be induced by triplet sensitization, but this usually gives a more complex product mixture because of facile photoisomerization of the alkene. However, the quantum efficiency often exceeds that for the unsensitized reaction. Copper salts are also known to catalyze the photodimerization of cyclic but not acyclic alkenes.

An important example of a [2 + 2] dimerization is shown in Eq. (41) involving the heterocyclic compound thymine.

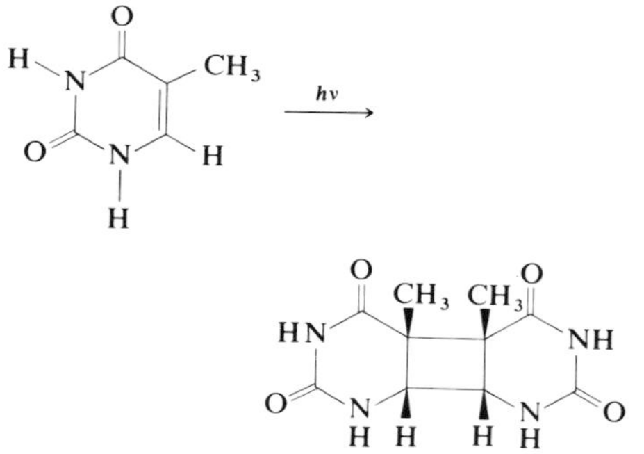

(41)

Thymine is one of the four bases found in DNA macromolecules which contain the genetic information in the cell nucleus and control the synthesis of proteins. Formation of thymine dimers between nearby thymine residues is known to be the primary reaction occurring on exposure of DNA to UV or gamma radiation. This process disrupts the normal replication of DNA and synthesis of RNA, and hence the synthesis of proteins, and can lead to chromosome defects, mutations, or even cell destruction. Breakdown of the thymine dimers to thymine monomers can be brought about by irradiation at 254 nm in vitro; in vivo, there is a reactivating enzyme that accomplishes the same result by absorption of light above 300 nm in a process that appears to be initiated by electron transfer.

Excitation of olefins in the crystalline state can also lead to cyclobutanes. The reaction in this case is controlled by the arrangement of the molecules in the crystal lattice, since only a limited amount of molecular movement is permissible. Since only groups that are near each other in the crystal can interact, the regio- and stereospecificity of solid-state reactions can differ greatly from that observed with the same compounds in solution. A classic example of such topochemical control is provided by *trans*-cinnamic acid (53), which exists in three crystalline forms. In the α form, there is a 3.6-Å separation of double bonds, and the molecules are lined up head to tail, while in the β form the separation is nearly identical, but the molecules are lined up in parallel. Irradiation of the α form of *trans*-cinnamic acid results in formation of α-truxillic acid (54), while the β form leads to β-truxinic acid (55). No reaction occurs with the third crystalline form because the intermolecular distance (4.7–5.1 Å) is too large to allow photodimeriza-

tion. Many other examples of topological control of alkene dimerizations in the solid state are known.

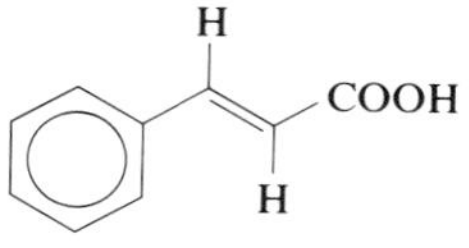

(53)

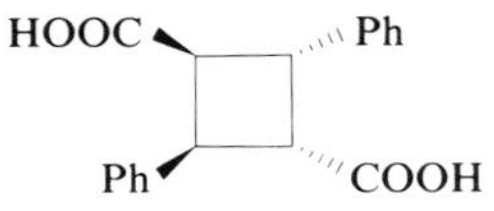

(54) 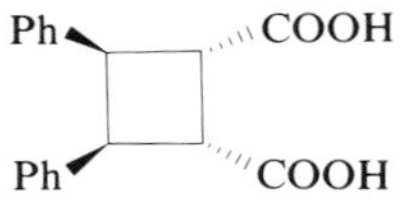   (55)

2. Intramolecular [2 + 2] Photocycloadditions

When two C=C bonds in the same molecule are spatially proximate, direct or triplet-sensitized excitation frequently gives rise to intramolecular cyclobutane formation. Notable examples are shown below involving direct excitation of norbornadiene [Eq. (42)] and hexamethyl Dewar benzene [Eq. (43)] and the formation of cage compounds on triplet sensitized excitation of tricyclic dienes [Eq. (44)]:

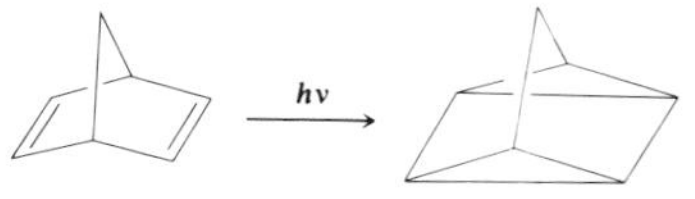

(42)

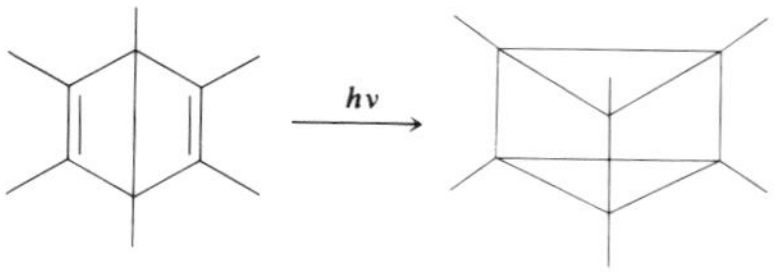

(43)

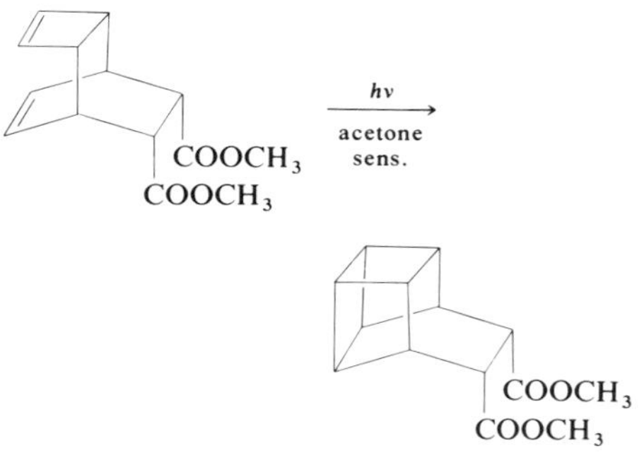

(44)

One can prepare such highly strained compounds photochemically because the reaction products are saturated and do not absorb the UV light used to drive the reaction, so there is no way to reverse the reaction under the reaction conditions. However, it is possible to convert quadricyclane back to norbornadiene using photosensitization. It would be very difficult to make such strained compounds by a ground-state (i.e., thermal) process, since the products are thermodynamically so unstable relative to the olefinic starting materials. The norbornadiene–quadricyclane system has been explored as a device for converting solar energy into heat, since the heat content stored in the highly strained quadricyclane can be released by conversion back to norbornadiene quantitatively in the presence of catalysts. There are, however, many practical problems to be solved before such organic systems can become economically viable as vehicles for solar energy conversion.

Another mode of intramolecular [2 + 2] cycloaddition is shown in Eq. (45) in the low-yield photoconversion of 1,3-butadiene into bicyclobutane, which accompanies formation of cyclobutene. If the 1,3-diene moiety can be constrained into an s-*trans* geometry, this type of reaction can be made to occur in good yield, as shown in the case of a heteroannular steroid diene in Eq. (46).

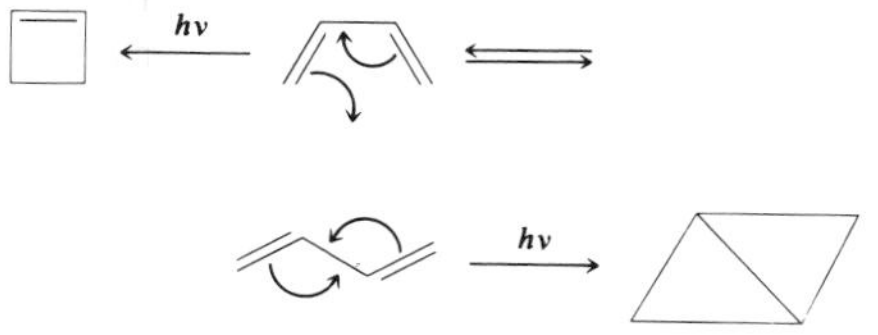

$$ \text{(45)} $$

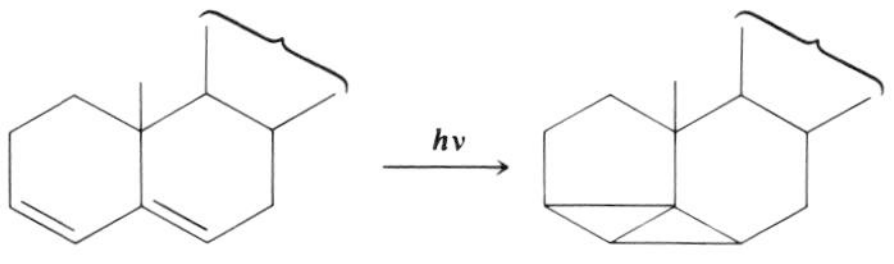

$$ \text{(46)} $$

Some examples of intramolecular [2 + 2] cycloadditions involving cyclopropanes are shown in Eqs. (47) and (48). Once again, this is a route to highly strained polycyclic structures.

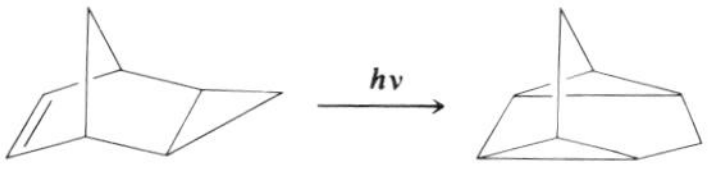

$$ \text{(47)} $$

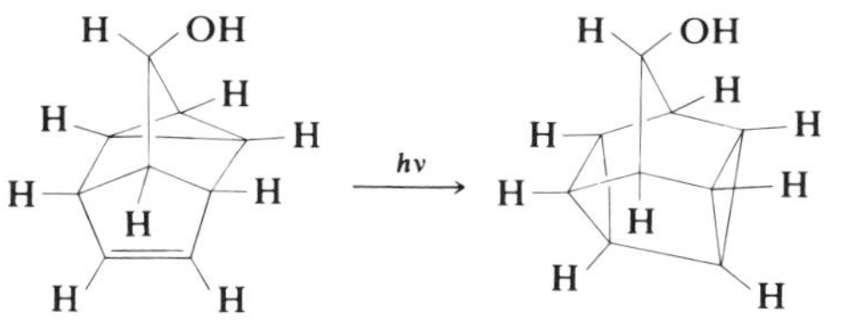

$$ \text{(48)} $$

3. Other Photocycloadditions Yielding Cyclobutanes

A large number of cross photocycloadditions to give cyclobutanes involving two different olefinic compounds are known. Since the singlet excited state lifetimes are very short and intersystem crossing is generally very inefficient, the chance that excited singlets of olefins can be effectively intercepted by another olefin in a collisional process to give a cyclobutane is small. Triplet states of olefins, generated by sensitization techniques, are much longer lived and therefore serve as effective precursors for cross cycloaddition reactions. Some examples of triplet-sensitized intermolecular [2 + 2] cycloaddition reactions of dienes are the following:

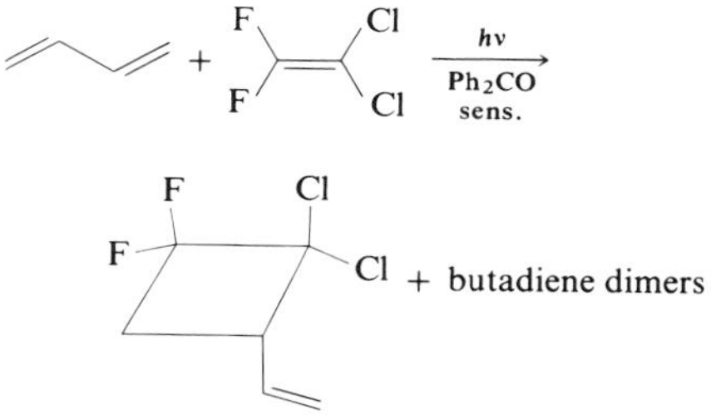

$$ \text{+ butadiene dimers} \qquad \text{(49)} $$

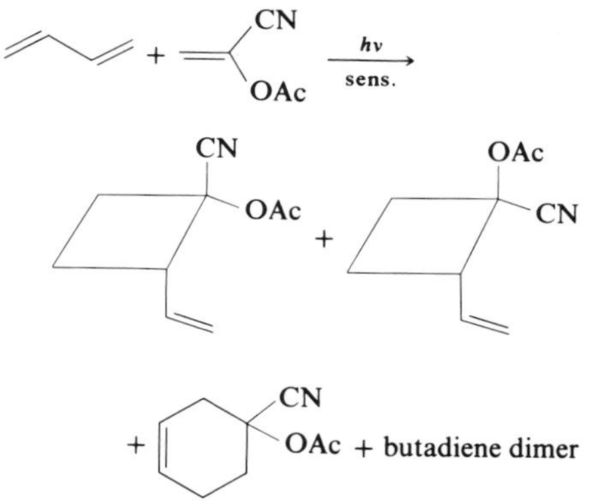

$$ \text{+ butadiene dimer} \qquad \text{(50)} $$

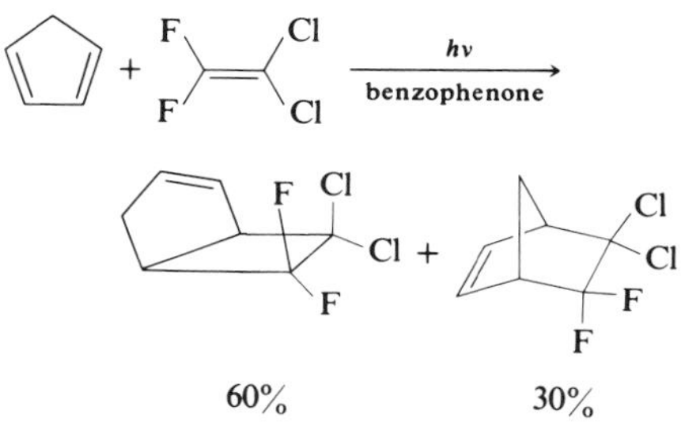

$$ \text{(51)} $$

60%   30%

It should be noted that most successful reactions of this type involve additions to an electron deficient alkene, suggesting that charge transfer or exciplex interactions may be important in such reactions.

A few examples of cross cycloadditions on direct excitation of olefins are known. For the reaction shown in Eq. (52) there is evidence that the triplet state of 1,1-diphenylethylene is probably the reactive intermediate. The observation of heavy atom effects on the efficiency of addition of acenaphthylene to acrylonitrile [Eq. (53)] also points to reaction via acenaphthylene triplets. However, the addition of *trans*-stilbene to tetramethylethylene [Eq. (54)] seems to involve stilbene singlets. In this case, kinetic studies indicate that a reversibly formed singlet exciplex is a precursor of the cycloadduct.

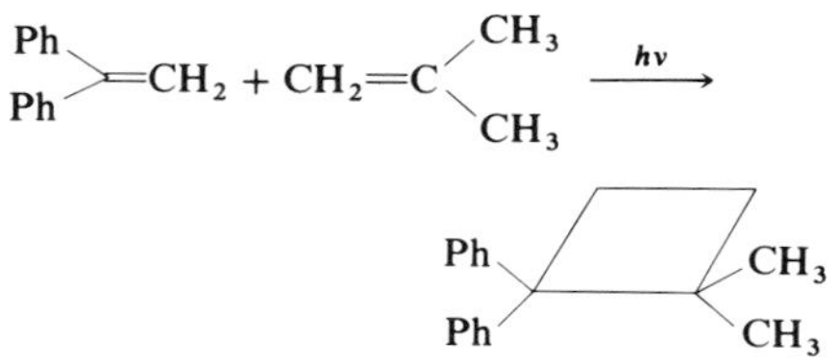

$$(52)$$

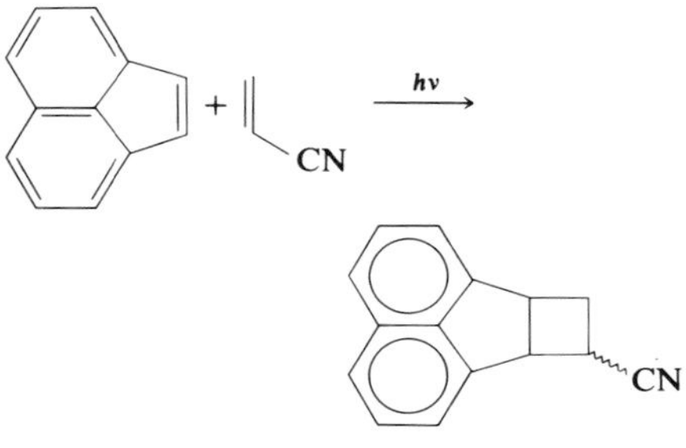

$$(53)$$

(syn + anti)

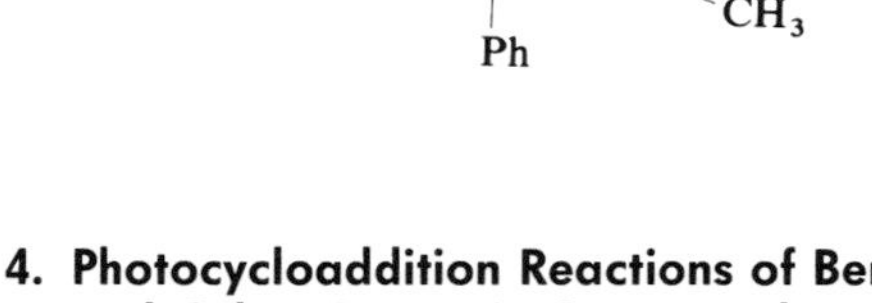

$$(54)$$

4. Photocycloaddition Reactions of Benzene and Other Aromatic Compounds

Benzene undergoes a variety of photocycloadditions to alkenes to produce 1,2, 1,3, and 1,4 adducts by [2 + 2], [2 + 3] and [2 + 4] processes, respectively. Specific examples of each of these are shown in Eqs. (55)–(57).

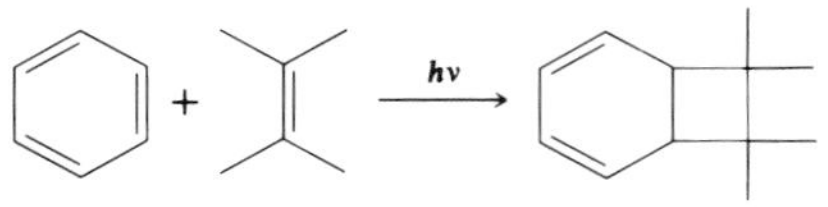

$$(55)$$

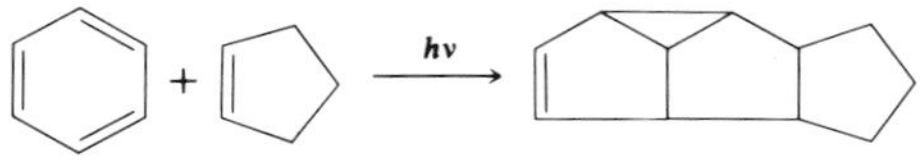

$$(56)$$

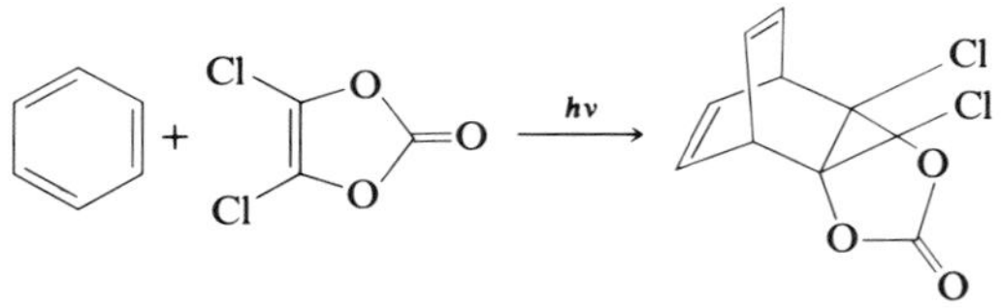

$$(57)$$

Since the olefin stereochemistry is usually retained in these reactions, it has been concluded that the S_1 state of benzene is the likely reactive intermediate. At least some, if not all, of these reactions seem to occur via singlet exciplexes. The formation of "ladder-type" cycloadducts as in Eq. (58) would seem to involve prior photoisomerization of benzene to the Dewar structure (51) followed by [2 + 2] cycloaddition. Structurally complex 2:1 and 2:2 adducts can also be formed photochemically.

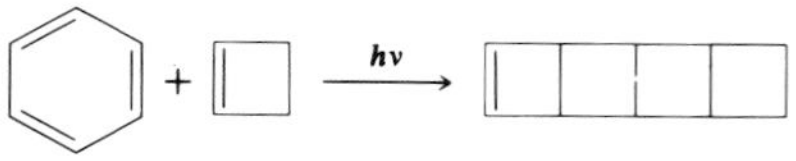

+ other products   (58)

Acrylonitrile adds to photoexcited naphthalene as shown in Eq. (59), probably via a singlet exciplex. Such exciplexes have been strongly implicated from spectroscopic and kinetic studies in the stereospecific [4 + 4] photoaddition of 1,3-dienes to naphthalene, illustrated in Eq. (60).

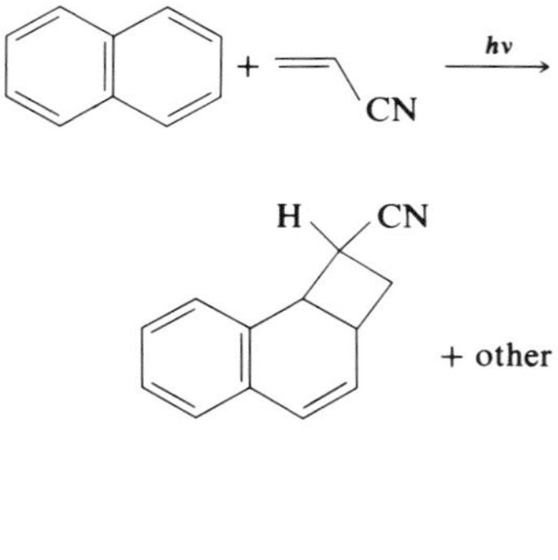

+ other products   (59)

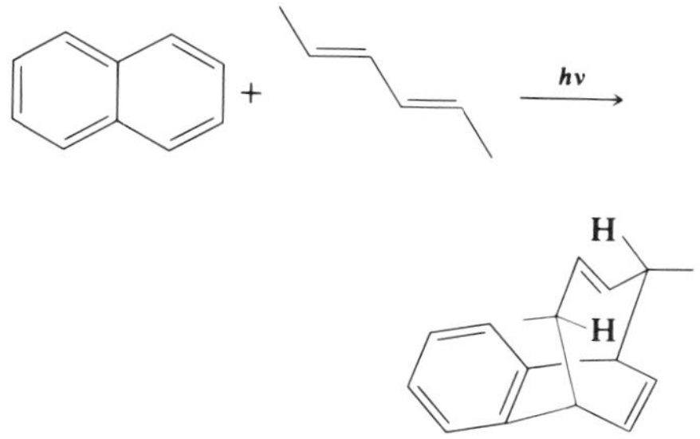

(60)

Formation of anthracene photodimers by [4 + 4] cycloaddition as shown in Eq. (61) is a general reaction, which occurs via the S_1 state of anthracene and probably involves excimeric intermediates. Although naphthalene does not undergo such a reaction, certain substituted naphthalenes and the bridged binaphthyl compound shown in Eq. (62) react in this manner.

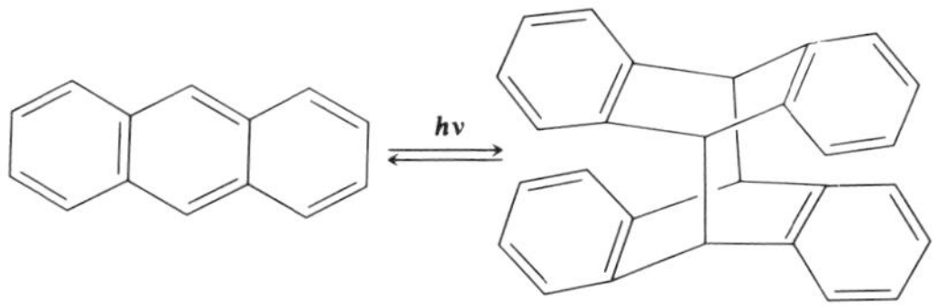

(61)

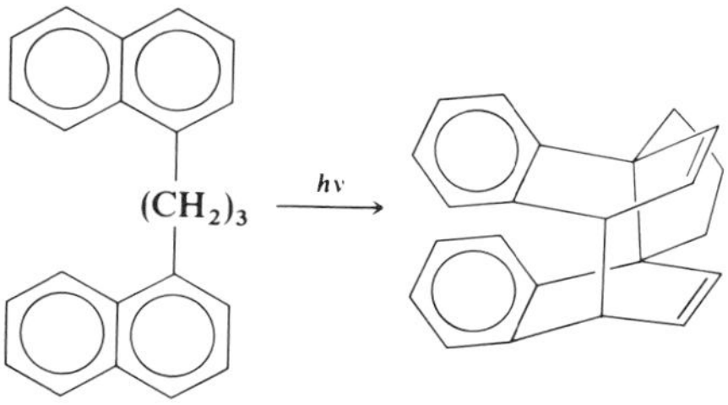

$(CH_2)_3$ (62)

5. Photoadditions of Carbonyl Compounds to Olefins: Formation of Oxetanes

The process shown in Eq. (63), involving [2 + 2] addition of an olefin to a carbonyl compound to give oxetanes, is a very general reaction.

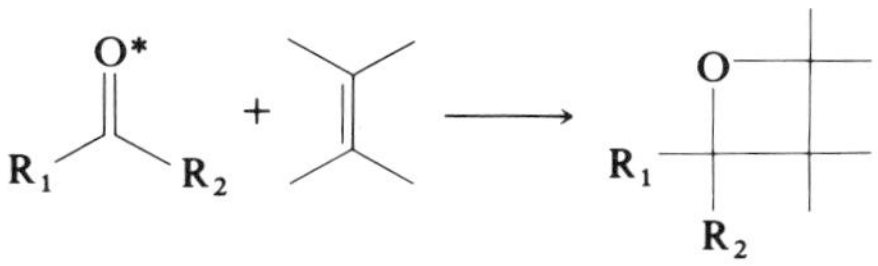

(63)

This reaction, known as the Paterno–Buchi reaction, can occur via either a singlet or triplet n, π^* state of the carbonyl component and involves attack of the half-filled oxygen n-orbital on the alkene to give a diradical, which then closes to

the product (a spin flip is of course necessary for closure of the diradical in the triplet reaction). The regiospecificity of such reactions is illustrated by the reaction of triplet benzophenone with isobutylene shown in Eq. (64),

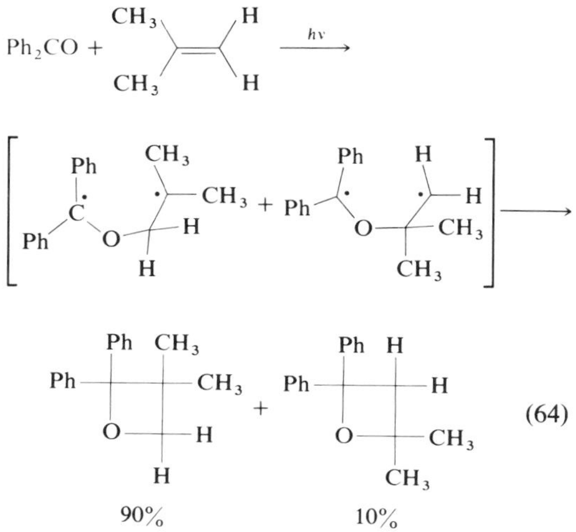

$$Ph_2CO +$$

(64)

$$90\% \quad\quad 10\%$$

which can be understood in terms of formation of the more stable diradical intermediate. The fact that reactions of aromatic ketones with electron-rich olefins are nonstereospecific, as seen in Eq. (65) for benzophenone with cis- or trans-2-butene, is consistent with the intermediacy of a triplet biradical in which rotation around the $C=C$ bond of the alkene is fast relative to ring closure.

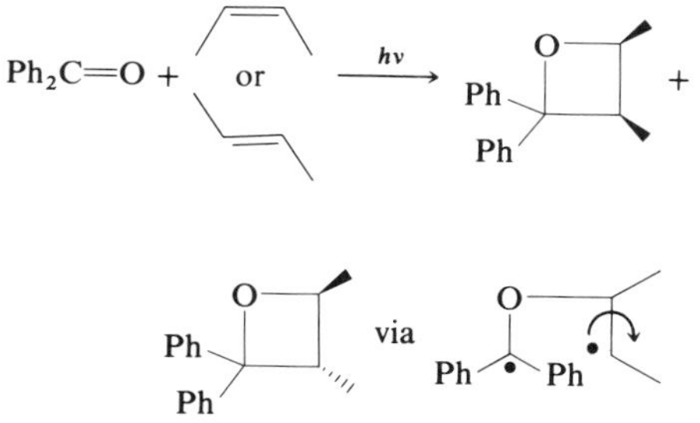

$$Ph_2C=O + \quad \text{or} \quad$$

(65)

Aromatic ketones react exclusively in bimolecular processes via their T_1 states since their singlet states are so short-lived.

Dialkyl ketones could in principle react via singlet or triplet n, π^* states. Indeed, quenching experiments indicate that reaction of acetone with electron-rich olefins such as enol ethers appears to involve both of these states. Thus, the ratio of the products in Eq. (66) depends on the concentration of the alkene component,

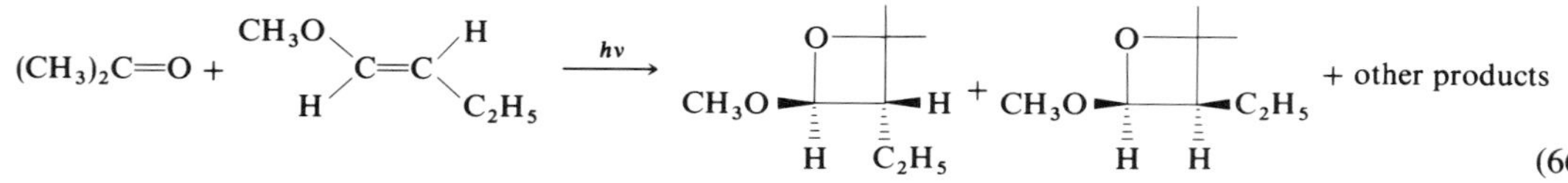

$$(CH_3)_2C=O + \quad \xrightarrow{h\nu} \quad + \text{ other products} \quad (66)$$

which suggests competition between a stereospecific singlet state cycloaddition and intersystem crossing to acetone triplets which react nonstereospecifically. Photocycloaddition of acetone to *cis*- or *trans*-dicyanoethylene, however, is completely stereospecific [Eq. (67)], suggesting reaction exclusively via the acetone S_1 state.

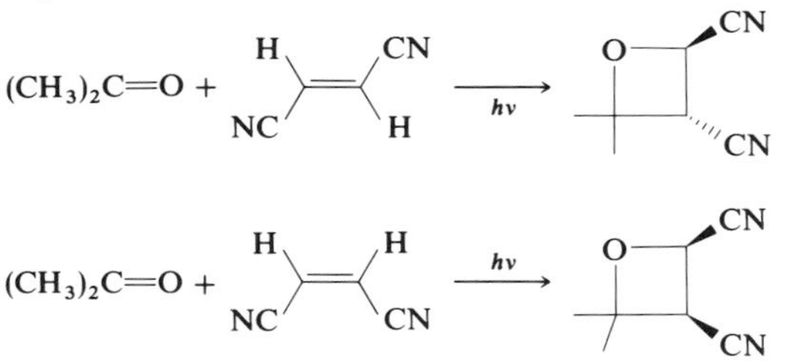

$$(CH_3)_2C=O + \quad \xrightarrow{h\nu}$$

$$(CH_3)_2C=O + \quad \xrightarrow{h\nu}$$

$$(67)$$

This is supported by the observation that electron-poor alkenes quench acetone fluorescence and that the formation of oxetanes in such systems is not affected by typical triplet quenchers. It has been suggested that these reactions occur by nucleophilic attack of the electron-rich π-system of excited acetone on the electron-poor π-system of the alkene. The accompanying side reaction involving isomerization of the alkene (i.e., formation of *cis*-dicyanoethylene) has been shown to occur via acetone triplets by fragmentation of the triplet diradical after bond rotation;

in these systems, closure of the triplet diradical to oxetane does not occur.

Formation of oxetanes followed by alternative bond cleavage has considerable utility in organic synthesis, as shown by the example in Eq. (68).

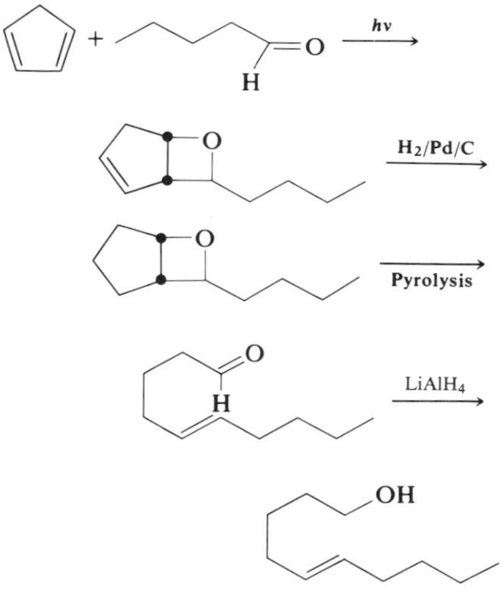

6. Photocycloaddition of Alkenes to Cyclic α, β-unsaturated Carbonyl Compounds

This photoreaction is general for cyclopentenones and cyclohexenones and is a good way of making bicyclic ketones in good chemical yields. Specific examples are given in Eqs. (69)–(72).

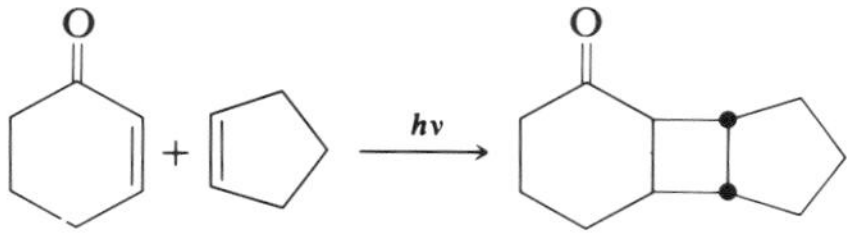

$$+ \quad \xrightarrow{h\nu} \quad (69)$$

$$+ \quad \xrightarrow{h\nu}$$

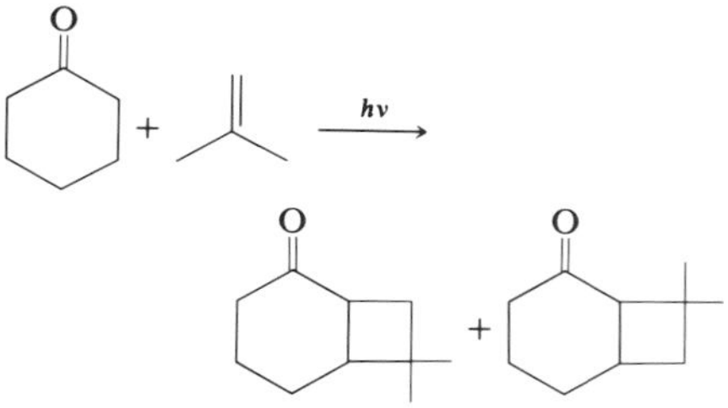

Major   Minor   (70)

$$+ \quad \xrightarrow{h\nu}$$

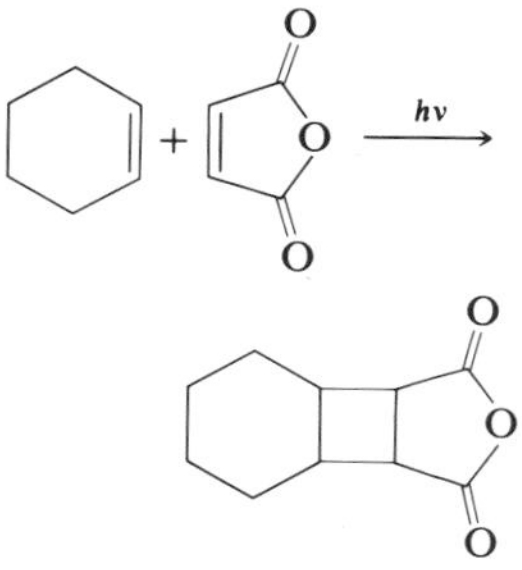

Mixture of Stereoisomers   (71)

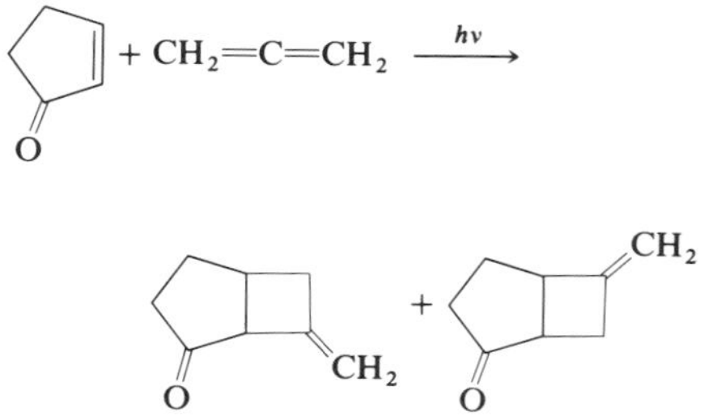

$$+ CH_2=C=CH_2 \quad \xrightarrow{h\nu}$$

$$\xrightarrow{H_2/Pd/C}$$

$$\xrightarrow{Pyrolysis}$$

$$\xrightarrow{LiAlH_4}$$

OH

(68)

Insect Pheromone

Major   Minor   (72)

Regioselectivity is observed, consistent again with formation of the more stable diradical intermediate, suggesting that the addition is stepwise. The stereochemistry of acyclic alkenes is lost in the reaction, indicating that 1,4-biradicals are probably involved. The T_1 state in these enones is most likely a $^3\pi, \pi^*$ state, and it has been proposed that the initial interaction with the olefin probably results in a triplet exciplex. Limited kinetic data indicate that the reaction rate in some systems is correlated by the electron density in the π-system of the alkene, although steric effects are also important. One of the most interesting observations is the preferred formation of cycloadducts with a highly strained trans ring fusion from cyclohexenones and electron-rich alkenes, as illustrated for a simple enone in Eq. (73) and for a steroid system in Eq. (74).

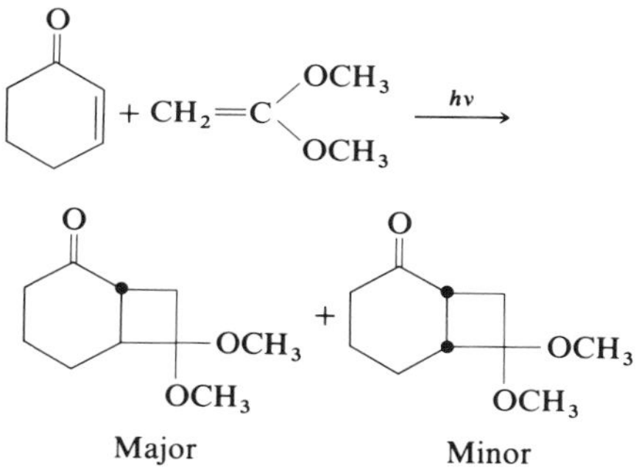

$$(73)$$

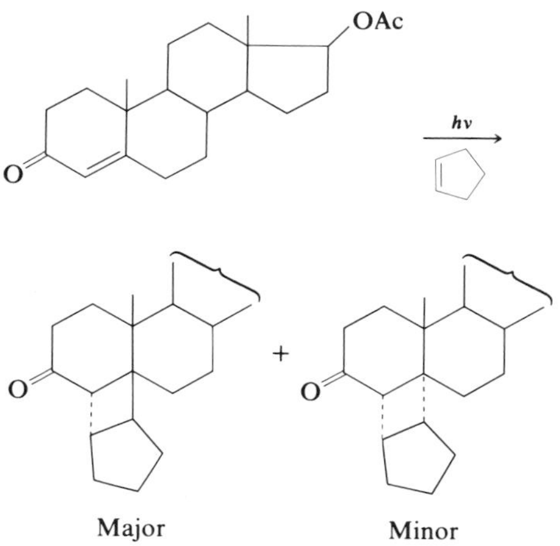

Major   Minor

$$(74)$$

Major   Minor

It has been proposed that the enone triplet excited state undergoes conformational relaxation around the C=C bond from a planar to a more stable twisted geometry prior to attack by the alkene and that formation of the two carbon–carbon bonds in an antarafacial manner to give

the trans-fused adduct is competitive with geometric relaxation of the 1,4-diradical on the way to the more stable cis-fused adduct. However, recent flash photolysis data indicate that for at least some cyclohexenones the "relaxed" twisted triplet does not react directly with alkenes but, rather, isomerizes to some other reactive species that then reacts. Larger ring enones, unless sterically constrained, undergo isomerization to the corresponding trans-enones systems rather than cycloaddition to electron-rich olefins. Recent data indicate that cycloheptenone undergoes photoaddition to electron-deficient alkenes in competition with photodimerization.

Intramolecular cycloadditions of this type are abundant. The classic conversion of carvone to carvomenthone is shown in Eq. (75).

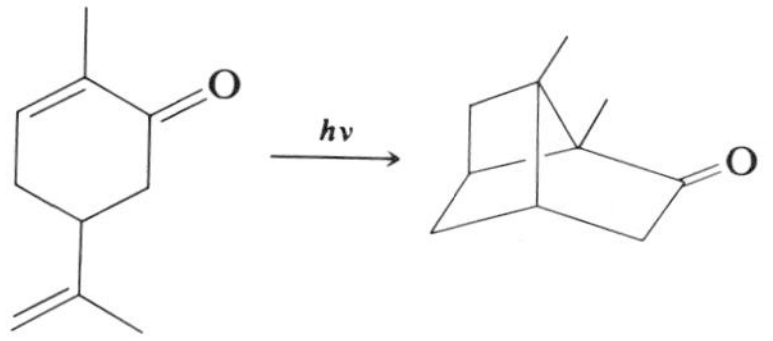

$$(75)$$

Formation of oxetanes competes with cyclobutane formation in some of these systems. Although in general the addition of electron-deficient alkenes to triplet excited cyclopentenones and cyclohexenones proceeds poorly, formation of [2 + 2] photodimers of enones occurs readily at high enone concentrations. This process, illustrated for cyclopentenone in Eq. (76), typically gives a mixture of head-to-head and head-to-tail isomers whose ratio is often sensitive to solvent polarity.

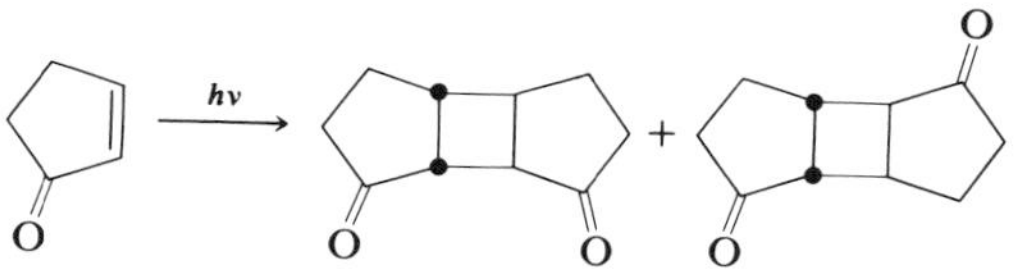

$$(76)$$

Cycloadditions of this type have found extensive use as a critical step in the synthesis of structurally complex organic molecules, a few of which are shown on page 408.

7. Intramolecular [4 + 2] Photocycloadditions

1,3,5-Trienes undergo intramolecular [4 + 2] photocycloadditions, the photochemical counterpart of the ground-state Diels–Alder reaction.

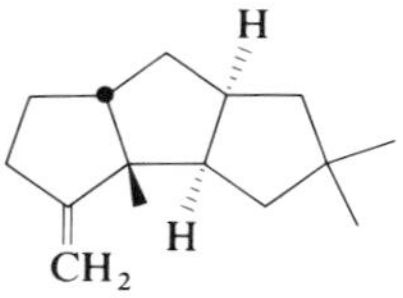

Hirsutene

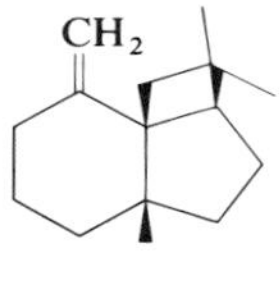

Panasinsene

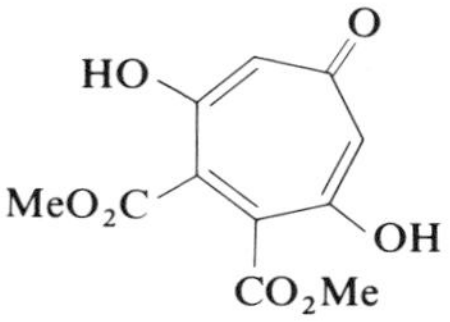

Stipitatonic Acid

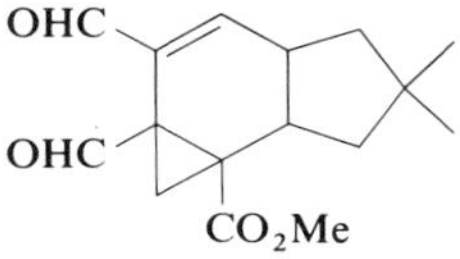

Methyl Isomarasmate

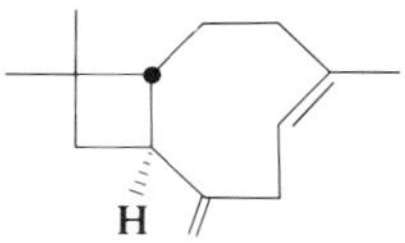

Caryophyllene

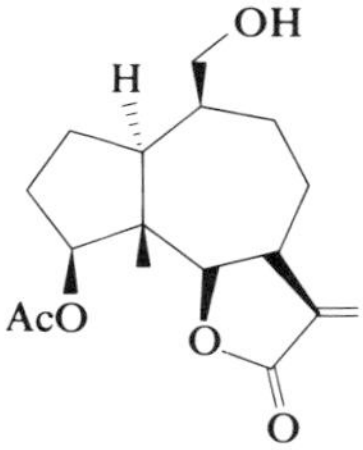

Hysterin

As illustrated in Eqs. (77) and (78), reaction occurs in a criss-cross manner to give bicyclo-[3.1.0]hexenes and represents a good synthesis of such compounds.

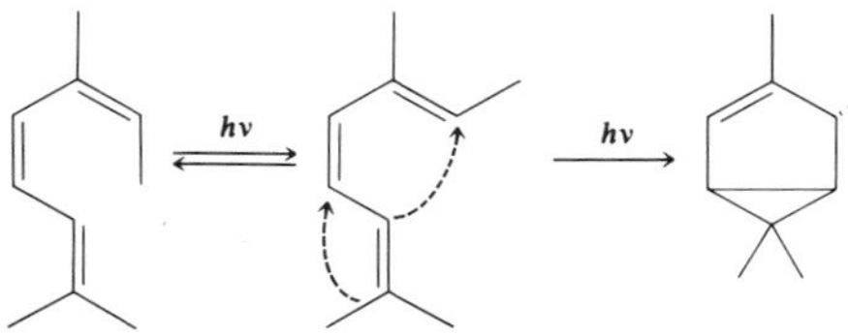

$$(77)$$

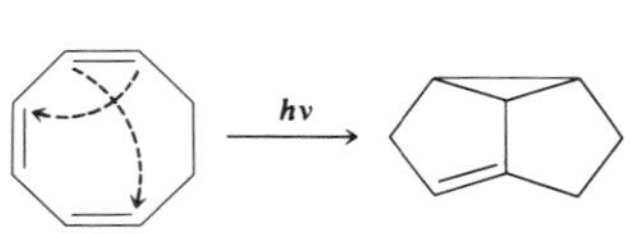

$$(78)$$

The reaction is usually stereospecific, following the [4s + 2a] or [4a + 2s] path predicted by the orbital symmetry theory of Woodward and Hoffmann, although exceptions to this rule are known. In the case of acyclic or large ring trienes, an s-*cis*, s-*trans* conformation is required for this reaction. An early example in which such a conformation is adopted is the photoconversion of vitamin D_2 into suprasterols I and II [Eq. (79)], which is one of the series of photoreactions occurring upon excitation of vitamin D precursors. The photoisomerization of benzene into benzvalene [Eq. (80)] is another example of

this reaction. The reactive excited state of benzene has been determined to be the S_1 state.

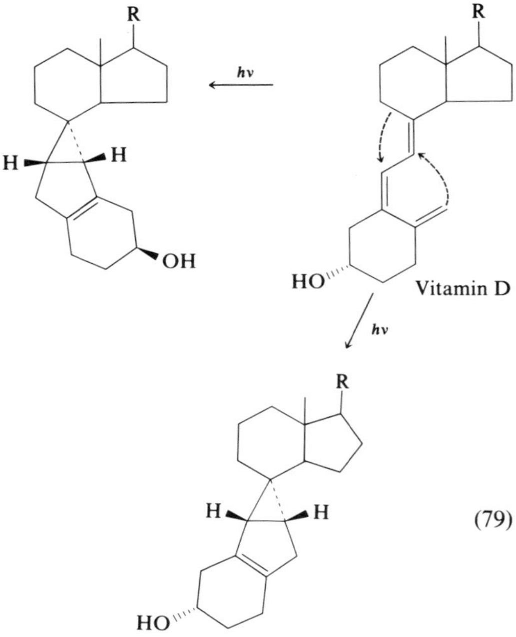

$$(79)$$

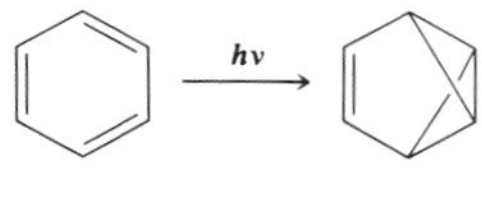

$$(80)$$

D. ELECTROCYCLIC REACTIONS

Electrocyclic reactions were defined by Woodward and Hoffmann as cyclization reactions involving formation of a sigma bond between the termini of a conjugated π-system and the reverse-ring cleavage processes. On the basis of the orbital symmetry rules developed for these reactions, it was predicted that the stereochemistry of such processes would be opposite in thermally and photochemically induced reactions. Thus, the ground-state (thermal) conversion of the $4\pi e$ system of 1,3-butadiene into cyclobutene was predicted to occur in a conrotatory fashion, while the corresponding excited state reaction was predicted to occur by a disrotatory path, as defined and illustrated in Eqs. (81) and (82).

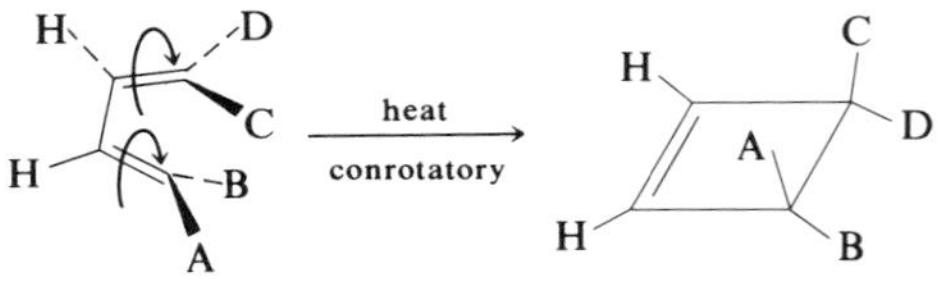

$$(81)$$

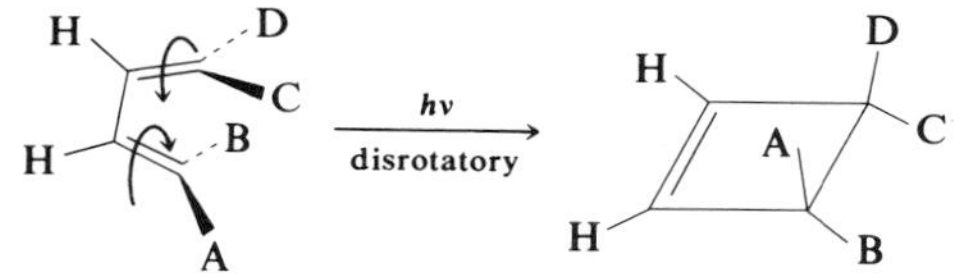

(82)

This was indeed confirmed experimentally. The same correlations hold for any $4n$ π-electron cyclization or the reverse ring cleavage, while the reverse stereochemical pathways (thermally disrotatory, photochemically conrotatory) are predicted for corresponding reactions of conjugated systems with $4n + 2$ π-electrons. Additional examples of photocyclizations of 4π systems are shown in Eqs. (83) and (84), while Eqs. (85) and (86) illustrate the photoinduced conrotatory ring openings of 1,3-cyclohexadienes. In the last two examples, the photoreaction is followed by a thermal disrotatory cyclization to give an isomer of the starting material.

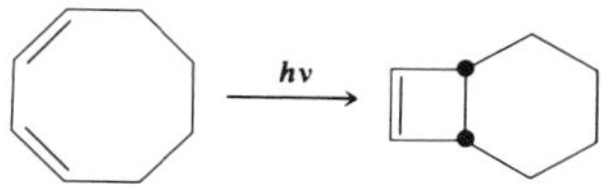

(83)

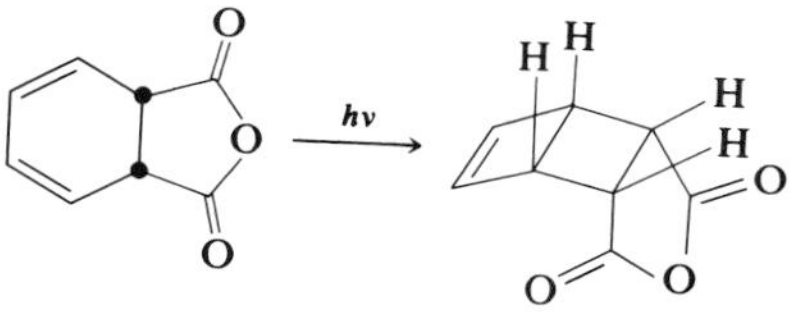

(84)

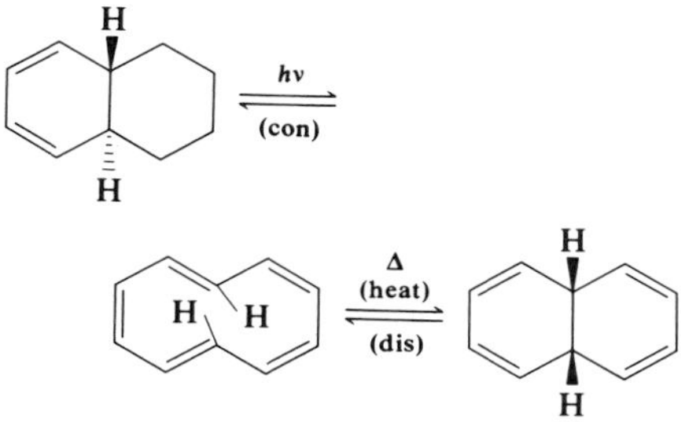

(85)

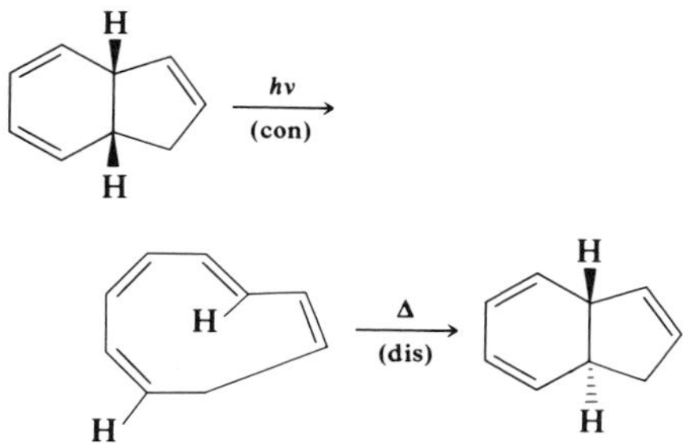

(86)

The stereochemical control associated with such reactions is dramatically illustrated by comparing reactions of the stereoisomeric steroids er-

gosterol (56) and isopyrocalciferol (57) shown in Eqs. (87) and (88).

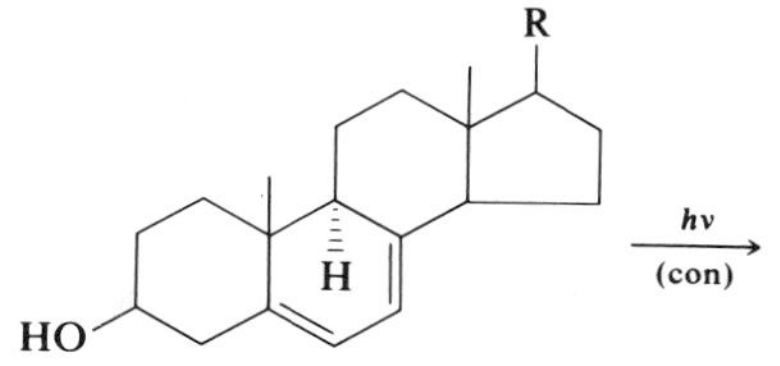

(87)

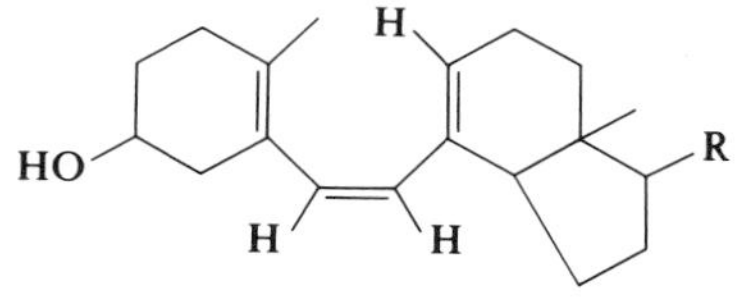

Previtamin D

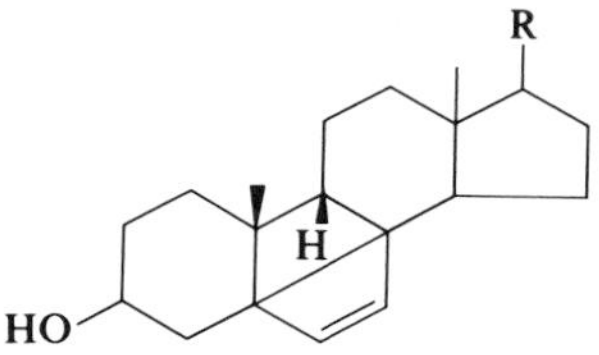

(88)

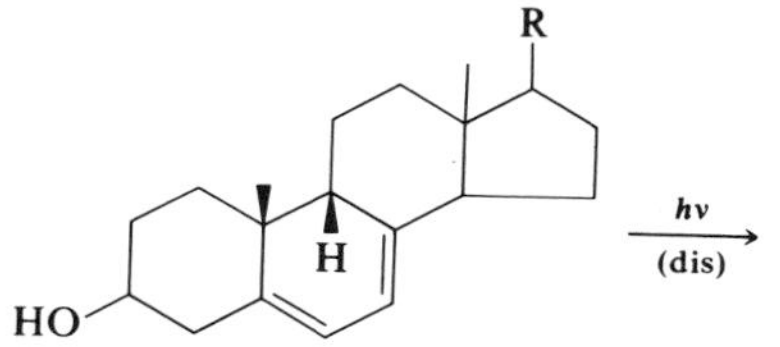

While (56) undergoes the expected conrotatory $6e$ ring opening to previtamin D, the formation of this same product from (57) would require a disrotatory ring opening, and this is not observed. Since conrotatory ring opening of (57) would necessarily result in formation of a *trans*-cyclohexene in ring A or ring C, which is energetically prohibitive, an alternative "allowed" reaction course of (57) is followed: namely, $4\pi e$ disrotatory ring closure to a highly strained bicyclo[2.2.0]hexene. The observation of these highly selective phototransformations was an important stimulus to formulation of the theory of orbital symmetry control of pericyclic reactions of π-electron systems. The theory in turn was an important stimulus to experiments designed to test the selection rules, resulting in the discovery of many new examples of such reactions.

A photoelectrocyclic reaction that has found considerable synthetic utility is shown in Eq.

(89), involving the conversion of *cis*-stilbene to the *trans*-dihydrophenanthrene (58).

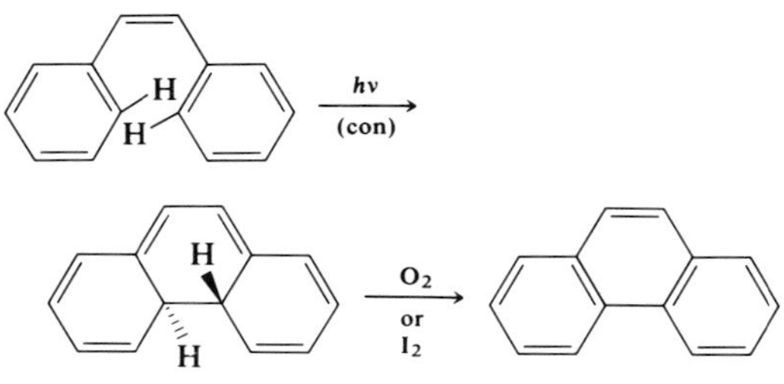

(58)

(89)

This reaction proceeds in a conrotatory fashion as expected for a $6\pi e$ excited state process. The intermediate, which has been directly detected spectroscopically, is readily oxidized by either oxygen or iodine to phenanthrene in high yields. Since trans-stilbenes photochemically isomerize to cis-stilbenes, either isomer can be converted in high chemical yields to phenanthrenes, and this route has found extensive utility in the preparation of substituted phenanthrenes. Similar reactions of condensed aromatic systems joined by a double bond (Ar—CH=CH—Ar) have been used for the preparation of helicenes. When such reactions are carried out using circularly polarized light, optically active products are produced, as in Eq. (90).

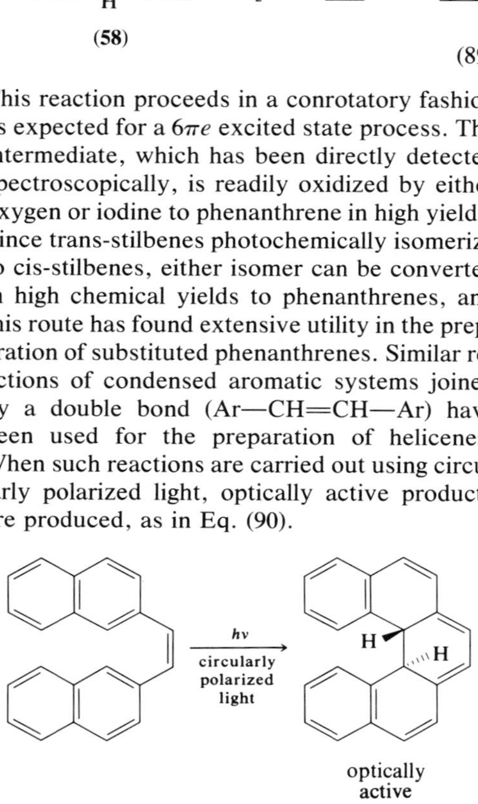

optically active

(90)

E. REACTIONS INVOLVING HYDROGEN ABSTRACTION

The prototypical example of photoreduction of ketones by H-abstraction from the solvent or other appropriate donors is the highly efficient conversion of benzophenone to benzpinacol [Eqs. (15) and (27)]. This type of reaction is general for ketones whose lowest triplet state is an n, π^* state, which is the case for most aromatic and aliphatic ketones. Aromatic ketones with strong electron-donating substituents undergo the reactions much less efficiently or not at all, which has been attributed to low-lying π, π^*, or

charge-transfer triplet states that have a low reactivity in this type of process. Photoreduction of ketones in the presence of amines appears to occur in two stages, with an initial electron transfer followed by proton transfer, to give the ketyl radical $R—\dot{C}(OH)—R'$, which then either dimerizes to pinacol or abstracts a second hydrogen to give a secondary alcohol.

The Norrish Type II reaction illustrated in Scheme I is the intramolecular counterpart of this reaction and is very general. Aromatic ketones undergo this reaction exclusively from triplet n, π^* states, while aliphatic ketones generally react from both singlet and triplet n, π^* states, since in these systems the rates of intersystem crossing and intramolecular H-abstraction are comparable. With appropriately substituted compounds, it has been shown that the singlet component of the reaction is largely stereospecific, indicating that a singlet 1,4-biradical is produced that cleaves to an alkene and an enol faster than it undergoes bond rotation. The opposite is true for the corresponding triplet reactions, which give mixtures of isomeric alkenes.

Other types of unsaturated compounds that undergo photoreduction in isopropyl alcohol and other good H-donor media include aromatic nitro compounds, azo compounds, and even certain alkenes, as illustrated in Eqs. (91)–(93). Intramolecular variants are shown in Eqs. (94)–(96). In all cases, reactions appear to occur via n, π^* excited states.

$$+ (CH_3)_2\dot{C}OH \longrightarrow \longrightarrow PhNH_2 \quad (91)$$

$$+ (CH_3)_2\dot{C}OH \longrightarrow PhNHNHPh \quad (92)$$

$$+ Ph_2\underset{\underset{CH_3}{|}}{C}—\underset{\underset{OH}{|}}{C}(CH_3)_2 + (CH_3)_2\underset{\underset{OH}{|}}{C}—\underset{\underset{OH}{|}}{C}(CH_3)_2 \quad (93)$$

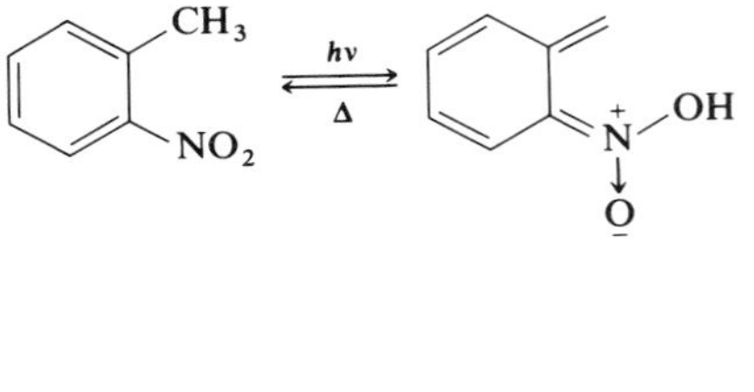

(94)

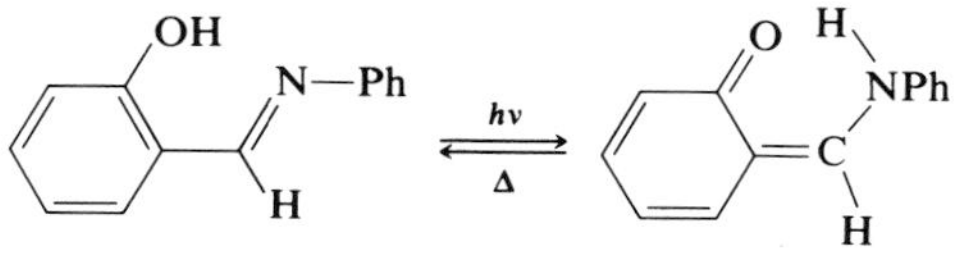

(95)

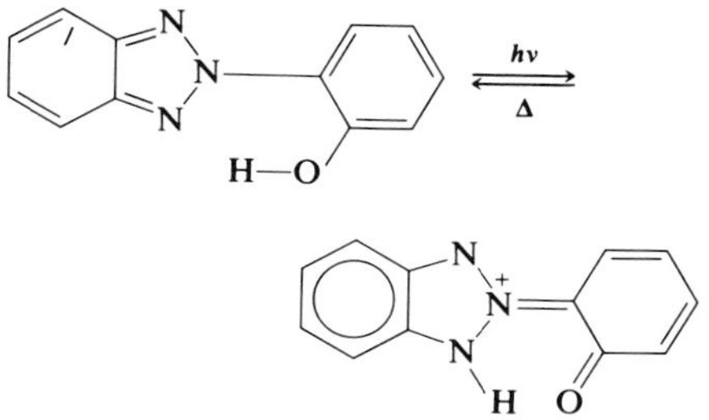

(96)

Another type of intramolecular hydrogen abstraction is shown by the conversion of the substituted cyclopentenone (59) into the bicyclic ketone (60) [Eq. (97)].

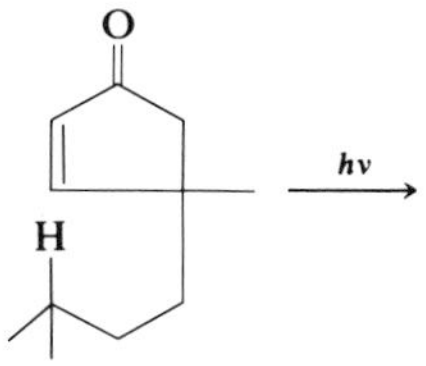

(59) $\xrightarrow{h\nu}$ (97)

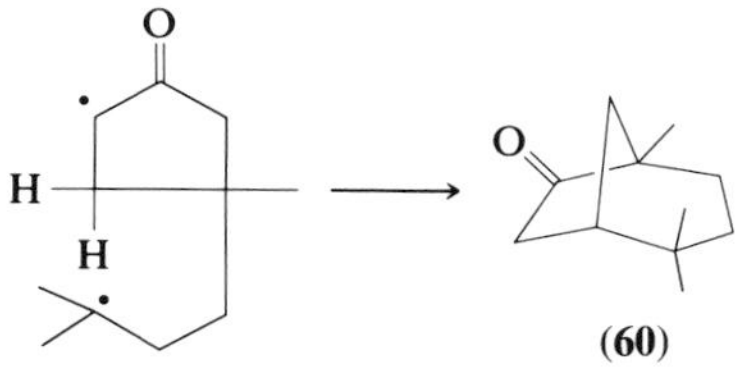

(60)

The reaction mechanism involves hydrogen abstraction from the tertiary center on the side chain by the β-carbon of the enone moiety, followed by coupling of the resulting biradical. The reaction has been shown to have considerable generality. A related observation is that cyclohexenones such as (16) and (27) undergo photoreduction in isopropyl alcohol to the corresponding saturated ketones by a process that appears to involve initial hydrogen abstraction by the β-carbon to give an enoxyl radical [Eq. (98)], followed by a second H-transfer to either the α-carbon or the oxygen.

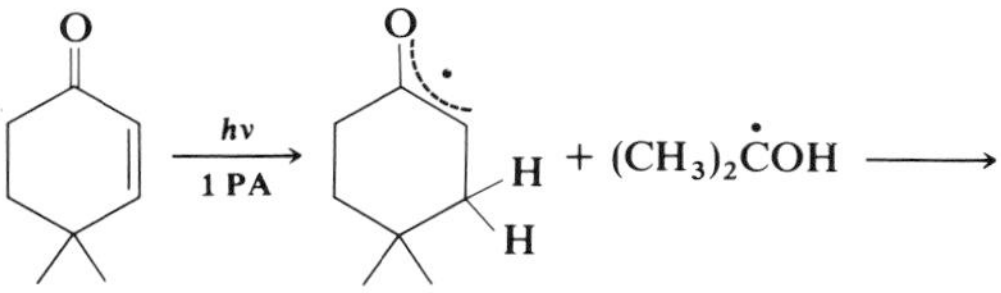

(27) $\xrightarrow[1\ PA]{h\nu}$ $+ (CH_3)_2\overset{\cdot}{C}OH \longrightarrow$

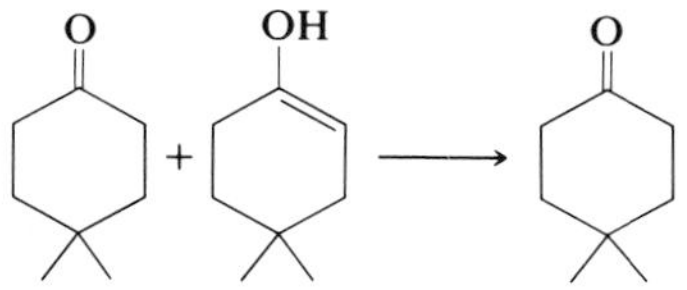

(98)

It is likely that this reaction, which is fundamentally distinct from the typical H-abstractions of $^3n, \pi^*$ states discussed above, occurs via a $^3\pi, \pi^*$ state. In most cases where inter- or intramolecular H-transfer to a conjugated enone can take place to the α or β carbon, the latter predominates. In some polycyclic systems where the topology of the system dictates the course of reaction (e.g., in reactions occurring in the crystalline state), both pathways are observed, depending on the distances between the reactive sites.

F. PHOTOFRAGMENTATION REACTIONS

A common type of photochemical reaction involves fragmentation into smaller molecular units, either free radicals or stable ground-state molecules. Several types of molecular fragmentations are discussed below.

1. Norrish Type I Cleavage of Ketones

Irradiation of ketones in the vapor phase leads most often to cleavage of one of the bonds to the carbonyl carbon, as shown in Eq. (99), to generate an acyl–alkyl radical pair.

$$R-\underset{O}{\underset{\|}{C}}-R' \rightarrow R\cdot + \underset{O}{\underset{\|}{C}}-R' \rightarrow CO + \cdot R' \quad (99)$$

The acyl radical usually subsequently loses CO to give a second alkyl radical. The final products arise by coupling and disproportionation of the various radicals. Similar reactions are seen for many ketones in solution. The radicals can be directly detected by ESR, and CIDNP experiments also provide evidence for the intermediacy of radicals in these reactions. A typical reaction in an acyclic system is shown in Eq. (100), in which a statistical mixture of coupling products is observed:

$$PhCH_2CO-CHPh_2 \rightarrow PhCH_2CO\cdot + \cdot CHPh_2$$
$$\rightarrow CO + PhCH_2\cdot + Ph_2CH\cdot$$
$$\rightarrow 25\%\ PhCH_2CH_2Ph$$
$$+ 50\%\ PhCH_2CHPh_2$$
$$+ 25\%\ Ph_2CHCHPh_2$$

(100)

The type I cleavage can take place from both singlet and triplet excited states and in solution frequently competes with other types of reactions, including the type II reaction when the system contains γ-hydrogens. The stability of the radicals plays a major role in determining the rate and efficiency of the cleavage. Inefficiency of the process in solution has been shown in some systems to be due to cage recombination of the initial radicals. Thus, optically active ketone (61) undergoes racemization attributed to recombination of the initial radical pair (ϕ = 0.56) competitive with disproportionation (ϕ = 0.44).

and ethylene, and (c) ring expansion to an oxacarbene that can be trapped in alcohol solvents. In the case of cyclohexanones, shown in Eq. (102), similar cleavage gives a 1,6-acyl–alkyl diradical that reacts further (competitive with recyclization) by intramolecular hydrogen migration from either the carbon alpha to the carbonyl (to give a ketene, path A) or the carbon adjacent to the alkyl radical site (to give an unsaturated open-chain aldehyde, path B).

The effects of structure on the competition between these competitive reaction pathways are reasonably well understood.

The Type I cleavage can also result in molecu-

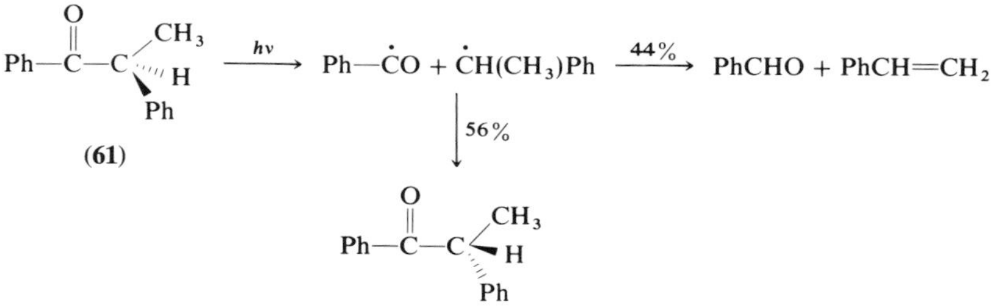

$$Ph-\overset{\overset{\displaystyle O}{\|}}{C}-\overset{\overset{\displaystyle CH_3}{|}}{\underset{\overset{|}{Ph}}{C}}{}_{\cdots H} \xrightarrow{h\nu} Ph-\overset{\cdot}{C}O + \overset{\cdot}{C}H(CH_3)Ph \xrightarrow{44\%} PhCHO + PhCH{=}CH_2$$

(61)

In cyclic systems containing small rings, another driving force for this reaction is relief of strain, as illustrated by the behavior of cyclobutanone [Eq. (101)].

lar rearrangements. The substituted cyclopentenone shown in Eq. (103) undergoes cleavage to a 1,5-diradical followed by ring closure to a cyclopropyl ketene that reacts with methanol to

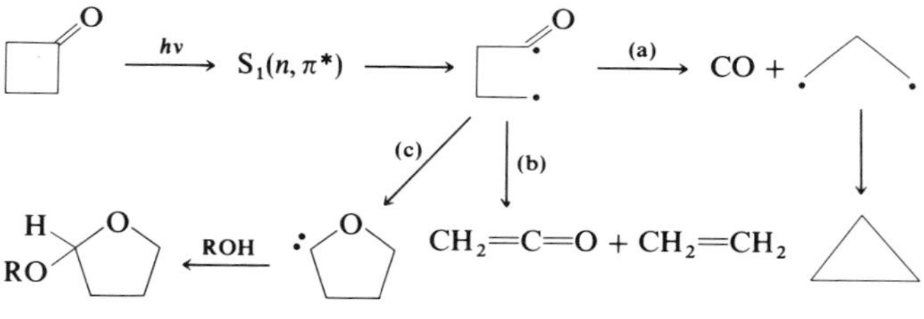

(101)

Here one sees initial cleavage to a 1,4-acyl–alkyl diradical that undergoes three competitive secondary reactions: (a) loss of CO to give a trimethylene 1,3-diradical that leads to cyclopropane, (b) cleavage into a mixture of ketene

give an ester.

In the case of β, γ-unsaturated ketones, cleavage gives an acyl and an allyl radical; recombination at the other allylic carbon gives a rearranged ketone, as shown for (62) in Eq. (104).

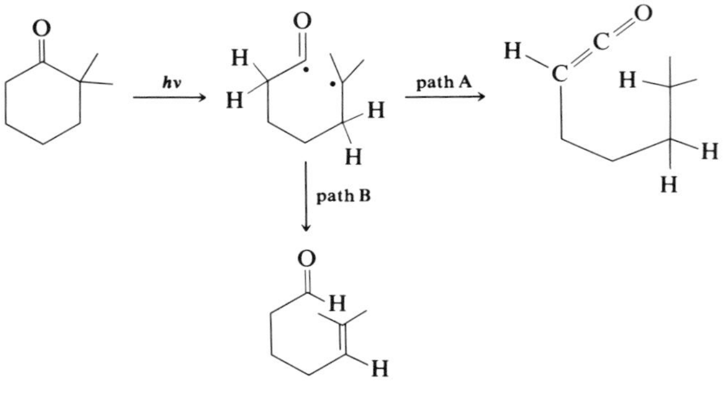

(102)

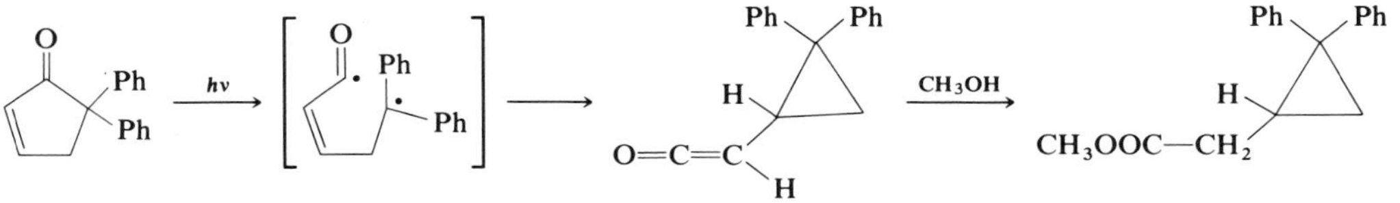

$$(103)$$

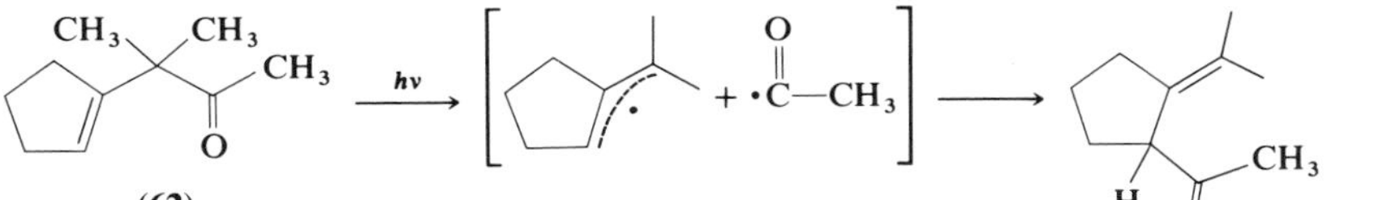

$$(104)$$

(**62**)

The overall process corresponds to a [1,3]-sigmatropic acyl shift, using the Woodward–Hoffmann terminology, and is the reaction usually observed on direct excitation of this class of compounds in solution. Gas phase studies demonstrate that there is a concerted pathway for the [1,3]-acyl shift which competes with the stepwise biradical route. Triplet-sensitized excitation of such ketones usually gives a different product resulting from an "oxa-di-pi-methane" rearrangement, to be discussed later.

2. Cleavage of Azo Compounds

Azo compounds, of structure R—N=N—R′, also readily lose nitrogen on UV irradiation to give a pair of radicals R· and R′·. The ease of cleavage again depends on the stability of the radicals, which can be detected using the ESR and CIDNP methods discussed earlier [see Eqs. (28)–(31)]. The reaction can occur in solution or the gas phase and by either direct or triplet-sensitized excitation. When the radicals R· and R′· are both stable (e.g., tertiary, benzylic, or allylic), cleavage of both C—N bonds occurs more or less simultaneously, but when the radical on one side is significantly less stable (as in compound (**18**), the reaction proceeds stepwise and diazo radicals RN₂· are discreet intermediates.

Fragmentation of cyclic azo compounds has found utility as a route to the synthesis of strained-ring systems, as illustrated in Eqs. (105) and (106). When 1,4-diradicals are formed, cyclization and cleavage are usually both observed, as in Eq. (107). Since the reaction can be carried out in matrices at very low temperature,

$$(105)$$

it has been used as a method of preparing a variety of diradicals for spectral studies under conditions where then do not react. An example [Eq. (108)] is the formation of trimethylene-methane diradicals of type (**63**) from the corresponding azo compound (**64**).

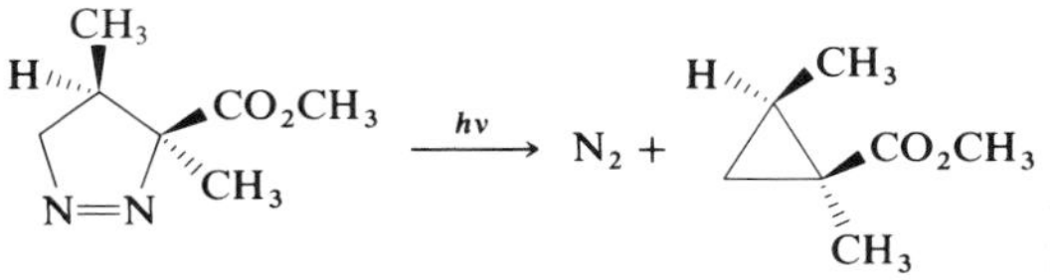

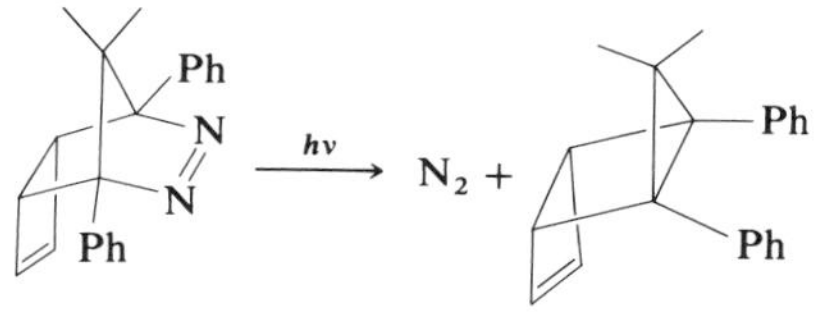

$$(106)$$

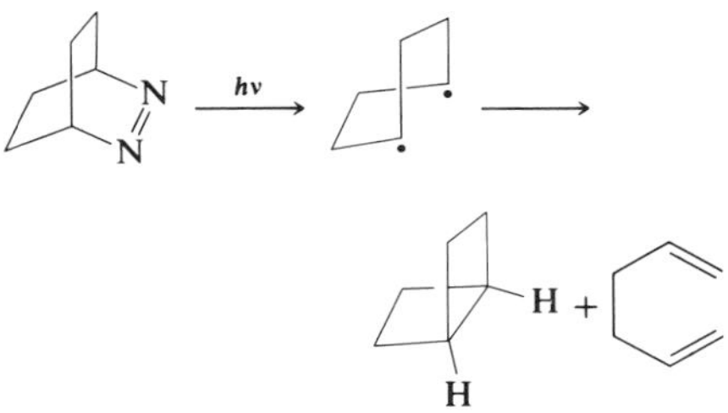

$$(107)$$

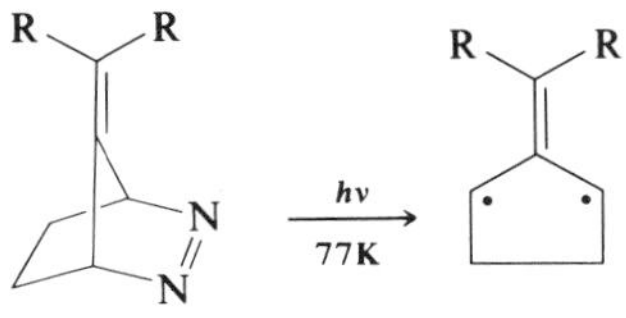

$$(108)$$

(**64**)   (**63**)

R = CH₃, Ph

3. Generation of Carbenes and Nitrenes from Diazo Compounds and Azides

Irradiation of diazocompound RR′CN₂ results in loss of nitrogen and formation of highly reactive carbenes RR′C:. Since the nonbonding pair of electrons can be either paired or unpaired, the carbene can exist as a singlet or a triplet, the latter usually being the more stable energetically. Direct excitation of diazo compounds gives initially singlet carbenes, which either react or undergo a spin flip to give the triplet carbene. The latter can be prepared directly using triplet photosensitization techniques.

The most common reactions of carbenes, typified by the parent CH_2:, involve addition to $C{=}C$ bonds to give cyclopropanes and insertion into $C{-}H$ bonds to give $C{-}CH_3$ groups. Both reactions are concerted stereospecific processes for the singlet carbene but occur stepwise via diradicals in the case of triplet carbene. Insertion into many other bonds (e.g., $C{-}C$, $O{-}H$, and $S{-}H$) is also observed. Substituted carbenes such as CH_3CH:, generated from CH_3CHN_2, usually undergo rearrangement (in this case to ethylene) at the expense of the above processes. The usual reaction path of singlet ketocarbenes, as illustrated in Eq. (109), is rearrangement to ketenes (Wolff rearrangement), which can have synthetic utility in ring condensations [see Eq. (110)]. In some cases, intramolecular addition to a $C{=}C$ bond competes effectively with the Wolff rearrangement and can be used to prepare highly strained ring systems, as in Eq. (111).

terization of biologically active macromolecules, in particular enzymes and receptors. The latter are sites of action of a variety of molecules, including neurotransmitters, hormones, cyclic nucleotides, natural and synthetic opiates, and so on, and are found throughout living systems. The technique involves synthesis of photochemically active analogs of known ligands for the enzyme or receptor and irreversible attachment of the photoaffinity label to the macromolecule (typically protein) by irradiation after the label has been incubated with appropriate preparations containing the active species. Inactivation of the site by covalent attachment of the label is demonstrated by showing alterations in biological activity and/or by direct detection of a covalently bound label containing a radioactive marker. With biochemical techniques, portions of the labeled enzyme or receptor can then be isolated and characterized. The most common functionalities used to date to create photoaffinity labels are diazo and azido moieties attached to ligands at sites such that the biological activity of the ligand is not altered. In recent years this photochemical tech-

$$R{-}CO{-}CHN_2 \xrightarrow[-N_2]{h\nu} R{-}CO{-}CH: \rightarrow R{-}CH{=}C{=}O \xrightarrow{MeOH} RCH_2COOMe \qquad (109)$$

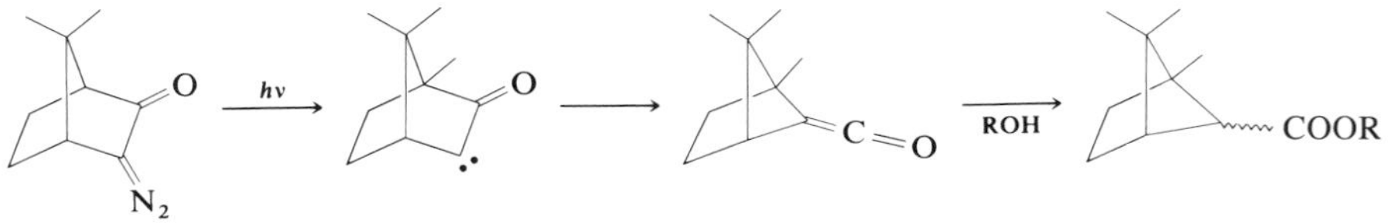

$$(110)$$

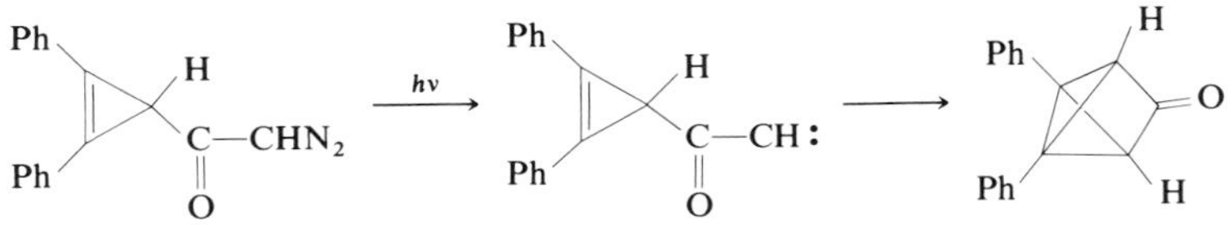

$$(111)$$

The formation of nitrenes R—N: from excitation of azides RN_3 is completely analogous to the formation of carbenes from diazo compounds. Again, both singlet and triplet nitrenes are possible, and direct and sensitized photolyses can be utilized. The reactions of nitrenes are also similar to those of carbenes, involving primarily formation of aziridines by addition to $C{=}C$ bonds and intra- and intermolecular insertions. Generally, nitrenes are not quite as reactive as carbons and react with somewhat greater discrimination.

Photoaffinity labeling is an important application of the photochemistry of azides and diazo compounds for the study and molecular charac-

nique has led to major advances in understanding the structure and mode of action of enzymes and receptors.

4. Photochemical Ring Openings of Oxiranes and Cyclopropanes

We have already discussed photochemical electrocyclic ring openings and ring closures and cleavage of cyclic ketones and azo compounds with generation of diradicals. Another group of related reactions involves photochemically induced ring openings of cyclopropanes and oxiranes to generate carbenes, which can be chemically trapped and deteced spectroscopically, as illustrated in Eq. (112). Cyclopropanes sometimes undergo ring opening to generate either zwitterions (direct excitation) or 1,3-diradicals (photosensitization) via excited singlets or triplets, respectively, as illustrated in Eq. (113). The zwitterions can be trapped nucleophilically, while the diradicals undergo ring closure to give an isomer of the starting material. The initial step in the photochemistry of oxiranes involves a 4e disrotatory ring opening to give a highly colored ylid [Eq. (114)], easily detected spectroscopically, which in a second step cleaves thermally or photochemically to a carbene and an aldehyde or ketone; ring closure of the zwitterion to an oxirane (starting material or an isomer) does not seem to be significant. Similar ring openings of aziridines (nitrogen analogs of oxiranes) are known.

Other types of photofragmentations are known that involve organic compounds such as halides, nitrites, peroxides, esters, amides, and nitro compounds. The interested reader is referred to monographs and the original literature for further details.

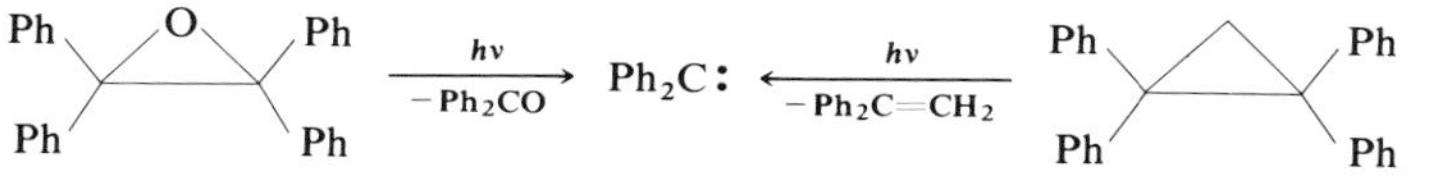

istry literature include benzene-sensitized extrusion of SO_2 from cyclic sulfones to form butadienes, illustrated in Eq. (115); this reaction shows only partial stereoselectivity that is opposite to that seen on thermal extrusion of SO_2 from these compounds.

The analogous photochemical ring opening of the isomeric ketones (65) and (66) gives a mix-

(65)   (66)

ture of butadienes, attributed to competitive triplet excited state pathways. In contrast, photodecarbonylation of 3,5-cycloheptadienone (67) and substituted analogs to 1,3,5-hexatrienes [Eq. (116)] is a clean singlet-state reaction that we have shown, by using appropriate substitu-

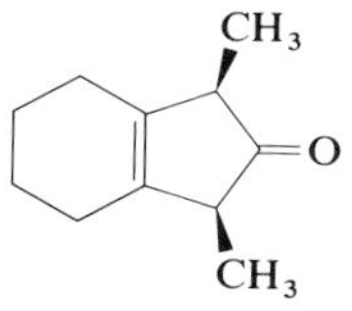

(112)

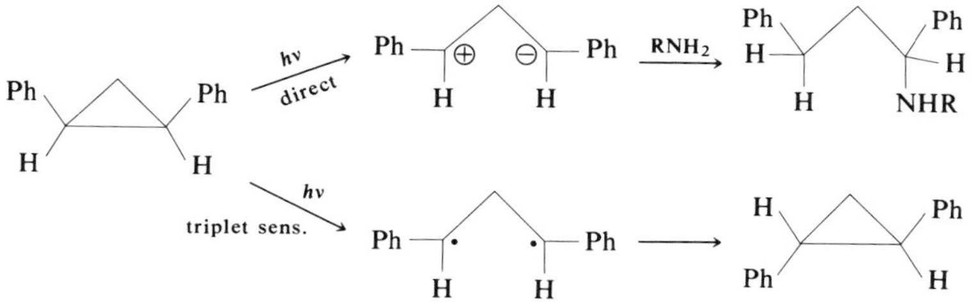

(113)

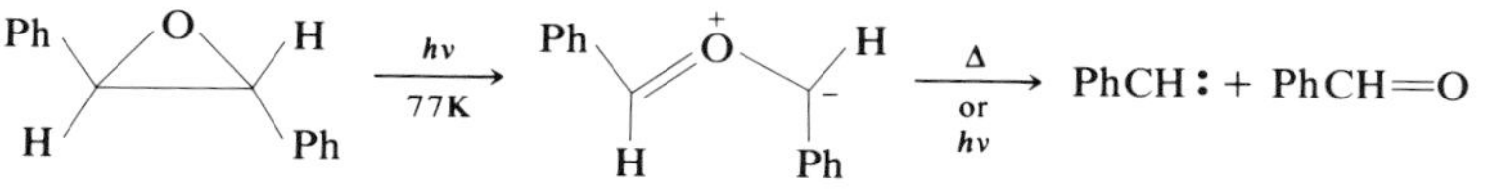

(114)

5. Cheletropic Reactions

Cheletropic reactions are fragmentations of cyclic systems that generate a linear conjugated system by extrusion of a small molecular fragment such as CO, N_2, O_2, or SO_2. Such reactions in the ground and excited states were predicted by Woodward and Hoffmann to occur by concerted processes with simultaneous cleavage of both bonds to the departing cheleofuge. Known examples of such reactions in the photochem-

Major   Minor

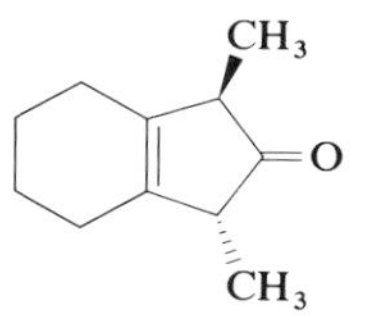

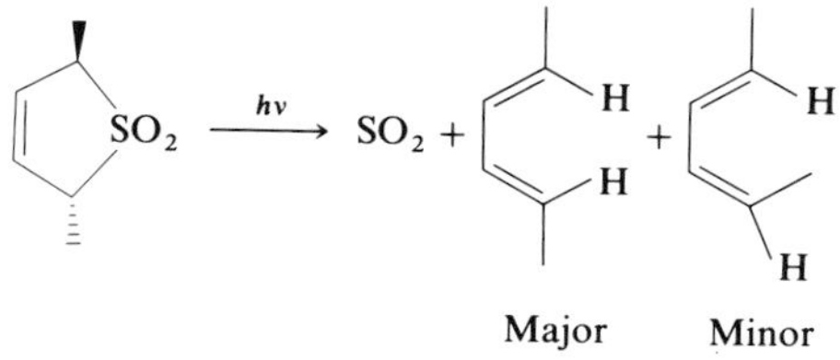

(115)

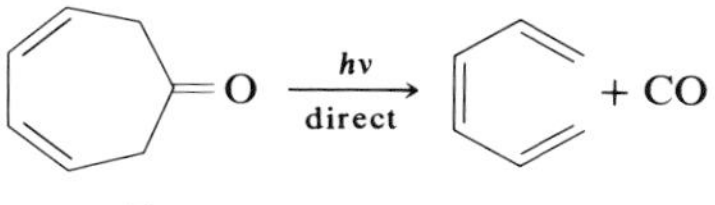

(67)

(116)

tion, to be stereospecific in a conrotatory sense. Although this reaction seems to be a true concerted cheletropic fragmentation, the reaction course followed is not that anticipated on the basis of application of the Woodward–Hoffmann rules but seems to be the path consistent with the principle of least nuclear motion. Specifically, the stereochemical course observed is the same (conrotatory) as in corresponding thermal extrusion of SO_2 from analogous sulfones.

G. PHOTOADDITION AND PHOTOSUBSTITUTION REACTIONS

It has already been mentioned (Sections VII,D,4 and VII,E) that alcohols (and water) undergo photoaddition to cyclic alkenes to give ethers (and alcohols). The reaction appears to involve carbocation intermediates generated by protonation of the alkene singlet excited state (in the case of small rings) or a ground-state trans-cycloalkene formed on photoisomerization of the starting material via the triplet state and therefore follows Markovnikov's rule. Two additional examples are shown in Eqs. (117) and (118) (see also Scheme II). Triplet-sensitized excitation of norbornene itself proceeds by a different course involving H-abstraction to give a norbornyl radical followed by characteristic radical coupling and disproportionation reactions

is a related process and is one of the most important reactions involved in photodeactivation of nucleic acids.

As discussed earlier (Section III), electron-transfer-induced reactions of water or alcohols with alkenes occur via radical ion intermediates and result in overall anti-Markovnikov addition. This is a very general reaction that has synthetic utility. Two additional examples are given in Eqs. (121) and (122).

Many other addition reactions to alkenes are known that involve both direct and sensitized excitation. Among the addends are simple amines, NH_2Cl, N-nitrosoamines, $HCONH_2$, aldehydes, ketones, halogens and halides, thiols, phosphines, and silanes. Some of these reactions occur via ionic mechanisms, but most seem to involve free-radical intermediates generated by initial abstraction of an H or a halogen atom by an alkene excited state. Addition of polar molecules to benzene involves 1,3-addition as in cycloaddition of alkenes (Section VIII,C,4). A typical example is shown in Eq. (123).

In nucleophilic aromatic photosubstitution reactions, orientation effects are quite different from those in analogous ground-state reactions. Thus, in the photochemical process, electron-withdrawing groups such as NO_2 direct the nucleophile to meta positions, while electron-do-

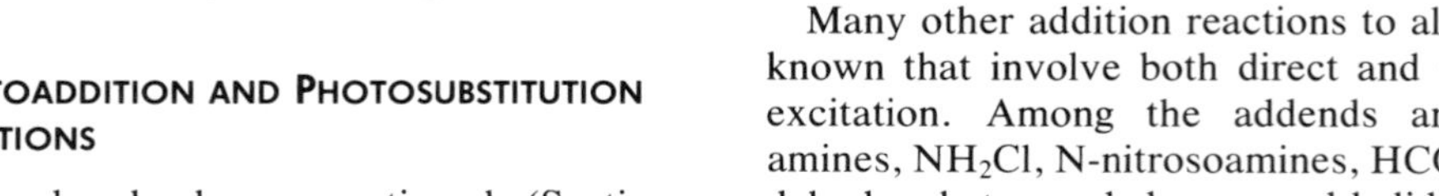

(117)

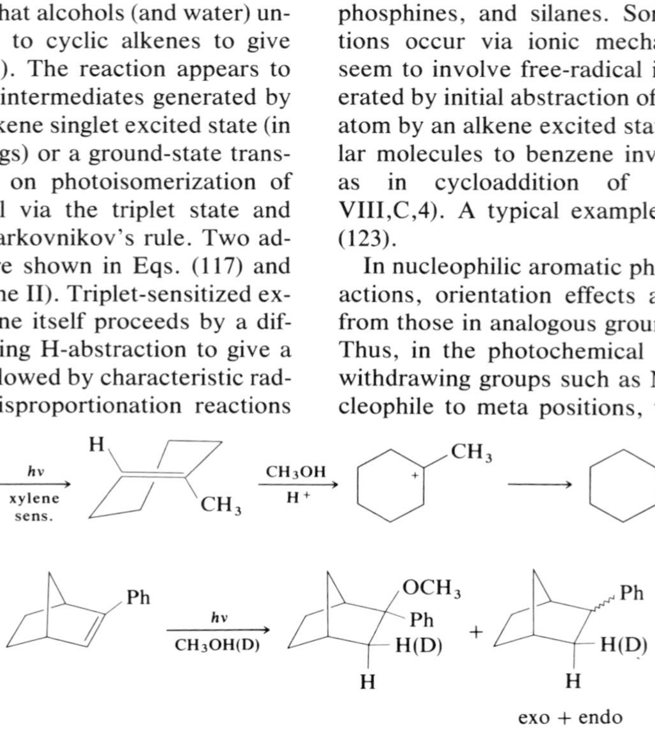

exo + endo

(118)

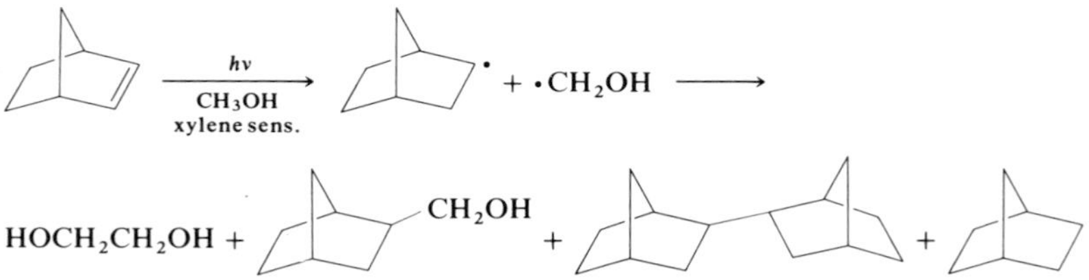

(119)

[Eq. (119)]. It is not at all clear why there is such a difference in behavior of norbornene and 2-phenylnorbornene on both direct and triplet-sensitized excitation. Photohydration of uracil [Eq. (120)] and cytosine and related nucleotides

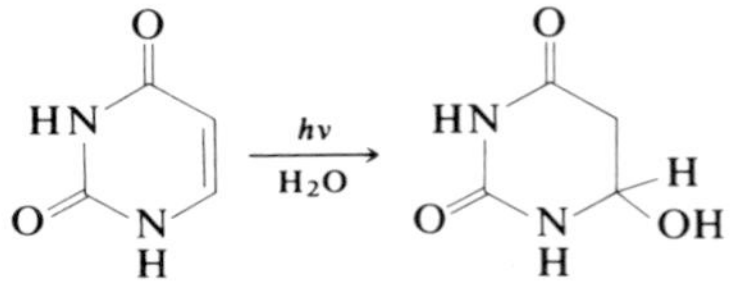

(120)

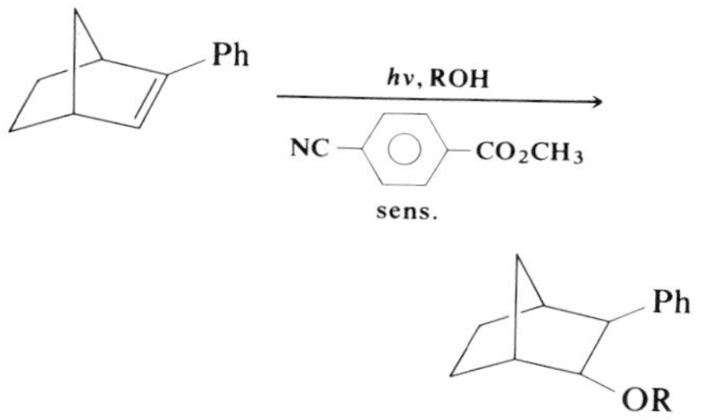

ortho and para positions. Some specific examples are shown in Eqs. (124)–(126).

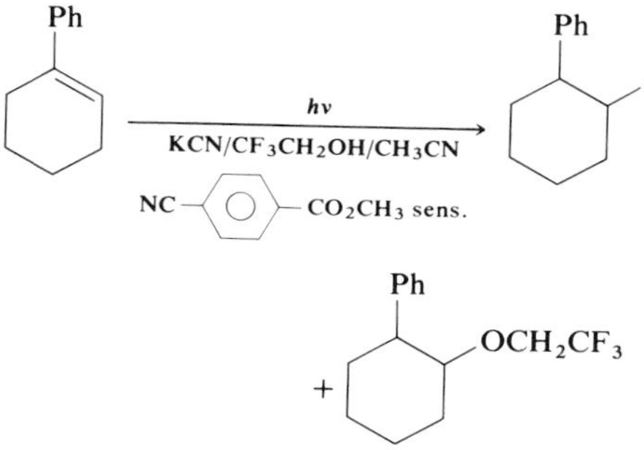

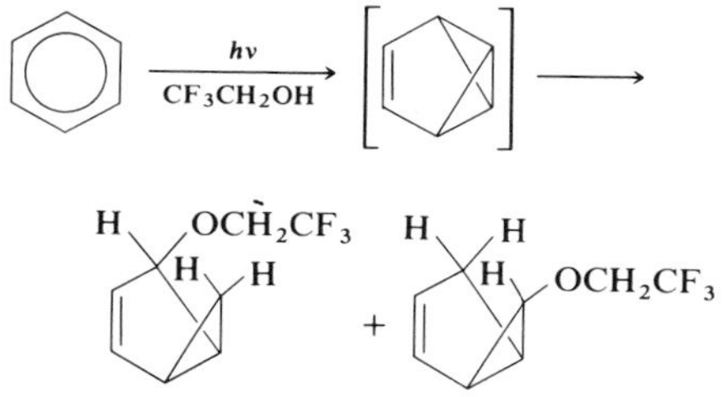

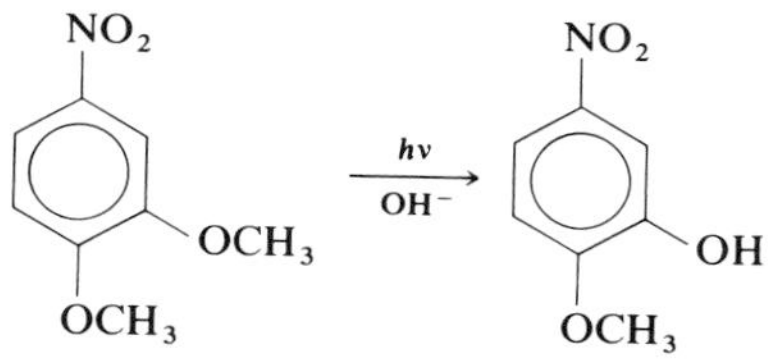

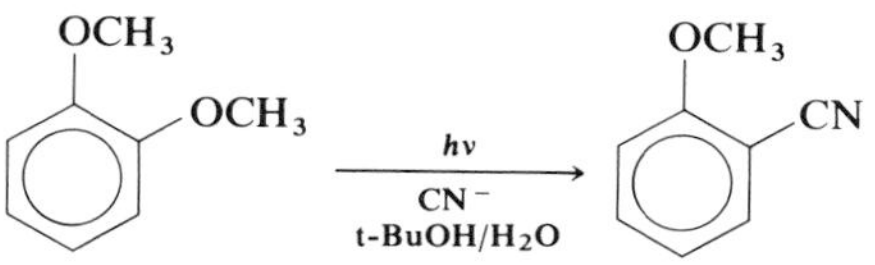

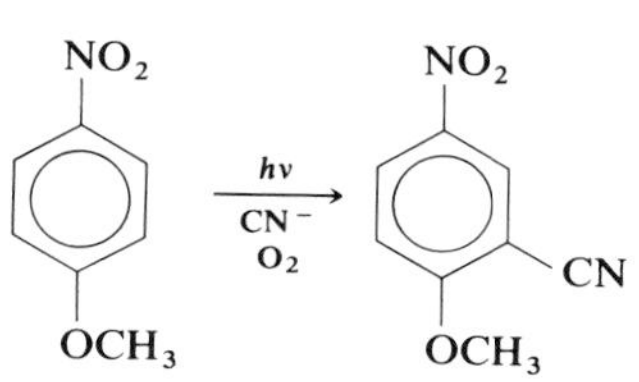

The course of these reactions has been rationalized in terms of calculated differences in charge densities of such molecules in the ground state and the S_1 state, leading in some cases to a charge reversal. Thus, in both singlet and triplet excited nitrobenzene, the position with greatest positive charge is meta to the nitro group, making this the preferred site for nucleophilic attack, the opposite of that in the ground state. However, in at least some of these reactions, electron transfer and even electron ejection (in the presence of strong electron-releasing substituents) with generation of radical ions can play an important role.

Photosolvolysis of compounds RX, where X is a suitable leaving group such as halide, acetate, sulfonate, or phosphate, frequently results in nucleophilic substitution. In some instances it seems clear that carbocations R^+ are involved, while in other cases the evidence is more ambiguous, and reactions may involve radical ions or even neutral radicals. In studies of compound (68) and a number of other benzobicyclic sys-

(68)

tems where the leaving group X is Cl or OSO_2CH_3, a number of reaction pathways not seen in the ground state have been uncovered, involving cationic as well as radical ion intermediates, leading to complex molecular rearrangements.

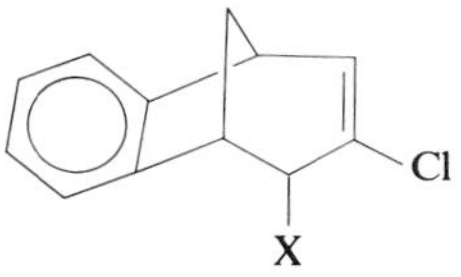

H. Photorearrangements

We have already seen a number of examples of reorganization of molecular structure under the influence of light. Relatively simple examples include *cis–trans* isomerization of alkenes and azo compounds (Section VIII,B) and electrocyclic reactions (Section VIII,D). The rearrangements of 2,5-cyclohexadienones discussed earlier (Section VII,E, Schemes V and VI), which appear to occur via photogenerated zwitterions, are examples of more deep-seated molecular reorganizations that have fascinated organic photochemists for many years. The ability to rationalize satisfactorily the course of these

nating groups such as CH_3O (which in the ground state are deactivating in nucleophilic aromatic substitution) direct the nucleophile to reactions was one of the early triumphs of mechanistic organic photochemistry. The discussion below will focus on these and some other complex photorearrangements.

1. Di–Pi-Methane Rearrangements

Direct irradiation of many acyclic 1,4-dienes results in rearrangement to cyclopropylethylenes, as shown in Eqs. (127)–(130).

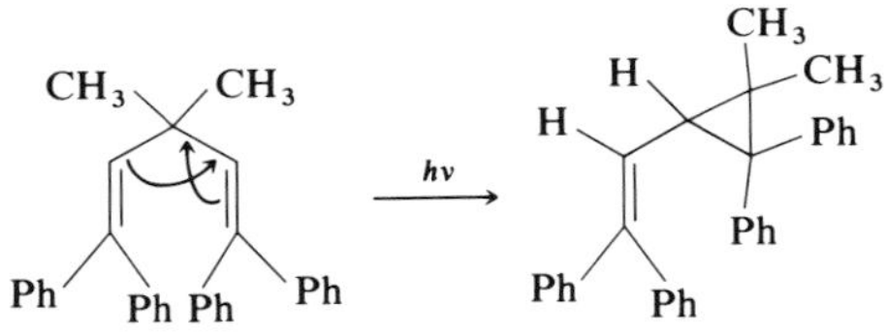

$$(127)$$

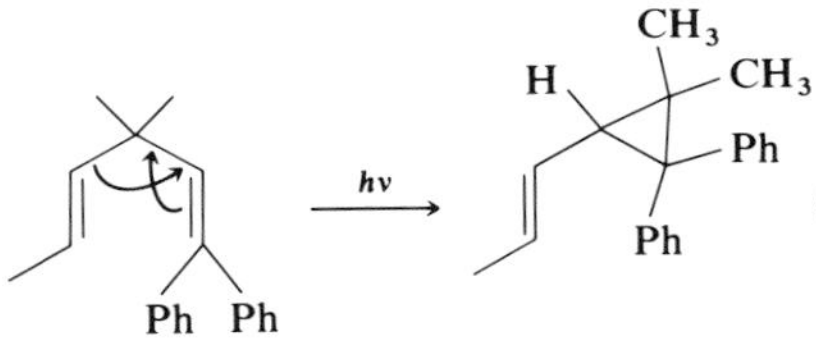

$$(128)$$

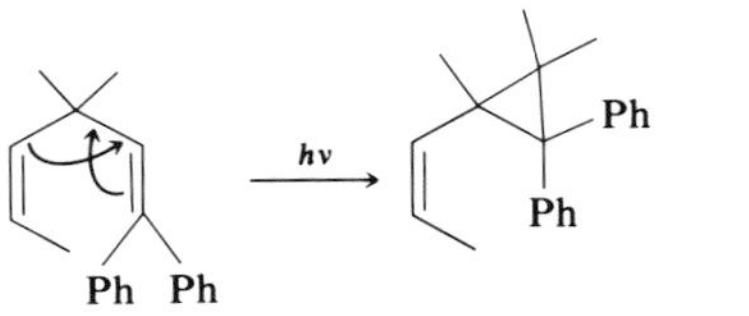

$$(129)$$

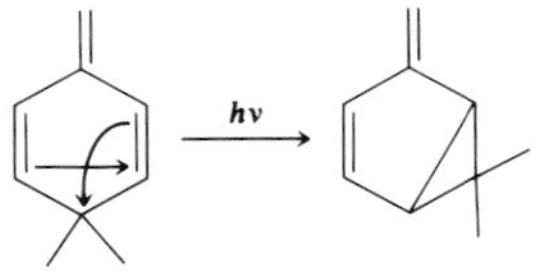

$$(130)$$

This so-called di–pi-methane (DPM) rearrangement shows regiospecificity [see Eq. (128)] as well as stereospecificity [see Eqs. (129) and (130)]. Generation of the diene T_1 state on triplet-sensitized excitation generally gives little or no rearrangement product. This is attributed to the preference of the triplet for undergoing isomerization around the double bond(s) (so-called free-rotor effect) rather than the bonding interaction required in the generally accepted rearrangement mechanism depicted in Scheme VII. Although the reaction may be concerted without the intervention of detectable intermediates, it is

SCHEME VIII.

convenient to think of the process in terms of initial bonding between C-2 and C-4 to give a 1,4-diradical that can cleave either to regenerate the starting diene or to give a bis-homoallylic 1,3-diradical that closes to give the final product. The regiospecificity can be understood in terms of reaction via the more stable 1,3-diradical intermediate, while the stereospecificity is indicative of a highly synchronous process.

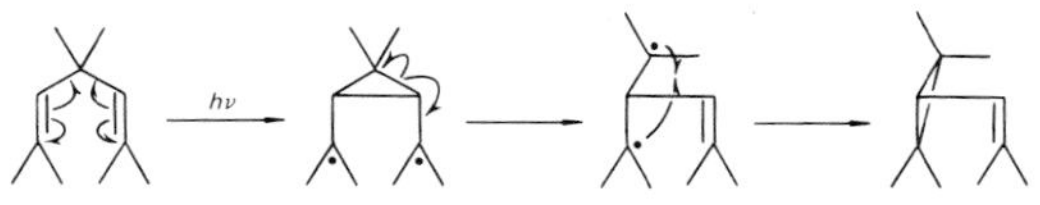

When one of the pi systems is incorporated in a benzene ring, the same rearrangement occurs, as seen in Eqs. (131) and (132).

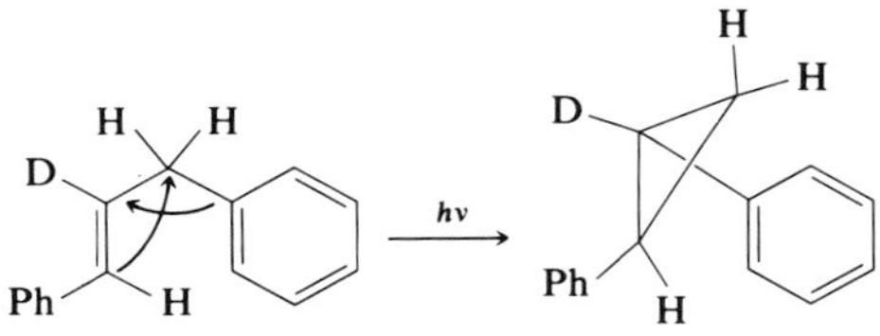

$$(131)$$

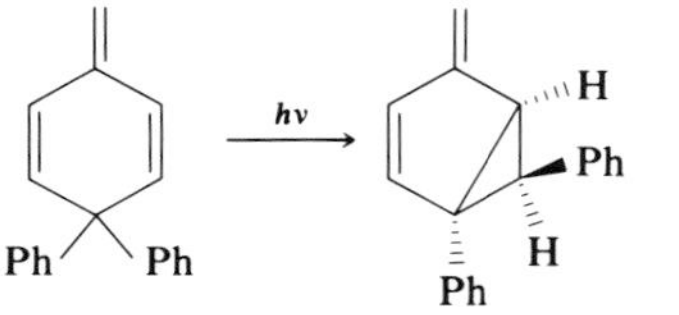

$$(132)$$

Here the reaction corresponds to a [1,2]-migration of a phenyl group. An instructive example is provided by the diphenylcyclohexadiene system in Eq. (133) which undergoes electrocyclic ring opening from S_1 and the DPM rearrangement from T_1.

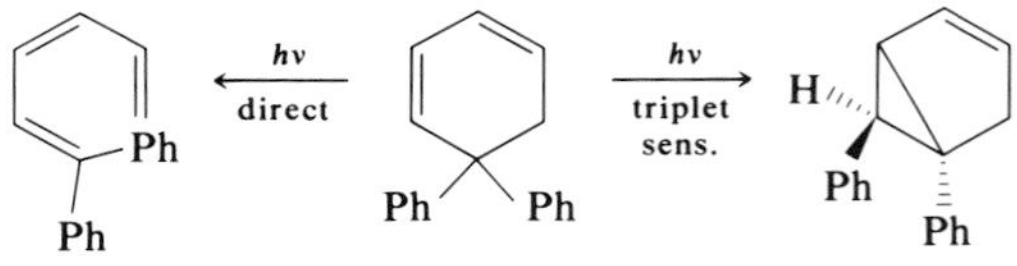

$$(133)$$

Similarly, direct excitation of barrelene (**69**) [Eq. (134)] leads to cyclooctatetraene while triplet sensitization gives semibullvalene (**70**).

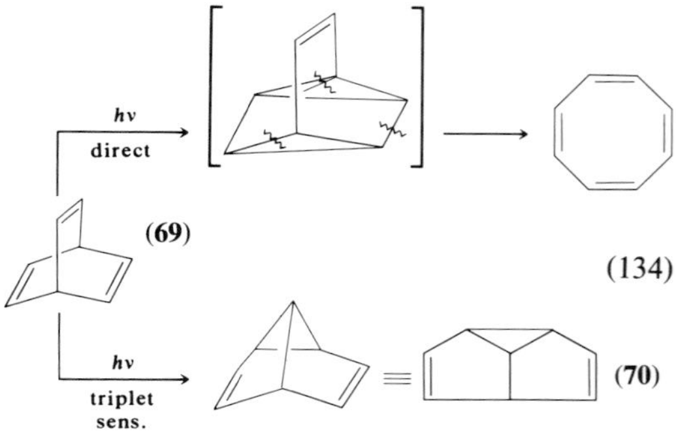

$$(134)$$

$$(70)$$

The former process may involve an initial intramolecular [2 + 2] cycloaddition, followed by the indicated retrocleavage, while labeling studies clearly show that semibullvalene (70) arises by a classic DPM process as given by Scheme VIII. As a rule, cyclic systems undergo the DPM rearrangement more efficiently on sensitization than on direct excitation although there are exceptions. Quantum efficiencies can run as high as 0.5.

A β, γ-unsaturated ketone is an oxa-analog of a 1,4-diene. As discussed earlier, direct irradiation of such compounds generally results in an allylic [1,3]-acyl shift [see compound (62), Eq. (104)], involving radical pairs at least in part. Triplet sensitization results in a molecular rearrangement to (71) [see Eq. (135)] structurally analogous to the DPM process.

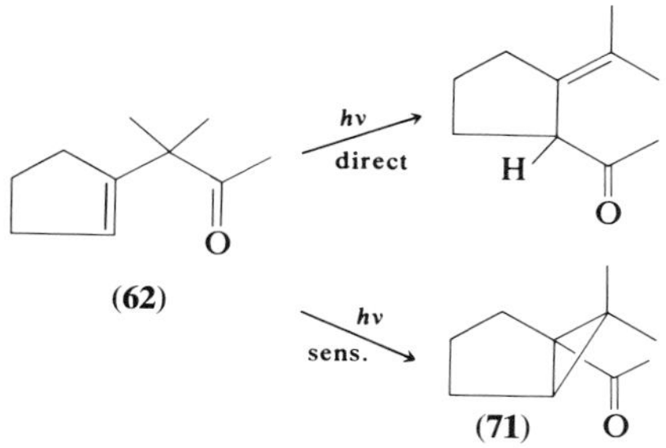

(135)

The proposed mechanism of this oxa-di-pi-methane rearrangement is analogous to that given in Scheme VIII, except that the intermediate 1,4- and 1,3-diradicals, being triplets, are relatively long lived, leading to loss of stereochemistry on the central carbon (α to the carbonyl). Such reactions of bicyclic ketones are synthetically useful as a route to highly strained tricyclic systems that are difficult to obtain by alternative methods, as shown by the examples in Eqs. (136), (137), and (138). The first of these, although it occurs on direct excitation, has been shown to occur via T_1; it is interesting to note that intersystem crossing in this compound occurs at the expense of the usual (1,3)-acyl shift, although this is not the case when methyl groups are substituted at the central carbon marked with an asterisk.

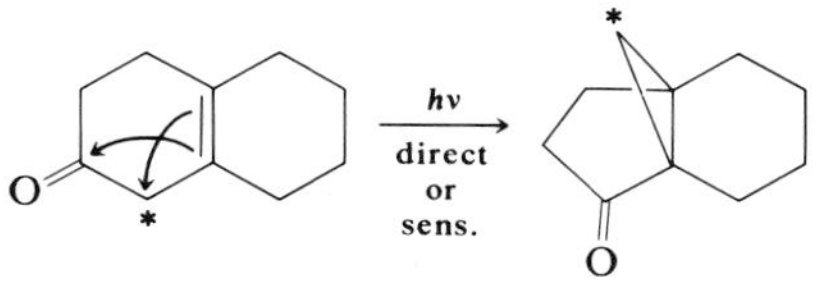

(136)

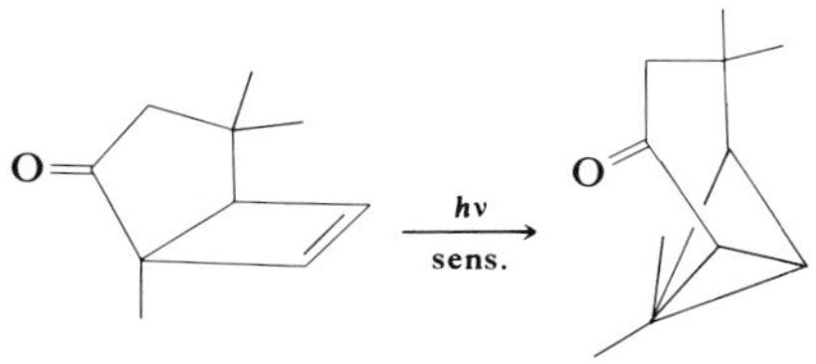

(137)

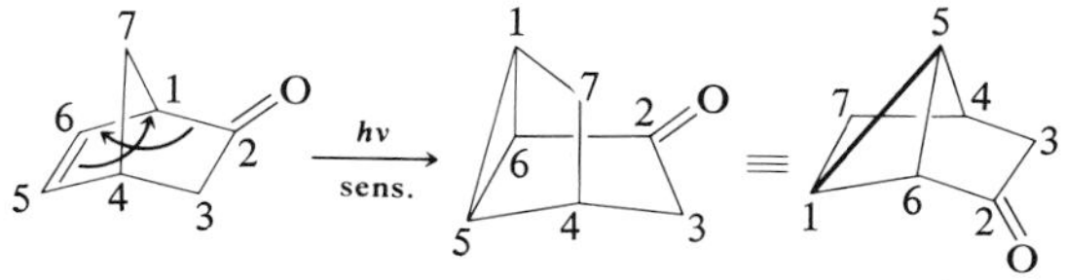

(138)

2. Rearrangements of Cyclohexadienones

The zwitterion mechanism for type A rearrangement of 2,5-cyclohexadienones (see Schemes V and VI) was one of the early successes of mechanistic organic photochemistry. This mechanism was able to account, for example, for the complex transformations of the sesquiterpene α-santonin (46) shown in Eq. (139) which had previously seemed to be beyond rational interpretation. Also, the different courses taken by the 2- and 4-methyl dienones in Eqs. (140) and (141) on irradiation in acidic solution could be rationalized in terms of stereoelectronic reactions of protonated zwitterions as indicated. In many of these systems, including santonin and some analogous steroid dienones, a consecutive series of photoreactions ensues, often leading to quite complex product mixtures.

The 2,4-cyclohexadienones usually undergo a ring-opening reaction to give a ketene, shown in Eq. (142), that is simply an electrocyclic ring opening of a 1,3-cyclohexadiene to a 1,3,5-hexatriene. In some cases, as here, the ring opening is photochemically (and sometimes thermally) reversible. The ketene (72) can undergo thermal ring closure to a bicyclic ketone (lumiketone, Section VII,E) of the type normally formed as the primary photoproduct of 2,5-cyclohexadienones or can be trapped nucleophilically by water (to give carboxylic acids), alcohols (to give esters), and amines (to give amides). Such ketenes have been directly observed spectroscopically under matrix isolation conditions at low temperature (Section VII,D,3). Thus, the formation of photosantonic acid (46a) on irradiation of santonin (46) [Eq. (139)] is due to secondary photoisomerization of lumisantonin (46b) to a linearly conjugated dienone (46c), followed by

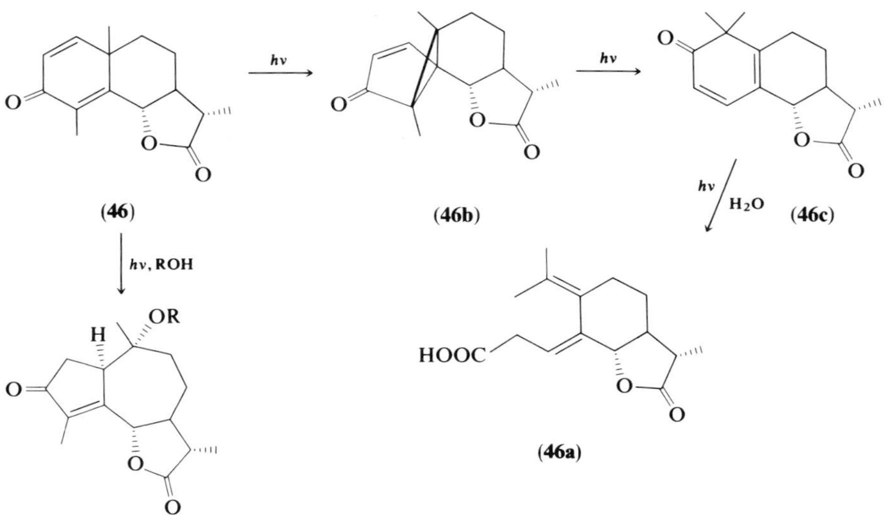

(139)

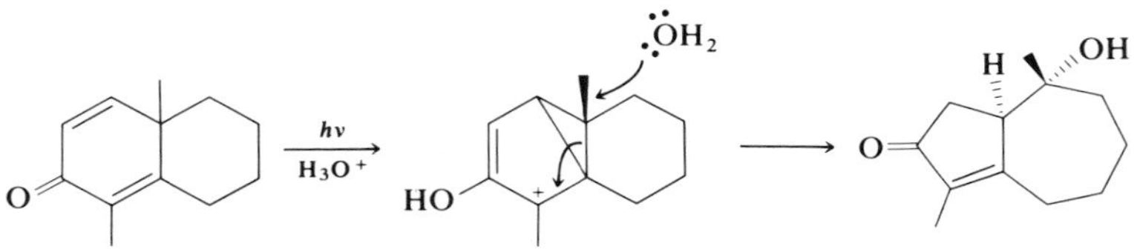

(46)   **(46b)**   **(46c)**

(46a)

(140)

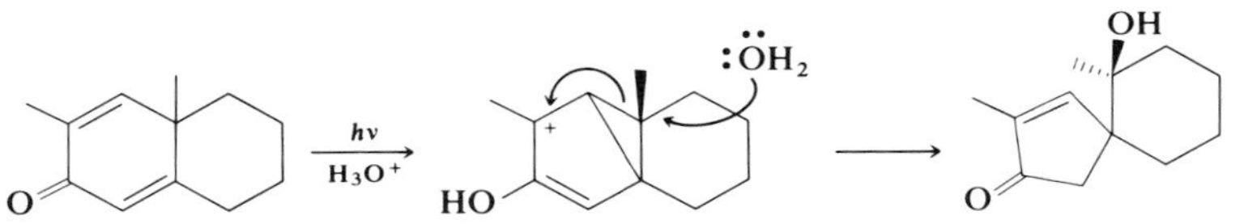

(141)

(72)

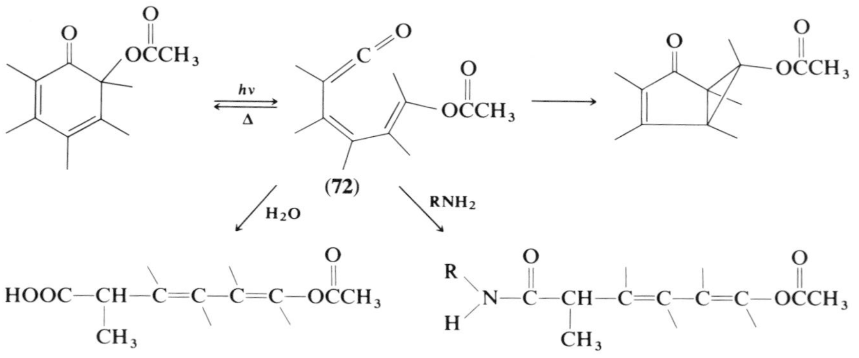

(142)

photoinduced ring opening to a ketene which is ultimately trapped by water.

3. Rearrangement of Substituted Cyclohexenones

The 4,4-disubstituted-2-cyclohexenones such as (**16**) and (**27**) (see Section VII,D,4) undergo a type A molecular photorearrangement to bicyclo[3.1.0]hex-3-en-2-ones (lumiketones) (**34**) and

(**35**) which is structurally analogous to the photorearrangement of corresponding 2,5-cyclohexadienones. There are significant differences in these reactions, however. Foremost is the fact that the quantum efficiency of the cyclohexenone reaction is very low, on the order of 0.01, while the cyclohexadienone rearrangement is very efficient, $\phi \sim 0.7$–1.0. Second, there is abundant evidence that the cyclohexadienone rearrangement proceeds via zwitterion interme- diates that in some cases can be trapped by nucleophiles, while there is no evidence for a corresponding intermediate in the cyclohex- enone rearrangement, although the efficiency seems to depend on the polarity of the medium. Third, while 4,4-diphenylcyclohexadienone (40) cleanly rearranges to (41), the corresponding cyclohexenone (73) undergoes phenyl migration as shown in Eq. (143) in what is essentially a

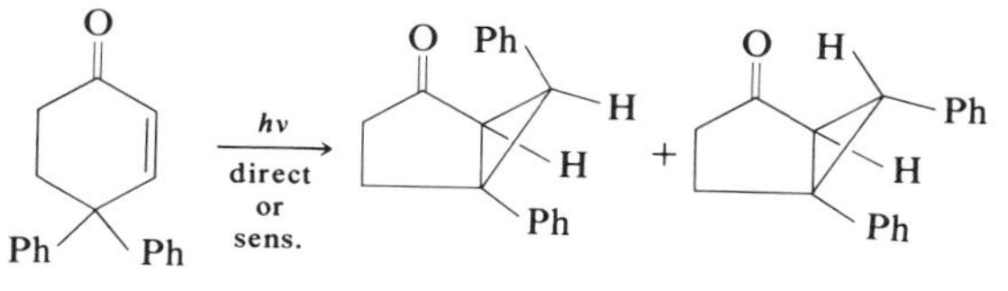

(73)   Major   Minor (143)

DPM rearrangement. The efficiency is again quite low (0.04). Thus, the second double bond in the cyclohexadienones participates directly in the reaction (as described in the previous sec- tion) in a very efficient process that has no anal- ogy in the photochemistry of cyclohexenones.

An important similarity in all of these reac- tions is that they clearly proceed via triplet ex- cited states. There is good evidence that the lowest triplet of cyclohexadienones (40) and (48) is an n, π^* state, and the classical mechanism for the type A rearrangement is built on that assign- ment, although there is some question about the triplet assignment to (45), (46), and related sys- tems. For cyclohexenones such as (27), the low- est triplet seems to be a π, π^* state. The pro- posed rearrangement in these systems is thought to originate from a twisted, $^3\pi, \pi^*$ state in which substantial torsion around the C=C bond has taken place in the direction of a trans-cyclo- hexenone, analogous to the formation of such trans-cyclic enones in larger ring systems (see Section VII,D,4). Flash data support this pro- posal, as well as observations that torsionally constrained cyclohexenones such as (74) do not

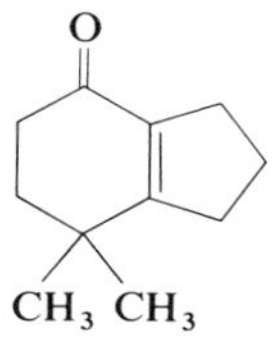

(74)

rearrange on photoexcitation. Furthermore, al- kenes quench the photorearrangement of (27) by intercepting either the twisted enone triplet or some subsequent intermediate to form trans-

fused [2 + 2] cycloadducts (Section VIII,D,6). With optically active chiral cyclohexenones (75), it was shown that the rearrangement occurs stereospecifically without loss of optical purity [see Eq. (144)], indicating that ring-opened di-

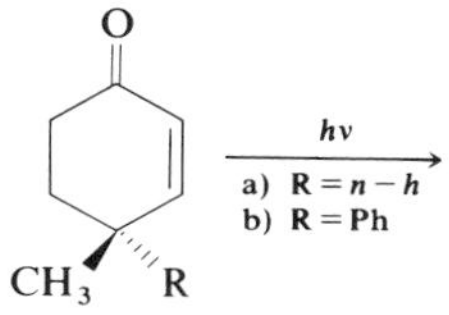

(75)

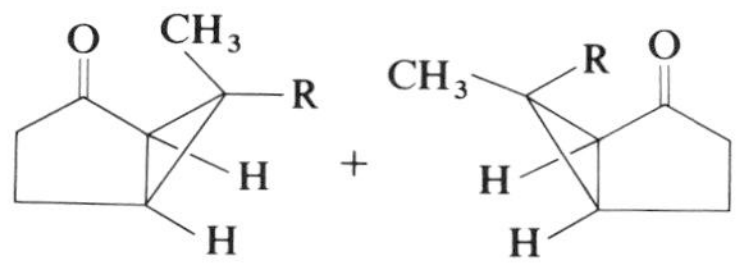

(144)

radicals are probably not intermediates in the reaction.

Since starting enones recovered after more than 100 hr had not lost optical activity, ineffi- ciency in these systems cannot be associated with reversible ring opening. The enone photo- rearrangement has been pictured as in Scheme IX as a concerted antarafacial (i.e., anti) addi- tion of the C_4—C_5 sigma bond to the C=C bond with inversion of configuration at C_4, which is consistent with theoretical predictions based on orbital symmetry. It is surprising that such a concerted process would occur through a triplet excited state with unpaired spins, implying that rearrangement and intersystem crossing may be synchronous in this case. An alternative mecha- nism, supported by flash-kinetic studies, is that the twisted triplet of (27) actually gives a ground state transoid enone, which rearranges ther- mally, in competition with isomerization back to the starting cis-enone.

One final note is that enone (75b), with methyl and phenyl groups at C_4, undergoes both the type A rearrangement to lumiketones and phenyl migration [see Eq. (145)], the former re-

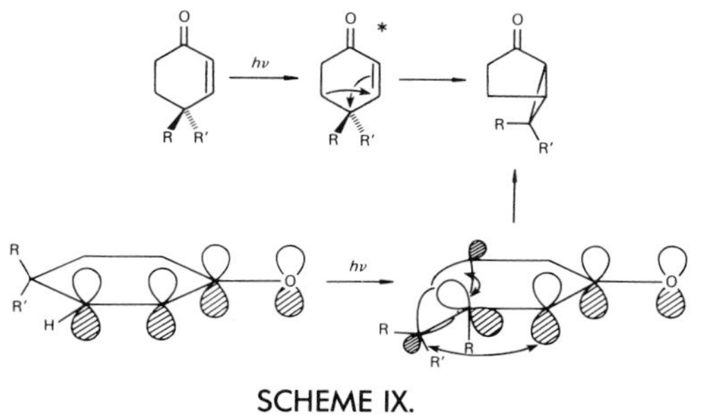

SCHEME IX.

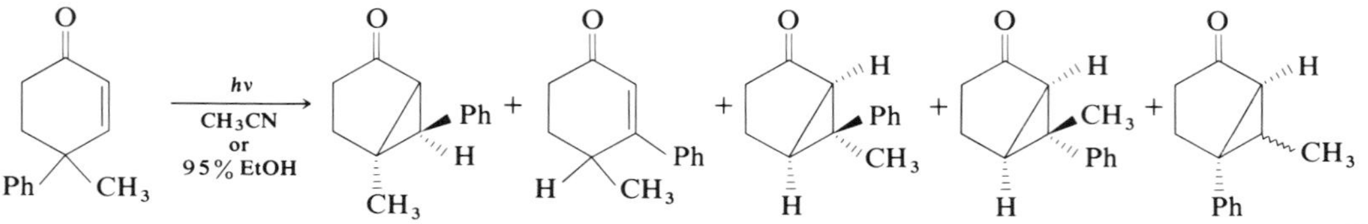

(75b)   (145)

action being favored in polar solvents and the latter in nonpolar solvents. This suggests that these reactions arise via two distinct triplet states, a π, π^* state and an n, π^* state, respectively. It is known that increasing solvent polarity lowers the energy of π, π^* states and raises the energy of n, π^* states. Indeed, the ordering of the states in these systems may well be a function of solvent polarity, which may explain why the type A rearrangement is usually observed only in polar solvents. See also results for (16) in Fig. 9.

IX. Photochemistry without Light; Chemiluminescence

In principle, electronic excited states of molecules could be generated by thermal excitation of ground states. In practice this is not possible because of the extreme temperatures that would be required (e.g., 8500°C to populate a state ~80 kcal above the ground state to ~1%). However, thermal energy provided by a chemical process can in some cases lead to population of an electronic excited state of one of the products. Light emission from the excited state under these circumstances is known as chemiluminescence.

A simple example of such a reaction is shown in Eq. (146). Heating the endoperoxide (76) causes a symmetry-allowed Diels–Alder reaction to give 9,10-diphenylanthracene and singlet oxygen, an excited state of oxygen.

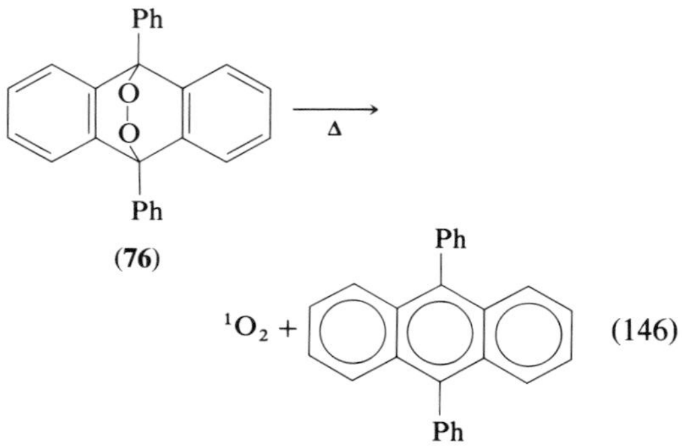

(76)

$$^1O_2 + \text{[9,10-diphenylanthracene]} \quad (146)$$

The activation energy for this reaction (28 kcal/mol) puts the energy of this transition state of this fragmentation above that of the $^1\Delta$ state of

O_2, which is 32 kcal/mol above the triplet ground state $^3\Sigma$. Chemiluminescence can be observed in this system, as well as in a variety of other thermal processes that generate $^1O_2$. This excited state of oxygen is relatively long lived (e.g., 25 μsec in benzene, 10 μsec in ethanol, and 0.7 msec in CF_3Cl), since decay to the ground state is a spin-forbidden reaction. It is also highly reactive. For example, heating (76) in the presence of tetramethylethylene leads to an allylic hydroperoxide, a characteristic product of reaction of $^1O_2$ with alkenes.

Another type of system that has been used to good effect for thermal generation of excited states is tetramethyldioxetane (7). This compound thermally decomposes to give two molecules of acetone [Eq. (147)]:

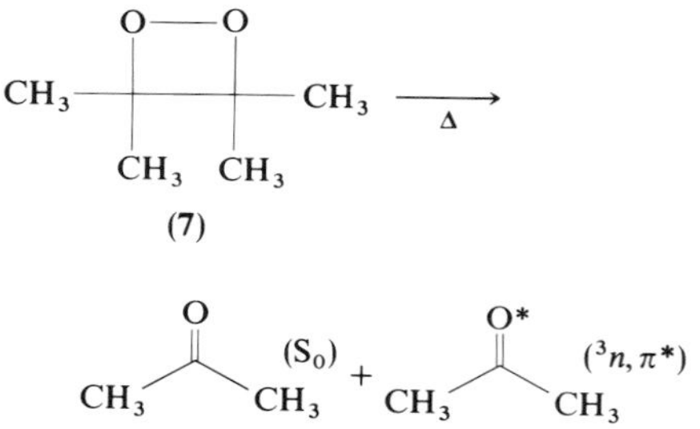

(7)

$$\text{[acetone]} (S_0) + \text{[acetone]}^* (^3n, \pi^*) \quad (147)$$

The enthalpy of reaction ΔH_0 is ~63 kcal/mol while the activation energy in acetonitrile is ~27 kcal/mol. This places the transition state for the fragmentation above the energies of both the S_1 and T_1 states of acetone, as shown in Fig. 11. The yields of both these states in the thermal reaction have been measured by thermal and spectroscopic methods and are about 0.5% for the singlet and 50% for the triplet. Thus, the chemiluminescence observed in deaerated solution is almost entirely acetone phosphorescence, while under the same conditions photoexcitation of acetone itself results in nearly pure acetone fluorescence. Thus, the remarkable conclusion emerges that in the thermal cleavage of (7), a crossing occurs with high efficiency from the ground-state potential surface to the triplet excited-state surface. A rationalization for the spin flip observed in this reaction has been given in terms of a comparison of the elec-

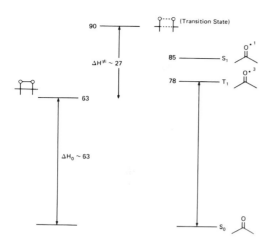

FIG. 11. Thermochemistry of tetramethyl-1,2-dioxetane.

tronic reorganization associated with fragmentation of dioxetanes and the spin–orbit coupling associated with $n \rightarrow \pi^*$ excitation of ketones. The effective spin–orbit coupling in these reactions facilitating intersystem crossing is in contrast to the poor coupling in endoperoxide decompositions [Eq. (146)] which does not give triplet excited states.

Thus, thermal decomposition of dioxetanes offers a route to triplet n, π^* states of ketones in the absence of light. Indeed, heating of dioxetane (**77**) results in lumiketone (**41**), the product

of the $^3n, \pi^*$ state of dienone (**40**), while dioxetane (**78**) on heating gives the typical Norrish

(78)

Type II fragmentation products expected from valerophenone triplet states.

The most efficient chemiluminescence system known is that associated with the firefly. In this system, the lumophore (**79**) undergoes biochemical oxidation promoted by the enzyme luciferase to give a hypothesized peroxylactone intermediate (**80**), which thermally decomposes to carbon dioxide and the electronically excited compound (**81**) that is the emitting species [see Eq. (148)]. The efficiency of the reaction affording chemiluminescence in this system is 100%. The similarity to dioxetanes is obvious.

Chemiluminescent systems not involving oxygen are rare. One notable example involves the electrocyclic ring opening of Dewar benzene to benzene [Eq. (149)].

$$(6) \quad \xrightarrow{\text{heat}} \quad (S_1) \quad \longrightarrow \quad (S_0) \quad + \; h\nu$$

(149)

In this case, the luminescence associated with the benzene S_1 excited state is very weak but can be enhanced by energy transfer to a suitable acceptor A, such as 9,10-dibromoanthracene. The highly exothermic ring opening in Eq. (149) is symmetry-forbidden and consequently has a high activation energy, which places the transition state above the lowest benzene excited state, much as in the dioxetane system shown in

(77)

(79) $\xrightarrow[\text{O}_2,\text{ATP},\text{Mg}^{2+}]{\text{Luciferase}}$

(148)

$$[\quad \mathbf{(80)} \quad] \quad \xrightarrow{-\text{CO}_2} \quad \mathbf{(81)}$$

Fig. 11. Thermal decomposition of the polycyclic azo compound (82) gives initially Dewar benzene (6), which under the reaction conditions affords benzene and chemiluminescence.

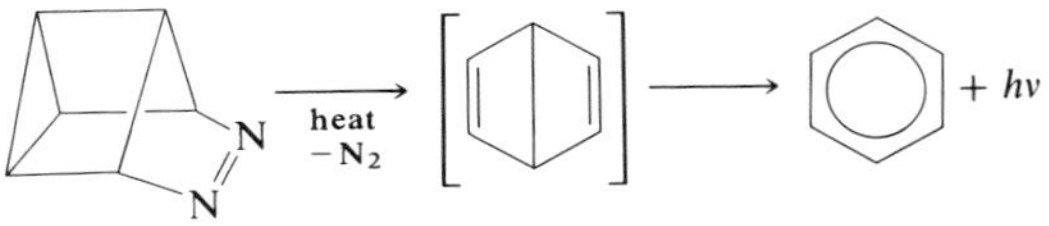

(82)

X. Conclusion

There are a number of photochemical processes that could not be included in this article because of space limitations. For example, photopolymerizations may be reasonable alternatives in some cases to free-radical or transition-metal catalyzed polymerization processes. The basic principles of organic photochemistry and the most commonly observed types of reactions have been presented and illustrated. For more details, the interested reader is referred to the basic texts listed in the bibliography.

BIBLIOGRAPHY

Bryce-Smith, D. (ed.) (ongoing series). "Photochemistry—Specialist Periodical Reports." The Royal Society of Chemistry, London.

Cowan, D. O., and Drisko, R. L. (1976). "Elements of Organic Photochemistry." Plenum, New York.

Coyle, J. D. (1986). "Introduction to Organic Chemistry." Wiley, New York.

de Mayo, P. (ed.) (1980). "Rearrangements in Ground and Excited States," Vol. 3. Academic Press, New York.

Hammond, G. S., et al. (eds.) (ongoing series). "Advances in Photochemistry." Wiley, New York.

Horspool, W. M. (ed.) (1984). "Synthetic Organic Photochemistry." Plenum, New York.

Padwa, A. (ed.) (ongoing series). "Organic Photochemistry." Marcel Dekker, New York.

Turro, N. J. (1978). "Modern Molecular Photochemistry." Benjamin/Cummings, Menlo Park, CA.

PHOTOCHROMIC GLASSES

R. J. Araujo   *Corning Glass Works*

GLOSSARY

Chemical strengthening: Strengthening of glasses by immersion in a bath of molten salt. An exchange of ions between the glass and the bath causes a thin layer of glass to be in compression and thus increases the physical strength of the glass.

Colloids: Small aggregates of matter having properties differing slightly from those of the bulk material because of the increased importance of surface effects.

Color coordinates: Numerical description of color determined by the ratios of transmittance at three different wavelengths.

CPF™ lenses: Ophthalmic lenses that because of their deep red color (strong blue absorption) provide a measure of comfort to some victims of *retinitis pigmentosa* and cataracts.

Dark adaptation: Process by which vision improves with time after the eye has been subjected to a dramatic decrease in the level of lighting. This phenomenon depends on the regeneration of rhodopsin in the eye as well as on complicated neurological processes.

Photolysis: Chemical or physical reaction induced by irradiation with light.

Photochromism is the phenomenon by which absorption of electromagnetic energy causes a reversible change in the color of a material. The word reversible implies that the system reverts to its initial state upon the cessation of irradiation. Certain glasses containing very small crystallites of silver halide exhibit this behavior.

I. Introduction

In 1964 the Corning Glass Works introduced to the marketplace a new family of ophthalmic lenses having the remarkable property that they become dark in sunlight and revert to their clear or colorless state indoors. Subsequently, both the Schott Glass Company in Germany and the Chance-Pilkington Glass Company in England began to manufacture these lenses. This family of glasses was invented by W. H. Armistead and S. D. Stookey, the latter being also the inventor of glass-ceramics, the material used in Pyroceram® nose cones and Corning Ware® cooking utensils. The phenomenon of changing color in response to sunlight irradiation is called photochromism, and materials showing such a property are said to be photochromic. Photochromism has been known since about 1880, and many hundreds of chemical compounds, most of them organic (that is, containing carbon), show this property. All the organic photochromic compounds share an unfortunate characteristic: They all fatigue, that is, they stop being photochromic after they have been darkened and faded a number of times. The photochromic glasses discovered by Armistead and Stookey stand as a dramatic exception: They never fatigue. Small but significant changes in the speed of darkening and fading do occur during the first few cycles, but the degree of darkening and fading possible does not change even after a half a million darkening and fading cycles. This family of photochromic glasses owes its remarkable properties to a suspension of very tiny (~1–2 millionth of an inch) crystallites of silver chloride. In the following sections, we discuss how these glasses are made and processed, the mechanism by which they darken and fade, and the manner in which conditions influence the performance characteristics of the glasses. [*See* GLASS.]

ENCYCLOPEDIA OF PHYSICAL SCIENCE
AND TECHNOLOGY, VOL. 10

II. Mechanism of Darkening and Fading

To a very good degree of approximation, the darkening and fading of photochromic glasses can be ascribed to the photolysis of the tiny crystallites of silver halide that are suspended in a vast ocean of inert glass. Because of their minute size, the crystallites do not appreciably scatter light, so the glass appears to be quite transparent (i.e., not milky).

Figure 1 is a two-dimensional representation of the way in which positively charged silver ions (Ag^+) and negatively charged chloride (or bromide) ions (Cl^-) are arranged in a periodic fashion in a silver halide crystal in photochromic glasses. Note that singly charged copper ions occur in a few of the positive ion sites in Fig. 1. In a perfect crystal, there would be no such impurity, and silver ions would occupy all positive-ion sites.

If light of sufficiently high energy (UV or blue) impinges on the crystal, an electron can be transferred from a chloride ion to a silver ion:

$$Cl^- + h\nu \rightarrow Cl^0 + e^- \qquad (1)$$

$$e^- + Ag^+ \rightarrow Ag^0 \qquad (2)$$

Unless the resulting chlorine atom (Cl^0) and the silver atom (Ag^0) are moved apart extremely quickly, the electron simply returns to the chlorine atom, restoring the original state of the crystal, and the absorption of the photon has caused no net change. In spite of the rather crowded condition depicted in Fig. 1, a silver ion can move rather quickly from one positive-ion site to an empty, neighboring positive-ion site, even though in doing so it is squeezed terribly as it passes through the very small spaces between the chloride ions. But although the silver ion can move with surprising speed, the silver atom cannot move fast enough to prevent the electron from returning to its original chlorine atom parent; but fortunately, it does not have to. Another mechanism is available for separating the silver atom from the chlorine atom. (The chlorine atom does not move either, since chlorine atoms are much too big to move at room temperature.) Since all positive-ion sites in the interior of the crystal are equivalent, the energy of the system is not changed by an electron moving from one silver site to another. This motion of the electron effectively separates the silver atom from the chlorine atom. If the electron reaches a silver atom that is not equivalent to those on interior lattice sites (e.g., those marked $\boxed{T}$ in Fig. 1), it becomes trapped there. This trapped electron (a Ag^0 on the surface) alternately attracts more electrons and silver ions until a speck of appreciable size forms, as can be seen in Fig. 1. This speck or colloid of silver is the entity that absorbs visible light in much the same way as does a metal and that causes the photochromic glass to appear dark. Note that the formation of the silver speck on the surface of the silver halide crystallite has caused some positive-ion sites to be vacant. Such sites are called vacancies and are designated by the symbol V in Fig. 2.

The process described so far is identical to that responsible for the formation of silver

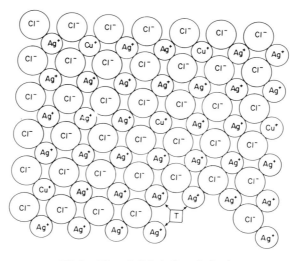

FIG. 1.   Silver halide before darkening.

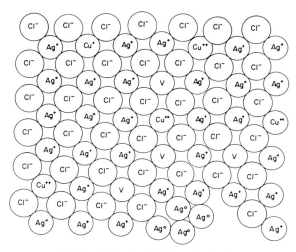

FIG. 2.   Silver halide after darkening.

specks in photographic film. That is to say, in response to irradiation, electrons and silver ions form silver specks in both photographic film and photochromic glasses. [*See* PHOTOGRAPHIC PROCESSES AND MATERIALS.]

Note that no Cl^0, formed according to Eq. (1), appears in Fig. 2. The reason is that since all chloride ions in the interior of a silver halide crystallite occupy equivalent sites, the energy of the system is not changed when the Cl^0 borrows an electron from a neighboring chloride ion, effectively moving the Cl^0 from the original lattice site to a neighboring site. The site of a missing electron—that is, the site that has one electron less than it had in the lowest energy state of the system—is called a hole. The hole is free to move from any lattice site to any equivalent neighboring site in the same manner as the electron. When the hole reaches a site next to a singly charged copper ion, it is at a site that is not equivalent to all the other negative-ion sites; so the hole, in effect, becomes trapped there. As a matter of fact, the singly charged copper ion gives an electron to its neighboring Cl^0 and becomes a doubly charged copper ion:

$$Cu^+ + Cl^0 \rightarrow Cl^- + Cu^{2+} \qquad (3)$$

By this process the hole is said to be trapped by the copper ion. The doubly charged copper ion formed by the process is sometimes called a trapped hole.

It is fortunate that this process of hole trapping occurs. Otherwise, the hole would continue to wander in a random way until it came near the trapped electron (Ag^0), whereupon it would re-trieve its electron, and the system would be restored to its initial state. Like the mobility (ease and speed of moving) of an electron, the mobility of a hole would be very high if it were not for the trapping; the process

$$Cl^0 + Ag^0 \rightarrow Cl^- + Ag^+ \qquad (4)$$

would then be so fast that colloid formation could not occur, and the photochromic glass would not darken. In effect, one vacancy and one doubly charged copper ion are formed for each silver ion that captures an electron and contributes to the growth of a silver speck. The net effect of Eqs. (1), (2), and (3) is given by

$$Cu^+ + Ag^+ \rightarrow Cu^{2+} + Ag^0 \qquad (5)$$

where Ag^0 is understood to be one of the atoms in the silver speck. Thus in Fig. 2, Cu^{2+} appears instead of Cl^0. Vacancies occur because the silver ions that previously resided on those sites are now part of the colloid.

In photographic film an analogous kind of hole trapping is required for colloid formation to occur. In this case, the Cl^0 chemically reacts with the gelatin in which the silver halide crystallite is suspended. Consequently, the hole is trapped in an infinitely deep trap. In no way can the process be reversed, and the silver speck or colloid is permanent. Thus, the darkening of a photographic film is said to be irreversible.

We have said that a photochromic glass returns to its initial clear state when irradiation ceases. That is, the darkening is said to be reversible. What accounts for the difference in behavior of a silver chloride crystallite in a photochromic glass and one in a photographic film is that in the glass there is no gelatin or any other substance with which the Cl^0 can react, other than with the singly charged copper ion. The doubly charged copper ion does not constitute an infinitely deep trap. Indeed, the energetics of the situation strongly favor a process such as

$$Cu^{2+} + Ag^0 \rightarrow Cu^+ + Ag^+ \qquad (6)$$

which is obviously the inverse of darkening and is, indeed, the mechanism by which photochromic glass fades. The limitation on the speed of process (6) is the requirement that the doubly charged copper ion must be extremely close to the silver speck (one or two lattice sites) or the electron cannot make the jump. One can see from Figs. 1 and 2 that the doubly charged copper ions are not, in general, close enough for the electron to jump, that is, for process (6) to occur. How, then, does the fading take place?

The entire doubly charged copper ion is able to move from one lattice site to another (which, of course, must be vacant) in spite of the tight squeeze involved in getting past the big chloride ions. In fact, the motion of the doubly charged copper ion is much slower than that of the silver ion, precisely because of its double charge. This accounts for the observation that fading is much slower than darkening in photochromic glasses. Moreover, the speed of the copper motion depends more strongly, by far, on the temperature of the system than does the speed of the silver ion. The speed of electron motion or hole motion is virtually independent of temperature. Consequently, the speed with which the colloids form in response to radiation (speed of darkening of the glass) does not depend much on temperature; but the speed with which the glass fades depends very strongly on temperature. Of course, the glass fades more rapidly as the temperature is increased.

The statement that the electron cannot jump a long distance from the silver speck to a doubly charged copper ion must be modified somewhat. When the colloid is irradiated with visible light, there is a small but finite probability that a single electron will acquire so much energy that it can make a very long jump. This infrequent broad jump by the electron has the result that a photochromic glass fades somewhat faster when it is irradiated by long wavelengths (with no UV present) than it does in the dark. This slight acceleration of the fade rate by long wavelength irradiation is called optical bleaching. The fading that occurs even in the dark is called natural or thermal fading. The latter term is an attractive one because the rate of natural fading depends so strongly on temperature.

III. Formation of Photochromic Glasses

If salts of silver, copper, and chloride are added to the batch materials that are used to make glass, the resulting glass contains silver ions (Ag^+), chloride ions (Cl^-), and copper ions (Cu^+ or Cu^{2+}) in solution. Such a system is analogous to a solution of sodium chloride in water. When the silver, copper, and halide ions are completely in solution, the glass is not photochromic.

Fortunately, the amount of silver and chloride ions that can stay dissolved (solubility) depends very strongly on temperature for some glasses. At the high temperature at which the glass is melted, the silver and chloride ions are quite soluble; but in suitable host glasses the solubility decreases markedly with decreasing temperature. The fraction of the silver and chloride ions that are no longer soluble at low temperatures precipitates out in the form of silver-halide crystals, and the material is then a photochromic glass. Of course, if the host glass is such that the solubility of silver and chloride ions does not depend strongly enough on temperature, no silver halide precipitation occurs, and the glass is not photochromic, even though it contains silver and chloride. Thus not all glasses are suitable hosts.

Simply cooling the glass quickly to room temperature does not make it photochromatic. The glass must be heated to a moderately elevated temperature (but a low temperature compared with that of melting) for photochromism to be observed.

At room temperature, even though the solubility of silver chloride is very low, the motion of the silver ions and chloride ions as they try to move through the glass looking for each other to form a silver-halide crystal is exceedingly slow. At such speeds the precipitation may take millions of years. At a moderately elevated temperature (but low enough so that the solubility of silver chloride is low) such as 600–700°C, the ions move fast enough to find each other and form droplets of molten silver chloride in about 20 min. (Silver chloride melts at about 455°C.) In most glasses, the mobility of chloride ions in the glass host is very slow at temperatures much lower than 500°C, and so growth of the silver-halide droplet or crystal is virtually nonexistent at such low temperatures. Nevertheless, heating the glass in a temperature range below 500°C and much above room temperature can alter the photochromic properties. The reasons are not well understood, but some relevant observations can be made. Above 450°C the silver halide is molten. The molten droplet can be considered to be an infinitely defective crystal (the periodicity is destroyed). At temperatures slightly below the melting point, a large number of defects such as vacancies and kinks exist, such as those at position $\boxed{T}$ in Fig. 1. If the temperature is decreased slowly enough, many of these defects are annealed out. If the temperature is decreased too fast, many of the defects that are stable at high temperature persist at the low temperature, because the ionic motion required to correct the defects is too slow. Such phenomena can influence the performance of a photochromic glass. A second kind of phenomena may also be influential on photochromic properties. Ions such as copper or lead ions can exist in a silver-halide crystal, or they can exist in the glass. The exact manner in which they partition themselves between the silver-halide droplets and the glass host depends on temperature. If the temperature is changed too rapidly, they may not have time to adjust to the partition they "want" to exist.

The reason the properties of a photochromic glass often change when the glass is chemically strengthened or tempered is probably related to phenomena similar to those just described. Ordinarily, nominal properties can be restored by repeating the heat treatment with more careful control of temperature and cooling rates.

IV. Performance of Photochromic Glasses

For our purposes, we may think of light as being comprised of small packets called photons. The energy of each of these packets is proportional to the frequency of the light, or equivalently, it is proportional to the reciprocal of the wavelength. Light of wavelength 350 nm is, therefore, more energetic than that of 550 nm wavelength. The reason green light does not cause typical photochromic glass to darken is that the energy of a photon of light with wavelength 550 nm is too low to remove an electron from a chloride ion and also give it enough energy to wander from silver site to silver site, as

described in Section II. A photon of UV light, on the other hand, is quite energetic enough for this purpose.

When a photon of light is absorbed by the material through which the light is traveling, the intensity of the light is said to decrease. The intensity is conveniently thought of as the number of photons that pass through some unit area in a plane per unit of time.

Figure 3 represents a beam of light traveling from left to right and passing through a transparent but absorbing material M; it has an intensity I_0 before passing through the material and a diminished intensity I after having passing through the material. The ratio of I to I_0 is called the transmittance. That is,

$$T = \frac{I}{I_0} \qquad (7)$$

The transmittance of undarkened photochromic glasses is routinely measured using light of wavelength 365 nm. This transmittance is often expressed in percentage units and is called UV transmittance. Typical values range from values like 30% in Photogray® lenses to values as low as 5% in Photobrown™ lenses. Another convenient number for expressing the fraction of the photons that pass through a material is called the absorption coefficient, which is defined as

$$\alpha = \frac{1}{L} \ln \frac{I_0}{I} \qquad (8)$$

where α is the absorption coefficient and L the thickness of the sample. A useful characteristic of the absorption coefficient is that it is proportional to the concentration of the species that is absorbing the light. The absorption coefficient of photochromic glasses rises very rapidly as the

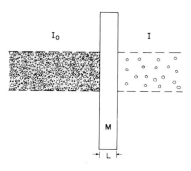

FIG. 3.   Representation of number of photons per unit area in a beam of light of intensity I_0 and the reduced number of photons in a beam of lower intensity I.

wavelength decreases below 350 nm. That is, photochromic glasses transmit very little UV.

The transmittance in the visible portion of the spectrum of an undarkened photochromic glass is almost 90% unless colorants such as cobalt or nickel are added. When the glass is irradiated, the transmittance decreases with time and approaches an asymptotic or steady-state value. The speed with which the glass darkens during the initial period of irradiation is proportional to the intensity of the UV radiation. The proportionality constant depends on the UV transmittance. The rate of fading decreases considerably as the glass fades toward complete clarity. For our present purposes, we can describe the time dependence of the darkening process by

$$A = \frac{k_d I A_0}{k_d I + k_f} (1 - e^{-(k_d I + k_f)t}) \qquad (9)$$

where A is the absorbance ($\alpha \times L$), $k_d I A_0$ represents the speed of the net process (5), and k_f represents the speed with which process (6) occurs. When t is set equal to infinity in Eq. (9), the resulting expression describes the steady-state absorption,

$$A_{ss} = \frac{k_d I A_0}{k_d I + k_f} \qquad (10)$$

As the intensity of the UV increases without limit, A_{ss} becomes equal to the limiting value A_0. The way in which A_{ss} varies with I at constant temperature is the following: If I is sufficiently small so that $k_d I \ll k_f$, A_{ss} increases proportionately to I; but if $k_d I \gg k_f$, A_{ss} virtually does not change at all with I.

In actual sunlight conditions, commercially available photochromic glasses are in the regime in which A_{ss} changes only slightly with the intensity of UV. Between the hours of noon and 4:00 p.m., the intensity of UV may change by more than a factor of 10, but the transmittance of the photochromic glass changes by less than a factor of 2.

Temperature plays an important role in both the rate of fading and the determination of A_{ss}, the steady-state absorbance (or transmittance). A_{ss} is determined by the balance of forces trying to darken the glass and the forces trying to fade it. The principle is illustrated by a simple analogy.

Consider a very, very deep bucket with small holes uniformly distributed over its walls. Consider what happens when one tries to fill the bucket with a water hose. The level of the water

rises until the rate at which the water leaks out the holes exactly equals the rate at which the hose delivers water to the bucket. Then it stays constant. If one uses a larger hose that delivers water at a faster rate, the steady-state level of the water will be higher. The rate at which water leaks out at steady state will also be larger, because it can leak out a larger number of holes since the water level is higher. As soon as the hose is turned off, the level begins to fall because the leaking is no longer compensated. As the water continues to leak and the water level falls, the rate at which the level falls slows down, because fewer and fewer holes are involved in the leaking.

In photochromic glass, A_{ss} is analogous to the water level, the hose is analogous to $k_d I$, and the holes correspond to k_f, the fade-rate constant.

If one raises the temperature, one increases the fade rate. In our analogy one can make the holes larger. This makes leaking faster and yields a lower steady-state water level for a given hose. In photochromic glass the speed of process (5) is virtually independent of temperature (like the hose). Process (6) depends strongly on temperature (as the leak rate depends on the size and number of holes); so A_{ss}, which depends on the balance, decreases with increasing temperature (just as the water level decreases with increasing hole size). That is, the higher the temperature, the less dark a photochromic glass can get. This is shown graphically in Fig. 4.

The fading rate in photochromic glass is not determined uniquely by how dark the glass becomes before irradiation was terminated (Fig. 5). If in two experiments performed at the same temperature, a glass is darkened (not to steady state) to the same value by UV irradiation of two different intensities, a faster fade rate is observed in the experiment in which the higher intensity is used. Of course, had the glasses been darkened to steady state, somewhat more darkening would be observed when the higher intensity is used. In that case, the fade rate may have been faster or slower. Figure 6 illustrates that as one increases A_{ss} by increasing I over a certain range, the fade rate increases. If one increases A_{ss} by increasing I beyond that range, the fade rate begins to decrease again.

Often the glass is darkened at one temperature, and the temperature is changed when the irradiation is terminated (and when the glass can fade). For example, consider glasses that are darkened while the wearer is skiing on a very cold day. Only when the skier goes indoors does

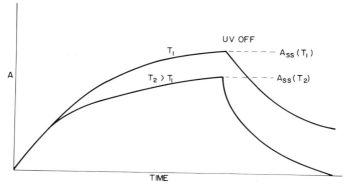

FIG. 4. Darkening and fading as a function of time at two different temperatures.

the sunlight irradiation of the glasses cease, and only then do the glasses fade. It is quite likely that the temperature indoors is higher than that out of doors. The magnitude of the effect of changing temperature during fading varies from glass to glass; but, in general, a glass darkened at a low temperature T_L and faded at a high temperature T_H fades faster than one darkened and faded at T_H. A glass darkened at T_H and faded at T_L fades more slowly than one darkened and faded at T_L.

The latter case is a common one. One wears sunglasses on a hot day. When one comes indoors, it is into an air-conditioned room. Because the situation is so common, Corning chooses to publish fade rates measured this way. Such published rates are the lowest likely to be observed. Other companies choose not to do this; the fade rates they publish would be higher than those published by Corning even if the glasses were identical.

Although the manner in which the absorption spectrum of a silver colloid changes with size and shape is reasonably well understood, the factors that determine the size and shape assumed by the colloid when a photochromic glass darkens are not understood at all. Nevertheless, some general observations about color as a performance characteristic are in order. The darkened color of most photochromic glasses changes after the first few cycles. Photogray Extra® lenses, for example, darken to a brownish-grey the first time they are exposed to light. After several darkening cycles (or a long period of continuous darkening), the color changes to the familiar blue–gray. An experimental glass known within Corning Glass Works as 15–15, darkens to a blue–gray when first exposed to

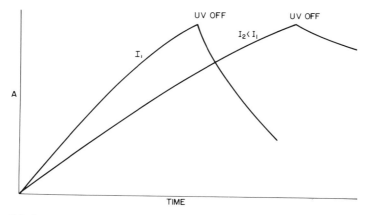

FIG. 5. Short-time darkening at two different intensities and the corresponding fading.

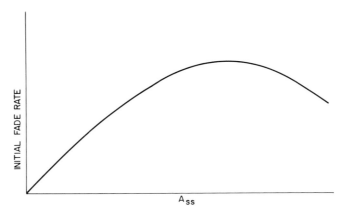

FIG. 6.   Initial fade rate as a function of the steady-state darkening level at constant temperature, A_{ss}, is varied by varying I.

UV. After several darkening cycles it darkens to a very bright red, similar to the color of Corning's CPF™ 550 lenses. A plot of the color coordinates measured at successive periods of time during the fading process may yield almost a straight line from the original color to the T_0 (original undarkened transmittance) color, but it may also yield a wild excursion in color-coordinate space. The darkened color and the color path during fading may both depend quite strongly on temperature. Photobrown™ lenses darkened to a brown color at temperature above 75°F, but at temperature below 32°F they were almost indistinguishable from Photogray® lenses.

Thickness also plays a key role in determining the color to which a photochromic glass darkens. In Photobrown™ lenses, since the brown color depends so much on the UV transmittance of the glass, thick lenses yield a far better brown color than do thin lenses.

V. Benefits and Alleged Detriments

Many tens of millions of pairs of sunglasses are sold in the United States every year, attesting to the desire of a large number of people for protection from very bright light. For wearers of prescription lenses, the advantages of a single pair of glasses that function as sunglasses outdoors and clear glasses indoors from the simple point of view of convenience is self evident.

There is no disagreement among vision researchers that UV and blue light can cause damage to both the anterior eye and the retina when the intensity reaches the level equivalent to

looking directly at the sun at its apex. There is some opinion that blue light is somewhat hazardous at considerably lower levels, but most researchers believe that the danger decreases sharply as the intensity decreases.

Because of the natural photophobia of the normal person, it is unusual for the eye to be exposed to those levels of blue light or UV that are hazardous by general consensus. Exceptions can occur, however. Reflections from snow fields or white sands can produce, at the cornea, intensities up to one-third the intensity produced by looking at the sun. Specular reflections (those observed on a mirror) from chrome ornaments or glass surfaces with reflecting coatings can lead to enormously high intensities. It is, of course, impossible to know the danger such random circumstances present to the average person. The "prudent man" concept, however, would suggest that glasses that absorb a substantial fraction (90%) of potentially harmful rays are a benefit.

There is a substantial body of thought that total deprivation of blue and UV constitutes a health hazard. While no general consensus exists on this point, it seems unwise as well as unnecessary to strive to remove 100% of the blue light from reaching the eye. Although further research may change our conclusions, at the present moment a dark grey glass (between 1 and 30% luminous transmittance) that absorbs more than 90% of all the radiation below 380 nm is a benefit. Although photochromic glasses normally do not darken below 10% transmittance, they are, nevertheless, remarkably well suited for eye protection as it is currently understood.

It is important to note that in this context Photo-brown Extra® lenses are to be regarded as gray. The most important advantage of the photochromic glasses over other glasses, which conceivably can be designed to have the same absorption spectrum, is that the convenience afforded by the automatically adjusting transmittance makes them more likely to be worn than glasses that must be exchanged when the environment changes.

A bright environment slows down the rate at which the eye can dark adapt. This effect is cumulative. A lifeguard, for example, who spends 8 hr on the beach every day for a week does not regain his normal nighttime vision even after 24 hr in the dark. A person who wears dark glasses in the sun dark adapts to low-level light more quickly than one wearing no glasses. If, in addition, the dark glasses are photochromic and begin to clear up when the light intensity is dramatically reduced, an added, albeit small, benefit exists.

Because photochromic glasses darken most in cold weather and because the fading rate observed at room temperature is maximal for a glass darkened at very low temperatures, and also because the intensity of light is so high on a snow field, the advantages of photochromic glasses are certainly maximized for a snowshoer or skier.

The question of color distortion of Photo-brown™ lenses has been occasionally raised. The absorption coefficient induced in this glass during darkening is really quite flat (the human eye is a remarkable detector of small differences); so although very slight color changes are perceivable when comparisons are made, absolutely no danger of color misidentification exists for the normal eye.

It has been suggested that the automatic response of photochromic glasses renders a changing pupil size unnecessary and, therefore, leads to atrophy of the iris. Virtually all the variations in size of the pupil at steady state occur at light levels considerably lower than those found in the typical office or classroom. Since no sunglasses of any kind are normally worn under such conditions, they cannot have an effect. Furthermore, the speed of response of photochromic glasses is much slower than the speed of pupillary response. Therefore, even if photochromic glasses were worn and even if they did respond to the same range of intensities to which the pupil responds, they would not replace pupil response because they are too slow.

Sunglasses yield comfort and perhaps a measure of eye protection. The eye protection is probably improved if the sunglass absorbs UV and short-wave blue rather strongly. Photochromic glasses do this very nicely, and they also provide the convenience of automatic adjustment. The wearer of photochromic glasses is not likely to forget to don his eye protectors when venturing into the outdoors.

BIBLIOGRAPHY

Araujo, R. J. (1968). *Appl. Opt.* **7,** 781.
Araujo, R. J. (1971). In *Photochromism,* G. H. Brown, ed. Wiley, New York, p. 680.
Araujo, R. J. (1977). In "Treatise on Materials Science," Vol. 12, M. Tomazawa and R. H. Doremus, eds. Academic, New York, p. 91.
Araujo, R. J. (1980). *Contemp. Phys.* **21,** 77.
Araujo, R. J. (1980). *J. Non-Cryst. Solids* **42,** 209.
Araujo, R. J. (1982). *J. Non-Cryst. Solids* **47,** 69.
Araujo, R. J., Borrelli, N. F., and Nolan, D. A. (1979). *Phil. Mag.* **B40,** 279.
Araujo, R. J., Borrelli, N. F., and Nolan, D. A. (1981). *Phil. Mag.* **B44,** 453.
Nolan, D. A., Borrelli, N. F., and Schreurs, J. W. H. (1980). *J. Am. Ceram. Soc.* **63,** 305.

PHOTOELECTRON SPECTROSCOPY

Georg Hohlneicher   *Cologne University*

GLOSSARY

Auger electron spectroscopy: Investigation of the kinetic energy distribution of electrons produced through the decay of a highly excited core hole state.

Characteristic x-ray emission: Emission of x rays with well-defined energy resulting from a different decay route of core hole states.

Core hole state: Excited state of an atom in which an electron has been removed from one of the inner shells.

Multiplet splitting: Substructure of bands in photoelectron and Auger electron spectra caused by spin–orbit interaction.

Photoelectron spectroscopy: Investigation of the kinetic energy distribution of photoemitted electrons.

Photoemission: Emission of electrons from a free molecule or a surface following irradiation with photons of sufficiently high energy.

Vibrational fine structure: Structure observed in highly resolved photoelectron spectra due to interaction of the electronic process with molecular vibrations.

Photoelectron spectroscopy is the investigation of the energy distribution of electrons released from a molecule or a solid by irradiation with sufficiently energetic radiation. It can be used as an analytical tool and has important applications in the investigation of the electronic structure of molecules and solids. Photoelectron spectroscopy has become especially important

in surface science because of its extraordinary surface sensitivity.

I. Fundamentals of Photoelectron Spectroscopy

A. BASIC DEFINITIONS

The subjects of photoelectron spectroscopy (PES) are atoms or molecules in the gas phase, solids, and, with very special technical requirements, liquids. As long as there is no need to specify the nature of the sample, we simply speak of the investigated system (M). Before photoionization takes place, the system is in a well-defined electronic state, usually the electronic ground state M_0, the initial state of the photoemission process. When the system is irradiated with radiation of sufficiently high energy $h\nu$, an electron can be removed from the system. The ion M^+ that is created by this process is again in a well-defined electronic state M_i^+ which is the final state of the photoemission process. In general, the lifetime of M_i^+ is long enough to prevent successive changes in the ion state to influence the kinetic energy of the photoelectron. The final state is either the electronic ground state M_0^+ or, provided $h\nu$ is sufficiently high, an electronically excited state of the ion. From conservation of energy it follows that

$$E(M_i^+) - E(M_0) + E_{kin}(e^-) + E_{kin}(M^+) = h\nu$$

(1)

where $E(M_0)$ and $E(M_i^+)$ are the energies of the initial and final states, respectively, and $E_{kin}(e^-)$ and $E_{kin}(M^+)$ the kinetic energies of the electron and of the ion. Since $E_{kin}(e^-)/E_{kin}(M^+)$ is determined by the mass ratio $m(M^+)/m(e^-)$, $E_{kin}(M^+)$ is much smaller than $E_{kin}(e^-)$ and therefore is neglected in most applications. This leads to

$$E_B(i) = E(M_i^+) - E(M_0) = h\nu - E_{kin}(e^-) \quad (2)$$

where $E_B(i)$ is called the "binding energy." To avoid confusion, this expression is used throughout this article. However, specifically in connection with the investigation of free molecules, the energy difference $E(M_i^+) - E(M_0)$ is also called the "ionization energy" or "ionization potential."

From Eq. (2) it is clear that it is possible to determine $E_B(i)$ by measuring the kinetic energy of the created photoelectrons, provided the exciting radiation has a well-defined energy $h\nu$. Photoelectron spectroscopy is the measurement of the kinetic energy of photoelectrons with the goal of deriving information about binding energies. This definition distinguishes PES from other methods in which photoionization is used mainly for detection (e.g., laser-induced multiphoton ionization) and information on the binding energies is not the primary goal of the investigation. [See MULTIPHOTON SPECTROSCOPY.]

A photoelectron spectrum is the number of photoelectrons with a specific kinetic energy observed per unit time displayed as a function of kinetic energy. Three examples of photoelectron spectra are shown: the gas phase spectrum of neon excited with $h\nu = 1253.6$ eV (Fig. 1a), the gas phase spectrum of $H_2C{=}O$ excited with $h\nu = 21.2$ eV (Fig. 1b), and the solid-state spectrum of copper excited with $h\nu = 1486.7$ eV (Fig. 1c). The meaning of the assignments given in these spectra is explained below. In all three cases two energy scales are shown corresponding to the IUPAC recommendations: the scale for the kinetic energy is given at the top of the spectrum and the scale for the binding energy at the bottom. The binding energy scale is obtained from the kinetic energy scale by means of Eq. (2). It must be kept in mind, however, that the quantity originally measured is $E_{kin}(e^-)$, even when only a scale for E_B is shown. The scales in Fig. 1 run in different directions for different examples. This is not a mistake but is due to different presentations of photoelectron spectra in the literature. Some researchers show values of the measured quantity $E_{kin}(e^-)$ increasing from left to right. The scale for E_B then runs from right to left. Others are interested mainly in E_B, so they draw the E_B scale with increasing energies from left to right. Sometimes only the $E_{kin}(e^-)$ scale or only the E_B scale is provided. As far as possible we will show both scales throughout this article, but when using spectra from the literature one

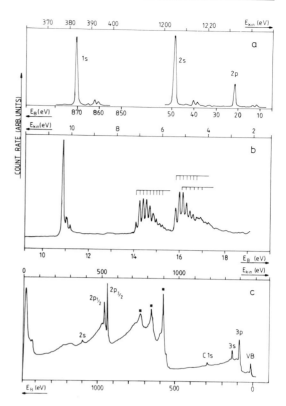

FIG. 1.   Examples of photoelectron spectra: (a) neon, $h\nu = 1253.6$ eV; (b) formaldehyde, $h\nu = 21.2$ eV; (c) metallic copper, $h\nu = 1486.7$ eV. [Spectrum of neon from Siegbahn, K., Nordling, C., Johansson, G., Hedman, J., Heden, P. F., Hamrin, K., Gelius, U., Bergmark, T., Werme, L. O., Manne, R., Baer, Y. (1971). "ESCA Applied to Free Molecules." Elsevier, Amsterdam.]

should always be careful about the scale used. [See ATOMIC SPECTROMETRY.]

Figure 1 gives a first impression of different types of PE spectra. From Eq. (2) we expect photoelectrons to appear only at kinetic energies that correspond to a certain final state M_i^+. We therefore expect a PE spectrum to consist of lines with widths defined by some experimental parameters. This is the result observed for neon (Fig. 1a). For molecules, vibrational and rotational states are coupled to the electronic states and, as in optical spectroscopy, band spectra are obtained (Fig. 1b). In this case, we speak of photoemission bands rather than photoemission lines. The resolution achievable in PES is, however, much lower than in optical spectroscopy (Section II). If the sample is a solid (Fig. 1c), each band or line is preceded by a tail extending toward lower kinetic energies. For low kinetic energies this leads to a considerable background. The tails are due to inelastic scattering of the photoelectrons within the solid.

The progress achieved with PES results from the fact that it is an energy-resolved method. Older methods for the determination of binding energies were mainly based on measurement of the photoionization current and depended on a variable excitation energy. For a given excitation energy E_a the photoionization current is proportional to the integral over the photoelectron spectrum from $E_B = 0$ to $E_B = E_a$. Even if a sufficiently variable source is available for excitation (which was not the case prior to the invention of synchrotron radiation), an energy-resolved method is always preferable to an integral one. In an energy-resolved measurement only electrons with a kinetic energy that falls in the window defined by the resolution of the analyzer contribute to the statistical noise, whereas in an integral method all electrons up to this energy contribute. The older methods therefore permitted determination of only the first or—under fortunate conditions—the first few binding energies. The photoemission processes leading to higher excited final states became accessible with PES. In Section III we will discuss in detail why and how these processes in particular contribute to a better understanding of the electronic structure of the systems investigated. First, however, we will clarify two terms frequently used in connection with PES: low- and high-energy PES. We will also discuss some historical aspects that will aid the reader in understanding the development of photoelectron spectroscopy.

In low-energy PES, usually termed ultraviolet PES or UPS, far-UV radiation is used for excitation. The most common source for UPS is the helium resonance lamp, which provides radiation with an energy of 21.2 eV and a half-width down to 10 meV (see Section II for further details). Only valence electrons can be photoionized with this energy. The development of UPS in its application to free molecules is mainly due to work by D. W. Turner and associates in Oxford, who reached a breakthrough in the early 1960s.

In high-energy PES, characteristic x radiation with energies between 100 and 2000 eV and half-widths of about 1 eV is used for excitation. Because of the higher excitation energy, valence electrons as well as electrons from inner shells

(core electrons) can be photoionized. The core electrons yield direct information on elemental composition and—when combined with chemical shift effects—on the electronic structure of solids and surfaces. Because of the lower resolution, the information obtained for valence electrons is more limited than in UPS. High-energy PES, which is usually termed electron spectroscopy for chemical analysis (ESCA) or x-ray photoelectron spectroscopy (XPS), was developed by K. Siegbahn and co-workers in Uppsala in the 1950s and early 1960s. In 1981 Siegbahn was honored with the Nobel Prize for this significant development.

B. COMPETITIVE PROCESSES

The photoemission processes $M_0 \rightarrow M_i^+$ are not the only processes that lead to the production of free electrons with discrete kinetic energies. Depending on the energy of the exciting radiation, some other processes also contribute to the PE spectrum. These processes are indicated by the circled numbers in Fig. 2, which shows a schematic diagram for the electronic states of the initial system (M) and the systems that have lost one (M^+) and two (M^{2+}) electrons.

In Fig. 2 process 1 is the normal photoemission leading to the photoionization of either a valence (1a) or an inner shell (1b) electron.

Process 2 represents an autoionization. If the energy $h\nu$ of the exciting radiation coincides with an electronic transition of the neutral system (which in the valence region is very likely for larger molecules), a photon can be absorbed by the neutral system. The final state M_k of this process is higher in energy than the ground state of M^+, and thus M_k can decompose into an electron and a low-lying state of M^+. Since both the

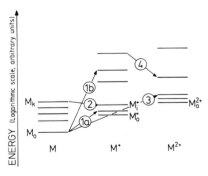

FIG. 2. Schematic representation of different processes that lead to the production of electrons with well-defined kinetic energy.

initial and the final state of the autoionization (AI) process are well-defined electronic states, the electron created in this process has the well-defined kinetic energy

$$E_{\text{kin}}^{\text{AI}}(e^-) = E(M_k) - E(M_i^+) \qquad (3)$$

In Eq. (3) we again neglected the rebound energy of the heavy particle. Unlike the kinetic energy of an electron produced in a normal photoemission process, $E_{\text{kin}}^{\text{AI}}(e^-)$ does not depend on $h\nu$. Therefore, the binding energy scale has no meaning for autoionization peaks. These peaks appear primarily at low kinetic energies since autoionization is usually efficient only when the initial and final states of the process are close in energy.

Process 3 in Fig. 2, which in principle contributes to the photoelectron spectrum, is a direct transition from M_0 to states of M^{2+} under simultaneous emission of two electrons. As discussed in Section III, such a process has much lower probability than process 1 and leads to a continuous energy distribution of the produced photoelectrons.

Process 4 represents an Auger transition. As discussed above, photoionization can lead to the creation of a core hole provided the energy of the exciting radiation is high enough. The resulting state of M^+ is highly excited. In about 10^{-16} s it relaxes to a lower excited state of M^+ by emission of a x-ray photon or to a lower-lying state of M^{2+} by emission of another electron. The latter transition, known as an Auger process, has a higher probability for light atoms, up to about $Z = 40$. Since the initial and final states of an Auger transition are well-defined electronic states of M^+ and M^{2+}, respectively, the emitted Auger electron has the well-defined kinetic energy.

$$E_{\text{kin}}^{\text{Au}}(e^-) = E(M_j^+) - E(M_k^{2+}) \qquad (4)$$

Auger transitions contribute strongly to high-energy PE spectra. For example, in Fig. 1c all peaks indicated with an asterisk result from Auger transitions. As in autoionization, the kinetic energy of the Auger electron does not depend on $h\nu$. Autoionization and Auger processes, therefore, can be separated from photoemission processes by variation of the excitation energy.

C. COMPARISON TO THE OPTICAL SPECTRUM OF THE ION

From inspection of Fig. 2, it is clear that the information obtained from a photoelectron spectrum of the system M is basically the same as the information obtained from an optical spectrum of M^+. The difference between the binding energy $E_B(i)$ corresponding to the photoemission process $M_0 \rightarrow M_i^+$ and the binding energy $E_B(0)$ corresponding to $M_0 \rightarrow M_0^+$ is equivalent to an optical transition between M_0^+ and M_i^+. If the 0–0 transition of the first photoemission band is used as the zero of a new energy scale, the energy of the higher-lying ionization bands (or lines) corresponds to an excitation energy within the system M^+ which in principle can be obtained by optical spectroscopy. This is shown in Fig. 3, where the upper and lower spectra are the PE spectrum of octafluornaphthalene and the absorption spectrum of the octafluornaphthalene cation, respectively. The two spectra are arranged so that the zero of the energy scale of the absorption spectrum is matched to the first peak of the PE spectrum. Considering that the PE spectrum was taken in the gas phase and the absorption spectrum in solution, the similarity of the spectra is striking. Most of the bands that are seen in the optical spectrum are also seen in the PE spectrum and vice versa. However, the intensities of the bands are different (note that the absorption spectrum has a logarithmic intensity scale). [*See* MOLECULAR OPTICAL SPECTROSCOPY.]

In spite of the similarities between the PE spectrum of a molecule and the optical spectrum of the corresponding ion, the spectra often yield different types of information:

1. The connection described above holds only for molecules. For extended systems like solids it is no longer valid.

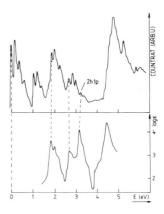

FIG. 3. Comparison of the PE spectrum of octafluornaphthalene to the absorption spectrum of octafluornaphthalene cation.

2.   Even for molecules it is often much more difficult to measure the optical spectrum of an ion than to measure the PE spectrum of the parent molecule.

3.   States with an excitation energy higher than about 6 eV are difficult to study in optical spectroscopy. This is particularly important for highly excited ion states that correspond to the removal of an inner shell electron.

4.   For reasons discussed in Section III, the selection rules are quite different for both types of spectroscopy. The number of transitions allowed in photoemission is usually much smaller than the number of allowed optical transitions, especially at high excitation energies. Therefore, PE spectra are considerably easier to interpret than optical spectra, as will be seen in the following sections.

D. SPECIAL ASPECTS OF SOLIDS

In the case of solids some special aspects, such as the very limited escape depth, must be considered. A photoelectron created inside a solid must escape into the vacuum to be measured. Since the probability of inelastic scattering is very high as long as the electron moves inside the solid, only electrons created close to the surface have a chance to escape without a secondary energy loss. Figure 4 shows the average escape depth as a function of the kinetic energy of the electron together with some of the most useful excitation lines (see Section II). For a kinetic energy of about 100 eV the escape depth is lowest, with an average value of only a few angstroms. Even for a kinetic energy of 10 or 1000 eV the escape depth is only on the order of 20 Å. PES probes only the few outermost atomic layers of a solid, which can be a disadvantage if we want to study the bulk material. First, the composition of the surface is often different from the composition of the bulk because of segregation effects or surface contamination (Section II,A). Even if there is no difference in composition, there is usually a strong contribution from the outermost layer, especially for kinetic energies around 100 eV. The outermost layer is chemically always different from the interior, since the atoms in this layer have fewer neighbors. The surface sensitivity of PES is advantageous, however, if we want to study the surface itself. Therefore, PES has become one of the most powerful tools in surface science. It allows us to study not only a surface, but also atoms or molecules sitting at the surface, with high sensitivity. It is possible to detect coverages down to a fraction of a monolayer. Thus, photoelectron spectroscopy is extremely useful for the investigation of adsorbates. [*See* SURFACE CHEMISTRY.]

Another special aspect of solids is the "reference problem." For an atom or molecule in the gas phase, ionization leads to the creation of an electron and a positive ion. The electron is either detected or lost at the walls of the instrument. The ion also leaves the ionization region rapidly. By calibration with accurately known binding energies (see Table I), the binding energies of the sample can be referred to the vacuum level that corresponds to an infinite separation of electron and ion.

In the case of solids it is necessary to distinguish between conductors and insulators. If the sample is a conductor and in electrical contact with the spectrometer, the Fermi levels ε_F equilibrate (Fig. 5a). The same is true for any metal that is used to calibrate the binding energy scale. The binding energy E_B^S of an arbitrary conducting sample can therefore be referred to the Fermi level of the spectrometer, which is the reference level used in most investigations. If the work functions of the reference material and sample are known, the binding energy of the sample can also be referred to the vacuum level of the sample.

The situation is more difficult for insulating samples. Photoionization creates positive charges within the sample that are not equilibrated immediately, and the sample becomes

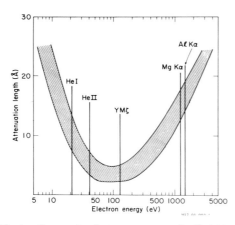

FIG. 4.   Connection between escape depth (attenuation length) and kinetic energy of the photoelectron. [From Cardona, M., and Ley, L., eds. (1978). "Topics in Applied Physics," Vol. 26, "Photoemission in Solids I," p. 193. Springer-Verlag, Berlin.]

TABLE I.   Useful Calibration Lines

Atom	Level	Compound/ phase	Energy (eV)[a]
Ne	$1s$	Gas	870.37
F	$1s$	CF_4/gas	695.52
O	$1s$	CO_2/gas	541.28
N	$1s$	N_2/gas	409.93
C	$1s$	CO_2/gas	297.69
Ar	$2p_{3/2}$	Gas	248.62
Kr	$3p_{3/2}$	Gas	214.55
Kr	$3d_{5/2}$	Gas	93.80
Ne	$2s$	Gas	48.47
Ne	$2p$	Gas	21.59
Ar	$3p$	Gas	15.81
Cu	$2p_{3/2}$	Metal	932.8
Ag	$3p_{3/2}$	Metal	573.0
Ag	$3d_{5/2}$	Metal	368.2
Cu	$3s$	Metal	122.9
Au	$4f_{7/2}$	Metal	83.8
Pt	$4f_{7/2}$	Metal	71.0

[a] For the metals the energies refer to the Fermi level instead of the vacuum level.

charged. At the same time there is usually a relatively high density of low-energy electrons close to the sample surface, which can neutralize the positive charges. The equilibrium between outgoing and incoming electrons depends on the measuring conditions, specifically on the intensity of the ionizing radiation and the cleanness of the surrounding metal parts. Therefore, the actual charging potential ϕ_{ch} (Fig. 5b) depends on the measuring conditions. The charging may not even be homogeneous over the surface area investigated (differential charging), resulting in broadening of the observed lines. Sample charging can be reduced by use of very thin samples or a separate source of low-energy electrons

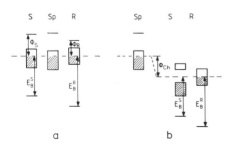

FIG. 5.   Reference schemes for solid samples; (a) conducting sample; (b) insulating sample with reference material on top. S, Sample; Sp, spectrometer; R, reference.

(flood gun). Alternatively, sample charging can be taken into account by evaporating small amounts of a reference material (usually gold) on the sample surface. Assuming that the reference material and the sample are at the same potential in the irradiated area (Fig. 5b), the binding energies for the sample are then referred to the binding energies of the reference material.

E. Angular Distribution of Photoelectrons

For a single atom or molecule, the probability of emission of an electron into a certain direction with respect to a molecule-fixed coordinate system is not isotropic. It depends on the initial and final states of the photoemission process, the orientation of the electric vector **E** of the ionizing radiation, and the energy $h\nu$. For example, if the electron is removed from an s orbital of an atom, the probability of finding the outgoing electron under an angle ϕ with respect to **E** is proportional to $\cos^2 \phi$. Thus, the probability distribution looks like an atomic p orbital.

In a gaseous sample the molecules are randomly oriented with respect to a laboratory-fixed coordinate system. To derive the angular distribution $I(\phi)$ in the laboratory system, we must integrate over all possible orientations of the molecules. If the ionizing radiation is plane-polarized, $I(\phi)$ can be expressed as

$$I(\phi) = (\sigma/4\pi)[1 + (\beta/2)(3 \cos^2 \phi - 1)] \quad (5)$$

where σ is the isotropic cross section and β the "asymmetry parameter." The possible range of β is -1 to $+2$. Both σ and β depend on the initial and final states of the photoemission process as well as the kinetic energy of the outgoing electron, with σ often showing strong variations at low kinetic energies and a smooth decrease at high kinetic energies (Fig. 6). At low kinetic energies pronounced maxima in σ, called "shape resonances," are often observed; these result from an interaction with quasi-bound states lying in the ionization continuum or from autoionization channels.

For unpolarized exciting radiation the emission is isotropic around the incident beam. In this case the angular distribution depends only on the angle θ between the propagation direction of the radiation and the direction of the outgoing electron. We obtain

$$I(\theta) = (\sigma/4\pi)[1 + (\beta/2)(\tfrac{3}{2} \sin^2 \theta - 1)] \quad (6)$$

Therefore, it is possible to measure σ and β even with unpolarized radiation.

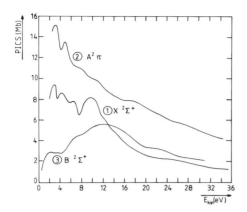

FIG. 6. Photoionization cross section for the first three ion states of CO. [From Plummer, E. W., Gustafson, T., Gudat, W., and Eastman, D. E. (1977). *Phys. Rev. A* **15**, 2339.]

If the emitted photoelectrons are observed under an angle of $\theta = 54°44'$ (the "magic angle"), the term in parentheses in Eq. (6) vanishes and the isotropic cross section σ is obtained directly. In normal gas phase PE spectrometers that are not designed for angle-resolved measurements, θ is commonly 90° and the intensity of the observed bands is influenced up to about 30% by the asymmetry parameter β.

The situation is changed completely if a molecule is adsorbed at a surface and thereby fixed in space. For example, consider a rodlike molecule (e.g., CO) which, for a given final state M_i^+, emits electrons preferably in the direction of the molecular axis. In addition, assume that the emission probability is proportional to the cosine between the molecular axis and the electric vector **E** of the ionizing radiation. The system is fully described by three angles and the polarization of the radiation. The three angles are shown in Fig. 7a: Ω is the angle between the surface normal **n** and the propagation direction **s** of the radiation, θ the angle between **n** and the direction **D** of the outgoing electron, and ϕ the angle between the **ns** plane and the **nD** plane. Usually Ω is called the "incidence angle," θ the "polar angle," and ϕ the "azimuthal angle."

If the molecule is standing on the surface (Fig. 7b) and the radiation is polarized perpendicular to the **ns** plane, there is no component of the electric vector in the direction of the molecular axis and no emission at all. If the radiation is polarized in the **ns** plane, there is emission in the direction of the surface normal which is strongest for grazing incidence (Ω close to 90°) and vanishes for normal incidence ($\Omega \approx 0°$). Now consider the situation where the molecule is lying flat on the surface with the long axis perpendicular to the **ns** plane (Fig. 7c). For **E** perpendicular to the **ns** plane the emission is strongest for $\phi = 90°$ and large polar angles (θ close to 90°) but there is almost no dependence on Ω. For molecules lying flat but randomly oriented on the surface and radiation polarized in the **ns** plane, there is still no emission in the direction of the surface normal. It is therefore possible to decide from angle-resolved PES (ARPES) whether a molecule is standing or lying on a surface.

The example above illustrates how ARPES can contribute to the study of adsorbate systems. It also shows the importance of the use of polarized radiation. Another field in which ARPES is important is the investigation of two- and three-dimensional periodic structures. Two-dimensional periodic structures are found in layer compounds or in well-ordered adsorbate layers formed at the surface of a single crystal. Three-dimensional periodic structures are found in all crystalline materials. To study ARPES it is necessary to have single crystals of at least a few millimeters in the two surface dimensions since the photoionization region cannot be made smaller than about 1 mm² without losing too much intensity or risking too much radiation damage.

The important difference between free or space-fixed single molecules and periodic structures is as follows. In the first case, the intensity of the peaks in the PE spectrum is angle-dependent but the position of the peaks is not. In the second case, the energy of the most prominent features in the spectrum of the valence region depends on the angles Ω, θ, and ϕ as well as the polarization of the ionizing radiation. One therefore speaks of energy distribution curves instead of PE spectra in such a case. We will come back to this type of investigation in Section IV,B.

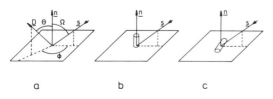

FIG. 7. Angle-resolved photoelectron spectroscopy (see text for details).

II. Instrumentation

A PE spectrum is measured with an instrument known as a photoelectron spectrometer. Figure 8 shows a sketch of the basic structure of this instrument, which is the same whether the instrument is designed for gases or solids. The technical details of the instrument, however, differ considerably depending on the use.

A. Vacuum Requirements

Since PES measures electrons, the whole path between the region where photoionization takes place and the detector must be kept at a pressure of about 10^{-5} torr or less to prevent collisions with the rest gas. For an exciting radiation with $h\nu > 11.3$ eV, no material exists that is sufficiently transmitting to serve as window material. Therefore, the path from the radiation source to the photoionization region (PIR) must also be kept under vacuum. [See VACUUM TECHNOLOGY.]

Photoelectron spectrometers are very sensitive to contamination of their inner surfaces. Deposition of the sample or its decomposition products on these surfaces can lead to local charges that strongly influence the path of the electrons. Therefore, the vacuum must not only provide a sufficient mean free path for the electrons but also prevent contamination as much as possible. Since even heavy pumping does not always prevent contamination, all vital parts of a PE spectrometer should be bakable to remove contamination. Modern PE spectrometers are built in high-vacuum (HV) technology and the necessary pumps are an important part of the spectrometer.

Photoelectron spectrometers designed for the study of solids require even better vacuum conditions. Because of the surface sensitivity of PES (see Section I,D and Fig. 4), any contamination of the sample surface contributes strongly to the measured PE spectrum. For a gas with a sticking coefficient of 1 (the sticking coefficient is the probability that a particle hitting the surface will remain on the surface), a clean surface is covered by a monolayer if 1 langmuir of gas is offered to the surface (1 langmuir is an exposure to 10^{-6} torr · sec). To maintain a reasonably clean surface for $\frac{1}{2}$ hr (about 10^3 sec), a vacuum better than 10^{-9} torr is necessary in the collision chamber. If well-defined adsorbates in the mono- or submonolayer range are being studied, the vacuum in the collision chamber must be even better (down to 10^{-11} torr). Spectrometers designed for this type of investigation are therefore built in ultrahigh-vacuum (UHV) technology.

B. Radiation Source

The source most commonly used in low-energy PES is a resonance lamp, usually operated with helium. The radiation emitted from this source results from the transitions $^1P(1snp) \rightarrow {}^1S(1s^2)$ (compare Table II). The radiation is produced by a high-voltage, direct-current discharge in a capillary, a high-current arc discharge with a heated cathode, or a microwave discharge. Examples of the actual design of the different types of lamps can be found in the bibli-

TABLE II.   Sources of Exciting Radiation

Source		Energy (eV)	Width (eV)
He	I	21.22	0.001
	I_α	21.2182	
	I_β	23.0848	
	I_λ	23.7423	
	II	40.81	0.01
	II_α	40.8140	
	II_β	48.3718	
	II_λ	51.0170	
	II_δ	52.2415	
Ne	I_α	16.6709	0.001
		16.8482	
	II_α	26.8141	0.01
		26.9110	
Ar	I_α	11.6237	
		11.8282	
Y	M_ζ	132.3	0.5
Zr	M_ζ	151.4	0.8
Ti	M_ζ	452.2	
Mg	K_α	1253.6	0.7
Al	K_α	1486.7	0.8
Si	K_α	1740.0	0.9
Cu	K_α	8047.8	2.5

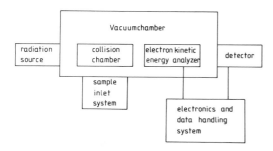

FIG. 8.   Scheme of a photoelectron spectrometer.

ography. The helium pressure is usually a few hundred millitorrs. Since this is higher than the pressure maintained in the collision chamber, differential pumping is necessary. The lamps have an intensity of about 10^{12} photons/cm · sec and a half-width of 10–15 meV, corresponding to about 100 cm^{-1}. The half-width, which determines the resolution obtainable in the spectrum, can be reduced by altering the operating conditions, but at the cost of rapid intensity loss. The numbers given above are a compromise between intensity and half-width. Under usual operating conditions about 98% of the emission consists of the He I$_\alpha$ line at 504 Å. Thus, the helium resonance lamp is a source of fairly monochromatic radiation. If the lamp is operated at lower helium pressure and higher voltage, it also emits He II radiation, which is the radiation emitted from He$^+$. The strongest emitted line is the He II$_\alpha$ line at 40.8 eV (compare Table II), and under optimal conditions up to 40% of the emitted radiation is He II$_\alpha$. However, He I radiation is also always present. Under the conditions for optimal He II emission, the total intensity of the emitted radiation is lower than the intensity under the standard conditions for He I operation.

Resonance lines from other atoms can also be used for excitation. Data for some of these lines are included in Table II. Since there is usually more than one intense line and since all the energies are lower than He I$_\alpha$, these lines are used only for special investigations. However, some of these lines can show up in the helium discharge if the helium is not of very high purity. Because of the lower energy of the corresponding excited states, the relative intensity of the impurity lines is much higher than the concentration of the impurity itself.

For higher excitation energies, the only easily available monochromatic radiations are characteristic x-ray emissions. Because of the short lifetime of the corresponding transitions (10^{-16} sec), the half-width of the emitted radiation is much larger than for the resonance lines discussed above and increases rapidly with the ordering number. The most commonly used sources of this kind are therefore the K_α lines of magnesium and aluminum. Other materials such as silicon, yttrium and silver have also been used with success. The energies and half-widths of these lines are given in Table II. Figure 9 shows the $4f$ doublet of polycrystalline gold measured with Mg, Al, and Si K_α radiation. The intensity obtained with characteristic x-ray lines is about two orders of magnitude lower than that

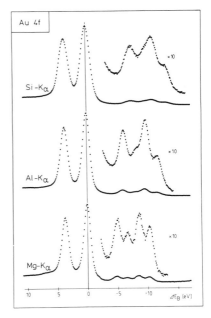

FIG. 9.   Au $4f$ doublet measured with different excitation sources.

of the resonance lines. High-power x-ray sources operating with a rotating anode have been designed for this reason; however, conventional x-ray tubes are still used in most cases. [See X-RAY ANALYSIS.]

Like the resonance lines, the x-ray lines are always accompanied by satellites with intensities up to 10% of that of the main line. This must be considered in the evaluation of spectra. For example, the small structures seen at the low binding energy side of the photoelectron lines in Fig. 1a and Fig. 9 are due to the satellites of the exciting radiation. In addition to the satellites, there is always a bremsstrahlung continuum underlying the characteristic x-ray lines. This continuum can be reduced by inserting a thin metal foil (the nature of which depends on the anode material) between the X-ray source and the collision chamber. The foil must be sufficiently thin and cannot be used to maintain a reasonable pressure difference between the source and the chamber.

To reduce the half-width of the characteristic x-ray lines, x-ray monochromators have been designed in which bent crystals are used to disperse the radiation. Line widths down to 0.3 eV have been reached for Al K_α with this technique. Use of the monochromator has the additional advantage that the K_α satellites and the bremsstrahlung continuum are removed. However, the intensity is strongly reduced. Thus, a high-power x-ray source must be used in connection with a monochromator. The spectrum shown in Fig. 15 was obtained with an x-ray monochromator.

Synchrotron radiation has become increasingly important as exciting radiation. When charged particles travel along a bent path with a velocity near the speed of light, as occurs in electron synchrotrons or storage rings, an intense beam of light is emitted tangentially to the path. This light has a smooth continuous energy distribution extending far into the vacuum UV and is strongly polarized in the plane of the ring. Figure 10 compares the brightness of the radiation (number of photons emitted per square centimeter, second, and steradian) from three synchrotron radiation facilities to that from several more conventional sources. Over a wide range synchrotron radiation is comparable or even superior to the line sources. The intensity of the synchrotron radiation can be increased further by two or three orders of magnitude by use of a "wiggler" or "undulator." Such devices are now available at most synchrotron radiation facilities. For the radiation to be used as an excitation source in PES, a certain wavelength must be selected by a monochromator. In the range

30–200 Å this can be achieved with grazing incidence monochromators; above this energy monochromatization becomes increasingly difficult. Synchrotron radiation is therefore used mainly below 200 eV, which is the energy range most suitable for surface investigations (see Fig. 4). Synchrotron radiation has the great advantage that the energy of the exciting radiation can be varied continuously over a wide range. In addition, it is the only source of polarized radiation practically useful for photoemission experiments. The disadvantage of synchrotron radiation is that it is not a laboratory source.

C. Collision Chamber and Sample Inlet System

For the measurement of gaseous samples the target gas is introduced through a capillary about 1 mm in diameter. The amount of gas entering the collision chamber is regulated by a leak valve. The beam of molecules coming from the capillary and the radiation coming from the source, which is usually a beam with a diameter of about 1 mm, can be either parallel or perpendicular to each other. The former arrangement leads to a rodlike photoionization region, the latter to a more pointlike one. Gas pressure in the PIR is about 10^{-3} torr. The photoelectrons created are usually observed perpendicular to the two incoming beams. This avoids having the beam of target gas point directly to the entrance of the kinetic energy analyzer and helps minimize contamination of the surfaces of the analyzer.

Vapors from liquids usually can be measured in the same way as gases. If the vapor pressure is lower than 10^{-3} torr, the liquid must be heated. In this case the inlet system should also be heated to prevent condensation.

To determine a gas phase spectrum from a sample that is solid under normal conditions, the sample must be volatile. If the volatility is fairly high, the same inlet can be used as for liquids. For samples with low vapor pressure, a direct inlet system can be used. A small capillary is filled with the sample and placed in a heatable sample holder, and the opening of the capillary is then brought close to the PIR. Molecules evaporating from the capillary reach the PIR directly and difficulties with deposition at narrow or cold parts of the inlet system do not arise. In addition, the molecules do not come into contact with heated metal parts, which often leads to catalytic decomposition. The volatility require-

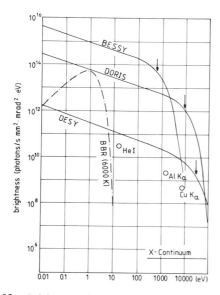

FIG. 10. Brightness of synchrotron radiation sources. The curves correspond to average operation conditions. For comparison, data for a blackbody radiator (BBR) at 6000 K and for several line sources are included. The arrows mark the cutoff energies. DESY and DORIS, Hamburg; BESSY, Berlin.

ments for the measurement of a gas phase PE spectrum are comparable to those for a mass spectrum. The amount of substance needed for the PE spectrum is about 20 mg, and it cannot be recovered.

To study photoemission from a solid, the sample must be brought directly to the ionization region. When the UHV requirements are not too high (up to 10^{-10} torr), the sample can be introduced through a lock. For measurements under extreme UHV conditions, which are necessary for most surface investigations, a fully metal-sealed vacuum system must be used. The system must be opened to insert the sample. In this case it is best to mount the sample on a manipulator that allows it to be positioned and turned from the outside.

If the sample is a conductor, it is brought in direct electrical contact with the sample holder. Metallic indium is often used to provide the necessary electrical and thermal contact between sample and sample holder. This is especially convenient because samples that are small particles can be pressed directly into the indium. If the sample is an insulator, even pressing it into indium does not always prevent charging. Powder samples can be mounted on double-stick tape or be pressed into a tablet, similar to the preparation used in infrared spectroscopy. Soluble samples can be dissolved and a drop of the solution brought to the surface of a metallic sample holder and evaporated. The latter method can yield a thin film, which is less sensitive to charging.

When a sample is brought into the vacuum, its surface is always contaminated, usually with O_2, H_2O, CO_2, and hydrocarbons. In many cases even the surface of the sample differs in chemical constitution from the bulk. The surface of most metals, for example, consists of oxides. In nearly all cases the sample surface must therefore be cleaned under UHV conditions. Among the techniques most widely used for the preparation of suitable surfaces are (1) bombardment of the surface with rare gas ions with a kinetic energy of a few hundred to a few thousand electron volts, (2) heating of the sample, sometimes in the presence of hydrogen gas (to reduce oxides), and (3) mechanical preparation of fresh surfaces.

Which preparation technique is most suitable always depends on the nature of the sample. For example, ion bombardment can change the surface constitution and heating can lead to phase transitions. To prevent contamination of the col-

lision chamber, the cleaning and preparation procedure is usually carried out in a separately pumped chamber. The sample is then transferred from the preparation chamber to the collision chamber.

D. KINETIC ENERGY ANALYZERS

The kinetic energy analyzer is the most important part of a photoelectron spectrometer. It is here that the electrons are discriminated with respect to their kinetic energy. The most important features of an electron kinetic energy analyzer (EKEA) are sensitivity and resolution. High resolution and high sensitivity always conflict with each other, and any analyzer is a compromise between them. Most EKEAs accept only electrons emitted into a certain solid angle. For usual work, this angle should be large to collect a great number of the electrons created. For angle-resolved work the acceptance angle must be limited to a few degrees; otherwise the angular resolution would be lost.

Because of the importance of EKEAs for PE and other kinds of electron spectroscopy, a great deal of attention has been devoted to their design and the theoretical analysis of their properties. This article can provide only a brief outline of the principles on which EKEAs are built and the design of those most widely used in PES. Further details can be found in the bibliography; Sevier's book provides more comprehensive information.

1. Retarding Field Analyzers

In a retarding field analyzer, an electron that is created in an area of well-defined electrostatic potential travels against a retarding field (Fig. 11a). If the kinetic energy of the electron is higher than the retarding potential U_R, the electron can pass the grid G and reach the collector C. The photoelectron spectrum is the first derivative of the photoelectron current I_{PE} mea-

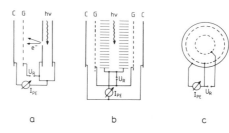

FIG. 11.   Different types of retarding field analyzers.

sured at the collector. For small values of U_R, nearly all electrons reach the collector. At low kinetic energies variations in I_{PE} are superimposed on a very high background. Since the statistical noise is proportional to the total number of electrons reaching the collector, retarding field analyzers have problems with the detection of structures in the low kinetic energy range. If an electron is not traveling in the direction of the electric field, only the component of the kinetic energy parallel to the electric field is measured. Therefore, to obtain good resolution it is necessary to ensure that the electrons travel in the direction of the electric field. Several designs have been used to reach this goal:

1. "Slotted grid" retarding field analyzers (Fig. 11b). The first grid consists of pinholes pointing in the direction of the electric field. To reach adequate collecting efficiency, the analyzer is designed to be cylindrical. This type of analyzer is very suitable for a rodlike ionization region.

2. "Spherical grid" analyzers (Fig. 11c). This type of analyzer has the highest possible collecting efficiency (4π); however, its resolution is strongly reduced if the collision area is not pointlike and has not been adjusted perfectly to the center of the spheres.

3. Electrostatic lenses used to form a parallel beam of electrons collected within a certain angle.

In spite of their high collecting efficiency and acceptable resolution, retarding field analyzers are no longer widely used in PES. The primary reason is high sensitivity to contamination. Contamination of the surface of the grids by sample material or decomposition products leads to a rapid loss of resolution.

2. Deflection Analyzers

These analyzers are based on the deflection of electrons in an electrostatic or magnetic field. Magnetic deflection analyzers, which were used in the early days of PES, are presently of no importance, because the weak magnetic fields necessary for the separation of low-energy electrons are difficult to control.

The operation of an electrostatic deflection analyzer is illustrated by the parallel-plate analyzer shown in Fig. 12a. This is a parallel-plate condenser with two holes in one plate. Between the two plates, a potential U_R is applied in such a manner that the resulting electric field is retarding for negatively charged particles. An electron

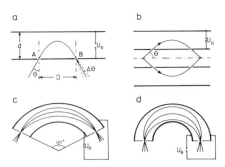

FIG. 12.   Different types of electrostatic deflection analyzers.

entering the field through hole A at angle θ follows a parabolic trajectory and leaves the field at hole B if the following "focusing condition" is fulfilled:

$$E_{kin} = e U_R D / 2d \sin 2\theta \qquad (7)$$

Varying U_R permits electrons with different kinetic energies to exit through hole B. Thus, measuring the number of electrons arriving at B as a function of U_R results in the PE spectrum. In this form the analyzer has little resolution because all electrons that fulfill the condition

$$E_{kin} \sin 2\theta = e U_R D / 2d \qquad (8)$$

reach hole B. If we accept only electrons that leave B under a small angle $\Delta\theta$, this problem can be solved. In such a case, however, an electron that has the correct kinetic energy but comes from an angle different from $\theta \pm \Delta\theta$ is lost. Therefore, the sensitivity is very poor. If the focusing conditions are fulfilled only for one angle and not for some range around a certain mean angle, the analyzer is said to have no focusing properties. For the parallel-plate analyzer the lack of focusing properties is no longer true for $\theta = 45°$. In this case the first-order dependence of E on θ vanishes. The analyzer is first-order focusing. However, the first-order focusing holds only for angular deviations $\Delta\theta$ in the plane of the drawing (Fig. 12a). Such an analyzer is called "single focusing." Any deflection analyzer should be at least first-order single focusing. If the angle θ that is acceptable without too much loss in resolution is small, we speak of "weak first-order focusing"; if the angle is large, we speak of "strong first-order focusing." For the 45° parallel-plate analyzer, $\Delta\theta$ is only a few degrees. If the ionization region is rodlike, entrance and exit slits can be used instead of holes to increase the sensitivity.

The cylindrical mirror analyzer (Fig. 12b) can be considered an extension of the parallel-plate analyzer with the two plates folded into two concentric cylinders. The entrance and exit are then ring slits in the inner cylinder. The cylindrical mirror analyzer is second-order focusing in θ and has no limitations with respect to the radial angle. Since focusing is good over a wide range of θ, this is the analyzer of highest sensitivity among the deflection analyzers. However, to reach good resolution the electrons must be emitted from a point source. For use in PES this means that the ionization region should be a relatively small spot. This is easy to achieve with solid samples but difficult with gases. For gas phase work, the cylindrical mirror analyzer has the advantage that its focusing properties are also good for the magic angle of $54°44'$.

The $127°$ cylindrical analyzer (Fig. 12c), a strong first-order single focusing analyzer, is the most easily built electrostatic deflection analyzer. Because it is single focusing, it is most successful with rodlike ionization regions and is therefore the most commonly used analyzer in gas phase PES.

The best matched analyzer for pointlike sources is the $180°$ spherical analyzer (Fig. 12d), which consists of two concentric hemispheres. This analyzer is first-order double focusing. It is the analyzer used in most instruments that are designed primarily for investigation of solids.

3. Hybrid Analyzers

Most analyzers now used in commercial photoelectron spectrometers are of a hybrid variety. Retarding and accelerating fields are used in connection with electrostatic lenses and deflection analyzers. The use of lenses makes it possible to increase the acceptance angle and achieve a geometric separation between the ionization region and the analyzer. Application of retarding fields in connection with deflection analyzers allows different modes of operation: the spectrum can be measured either by sweeping the retarding potential while keeping the deflection potential constant, or by sweeping the deflection potential while keeping the retarding potential constant, or by a combination of both methods. With this type of instrument it is possible to measure a spectrum with constant absolute resolution (ΔE = constant) or constant relative resolution ($\Delta E/E$ = constant). An example of the design of a hybrid analyzer is shown in Fig. 13.

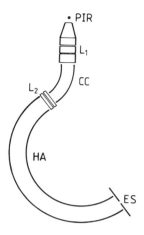

FIG. 13.   Sketch of a hydrid analyzer as used in the Leybold–Heraeus UPS spectrometer. PIR, Photoionization region; L, lenses; CC, cylindrical condenser; HA, hemispherical analyzer; ES, exit slit.

E. Detectors

Direct measurement of the current provided by the electrons passing the analyzer is possible only for retarding field analyzers with their high collecting efficiency. For currents between 10^{-14} and 10^{-16} A, corresponding to 10^5 to 10^3 electrons/sec, an electrometer can be used for detection. In an electrometer the collected electrons develop a voltage over a very high resistor and this voltage is then amplified and measured. The limit for use of an electrometer is set at about 10^{-16} A by the random noise of the electrons in the input resistor.

In PE spectrometers with deflection analyzers, electron fluxes down to a few electrons per second must be recorded. In this case electron multipliers are most often used for detection. Since windows cannot transmit electrons, an open multiplier is used; the electrons come from the analyzer and directly hit the first electrode (dynode) of the multiplier. A major disadvantage of open multipliers is their sensitivity to contamination. Decomposition products from the sample can be deposited on the surface of the dynodes and chemically aggressive gases can react with the dynode material. Such contamination drastically reduces the efficiency of the multiplier.

Measurement of a PE spectrum with count rates down to a few electrons per second is time-consuming. Statistical noise is always proportional to the square root of the number of electrons counted per unit time. If only 10 electrons are counted for a given kinetic energy, the uncertainty is about 30%. If 90% of the electrons are due to background (see Fig. 1c), the signal is lost in the noise. To detect the signal, the noise must be reduced at least to 5%, corresponding to a collection of about 400 electrons. A count rate of 10 per second leads to an observation time of 40 sec at this single kinetic energy. It is obvious that such lengthy measuring times lead to specific complications. The important complications are the stability of the sample, the stability of the electrical potentials in the analyzer, and the stability of the intensity of the exciting radiation. Variation of the target gas pressure, for example, influences the energy and the relative intensity of different lines. Therefore, most modern instruments use a sweep technique in which the investigated energy range is measured within a few minutes and the measurement is repeated at intervals. The data from each sweep are accumulated and stored in a data handling system. The development of the spectrum can be watched on a screen and the measurement can be terminated when the signal-to-noise ratio has reached an acceptable limit.

A new development that leads to reduced measuring times is the use of multichannel plates as detectors. A multichannel plate is an array of small continuous electron multipliers (channeltrons). This device permits simultaneous measurement of electrons within a certain energy range, not just at a single kinetic energy. To make use of this multiplex advantage a specifically designed electron kinetic energy analyzer is necessary to match the focusing requirements set by the multichannel plate.

F. Magnetic Shielding

Low-energy electrons are very sensitive to the stray magnetic fields that exist in any laboratory, such as the earth's magnetic field. These stray fields strongly influence the trajectories of the traveling electrons. As a result, all parts of a PE spectrometer between ionization region and detector must be carefully shielded. This can be done with Helmholtz coils, which compensate the external fields, or with magnetically shielding materials like mu-metal. The latter type of shielding is used in most modern instruments, because shielding with Helmholtz coils is difficult for the dynamic stray fields produced by many types of laboratory equipment. Improper shielding leads to reduced sensitivity and resolution, as well as to asymmetric skewed lines.

G. Instruments for Angle-Resolved Photoelectron Spectroscopy

Instruments designed for angle-resolved measurements must meet two specific requirements: (1) the analyzer must be movable around the ionization region, and (2) the acceptance angle of the analyzer must be small enough (usually a few degrees) to achieve angular resolution. The requirements for an angle-resolving spectrometer differ if the sample is a gas or a solid.

The angular distribution for atoms or molecules in the gas phase is fully described by the asymmetry parameter β (Section I,D). To measure β, the angle θ between the incoming radiation and the outgoing electrons must be varied [Eq. (6)]. It is sufficient for this type of measurement to move the analyzer in one plane.

For solid samples there are three degrees of freedom, the angles Ω, θ, and ϕ defined in Section I,D. If the incident radiation is polarized, an additional degree of freedom is obtained because of the orientation of the polarization vector. Angle-resolved measurements with unpolarized radiation can be made with an instrument in which the analyzer can be moved in one plane, provided the sample is mounted on a manipulator that allows adjustment of the direction of the surface normal. With this setting, Ω, θ, and ϕ can be varied over a fairly wide range. However, a better range for all three angles is achieved if the analyzer can be moved in two planes that are perpendicular to each other. Because of the additional degree of freedom obtained with polarized radiation it is necessary to use an instrument that allows movement of the analyzer in two dimensions in order to obtain all the available information.

Some limited information on the angular distribution of the electrons can be derived even with an instrument in which the angle between excitation source and analyzer is fixed: if the sample is mounted on a manipulator, it is possible to vary the orientation of the surface normal.

III. Electronic Structure of Atoms, Molecules, and Solids

A. Interpretation of PE Spectra in the Single-Particle Approximation

To discuss the information on electronic structure obtainable by PES, we will start with the independent-particle model. This model is best known for atoms. The electronic structure of an atom is described by orbitals $\varphi(q_i)$, which are functions of the spatial coordinates q_i. Each orbital can hold a maximum of two electrons if these electrons have antiparallel spin. With each orbital we can associate an orbital energy ε_i. According to the aufbau principle, the state of lowest total energy is reached if the orbitals are filled in order of increasing orbital energy. Some of the orbitals are degenerate with respect to orbital energy, and Hund's rule states that degenerate orbitals are first filled singly with electrons of parallel spin. The orbitals holding the inner shell electrons are energetically well separated from the orbitals of the outermost or "valence" electrons (Table III). For most atoms the valence electrons have binding energies of less than 40 eV. The inner shell electrons are usually called "core electrons." If the state of lowest energy consists only of doubly occupied orbitals, it is called a "closed shell ground state." This situation is depicted in Fig. 14a. The rest of this article will be restricted to the discussion of systems with a closed shell ground state. This state is an exception for atoms but is very common for molecules. Apart from the orbitals that are occupied in the ground state, there are unoccupied or "virtual" orbitals. For example, for the neon atom, with ground state $1s^2 2s^2 2p^6$, the lowest unoccupied orbital is $3s$. In the following discussion, it is important to keep in mind that "occupied" always refers to the set of orbitals occupied in the ground state M_0. Correspondingly, "unoccupied" always refers to the set of orbitals not occupied in the ground state.

When we turn from atoms to molecules, the

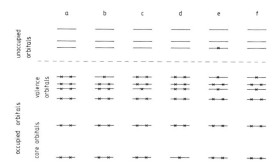

FIG. 14. Orbital representation of (a) the initial and (b–f) different final states of the photoemission process.

situation changes only slightly. The major difference is that the valence orbitals no longer belong to a single atom. To a greater or lesser extent, they are extended over the whole molecule. The same is true of the unoccupied orbitals. The core orbitals are still localized at their respective atoms. Degenerate valence orbitals are less common in molecules than in atoms; they are usually found in molecules of higher symmetry, with at least one threefold axis.

For solids the number of atoms and therefore the number of orbitals goes to infinity. The valence orbitals develop into continuous energy bands. The details of the electronic structure of a crystalline solid are described in terms of the "band structure." The energy up to which the bands are occupied is called the Fermi energy (E_F). When the Fermi energy lies in a band, the solid is a metal; when it lies in a gap between

TABLE III. Calculated Orbital Energies (ε_i), Experimental Binding Energies (E_B), and Relaxation Energies (R_i) for Formaldehyde and Water[a]

Molecule	No.	$-\varepsilon_i$	E_B	R_i	Symmetry	Type
$H_2C{=}O$	8	11.96	10.80	1.16	$2b_1$	n(O)
	7	14.81	14.5	0.3	$1b_2$	π(CO)
	6	17.91	16.0	1.9	$5a_1$	σ(CH)
	5	19.07	16.6	2.5	$1b_1$	σ(CH)
	4	23.67	20.5	3.17	$4a_1$	σ(CO)
	3	38.68			$3a_1$	O 2s
	2	308.45	294.47	13.98	$2a_1$	C 1s
	1	559.86	539.44	20.43	$1a_1$	O 1s
H_2O	5	13.59	12.6	0.99	$1b_2$	n(O)
	4	15.26	14.7	0.56	$1b_1$	σ(OH)
	3	19.37	18.4	0.97	$3a_1$	σ(OH)
	2	36.27	32.2	4.07	$2a_1$	O 2s
	1	559.27	539.7	19.57	$1a_1$	O 1s

[a] Orbital energies courtesy of D. Cremer and J. Gauss.

two bands the solid is a semiconductor (small gap) or an insulator (large gap). As in molecules, the core orbitals in solids still behave like those in atoms.

We now return to photoemission. For a first approximation, we assume that we can remove a single electron from one of the occupied orbitals without disturbing the remaining electrons. This is called the "frozen orbital approximation" (FOA). This process creates a hole in the manifold of the occupied orbitals and we call the resulting states "hole states." If the electron is removed from the highest occupied orbital, the final state of the photoemission process is the ground state M_0^+ of the ion (Fig. 14b). If the electron is removed from a lower-lying orbital, an excited state of the ion is reached which we call a "hole excited state" because moving a hole downward is the same as moving an electron upward (Fig. 14c). When the electron is removed from a core orbital, we speak of a "core hole state" (Fig. 14d). Such a state is usually labeled by the chemical symbol of the atom and the orbital from which the electron has been removed (e.g., C 1s, O 1s, P 2p).

If we start from M_0^+ (Fig. 14b) and excite one of the remaining electrons to an unoccupied orbital (Fig. 14e), a new type of excited ion state is reached. Compared to the ground state M_0 of the initial system, this state has two holes in the occupied orbitals and one electron in the unoccupied orbitals. Therefore, it is called a "two-hole one-particle (2h1p) state." Correspondingly, we can define three-hole two-particle states and so on. These states are frequently called "shake-up states," based on the idea that

photoionization is such a strong perturbation that the whole electron system is "shaken" and one or more of the remaining electrons are "shaken up" to unoccupied orbitals. If the "shake-up electron" receives enough energy to leave the system, the final state of the photoionization is a state of the doubly ionized system (Fig. 14f). This type of transition was shown as process 3 in Fig. 2 and, in a further extension of the above ideas, is called a "shake-off" process. Since the second electron is no longer bound, shake-off excitations lead to a continuum. This corresponds to the ionization continuum following discrete excitation in an optical spectrum.

Examples of the processes discussed above are seen in the photoemission spectrum of neon (Fig. 1a). The first line appears at 21.6 eV, corresponding to photoionization of one of the six 2p electrons. The second line, at 49 eV, corresponds to photoionization of a 2s electron and the third line, at 870 eV, to photoionization of a 1s electron. On an enlarged intensity scale (Fig. 15) shake-up satellites are seen at the high binding energy side of the Ne 1s line which result from transitions into 2h1p states. The lines indicated—3, 5, 6, and 7, for example—have been identified as transitions into states where, in addition to the creation of the 1s hole, a 2p electron is excited to 3p, 4p, 5p, and so on. In addition, a shake-off continuum underlies the high-energy part of the 1s shake-up spectrum.

The appearance of shake-up structures is not restricted to the core region. Because of the close spacing of the valence orbitals and the vibrational structure of the corresponding photo-

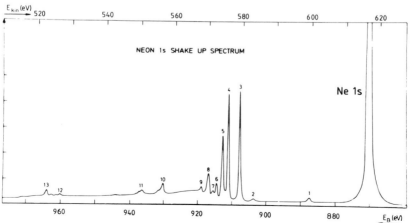

FIG. 15.   Shake-up satellites accompanying Ne 1s photoionization; $h\nu = 1486.7$ eV. [From Siegbahn, K. (1974). *J. Electron Spectrosc.* **5**, 34.]

emission bands (Section III,C), shake-up structures in the valence region are more difficult to observe. Figure 3 shows an example where it was possible to detect a shake-up structure (indicated 2h1p) by comparison with the optical spectrum of the ion. A transition from the ion ground state M_0^+ to this particular 2h1p state is allowed and leads to a strong band in the absorption spectrum of the ion. In the PE spectrum, which starts from M_0^+ instead of M_0, the 2h1p final state gives rise only to a weak structure. From the PE spectrum alone it would be difficult to detect this weak structure as a result of an independent transition. Comparison with the UV spectrum of the ion, however, allows an unambiguous assignment.

In both examples shake-up transitions lead to weak structures in the PE spectrum. This is a very general observation. Only in special cases can shake-up satellites reach an intensity comparable to that of the main line (see Section IV,C and Fig. 25). Why are shake-up transitions usually weak? In the FOA, transitions are allowed only when the initial and final states differ by one electron. Transitions that involve more than one electron are forbidden. Since all 2h1p states differ from the initial state by at least two electrons (Fig. 14), transitions from M_0 to 2h1p states (and also to shake-off states) are strictly forbidden in the FOA. Only transitions to final single-hole states are allowed. When degenerate orbitals are counted only once, the number of lines or bands observable in a PE spectrum should be equal to the number of occupied orbitals. Therefore, all the prominent bands in the valence region of free molecules and all main lines in the inner shell region of free molecules and solids are usually assignable in the single-particle picture. Basically, this type of interpretation also holds for photoemission from solids. However, in this case some special aspects must be considered, which we will discuss in part in Section IV,B.

The correspondence between orbital picture and PE spectrum goes even further. In the FOA the binding energies referred to the vacuum level are directly connected with the orbital energy of the occupied orbitals

$$E_B(i) = -\varepsilon_i \qquad (9)$$

This relation was derived by T. Koopmans in 1932 and is now known as Koopmans' theorem (KT). It is the basis of most applications of PES in electronic structure elucidation. If KT were strictly valid, we could experimentally observe orbital or single-particle energies, which in reality exist only in the theoretical framework of the independent-particle model. It must be clearly understood that orbital energies are not observable in the sense of quantum mechanics. However, they can be calculated by a variety of different methods, and these calculations can be performed with a high degree of accuracy for small molecules. A comparison of orbital energies from such calculations with experimentally observed binding energies shows that KT is usually fulfilled within a margin of about 10% (see Table III). Therefore, KT is a close approximation.

However, one point must be kept in mind in relation to the application of Koopmans' theorem: orbital energies are quantities solely related to the initial state of the photoemission process. Thus, applying KT is equivalent to neglecting any influence of the final state on the binding energies. Obviously, this cannot be true. To take final state effects into account, we must go beyond the FOA. We return to this point in the following section.

In the application of PES to the study of molecular electronic structure, KT is seldom used in the form of Eq. (9). In most cases a direct comparison of calculated ε_i and measured $E_B(i)$ is not the main objective. Often it is more interesting to investigate how a certain ε_i and correspondingly a certain E_B vary with alterations in chemical constitution. In this case we use a "weaker" form of KT:

$$\Delta E_B(i) = \Delta \varepsilon_i \qquad (10)$$

where we connect binding energy shifts (ΔE_B) with orbital energy shifts. In this case we do not completely neglect final state effects. We only assume that they are approximately constant within a certain class of compounds. The chemically appealing feature of this approach lies in the great variety of models available to estimate orbital energy shifts. For example, all the models that have been developed to describe the influence of different substituents on physical properties and chemical reactivities of molecules can be applied in this case, and frequently a more direct proof of a given model is possible with PES. Two examples of this type of application are discussed in Section IV,A.

One model that applies to core electron binding energy shifts will be described here in some detail. Chemists usually attribute partial charges q_A to the different atoms of a molecule, even though this concept is problematic from a strictly theoretical point of view. We now consider a core orbital at atom A. In the "point charge approximation" the orbital energy $\varepsilon_i(A)$ can be expressed in terms of partial charges by

$$\varepsilon_i(A) = k(i, A)q_A + V(q_B) + k_0(i, A) \quad (11)$$

where q_A is the charge at the considered atom, $V(q_B)$ the "off-atom potential" created by the charges at all other atoms, and k_0 and k are parameters specific for atom A and orbital i. Within the limits of applicability of Eq. (10), the model makes it possible to derive information on differences in atomic charges from core electron binding energy shifts.

B. FINAL STATE EFFECTS

In the preceding section we saw that the most prominent structures in PE spectra can be explained adequately in the single-particle approximation. Two facts, however, clearly show the limitations of this type of interpretation: (1) the deviations from Koopmans' theorem [Eq. (9)] and (2) the appearance of shake-up and shake-off satellites.

From these two observations, it is obvious that we cannot completely neglect final state effects. Using Koopmans' theorem as a starting point, we can write

$$E_B(i) = -\varepsilon_i - R_i \quad (12)$$

where R_i is called the "relaxation energy." The meaning of Eq. (12) can be understood by the following gedankenexperiment. We separate the photoemission process into two steps. In the first step, we use the frozen orbital approximation and remove one electron from a given orbital i, leaving all other electrons unperturbed. To remove the electron we need energy ε_i. In the second step, we consider the reaction of the remaining electrons to the presence of the hole created in the first step. The system will "relax" to a new energetically favorable situation. Therefore, the relaxation energy is usually positive. Only in rare cases can specific quantum mechanical effects (so-called correlation effects) lead to small negative values of R. In Table III we have compared experimental binding energies to calculated orbital energies. The relaxation energies derived from these two sets of data vary from orbital to orbital. R is much larger for core electrons than for valence electrons; however, relative to the magnitude of the binding energy, R is similar for all shells.

From our gedankenexperiment we suggest that R_i will be connected strongly to the mobility of the electrons in the system. If the orbital from which the electron is removed is highly localized, as in a core orbital or a lone pair orbital, the most efficient stabilization of the hole state will be achieved by transfer of negative charge to the vicinity of the hole. If a direct transfer is not possible, the stabilization can be achieved only by polarization of the surroundings. From this consideration we expect larger relaxation energies for the core ionization of metals, where the electrons in the valence band move almost freely, than for the core ionization of insulators, where the polarization of the nearest neighbor atoms yields the most important contribution to the relaxation. For delocalized holes that result from the photoionization of π electrons of unsaturated molecules or valence electrons of solids, the relaxation contribution is expected to be smaller and less dependent on the individual orbital. This is in accordance with the data shown in Table III; for formaldehyde, for example, the smallest relaxation energy is found for the π orbital $1b_1$.

The considerations discussed above can also be viewed in a somewhat different manner. The ion state $(M_i^+)^*$ formed in the frozen orbital approximation is not a real state (eigenstate) of the investigated system. It exists only in the framework of the theoretical model. It can be described, however, by a superposition of the eigenstates M_k^+ of the ion, where each of these states contributes with a certain weight factor g_k:

$$(M_i^+)^* = \sum_k g_k M_k^+ \quad (13)$$

In the high-energy limit where the electron leaves the ionized system very quickly (sudden limit), the probability of a transition to a final state M_k^+ is equal to g_k^2. The main contribution usually comes from a state M_i^+ that has the same orbital occupation as the hypothetical frozen orbital state $(M_i^+)^*$. However, because of the presence of the hole, the orbitals of M_i^+ differ somewhat from the orbitals of the initial state M_0. Since the weight factor g_i is usually greater than 0.5, we normally can give an assignment of the main bands or main lines of a PE spectrum in terms of Koopmans' theorem, as discussed in the previous section. The remainder of the weights is frequently distributed over a variety of final states M_k^+ including the continuum states, thus explaining shake-up satellites and shake-off continuum. When we are able to describe M_k^+ to a good approximation by a single 2h1p state, we reach a situation as discussed in the previous section in connection with Fig. 15, where several of the observed shake-up satellites could be assigned to specific electronic excitations in the core ionized system. For larger systems, where the number of possible electronic excitations becomes very large, we will be able to identify the shake-up satellites only in cases where the remaining weight is not distributed more or less equally over a large number of final states. When g_k has a somewhat larger value for a specific final state M_k^+, we will observe a characteristic shake-up satellite even for an extended system. An example of such a situation will be discussed in Section IV,C.

We now turn to binding energy shifts. From Eq. (12) we obtain

$$\Delta E_B(i) = -\Delta\varepsilon_i - \Delta R_i \qquad (14)$$

Binding energy shifts depend as much on initial state effects (via $\Delta\varepsilon_i$) as on final state effects (via ΔR_i). Often we are mainly interested in initial state effects, because we want to derive information on the electronic structure of the initial system M_0 and its dependence on variations in chemical constitution. This information, however, can be derived only if ΔR is negligibly small or we are able to obtain independent information about ΔR. The relaxation contribution itself also contains valuable information, since it is connected with electronic relaxation processes that can take place during a chemical transformation. In a wide variety of chemical reactions the transition state is charged. The better this charge can be screened by a relaxation of the whole electronic system, the lower the energy of the transition state.

A variety of theoretical models have been developed in which relaxation is taken into account (transition state models, relaxed potential models, equivalent core models). A discussion of these models is far beyond the scope of this article. Here, we will only add some comments on methods by which it is possible to separate initial and final state effects with the use of experimentally available data. These methods are based on a combination of photoelectron and Auger electron spectroscopy. We consider an Auger transition from an initial state that has a single hole in the inner shell k to a final state that has two holes in another inner shell i. This Auger transition is combined with photoemission processes that correspond to the photoionization of an electron from orbital k and from orbital i. This yields

$$\Delta\beta(i) = \Delta E_{kin}^{Au}(kii) + 2\,\Delta E_B(i) - \Delta E_B(k) \qquad (15)$$
$$= 2\,\Delta R_i - \Delta R_{ii}$$

where ΔR_{ii} is the relaxation contribution of the double-hole final state of the Auger transition. $\Delta\beta(i)$ is independent of the reference level. Therefore, it can be obtained for molecules in the gas phase as well as for solids. Since it is independent of the reference level, it is also independent of sample charging if the Auger kinetic energy $E_{kin}^{Au}(kii)$ and the binding energies are derived from the same measurement. When we introduce the approximation that the relaxation energy is mainly a result of classical electrostatic contributions, ΔR_{ii} should be four times as large as ΔR_i. We then obtain

$$\Delta\beta(i) = -2\,\Delta R_i \qquad (16)$$

If experimental constraints allow only the measurement of one of the binding energies, we can use the cruder approximation

$$\Delta\alpha(i) = \Delta E_{kin}^{Au}(kii) + \Delta E_B(i) = -2\,\Delta R_i \qquad (17)$$

where $\Delta\alpha$ is the Auger parameter shift, introduced by C. Wagner.

The drawback of these experimental methods for disentangling initial and final state effects is that such methods are applicable only to core ionizations and heavy atoms. An Auger transition that does not involve valence electrons can be observed only for atoms with at least two inner shells. Thus, an experimental estimate of final state relaxation effects can be derived only for third row and higher elements. For the light elements, however, highly accurate calculations are increasingly available that allow us to derive ΔR from a comparison of experimental and theoretical data as shown in Table III.

C. VIBRATIONAL FINE STRUCTURE

The low-energy PE spectra of small and medium-size molecules often contain bands with well-resolved vibrational fine structure. Since the resolution in UPS is usually limited to about 150 cm^{-1} (≈ 15 meV) only excitations into well-separated vibrational states of the final ion state can be observed. As long as the temperature of the sample is not much higher than 300 K, vibrational excitations in the initial states do not perturb the observed spectra; low-frequency vibrations excited at these temperatures are covered by the limited resolution.

To first order, the intensity distribution of the different lines of a vibrationally structured PE band can be interpreted in terms of the Franck–Condon principle (Fig. 16). The most probable transition is the "vertical transition." In a potential diagram, where the electronic energy is drawn as a function of interatomic distances, the vertical transition is best approximated by a vertical line drawn from the minimum in the potential of the initial state to the potential curve of the final state. It is the binding energy that corresponds to the vertical transition [the "vertical ionization potential" (VIP)] that must be used in connection with Koopmans' theorem [Eq. (9)]. The transition from the zero vibrational level of the initial state to the zero vibrational level of the final state is called an "adiabatic transition." If the minimum of the final state is not displaced with respect to the initial state, vertical and adiabatic transitions fall together (Fig. 16). The difference between adiabatic and vertical ionization potentials is therefore a measure of the change in equilibrium geometry between ion state and initial state. For example, the first ionization band of formaldehyde (Fig. 1b) shows a very intense 0–0 transition and little vibrational fine structure. From this we can conclude that the electronic ground state of the formaldehyde cation is very similar in geometry to the electronic ground state of formaldehyde. For the

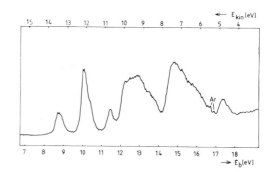

FIG. 17.   UP spectrum of *trans*-stilbene.

second photoionization band the vertical transition corresponds to the fifth vibrational level, telling us that the equilibrium geometry of the first electronically excited ion state differs considerably from the ground state of the neutral system. In the fourth ionization band of formaldehyde we observe a rapid loss of vibrational fine structure after the fifth vibrational excitation. This is a typical pattern for an excitation into a dissociative final state (see Fig. 16). Excitation with an energy greater than the dissociation limit leads to fragmentation of the ion. The vibrational structure becomes broadened in the region of the dissociation limit and then continuous. From the examples discussed it is clear that the vibrational fine structure of a photoionization band contains information on the geometry and the potential surface of the final state.

For larger molecules we usually do not observe vibrationally structured bands (Figs. 17, 18, and 19). The number of possible vibrations that can be excited in the final ion state increases rapidly with the size of the system, and the superposition of the different vibrational excitations leads to more or less continuous bands. For these unstructured bands the VIP is attributed to the band maximum. Some bands show vibrational fine structure even for larger molecules (Figs. 4 and 20). In planar unsaturated compounds, for example, an ionization from the π system usually couples to skeleton modes that lie near 1400 cm^{-1}, resulting in a well-developed vibrational structure. However, for the same molecules the bands that result from ionizations from CH σ bonds are mostly broad and unstructured.

Analysis of the vibrational fine structure can be carried even further when we combine it with the assignment of the main bands in the single-particle approximation. From theoretical calcu-

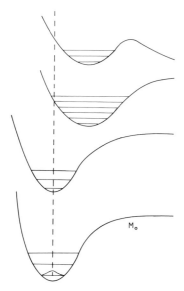

FIG. 16.   Schematic representation of potential curves for the ground state of the initial system (M_0) and different ion states M_k^+.

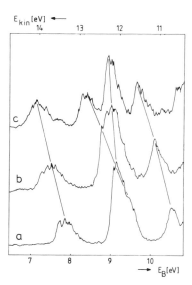

FIG. 18. UPE spectra for (a) *trans*-stilbene, (b) *trans*-*p*-methoxystilbene, and (c) *trans*-*p*,*p*′-dimethoxy-stilbene.

lations, for example, the highest occupied orbital of formaldehyde is a nonbonding orbital (*n* orbital) that is mainly localized at the oxygen atom. When an electron is removed from such

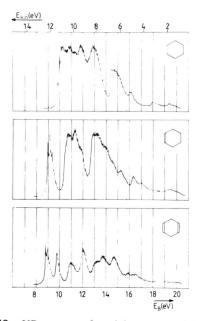

FIG. 19. UP spectra of cyclohexane, cyclohexene, and 1,4-cyclohexadiene. [From Bischof, P., Hashmall, J. A., Heilbronner, E., and Hornung, V., (1969). *Helv. Chim. Acta* **52**, 1745.]

an orbital the geometry is expected to change only little, in agreement with what we concluded from the fine structure of the first band. A detailed analysis of this structure reveals that the CO stretching vibration, which is 1744 cm^{-1} in the ground state of the neutral molecule, is only slightly reduced to 1590 cm^{-1} in the lowest ion state. The second highest orbital is the π orbital of the CO double bond, an orbital that is strongly CO-bonding. When an electron is removed from this orbital, the bond is considerably weakened. This should lead to an increased CO bond length, in accordance with the difference between vertical and adiabatic ionization potential observed for the second band. The weakening of the CO bond is also seen in the lowering of the CO stretching frequency, which is reduced to 1210 cm^{-1} in the first excited state of the ion. These examples show how careful analysis of the vibrational structure of the photoionization bands and comparison of the observed vibration frequencies to the vibration frequencies known for the molecular ground state can be used to derive information on the bonding characteristics of molecular orbitals.

D. MULTIPLET SPLITTING

As mentioned earlier, core electrons behave in an atomlike way, even when the respective atom is part of a molecule or a solid. As a consequence, the angular quantum number is always a good quantum number for inner shell electrons. We can denote such an electron by its principal quantum number *n* and its angular quantum number *l*. We therefore speak of 2*s*, 2*p*, or 3*d* electrons and correspondingly of a 2*s*, 2*p*, or 3*d* hole. However, an additional effect must be taken into account for core holes: the spin–orbit coupling. Since we can treat a hole in an otherwise completely filled shell in the same way as a single electron in an otherwise empty shell, we can attribute a spin $+1/2$ or $-1/2$ to the hole. Under the action of spin–orbit coupling the angular momentum resulting from the spin couples to the angular momentum corresponding to the spatial motion of the hole to form a total angular momentum *J*. The two possible *J* values that can be formed are

$$J = l \pm 1/2 \tag{18}$$

Only $J = +1/2$ is possible for $l = 0$ (*s* orbitals). Photoionization of a core electron with angular quantum number $l > 0$ therefore leads to two possible final states that correspond to the two

TABLE IV. Spin–Orbit Splitting in Xenon

	Splitting (eV)	
n	$l = 1$	$l = 2$
5	1.3	
4		2.0
3	61.5	12.6
2	321.5	

possible J values. Because of spin–orbit coupling, the two final states have different energies. For different inner shells spin–orbit coupling decreases with increasing n and within the same shell it decreases with increasing l. To give an impression of the magnitude of spin–orbit splitting, data for xenon are provided as an example in Table IV. If the splitting exceeds 1 eV, we observe a resolved doublet in the XPE spectrum. In the case of resolved doublets the individual lines are labeled nl_J. Table V compares this type of labeling with the notation commonly used in x-ray spectroscopy.

The intensity ratio of the two doublet lines is given by

$$I_{l+1/2}/I_{l-1/2} = (l + 1)/l \qquad (19)$$

The line with the higher intensity is always found at lower binding energies (see Fig. 9). Deviations from this ideal intensity ratio exist because of the effects discussed in Section III,B: one of the lines can lose more intensity to satellites than the other.

Spin–orbit coupling is less important for the valence electrons because l is no longer a good quantum number. Only for molecules that contain at least one symmetry axis that is threefold or higher can the angular momentum around this axis be described by a good quantum number. If the angular momentum is not zero, the angular motion can still couple to the spin. The intensity ratio in this case is 1 : 1 for the two doublet lines. In the valence region spin–orbit interaction is observed only when at least one atom with $Z > 10$ is present in the molecule. For light atoms, the doublet splitting due to spin–orbit coupling is so small that it is not resolved in UPE spectra.

Spin–orbit interaction also influences the band structure of solids.

IV. Some Examples

In this section we will discuss a few examples of the application of PES in different fields of research. The number of examples is too limited to allow more than a glimpse of the broad and still rapidly evolving field of photoelectron spectroscopy. Further examples and other fields of application can be found in the references cited in the bibliography. Current work in PES is published in a variety of scientific journals. However, a journal specifically devoted to this field is the *Journal of Electron Spectroscopy and Related Phenomena*, published by Elsevier, Amsterdam.

A. FREE MOLECULES

The type of investigation that can be performed with free molecules has been illustrated throughout this article by the example of formaldehyde: experimental binding energies can be compared to calculated orbital energies, we can try to understand differences in relaxation contributions, and, by inspection of the shape of the PE bands, we derive information on the bonding characteristics of different orbitals. This type of analysis is, however, limited to smaller molecules. When the molecules become larger, the number of orbitals in the valence region increases and the different bands overlap so strongly that a separate assignment is often impossible. An example of such a case is shown in Fig. 17. The bands between 11.8 and 18 eV are caused mainly by photoemission of electrons from CH bonds. Because of the large number of CH bonds in stilbene and the vibrational broadening of the single bands, this energy region of the spectrum can no longer be analyzed in detail. In the low-energy region, however, three well-separated bands are observed with an intensity ratio of approximately 1 : 3 : 1. These bands can be correlated to the five highest occupied orbitals of *trans*-stilbene, which are all π orbitals. This situation, where it is possible to assign several bands at the beginning of the PE

TABLE V. Labeling Schemes Used in XPS and X-ray Spectroscopy

XPS	$1s$	$2s$	$2p_{1/2}$	$2p_{3/2}$	$3s$	$3p_{1/2}$	$3p_{3/2}$	$3d_{3/2}$	$3d_{5/2}$
X-ray	K	L_I	L_{II}	L_{III}	M_I	M_{II}	M_{III}	M_{IV}	M_V

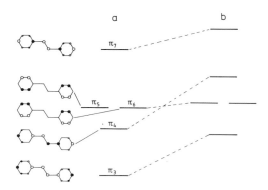

FIG. 20. Orbital diagram for stilbene: (a) ordering as in *trans*-stilbene; (b) ordering as in *trans-p,p'*-dimethoxystilbene.

spectrum of a larger molecule, is usually met in cases where the uppermost occupied orbitals are energetically well separated from the main body of the valence orbitals. Among these special orbitals are usually n orbitals, π orbitals, or the d orbitals of transition metal complexes.

Very often it is not the electronic structure of a single compound that is studied by PES. In most cases it is the change in electronic structure caused by some variations in chemical structure that one wants to understand. This type of investigation is illustrated by the following two examples.

When stilbene is substituted in the para position with an electron-donating substituent such as a methoxy group, the π orbitals are expected to rise in energy, which in turn should lead to smaller binding energies. The influence of a substituent on a given orbital is proportional to the contribution to this orbital of the atom on which the substituent is located. The contribution of the different carbon atoms to the five highest occupied π orbitals of *trans*-stilbene are indicated in Fig. 20. Orbitals π_5 and π_6 have negligible contributions are the para atoms and are expected to be relatively insensitive to para substitution. Comparison of the observed spectra (Fig. 18) to the expected band shifts leads to an unambiguous assignment, which cannot be derived from one of the individual spectra. For example, it is not directly obvious that the shoulder observed in the *trans*-stilbene spectrum at 9.5 eV corresponds to an ionization from π_4. However, if the series stilbene, p-methoxystilbene, and p,p'-dimethoxystilbene is considered, it is clear that the orbitals that have contributions at the para atoms (π_3, π_4, and π_7) are shifted upward by nearly equal amounts. In contrast, the band corresponding to the nearly degenerate orbitals π_5 and π_6 moves very little.

The spectra of our second example are shown in Fig. 19. Cyclohexene shows a well-separated band at the onset of the PE spectrum with a vertical ionization potential of 9.1 eV. This band, which is not seen in cyclohexane, is undoubtedly related to the π orbital of the double bond. In contrast to the bands at higher energies, the π ionization band exhibits vibrational fine structure due to a strong coupling of the ionization process to the C–C stretching vibration of the double bond. In 1,4-cyclohexadiene we find two bands at the beginning of the spectrum (VIPs 8.8 and 9.8 eV) that must be attributed to ionizations from the π system. At first thought, one expects only one band in 1,4-cyclohexadiene, since the two double bonds are not conjugated. The splitting tells us that there is some interaction between the two nonconjugated π bonds. This interaction can be caused either by direct overlap, resulting from the nonplanar conformation of the system, or by an interaction with σ bonds of appropriate symmetry. The first type of interaction is called a "through-space" and the second a "through-bond" interaction. In order to distinguish between these two types of interactions a careful theoretical analysis is necessary.

B. ENERGY BAND MAPPING

The quadratic relation between energy and momentum that holds for a free electron is no longer valid in a crystalline solid, where the electron moves under the influence of the periodic lattice potential. The relation between electron energy and electron momentum along certain directions within the Brillouin zone is called the "band structure" of the solid. Angle-resolved PE spectroscopy is presently the only method that allows experimental determination of the band dispersion (the relation between energy and momentum) of occupied bands that do not lie close to the Fermi energy. This capability is due to the fact that the photon adds only negligibly to the electron momentum. Because of momentum conservation, the momentum of the created photoelectron (p_{PE}) is the same as that of the electron in the initial state band (p_i). The simultaneous fulfillment of energy conservation

$$E_{kin} = h\nu - E_B \qquad (20)$$

and momentum conservation

$$\mathbf{p}_{PE} = \mathbf{p}_i \qquad (21)$$

has consequences that can be understood from the simplified band structure shown in Fig. 21. As usual for band structures, the wave vector $\mathbf{k} = \mathbf{p}/\pi$ instead of the momentum is drawn at the abscissa. To a first approximation the created photoelectron can be treated as a free electron, which yields

$$E_{kin} = \mathbf{p}_{PE}^2/2m = \mathbf{p}_i^2/2m = \pi^2\mathbf{k}^2/2m \quad (22)$$

Thus, selecting a specific kinetic energy corresponds to selecting a specific k value. All processes that lead to the proper momentum must lie on a vertical line defined by this k value. Conservation of energy [Eq. (20)] allows photoemission to occur only in cases where the binding energy with respect to the vacuum level and the kinetic energy of the photoelectron add up to the excitation energy $h\nu$ (see Fig. 21). For a given $h\nu$, photoemission is possible only for selected binding energies corresponding to a point within a given band. When the excitation energy is changed, the simultaneous fulfillment of energy and momentum conservation selects other points within the band and peaks in the energy distribution curve will occur at other binding energies. An example of such a measurement is shown in Fig. 22, where photoemission from a Pt(111) surface is studied in the valence region with excitation energies between 12.7 and 23.9 eV.

The analysis outlined above is complicated by the fact that for kinetic energies up to about 50 eV the created photoelectron is not really a free

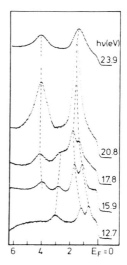

FIG. 22. Energy distribution curves obtained with different excitation energies from a Pt(111) surface under normal emission ($\theta = 0°$). The binding energy is given with respect to the Fermi energy. [From Leschik, G., Courths, R., and Baalmann, A. (1984). Annual report of the Berlin synchrotron radiation facility (BESSY).]

electron. It is excited into a final state whose dispersion may deviate considerably from the parabolic form of Eq. (22). A second type of complication results from the fact that to be measured the electron must escape into the vacuum, and in going from the interior of the crystal through the surface into the vacuum only the parallel component $k_\parallel$ of the momentum is conserved and not the component perpendicular to the surface (see insert in Fig. 23). A variation of the polar angle θ selects different parallel components

$$k_\parallel = (2mE_{kin}/\pi^2)^{-1/2} \sin\theta \quad (23)$$

Consequently, energy distribution curves measured under different polar angles, but using the same excitation energy $h\nu$, strongly depend on θ (Fig. 23). For two-dimensional lattices as they occur in well-ordered adsorbates or in layer compounds the band dispersion is only a function of $k_\parallel$, and the unknown change in the vertical component $k_\perp$ is of little importance in the analysis of the energy distribution curves. For three-dimensional lattices a variety of methods have been developed to overcome the problems resulting from the change in $k_\perp$ (see Himpsel and Plummer and Eberhardt for further details). An experimentally determined band structure of

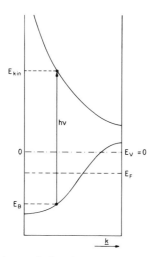

FIG. 21. Scheme of a band structure showing the constraints imposed by simultaneous fulfillment of energy and momentum conservation.

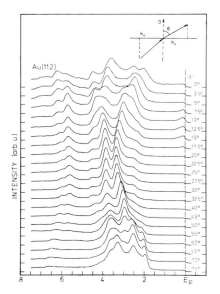

FIG. 23. Energy distribution curves measured under different polar angles θ from an Au(112) surface. The binding energy is given with respect to the Fermi energy. The excitation energy is 16.85 eV. [From Heimann, P., Miusga, H., and Neddermeyer, H. (1979). *Solid State Commun.* **29**, 463.]

copper is shown in Fig. 24. The dashed lines show the results of theoretical predictions. To derive the experimental band structure the polar angle θ and excitation energy $h\nu$ have been varied.

C. Investigation of Adsorbates

When a molecule is adsorbed at the surface of a solid it can be either physisorbed or chemi-

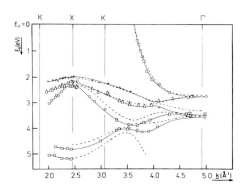

FIG. 24. Experimentally determined band structure along the Σ direction. [From Baalmann, A., Neumann, M., Braun, W., and Radlik, W. (1985). *Solid State Commun.* **54**, 583]. The dashed lines represent theoretical results derived by H. Eckardt, L. Fritsche, and J. Noffke [(1983). *J. Phys. F* **14**, 97].

sorbed. We speak of "physisorption" when the bonding is mainly caused by van der Waals interactions and of "chemisorption" when we have some type of a chemical bond between adsorbate and substrate. However, there is a more or less continuous transition between both types of bonding. Understanding of the bonding between adsorbate and substrate is very important for the understanding of reactions that take place in heterogeneous catalysis. Catalytic processes of this type have great technical importance but often are not fully understood on a microscopic level. Because of its extreme surface sensitivity, PES is a powerful tool in achieving such an understanding. The applications of PES to the study of adsorbates are too widespread to be reviewed here. We show only a single example to give an impression of how these studies can be done.

For a variety of adsorbate systems, such as the adsorption of small molecules (CO, N_2, H_2O, etc.) on transition metal surfaces, a "molecular view" of the bonding seems to give a useful description. In a molecular view we do not look at the substrate as a metal with band structure and all the typical solid-state properties. Instead, we look primarily at a few metal atoms to which the adsorbed molecule is bound. Such a view immediately connects the bonding in an adsorbate to the bonding in a metal complex. Thus, studying adsorbates together with related complexes can contribute strongly to the understanding of the chemisorption bond. Figure 25 shows an example. The C1s and O1s spectra of gaseous CO are shown at the bottom. A weak satellite structure at the high binding energy side of the main line in both cases is blown up in the figure in order to reveal its structure. The indicated satellites are connected with final ion states in which, in addition to the removal of a core electron, a second electron is excited from the π to the π^* orbital. When CO is adsorbed on the 110 surface of tungsten a new satellite is observed closer to the main line. This new satellite increases strongly in intensity with decreasing strength of the chemisorption bond. For the system CO on copper (100) (at the top in Fig. 25), where CO is only weakly bound, this new satellite becomes very strong. What is the nature of this satellite and what does it tell us about bonding? The fact that the same satellite observed for CO on W(110) is also observed in the metal complex W(CO)$_6$ reveals that this satellite cannot be connected with the special properties of metallic tungsten. It must be connected

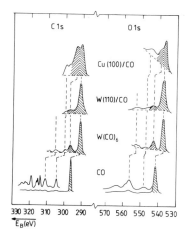

FIG. 25. Comparison of C $1s$ and O $1s$ spectra for free CO, W(CO)$_6$, and CO adsorbed on W(110) and Cu(100). [From Freund, H. J., and Plummer, E. W. (1981). *Phys. Rev. B* **23**, 4859.]

with the local bonding between CO and a tungsten atom. The satellite is connected with a final state where, in addition to the core ionization, an electron is transferred from an occupied d orbital of the tungsten atom to the π^* orbital of CO (charge transfer excitation). A detailed analysis reveals that the intensity of the satellite is a measure of the "backbonding" caused by the overlap of occupied d orbitals of the metal atom with empty π^* orbitals of the ligand. The similarities and differences observed in the satellite structure of the C$1s$ and O$1s$ photoemission line can be used to derive additional information on the chemisorption bond.

D. ATOMIC CHARGES

In Section III,A we showed how core orbital energies can be expressed in the point charge approximation [Eq. (11)]. By combining Eq. (11) with Eq. (10) we obtain

$$\Delta E_B(i) = k(A, i)\, \Delta q_A + \Delta V(q_B) \quad (24)$$

which connects core electron binding energy shifts ΔE_B with variations in atomic charges. The most drastic approximation is to correlate binding energy shifts only with the charge at the atom that holds the orbital. This simple approximation was widely used in the early days of XPS. However, it is an oversimplification because it neglects not only the off-atom potential $V(q_B)$ but also final state effects. A much better result is obtained when we use the method outlined at the end of Section III,B. From the

proper combination of Auger kinetic energies and binding energies we can derive a close approximation to the relaxation contribution ΔR. Combination of Eqs. (16) and (14) allows us to take final state effects into account. We then derive quasi-experimental $\Delta\varepsilon_i$ values that can be used in connection with Eq. (11) and theoretically calculated atomic charges. An example of such a study is shown in Fig. 26. The numbers in the figure refer to the phosphorus compounds listed in Table VI, which shows the measured data (Auger kinetic energy shift and binding energy shifts relative to PH$_3$), the final state relaxation contribution ΔR, and the "experimentally" derived orbital energy shift $\Delta\varepsilon$ for the $2p$ orbital of phosphorus. Nearly all the ΔR's are positive, indicating that in almost all investigated compounds the final state relaxation is larger than in PH$_3$. Only PF$_3$ and OPF$_3$ exhibit small negative relaxation shifts. As suggested in Section III,B, the final state relaxation is largest for the molecules that contain easily polarizable substituents. In Fig. 26 the $\Delta\varepsilon$ values obtained from the experimental binding energy shifts and the experimentally estimated relaxation contribution are compared to the charge at the phosphorus atom and the off-atom potential, both calculated by a semiempirical quantum chemical method. The excellent correlation obtained for this example tells us that the theoretical method used is suitable for predicting reasonable atomic charges.

The energies shown in Table VI were measured for gaseous samples. For solids an addi-

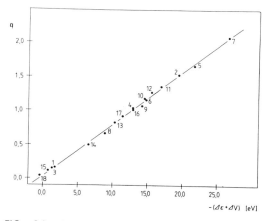

FIG. 26. Correlation between calculated atomic charges and experimentally derived initial state contributions (see text for details). The numbers refer to the compounds listed in Table VI.

TABLE VI. Evaluation of Initial ($\Delta\varepsilon$) and Final State (ΔR) Effects from Experimental Binding Energy (P $1s$ and P $2p$) and Auger Kinetic Energy ($KL_{2,3}L_{2,3}$) Shifts[a,b]

Compound	No.	$(1s)$	$(2p)$	E_{kin}^{Au}	ΔR	$\Delta\varepsilon$
PH_3	1	0.00	0.00	0.00	0.00	0.00
SPF_3	2	6.20	5.15	-2.98	0.56	-5.71
PMe_3	3	-0.98	-1.10	3.30	1.04	0.06
$SPCl_3$	4	4.21	3.57	0.47	1.70	-5.27
OPF_3	5	6.96	5.91	-5.14	-0.14	-5.77
$OPCl_3$	6	4.70	4.01	-0.72	1.30	-5.31
PF_5	7	8.55	7.33	-5.93	0.09	-7.42
PCl_3	8	3.28	2.79	0.39	1.35	-4.14
PF_3	9	5.48	4.72	-4.62	-0.33	-4.39
$(CH_3O)_2PSCl_2$	10	3.60	2.78	0.95	1.46	-4.24
$(CH_3O)_3PO$	11	3.34	2.54	0.11	0.92	-3.46
$(CH_3O)_3PS$	12	3.26	2.39	0.95	1.24	-3.63
$(CH_3O)_3P$	13	1.82	1.19	1.03	0.80	-1.98
CH_3PCl_2	14	1.89	1.53	1.33	1.25	-2.78
PEt_3	15	-1.48	-1.53	4.48	1.45	0.08
$(CH_2Cl)POCl_2$	16	3.77	3.18	0.10	1.34	-4.52
CH_3PSCl_2	17	3.19	2.59	1.44	1.71	-4.30
$P(CF_3)_3$	18	1.74	1.70	1.16	1.41	-3.11

[a] Experimental data from Sodhi, R. N. and Cavell, R. G. (1983). *J. Electron Spectrosc.* **32**, 283.
[b] All energies are in electron volts.

tional problem arises that is connected with the reference problem (Section I,D). For solids the binding energies are usually referred to the Fermi level and not to the vacuum level. The binding energy shifts appearing in Eq. (14), however, refer to the vacuum level. Binding energy shifts referred to the Fermi level and those referred to the vacuum level are equal only when the work function is the same for both samples. If this is not true the work function difference, which in principle is also an experimentally obtainable quantity, must be taken into consideration.

As mentioned at the beginning of this section, these few examples can provide only a glimpse of a very broad field. We hope, however, they can help the reader to gain an idea of how and where PES can be applied.

BIBLIOGRAPHY

Briggs, D., and Seah, M. P. (Eds.). (1983). "Practical Surface Analysis by Auger and X-ray Photoelectron Spectroscopy." Wiley, New York.

Brundle, C. R., and Baker, H. D. (Eds.). (1977). "Electron Spectroscopy: Theory, Techniques and Applications," Vols. I–IV. Academic Press, New York.

Cardona, M., and Ley, L. (1978, 1979). "Photoemission in Solids," Vols. I and II. Springer, Berlin.

Eland, J. H. D. (1984). "Photoelectron Spectroscopy," 2nd ed. Butterworths, London.

Himpsel, F. J. (1980). *Appl. Opt.* **19**, 3964.

Kimura, K., Katsumata, S., Achiba, Y., Yamazaki, T., and Iwata, S. (1981). "Handbook of HeI Photoelectron Spectra of Fundamental Organic Molecules." Japan Scientific Societies Press, Tokyo; Halsted, New York.

Plummer, E. W., and Eberhardt, W. (1982). "Angle-Resolved Photoemission as a Tool for the Study of Surfaces." Wiley, New York.

Rabalais, J. W. (1977). "Principles of Ultraviolet Photoelectron Spectroscopy." Wiley, New York.

Sevier, K. D. (1972). "Low Energy Electron Spectroscopy." Wiley, New York.

Siegbahn, K., Nordling, C., Fahlmann, A., Nordberg, R., Hamrin, K., Hedman, J., Johansson, G., Bergmark, T., Karlsson, S.-E., Lindgren, I., Lindberg, B. (1967). ESCA, atomic, molecular and solid state structure studied by means of electron spectroscopy. *Nova Acta Regiae Soc. Sci. Ups. Ser. IV* **20**.

Turner, D. W., Baker, C., Baker, A. D., and Brundle, C. R. (1970). "Molecular Photoelectron Spectroscopy." Wiley-Interscience, London.

PHOTOGRAPHIC PROCESSES AND MATERIALS

P. S. Vincett   *Xerox Research Centre of Canada*

GLOSSARY

Amplification: Extent to which the light-sensitivity of a photographic process exceeds that of a hypothetical process in which light *directly* produces or destroys the light-absorbing material of an image. Amplification usually occurs in the development step.

Density: In photography, the ability of a material to absorb light, usually as a result of exposure and development. Accurately, density refers to $\log_{10}(I_0/I)$, where I_0 and I are the light incident on, and transmitted by, the material. Often referred to as optical density.

Development: Process of making the latent image visible by providing appropriate energy or matter to the imaging material; this couples to the latent image differently from the rest of the material.

Emulsion: As (incorrectly) used in photography, the active (light-sensitive) layer of the photographic material.

Fixing: Removal of the light-sensitivity of the emulsion after exposure (usually after development and usually by removing the light-sensitive material).

Gamma: Measure of the rate at which the final image density increases or decreases as a function of the exposure. More accurately, gamma is the slope of the nearly linear region that usually occurs in the plot of density versus the logarithm of exposure. A high gamma allows a full-density-range image from a low-contrast input, but (because there is a reduced range of exposures over which good output density variations can be obtained) it tends to reduce the latitude and the ability to reproduce a wide tonal range.

Hole: Object having many of the characteristics of an electron but with positive charge. (This is the sense commonly used in discussions of photographic mechanisms.) See also Section I,B,1.

Latent image: Invisible image-precursor formed by the imaging light; it makes a region of the material developable.

Latitude: Ability of a material to provide an acceptable image when various imaging parameters, particularly the exposure, are varied.

Oxidation: Reverse of reduction, particularly when an element or compound combines with oxygen.

Photoconductor: Insulator or semiinsulator whose electrical conductivity is changed (usually increased) when light is shone on it.

Photon: Minimum packet of light, having some of the characteristics of a discrete particle.

Positive/negative and positive/negative-working: Positive-working process or material renders the lighter and darker objects of an original lighter and darker, respectively, in the image. A negative-working system does the opposite. More correctly referred to as sign-retaining and sign-reversing. A positive is an image whose lighter and darker

ENCYCLOPEDIA OF PHYSICAL SCIENCE
AND TECHNOLOGY, VOL. 10

areas correspond to the lighter and darker areas of a typical original, such as a scene or a document; a negative is the opposite.

Reciprocity failure: Failure of a photographic response to be dependent on the illumination energy but not on the rate at which the energy is supplied. There is then no longer a reciprocal relationship between the exposure intensity and exposure time for a given response.

Reduction: Chemical process of providing one or more electrons to the metal ion in a chemical compound so as to reduce its oxidation state, often returning it to the metal by removing nonmetallic elements. For example, a silver halide can be reduced to silver.

Sharpness: Used in this article to include the various factors related to image definition of a material. These include granularity, a measure of the random density fluctuations in an image, acutance, a measure of the faithfulness with which the edge of an object is rendered, and resolution, a measure of the ability of the material to keep closely-spaced objects separate in the image. These factors are related but do not correlate perfectly.

Silver halide: Compounds or mixed compounds of silver with chlorine, bromine, and/or iodine. Strictly speaking, the term would include silver fluoride, but this compound has little, if any, photographic importance.

Updating: Addition of new images to a photographic material after development and use of the original image(s).

For our purposes, photography may be defined as the creation of a chemical or physical spatial-image pattern by the direct action of a corresponding pattern of visible light (or other radiation) on a sensitive material; the pattern is either visible and more or less permanent or, much more commonly, is made so by a separate development process applied to the pattern. As such, it does not include processes such as electronic printing, in which an electronic information stream is converted to a visible image, even though the electronic signal may have come from scanning a visible image; nevertheless, such processes often use processes and materials similar to those used in certain branches of photography. The present article discusses the processes and materials used in photography, with an emphasis on the fundamental mechanisms, but does not describe the corresponding hardware (cameras etc.), except where this is necessary to understand the process. The creative aspects of photography are also not discussed. While silver halide materials dominate, much more attention than usual will be given to electrophotography (now one of the largest branches of photography), other nonsilver processes, digital optical recording materials, and recent commercial and research directions.

I. Silver Halide (AgX)—Black-and-White

Most commercially available photographic materials employ silver halide (AgX) "emulsions" as the light-sensitive element. A black-and-white (b/w) "emulsion," which is actually a solid dispersion, consists mainly of small AgX crystals ("grains") and a protective matrix, usually gelatin (Fig. 1). The end product of exposure and development of the AgX crystals is an image whose dark areas usually consist of a large number of small grains of silver (Ag). Exposure, in a camera or other suitable device, gives rise to an initial, minute, invisible latent image (LI) in some or all of the exposed crystals, the LI consisting typically of just a few atoms of Ag per crystal. Subsequent development with chemical reducing agents can convert such crystals to a mass of opaque metallic Ag, thus providing amplification factors up to around a billion (10^9). The visible darkening is determined by the exposure and the extent of development and depends on the number of crystals devel-

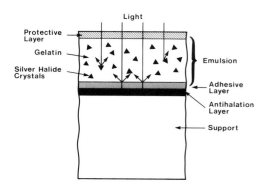

FIG. 1. Schematic cross section of a typical black-and-white AgX emulsion.

oped in a particular area and their size. Nonexposed crystals are left largely unreduced. Typically, the unchanged AgX is then dissolved away ("fixing"), then the emulsion is washed to remove the remaining chemicals and is dried. Since more density is thus usually produced where the original was lighter, the process is usually negative-working. A positive can then be made by exposing another AgX emulsion (often on paper) through the negative and again developing, fixing, and washing. Some materials, however, give rise to a direct positive, and others give a positive by modified processing. In conventional color processes, the developer which is oxidized in reducing the AgX is generally used to form or destroy organic dyes, and the Ag itself is usually removed. Instant color processes typically rely on the oxidation of the developer to cause changes in the mobility of dyes, which are then captured by a receiver layer.

A. Conventional B/W Silver Halide Emulsions

Typical b/w emulsions contain 30–40% by weight (about 7–12% by volume) of AgX in gelatin; smaller amounts of other compounds are added for specific purposes as described later. The AgX crystals are typically mixed ones, usually containing silver chlorobromide or iodobromide, depending on the intended use of the emulsion. The proportion of chloride is higher where rapid processing (reflecting its higher solubility) is important and its lower sensitivity can be tolerated, as in printing papers; more sensitive emulsions use iodobromide with about 1–10% iodide.

1. AgX Crystals

To a first approximation, the AgX crystals act as independent units during exposure and development, and so their size and size-distribution exert a profound influence on the photographic properties of the emulsion. An entire crystal can be developed from one LI speck, which must contain a certain minimum number of Ag atoms; this number (and the number of absorbed photons required to produce it) depends very little on crystal size, at least up to a point. Other things being equal, the number of photons absorbed by a given crystal during a given exposure increases rapidly with its size; the number of incident photons per unit area required to form the minimum LI therefore falls rapidly as

the crystal size increases, so the sensitivity of the material increases. Its sharpness will also fall as the crystal size is increased, although sharpness is also determined by light scattering within the emulsion, which depends on factors such as its thickness. Furthermore, as the spread of crystal sizes is increased, the variation of sensitivity between crystals increases; other things being equal, the gamma of the emulsion is thus decreased and its exposure latitude and tonal range increased. Other factors reducing the gamma include crystal-to-crystal variations in the number of absorbed photons required to render the crystal developable, changes in the intensity of the light as it passes through the emulsion (in which the AgX is usually distributed randomly), and to a small extent even statistical variations in light absorption caused by the random arrival of the photons themselves. The gamma is also affected by the development conditions. The mean crystal size in high-sensitivity emulsions is in the region of 1 μm, while for an ultrahigh-sharpness emulsion the mean may be about 0.05 μm. The spread of crystal sizes is typically of the same order as the mean size. A high-sensitivity emulsion is typically about 7 μm thick, and it may contain roughly 5 × 10^8 crystals per cm^2.

2. Gelatin

Gelatin is a high-molecular-weight natural protein made by the hydrolysis of collagen, which in turn is extracted from the hides and bones of animals, particularly cows. When sufficiently concentrated aqueous solutions of gelatin are cooled, they set reversibly to form gels, from which the water can subsequently be removed; this layer can later be penetrated and swelled by developing solutions, although the swelling can be prevented from becoming excessive, especially at the higher processing temperatures used for rapid development, by the addition of hardeners to induce permanent cross-links between the gelatin chains. The gelatin acts as a mechanical support and also fulfills various other functions: it attaches strongly to the AgX crystals during emulsion-making, permitting their growth while preventing coagulation; during development, it permits chemicals to penetrate the emulsion, but it retards the reduction of the AgX, thus facilitating the selective development of the exposed crystals; it also stabilizes the LI against oxidation (without gelatin, the LI can decay in seconds). It used to serve as a source of necessary trace compounds:

one famous photographic scientist, referring to an important trace compound, remarked that if cows didn't like mustard, we couldn't go to the movies! However, nowadays the gelatin is generally purified as far as possible before use and controlled amounts of trace compounds are then added. Gelatin absorbs water vapor, so the stability of the emulsion depends on the relative humidity, and fungus and bacteria growth can occur under appropriate conditions.

3. Emulsion Production and Additives

The emulsion is formed by mixing solutions of a soluble silver salt and of the appropriate halide(s) in the presence of gelatin. Fine particles of the AgX are precipitated. The precipitation is followed by a ripening stage, during which larger particles grow at the expense of smaller ones, and the average particle size increases; this occurs principally because the solubility of small particles is larger than that of large ones, owing to the significant surface energy of the former. If the composition of the bulk solution is held fairly constant via continuous addition of the original solutions (double-jetting), a quasi-equilibrium situation can arise and all the AgX crystals grow to a similar size. If one solution (usually the silver salt) is added continuously to the other (single-jetting), however, conditions change throughout the precipitation, a range of crystal sizes is formed, and the final emulsion will have a lower gamma. After removal of by-products, the emulsion is treated with minute amounts of sensitizers and other additives such as coating agents, antistatic materials, hardeners, plasticizers, stabilizers, disinfectants (to prevent bacterial growth), and development modifiers; it is then coated onto a base and dried. A more sensitive layer may be coated over one or more less sensitive ones, so that reasonable exposure latitude is obtained; alternatively, the high- and low-sensitivity components may be blended. Subsequent aging leads to significant sensitivity changes for at least several days; in fact, it is said that one cannot predict the exact characteristics of a given emulsion in advance of this aging process.

4. Emulsion Substrates

The most common film substrates (bases) are cellulose triacetate and polyester, typically a few thousandths of an inch thick; the latter is used principally when dimensional stability or high strength are important. An adhesive sub-coating is used between the emulsion and the base, and an anticurl layer may be placed on the rear of the base. Protection is often provided against halation (reflection at the base/air interface of light that has passed through the emulsion, leading to a spurious image where it reenters the emulsion). This can be done by coating one side of the base with a light-absorbing material or by incorporating it in the emulsion; it is removed or bleached during processing of the film. If an increase in background density is acceptable, a gray base may be used instead. An antistatic layer and an abrasion-resistant topcoat may also be incorporated. A paper substrate (with high wet-strength and appropriate chemical inertness) is generally used when opacity or low cost is needed; barium sulfate (baryta) in a gelatin binder may be applied first to improve opacity and brightness, and colorless blue-fluorescent brighteners may also be added to the paper to improve "whiteness." Recently, "resin-coated" (RC) paper, which is coated on both sides with a thin layer of polyethylene, has gained wide acceptance, primarily because it absorbs far less of the developing solutions and thus reduces the processing time, making machine processing rapid and easy. There is also much less tendency for the paper to tear when wet; furthermore, the curl of the paper can be controlled by the polyethylene, the dimensional stability is improved, and desired gloss levels can be obtained without separate postdevelopment processes. Pigments, such as titanium dioxide, in the polyethylene can be used to control the background color and brightness and may also improve the image sharpness. Finally, when extreme dimensional stability or flatness is required, glass-plate substrates are used.

B. Latent Image (LI) Formation

On exposure of the emulsion to low-intensity light such as that transmitted by a camera lens, an LI consisting of a very minute speck of metallic Ag, containing just a few Ag atoms, is formed in some or all of the exposed AgX crystals. This LI is generally located on the surface of the crystals; although it may contain only 1 part in 10^7–10^{10} of a crystal's Ag, it renders the whole crystal developable, that is, during subsequent chemical treatment, reduction of the crystal to metallic Ag (commencing at the Ag speck) is significantly faster than it is in the absence of the LI. The number of photons that must be absorbed by a crystal to make it developable is of the order of 15 for sensitive emulsions, although a significant portion of the crystals may be made developable by four photons each; when certain special sensitizers are used under special conditions, every crystal may be made developable by only two or three absorbed photons each. However, the sensitivity of AgX is limited by the fairly low absorption (of order 10%) of incident photons by each small crystal; furthermore, the sharpness is reduced by the scattering which can therefore take place off of the crystals before a photon is absorbed.

Remarkably, the most basic mechanisms of LI formation are still highly controversial and the subject of much research. We shall describe what is perhaps the most widely held theory and later mention the other proposed mechanisms that are still current. First, however, we must give some background.

1. AgX Electronic Structure

Silver halides are ionic crystals consisting of a regular cubic lattice of Ag and halide ions together with a small proportion of defects, such as Ag ions that have been displaced from their regular lattice position to another "interstitial" position (the Ag ions are much smaller than the halide ions), and the corresponding vacancy in the lattice. Although the lattice itself is fairly rigid, such interstitials and vacancies are fairly mobile, because they can jump through the crystal, one lattice spacing at a time, without other ions having to move substantially to make room for them. When ions are brought together in a crystal, their sharply defined electronic energy levels, corresponding to the energy states of the outer-shell electrons, are broadened by the interactions between them into a series of bands; electrons within these bands are no longer each localized on one ion, belonging instead to the crystal as a whole.

In AgXs, all the energy levels of the highest occupied energy band (the valence band, VB) are occupied by electrons. Although the electrons are delocalized, any increase of one electron's momentum under the action of an electric field must be balanced by a reduction of momentum of another one, because no spare energy states are available. Thus, in a perfect crystal, essentially no electronic conduction occurs. The next higher energy band (the conduction band, CB) is essentially empty in a perfect AgX crystal, being too far (2.5 eV or so) above the VB for significant thermally excited transitions to take

place from the VB. In a perfect crystal, there are no energy levels in the gap between the VB and CB, but in practice there are a few such levels associated with crystal defects.

When a photon is absorbed by an AgX, there is a high probability of formation of an "electron–hole" pair: an electron is raised from the VB to the nearly empty CB, in which it can move more freely, while the "hole" is a way of conceptualizing the behavior of the previously full VB from which the one electron has been removed. Since the depleted band of electrons responds to an electric field, for example, like the full band less one electron, and since the full band gives no net response, the overall effect is like that of minus one electron moving in the VB. This "hole" can thus be considered as somewhat like a positively charged electron. Such electron–hole formation occurs in all photoconductors upon light absorption, but in most cases the electrons and holes recombine rapidly in the absence of an external electric field; in AgXs, however, the combination of electrons with Ag ions to form Ag atoms can dominate over recombination. Electron–ion combination cannot occur if both the electron and ion are mobile, since the total energy of the ion–electron system would make them separate again, nor can it occur if the Ag ion is in a lattice position, since there is not enough energy to overcome the force holding the ion in the lattice. However, it can happen if the electron or ion is at a trap; an electron-trap, for example, is a position in the crystal lattice (usually associated with a defect) where the electron's energy is lower than it is when free and where it may consequently reside for a time.

2. Step-by-Step Mechanism

One of the most important points that any theory of the LI formation process must explain is how it is that photons that are absorbed throughout each AgX crystal become concentrated, in the sense that the Ag formed from them becomes concentrated in one or just a few LI specks. The most widely accepted theory is the "step-by-step" mechanism, which had its origins with Gurney and Mott as long ago as 1938. The electron is considered to move through the crystal in the CB until it reaches a trap, where it remains for long enough that a mobile Ag ion can move to it because of the attraction of their opposite charges, and combine with it to form an Ag atom. Once this happens, the same thing can occur again at the same place, until a tiny speck of Ag is built up to form an LI. In fact, a stable Ag nucleus of more than a certain size is itself a deep electron trap, so beyond this size further electrons are even more likely to be trapped at the growing nucleus rather than elsewhere. Thus, the mobility of the electrons and their subsequent trapping provide the concentration mechanism. In practice, the all-important trap or "sensitivity center" is usually associated with areas of the surface of the crystal that have been chemically sensitized as discussed below, so that usually the LI is formed primarily at the surface. In fact, unsensitized crystals are fairly insensitive, especially when conventional developers (which do not dissolve the AgX and therefore act primarily on surface LIs) are used. Of course, there may be more than one such trap per crystal, although (since one LI center can render a whole crystal developable) it is clearly more efficient to have as few LI's as possible.

It is important to note that one Ag atom, formed as above, is not completely stable against redissociation to the ion and electron, and in fact there is a minimum size (probably two atoms) of the Ag speck for stability (as opposed to developability which requires further growth). Thus, the probability of formation of an LI depends not only on the total number of photons absorbed but also on the time between absorption of successive photons: if the time is too long, the chance of two Ag atoms being formed at the same place at the same time drops. This gives rise to low-intensity reciprocity failure, an effect well-known to anyone who has attempted photography in very low light levels, especially with color films, and particularly to photographers of astronomical objects. In addition, especially in the absence of chemical sensitization or in the presence of oxygen or moisture, electron–hole recombination is a significant factor limiting the photographic sensitivity. It is not entirely understood why such recombination is generally less important with AgXs than with most other materials, but suggested mechanisms include trapping of holes at certain chemical impurities, and complexing of the holes with Ag ion vacancies (which are effectively negatively charged). There is also a high concentration of interstitial Ag ions near the surface of the crystals, probably related to imperfections there. Support for many aspects of the above theory comes from the change in physical position of the Ag image when electric fields, and even combinations of electric and magnetic fields, are applied to the

AgX at various times relative to the light exposure.

3. Chemical Sensitization

In practical emulsions, the sensitivity and color response are usually greatly increased by sensitizers, compounds that are added during emulsion-making and are adsorbed on the surface of the crystals. "Chemical sensitization" produces a major decrease in the number of absorbed photons needed to make a crystal developable and can greatly decrease low-intensity reciprocity failure. Care must be exercised, however, since too much sensitization can cause the formation of chemical nuclei that can act like LI during development, giving rise to fog (spurious density in nonexposed areas).

The three most common forms of chemical sensitization are based on reaction of the AgX with very small quantities of sulfur compounds, salts of gold or other noble metals, and reducing agents. The premier commercial technique for high sensitivity is a combination of the first two. The sulfur compounds react with the AgX to form some silver sulfide on the surface of the crystals; while catalytic mechanisms have been proposed and enhanced hole trapping is also possible, it is generally believed that sulfur sensitization acts by formation of electron traps, thus diminishing recombination, perhaps by increasing the electron trap depth of a single Ag atom. Moreover, it may stabilize single Ag atoms somewhat against redissociation and leads to less formation of LI in the interior of the crystal where most developers cannot reach it. If too much sulfur is added, the sensitivity decreases again; this has been attributed to formation of multiple LI centers, which is an inefficient use of photons. Gold sensitization causes gold to become part of the LI (by formation of a mixed silver–gold cluster, probably during exposure but possibly on immersion in the developer). This decreases the size at which the LI becomes developable: a three-atom (or possibly two-atom) gold-containing cluster is developable (e.g., Ag_2Au or $AgAu_2$), compared with four Ag atoms. Gold sensitization may also improve the stability of Ag atoms and increase the depth of associated electron traps. Reduction sensitization is produced by heating the emulsion with reducing agents such as stannous salts; they are believed to act by forming very small Ag centers. While these could act as electron traps and thus constitute LI nuclei, it is believed more likely that they capture photogenerated holes, reducing the hole concentration and hence recombination. Indeed, if a photon produces an electron and a hole, and the hole attacks a reduction center consisting of two Ag atoms, the Ag_2^+ first formed will dissociate spontaneously into two Ag ions and another electron:

$$Ag_2 + (hole)^+ \rightarrow Ag_2^+ \rightarrow 2\ Ag^+ + e^-$$

The net result is two electrons for one photon!

4. Other LI Formation Theories

One alternative viewpoint of the LI formation process considers that single Ag atoms are formed separately by electron–ion combination, leading to the AgX becoming supersaturated with Ag. This Ag then precipitates (starting as a two-atom cluster) and grows as in any supersaturation–nucleation situation, the required diffusion (probably of single Ag atoms) in this case occurring on the AgX crystal surface or possibly under the action of crystalline elastic energy forces. In this viewpoint, the development process (which we discuss later from a more conventional perspective) is also viewed as a supersaturation situation: the reducing agent in the developer maintains a supersaturation of Ag in the crystals, since it provides electrons to push the equilibrium

$$Ag \rightleftharpoons Ag^+ + e^-$$

toward the left-hand side; this causes precipitation (development) onto the existing Ag nuclei if they are large enough to be stable under these conditions. In this model, sensitivity centers (e.g., silver sulfide or sulfide plus gold) are considered to act as foreign condensation centers for the Ag supersaturation produced by light exposure, decreasing the energy required for Ag nucleation.

Another alternative viewpoint objects to earlier theories on the basis of trap-depth calculations. To overcome this problem, with sulfur sensitization, for example, a large number of Ag_2 molecules is considered to be present around the edges of silver sulfide monolayer islands on the crystals. Ag_2 can trap holes, leading as above to the formation of two Ag ions and a second electron. The electron and an Ag^+ ion can recombine at another Ag_2 to form Ag_3. This can then absorb an Ag^+ ion to form Ag_4^+, which now provides a deep trap for other photoelectrons. After formation of Ag_4, another ion can be captured, followed by another electron, and so on. Thus, a center that gets this far continues

to grow and can concentrate all the Ag. Once the size of the nucleus is Ag_5^+ or larger, the speck can be shown to be stable in aqueous environments, and developers can therefore continue the growth process (i.e., the speck becomes developable).

Even now, new mechanisms of LI formation and development, either combinations of existing proposals or new ones such as a recently proposed mechanism based on the redox potential of the AgX crystals, are still being introduced.

C. Color Sensitization and Hypersensitization

1. Color Sensitization

A pure or chemically sensitized AgX emulsion absorbs significantly (and thus produces electron–hole pairs) only up to the wavelength of its absorption edge, corresponding to the gap between the valence band (VB) and the conduction band (CB). For silver bromide and silver chloride, the absorption falls to very low levels by about 490 nm (with a slight extension for iodobromide) and 420 nm, respectively, leading to sensitivity only in the blue region. To extend the sensitivity to other regions, emulsions are often doped with dyes that absorb in the wavelength region desired. Emulsions sensitive from the blue to the green–yellow are often termed orthochromatic, while those with reasonably constant sensitivity throughout the visible region (~400–700 nm) are said to be panchromatic. Good sensitivity out to the 900-nm region of the near infrared can be obtained, and some response to around 1300 nm. (Beyond this, either the stability of the emulsions becomes very low or indirect methods must be used, e.g., detection by an electrical infrared cell, which then modulates a scanning light source.)

Sensitizing dyes are usually from the polymethine class, characterized by the presence of either the $=N-$ or $=CO$ group in a conjugated electron system and most commonly part of a heterocyclic nucleus. The three major subclasses of these dyes are the cyanines, the merocyanines, and the oxonols. These dyes have high absorption coefficients, a wide range of structures and absorptions, and are easily purified. Generally, the sensitizing dye molecules are effective only when in close contact with the AgX crystals. The absorption of appropriate visible (or near-infrared) radiation by the dye results in the promotion of an electron from the highest filled electronic level of the dye molecule to one of the lowest unoccupied levels. (Vibrational and rotational states in the dye combine with the main electronic transition to produce a broad absorption band rather than a single-wavelength absorption.) As explained below, the result is an electron in the CB of the AgX that then undergoes the same processes we have described for directly produced photoelectrons. While this effect increases with increasing light absorption by a given dye, the dye may also desensitize the emulsion by acting on some of the secondary processes of LI formation; since the increase in absorption can become slower than the increase in desensitization, there is an optimum dye concentration, which generally corresponds to less than a complete molecular monolayer coverage of the crystals. This is very important, since it limits the sensitivity obtainable. Because of this, dye-enhanced sensitivities tend to be proportional to the area of the crystals, while intrinsic sensitivities (which depend on bulk absorption by the AgX) tend to be proportional to crystal volume. Both proportionalities tend to saturate at about 1 μm crystal diameter: above this, imperfections in the crystals tend to stop them from acting as one unit.

2. Color Sensitization Mechanisms

The mechanism of sensitization is generally believed to be direct transfer of an electron (Fig. 2a) from the excited electronic state of the dye into the CB of the AgX (by quantum-mechanical tunneling through the energy barrier between them and, if necessary, by slight thermal excitation), possibly followed by regeneration of the dye by transfer of an electron from a defect state in the AgX bandgap, leaving a trapped hole. As would therefore be expected, there is a reasonably good correlation between the energy levels of otherwise similar dyes and their sensitizing efficiency. Another possible sensitizing mechanism is indirect transfer of the excited electron by an energy transfer mechanism: here the excitation energy of the excited dye is transferred by resonance to a defect state lying in the bandgap of the AgX, promoting an electron into the CB as before (Fig. 2b). This mechanism has been shown to operate almost certainly under special circumstances but is generally assumed to be of secondary importance in practical situations.

Because of local variations in the relevant energy levels, sensitizing dyes could also desensi-

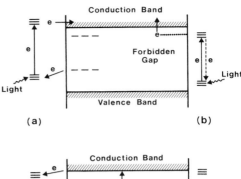

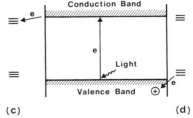

FIG. 2.   Energy levels of AgX and of color-sensitizing dyes, showing sensitization by (a) electron- and (b) energy-transfer, and desensitization by (c) electron trapping and (d) hole trapping. "*e*" represents electron pathways.

tize when their excited energy level is below the CB of the AgX, causing them to behave as an electron trap (Fig. 2c); this would possibly allow the electron to combine with oxygen and prevent it from taking part in the formation of metallic Ag. Also, a dye may trap a hole (Fig. 2d) by transferring an electron from its ground state to a hole in the VB, or (if the ground state of the dye is near or below the VB) an electron may be injected from the VB to the photoexcited dye, creating a hole in the VB. Such trapped or free holes may then recombine with photoelectrons. On the other hand, the additional electron captured from the VB by an excited dye (whose excited energy level is normally too low for electron transfer into the CB) may change the relative energy levels of the dye and the AgX enough to allow such transfer.

3. Supersensitization

A second sensitizing dye adsorbed on the crystal may have a major effect on dye sensitization, and the improvement may be greater than merely an additivity of sensitizing effects. Such "supersensitization" can arise from increased absorption caused by a change in the structure of the dyes on the surface, but the more important form involves an actual increase of the efficiency of transfer of electrons to the CB. A number of theories have been proposed for this. For example, transfer of an electron from the second dye to the first, so that the excited electron and the corresponding hole are separated, may reduce the possibility of recombination and may also change the energy levels favorably. The supersensitizing dye may also act as a discontinuity in the dye layer, trapping the excitation long enough at a given point to give time for transfer to the crystal. In addition, the supersensitizer may provide an energetically favorable site for transfer because of electrostatic interactions with the first dye.

4. Hypersensitization. Number of Photons Required per Crystal

For applications such as low-light-level astronomical photography, hypersensitization procedures are often applied to the emulsion, which is exposed shortly thereafter to avoid the formation of fog. Such procedures can greatly reduce low-intensity reciprocity failure as well as increasing overall sensitivity, and can allow the recording of images from light that is one-millionth of the intensity required to be visible to the naked eye. Hydrogen, for example, has been used as a reduction hypersensitizer on sulfur-plus-gold sensitized emulsions. In favorable circumstances, *all* the AgX crystals have then been made developable by absorption of only two or three photons each (a higher *average* number of photons was needed because of the random statistics of photon impingement on the crystals, but half the crystals were developable for an *average* of only two absorbed photons each, and 10% were developable for an average of only one photon each). This is to be compared with about 15 photons for 50% developability in one sulfur-plus-gold sensitized silver bromide emulsion that was studied under more typical conditions, about 50 for sulfur sensitization alone, and about 150 for the unsensitized emulsion. Part, but not all, of such sensitivity increases is due to initial vacuum-outgassing of the emulsion, since oxygen or moisture causes significant desensitization of AgX emulsions, especially at low exposures; this is probably because of combination of electrons with the oxygen, followed by reaction of the O_2^- with a hole or with water. Other forms of hypersensitization include soaking the emulsion in water, ammonia, or silver nitrate solution. These treatments are believed to work by mechanisms involving increased Ag ion con-

centration in the vicinity of the crystals. The sensitivity gains can be spectacular: one infrared emulsion increased in sensitivity by a factor of 18,000.

5. UV and X-Ray Sensitivity

For decreasing wavelengths below that of the absorption edge of the AgXs, the practical limitation on sensitivity is absorption by the gelatin, and below 200 nm special emulsions must be used; they may be very low in gelatin, may have AgX crystals projecting above the top surface of the gelatin, or may contain materials that give longer-wavelength fluorescence. Gelatin becomes transparent again in the soft X-ray region. The electrons first liberated by an X-ray photon (or by a high-energy particle) have too great a velocity to be trapped and to combine with Ag ions. Their collisions with ions in the crystal, however, release secondary electrons of decreasing velocity. For energies of the order of tens of kilovolts, the secondary electrons do not leave the crystal, but it becomes developable. Higher energies may expose other crystals. However, the sensitivity in terms of *incident* energy or incident photons is low, because of the weak absorption of X-rays in the emulsion and the inefficient use of their energy. AgX emulsions can be used directly for some X-ray applications, but when a minimum X-ray dose is required (as in medical applications), the film (emulsion-coated on both sides) is usually sandwiched between two screens that absorb the X-rays efficiently and fluoresce in color regions to which the emulsion is highly sensitive. Alternatively, the screen can be photographed using a camera. Medical films need high gamma, because the contrast of the X-ray image is low; various contrast-enhancement techniques can also be used. Recently, xerography has been used for certain medical applications, because of its ability to enhance the edges of faint lines in the image (Section VI,A,2).

D. Special Exposure Effects

The mechanisms discussed above can be used to interpret a number of interesting exposure effects observed in AgX. Low-intensity reciprocity failure, as previously mentioned, is largely due to dissociation of Ag atoms before two atoms have time to be formed at the same site to make the Ag nucleus stable. Another cause is slow diffusion of halogen to surrounding crystals, where it attacks image Ag. Reciprocity failure can also occur at high intensities, because the probability of two photons being absorbed within the lifetime of an Ag atom becomes high; stable nuclei can then be formed at relatively shallow traps, so that multiple, competing nuclei are formed within a single crystal, and each nucleus is less likely to grow into a developable size during a given exposure.

A wide range of other exposure effects can occur. Uniform (i.e., nonimagewise) exposure of the emulsion close to the threshold of developability, before, during, or after the image exposure, can produce a simple additive effect: the effective sensitivity of the emulsion is increased for the imagewise exposure, although there is also a reduction of gamma. The effects of multiple exposure may also be better than additive, especially if one exposure is to a brief flash of high-*intensity* (i.e., bright) light. For example, a high-intensity uniform exposure can significantly increase the effect of a subsequent low-intensity imagewise exposure ("hypersensitization," a term also used for the chemical treatments mentioned previously); similarly, a low-intensity uniform exposure can increase the densities formed from a previous high-intensity imagewise exposure ("latensification," again a term also used for certain chemical treatments). Both techniques work because nucleation of a stable Ag speck (smaller than that required for developability) is most efficient at high intensities (because unstable single Ag atoms are more quickly stabilized by a second atom), while growth of the stable nucleus is efficient at both high and low intensity. The latter technique can be used without chemical development to produce direct Ag ("print-out") images from high-intensity oscillograph traces.

Very high-intensity exposures followed by low-intensity ones can also cause less-than-additive effects. In the Clayden effect, the first exposure desensitizes the emulsion for the second one, while with higher intensities still, the second exposure can destroy the image produced by the first (the Villard effect). The Clayden effect is believed to occur because the first exposure causes such a high concentration of electrons that many Ag nuclei (many too small to be developable) are formed at shallow traps, many of them internal to the crystal; these then compete with the surface for electrons liberated during the second exposure. The Villard effect probably arises from the large number of internal nuclei formed during the first exposure: the second exposure increases their size, releasing

halogen, which reacts with and destroys the surface Ag.

In the Herschel effect, long-wavelength light, which is not absorbed by AgX but is absorbed by the Ag LI, can cause effective destruction of the image: the Ag is dissociated into ions and electrons, and when the ions are subsequently again reduced, they may be in the interior of the crystal where conventional developers are ineffective; under some circumstances, rehalogenation (somewhat analogous to the Villard effect) has also been implicated. A closely related effect can be used on a deliberately prefogged image to give a "reversal" or positive-working process. In "solarization," the developed density formed as a function of very high exposure goes through a maximum and then starts to reduce; this is attributed primarily to attack of the developable surface Ag LI by halogen liberated via formation of internal LI. At similar exposures, printout Ag formation can occur. The term "solarization" is sometimes applied incorrectly to the Sebatier effect, in which partially positive images are produced for pictorial purposes by partial development followed by reexposure to uniform light and redevelopment. The reversal may be partly caused by the screening action of the first image during the second exposure, but more complex mechanisms are also involved.

E. Development

During development, exposed crystals are chemically reduced to Ag much more rapidly than nonexposed crystals, while the developing agent is oxidized. Since more silver is usually produced at exposed crystals, most b/w development is negative-working. Dye image formation, which is necessary for color photography, depends on the oxidation of the developing agent and will be discussed in a later section.

1. Developers

Most of the known developers are organic compounds, although some use certain nonmetallic inorganics or metal ions that can be converted to a higher oxidation state by Ag. Most are aromatic, usually derivatives of benzene, and are typically phenols or amines (or simple derivatives) with at least two hydroxy groups or two amine groups, or one of each arranged ortho or para to each other (adjacent or opposite, respectively) on the benzene ring. Examples of such developers are hydroquinone (1,4-dihydroxybenzene), N-methyl-p-aminophenol sulfate (Elon, Metol), Amidol (the dihydrochloride of 2,4 diaminophenol), and derivatives of p-phenylene-diamine. Other important organic developers are the reductones (including ascorbic acid) and the pyrazolidones (including 1-phenyl-3-pyrazolidone or Phenidone). Combinations of hydroquinone with Metol and with Phenidone are particularly important.

In addition, practical developers usually contain (i) an alkali (commonly a hydroxide or carbonate of sodium or potassium, or sodium metaborate or tetraborate) to adjust and maintain the hydrogen ion concentration (acidity or pH) to a value appropriate for sufficient development activity; (ii) an easily oxidized sulfite, to react with oxidized developing agent (maintaining the reaction kinetics reasonably constant and removing unwanted colored products that may cause staining) and to act as a preservative against oxidation of the developer by air; and (iii) one or more antifoggants, such as potassium bromide and certain organic heterocyclics, that retard the development of nonexposed regions of the emulsion more than that of exposed regions and thus improve the discrimination between the two.

The sensitivity of many developers to pH is very important, since in some situations it can be used to "turn on" the development when required. Many developers have one or more ionizable hydroxy groups and become far more active when this group is ionized. For hydroquinone, for example, the equilibrium between nonionized and ionized forms, and the subsequent reduction of the silver ions of the AgX, can be represented thus:

$$C_6H_4(OH)_2 \rightleftharpoons C_6H_4O_2^{2-} + 2H^+$$

and

$$C_6H_4O_2^{2-} + 2Ag^+ \rightleftharpoons C_6H_4O_2 + 2Ag$$

or, overall,

$$C_6H_4(OH)_2 + 2Ag^+ \rightleftharpoons C_6H_4O_2 + 2Ag + 2H^+$$

The large concentration of OH^- ions present in an alkaline solution tends to remove the H^+ ions (forming water), thus pushing the equilibria towards the right and favoring development.

2. Types of Development

If, as is most common, essentially all the Ag for the growing image comes from the AgX crystals, the process is often called "chemical" development; most of such development is from the "direct" process in which the Ag ions are reduced directly at the Ag/AgX interface. If the developer contains a soluble silver salt to supply Ag to the image, this is called "physical" development, a misnomer since the process is still chemical. Physical development as such is of little practical importance, but "solution physical development," in which Ag ions from exposed or nonexposed AgX crystals pass into solution and are then reduced elsewhere at an LI or a growing visible image, is very important in some processes and plays some part in normal "chemical" development. The balance between these various processes can be changed by the concentration of AgX solvents, such as sodium sulfite or sodium thiosulfate, in the developer.

3. Mechanisms

In conventional negative-working development, reduction of the exposed crystal starts at the one or more LI nuclei on the crystal surface and proceeds until the entire crystal is reduced or the developer is removed. The electron microscope reveals, however, that the final black Ag image usually does not have the same shape as the crystal, but rather is a tangled mass of 15–25-nm-diameter filaments of Ag, roughly resembling a piece of steel wool. Very fine crystals may give only one or a few filaments each; such images tend to be somewhat colored, because of wavelength-dependent light scattering. Exposed crystals are reduced to Ag much more rapidly than nonexposed crystals, because the LI nuclei in the former act as catalytic centers for reduction of further AgX (i.e., it is easier to form Ag where Ag already exists). The new Ag in turn accelerates the reduction of further Ag ions, and the process of reduction thus accelerates as an autocatalytic reaction. The presence of Ag or LI has been shown to reduce the activation energy for the reduction process by several kilocalories per mole, but the exact mechanism of the catalysis is not fully understood. An electrochemical mechanism has been proposed, in which the growing Ag nucleus (or the LI) acts as an electrode to permit both an anodic partial process, in which the nucleus (Ag_n) gains an electron:

$$\text{developer)} + Ag_n \rightarrow$$
$$\text{(oxidized developer)}^+ + Ag_n^-$$

and a cathodic partial process, in which Ag ions are neutralized at the nucleus:

$$Ag^+ + Ag_n^- \rightarrow Ag_{n+1}$$

A second possible mechanism is adsorption catalysis, in which one or more of the reactants is first adsorbed by the Ag LI or at the Ag/AgX interface. An intermediary complex is formed between the developing agent and the Ag ion. This complex interacts with the Ag (or the adsorption site on the Ag surface may form part of the complex), and the complex is distorted, lowering the activation energy required for the reduction reaction. Reaction then occurs by electronic rearrangement, and the complex decomposes to Ag and oxidized developer. Of course, adsorption could also be involved in the electrochemical mechanism. Adsorption by Ag, AgX, or both has actually been demonstrated for many developers. It is possible that the mechanism changes from one to the other during development, especially since a very small LI might not be able to act as an electrode in the early stages. It has also been suggested that, in the very early stages of development, developers may be able to transfer an electron indirectly to the conduction band of the AgX, giving rise to growth of the LI just as if the electron had been produced by light absorption.

The formation of Ag filaments is also not fully understood, but elongation of the initial spherical nuclei may start when new Ag atoms do not have time to diffuse far enough to maintain a spherical form. This elongation may be extended by the electrostatic repulsion of electrons in the Ag to the ends of the spheroid, giving rise to a substantial electric field in the AgX and concentrating further growth at the ends. Physical development often gives rise to much less filament formation than chemical development.

4. Rate of Development

The rate of development (and of subsequent processing) of the exposed crystals is an important practical consideration and is influenced by the rate of the primary processes, the rate of swelling of the gelatin when it is immersed in the developer, and the rates of diffusion of the processing chemicals in the swelled gelatin. Rates are increased by agitation of the solutions and by increase of temperature. Too high a temperature, however, can cause excessive swelling of the gelatin and consequent physical damage to the emulsion; this and subsequent damage can be decreased by hardening (crosslinking) the gelatin during manufacture or processing. Alternatively, a high concentration of a salt that restrains swelling, such as sodium sulfate, can be added. Certain development accelerators can increase the rate of development, sometimes by neutralizing the negative charge found on AgX crystals to permit the easier access of negatively charged developer ions. Some combinations of developers (e.g., hydroquinone with Metol or Phenidone) show superadditive rates of development. In these cases, one agent is active as the undissociated molecule or as the singly charged ion and is therefore not too affected by the charge on the crystal; it also typically has a chemical affinity for the AgX and/or the Ag and adsorbs on it. This material therefore rapidly initiates development, forming a fairly stable oxidation product that probably remains adsorbed. The second developer is doubly or triply charged in its most active form when used alone but can rapidly reduce the oxidation product, regenerating the first developer and removing any tendency for the oxidation product to retard development. In effect, the first developer finds the LI and acts as a "wire" for electrons from the second developer.

The effective sensitivity of AgX emulsions can be increased by "pushing" (developing for longer times, at higher temperatures, or with more active developers). However, such processes are limited by fog formation, by changes in the characteristics of the emulsion (for example in the gamma), and by large increases in the graininess of the image.

5. Fine-Grain Development

When high magnification will be used to view or print the finished image, high sharpness is required, and special "fine-grain" developers can be used. These contain AgX solvents such as sodium sulfite and use low to medium rates of development, so that solution physical development is significant. Fine-grain development results in a decrease of the average size of the Ag grains, perhaps by separating several LI centers from the AgX crystal and allowing them to grow separately by solution physical development; more importantly, there is a decrease in the size and probability of formation of randomly distributed clumps of Ag grains. Fine grain is particularly important for microfilm applications, in which documents or computer-output displays are photographed at reduction ratios of typically about 50; this is still (and will probably remain) a widely used and inexpensive way to store large quantities of data.

F. Post-Development Processing

1. Fixing and Washing

After development, other processes are usually employed to render the image stable, and rinsing and washing steps may be used at various points. Preferably the image is first rinsed in water or in a stop bath or clearing bath to lower the pH in the emulsion (make it less alkaline) and to remove developer chemicals, including oxidized developer, which can cause staining, especially in color materials. The material is then immersed in a "fixing" bath to dissolve out undeveloped AgX. This bath usually contains sodium thiosulfate ("hypo"), or sometimes ammonium thiosulfate for faster results. It may be buffered acidic to neutralize any alkaline developer carried into it, in which case a preservative (e.g., bisulfite) can also be added to retard decomposition of the thiosulfate to bisulfite and sulfur. In recent years, with the increasing price of Ag and more emphasis on pollution control, there has been increasing interest in recovering the Ag from the fixing bath, which can also increase the bath's useful life. Finally, washing removes fixing-bath chemicals and soluble Ag ion complexes, and the material is dried as uniformly as possible.

2. Image Stability and Modification

If residual thiosulfate is not removed sufficiently, the image will be more or less unstable. In fact, the likely keeping (archival) properties of a given Ag image are determined by measuring this residue. If washing is incomplete, thiosulfate or its decomposition products can react with finely divided Ag, or silver thiosulfate complexes already adsorbed to the image Ag can decompose to give yellow or brown silver sulfide; thiosulfate can also enhance atmospheric oxidation of the Ag. Complete washing of a film can take as long as 30 min, and porous-paper prints (as opposed to those using the more recent resin-coated papers described earlier) can take much longer. Washing can be speeded-up by initially using salt solutions to displace the thiosulfate. Even after proper fixing and washing, Ag images can be physically or chemically unstable under some environmental conditions.

Various chemical treatments can be applied to the final image to change its color, or to increase or reduce its density, or to improve stability (e.g., by converting the image Ag to the sulfide or selenide). Density-enhancement techniques, such as using Ag-catalyzed dye formation reactions, may become more important in the future with the growing need to conserve finite supplies of Ag. In special circumstances, Ag images have even been rendered radioactive, and an image with enhanced density has been formed by the absorption of radioactive emanations in an adjacent emulsion or glass plate.

3. Modified Development Processes

"Monobaths" combine the development and fixing steps in one operation; competition between development and fixing then occurs, and the final image depends on the relative rates of the two processes. Photographic papers often contain the developer and are activated by alkaline solutions, but need normal fixing and washing steps. When images are required in a minimum time, and reduced permanence is acceptable, fixing and washing are sometimes replaced by a single step called stabilization. The stabilizing agent (e.g., sodium bisulfite–thiosulfate, thiocyanates, thioureas, or certain organic sulfur compounds) forms reasonably light-stable complexes with undeveloped Ag ions. Stabilization papers contain a developer in the emulsion layer or in an overcoat. They are developed by immersion in an alkaline–sulfite solution and are then placed in the stabilization bath; a print can be obtained in as little as 10 sec.

G. Obtaining a Positive

Negative-working development is used in most b/w work, often followed by exposure of another emulsion (e.g., an AgX printing paper) through the negative to reverse the sign again and form a positive. Papers of various "hardness" (gammas) can be used to correct nonoptimum density range in the negative or to achieve desired esthetic effects. Variable-gamma papers contain a mixture of emulsions having different gammas and different color sensitivities; the color of the printing light controls their effective gamma.

For some applications, such as b/w motion picture film, it is useful to produce a positive at once. This is known, somewhat confusingly, as reversal processing (since the sign of the image is reversed during processing), although the process is in fact positive-working. Most commonly, the initial unfixed negative Ag image is removed by a solvent (e.g., a solution of potassium dichromate and sulfuric acid to convert the Ag to the soluble sulfate) that does not dissolve the nonexposed AgX crystals. A clearing bath then stops the bleaching action and prevents staining. The resulting image pattern of AgX is exposed uniformly to light or to the action of fogging (reducing) materials such as stannous compounds or amine boranes and is then developed and fixed in a conventional developer. Direct positive materials, which are positive-working with more-or-less normal processing, are also available; one is discussed in Section III,B.

II. Silver Halide—Color

A. BASIC PRINCIPLES

Present-day color emulsions are tricolor systems: they can reproduce most natural colors, within a limited color-saturation range, by adding together suitable proportions of the three primary colors, blue, green, and red. Tricolor systems can work by combining the actual primary lights; this is known as additive synthesis and is used in color TV for example. Most color photographs, however, are subtractive: white light (which contains all colors) passes successively through three single-color image layers, each of which absorbs some or all of one of the primary colors: the colors of these dye images are cyan (which absorbs red and transmits blue and green), yellow (which absorbs blue and transmits red and green, which is why yellow filters are often used in b/w photography to darken the rendering of the sky), and magenta (which absorbs green and transmits red and blue, appear-

ing as a purplish pink). In principle, then, appropriate optical densities in the three layers can produce any proportion of the three additive primaries (blue, green, and red) in the transmitted light. With transparencies, the colored light is transmitted directly to the eye, while in photographic prints the light must pass through the image twice and is reflected from the underlying substrate, usually paper.

Practical color photographic materials are usually multilayer structures with separate but superimposed red-, green-, and blue-sensitive AgX layers. Most color development processes are chromogenic: the dye images are formed during development by reactions between the oxidation products of the developer and color-couplers; the latter are special chemicals (which may or may not be colored themselves) contained either in the developing solution or within the layers of the emulsion. It is important to note that these color-couplers and the dyes resulting from their reaction with the oxidized developer are quite separate and distinct from the dyes used to sensitize the AgX to the appropriate exposure colors. A typical chromogenic color film is shown schematically in Fig. 3 together with the colors that can form in each layer. A cyan image (which transmits red) develops where the red-sensitive layer is exposed, a yellow image where the blue-sensitive layer is exposed, and a magenta image where the green-sensitive layer is exposed. Thus, with direct development, a pure blue area (for example) in the original scene causes development only in the blue-sensitive layer, yielding a yellow area in the final image;

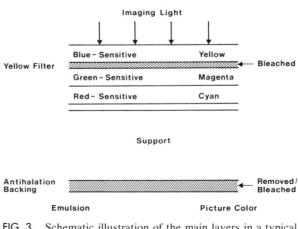

FIG. 3. Schematic illustration of the main layers in a typical chromogenic color emulsion. The color sensitivity of the active layers is shown on the left-hand side, and the colors formed after development are given on the right-hand side.

this absorbs blue, so the image is a negative of the original, analogous to a b/w negative. Similarly, a white area in the original scene causes development and dye formation in all the layers, yielding black.

Reversal (positive-working) color materials usually employ b/w (nonchromogenic) development of exposed crystals, followed by reexposure of the remaining AgX and chromogenic development. In the above example, the blue-sensitive layer is then not available for chromogenic development but the other two are; cyan and magenta therefore develop. Only blue is transmitted by both of them, so a positive results. In the area exposed to white light, none of the layers is available for chromogenic development, leaving a clear area; this appears white when viewed normally.

The yellow filter of colloidal Ag (which is bleached during processing) protects the underlying layers (Fig. 3), which are intrinsically sensitive to blue as well as to the colors to which they are sensitized, from blue light transmitted by the blue-sensitive layer. It is crucial that the various layers stay separate and do not undergo unwanted interlayer diffusion of some of the chemicals; this was one of the key problems that delayed the introduction of color films. The dyes used are never perfect matches to the primaries (especially given the other criteria, such as stability and chemical properties, that they must fulfill), and exposure and development errors can cause additional color distortion. Nevertheless, in practice, a reasonable representation of most originals can be obtained. This is done by paying particular attention, both in selecting the dyes and during the photofinishing process, to various colors to which the eye is especially critical, particularly skin-tones.

B. Reversal Processes

Kodachrome was the first commercially successful color film. It forms reversal (positive) dye transparency images in successive color development steps, using soluble couplers in the developer. Typical development steps with such materials are as follows. After b/w (nonchromogenic) development of exposed crystals, the remaining AgX is uniformly exposed to red light to make the remaining parts of the red-sensitive layer developable; this layer is then developed (using a cyan-forming coupler in the developer) to form the positive cyan image. The emulsion is then exposed to blue light to make the remaining

parts of the blue-sensitive layer developable, and this layer is developed to form the yellow image. Finally, the emulsion is exposed to white light (or to a chemical fogging compound) to make the remaining AgX (in the green-sensitive layer) developable, and the corresponding magenta image is developed. (Any unused AgX in the red- or blue-sensitive layers is removed by nonchromogenic "auxiliary" development immediately after the respective color development steps, to prevent color contamination by magenta in the yellow and cyan layers.) Processing finishes with bleaching (to remove the Ag), fixing, washing, and drying steps. Water rinses, stop baths, and gelatin-hardening baths may be used at appropriate points in the process.

Processing of reversal emulsions is simpler if colorless couplers are incorporated into the emulsion in such a way that there is little interlayer diffusion; this may be done, for example, by attaching long-chain substituents to the molecules or by dispersing them first in oily globules. Materials of this type include Agfachrome and Ektachrome. After nonchromogenic development and reexposure to white light (or to a fogging compound), the three dye images can then be developed in a single step, followed by bleaching, fixing, and so on. An additional step may be included to stabilize the unused couplers.

C. Negative Processes

Negative-working color materials work similarly, usually with the couplers in the emulsion, but yield a negative dye image by chromogenic development of the original exposed image. A positive is later made by printing onto a negative color paper. The papers work by principles similar to the films, although they may have a different emulsion sequence, sometimes to put the most light-stable dyes at the top and sometimes to put the magenta layer (which carries most of the sharpness information) there to reduce the effects of light scattering. Their color sensitivity may be different from those of films to compensate for the characteristics of the negative dyes. The couplers in negative-working films are often not colorless but are colored red or yellow or both; this color is destroyed when the couplers react to form the image dye, but the unreacted couplers remain as positive images and compensate for unwanted absorptions of the image dyes. Such couplers give the negative the familiar overall red or orange tint. Of course, since the materials controlling the color sensitivity of a layer and its color formation process are quite independent, one or more layers can deliberately be made to form a "wrong" color; some films, especially those sensitive to infrared, use such false color for special applications, for example, to distinguish visually between items of similar color that reflect infrared differently, such as vegetation and camouflage paint, or live and dead vegetation. (Infrared films also tend to "cut through" haze, because of its low infrared scattering.)

D. Further Details

Because of the different color compositions of light sources, especially sunlight compared with tungsten illumination, color films are available "balanced" for the effective temperatures of various light sources, which determine the composition of the light; alternatively, filters can be used to make the correction. This is especially important for reversal films which cannot be corrected at the printing stage by interposing various densities of cyan, magenta, and yellow filters or by using successive red, green, and blue exposures of appropriate lengths. Since the blue-sensitive layer of color films is also sensitive to the ultraviolet (UV), a UV-absorbing filter is often used on the camera, especially for distant outdoor views which may be comparatively rich in UV. Polarizing filters are also often used to cut down on glare from objects such as water and glass and to darken the rendering of the sky.

In practice, there may be a large number of layers in color emulsions. In addition to antihalation, abrasion-resistant, and UV-absorbing layers, for example, gelatin interlayers can be used to cut down interlayer interactions and may contain chemical scavengers to prevent unwanted interlayer migration of oxidized developer; also, as with b/w emulsions, there may be more than one light-sensitive layer for each color. Good exposure latitude is more difficult to obtain with color than with b/w emulsions, because too low a gamma can reduce the saturation of the colors obtained.

Some b/w emulsions (e.g., Ilford's XP1 400 and Agfa's Vario-XL) also use chromogenic dye chemistry. The Ag can be recovered, good densities can be obtained, and there is less visible grain and light scattering than with an Ag image. As a result, excellent sensitivity and sharpness

characteristics, latitude, and tonal range can be achieved.

The developing agents used for chromogenic development are usually members of the p-phenylenediamine class, in which the hydrogens of one amino group are replaced by organic groups such as $-C_2H_5$, $-C_2H_4OH$, and $-C_2H_4NHSO_2CH_3$. Most couplers are either (i) of the phenol or naphthol class, which form cyan-colored indoaniline dyes, or (ii) compounds with an active methylene group, which form the magenta and yellow dyes. Quinonediimines formed by oxidation of the developer react with the coupler, usually in its ionized form. Although removal of the oxidation products of the developer by sulfite is not desired in chromogenic development, a little sulfite is still usually added to protect the developer from oxidation by air. Bleaching of the Ag consists of oxidizing it to a sparingly soluble salt, using a mild oxidizing agent that does not attack the dye image, and then removing the salt in a thiosulfate fixing bath. A bleach of ferricyanide and bromide is often used; this converts the Ag to silver bromide.

Remarkable improvements have been made to color emulsions and processes over the years. Many of the individual processing steps have been combined, and the times required have been much reduced, partly by using high processing temperatures while still maintaining adequate gelatin hardness. Dyes with more ideal color absorption characteristics and much better stability have been introduced, and a UV-absorbing layer may be used. Nevertheless, even more than with b/w materials, it is still important to store color images at moderate temperatures and humidities and, if possible, in the dark. Sharpness has been improved by a variety of methods. One of the most important was to reduce light scattering by reducing the thickness of the layers. This involved better control over the mechanics of coating and also use of the minimum amount of imaging material consistent with good photographic sensitivity. The quantity of couplers and of AgX was reduced by using couplers of maximum efficiency. The softening of sharp edges has also been minimized by using edge-enhancement techniques. For example, certain developers can release a development-inhibiting compound during the coupling reaction; since this inhibitor tends to migrate laterally, it is less concentrated at the edges of any object in the image, thus allowing enhanced development there.

Advances of this kind can lead to major changes in the way photography is practiced. In the disc system of photography, for example, it was found possible to design small, nonadjustable cameras able to give satisfactory results over a wider range of light-level and distance conditions than previously (as well as allowing easier photofinishing), so long as the size of the image could be reduced substantially compared with earlier cameras. The reduced size obviously imposes increased sharpness requirements on the film, since the magnification during printing is greater. Films with enhanced sharpness, together with the excellent film flatness achievable with the disc system, permitted this reduced image size.

E. T-Grains

Recently, a major change in emulsion-making technology, the T-grain (T for tabular) system was introduced and used for the first time in the Kodak VR1000 color negative-working film, and more recently in disc films. The AgX crystals are made flat and thin (i.e., tablet-shaped) in contrast to the more symmetrical crystals of conventional emulsions, and tend to lie parallel to the layers.

While little appears to have been published on this system, it probably works as follows. The increased surface-to-volume ratio enables better dye sensitization than heretofore, because as discussed previously such sensitization is limited by the quantity of dye that can be incorporated into somewhat less than a monolayer on the crystals. Thus, increased sensitivity can be attained without increasing the "depth" of the crystals, which would increase the thickness of the emulsion and degrade the image sharpness via light scattering. Furthermore, the tendency for the crystals to lie parallel to the layers also tends to cut down on lateral light scattering. Since the sensitization is so good, the native blue-sensitivity of the green- and red-recording layers is nearly negligible by comparison, and the yellow-absorbing layer normally used in color films can be omitted. Moreover, these two layers (which make the most important contribution to the visual sharpness) can now be placed at the top, thereby avoiding scattering from the blue-sensitive layer. Furthermore, the shape of the crystals improves the covering power of b/w images, and the high surface area improves processing speed. Dye-sensitization in the blue (where the native AgX sensitivity is normally relied upon) can also produce substantial sensitivity advantages with these crystals.

F. Dye Destruction Systems

A different method of forming tricolor images uses dye destruction instead of dye creation. The emulsion starts with the final dyes; these are destroyed during development (by reactions involving or catalyzed by Ag) in proportion to the Ag image formed. Straightforward processing is positive-working, but more elaborate methods (analogous to normal reversal processing) yield a negative. Dye-bleach materials (e.g., Cibachrome) are claimed to use far more stable dyes (generally of the azo class) than those produced chromogenically. The dyes also act as a barrier to light scattering during exposure, thus reducing the loss of sharpness in the image.

III. One-Step and Other Diffusion-Transfer Silver Halide Processes

One-step photography (which is one-step only from the point of view of the user) is normally associated with the name of the Polaroid Corporation, although other systems now exist. Such photography depends on diffusion processes (the tendency of matter to move so as to even-out concentration differences), and these processes are also the basis of other forms of photography. One-step photography has a great advantage in applications that need quick access to a finished image, coupled with high sensitivity (e.g., in amateur photography, exposure testing, and scientific photography). However, sharpness is usually not high (because diffusion takes place over significant distances, leading to sideways "spread"), duplication is difficult, and manufacturing is not inexpensive.

A. One-Step Black-and-White Photography

In the Polaroid one-step b/w systems, a solvent (e.g., "hypo") for the AgX is included in the developing solution, which is also made viscous ("thick") by incorporation of soluble plastics. A small quantity of this solution is contained in an impervious pod in the film pack. After exposure, the emulsion (the "donor") and a "receiver" sheet containing development nuclei (particles that can catalyze the reduction of silver, analogous to LIs in conventional materials) are drawn together through a pair of pressure rollers; this action ruptures the pod and releases the processing reagent, which spreads between the sheets. AgX that is not developed in the emulsion dissolves in the processing solution, and the Ag ion complex so formed diffuses to the receiver where the Ag ions are reduced by solution physical development at the development nuclei. The process is positive-working, since the amount of Ag complex reaching the receiver varies inversely with the amount of AgX developed in the donor. After an appropriate time, the donor and receiver sheets are stripped apart, and a stabilizing coating is often applied to the surface of the positive print. The negative on the donor may be discarded or, with appropriate films, used after washing as a negative for making more prints.

Silver formed by solution physical development tends not to form filaments, so the original images tended to be brownish (e.g., sepia) rather than black: the small, fairly compact Ag particles scattered blue light preferentially. Special techniques are therefore used to make the color more neutral, by making the particles grow in small regions where they clump together to form larger aggregates of order 100 nm in diameter, still much smaller than the original AgX crystals and of a very high covering power. This can be done by precipitating metallic sulfide catalytic nuclei in the interstices of a very fine colloidal silica suspension. Most of the developing reagent is removed by adherence to the negative when it is stripped from the positive, but some reagents are left in the very thin ($\sim$0.4 μm) image layer in images formed in colloidal silica; moreover, the image is susceptible to abrasion and to attack by the atmosphere. Such images are therefore coated with an acidic aqueous solution of a film-forming plastic; the acid tends to deactivate the developer, and the coating action washes away the reagent and provides the required protection. Coaterless Polaroid emulsions deposit the Ag in depth within a receiving layer of regenerated cellulose, where it is protected from abrasion and atmospheric attack. The developing reagents employed have colorless oxidation products and leave no residues, and image stabilizers and immobilized polymeric acids (to neutralize excess alkali) may be incorporated into the receiver.

Naturally, the various processes occurring during development must be properly balanced for this system to work. However, as a result of the tendency to remove any AgX that is not quickly developed, very active developers can be used without serious fogging and (since they are sealed in the pod prior to use) without attack by air; because of this, and perhaps because these images are rarely enlarged, very high effective sensitivities [even ISO (ASA) 20,000] can be attained. The high covering power (high density) of the small Ag image particles assists the attainment of such sensitivities and also leads to a reduction in the quantity of Ag necessary in the original emulsion.

B. COLOR ONE-STEP PROCESSES

1. Polacolor Process

In the Polacolor process, the donor consists of three AgX emulsion layers separated by spacer layers, each associated with a layer that contains a dye complementary in color to that of the light to which the emulsion layer is sensitive. That is, there are six active layers: (i) a blue-sensitive layer backed by (ii) a layer having a yellow dye and developer compound, (iii) a green-sensitive layer backed by (iv) a magenta dye and developer, and (v) a red-sensitive layer backed by (vi) a cyan dye and developer. The dye-developer in each case consists of actual dye molecules (not couplers) chemically linked to a hydroquinone developer molecule. The dye-developer molecules are insoluble in acid media but are made soluble by the alkaline processing solution (from the pod) that ionizes the hydroquinone portion of the molecule, making it an active developing agent. Dye-developer molecules that react with exposed AgX are oxidized to the quinone, which is fairly insoluble and may react with gelatin, thus remaining in the donor layer. Unreacted dye-developer molecules diffuse into the receiver layer and are immobilized by reaction with compounds contained in it, thus forming the image. This is then stripped apart from the donor. The receiver also contains an immobile polymeric acid to neutralize the alkali of the processing solution (once it has passed through a "timing" spacer layer between it and the acid) and to stabilize the dye image.

In order that the oxidation state of the developer shall not affect the color of the dye part of the molecule, the two are separated by "insulating" (nonconjugated) chemical links. The efficiency of development can be increased by adding small auxiliary "messenger" developers of greater alkali-solubility and greater diffusion rate than the dye-developers. Such a developer reaches the AgX crystals quickly to initiate development, and its oxidation product can then oxidize and immobilize the slower moving dye developer, regenerating the messenger. Phenidone and Metol were among the first messengers used; later systems included substituted hydroquinones, alone or combined with Phenidone. The timing of the processes is critical in systems of this kind: for example, development in the individual layers must be near completion before the dyes diffuse significantly, otherwise the "wrong" dye could develop the crystals; this timing is assisted by the spacer layers, and steps are also taken to deactivate the emulsions after a specific time.

2. The SX-70 System

The Polaroid SX-70 system was the first fully integral one-step color film, in the sense that no strip-apart step is involved. One of the film packs used is shown in Fig. 4. After exposure, the film is ejected from the camera through rollers to rupture the pod containing the processing fluid. The very high optical density of the new reagent layer, due to the presence of white TiO_2 and of opacifying (light-blocking) dyes, protects the emulsion from further exposure. The development process is in principle broadly similar to the Polacolor system, except that the donor and receiver are integral and remain so, the dyes that remain in the donor being hidden behind the TiO_2 layer; the donor as well as the receiver must therefore be stable after completion of the image formation process. The opacifying dyes (of the phthalein pH indicator

class) of the processing fluid are colored only in alkaline media and become colorless as the alkali is finally neutralized by the polymeric acid, leaving the TiO_2 to provide a white background for the image.

3. The Kodak System

The Kodak instant print scheme (Fig. 5) introduced in 1976 uses a different chemistry. The dye releasers are sulfonamidophenols, rendered immobile (ballasted) by the incorporation of long-chains into the molecules; cyan, magenta, or yellow dyes are attached to the sulfonamido group. Upon oxidation by the developing agent (a derivative of 1-phenyl-3-pyrazolidone) and reaction of the oxidized form with hydroxide ions, the dye is split off, leaving an immobile ballasted quinone. The pyrazolidone is regenerated by its reaction with the dye-releasers and can thus act repeatedly as a messenger. The split-off dye can diffuse to the receiving layer, where it is immobilized. To make the process positive-working, the AgX emulsion has to be one that develops directly, by special mechanisms, to a positive. A "core–shell" type of AgX crystal is apparently used: sensitivity centers are incorporated onto the surface of an initial crystal, and an outer shell with little surface sensitivity is then grown on this core. The developer contains a fogging agent, which provides electrons to the crystals, causing the nonexposed crystals to become developable via the formation of surface Ag nuclei. The internal LI centers in the exposed crystals, however, act as very good traps for these elec-

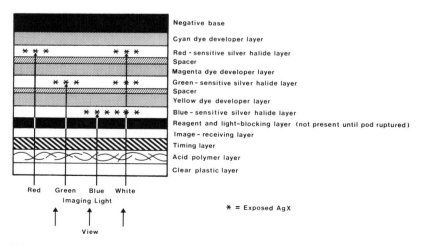

FIG. 4. Schematic cross section of a Polaroid SX-70 integral one-step color film. Imaging and eventual viewing take place from the same side of the film.

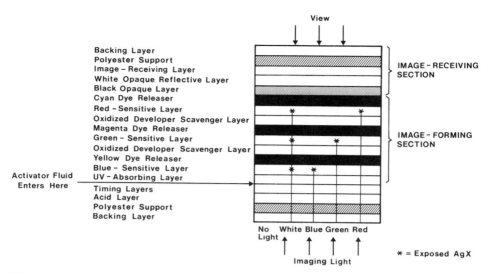

FIG. 5.   Schematic cross section of a Kodak integral one-step color film. Imaging and eventual viewing take place from opposite sides of the film.

trons, so no surface Ag nuclei form, and (since the developer, as usual, cannot reach the internal LI) no development takes place.

4. Color Printing Systems

Instant photography systems can also be used for making prints from existing negatives or transparencies. The Kodak Ektaflex system, for example, is based on the chemistry described above but uses donor and receiver sheets and a tray of processing solution, rather than an integral structure and a pod of solution. The Agfachrome-Speed material, on the other hand, is integral and is developed simply by immersion for $1\frac{1}{2}$ minutes in a tray of activating solution (mainly alkali), followed by a rinse. Each color layer contains appropriately sensitized AgX, an immobile (ballasted) reducing agent, and an immobile color-providing compound (CPC); the CPC itself contains various groups, including a dye and a ballast group. Also present is a mobile developer to act as a messenger. In exposed areas the AgX is reduced by the developer, which then oxidizes the reducing agent. In nonexposed areas, the reducing agent is not used up in this way and instead reduces a group in the CPC; this reduced group, in the presence of alkali, then initiates the cleavage of the dye from its ballast, allowing the dye to diffuse to a receiving layer. The resulting system is positive-working and suitable for printing transparencies. Since it depends on a competition between the rates of

two reduction reactions, it is fairly insensitive to temperature, which affects both reactions similarly; however, the gamma can be varied by changing the relative rates, for example, by simply diluting the activator.

5. Additive System

An instant color (and b/w) slide transparency process has recently been introduced by Polaroid, using *additive* color principles similar to those used in an earlier instant movie film. The emulsion (Fig. 6) is exposed through a microscopically fine-lined integral color screen; nearly 400 red, blue, and green triplets (the additive primaries) are used per cm. Green light, for example, passes through only the green parts of the screen and exposes the AgX behind it. The entire film cartridge is developed at one time in a semiautomatic processor that applies processing fluid (alkali, developer, AgX solvents, etc.) to the back of the film. In the above example the AgX areas behind the green screen are developed, while the nonexposed areas (behind the red and blue parts of the screen) are dissolved; these transfer to the image-receiving layer between the original AgX and the screen, where an Ag image of good covering power is formed in about a minute. The film is then rewound, which strips away the top layers. Light from subsequent illumination of the film is transmitted through the green regions but is blocked by the Ag image in the blue and red areas. The original

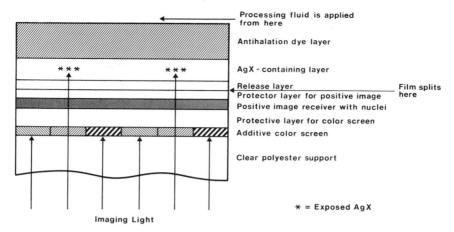

FIG. 6.   Schematic cross section of a Polaroid instant additive color transparency film. After development, the top portion of the film is stripped away, as indicated.

color is thus reproduced, although the continued presence of the additive screen confers a fairly high background density on the transparencies and its use limits their sharpness.

C. Other Diffusion-Transfer Processes

The b/w diffusion transfer system (using a nonviscous developer and, typically, separate donor and receiver sheets) has also been used in other processes to produce a positive image with improved processing convenience, quick access to the image, and moderate use of Ag and thiosulfate. Applications have included reprography (e.g., office copying, often by direct contact with the document original and illumination through the back of the donor sheet) and more conventional photographic areas such as printing-plate preparation, one well-known system, for example, being known as "Photomechanical Transfer."

The developing agents can be incorporated into the donor or the receiver, so the processing solution is really an activator containing mainly alkali. The nuclei in the receiver, typically heavy-metal sulfide or selenide particles about 5 nm in diameter, are usually coated in gelatin onto a paper base; various chemicals can be added to improve the color of such images. A variant of this system eliminated the processing solution by incorporating the developing agents, together with a dry, fairly weak alkali, into the donor; development was initiated by running the sheets between heated rollers. The heat released

moisture from a water-containing inert salt, which activated the developer. The Kodak Bi-Mat system eliminated the processing solution as a separate component by soaking it into the gelatin of the receiver, leaving it dry to the touch. The final positive could be stabilized by subsequent washing.

Diffusion-transfer products have been marketed with both donor and receiver on the same support, the unhardened negative layer being washed off after processing; such a system was used in "photostat" copying machines, which earlier used other processes. Alternatively, the transferred Ag (at the outer surface of the receiver) has been used, after reaction to make it ink-receptive, as a printing plate. Other diffusion-transfer systems using various substrates (e.g., aluminum) have been used as offset plates for printing. A negative-working diffusion-transfer material has been described that uses a direct-positive emulsion as the light-sensitive element. Single-color images can be formed with diffusion-transfer techniques, and these are suitable for color proofing by overlaying three appropriately colored images; a pigment is incorporated into the receiver, and the Ag deposited there reacts with additional processing chemicals to degrade the gelatin, which is then washed away.

The "physical transfer" process uses the gelatin-hardening action of some oxidized developers to form an image pattern ("tanning" development). The amino groups of the gelatin can react with the oxidized developer forming a compound that can react with other gelatin amino groups, thus crosslinking and hardening the gelatin. The unhardened gelatin can be transferred to a receiver to form a positive image (the basic idea of the Verifax process), or an image of hardened gelatin (which is more oil-receptive and less water-receptive than unhardened gelatin) can be used as a printing plate. If the unhardened gelatin is removed, the relief image can be dyed and used to transfer dye to a receiver, the basis of the Technicolor and dye-transfer processes.

IV. Other Processes Related to Silver Halide

A. PRINTING

Books, magazines, and newspapers, including the illustrations, are produced by one of several common printing processes, using a printing plate usually made from a photographic original. We shall briefly describe these processes, since most of the photographic images we see every day are made in this way and because we shall later describe some of the photographic materials used in printing plates themselves.

1. Principles of Printing

Printing plates may use a raised or depressed relief pattern to carry the ink (letterpress and gravure, respectively), or the inked and non-inked areas may be in essentially the same plane but differ in their receptivity to ink (lithography). Printing plates are typically made by contact exposure to a photographic film, using one of several photoactive systems on the plate (see Sections VIII,A and D).

Most printing systems are "on–off" in nature, so they cannot directly reproduce a range of tones. In a typical system, therefore, pictorial input is rephotographed at the full final size through a screen consisting of a dense array of dots, each of whose optical density is highest in the center and reduces gradually towards the edge; this converts tone gradations of the original to variable-sized dots in the copy when very high-gamma emulsions are used. The spatial frequency of the dots is high enough that the eye perceives such gradations as a nearly continuous tone. A uniform exposure through the dots alone may be added, to change the effective gamma of the process. Text material, often produced electronically, is printed out, pasted-up

into the desired layout, and photographed at full-size. The pictorial and text material is combined and contact exposed to the plate (which has very low sensitivity), and the plate is developed. Colored pictures must be color-separated by photographing them through appropriate filters and making multiple plates for printing with cyan, magenta, yellow, and black inks, the last to ensure a good neutral black.

Nowadays, there is a strong tendency to move to more electronic systems in which scanners convert pictorial material into electrical signals, and various manipulations are performed on the signals; they are then output, typically by a laser-scanner, to the photographic film, which is subsequently contact-exposed to the plate.

2. Lith Development

Photographic AgX films used in the above processes must generally have very high gamma, so that the edges of text characters, and especially the dot edges, are sharp. This is accomplished by using films with a very high Ag content to permit high developed densities, and by using special "lith" developers: in these, the oxidation products of the special hydroquinone developer are not removed as efficiently as they would be by the higher sulfite content of most developers. The oxidation products, as they are formed during development, therefore build up in the vicinity of exposed crystals. They are more active as developers than the original hydroquinone or may form superadditive combinations with it, so that development is autoaccelerating. Furthermore, the oxidation products, or Ag ions released from the crystals, can diffuse to neighboring exposed crystals and start development there, thus increasing the developed density still further. The oxidation products may also retard development of crystals with small exposures, further increasing the sharpness of the dot edges.

B. STEREOSCOPIC PHOTOGRAPHY

Stereoscopic effects can be obtained by making two photographs from viewpoints separated by about the distance between the eyes, and presenting them independently to the two eyes, for example, by using two independent viewing boxes for transparencies. Alternatively, monochrome images can be printed in two different colors and viewed through spectacles of similar colors. Color stereo pairs can be placed over one another in polarizing layers and viewed with polarizing spectacles, or the stereo pair (or more than two images) can be split into narrow interlaced strips and viewed more or less independently through a carefully aligned grid of microlenses placed on the composite.

C. THERMALLY PROCESSED SILVER HALIDE

1. Dry Silver

Thermally processed AgX, commonly known as dry silver, is somewhat similar in principle to conventional (wet-processed) AgX except that development simply uses heat. The emulsion contains the AgX crystals and a much larger quantity of organic silver compounds that are insensitive to light; also present are a developing agent, additives to promote stability, and a plastic. The organic silver compounds are silver salts of long-chain organic acids (e.g., behenic acid), and the AgX can be mixed with them or formed from them by treatment with a halide. The developers are much less active than those used for wet processing, but on being heated to the region of 125°C for a few seconds, they reduce the organic silver compounds to silver in those regions where an LI catalyzes the process.

Dry silver materials are easy and quick to develop, but their sensitivity, shelf-life, and image stability are somewhat in conflict; this is because the processes occurring during development have some tendency to occur during storage also, especially the storage of finished images in the light. Stable dry silver materials could be said to represent a half-way house between high-sensitivity (camera-speed) materials and the much less sensitive duplicating materials, discussed elsewhere, whose sensitivity they exceed by several orders of magnitude. Unlike some such materials, they are sensitive enough that they must be kept in the dark prior to imaging. They are largely limited to negative-working applications and tend to have a significant background density. Nevertheless, they have good sharpness and can be made sensitive throughout the visible region. Dry silver papers and films (generally on a polyester base) have achieved considerable acceptance in areas such as microfilm, duplication film, and recording paper.

2. Other Photothermographic Systems

Dry silver is a special case of photothermographic systems. Another example, the 3M Dual Spectrum process, was widely used in office copying. An intermediate sheet is placed in direct contact with an original, and light is shone from the intermediate side. Light reflected from the white areas of the original decomposes a volatile material, 4-methoxy-1-naphthol, in the intermediate. This sheet is then heated in contact with a receiver sheet containing silver behenate/behenic acid, which is chemically reduced by the material remaining in the intermediate.

V. Nonsilver Processes: Overview

A. SENSITIVITY UNITS

The quantity of light required for exposure of a photographic system may be expressed in terms of ergs/cm^2; this is numerically equal to mJ/m^2. The higher the number, the less sensitive the material. For comparison, a *very* rough rule of thumb would be that arithmetic ISO sensitivities (ASA) are equal to the reciprocal of the exposure in ergs/cm^2 required to obtain slight development; reasonably complete development, however, may need roughly 10 to 100 times more exposure. (This correlation is a rough one, since ISO sensitivities are most appropriate for white-light sensitivity, while ergs/cm^2 numbers usually refer to the exposure required at the most sensitive wavelength.) Sensitive AgX emulsions (say ISO 1000–10,000) thus require exposures *for slight development* in the region of 10^{-3} to 10^{-4} ergs/cm^2 while the very highest-sharpness AgX materials need as much as 10^3 ergs/cm^2. A typical xerographic office-copying photoreceptor needs about 0.5 and 5 ergs/cm^2 for threshold and complete development, respectively, and the XDM electrophotographic film system is presently several times more sensitive. A hypothetical system that formed one molecule of an efficiently absorbing dye for every photon absorbed but had no amplification and no sources of inefficiency would need energy in the 10^5 ergs/cm^2 region. Many insensitive systems with little or no amplification are actually in the 10^6–10^7 ergs/cm^2 range.

B. STATUS OF NONSILVER SYSTEMS

While AgX materials obviously possess enormous advantages, especially their outstanding sensitivity, there are also serious shortcomings for many applications. These include their cost, the use of a limited resource (silver) which may greatly increase in price one day, the normal use of wet-processing (or jellies), the long time usually required after exposure to obtain a finished image, the difficulty of updating them, their limited shelf-life, their fairly complex manufacturing process, the need to handle them in the dark, and the reciprocal limitations on their sharpness and sensitivity. In the next sections, we describe the various materials that have been developed to overcome some of these and other disadvantages for specific applications.

The success and fairly high sensitivity of electrophotography is well-known, while many of the nonelectrophotographic approaches have also been reasonably successful and have filled critically important niches in the realm of duplication-speed rather than camera-speed applications. (In duplication processes, the original and the photosensitive material are placed in contact; the absence of a lens increases the transmitted light intensity by a large factor.) However, there is in general a reciprocal relationship between the simplicity, cost, and convenience of nonsilver nonelectrophotographic systems on the one hand, and many of their imaging properties, especially sensitivity and stability, on the other.

In particular, leaving aside electrophotographic systems, it does not seem to have been possible to obtain a stable nonsilver system needing less than about 10^2–10^3 ergs/cm^2 for reasonably complete development, or to obtain a stable, instant, heat-developed material (even if silver systems are included) better than the roughly 10^2 ergs/cm^2 available with dry silver. Moreover, most nonsilver nonelectrophotographic systems are very difficult to sensitize into the red without incurring serious stability problems. Techniques for overcoming the sensitivity/stability problem have been suggested (e.g., to build in a slow back-reaction to erase unwanted latent images appearing on storage) but do not appear to have been successfully applied.

Electrophotographic systems of the well-known xerographic toner type have overcome many of these limitations for reprographic and other applications. In addition, the heat-developed, <1 erg/cm^2 XDM electrophotographic material overcomes almost all the above limitations for photographic and information-recording-type applications and could also eventually even exceed the sensitivity of AgX.

VI. Electrophotography

Electrophotography is based on the formation of images by the movement of matter, usually charged particles, in electric fields. Most commonly, the fields are reduced or destroyed in exposed areas via the action of a photoconductive insulator. Amplification of at least ~10^5 is achievable, because previously stored charge at high voltages can be discharged by photons that each carry the energy equivalent of only a few volts, so that comparatively large particles can be moved by the attraction of charges equivalent to relatively few electrons. The exposure required for complete development by typical, efficient electrophotographic processes is thus in the several ergs/cm^2 region, easily good enough for photocopying applications (a lens-coupled camera-exposure situation in which fairly high illumination intensities can be supplied) and also for a number of more conventional photographic applications. Furthermore, conventional electrophotography is capable of much higher sensitivities; for example, very experimental CdSe systems have exhibited complete discharge for exposures as low as 0.3 ergs/cm^2. The practical XDM film system (Section VII) can be more sensitive than this and appears to offer the eventual possibility of exceeding the sensitivity of AgX.

The most important electrophotographic process is transfer xerography, which is the basis of the majority of modern office copiers. A photoconductive material, which is the active part of the photoreceptor (P/R), is electrically charged with ions produced from the air and is exposed to the imaging light; in exposed areas, the voltage across the P/R is largely discharged, leaving an imagewise pattern of charge on its surface. This is the latent image. Development is achieved by the differential attraction to the charged or uncharged areas of charged pigmented plastic toner particles; the toner is usually a dry powder and the pigment is commonly carbon black. The resulting toner image is then transferred to the final substrate, usually paper, to which it is fixed. After discharge and cleaning steps, the P/R is ready for reuse for as many as 300,000–400,000 cycles.

Other important processes include direct xerography (in which the P/R is coated on the paper), liquid-development processes using liquid-borne toners, and photoactive-particle processes in which the electric field can be the same in exposed and nonexposed areas but the exposed particles themselves change their charges. Yet further development processes rely on the wrinkling of a charged surface yielding a light-scattering image, and changes in electrostatic adhesion.

Electrophotographic development is rapid

and usually involves no liquids. With reusable P/Rs, the main consumables are plain paper and a toner system so the materials costs per unit area are extremely low by photographic standards. The final paper image is very stable. Electrophotographic *films* can be handled in the light before use and can be updated after development and use; their shelf life and image stability can be very good, because the sensitivity to light is "gated" by application and later removal of the electric field. Although the sharpness of many electrophotographic processes is fairly low, as is appropriate for document copying applications and the like, there is nothing fundamental about this, and excellent sharpness can be attained under appropriate circumstances.

Both positive- and negative-working systems are available. Most electrophotographic processes show little, if any, reciprocity failure.

A. TRANSFER XEROGRAPHY: PROCESSES

1. Charging and Exposure

The basic steps in transfer xerography are illustrated in Fig. 7. They are generally carried out around the periphery of a continuously rotating P/R drum or flexible belt. The charging step normally employs a corotron, a device with one or several thin wires largely surrounded by a grounded shield that stabilizes the process; the wires are held at a high enough voltage (typically

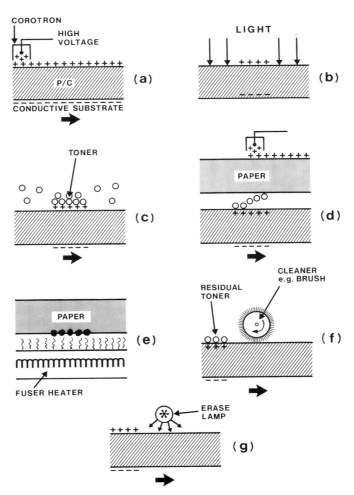

FIG. 7. The basic steps in transfer xerography: (a) charging; (b) light exposure; (c) toning; (d) toner transfer to paper; (e) toner fusing; (f) P/R cleaning; (g) P/R discharge.

2–10 kV) to ionize the air in the vicinity of the wire. Because the field is high only in the vicinity of the thin wire, no sparking occurs, but ion multiplication processes near the wire form a stream of positive or negative ions, such as hydrated protons and CO_3^-, that flow to and thus charge the nearby P/R. Scorotrons use a control grid between the wire and the P/R, somewhat analogous to a triode: the ion flow tends to stop when the P/R surface voltage is similar to that connected to the grid. The ions are reasonably stable on the surface of appropriate P/Rs, such as noncrystalline (glassy) selenium, and the resulting field across the P/R is reasonably constant for times typically of the order of seconds. The surface charge induces an equal and opposite charge in the grounded base electrode to maintain overall charge neutrality. Exposure of the P/R to light produces electron–hole pairs that separate under the action of the applied field. With a selenium P/R, for example, the light is absorbed near the surface, but if the ionic charge is positive, the holes can travel right to the bottom of the selenium in a fraction of a second, with little deep trapping. Thus, virtually total discharge is achieved in the exposed areas.

2. Development

To achieve a photographically positive image when the ionic charge is positive, negatively charged toner particles must be attracted as uniformly as possible to the nondischarged P/R areas. However, in the absence of another grounded surface, a toner particle near the surface of the P/R "sees" as much negative charge on the electrode as positive on the surface, even though the toner may be nearer to the positive charge: the effect of an "infinitely" extended sheet of uniform charge does not depend on the distance. However, if the toner is near an edge of a charged area on the P/R, the charge does not appear infinite in extent, and then the toner's closer proximity to the positive charge leads to the intuitively expected attractive force. Expressing the situation another way, the electric fields associated with the surface charges extend almost entirely to the opposite charges induced in the conducting substrate; only near the edges of a charged area do significant fields extend above the P/R surface where a charged toner particle can experience the resulting force. Figure 8a shows this situation, expressed in terms of the direction and density of the lines of force that a negatively charged toner tends to follow

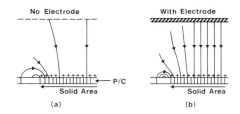

FIG. 8. Illustration of the direction and density of the electrical lines of force "seen" by a negatively charged toner near the edge of a uniform positively charged area, (a) without and (b) with a development electrode.

toward a large positively charged area. In these circumstances, then, toner is attracted primarily to the edges of charged areas, and this leads to the well-known edge effects of some xerographic processes: thin objects such as typewritten characters are well reproduced, but solid black areas are not. This is useful up to a point in enhancing the reproduction of documents and has been used to great effect in xeroradiography, in which medical X-rays are detected xerographically; images that would otherwise have very low density variations can be greatly enhanced by the edge effects. The effect also confers process latitude, since development is purely a function of voltage differences between adjacent areas rather than depending on absolute voltages.

For most applications, however, it is desirable to reduce these edge effects or enhance solid-area development. Screening techniques (see Section IV, A) can assist in this regard. Usually, however, a development electrode is placed in close proximity to the P/R (Fig. 8b): negative charges are then induced on it at the expense of the P/R substrate electrode, and the toner now sees a net force towards the P/R even in solid areas. The difficulty that must be overcome with this is that for maximum effect the spacing between this electrode and the P/R must be small, so that the electrode can interfere with access of the toner.

The toning process can be made negative-working by using a development electrode biased with the same sign of voltage as the original charge on the P/R. Alternatively, if only small lines are to be developed (using edge fields), then simply reversing the charge of the toner is sufficient, because the edge fields reverse direction (Fig. 8) a small distance from a charged area.

3. Toner Transfer, Fixing, and Cleaning

After the P/R has been toned, the toner is transferred to the paper (or other substrate such as a coated polyester for transparencies), usually assisted by a bias field applied by a corotron to the rear of the paper. An electrically biased roller can be used in place of the corotron, and this can prevent ionization effects occurring on separation of the P/R and the paper; the electrostatic image on the P/R can then be reused to produce more than 100 copies, although this is rarely done commercially. Steps can be taken to suppress the transfer of some of the toner on the P/R, thus reducing background (toner deposition in areas of the paper that should be toner-free). To facilitate paper stripping, a second corotron can discharge the paper after transfer charge has been applied. The toner is then made to flow partly into the paper to fix it in position. (Note the somewhat different usage of the term "fix" in electrophotography and conventional photography). This may be done by fusing (melting) the toner via contact with a heated roll that has a fluorocarbon or silicone release surface or a thin surface layer of a suitable liquid. Alternatively, radiant heat (e.g., from a flash lamp) can be used; this has the advantage of reducing the copier warm-up time and eliminating the energy used to keep a roll hot. Cold pressure alone can be employed, but the high pressures required may give the paper an objectionably glossy appearance and feel; also, a careful balance must be struck in terms of toner softness, or toners of this kind may tend not to stick well enough to the paper or to stick too well to the P/R, making cleaning difficult. Solvent vapor has also been used for fixing.

Residual charge on the P/R is removed by uniform illumination. The last toner residues are neutralized by a corotron driven by an alternating voltage and are removed by mechanical means such as mechanical brushes, wiper blades, disposable sheets, or developers working in reverse (often with an electrical bias); removal of the toner may be assisted with a vacuum, and solid lubricant can be incorporated in the toner to assist transfer and cleaning.

4. Electronic Printing

In addition to the familiar lens-coupled copying of existing documents, xerography is becoming widely used as an electronic printing device. In such applications (which strictly are not photography), the original is not a light pattern but a stream of electronic signals. Usually, a laser beam is scanned repeatedly back and forth across the surface of a P/R, using a rapidly spinning multifaceted mirror, while the P/R moves slowly at right-angles to the scan direction. The laser is turned on and off by the electronic bit stream as necessary to generate a pattern on the P/R corresponding to the print desired. Such systems, using visible lasers, have become very important and widely available recently and may become more so as inexpensive and easily switched infrared diode lasers become more widely used. Other techniques of coupling electronic input into xerographic systems may eventually become important; these include active (e.g., electroluminescent) and passive (e.g., liquid crystal) linear displays. The latter would be used as a line of "shutters" in conjunction with a continuous light source such as a fluorescent tube.

5. Color

Full-color transfer xerographic systems typically use three full charge, exposure, development, transfer, and cleaning passes of the photoreceptor for each print; the exposures are through appropriate color filters, and development is done with three or four different colored toners. Colored slide originals can be used by projection of the image onto a special Fresnel lens placed on the document platen to redirect the projected light towards the copier lens; a screening system (see Section IV,A) is used to reduce the effect of the high gamma of the xerographic process, and good pictorial reproductions can be made.

6. Process Variants

Transfer xerography can also use somewhat different electrostatic procedures. Figure 9, for example, shows one of several related photodielectric processes. Charge is first deposited onto the insulating overcoat; if the photoconductor (P/C) permits charge flow in one direction in the dark (as many do), the normal countercharge induced on the ground electrode can be injected into the P/C and can travel to the insulator interface. Image illumination is applied through a transparent corotron (e.g., a grid of wires) that is driven with an alternating voltage so as to produce roughly zero voltage on the whole P/R surface, the charge varying imagewise. (In the dark the negative charges in the P/C cannot move, and so a smaller positive charge

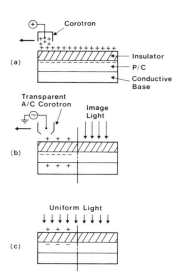

FIG. 9. The charging and exposure steps in a typical photodielectric process: (a) charging; (b) light exposure through transparent A/C corotron; (c) uniform light exposure.

is needed on the surface to neutralize their effect there and to give zero voltage. Some positive charge also resides on the substrate electrode to preserve overall charge neutrality). Uniform illumination then collapses the field across the P/C, reducing the negative charge and producing a stable variable-voltage image that can then be toned. Processes of this kind allow the P/C to be protected against damage and can use P/Cs with higher dark conductivity than the more conventional approaches; process speed, however, is probably limited.

Multiple charging and exposure processes can also be used with multilayers of P/Cs that respond to different colors, to produce electrostatic images of two different charge polarities in the same P/R. These can then be toned with toners of different colors (e.g., black and red) and opposite charge to produce two-color copies.

An interesting development variant, which does not seem to have been used commercially, employs uniform toning of the P/R before charging and exposure through a transparent conducting substrate. In exposed areas, charge moves from the electrode to the P/R surface and holds the toner more tightly. After toner transfer from the nonexposed areas and erasure of the electrostatic image, the removed toner is replaced, and no cleaning step is necessary.

Modern copiers continuously monitor many

of the parameters of the entire process and use microprocessor control to effect appropriate changes to the system.

B. TRANSFER XEROGRAPHY: MATERIALS

1. P/R Materials

A wide range of P/C materials has been used in transfer xerography. Noncrystalline selenium (produced by vacuum deposition onto a cylindrical drum held at a special temperature) has been widely used but responds primarily to blue light; green illumination can weight the response towards longer-wavelengths well enough for most applications, but the insensitivity to pure blue lines (which seem like white to the selenium) is well-known. Alloys of selenium with tellurium, or compounds with arsenic, can give panchromatic response, and the latter can improve stability against crystallization. The P/C is often deposited on a thin barrier layer to prevent unwanted charge injection from the base electrode, which is typically aluminum. Dispersions of photoconductive pigments such as cadmium sulfide, zinc oxide, phthalocyanine, or thiopyryliums in insulating (usually plastic) binders may be used, provided the volume-loading of pigment is high enough for good charge transport or a charge-transport material is used in the insulator; many such materials can be dye sensitized, analogous to AgX (Section I,C). Near-infrared-active materials such as phthalocyanines may become very important, because they would permit the use of long-wavelength diode lasers for electronic printing applications. Special plastic layers have been used as the P/C itself, a well-known example being the charge-transfer complex between poly(N-vinylcarbazole) and trinitrofluorenone; this has good photoconductivity and color response and transports both electrons and holes. Recently, there has been much interest in using noncrystalline silicon in P/Rs. This has a very hard and durable surface and is very photoactive but may be expensive to produce.

It is often desirable to coat a very stable, abrasion-resistant material such as a tough plastic at the top surface of the P/R, so as to protect a softer photoconductive material from abrasion, attack by ions and so on. Of course, a simple overcoating would tend to trap charge, although a corotron erase step may prevent this; alternatively, photodielectric processes may be used. More recently, however, the functions of photo-activity and charge transport have been separated using multiple-layer systems. For example, trigonal selenium (which has better color response than noncrystalline selenium) or organic P/Cs with or without sensitization have been used in plastic dispersions to form a thin photosensitive layer at the base of the P/R; a much thicker overlayer of a tough plastic such as a polycarbonate provides the charge transport and hence the insulating ability and voltage contrast. Photogenerated carriers are injected into the overlayer and are transported therein by virtue of a special charge-transport material dissolved in the layer. In the case of a negatively charged P/R this material needs an unusually small amount of energy to lose an electron (i.e., it has a low ionization potential) and can therefore easily donate an electron to neutralize the hole left in the photosensitive layer after photoexcitation; after this, the charged molecule can accept another electron from a neighboring molecule, and so on. The initial photogenerated hole is thus effectively transported through the insulator to neutralize the negative charge at the surface.

Such P/Rs can have the additional advantage of flexibility, so that flexible P/R belts (typically on aluminized polyester) are more easily used in place of the conventional P/R drum; selenium on special, flexible nickel belts has also been used. Such belts need not move in a circular path and can be made flat in the exposure region. This enables rapid flash exposure of the entire document at once (without a curved document platen) rather than the usual slit-scanning procedure in which only a small portion of the document is illuminated and imaged onto the moving P/R at any instant; slit scanning is optically less efficient and can produce slight distortion of the copy. Furthermore, the various process stations can be placed more conveniently with a flexible belt, high copy speed is easier since several images can be made for each rotation of the P/R without requiring a very large drum, and the belt can be bent round a sharp corner to assist separation of the paper after toner transfer. Alternatively, the belt can be rolled up and placed inside a drum around which part of the belt is wrapped for use; when this part of the belt is worn-out, new material is advanced from inside the drum.

The P/R must meet a number of requirements, some of them in conflict. The sensitivity must be high; this requires good absorption of the appropriate light, high efficiency of electron–hole generation and transport (which tends to imply high fields), and reasonably low surface charge (which is proportional to V/d, where V is the charging voltage and d the P/R thickness). V must be high enough for good development. The thickness d must not become so large that the carriers start to be deeply trapped before they traverse the whole P/R thickness, and economic and adhesive factors speak for lower values of d. The field in the P/R must not be so large that breakdown or excessive dark conduction occurs. In addition, the P/R must cycle reproducibly many times without excessive trapped charge (which can prevent proper photodischarge or can effectively increase the P/R dark conductivity because of an increase in field at the electrode), without other increases in conductivity, and without mechanical damage or physical changes such as crystallization.

Typical charging voltages are in the region of 1000 V, although excellent results have been obtained with only a hundred volts or so; selenium-based P/Rs are usually about 50 μm thick while organic layers tend to be about 25 μm. A typical selenium P/R therefore acquires an initial surface charge density of about 10^{-7} C/cm^2 or the equivalent of about 7×10^{11} electronic charges/cm^2. Substantial photodischarge with a efficiency of, say, 0.7 electron–hole pairs per absorbed photon therefore requires about 10^{12} photons/cm^2. If each photon is blue and therefore carries about 3 eV or about 5×10^{-12} ergs, the required exposure for full development is about 5 ergs/cm^2. The charge density in the nonexposed areas is enough to attract a dense black layer of toner particles. Xerography may be capable of much higher sensitivities (ISO 500 has been suggested) if P/Cs having few defects would permit the use of much lower fields, thus requiring neutralization of fewer charges; for example, very experimental CdSe systems can be completely discharged by exposures as low as about 0.3 ergs/cm^2, and ISO sensitivities up to 100 have been claimed.

2. Dry Toners

Typical dry xerographic toners are plastic particles (often styrene–acrylic copolymers), typically about 10–15 μm in size, colored black with about 5–10% of submicrometer carbon-black particles, or (for color processes) colored with cyan, magenta, or yellow colorants. Early toners were charged by passing insulating particles through electrically active nozzles, thus forming a charged powder cloud. This process can give excellent continuous-tone images and can allow a close-spaced development electrode, but is difficult to control, can tend to give background, and is difficult to scale to high process speeds. Most modern dry toners are charged via contact electrification caused by contact with relatively massive carrier particles (Fig. 10): if different materials are brought into intimate contact, they tend to charge each other via electron flow under the influence of their different electronic energy levels. Since the toner particles, which constitute about 1–2% of the total mass, are attracted to the carrier as well as to the P/R, there is less tendency for them to be deposited in uncharged areas, and if they are deposited there, they tend to be cleaned up later by reattachment to other carriers. In cascade development (Fig. 10), this mixture is poured over the exposed P/R, and toner is knocked loose, redistributing itself between the carrier and charged areas of the P/R. Inverted cascade development uses a similar concept, but the carrier is cascaded down the development electrode with the P/R above it: gravity thus acts against transfer in this case. With insulating carriers such as small glass beads, edge development is prominent, because the size of the carrier beads (typically 250–600 μm) makes it impractical to bring a development electrode close to the P/R.

Conductive carrier beads, however, can act as their own development electrode. A widely used technique is the magnetic brush system. A magnet carries a mixture of fine toner with coarse (typically 100 μm but sometimes down to about 20 μm) magnetized carrier beads; the beads may be of ferrite or steel, usually overcoated with an insulator such as PTFE to confer the required charging properties and able to release the toner easily. The system is in the form of a soft brushlike structure (Fig. 11) which is moved

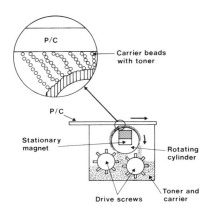

FIG. 11.   Schematic illustration of the magnetic brush development process. For simplicity a flat P/R is assumed.

against the P/R. The brush electrode may be electrically biased to compensate for incomplete P/R discharge or to change the gamma of the process. By adjustment of the conductivity of the developer mixture either edge-contrast or solid-area development can be favored. The magnetic brush system offers good process control, is capable of high process speed, gives low background, can be fitted into a smaller space than cascade systems, and does not depend on gravity, giving greater process freedom.

Single-component development systems dispense with the carrier and may use toner particles that themselves carry magnetic material. Such toner may be applied, for example, from a magnetic roller. Single-component developers have the advantages that there is no carrier to wear out and no necessity to keep a reasonably constant balance of toner and carrier, there may be less P/R abrasion, and the development unit may be simpler. The toner-bearing surface may still be moved against the P/R to give a good supply of toner. Alternatively, "touchdown" or "impression" development may be used: a toner layer, typically on a soft donor surface (such as a conductive brush fabric or an elastomer layer) is pressed into contact with the P/R without shearing motion. Close spacing between the roller or sheet and the P/R can provide solid-area coverage, and an electrical bias may be applied. However, close control of the P/R voltages is required, and care must be taken to avoid background.

Single-component systems may use relatively conductive toner, so that the charged P/R itself induces charge separation in them, giving the

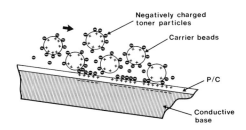

FIG. 10.   Schematic illustration of the cascade development process, showing toners and carrier particles.

required attraction to the P/R. However, toner transfer to the paper can then be more difficult, since the conductive toner may exchange charge with the paper and be repelled; moreover, the conductivity may be humidity-dependent. Other systems contact-charge an insulating single-component toner by contacting it to a blade held at a suitable voltage or by applying it to the P/R from a roller of an appropriate insulating material. One interesting insulating single-component magnetic system first forms a uniform layer of toner on a magnetic roller. The toner then jumps a 300-μm gap to the P/R, propelled by a combination of an ac and dc bias between the rotating roller and the P/R. As the distance between any given region of the toner layer and the drum reduces and then increases again, toner transfers, oscillates between the two, and finally is back-transferred from the light areas.

Toner plastics are chosen to melt at the lowest temperature possible (since fuser power and warm-up time are important practical considerations) but must not stick to each other, to the carrier during storage or in the photocopier prior to use, or too well to the P/R. Tacky toner encapsulated in a higher-melting coat may be a way around this conflict.

3. Liquid Toners

A different form of development uses micrometer or submicrometer pigment or pigmented plastic particles in a kerosene-type liquid. The particles may become charged by selective absorption of ionic species that are added as charge-control agents; reasonable stability against aggregation is then achieved by the resulting electrostatic repulsion. Alternatively, long-chain molecules may be attached to the particles to charge them and to form a diffuse shell around them to keep them apart. Very sharp images can be obtained with these small particles, which are carried very close to the latent image by the liquid and thus sample the field very close to it. (Dry toners in this size range are difficult to handle and may be a health risk). Heat fixing may be avoided, since the image may fix itself as the liquid dries, especially if a soluble film-forming plastic is contained in the liquid. Major disadvantages include the tendency of conventional toner suspensions to aggregate and settle out, and carry out of liquid on the paper, often with absorption of toner-bearing liquid in the background areas. Moreover, reusable P/Rs can be hard to clean when such developers are used, and some P/Rs are incompatible with the toner liquid over long periods. Fine lines may be smeared as the liquid image is transferred to paper. Special paper may be used to prevent undue spread of the liquid image by a blotting-paper effect.

Various ways of avoiding or reducing the liquid carry-out problem have been proposed. For example, a counter-rotating roller may be brought close to the P/R to remove excess liquid. Alternatively, much of the liquid can be squeezed from the toner suspension by means of a charge applied to it on the P/R; at the transfer stage, the toner can be made to jump to the paper under the influence of a field (the required separation may be maintained by large particles in the liquid), and enough solvent is transferred that no fixing is necessary. Another system applies the liquid toner on a sponge-rubber roller covered with a thin insulating mesh. This is pushed against the P/R and is thus squeezed. Excess toner and liquid are removed by the sponge as it expands on separating from the P/R. Yet another approach is to use a P/R that is not wetted by the developer except with the aid of the field from the surface charge. In a different liquid-development method, conductive ink has been electrostatically drawn out of the recesses of a finely patterned applicator onto the P/R only in charged, nonexposed areas.

C. Direct Xerography

In direct xerography, the P/R is factory-coated onto the image substrate, usually paper, and remains there after imaging. There is no requirement for image transfer or P/R cleaning and recycling, which significantly reduces the mechanical complexity of the copier. However, the paper no longer appears "plain" (apart from its greasy feel, it cannot be perfectly white if it is to absorb visible light), and its cost is significantly increased. Typically, zinc oxide, incorporated in a binder and with appropriate dye sensitizers, is used as the P/C. The paper is usually treated to make it reliably conductive (reasonably independent of humidity), so as to form the required underlying electrode. Development may be with dry or liquid toners, and repeated exposures through appropriate filters can give color images. Direct xerography with no charging step has been reported from development of the small "Dember" voltages arising from independent migration of photogenerated electrons and holes in the absence of a field.

D. PICTORIAL XEROGRAPHY

Xerography is typically a fairly high-gamma process, which is appropriate for applications such as document copying. The high gamma arises principally because the discharge of most P/Rs occurs over a fairly narrow range of light exposures, the sharply defined voltages on them then being converted nearly proportionally to density with most modern developers. For pictorial applications, however, it has been demonstrated that multiple screening techniques can reduce the effective gamma to levels acceptable for pictorial purposes, and it appears that the other factors that militate against high-quality color or b/w pictorial reproduction (such as sharpness and process reproducibility) can be dealt with using liquid or maybe even dry toners. Color fidelity can probably be controlled either with special P/Rs and multiexposure processes or by appropriate correction of electronic input. Coated paper is quite acceptable for many pictorial applications, so the degradation that can accompany toner transfer may be avoided with some processes. Some liquid-developed processes have already been claimed to give photographic quality, although processing can be slow; liquid developers may also need to be understood better if stability and reproducibility is to be obtained, especially without electronic correction systems. Some of the PAPE systems described below also offer serious possibilities for pictorial applications.

E. PHOTOACTIVE PARTICLE ELECTROPHOTOGRAPHY (PAPE)

In PAPE, as in xerography, development occurs via the difference in Coulombic (charge multiplied by field) forces between exposed and nonexposed areas. Unlike xerography, however, in which the toners have a constant charge and are differentially attracted to the P/R via differences in the fields that they see, in PAPE the "toners" themselves undergo a change (usually a reversal) of charge, and the developing field is usually constant across the image. The related XDM process is discussed in Section VII.

Conceptually, the simplest PAPE process involves dusting a uniform layer of photoconductive particles across a receiver electrode, charging and exposing the powder, blowing away the discharged powder in the exposed areas, and fixing. This process has been used extensively in special applications, for example, in marking metal plates during shipbuilding.

1. Color PAPE

One of the most extensively investigated PAPE processes is the one-step full-color photoelectrophoresis (PEP) system. An oil suspension of roughly equal parts of photosensitive cyan, magenta, and yellow particles is subjected to a high electric field (10^5–10^6 V/cm), using a transparent electrode on one side and a metallic electrode coated with an insulating layer (to prevent short-circuits and charge exchange at the metal) on the other; it is exposed once to a full-color image. The particles acquire an initial charge in the dark. In PAPE systems generally, this may be deposited by a corotron or may come from charge injected into the liquid by one of the electrodes, generally the noninsulated one; this is often controlled by very small concentrations of impurities deliberately added to the liquid, and the sensitivity of the systems is determined partly by the minimum dark charge that can be used without some of the particles ending-up with the "wrong" charge. When the PEP particle absorbs light, electron–hole separation occurs, and one sign of charge may be injected out, probably into an impurity in the liquid. When the other sign of charge remaining in the particle becomes large enough, it may neutralize, and eventually reverse, the dark charge. Such particles therefore move from one electrode to the other. The receiver electrode thus becomes a negative of the original, while a positive is left on the other electrode. For example, red light is absorbed by the cyan pigment, which therefore migrates; the donor electrode (which originally appeared black) is now deficient in red-absorbing pigment and therefore appears red. The positive, for example, is transferred to paper, and the electrode reinked.

The system needs exposures in the 100–1000 ergs/cm² range and can give good images, but there are some problems. For example, charge exchange may occur between a particle that has absorbed light and one that has not, especially if one is located above the other; moreover, some particles tend to remain on the donor electrode, giving background, and most available solid-state photoactive pigments have nonideal absorption.

To get around some of these problems and to

remove the necessity for an ink, another system (known as charge injection mosaic, CHIMO, or color PAPE, COPAPE) uses a precoated and dried *monolayer* (thus reducing interparticle injection) of plastic particles on a substrate such as a polyester, each particle containing one of the required three photoactive pigments and also a solid dye of the appropriate color. The mosaic is wetted with an insulating liquid and is exposed to light in the presence of a high electric field. If a pigment is on the surface of the particle, either the electron or hole produced by illumination may be injected into the liquid as before, and again the existing dark charge can be reversed. The resulting image is further developed by heating, so that the dye diffuses into the plastic-coated receiver and is diluted, giving more ideal color; the remaining insoluble photoactive pigment is washed away. Alternatively, the dye could be dissolved in the particle to start with. Although the images are somewhat grainy, such a process holds promise for remarkably simple functional color reproduction. Exposures in the 100–1000 ergs/cm^2 range are needed.

2. B/W PAPE

Black and white PAPE systems are often driven by the sensitive X-phthalocyanine material and therefore require only about 10 ergs/cm^2 exposure. They include ink-based processes ("MINK"), that use plastic particles with appropriate photoactive materials and colorants; the ink is charged by a corotron so that it sticks to a polyester web, after which the field is applied by contacting the ink to a metal roller and charging the back of the polyester. After exposure and separation of the polyester from the metal, the ink layer splits imagewise. The process has high gamma and process speed and sufficient sharpness for electronic printing. It works best in the negative mode, because some ink is left on the polyester in illuminated areas. Other systems use a continuous layer ("Manifold") of agglomerated particles containing similar materials: this is softened by gently heating to the melting point of a wax that is also present in the layer, whereupon the photogenerated charge permits fracturing of the layer to form the image: the exposed areas stick to one electrode and the nonexposed to the other electrode. A process related to MINK and PEP uses either photoactive particles mixed with inert coloring particles, or a photoactive layer in contact with

a suspension of inert particles: the colored particles migrate because of charge exchange with the photoactive material. Such a system could be used with light of a single wavelength to form color reproductions from color-separation negatives or from electronic input.

F. PHOTOPLASTIC IMAGING

A plastic layer with a very high field applied across it, for example, a layer with an ionic charge on one surface and an electrode on the other, is unstable when heated or otherwise softened and tends to form a pattern of wrinkles in the surface. This is related to a mutual repulsion of surface charges, which tends to increase the surface area, and to attraction between this charge and the counter charge in the electrode, which enhances edge contrast. If the plastic is also a P/C or if it is in contact with another photoconductive layer, this phenomenon can be used to form visible images. For example, a commercial microfilm uses a plastic containing a dispersion of fine P/C particles. After charging and exposure the material is developed rapidly by heat alone. However, to keep the finished image reasonably transparent, the selection and concentration of the P/C is limited, and exposures in the 100 ergs/cm^2 region are needed. Moreover, special equipment must be used for viewing the scattered-light images, the sharpness is limited by the minimum wrinkle spacing (which is in turn directly related to the plastic thickness), solid areas are not well reproduced, and the material (since it relies on a surface effect) is very sensitive to surface contamination or damage. On the other hand, it is possible to update old images and also to erase them (which may or may not be an advantage) by heating in the absence of a field. The material can be panchromatic and (like most electrophotographic materials) can be handled in reasonable light prior to use; it is more or less stable before and after imaging. Systems with much higher sensitivity have been described, for example, using a layer of selenium as the P/C in contact with a plastic.

The change in fields produced in a sandwich structure, containing a layer of a P/C and another layer, can also be used to change optical properties in other ways. Second-layer materials have included liquid crystals, ferroelectrics, or dispersions of flat particles that reorient like a venetian blind in a field.

G. Microfilms Using Liquid Toner Development

Several direct-xerographic microfilms are now available. To obtain high sharpness, these are developed with liquid toners, and no transfer step is used. Since the P/R is visible in the final image, it is typically an organic P/C in a transparent plastic with dye-sensitizers also present. Such films are charged, exposed, toned, and dried, and the toner is fused onto the P/R, usually by heat. Drying usually employs an air-knife (a long thin jet of air), a vacuum, or warm air. As with photoplastic films, the need to keep the P/R reasonably transparent dictates typical exposures in the 100 ergs/cm^2 range. A material using a crystalline solid P/R layer such as cadmium sulfide has also been described. Handling liquid toners and keeping them stable is inconvenient, but processing is rapid; the sharpness can be very high with fine toners, and color response can be reasonably good. Like photoplastic materials, these films can be handled in reasonable light prior to charging, are more or less stable before and after imaging, and can be updated after use by addition of new images next to old ones. The lifetime of the charge and of the latent image is not very long, and camera systems must be designed with this in mind. Positive- and negative-working systems can be designed using appropriate ionic-charging and toner-charge polarities, and (especially with a development bias) the gamma can be made low enough for continuous tone reproduction.

H. Some Other Electrophotographic Processes

The TESI (transfer of electrostatic image) processes transfer the electrostatic image (during or after exposure), rather than the toned image, from the P/R to a receiver such as insulator-coated paper placed in contact with the P/R. A bias field is generally applied. Such processes have been used commercially, but disadvantages include the need for specially coated paper and the dependence of the process on the contact geometry and the humidity.

Instead of charging uniformly and imagewise exposing, it is possible to charge an insulating material imagewise and then develop it conventionally. A P/C-coated metal screen is interposed between the ion source and the ion-receiving layer and is exposed before or during the charging step. Where there is no light, the insulating screen becomes charged by the incident ions and therefore soon repels them, stopping their transmission to the receiving layer. The resulting latent image formed on the receiver is toned and can be transferred and cleaned conventionally. This process has the advantage that the toning and cleaning steps do not have the opportunity to damage the P/C. If another insulating layer is used in the screen itself, the electrostatic image can be trapped, and many latent images can be made from the same exposure. Electronic signals rather than light can be used to control ion flow, for example, using an array of styli that are scanned past an insulator (typically paper with an insulating layer on it) and are raised to a high voltage at the appropriate time. Liquids are often used to develop such images. Processes of this type are known as electrography.

Other techniques have been used to allow many copies to be made from the charge pattern produced by one exposure. For example, some materials become significantly more conductive for some time after light exposure, because of trapping and slow release of electrons or holes. The material can then be repeatedly charged and toned, the charge in the once-exposed areas moving rapidly to ground without further exposure. Various "xeroprinting" processes have also been tried in which an insulating image (e.g., a toner image) on a conducting layer is itself repeatedly charged and toned, followed by transfer as usual.

A widely used approach to allow many copies from one exposure is to use a toner image (if necessary with appropriate post-toning treatment to give the required liquid attracting and repelling properties) as a printing master for conventional printing processes such as lithography. In fact, the first xerographic machines were used for this purpose and were probably crucial in ensuring the economic survival of the technology. Such masters are widely used, for example on paper-based substrates such as ZnO paper. Metal or other substrates may be used to improve on the fairly short print life of paper systems, but imperfect toner adhesion can still limit the life. One approach to avoid this is to form the toner image on a P/R that is itself coated onto or into a conventional (e.g., photopolymer) plate and to use the toner image as an *in situ* mask for subsequent conventional exposure of the plate; the toner image and the P/R are removed during the normal processing of the plate.

The charge transport in P/Cs has been used to trigger electrolytic development of silver ions or of ionic dye precursors. Materials so deposited have also been used as catalytic nuclei for subsequent silver or photopolymerization development. The gating of the sensitivity by the electric field in electrophotography has been used to make infrared-sensitive systems using development processes of this and other types; sensitivity has been reported in wavelength regions where the shelf-life of a conventional material would be so short (because of thermally generated fog) as to make it virtually useless.

Electrostatic means for imaging X-rays, in addition to the previously mentioned xeroradiography technique, include absorbing the X-rays in a liquid or gas placed between two electrodes, collecting the ions so generated on an insulator, and toning it.

Toning techniques somewhat similar to those used in electrophotography can be used with magnetic toner to visualize magnetic images formed either by writing with a magnetic head system, or by magnetic changes brought about by the heating effect of a bright image.

Many of the above processes have been applied to color as well as b/w reproduction. Various other color processes have been used. For example, three separate black toner images corresponding to the required three colors may be made; these are then used to absorb heat and thus transfer dyes off of the back surfaces of the disposable P/Rs to a single receiver sheet.

VII. Electrophotographic Migration Imaging: XDM and AMEN

XDM, and its recent outgrowth AMEN, are electrophotographic, silverless, nonpolluting photographic films and information-storage media. They are perhaps the first technologies that offer (and in some ways even improve upon) the high sensitivity/sharpness and stability, and the reasonable hardware/materials costs, that are usually associated with the former photographic term, while at the same time offering the convenience, office-compatibility, and user-friendliness that one has come to associate with the computer world, which is conjured-up by the latter term. They are the only nonsilver photographic films to approach AgX sensitivities (they could eventually exceed them) and are easily the most sensitive stable dry-developed films, even if AgX systems are included. They work by

unique mechanisms and exhibit high sharpness (because the photosensitive particles are very small and the emulsion is very thin), instant dry development, and excellent stability before and after imaging; they can also be panchromatic, can be handled in room light before exposure and sometimes even during development, and can be updated after use.

A. STRUCTURE

XDM and AMEN possess a unique, inexpensive structure, consisting (Fig. 12) of a nearly-close-packed monolayer of submicrometer photosensitive particles embedded a few tens of nanometers under the surface of a softenable plastic. This matrix polymer is usually about 1.5 μm thick and is supported on a transparent conductive substrate. The particles are typically selenium (Se) but can also be (for example) panchromatic Se alloys or infrared-sensitive phthalocyanines. The films, which do not contain any rare materials, are fabricated by solution-coating the conductive base with the matrix polymer (often with an adhesive underlayer) using conventional coating equipment, and then vacuum-evaporating the Se onto it. Immediately before passing over the selenium vapor stream, the films are heated to soften the matrix polymer. Rather than forming a normal uniform above-surface layer, the selenium then forms the extraordinary monolayer structure. This

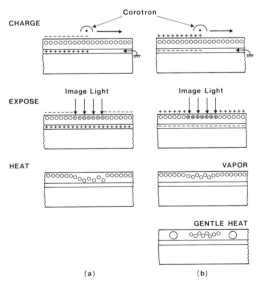

FIG. 12. The basic steps in the XDM (heat-development) and AMEN processes.

happens because the polymer has a greater affinity for the selenium than it does for itself, so there is a force driving the selenium molecules in; because of the softened polymer, the kinetic limitations that would normally cause a surface layer to form do not apply. The particles stay in a monolayer because of an equilibrium between a novel entropic force that pushes them towards the surface and a van der Waals force that tends to pull them deeper in. An abrasion-resistant polymer overcoat is usually solution-coated onto the finished structure. Because the optical density of the nonexposed films is high, no anti-halation layer is needed.

B. XDM

After charging (Fig. 12), the field in the films (typically about 10^6 V/cm) causes electron–hole separation with high efficiency in those selenium spheres that are exposed to light. Development consists of softening the matrix polymer for a few seconds, whereupon one sign of this charge annihilates the ionic charge on the adjacent film surface, leaving the spheres electronically charged to the same polarity as the original ionic charge. Most of the ionic charge in the nonexposed regions simply migrates through the matrix polymer to ground. The exposed spheres are attracted towards the electrode, and the other function of the XDM (migration) development process is to allow them to migrate towards it, leaving them dispersed in depth. The finished image is available a few seconds after exposure. No *removal* of material has occurred, merely *migration* (giving no effluent) in the exposed areas. However, the size of the spheres is similar to the wavelength of light, and so the exposed dispersed regions actually look quite different from the nonexposed close-packed areas, yielding a very good image. The small dispersed particles are independent, fairly-inefficient light-scatterers. Their optical density is therefore fairly low, certainly lower than a continuous layer containing the same amount of Se, because light can diffract (bend) around the individual particles. The nonmigrated monolayer particles, however, interact strongly in an optical sense and can be regarded as a continuous layer of Se (which absorbs strongly, mainly in the blue), with the small spaces between the particles producing an additional scattering peak in the green.

By far the most common development techniques for conventional XDM use heat alone (obviously a totally dry process) *or* the vapor of a widely used organic solvent. From the user's viewpoint vapor development is also essentially dry, since it involves only contacting the film with a small quantity of a simple vapor: the liquid remains in a small sump or sponge, never touching the film; no material is removed from the film, and no effluent is produced. Various forms of liquid contact development are also possible. XDM (and AMEN) exhibit excellent stability both before and after imaging, mainly because they then consist of almost uncharged particles of an inert material in an inert matrix: sensitivity to light and heat/vapor is "gated" when required by the application of the electric field.

C. AMEN

AMEN is developed (Fig. 12) by vapor followed by gentle heating. This causes agglomeration of nonmigrated uncharged particles in the nonexposed areas, resulting in a very low background-density. An electrically active additive in the film ensures that only slight migration occurs in the exposed areas, barely changing the optical density. This additive works by limiting the photogenerated positive charge on the exposed particles to a reproducible but small level. The slight migration in the exposed areas and the slight charge are enough to prevent the agglomeration that occurs elsewhere. The agglomerated particles are so widely separated and so much larger than the wavelengths of visible light that they become essentially invisible, leaving the film with virtually no background density. The processing latitudes of AMEN are very high, because most of the development processes are largely self-limiting. Unlike most XDM processes, AMEN is usually negative-working, although positive output can also be obtained. A simple additional negative ionic recharge step between exposure and development increases the sensitivity and optical density range of the film and allows development in normal room light!

D. Applications

XDM and AMEN possesses unique advantages in a variety of typically photographic areas including the printing, microfilm, and instrumentation recording businesses. Looked at as optical information storage media, they promise to simplify and reduce the cost of many information storage systems, both analog and digital, especially for the smaller user. Possible future improvements, or increases in the price of silver, could greatly expand the uses. In addition to light exposure, images can be formed on XDM by exposure to electron beams or to high-voltage styli. Migration images, especially if the particles are driven right to the base by liquid development, can be used for purposes other than simple viewing; resist-type applications, for example, have been described.

E. ULTIMATE SENSITIVITY

Typical XDM materials under appropriate conditions need exposures in only the 0.2–0.4 ergs/cm^2 region for 10% and 50% development, respectively, corresponding to only about 45 and 90 incident photons per 0.3-μm particle. (Since XDM and AMEN can absorb almost all the incident light, the number of absorbed photons is similar.) Optimally sensitized AgX crystals under normal atmospheric conditions need to absorb about one-fifth of these numbers of photons for similar development. However, AgX crystals of this size range typically absorb only of order 10% of the incident photons; thus, the required number of *incident* photons per particle is probably actually lower for XDM than for AgX. Experimental forms of XDM using different matrix polymers have been observed to have sensitivities higher by about another factor of five or so. It is the required number of incident photons per particle that really determines the ultimate sensitivity/particle-size trade-off. Thus, XDM has already exceeded this fundamental AgX trade-off, possibly the first practical material to do so. Nonoptimized panchromatic XDM materials have shown ISO sensitivities in the region of 20. The photogenerated charge in XDM and AMEN need not be nearly as large as the original ionic charge, since migration efficiency can be varied by changing the matrix-polymer viscosity, but it must not be swamped by the small amount of dark charge picked up by the nonexposed particles from the surface or from thermal generation. There is no reason to suppose that dark charging could not be further minimized, so still higher sensitivity appears possible even with similar particle size: the fundamental adhesive-force factors that normally limit electrophotographic sensitivity do not seem to apply to XDM. With larger particles, absolute sensitivities should be capable of great improvement; the inefficiencies that occur in AgX for particles larger than about a micrometer

probably have no counterpart in XDM, and ultimate sensitivities well in excess of the highest-sensitivity AgX materials appear possible in principle.

VIII. Other Processes

A. DIAZO MATERIALS

1. Principles

The diazo process is based on the ability of certain diazonium salts such as 2,5-dialkoxy-4-morpholino benzene diazonium fluoroborate to combine with dye couplers to form strongly colored dyes. When exposed to near-UV radiation the salts are decomposed, destroying their dye-forming ability, leaving an inert product and liberating gaseous nitrogen.

Examples of the reactions are the following (R and R$'$ being aromatic groups):

(No light) $\underset{\text{diazonium salt}}{R\!-\!N\!\!=\!\!NCl} + \underset{\text{coupler}}{R'OH} \rightarrow$

$\underset{\text{diazo dye}}{R\!-\!N\!\!=\!\!N\!-\!R'OH} + \text{other products}$

(UV) $R\!-\!N\!\!=\!\!NCl + H_2O \rightarrow$

$ROH + N_2 \uparrow + HCl$

The diazo material and the coupler are dissolved, each at about 10% loading, in a plastic matrix (usually a cellulose ester) about 4–8 μm thick. A few percent of an acid stabilizer is incorporated to prevent premature coupling and dye formation. After exposure, the "emulsion" is treated with an alkali such as ammonia to neutralize the stabilizer and permit coupling. Since the dye is formed in the nonexposed areas, the imaging process is positive-working. The material is generally heated during development to speed up the coupling process and keep the development time down to a second or so. Heat alone can be used to develop some diazo systems: for example, a material that promotes an alkaline environment can be incorporated into a separate layer of the emulsion, and when heated it diffuses into the layer containing the diazo salt and coupler. The problem with this, of course, as with most nonelectrophotographic heat-development processes, is that the process desired during development can also occur slowly during storage, leading to instability.

2. Status

During exposure, sufficient energy must be provided to destroy the diazonium salts by photolysis. There is thus no amplification, and the material is very insensitive, needing exposures in the 10^6 ergs/cm^2 region. (Related experimental, thermally developed systems incorporating leuco dyes have been described having significant amplification and thus higher sensitivity; other systems have produced minute particles of mercury by reaction of mercurous salts with the photoproducts, followed by amplification by physical development with a silver salt and a reducing agent.) Very often, diazo images are blue, but a range of colors can be formed using different dyes, and several dyes can be combined to produce black. Diazo emulsions are coated onto a variety of base materials including papers, cloths, and films. Negative-working systems have been reported, for example, using coupler precursors that are activated by the acid produced as a by-product by some of the exposed diazo compound.

While the sensitivity of diazo systems is very low, their sharpness is very high since the image is formed on a molecular scale. They normally have little sensitivity in the visible and can therefore be handled with care in room light, especially if steps are taken to limit UV exposure. Processing is very rapid, but the smell and handling of ammonia is a significant disadvantage. The shelf-life is limited, because the stabilizer can decompose allowing premature coupling; increasing the acid content will stabilize the material but at the cost of increasing the amount of ammonia needed and hence increasing the development time. The image lifetime can be good, and the material and processing costs are low.

Diazo materials are widely used as duplicating media, for example, in the microfilm industry: most people have seen roll-film or 6 × 4-inch microfiche with the typical diazo-blue microimages. Diazo systems are also used for duplication of engineering drawings and as the light-sensitive elements for certain photopolymer-type materials (Section VIII,D), for example, printing plates and resists. Certain diazo compounds can be imagewise decomposed by light to yield materials that absorb moisture from the air and thus become tacky; this pattern can then, for example, be dusted with phosphor powder in the production of color cathode-ray tubes.

B. Vesicular Materials

1. Principles

The vesicular process also uses diazo materials, but it works on a different principle, is developed by heat alone, is somewhat more sensitive than normal diazo, and is usually negative-working.

As in conventional diazo materials, the diazo salts decompose under the action of UV light, releasing nitrogen. However, the plastic matrix has a low permeability to nitrogen by virtue of a high density of very polar groups; for a short period, therefore, the gas remains largely trapped. This latent image is developed soon after exposure by heating the material to around 130°C for a second or so to soften the plastic, which allows the nitrogen to expand and form microscopic (~0.5-μm) bubbles or vesicles. The bubbles become firm and stable when the film cools, and they scatter light efficiently enough to form a visible image; it has been suggested that the plastic crystallites in partially crystalline matrices also tend to localize preferentially in a shell around the bubble, increasing the scattering effect. A dye may be present to improve the appearance and density range of the final image: because the path length of light passing through the vesiculated areas is increased by multiple reflections and refractions, the dye produces more density increase there than in the background regions. After development, the film is fixed by uniform reexposure to decompose the remaining diazonium salts; the resulting nitrogen escapes harmlessly over the next few hours as long as the film is not reheated.

Vesicular materials are generally coated on a polyester base that is dimensionally stable at the development temperature. The plastic matrix may contain additives such as other plastics or waxes to render it nonuniform and create formation sites for the bubbles; alternatively, it is treated, usually with hot water or steam, to produce microscopic cracks for the same purpose. The process may be made positive-working. For example, the initial nitrogen latent image may be allowed to diffuse away; subsequent uniform reexposure followed by development produces vesicles from the diazo remaining in the originally-nonexposed areas. However, the time taken for the first diffusion step can become excessive. Alternatively, a short, intense overall exposure after the image exposure can cause enough heat absorption in the originally-nonexposed areas that bubbles are selectively formed there.

2. Status

Because the bubble formation gives slight amplification of the image, the sensitivity is about 10 times better than diazo. Sharpness is fairly high, and like diazo the material has little sensitivity in the visible and can therefore be handled with care in room light. Processing is very simple and quick, and the material is inexpensive. The gamma is typically high, although (as in most materials) this can be reduced by a uniform preexposure; this reinforces the image exposure, relatively more in the lightly exposed areas. The shelf-life is long, and the images are stable so long as excessive temperatures are avoided. The images appear almost black-and-white in normal viewing situations, but it is usually difficult to duplicate them by contact to another material, because a wide-angle optical system collects the light scattered by the bubbles, and there is then little effective difference between the image and the clear background. Vesicular films have found wide use for microfilm duplication, especially for duplication of microimages generated via computer-output displays, where vesicular's negative-working characteristics can be convenient.

C. FREE-RADICAL DYE-FORMING PROCESSES

1. Principles

In free-radical photography, dyes are formed or destroyed via the action of a free-radical (a reactive intermediate whose molecules have an unpaired electron) on a dye intermediate, the free radical being formed via the absorption of light.

2. Examples

A very wide variety of chemical reactions falls under the above definition. The light sensitivity of halogenated hydrocarbons forms the basis for many of them. UV irradiation of carbon tetrabromide forms the $CBr_3^{\cdot}$ and $Br^{\cdot}$ free radicals, which can initiate many reactions. For example, a mixture of CBr_4 and diphenylamine in polystyrene gives a blue image on irradiation with UV, probably via a chain reaction that starts from the initial free-radical; the image can be fixed by solvent extraction of the unreacted materials or by heating to drive off volatile reactants. Various color images can be formed by different amines

or different halogenated hydrocarbons, stabilizers can be added to reduce premature dye formation, and sensitizers can extend the sensitivity well into the visible.

Another system uses iodoform as the light-sensitive material. This photodecomposes to release hydrogen iodide, which cleaves an acid-sensitive furfurylidene compound, one of whose reaction products reacts with an aromatic amine in the emulsion to form a dye. Development of the full optical density is accomplished by a heating step after exposure, which also drives off the volatile amine. A related system uses the difference in hydrocarbon solubility between the dye and its precursor either to fix the image by solvent extraction or to solvent-transfer unconsumed precursor to a receiving sheet. By using different dyes and multiple-layer films (or multiple single-layer films), full-color images can be built up on a single receptor sheet in this way. The DuPont Dylux system uses a photooxidizer, sensitive to light in the UV region, a leuco dye that is mainly colorless until it reacts with the photooxidizer, and also compounds that (when subjected to a uniform exposure to *visible* light after the image exposure) produce materials that interfere with the color-forming chemistry, thus photodeactivating and fixing the system. Deactivation can also be achieved by a higher-sensitivity photopolymerization (Section VIII,D) reaction prior to or during the color-forming reaction; this increases the system's viscosity enough to stop further color formation. A low-intensity imagewise photopolymerization exposure followed by a high-intensity blanket color-forming exposure is positive-working, while a high-intensity image exposure gives a negative from color formed before photopolymerization is well-advanced.

3. Optical Amplification

Blanket exposure to red light after the initial imaging exposure has been used in some systems to intensify the image and to provide some amplification: the dye image (but not the original photosensitive material) absorbs red light and transfers energy to lower-lying energy levels of the original material; these are energetic enough to initiate the color-forming reaction. An interesting variation of this, giving a dry-developed system with fairly high sensitivity, involves a dye base that responds to blue light to produce a material that sensitizes a subsequent photopolymerization process to occur under the influence of uniform red light; the unreacted monomer can then form a colored compound by a subsequent free-radical reaction with CBr_4 under the influence of uniform light and heat.

4. Status

While some amplification may be achieved in typical free-radical systems, for example, by chain reactions, the exposures required are usually in the 10^4 ergs/cm² range and often much more. Optical amplification can improve this to the 10^2 ergs/cm² region or better, while the described combination of this with photopolymerization can push it to about 10^{-1} ergs/cm². However, these higher-sensitivity materials, and to a lesser extent many of the others, have serious stability problems, because many of the processes that are intended to occur on exposure and fixing also have a tendency to occur in storage. The lifetime of the image dyes in free-radical systems can also be limited, and objectionable vapors can be produced on heat development. The sensitivities are often well into the blue or UV end of the spectrum. Maximum optical densities are often not high, and the more sensitive materials may not give black images. These materials can, however, be convenient and reasonably rapid to use, often with no liquids being involved, and room-light handling may be possible. Their sharpness can be high and their cost reasonable. They have achieved acceptance for applications such as duplicating, proofing (rapid production of preliminary prints for inspection purposes in the printing industry), and the like.

A number of low-sensitivity dye bleach processes have also been described.

D. Photopolymers

1. Principles

A polymer is a large molecule built up by the repeated linking of smaller units called monomers. "Plastics" are generally polymers. In the broad sense, photopolymerization processes may be defined as those involving a light-induced change of molecular weight in a polymeric system. As such, they may involve actual photopolymerization (polymerization of a monomer on exposure to light), and/or photocrosslinking (linking of polymer chains to form higher molecular weight, three-dimensional networks with greatly decreased solubility), or photodegradation of a polymer to lower molecular

weight fragments via the breaking of chemical bonds. In addition, photopolymers may work by a change of properties, such as solubility, occurring without major change in molecular weight, either because of light-induced changes in some or all of the repeat units or because of a change in an additive incorporated in the polymer.

2. Processes

Very often an initiator such as benzoyl peroxide is added to a monomer to start polymerization. In photopolymerization, this initiator is activated by light to form a material such as a free radical or an ion radical; this starts a chain reaction of polymerization, the radical reaction center propagating by successive addition of large numbers of monomer molecules. Photoreducible dyes may be used to extend the sensitivity from the UV into the visible; the dye in its excited state reacts with a weak reducing agent to generate free radicals. The chain reaction, which can be controlled and inhibited to achieve good shelf life, builds an amplification step into the process that is limited by chain termination arising from unwanted reactions or from the ends of the growing chain locally running out of monomer. Photocrosslinking also gives amplification, because large areas can be crosslinked with comparatively few links; chain reactions can also occur. Amplification can be achieved with photodegradation if, for example, a degradation product initiates another reaction or if a polymer chain, cleaved at one point, is unstable and then undergoes partial or complete depolymerization. Nonpolymeric materials, apart from initiators, may undergo light-induced changes that may have a variety of effects on the polymers with which they are in contact. The new material may, for example, take part in reactions such as crosslinking, or it may be a catalyst (e.g., an acid) for polymerization, depolymerization, changes to some or all of the repeat units, and so on; alternatively, it may inhibit the solubility of a polymer more or less than the original material.

3. Materials

Actual photopolymerization materials are typically in the form of a monomer liquid, alone or dispersed in a film-forming polymer; in the latter case, the photopolymerization of the monomer also causes linkages to the existing polymer and thus changes the mixture's solubility and other properties. In photocrosslinking, the reaction may occur via excitation of groups in the polymer or of an additive that can link to two chains. An example of the former is the cinnamoyl group which undergoes cyclodimerization to form a cyclobutane ring between chains containing the group. Again, sensitizers may extend the wavelength response. Azide additives photodecompose to nitrogen and imidogen radicals or nitrenes which can crosslink polymer chains, such as cyclized rubbers, that contain residual bond-unsaturation or other suitable groups. A well-known older system uses bichromate ions as the light-sensitive element; a change in oxidation state of the chromium during exposure produces ions that crosslink many natural and synthetic polymers, including proteins, gelatin, albumen, certain glues, shellac, and poly(vinyl alcohol). However, the sensitivity of bichromate systems is low and depends on a number of factors that are difficult to control, and their use has declined greatly. A common photopolymer system employs a diazo-oxide dissolved in a phenol-formaldehyde-type acidic polymer. The diazo-oxide inhibits the dissolution of the polymer in aqueous alkali solutions, but light-exposure converts the inhibitor to an acid, permitting the polymer to dissolve.

4. Development

The development of photopolymers generally makes use of one the changes in properties that occurs on exposure, most typically by washing away the more soluble regions, or more recently by removing one component by vacuum processes such as etching with reactive ions. Other techniques include transfer of the softer regions to another support, dusting-on of a pigment (using differences in tackiness) followed by transfer of the pigmented areas to another support, delamination or peeling apart of a film sandwich containing the photopolymer (using changes in the adhesive properties), or making use of changes in the transmissibility of inks or chemicals (which, for example, can permit color-coupling reactions to proceed imagewise, analogous to AgX chromogenic materials) or of differences in light scattering. In many cases, as we shall see, the visibility of a photopolymer "image" is not the prime concern, although the material may be specially pigmented where it would not otherwise be visible. When nonexposed material is not removed, for example, in the case of certain light-scattering images, appropriate materials may be included in the formulation so that

the primary reactants can be desensitized to form nonreactive compounds after exposure, for example, by heating or by irradiation with light of a wavelength below that used for exposure.

5. Status

Photopolymers can be more sensitive than many nonsilver systems, often needing exposures in the $10^4–10^5$ ergs/cm^2 region. (Experimental systems have improved this to the 100 ergs/cm^2 region; methods include minimizing chain termination effects, for example, those caused by oxygen, and using largely-crystalline monomers or other matrices in which photopolymerization may occur only in small liquid-like areas.) Considerable high- and low-intensity reciprocity failure may occur. UV (or shorter-wavelength) exposure is usually employed. Photopolymers find extremely important use in printing plates and as resists (e.g., in the fabrication of microcircuits, printed circuit boards, color cathode-ray tubes, and various other precision parts), where their good mechanical and film-forming properties, solvent resistance, image sharpness, and flexibility are critical. Other uses include holographic recording, replication of video discs, and three-dimensional object generation using a focused laser beam.

6. Printing Plate Applications

For printing plates, photopolymers may be applied to the baseplate during manufacture, or they may be coated by the user. A wide variety of photopolymerization, photocrosslinking, and other materials is available. Removal of the more soluble part of the image after exposure, leaving the underlying material, can give rise to a more-or-less planar plate, part of which is hydrophilic (water loving) and part oleophilic (oil loving) as needed for lithography. With thicker polymer layers the relief pattern needed by letterpress printing may be created.

7. Electronic Applications

During the microfabrication of electronic integrated circuits, a resist is coated onto an Si/SiO$_2$ wafer and exposed to radiation through a patterned mask. (The mask itself was conventionally made via photographic reduction processes onto high-sharpness AgX film, although electron-beam methods are now widely used.) After solvent development, the mask pattern formed by the resist is left on the wafer. Subsequent chemical etching of the SiO$_2$ along the pattern and removal of the remaining resist in a strong solvent generates a pattern of bared Si on the wafer, into which dopants and the like can be diffused. The whole process may be repeated several times.

Photopolymer resists may be applied from a solvent using techniques such as spin coating, in which the substrate is rapidly rotated as the solution is applied, leading to a uniform layer roughly 1 μm in thickness. Very high sharpness is obtainable with these systems, and they are used to photofabricate lines down to the 1–2 μm range. Positive-working photoresists (in which the irradiated regions are dissolved) commonly employ the diazo–oxide/phenol–formaldehyde solubility-inhibition system described previously, while negative-working resists have generally used photocrosslinking, often employing the cyclized rubber/bisazide system. To reduce the effects of diffraction on the sharpness of the lines obtainable, photopolymers may be exposed with deep-UV (towards 200 nm), electron beams, X-rays, or ion beams. (Experimentally, objects down to about 10 nm have been made with electron beams.) For these applications, different resists are needed. In the case of positive-working deep-UV resists, for example, chain scission of poly(methyl methacrylate) has been widely used. It is desirable to eliminate the wet-development step after exposure of the resist. Systems have been reported in which the light exposure degraded the polymer to volatile products; another approach is to use a dry-development technique such as plasma etching. Multiple-layer systems are being used in which a thin, high-sharpness resists acts (after development) as an *in situ* mask for a thicker one that smooths-over steps in the substrate, and optically bleachable overlayers with nonlinear optical properties can enhance contrast and hence sharpness. Inorganic resists are under investigation, since they may give higher effective sharpness as well as being plasma-developable.

Resists for lower-sharpness applications, such as printed-circuit boards, use much thicker emulsions than those for fine-line applications, and dry, self-supporting, adhesive photopolymer films are available.

E. PHOTOCHROMISM

Photochromic imaging processes are based on the ability of certain compounds to change their molecular arrangement (for example, between the *cis* and *trans* forms) and hence their color on exposure to light. The rearrangement is generally reversible, usually under the influence of thermal processes, although the reversal time may be long in appropriate environments. The erasability of photochromic images is of interest, although a number of approaches, for example, reduction of the temperature or various thermal or chemical treatments, has also been explored to render them more permanent. Photochromic image formation is instantaneous and needs no development steps, and the sharpness is high. However, given the lack of amplification, the sensitivity is generally very low. Furthermore, the color response is narrow, and the images are usually intrinsically impermanent. For transient imaging, the number of cycles is generally limited. A photochromic material has been used for making short-lived photographic intermediates during microfilm duplication, and certain metal compounds have been used in light-control devices such as sun glasses. AgX crystals dispersed in glass, for example, can be made to decompose to visible Ag and halogen in the light; because the halogen cannot escape, it recombines with the Ag when the light is removed.

F. TRUE ONE-STEP PROCESSES: OPTICAL DISKS

1. Principles

The rapidly increasing need for quick, convenient, and high-capacity storage and retrieval of information, much of it in digital form, and the wide availability of lasers has prompted much recent activity on photographic materials capable of true one-step instant information storage. The advent of consumer video discs, some of whose mastermaking and playback systems have aspects in common with the proposed digital "optical disks," has assisted the development of the latter, and such systems are becoming available.

In a typical optical disk system, a laser beam is switched on and off at rates approaching the ten MHz range by the digital information to be stored and is focused onto a spinning disk several inches to about a foot in diameter. An image consisting typically of spots in the 1-μm range is formed within nanoseconds with no processing; this pattern of spots can be instantly verified by a reading laser beam (usually reading by reflection) and constitutes the digital recording, each spot (or its absence) most simply corresponding to one binary bit of information. Thus the user can write and read information, but can typically not erase or alter it (except by recording updated information elsewhere on the disk).

For reasonable laser powers the materials should need exposures in the 10^5 to 10^6 ergs/cm^2 range; higher sensitivity could actually be detrimental, since reasonable signal-to-noise ratio in the reading system requires that the reading beam be able to apply exposures of order ten times less than the writing energy as often as desired, without causing unwanted writing. Clearly, therefore, the imaging material must possess a threshold energy level below which unlimited exposures do not cause imaging. That is, it must show extreme reciprocity failure for exposures extending over periods long compared with, say, a microsecond. Obviously the material must also show extreme sharpness, have very few blemishes (although software can correct for a few errors), and be at least reasonably stable before and after imaging. To keep surface dust out of focus, many schemes rely on exposure through an ~1-mm substrate. This must be very flat and smooth if proper focus is to be maintained, and can easily be the dominant factor in the very high cost per unit area of the recording materials. The other surface of the photosensitive material may also need protection from contamination. Alternatively, various means for keeping the recording surface permanently sealed from its environment have been used. Moreover, the small spot size implies a very small depth of focus, and the writing/reading hardware appears to be intrinsically rather expensive, at least for large disks.

2. Materials

The most widely studied one-step processes involve ablation of a thin solid layer, in which the light energy melts (and probably partly evaporates) the material, which pulls back by surface tension forces to form a pit, revealing the underlying layer. Thin (~5–30-nm) metal layers such as tellurium alloys have been widely studied. Their melting point is fairly low and their thermal conductivity is low enough that little sideways spreading of the heat occurs on timescales of a fraction of a microsecond; lower exposures, however, even if repeated many times, do not approach the threshold required for melting. Dyed plastics are another very important ablative system, pioneered by Xerox and Eastman Kodak; they can be efficiently sensitized in the near infrared for exposure by low-cost diode lasers, can be stable, can be fabricated very simply, and have the desired low thermal conductivity.

Quarter-wavelength transparent coatings may be applied between metal photosensitive layers and an aluminum underlayer; optical interference effects then minimize reflection by the nonexposed material, increasing sensitivity. The dye systems can be thick enough to act as their own quarter-wavelength layers. The dye-in-plastic systems have been read either by a change in reflectivity like the metal devices, or by optical interference effects resulting from phase changes between the light reflected from the imaged and nonimaged areas (perhaps assisted by the ridges around the holes); with the latter technique, they can be read with high laser powers at wavelengths outside their fairly narrow absorption bands, thus allowing better signal-to-noise ratio.

Other ablative materials include colloidal Ag (produced photographically) or other metals dispersed in a plastic matrix. Bubble-forming media have also been employed. For example, a layer of gold may absorb the laser light and transfer the heat to an underlying plastic layer that produces gases on heating. A "blister" is thus formed in the metal layer, changing its reflectivity. The top layer then does not need a low melting point and can be chosen partly for good stability; it can also absorb over a wide wavelength range. Yet another system involves the fusing together of separate metal layers, changing their reflectivity. Various other systems have been investigated.

Although such media may have information packing densities in the range of 10^{10} bits on a 12-in. disk, it may be desirable for applications not involving archival data to be able to erase and reuse some or all of the disk. Various systems are under investigation. They include (i) photochromics (Section VIII,E), (ii) special chalcogenide materials (i.e., materials containing tellurium, selenium, and/or sulfur) that can undergo a heat-reversible noncrystalline-to-crystalline transition that changes their optical properties, and (iii) magnetooptic systems in which, for example, a laser heats the surface of a noncrystalline magnetic layer enough to allow its direction of magnetization to be changed by a magnetic field; writing or erasure depends on the direction of the applied magnetic field and readout is via the change in polarization direction of a laser beam occurring on reflection from the material.

3. Prognosis

While optical disk systems will probably become important for some applications, there are serious concerns that may restrict their uses, particularly for write-once applications. Digital storage generally may be questionable for many applications, especially where the information starts in analog (human-readable) form, because of the cost, time, and complexity associated with conversion of the information to and from digital form and the access problems that may arise if digital formatting practices change. Optical disk systems themselves appear to be intrinsically rather expensive for both hardware and recording media, although their large storage density reduces the cost per bit to a reasonable level. There are serious unanswered questions about their archival stability, and exposed disks may often be difficult to duplicate conveniently. Once the capacity of a single disk is exceeded, one must resort to complex "juke-box" or multiple-disk systems if rapid retrieval is to be maintained. These limitations suggest that, for information which is not accessed very frequently, flexible roll-film media may have major advantages, since the very high spooling rates which are now becoming available allow a much larger area to be quickly retrieved than is possible with a disk, and could therefore permit somewhat larger spot size and much simpler optical systems; a combination of such a system with an inexpensive, easily developed, stable, and sensitive photographic medium like XDM (Section VII) may offer major hardware and material cost advantages for both digital and analog storage, especially for the smaller user. On the other hand, if reliable, low cost, erasable optical-disk systems can be produced, they may prove to be very competitive with existing means, primarily magnetic, of storing frequently-erased data.

4. Other Applications

In addition to digital applications, an ablative system on a flexible substrate has been used as an instant no-processing photographic film for recording computer-generated human-readable images ("Computer Output Microfilm" recording) using a high-power laser-scanning exposure system; ablative materials can also be used as the light-sensitive element of printing plates. A microfilm duplication film is also based on the structural changes (not ablation) occurring when a thin metal layer is momentarily melted by a short pulse of light. There are many older systems that make use of heat to produce images of various kinds, often by direct contact with an image to be copied and overall exposure to infrared radiation. Scanning lasers can also be used to transfer volatile dyes, and the like, imagewise from a donor sheet to a receiver.

G. Miscellaneous

Various other inorganic processes have been used commercially for special products. They include the blueprint process, which is based on the sensitivity of certain iron compounds such as ferric ammonium citrate and ferric oxalate. On light exposure, ferrous ions are produced, which react with ferricyanide ions during aqueous development to form insoluble, blue ferrous ferricyanide. Modern blueprints generally use diazo materials. The brownprint process is also based on the sensitivity of iron compounds (e.g., ferric ammonium oxalate). Again, the ferric ions are reduced to ferrous on exposure; these act as a reducing agent for various metal ions, such as Ag which produces metallic Ag images. These processes have very low sensitivity. Various other inorganic salts also have light sensitivity, including certain salts of mercury, copper, thallium, and lead.

Apart from the processes discussed elsewhere in this article, many attempts have been and are still being made to mimic the amplification capability of AgX while avoiding some of its problems. Various inorganic and organic systems are being used, including the following, most of which are experimental systems that have not been commercialized.

A number of "physical development" processes (where the developer, rather than the photosensitive medium, provides the reducible material for the visible image in addition to its normal role of providing reducing agent) have been used with novel photosensitive materials such as titanium dioxide and various organic materials. In the Itek RS process, photoelectrons produced in TiO_2 (with or without dye sensitization) were briefly trapped at the surface, long enough to form Ag nuclei from a silver nitrate solution in an activation stage of development; physical development, using the Ag nuclei to catalyze reduction of Ag or other metal ions, or of certain organic dyes, yielded the final image. Exposure requirements down to 1 erg/cm^2 were claimed, but the lifetime of the latent image was very short. While this was a major problem for many applications, it did mean that the image was reversible and updatable, and manufacture did not need to be in the dark. Image access time was short, and sharpness could be high. Most other systems of this type have been limited to the 10^3 ergs/cm^2 region.

Tellurium compounds have been used in recent nonsilver physical development processes. They are reducible to tellurium in the presence of photoreduced species such as hydroquinone, most efficiently if palladium compounds are included to form catalytic palladium nuclei; development can be by heat. Systems based on the *photo*reduction of cobalt(III) compounds to cobalt(II), followed by certain catalytic reactions, have also been extensively studied as heat-developed systems. Biochemical systems have been proposed: for example, vesicles of a membrane containing rhodopsin (obtained from dark-adapted bovine retinae) become permeable on exposure because of a *cis–trans* isomerization in the rhodopsin; this allows a color reaction to take place between reactants that were separated by the membrane, and high amplification can be achieved.

An interesting non-dye color process known as zero-order-diffraction has been described. Periodic surface-relief structures in a transparent medium can act as diffraction gratings, giving wavelength-dependent deflection of light because of its wave character. They can be made so that only light of one color is zero-order diffracted (i.e., is not deflected). Three such layers, each containing a grating (in an image pattern) tuned to one of the primary colors, can form a full-color image. While the cost of making such imaged diffraction gratings is high, they are easy to replicate by embossing techniques and therefore offer an inexpensive way of making multiple, very high-sharpness images that do not bleach under strong illumination (because they absorb little light). Possible applications include micropublishing and archival storage.

IX. Conclusion

We have tried to show that, despite its enormous achievements and its major role in our lives, the field of photographic processes and materials is very far from being a static, closed one. We can continue to expect significant advances in most of the areas we have covered.

In AgX photography, in addition to the normal steady improvements in such areas as dye stability and color characteristics, we may for example see attacks on many of the existing in-

efficiencies of the process. The low light absorption per AgX crystal may be addressed with new techniques involving the funneling of energy to the crystals via "antenna" multilayers of dyes. The structure of the crystals themselves may become more sophisticated, for example, by incorporation of dyes right into the crystals or by formation of hybridized crystals, a part of which absorbs the light efficiently while another part forms the latent image. Other types of two-component crystals may reduce the overall desensitization that can be caused by electron–hole recombination at sensitizing dyes, by trapping the latent image on a core inside the crystal and keeping it away from the dyes. Larger crystals may perhaps be made without the present inefficiencies that prevent the expected increase of sensitivity, and crystals of special shapes have already made their commercial appearance (T-grains). Better understanding of the latent-image formation process may make it possible to reduce recombination without resorting to hydrogen atmospheres and the like, and different size distributions, together with some of the above improvements, may increase the fairly low efficiency of AgX systems when they are judged as photon detectors.

The trend to photographic systems that are easier to process can be expected to continue both in AgX and other systems. Processes using less or no silver may become even more important if the price of silver eventually inflates substantially. Various electrophotographic processes (perhaps especially XDM-type systems) may find wider use in what are now regarded as camera-sensitivity photographic areas, and xerography and related processes will move more and more into mainstream printing. Electrophotography, especially with the help of electronic image-correction systems, may become a serious contender in areas such as photofinishing which require high pictorial quality. Electrophotographic and/or true one-step processes such as ablation will probably become extremely important in digital mass information storage, and systems like XDM offer great promise for this and for direct storage of existing human-readable information. Photopolymers and related systems will continue to make great strides as the requirements for electronic devices become more and more stringent.

Electronics will come to play a major role in photography, but no one can really predict whether or how far it will revolutionize the field. Already, to give just two examples, it is making cameras easier to use, for example, via intelligent exposure control systems, and it is improving the output of mass photofinishing processes by means such as intelligent color-correction systems. Electronic scanning (using a solid-state detector array) of photographic negatives in a camera store, projection of the resulting digital image onto a TV screen, color correction, cropping and other editing of the image by the consumer, followed by an immediate electronic printing process such as laser xerography, is one of many scenarios that are discussed. Alternatively, a consumer might own a similar device, which would record the user's editing actions, for example, onto magnetic tape; the tape and the image would then be sent to the photofinisher.

More radically, electronic cameras have been developed that scan the image transmitted by the lens and store it in digital form on a magnetic disk in the camera. The "final" image can then be a TV display of the original scene. While such devices are largely experimental at present, they could offer obvious advantages to specialized users such as press cameramen who wish to transmit photographs immediately via a phone line or similar system to their newspapers; indeed, a system of this kind was in use at the 1984 Los Angeles Olympics. It is much less obvious that such a system will ever supplant conventional camera films, especially when the latter are supplemented by systems such as those mentioned above. To record a series of really high-quality color images (analogous to the capability of a film) requires a very large amount of digital storage space, and the recording system is usually heavy and bulky. Even with expected improvements in the various technologies, it is hard to see an overriding force to drive a change-over from photographic materials.

As the ability to digitize other kinds of existing images becomes greater, nonpictorial photographic (i.e., direct visible recording of light) systems such as copiers can also *in principle* be replaced by combining the scanning system with an electronic printing device to visualize the electronic bit stream. Sometimes the advantages of this (such as the ability to modify and/or electronically transmit the image) may outweigh the possible disadvantages such as cost, complexity, memory limitations, and lowered quality; often they do not. In any event, the printing system is very often based directly on the corresponding photographic device and may

even use light to do the recording. A particularly important example is laser xerography.

In summary, we expect to see major advances in the very active field of photographic processes themselves and to see exciting improvements and simplifications based on the marriage of such processes with electronic systems. However, we see no reason to expect replacement of the photographic processes by electronic systems, except in specialized cases. Indeed, the uses of photographic processes may well become broader as major new applications are developed in areas such as printing and information/data storage. We believe that there are still major advances to be made in photographic processes, materials, and applications, and that they may well come from directions that will surprise most of us.

BIBLIOGRAPHY

Anonymous (1983). "Sensitized High Aspect Ratio Silver Halide Emulsions and Photographic Elements." Research Disclosure #22534, pp. 20–58, January.

Bartoli, R., Axelrod, D. J., Kurttila, K. R., McCormick-Goodhart, M. H., and Wolf, D. R. (1983). "Micrographic Film Technology." Association for Information and Image Management, Silver Spring, Maryland.

Bartolini, R. A., Weakliem, H. A., and Williams, B. F. (1976). *Optical Engineering* **15**, 99–108.

Carroll, B. H., Higgins, G. C., and James, T. H. (1980). "Introduction to Photographic Theory. The Silver Halide Process." Wiley (Interscience), New York.

Dessauer, J. H., and Clark, H. E., eds. (1965). "Xerography and Related Processes." Focal Press, London.

Jacobson, R. E. (1983). *J. Photographic Sci.* **31**, 1–12.

Schaffert, R. M. (1975). "Electrophotography." Focal Press, London.

Sturge, J. M. (ed.) (1977). "Neblette's Handbook of Photography and Reprography. Materials, Processes and Systems." Van Nostrand Reinhold, New York.

Thomas, W. Jr., ed. (1973). "SPSE Handbook of Photographic Science and Engineering." Wiley (Interscience), New York.

Thompson L. F., Willson, C. G., and Frechet, J. M. J., eds. (1984). "Materials for Microlithography." American Chemical Society, Washington, D.C.

Vincett, P. S., Kovacs, G. J., Tam, M. C., Pundsack, A. L., and Soden, P. H. (1986). *J. Imaging Sci.* **30**, 183–191.

Weigl, J. W. (1977). *Angewandte Chemie*, International ed. in English **16**, 374–392.

PHOTOVOLTAIC DEVICES

Stephen J. Fonash   *Pennsylvania State University*

GLOSSARY

Bucking current: Current produced in a photovoltaic device to oppose the photocurrent; its origins lie in recombination.

Photocurrent: Current produced in a photovoltaic device by the sweeping out of photogenerated electrons and holes.

Photo-generation rate: Number of electron–hole pairs created per volume per second in a semiconductor due to the absorption of light.

Photovoltaics: Direct conversion of light energy into electrical energy.

Recombination rate: Number of free electron–hole pairs that are annihilated per volume per second in a semiconductor.

Solar cells: Photovoltaic devices designed specifically to convert sunlight into electrical energy.

Photovoltaic devices are all solid-state or, in some cases, solid/liquid materials systems that are structured so as to convert light directly into electrical energy. These devices can be designed to optimize this energy conversion capability or they can be designed to optimize the obvious light detection capability. When photovoltaic structures are designed specifically to convert sunlight directly into electrical energy, they are termed solar cells.

I. Introduction

Photovoltaic devices are very closely akin to the transistors and diodes of microelectronics. Both technologies are based on semiconducting materials. The device physics and much of the processing used to produce photovoltaic devices is the same as that of microelectronic devices. However, in microelectronics one usually wants to attain as many devices as possible per unit area; in photovoltaics—at least in photovoltaics used for energy conversion—one usually wants to attain as much area as possible per device. The requirement of very small devices in microelectronics is driven by the need for low cost, high-speed complex circuits. The requirement of very large devices in photovoltaic energy conversion is driven by the need to produce as much energy as possible while minimizing assembly and interconnection costs. [*See* MICROELECTRONICS.]

II. Origins of Photovoltaic Behavior

Since light interacts with materials through energy quanta called photons, photovoltaic action is based on having materials that absorb photons. Since photovoltaic devices produce voltage and current, these materials, which absorb the photons, must do so by creating charged particles that are capable of moving through the material and carrying a current. The class of materials that absorbs photons by producing free, charged particles is semiconductors.

Photovoltaic action requires that the free, excited-charged particles created by light absorption be swept out of the photovoltaic device before they lose their energy. That is, a "check value" is needed that forces the excited charged particles to exit the photovoltaic device and to dissipate their energy as useful work in an external circuit. This sweeping out must occur before the free carriers can dissipate their energy uselessly within the device. The class of materials in which such a check valve, or sweeping action, can be achieved is again semiconductors. The very effective check valve that can be built into a semiconductor to sweep out the charge carriers is an electric field.

Figure 1 shows schematically these basic ingredients required for photovoltaic action. The photon of energy $h\nu$ (where h is Planck's constant and ν is the frequency of the light that is assumed for simplicity to be monochromatic) is seen to be absorbed due to an electron transition in the material from a ground state to an excited state. This transition caused by the absorption of light (light in the general sense, not just visible light) produces a free negatively charged particle (the free electron e) and a free positively charged particle (the resulting empty ground state or free hole h) noted in the schematic. A transport mechanism (drift in an electric field) is seen to sweep out the free electron and free hole from the device in opposite directions. This drift field also serves as a check valve preventing the mutual annihilation of the electron–hole pair seen in Fig. 1 and preventing the concomitant energy dissipation within the device. The electron is seen to be allowed to fill in the hole (recombine) only by traveling through an external circuit where it must do useful work.

In a semiconductor the ground state is actually a band of states that exist over a range of energies called the valence band. These delocalized states, capable of allowing hole motion, are not necessarily uniformly distributed in energy. Hence it is convenient to define a density of states $N_v(E)$ such that $N_v(E)\, dE$ is the number of valence band states between the energies E and $E + dE$. Likewise the excited state in a semiconductor is actually a band of states existing over a range of energies called the conduction band. Since these delocalized states, capable of allowing electron motion, are also not necessarily uniformly distributed in energy, there is a density of states $N_c(E)$ for the conduction band. The difference between the highest energy available in the valence band E_v and the lowest energy available in the conduction band E_c is termed the energy gap E_G. The difference between the lowest energy available in the conduction band E_c and the highest energy available in the material E_{LV} is the electron affinity χ. Electrons with energies greater than E_{LV} are able to escape from the material. In general, E_{LV}, χ, and E_G may be functions of position in a semiconductor. Since a range of ground states and excited states are available in a semiconductor, all photons obeying

$$h\nu > E_G \qquad (1)$$

can be absorbed by a semiconductor with an energy gap E_G.

Light of a frequency ν obeying Eq. (1) is absorbed according to

$$\phi = \phi_0(\nu)e^{-\alpha(\nu)x} \qquad (2)$$

Here $\phi(\nu)$ is the flux density (photons of frequency ν per area per time) present in the device at $x = 0$ and α is the absorption coefficient for these photons of frequency ν. The number of electron–hole pairs created per volume per time $G(X, \nu)$ due to the absorption of this light is the magnitude of the derivative of Eq. (2); namely,

$$G = \alpha\phi_0 e^{-\alpha x} \qquad (3)$$

This quantity G is also known as the optical generation rate. If the light impinging on a photovoltaic device has a spectrum of frequencies present, then Eq. (3) must be summed over the spectrum. Those photovoltaic devices that use sunlight as their source of light energy have the photon spectra seen in Fig. 2 as their basic energy source.

Actually Fig. 2 shows two examples of the many solar spectra possible. These examples are air mass zero (AM0) and air mass two (AM2) spectra. The AM0 spectrum is the solar energy distribution available just outside the atmosphere of the earth whereas the AM2 spectrum is essentially a representation of the solar energy distribution available at the surface of the earth for average weather conditions. If one were to take a semiconductor with $E_G = 1.24$ eV, for example, then Fig. 2 shows that a substantial portion of the spectrum of solar photons (all those with energies ≥ 1.24 eV) would be available for energy conversion. Each of these pho-

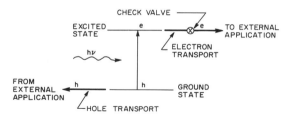

FIG. 1.   Basic features of photovoltaic action.

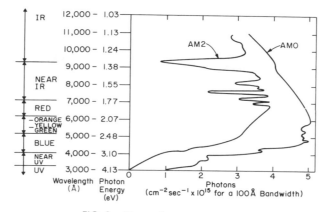

FIG. 2.   Two solar energy spectra.

tons with $h\nu \geq 1.24$ eV would produce an electron–hole pair as seen in Fig. 1. To achieve a useful solar cell it remains to design a structure, based on this semiconductor, which would sweep out these electron–hole pairs thus setting up a current and giving rise to photovoltaic action.

III. Origins of Photovoltaic Behavior—A Closer Look

As we have noted, light impinging on a semiconductor produces an electron–hole pair when a photon obeying Eq. (1) is absorbed. If these electrons and holes are created in, or diffuse to, a built-in electrostatic field region in the structure, then this built-in electric field can separate the carriers according to charge and sweep them out, setting up a current density J. If the photovoltaic device is under short-circuit conditions (terminals shorted together), this current density J would take on its maximum value J_{sc} and the voltage developed across the device would be zero. If the photovoltaic device is under open-circuit conditions (terminals open), this current density J would be zero. However, in this case the voltage developed across the device would be at its maximum value termed the open-circuit voltage V_{oc}.

It is interesting to note that at open circuit the device must develop a bucking current that is just able to counter the light-caused current yielding a net zero current. This bucking current has some value at all voltages but at short circuit it is zero. Hence, the full light-caused current J_{sc} is collected at short circuit unopposed by a bucking current. Expressed mathematically the net current density J under illumination is re-

lated to the voltage developed by the device according to

$$J(V) = J_{sc} - J_{BK}(V) \qquad (4)$$

In the idealization of Eq. (4) J_{sc} is the constant photocurrent arising from the presence of light, J_{BK} is the voltage-dependent bucking current opposing this photocurrent, and V is the voltage developed. It is seen from Eq. (4) that the open-circuit voltage is such that

$$J_{BK}(V_{oc}) = J_{sc} \qquad (5)$$

whereas

$$J_{BK}(0) = 0 \qquad (6)$$

and

$$J(0) = J_{sc} \qquad (7)$$

A sketch of the idealization represented by Eq. (4) seen in Fig. 3. Figure 3 shows that there is some current density J_{mp} and some corresponding voltage V_{mp} such that the product $J_{mp}V_{mp}$ is the maximum power density that can be produced by the device.

We have stressed that a built-in electric field is the source of photovoltaic action in a semiconductor due to the sweeping out and check-value effect it has on photogenerated elections and holes. However, there are two other possible sources of photovoltaic action that may be present in a device structure and which, if present, will add to the photovoltaic effect caused by built-in electrostatic fields. These two additional sources are effective fields due to material property variations and the Dember potential.

In a very general derivation of the sources of photovoltaic action, formulated in terms of con-

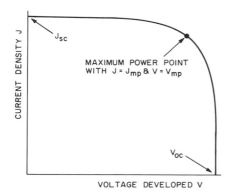

FIG. 3.   The current–voltage (J–V) characteristic of a photovoltaic structure under illumination. The device is producing power since the current density J is leaving the positive terminal.

tributions to the open-circuit voltage V_{oc}, it has been shown that the magnitude of the open-circuit voltage that is due to the presence of the built-in electric field is given by

$$\left| \int_0^L \left[\frac{e\mu_n \, \Delta n + e\mu_p \, \Delta p}{\sigma} \right] \xi_0 \, dx \right| \quad (8)$$

In this expression μ_n is the mobility of the free electrons in the conduction band and μ_p is the mobility of the free holes in the valence band. If n_0 is the population of free electrons per volume in the conduction band in thermodynamic equilibrium (no voltage and no illumination), then Δn in Eq. (8) is defined by $n = \Delta n + n_0$, where n is the population of free elections per volume under illumination. This Δn is a positive quantity since additional free elections are created by the photon absorption. Correspondingly Δp is defined by $p = \Delta p + p_0$ where p_0 is the free hole population per volume in thermodynamic equilibrium.

The quantity σ appearing in Eq. (8) is the conductivity present under illumination. This conductivity is given by

$$\sigma = e\mu_n(n_0 + \Delta n) + e\mu_p(p_0 + \Delta p) \quad (9)$$

The quantity ξ_0 in Eq. (8) is the electrostatic field present in the device at thermodynamic equilibrium; it is what we have already termed the built-in electrostatic field. This built-in field exists in the region $0 < x < L$ where $x = 0$ is one terminal and $x = L$ is the other. Since the quantity in the square brackets is ≤ 1, Eq. (8) allows us to note that the upper bound on the contribution to V_{oc} from the built-in electrostatic field is

$$\left| \int_0^L \xi_0 \, dx \right| \quad (10)$$

This expression stresses mathematically the importance of the built-in electric field to photovoltaic action.

A very general derivation of the sources of photovoltaic action also shows that the magnitude of the additional contribution to V_{oc} coming from the effective fields due to material property variations is given by

$$\left| \int_0^L \left[\left(\frac{e\mu_n \, \Delta n}{\sigma} \right) \frac{d\chi}{dx} \, dx \right] + \int_0^L \left[\left(\frac{e\mu_p \, \Delta p}{\sigma} \right) \right. \right.$$
$$\times \left(\frac{d\chi}{dx} + \frac{dE_G}{dx} \right) dx \right] - kT \int_0^L \left[\left(\frac{e\mu_p \, \Delta p}{\sigma} \frac{d}{dx} \right. \right.$$
$$\left. \left. \times \ln N_v^{\text{eff}} - \frac{e\mu_n \, \Delta n}{\sigma} \frac{d}{dx} \ln N_c^{\text{eff}} \right) dx \right] \right| \quad (11)$$

The quantity kT is a measure of thermal energy at some temperature T and k is Boltzmann's constant. All the other quantities in this expression have been defined with the exception of N_v^{eff} and N_c^{eff}. These are effective valence and conduction band densities of states, respectively. These effective densities of states weight the actual densities of states $N_v(E)$ and $N_c(E)$ toward the states that are more likely to be populated (i.e., toward the states near E_c for the conduction band and toward the states near E_v for the valence band).

The quantities $d\chi/dx$, $d/dx(\chi + E_G)$, $d/dx(\ln N_v^{\text{eff}})$, and $d/dx(\ln N_c^{\text{eff}})$ play the role of effective fields, as may be seen by comparing Eq. (11) with Eq. (8). Equation (11) shows that varying material parameters, if present, can be used to give an effective fields component to V_{oc} that may augment the electrostatic field component. The importance of this effective fields component to V_{oc} is seen to depend on the ratios $(e\mu_n \, \Delta n)/\sigma$ and $(e\mu_p \, \Delta p)/\sigma$ in the regions of material property variation. Equation (11) shows, for example, that a properly oriented change in the semiconductor conduction bandwidth $\Delta\chi$ could add $\Delta\chi$ to the V_{oc} arising from the electrostatic field, if $(e\mu_n \, \Delta n)/\sigma = 1$ in the region where χ is changing.

A very general derivation of the sources of photovoltaic action also shows that the third and final, possible additive contribution to V_{oc} is the Dember potential whose magnitude is given by

$$\left| kT \int \frac{1}{\sigma} \left(e\mu_p \frac{d}{dx} \Delta p - e\mu_n \frac{d}{dx} \Delta n \right) dx \right| \quad (12)$$

Generally it has been argued that this Dember contribution to V_{oc} should be small compared to the built-in electrostate field term. The basis of this argument has been to assume that $\Delta n \simeq \Delta p$. Then Eq. (12) shows the Dember potential is zero unless $\mu_n \neq \mu_p$.

It would seem from Fig. 1 that Δn should be equal to Δp. However, carriers move away from their point of generation with different effectiveness (mobilities). Furthermore, there can be another very important reason why $\Delta n \neq \Delta p$: in amorphous solids and in organics there can be energy levels in the gap between E_v and E_c. Unlike the delocalized states of the valence and conduction bands, these are localized (atomiclike) states that do not allow carrier motion and, therefore, can not contribute to photovoltaic action. These gaps states, as they are called, may trap, however, some of the photoexcited electrons or holes or both. Consequently, in these materials, the steady-state photoinduced Δn in the conduction band and Δp in the valence band may be very different. The difference would depend on how many photocreated electrons and holes get trapped in the localized gap states. Hence, it is concluded that the Dember component to V_{oc} could be significant in materials with a large difference between the free-electron and free-hole mobilities and in materials with large densities of gap states. [See BONDING AND STRUCTURE IN SOLIDS.]

Photovoltaic devices are designed so that the photovoltaic action arising from any spatial variations in semiconductor properties (effective fields) is oriented so as to add to the photovoltaic action of the built-in electrostatic field. Similarly any Dember effect must be designed to be minimal or oriented in the sense of the photovoltaic action of the built-in electric field.

IV. Photovoltaic Structures

The question now arises of how one creates the all-important built-in electrostatic field somewhere in the light absorbing semiconductor. The establishment of a built-in electrostatic field region necessitates the creation of an interface between dissimilar materials. Since we already know a semiconductor is the key to photovoltaic action, the other required (dissimilar) materials could possibly be other semiconductors, metals, or liquids. Whatever is used, it must be capable of carrying a current so that photocurrent can be drawn from the device.

Electrostatic fields are created in materials during the exchange of particles and energy, which takes place whenever dissimilar materials are first placed into intimate contact. This exchange continues until thermodynamic equilibrium is established. This means that when materials are placed in contact to form a materials system, a flow of electrons, holes, and energy occurs across the interface until one electrochemical potential (Fermi level E_F) and one temperature T are established for this materials system. When one Fermi level and temperature are achieved, the materials system has reached thermodynamic equilibrium. Fermi level (or Fermi energy) E_F and temperature T are the two thermodynamic parameters needed to characterize a materials system in thermodynamic equilibrium.

Since Fermi levels generally lie at different energies in isolated materials, an electrostatic potential energy develops at an interface between dissimilar materials to insure that there is one Fermi level for the materials after a junction is formed. That is, an electrostatic potential energy, with its concomitant electrostatic field, develops across a space charge (or barrier or junction) layer at the interface between dissimilar materials. This potential energy develops to shift the energy bands of the constituent materials until their Fermi levels coincide, as required in thermodynamic equilibrium (no bias, light, or temperature gradients present). It follows that electrostatic fields can be built into photovoltaic devices by forming a junction between a light absorbing semiconductor and another conductor such as some other semiconductor (termed a homojunctions or heterojunctions), a metal (termed a Schottky barrier or surface barrier structure), or a conducting liquid (also termed a surface barrier structure).

Homojunction photovoltaic devices are materials systems that use the same basic semiconductor on both sides of the junction to form the electrostatic field, barrier region. On one side of such a homojunction the semiconductor is doped n type and on the other side it is doped p type. Doping a semiconductor n type means that, if this semiconductor were isolated, its Fermi level would lie in the upper part of the energy gap. This n-type doping is achieved by adding carefully chosen impurities (donors) that cause electron-donating localized gap states just below E_c in energy. These donor states cause the semiconductor to become n type because they have the attribute of being able to donate free electrons to the conduction band without

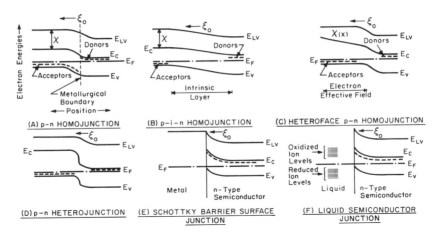

FIG. 4.   Types of photovoltaic structures. Shown in thermodynamic equilibrium (no light, no bias).

creating the corresponding free hole in the valence band. That is, when an electron leaves the donor site and becomes a free electron with an energy above E_c, a fix positive charge remains behind at the donor site and there is no corresponding free hole in the valence band. This results in a shifting of the Fermi level towards the top part of the gap.

On the other side of the homojunction we have noted that the semiconductor is doped p type. Doping a semiconductor p type means that, if this semiconductor were isolated, its Fermi level would lie in the lower part of the energy gap. This p-type doping is achieved by adding carefully chosen impurities (acceptors) that cause electron-accepting localized gap states just above E_v energy. Clearly the presence of these states results in the creation of free holes in the valence band without the creation of the corresponding free electron in the conduction band. That is, when an electron leaves the valence band and is captured by an acceptor state, a free hole is created in the valence band but the electron is immobilized at the acceptor, localized gap state.

When a p-n homojunction is formed between an n-type region and a p-type region, the necessity of establishing one Fermi level sets up the electrostatic field, barrier region. This is seen schematically in Fig. 4A. As may be noted from this figure, there is one Fermi level E_F for the structure in thermodynamic equilibrium; however, E_c, E_v, and E_{LV} have become functions of position. The region in which they vary is the space charge or barrier region (i.e., it is the location of the electrostatic field). In fact the electro-

static field ξ_0 present in thermodynamic equilibrium is given by

$$\xi_0 = \left.\frac{dE_{LV}}{dx}\right|_{\substack{\text{Thermo.}\\\text{Equilib.}}} \qquad (13)$$

which for a homojunction (same χ and E_G throughout the device) is also given by

$$\xi_0 = \left.\frac{dE_c}{dx}\right|_{\substack{\text{Thermo.}\\\text{Equilib.}}} = \left.\frac{dE_v}{dx}\right|_{\substack{\text{Thermo.}\\\text{Equilib.}}} \qquad (14)$$

If light were impinging on the p-n homojunction seen in Fig. 4A, the device would be driven out of thermodynamic equilibrium. Electron–hole pairs created in the electrostatic field region due to absorption of the photons of the light would have the free electrons swept to the right and the free holes swept to the left by the built-in electrostatic field. This would set up the photocurrent J_{sc} flowing out the left side. Free electrons and holes that are created outside the electrostatic field region could also contribute to J_{sc}, if they did not recombine and if they diffused into the field region for subsequent sorting out by the electrostatic field.

When light impinges on the device seen in Fig. 4A, a photovoltage is also developed. The voltage ranges in value from zero at short circuit to its maximum value V_{oc} at open circuit, as we have noted. This photovoltage sets up the bucking current J_{BK} that opposes J_{sc} and just balances it at V_{oc}. Consequently, this photovoltage is oriented to set up J_{BK} in opposition to J_{sc} as we have pointed out in Eq. (4). Hence the photovoltage in Fig. 4A is such that the left terminal is at a higher voltage than the right terminal. The device is producing power under light since the current is leaving the higher voltage terminal.

In semiconductor materials with poor diffusion lengths such as amorphous semiconductors photocarriers generated outside the electrostatic field region may not be able to get to that barrier region to be swept out. In this situation one would want to extend the electrostatic field over as much of the device as possible. In addition, in some semiconductor materials, such as amorphous semiconductors, there can be high densities of localized gap states, which can store charge and, therefore, can cause the development of electric fields. In such a situation one would again want a strong field built in across the device that would dominate over any fields developed by the trapping of charge in localized states. For these reasons the $p-i-n$ variation of the homojunction seen in Fig. 4B has been developed. It has also proven to be highly effective in single crystal silicon solar cells even though the diffusion length is long and the gap-state density is low. The i layer in a $p-i-n$ device is undoped and, therefore, relatively carrier free in thermodynamic equilibrium. Thus electrostatic flux lines have no charge to terminate on in this region and, consequently, extend from the p layer to the n layer as may be seen from the figure by noting that $\xi = dE_c/dx \neq 0$ over this region.

Another variation in the $p-n$ homojunction is shown in Fig. 4C. In this case a $p-n$ homojunction is the principal charge separating region. However, as may be noted from the figure, an effective field region has been added to the device forming what is termed a heteroface structure. This has been achieved in this example by adding a region where the electron affinity χ has been made a function of position. Note that the effective force [see Eq. (11)] in this situation acts only on elections and is oriented so as to aid the electrostatic field region.

As we have mentioned, photovoltaic devices that use junctions between two different semiconductors to set up the built-in electrostatic field region are termed heterojunctions. A heterojunction between a semiconductor doped p type and another semiconductor doped n type is seen in Fig. 4D. This is an abrupt $p-n$ heterojunction in that the metallurgical boundary between the two materials is seen to be sharp. If we exclude $n-n$ and $p-p$ heterojunctions, since they do not have large enough Fermi level differences to cause extensive built-in field regions, we find that there are 24 different types of heterojunctions; however, only 8 of these are useful since the others involve effective force fields that oppose the built-in electrostatic field. These eight, of which Fig. 4D is one, can also be fabricated in the $p-i-n$ variation.

Heterojunctions are useful photovoltaic structures for a number of reasons. They allow (1) the use of semiconductors that can only be effectively doped either n- or p-type and yet may have attractive property or cost considerations, (2) the use of wide-gap semiconductors as barrier formers, (3) the use of effective fields to aid in the collection of photocarriers, and (4) the use of effective fields to give rise to additional photovoltage.

A third type of interface structure used for photovoltaic devices is the metal–semiconductor junction. This is a surface barrier device in that the barrier is formed at the surface of the semiconductor. Such devices are also referred to as Schottky barrier photovoltaic structures. A metal–semiconductor surface barrier device is seen in Fig. 4E. Note that a metal has a continuum of energy levels available to electrons; hence, there is no E_c nor E_v for the metal in Fig. 4E. Since a metal has an enormous number of charge carriers on which elective flux lines may terminate, Fig. 4E and Eq. (14) show that the built-in electrostatic field region exists only in the semiconductor in a metal–semiconductor photovoltaic structure.

The metal–semiconductor (M–S) photovoltaic structure of Fig. 4E is seen to be very simple. It involves only one semiconductor and the barrier is formed by depositing a metal on the surface of that semiconductor. If light impinges on the structure through the metal, then the metal thickness must be <100 Å to allow light to pass through. To enhance photovoltaic response (mainly to enhance V_{oc}) a thin insulator layer is often inserted between the metal and the semiconductor forming the M–I–S device. This insulator layer must be thin enough so that electrons and holes can quantum mechanically tunnel across it, or it must support hopping transport, in order that the photocurrent can be collected.

The second type of surface barrier photovoltaic device is the conducting liquid–semiconductor junction seen in Fig. 4F. The electrochemical potential (Fermi level) difference between the isolated ion bearing liquid and the isolated semiconductor (shown to be n type in Fig. 4F) leads to the built-in electric field region, when they contact, seen in the figure. In properly designed liquid–semiconductor junctions much of the electric field region exists in the semiconductor, as shown in Fig. 4F. The built-in electric field in this type of photovoltaic structure is seen to collect photocurrent by sweeping holes to the reduced ion species and electrons away from the oxidized ion species in solution.

V. Optimization of Photovoltaic Response

Equation (4) shows that, to optimize photovoltaic action, one wants to enhance J_{sc} and suppress $J_{BK}(V)$ as much as possible. Doing this will, of course, assure the enhancement of V_{oc} and will, in general, allow a larger current density J to be collected at a voltage V. The ultimate cause of J_{BK} may be viewed as recombination (a free electron filling in a free hole) whether it is recombination in the semiconductor materials or recombination at contacts. In fact, at V_{oc} this recombination just balances the photogeneration caused by the light resulting in no net current. At voltages less than V_{oc} (see Fig. 3) generation dominates over recombination. In general, recombination decreases as the energy gap E_G increases.

Unfortunately, for an uniform photon spectrum J_{sc} would decrease as E_G is increased. Hence, there is a trade-off between increasing E_G to decrease $J_{BK}(V)$ and decreasing E_G to increase J_{sc}. The presence of such a trade-off means that there is some optimum value of E_G (hence, some optimum semiconductor) that will maximize the JV (power density) product and, therefore, maximize device performance. This optimum E_G will depend on the light spectrum, which will not necessarily be uniform with frequency (see Fig. 2), and on the detailed relationship between $J_{BK}(V)$ and E_G. For photovoltaic devices used as solar cells, the optimum band gap for the AM0 solar spectrum is found to be between 1.4 and 1.6 eV. This range in E_G is caused by the dependence assumed for J_{BK} on E_G.

VI. Device Performance

Homojunctions based on semiconductors such as silicon and GaAs have been developed and optimized as solar cells to the point where energy conversion efficiencies for concentrated sunlight in the 20 to 30% range and for one sun (no concentration) energy conversion efficiencies in the 20 to 25% range are attainable. These high efficiency devices use single-crystal semiconductors, which causes them to be expensive; hence, their use must be justified by space vehi-

cle or other applications where high costs can be accepted. In the case of silicon, the most efficient of these single-crystal devices uses the $p–i–n$ configuration. In the case of GaAs, heteroface structures have proven very effective.

Homojunction solar cells based on amorphous silicon, also in a $p–i–n$ configuration, have yielded solar energy conversion efficiencies in the 10% range. These amorphous silicon devices suffer from a degradation mechanism that causes more sites for recombination, and therefore, enhanced recombination, as the cell is exposed to light. In some versions of the amorphous silicon $p–i–n$ solar cell, the $p–i–n$ structure has evolved to a $p–i–n$ heterojunction in efforts to reduce this degradation and to enhance efficiency. In these devices the p or n layers, or both, can be other amorphous semiconductors or polycrystalline semiconductors.

The use of amorphous-semiconductor-based solar cells offers the hope of reduced cost since expensive, single-crystal materials are not employed. Also of paramount importance is the fact that amorphous materials really have different properties than their single-crystal counterparts and generally absorb light more strongly than the corresponding single-crystal material. Thus the α in Eq. (2) is larger for a given frequency for amorphous silicon, for example, than it is for crystalline silicon. The result is that thinner amorphous silicon films are needed to absorb the same amount of light. This again results in savings in material costs.

Besides amorphous-silicon-based-heterojunction solar cells there are already a number of other heterojunction photovoltaic devices that have surpassed the 10% solar energy conversion efficiency plateau. The list includes the $n–p$ device $(n)CdS/(p)CdTe$, the $n–i–p$ device $(n)Cd_xZn_{1-x}S/(i)CuInSe_2/(p)CuInSe_2$, and the $n–p$ device $(n)Cd_xZn_{1-x}S/(p)Cu_2S$. All three of these devices are thin-film polycrystalline structures.

Surface barrier photovoltaic devices of the M–S or M–I–S type have been fabricated on a variety of semiconductors. Among them are crystalline silicon, amorphous silicon, single-crystal GaAs, InP, and polycrystalline CdTe. Efficiencies greater than 10% for solar energy conversion have been achieved in a number of these structures. Attaining these conversion efficiencies with metal–semiconductor-type surface barrier devices has necessitated employing the M–I–S configuration to enhance the voltage. In general, M–S and M–I–S devices are found to lack the stability of homojunction or heterojunction solar cells due to interface instability problems.

Surface barrier photovoltaic devices based on conducting liquid–semiconductor interfaces have also been fabricated on a variety of semiconductors. These devices, often referred to as photoelectrochemical cells (PECs), have yielded efficiencies >10% for solar energy conversion using single-crystal semiconductors and comparable efficiencies using polycrystalline materials. This type of photovoltaic device has the unique attribute of spontaneous junction formation on immersing a semiconductor in a conducting liquid. However, these devices can be unstable under illumination due to photocorrosion, electrolyte impurity effects, ion exchange processes, and electrolyte degradation. Although extremely interesting for basic photochemistry and interface studies, it remains to be seen if this particular type of photovoltaic device will have any technological impact.

BIBLIOGRAPHY

All, R. H., Birkmire, R. W., Phillips, J. E., and Meakin, J. D. (1981). *Appl. Phys. Lett.* **38,** 925.

Baer, T. (1986). *High Technol.* (July), 26.

Emtage, P. R. (1962). *J. Appl. Phys.* **33,** 1950.

Fonash, S. J. (1981). "Solar Cell Device Physics," Academic Press, New York.

Fonash, S. J., and Rothwarf, A. (1985). Heterojunction solar cells, *in* "Current Topics in Photovoltaics" (T. Coutts and J. Meakin, eds.), Academic Press, New York.

Hodes, G., Fonash, S. J., Heller, A., and Miller, B. (1985). *In* "Advances in Electrochemistry and Electrochemical Engineering," (H. Gerischer, ed.), Vol. 13, Wiley (Interscience), New York.

Hovel, H. J. (1976). *In* "Semiconductors and Semimetals" (R. K. Willardson and A. C. Beer, eds.), Vol. 11, Academic Press, New York.

Kromer, H. (1957). *RCA Rev.* **18,** 332.

Marfaing, Y., and Chevallier, J. (1971). *IEEE Trans. Electron Devices* **ED-18,** 465.

Mickelsen, R. A., and Chen, W. S. (1982). *Conf. Rec. IEEE Photovoltaic Spec. Conf.* **16,** 781.

Rothwarf, A., and Bower, K. W. (1975). *Prog. Solid State Chem.* **10,** 71.

Tauc, J. (1957). *Rev. Mod. Phys.* **29,** 308.

Tyan, Y. S., and Peres-Albuerne, E. A. (1982). *Conf. Rec. IEEE Photovoltaic Spec. Conf.* **16,** 794.

Van Ryuven, L. J., and Williams, F. E. (1967). *Am. J. Phys.* **35,** 705.

PHYSICAL CHEMISTRY

Douglas J. Henderson   *IBM Almaden Research Center*
Charles T. Rettner

GLOSSARY

Activated complex: Short-lived transition state that occurs at the point of maximum energy along a reaction path when the molecules in a chemical reaction can no longer be considered as reactants or products.

Adsorption: Adhesion of a gas or liquid at a surface resulting in an increased concentration of the gas in the vicinity of the surface; to be distinguished from absorption which occurs throughout the solid or liquid.

Critical point: Point where two phases become identical and form one phase.

Degrees of freedom: Variables which must be determined to specify the state of a system.

Elementary reaction: Reaction concerning a single chemical step, such as dissociation or recombination, as distinct from complex reactions which occur through a series of separate elementary reactions.

Equation of state: Relation between the thermodynamic properties of a system.

Equilibrium: State of an isolated system which is specified by quantities which are independent of time.

Isotherm: Curve joining states for which the temperature is constant.

Kinetics: Study of how chemical systems change, concerning the rate at which change occurs and the factors upon which this rate depends. Also used to refer to the sequence of reactions by which a complex reaction occurs.

Molecular beam: Stream of molecules all traveling in the same direction in vacuum, used in studies of isolated molecules and to examine the dynamics of single molecular collisions.

Normal mode: One of a set of coordinates of a system that can be excited while the others remain at rest.

Order of a transition: Transition from one thermodynamic phase to another is of order n if the first discontinuous derivative of the free energy with respect to the thermodynamic variables is of order n.

Phase, thermodynamic: Region of the space specified by the thermodynamic degrees of freedom of system separated from the remainder by a clearly defined surface and within which the thermodynamic properties differ from those of the remainder.

Rate constant: Constant that gives a measure of the rate of a chemical reaction; the proportionality constant between the rate of product formation and the product of the reagent concentrations. If the rate expression involves N molecules of the same reagent, the concentration must be raised to the power of N.

Reversible process: A process in which a system changes from one thermodynamic state to another is reversible if the thermodynamic variables in the inverse process pass through the same values but in the inverse order and in which all exchanges of heat, work, etc., with the surroundings occur with reverse sign and in inverse order.

Spectroscopy: Analytical technique concerned with the measurement of the interaction of energy and matter, the development of instruments for such measurements, and the interpretation of such information for analysis of the structure or constituents of a system. Techniques, such as mass spectrometry which do not involve energy are often called spectroscopic because they also yield output scans in the form of spectra.

Spectrum: Intensity of a signal due to a process such as optical absorption or emission displayed as a function of some varying characteristic such as wavelength, energy, or mass. Also used in quantum mechanics and applied mathematics to specify the pattern of eigenvalues of a linear operator and in electrodynamics to specify the range of frequencies of electromagnetic radiation.

State function: In thermodynamics a variable is a state function if, when all the thermodynamic variables are specified, it has a unique value. As a result, the change in any state function in a reversible cyclic process must be zero.

Thermodynamics: Study of the changes in the properties of a system, usually as a result of changes in temperature or pressure.

Physical chemistry is the branch of chemistry in which experimental and theoretical techniques of physics are used to investigate and interpret chemical phenomena. Physical chemistry has its origins in the late nineteenth century, where it was largely concerned with the application of classical thermodynamics to chemistry. Modern physical chemistry is based more on quantum and statistical mechanics, which were developed only during the twentieth century. The branch of physical chemistry that employs twentieth century physical techniques is sometimes called chemical physics, with physical chemistry being regarded as concerned only with classical techniques. However, the distinction is artificial. Physical chemistry and chemical physics are really the same field and are considered as such here. Experimental physical chemistry has been revolutionized by relatively recent advances in electronic instrumentation, vacuum technology, and by the introduction of lasers. Equally, advances in computer power have had a great impact on theoretical studies, with an increasing emphasis on computer simulations and the detailed modeling of chemical systems.

I. Classical Mechanics

The dynamics (i.e., motion and energetics) of molecules and atoms and, at a more fundamental level, electrons, are the origin of chemical phenomena. Prior to the twentieth century it

was believed that all of the dynamics of a system, whether astronomical or molecular, were described by Newton's equation of motion,

$$\mathbf{F} = m \frac{d\mathbf{v}}{dt} \tag{1}$$

where $\mathbf{F}$ is the force, $\mathbf{v}$ the velocity, t the time, and the proportionality factor, m, the mass of the particle or object.

The force and velocity are vectors, whose direction and magnitude are both of importance. In complex problems it is often preferable to reformulate classical mechanics in terms of a scalar, such as the energy, which is characterized only by its magnitude. This gives rise to the Lagrangian and Hamiltonian equations of motion. The latter equations are of most interest here and are

$$\frac{\partial q_i}{\partial t} = \frac{\partial \mathcal{H}}{\partial p_i}$$
$$\frac{\partial p_i}{\partial t} = -\frac{\partial \mathcal{H}}{\partial q_i} \tag{2}$$

where p_i and q_i are generalized momenta and positions, respectively, and $\mathcal{H}$, the Hamiltonian, is the total energy of the system using momenta and position as variables. The space spanned by the p_i and q_i is called phase space. The dynamics of a system are described by a path in phase space. If the system is periodic, as is the case for electrons in an atom or molecule, then the path is a closed orbit in phase space. The advantage of the Hamiltonian formulation in physical chemistry is the fact that all variables are treated on an equal footing. However, the Hamiltonian and Newtonian formulations of classical mechanics are completely equivalent.

II. Quantum Mechanics

A. DUALITY OF MATTER AND ENERGY; UNCERTAINTY PRINCIPLE

During the nineteenth century it was established that matter consists of atoms and chemically bound aggregates of atoms called molecules. At first, it was thought that atoms were structureless. However, by about the turn of the century it was shown that atoms were miniature solar systems consisting of a positively charged nucleus, whose structure is irrelevant for chemical phenomena, which contains nearly all the atomic mass, and negatively charged "planetary" electrons which orbit the nucleus. Because of this, nuclei are slow moving, virtually motionless, on the time scale of electronic motions. For many chemical phenomena, the nuclei can be regarded as fixed in space. Only the electronic dynamics need be considered. This simplification is called the Born–Oppenheimer approximation. [See QUANTUM CHEMISTRY; QUANTUM MECHANICS.]

At first it was thought that the electronic motions could be described by classical mechanics. However, it is impossible to describe the microscopic world by classical mechanics. It became apparent, through for example the photoelectric effect, that electromagnetic radiation was not always wavelike, but, in some circumstances, consisted of discrete, particlelike, units of magnitude called quanta,

$$E = h\nu, \tag{3}$$

where ν is the frequency of the radiation and $h = 6.626 \times 10^{-34}$ J s is Planck's constant. Conversely, it became apparent, from for example the diffraction of electrons, that matter was not always particlelike but, in some circumstances consisted of waves of wavelength

$$\lambda = h/p \tag{4}$$

In other words, there is a duality of matter and energy. Whether the particlelike or the wavelike character of matter/energy is dominant depends on the experiment. In fact, the experiment itself interacts with the matter/energy and defines some aspect of the system at the cost of indefiniteness of some other aspect. This uncertainty principle was made precise by Heisenberg who showed that even under the most ideal circumstances

$$\Delta p_i \, \Delta q_i = h/4\pi \tag{5}$$

If the experiment defines the particle character of the system, the uncertainty of the positions, Δq_i, is small and the uncertainty in momenta, Δp_i, or frequency is large. However, if the experiment defines the wave character of the system, the reverse is true. The momenta p_i and positions q_i are called conjugate variables. Energy and time are also conjugate variables so that

$$\Delta E \, \Delta t = h/4\pi \tag{6}$$

Classical mechanics, where there is no uncertainty, is a limiting case in which the magnitudes of the variables are large compared to h. As a result, classical mechanics is appropriate for large macroscopic bodies.

B. Wave Equation

In the earliest formulation of quantum mechanics, classical mechanics was assumed valid with the exception that some periodic variables were quantized (i.e., had discrete values). Their values could be determined by integrating the p_i over their orbits in phase space, according to

$$\int p_i \, dq_i = n_i h \tag{7}$$

where n_i is an integer called a quantum number. The integral in Eq. (7) over a closed path is called a phase integral.

However, as the implications of the duality of matter and energy and the uncertainty principle were accepted, it became apparent that one could refer only to the probability of finding the system in some configuration. Just as the wave nature of radiation meant that there was a wave equation for radiation, the wave nature of matter implied the existence of a new wave equation. This wave equation, called the Schrödinger equation, is formulated as an eigenvalue equation (eigen ≡ characteristic or proper) where the Hamiltonian operator "operates" on the probability function or wave function or eigenfunction, ψ, to give the energy eigenvalue, E, times ψ. Thus, the wave equation is

$$\mathcal{H}\psi = E\psi \tag{8}$$

The wave function has the property that $|\psi|^2$ gives the probability of the system having the eigenstate whose energy is E. The Hamiltonian operator is formed by replacing p_j in the classical Hamiltonian by the operator $-(\hbar/i)(\partial/\partial q_j)$, where $\hbar = h/2\pi$ and $i = \sqrt{-1}$. The q_j remain unchanged. Interestingly, the earlier phase integral formulation [Eq. (7)] becomes the approximate Wentzel–Kramers–Brillouin (WKB) method of solution of Schrödinger's equation and remains useful in many problems in the sense that differences between quantized values of a phase integral are integral multiples of h except that there may be a zero point value of the phase integral given by a fractional value of h.

C. Hydrogenlike Atom; Electronic Transitions

One of the first systems to which quantum mechanics was applied was the hydrogenlike atom consisting of a single electron orbiting a nucleus of charge Ze_0. The energy eigenvalues or levels are obtained by solving Schrödinger's equation and are given by

$$E = -\frac{2\pi^2 m_0 Z^2 e_0^4}{n^2 h^2} = -\frac{hcRZ^2}{n^2}, \qquad (9)$$

where e_0 is the charge of an electron, n is an integer, c is the velocity of light, and R is called Rydberg's constant. Strictly speaking we should not use the electronic mass m_0 in this formula but the reduced electronic mass. However, the effect is small.

This means that electronic transitions between states characterized by n_1 and n_2 emit or adsorb energy or radiation whose wavelength is given by

$$\frac{1}{\lambda} = Z^2 R \left(\frac{1}{n_1^2} - \frac{1}{n_2^2} \right) \qquad (10)$$

An energy level diagram for the hydrogen atom is shown in Fig. 1 which also displays some of the allowed transitions.

The energy levels of the hydrogenlike atom are degenerate because more than one state corresponds to a specific value of n (the principal quantum number). These degenerate states are characterized by the quantum numbers l and m which characterize the spherical harmonics of the wave function. For each value of n there are n values of $l(l = 0, ..., n - 1)$ and for each value of l there are $2l + 1$ values of $m(m = -l, -l + 1, ..., 0, ..., l - 1, l)$ giving n^2 values of l and n for each value of n.

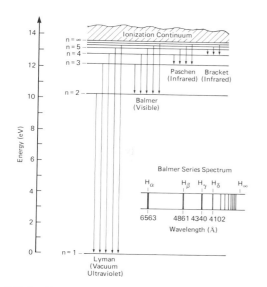

FIG. 1. Energy level diagram for the hydrogen atom showing allowed transitions. Insert displays the Balmer Series spectrum that can be observed in the visible.

D. MANY-ELECTRON ATOMS; PAULI PRINCIPLE; ELECTRON SPIN

To understand more complex atoms containing many electrons, we must solve the many-electron Schrödinger equation. Even in classical mechanics, many-body problems are difficult, so it is not surprising that many-electron quantum mechanics, usually called quantum chemistry, is an active research field today. However, an understanding of the electronic structure of atoms can be understood in terms of the *aufbau* (building up) principle whereby electrons are added one at a time to the atom.

However, two additional facts should be mentioned. First, the quantum numbers n, l, and m are not sufficient to specify the state of an electron. The spin of an electron must also be specified. Electrons can have one of two spins (say, up or down). This is specified by the spin quantum number, $s = \pm\frac{1}{2}$ (so that there are $2|s| + 1 = 2$ spin states). Thus, the state of an electron is specified by n, l, m, and s. For historical reasons the values $l = 0, 1, 2, 3, 4, ...$ are specified by the spectroscopic notation s, p, d, e, f, ... Thus, an electron might be said to be in a $1s(n = 1, l = 0)$ or a $2p(n = 2, l = 1)$, state. Similarly, states for the whole atom are termed, S, P, D, E, ... corresponding to $L = 0, 1, 2, 3, ...$, where L is the total orbital angular momentum for the atom, which is arrived at by combining the orbital angular momenta of the individual electrons.

Second, electrons obey the Pauli exclusion principle. This means that only one electron can occupy a quantum state. Thus, as the *aufbau* principle is employed, the electrons are added one at a time to the state of lowest energy, each state being filled by one electron.

From these principles a simple understanding of the periodic table is gained. Each electronic shell is specified by n and contains $2n^2$ states. Thus, the first row of the periodic table corresponds to $n = 1$ and contains 2 elements (H and He). The electronic configurations of these elements are denoted $1s$ (H) and $1s^2$ (He). The superscript indicates the number of electrons with the given value of l. The second row corresponds to $n = 2$ and contains 8 elements with configurations $1s^2 2s$, $1s^2 2s^2$, $1s^2 2s^2 2p$, ..., $1s^2 2s^2 2p^6$. The subsequent rows contain 8 columns even though $2n^2$ exceeds 8 because the energy of the electronic states is ordered (approximately) $1s/2s\ 2p/3s\ 3p/4s\ 3d\ 4p/5s\ 4d\ 5p/6s\ 4f\ 5d\ 6p/7s$ ... so that the third row is still filled with 8 elements. With potassium ($Z = 19$) the nineteenth electron goes into a 4s rather than a 3d level. The transition elements are regarded as occupying one position in the table since the outer shell configuration does not change as the d electrons are added and, as a result, they have similar chemical properties. As n increases, not only must the d electrons be accommodated in single positions, but the f electrons must also be accommodated in a single position, so that the table becomes more complex. However, the underlying principles are simple. [See PERIODIC TABLE (CHEMISTRY).]

E. MOLECULAR SYSTEMS: CHEMICAL BOND

Many-electron molecular systems are even more complex than atomic systems. The theory of such systems is an active field of research. The potential energy terms in the wave equation for the molecule involve not only the Coulomb repulsions between the electrons and the Coulomb attractions of the electrons and nuclei but also the Coulomb repulsion between the nuclei.

Given the repulsion between the nuclei, we are inclined to ask how atoms form a chemical bond in a molecule. A simple answer can be obtained by considering the one electron hydrogen molecule ion H_2^+. The wave functions of each H atom separately are $\psi_{1s}(A)$ and $\psi_{1s}(B)$. An approximation to the H_2^+ wave function can be obtained by forming a molecular orbital from a linear combination of these two wave functions

$$\psi_b = \psi_{1s}(A) + \psi_{1s}(B)$$

or

$$\psi_a = \psi_{1s}(A) - \psi_{1s}(B) \qquad (11)$$

In the first (bonding) orbital, the electron is concentrated between the nuclei, and is simultaneously attracted by both nuclei resulting in a lower electronic energy which more than offsets the repulsion of the nuclei. In other words when the electron is between the nuclei it acts as a cement holding them together. There is zero probability of finding the electron between the nuclei in the second (antibonding) orbital. As a result a chemical bond is not formed by this orbital.

F. MOLECULAR SYSTEMS; ENERGY LEVELS

When atoms combine to form molecules, the individual atomic energy levels give rise to discrete electronic energy levels or states of the molecule. The number of these molecular electronic states far exceeds those of the individual

atoms because of the many different ways in which the atomic states can be combined. Electronic states of molecules are classified in terms of several molecular quantum numbers in a manner analogous to atomic electronic states. For a diatomic molecule these include the electronic orbital angular momentum, l, its component along the internuclear axis, λ, and the corresponding quantities for the molecule as a whole, L and Λ. Just as $l = 0, 1, 2, \ldots$ gives rise to s, p, d, ... electron states and $L = 0, 1, 2 \ldots$ gives S, P, D, ... atomic states, so $\lambda = 0, 1, 2 \ldots$ yield σ, π, δ ... electron states and $\Lambda = 0, 1, 2 \ldots$ correspond to Σ, Π, Δ ... molecular states. The component of the total electronic angular momentum along the internuclear axis, Ω, is also of importance.

The ground electronic states of H_2, O_2, and N_2 are Σ states, while those of OH and NO are Π states. These latter two molecules have open valence shells with net spin of $\frac{1}{2}$, so that the multiplicity, S, of these states is 2, ($S = 2S + 1$), and they are termed doublet states. The former three have zero spin, a multiplicity of 1, and are termed singlet states, while states with a multiplicity of 3 and 4 are termed triplet and quartet states. Thus, the ground electronic state of NO is written as $^2\Pi_{1/2,3/2}$, where the 2 refers to the doublet nature of the state, the Π to $\Lambda = 1$, and the $\frac{1}{2}$ and $\frac{3}{2}$ refer to values of Ω which are arise from the two possible ways that the spin and orbital angular momentum can add.

In addition to electronic energy states, molecules posses both rotational and vibrational energy levels. Assuming a fixed distance between two atoms (rigid rotor approximation), the Schrödinger equation yields for the allowed rotational energy levels of a diatomic molecule

$$E_r \approx \frac{h^2}{8\pi^2 I} J(J + 1) \qquad (12)$$

where I is the moment of inertia of the molecule, J the rotational quantum number and the quantity $(h^2/8\pi^2 I)$ is termed the rotational constant for that particular electronic state of the molecule (usually given the symbol B).

Vibrational energy levels can be estimated by inserting the Hooke's law potential energy, $U(r) = 0.5k(r - r_e)^2$, in the Schrödinger equation (harmonic oscillator approximation). This yields eigenvalues, E_v, for the permissible energy levels, of

$$E_v = (h/2\pi)\sqrt{k/\mu}\,(v + \tfrac{1}{2}) = h\nu_0(v + \tfrac{1}{2}) \qquad (13)$$

where μ is the reduced mass of the system [$\mu = m_1 m_2/(m_1 + m_2)$] and ν_0 is known as the fundamental vibrational frequency. The smallest amount of vibrational energy a molecule can possess is thus $\nu_0/2$, termed the zero-point energy. Rather than the simple Hooke's law potential we may consider more realistic molecular potential energy curves such as a Morse potential given by

$$U(r) = D_e\{1 - \exp[-\beta(r - r_e)]\}^2 \quad (14)$$

where D_e is the dissociation energy of the molecule and β is related to D_e and ν_0. This leads to a similar expression for E_v, but with an additional quadratic term in $(v + \frac{1}{2})$, which is negligible for low vibrational energies. Similarly, an accurate treatment of molecular rotation leads to additional terms in higher powers of the quantity $\{J(J + 1)\}$.

G. Spectroscopy

Atoms and molecules can adsorb and emit radiation to change their internal energy states. The electronic transitions of the hydrogenlike atoms have already been mentioned. The quantization of the energy levels restricts the possible wavelengths of the radiation to discrete spectral lines. Only certain transitions are allowed and these are given by separate selection rules for electronic, vibrational and rotational transitions. For example in atomic transitions Δl can only take values of ± 1. In addition, each allowed transition has an associated transition linestrength. In electronic spectroscopy involving transitions between electronic states, these are termed transition moments, Franck–Condon factors and Hönl–London factors for electronic, vibrational, and rotational transitions, respectively.

The spectrum of absorption lines associated with a given species can be used as a fingerprint for identification and quantification in the process of absorption spectroscopy. An absorption spectrum is usually recorded with an absorption spectrometer or, more recently, with a laser. Similarly, emission spectroscopy is concerned with the radiation that is emitted by an excited atom or molecule. Here radiation is generally spectrally resolved using a spectrograph. In the various spectral regions, these techniques may be referred to as vacuum ultraviolet (VUV), ultraviolet (UV), infrared (IR), and microwave spectroscopies. Microwave photons have energies of less than 1 meV and are associated with rotational transitions, while infrared photons have energies of up to 1 eV and are involved with vibrational transitions. More energetic ra-

diation from the visible to VUV is usually associated with electronic transitions. In addition to identifying species, these and related spectroscopic techniques are frequently employed to obtain detailed structural information, such as bond lengths and bond angles, and to study the flow of energy in chemical reactions.

Other spectroscopic techniques include:

1. Raman spectroscopy, which concerns the spectral analysis of radiation scattered by an atom or molecule. Recent developments are coherent anti-Stokes Raman spectroscopy (CARS) and surface enhanced Raman spectroscopy (SERS), both of which are sensitive laser-based techniques, and examples of laser spectroscopy.

2. Photoelectron spectroscopy, which is based on kinetic energy analysis of electrons ejected from an atom or molecule by an enegetic photon and provides information on the binding energies or ionization potentials of the ejected electrons. Recent developments include X-ray photoelectron spectroscopy (XPS) of surfaces, and the use of lasers as radiation sources.

3. Auger electron spectroscopy (AES), which involves the electron impact ionization of an atom to give an exited electronic state that decays by emission of a second electron whose energy is characteristic of the atom. This is most commonly used in surface analysis.

4. Spin resonance spectroscopy, which concerns the application of a magnetic field to split energy states associated with electron or nuclear spin orientations. Electron spin resonance (ESR) involves absorption of microwave radiation, while nuclear magnetic resonance (NMR) is a radiofrequency technique.

5. Mössbauer spectroscopy, which involves the resonant absorption of a γ-ray photon by a nucleus. The resonant condition is achieved via the Doppler effect, by sweeping the velocity of a sample relative to the source. The chemical environment of the nucleus causes characteristic frequency shifts. [See ATOMIC SPECTROMETRY.]

III. Statistical Thermodynamics

A. First and Second Laws of Thermodynamics; Entropy

The first law of thermodynamics states the conservation of energy,

$$\delta Q = dU + \delta W \quad (15)$$

where δQ is the heat absorbed by the system, dU is the change in internal energy of the system, and δW is the work done by the system. The second law of thermodynamics states that heat cannot pass from a cold reservoir to a hot reservoir without the application of work. [See CHEMICAL THERMODYNAMICS.]

The change in entropy, dS, is just $\delta Q/T$, where T is the temperature. The factor $1/T$ is an integrating factor which transforms δQ into an exact differential just as $1/v^2$ transforms $v\,du - u\,dv$ into the exact differential $d(u/v)$. Because the change in entropy, $dS = \delta Q/T$ is an exact differential, the change in entropy in a reversible cyclic process is zero. The entropy of a thermodynamic state is a well-defined single-valued function and the entropy is said to be a state function. An equivalent statement of the second law of thermodynamics is

$$\Delta S \geq 0 \qquad (16)$$

where the change in entropy is zero for a reversible cyclic process. The entropy increases in an irreversible process.

B. Free Energy; Experimental Measurements

The first and second laws of thermodynamics can be combined to give

$$T\,dS = dE + \delta W \qquad (17)$$

If the only work done by the system results from an expansion dV or a change in the amount dN_i of the constituents, then Eq. (14) becomes

$$T\,dS = dU + p\,dV - \sum_{i=1}^{m} \mu_i \, dN_i \qquad (18)$$

where p is the pressure, V the volume, μ_i the chemical potential of constituent i, and N_i the amount or concentration of constituent i. The chemical potential of a gas is the value of the Gibbs function for one mole. In a purely mechanical system, equilibrium is achieved when the energy is a minimum. In a thermodynamic system, entropy changes as well as energy changes must be considered. At constant temperature, volume, and concentration, the Helmholtz free energy,

$$A = U - TS \qquad (19)$$

is a minimum at equilibrium whereas at constant pressure and temperature the Gibbs free energy

$$G = A + pV = \sum_{i=1}^{m} \mu_i N_i \qquad (20)$$

is a minimum at equilibrium.

For a dilute gas, where the perfect gas law ($pV = nRT$) applies, the value of μ per mole at a pressure p is

$$\mu(p) = \mu° + RT \ln(p/\text{atm}) \qquad (21)$$

where $\mu°$ is the value of μ at 1 atmosphere, which is the pressure at which the standard state is established. The value of $\mu°$ is often taken as zero for the elements. In Eqs. (20) and (21) R ($=8.3144$ J mole^{-1} K^{-1}) is the so-called gas constant. A mole of any substances is the amount with a mass in grams equal to its molecular weight in atomic mass units, so that a mole of molecular hydrogen has a mass of 0.002 kg. The value of G or μ can be determined for some pressure p by measuring the volume of the gas as a function of pressure up to p and integrating. Thus,

$$\mu(p) = \mu° + \frac{1}{n} \int_{p_0}^{p} V(p)\,dp \qquad (22)$$

where p_0 is the pressure of the standard state (usually 1 atm). The free energy of a condensed phase can be related to that of a dilute gas through the vapor pressure, the pressure of the gas in equilibrium with the condensed phase. Once the free energy of the condensed phase has been established, values for other states can be obtained by measuring pressure or energy through a sequence of states leading to the desired state.

C. Statistical Mechanics; Partition Function

The thermodynamic properties of a system result from the dynamics of its molecules. Since even a three-body system is difficult, statistical methods must be employed to treat the large number of molecules in a thermodynamic system. [See STATISTICAL MECHANICS.]

The fundamental result in statistical mechanics is the fact that the probability of a system occupying the energy level E_i is proportional to the Boltzmann factor, $\exp(-\beta E_i)$, where $\beta = 1/kT$ and $k = 1.3804 \times 10^{-23}$ J K^{-1} is the Boltzmann constant. The Boltzmann constant is the so-called gas constant (R) divided by Avogadro's number, $N_A = 6.022 \times 10^{23}$ mole^{-1}, the number of molecules in a mole.

The thermodynamic properties are related to the energy levels E_i through the partition function Z defined by

$$Z = e^{-\beta A} = \sum_{i} \exp(-\beta E_i), \qquad (23)$$

where the sum is over all the energy levels of the system. Thus, to obtain the thermodynamics of a system, all that is required is that Schrödinger's equation for the system be solved and the partition function summed. For most systems this is a difficult task, often impossible without some approximations.

There is also a relation between the entropy and the microscopic configurations of the system. The entropy is proportional to the logarithm of the number of accessible states, Ω, of the system. Thus,

$$S = k \ln \Omega \qquad (24)$$

Equation (24) is called Boltzmann's relation. At absolute zero, the system is in its ground state, and the number of accessible states is unity. Thus, the entropy of a system tends to zero as the temperature goes to zero. This is called the third law of thermodynamics.

The Boltzmann relation provides a statistical interpretation of the entropy. The greater the number of accessible states, the less our knowledge of the system and the more randomness or disorder in the system. This entropy is a measure of disorder. The tendency of the entropy to increase reflects the tendency of thermodynamic systems to increase in disorder just as an initially ordered deck of cards increases in disorder during a game of cards.

If the system is classical, the energy levels merge into a continuum and an important simplification results. The sum in the partition function becomes an integral. Moreover, if the kinetic energy degrees of freedom (i.e., the momenta) are independent of the potential energy or internal degrees of freedom (i.e., the generalized positions) then the momenta can be integrated immediately. For the particular case in which there is only translational motion,

$$Z = \frac{\lambda^{-3N}}{N!} \int \exp(-\beta\Phi) \, d\mathbf{r}_1 \cdots d\mathbf{r}_N \qquad (25)$$

where $\lambda = h/(2\pi mkT)^{1/2}$, N is the number of molecules in the system, and $\Phi = \Phi(\mathbf{r}_1, \ldots, \mathbf{r}_N)$ is the potential energy. The factor $N!$ is required because states that differ only by an interchange of molecules are not distinguishable.

From Eq. (20), it follows that the average kinetic energy of the system is

$$\langle \text{KE} \rangle = \tfrac{3}{2}NkT \qquad (26)$$

In other words, there is a statistical relation between the temperature and the average motion of the molecules. The greater the temperature, the more rapidly the molecules move and the greater their kinetic energy.

The problem of predicting the thermodynamic properties of such a classical system becomes the problem of evaluation of the configuration integral, the integral over $\exp(-\beta\Phi)$. This is still a difficult task. In general, it can be done only through computer simulations (Monte Carlo and molecular dynamics methods). However, there are a few simple approximations which are helpful.

D. Perfect and Imperfect Gases

The simplest system is the perfect gas in which the molecules do not interact, i.e., $\Phi = 0$. Thus, the configuration integral is just the volume raised to the power N. Using Stirling's approximation, $N! = (N/e)^N$,

$$Z = \lambda^{-3N}(eV/N)^N \qquad (27)$$

and

$$pV = NkT \qquad (28)$$

If the molecules interact, then the problem is more complex. The gas is called imperfect because there are deviations from the perfect gas result. These deviations can be written as a power series in the density, $\rho = N/V$, called a virial series. For example, if the molecules are hard spheres such that the molecules collide elastically but exert no attractive forces on each other, then

$$\beta p/\rho = 1 + \rho b + \tfrac{5}{8}(\rho b)^2 + \cdots \qquad (29)$$

For hard spheres, the coefficients of ρ^n, called virial coefficients, are independent of the temperature. For more complex gases the virial coefficients are temperature dependent.

The virial coefficients can be related to the forces between the molecules. However, both the relation itself and the evaluation of the resultant integrals rapidly become complex as the power n of ρ^n increases. In general, it is difficult to go beyond $n = 4$.

The pressure of the hard-sphere gas exceeds that of the perfect gas at the same temperature and density. To a first approximation, this can be thought to be a result of a reduction in the volume available to the molecules because of the volume occupied by the molecules themselves. The hard spheres can be said to have less free volume than the perfect gas.

The hard-sphere gas cannot be liquified. Liquification requires attractive forces. Attractive forces can also cause the pressure to be less than the perfect gas result. Interestingly, attractive forces are not required for the existence of a solid phase. If the hard sphere gas is compressed, computer simulations show that it will freeze and exist as a close-packed solid.

E. Liquids; van der Waals Theory; Critical Point; Renormalization Group

In contrast to a gas, a liquid need not fill space but can exist in equilibrium with its vapor with a surface separating the liquid and vapor. The pressure at which the equilibrium occurs is called the vapor pressure. Below the vapor pressure, liquid will evaporate until equilibrium is reached. For pressures greater than the vapor pressure, there is no interface between liquid and vapor. The liquid fills the container and there is no clear distinction between liquid and gas. The liquid under pressure can be heated at constant volume to a temperature greater than the critical temperature (the highest temperature at which liquid–vapor coexistence can occur), then allowed to expand and cool to the original temperature and pressure without any transition from liquid to gas being observed. A continuity of states between liquid and gas is said to exist. This is illustrated in Fig. 2. [See LIQUIDS, STRUCTURE AND DYNAMICS.]

The liquid–gas phase can be referred to by the single term fluid. Thus, a theory of the liquid state is of necessity also a theory of an imperfect gas. The earliest theory of the liquid state is that of van der Waals. Although more than a century old, with slight modifications it is viable today. The idea of van der Waals was that a liquid be-

haved as a hard sphere gas except that the pressure must include the internal pressure due to the attractive forces of the molecules in the liquid. It is reasonable to assume that the contribution of the internal pressure to the free energy is proportional to the density. Thus,

$$p = p_0 - \rho^2 a \tag{30}$$

where p_0 is the pressure of the hard-sphere gas and a is a constant depending on the nature of the attractive forces. In its original formulation, the van der Waals theory was only qualitatively successful because van der Waals approximated p_0 by the perfect gas expression with a reduced free volume, i.e.,

$$\frac{\beta p_0}{\rho} = \frac{1}{1 - \rho b} \tag{31}$$

This expression gives only the second hard-sphere virial coefficient correctly and seriously overestimates p_0. Much more satisfactory results can be obtained from the approximation

$$\frac{\beta p_0}{\rho} = \frac{1}{(1 - \eta)^4} \tag{32}$$

where $\eta = \frac{1}{4}\rho b$. The van der Waals theory predicts that the equation of state of a liquid can be expressed in a universal form if the following reduced variables, $T^* = bkT/a$, $p^* = b^2 p/a$, and $\rho^* = \rho b$, are used. This is called the law of corresponding states. As is illustrated in Fig. 3, the theory also predicts that below the critical temperature there is a first-order phase transition between the liquid and vapor accompanied by a discontinuous change in the density ρ. At the critical temperature the transition becomes second order since the liquid and vapor have become identical. For temperatures above the critical temperature, there is no phase transition. In the van der Waals theory, the critical point occurs when

$$\left(\frac{\partial p}{\partial \rho}\right)_T = \left(\frac{\partial^2 p}{\partial \rho^2}\right)_T = 0, \tag{33}$$

i.e., the pressure isotherms have a point of inflection at the critical point.

Modern theories show that the van der Waals theory is a first approximation to a systematic approach, called perturbation theory, in which the pressure is obtained as a power series in $1/T$.

In the van der Waals theory, the first two derivatives of p at constant T with respect to the density vanish at the critical point. This is not just a prediction of the van der Waals theory. Any theory in which the equation of state is ana-

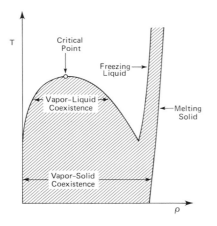

FIG. 2. Phase diagram of a typical simple liquid. The shaded region is not thermodynamically stable.

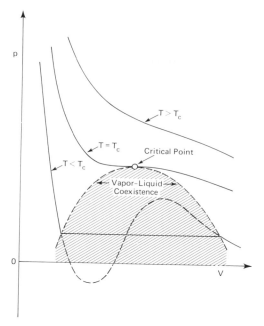

FIG. 3. Typical pressure isotherms as a function of the volume V in the van der Waals theory. The shaded region is not thermodynamically stable. Here T_c is the critical temperature.

lytic at the critical point will yield this result. By analytic, it is meant that the pressure can be expanded as a power series about the critical point. Experimentally, the equation of state is not analytic at the critical point. The exponents in an expansion near the critical point are generally not integers. At least one, and possibly two, more derivatives of p with respect to the density at constant T vanish near the critical point. There has been a great deal of work on the fascinating properties of the equation of state in the vicinity of the critical point. The most far reaching is the renormalization group approach in which a group of successive transformations is applied to the liquid, yielding ultimately a renormalized system in which only the long-range correlations typical of the critical point remain. In this system the critical point properties can be examined.

F. MIXTURES

Mixtures of two gases or liquids can be treated by the same techniques as liquids. The analog of a perfect gas is the ideal mixture, where molecules of the components are very similar so that the partition function can be writ-ten (for the two-component case) as

$$Z = \frac{N!}{N_1! N_2!} Z_1 Z_2$$

where $N = N_1 + N_2$ and N_i is the number of molecules of species i. From this, it can be deduced that the partial pressures of the components are proportional to their concentrations. This result is known as Raoult's law. The factor $N!/N_1! N_2!$ gives rise to the entropy of mixing

$$\Delta S = -Nk[x_1 \ln x_1 + x_2 \ln x_2], \quad (34)$$

where $x_i = N_i/N$.

For a nonideal mixture, the original approach of van der Waals is still useful. In this approach

$$Z = \frac{N!}{N_1! N_2!} \langle Z \rangle \quad (35)$$

where $\langle Z \rangle$ is obtained from the van der Waals equation of state of a liquid, preferably with the more satisfactory expression for p_0, with a and b replaced by the concentration-dependent quantities

$$\langle a \rangle = \sum_{ij} x_i x_j a_{ij} \quad (36)$$

and

$$\langle b \rangle = \sum_{ij} x_i x_j b_{ij} \quad (37)$$

with $a_{ij} \simeq \sqrt{a_{ii} a_{jj}}$ taken as a parameter. Assuming that the spherical cores of the molecules do not overlap,

$$b_{ij} = \left(\frac{b_{ii}^{1/3} + b_{jj}^{1/3}}{2} \right)^3 \quad (38)$$

This approach is satisfactory for mixtures of nonelectrolytes. The situation is a little more complex for mixtures of electrolytes because of the long range of the Coulomb potential. However, each ion in the mixture tends to be surrounded by ions of opposite charge which causes the potential to decay exponentially with a decay factor, κ, the Debye parameter, which is proportional to the square root of the product of the density and T^{-1}. As a result, an appropriate expansion parameter for electrolytes is κ, or $T^{-1/2}$, which is different from nonelectrolytes, where T^{-1} is the expansion parameter. These ideas become quantitative in the Debye–Hückel theory.

G. SOLIDS

In contrast to the disorder of gases and liquids, there is translational order in crystals. Disordered or amorphous solids (i.e., glasses) exist which lack this order. However, they are really highly viscous liquids. This translational order is such that the entire structure, or lattice, can be generated by repeated replication of a small regular figure, termed the unit cell. The planes of any crystalline structure can be specified using Miller indices, which also serve to identify single crystal faces. Miller indices are obtained by determining the intercepts of the plane with the unit cell axes in terms of the length of the cell in that direction, taking the reciprocal, and normalizing so the indices are all integers. [*See* CRYSTALLOGRAPHY.]

The ordered structure, or lattice, of a solid can be determined by X-ray or neutron diffraction studies, in which a beam of X-rays or neutrons is scattered from the sample to produce a diffraction pattern, which can be analyzed to reveal the crystal structure of the sample. All crystal lattices can be classified into fourteen Bravais lattices belonging to seven systems. For example, the simple cubic, face-centered cubic and body-centered cubic lattices are the three lattices of the cubic system. Cubic and hexagonal close packed structures have the structure of tightly packed spheres where each sphere touches twelve neighbors, six in the same plane and three above and three below. These two close-packed structures differ in the placement of successive planes or layers. For the cubic case, a third layer is laid down to reproduce the first layer, so that the structure could be represented by ABABAB.... For hexagonal close packing, the third layer is again displaced, corresponding to ABCABC. ...

No theory of freezing exists. That is, there is no partition function that encompasses both the solid and fluid phases. However, separate theories of solids and fluids can be developed and their solid–fluid coexistence examined. To that extent theories of melting or freezing exist. Since freezing can occur in the hard-sphere system, no critical point is expected for freezing. This transition is expected to be first order at all temperatures, as illustrated in Fig. 1.

If a solid were classical, the heat capacity would be $3Nk$. This is indeed the case at high temperatures and is called the law of Dulong and Petit. However, the experimental heat capacity goes to zero at low temperatures. This can be explained by regarding the solid as a collection of quantized oscillators. The only difficulty is to determine the spectrum of frequencies of the oscillators. For many purposes, the solid can be regarded as an elastic continuum. The result is the Debye theory. If something more sophisticated is needed one must solve for the normal modes of the crystal, i.e., the method of lattice dynamics.

The conduction of electricity in a metal is due to the presence of free or quasi-free electrons in the metal. Classically, free electrons would contribute $3nk/2$ to the heat capacity, n being the number of free electrons. However, experimental evidence indicates that the electrons do not contribute significantly to the heat capacity of a metal. The reason for this is the exclusion principle. Although the electronic gas is in its ground state, because of the exclusion principle the electrons can each occupy one energy level. The electrons occupy the levels up to a maximum energy, called the Fermi energy, ε_F. Only the small number of electrons with energies near ε_F can be thermally excited and, as a result, the electronic heat capacity is small. If the exclusion principle is taken into account, treating the conduction electrons as free describes many of the electronic properties of a metal. To treat metals in a more sophisticated manner and to account for semiconductors, the structure of the solid must be included. If this is done, the electrons are not free but are restricted to bands of energy.

H. Interfaces

The study of interfaces is becoming an increasingly important area of modern physical chemistry. Of particular interest are the interfaces between gases and solids, liquids and vapors, between two immiscible liquids.

Consider, for example, the physical adsorption of a gas by a solid. If the solid is regarded as a giant sphere, the adsorption of the gas can be regarded as the interaction of a gas with a single infinitely large molecule dissolved in that gas. If the simplest form of the van der Waals theory of mixtures is applied to that system, then the adsorption isotherm is just

$$\Gamma = \rho\beta \, \frac{a'_{12}}{\beta(\partial p/\partial\rho)_T}, \tag{39}$$

where a'_{12} is a constant. At low densities $\beta(\partial p/\partial\rho)_T = 1$ and the adsorption is proportional to the density (Henry's law). However, at higher densities $\beta\partial p/\partial\rho$ is a function of the density, and deviations from Henry's law are observed. Especially interesting is the region near the critical point of the gas where $(\partial p/\partial\rho)_T \to 0$ and singularities in the adsorption are observed.

Interfaces between dissimilar materials may also become electrically polarized, with a separation of charge occurring at the interface. When a metal is placed into an electrolyte for example, electrons may leave it to reduce cations in the solution, giving it a net positive charge and making the solution slightly negative. These charges arrange themselves at the interface in two layers, known as the double layer. The most important property of this double layer is the variation of potential in its vicinity. The potential governs the rate at which ions can be transported through the interface, and so controls the rate of electrochemical processes.

IV. Kinetics and Dynamics

The previous sections have dealt only with the equilibrium properties of a system of molecules. Such properties tell us nothing of the time required for equilibrium to be reached or about the dynamical properties of these systems. The rate at which change occurs is the province of kinetics. The detailed manner in which chemical forces act to bring about atomic and molecular motion is the province of chemical dynamics. This section deals with the motions of atoms and molecules and the processes associated with chemical change. [*See* KINETICS (CHEMISTRY).]

A. KINETIC THEORY OF GASES

The kinetic theory of gases assumes that molecules have negligible size compared to their separation, are in continuous random motion, and interact only via elastic scattering. These postulates permit the calculation of molecular speed and velocity distributions. The probability that a molecule has a speed between v and $v + dv$ is found to be

$$dF(v) = 4\pi(m/2\pi kT)^{3/2} v^2 \exp(-mv^2/2kT)\, dv$$

$$(40)$$

where T is the gas temperature and m the molecular mass. This is the Maxwell distribution of molecular speeds. Figure 4 displays this distribution for nitrogen gas at 25 and 500°C. This distribution permits the evaluation of such important quantities as the pressure p exerted by a dilute gas and the collision frequency Z in the gas under given conditions.

The pressure is then given by

$$p = \tfrac{1}{3}\rho m v_{rms}^2 \qquad (41)$$

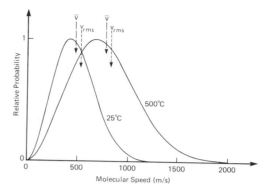

FIG. 4.   Maxwell distributions of speeds for molecular nitrogen at 25°C (298 K) and 500°C (773 K). Arrows indicate $\bar{v}$ and v_{rms} for each case. The most probable velocity has been arbitrarily scaled to unity in each case.

while the collision frequency for a one-component gas is given by

$$Z = \sqrt{2}\sigma\bar{v}\rho \qquad (42)$$

where ρ is the number of molecules per unit volume and σ the collision cross section. The quantity v_{rms} is the average square speed and is related to the average kinetic energy and thus to the temperature, so that

$$v_{rms} = \sqrt{\langle v^2 \rangle} = \sqrt{\frac{3kT}{m}} \qquad (43)$$

where m is the molecular mass. This quantity differs from the average speed,

$$\bar{v} = \langle v \rangle = \sqrt{\frac{8kT}{\pi m}}$$

Equation (43) is equivalent to Eq. (26).

Knowing the collision frequency and the molecular speed, it is possible to estimate the mean free path between collisions, $\lambda = v/Z$, so that

$$\lambda = 1/\sqrt{2}\rho\sigma \qquad (44)$$

which shows the expected behavior that λ must decrease as the diameter of the molecule increases or as the density of the gas increases.

Helium and nitrogen have estimated self-collision cross sections of 0.13 and 0.31 nm² and at a pressure of 1 Torr (= 133.3 N m⁻² = 1.32×10^{-3} atm) there are about 3×10^{22} molecules m⁻³, giving mean free paths of ~1.8×10^{-4} and 7.6×10^{-5} m for helium and nitrogen, respectively. At 25°C, these species have respective velocities of $\bar{v} = 1254$ and 474 m/sec, giving collision frequencies of 6.9 and 6.2×10^6 sec⁻¹ per molecule.

The kinetic theory of gases can also be applied to the free-electron gas to describe the transport properties of metals, obtaining the result that the ratio of the thermal and electrical conductivities is a universal constant times the temperature (the Wiedemann–Franz law).

B. Transport Properties

The kinetic motion of molecules may cause them to change their spatial distribution through successive random movements. This is the process of diffusion, which is a transport property. Other transport properties include viscosity, electrical conductivity, and thermal conductivity. While diffusion is concerned with the transport of matter, these are associated with the transport of momentum, electrical charge and heat energy, respectively. Transport is driven in each case by a gradient in the respective property. Thus, the diffusion rate of species A is given by Fick's Law,

$$J_z(A) = -D[d\rho(A)/dz] \qquad (45)$$

where $J(A)$ is the net flux of A molecules crossing unit area in the z direction and D is the diffusion coefficient; simply kinetic theory leads to

$$D \approx \tfrac{1}{3}\bar{v}\lambda \qquad (46)$$

Derivation of other transport properties follow from similar relationships.
The viscosity coefficient or viscosity of a gas is given by

$$\eta \approx \tfrac{1}{3}\bar{v}\rho m\lambda \qquad (47)$$

while the thermal conductivity coefficient κ is given by

$$\kappa \approx \tfrac{1}{3}\bar{v}\rho\lambda C_v = \eta C_v/m \qquad (48)$$

where C_v is the molar heat capacity of the gas at constant volume. Notice that since λ is inversely proportional to ρ, both η and κ are independent of the gas density. This will be true so long as λ is small compared to the dimensions of the apparatus.
In solution, D is given by the Stokes–Einstein relation which relates D to the viscosity coefficient of the solution, η, and the effective hydrodynamic radius a, where

$$D = kT/6\pi\eta a \qquad (49)$$

and by the Einstein–Smoluchowski relation:

$$D = d^2/2\tau \qquad (50)$$

where τ is the characteristic time between jumps of distance d.

More elaborate theories of transport phenomena make use of the Boltzmann transport equation or computer simulations.

C. Chemical Kinetics; Activated Complex Theory

The quantitative study of chemical reaction rates and the factors upon which these rates depend constitutes the field of chemical kinetics. Chemical reaction rates can be expressed in terms of the rate of production of any reaction product or as the rate of decrease in the concentration of any reactant. The individual steps in a chemical reaction sequence are termed elementary reactions. A number of consecutive elementary reactions may be responsible for a given chemical change. If the observed rate is found to be proportional to the concentration of a given reactant raised to some power, α, then α is said to be the order of the reaction with respect to that reagent. The sum of the orders over all reagents gives the overall order of the reaction. A complex reaction, involving a number of elementary steps may have a noninteger order. Thus the order should not be confused with the molecularity of an elementary reaction, which is the number of reagent molecules involved in a single reaction step. The sequence of elementary reactions by which a reaction proceeds is termed the reaction mechanism, a term also used to describe the detailed process of bond breaking and formation in a single reactive collision.
By way of illustration, consider the formation of nitrogen dioxide from nitric oxide and oxygen. This reaction is found to be third order, corresponding to

$$2NO + O_2 = 2NO_2 \qquad (51)$$

with a third-order rate law corresponding to

$$\frac{d[NO_2]}{dt} = k_{eff}[NO]^2[O_2]. \qquad (52)$$

Here the constant k_{eff} is the reaction rate constant and square brackets indicate concentrations. If concentrations are given in moles per liter, the rate constants will have units of (moles/liter)$^{1-n}$ sec^{-1}, where n is the order of the reaction. A likely mechanism for this process can be written in terms of the elementary steps:

$$NO + NO \xrightarrow{k_1} N_2O_2 \qquad (53a)$$

$$N_2O_2 \xrightarrow{k_{-1}} NO + NO \qquad (53b)$$

$$N_2O_2 + O_2 \xrightarrow{k_2} 2NO_2 \qquad (53c)$$

which leads to the observed rate law if the first two steps are assumed to come to equilibrium prior to the third reaction, or if the steady-state approximation, which assumes that the rate of change of all concentrations is zero, is invoked. Reaction (53a) has a molecularity of 2 and is a bimolecular reaction, while reaction (53b) is an example of a unimolecular reaction, involving a single species.

An important class of reaction mechanisms are those in which a reaction product from one step is a reagent in a prior step. The species concerned is often a highly reactive molecule with a vacancy in its outermost shell of electrons, termed a free radical. Such processes are termed chain reactions. Chain reactions are very important in polymerization reactions, where a radical may add to another reactant to form another (larger) radical. In cases where more than one reagent species is formed as a product of a later step, the chain is said to be branched, and such branching chain reactions often lead to explosions. In other cases, explosions may occur as a result of a fast exothermic reaction which yields a net excess of energy in the form of heat and in a time too short for the energy to be dissipated. The increase in temperature then causes an increase in rate, and the cycle ends in a thermal explosion.

In some mechanisms a species may be consumed in one step of a reaction only to be regenerated in a subsequent step. In cases where the presence of this species increases the overall reaction rate, it is termed a catalyst, which is defined as a species that increases the rate of a reaction without being consumed or changing the reaction products. A catalyst must increase the rate of both forward and backward reactions in any system at equilibrium and can be thought of as lowering ΔE_0 (see below).

An expression for bimolecular rate constants can be obtained by observing that along a reaction coordinate the energy surface consists of two wells, representing the reactants and products, separated by a saddle point, representing the maximum energy required to pass along the minimum energy path between reactants and products. If the height of this maximum, relative to the reactant well, is ΔE_0 then only collisions where the energy exceeds ΔE_0 can lead to reaction. Integrating the Boltzmann distribution of energies over all energies exceeding ΔE_0, shows that probability of a collision with energy in excess of ΔE_0 is proportional to $\exp(-\beta \Delta E_0)$. This is consistent with the rate law of Arrhenius

$$k = A \exp(-\Delta E_0 / k_B T) \tag{54}$$

where A is known as the pre-exponential factor, and the Boltzmann constant is written as k_B here to avoid confusion; A can readily be estimated from collision theory, using the expression for the collision frequency for one reagent with another, Z_{12}

$$Z_{12} = \rho_1 \rho_2 \sigma_{12} [8 k_B T / \mu \pi]^{1/2} \tag{55}$$

which leads to

$$A = P \sigma_{12} N_A [8 k_B T / \mu \pi]^{1/2} \tag{56}$$

where N_A is Avogadro's number, which converts ρ to molar units, and P is the so-called steric factor, which accounts for the fact that not all collisions lead to reaction. Alternatively, we can replace the product $P\sigma_{12}$ with σ_{reac}, where σ_{reac} is termed the reactive cross section.

A more general treatment of detailed reaction rates is available in the activated complex theory of Eyring, which assumes that there is an intermediate state between the reactants and the products, called the activated complex or transition state which can be regarded as at least somewhat stable and which is in thermodynamic equilibrium with the reactants, thus permitting thermodynamics to be applied. Instead of an energy, we must use the free energy G (because the pressure is constant) in the exponential. This treatment yields

$$k = \kappa \frac{kT}{h} \exp\left(-\frac{\Delta G_0^\ddagger}{RT}\right)$$
$$= \kappa \frac{kT}{h} \exp\left(\frac{\Delta S_0^\ddagger}{R}\right) \exp\left(-\frac{\Delta H_0^\ddagger}{RT}\right) \tag{57}$$

where R is the Gas constant, κ is the transmission coefficient and $\Delta G_0^\ddagger$, $\Delta S_0^\ddagger$, and $\Delta H_0^\ddagger$ refer to differences between the activated complex and the reactants.

Extensions of this statistical thermodynamical approach to estimating reaction rates include the *RRK* and *RRKM* theories of unimolecular decay rates, and the information theoretic formulation of reaction dynamics. These theories are remarkably successful, although generally more successful at interpreting experimental data and correlating results than at deriving results *a priori*.

D. REACTION DYNAMICS; INELASTIC COLLISIONS

Kinetic measurements and knowledge of reaction products and yields can provide only rather limited insight into the molecular dynamics of chemical reactions. To understand the detailed manner in which atoms and molecules move together and come apart in the process of a chemical reaction, it is necessary to study the isolated elementary reactions in as much detail as possible. Such isolation is most often provided by a dilute gas environment. Ultimately, the hope is to understand reaction dynamics in terms of electronic structure and to be able to calculate this for a chosen system. The electronic structure or potential energy surface is the meeting ground between theory and experiment. Currently most studies seek to probe those factors, or states, which influence the rate of chemical reactions, such as vibration and translational energy, and to examine the manner in which energy and angular momentum are disposed among the product states for various processes. This is the area known as state-to-state chemistry.

Molecular photodissociation is an ideal process for such studies and has been examined in considerable detail. This unimolecular event is sometimes considered as a "half collision," where the absorption of a photon excites the system to a repulsive state that flies apart. A number of radiation sources have been employed for such photolysis experiments, including discharge lamps, flash lamps and synchrotrons. However, most recent studies have concerned laser photolysis. The photofragments are detected, for example, by emission or laser spectroscopy, which provides information on the velocity and quantum-state distribution of the fragments, with respect to rotational, vibrational, and electronic states. Such measurements can provide information on the shape of the excited state potential energy surface.

Bimolecular reactions are often studied by firing collimated streams of reagents at each other in the form of crossed molecular beams. The scattered reagents and products can be detected by a rotatable mass spectrometer in order to measure angular distributions. Such experiments have shown that many reactions occur in essentially a single encounter in a direct mechanism, while others proceed through a long-lived complex mechanism. In other experiments, spontaneous light emission from the unrelaxed, or nascent, products, termed chemiluminescence, has been analyzed to yield quantum-state distributions. Lasers are often used to probe internal states of products, for example by inducing emission as in laser-induced fluorescence (LIF) detection, and to prepare molecules in specific states and with chosen orientations. Vibrational energy is often found to be more efficacious in promoting reaction than is translational

or rotational energy, since it is more strongly coupled to the reaction coordinate, or path in phase space along which reaction takes place. Product distributions are frequently observed to be far from equilibrium. For example, in direct reactions, high vibrational levels are often found to be more populated than low ones. This so-called population inversion forms the basis of the chemical laser.

Reaction rate constants cannot be used to describe such detailed processes. Instead the differential reaction cross section, σ_{react} (n_1, n_2, n_3, ... n'_1, n'_2, n'_3), is employed, where n_i are various quantum numbers and the primed quantities refer to reaction products. Such cross sections represent the effective collision area for reagents with given n_1, n_2, ..., to give specific products. Rate constants represent the effective average of the product of the cross section with the approach velocity taken over the calculated distribution of reagent quantum states.

Cross sections can be predicted from semi-classical trajectory calculations, in which equations of motion are solved by numerical integration, or they can be obtained from quantum calculations via the time-dependent Schrödinger equation. Both approaches require a previously calculated potential energy surface. However, accurate potentials are currently available only for the $H + H_2$ reaction and its isotopic analogues, for which precise quantum calculations can be made. For other systems approximate surfaces can be obtained either semi-empirically (e.g., using the LEPS or BEBO functions), or from approximate a priori calculations. Statistical theories are also employed. These are based on the assumption that different reaction channels are populated in proportion to the volume of phase space with which they are associated, which is consistent with conservation of energy and linear and angular momentum. Dynamical factors may cause deviations from such statistical behavior, providing information on the reaction mechanism. This is recognized in the information theoretic approach where the fully statistical outcome, or prior distribution, is compared with observations in so-called surprisal plots which indicate the degree to which the data deviate from statistical behavior. This approach has proven very valuable in the correlation and extension of a wide body of data.

Since reaction rates can depend not only on reagent energy, but also on the form in which it is available, a full understanding of chemical behavior requires knowledge of the manner in which energy in various forms is redistributed by collisions. This information must be obtained by studies of energy transfer associated with inelastic collisions. Experimental studies vary from kinetic measurements of decay rates, to full state-to-state studies. It is found that rotational energy is readily transferred from one molecule to another, occurring on almost every collision. Transfer from rotation to translation can be 10^2 times slower, while transfer to vibration may be 10^4 times slower. Transfer between translation and vibration occurs only about once in a million collisions at room temperature. In general, the rate of energy transfer decreases rapidly as the amount of energy transferred increases, following an approximate exponential gap rule.

E. REACTIONS IN SOLUTION

In principle, reactions in solution occur in a similar manner to those in the gas phase and in some favorable cases the observed rate constant is the same in both phases. For example, the unimolecular decomposition of N_2O_5 yields similar A and ΔE_0 values in the gas phase and in a large range of solvents. However, there are many important differences. Reactions of ionic species and of large molecules such as proteins and polymers are rare in gas-phase studies but are common in solution. Reactions in solution are often catalyzed, for example by protons in acid catalysis and by enzymes in many biological systems. Moreover, interactions with solvent molecules may grossly alter the potential energy surface on which reaction occurs, compared to the isolated gas-phase system. Such interactions are strongest for polar reagents and solvents.

Reactions in solution are often diffusion controlled, where the limiting step is the rate at which reagents can find each other. In the absence of strong interactions such as those between ions, the rate constant may be estimated from Fick's Law [Eq. (45)] together with the Stokes–Einstein relation [Eq. (49)] giving:

$$k = 8RT/3\eta \qquad (58)$$

where R is the gas constant. Since reactants are also slow to drift apart, the time-averaged collision frequency per molecule may be close to the gas-phase value, so that reactions with rate constants much smaller than given by Eq. (58) may be relatively insensitive to diffusion effects (in practice this means $\Delta E_0 \geq 40$ kJ/mole). Reaction

products may also be slow to move apart, thus in liquid-phase photodissociation, where the adsorption of light causes a molecule to fall apart, the surrounding solvent cage may hold the products together long enough for recombination to occur.

If two reactions are in equilibrium with an equilibrium constant K, and the back reaction is held constant at the diffusion limit, k_d, then the forward rate constant will be equal to Kk_d. More generally, it is often found that for a given reaction involving a series of similar reagents,

$$k \propto K^\alpha \qquad (59)$$

$$\log k = b + \alpha \log K \qquad (60)$$

this is the Brønstead equation, and is an example of a linear free energy relationship, since log $K \propto \Delta G_0$ and log $k \propto \Delta G_0^\ddagger$, then Eq. (60) could be written as

$$\Delta G_0^\ddagger = b' + \alpha \Delta G_0 \qquad (61)$$

Related to the Brønstead equation is the Hammett equation, which expresses the rate constant k of one of a series of related reactions in terms of a specific reference reaction with rate k_0, giving

$$\log (k/k_0) = \rho\sigma, \qquad (62)$$

where ρ is a characteristic of the type of reaction and σ is characteristic of the specific system. Expressions such as the Brønstead and Hammett equations are particularly useful since the complex nature of the environment makes absolute rate theories such as the activated complex theory difficult or impossible to apply in solution.

The rate constant for a bimolecular reaction in solution can be expressed in terms of the activity coefficients of the reagents, $\gamma(A)$ and $\gamma(B)$, and of the transition state, $\gamma(AB)^\ddagger$, by

$$k = k_0[\gamma(A)\gamma(B)/\gamma(AB)^\ddagger] \qquad (63)$$

where an activity coefficient is defined as the ratio of the effective concentration to the actual concentration, as required to obey Raoult's Law. The Debye–Hückel limiting law, gives γ in terms of the ionic charges z of the species:

$$\log \gamma(i) = -sz_i^2 \sqrt{I} \qquad (64)$$

where I is the ionic strength:

$$I = \frac{1}{2} \sum_{i=0}^{i=\infty} x_i z_i^2, \qquad (65)$$

in which x_i indicates molar concentration, giving:

$$\log k = \log k_0 + 2sz_A z_B \sqrt{I}. \quad (66)$$

This indicates that the rate constant for reactions in solution depends on the ionic strength of the solution, behavior referred to as the kinetic salt effect. Thus, addition of inert ions will increase the rate of reaction between ions of like charge and vice-versa.

F. Reactions at Surfaces; Heterogeneous Catalysis; Corrosion

Atoms and molecules frequently adsorb on surfaces, where they may decompose and/or react with other adsorbed species. Modern technology is increasingly dependent on surface chemistry which underlies many industrially important processes as well as destructive processes such as corrosion. It is useful to distinguish two types of adsorption: physical adsorption, or physisorption, and chemical adsorption, or chemisorption. Physisorption is similar in nature to condensation and involves little chemical interaction with the surface, being associated with van der Waals forces. Chemisorption involves a true chemical interaction with the surface, with the formation of a chemical bond. Thus, the enthalpy of physisorption is usually of the order of 20 kJ mole^{-1}, while for chemisorption values are in the region of 200 kJ mole^{-1}. A chemisorbed molecule may either remain intact in molecular chemisorption, or fall apart in dissociative chemisorption. [*See* SURFACE CHEMISTRY.]

The fraction of a surface covered by a gas, Θ, is given in terms of monolayers, (ML), where 1.0 ML represents complete coverage by a single layer. The pressure dependence of Θ at a given temperature is termed an adsorption isotherm. If the rate of adsorption is $k_a p(1 - \Theta)$, and the rate of desorption is $k_d \Theta$, we obtain the Langmuir isotherm:

$$\Theta = \frac{Kp}{1 + Kp} \quad (67)$$

where $K = k_a/k_d$. Thus, when Kp is small, Θ is simply proportional to the pressure. An adsorption dependence on $1 - \Theta$ arises in the ideal case in which each molecule adsorbs at and occupies a single surface site. If two adjacent sites are required for adsorption, a $(1 - \Theta)^2$ dependence might hold. The probability of adsorption for a single collision is termed the sticking probability, which usually implies chemisorption and may range from unity, for say oxygen on a clean metal surface, to close to zero for an inert system. At low temperatures, even inert gases may trap into a physisorption state, in proportion to their trapping probability.

With a sticking or trapping probability of unity, a monolayer will form in about one second at 10^{-6} Torr, which means that studies of clean surfaces must be carried out under conditions of ultra high vacuum (UHV), or below $\sim 10^{-9}$ Torr.

The ability of surfaces to promote chemical reactions stems largely from their ability to cause dissociation. Consider the decomposition of ammonia to nitrogen and hydrogen:

$$2NH_3 \rightarrow N_2 + 3H_2 \quad (68)$$

In the gas phase, this process has an activation energy of ~ 330 kJ mol^{-1}, whereas in the presence of on a tungsten surface this falls to ~ 160 kJ mol^{-1}. By providing an alternative low-energy path for reaction, the surface causes a large increase in the reaction rate. This is an example of heterogeneous catalysis. (See Section IV,C for a definition of catalysis.) For such a reaction to occur at a surface, the ammonia must first diffuse to the surface, it must dissociatively chemisorb, the hydrogen and nitrogen atoms must then recombine and they must desorb and diffuse away from the surface. The recombination and desorption may actually occur as one step, as the reverse of dissociative adsorption. Either the adsorption or desorption steps are rate-limiting in gas–surface reactions, although fast liquid–surface reactions may be diffusion limited. This is consistent with the fact that dissociative chemisorption and recombinative desorption are often activated processes.

Chemical reactions at the gas–surface interface can be followed by monitoring gas-phase products with, for example, a mass spectrometer, or by directly analyzing the surface with a spectroscopic technique such as Auger electron spectroscopy (AES), photoelectron spectroscopy (PES) or electron energy loss spectroscopy (EELS), all of which involve energy analysis of electrons, or by secondary ionization mass spectrometry (SIMS), which examines the masses of ions ejected by ion bombardment. Another widely used surface probe is low-energy electron diffraction (LEED), which can provide structural information via electron diffraction patterns. At the gas–liquid interface, optical reflection ellipsometry and optical spectroscopies are employed, such as Fourier transform infrared (FTIR) and laser Raman spectroscopies.

Elastic and inelastic collisions of atoms and molecules at surfaces are also of importance. The scattering of hydrogen and helium from surfaces leads to diffraction patterns in the same manner as with LEED, but since the atoms penetrate the surface far less deeply than even low-energy electrons, the structures obtained reflect the very surface of the sample. The inelastic surface scattering of molecules can be examined in detail using laser and mass spectrometric detection for the scattered molecules. Such measurements can be used to model the form of the gas–surface interaction potential, knowledge of which is a prerequisite for any detailed picture of gas–surface reaction dynamics.

Not all surface chemistry is catalytic. In many cases the surface itself may be consumed, in processes such as etching and corrosion. Etching is employed to fabricate devices where it provides for the controlled removal of material. In the semiconductor industry, for example, discharges containing fluorine are used to etch silicon by volatilization as silicon fluorides. Corrosion is generally an unwanted process whereby items are destroyed through dissolution and/or oxidation. Metals may corrode through many different (usually electrochemical) processes. For example, in the presence of oxygen, a metal may displace protons as water or reduce oxygen to OH^-, in acid and alkaline environments, respectively. In principle this process requires the additional presence of a second metal, with a lower electrochemical potential. However, all samples have regions of high and low strain, which will have slightly different potentials. A given metal can be protected by contact with another (sacrificial) metal with a more negative potential, which will be preferentially corroded. This is applied in the galvanizing of iron by zinc.

BIBLIOGRAPHY

Adamson, A. W. (1980). "Physical Chemistry of Surfaces." Wiley, New York.

Adamson, A. W. (1979). "A Textbook of Physical Chemistry," 2ed. Academic Press, New York.

Atkins, P. W. (1982). "Physical Chemistry," 2ed. Freeman, San Francisco.

Bernstein, R. B. (1982). "Chemical Dynamics via Molecular Beam and Laser Techniques." Oxford Univ. Press, New York.

Berry, R. S., Rice, S. A., and Ross, J. (1980). "Physical Chemistry." Wiley, New York.

Eyring, H. (1944). "Quantum Chemistry." Wiley, New York.

Eyring, H., Henderson, D., and Jost, W. (1967). "Physical Chemistry–An Advanced Treatise," 15 Vol. Academic Press, New York.

Eyring, H., Henderson, D., Stover, B. J., and Eyring, E. M. (1982). "Statistical Mechanics and Dynamics," 2ed. Wiley, New York.

Herzberg, G. (1950). "Molecular Spectra and Molecular Structure: I. Spectra of Diatomic Molecules," 2ed. Van Nostrand-Reinhold, New York.

Kauzmann, W. (1957). "Quantum Chemistry." Academic Press, New York.

Levine, I. N., and Bernstein, R. B. (1974). "Molecular Reaction Dynamics," Oxford Univ. Press, New York.

Laidler, K. J. (1965). "Chemical Kinetics," 2ed. McGraw-Hill, New York.

Linnett, J. W. (1960). "Wave Mechanics and Valency." Methuen, London.

McQuarrie, D. A. (1976). "Statistical Mechanics." Harper and Row, New York.

Moore, W. J. (1983). "Basic Physical Chemistry." Prentice Hall, New York.

Partington, J. R. (1954). "An Advanced Treatise on Physical Chemistry," 5 Vols. Wiley, New York.

Rowlinson, J. S. (1969). "Liquids and Liquid Mixtures," 2ed. Butterworths, London.

Smith, I. W. M. (1980). "Kinetics and Dynamics of Elementary Gas Reactions," Butterworths, London.

Smith, R. A. (1961). "Wave Mechanics of Crystalline Solids." Wiley, New York.

Steinfeld, J. I. (1974). "Molecules and Radiation: An Introduction to Modern Molecular Spectroscopy." MIT Press, Cambridge, Massachusetts.

PHYSICAL OCEANOGRAPHY

Arnold Gordon   *Columbia University*

GLOSSARY

Antarctic bottom water: Dense (cold) water mass formed around Antarctica. This water mass spreads along the sea floor well into the northern hemisphere.

Antarctic intermediate water: Low-salinity water mass formed in the Antarctic circumpolar belt. Spreads northward immediately below the thermocline.

Ekman transport: Transport of surface water by the direct stress of the wind on the sea surface, modified by the rotation of the earth. Directed at right angles to the wind: to the right in the northern hemisphere and to the left in the southern hemisphere.

El Niño: Interannual change in the temperature and circulation of the equatorial Pacific ocean. Occurs every 3–8 years as warm water spreads eastward in response to relaxation of the trade winds.

Mixed layer: Surface layer of the ocean in which water characteristics are more-or-less uniform. This layer is produced mainly by wind-induced mixing and is 10–100 m thick, although thicker mixed layers occur at some sites.

North Atlantic deep water: Relatively salty water mass formed in the northern North Atlantic that subsequently spreads throughout the world ocean at a depth near 3000 m.

Potential temperature: Temperature of a parcel of water if moved vertically to the sea surface without exchange of heat with its surroundings (adiabatic displacement). The potential temperature is slightly cooler (tenths of a degree) than the temperature at the initial depth of the water.

Pycnocline: Layer in the ocean in which there is a rapid increase in density with increasing depth. Usually coincides with the thermocline.

Salinity: Dissolved salts in seawater expressed in parts per thousand.

Thermocline: Layer in the ocean in which there is a rapid decrease of temperature with increasing depth. Occurs within the upper kilometer of the ocean.

Thermohaline circulation: Ocean currents produced by the large-scale buoyancy forces (heat/fresh–water exchange with the atmosphere) acting on the ocean.

Tides: Variations in sea level and oscillatory horizontal flow induced by the gravitational interaction of the earth–moon–sun system.

Upwelling: Upward movement of water to the sea surface induced by Ekman transport.

Water mass: Significant volume of seawater with specific temperature and salinity characteristics.

Waves: Propagating oscillation of sea surface density interfaces of the ocean (internal waves) induced by wind, tides, or seismic activity of the sea floor.

The study of ocean circulation, waves, and tides, horizontal and vertical mixing in the ocean, and the exchange of energy and properties with the atmosphere are all included in the field of physical oceanography. Physical oceanography is in the most general sense the application of the science of physics (e.g., mechanics, thermodynamics, and optics) to the study of the ocean.

ENCYCLOPEDIA OF PHYSICAL SCIENCE
AND TECHNOLOGY, VOL. 10

I. Introduction

A. OCEAN DIMENSIONS

The ocean covers 361.25 million km², 70.8% of the earth's surface, to a mean depth of 3790 m; if the continental margins are neglected, the mean depth is 4100 m, 84% is deeper than 2000 m, and 53.6% is deeper than 3000 m. At all latitude belts, except for the interval from 45 to 70°N and south of 70°S (Antarctica), the ocean accounts for more than 50% of the surface area. Between 35 and 65°S the ocean cover is 97.5% of the total area. The total volume of the ocean is 1368 million km³ accounting for 93% of the free water on earth. Terrestrial water represents 5% of the earth's water, 2% is locked in glacial ice. Once a water molecule is in the ocean, it's expected residence time is 3600 yr, compared with a residence time of only 10 days in the atmosphere (before it is rained out) and 10 yr in the larger of the earth's lakes.

The ocean is divided into basins of varied sizes, from the major basins called oceans (Table I) to smaller seas along the ocean margins. There are three oceans: Pacific, the largest ocean, is 46% of the water surface; atlantic and indian, are 23% and 20% of ocean area, respectively. The polar ocean areas are not considered as separate oceans: the Atlantic Ocean includes the Arctic Sea, and the Antarctic or Southern Ocean is included in the southern limits of the three oceans. The largest of the marginal seas (greater than 2 million km²) are the Mediterranean Sea, Arctic Sea, Caribbean Sea of the Atlantic Ocean; the Bering Sea of the Pacific Ocean, and the numerous seas that separate the Pacific and Indian Oceans in the Indonesian region. The separation of the Pacific and Indian Oceans south of Australia is at 120°E longitude. The division between the Pacific and Atlantic Oceans is taken as the Bering Strait in the north and the Drake Passage in the south. The Atlantic and Indian Oceans are divided at 20°E longitude.

B. OCEAN IMPORTANCE

The ocean provides food, recreational activity, transportation, and an arena for military action. The ocean also plays a critical role in stabilizing the earth's climate, maintaining a habitable range of temperature and providing a source of water to feed the moisture in the atmosphere. The ocean water stores the excess heat of summer, which is derived from solar radia-

TABLE I.   Ocean Areas and Depths

Ocean	Area (10⁶ km²)	Average depth (m)
Pacific	180	3940
Atlantic	107	3310
Indian	74	3840

tion, and releases this heat during winter when solar heating is diminished. This attenuates the seasonal range of temperatures of the atmosphere. One needs only to compare the annual extremes of air temperature in the interior of the large continental blocks with the air temperature range of typical ocean environments to see the power of this modification. The air temperature range in central Asia is over 120°F while the air temperature range over the ocean at a similar latitude is only 10–20°F.

Additionally, the ocean circulation, working with that of the atmosphere, carries the excess heat of the tropics to higher latitudes which receive lesser amounts of solar heat. This meridional heat flux is required to maintain a steady-state thermal pattern on the earth. In general, the ocean poleward heat transfer is believed to be greater than that of the atmosphere for the low latitudes (equator to 30° latitude), while the atmosphere dominates from there to the poles. Conceptually, the ocean moves heat away from the tropics, transferring it to the atmosphere in the midlatitudes, primarily from the western parts of the oceans where warm tropical water is carried northward below a cooler atmosphere. The heated air is carried to higher latitudes by atmospheric circulation.

The ocean also plays an important role in governing the atmospheric chemistry in that gases are exchanged between the ocean and atmosphere. For example, much of the carbon dioxide introduced to the atmosphere in the last century by the burning of fossil fuels, cement making, and forest clearing no longer resides in the atmosphere, having been taken up by the ocean. This is important because carbon dioxide in the atmosphere warms the air by the greenhouse effect. Removal of this gas by the ocean thus acts to reduce the climatic impact of these human activities. However, the ocean has absorbed only 50% of the excess carbon dioxide. [See AIR POLLUTION (METEOROLOGY).]

In the ocean there is a complex interdependent network of phenomena involving both biological and physical processes that act to stabilize life and climate on the earth. Human activities many now be perturbing the ocean system, which may have a great impact on the course of civilization. This is particularly relevant as we place more demands on the ocean for materials and food while continuing to stress the natural system. Human and natural perturbations and internal cycles raise important concerns for future trends and the stability of the earth's life and climate. As we learn more about the ocean, we can provide more accurate assessment as to its contribution to these changes. [See POLLUTION, ENVIRONMENTAL.]

II. Seawater

The ocean basins are filled with a saline water solution, roughly 3.5% salt. Seawater density depends on its temperature, concentration of salt in solution, and the hydrostatic pressure exerted by the overlying water. [See CHEMICAL OCEANOGRAPHY.]

The properties of seawater can best be appreciated by looking at the unique properties of its basic ingredient: the water molecule H_2O, which is most unusual. Despite its low molecular weight of 18, it remains in a lower (denser) physical state than expected. Pure water freezes at 0°C and boils at 100°C, whereas by virtue of its molecular weight it should freeze at −90°C and boil at −70°C. It also has the unusual characteristic of expanding as its temperature decreases below 3.98°C, and it experiences a sudden expansion of 8% upon freezing. These properties are the consequence of the polar structure of the water molecule: The positively charged hydrogen ions are both to one side of the molecule, which induces a charge separation within the molecule. The molecules attract one another in what is called hydrogen bonding, so more energy is needed to thermally agitate or separate water molecules and change the temperature or physical phase. Water's heat capacity is the highest of all solids and liquids except for liquid ammonia (which also experiences hydrogen bonding), and the latent heat of vaporization (vapor to liquid state) is about 570–590 cal/gm. The latent heat of freezing (liquid to ice state) is about 80 cal/gm.

Hydrogen bonding is also responsible for the great dissolving power of water. It is able to hold in solution much material, such as salts. While the presence of salt in solution slightly attenuates these unusual powers, seawater retains most of these properties. For example, the freezing point of seawater is reduced to −1.9°C

(for average concentrations of salt), and the heat capacity is reduced by 6%.

The salt in solution, referred to as salinity, is composed of all the naturally occurring elements, though most are found only in trace amounts (Table II). Six constituents account for 99% of the total salt content: chloride (55%), sodium (30.6%), sulfate (7.7%), magnesium (3.7%), calcium (1.2%), and potassium (1.1%). The ocean is well mixed, so the ratio of the various constituents of the salinity to one another is approximately fixed. Only near river outflows or the recently discovered submarine vents do the relative proportions significantly differ. Salinity is expressed in parts per thousand; thus a saline solution of 3.5% becomes 35 parts per thousand (or salinity units).

The density of fresh water is 1.0 gm/cm³ at 3.98°C and is somewhat lower at all other temperatures (0.99987 gm/cm³ at the freezing point and then again at 8.2°C). Salinity increases water density; for example, seawater at 0°C with average salinity would have a density of 1.02811, while at 4°C the density would be 1.02779. Oceanographers have adopted the convention of expressing density as the difference from the density of fresh water at 3.98°C, multiplied by 1000. Hence the density of seawater for the two examples given above becomes 28.11 and 27.79. This notation is called sigma density.

Density is also a function of temperature, increasing as temperature decreases. For seawater the temperature of maximum density occurs below its freezing point, so its most dense state is achieved at the freezing point. Density increases rapidly with pressure. For example, for sea water at 0°C with a salinity of 3.5%, the density in sigma notation is 28.11 at the sea surface where the pressure is equal to 1 atm (about 10^6 dyn/cm² or 14.7 lb/in²) but increases to 50.68 at the pressure found at 5000 depth in the ocean where the pressure is about 500 atm.

TABLE II. Major and Minor Constituents of Seawater[a]

Constituent	Amount (g/kg seawater)
Water (H_2O)	965.1
Chloride (Cl^-)	19.215
Sodium (Na^+)	10.685
Sulfate (SO_4^{2-})	2.511
Magnesium (Mg^{2+})	1.287
Other elements[b]	1.204

[a] Salinity is 34.7%.
[b] Calcium, potassium, bromide, and bicarbonate compose about 90% of the other elements.

Pressure increases with depth following the hydrostatic relationship. With the range of density encountered, a 1-m vertical movement is equivalent to 0.1 atm, so for every 10 m of depth, the pressure changes by 1 atm. The temperature change associated with pressure change is an adiabatic process. As pressure is exerted on a parcel of seawater that is isolated thermally from the environment, the temperature in the parcel changes. An increase of 100 atm (1000 m of ocean depth) induces a temperature change of about 0.1°C. The adiabatic effect is often removed in oceanographic studies by altering the actual temperature, as measured at some depth in the ocean (in situ temperature) to the value it would have at the sea surface if it were moved there adiabatically. This is called the potential temperature. Using the potential temperature for the determination of density yields the potential density, expressed as sigma-0. Potential temperature and potential density are used mainly for tracing ocean water movements, which is made easier if the pressure effect is removed.

Often, tracing water mass movements at depth can be achieved if the density and adiabatically corrected temperature are determined at a reference level below the ocean surface where the pressure is greater than 1 atm. For example, the density at 2000 m (approximately 200 atm) may be determined; it is denoted as sigma-2 (the 2 standing for the depth in kilometers).

III. Air–Sea Interaction

The ocean and atmosphere are well coupled, meaning that heat and water vapor are exchanged freely, and it is not possible to consider the heat/water budgets of one without the other. In view of water's high heat capacity and hence the ocean's great ability to store heat relative to that of the atmosphere, the ocean plays an effective role in controlling the range of atmospheric temperature and humidity. The atmosphere in contact with the ocean can experience only minor changes in temperature and generally remains humid, holding nearly its maximum load of water vapor; the typical difference between air and ocean temperature is 1–2°C. While the marine effect of the atmosphere is most obvious over the ocean, air influenced by the ocean invades the continents, carrying with it many of the marine characteristics. The entire atmosphere of the earth is influenced by the ocean.

The exchange of heat and water between the ocean and atmosphere not only has a strong impact on the global climate but also determines the temperature and salinity of the ocean. Surface water attains certain salinity characteristics and then spreads by mixing and circulation into the interior of the ocean volume. In this way water at every depth and position in the ocean attains thermal and salinity characteristics from the ocean surface, often from the surface of a remote region.

The heat exchange between ocean and atmosphere is accomplished by radiation and other processes. Solar radiation passes through the atmosphere without much diminution, but on entering the ocean it is effectively absorbed and thus converted to heat. Nearly all solar radiation is attenuated in the upper 100 m of the ocean, leaving the bulk of the ocean in darkness. The amount of solar radiation reaching the ocean is primarily a function of latitude and season, the higher the sun above the horizon the greater the solar heating of the ocean. A modifying influence is cloud cover, which reflects the solar radiation back to space. On the average the ocean receives 150 W/m² of energy from the sun.

Heat is lost by three methods (though each of these processes can also lead to heating of the ocean, it is not common): back radiation, a process in which the ocean radiates energy back to the atmosphere; sensible heat exchange (direct transfer of heat to the atmosphere); and evaporation of water into the atmosphere. The most significant of these is evaporation, which has the added effect of introducing water vapor to the air. On a global average, evaporation transfers 90 W/m² to the atmosphere (corresponding to slightly over a meter of water, which is compensated by precipitation over the ocean and by continental runoff). The back radiation transfers on average 50 W/m²; and the sensibly heat transfer is the least important, amounting to an average of 10 W/m².

All three processes are driven by the temperature difference between the ocean and atmosphere. The greater this difference the larger is the heat exchange, which is the essence of the high degree of coupling between the ocean and the atmosphere.

IV. Distribution of Properties

A. SURFACE DISTRIBUTION

The surface layer of the ocean is in direct contact with the atmosphere and hence reflects a near equilibrium condition with the local atmosphere (Fig. 1). The highest surface tempera-

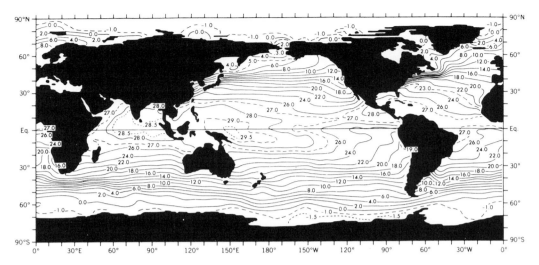

FIG. 1.   Surface temperature (°C). [From Levitus, S. (1982). "Climatological Atlas of the World Ocean," NOAA Professional Paper #13. U.S. Superintendent of Documents, Washington, D.C.]

tures are reached near, but not exactly at, the equator. The ocean thermal equator lies slightly north of the geographic equator, approximately at 5°N. The surface temperature drops off rapidly across the midlatitudes, attaining freezing point values in the polar regions where it forms sea ice. Sea ice has about 20–50% of the salinity of the seawater and is about 2–3 m thick in the Arctic and 1 m in the Southern Ocean. The mean cover of sea ice on the ocean is about 4.5×10^4 km^2, about 7% of the earth surface area. The sea ice cover of the Southern Ocean varies from

about 2 million km^2 in summer to 20 million km^2 at the end of winter.

The surface salinity values more or less reflect the differences of precipitation and evaporation (Fig. 2). Salinity is highest in the subtropical regions where evaporation is greater than precipitation and is lowest in the polar regions where evaporation is relatively low. Near the ocean's thermal equator there is excess precipitation due to the presence of the atmospheric intertropical convergence zone, which is associated with decreased salinity.

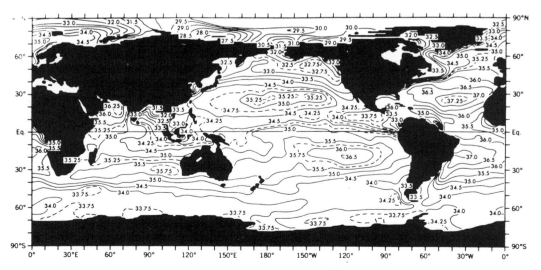

FIG. 2.   Surface salinity (‰). [From Levitus, S. (1986). "Annual Cycle of Salinity and Salt Storage in the World Ocean." *J. Phys. Oceanography* **16**, 322–343.

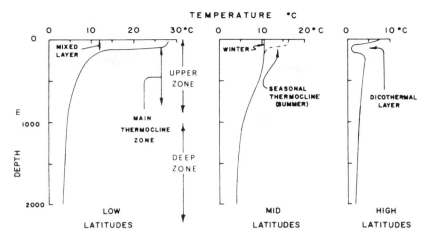

FIG. 3. Typical mean temperature–depth profiles for the open ocean. [From Pickard, G. L., and Emery, W. J. (1982). "Descriptive Physical Oceanography," 4th ed. Pergamon, New York.]

The distribution of surface temperature and salinity indicates some longitudinal dependence. Warmer water reaches further poleward along the western margins of the subtropical ocean and along the eastern margins of the subpolar regions. This characteristic is brought about by ocean circulation (see below), as warm water flows from the equator to higher latitudes. In the western tropical Pacific is a pool of very warm low-salinity water and in the eastern tropical Pacific is a tongue of colder water, induced by the local atmosphere. The large-scale ocean circulation redistributes heat and salinity to further shape the patterns initiated by the local ocean–atmosphere heat exchanges.

B. Vertical Stratification

As depth increases, water characteristics change (Figs. 3 and 4). In general, as depth increases, the water temperature and salinity are similar to the surface water properties of increasingly higher latitudes, reflecting the origin of the subsurface water characteristics. For example, the bottom water in the tropical regions is near 0°C and is similar in temperature and

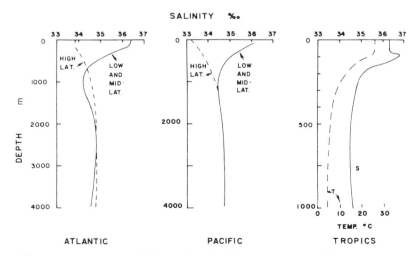

FIG. 4. Typical mean salinity profiles for the open ocean and the temperature profile for the tropics. [From Pickard, G. L., and Emery, W. J. (1982). "Descriptive Physical Oceanography," 4th ed. Pergamon, New York.]

salinity to the water at the sea surface in the polar regions.

The surface layer of the ocean (10–100 m thick, on average) is well mixed by wind action. This is called the mixed layer. Below this layer the change with depth is not uniform, rather the temperature and salinity decrease more rapidly in the upper kilometer, in what is called the thermocline and halocline, respectively. The density increases rapidly with depth across this layer (pycnocline). Below this feature the vertical gradients are much reduced. In the polar regions the usual trend in the vertical profile is altered as the surface water is colder and fresher than the water at depth. During the polar summer the remnant of the winter cold mixed layer forms a subsurface temperature minimum, the dicothermal layer (Fig. 3).

The density field in the ocean is for the most part stable, that is, density increases with depth. A small vertical displacement of a water parcel leads to a restoring force that tends to return the parcel back to its original depth. The parcel oscillates around this initial position, gradually coming to rest as friction removes the oscillatory energy. The period of oscillation depends on the stability of the water column, being more stable as the rate of density increase with depth increases. The period is called the Brunt–Väisälä period (or its inverse, the Brunt–Väisälä frequency or buoyancy frequency). This period is minutes in the pycnocline and close to an hour in the more homogeneous deep water. In homogeneous water the buoyancy frequency is infinitely long.

The small reversals in the density versus depth gradient that may occur in the surface layer as buoyancy is removed through cooling or evaporation are quickly corrected by overturning or convection of the water. As the surface water is made denser, the surface mixed layer deepens. At some ocean sites, deep mixed layers occur frequently, and a specific type of water mass is formed that spreads for great distances within the ocean.

The distribution of temperature and salinity along a meridional plane in the ocean (Figs. 5 and 6) reveals horizontal variations in water characteristics. These are related to the distance from the surface water source for the various layers and to ocean mixing.

V. Water Masses

Water masses are volumes of ocean water with similar temperature and salinity character-

istics. A water mass is defined by the relationship of these two parameters in a potential temperature–salinity diagram (T–S diagram).

The distribution of water masses in the world ocean is a reflection of atmospheric heat and water transfers by the ocean circulation and mixing processes. There are significant differences between the oceans, which diminish as the interocean mixing conduit afforded by the Antarctic circumpolar current and the more or less radially symmetric climate belts of the Southern Ocean are approached. Layers of minimum and maximum salinity identify the water masses core layers in the meridional sections (Figs. 5 and 6). The water-column potential temperature–salinity relation for the oceans (Fig. 7) is discussed in the order of decreasing distance towards Antarctica for each ocean.

A. Northern Subtropical Oceans

Northern oceans are distinctly different from each other. Presumably this is forced by their relative isolation from interocean connections and different atmospheric wind and thermohaline stresses. The Atlantic is the saltiest at all temperatures, with the exception of the 5–9°C region, in which the Indian ocean is most saline. This break in the salinity dominance of the Atlantic is induced by the effective transfer into the North Atlantic of low-salinity Antarctic intermediate water formed in the subpolar circumsolar belt near 50°S.

The abyssal salinity maximum in the Atlantic near 4–5°C is traced to the Mediterranean outflow. Inspection of the characteristics of the Mediterranean outflow, once it begins its lateral spread into the Atlantic interior, reveals that it is a mixture of subsurface Mediterranean Sea water with entrained middepth North Atlantic thermocline water (near the 11°C level) in an approximate 1 : 1 ratio. Below this layer are the components of North Atlantic deep water derived from the northern North Atlantic: from the Labrador Sea and overflow from the Greenland and Norwegian Seas. The sinking of relatively salty water in the northern North Atlantic during the production of North Atlantic deep water is a major feature of the world ocean. It effectively transfers upper-layer water (thermocline water) to deep levels. Compensating upward transfer occurs in the Southern Ocean region and perhaps in the main thermoclines.

Below 2°C the potential temperature–salinity curve extends into colder water that can be traced to Southern Ocean sources for bottom

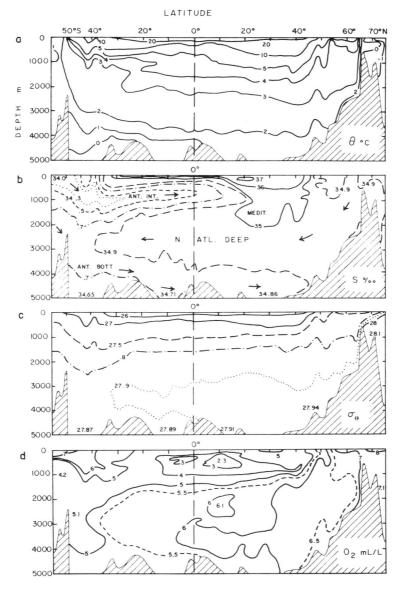

FIG. 5. Atlantic Ocean. S–N vertical sections of water properties along the western trough: (a) temperature, (b) salinity, (c) sigma-0, and (d) dissolved oxygen. [From Pickard, G. L., and Emery, W. J. (1982). "Descriptive Physical Oceanography," 4th ed. Pergamon, New York.]

water, the Antarctic bottom water. This water mass forms along the continental margins of Antarctica, mainly the Weddell Sea.

The entire water column of the immense North Pacific Ocean is most unique by its freshness. The lowest salinity occurs at 5–6°C and is identified as the North Pacific intermediate water, believed to form off the coast of Kamchatka by a process more akin to a vertical mixing process rather than convection. The origin of Pacific deep water (deeper than 600 m for the most part) is not clear, since no North Pacific surface water attains densities sufficient to replace the deep water, yet a low salinity input is required to alter the more saline cold water from the south.

The North Indian Ocean deep water may also

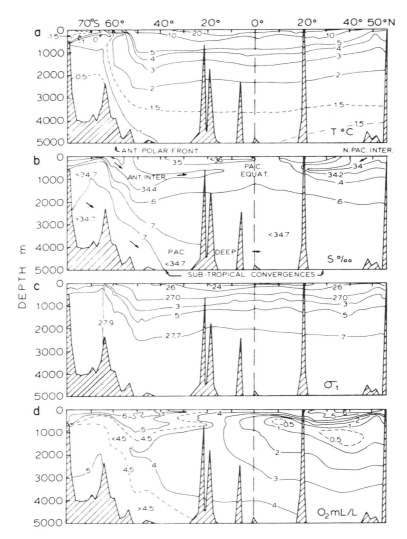

FIG. 6.   Pacific Ocean. S–N vertical sections at 160°W of water properties: (a) temperature, (b) salinity, (c) sigma-0, and (d) dissolved oxygen. [From Pickard, G. L., and Emery, W. J. (1982). "Descriptive Physical Oceanography," 4th ed. Pergamon, New York.]

be formed to some extent by vertical mixing processes, since the potential temperature–salinity curve is nearly linear between the salinity maximum near 10°C, which marks outflow from the salty Red Sea and the cold end-member near the sea floor.

The northern thermoclines also display significant differences. The Atlantic thermocline is about 1.5 salinity units saltier than the Pacific thermocline, while the Indian thermocline is nearly isohaline. The cause of the saltiness of the Atlantic relative to the Pacific is be-

be a result of global-scale circulation patterns.

The North Indian Ocean is unique in its lack of a subpolar segment. The lack of a subpolar segment forces tropical heating to be transferred to the southern subpolar region, which may account for the Indian heat source for the Atlantic and Pacific.

B. SOUTHERN SUBTROPICAL OCEANS

The potential temperature–salinity curves for the three oceans in the southern hemisphere subtropical region are more similar to each other than are those of the northern hemisphere. The Atlantic is the saltiest with the exception of the 3–8°C layer, where the Indian is saltier. In fact, near 4°C the Atlantic is the freshest of the three oceans, due to the direct input of low-salinity Antarctic intermediate water. The Pacific receives a slightly higher salinity Antarctic intermediate water from the southeast Pacific, while the Indian has the weakest intermediate water input.

The south Indian and Pacific deep water potential temperature–salinity curves are similar to those of the northern deep water, and the Indian and Pacific waters are believed derived from the northern deep water reservoirs. The near bottom stratification for the Atlantic becomes more complex as the potential temperature–salinity curve reflects a strong influence of Antarctic bottom water characteristics derived from the Weddell Sea. The Pacific potential temperature–salinity structure is the same as that of

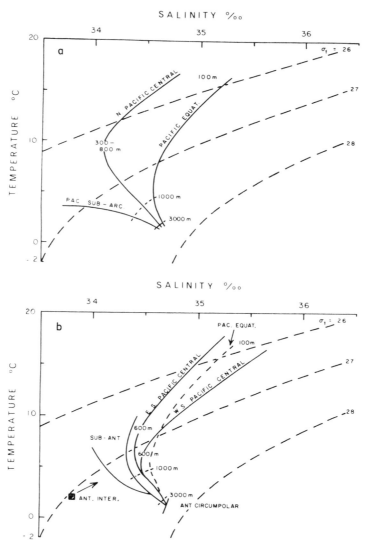

FIG. 7. Average temperature–salinity diagrams for the main water masses of (a) North Pacific Ocean, (b) South Pacific Ocean, (c) Atlantic Ocean, and (d) Indian Ocean. [From Pickard, G. L., and Emery, W. J. (1982). "Descriptive Physical Oceanography," 4th ed. Pergamon, New York.]

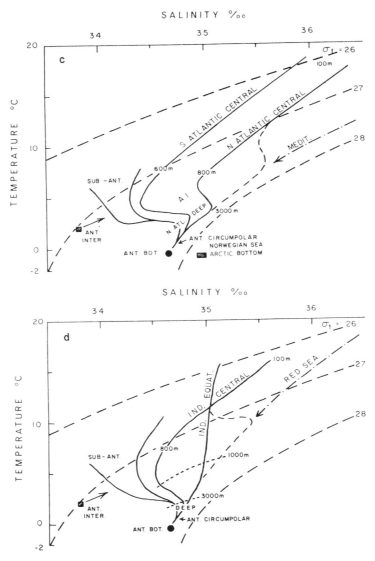

FIG. 7. (*Continued*)

the Indian below 1.2°C, both extended to colder water are directed towards a mixture of the cold Antarctic bottom waters with warmer, saltier North Atlantic deep water.

C. Subantarctic Zone

The deep salinity-maximum, induced by the North Atlantic is clearly seen attenuating with distance from the Atlantic. The attenuation is due to lateral mixing with lower salinity water in the Indian and Pacific sectors of the Southern Ocean.

D. Antarctic Zone

The Atlantic (Weddell Basin) stands out as the coldest Antarctic water column. This fact is associated with the size of the cyclonic subpolar gyre (see below) of the Weddell Basin, which must prolong the residence of water in the regions of strong heat flux to the atmosphere. The Pacific, representative of a smaller cyclonic gyre adjacent to the Ross Sea, is warmer, though the transition from the winter water temperature minimum to the deeper temperature maximum is similarly stable. In the bottom layer (the lower

1–2 km) a salty variety of Antarctic bottom water forming in the Ross Sea influence is clearly seen in the Pacific.

VI. Circulation, Tides, and Waves

The mean or climatic averaged movement of sea water is called the general circulation, which, like the atmosphere, follows a specific pattern (Fig. 8). Circulation derives its energy from the wind, which exerts stress on the sea surface, and from the density variations imposed at the sea surface by buoyance exchange (accomplished by heat and water flux) between the ocean and atmosphere. These two types of circulation are known as wind-driven and thermohaline circulation, respectively.

The sea-surface is not flat relative to a surface of equal gravitational potential or horizontal surface. It has hills and valleys at the meter scale, some of which change rapidly with time as the sea surface oscillates up and down (Fig. 9), at the 10-m scale in extreme cases, with the waves and tides. There are also slower-changing variations associated with transient components of ocean circulation, such as meanders of a current or large eddies which may be generated from larger-scale currents. All this is superimposed

on the mean or climatic topography of the ocean's surface relative to the horizontal. It is this climatic topography that is associated with the general circulation of the ocean; it has a total relief of about 2 m. The ocean's mean sea surface topography is expressed in energy units with a factor, called dynamic meters, that makes the numbers numerically similar to the meter-length scale.

The mean ocean surface topography is due to the nonuniform distribution of mass induced by the wind-driven and nonuniform thermohaline alterations at the air–sea interface.

The large-scale currents composing the general circulation of the ocean are basically a balance between two horizonal directed forces: (a) pressure gradients directed from areas of high to low sea level and (b) the Coriolis force, which depends on the current's speed and the latitude. The Coriolis force is an apparent force in that an observer standing on the surface of the rotating earth must invoke such a force to explain the dynamics of moving objects (or wind or currents). This simple linear balance is called the geostrophic relation. Most ocean currents are in geostrophic balance, only for high-velocity currents or for currents that execute sharp bends or meanders do other factors in ocean dynamics become important. The direction and speed of

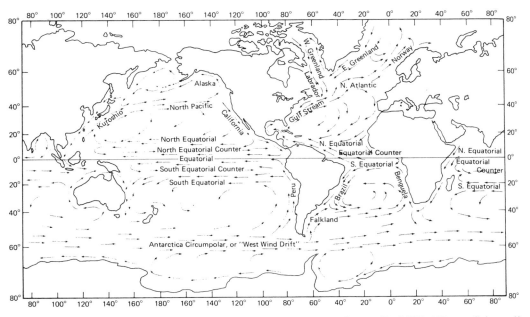

FIG. 8. Ocean circulation, with some major surface currents. [From Stowe, K. (1983). "Ocean Science." Wiley, New York.]

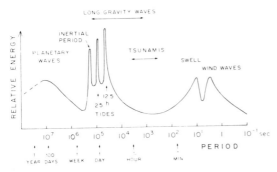

FIG. 9. Schematic energy spectrum of oceanic variability, showing approximate relative energy levels. [From Pond, S., and Pickard, G. L. (1983). "Introductory Dynamic Oceanography," 2nd ed. Pergamon, New York.]

the geostrophic circulation is determined by the relief of the sea surface and the latitude.

The water on a nonrotating globe would spread away from a hill in the sea surface, tending to flatten it out. The coriolis force diverts the flow to the right of its motion in the northern hemisphere and to the left in the southern hemisphere. A balance is established when the pressure forces that are directed away from the hill are matched by the Coriolis force directed toward the hill. This requires a current perpendicular to the pressure and Coriolis forces and directed to the right of a line pointed from the top of the hill to the surrounding low sea surface.

A. THERMOHALINE CIRCULATION

While the regional elements in the general circulation pattern are generally thought to be wind-driven features, the global-scale circulation is of the more sluggish thermohaline type, driven by large-scale buoyancy (heat and fresh water) fluxes between ocean and atmosphere. These fluxes, which induce a circulation pattern, are concentrated in the ocean polar regions. Buoyancy fluxes are large in various locations in lower latitudes, but the resultant thermohaline circulation is embedded in the strong wind-driven flow and cannot be distinguished.

Perhaps the most influential global-scale thermohaline circulation theory is that for the deep water circulation proposed by Stommel and Arons (1960). In this concept, sinking of water in rather limited sites in the northern North Atlantic (during production of North Atlantic deep water) and in the Weddell Sea of the southern South Atlantic (by production of Antarctic bottom water) is balanced by slow upwelling of equal volumes of deep water into the layer from which the initial sinking was drawn.

There are two global-scale thermohaline circulation cells: one composed of Antarctic bottom water compensated by return to the southern ocean as the somewhat warmer saltier circumpolar deep water and the other driven by North Atlantic events, as upper layer or thermocline water is converted to North Atlantic deep water. As North Atlantic deep water spreads throughout the ocean, slow upwelling returns water to the upper layer, which initially fed the production of the North Atlantic deep water. Naturally, the upper layer water must flow back to the North Atlantic to balance the export of deep water. This thermohaline cell is so well developed that it is clearly seen in the North Atlantic transport projected along the meridional plane: Upper layer water in the thermocline flows northward to balance approximately 17 million m³/sec, or 17 sv [where 1 million m³/sec equals 1 Sverdrup (sv)] of southward North Atlantic deep water transport within the deep water.

Intense water mass alteration around Antarctica occurs as the relatively warm circumpolar deep water upwells to interact with the polar atmosphere, as well as sea ice and glacial ice, to form dense Antarctic bottom water. This water mass creeps northward within the western boundary of the oceans, reaching well into the northern hemisphere. The estimates for Antarctic bottom water production rate are 30 sv. At the Antarctic circumpolar current there is sinking of low-salinity water, that feeds the salinity minimum layer immediately below the thermocline, called Antarctic intermediate water.

The thermohaline circulation pattern is one of strong meridional flow (typical speeds of 10 cm/sec or 0.2 knots) in the western boundaries of the ocean, which is then compensated by sluggish interior flow (typical speeds of less than 1 cm/sec). The Stommel–Arons concept has been well verified by observations and has been extended, using more realistic ocean bathymetry, and its validity to ocean basin-ridge scale has been achieved.

B. WIND-DRIVEN CIRCULATION

The wind exerts a stress on the ocean surface proportional to the square of the wind speed in the direction of the wind. Transfer of momentum from the wind to the ocean results, and the ocean responds in what is called the wind-driven circulation.

The upper layer of the ocean (typically the upper 100 m or the homogeneous mixed layer) that feels the direct stress of the wind is called the planetary boundary layer. In this layer the wind stress is balanced by the velocity-dependent horizontal Coriolis force. The resultant transport in this layer is directed perpendicular to the wind direction: to the right of the wind vector in the northern hemisphere and to the left in the southern hemisphere. This transport is called the Ekman transport. The velocity pattern in the planetary boundary layer, under certain conditions, may take on the structure of a spiral in which the surface water is directed at an angle of 45° to the wind (to the right in the northern hemisphere and to the left in the southern hemisphere). With increasing depth in the boundary layer the current vector rotates further away from the wind direction, becoming antiparallel to the surface flow at the base of the boundary layer. However, the Ekman Spiral may be the exception rather than the rule, as the conditions are not often met. It is more likely that the velocity structure in the mixed layer that is associated with Ekman transport is close to uniform with depth.

Where the wind field is not uniform or where the wind field has a component parallel to a coast line, the Ekman transport induces vertical motion, either upwelling or downwelling. The principal upwelling or divergence regions are along the eastern boundary of the subtropical ocean (e.g., the coastal region of Peru and northwestern Africa) and around Antarctica. Upwelling within the subtropical regions carries cold subsurface waters which are high in nutrient concentration to the sea surface, where in the presence of sunlight they can support high levels of biological activity. The cold water surface in these Ekman divergence regions induces cooler climates than might be expected for the latitude and suppresses rainfall and encourages fog conditions.

The extensive area of Ekman-induced upwelling in the Southern Ocean introduces about 45 sv of deeper, slightly warmer circumpolar deep water into the surface layer. Replacement of the deep water mass is derived from the north. This water cools and is converted into Antarctic bottom water and Antarctic intermediate water, which then spread northward.

Wind-induced downwelling or convergence regions force surface water to subsurface layers. These regions are low in nutrient concentration and hence of low biological productivity. The principal site of convergence is in the subtropical latitudes which separate the westerlies from the trade wind regimes.

C. Wind-Driven Currents

The vertical motion associated with divergence and convergence of surface water, resulting from action of the Ekman transport, forces a deeper reaching circulation pattern which is called the wind-driven circulation. The pattern and transport of the wind-driven circulation is related to the pattern of the wind field and structured by the coriolis force. The depth penetration of these currents depends on the intensity of ocean stratification: For those regions of strong stratification such as the tropics, the surface currents extend to less than 1000 m. Within the low stratification polar regions the wind-driven circulation reaches the sea floor.

1. Southern Ocean

The southern ocean, where the major oceans are linked by the deep reaching circumantarctic belt in the 50–60°S range, there is the most extensive current: the Antarctic circumpolar current. It flows to the east and can be considered to be the only global current since it encircles the longitudes at high latitudes. With a path of 24,000 km and transport of 125 sv, it is the strongest, most dominant current of the ocean. It is the most important factor in diminishing the differences among oceans. The Antarctic circumpolar current is composed of a series of individual filaments separated by frontal zones, rather than a well-defined single-axis current. Superimposed on the mean transport is a variable component amounting to 25% of the mean, which can be related to the variability of the wind acting on the circumpolar belt. The Antarctic current reaches the seafloor and hence is guided by the bottom topography. Large meanders and eddies develop in the current. These features induce poleward transfer of heat, which may be significant in balancing the oceanic heat loss to the Antarctic atmosphere further south.

The rest of wind-driven circulation is divided into large gyres: the subtropical gyres extending from the equatorial regions to the maximum westerlies in the wind field at midlatitude, and the subpolar gyres poleward of the maximum westerlies.

2. Equatorial Circulation

At the equator the Coriolis force is zero. For a wind blowing from east to west across the whole equatorial zone (say from 15°N to 15°S) the Ekman transport would induce a divergence right at the equator; the Ekman transport would move surface water away from the equator for both hemispheres. Since the trade winds of both hemispheres meet at the equator, the wind is in general from east to west, so equatorial upwelling is indeed expected. The sea level trough forming along the equator as a consequence of the equatorial upwelling corresponds to a crest in the thermocline, and it produces a geostrophic ocean current to the west. The component in the northern hemisphere is the north equatorial current, and the current to the south is the south equatorial current. For most of the equatorial ocean the trades actually meet slightly north of the geographic equator, and where they meet there is a zone of calms or west to east wind; this is the intertropical convergence zone. The wind asymmetry with the equator also induces an asymmetrical ocean circulation. This induces two regions of Ekman divergence (one at the geographic equator and the other at 10°N) and one of Ekman convergence (5°N). The geostrophic flow is to the west except for the band between 5 and 10°N, where an eastward current occurs; this is called the equatorial counter current.

There is also an east–west trend of ocean characteristics in the equatorial zone. This is due to piling up of the low-density surface water in the west. This is due to the direct action of the wind field at the equator where there is no Coriolis force or Ekman transport, and also the Ekman transport off the equator is not perfectly aligned with the meridians. Sea level stands higher in the west, and the thermocline is depressed in the west. This tilt in sea level up to the west advects water away from the equator in a geostrophic current for latitudes off the equator (further than 1° of latitude off the equator); these currents feed the western boundary currents of the subtropical gyres.

At the geographic equator the sea surface tilt initiates a west-to-east pressure gradient, where in the absence of the Coriolis force, forces water to flow back to the east along the equator. Within the surface layer this eastward flow is retarded by the direct stress of the wind, but below the surface layer (below a few tends of meters) within the upper thermocline, the flow is

to the east. This subsurface current is called the equatorial undercurrent.

The equatorial undercurrent attains speeds of over 1 m/sec. It shallows somewhat as it flows eastward and actually injects colder water into the surface layer along the equator. This is seen in a band of colder surface water along the equator. This band also is cooled by colder water entering the equatorial zone from the eastern boundary upwelling regions.

The intensity of the equatorial circulation directly depends on the wind intensity. The surface water temperature pattern (which is a consequence of the ocean circulation and upwelling) determines the strength of the tropical Hadley cell and so has an influence on the tropical wind field. In this way the equatorial ocean and atmosphere are well coupled.

In addition to the Hadley cell circulation in the atmosphere is a circulation cell aligned along the equator. This is called the Walker circulation. Basically there is net upward motion of air and low atmospheric pressure over the Indonesian Seas and sinking with high pressure over the eastern Pacific. Similar circulation cells occur over the rest of the global equatorial belt, but they are weaker. The strength of the Walker circulation varies. This variation is called the southern oscillation: The low pressure over the Indonesian gets weaker (higher in pressure) as the high over the eastern Pacific gets weaker (lower in pressure). In this way the intensity of the Walker circulation weakens. During this weak phase, rainfall in the western Pacific decreases while rainfall in the eastern Pacific increases. The southern oscillation has irregular periods but in general occurs every 3–8 yr. As the Walker circulation weakens, the wind directed from east to west gets weaker, and the forces responsible for piling of warm water in the west decreases. When this occurs, the warm water spreads back to the east, and surface water temperatures and sea levels decrease in the west and increase in the east. This event is called El Niño. The combined southern oscillation/El Niño effect has received much attention because it is associated with global-scale climatic variability. During such events the equatorial currents that direct water to the east (the equatorial counter current and equatorial undercurrent) are stronger, while the west-flowing currents weaken. Normally, the first signs of an anomaly in surface temperature occurs in the eastern Pacific; then the anomaly spreads westward. In 1982–1983 a giant El Niño event occurred. It was anomalous in two other respects: It began in late spring 1982, and the temperature anomaly was initiated in the central Pacific and then moved eastward. It was associated with anomalous weather conditions at rather remote regions of the globe.

3. Subtropical Gyres

The subtropical gyres are anticyclonic circulation features (meaning sea level stands higher within the central part of the circulation cell); the geostrophic flow is clockwise in the northern hemisphere and counterclockwise in the southern hemisphere. The flow pattern of the subtropical gyres is not symmetric but displays a western intensification: The western boundary currents are jetlike, high-velocity features carrying water poleward, while the flows within the interior and eastern boundary are directed toward the equator and are more sluggish. The western intensification arises from the changing Coriolis force with latitude. In the most general sense a balance is achieved between the wind stress acting on the sea surface over the ocean interior, which introduces the anticyclonic momentum into the ocean, and the friction exerted at the western boundary, which removes it by inducing an equal amount of cyclonic momentum at the western boundary.

Western boundary currents transport the excess heat of the low latitudes to higher latitudes, hence playing an important role in diminishing north–south temperature gradients. The strongest member of the western boundary currents is the Gulf Stream, a component of the North Atlantic subtropical gyre; it carries about 30 sv through the Florida Straits and about 100 sv after its separation from the western boundary at Cape Hatteras. The Gulf Stream carries warm water from the north equatorial current to the North Atlantic drift. This is evident in the northward deflection of surface water temperature in the western North Atlantic and in the associated large transfer of heat from the ocean to the atmosphere within the Gulf Stream region, off the east coast of the United States. This heat exchange is driven by the large contrast of ocean and atmospheric temperatures of the region (ocean being warmer and thus driving heat into the atmosphere). As the Gulf Stream leaves the western boundary near Cape Hatteras, it begins to form waves or meanders and to generate eddies and is quite similar to the atmospheric jet stream. The warm and cold eddies that form drift back to the southwest and reenter the Gulf Stream. The warm eddies are formed north of the Gulf Stream, the cold eddies to the south.

The Kuroshio current of the North Pacific subtropical gyre is about the same magnitude as the Gulf Stream. Both currents form many meanders and eddies after separation from the western boundary.

In the southern hemisphere the western boundary currents are weaker, the strongest being the Agulhas current of the south Indian Ocean, with a maximum transport of 70 sv. The Agulhas is unique in that it turns westward along the southern terminus of Africa and then abruptly curls back to the east, returning water to the subtropical gyre of the south Indian Ocean. The Brazil current is the weakest of the western boundary currents. This is a product of the strong thermohaline circulation in the Atlantic Ocean which directs upper-layer water toward the north in response to North Atlantic deep water formation.

4. Subpolar Gyres

These circulation cells are driven by the cyclonic wind systems poleward of the maximum westerlies. Due to the low degree of stratification of these regions, the circulation reaches well into the water column, attaining the sea floor in most sites. The Ekman divergence induced on the ocean surface layer causes a slight depression in sea level, thus forcing the cyclonic ocean circulation of the subpolar gyres.

In the North Atlantic the subpolar gyre consists of the North Atlantic drift at its equatorward side and the Norwegian current, carrying relatively warm water northward along the coast of Norway. It is this warm water that, on releasing heat into the atmosphere, maintains a moderate climate at these northern extremes. Along the east coast of Greenland is the southward-flowing, very cold, East Greenland current. It loops around the southern tip of Greenland and continues flowing in a counterclockwise fashion around the Labrador Sea. The continued southward flow off the coast of Canada is called the Labrador current. This current for the most part separates from the coast near Newfoundland to complete the subpolar gyre of the North Atlantic. Some of the cold Labrador current water, however, continues to flow south and contributes to the waters off the northeast coast of the United States, eventually meeting the Gulf Stream at its separation at Cape Hatteras.

In the North Pacific the cyclonic (counter-clockwise) flowing subpolar gyres are composed of the northward flowing Alaska current, the subarctic or Aleutian current, and the southward-flowing cold Oyashio current, with the North Pacific drift forming the separation between the subpolar and subtropical gyres of the north Pacific.

In the southern hemisphere, the subpolar gyres are less defined. They lie polarward of the Antarctic circumpolar current, in the cyclonic wind field. The best formed is the Weddell gyre of the South Atlantic sector of the Southern Ocean. The northward-flowing current off the east coast of the Antarctic Peninsula carries very cold Antarctic coastal water into the circumpolar belt.

D. Tides and Waves

Sea level oscillates up and down on a variety of time scales and for a variety of reasons (Fig. 9). The wind stress in the sea surface produces wind waves. These are high-frequency oscillations of seconds to minutes. The gravitational interactions of the earth–moon–sun system drives the lower-frequency tidal waves (half-day to long periods) and earth movements, as undersea earthquakes induce intermediate-frequency or long waves called tsunami or seismic sea waves.

The most obvious waves are those at the sea surface; however, in the ocean interior, where there are strong vertical changes in density (e.g., the pycnocline) oscillation of water occurs. These are internal waves. Internal waves have larger amplitudes and longer time scales than do surface waves. The Brunt–Väisälä frequency (which is measured in minutes) marks the most rapidly oscillating internal wave field. The energy driving internal waves comes from surface waves, and hence from the wind, and from shear in the water column.

1. Wind Waves

The wind ripples the surface of a smooth ocean by forming centimeter-scale waves called capillary waves. These waves increase the energy transfer from wind to ocean, which increasingly enters into longer and higher-amplitude waves. The size of these waves (wavelength and amplitude) is proportional to the duration of the wind and its fetch (or distance) over the ocean. The waves grow for any given wind condition

until an equilibrium is reached where wave energy is dissipated by friction at the rate of energy input from the wind. This is called the fully developed sea state. White caps are common in a developing and fully developed sea. Wave amplitudes of 1–5 m with wavelengths of around 10 m are fairly typical. However, in storms, amplitudes of 10–15 m occur.

When the wind ceases, the waves continue to progress and can travel tens of thousands of kilometers, hitting coast lines. These are called ocean swells. They are long wavelength features (10–100 m). While waves progress over the ocean surface, the water particles associated with the wave do not. They move in vertically oriented circles of decreasing diameter with increasing depth. Near the surface the diameter of these circles equals the wave amplitude, at a depth of 1.5 times the wavelength, their diameter is nearly zero.

As waves move into shallow water, the frictional effects of the sea floor slow the waves. This occurs when the water depth is near 1.5 times the wavelength. The waves essentially pile up, becoming steeper as their wavelength decreases. Eventually they break and form the surf. Waves in shallow water have speeds proportional to the water depth. They refract (propagation direction of the wave changes as the depth varies). The refraction focuses the wave energy to the shallows and headlands along the coast.

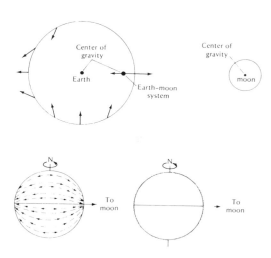

FIG. 10.   Forces producing tides in the earth's bodies of water. [From Weisberg, J., and Parish, H. (1974). "Oceanography." McGraw-Hill, New York.

TABLE III.   Characteristics of Some of the Principal Tide-Producing Force Constituents[a]

Species and name	Symbol	Period (solar hours)	Relative size
Semidiurnal			
Principal lunar	M_2	12.42	100
Principal solar	S_2	12.00	47
Larger lunar elliptic	N_2	12.66	19
Luni-solar semidiurnal	K_2	11.97	13
Diurnal			
Luni-solar diurnal	K_1	23.93	58
Principal lunar diurnal	O_1	25.82	42
Principal solar diurnal	P_1	24.07	19
Larger lunar elliptic	Q_1	26.87	8
Long period			
Lunar fortnightly	M_f	327.9	17
Lunar monthly	M_m	661.3	9
Solar semiannual	S_{sa}	4383	8

[a] From Pond, S., and Pickard, G. L. (1983). "Introductory Dynamic Oceanography," 2nd ed., p. 260. Pergamon, New York.

2. Tidal Waves

The earth and moon revolve around a common center of mass, which is situated 4700 km from the earth's center (the point lies within the earth). The balance between the gravitational attraction between the two bodies and the centrifugal forces tends to stretch the earth, and the bodies move along the line passing through their centers. The solid component of the earth resists this, though earth tides do occur; but the fluid parts, atmosphere and ocean, move in response to these tide-generating forces. The motion is mainly due to the horizontal or traction forces, rather than the vertical forces, directly on a line with the moon portion (Fig. 10). The rotation of the earth induces variations in these forces that result in a time-varying tidal generating force.

The earth and sun also revolve around a common center of mass close to the sun. The greater distance of the sun from the earth reduces the tidal generating force due to the earth–sun system. However, the great mass of the sun still produces a significant tidal force on the earth, amounting to 47% of the lunar force.

The tide-generating forces from the complete earth–moon–sun system are separated into various constituents of specific periods and amplitudes (Table III).

The oscillation of the sea surface height at a specific site along the coastal ocean depends on the relative configuration of the earth–moon–sun position and in the local geometry and mathematical depth of the region. When the natural period of oscillation of a basin is close to one of the major tide-generating forces, resonance occurs, and the tidal amplitude is enhanced.

3. Seismic Sea Waves

Sudden movement of the sea floor by seismic activity induces waves. These waves, often called tsunami, have amplitudes of only a few centimeters but 100-km wavelengths. Tsunamis, because of their long wavelength, are associated with particle motion down to the sea floor. When they reach the shallow regions of the continental shelves, they slow and pile up, and their amplitudes increase to tens of meters, causing great damage along the coast in a tsunami surf.

BIBLIOGRAPHY

Gordon, A. L. (1986). "Inter-Ocean Exchange of Thermocline Water." *J. Geophys. Research* **91**, 5037–5046.

Killworth, P. D. (1983). "Deep Convection in the World Ocean." *Reviews of Geophysics and Space Physics* **21**(1), 1–26.

Pickard, G. L., and Emery, W. J. (1982). "Descriptive Physical Oceanography," 4th ed. Pergamon, New York.

Pond, S., and Pickard, G. L. (1983). "Introductory Dynamic Oceanography," 2nd ed. Pergamon, New York.

Warren, B. A. (1981). In "Evolution of Physical Oceanography, Scientific Surveys in Honor of Henry Stommel," B. A. Warren and C. Wunsch, eds., pp. 6–41. MIT Press, Cambridge, MA.

Worthington, L. V. (1981). In "Evolution of Physical Oceanography, Scientific Surveys in Honor of Henry Stommel," B. A. Warren and C. Wunsch, eds., pp. 42–69. MIT Press, Cambridge, MA.

PHYSICAL ORGANIC CHEMISTRY

Gary A. Epling   *University of Connecticut*

GLOSSARY

Carbanions: Compounds containing a trivalent carbon bearing a negative charge.

Carbenes: Compounds containing an electrically neutral, divalent carbon.

Carbocations: Compounds containing a trivalent carbon bearing a positive charge; a "carbonium ion."

Concerted reaction: Chemical transformation in which bond formation and bond cleavage proceed simultaneously, and the product forms without proceeding through a discrete intermediate.

Conformations: Stereoisomers that may be interconverted by rotation around single bonds between atoms.

Diastereomers: Stereoisomers that are not mirror images of one another.

Electron pushing: Formalism used to describe the movement of electrons during reaction using a curved arrow ($\frown$).

Enantiomers: Stereoisomers that are mirror images of one another.

Intermediate: Structure that during a reaction is a "local minimum" on the potential energy surface. It is more stable than any species formed by stretching or shortening bonds on the structure.

Intermolecular: Chemical changes that involve two species as reactants.

Intramolecular: Reaction in which changes occur entirely within a single molecule.

Nonclassical carbonium ion: A delocalized carbocation in which multicenter bonding leads to carbons attached formally to five atoms.

Pericyclic reactions: Concerted reactions that involve a closed loop of interacting orbitals.

Photosensitization: Reaction proceeding by light absorption by the sensitizer, then energy transfer to another species, which undergoes reaction.

Reaction mechanism: Detailed step-by-step description of the bond changes taking place during a chemical transformation.

Stereoisomers: Compounds with the same bonds, but differing in spatial orientation of some bonds.

Transition state: Point of highest energy during chemical reaction, a local maximum on the potential energy surface for the reaction.

Physical organic chemistry is the study of the physical and chemical properties of organic molecules, using physical laws and relationships to understand and predict reactivity and reaction products. To accomplish these goals, a physical organic chemist often relies upon sophisticated instrumentation for the measurement of properties of molecules, rates of their reaction, detection of short-lived reaction intermediates, and products of reaction. A variety of spectroscopic techniques are well suited for these goals, and are widely utilized by physical organic chemists. Computers are increasingly utilized—both in the analysis of data and in the prediction of properties of compounds, often before they have been synthesized in the laboratory.

I. Structural Concepts

A. BONDING

Although most organic compounds have complex structures, one generally finds that each atom is attached to others by sharing a pair of electrons (a covalent bond). In organic compounds most atoms share electrons in such a way that an octet of electrons around each atom is achieved. Examples are easily shown using a Lewis dot formalism, where a dot represents an electron:

$$
\begin{matrix}
& :\ddot{\text{C}}\text{l}: & & \text{H} & \text{H} & & :\ddot{\text{O}}: & \\
:\ddot{\text{C}}\text{l}:\text{C}:\ddot{\text{C}}\text{l}: & & \text{H}:\ddot{\text{O}}:\text{C}:\text{C}:\text{H} & & \text{H}:\ddot{\text{C}}:\ddot{\text{O}}:\text{H} & \text{H}:\text{C}:::\text{C}:\text{H} \\
& :\ddot{\text{C}}\text{l}: & & \text{H} & \text{H} & & &
\end{matrix}
$$

$$\text{1} \qquad\qquad \text{2} \qquad\qquad \text{3} \qquad\qquad \text{4}$$

Only the valence shell electrons are shown in such representations. Achievement of a complete shell results in the maximum stability for an atom. For hydrogen two electrons would complete the first shell. For the main row elements in the periodic table (Li, Be, B, C, N, O, F, Ne) a completed outer shell requires eight electrons. Observation of the exceptional stability of compounds in which every atom has a completed outer shell led to the concept of "magic numbers." This concept can be relied on to predict whether a molecule is more stable than a similar one that has an incomplete outer shell, or in which too many electrons have been forced. In structure **1,** each chlorine has provided seven electrons to the molecule (the number of electrons in the incomplete second shell of a neutral chlorine atom) and carbon has provided four electrons. The sharing of electrons, as shown, allows every chlorine atom and the carbon atom to have a complete shell of eight electrons. The structure indicates that carbon forms four bonds, one to each chlorine atom, by sharing a pair of electrons. Each chlorine has six electrons that are not shared with any other atom. These are referred to as three "unshared pairs" of electrons. [*See* PERIODIC TABLE (CHEMISTRY).]

In structure **2,** ethanol, the oxygen has two unshared pairs of electrons. There are shared pairs of electrons between hydrogen and oxygen, oxygen and carbon, carbon and carbon, and carbon and hydrogen. These are all single bonds because they result from the sharing of one pair of electrons. In structure **3,** acetic acid, there is a double bond, with two pairs of electrons being shared by the carbon and one of the two oxygens. Several types of single bonds also exist in this compound. Even triple bonds are possible, with three shared pairs of electrons between two atoms as in structure **4,** acetylene, which has a triple bond between the two carbon atoms.

The most common elements in organic compounds are carbon, hydrogen, oxygen, nitrogen, sulfur, and halogens. Three traits of carbon are responsible for the great complexity of structures, which are possible with such a small number of elements:

1. Carbon needs to share four more electrons for a complete octet, which could be achieved by forming four bonds with other elements. In other words, the valence of carbon is four.
2. The electronegativity of carbon is intermediate, being neither a strong metal nor a strong nonmetal.
3. Carbon forms strong bonds to itself.

The first feature of carbon means that a great variety of structures can result with a small number of atoms. The second feature, middling electronegativity, means carbon usually forms covalent bonds since there is not a great deal of difference in electronegativity between carbon and other atoms. Electrons are shared between the two atoms rather than donated as in the interaction between two atoms of greatly differing electronegativity, such as sodium chloride, **5:**

$$\overset{\oplus}{\text{Na}} \qquad :\ddot{\text{C}}\overset{\ominus}{\text{l}}:$$

5

Since there is a great difference in electronegativity between the two atoms, the sodium achieves a complete octet by donating its outer-shell electron to chlorine, forming Na^+ and Cl^-. In a crystal there is an electrostatic attraction, an ionic bond, between these two ions. However, in solution (the compound dissolved in water) these ions become separated by solvation, and the complex structure of the crystal is lost as the constituent ions are separated. In contrast, the covalent bonds of an organic compound do not become broken when the compound dissolves in a solvent. Hence, even complex compounds comprising thousands of atoms (such as in DNA) retain the exact structure that existed in the crystal form of the compound.

The third feature of carbon, its ability to form strong bonds with itself, is important because such behavior is not common for any other element. Thus, carbon can form a skeleton of carbon–carbon bonds in a continuous chain to which other atoms can be attached by other co-valent bonds.

Although Lewis structures have great value in understanding simple structures, they are not widely used because of two major deficiencies:

1. They are cumbersome and slow to use.
2. They do not accurately depict the three-dimensional structure of the compound.

To overcome the first deficiency it is common to show a shared pair of electrons (a covalent bond) as a line; unshared pairs are frequently omitted unless crucial in understanding a particular reaction. However, their existence is often extremely important, and much can be overlooked if they are forgotten. Nevertheless, one would usually see carbon tetrachloride represented in one of the ways shown in **6** and **7**:

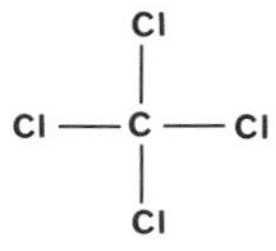

6 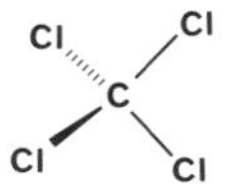   7

Structure **7** is an attempt to represent molecules in such a way that the three-dimensional structure can be visualized. Structure **6** implies that all five atoms lie in a single plane, which is erroneous; structure **7** uses a dotted line to depict a bond that goes back behind the plane of the paper and a wedged line to indicate a bond that projects in front of the paper. The actual bond angles in CCl₄ (the angle formed by any two bonds, Cl—C—Cl) are all 109.5°, rather than the 90° angles implied by structure **6**. Structures **8** and **9** are accurate depictions of the three-dimensional structure indicating the characteristic bond angles of 120° when a double bond to carbon exists, and 180° when a triple bond exists:

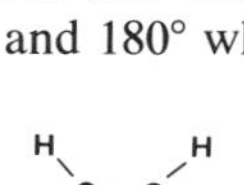

8   9

The bond angles between atoms in organic compounds are always very close to these three characteristic values (109.5, 120, and 180°) because the covalent bonds result from electron-sharing overlap involving electrons in hybridized atomic orbitals, rather than pure hydrogenic orbitals. Thus, the four electrons in the valence shell of carbon, which in a simple process of filling the lowest energy orbitals would be $1s^2 2s^2 2p_x^1 2p_y^1$, are physically mixed to produce

four sp^3 hybrids that are mathematically orthogonal. These make it possible to form four covalent bonds, which have angles of 109.5°, between any three atoms involved in the bonding.

Of somewhat lesser stability are the molecules formed by carbons utilizing two other types of hybrid orbitals. Double bonds to carbon are possible when the carbon is sp^2 hybridized resulting in an unhybridized p_z orbital that can overlap in a sideways (π) fashion to form a second bond in addition to the end on overlap of the sp^2 hybrids that form the σ bond, as shown in structures **10** and **11**:

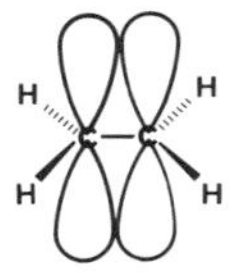

10 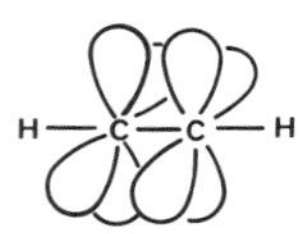   11

For simplicity, the C—H bonds involving the other two sp^2 orbitals are shown as a dotted line and wedge rather than an orbital. Similarly, a triple bond is possible if a molecule uses sp hybrids, because two p orbitals (p_y and p_z) are not used for σ bonding. The orbitals are available for sideways (π) overlap to form a total of three bonds between a pair of atoms. It is important to note that the sp^2 and sp hybrids are not quite as stable as those with sp^3 hybridized carbon, causing a molecule to be susceptible to react with a great variety of reagents. For example, H₂ in the presence of a Pt catalyst would readily react with both of the above structures to form ethane (CH₃—CH₃), but the ethane, having only sp^3 hybridized carbons, will not react under the same conditions. Hence, it is called a "saturated" hydrocarbon, while those compounds of carbon and hydrogen which have double or triple bonds are called "unsaturated" hydrocarbons. [See ORGANIC CHEMISTRY, SYNTHESIS.]

B. Delocalized Pi Bonding

In a few instances we encounter compounds that are unusual because there seems to be more than one reasonable structural representation. Two examples of such a situation are the acetate ion (CH₃CO₂⁻), formed when a base removes a proton from acetic acid (structures **12** and **13**)

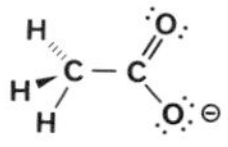

12 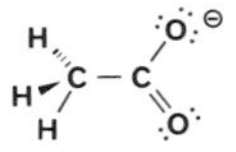   13

and benzene (structures **14** and **15**):

14   15

When this situation exists, one generally finds that the particular molecule in question is more stable than would be expected based on examination of either of the structures. For example, benzene has three double bonds and is an unsaturated hydrocarbon, but reacts very slowly with H_2 even in the presence of a platinum catalyst. High temperatures and pressures are used to make the reaction proceed at a reasonable rate. This behavior is readily explicable by advanced theoretical treatments of structure that involve mathematical computation of the stability, structure, and reactivity of such molecules. These calculations show that neither of the two structures is a very good picture of the true molecular structure. The acetate ion, for example, does not have a -1 charge on either of the oxygen atoms, but instead has a $-\frac{1}{2}$ charge on each oxygen atom. Similarly, the bond distance between any pair of carbon atoms in benzene is neither the normal single bond distance, nor the shorter double bond distance. Instead, it is intermediate between these two, and all of the C—C bond distances are identical. The bonding in such molecules is described as delocalized π bonding. No picture can be readily drawn to represent such delocalization, so either one of the inaccurate pictures is accepted. However, with benzene one common way of depicting this delocalization is by using a circle to show the delocalization of the π bonds:

16   17   18   19   20

It is also common (particularly in cyclic compounds) to abbreviate structures by omission of the specific depiction of a carbon atom. The intersection of two lines, and the terminus of a line, implies a carbon atom at each site. In another valuable abbreviation, the hydrogens attached to these carbons do not have to be written at all. Thus, structures **14, 15, 16, 17, 18, 19,**

or **20** as representations of benzene would be acceptable. This may be confusing at first because the reader must be able to realize what missing elements are implied. However, the knowledge that four bonds must exist to every carbon allows one to readily count how many hydrogens have been omitted for simplicity.

The additional stability of such π-delocalized compounds was recognized long before it was possible to explain such stability mathematically. The stabilization was explained using a nonmathematical concept called "resonance." When it is possible to depict the structure of a molecule using two or more acceptable Lewis structures and the differences between the structures are solely in the positions of the π electrons (i.e., the positions of the atoms and the σ bonds do not change), the molecule is said to be "resonance stabilized." The true molecular structure would be a hybrid of the two (or more) Lewis structures. Thus, the bond order between all adjacent carbons in benzene is $1\frac{1}{2}$, rather than one or two. Any formal charges that might exist are averaged over the atoms indicated by the various Lewis structures. Finally, the compound will be more stable than one would anticipate by examination of any one of the structures alone. By application of this nonmathematical generalization it is possible to quickly identify structures of unexpected stability.

C. AROMATICITY

In many resonance-stabilized compounds the extent of stabilization is so great many physical and chemical properties are altered. Traditionally, unsaturated compounds that were so greatly stabilized by resonance were called "aromatic," while unsaturated compounds that were not significantly stabilized by resonance were termed "aliphatic." Benzene is the best-known example of an aromatic compound, and its sluggish reactivity with H_2 (**23** to **24**) in comparison with aliphatic compounds (**21** to **22**) has already been mentioned.

$H_2C=CH_2 \xrightarrow[Pt]{H_2} H_3C-CH_3$ FAST

21   22

23   24   SLOW

Another important difference of aromatic compounds is reaction with electrophiles, such as

Br₂, usually requires a catalyst, and the product is different [being a substitution product (**27** to **28**) rather than an addition product (**25** to **26**)].

aliphatic: addition
$H_2C=CH_2$ →[Br₂] $BrCH_2-CH_2Br$
  25   **26**

aromatic: substitution
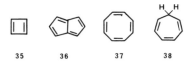
  27   **28** + HBr

Because of their differing chemical and physical properties, it is important to be able to predict if a compound (perhaps newly synthesized) will have aromatic character. Although molecular orbital calculations can accurately predict aromaticity, these are time consuming. A simple, but reliable, predictor of aromaticity is the Huckel rule, which stipulates that if there are $4N + 2\pi$ electrons in a circle of continuously overlapping p orbitals the molecule will be aromatic (N is any integer). Thus, the "magic numbers" for aromaticity are 2, 6, 10, 14, ..., π electrons. Examples of compounds (or ions) whose aromaticity is accurately predicted by the Huckel rule are **29** to **34**:

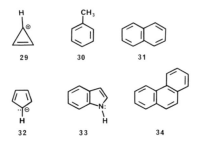

29   **30**   **31**

32   **33**   **34**

(Note that unshared electrons may be part of the π system.) A compound will be aliphatic (structures **35–38**) when the π electrons do not total one of these magic numbers or whenever there is a break in the conjugation by the existence of an intervening sp^3 hybridized carbon:

35   **36**   **37**   **38**

In these cases resonance stabilization is not as important, and the compounds will be aliphatic. In general these will be more reactive than an aromatic compound with a given reagent.

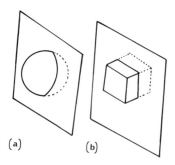

FIG. 1.   Symmetry in familiar objects: (a) ball and (b) cube.

D. Symmetry

Many objects have an inherent property referred to as symmetry. For such objects a symmetry operation (a particular kind of manipulation) transforms the object into an identical species. Common symmetry elements are lines, points, or planes of symmetry. For example, a plane of symmetry, or mirror plane, means that a plane passing the center of the object will divide into two equal parts. Thus, if the plane were to be perceived as a mirror the reflection of the top half in the mirror would be identical with the actual shape of the bottom half of the object. A ball, cube, or doughnut all have at least one plane of symmetry (Fig. 1). An irregular object like a rock will rarely have a plane of symmetry.

Many organic compounds have a plane of symmetry. One mirror plane for CCl_4 and $CH_2=CH_2$ is shown in Fig. 2. The existence of a mirror plane is the most important element of symmetry in an organic molecule because it is the key to predicting the existence of some types of stereoisomers.

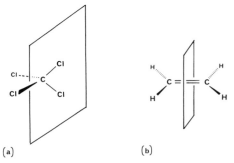

FIG. 2.   Planes of symmetry in organic compounds: (a) carbon tetrachloride and (b) ethylene.

E. Stereochemistry

The study of the three-dimensional structure of molecules is called stereochemistry. A good knowledge of stereochemistry is essential to fully understand isomerism and reactivity. Isomers are compounds that have the same molecular formula but different molecular structure. For example, both ethanol (structure **39**) and dimethyl ether (structure **40**) have the molecular formula C_2H_6O,

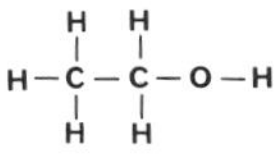

but they have different connectivity (i.e., the atoms are connected in a different pattern). Ethanol has an alcohol functional group (C–O–H bonding), whereas the ether does not. Isomers having different connectivity are constitutional isomers. More difficult to perceive are stereoisomers—isomers that have exactly the same connectivity but differ in the way the atoms are oriented in space.

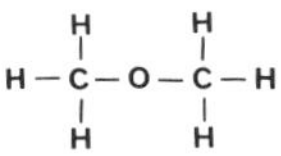

The most common stereoisomers are the geometric (or cis/trans) isomers. A pair of such compounds are *cis*- and *trans*-2-butene (structures **41** and **42**). Because the geometry of the molecule is critically determined by the 120° bond angles around the central carbons, the two methyl (CH_3) substituents on the double bond can be either on the same side (cis) or on opposite sides (trans).

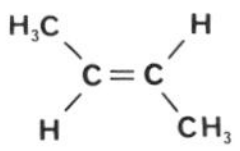

When they are on the same side they are closer together, causing an unfavorable interaction that makes the cis compound slightly less stable than the trans.

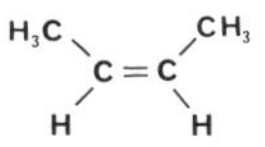

Geometric isomerism is also seen in cyclic compounds, for example in *cis*- and *trans*-1,2-dichlorocyclobutane (structures **43** and **44**).

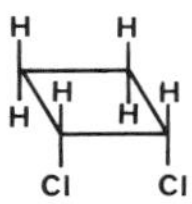

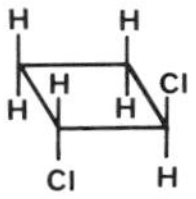

Geometric isomers do not readily interconvert. To transform one into the other would require the cleavage of a bond. One can often purchase

both the cis and trans stereoisomers of a particular compound.

The type of stereoisomerism that is the most difficult to visualize is enantiomerism. Enantiomers are mirror-image stereoisomers (i.e., they are molecules that have the relationship of being mirror images of one another). In most cases an object is identical (superimposable) on its mirror image. In rare cases it is not. A glove is not superimposable on its mirror image, and there exists both left-handed and right-handed gloves. Similarly, compound **45** is not identical to its mirror image, structure **46**. Construction of accurate molecular models for both would prove it is impossible to line up all the atoms so they are in identical positions in space. Since they are mirror images of one another, they are enantiomers:

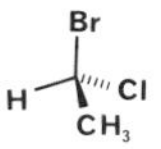

It can be shown mathematically that a substance with a plane of symmetry passing through it will be superimposable on its mirror image. Thus, a molecule with a plane of symmetry will not have an enantiomer. However, if any carbon in a molecule has four different substituents attached to it, the compound will not have a plane of symmetry, and two enantiomers for that substance can exist. Such a compound is called "chiral" ("handedness"). The enantiomers will have virtually identical chemical and physical properties. The only difference in physical properties will be in the opposite direction of rotation of polarized light. The only difference in chemical properties will be in reaction with chiral reagents. A "left-handed" glove will fit more easily on a left hand than a right hand, and the reaction of one enantiomer with a chiral reagent will always be at least a little easier. In proteins only one enantiomer of the constituent amino acids is found, and a significant difference of reaction rate with chiral molecules is seen.

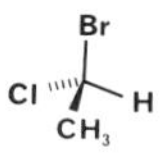

When more than one chiral carbon exists in a molecule, the situation is more complex. In general, for N chiral carbons there will exist 2^N stereoisomers. Compound **47** contains 2 chiral carbons, indicated by an asterisk. Its mirror image is compound **48**; these nonsuperimposable compounds are enantiomers. If only one chiral center is changed compound **49** results, which is neither identical to compound **47** nor **48**. It is a diastereomer (a nonmirror image stereoisomer)

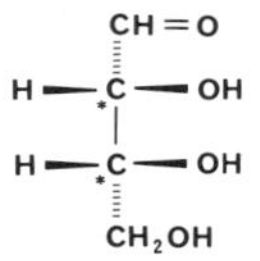

CH=O
H—C—OH
H—C—OH
CH₂OH

47

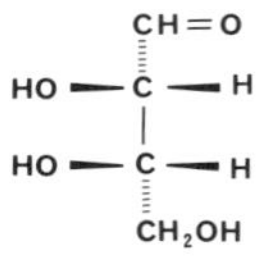

CH=O
HO—C—H
HO—C—H
CH₂OH

48

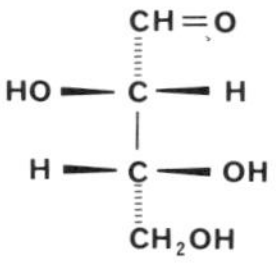

CH=O
HO—C—H
H—C—OH
CH₂OH

49

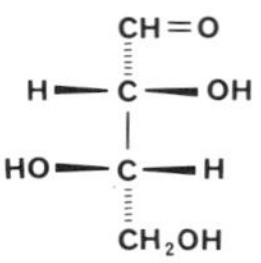

CH=O
H—C—OH
HO—C—H
CH₂OH

50

of both of these compounds. Its mirror image is a fourth stereoisomer, compound **50**.

In some cases a molecule with chiral carbons has a mirror image that is identical. Structures **51** and **52** represent the *same* compound. Such a compound is called a meso compound. The superimposability of the mirror image results from the existence of a plane of symmetry passing through the molecule, as shown in Fig. 3. This makes the actual number of stereoisomers that exist less than predicted by the 2^N rule.

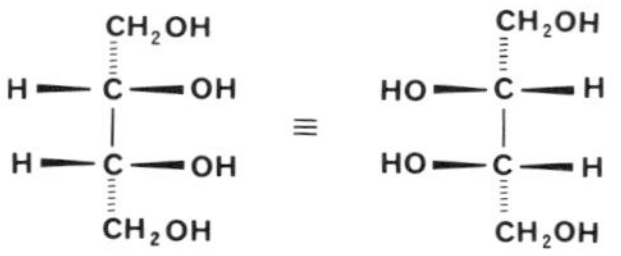

CH₂OH
H—C—OH
H—C—OH
CH₂OH

51

≡

CH₂OH
HO—C—H
HO—C—H
CH₂OH

52

Conformational isomers are another important type of stereoisomers. Such compounds can be interconverted by rotations about single bonds in a molecule. Interconversion is generally very rapid at room temperature. This is different from enantiomers or diastereomers, which do not readily interconvert. In fact, an infinite number of conformations for most molecules exists, although only the most important ones have been given names.

For acyclic molecules, such as ethane, rotation around the central C—C bond causes

CH₂OH
H—C—OH
H—C—OH
CH₂OH

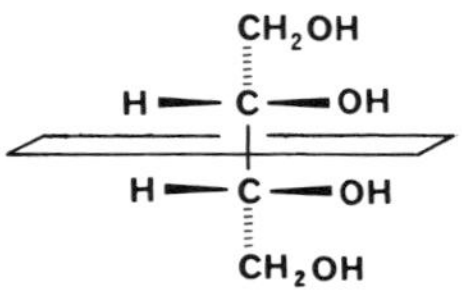

FIG. 3.   Plane of symmetry in a meso compound.

"eclipsing" of the attached hydrogens. The most stable conformation is the "staggered" arrangement (structure **53**) and the least stable is the fully eclipsed arrangement (structure **54**):

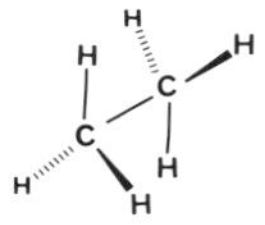

53 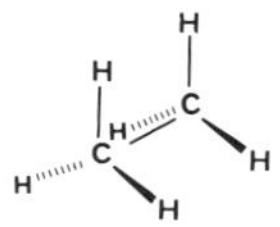   **54**

In cyclic compounds the most important, and best-studied, system is the six-membered ring—cyclohexane. Two "chair" conformations (**55** and **56**) for a substituted cyclohexane exist. In structure **55**, the most stable conformation, the methyl substituent is "equatorial" (lying roughly in the plane of the molecule). In structure **56**, the methyl substituent is "axial" (lying parallel to the axis of the plane). Substituents in axial positions are more crowded, making structure **56** less stable. The larger the substituent, the greater the energy difference between the two. Structure **57** is a "boat" conformation, which is particularly unstable because of "flagpole" interactions between hydrogens on carbons 1 and 4, and eclipsing interactions for the hydrogens on the four carbon atoms on the base of the boat:

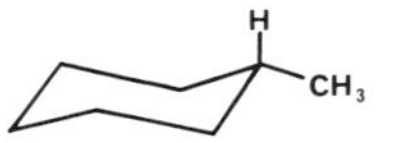

55    **56**

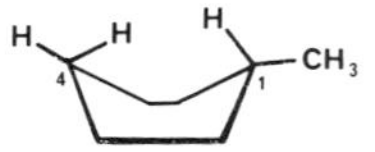

57

F. Acids and Bases

The first step in most chemical reactions is the interaction of a pair of nonbonding electrons (an "unshared pair") of one molecule with a center of electron deficiency (a partial or complete positive charge) in another species. One of the simplest reactions of this type is an acid–base reaction. The principles controlling such reactions are applicable to other reactions, and therefore is a convenient beginning for discussion of reactivity.

A typical acid–base reaction would be the reaction of sodium hydroxide with acetic acid:

$$HO^- + CH_3CO_2H \rightarrow HOH + CH_3CO_2^- \quad (1)$$

In this reaction hydroxide is acting as a base, removing a proton from acetic acid to produce the acetate ion. The Bronsted definition of an acid is a substance that is a proton donor, while a base is a proton acceptor. Clearly, this definition is appropriate in this case, with acetic acid being the substance that is donating a proton (and therefore is the acid) to the hydroxide (the base). The reaction is initiated by the unshared pair of electrons on the hydroxide starting to form a bond to the proton on the acetic acid (which is the aforementioned electron-deficient center).

Such acid–base reactions are generally ones in which the position of equilibrium must be carefully considered, because the reverse reaction is also a feasible reaction. In the reverse reaction the acetate ion would act as a base accepting a proton from water, the acid. To predict whether the forward reaction or the reverse reaction would be more important requires consideration of many of the same elements governing reactivity, which are important in the more general case. For acid–base reactions of the Bronsted type the most important consideration is the relative stability of the various reactants and products. The stability that is most important is that of the resultant ions from proton transfer. The equilibrium favors the products as shown, and one can say hydroxide is a stronger base than acetate ion, or acetic acid is a stronger acid than is water. In each case the conclusion could be achieved by noting that a negative charge is being placed on an oxygen atom, but in the case of the acetate ion the aforementioned resonance stabilization of the anion will delocalize the charge over two oxygen atoms. Thus, structural features that stabilize the negative charge make it more likely that the reaction will favor the formation of that ion.

Three properties of an atom are important in determining whether the formation of a charge on that atom will be facile: (1) the electronegativity of the atom, (2) whether resonance stabilization can delocalize the charge; and (3) the size of the atom (atomic radius).

The electronegativity of the atom is usually the most important consideration, with higher electronegativity giving a more stable ion, and thus a weaker base. In a given row of the periodic table, the basicity decreases as one goes across to the right. For example, H_2N^- is a stronger base than HO^-, which in turn is stronger than F^-.

The second important factor, resonance stabilization, has been exemplified with the hydroxide/acetate example. Similarly, nitrate, phosphate, and sulfate are well-stabilized anions by resonance, and are very weak bases that form readily. This makes their conjugate acid (the protonated form of these anions) a strong acid because the molecule is willing to give up the proton and form the very stable anion.

The third important factor is more subtle—as one increases the size of an anion the stability increases. The charge is dispersed over a larger area, so it is stabilized in somewhat the same way that resonance delocalization stabilizes an anion. The charge is in a more diffuse cloud of electrons, which does not as readily form a bond to a proton, causing a weaker base. For this reason, if one goes down a family in the periodic table, the basicity becomes weaker. Thus, HS^- is a weaker base than HO^-. Similarly, F^- is the strongest base among the halogens, while I^- is the weakest.

The Bronsted definition of acids and bases provides a framework for understanding many reactions, if one realizes that a reaction will always tend to form the weaker (more stable) base. The general principles just outlined explain reactivity of many other systems if one uses a more general definition of acids and bases. In the Lewis formalism, a base is an electron pair donor, while an acid is an electron pair acceptor. It is easy to see that a Bronsted base is also a Lewis base and that a Bronsted acid is also a Lewis acid. However, a reaction such as that of BF_3 with NH_3 to give compound **58** can be classified as a simple acid–base reaction according to the Lewis definition, although no proton is being transferred, as required by the Bronsted definition:

$$(2)$$

58

Many substances such as BF_3 or $AlCl_3$ are excellent Lewis acids because they are electron-deficient species, lacking a complete octet of electrons. Their reactivity can be readily understood as a desire to form a more stable product by completion of the octet. These same considerations explain reactivity of a number of highly reactive organic compounds that are often called "reactive intermediates." Their reaction, and the preferred sites of reaction, can be understood by recognizing that they are Lewis acids, which will seek electron pairs from other reactants.

II. Techniques Used in the Study of Reactions

A. WHY MECHANISMS ARE IMPORTANT

Reaction mechanisms (and physical properties of individual molecules) are studied in order to apply the resulting knowledge in a predictive manner to reactions yet to be examined experimentally. When considering a reaction one believes might be feasible, two questions must be answered to resolve whether the reaction will be "good:"

1. How far will the reaction go? (What is the "equilibrium constant" for the reaction?)
2. How fast will the reaction proceed? (What is the "rate" of the reaction?)

The answers to both these questions must be favorable for a proposed new reaction to be a valuable transformation.

The first question asks how much product ultimately forms with specified reactants. The equilibrium constant for a reaction is the mathematical expression of the ratio of products to reactants. In organic chemistry a reaction is not generally useful unless the equilibrium constant K is favorable (i.e., much greater than unity). Constant K is directly related to the difference of free energies of starting materials and products:

$$\Delta G^0 = -2.303 RT \log K \qquad (3)$$

where T is the absolute temperature and R the gas constant. When the products are more stable (have a lower free energy) than the starting materials, the equilibrium constant will be large and the reaction is said to be one that would "go to completion"—very little unreacted starting materials remains after the reaction ceases. To predict when the equilibrium constant will be favorable is a goal achieved by accurately predicting the relative stability of the various compounds. A close examination of the logic utilized in the discussion of acids and bases reveals that the key element of the reaction is the relative stability of the various ions involved. An elegant way of determining relative stabilities for nonionic

reactions is by molecular orbital calculations, but this is frequently too time consuming when dealing with complex molecules. A simpler analytical method, which is often reasonably accurate, is comparison of the relative strengths of bonds broken and formed during the course of the reaction. When stronger bonds are formed than are broken, more stable molecules are formed and the reaction will be exothermic. The equilibrium constant is favorable. Another important consideration is that maximization of entropy is desired and is achieved if greater disorder (randomness) results. This factor is particularly important at very high temperatures, where the formation of smaller molecules becomes increasingly more likely entropically because the process leads to greater disorder. The mathematical evaluation of the changes of free energy during a reaction is referred to as the thermodynamics of the reaction, and it is crucial that this be favorable for a successful reaction.

Although favorable thermodynamics are an essential element in successful reactivity, the kinetics of the reaction are equally important. Kinetics deals with how fast the reaction will proceed, and a well-known example will illustrate its importance. Consider the "hypothetical" reaction of air oxidation (combustion) of cellulose, a polymer of molecular formula $(C_6H_{10}O_5)_n$:

$$(C_6H_{10}O_5)_n + 6nO_2 \rightarrow 6nCO_2 + 5nH_2O \quad (4)$$

Even simple calculations (and personal experience) show this reaction to be thermodynamically favorable—being a highly exothermic reaction. However, we do not fear that this page will burst into flame, or slowly disappear by conversion to the gaseous products of oxidation (carbon dioxide and water). For this reaction we are keenly aware that the rate of such oxidation at normal temperatures is very slow, and books can survive for centuries without disappearing into gaseous products. This example focuses on the second important consideration for a "good" reaction—whether it will have a favorable reaction rate. Unlike the equilibrium constant, which depends on the relative stability of products and reactants, the rate of the reaction depends on the existence of a favorable mechanism for reaction. The oxidation of cellulose is such a complex molecular reorganization that the activation energy for reaction is very high, and only by applying considerable heat can the reaction be forced to proceed at a rapid rate. Once the reaction has begun, the exothermicity of the reaction causes it to continue, making auxiliary heating no longer necessary. The consideration of the kinetics of a reaction is whether a reaction can proceed at a reasonable rate, perhaps by supplying a reasonable amount of heating to prompt the reaction to begin. [*See* KINETICS (CHEMISTRY).]

The study of a reaction mechanism is an examination of the details of each step of a successful reaction so one can understand the types of changes that require only a modest input of energy before they proceed. Understanding a small number of reaction mechanisms allows one to begin to predict when reactions to be discovered might proceed at reasonable rates. [*See* CHEMICAL KINETICS, EXPERIMENTATION.]

B. KINETICS

The most important observation of kinetics is that every reaction will proceed at a rate directly dependent on the concentration of some or all of the reactants. The most common case of two reactants, giving two products

$$A + B \rightarrow C + D \qquad (5)$$

gives Eq. (6) as the rate expression:

$$\text{Rate} = k[A]^x[B]^y \qquad (6)$$

The progress of the reaction could be observed by monitoring either the formation of product (C or D) or the disappearance of starting material (A or B). Graphing the changes in the concentration of one of these species would give a curve somewhat like the ones shown in Fig. 4. The rate of reaction at any point in time is the tangent to the curve drawn, and may be equated to any of the reactant or product species by using the balanced equation of reaction. Thus, $-dA/dt = dC/dt$ if the coefficients are unity as shown in our example.

The most common kinetic relationship for such a reaction is that both exponents (x and y) in the rate expression are unity, and the *order* of the reaction (the sum of the exponents in the differential rate expression) is two. Such a reaction is said to be first order with respect to each

of the two reactants, and second order overall. Unfortunately, not every reaction is second order. It is common to have first-order reactions, zero-order reactions, third-order reactions, and even reactions with nonintegral exponents. The determination of the order of a reaction is important because a proposed mechanism must be consistent with this number. A reaction that takes place in one step, brought about solely by collision of two reactants, must have second-order kinetics. Observation of a different kinetic order is a clue that the reaction mechanism is more complex than one step.

Examination of Eq. (6) reveals there are essentially only two ways in which one can experimentally alter the rate of a reaction—by changing the concentration of the reactants, which appear in the rate expression, and by changing the rate constant k, for the reaction. In the example of a second-order reaction, one could double the rate of reaction by doubling the concentration of either one of the two reactants. A one-step reaction induced by collision of two reactants is more rapid as collisions increase by higher concentration of either reactant. How one could change the rate constant in the differential rate expression is less apparent. The rate constant for a reaction is a number (which can be experimentally determined) that is invariant with changes in the concentration of reactants. It does change if either reactant is changed. For example, the reaction of HCl with cyclohexene will have a different rate constant than the reaction of HBr with cyclohexene:

$$k_{HCl} < k_{HBr} \qquad (9)$$

Furthermore, the rate constant will vary with the temperature in a manner predictable by the Arrhenius relationship:

$$k = A e^{-E_A/RT} \qquad (10)$$

In this relationship, increasing the temperature is seen to exponentially accelerate the reaction by affecting the rate constant. The pre-exponential factor A is dependent on such factors as solvent and entropy and is not easily altered in a predictable way. However, an even greater de-

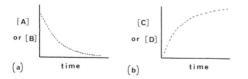

FIG. 4.   Concentration changes a typical reaction: (a) starting materials and (b) products.

pendence (exponential) of the rate constant arises from variation of E_A, the activation energy for the reaction. The activation energy can be illustrated by a graph of the overall energy of all reacting species as the reaction begins. Since every reactant is likely to have a full complement of bonds attached to every atom, it is impossible for new bonds to form before some existing bonds have broken. Thus, the beginning point of every reaction represents a local minimum, and the energy must first increase in order to cause a reaction. As the reaction progresses, there will come a point at which bond formation can begin, and the overall energy decreases. For an exothermic reaction the energy will continue to decrease until products of lower energy than the reactants are formed. As shown in Fig. 5, the height of the energy barrier that must be traversed from starting materials is the activation energy.

A good reaction is one that has a reasonable rate constant at ordinary temperatures. A critical factor is the activation energy. A reaction with a very high activation energy will proceed at an extremely slow rate; one with a low activation energy will be rapid. Although other mechanisms of reaction may be more complex than the example, these conclusions will still be appropriate. The second crucial question in predicting reactivity of two compounds becomes a question of whether a reasonable mechanism of reaction exists—one which smoothly converts starting materials to products without going over a large energy barrier of a high activation energy.

Before discussing the evidence that led to acceptance of some well-known mechanisms of reaction, a brief summary of the major mathematical relationships of first- and second-order reactions will highlight differing behavior. More exhaustive treatments of these and other cases will be found in Kinetics (Chemistry).

First-order reactions, which follow the differential rate expression of Eq. (11), are quite common. For example, the decay of radioactive isotopes follows this rate law:

$$-d[A]/dt = k[A] \tag{11}$$

$$\ln[A/A_0] = -kt \tag{12}$$

Integration of Eq. (11) gives Eq. (12), which describes how the concentration of A varies with time. A plot of $\ln A/A_0$ versus time (where A_0 is the initial concentration of species A) is linear, giving a slope of $-k$. An important concept is the phrase half-life, which is the time required for half of the material to react (or decay, in the case of radioisotopes). For a first-order reaction one can easily verify that the half-life ($t_{1/2}$) equals $0.693/k$. This relationship is *not* true for reactions of any order other than first order. The determination of the dependence of half-life on the initial concentration of the reactants is a useful technique for establishing reaction order.

Second-order reactions, as mentioned earlier, are the most common in organic chemistry. These often arise from one-step mechanisms induced by the collision of two reactants, so-called bimolecular reactions. The differential rate expression

$$-d[A]/dt = k[A][B] \tag{13}$$

can be integrated to give

$$\frac{1}{A_0 - B_0} \ln \frac{[B]_0[A]}{[A]_0[B]} = kt \tag{14}$$

For a second-order reaction a plot of $\ln(A/A_0)$ versus time will not be linear, but a plot of the left-hand side of Eq. (14) versus time will be, with a slope of k. Unlike a first-order reaction, the half-life is inversely proportional to the initial concentrations and will vary during the course of the reaction as the concentration of the reactants changes.

A surprising number of reactions are observed to follow either first- or second-order kinetics. With this background one can examine how mechanisms can be probed using kinetic tools.

C. ISOTOPES

The commercial availability of a number of unusual isotopes of common elements has provided an important tool for probing reaction mechanisms, using both kinetic and nonkinetic methods. These isotopes, which have different atomic weight because of a greater number of neutrons, will react in the same manner as the ordinary isotopes. However, they may be used as labels to detect what atoms become incorpo-

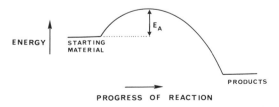

FIG. 5.   Energy change during a one-step exothermic reaction.

rated, or may serve as a way of subtly altering the rate of a reaction. Some isotopes that have been utilized in such ways are

$$^2H, {}^3H, {}^{13}C, {}^{14}C, {}^{15}N, {}^{18}O, {}^{32}P, {}^{35}S, {}^{37}Cl, {}^{131}I$$

Some of these isotopes are radioactive and allow sensitive detection of small differences in reactions. Others are nonradioactive and their successful use relies on either mass spectrometric or spectroscopic detection methods. The first two isotopes mentioned, deuterium and tritium, are so commonly used in mechanistic studies they have been given symbols as if they were separate elements (D and T). It is more convenient to use these single letters than the atomic weight superscripts.

The nonkinetic applications of isotopes can be readily exemplified with the important reaction of hydrolysis of an ester, a reaction that will be discussed in more detail later. The hydrolysis of amyl acetate by hydroxide in water could occur in one of two ways—either the oxygen from the hydroxide becomes attached to the carbonyl of the acetate, causing cleavage of bond 1 [Eq. (15)], or the oxygen from the hydroxide becomes attached to the terminal carbon of the amyl group, causing cleavage of bond 2 [Eq. (16)]

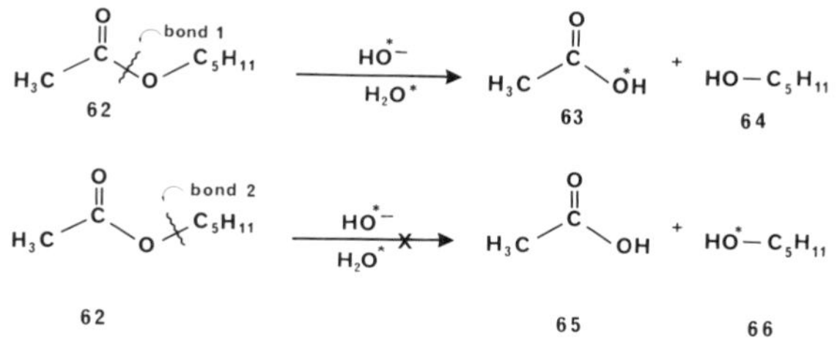

gen and carbon bonded to deuterium. The bond to deuterium appears to be a little stronger. This is because the larger mass of the deuterium leads to a lower frequency of vibration of the C—D bond than the C—H bond. Thus, the vibrational energy is somewhat smaller in the starting material, and if this bond is broken during the reaction about 5 kJ/mol more energy must be supplied to break the bond when deuterium is involved. Both theoretical calculations and experiments show that substitution of deuterium for hydrogen in a molecule can cause the reaction to slow down by about a factor of 8 if the C—H bond is broken during the course of reaction. A measurement of the ratio of the rate of reaction of the undeuterated and deuterated material (k_H/k_D) is reported as supporting, or rebutting, the cleavage of a C—H bond in the rate-determining step of a reaction. An experimental value of 8 or so for this ratio is considered a primary kinetic isotope effect, and is strong evidence for rate-determining cleavage of a C—H bond.

The concept of a rate-determining step emerges from mathematical treatment of more complex kinetic relationships than this section has discussed. Calculations involving multistep reactions lead to some simplifying conclusions

$$(15)$$

$$(16)$$

By using water (and hydroxide) that has been enriched in ^{18}O (indicated by an asterisk), one can immediately determine which of these two paths is followed by determining whether the ^{18}O becomes attached to the acetate ion or the amyl alcohol product. Unequivically, the ^{18}O from the hydroxide becomes attached only to the acetate ion, so bond 1 is the carbon–oxygen bond broken during this reaction, not bond 2.

A more subtle application of isotopes as mechanistic probes is their application in kinetic studies. Two important concepts are utilized in such studies—the kinetic isotope effect and the concept of a rate-determining step.

The existence of a kinetic isotope effect is due to a small difference in the strength of a bond between, say, carbon bonded to ordinary hydro-

that have great value in mechanistic studies. Consider a two-step reaction, for which two "limiting" cases might apply:

$$A \xrightarrow{slow} B \xrightarrow{fast} C \qquad (17)$$

$$A \xrightarrow{fast} B \xrightarrow{slow} C \qquad (18)$$

In Eq. (17), unlike Eq. (18), the first step can be shown to be rate determining. It represents a situation in which a large energy barrier must be overcome to get to species B, and the overall rate of reaction is found not to depend on the concentration of any species that reacts with B, nor to isotopic substitution on any bonds broken in the conversion of B to C. However, in Eq. (18) the second step does affect the overall rate of the reaction, and isotopic substitution of a bond being broken in the second step of this case would lead to a primary kinetic isotope effect. A more detailed discussion involving the S_N1 reaction, which appears in Section IV,B, will illustrate how the conclusions described for Eq. (17) can exist.

$$
\underset{70}{\text{C}-\text{Cl}} \xrightarrow{\text{H}_2\text{O}} \underset{71}{\text{C}-\text{OH}} + \underset{72}{\text{HO}-\text{C}}
$$

(20)

D. Chirality and Stereoisomerism

The structural complexity arising from stereoisomerism provides tools for mechanistic study. Particularly informative is the study of reactions by choosing starting materials that form geometric isomers or chiral products.

The reaction of cyclohexene with Br_2 exemplifies a reaction conceivably forming two addition products, *cis-* or *trans*-1,2-dibromocyclohexane. Yet only the trans product is formed:

(19)

This is mechanistically significant, revealing the absence of a carbocation intermediate (as shall be apparent later). Furthermore, it negates a concerted reaction in which π-bond cleavage, σ-bond cleavage (of the Br—Br), and σ-bond formation (C—Br bonds) occur simultaneously.

The reaction of a chiral molecule can be mechanistically informative because three outcomes are possible:

1. The retention of configuration at the chiral carbon.
2. The inversion of configuration at the chiral carbon.
3. Racemization, or the formation of both enantiomers of the product.

This third possibility is particularly revealing, since it establishes the presence of an intermediate with a plane of symmetry. The reaction of structure **70** in Eq. (20) with water in 80% acetone to water gives virtually equal amounts of structures **71** and **72**. The proposed formation of a planar intermediate (which is achiral) in the reaction was largely because of this stereochemical observation.

E. Direct Detection of Intermediates

In a multistep transformation, verification of the existence of an intermediate significantly improves confidence in a proposed mechanism. At other times an intermediate may be isolated (when stable enough), and shown to lead to the same product at a faster rate than the starting material. Sometimes the intermediate can be trapped by reaction with another component whose reaction is well known to verify such intermediate's existence. When the lifetime of an intermediate is fleeting, these methods may not be feasible. In such instances direct detection by spectroscopy is convincing evidence for an intermediate's existence.

The most common spectroscopic tools for detection of intermediates are infrared (IR), nuclear magnetic resonance (NMR), and electron spin resonance (ESR) spectroscopy. Infrared can detect the presence of an unusual bond (such as an enol) in an intermediate. The recent development of Fourier Transform (FT–IR) instrumentation will make this an increasingly used method for such studies. Nuclear magnetic resonance, particularly FT–NMR, is a sensitive probe for many intermediates. Carbocations have been well studied by this technique. Electron spin resonance is the most sensitive tool, detecting the existence of extremely small amounts of free radicals (species with an unpaired electron). Another common spectroscopic method that detects radical intermediates is the observation of CIDNP (altered NMR spectra of products) due to radical involvement during product formation.

III. Reactive Intermediates

A. Radicals

A species with an unpaired electron is called a *radical* or a *free radical*. Methyl radical, **73,** is an example:
Organic chemists generally show this unpaired electron explicitly by a dot; other disciplines sometimes do not. Inspection of structure **73**

73

will verify that the electrical charge is neutral, but the carbon has an incomplete octet. Thus, as an electron-deficient species a radical is very reactive, needing another electron to complete the octet. Commonly, a radical will react by abstraction of a hydrogen atom from another species, sometimes forming another radical, though perhaps a more stable one:

$$H_3C\cdot \;+\; H-R \;\longrightarrow\; H_3C-H \;+\; R\cdot \qquad (21)$$

Another common reaction of a radical is addition to a π bond, a commonly used reaction for initiation of polymerization of alkenes:

$$R\cdot \;+\; H_2C=CH_2 \;\longrightarrow\; R-CH_2\dot{C}H_2 \qquad (22)$$

Radicals may be detected spectroscopically, or their existence verified indirectly using radical scavengers. Some compounds (e.g., hydroquinone, thiols, BHA, or BHT), donate a hydrogen atom readily to form a relatively stable free radical. Thus, any radicals present would react rapidly with such reagents, and other reactions of those radicals would be halted.

$$R\cdot \;+\;
\begin{array}{c} OH \\ \bigcirc \\ OH \end{array}
\;\longrightarrow\; R-H \;+\;
\begin{array}{c} O\cdot \\ \bigcirc \\ OH \end{array}
\qquad (23)$$

74

Hydroquinone (structure **74**), for example, would halt polymerization initiated by radicals, and is a common stabilizer. Similarly, BHA and BHT, familiar food additives, are antioxidants because they readily scavenge free radicals. The thiol-containing tripeptide glutathione is found in high concentrations throughout our body, and is believed to be synthesized as a protection against cellular damage that would be induced by free radicals.

Several factors will affect the stability of radicals. First, a radical will be more stable if alkyl groups instead of hydrogen is attached to the radical center, so the order of decreasing stability in some common radicals would be

$$H_3C-\underset{H_3C}{\overset{H_3C}{\underset{|}{C}}}\cdot \quad H_3C-\underset{H}{\overset{H_3C}{\underset{|}{C}}}\cdot \quad H-\underset{H}{\overset{H_3C}{\underset{|}{C}}}\cdot \quad H-\underset{H}{\overset{H}{\underset{|}{C}}}\cdot$$

$$\xrightarrow{\text{decreasing stability}}$$

Second, resonance delocalization of the unpaired electron is an important stabilizing feature:

$$H_2C=CH-\dot{C}H_2 \;\longleftrightarrow\; H_2\dot{C}-CH=CH_2 \;>\; CH_3CH_2\dot{C}H_2$$

Third, an adjacent pair of electrons will help stabilize a radical:

$$H_3C-\underset{\ddot{O}H}{\overset{|}{\dot{C}}}-CH_3 \;>\; H_3C-\underset{}{\overset{H}{\underset{|}{\dot{C}}}}-CH_3$$

This latter stabilization is the reason ethers are so prone to air oxidation, readily forming radicals which attack O_2 to ultimately yield explosive peroxides.

B. CARBOCATIONS

In organic chemistry the most common cation reactive intermediate is a carbocation (carbonium ion), for example, **75**.

$$\overset{\oplus}{C}$$

75

Reactions proceeding through formation of carbocations are now known to be common. Such ions have a net positive charge on the carbon, and are species with only six electrons in the valence shell. Therefore, they are very reactive species. Carbocations react in many ways to relieve this electron deficiency.

1. The elimination of a proton from an α carbon to form an alkene.
2. They abstract a hydride from another compound to form an alkane.
3. They rearrange by migration of a hydride or alkyl group to form a more stable carbocation.
4. They attach a nucleophilic species such as a water (forming an alcohol) or halide ion.
5. They attack an alkene to form a new C—C bond (and higher molecular weight material).

The great variety of reactions available frequently leads to complicated mixtures of products whenever a carbocation is involved as a reaction intermediate.

Polymerization of alkenes is feasible via carbocations, initiated by protonation of the alkene by a strong acid. Subsequent electrophilic attack on another alkene by the resulting carbocation gives an increasingly longer carbon chain:

$$H_2C=CH_2 \xrightarrow{H^\oplus} CH_3\overset{\oplus}{C}H_2 \xrightarrow{H_2C=CH_2} CH_3CH_2CH_2\overset{\oplus}{C}H_2$$

$$(24)$$

The stability of carbocations is greatly increased by substituents that alleviate the electron deficiency, most commonly alkyl groups. The more highly substituted carbocations are the most stable. Carbocations (and radicals) are frequently classified according to the number of alkyl groups attached to the carbon bearing the incomplete octet—primary for one alkyl substituent, secondary for two alkyl substituents, and tertiary if three alkyl substituents are attached:

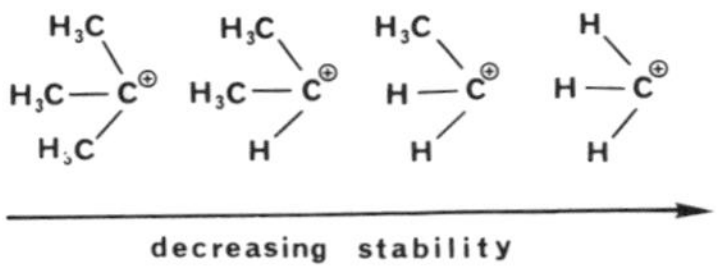

decreasing stability

Like radicals, carbocations are stabilized if the positive charge can be delocalized by resonance. For this reason an allyl (structure **76**) or benzyl (structure **77**) carbocation is very stable, being comparable in stability to a tertiary carbocation:

76   77

The previously mentioned tendency for a carbocation to undergo a rearrangement by migration of a hydride or an alkyl group can be understood on the basis of preferential formation of a more stable product (the rearranged carbocation). A secondary carbocation will tend to rearrange (reaction 25) to the more stable tertiary carbocation when structurally possible unless other reactants are present to capture the carbocation:

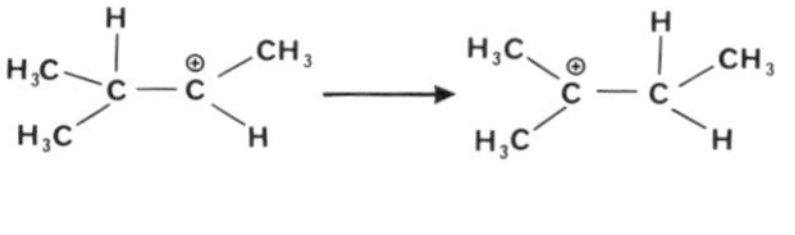

$$(25)$$

C. CARBANIONS

Unlike carbocations, which often undergo myriad reactions to give complex product mixtures, carbanions are often useful in synthesis. They vary widely in their stability, depending on the ability of substituent groups to stabilize a negative charge.

Carbanions are usually produced directly by the removal of an acidic hydrogen by a strong base (reaction 26):

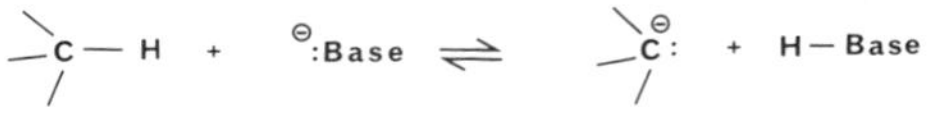

$$(26)$$

The acidity of the precursor, as measured by its pK_a, is a direct measure of the stability of the resulting carbanion. The more acidic compounds are more acidic because the resulting anions are more stable. Table I shows the relative stability of representative carbanions, and the pK_a of their neutral precursors, their "conjugate acids."

Resonance delocalization is clearly important, particularly when a "magic number" of π electrons is produced by inclusion of the unshared

TABLE I. Comparative Carbanion Stabilities[a]

Carbanion	pK_a of conjugate acid
CH_3-	47
$H_2C=CH-CH_2-$	35
$(Ph_2)CH-$	34
$N\equiv C-CH_2-$	31
$-CH(CO_2CH_2CH_3)_2$	30
$CH_3-\overset{\overset{\displaystyle O}{\|\|}}{C}-CH_2-$	26
$Ph-\overset{\overset{\displaystyle O}{\|\|}}{C}-CH_2-$	25
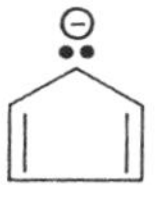	18
$-CH_2NO_2$	17
$-CH(CO_2CH_3)_2$	13
$-CH(CN)_2$	11
$CH_3-\overset{\overset{\displaystyle O}{\|\|}}{C}-\overset{\ominus}{C}H-CO_2-CH_2-CH_3$	11

[a] Listed from lowest to highest stability.

pair of the anion. Electron-withdrawing substituents (CN, NO$_2$, CO$_2$R, COR) are seen to stabilize an adjacent anion—two being even more effective than one in this regard. Compounds bearing such substituents are referred to as active methylene compounds. Carbanions can be quantitatively obtained from them by using a strong base such as lithium diisopropylamide (LDA):

$$H_2C(CO_2C_2H_5)_2 + \overset{\oplus}{Li} \overset{\ominus}{:}N\overset{CH(CH_3)}{\underset{CH(CH_3)_2}{\diagup}} \longrightarrow$$

$$H\overset{\ominus}{\underset{}{C}}\overset{Li^\oplus}{(CO_2C_2H_5)_2} + HN\overset{CH(CH_3)_2}{\underset{CH(CH_3)_2}{\diagup}}$$

Subsequent reaction of the carbanion with an alkyl halide (or other electrophile) leads to C—C bond formation and halide cleavage. In each step the driving force for the reaction is formation of a weaker base. From the strongest base (LDA) a weaker one (the carbanion) is formed, and finally the weakest of all—the halide ion. Synthetic organic chemistry utilizes such reactivity often; more examples of the use of carbanions in synthesis are provided in Organic Chemistry, Synthesis.

D. CARBENES

Divalent compounds of carbon, carbenes, are particularly unusual. Examination of methylene (structure 78) shows the electrical charge on carbon to be neutral, yet an incomplete octet exists:

$$H\diagdown \\ \quad \overset{..}{C}: \\ H\diagup$$

78

A carbene will seek electron-rich species for reaction so that its octet may be completed. Typically, carbenes react by addition to π bonds (e.g., compound 79):

$$Cl\diagdown \overset{..}{C}: + \overset{H\diagdown \;/CH_3}{\underset{H\diagup \;\backslash CH_3}{\overset{C}{\underset{C}{\parallel}}}} \longrightarrow \text{(27)}$$

79

In addition, very reactive carbenes like methylene will even react with σ-bonded electrons by insertion:

$$H\diagdown \overset{..}{C}: + H_3C-\overset{H}{\underset{}{CH_2}} \longrightarrow H_3C-\overset{CH_3}{\underset{}{CH_2}} \quad \text{(28)}$$

80

In spite of their high reactivity, carbenes have been generated in a variety of ways: α elimination from carbanions [Eq. (29)]; photolysis of diazoalkanes [Eq. (30)] or ketenes; and by Simmons–Smith reaction:

$$HO^\ominus + CHCl_3 \rightleftharpoons H_2O + {}^\ominus:CCl_3 \longrightarrow :CCl_2 + {}^\ominus Cl \quad \text{(29)}$$

$$CH_2N_2 \overset{h\nu}{\longrightarrow} :CH_2 + N_2 \quad \text{(30)}$$

The Simmons–Smith reaction of CH$_2$I$_2$ with Zn(Cu) leads to a carbenoid rather than a true carbene, because its reactivity is somewhat different—reacting more selectively and less rapidly.

In reality two electronic arrangements of carbenes are possible: a singlet (structure 81) and a triplet (structure 82) state:

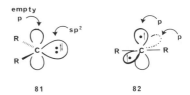

In 81, the singlet both electrons are in the same sp^2 orbital and have opposite spins. In 82, the triplet the electrons are in different orbitals and have the same spin. The most stable state of a carbene depends on its substituents. In methylene the triplet is about 8–11 kcal/mol more stable than the singlet. In contrast, halomethylenes are more stable as the singlet. The electronic configuration is important in determining mechanism, because the stereospecific reaction shown in Eq. (27) (cis-2-butene leading to the cis-dimethylcyclopropyl product) is only possible with the singlet. The triplet must react stepwise and less selectivity is observed.

E. ELECTRONICALLY EXCITED STATES

The last type of reactive intermediate to be considered is the easiest to generate. In every molecule there are bonding orbitals that are filled and antibonding orbitals that are empty. Irradiating the molecule with a photon of light that has precisely the energy content of the separation between the bonding and antibonding orbital (indicated by the arrow) will cause absorption of the light and the promotion of an electron from the π orbital to a π^* orbital. The resulting electronic configuration is shown in Fig. 6b, and would be referred to as a (π, π^*) singlet state, or $^1(\pi, \pi^*)$. A change in spin of an unpaired electron would give the triplet state, $^3(\pi, \pi^*)$ (Fig. 6c).

An accurate representation of a molecule in an excited state is even more problematical than one for benzene. The π bond has been half-broken. Such an excited state is sometimes written as a biradical:

$$H_2C = CH_2 \xrightarrow{h\nu} \left[H_2\dot{C} \overset{\nearrow}{-} \dot{C}H_2 \right]^* \qquad (31)$$

There is some merit to this picture since it suggests the possibility of a reaction of excited alkenes—cis/trans isomerization due to free rotation around the single bond—which in fact is common:

(32)

83   84

The regeneration of the alkene in the last step by demotion of the excited electron back into the bonding orbital [the reverse of Eq. (31)] is another possible reaction of an electronically excited state, deactivation, which liberates the excess energy as heat. Other common events for excited states include emission of a photon of light (called fluorescence if the excited state was a singlet—phosphorescence if it was a triplet) and a variety of photochemical reactions.

Photochemistry (study of the chemical reactivity of electronically excited states) is an ac-

tive area of investigation. Some important types of reactions that have been seen are photoreduction [Eq. (33)], photorearrangement [Eq. (34)], photodimerization [Eq. (35)], and photofission [Eq. (36)] (note: Ph is the abbreviation for a Phenyl substituent):

(33)

(34)

(35)

(36)

Modern photochemistry focuses not only on discovering what reactions occur from excited states but also on the reasons for those changes. Some photoreactions are so efficient they have been used preparatively on an industrial scale. [*See* PHOTOCHEMISTRY, ORGANIC.]

IV. Case Studies

A. ESTER HYDROLYSIS

Esters can be hydrolyzed in either acidic or basic solution. In both cases second-order kinetics are observed—first order in the ester and first order in either the acid or base. The simpler mechanism is the base-promoted hydrolysis (saponification). The accepted mechanism is shown in Fig. 7. The direct attack of hydroxide on R' was ruled out by isotopic labeling, as described in Section II,B:

In acid hydrolysis is reversible, and the mechanism (Fig. 8) is the reverse of the mechanism of acid-catalyzed esterification. The accepted mechanism must account for the partial exchange of isotopically labeled oxygen from the

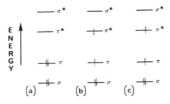

FIG. 6. Electronic configurations of ethylene: (a) ground state, (b) singlet excited state, and (c) triplet excited state.

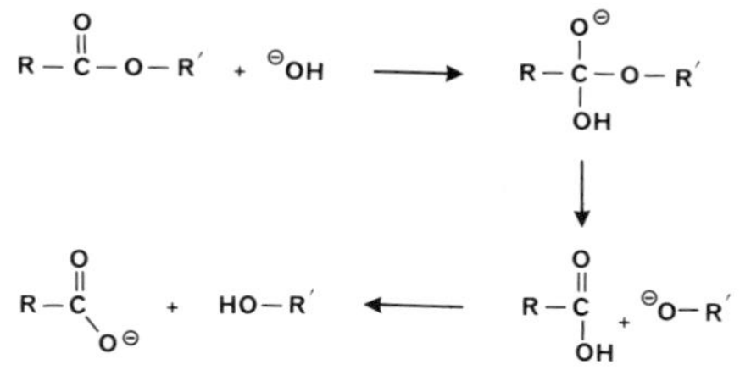

$$R-\overset{\overset{\displaystyle O}{\|}}{C}-O-R' \;+\; {}^{\ominus}OH \;\longrightarrow\; R-\overset{\overset{\displaystyle O^{\ominus}}{\|}}{\underset{OH}{C}}-O-R'$$

$$R-\overset{\overset{\displaystyle O}{\|}}{C}\underset{O^{\ominus}}{} \;+\; HO-R' \;\longleftarrow\; R-\overset{\overset{\displaystyle O}{\|}}{\underset{OH}{C}} \;+\; {}^{\ominus}O-R'$$

FIG. 7. Mechanism of base-promoted ester hydrolysis.

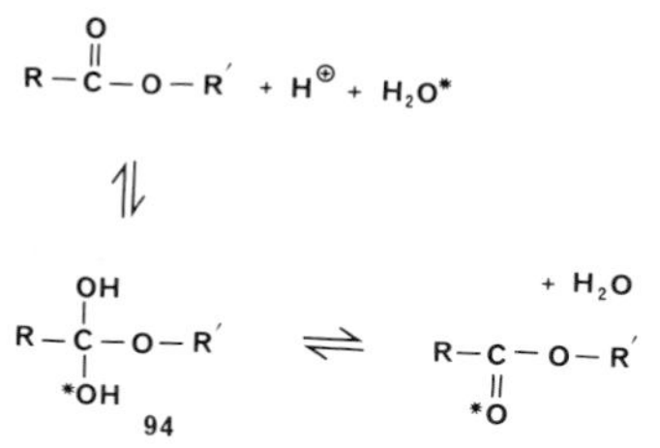

$$R-\overset{\overset{\displaystyle O}{\|}}{C}-O-R' \;+\; H^{\oplus} \;+\; H_2O^*$$

$$\Updownarrow$$

$$R-\overset{OH}{\underset{{}^*OH}{C}}-O-R' \;\rightleftharpoons\; R-\overset{+\,H_2O}{\underset{{}^*O}{C}}-O-R'$$
$$94$$

FIG. 9. Exchange of ${}^{18}O$ during ester hydrolysis.

solvent. If the starting material is recovered before total reaction it is found to have incorporated some ${}^{18}O$ from ${}^{18}O$ labeled water as solvent. This can be explained from the above mechanism, as shown in Fig. 9. From the tetrahedral intermediate (structure 94 in Fig. 9) loss of either labeled (*) or ordinary oxygen is possible, and sometimes this intermediate reverts to starting material by loss of the ordinary oxygen, leaving the ${}^{18}O$ incorporated in the starting material.

B. NUCLEOPHILIC SUBSTITUTION REACTIONS

A substitution reaction usually involves the attachment of an ion bearing a negative charge (the nucleophile) to a carbon that simultaneously detaches a substituent halide:

$$\ddot{B}r-CH_3 \;+\; {}^{\ominus}:\ddot{O}H \;\longrightarrow\; :\ddot{B}r:^{\ominus} \;+\; CH_3-\ddot{O}H \quad (37)$$

A variety of nucleophiles react in this manner (HO—, RO—, HS—, CN—, etc.), and groups other than halides can serve as the "leaving group." The driving force for the reaction is the formation of a more stable anion (a weaker base). This reaction is so important in synthetic organic chemistry that its mechanism has been carefully examined. Studies show this simple reaction can proceed by two very different paths called the S_N1 and S_N2 mechanisms, written as shorthand for substitution-nucleophilic-unimolecular and substitution-nucleophilic-bimolecular mechanisms, respectively.

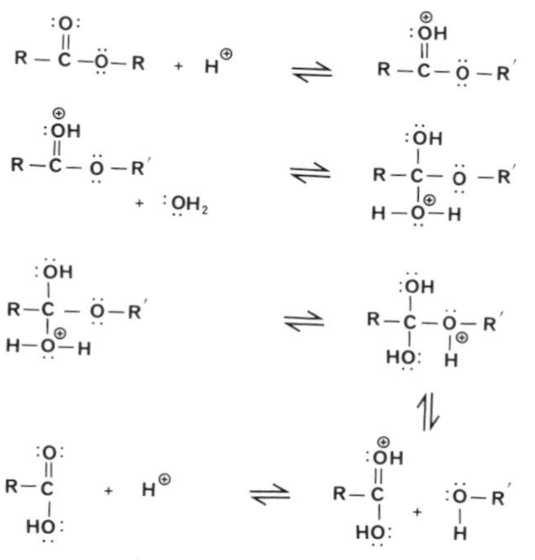

The bimolecular mechanism is the most straightforward. Consistent with its name, it follows second-order kinetics. The accepted mechanism is a concerted reaction where bond formation and bond cleavage occur simultaneously. There is no intermediate at all, but a "half-reacted" state called a transition state (structure 95) which represents the most unstable species produced during the course of the reaction:

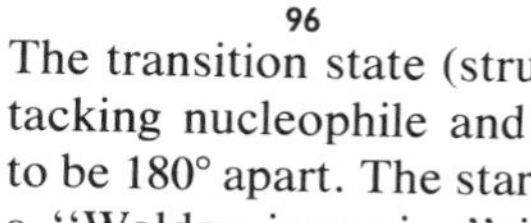

$$HO^{\ominus} \quad \overset{H\;\;CH_2CH_2CH_3}{\underset{CH_3}{C}}-Br \;\longrightarrow$$
$$95$$

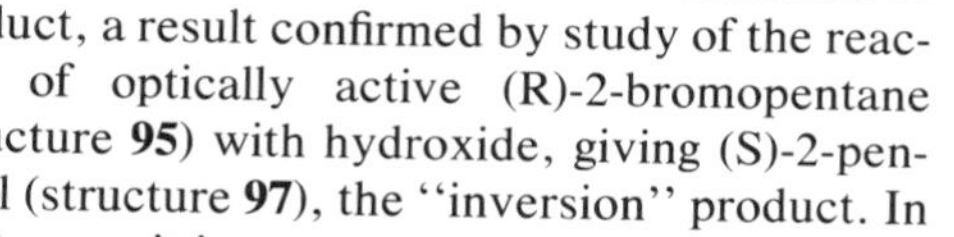

$$\left[HO^{\delta-}\cdots\cdots\overset{H\;CH_2CH_2CH_3}{\underset{CH_3}{C}}\cdots\cdots Br^{\delta-} \right] \;\longrightarrow\; HO-\overset{CH_2CH_2CH_3}{\underset{CH_3}{C}}{\underset{}{\overset{H}{\diagup}}} \quad Br^{\ominus}$$
$$96 \hspace{6cm} 97$$

The transition state (structure 96) shows the attacking nucleophile and the departing bromide to be 180° apart. The starting material undergoes a "Walden inversion" in its transformation to product, a result confirmed by study of the reaction of optically active (R)-2-bromopentane (structure 95) with hydroxide, giving (S)-2-pentanol (structure 97), the "inversion" product. In most cases it is not apparent that such an inversion has occurred because of symmetry in the reactant, but in cyclic compounds or chiral compounds this behavior is noted for every S_N2 reaction.

The alternate mechanism of substitution, S_N1, is less common, but is not rare. In dilute aque-

$$R-\overset{\overset{\displaystyle O}{\|}}{C}-\ddot{O}-R' \;+\; H^{\oplus} \;\rightleftharpoons\; R-\overset{\overset{\displaystyle \oplus OH}{\|}}{C}-\ddot{O}-R'$$

$$R-\overset{\overset{\displaystyle \oplus OH}{\|}}{C}-\ddot{O}-R' \;+\; :\ddot{O}H_2 \;\rightleftharpoons\; R-\overset{\overset{\displaystyle :\ddot{O}H}{|}}{\underset{H-\ddot{O}-H}{C}}-\ddot{O}-R'$$

$$R-\overset{\overset{\displaystyle :\ddot{O}H}{|}}{\underset{H-\overset{\oplus}{\ddot{O}}-H}{C}}-\ddot{O}-R' \;\rightleftharpoons\; R-\overset{\overset{\displaystyle :\ddot{O}H}{|}}{\underset{HO:\;H}{C}}-\overset{\oplus}{O}-R'$$

$$\Updownarrow$$

$$R-\overset{\overset{\displaystyle :O:}{\|}}{\underset{HO:}{C}} \;+\; H^{\oplus} \;\rightleftharpoons\; R-\overset{\overset{\displaystyle \oplus OH}{\|}}{\underset{HO:}{C}} \;+\; :\ddot{O}-R'$$

FIG. 8. Mechanism of acid-catalyzed ester hydrolysis.

ous potassium hydroxide *t*-butyl bromide (structure **98**) forms *t*-butyl alcohol (structure **99**), and the rate of the reaction is found to be independent of hydroxide concentration:

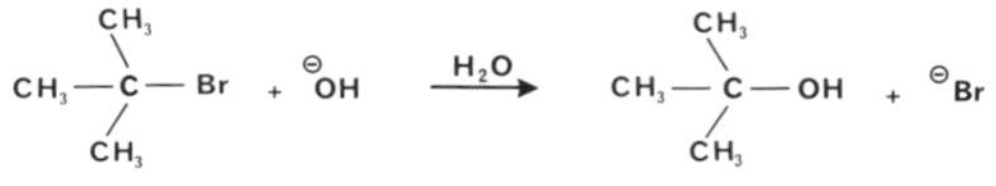

$$CH_3-\underset{\underset{CH_3}{|}}{\overset{\overset{CH_3}{|}}{C}}-Br + {}^{\ominus}OH \xrightarrow{H_2O} CH_3-\underset{\underset{CH_3}{|}}{\overset{\overset{CH_3}{|}}{C}}-OH + {}^{\ominus}Br$$

$$\quad\quad 98 \quad\quad\quad\quad\quad\quad\quad\quad\quad 99$$

It is difficult to visualize how first-order kinetics might be possible when two reactants are clearly involved. The accepted mechanism of reaction is two steps, with the first step being rate determining (see Fig. 10). Equation (38) is the differential rate expression that would apply to such a scheme:

$$-\frac{d[RX]}{dt} = k_f[RX]\left[\frac{k_2[OH^-]}{k_r[X^-] + k_2[OH^-]}\right] \quad (38)$$

This relationship is complex in general, but if $k_r[X^-] \ll k_2[HO^-]$ the second term becomes approximately unity, and the expression reduces to

$$-d[RX]/dt = k_f[RX] \quad (39)$$

First-order kinetics would be observed in this situation. Qualitatively, this situation exists if the second step is much faster than alternatives.

The stereochemical outcome of an S_N1 reaction is different from that of an S_N2 reaction because of the intermediacy of a carbocation. Because a carbocation is planar, a chiral reactant would lose its chirality during reaction and mixtures of both enantiomers of product would result. This has been experimentally verified, and S_N1 reactions are said to proceed with racemization, implying loss of some (or all) of the optical activity originally due to the reacting carbon.

Since an S_N1 reaction involves a carbocation intermediate it is a less desirable mechanism than the S_N2. The S_N1 mechanism will usually be followed by all tertiary halides, and secondary halides in polar solvents or in reaction with poor nucleophiles. In other cases the S_N2 reaction predominates. The shift in mechanism because of halide structure is entirely due to the

$$R-X \underset{k_r}{\overset{k_f}{\rightleftharpoons}} R^{\oplus} + X^{\ominus}$$

$$R^{\oplus} + {}^{\ominus}OH \xrightarrow{k_2} R-OH$$

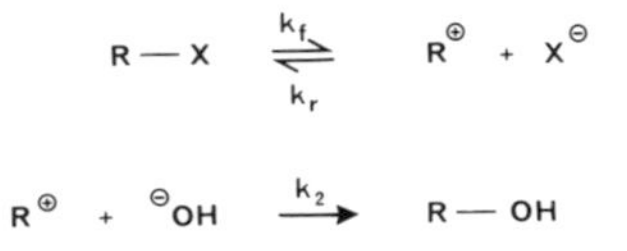

FIG. 10.   Mechanism of S_N1 reaction.

stability of the intermediate carbocation. A tertiary carbocation is the most stable, so the S_N1 reaction is the fastest with a tertiary halide. On the other hand, an S_N2 reaction leads to a very crowded transition state if a highly substituted halide is the reactant, and this path is the fastest with a smaller number of alkyl substituents. With two alkyl substituents (a secondary halide) the rate of reaction by both pathways is very comparable, and the polarity of the solvent (increased polarity stabilizes carbocations better) may be sufficiently important to determine the mechanism.

C. ELIMINATION REACTIONS

The most common elimination reaction is that of an alkyl halide with a base to give an alkene:

$$HO^{\ominus} + H_3C-CH_2-Br \longrightarrow H_2O + H_2C{=}CH_2 + {}^{\ominus}Br$$

As with substitution reactions, two paths of reaction are common: E2 (elimination–bimolecular) and E1 (elimination–unimolecular). These mechanisms are supported by considerable mechanistic evidence.

The E2 mechanism is described as a concerted reaction (simultaneous bond formation and cleavage) through a well-defined geometry described as trans-antiperiplanar. The proton removed by the base and the departing halide are oriented 180° apart from one another. If they cannot be so oriented because the reactants are in a ring, elimination becomes very slow:

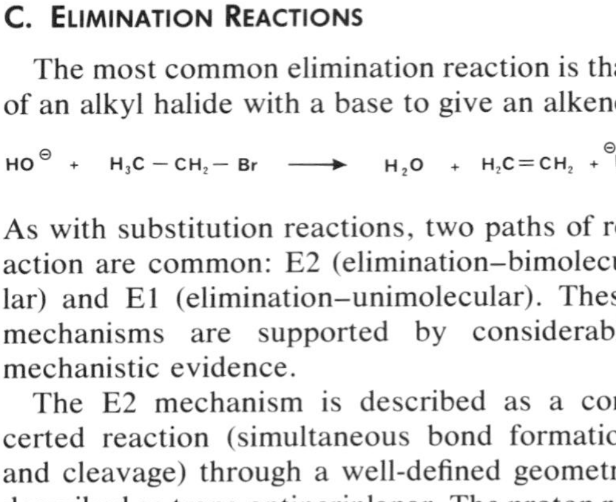

A one-step process is supported by second-order kinetics and by observation of a kinetic isotope effect. The reaction of hydroxide with Br-$CH(CH_3)_2$ is seven times faster than that of Br-$CH(CD_3)_2$, supporting C—H bond cleavage in the rate-determining step.

The E1 mechanism contrasts with the E2 in a number of ways. Analogous to the S_N1 reaction in being a two-step process, it begins by formation of a carbocation (Fig. 11). The first-order kinetic behavior is consistent with a slow ionization step followed by a rapid second step of proton abstraction by the base. The E1 reaction does not exhibit a kinetic isotope effect, nor

$$(CH_3)_3C-Br \underset{slow}{\rightleftharpoons} (CH_3)_3C^{\oplus} + Br^{\ominus}$$

$$(CH_3)_3C^{\oplus} + {}^{\ominus}OH \xrightarrow{fast} (CH_3)_2C=CH_2 + H_2O$$

FIG. 11. Mechanism of E1 reaction.

does it have a strong stereochemical requirement like the E2 reaction.

Like the S_N1 reaction, the E1 is important with tertiary halides, and secondary halides in polar solvents or stereochemically restricted so that the E2 reaction is impossible. Competing S_N1 reaction can occur, and is important at low concentrations of base.

D. DIELS–ALDER ADDITION

The addition of a diene to a second unsaturated molecule (generally called a dienophile) was discovered by Otto Diels and Kurt Alder. The stereospecificity of this reaction, and its formation of the synthetically important six-membered ring, makes it one of the most important reactions of organic chemistry:

The kinetics of the reaction (second-order) and its stereospecificity are suggestive of a concerted, one-step reaction without any intermediates (path 1 of Fig. 12). However, the rate acceleration by electron-withdrawing substituents makes a zwitterion route attractive (path 2). This two-step mechanism and a biradical mechanism (path 3) are both consistent with all these observations, as long as the ring closure step is faster than bond rotations in the intermediates.

Figure 13 outlines how the stereospecificity of the reaction of *cis* and *trans*-1,2-dichloroethylene with cyclopentadiene makes path 3 un-

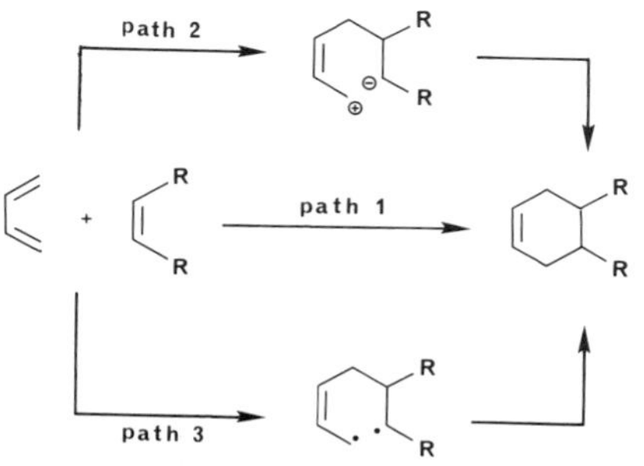

FIG. 12. Potential paths in the Diels–Alder reaction.

likely (and by the same logic, path 2 as well). Only if k_5 and k_6 are much faster than k_3 or k_4 could the reaction appear to be stereospecific. Careful examination of this reaction showed no evidence of the "wrong" stereoisomer (to the limits of sensitivity of the chromatographic method—~0.5%), making it unlikely that path 2 or path 3 is followed because such stereospecificity would be unexpected. To further establish path 1 as the mechanism requires that the relative rates of bond rotation (k_5 and k_6) be shown to be at least as fast as k_3 and k_4. Other workers have focused on independent generation of intermediates to compare the rate of bond rotation with that of ring closure. Their observations make biradical intermediates in the Diels–Alder reaction unlikely since bond rotation in the biradicals appears easier than required by the two-step mechanisms.

E. FAVORSKII REARRANGEMENT

In the Favorskii rearrangement attack of base on an α haloketone converts it into a carboxylic acid:

$$RCH_2 \overset{O}{\underset{\parallel}{C}} CH_2Cl \xrightarrow{HO^{\ominus}} RCH_2CH_2CO_2H$$

In this complex reaction it is difficult to even guess which carbon was transformed into the carboxylic group of the product. Clearly a multistep mechanism is involved, and possibilities abound. One imaginative proposal rationalizes the transformation and is also consistent with the observed second-order kinetics:

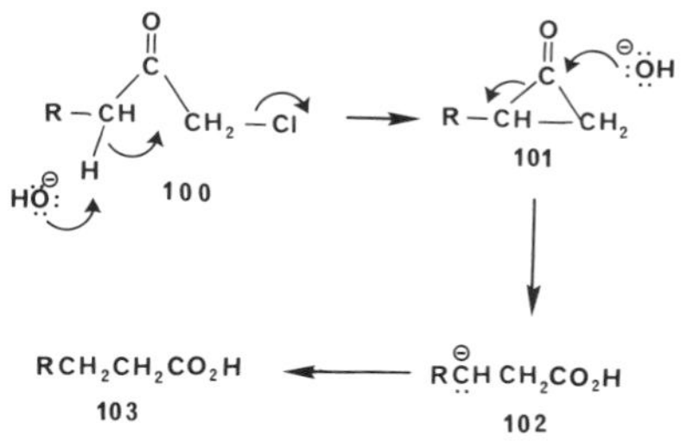

RCH₂CH₂CO₂H ← RĊHCH₂CO₂H
103   102

The key element of this proposal is the formation of the cyclopropanone intermediate (structure **101**). Studies involving isotopic labeling (indicated by an asterisk on the labeled carbon) provided additional support for this mechanism. First, labeling the carbonyl carbon showed it was the carbon transformed into the carboxyl:

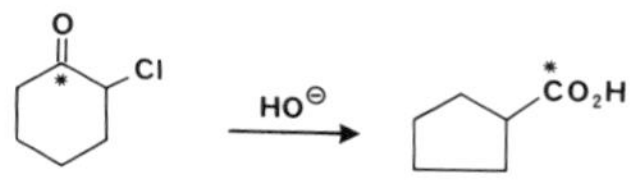

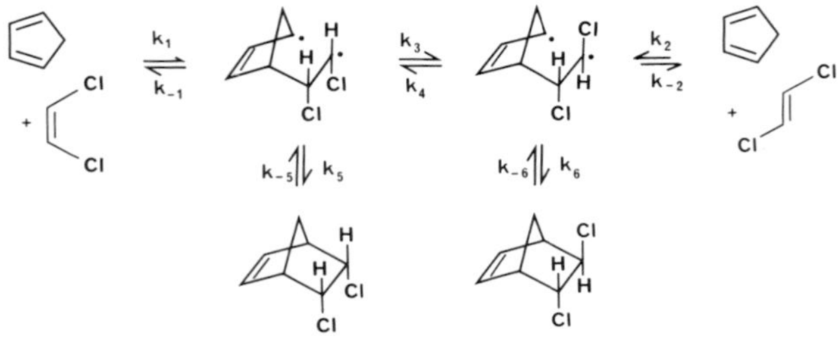

FIG. 13. Kinetic evidence for concerted Diels–Alder reaction.

Second, labeling the adjacent carbon showed it was transformed with equal probability into *two* other positions of the product:

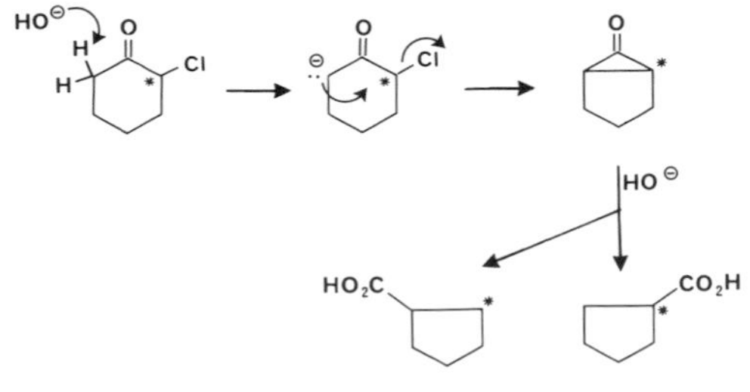

The symmetrical cyclopropanone is equally likely to be attacked from either side, and the label becomes distributed to two locations in the product.

Another strong piece of supporting evidence for the cyclopropanone mechanism was the successful trapping of the intermediate as a Diels–Alder addition product with furan:

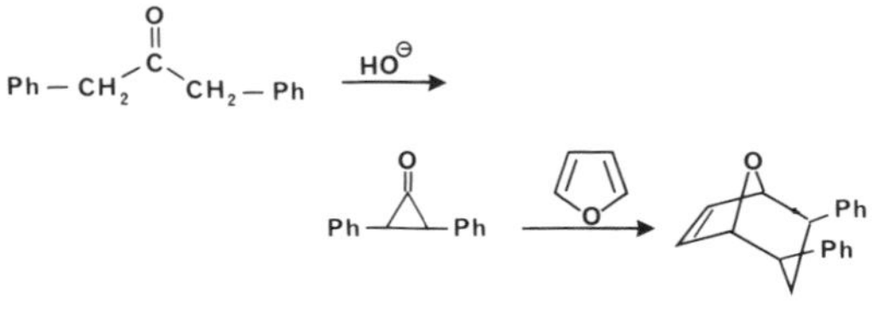

Other evidence supporting this mechanism has been obtained, but the labeling, trapping, and kinetic studies are sufficiently compelling to establish the key elements of the mechanism, especially the cyclopropanone intermediate.

V. Structure–Reactivity Correlations

A. Linear Free-Energy Relationships

In ideal circumstances, if one knew the rate of a particular reaction it would be possible to quantitatively predict how the rate would change as the structure of the reactants were changed. Considerable progress has been made in accomplishing this goal, particularly in predicting the effect of changing a substituent on an aromatic ring that undergoes reaction elsewhere in the molecule:

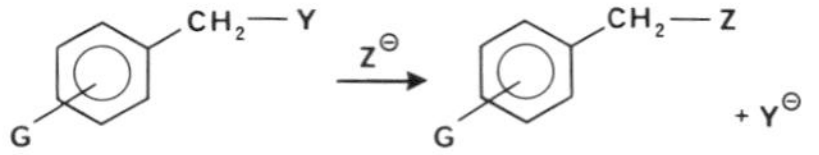

An attacking reagent will present a specific electronic demand that may be provided by substituent G. On the other hand, some substituents will *retard* the reaction because they have the wrong electron-withdrawing or electron-supplying properties. Agreement with the Hammett relationship

$$\log(k/k_0) = \sigma\rho \qquad (40)$$

is common. Here, the symbols k and k_0 refer to the rate constant for the substituted and unsubstituted compounds, respectively. The substituent constant σ depends only on the structure of substituent G (see Table II). The sign of σ indicates if a substituent is electron-supplying (−) or electron-withdrawing (+). The reaction constant ρ depends in sign and magnitude on whether a reaction develops cationic character (− value of ρ) or anionic character (+ value). Representative values of ρ appear in Table III.

One can predict the rate of reaction k of a substituted compound in comparison with an unsubstituted compound (k_0) using Eq. (40). For example, the rate of S_N2 reaction of ArO^- with ethyl iodide ($\rho = -0.991$) will be accelerated by a factor of about $10^{0.099}$ (about a factor of 1.26) by attachment of a t-butyl group to the meta position ($\sigma = -0.10$):

$$\log(k/k_0) = (-0.10)(-0.991) = +0.0991$$

For a new reaction one would need to determine the ρ experimentally by measuring the

TABLE II. Substituent Constants in the Hammett Relationship

Substituent	σ_m	σ_p
NH$_2$	−0.16	−0.66
CH$_3$	−0.07	−0.17
C(CH$_3$)$_3$	−0.09	−0.15
C$_6$H$_5$	0.06	−0.01
OH	0.12	−0.37
OCH$_3$	0.12	−0.27
F	0.34	0.06
I	0.35	0.18
CO$_2$H	0.37	0.45
Cl	0.37	0.23
COCH$_3$	0.38	0.50
Br	0.39	0.23
CO$_2$R	0.37	0.45
CF$_3$	0.43	0.54
CN	0.56	0.66
NO$_2$	0.71	0.78
+N(CH$_3$)$_3$	0.99	0.96

rates of two compounds of known σ's, and then could quantitatively predict reactivity of a host of other compounds.

B. SOLVENT EFFECTS

One of the most important factors affecting reaction rates is the solvent. The reaction constants ρ of the Hammett relation will change if the solvent is changed. For example, the base promoted hydrolysis of ArCO$_2$Et at 25° in 70% aqueous dioxane is 1.828, while in 85% aqueous ethanol the value is 2.537. Instead of measuring how these reaction constants would vary, a good knowledge of the manner in which solvents will affect reactions would allow prediction of these changes.

Much effort has focused on correlation of solvent effects on reaction rate with a readily measurable property of the solvent. Unfortunately,

these efforts have shown that solvent effects are so complex this may not be possible. A reaction that has a separation of charge in the rate-determining step is the S_N1 reaction of halides, where the transition state resembles an ion pair, R$^+$X$^-$. Increased "polarity" of solvent will help stabilize the developing charges, and accelerate the reaction. In general, this is the observed behavior. The dielectric constant ε (Table IV) is easily measured and ranks polarity of solvents. Unfortunately, there is only a qualitative correlation between rate of reaction and ε within a series of alcohols, for example. Furthermore, reaction in a nitrile is more greatly misplaced by its ε. Similarly, other simple parameters have failed to be universally applicable to a great variety of reactions.

In a manner reminiscent of the Hammett relationship, efforts have focused on measurement of a solvent parameter that would accurately predict changes of reactivity of a particular type of reaction. The Grunwald–Winstein relationship for the solvolysis (S_N1 displacement by solvent) of t-butyl chloride was developed to measure solvent effects quantitatively:

$$\log(k/k_0) = mY \qquad (41)$$

Here, k is the rate of reaction in the new solvent, while k_0 is the rate of reaction in the standard solvent—80% aqueous ethanol. The compound parameter m varies with the reactant (it is 1.00 for t-butyl chloride), while Y is the solvent pa-

TABLE III. Representative Values of Reaction Constants

Reaction	ρ value
ArCO$_2$Et + HO$^-$ → ArCO$_2^-$ + EtOH	2.61
ArOH → ArO$^-$ + H$^+$ (in water)	2.26
ArCO$_2$H → ArCO$_2^-$ + H$^+$ (in water)	1.00
ArCH$_2$CO$_2$Et + HO$^-$ → ArCH$_2$CO$_2^-$ + EtOH	1.00
ArCH$_2$Cl + H$_2$O → ArCH$_2$OH + HCl	−1.31
ArNH$_2$ + PhCOCl → ArNHCOPh + HCl	−3.21
ArC(Me)$_2$Cl + H$_2$O → ArC(Me)$_2$OH + HCl	−4.48

TABLE IV. Dielectric Constants of Common Solvents

Substance	ε value
Water	78
Dimethyl sulfoxide	47
Acetonitrile	38
Dimethylformamide	37
Methanol	32.7
Ethanol	24.5
Acetone	21
t-Butyl alcohol	12.5
Pyridine	12
Tetrahydrofuran	7.6
Acetic acid	6.1
Chloroform	4.8
Diethyl ether	4.3
Benzene	2.3
Dioxane	2.2
Carbon tetrachloride	2.2
n-Hexane	1.9

rameter that reflects the ionizing power of a particular solvent. Since different reactants will have a different charge buildup in the transition state, the m term is necessary, and is the counterpart of the ρ in the Hammett relationship. The m values (Table V) do not vary as much as ρ values do, though they are only applicable to a much more limited set of reactants (the reactant must undergo S_N1 reaction). Although the Y values have merit for prediction of solvent effects, there are many cases where the predictions are quantitatively and qualitatively very inaccurate. Only very polar solvents may be ranked because the solvent must be polar enough to cause solvolysis of t-butyl chloride. Other parameters for measuring solvent effects have been derived and measured, such as the observed λ_{max} of a pyridinium salt (Z value) or a stable zwitterion (E_T value) (see Table VI). These correlate with one another very well, and frequently are qualitatively accurate in predicting solvent effects on reactions. There are instances in which the predictions are poor. It is apparent there may never be a single model for solvent polarity that will accurately predict solvent effects on the whole range of organic reactivity.

C. NUCLEOPHILICITY

The synthetic importance of substitution reactions has made them the subject of extensive quantitative study. Reaction constants for many reactants have been measured. The rate of such reactions depends on the attacking nucleophile, the departing substituent, and the solvent. An attempt has been made to examine these factors independently of one another for maximum understanding and simplicity. The concept of nucleophilicity reflects the effect on reaction rate of varying the structure of the attacking species, the nucleophile. Empirical measures of nucleophilicity may be obtained by choosing a standard reaction (the S_N2 substitution reaction of methyl iodide in methanol at 25°). It is con-

TABLE VI. Quantitation of Solvent Polarity: Z, E_T, and Y Values

Solvent	Z	E_T	Y
Water	94.6	63.1	3.49
Formic acid			2.05
Methanol	83.6	55.5	−1.09
Ethanol	79.6	51.9	
Acetic acid	79.2	51.7	−1.68
Isopropyl alcohol	76.3	48.6	
t-Butyl alcohol	71.3	43.9	−3.3
Acetonitrile	71.3	46.0	
Dimethyl sulfoxide	71.1	45.0	
Dimethylformamide	68.5	43.8	
Acetone	65.7	42.2	
Chloroform	63.2	39.1	
Tetrahydrofuran		37.4	
Ether		34.6	
Benzene	54.0	34.8	
Carbon tetrachloride		33.6	
n-Hexane		33.1	

venient to express the change in the reaction rates in a manner similar to the tabulation of reaction constants in the Hammett equation:

$$n = \log(k_n/k_0) \qquad (42)$$

Here n is the nucleophilic constant for a given nucleophile; k_n and k_0 are the rate of reaction of the "test" nucleophile and methyl alcohol, respectively, with methyl iodide. Table VII gives

TABLE VII. Nucleophilic Constants of Common Nucleophiles

Nucleophile	n value
CH_3OH	0.0
NO_3^-	1.5
F^-	2.7
$CH_3CO_2^-$	4.3
Cl^-	4.4
$(CH_3)_2S$	5.3
NH_3	5.5
N_3^-	5.8
PhO^-	5.8
Br^-	5.8
CH_3O^-	6.3
HO^-	6.5
NH_2OH	6.6
NH_2NH_2	6.6
$(CH_3CH_2)_3N$	6.7
CN^-	6.7
I^-	7.4
HO_2^-	7.8

TABLE V. Substrate Parameters for the Winstein–Grunwald Relationship

Substrate	m value
1-Adamantyl chloride	1.20
1-Phenylethyl chloride	1.12
t-Butyl chloride	1.00
t-Butyl bromide	0.90
$(Ph)_2CHCl$	0.76

nucleophilic constants for many common nucleophiles.

Weak nucleophiles (slow reaction rates) have small n values; strong nucleophiles have fast reaction rates and large n values. Careful scrutiny of such data has failed to reveal a direct correlation with any single measurable property of the nucleophile. Rather, it appears a number of factors are important including size, electronegativity, polarizability, strength of its bond to carbon, and solvation energy. Nevertheless, some generalizations are apparent. If one holds the attacking atom constant, the nucleophilicity correlates very well with the basicity (e.g., $CH_3O^- >$ $C_6H_5O^- > CH_3CO_2^- > NO_3^-$). Nucleophilicity usually decreases in going across a row in the periodic table, probably due to changes in electronegativity ($HO^- > F^-$). Nucleophilicity increases in going down a column ($I^- > Br^- >$ $Cl^- > F^-$), apparently due to the greater polarizability and decreased solvation of the heavier atoms. These generalizations provide a qualitative framework for assessing the effect of structure on reaction rate.

D. Hard and Soft Acids and Bases

The limited correlation of nucleophilicity with basicity shows there is a conceptual relationship between the two. Both involve the formation of a new bond by donation of an electron pair to an electrophilic species. However, as the atoms are changed, the quantitative relationship breaks down as such considerations as polarizability become important. Many compounds might react with a base/nucleophile in two entirely different ways. The E2 and S_N2 reactions are an example. A particular nucleophile, such as ethoxide ion, might displace a halide (S_N2 reaction) or remove a proton from a β carbon (E2 reaction, where basicity of ethoxide is the important

parameter). It would be valuable to predict which chemical species Y^- would react with a given compound as a base, and which ones would react as nucleophiles.

The complexity of this problem has precluded a concise quantitative treatment. However, Pearson has presented a useful qualitative concept. He defined acids and bases as being either hard or soft. Hard acids are small atoms that have outer electrons which bear considerable positive charge. Soft acids have atoms of lower positive charge and large size. Hard bases contain highly electronegative atoms of low polarizability, which are difficult to oxidize. Soft bases are polarizable, have less electronegative donor atoms and are easily oxidized. Table VIII classifies several common ions according to this scheme.

The utility of this concept arises from its application to reactions. The key concept is that reactions will occur most readily between species matched with respect to hardness and softness. Hard acids prefer to react with hard bases, while soft acids prefer to react with soft bases. To apply this concept to the particular problem of nucleophilic substitution versus elimination (caused by deprotonation), we note that the proton is a hard acid, whereas an sp^3 carbon is a soft acid. As a result, the soft bases are more likely to effect substitution, whereas hard bases are more likely to abstract a proton, act as a base, and initiate an E2 reaction.

E. Leaving Group Effects

A final factor affecting reactivity in substitution reactions is the nature of the leaving group. Although the rate of reaction is strongly dependent on the nature of the group being cleaved, no general relationship has been developed to quantitatively predict such effects. Since a leav-

TABLE VIII. Hardness and Softness of Common Reactants

Bases	Acids
Soft	
H^-, RSH, RS^-, I^-,	Br_2, RCH_2X, RSX
R_3P, CN^-	I_2, Hg^{2+}
Borderline	
Br^-, N_3^-, $ArNH_2$	R_3C_+, R_3B
Hard	
H_2O, HO^-, ROH, RO^-	$H—X$, H^+, Li^+, Na^+, K^+
RCO_2^-, F^-, Cl^-, NO_3^-	Mg^{2+}, Ca^{2+}, Al^{3+}
RNH_2, NH_3	$R_3Si—X$

ing group departs with a pair of electrons (and usually a negative charge) it would seem that electronegativity would be the critical factor. However, the situation is much more complex. This is evident from Table IX, which compares the rate of solvolysis (S_N1 reaction) of several 1-phenylethyl esters and halides at 75° in 80% aqueous ethanol. This table compares the rate of reaction of a specified leaving group (k) with that of the chloride (k_0). Enormous rate differences are apparent from the table. Trifluoroacetate is 10^6 times more reactive than acetate, and iodide is 10^7 times more reactive than fluoride. The order of reactivity of the halide leaving groups ($I^- > Br^- > Cl^- \gg F^-$) is interesting; it is the opposite of the order of electronegativity. This is mainly because of the relative carbon–halogen bond strengths and the polarizability of the respective anions. The C—I bond is much weaker than the C—F bond, and the larger ions are the more polarizable.

In the family of esters, it is apparent the weaker bases (the more stable anions) are better leaving groups. The electron-withdrawing substituents on the trifluoroacetate stabilize the anion better than the unsubstituted acetate. This is clearly reflected in the rates.

The sulfonate esters are important leaving groups. Weak basicity makes their reaction rapid, and they are readily prepared. An alcohol is unreactive in substitution reactions because of the strong basicity of the HO^- that would be lost. Transformation of the OH group into a sulfonate by a sulfonyl halide transforms the hydroxide into a weak base when the C—O bond is cleaved.

TABLE IX.   Effect of Leaving Group on Reaction Rates of 1-Phenylethyl Esters and Halides

Leaving group	log (k/k_0)
$CF_3SO_3^-$	8.15
p-Nitrobenzenesulfonate	5.64
p-Toluenesulfonate	4.57
$CH_3SO_3^-$	4.48
I^-	1.96
Br^-	1.15
$CF_3CO_2^-$	0.32
Cl^-	(0.00)
F^-	−5.05
p-Nitrobenzoate	−5.26
$CH_3CO_2^-$	−5.85

One of the most unusual leaving groups is molecular nitrogen from diazonium salts. Diazonium ions can be generated by nitrosation of primary amines; their cleavage gives a neutral molecule (N_2), so ion pairing after cleavage is not a complication as when an anion is the leaving group.

VI. Theoretical Concepts and Treatments

A. EMPIRICAL, SEMIEMPIRICAL, AND *Ab Initio* METHODS

There are many approaches to the development of predictive ability in chemistry. Empirical approaches, such as resonance, organize data, and seek patterns of reactivity from which predictions are drawn. *Ab initio* theoretical methods begin with the physical laws controlling the reactants, calculate forces of interaction, and derive information about stability and reactivity. A semi-empirical method will choose which interactions are important and which are negligible. Certain experimentally measured interaction energies will be assumed appropriate, and probable results of reaction calculated on the basis of the important interactions alone. Each of these approaches is important in its own right, and has advantages and disadvantages. The *ab initio* approaches are the most intellectually satisfying, yet require such effort only the simplest molecules are amenable to consideration. Empirical methods are the most direct, but yield valid predictions only with wisdom and insight by the analyst. More effort now focuses on semi-empirical methods that utilize simplifying assumptions, yet retain enough theoretical validity to have predictive merit. Some of these approaches will be briefly described to illustrate utility. [*See* ORGANIC CHEMICAL SYSTEMS, THEORY; POTENTIAL ENERGY SURFACES; QUANTUM CHEMISTRY.]

B. PERICYCLIC AND CONCERTED REACTIONS

Perhaps the most puzzling type of reactions historically has been the "no mechanism" or concerted reactions where bond formation and bond cleavage occurred simultaneously. While easy to visualize the resulting connections of atoms, the existence of these reactions was difficult to forecast. For example, the Diels—Alder reaction [Eq. (43)] was clearly facile, yet conceptually similar reactions [Eqs. (44) and (45)] were not:

$$\langle\!\!\langle \quad + \quad \| \quad \longrightarrow \quad \bigcirc \quad\quad (43)$$

$$\langle\!\!\langle \quad + \quad \rangle\!\!\rangle \quad \xrightarrow{\;\;\times\;\;} \quad \bigcirc\!\!\bigcirc \quad (44)$$

$$\| \quad + \quad \| \quad \xrightarrow{\;\;\times\;\;} \quad \square \quad\quad (45)$$

One of the most exciting developments of physical organic chemistry has been the development of the ability to predict when such reactions will be practical. The cycloaddition reactions (**43–45**), sigmatropic reactions (a σ bond migrates over a system to a new location), cheletropic reactions (expulsion or addition of a fragment), and electrocyclic reactions (ring opening or ring closure of unsaturated compounds) are readily analyzed. Several theoretically equivalent methods for analysis of such problems were described within a relatively short period. The Huckel–Mobius analysis popularized by Zimmerman is probably the easiest to visualize; the Woodward–Hoffmann rules are so widely used they are indispensable; and frontier molecular orbital methods pioneered by Fukui are easier to extend into entirely new systems. Each of these methods determines the practicality of a reaction by determining whether the transition state will be of low energy or high energy. They focus on changes in the energy of the component molecular orbitals upon reaction by observation of the symmetry properties of the orbitals to determine if the interactions will be stabilizing or destabilizing. They are referred to as the orbital symmetry methods of analysis as a reflection of the underlying hypothesis that these will constitute the most important interactions in determining the practicality of a reaction.

C. HUCKEL–MOBIUS ANALYSIS

Zimmerman's approach relies on examination of the interaction of orbitals involved in the reaction and comparison of the total number of electrons involved in the transition state of the reaction with the magic numbers ($4N + 2$) of the Huckel rule. The Diels–Alder reaction, for example, arises from the interaction of six overlapping p orbitals (constituting the π bonds of the two reactants):

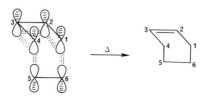

The overlap of the p orbitals on carbon 1 and 4 of the butadiene with the ethylene (that will become carbon 5 and 6 of the product) causes formation of two new σ bonds. The remaining two p orbitals overlap with each other to form the π bond of the product. The predictive power comes into play when the total number of electrons involved is added. Each p orbital contributes one electron, for a total of six electrons. This is a (4 + 2) cycloaddition. The total, six, is equal to one of the magic numbers, and this reaction would be termed *allowed* or *favored*. On the other hand the (4 + 4) cycloaddition [Eq. (44)] and the (2 + 2) cycloaddition [Eq. (45)] do not sum electrons to a magic number (eight and four electrons being involved, respectively). Such reactions would be called *forbidden* or *disallowed*.

Two additional observations complete the rules for application of this method to the host of possible applications. First, a disallowed reaction, one having $4N$ instead of $4N + 2\pi$ electrons, becomes allowed if the reaction is photolyzed. Thus, the (2 + 2) cycloaddition is allowed in the excited state, consistent with experimental observation. Finally, one occasionally finds that a Mobius strip of orbitals could be involved, where the top of one p orbital (arbitrarily shaded and called a + lobe) could overlap with the bottom of another orbital (a − lobe). This is exemplified by the ring closure of a trans, trans-disubstituted octatetraene (an eight electron system):

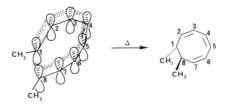

Here, $4N$ electrons are involved in a Mobius interaction, and the conclusion would be that this would be an allowed reaction, consistent with experimental observation. Note that as a bond forms between carbons 1 and 8 the orbitals must rehybridize from p to sp^3. In so doing they twist to overlap better. Here, the twisting is such that one methyl moves down while the other moves up. One *p* orbital tilts to turn its bottom lobe to touch the top of the other *p* orbital. Such motion is conrotatory. Had the methyls moved so the tops of both overlap the closure would be called disrotatory.

The simplicity of this approach allows a short summary to encompass all the rules needed to analyze any concerted reaction. The allowed combinations are summarized as

	Huckel overlap	Mobius overlap
Thermal reactions	$4N + 2$ electrons	$4N$ electrons
Photochemical reactions	$4N$ electrons	$4N + 2$ electrons

In cases that do not involve these magic numbers of electrons the reactions will be forbidden and are unlikely to ever be observed.

D. WOODWARD–HOFFMANN RULES

Familiarity with the Zimmerman analysis makes the Woodward–Hoffmann methods easy. This approach also analyzes the overlap in the transition state to determine if a transformation will be allowed or forbidden. There are two ways in which orbitals can interact during reaction—a suprafacial overlap in which the orbitals interact from the same face of the molecule, and an antarafacial overlap in which they form bonds from opposite faces:

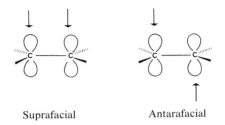

Suprafacial   Antarafacial

Sigma orbitals are analogously treated.

For pericyclic reactions ("around the ring") such as cycloadditions, the reaction is *favored* when $4N + 2$ electrons are suprafacial and $4N$ electrons are antarafacial. This principle can be applied to previously mentioned examples:

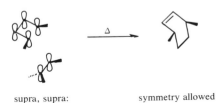

supra, supra: symmetry forbidden

With four electrons an antarafacial process would be needed for an allowed reaction, a geometry that would be sterically difficult:

supra, supra: symmetry allowed

With six electrons the suprafacial interaction is favored. Here, an important stereochemical consequence is observed if the transformation of the *p* orbitals into σ bonds is followed using molecular models. The cis product from reaction of a trans, trans-disubstituted 1,4-diene is correctly predicted by this theoretical model. Such stereochemical predictions are consistently accurate, leading the original authors to boast "Violations: There are none!"

The underlying assumption of the Woodward–Hoffmann approach is conservation of orbital symmetry; the symmetry of the molecular orbitals will affect whether an interaction will be stabilizing or destabilizing. The underlying theory is sound and properly applied, the method can be used to make predictions with confidence.

E. FRONTIER ORBITALS

The roots of the Woodward–Hoffmann, Huckel–Mobius, and similar approaches lie in the same energetic considerations that constitute the frontier orbital approach. One chooses between the methods not on the basis of which is most theoretically sound, but on the one that can provide the most straightforward analysis of the problem. A greater variety of problems can be tackled by frontier orbital analysis.

The frontier orbital method treats the interactions of one molecule with another as a perturbation on the respective orbitals, and identifies whether the most significant energy terms will be favorable or unfavorable. When two reactants are brought together it is the interaction of the electrons that gives rise to stabilization or destabilization, because it is the electrons which are involved in bonding. Each molecule has its electrons in certain occupied molecular orbitals, and each molecule will have some orbitals of higher energy that are not occupied. All orbitals, occupied or unoccupied, will interact with one another as reaction occurs. When two orbitals interact, the two new orbitals are always produced—one of lower energy than the original pair and one of greater energy (Fig. 14). Calculations show that when the two orbitals are of greatly differing energies, the interaction will cause little change in the energy of the two original orbitals. When two orbitals of similar energy interact, the resulting changes are significant. In Fig. 14 the original orbitals 1 and 2 change energy to 1' and 2'.

If both original orbitals were occupied there will be no improvement in energy due to the interaction, because the drop in energy of one orbital will be counterbalanced by rise in the energy of the other. However, if the orbital of higher energy (orbital number 2) is unoccupied, the molecule will not suffer destabilization as the energy of this orbital is raised, but will have a stabilization based on a decrease in energy of the occupied orbital (number 1). The key element of frontier orbital theory is to select the highest occupied molecular orbital (HOMO) of one reactant and the lowest unoccupied molecular orbital (LUMO) of the other as a pair of interactions of greatest importance in determining the overall interaction energy. When the HOMO of one reactant interacts with the LUMO of the other, a favorable interaction results, and attraction between the molecules.

The frontier orbital method recognizes that there are other interactions that may be important, principally the coulombic attraction if the reactants are charged. This favorable interaction, when the charges are opposite, may be larger than the HOMO–LUMO contribution.

The previous examples are easy to describe using the frontier orbital method. For example, consider the dimerization of ethylene (2 + 2 cycloaddition). The two orbitals of closest energy are the π orbitals, and these orbitals are occupied in both reactants. The interaction of these two orbitals (Fig. 15) will create counterbalancing energetic terms as the energy of one occupied orbital is raised while the other is lowered. The HOMO–LUMO interaction will be small because of the large energy separation, and the reaction (correctly) is considered disfavored.

Cases difficult to treat by the other approaches may be considered. The principle of hard and soft acids and bases (HSAB), for example, can be summarized in frontier orbital terminology. Hard bases have a low energy HOMO (and usually a negative charge). Soft bases have a high energy HOMO but do not necessarily have a negative charge. Hard acids have a high energy LUMO and usually have a positive charge. Soft acids have a low energy LUMO but do not necessarily have a positive charge. The conclusions of the HSAB treatment are reasonably derived:

1. A hard acid/hard base interaction is favorable because of a large coulombic attraction.
2. A soft acid/soft base interaction is favorable because of a large interaction between the HOMO of the base and the LUMO of the electrophile.

Other combinations do not have an "unexpectedly large" reaction rate because there is no unusually favorable energy of interaction as there is in these combinations.

A last example will focus on explaining one of the mysteries of nucleophilic substitution—the α effect. Some nucleophiles (HO_2^-, ClO^-, $HONH_2$, N_2H_4, and R_2S_2) stand out because they are much more nucleophilic than would be expected from considerations described earlier, such as pK_a. In each case the nucleophilic atom

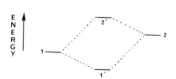

FIG. 14.   Energy changes in orbitals due to interaction.

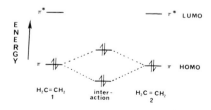

FIG. 15.   HOMO–LUMO interactions in ethylene dimerization.

is flanked by an additional atom that has a lone pair of electrons. If the orbital containing the electrons of this atom interact with the orbital bearing the electrons on the nucleophilic atom, the net result will be to raise the energy of the HOMO, as shown by the interaction for HOO⁻ (Fig. 16). This will cause the base to be softer than if no lone pair was present. The reaction of such a base with a soft acid will be much more favored because of the existence of the lone pair, while the reaction with hard acids will not be favored. These conclusions precisely reflect observation. The approximate relative reactivity of HOO⁻ and HO⁻ with different substrates is compared in Table X. With soft electrophiles HOO⁻ is substantially more reactive, but with a hard electrophile (H^+) the reactivity of HO⁻ is superior.

F. MOLECULAR ORBITAL THEORY

Molecular orbital (MO) theory is doubtless the most promising method for determination of molecular properties using physical principles rather than empirical observation. The MO theory is extensively treated in the article on "Organic Chemical Systems, Theoretical Interpretation," and many of the results of this method have been incorporated into this chapter. A brief introduction will summarize the origin of some of these conclusions.

The basic challenge of an MO calculation is the solution of the Schrodinger equation, which describes the particular molecule or molecules undergoing reaction:

$$\mathcal{H}\psi = E\psi \qquad (46)$$

The Hamiltonian operator $\mathcal{H}$ is a term that describes all the physical interactions, including potential and kinetic energy terms, expressed in a differential notation. The wave functions ψ are the descriptions of the location of electrons, and are the "solutions" to the differential equation defined by the Schrodinger equation. The energy E is a quantity that emerges from operation on the wave functions that are the solutions.

TABLE X.   Comparative Reactivity of Peroxide (HOO⁻) and Hydroxide (HO⁻) with Different Electrophiles

Electrophile	k_{HOO^-}/k_{HO^-}
PhCN	10^5
p-$NO_2C_6H_4CO_2CH_3$	10^3
$PhCH_2Br$	50
H_3O^+	10^4

Each solution ψ to the differential equation has an associated energy, and the energy is readily obtained by the above relationship.

The solution of such differential equations is difficult when complex molecules are involved, and theoreticians make advances by developing techniques to obtain approximate solutions using simplifying assumptions. For example, one of the most approximate MO methods—Huckel MO calculations—uses the assumption that only the π electrons will be important. Such assumptions are intellectually dissatisfying, but allow rapid solutions. In many cases the level of accuracy obtained is sufficient to answer the question being posed.

The MO methods produce much data of interest and possible utility. Most fundamentally, they yield the description of the possible molecular orbitals which the electrons can occupy, and the relative energy of these orbitals. For example, a Huckel calculation (which focuses only on the π electrons) on hexatriene would yield the relative ordering of orbitals illustrated in Fig. 17. Perturbational methods, such as frontier orbital analysis, can often provide insight by analysis of the energy and orbital descriptions derived from these simple MO calculations. As computational methods improve and computational power increases by advances in computer technology, the sophistication and accuracy of practical MO methods will increase. Physical organic chemistry will rely increasingly on such methods both for understanding and for their predictive power.

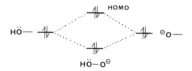

FIG. 16.   Energy changes in RO⁻ due to an adjacent lone pair.

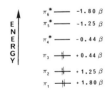

FIG. 17.   Energy levels in hexatriene.

BIBLIOGRAPHY

Carey, F. A., and Sundberg, R. J. (1984). "Advanced Organic Chemistry. Part A. Structure and Mechanisms," 2nd ed., Plenum, New York.

Carpenter, B. K. (1984). "Determination of Organic Reaction Mechanisms," Wiley (Interscience), New York.

Drenth, W., and Kwart, H. (1980). "Kinetics Applied to Organic Reactions," Marcel Dekker, New York.

Fleming, I. (1976). "Frontier Orbitals and Organic Chemical Reactions," Wiley (Interscience), New York.

Lowry, T. H., and Richardson, K. S. (1981). "Mechanism and Theory in Organic Chemistry," 2nd ed., Harper & Row, New York.

March, J. (1985). "Advanced Organic Chemistry," 3rd ed., Wiley (Interscience), New York.

Maskill, H. (1985). "The Physical Basis of Organic Chemistry," Oxford Univ. Press, London and New York.

Orchin, M., Kaplan, F., Macomber, R. S., Wilson, R. M., and Zimmer, H. (1980). "The Vocabulary of Organic Chemistry," Wiley (Interscience), New York.

Wentrup, C. (1984). "Reactive Molecules," Wiley (Interscience), New York.

Zuman, P. and Patel, R. (1984). "Techniques in Organic Reaction Kinetics," Wiley (Interscience), New York.

PLANETARY ATMOSPHERES

Joel S. Levine *NASA Langley Research Center*

GLOSSARY

Atmospheric pressure: Weight of the atmosphere in a vertical column, 1 cm^2 in cross section, above the surface of a planet. On earth, the average value of atmospheric pressure at sea level is 1.013×10^6 dyne cm^{-2}, or 1013 mbar, which is equivalent to a pressure of 1 atmosphere.

Cosmic abundance of the elements: Relative proportion of the elements in the cosmos based on abundances deduced from astronomical spectroscopy of the sun, the stars, and interstellar gas clouds and chemical analyses of meteorites, rocks, and minerals.

Gravitational escape: Loss of atmospheric gases from a planetary atmosphere to space. If an upward-moving atmospheric atom or molecule is to escape the gravitational field of a planet, its kinetic energy must exceed its gravitational potential energy. The two lightest atmospheric gases, hydrogen and helium, usually possess enough kinetic energy to escape from the atmospheres of the terrestrial planets. In photochemical escape, some heavier atmospheric species, such as atomic nitrogen and atomic oxygen, are imparted with sufficient kinetic energy from certain photochemical and chemical reactions to escape from planetary gravitational fields. Over geological time, gravitational escape has been an important process in the evolution of the atmospheres of the terrestrial planets.

Greenhouse effect: Increase in the infrared opacity of an atmosphere which leads to an increase in the lower atmospheric and surface temperature. For example, water vapor and carbon dioxide, the two most abundant outgassed volatiles, increase the infrared opacity of an atmosphere by absorbing outgoing infrared radiation emitted by the surface and lower atmosphere. The absorbed infrared radiation is then re-emitted by the absorbing molecule. The downward directed component of the re-emitted radiation heats the surface and lower atmosphere.

Magnetosphere: Region in upper atmosphere of planet possessing magnetic field where ions and electrons are contained by magnetic lines of force. The earth and Jupiter are surrounded by magnetospheres.

Mantle: One of the three major subdivisions of the earth's interior (the core and the crust being the other two). The mantle contains about 70% of the mass of the earth and is iron-deficient. The mantle surrounds the core, which is believed to consist mainly of iron. Surrounding the mantle is the relatively thin-layered crust. The core, mantle, and crust are composed of refractory elements and their compounds.

Mixing ratio: Ratio of the number of atoms or molecules of a particular species per cm^3 to the total number of atmospheric atoms or molecules per cm^3. At the earth's surface, at standard temperature and pressure, there are about 2.55×10^{19} molecules per cm^{-3}. The mixing ratio is a dimensionless quantity, usually expressed in parts per million by volume (ppmv = 10^{-6}), parts per billion by volume (ppbv = 10^{-9}), or parts per trillion by volume (pptv = 10^{-12}).

Photodissociation: Absorption of incoming solar radiation, usually radiation of visible wavelengths or shorter, that leads to the dissociation of atmospheric molecules to

ENCYCLOPEDIA OF PHYSICAL SCIENCE
AND TECHNOLOGY, VOL. 10

their constituent molecules, atoms, or radicals. For example, the photodissociation of water vapor leads to the formation atomic hydrogen (H) and the hydroxyl radical (OH).

Primary or primordial atmosphere: Atmosphere resulting from capture of the gaseous material in the primordial solar nebula from which the solar system condensed about 4.6 billion years ago. The atmospheres of Jupiter, Saturn, Uranus, and Neptune are believed to be remnants of the primordial solar nebula and, hence, contain atoms of hydrogen, helium, nitrogen, oxygen, carbon, and so on in the same elemental proportion as the sun. The atmospheres of these planets are composed of molecular hydrogen and helium, with smaller amounts of methane, ammonia, and water vapor, and their photodissociation products.

Primordial solar nebula: Interstellar cloud of gas, dust, and ice of a few solar masses, at a temperature of about 10 K, that collapsed under its own gravitational attraction to form the sun, the planets, and the rest of the solar system about 4.6 billion years ago. Compression caused the temperature of the contracting cloud to increase to several thousand degrees, vaporizing all but the most refractory compounds, while conservation of angular momentum flattened the cloud into a disk. The refractory elements in the equatorial plane began to accumulate into large bodies, eventually forming the planets by accretion and coalescence. The bulk of the mass of the primordial solar nebula, composed primarily of hydrogen and helium, formed the sun.

Refractory elements: Elements or their compounds that volatilize only at very high temperatures, such as silicon, magnesium, and aluminum. Refractory elements and their compounds formed the terrestrial planets through the processes of accretion and coalescence.

Secondary atmosphere: Atmosphere resulting from the outgassing of trapped volatiles, that is, the atmospheres surrounding earth, Venus, and Mars.

Troposphere: Lowest region of the earth's atmosphere, which extends from the surface to about 15 km in the tropics and to 10 km at high latitudes. About 80% of the total mass of the atmosphere is found in the troposphere (the rest of the total mass of the atmosphere is found in the stratosphere, which extends to about 50 km, with only a fraction of a percent of the total mass of the atmosphere found in the atmospheric regions above the troposphere and stratosphere: the mesosphere, thermosphere, exosphere, ionosphere, and magnetosphere).

Volatile elements: Elements that are either gaseous or form gaseous compounds at relatively low temperatures.

Volatile outgassing: Release of volatiles trapped in the solid earth during the planetary formation process. The release of the trapped volatiles led to the formation of the atmosphere and ocean.

Gravitationally bound to the planets are atmospheres, gaseous envelopes of widely differing masses and chemical compositions. The origin of the atmospheres of the planets is directly related to the origin of the planets some 4.6 billion years ago. Much of our knowledge and understanding of the origin, evolution, structure, composition, and meteorology of planetary atmospheres has resulted from the exploration of the planets and their atmospheres by a series of planetary fly-bys, orbiters, and landers. The atmospheres of the terrestrial planets (earth, Venus, and Mars) most probably resulted from the release of gases originally trapped in the solid planet during the planetary formation process. Water vapor, carbon dioxide, and molecular nitrogen outgassed from the terrestrial planets to form their atmospheres. By contrast, it is generally thought that the very dense hydrogen and helium atmospheres of the outer planets (Jupiter, Saturn, Uranus, and Neptune) are the gaseous remnants of the primordial solar nebula that condensed to form the sun and the planets. Of all of the planets in the solar system, the atmosphere of the earth has probably changed the most over geological time in response to both the geochemical cycling of a geologically active planet and the biochemical cycling of a biologically active planet.

I. Formation of the Planets and Their Atmospheres

The sun, earth, and the other planets condensed out of the primordial solar nebula, an interstellar cloud of gas and dust, some 4.6 billion years ago (orbital information and the physical characteristics of the planets are summarized in Table I). The chemical composition of the primordial solar nebula most probably reflected the cosmic abundance of the elements (see Table II). Volatiles, elements that were either gaseous or that formed gaseous compounds at the relatively low temperature of the solar nebula, were the major constituents. The overwhelmingly prevalent volatile element was hydrogen, followed by helium, oxygen, nitrogen, and carbon (see Table II). Considerably less abundant in the solar nebula, but key elements in the formation of the solid planets, were the nonvolatile refractory elements, such as silicon, iron, magnesium, nickel, and aluminum, which formed solid elements and compounds at the relatively low temperature of the solar nebula. The terrestrial planets (Mercury, Venus, earth, and Mars) formed through the processes of coalescence and accretion of the refractory elements and their compounds, beginning with grains the size of dust, to boulder-sized "planetesimals", to planetary-sized bodies. The terrestrial planets may have grown to their full size and mass in as little as 10 million years. Volatiles incorporated in a late-accreting, low-temperature condensate may have formed as a veneer surrounding the newly formed terrestrial planets. The chemical composition of this volatile-rich veneer resembled that of carbonaceous chondritic meteorites, which contain relatively large amounts of water (H_2O) and other volatiles. The collisional impact of the refractory material during the coalescence and accretion phase caused widespread heating in the forming planets. The heating was accompanied by the release of the trapped volatiles through a process termed volatile outgassing. The oxidation state and, hence, the chemical composition of the outgassed volatiles depended on the structure and composition of the solid planet and, in particular, on the presence or absence of free iron in the upper layers of the solid planet. If the terrestrial planets formed as geologically differentiated bodies, i.e., with free iron having already migrated to the core (as a result of the heating and high temperature accompanying planetary accretion), surrounded by an iron-free mantle of silicates, the outgassed volatiles would have been composed of water vapor, carbon dioxide (CO_2), and molecular nitrogen (N_2), not unlike the chemical composition of present-day volcanic emissions. Current theories of planetary formation suggest that the earth, Venus, and Mars formed as geologically differentiated objects. Some volatile outgassing may have also been associated with the impact heating during the final stages of planetary formation. This outgassing would have resulted in an almost instantaneous formation of the atmosphere, coincident with the final stages of planetary formation. As a result of planetary accretion and volatile outgassing, the terrestrial planets are characterized by iron–silicate interiors with atmospheres composed primarily of carbon dioxide (Venus and Mars) or molecular nitrogen (earth), with surface pressures that

TABLE I. The Planets: Orbital Information and Physical Characteristics

	Mercury	Venus	Earth
Mean distance from sun (millions of km)	57.9	108.2	149.6
Period of revolution	88 days	224.7 days	365.26 days
Rotation period	59 days	−243 days Retrograde	23 hr 56 min 4 sec
Inclination of axis	2°	3°	23°27′
Inclination of orbit to ecliptic	7°	3.4°	0°
Eccentricity of orbit	0.206	0.007	0.017
Equatorial diameter (km)	4880	12,104	12,756
Atmosphere (main components)	Virtually none	Carbon dioxide	Nitrogen Oxygen
Known satellites	0	0	1
Rings	—	—	—

range from about 1/200 atm (Mars) to about 90 atm (Venus) (the surface pressure of the earth's atmosphere is 1 atmosphere).

Since Mercury does not possess an appreciable atmosphere, it is not discussed in any detail in this article, which concentrates on the chemical composition of planetary atmospheres. Measurements obtained by Mariner 10, which encountered Mercury three times in 1974–75 after a 1974 Venus fly-by, indicated that the surface pressure of the atmosphere of Mercury is less than a thousandth of a trillionth of the earth's, with helium resulting from radiogenic decay and subsequent outgassing as a possible constituent. Mercury was found to possess an internal magnetic field, similar to but weaker than the earth's. Mariner 10 photographs indicated that the surface of Mercury is very heavily cratered, resembling the highlands on the moon (Fig. 1). A large impact basin (Caloris), about 1300 km in diameter, was discovered. Long scarps of cliffs, apparently produced by crustal compression, were also found.

In direct contrast to the terrestrial planets, the outer planets (Jupiter, Saturn, Uranus, and Neptune) are more massive (15–318 earth masses), larger (4–11 earth radii), and possess multiple satellites and ring systems (see Table I). The atmospheres of the outer planets are very dense and contain thick clouds and haze layers. These atmospheres are composed primarily (85–95% by volume) of molecular hydrogen (H_2) and helium (He) (5–15%) with smaller amounts of com-

pounds of carbon, nitrogen, and oxygen, primarily present in the form of saturated hydrides [methane (CH_4), ammonia (NH_3), and water vapor] at approximately the solar ratio of carbon, nitrogen, and oxygen. The composition of the atmospheres of the outer planets suggests that they are captured remnants of the primordial solar nebula that condensed to form the solar system, as opposed to having formed as a result of the outgassing of volatiles trapped in the interior, as did the atmospheres of the terrestrial planets. It has been suggested that a thick atmosphere of molecular hydrogen and helium, the overwhelming constituents of the primordial solar nebula, may have surrounded the terrestrial planets very early in their history (during the final stages of planetary accretion). However, such a primordial solar nebula remnant atmosphere surrounding the terrestrial planets would have dissipiated very quickly, due to the low mass of these planets and, hence, their weak gravitational attraction, coupled with the rapid gravitational escape of hydrogen and helium, the two lightest gases, from the "warm" terrestrial planets. Therefore, an early atmosphere composed of hydrogen and helium surrounding the terrestrial planets would have been extremely short-lived, if it ever existed at all. The large masses of the outer planets and their great distances from the sun (and colder temperatures) have enabled them to gravitationally retain their primordial solar nebula remnant atmospheres. The colder temperatures resulted in a "freezing

	Mars	Jupiter	Saturn	Uranus	Neptune	Pluto
	227.9	778.3	1427	2869	4496	5900
	687 days	11.86 yr	29.46 yr	84.01 yr	164.1	247.7 yr
	24 hr	9 hr	10 hr	17.24 hr	22 hr	−6 days
	37 min	55 min	39 min		or less	9 hr
	23 sec	30 sec	20 sec			18 min
						Retrograde
	25°12′	3°5′	26°44′	97°55′	28°48′	60°?
	1.9°	1.3°	2.5°	.8°	1.8°	17.2°
	0.093	0.048	0.056	0.047	0.009	0.25
	6787	142,800	120,400	51,800	49,500	3500
	Carbon dioxide	Hydrogen	Hydrogen	Hydrogen	Hydrogen	Methane (?)
		Helium	Helium	Helium	Helium	
		Methane	Methane	Methane	Methane	
	2	16	21	15	2	1
	—	Yes	Yes	Yes	—	—

TABLE II. Cosmic Abundance of the Elements[a]

Element	Abundance[b]	Element	Abundance[b]
$_1$H	2.6×10^{10}	$_{44}$Ru	1.6
$_2$He	2.1×10^9	$_{45}$Rh	0.33
$_3$Li	45	$_{46}$Pd	1.5
$_4$Be	0.69	$_{47}$Ag	0.5
$_5$B	6.2	$_{48}$Cd	2.12
$_6$C	1.35×10^7	$_{49}$In	2.217
$_7$N	2.44×10^6	$_{50}$Sn	4.22
$_8$O	2.36×10^7	$_{51}$Sb	0.381
$_9$F	3630	$_{52}$Te	6.76
$_{10}$Ne	2.36×10^6	$_{53}$I	1.41
$_{11}$Na	6.32×10^4	$_{54}$Xe	7.10
$_{12}$Mg	1.050×10^6	$_{55}$Cs	0.367
$_{13}$Al	8.51×10^4	$_{56}$Ba	4.7
$_{14}$Si	1.00×10^6	$_{57}$La	0.36
$_{15}$P	1.27×10^4	$_{58}$Ce	1.17
$_{16}$S	5.06×10^5	$_{59}$Pr	0.17
$_{17}$Cl	1970	$_{60}$Nd	0.77
$_{18}$Ar	2.28×10^5	$_{62}$Sm	0.23
$_{19}$K	3240	$_{63}$Eu	0.091
$_{20}$Ca	7.36×10^4	$_{64}$Gd	0.34
$_{21}$Sc	33	$_{65}$Tb	0.052
$_{22}$Ti	2300	$_{66}$Dy	0.36
$_{23}$V	900	$_{67}$Ho	0.090
$_{24}$Cr	1.24×10^4	$_{68}$Er	0.22
$_{25}$Mn	8800	$_{69}$Tm	0.035
$_{26}$Fe	8.90×10^5	$_{70}$Yb	0.21
$_{27}$Co	2300	$_{71}$Lu	0.035
$_{28}$Ni	4.57×10^4	$_{72}$Hf	0.16
$_{29}$Cu	919	$_{73}$Ta	0.022
$_{30}$Zn	1500	$_{74}$W	0.16
$_{31}$Ga	45.5	$_{75}$Re	0.055
$_{32}$Ge	126	$_{76}$Os	0.71
$_{33}$As	7.2	$_{77}$Ir	0.43
$_{34}$Se	70.1	$_{78}$Pt	1.13
$_{35}$Br	20.6	$_{79}$Au	0.20
$_{36}$Kr	64.4	$_{80}$Hg	0.75
$_{37}$Rb	5.95	$_{81}$Tl	0.182
$_{38}$Sr	58.4	$_{82}$Pb	2.90
$_{39}$Y	4.6	$_{83}$Bi	0.164
$_{40}$Zr	30	$_{90}$Th	0.034
$_{41}$Nb	1.15	$_{92}$U	0.0234
$_{42}$Mo	2.52		

[a] From Cameron, A. G. W. (1968). In "Origin and Distribution of the Elements." L. H. Ahrens, ed. Pergamon, New York. Copyright 1968 Pergamon Press.
[b] Abundance normalized to silicon (Si) = 1.00×10^6.

out" or condensation of several atmospheric gases, such as water vapor, ammonia, and methane forming cloud and haze layers in the atmospheres of the outer planets.

The most distant planet in the solar system, Pluto, has a very eccentric orbit, which at times brings it closer to the sun than Neptune's orbit.

By virtue of its great orbital eccentricity and small mass, it is suspected that Pluto may have originally been a satellite of another planet. Methane has been detected on Pluto. Very little is known about Pluto. Most of what we know about Pluto is summarized in Table I. Of the numerous smaller bodies in the solar system, including satellites and asteroids, only Saturn's satellite, Titan, has an appreciable atmosphere (surface pressure about 1.5 atm), composed of molecular nitrogen and a small amount of methane. [See PRIMITIVE SOLAR SYSTEM OBJECTS: ASTEROIDS AND COMETS.]

II. Earth

The atmospheres of the earth, Venus, and Mars resulted from the outgassing of volatiles originally trapped in their interiors. The chemical composition of the outgassed volatiles was not unlike that of present-day volcanic emissions: water vapor = 79.31% by volume; carbon dioxide = 11.61%; sulfur dioxide (SO_2) = 6.48%; and molecular nitrogen = 1.29%. On earth, the bulk of the outgassed water vapor condensed out of the atmosphere, forming the earth's vast ocean. Only small amounts of water vapor remained in the atmosphere, with almost all of it confined to the troposphere. Some atmospheric water is in the condensed state, found in the form of cloud droplets. Water clouds cover about 50% of the earth's surface at any given time and are a regular feature of the atmosphere (Fig. 2). Near the ground, the water vapor concentration is variable, ranging from a fraction of a percent to a maximum of several percent by volume. Once the ocean formed, outgassed carbon dioxide, the second most abundant volatile, which is very water soluble, dissolved into the ocean. Once dissolved in the ocean, carbon dioxide chemically reacted with ions of calcium and magnesium, also in the ocean, and precipitated out in the form of sedimentary carbonate rocks such as calcite ($CaCO_3$), and dolomite [$CaMg(CO_3)_2$].

The concentration of carbon dioxide in the atmosphere is about 0.034% by volume, which is equivalent to 340 parts per million by volume (ppmv). It has been estimated that the preindustrial (ca. 1860) level of atmospheric carbon dioxide was about 280 ppmv, with the increase to the present level attributable to the burning of fossil fuels, notably coal. For each carbon dioxide molecule in the present-day atmosphere, there are approximately 50 carbon dioxide molecules

FIG. 1.   Mosaic of Mariner 10 photographs of Mercury. With no appreciable atmosphere, we can see right down to the cratered surface of Mercury, which is similar to the cratered highlands on the moon. The largest craters in the photograph are about 200 km in diameter.

FIG. 2.   Planet earth as photographed by Apollo 17 astronauts on their journey to the moon. Scattered clouds, which cover only about 50% of the earth at any given time, permit viewing the surface of the earth. Earth is a unique planet in many respects, including the presence of life, the existence of liquid water on the surface, and the presence of large amounts of oxygen in the atmosphere.

physically dissolved in the ocean and almost 30,000 carbon dioxide molecules incorporated in sedimentary carbonate rocks. All of the carbon dioxide presently incorporated in carbonate rocks originally outgassed from the interior and was at one time in the atmosphere. Hence, the early atmosphere may have contained 100 to 1000 times or more carbon dioxide than it presently contains. Sulfur dioxide, the third most abundant gas in volcanic emissions, is chemically unstable in the atmosphere. Sulfur dioxide is rapidly chemically transformed to water-soluble sulfuric acid (H_2SO_4), which readily rains out of the atmosphere. Hence, the atmospheric lifetime of sulfur dioxide is very short. [*See* AIR POLLUTION (METEOROLOGY).]

Molecular nitrogen is the fourth most abundant gas in volcanic emissions. Nitrogen does not condense out of the atmosphere (as does water vapor), is not water soluble (as is carbon dioxide), and is not chemically active (as is sulfur dioxide). As a result, the bulk of the outgassed molecular nitrogen accumulated in the atmosphere, and over geological time became the most abundant atmospheric constituent (about 78% by volume). Molecular oxygen (O_2), produced as a by-product of photosynthetic activity, built up in the atmosphere to become the second most abundant species (about 21% by volume). Argon (isotope 40), the third most abundant atmospheric species (about 1% by volume), is a chemically inert gas resulting from the radiogenic decay of potassium (isotope 40) in the crust. Hence, the bulk chemical composition of the earth's atmosphere can be explained in terms of volatile outgassing and the ultimate sinks of the outgassed volatiles, including condensation/precipitation, dissolution and carbonate formation in the ocean, biogenic activity, and radiogenic decay. These and other processes, including photochemical and chemical reactions, have resulted in a myriad of other trace atmospheric gases in the atmosphere, which are listed in Table III.

TABLE III.   Composition of the Earth's Atmosphere[a]

	Surface concentration[b]	Source
Major and minor Gases		
Nitrogen (N_2)	78.08%	Volcanic, biogenic
Oxygen (O_2)	20.95%	Biogenic
Argon (Ar)	0.93%	Radiogenic
Water vapor (H_2O)	Variable, up to 4%	Volcanic, evaporation
Carbon dioxide (CO_2)	0.034%	Volcanic, biogenic, anthropogenic
Trace gases		
Oxygen species		
Ozone (O_3)	10–100 ppbv	Photochemical
Atomic oxygen (O) (ground state)	10^3 cm^{-3}	Photochemical
Atomic oxygen [O(^{1}D)] (excited state)	10^{-2} cm^{-3}	Photochemical
Hydrogen species		
Hydrogen (H_2)	0.5 ppmv	Photochemical, biogenic
Hydrogen peroxide (H_2O_2)	10^9 cm^{-3}	Photochemical
Hydroperoxyl radical (HO_2)	10^8 cm^{-3}	Photochemical
Hydroxyl radical (OH)	10^6 cm^{-3}	Photochemical
Atomic hydrogen (H)	1 cm^{-3}	Photochemical
Nitrogen species		
Nitrous oxide (N_2O)	330 ppbv	Biogenic, anthropogenic
Ammonia (NH_3)	0.1–1 ppbv	Biogenic, anthropogenic
Nitric acid (HNO_3)	50–1000 pptv	Photochemical
Hydrogen cyanide (HCN)	~200 pptv	Anthropogenic(?)
Nitrogen dioxide (NO_2)	10–300 pptv	Photochemical
Nitric oxide (NO)	5–100 pptv	Anthropogenic, biogenic, lightning, photochemical
Nitrogen trioxide (NO_3)	100 pptv	Photochemical
Peroxyacetylnitrate ($CH_3CO_3NO_2$)	50 pptv	Photochemical
Dinitrogen pentoxide (N_2O_5)	1 pptv	Photochemical
Pernitric acid (HO_2NO_2)	0.5 pptv	Photochemical
Nitrous acid (HNO_3)	0.1 pptv	Photochemical

(continued)

TABLE III.   (*Continued*)

	Surface concentration[b]	Source
Nitrogen aerosols		
Ammonium nitrate (NH_4NO_3)	~100 pptv	Photochemical
Ammonium chloride (NH_4Cl)	~0.1 pptv	Photochemical
Ammonium sulfate [$(NH_4)_2SO_4$]	~0.1 pptv(?)	Photochemical
Carbon species		
Methane (CH_4)	1.7 ppmv	Biogenic, anthropogenic
Carbon monoxide (CO)	70–200 ppbv (N hemis.) 40–60 ppbv (S hemis.)	Anthropogenic, biogenic, photochemical
Formaldehyde (H_2CO)	0.1 ppbv	Photochemical
Methylhydroperoxide (CH_3OOH)	10^{11} cm^{-3}	Photochemical
Methylperoxyl radical (CH_3O_2)	10^8 cm^{-3}	Photochemical
Methyl radical (CH_3)	10^{-1} cm^{-3}	Photochemical
Sulfur species		
Carbonyl sulfide (COS)	0.5 ppbv	Volcanic, anthropogenic
Dimethyl sulfide [$(CH_3)_2S$]	0.4 ppbv	Biogenic
Hydrogen sulfide (H_2S)	0.2 ppbv	Biogenic, anthropogenic
Sulfur dioxide (SO_2)	0.2 ppbv	Volcanic, anthropogenic, photochemical
Dimethyl disulfide [$(CH_3)_2S_2$]	100 pptv	Biogenic
Carbon disulfide (CS_2)	50 pptv	Volcanic, anthropogenic
Sulfuric acid (H_2SO_4)	20 pptv	Photochemical
Sulfurous acid (H_2SO_3)	20 pptv	Photochemical
Sulfoxyl radical (SO)	10^3 cm^{-3}	Photochemical
Thiohydroxyl radical (HS)	1 cm^{-3}	Photochemical
Sulfur trioxide (SO_3)	10^{-2} cm^{-3}	Photochemical
Halogen species		
Hydrogen chloride (HCl)	1 ppbv	Sea salt, volcanic
Methyl chloride (CH_3Cl)	0.5 ppbv	Biogenic, anthropogenic
Methyl bromide (CH_3Br)	10 pptv	Biogenic, anthropogenic
Methyl iodide (CH_3I)	1 pptv	Biogenic, anthropogenic
Noble gases (chemically inert)		
Neon (Ne)	18 ppmv	Volcanic
Helium (He)	5.2 ppmv	Radiogenic
Krypton (Kr)	1 ppmv	Radiogenic
Xenon (Xe)	90 ppbv	Radiogenic

[a] From Levine, J. S. (1985). In "The Photochemistry of Atmospheres: Earth, the Other Planets, and Comets," J. S. Levine, ed. Academic, Orlando, FL.

[b] Species concentrations are given in percentage by volume, in terms of surface mixing ratio, parts per million by volume (ppmv $\equiv 10^{-6}$), parts per billion by volume (ppbv $\equiv 10^{-9}$), parts per trillion by volume (pptv $\equiv 10^{-12}$), or in terms of surface number density (cm^{-3}). The species mixing ratio is defined as the ratio of the number density of the species to the total atmospheric number density (2.55×10^{19} molec cm^{-3}). There is some uncertainty in the concentrations of species at the ppbv level or less. The species concentrations given in molec cm^{-3} are generally based on photochemical calculations, and species concentrations in mixing ratios are generally based on measurements.

III. Venus

Venus has been described as the earth's twin because of its similar mass (0.81 earth masses), radius (0.95 earth radii), mean density (95% that of earth), and gravity (90% that of earth) (see Table I). However, in terms of atmospheric structure and chemical composition, Venus is anything but a twin of earth. The mean planetary surface temperature of Venus is about 750 K, compared with about 300 K for earth; the surface pressure on Venus is about 90 atm, compared with 1 atm for earth; carbon dioxide at 96% by volume, is the overwhelming constituent in the atmosphere of Venus, while it is only a trace constituent in the earth's atmosphere (0.034% by volume). In addition, Venus does not have an ocean or a biosphere and is completely covered by thick clouds, probably composed of sulfuric acid (Fig. 3). Hence, the atmosphere is inhospitable and very unlike the earth's atmosphere.

Much of our information on the structure and composition of the atmosphere of Venus has been obtained through a series of U.S. and U.S.S.R. Venus flybys, orbiters, and landers, which are summarized in Table IV.

The clouds on Venus are thick and contain no holes, hence, we have never directly observed the surface of Venus from earth. These clouds resemble a stratified low-density haze extending from about 45 to about 65 km. The total extinc-

FIG. 3.   Venus as photographed by the Pioneer Venus Orbiter. The thick, cloud-covered atmosphere continually and completely hides the surface of Venus, which at very high temperature and atmospheric pressure is not a hospitable environment.

TABLE IV. U.S. and U.S.S.R. Missions to Venus[a]

Name	Designator (U.S.S.R.)	Launch date	Mission remarks
Venera 1	61-Gamma 1	Feb. 12, 1961	Passed Venus at 100,000 km May 19–21, 1961; contact lost Feb. 27, 1961.
Mariner II	—	Aug. 27, 1962	Planetary exploration: First successful interplanetary probe. Found no magnetic field; high surface temperatures of approximately 800°F. Passed Venus Dec. 14, 1962 at 21,600 miles, 109 days after launch.
Zond 1	64-16D	April 2, 1964	Passed Venus at 100,000 km July 19, 1964; communications failed after May 14, 1964.
Venera 2	65-91A	Nov. 12, 1965	Passed Venus at 24,000 km, Feb. 27, 1966; communications failed.
Venera 3	65-92A	Nov. 16, 1965	Struck Venus March 1, 1966; communications failed earlier.
Venera 4	67-58A	June 12, 1967	Probed atmosphere.
Venera 5	69-1A	Jan. 5, 1969	Entered Venus atmosphere May 16, 1969.
Mariner V	—	June 14, 1967	Planetary exploration: All science and engineering subsystems nominal through encounter with Venus; data indicate Venus has a moonlike effect on solar plasma and strong H_2 corona comparable to Earth's, 72 to 87% CO_2 atmosphere with balance probably nitrogen and O_2. Closest approach, 3,900 km on Oct. 19, 1967.
Venera 6	69-2A	Jan. 10, 1969	Entered Venus atmosphere May 17, 1969.
Venera 7	70-60A	Aug. 17, 1970	Soft landed on Venus; signal from surface.
Venera 8	72-21A	Mar. 27, 1972	Soft landed on Venus; sent data from surface.
Mariner 10 Venus/Mercury	—	Nov. 3, 1973	Conducted exploratory investigations of planet Mercury during three flybys by obtaining measurements of its environment, atmosphere, surface, and body characteristics and conducted similar investigations of Venus. Mariner 10 encountered Venus on Feb. 5, 1974 and Mercury on Mar. 29 and Sept. 21, 1974 and Mar. 16, 1975. Resolution of the photographs was 100 m, 7000 times greater than that achieved by earth-based telescopes.
Venera 9-orbiter	75-50A	June 8, 1975	Orbited Venus Oct. 22, 1975. Orbiter and lander launched from single D-class vehicle (Proton), 4650 kg thrust.
Venera 9-lander	75-50D	June, 1975	Soft landed; returned picture.
Venera 10-orbiter	75-54A	June 14, 1975	Orbited Venus Oct. 25, 1975. Orbiter and lander launched from single D-class vehicle (Proton), 4659 kg thrust.
Venera 10-lander	75-54D	June 14, 1975	Soft landed; returned picture.
Pioneer 12 Pioneer 13 Pioneer Venus	—	May 20, 1978 Aug. 8, 1978	Orbiter launched in May studied interaction of atmosphere and solar wind and made radar and gravity maps of the planet. The multiprobe spacecraft launched in August returned information on Venus' wind and circulation patterns as well as atmospheric composition, temperature and pressure readings. Pioneer 12 entered Venus orbit Dec. 4, 1978; Pioneer 13 encountered Venus Dec. 9, 1978.
Venera 11-orbiter	78-84A	Sept. 9, 1978	Passed Venus as 35,000 km Dec. 25, 1978; served as relay station. Orbiter and lander launched from single D-class vehicle (Proton), 4650 kg thrust.
Venera 11-lander	78-84E	Sept. 9, 1978	Soft-landed on Venus.

(continued)

TABLE IV.   (*Continued*)

Name	Designator (U.S.S.R.)	Launch date	Mission remarks
Venera 12-orbiter	78-86A	Sept. 14, 1978	Passed Venus at 35,000 km Dec. 21, 1978; served as relay station. Orbiter and lander launched from single D-class vehicle (Proton), 4650 kg thrust.
Venera 12-lander	78-86E	Sept. 14, 1978	Soft-landed on Venus.
Venera 13-orbiter	1981-106A	Oct. 30, 1981	Both orbiter and lander launched from single D-class vehicle (Proton), 4650 kg thrust.
Venera 13-lander	None	Oct. 30, 1981	Soft-landed on Venus Mar. 3, 1982; returned color picture.
Venera 14-orbiter	1981-110A	Nov. 4, 1981	Both orbiter and lander launched from single D-class vehicle (Proton), 4650 kg thrust.
Venera 14-lander	None	Nov. 4, 1981	Soft-landed on Venus Mar. 5, 1982; returned color picture.

[a] From NASA (1983). "Planetary Exploration Through Year 2000: A Core Program." Solar System Exploration Committee of NASA Advisory Council, Washington, DC.

tion optical depth of the clouds in visible light is about 29. The extinction of visible light is due almost totally to scattering. The lower clouds are found between 45 and 50 km; the middle clouds from 50 to 55 km; and the upper clouds from 55 to 65 km. The tops of the upper clouds, which are the ones visible from earth, appear to be composed of concentrated sulfuric acid droplets (see Fig. 3).

As already noted, carbon dioxide at 96% by volume, is the overwhelming constituent of the atmosphere of Venus. The next most abundant atmospheric gas is molecular nitrogen at 4% by volume. The relative proportion by volume of carbon dioxide and molecular nitrogen in the atmospheres of Venus and Mars is almost identical. The chemical composition of the atmosphere of Venus is summarized in Table V. At the surface of Venus, the partial pressure of carbon dioxide is about 90 bar, molecular nitrogen is about 3.2 bar, and water vapor is only about 0.01 bar (more about water on Venus later). For comparison, if the earth were heated to the surface temperature of Venus (about 750 K), we would have a massive atmosphere composed of water vapor at a surface partial pressure of about 300 bar (resulting from the evaporation of the ocean), a carbon dioxide partial pressure of about 55 bar (resulting from the thermal composition of crustal carbonates), and a molecular nitrogen pressure of about 1 to 3 bar (resulting from the present atmosphere plus the outgassing of crustal nitrogen).

A major puzzle concerning the chemical composition of the atmosphere of Venus (as well as

the atmosphere of Mars) is the stability of carbon dioxide and the very low atmospheric concentrations of carbon monoxide (CO) and oxygen [atomic (O) and molecular], which are the photodissociation products of carbon dioxide. In the daytime upper atmosphere (above 100 km), carbon dioxide is readily photodissociated with a photochemical atmospheric lifetime of only about one week. The recombination of carbon monoxide and atomic oxygen in the pres-

TABLE V.   Composition of the Atmosphere of Venus[a]

Gas	Volume mixing ratio	
	Troposphere (below clouds)	Stratosphere (above clouds)
CO_2	9.6×10^{-1}	9.6×10^{-1}
N_2	4×10^{-2}	4×10^{-2}
H_2O	10^{-4}–10^{-3}	10^{-6}–10^{-5}
CO	$(2-3) \times 10^{-5}$	5×10^{-5}–10^{-3}
HCl	$<10^{-5}$	10^{-6}
HF	?	10^{-8}
SO_2	1.5×10^{-4}	5×10^{-8}–8×10^{-7}
S_3	$\sim 10^{-10b}$	?
H_2S	$(1-3) \times 10^{-6b}$	?
COS	$<2 \times 10^{-6}$	?
O_2	$(2-4) \times 10^{-5b}$	$<10^{-6b}$
H_2	?	2×10^{-5b}
^4He	10^{-5}	10^{-5}
$^{20,22}Ne$	$(5-13) \times 10^{-6}$	$(5-13) \times 10^{-6}$
$^{36,38,40}Ar$	$(5-12) \times 10^{-5}$	$(5-12) \times 10^{-5}$
^{84}Kr	$<2 \times 10^{-8}$–4×10^{-7}	$<2 \times 10^{-8}$–4×10^{-7}

[a] From Lewis, J. S., and Prinn, R. G. (1986). "Planets and Their Atmospheres." Academic, New York. Copyright 1984 Academic Press.

[b] Single experiment; corroboration required.

ence of a third body to reform carbon dioxide is efficient only at the higher atmospheric pressures occurring at and below 100 km. However, at these lower altitudes, atomic oxygen recombines with itself in the presence of a third body to form molecular oxygen considerably faster than the three-body reaction that leads to the recombination of carbon dioxide. Thus, essentially all of the photolyzed carbon dioxide produces carbon monoxide and molecular oxygen. Yet, the observed upper-limit atmospheric concentration of molecular oxygen above the cloud tops could be produced in only about one day, and the observed abundance of carbon monoxide could be produced in only about three months. Photodissociation could easily convert the entire concentration of carbon dioxide in the atmosphere to carbon monoxide and molecular oxygen in only about 4 million years, geologically a short time period.

This dilemma also applies to carbon dioxide on Mars. Considerable research has centered around the recombination of carbon monoxide and molecular oxygen back to carbon dioxide. It became apparent that the only way to maintain low carbon monoxide and oxygen concentrations and high carbon dioxide concentrations in the 100–150-km region is by the rapid downward transport of carbon monoxide and oxygen, balanced by the upward transport of carbon dioxide. It is believed that carbon dioxide is reformed from carbon monoxide and oxygen at an altitude of about 70 km through various chemical reactions and catalytic cycles involving chemically active compounds of hydrogen and chlorine.

If Venus and the earth contained comparable levels of volatiles and outgassed them at comparable rates, then Venus must have somehow lost about 300 bar of water vapor. This may have been accomplished by a runaway greenhouse. In the runaway greenhouse on Venus, outgassed water vapor and carbon dioxide entered the atmosphere, contributing to steadily increasing atmospheric opacity and thus to increasing surface and atmospheric temperatures via the greenhouse effect. On earth, water vapor condensed out of the atmosphere forming the ocean, and the oceans then removed atmospheric carbon dioxide via dissolution and subsequent incorporation into carbonates. The greater proximity of Venus to the sun and its higher initial surface temperature appear to be the simple explanation for the divergent fates of water vapor and carbon dioxide on Venus and

earth. In the runaway greenhouse scenario, the photodissociation of massive amounts of outgassed water vapor in the atmosphere of Venus would have led to the production of large amounts of hydrogen and oxygen. Hydrogen could have gravitationally escaped from Venus, and oxygen could have reacted with crustal material. The runaway greenhouse and the accompanying high surface and atmospheric temperatures, too hot for the condensation of outgassed water on Venus, would explain the present water vapor-deficient and carbon dioxide-rich atmosphere of Venus. An alternative suggestion is that Venus may have originally accreted without the levels of water that the earth contained, resulting in a much drier Venus.

IV. Mars

The atmosphere of Mars is very thin (mean surface pressure only about 6.36 mbar), cold (mean surface temperature about 220 K, with the temperature varying from about 290 K in the southern summer to about 150 K in the polar winter), and cloud-free, making the surface of Mars readily visible from the earth (Fig. 4). Much of our information on the structure and composition of the atmosphere of Mars has been obtained through a series of flybys, orbiters, and landers, which are summarized in Table VI. As already noted, the composition of the atmosphere of Mars is comparable to that of Venus. Carbon dioxide is the overwhelming constituent (95.3% by volume), with smaller amounts of molecular nitrogen (2.7%) and argon (1.6%), and trace amounts of molecular oxygen (0.13%) and carbon monoxide (0.08%), resulting from the photodissociation of carbon dioxide (the composition of the atmosphere of Mars is summarized in Table VII). Water vapor and ozone (O_3) are also present, although their abundances vary with season and latitude. The annual sublimation and precipitation of carbon dioxide out of and into the polar cap produce a planet-wide pressure change of 2.4 mbar, or 37% of the mean atmospheric pressure of 6.36 mbar.

The amount and location of water vapor in the atmosphere of Mars are controlled by the temperature of the surface and the atmosphere. The northern polar cap is a source of water vapor during the northern summer. The surface of Mars is also a source of water vapor, depending on the location and season. The total amount of water vapor in the atmosphere varies seasonally between the equivalent of 1 and 2 km^3 of liquid

FIG. 4.   Mars as photographed by the Viking 2 orbiter. The thin, cloud-free atmosphere of Mars permits direct observation of the Martian surface from space.

TABLE VI.   U.S. and U.S.S.R. Missions to Mars[a]

Name	Designator (U.S.S.R.)	Launch date	Mission remarks
Mars 1	62-Beta Nu 3	Nov. 1, 1962	Passed Mars June 19, 1963 at 193,000 km; communications failed March 21, 1963.
Mariner IV	—	Nov. 28, 1964	Planetary and interplanetary exploration: Encounter occurred July 14, 1965 with closest approach 6100 miles. Twenty-two pictures taken.
Zond 2	64-78C	Nov. 30, 1964	Passed Mars at 1500 km Aug. 6, 1965; communications failed earlier.
Mariner VI	—	Feb. 25, 1969	Planetary exploration: Mid-course correction successfully executed to achieve a Mars flyby within 3330 km on July 31, 1969. Designed to perform investigations of atmospheric structures and compositions and to return TV photos of surface topography.
Mariner VII	—	Mar. 27, 1969	Planetary exploration: Spacecraft identical to Mariner VI. Mid-course correction successful for 3518 km flyby on Aug. 5, 1969.
Kosmos 419	71-42A	May 10, 1971	Failed to separate.
Mars 2-orbiter	71-45A	May 19, 1971	Orbited Mars Nov. 27, 1971. Mars 2 orbiter and lander launched from single D-class vehicle (Proton), 4650 kg thrust.
Mars-lander	71-45E	May 19, 1971	Landed 47°E.
Mars 3-orbiter	71-49A	May 28, 1971	Orbited Mars Dec. 2, 1971. Mars 3 orbiter and lander launched from single D-class vehicle (Proton), 4650 kg thrust.
Mars 3-lander	71-49F	May 28, 1971	Landed 45°S, 158°W.
Mariner IX	—	May 30, 1971	Entered Mars orbit on Nov. 13, 1971. Spacecraft responsed to 38,000 commands and transmitted 6900 pictures of the Martian surface. All scientific instruments operated successfully. Mission terminated on Oct. 27, 1972.
Mars 4	73-47A	July 21, 1973	Passed Mars at 2200 km Feb. 10, 1974, but failed to enter Mars' orbit as planned.
Mars 5	73-49A	July 25, 1973	Orbited Mars Feb. 2, 1974 to gather Mars data and to serve as relay station.
Mars 6-orbiter	73-52A	Aug. 5, 1973	Mars 6 orbiter and lander launched from single D-class vehicle (Proton), 4650 kg thrust.
Mars 6-lander	73-52E	Aug. 5, 1973	Soft landed at 24°S, 25°W; returned atmospheric data during descent.
Mars 7-orbiter	73-53A	Aug. 9, 1973	Mars 7 orbiter and lander launched from single D-class vehicle (Proton), 4650 kg thrust.
Mars 7-lander	73-53E	Aug. 9, 1973	Missed Mars by 1300 km (aimed at 50°S, 28°W).
Viking 1 Lander and orbiter		Aug. 20, 1975	Scientific investigation of Mars. United States' first attempt to soft land a spacecraft on another planet. Successfully soft landed on July 20, 1976. First in situ analysis of surface material on another planet.
Viking 2 Lander and orbiter		Sept. 9, 1975	Scientific investigation of Mars. United States' second attempt to soft land on Mars. Successfully soft landed on Sept. 3, 1976 and returned scientific data. Orbiter from both missions returned over 40,000 high resolution photographs showing surface details as small as 10 m in diameter. Orbiter also collected gravity field data, monitored atmospheric water levels, thermally mapped selected surface sites.

[a] From NASA (1983). "Planetary Exploration Through Year 2000: A Core Program." Solar System Exploration Committee of NASA Advisory Council, Washington, DC.

TABLE VII. Composition of the Atmosphere of Mars[a]

Species	Abundance (mole fraction)
CO_2	0.953
N_2	0.027
^{40}Ar	0.016
O_2	0.13%
CO	0.08%
	0.27%
H_2O	$(0.03\%)^b$
Ne	2.5 ppm
^{36}Ar	0.5 ppm
Kr	0.3 ppm
Xe	0.08 ppm
O_3	$(0.03 \text{ ppm})^b$
	$(0.003 \text{ ppm})^b$

Species	Upper limit (ppm)
H_2S	<400
C_2H_2, HCN, PH_3, etc.	50
N_2O	18
C_2H_4, CS_2, C_2H_6, etc.	6
CH_4	3.7
N_2O_4	3.3
SF_6, SiF_4, etc.	1.0
HCOOH	0.9
CH_2O	0.7
NO	0.7
COS	0.6
SO_2	0.5
C_3O_2	0.4
NH_3	0.4
NO_2	0.2
HCl	0.1
NO_2	0.1

[a] From Lewis, J. S., and Prinn, R. G. (1984). "Planets and Their Atmospheres." Academic, New York. Copyright 1984 Academic Press, New York.

[b] Very variable.

water, with the maximum occurring in the northern summer and the minimum in the northern winter. Ozone is also a highly variable constituent of the atmosphere of Mars. Ozone is present only when the atmosphere is cold and dry.

There is evidence to suggest that significant quantities of outgassed carbon dioxide and water vapor may reside on the surface and in the subsurface of Mars. In addition to the polar caps, which contain large concentrations of frozen carbon dioxide and, in the case of the northern polar cap, of frozen water, there may be considerable quantities of these gases physically adsorbed to the surface and subsurface material. It has been estimated that if the equilibrium temperature of the winter polar cap would increase from its present value of about 150 K to 160 K, sublimation of frozen carbon dioxide would increase the atmospheric pressure to more than 50 mbar. This in turn would cause more water vapor to leave the polar cap and enter the atmosphere. Mariner and Viking photographs indicate the existence of channels widely distributed over the Martian surface. These photographs show runoff channels, tributary networks, and streamlined islands, all very suggestive of widespread fluid erosion. Yet, there is no evidence for the existence of liquid water on the surface of Mars today. In addition, a significant quantity of water vapor may have escaped from Mars in the form of hydrogen and oxygen atoms, resulting from the photolysis of water vapor in the atmosphere of Mars. If the present gravitational escape rate of atoms of hydrogen and oxygen has been operating over the history of Mars, then an amount of liquid water covering the entire planet about 2.5 m high may have escaped from Mars. Viking measurements of argon and neon in the atmosphere of Mars suggest that Mars may have formed with a lower volatile content then either earth or Venus. This is consistent with ideas concerning the capture and incorporation of volatiles in accreting material and how volatile incorporation varies with temperature, which is a function of the distance of the accreting terrestrial planets from the sun.

Unlike the very thick atmosphere of Venus, where the photolysis of carbon dioxide occurs only in the upper atmosphere (above 100 km), on Mars the photodissociation of carbon dioxide occurs throughout the entire atmosphere, right down to the surface. For comparison, the 6.36-mbar surface pressure of the atmosphere of Mars corresponds to an atmospheric pressure at an altitude of about 33 km in the earth's atmosphere. On Mars, carbon dioxide is reformed from its photodissociation products, carbon monoxide and oxygen, by reactions involving atomic hydrogen (H) and the oxides of hydrogen.

Viking photographs indicate that the surface rocks on Mars resemble basalt lava (see Fig. 5). The red color of the surface is probably due to oxidized iron. The soil is fine-grained and cohesive, like firm sand or soil on earth. Viking experiments gave no evidence for organic molecules or for biological activity in the Martian soil, despite unusual chemical reactions pro-

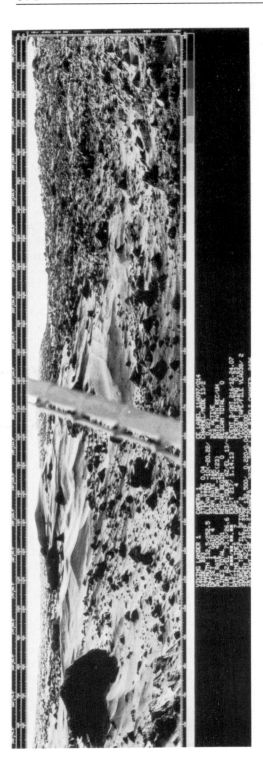

FIG. 5.   Surface of Mars as photographed by the Viking I lander. Upper photograph shows the Martian landscape with features very similar to those seen in the deserts of earth. Lower photograph, the first ever taken on the surface of Mars, was obtained just minutes after the Viking I landed on July 20, 1976.

duced by the soil and measured by the life detection experiments.

V. The Outer Planets

Jupiter, Saturn, Uranus, and Neptune are giant gas planets—great globes of dense gas, mostly molecular hydrogen and helium, with smaller amounts of methane, ammonia, water vapor, and various hydrocarbons produced from the photochemical and chemical reactions of these gases. They formed in the cooler parts of the primordial solar nebula, so gases and ices were preserved. These gas giants have ring systems and numerous satellites orbiting them. As already noted, the outer planets are more massive and larger and have very dense atmospheres that contain thick clouds and layers of aerosol and haze. The solid surfaces of the outer planets have never been observed, and we have only observed the top of the cloud and haze layers.

In many ways, Jupiter and Saturn are a matched pair, as are Uranus and Neptune. Jupiter and Saturn appear to have cores of silicate rocks and other heavy compounds comprising about 25 earth masses, surrounded by thick atmospheres of molecular hydrogen and helium. The total mass of Jupiter and Saturn are 318 and 95 earth masses, respectively. Uranus and Neptune appear to possess much less massive hydrogen/helium atmospheres relative to their cores. The total mass of Uranus and Neptune are only 14.5 and 17 earth masses, respectively.

The large satellites of Jupiter, Saturn, and Neptune are all larger than the earth's moon, with several comparable to the size of Mercury (see Table VIII for a summary of the orbital information and physical characteristics of these satellites). One of these satellites, Titan, the largest satellite of Saturn, has an appreciable atmosphere. Much of the new information about the atmospheres of Jupiter, Saturn, Uranus, their rings and satellites was obtained by the Voyager encounters of these planets (see Table IX for a summary of the missions to Jupiter and Saturn). [See PLANETARY SATELLITES, NATURAL.]

The Voyager spacecraft obtained high-resolution images of Jupiter, Saturn, and Uranus, their rings and satellites. Voyager instrumentation gathered new information on the chemical composition of their atmospheres. Jupiter, with its colorful banded, turbulent atmosphere, photographed by Voyager 1, is shown in Fig. 6. The Great Red Spot of Jupiter can be clearly seen in this photograph. The helium abundance in the atmosphere of Jupiter was found to be 11% by volume (with molecular hydrogen at 89% by volume), very close to that of the sun. The presence of methane, ammonia, water vapor, ethylene (C_2H_4), ethane (C_2H_6), acetylene (C_2H_2), benzene (C_6H_6), phosphine (PH_3), hydrogen cyanide (HCN), and germanium tetrahydride (GeH_4) in the atmosphere of Jupiter was confirmed (see Table X). The magnetosphere of Jupiter was found to be the largest object in the solar system, about 15 million km across (10 times the diameter of the sun). In addition to hydrogen ions, the magnetosphere was found to contain ions of oxygen and sulfur. A much denser region of ions was found in a torus surrounding the orbit of Jupiter's satellite, Io. The Io torus emits intense ultraviolet radiation and also generates aurora at high latitudes on Jupiter. In addition to a Jovian aurora, huge lightning flashes and meteors were photographed by Voyager on the nightside of Jupiter. A thin ring

TABLE VIII. The Large Satellites of the Outer Planets: Physical Characteristics[a]

Planet	Satellite	Distance from planet (10^3 km)	Sidereal period (days)	Radius (km)	Density (g cm^{-3})	Surface gravity (cm sec^{-2})
Earth	Moon	384	27.3	1738	3.34	162
Jupiter	Io	422	1.77	1815	3.55	181
	Europa	671	3.55	1569	3.04	132
	Ganymede	1070	7.15	2631	1.93	144
	Callisto	1883	16.7	2400	1.83	125
Saturn	Titan	1222	16	2575	1.89	136
Neptune	Triton	355	6	~2500	~2.1	~150

[a] From Strobel, D. F. (1985). In "The Photochemistry of Atmospheres: Earth, the Other Planets, and Comets," J. S. Levine, ed. Academic, Orlando, FL. Copyright 1985 Academic Press.

TABLE IX. U.S. and U.S.S.R. Missions to Jupiter and Saturn[a]

Name	Launch date	Mission remarks
Pioneer 10	Mar. 3, 1972	Investigation of the interplanetary medium, the asteroid belt, and the exploration of Jupiter and its environment. Closest approach to Jupiter 130,000 km on Dec. 3, 1973. Exited solar system June 14, 1983; still active.
Pioneer 11 Jupiter/Saturn	Apr. 6, 1973	Obtained scientific information beyond the orbit of Mars with the following emphasis: (a) investigation of the interplanetary medium; (b) investigation of the nature of the asteroid belt; (c) exploration of Jupiter and its environment. Closest approach to Jupiter on Dec. 2, 1974; Saturn encounter: Sept. 1, 1979.
Voyager II Voyager I	Aug. 20, 1977 Sept. 5, 1977	Voyager II encountered Jupiter July 9, 1979, Saturn Aug. 26, 1981, and Uranus Jan. 24, 1986 and is targeted for a Neptune encounter in Sept. 1989. Voyager I encountered Jupiter Mar. 5, 1979 and Saturn Nov. 13, 1980. Both returned a wealth of information about these two giant planets and their satellites including documentation of active volcanism on Io, one of the Galilean satellites.

[a] From NASA (1983). "Planetary Exploration Through Year 2000: A Core Program." Solar System Exploration Committee of NASA Advisory Council, Washington, DC.

surrounding Jupiter, much narrower than Saturn's, was discovered by Voyager. The four large Galilean satellites (Ganymede, Callisto, Europa, and Io) were studied in detail (see Figs. 7–10 for Voyager photographs of these geologically varied satellites of Jupiter). Io was found to have at least 10 active volcanos (see Fig. 10). Sulfur resulting from the volcanic emissions on Io is responsible for the orange color of its surface, as well as the presence of sulfur dioxide in its atmosphere (at a partial pressure of only about one ten millionth of a bar). Io's volcanic emissions are also responsible for the ions of oxygen and sulfur in Jupiter's magnetosphere.

After encountering Jupiter and its satellites, both Voyager spacecraft visited Saturn and its satellite system (see Figure 11). The six previously known rings were found to be composed of innumerable, individual ringlets with very few gaps observed anywhere in the ring system. Complex dynamical effects were photographed in the ring system, including spiral density waves similar to those believed to generate spiral structure in galaxies. The helium content of the atmosphere of Saturn was found to be about 6% by volume (with molecular hydrogen at about 94% by volume), compared with about 11% for Jupiter. The trace gases in the atmosphere of Saturn are similar to those in the atmosphere of Jupiter and include methane, acetylene, ethane, phosphene, and propane (C_3H_8) (see Table XI).

Titan, the largest satellite of Saturn, was found to have a diameter slightly smaller than that of Jupiter's largest satellite, Ganymede (see Table VIII). The atmosphere of Titan is covered by clouds and layers of aerosols and haze and has a surface pressure of about 1.5 bar, which makes it about 50% more massive than the

TABLE X. Composition of the Atmosphere of Jupiter[a]

Constituent	Volume mixing ratio[b]
H_2	0.89
He	0.11
CH_4	0.00175
C_2H_2	0.02 ppm
C_2H_4[c]	7 ppb
C_2H_6	5 ppm
CH_3C_2H[c]	2.5 ppb
C_6H_6[c]	2 ppb
CH_3D	0.35 ppm
NH_3[d]	180 ppm
PH_3	0.6 ppm
H_2O[d]	1–30 ppm
GeH_4	0.7 ppb
CO	1–10 ppb
HCN	2 ppb

[a] From Strobel, D. F. (1985). In "The Photochemistry of Atmospheres: Earth, the Other Planets, and Comets," J. S. Levine, ed. Academic, Orlando, FL. Copyright 1985 Academic Press.
[b] ppm ≡ parts per million; ppb ≡ parts per billion.
[c] Tentative identification, polar region.
[d] Value at 1 to 4 bar.

FIG. 6. Jupiter as photographed by Voyager 1. The Great Red Spot can be seen in the lower center of the photograph. The atmosphere of Jupiter is massive and completely covered with clouds and aerosol haze layers.

earth's atmosphere. The surface temperature of Titan is a cold 100 K. The cloud- and haze-covered Titan is shown in Figs. 12 and 13, obtained by Voyager. Titan's atmosphere is mostly molecular nitrogen, with smaller amounts of methane and trace amounts of carbon monoxide, carbon dioxide, and various hydrocarbons (see Table XII for the chemical composition of Titan in different regions of its atmosphere). The surface of Titan may hold a large accumulation of liquid methane.

After encountering Saturn, Voyager 2 was targeted for Uranus. On January 24, 1986, Voyager 2 had its closest approach to Uranus. Prior to this encounter, very little was known about Uranus, one of the three planets (Neptune and Pluto, the other two) not known to the ancients. Uranus was discovered accidentally by Sir William Herschel in March 13, 1787. Uranus is so far away from the sun (2,869.6 million km) it only receives about 1/400 of the incident solar radiation that the earth receives. Voyager's ra-

TABLE XI. Composition of the Atmosphere of Saturn[a]

Constituent	Volume mixing ratio
H_2	0.94
He	0.06
CH_4	0.0045
C_2H_2	0.11 ppm
C_2H_6	4.8 ppm
CH_3C_2H[b]	No estimate
C_3H_8[b]	No estimate
CH_3D	0.23 ppm
PH_3	2 ppm

[a] From Strobel, D. F. (1985). In "The Photochemistry of Atmospheres: Earth, the Other Planets, and Comets," J. S. Levine, ed. Academic, Orlando, FL. Copyright 1985 Academic Press.
[b] Tentative identification.

dio signals took 2 hr and 45 min to reach the earth from Uranus.

As Voyager approached Uranus, its cameras indicated that Uranus did not exhibit the colorful and very turbulent cloud structure of Jupiter or the more subdued cloud banding and blending of Saturn. The very low contrast face of Uranus exhibited virtually no detail (see Fig. 14). The atmosphere of Uranus, like those of Jupiter and Saturn, is composed primarily of molecular hydrogen (about 85%) and helium (15 ± 5%). Methane is present in the upper atmosphere and is also frozen out in the form of ice in the cloud layer. The methane in the upper atmosphere selectively absorbs the red portion of the spectrum and gives Uranus its blue-green appearance. The volume percentage of methane may be as much as 2% deep in the atmosphere. Acetylene (C_2H_2) with a mixing ratio of about 2×10^{-7} was also detected in the atmosphere of Uranus. The temperature of the atmosphere was found to drop to a minimum of about 52 K (at the 100 mbar pressure level) before increasing to about 750 K in the extreme upper atmosphere.

Uranus has a ring system (as do Jupiter and Saturn). Two new rings (designated 1986 U1R and 1986 U2R) were discovered in Voyager 2 images of Uranus. The ring system of Uranus consists of 11 distinct rings that range in distance from about 37,000 to 51,000 from the center of Uranus.

Prior to the Uranus encounter, five satellites were known to be orbiting Uranus: Miranda

TABLE XII. Composition of the Atmosphere of Titan[a]

Constituent	Volume mixing ratio		
N_2	0.76–0.98[b]		
	Surface	Stratosphere	Thermosphere (3900 km)
CH_4	0.02–0.08	≤0.026	0.08 ± 0.03
Ar	<0.16		<0.06
Ne	<0.002		<0.01
CO	60 ppm		<0.05
H_2	0.002 ± 0.001		
C_2H_6		20 ppm	
C_3H_8		1–5 ppm	
C_2H_2		3 ppm	~0.0015 (3400 km)
C_2H_4		0.4 ppm	
HCN		0.2 ppm	<0.0005 (3500 km)
C_2N_2		0.01–0.1 ppm	
HC_3N		0.01–0.1 ppm	
C_4H_2		0.01–0.1 ppm	
CH_3C_2H		0.03 ppm	
CO_2		1–5 ppb	

[a] From Strobel, D. F. (1985). In "The Photochemistry of Atmospheres: Earth, the Other Planets, and Comets," J. S. Levine, ed. Academic, Orlando, FL. Copyright 1985 Academic Press.
[b] Preferred value.

FIG. 7. Ganymede, the largest satellite of Jupiter, as photographed by Voyager 2. The photograph shows a large, dark circular feature about 3200 km in diameter. The bright spots dotting the surface are relatively recent impact craters, while lighter circular areas may be older impact areas.

(distance = 129,000 km from the center of Uranus; diameter = 484 ± 10 km), Ariel (distance = 190,900 km; diameter = 1160 ± 10 km), Umbriel (distance = 266,000 km; diameter = 1190 ± 20 km), Titania (distance = 436,300 km; diameter = 1610 ± 10 km), and Oberon (distance = 583,400 km; diameter = 1550 ± 20 km). Ten new satellites were discovered on Voyager 2 images. All 10 satellites orbit Uranus within the orbit of Miranda (at distances that range from 49,700 to 86,000 km from the center of Uranus) and have diameters that range from about 40 to 80 km. Two of the newly discovered satellites are located within the ring system of Uranus and "shepherd" one of the rings.

After its encounter with Uranus in January 1986, Voyager 2 was targeted for an encounter with Neptune in September 1989. After its encounter with Neptune, Voyager 2 will join Voyager 1 and Pioneer 10 and 11 and escape the gravitational pull of the sun and head for the stars.

FIG. 8.   Callisto, the second largest satellite of Jupiter, as photographed by Voyager 1. Far more craters appear on the surface of Callisto than on the surface of Ganymede, suggesting that Callisto may be the oldest satellite of Jupiter.

FIG. 9. Europa, smallest of Jupiter's Galilean satellites, as photographed by Voyager 2. It is believed that Europa has a reasonable quantity of water in the form of a mantle of ice with interior slush, perhaps 100 km thick.

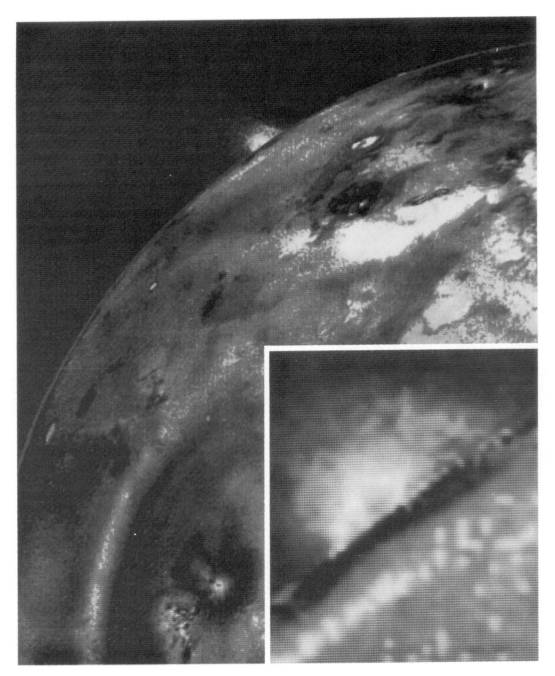

FIG. 10.   Io, satellite of Jupiter photographed by Voyager 1. An enormous volcanic eruption can be seen silhouetted against space over Io's bright limb. Solid volcanic material has been ejected up to an altitude of about 160 km.

FIG. 11. Saturn photographed by Voyager 1. Like Jupiter, the atmosphere of Saturn is massive and completely covered with clouds and aerosol and haze layers. The projected width of the rings at the center of the disk is 10,000 km, which provides a scale for estimating feature sizes on the image.

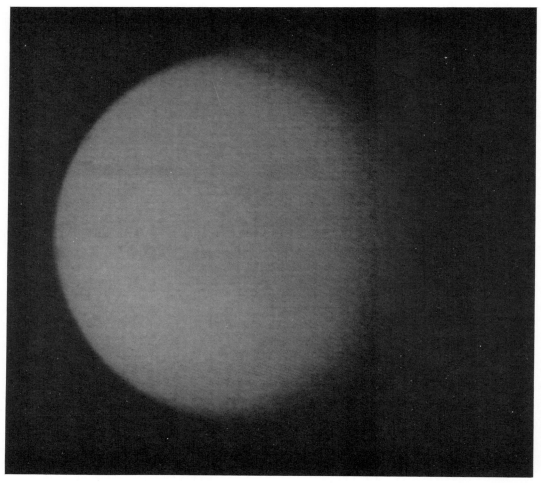

FIG. 12.   Titan, satellite of Saturn photographed by Voyager 1. Titan is the only satellite in the solar system with an appreciable atmosphere. The brownish-orange atmosphere of Titan contains clouds and haze and aerosol layers.

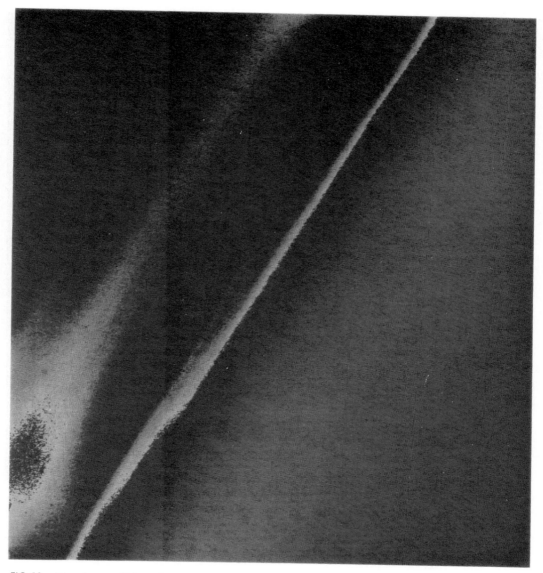

FIG. 13.   Haze layers of Titan photographed by Voyager 1. The upper level of the thick aerosol layer above the satellite's limb appears orange. The divisions in the haze occur at altitudes of 200, 375, and 500 km above the limb.

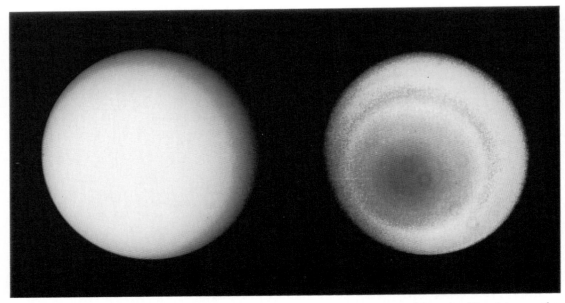

FIG. 14.   Two Voyager 2 views of Uranus—one in true color (left) and the other in false color. The picture on the left shows how Uranus would appear to human eyes from the vantage point of the spacecraft. The picture on the right uses contrast enhancement to bring out subtle details in the polar region of Uranus.

BIBLIOGRAPHY

Barth, C. A. (1985). In "The Photochemistry of Atmospheres: Earth, the Other Planets, and Comets," J. S. Levine, ed. Academic, Orlando, FL, pp. 337–392.

Cameron, A. G. W. (1968). In "Origin and Distribution of the Elements," L. H. Ahrens, ed., Pergamon, New York, pp. 125–143.

Levine, J. S., ed. (1985a). "The Photochemistry of Atmospheres: Earth, the Other Planets, and Comets." Academic, Orlando, FL.

Levine, J. S. (1985b). In "The Photochemistry of Atmospheres: Earth, the Other Planets, and Comets," J. S. Levine, ed. Academic, Orlando, FL, pp. 3–38.

Lewis, J. S., and Prinn, R. G. (1984). "Planets and Their Atmospheres: Origin and Evolution." Academic, New York.

NASA (1983). "Planetary Exploration Through Year 2000: A Core Program." Solar System Exploration Committee of the NASA Advisory Council, Washington, DC.

Prinn, R. G. (1985). In "The Photochemistry of Atmospheres: Earth, the Other Planets, and Comets," J. S. Levine, ed. Academic, Orlando, FL, pp. 281–336.

Strobel, D. F. (1985). In "The Photochemistry of Atmospheres: Earth, the Other Planets, and Comets," J. S. Levine, ed. Academic, Orlando, FL, pp. 393–434.

Walker, J. C. G. (1977). "Evolution of the Atmosphere." Macmillan, New York.

PLANETARY RADAR ASTRONOMY

Steven J. Ostro *Jet Propulsion Laboratory*[†]

GLOSSARY

Aliasing: Overlapping of echo at different frequencies or at different time delays.

Antenna gain: Ratio of an antenna's sensitivity in the direction it is pointed to its average sensitivity in all directions.

Circular polarization ratio: Ratio of echo power received in the same sense of circular polarization as transmitted (the SC sense) to that received in the opposite (OC) sense.

Doppler shift: Difference between the frequencies of the radar echo and the transmission, caused by the relative velocity of the target with respect to the radar.

Echo bandwidth: Dispersion in Doppler frequency of an echo, that is, the width of the echo power spectrum.

Ephemeris: Table of planetary positions as a function of time (plural: ephemerides).

Klystron: Vacuum-tube amplifier used in planetary radar transmitters.

Radar albedo: Ratio of a target's radar cross section in a specified polarization to its projected area; hence, a measure of the target's radar reflectivity.

Radar cross section: Most common measure of the intensity of a target's echo, equal to the projected area of the perfect metal sphere that would give the same echo power as the target if observed at the target's location.

Scattering law: Function giving the dependence of a surface element's radar cross section on viewing angle.

Synodic rotation period: Rotation period of a target as viewed by a moving observer, to be distinguished from the sidereal rotation period measured with respect to the fixed stars.

Time delay: Time between transmission of a radar signal and reception of the echo.

Planetary radar astronomy is the study of solar system entities (the moon, asteroids, and comets, as well as the major planets and their satellites and ring systems) by transmitting radio signals toward a target and receiving and analyzing the echoes. This field of research has primarily involved observations with earth-based radar telescopes but is now moving increasingly to include certain experiments with the transmitter and/or the receiver on board spacecraft orbiting or passing near planetary objects. Radar studies of the earth's surface, atmosphere, and ionosphere from spacecraft, aircraft, and the ground are not usually considered part of planetary radar astronomy. Radar studies of the sun involve such distinctly individual methodologies and physical considerations that solar radar astronomy is considered a field separate from planetary radar astronomy.

I. Introduction

A. SCIENTIFIC CONTEXT

Planetary radar astronomy is a field of science at the intersection of planetology, radio astronomy, and radar engineering. A radar telescope is essentially a radio telescope equipped with a high-power radio transmitter and specialized

[†] This research was conducted at the Jet Propulsion Laboratory, California Institute of Technology, under contract with the National Aeronautics and Space Administration.

electronic instrumentation designed to link transmitter, receiver, data-acquisition, and telescope-pointing components together in an integrated radar system. The principles underlying operation of this system are not fundamentally very different from those involved in radars used, for example, in marine and aircraft navigation, measurement of automobile speeds, and satellite surveillance. However, planetary radars must detect echoes from targets at interplanetary distances ($\sim 10^5-10^9$ km) and therefore are the largest and most powerful radar systems in existence. [See RADAR; RADIO ASTRONOMY, PLANETARY.]

The advantages of radar observations in astronomy stem from the high degree of control exercised by the observer on the signal transmitted to illuminate the target. Whereas virtually every other astronomical technique relies on passive measurement of reflected sunlight or naturally emitted radiation, the radar astronomer controls all the properties of the illumination including its intensity, direction, polarization, and time/frequency structure.

The properties of the transmitted waveform are selected to achieve particular scientific objectives. By comparing the properties of the echo to the very well-known properties of the transmission, one can deduce the target's properties. Hence, the observer is intimately involved in an active astronomical observation and, in a very real sense, performs a controlled laboratory experiment on the planetary target.

Computer analysis of the echo enables the target to be resolved spatially in a manner independent of its apparent angular extent, thereby bestowing a considerable advantage on radar over optical techniques in the study of asteroids, which appear as point sources through ground-based optical telescopes. Furthermore, by virtue of the centimeter–meter wavelengths employed, radar is sensitive to scales of surface structure many orders of magnitude larger than those probed in visible or infrared regions of the spectrum. Radar is also unique in its ability to "see through" the dense clouds that enshroud Venus and the glowing gaseous atmosphere or coma that conceals the nucleus of a comet. Because of its unique capabilities radar astronomy has made essential contributions to planetary exploration for a quarter of a century.

B. History

Radar technology was developed rapidly to meet military needs during World War II. In 1946, soon after the war, groups in the United States and Hungary obtained echoes from the moon, giving birth to planetary radar astronomy. These early efforts were motivated primarily by interest in electromagnetic propagation through the ionosphere and the possibility for using the moon as a relay for radio communication.

During the next two decades the development of nuclear weaponry and the need for ballistic missile warning systems prompted enormous improvements in radar capabilities. This period also saw rapid growth in radio astronomy and the construction of huge radio telescopes. In 1957, the Soviet Union launched Sputnik and with it the Space Age; and in 1958, with the formation by the U.S. congress of the National Aeronautics and Space Administration (NASA), a great deal of scientific attention turned to the moon and to planetary exploration in general. During the ensuing years exhaustive radar investigations of the moon were conducted at wavelengths from 0.9 cm to 20 m, and the results generated theories of radar scattering from natural surfaces that still see wide application.

By 1963, improvements in the sensitivity of planetary radars had permitted the initial detections of echoes from the terrestrial planets (Venus, Mercury, and Mars). During this period radar investigations provided the first accurate determinations of the rotations of Venus and Mercury and the earliest indications for the extreme geologic diversity of Mars. Radar images of Venus have revealed small portions of that planet's surface at increasingly fine resolution since the late 1960s; and in 1979, the Pioneer Venus Spacecraft Radar Experiment gave us our first look at Venus's global distributions of topography, radar reflectivity, and surface slopes.

The first echoes from a near-earth asteroid (1566 Icarus) were detected in 1968; it was nearly another decade before the first radar detection of a mainbelt asteroid (1 Ceres in 1977), followed in 1980 by the first detection of echoes from a comet (Encke).

During 1972 and 1973, detection of 13-cm wavelength (λ13 cm) radar echoes from Saturn's rings shattered prevailing notions that typical ring particles were 0.1–1.0 mm in size; the fact that decimeter-scale radio waves are backscattered efficiently requires that a large fraction of the particles be larger than a centimeter. Observations by the Voyager spacecraft have confirmed this fact and further suggest that particle sizes extend to at least 10 m.

TABLE I. Radar-Detected Planetary Targets

Year of first detection	Terrestrial objects	Jupiter's satellites	Ring system	Mainbelt asteroids	Near-earth asteroids	Comets
1946	Moon					
1961	Venus					
1962	Mercury					
1963	Mars					
1968					1566 Icarus	
1972					1685 Toro	
1973			Saturn's rings			
1974		Ganymede				
1975		Callisto			433 Eros	
		Europa				
1976	Io				1580 Betulia	
1977				1 Ceres		
1979				4 Vesta		
1980				7 Iris	1862 Apollo	Encke
				16 Psyche		
1981				97 Klotho	1915 Quetzalcoatl	
				8 Flora	2100 Ra–Shalom	
1982				2 Pallas		Grigg–Skjellerup
				12 Victoria		
				19 Fortuna		
				46 Hestia		
1983				5 Astraea	1620 Geographos	IRAS–Araki–Alcock
				139 Juewa	2201 Oljato	Sugano–Saigusa–Fujikawa
				356 Liguria		
				80 Sappho		
				694 Ekard		
1984				9 Metis	2101 Adonis	
				554 Peraga		
				144 Vibilia		
1985				6 Hebe	1627 Ivar	
				41 Daphne	1036 Ganymed	
				21 Lutetia	1866 Sisyphus	
				33 Polyhymnia		
				84 Klio		
				192 Nausikaa		
				230 Athamantis		
				216 Kleopatra		
				18 Melpomene		

In the mid-1970s, echoes from Jupiter's Galilean satellites Europa, Ganymede, and Callisto revealed the manner in which these icy moons backscatter circularly polarized waves to be extraordinarily strange and totally outside the realm of previous radar experience. In 1984, it was discovered that the echo polarization properties of the near-earth asteroid 2101 Adonis resemble those of Callisto more than those of any other observed asteroid or comet. Our theoretical understanding of the anomalous radar behavior is still not complete, but it has improved substantially during the past decade.

By 1985, the list of radar-detected planetary objects included four comets, 13 near-earth asteroids, and 27 mainbelt asteroids (Table I). Comets and asteroids comprise an enormous, diverse population of bodies containing important clues to the origin and evolution of the solar system, and radar observations are now providing new information about these objects' sizes, shapes, spin vectors, surface structure, and composition (e.g., metal concentration). [See PRIMITIVE SOLAR SYSTEM OBJECTS: COMETS AND ASTEROIDS.]

II. Techniques and Instrumentation

A. ECHO DETECTABILITY

How close must a planetary target be for its radar echo to be detectable? For a given transmitted power P_T and antenna gain G, the power flux at distance R from the radar is $P_T G/4\pi R^2$. We define the target's radar cross section σ as 4π times the backscattered power per steradian per unit incident flux at the target. Then, letting λ be the radar wavelength and defining the antenna's effective aperture as $A_e = G\lambda^2/4\pi$, we find the received power to be

$$P_R = \frac{P_T G A_e \sigma}{(4\pi)^2 R^4} \quad (1)$$

This power might be much less than the receiver noise power $P_N = kT_S \, \Delta f$, where k is Boltzmann's constant, T_S the receiver system temperature, and Δf the frequency resolution of the data. However, the mean level of P_N constitutes a background that can be determined and removed, so P_R is detectable as long as it is at least several times larger than the standard deviation of the random fluctuations in P_N. These fluctuations can be shown to have a distribution that for usual values of Δf and the integration time Δt is nearly Gaussian with standard deviation $\Delta P_N = P_N/(\Delta f \, \Delta t)^{1/2}$. The highest signal-to-noise ratio (SNR), $P_R/\Delta P_N$, is achieved for a frequency resolution equal to the intrinsic bandwidth of the echo. That bandwidth is proportional to $D/\lambda P$, where D is the target's diameter and P the target's rotation period, so let us assume that $\Delta f \sim D/\lambda P$. By writing $\sigma = \hat{\sigma}\pi D^2/4$, where the radar albedo $\hat{\sigma}$ is a measure of the target's radar reflectivity, we arrive at the following expression for the echo's SNR:

$$\text{SNR} \sim (\text{system factor})(\text{target factor}) \, (\Delta t)^{1/2} \quad (2)$$

where

$$\text{system factor} \sim \frac{P_T A_e^2}{\lambda^{3/2} T_S} \quad (3)$$

$$\sim \frac{P_T \, G^2 \lambda^{5/2}}{T_S}$$

and

$$\text{target factor} \sim \frac{\hat{\sigma} D^{3/2} P^{1/2}}{R^4} \quad (4)$$

The inverse-fourth-power dependence of SNR on target distance is a severe limitation in ground-based observations but can be overcome by constructing very powerful radar systems.

B. Radar Systems

Two active planetary radar facilities exist: the National Astronomy and Ionosphere Center's Arecibo Observatory in Puerto Rico and the Jet Propulsion Laboratory's Goldstone Solar System Radar in California. The Arecibo telescope (Fig. 1) consists of a 305 m diameter, fixed reflector whose surface is a section of a 265 m radius sphere. Movable line feeds designed to correct for spherical aberration are suspended from a triangular platform ~130 m above the

reflector and can be aimed toward various positions on the reflector, enabling the telescope to point to within about 20° of the overhead direction (declination 18.3°). The Goldstone main antenna is a fully steerable, 65 m parabolic reflector with horn feeds (Fig. 2). Radar wavelengths are 13 and 70 cm for Arecibo and 3.5 and 13 cm for Goldstone; with each instrument, greater sensitivity is achievable with the shorter wavelength. Order-of-magnitude values of optimum system characteristics for each radar are $P_T \approx$ 400 kw, $G \approx 10^{7.1}$, and $T_S \approx 25$ K. Arecibo is about an order of magnitude more sensitive at $\lambda 13$ cm than Goldstone is at $\lambda 3.5$ cm. However, Goldstone has access to the entire sky north of declination $-50°$ and can track targets continuously for longer periods.

Figure 3 is a simplified block diagram of a planetary radar system. A waveguide switch is used to connect the antenna to the transmitter or to the receiver. In most observations one transmits for a duration near the round-trip propagation time to the target (that is, until the echo from the beginning of the transmission is about to arrive) and then receives for a similar duration.

The heart of the planetary radar transmitter is one or more klystron vacuum-tube amplifiers. In these tubes, magnets focus electrons falling through a potential drop of some 60 kV and modulate their velocities, and hence the electron density and energy flux, at radio frequencies (RF). Internal resonant cavities enhance this modulation, and about half of the nearly one megawatt of input dc power is converted to RF power, sent out through a waveguide to the antenna feed, and radiated toward the target. The other half of the input power is waste heat that must be transported away from the klystron by cooling water. The impact of the electrons on the collector anode generates dangerous X-rays that must be contained by heavy metal shielding surrounding the tube, a requirement that further boosts the weight, complexity, and cost of the klystron.

In the receiving system, the maser-amplified echo signal is converted in a superheterodyne mixer from RF frequencies (e.g., ~2380 MHz for Arecibo at $\lambda 13$ cm) down to intermediate frequencies (IF, e.g., ~30 MHz), for which transmission line losses are small, and passed from the proximity of the antenna feed to a remote control room containing additional stages of signal-processing equipment, computers, and digital tape recorders. The signal is converted to very low (baseband) frequencies and is filtered

FIG. 1. (a) Arecibo Observatory's radio/radar telescope in Puerto Rico. The diameter of the spherical reflector is 305 m. (b) Close-up view of the structure suspended above the reflector. Antenna feeds extend from the bottom of the two "carriage houses" that contain receiver and transmitter equipment.

FIG. 2.   The 65-m Goldstone Solar System Radar antenna in California.

and amplified, and samples of the signal's voltage are converted from analog to digital form. The nature of the final processing prior to recording data on magnetic tape depends on the nature of the radar experiment and particularly on the time/frequency structure of the transmitted waveform.

C. ECHO TIME-DELAY AND DOPPLER SHIFT

The time between transmission of a radar signal and reception of the echo is called the echo's round-trip time delay τ and is of order $2R/c$, where c is the speed of light. Since planetary targets are not points, even an infinitesimally short transmitted pulse would be dispersed in time delay, and the total extent $\Delta\tau_{TARGET}$ of the distribution $\sigma(\tau)$ of echo power (in units of radar cross section) would be D/c for a sphere of diameter D and in general would depend on the target's size and shape.

The translational motion of the target with respect to the radar introduces a Doppler shift ν in the frequency of the transmission. Both the time delay and the Doppler shift of the echo can be predicted in advance from the target's ephemeris, which is calculated using the geodetic position of the radar and the orbital elements of the earth and the target. The predicted Doppler shift can be removed electronically by continuously tuning the local oscillator used for RF to IF frequency conversion (Fig. 3). The predicted Doppler must be accurate enough to avoid smearing out the echo in frequency, and this requirement places stringent demands on the quality of the observing ephemeris.

Because different parts of the rotating target have different velocities relative to the radar, the echo is dispersed in Doppler frequency as well as in time delay. The basic strategy of any radar experiment always involves measurement of some characteristic(s) of the function $\sigma(\tau, \nu)$, perhaps as a function of time and perhaps using more than one combination of transmitted and received polarizations. Ideally, one would like to obtain $\sigma(\tau, \nu)$ with very fine resolution, sampling that function within intervals whose dimensions $\Delta\tau \times \Delta\nu$ are minute compared with the echo dispersions $\Delta\tau_{TARGET}$ and $\Delta\nu_{TARGET}$. Unfortunately, SNR is proportional to $(\Delta\nu)^{1/2}$, so

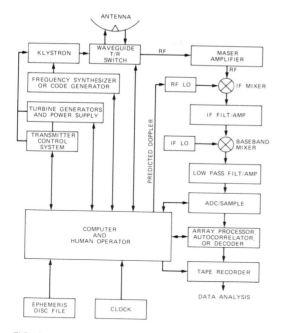

FIG. 3.   Block diagram of a planetary radar system. RF LO and IF LO denote radio frequency and intermediate frequency local oscillators, and ADC denotes anolog-to-digital converter.

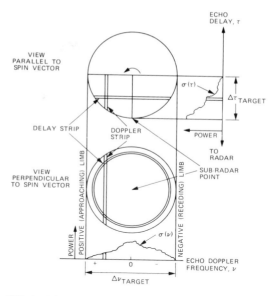

FIG. 4.   Time-delay and Doppler-frequency resolution of the radar echo from a rotating spherical target.

one's ability to resolve $\sigma(\tau, \nu)$ is necessarily limited by the available echo strength. Furthermore, an intrinsic upper bound on the product $\Delta\tau\,\Delta\nu$ forces a trade-off between delay resolution and Doppler resolution. Under these constraints, many planetary radar experiments employ waveforms aimed at providing estimates of one of the marginal distributions, $\sigma(\tau)$ or $\sigma(\nu)$. Figure 4 shows the geometry of delay-resolution cells and Doppler-resolution cells for a spherical target and sketches the relation between these cells and $\sigma(\nu)$ and $\sigma(\tau)$. Delay–Doppler measurements are explored further below.

D. Radar Waveforms

In the simplest radar experiment, the transmitted signal is unmodulated (continuous wave or cw) and highly monochromatic. Analysis of the received signal comprises Fourier transformation of a series of time samples and yields an estimate of the echo power spectrum $\sigma(\nu)$, but it contains no information about the distance to the target or $\sigma(\tau)$. To avoid aliasing, the sampling rate must be at least as large as the bandwidth of the low-pass filter (Fig. 3), and usually

is comparable to or larger than the echo's intrinsic dispersion $\Delta\nu_{\text{TARGET}}$ from Doppler broadening.

Fast-Fourier transform (FFT) algorithms, implemented via software or hard-wired in an array processor, greatly speed the calculation of discrete spectra from time series and are ubiquitous in radar astronomy. In a single FFT operation, a string of N time samples taken at intervals of Δt sec is transformed into a string of N spectral elements with frequency resolution $\Delta f = 1/N\,\Delta t$. Many thousands of FFTs may be needed to reduce the data from a single transmit/receive cycle to echo spectra, and data reduction times can drastically exceed data acquisition times even with the most efficient FFT algorithms. (Needless to say, the high-speed computer is utterly essential in radar astronomy.) Most planetary radar targets are sufficiently narrowband ($\Delta\nu_{\text{TARGET}} \lesssim$ a few tens of kilohertz) for power spectra to be computed and accumulated in an array processor (Fig. 3) and recorded directly on magnetic tape at convenient intervals. In some situations it is desirable to record time samples on tape and Fourier analyze the data later, perhaps using FFTs of different lengths to obtain spectra at a variety of frequency resolutions. For wideband echoes (e.g., $\sim 10^6$ Hz for Saturn's rings) the sampling-rate requirements are met most readily by passing the signal through an autocorrelator, recording autocorrelation functions, and then applying FFTs to extract spectra.

To obtain delay resolution one must apply some sort of time modulation to the transmitted waveform. For example, a short-duration pulse of cw signal lasting 1 μsec would provide delay resolution of 150 m. However, the echo would have to compete with the noise power in a bandwidth of order 1 MHz, so the echo power from many consecutive pulses would probably have to be summed to yield a detection. One would not want these pulses to be too close together, however, or there would be more than one pulse incident on the target at once and interpretation of echoes would be insufferably ambiguous. Thus, one arranges the interpulse period t_{IPP} to exceed the target's intrinsic delay dispersion $\Delta\tau_{TARGET}$, ensuring that the echo consists of successive, nonoverlapping replicas of $\sigma(\tau)$ that are separated from each other by t_{IPP}. To generate this pulsed cw waveform, the transmitter is switched on and off while the frequency synthesizer (Fig. 3) maintains phase coherence from pulse to pulse. Then Fourier transformation of time samples, taken at the same position within each of N successive replicas of $\sigma(\tau)$, yields the power spectrum of echo from a certain delay resolution cell on the target. This spectrum has

an unaliased bandwidth of $1/t_{IPP}$ and a frequency resolution of $1/Nt_{IPP}$. Repeating this process for a different position within each replica of $\sigma(\tau)$ yields the power spectrum for echo from a different delay resolution cell, and in this manner one obtains $\sigma(\tau, \nu)$.

In practice, instead of pulsing the transmitter, one usually codes a cw signal with a sequence of 180° phase reversals and cross-correlates the echo with a representation of the code (e.g., using the decoder in Fig. 3), thereby synthesizing a pulse train with the desired values of Δt and t_{IPP}. With this approach, one optimizes SNR by transmitting power continuously and extends the klystron's lifetime by avoiding on–off switching intervals comparable to the tube's internal thermal time constants.

A limitation of coherent-pulsed or phase-coded cw waveforms follows from combining the requirement that there never be more than one echo received from the target at any instant (i.e., that $t_{IPP} > \Delta\tau_{TARGET}$) with the antialiasing frequency requirement that the rate $1/t_{IPP}$ at which echo from a given delay resolution cell is sampled be no less than the target bandwidth $\Delta\nu_{TARGET}$. Therefore, a target must satisfy $\Delta\tau_{TARGET} \Delta\nu_{TARGET} \leq 1$ or it is overspread (Table II) and cannot be investigated completely and simultaneously in delay and Doppler without

TABLE II. Characteristics of Selected Planetary Radar Targets[a]

Target	Minimum echo delay[b] (min)	Radar cross section, σ_{OC} (km²)	Radar albedo $\hat{\sigma}_{OC}$	Circular polarization ratio, μ_C	Maximum dispersions[c] Delay (ms)	Maximum dispersions[c] Doppler (Hz)	Product
Moon	0.04	6.6×10^5	0.07	0.1	12	60	0.7
Mercury	9.1	1.1×10^6	0.06	0.1	16	110	2
Venus	4.5	1.3×10^7	0.11	0.1	40	110	4
Mars	6.2	2.9×10^6	0.08	0.3	23	7600	200
1 Ceres	27	3.3×10^4	0.05	0.0	3	3100	9
2 Pallas	25	2.1×10^4	0.09	0.0	2	2000	4
12 Victoria	17	2.1×10^3	0.15	0.1	0.5	590	3
16 Psyche	28	1.4×10^4	0.29	0.1	0.8	2200	2
1685 Toro	2.3	1.7	0.1	0.2	0.02	14	10^{-4}
1862 Apollo	0.9	0.2	0.1	0.4	0.01	16	10^{-3}
2100 Ra–Shalom	3.0	1.0	0.1	0.2	0.01	5	10^{-4}
2201 Adonis	1.5	0.02	?	1.0	?	2	?
Comet IRAS–Araki–Alcock	0.5	2.2	?	0.1	?	300	?
Io	73	1.9×10^6	0.2	?	12	2400	30
Europa	73	7.7×10^6	1.0	1.6	10	1000	10
Ganymede	73	1.3×10^7	0.6	1.6	18	850	15
Callisto	73	5.3×10^6	0.3	1.2	16	330	5
Saturn's rings	134	10^8–10^9	0.7	0.5	1600	6×10^5	10^6

[a] Typical 13-cm-wavelength values. Question marks denote absence of radar data (μ_C) or of prior information about target dimensions.

[b] For asteroids this is the minimum delay for observations to date.

[c] Doppler dispersion for transmitter frequency of 2380 MHz ($\lambda = 12.6$ cm). The product of the dispersions in delay and Doppler is the overspread factor at 2380 MHz.

aliasing, at least with the waveforms discussed so far. Various degrees of aliasing may be acceptable for overspread factors $\lesssim 10$, depending on the precise experimental objectives and the exact properties of the echo. One can deal with extremely overspread targets (e.g., Saturn's rings with an overspread factor $\sim 10^6$ at $\lambda 13$ cm) by using frequency-stepped and frequency-swept waveforms, but only at the expense of more complex data acquisition and reduction procedures. Virtually all modern ground-based radar observations of planetary targets employ cw or phase-coded cw waveforms.

III. Radar Measurements and Target Properties

A. ALBEDO AND POLARIZATION RATIO

A primary goal of the initial radar investigation of any planetary target is estimation of the target's radar cross section σ and its normalized radar cross section or radar albedo $\hat{\sigma} \equiv \sigma/A_p$, where A_p is the target's geometric projected area. Since the radar astronomer selects the transmitted and received polarizations, any estimate of σ or $\hat{\sigma}$ must be identified accordingly. The most common approach is to transmit a circularly polarized wave and to use separate receiving systems for simultaneous reception of the same sense of circular polarization as transmitted (i.e., the SC sense) and the opposite (OC) sense. The handedness of a circularly polarized wave is reversed on normal reflection from a smooth dielectric interface, so the OC sense dominates echoes from targets that look smooth at the radar wavelength. In this context a surface with minimum radius of curvature very much larger than λ would look smooth. SC echo can arise from single scattering from rough surfaces, multiple scattering from smooth surfaces, or certain subsurface refraction effects. The circular polarization ratio $\mu_C \equiv \sigma_{SC}/\sigma_{OC}$ is thus a useful measure of near-surface roughness.

When linear polarizations are used, it is convenient to define the ratio $\mu_L = \sigma_{OL}/\sigma_{SL}$, which would be close to zero for normal reflection from a smooth dielectric interface. For all radar-detected planetary targets, $\mu_L < 1$ and $\mu_L < \mu_C$. Although the OC radar albedo $\hat{\sigma}_{OC}$ is the most widely used gauge of radar reflectivity, some radar measurements are reported in terms of the geometric albedo, equal to $(\hat{\sigma}_{OC} + \hat{\sigma}_{SC})/4 = (\hat{\sigma}_{SL} + \hat{\sigma}_{OL})/4$. A perfectly smooth metallic sphere would have $\hat{\sigma}_{OC} = \hat{\sigma}_{SL} = 1$, a geometric albedo of 0.25, and $\mu_C = \mu_L = 0$.

If μ_C is close to zero (as for the planet Venus and the mainbelt asteroid 2 Pallas), its physical interpretation is unique, because the surface must be smooth at all scales within about an order of magnitude of λ, and there can be no subsurface structure at those scales within several $1/e$ power absorption lengths L of the surface proper. In this special situation, we can interpret the radar albedo as the product $g\rho$ where ρ is the Fresnel power-reflection coefficient at normal incidence, and the backscatter gain g depends on target shape, the distribution of surface slopes with respect to that shape, and target orientation. For most applications to date, g is $\lesssim 10\%$ larger than unity, so the radar albedo provides a reasonable first approximation to ρ. Both ρ and L depend on very interesting characteristics of the surface material, including bulk density, porosity, particle size distribution, and metal abundance.

If μ_C is $\gtrsim 0.3$ (e.g., Mars and some near-earth asteroids), then much of the echo arises from some backscattering mechanism other than single coherent reflections from large, smooth surface elements. Possibilities include multiple scattering from buried rocks or the interiors of concave surface features such as craters, or reflections from very jagged surfaces with radii of curvature much less than a wavelength. Most planetary targets have values of $\mu_C \lesssim 0.2$ at decimeter wavelengths, so their surfaces are dominated by a component that is smooth at centimeter-to-meter scales.

The observables $\hat{\sigma}_{OC}$ and μ_C are disk-integrated quantities derived from integrals of $\sigma(\nu)$ or $\sigma(\tau)$ in specific polarizations. Later, we shall see how their physical interpretation profits from knowledge of the functional forms of $\sigma(\nu)$ and $\sigma(\tau)$.

B. DYNAMICAL PROPERTIES FROM MEASUREMENT OF DELAY AND DOPPLER

Consider radar observation of a point target a distance R from the radar. As noted earlier, the round-trip time delay between transmission of a pulse toward the target and reception of the echo would be $\tau = 2R/c$. It is possible to measure time delays to within 10^{-6} sec. Actual delays range from $2\frac{1}{2}$ sec for the moon to $2\frac{1}{2}$ hr for Saturn's rings. For a typical target distance ~ 1 AU, the time delay is ~ 1000 sec and can be measured within a fractional timing uncertainty of 10^{-9}. Since the fractional error in our knowledge of the speed of light is $\sim 10^{-7}$, we know the light-second equivalent of a radar-determined distance better than the distance in, say, meters. Consequently, results of very precise planetary radar range measurements are generally reported in units of time delay instead of distance, which is a more poorly known, derived quantity.

If the target is in motion and has a line-of-sight component of velocity toward the radar of v_{LOS}, the target sees a frequency that, to first order in v_{LOS}/c, equals $f_{TX} + (v_{LOS}/c)f_{TX}$, where f_{TX} is the transmitter frequency. The target reradiates the Doppler-shifted signal, and the radar receives echo whose frequency is, again to first order, given by

$$f_{TX} + 2(v_{LOS}/c)f_{TX}$$

That is, the total Doppler shift in the received echo is

$$2v_{LOS}f_{TX}/c = v_{LOS}/(\lambda/2)$$

so a 1-Hz Doppler shift corresponds to a velocity of half a wavelength per second (e.g., 6.5 cm sec^{-1} for λ13 cm). It is not difficult to measure echo frequencies to within 0.01 Hz, so v_{LOS} can be estimated with a precision ~1 mm sec^{-1}. Actual values of v_{LOS} for planetary radar targets can be as large as several tens of kilometers per second, so radar velocity measurements have fractional errors as low as ~10^{-8}. At this level, the second-order (special relativistic) contribution to the Doppler shift becomes measurable; in fact, planetary radar observations have provided the initial experimental verification of the second-order term.

By virtue of their high precision, radar measurements of time delay and Doppler frequency are very useful in refining our knowledge of various dynamical quantities. The first delay-resolved radar observations of Venus during 1961 and 1962 yielded an estimate of the light-second equivalent of the astronomical unit (AU) that was accurate to one part in 10^6, constituting a thousand-fold improvement in the best results achieved with optical observations alone. Subsequent radar observations provided additional refinements of nearly two more orders of magnitude. In addition to determining the scale of the solar system precisely, these observations greatly improved our knowledge of the orbits of earth, Venus, Mercury, and Mars, and were essential for the success of the first interplanetary missions. Radar observations still contribute to maintaining the accuracy of planetary ephemerides for objects in the inner solar system and have played an important role in dynamical studies of Jupiter's Galilean satellites.

Precise interplanetary time-delay measurements have also been used to test Einstein's theory of general relativity. Radar observations verify the theory's prediction that for radar waves passing near the sun, echo time delays are increased because of the distortion of space by the sun's gravity. The extra delay would be ~100 μsec if the angular separation of the target from the sun were several degrees. (The sun's angular diameter is about half a degree.)

Since planets are not point targets, their echoes are dispersed in delay and Doppler, and the refinement of dynamical quantities and the testing of physical theories are tightly coupled to estimation of the mean radii, the topographic relief, and the radar scattering behavior of the targets. The key to this entire process is resolution of the distributions of echo power in delay and Doppler. In the next section we shall consider inferences about a target's dimensions and spin vector from measurements of the dispersions ($\Delta\tau_{TARGET}$, $\Delta\nu_{TARGET}$) of the echo in delay and Doppler. Then we shall examine the physical information contained in the functional forms of the distributions $\sigma(\tau)$, $\sigma(\nu)$, and $\sigma(\tau, \nu)$.

C. DISPERSION OF ECHO POWER IN DELAY AND DOPPLER

Each backscattering element on a target's surface returns echo with a certain time delay and Doppler frequency (Fig. 4). Since parallax effects and the curvature of the incident wave front are negligible for most ground-based observations (but not necessarily for observations with spacecraft), contours of constant delay are intersections of the surface with planes perpendicular to the line of sight. The point on the surface with the shortest echo time delay is called the subradar point; the longest delays generally correspond to echoes from the planetary limbs. As noted already, the difference between these extreme delays is called the dispersion $\Delta\tau_{TARGET}$ in $\sigma(\tau)$, or simply the delay depth of the target.

If the target appears to be rotating, the echo is dispersed in Doppler frequency. For example, if the radar has an equatorial view of a spherical target with diameter D and apparent rotation period P, the difference between the line-of-sight velocities of points on the equator at the approaching and receding limbs is $2\pi D/P$. Thus the dispersion of $\sigma(\nu)$ is $\Delta\nu_{TARGET} = 4\pi D/\lambda P$. This quantity is called the bandwidth B of the echo power spectrum. If the view is not equatorial, the bandwidth is simply $(4\pi D \sin \alpha)/\lambda P$, where the aspect angle α is the acute angle between the spin vector and the line of sight. Thus, a radar bandwidth measurement furnishes a joint constraint on the target's size, rotation period, and pole direction.

In principle, echo bandwidth measurements obtained for a sufficiently wide variety of line-of-sight directions can yield all three scalar coordinates of the target's intrinsic (i.e., sidereal) spin vector $\boldsymbol{\omega}_s$. This capability follows from the fact that the apparent spin vector $\boldsymbol{\omega}$ is the vector sum of $\boldsymbol{\omega}_s$ and the contribution ($\boldsymbol{\omega}_0 = \dot{\mathbf{e}} \times \mathbf{e}$, where the unit vector $\mathbf{e}$ points from the target to the radar) arising from the changing position of the radar on the celestial sphere as seen from the target's center of mass. Variations in $\mathbf{e}$, $\dot{\mathbf{e}}$, and hence $\boldsymbol{\omega}_0$, all of which are known, lead to measurement of different values of $\boldsymbol{\omega} = \boldsymbol{\omega}_s + \boldsymbol{\omega}_0$, permitting unique determination of $\boldsymbol{\omega}_s$.

These principles were applied in the early 1960s to yield the first accurate determination of the rotations of Venus and Mercury (Fig. 5). Venus's rotation is retrograde with a 243-d sidereal period. The period is close to the value (243.16 d) characterizing a resonance with the relative orbits of earth and Venus, wherein Venus would appear from earth to rotate exactly four times between successive inferior conjunctions with the sun. However, two decades of Venus observations yield a refined value (243.01 ± 0.03 d) for the period that seems to establish nonresonance rotation. To date, a satisfactory explanation for Venus's curious spin state is lacking.

For Mercury, long imagined on the basis of optical observations to rotate once per 88-day revolution around the sun, radar bandwidth measurements (Fig. 5) demonstrated direct rotation with a period of 59 days, equal to $\tfrac{2}{3}$ of the orbital period. This spin–orbit coupling is such that during two Mercury years, the planet rotates three times with respect to the stars but only once with respect to the sun, so a Mercury-bound observer would experience alternating years of daylight and darkness.

What if the target is not a sphere but is irregular and nonconvex? In this situation, which is most applicable to small asteroids and cometary nuclei, the relationship between the echo power spectrum and the derivable information about the target's dimensions is shown in Fig. 6. We must interpret D as the sum of the distances r_+ and r_- from the plane ψ_0 containing the line of sight and the spin vector to the surface elements with the greatest positive (approaching) and negative (receding) line-of-sight velocities. In differ-

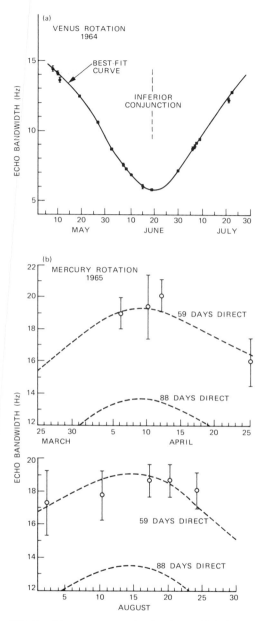

FIG. 5. Measurements of echo bandwidth (i.e., the dispersion of echo power in Doppler frequency) used to determine the rotations of Venus and Mercury.

ent words, if the planes ψ_+ and ψ_- are defined as being parallel to ψ_0 and tangent to the target's approaching and receding limbs, then ψ_+ and ψ_- are at distances r_+ and r_- from ψ_0. Letting f_0, f_+, and f_- be the frequencies of echoes from portions of the target intersecting ψ_0, ψ_+, and ψ_-, we have $B = f_+ - f_-$. Any constant-Doppler contour lies in a plane parallel to ψ_0.

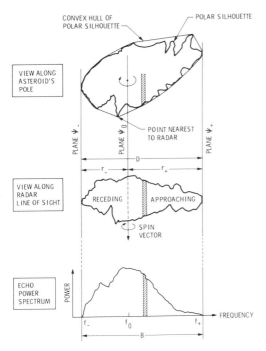

FIG. 6. Geometric relations between an irregular, nonconvex rotating asteroid and its echo power spectrum. The plane ψ_0 contains the asteroid's spin vector and the asteroid–radar line. The cross-hatched strip of power in the spectrum corresponds to echoes from the cross-hatched strip on the asteroid.

It is useful to imagine looking along the target's pole at the target's projected shape (i.e., its polar silhouette). D is simply the width, or breadth, of this silhouette (or, equivalently, of

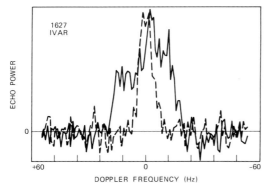

FIG. 7. Two λ13-cm, OC radar echo spectra obtained for the near-earth asteroid 1627 Ivar near rotational phases ~90° apart. The asteroid evidently is two to three times longer than it is wide.

the silhouette's convex envelope or hull H_p) measured normal to the line of sight (Fig. 6). In general, r_+ and r_- are periodic functions of rotation phase ϕ and depend on the shape of H_p as well as on the target's mass distribution (i.e., the point about which H_p rotates). If the radar data sample all rotational phases modulo 180°, then in principle one can determine $f_+(\phi)$ and $f_-(\phi)$ completely and can recover H_p. For many small, near-earth asteroids, pronounced variations in $B(\phi)$ reveal highly noncircular polar silhouettes. For example, echo spectra obtained for the ~7-km object 1627 Ivar (Fig. 7) indicate a polar silhouette two to three times longer than it is wide.

D. Delay and Doppler Distributions of Echo Power

1. Angular Scattering Law

The functional forms of the distributions $\sigma(\tau)$ and $\sigma(\nu)$ contain information about the radar-scattering process and about the structural characteristics of the target's surface. Suppose the target is a large, smooth, spherical planet. Then echoes from the subradar region (near the center of the visible disk; see Fig. 4), where the surface elements are nearly perpendicular to the line of sight, would be much stronger than those from the limb regions (near the disk's periphery). This effect is seen visually when one shines a flashlight on a smooth, shiny ball: A bright glint appears where the geometry is right for backscattering. If the ball is roughened, the glint is spread out over a wider area, and in the case of extreme roughness, the scattering is described as diffuse rather than specular.

For a specular target $\sigma(\tau)$ would have a steep leading edge followed by a rapid drop. The power spectrum $\sigma(\nu)$ would be sharply peaked at central frequencies, falling off rapidly toward the spectral edges. If, instead, the spectrum were very broad, severe roughness at some scale(s) $\gtrsim \lambda$ would be indicated. In this case, knowledge of the echo's polarization properties would help to ascertain the particular roughness scale(s) responsible for the absence of the sharply peaked spectral signature of specular scattering.

By inverting the delay or Doppler distribution of echo power one can estimate the target's average angular scattering law, $\sigma_0(\theta) \equiv d\sigma/dA$, where dA is an element of surface area and θ the incidence angle between the line of sight and the normal to dA. For the echo's polarized (i.e., OC or SL) components, $\sigma_0(\theta)$ can be related to statistics describing the probability distribution for the slopes of surface elements. Examples of scattering laws applied in planetary radar astronomy are the Hagfors law,

$$\sigma_0(\theta) \sim s_0^{-2}(\cos^4\theta + s_0^{-2}\sin^2\theta)^{-3/2} \quad (5)$$

and the Gaussian law,

$$\sigma_0(\theta) \sim \frac{s_0^{-2}\exp(-s_0^{-2}\tan^2\theta)}{\cos^4\theta} \quad (6)$$

where s_0 is the rms slope in radians. In the following paragraphs, we shall explore the diversity of radar signatures encountered for solar system targets.

2. Radar Signatures of the Moon and the Inner Planets

Echoes from the moon, Mercury, Venus, and Mars are characterized by sharply peaked OC echo spectra (Fig. 8). Although these objects are collectively referred to as quasi-specular radar targets, their echoes also contain a diffusely scattered component and have circular polariza-

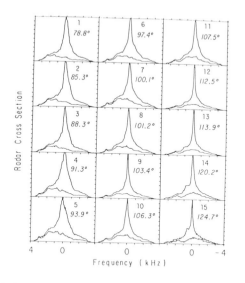

FIG. 8. Mars λ13-cm radar echo spectra for subradar points along 22° north latitude at the indicated west longitudes obtained in the OC (upper curves) and SC (lower curves) polarizations. In each box, spectra are normalized to the peak OC cross section. The echo bandwidth is 7.1 kHz. Very rough regions on the planet are revealed as bumps in the SC spectra that move from positive to negative Doppler frequencies as Mars rotates. (Courtesy of J. K. Harmon, D. B. Campbell and S. J. Ostro.)

tion ratios averaging about 0.07 for the moon, Mercury, and Venus, but ranging from 0.1 to 0.4 for Mars, as discussed below.

Typical rms slopes obtained at decimeter wavelengths for the four quasi-specular targets are around 7°, and consequently these objects' surfaces have been described as gently undulating. As might be expected, values estimated for s_0 increase as the observing wavelength decreases. For instance, for the moon s_0 increases from ~4° at $\lambda \sim 20$ m to ~8° at $\lambda \sim 10$ cm, and to ~33° at $\lambda \sim 1$ cm. At optical wavelengths ($\lambda \sim 0.5\ \mu$m), the moon shows no trace of a central glint, and the scattering is entirely diffuse. This phenomenon arises because the lunar surface (Fig. 9) consists of a regolith (a layer of loose, fine-grained particles) with much intricate structure at the scale of visible wavelengths. Surface structure with radii of curvature near radar wavelengths $\geqslant 1$ cm is relatively rare, so the diffuse component of lunar radar echoes is much weaker than the quasi-specular component, especially at very long wavelengths. At decimeter wavelengths the ratio of diffusely scattered power to quasi-specularly scattered power is ~1/3 for the moon, Mercury, and Venus. This ratio can be determined by assuming that all the SC echo is diffuse and calculating the diffusely scattered fraction (x) of OC echo by fitting to the OC spectrum a model employing a composite scattering law (e.g., $\sigma_0(\theta) = x\sigma_{\text{DIF}}(\theta) + (1 - x)\sigma_{\text{QS}}(\theta)$). Here $\sigma_{\text{QS}}(\theta)$ might be the Hagfors law, and usually $\sigma_{\text{DIF}}(\theta) \sim \cos^m\theta$. Estimated values of m usually fall between one (i.e., geometric scattering, which describes the optical appearance of the full moon) and two (Lambert scattering). Physical interpretations of the diffusely scattered echo employ information about albedo, scattering law, and polarization to constrain the size distributions, spatial densities, and electrical properties of wavelength-scale rocks near the surface.

3. The Radar Heterogeneity of Mars

Diffuse scattering from Mars is much more substantial than that from the other quasi-specular targets and often accounts for most of the echo power. It seems, therefore, that Mars possesses an unusually high concentration of near-surface, centimeter-to-meter-scale rocks. Furthermore, features in this planet's SC spectra reveal the existence of regions of extreme small-scale roughness (Fig. 8). The precise geographic location of the source of a spectral feature can-

FIG. 9. Structure on the lunar surface near the Apollo 17 landing site. Most of the surface is smooth and gently undulating at scales much larger than a centimeter. This smooth component of the surface is responsible for the predominantly quasi-specular character of the moon's radar echo at $\lambda \gg 1$ cm. Wavelength-scale structure produces a diffuse contribution to the echo. Wavelength-sized rocks are much more abundant at $\lambda \sim 4$ cm than at $\lambda \sim 10$ m (the scale of the boulder being inspected by Astronaut H. Schmitt), and hence diffuse echo is more substantial at shorter wavelengths.

not be ascertained from a single spectrum, since echo from anywhere along a constant-Doppler contour contributes to a given spectral element (Fig. 4). However, intersections between the contours corresponding to the feature's Doppler frequency in spectra obtained at different rotational phases can yield the longitude and the absolute value of the latitude of the source region. (The ambiguity in the sign of the latitude is discussed later.) For Mars the sources of SC spectral features evident in Fig. 8 apparently are volcanic regions, and the best terrestrial analog for this extremely rough terrain might be young lava flows (Fig. 10).

Since the motion in longitude of the subradar point on Mars (whose rotation period is only 24.6 hr) is rapid compared with that on the moon, Venus, or Mercury, and since the geometry of Mars' orbit and spin vector permits subradar tracks throughout the Martian tropics,

ground-based investigations of Mars have achieved more global coverage than those of the other terrestrial targets. Bistatic (i.e., two-station) radar observations of Mars, consisting of transmissions from an orbiting Viking spacecraft and ground-based reception of echoes from the Martian surface, have added information about surface properties near polar regions. The existing body of Mars radar data reveals extraordinary diversity in the degree of small-scale roughness as well as in the rms slope of smooth surface elements. For example, Fig. 11 shows the variation in OC echo spectral shape as a function of longitude for a subradar track along $\sim 16°$S latitude. Surface slopes on Mars have rms values from less than 0.5° to more than 10°. Chryse Planitia, site of the first Viking Lander, has fairly shallow slopes ($s_0 \sim 4$–5°), and in fact, radar rms slope estimates were utilized in selection of the Viking Lander sites.

FIG. 10.   This lava flow near Mount Shasta in California is an example of an extremely rough surface at decimeter radar wavelengths, possibly analogous to regions on Mars with very high circular polarization ratios. (Courtesy of D. Evans.)

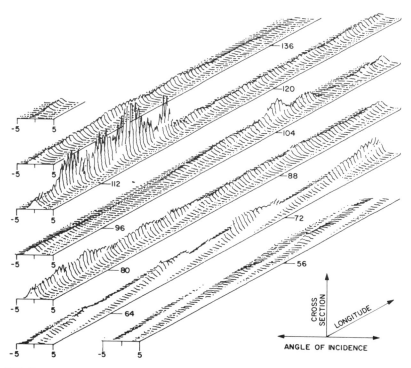

FIG. 11.   Mars echo power spectra as a function of longitude obtained along a subradar track at 16° south latitude. The most sharply peaked spectra correspond to the smoothest regions (i.e., smallest rms slopes). (Courtesy of G. S. Downs, P. E. Reichley, and R. R. Green.)

4. Radar Signatures of Asteroids

Echo spectra for the several dozen radar-detected asteroids show no hint of the sharply peaked signature of quasi-specular scattering seen for the terrestrial planets. However, the circular polarization ratios for mainbelt asteroids average about 0.1, so the echoes arise primarily from single-reflection backscattering from smooth surface elements. The broad spectra indicate that the rms slopes of these elements must be steeper than, say, the typical lunar value of 7°. For example, the nearly spherical, ~540-km-diameter asteroid 2 Pallas (Fig. 12) has $\mu_C = 0.05 \pm 0.02$, and fitting a spectral model based on a Gaussian scattering law indicates that $s_0 \gtrsim 20°$. This result suggests rather dramatic topographic relief, perhaps the outcome of impact cratering.

The much smaller near-earth asteroid 433 Eros (mean diameter ~20 km) has λ3.5-cm and λ13-cm values of μ_C about six times larger than Pallas's value, but not very different from the mean value for Mars. (Of course, very different physical processes are responsible for the small-scale roughness on Mars and Eros.) The Eros OC spectra vary in bandwidth and spectral

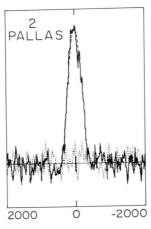

2
PALLAS

2000   0   -2000

FIG. 12. Radar echo power spectra obtained at λ13 cm for the asteroid 2 Pallas. Echo power is plotted against Doppler frequency (Hz). The circular polarization ratio μ_C of power received in the same sense of circular polarization as transmitted (i.e., the SC sense; dotted curve) to that in the opposite or OC sense (solid curve) is 0.05 ± 0.02, indicating an extremely smooth surface at decimeter scales. However, the model (dashed curve) fit to the OC spectrum indicates that the surface is very rough at some scale(s) no smaller than several meters and possibly as large as many kilometers.

shape as the asteroid rotates (Fig. 13). Some of the spectra are noticeably asymmetrical, and the mean frequency of the distribution $\sigma(\nu)$, labeled center frequency in Fig. 13, also oscillates with rotational phase. As indicated in that figure, the Eros data can be represented by a triaxial ellipsoid scattering according to a Lambert law, although the postfit residuals reveal significant departures from this model. In general, radar data obtained to date for near-earth asteroids seem very difficult to reconcile with models invoking homogeneous scattering and axisymmetric shapes.

E. TOPOGRAPHIC RELIEF

Topography along the subradar track superimposes a modulation on the echo delay above or below that predicted by ephemerides, which generally are calculated for a sphere with the target's a priori mean radius. There frequently are at least small errors in the radius estimate as well as in the target's predicted orbit. These circumstances generally require that an extended series of measurements of the time delay of the echo's leading edge be folded into a computer program designed to estimate simultaneously parameters describing the target's orbit, mean radius, and topography. The analysis program might also contain parameters from models of wave propagation through the interplanetary medium or the solar corona as well as parameters used to test general relativity, as noted earlier.

Radar has been used to measure topography on the moon and on the inner planets. For example, Fig. 14 shows a three-dimensional reconstruction of topography derived from altimetric profiles obtained for Mars in the vicinity of the giant shield volcano Arsia Mons. The altimetric resolution of the profiles is about 150 m, corresponding to 1-μsec delay resolution, but the surface resolution, or footprint, is very coarse (~75 km). Radar altimetry is best carried out from orbiting spacecraft, and the next generation of space missions to inner-solar-system bodies are likely to employ radar altimeters to make topographic measurements with footprints ~1 km and altimetric resolution as fine as ~10 m.

F. DELAY-DOPPLER RADAR MAPS

As illustrated in Fig. 4, intersections between constant-delay contours and constant-Doppler contours constitute a two-to-one mapping from

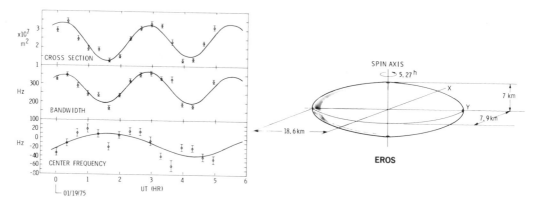

FIG. 13. Measurements of the radar cross section, spectral bandwidth, and apparent center frequency for λ3.5-cm OC echoes from the near-earth asteroid 433 Eros. The solid curves correspond to the sketched ellipsoid. (Courtesy R. F. Jurgens and R. M. Goldstein.)

the target's surface to delay-Doppler space: For any point in the northern hemisphere there is a conjugate point in the southern hemisphere at the same delay and Doppler. Therefore, given $\sigma(\tau, \nu)$, the source of echo in any delay-Doppler resolution cell can be located only to within a two-fold ambiguity. This north–south ambiguity can be avoided if the radar beamwidth is comparable to or smaller than the target's apparent angular radius, as in the case of observations of the moon (angular radius ~15 arcminutes) with the Arecibo λ13-cm system (beamwidth ~2 arcminutes). Similarly, no such ambiguity arises in the case of side-looking radar observations from spacecraft (e.g., the Pioneer Venus Orbiter), for which the geometry of delay-Doppler surface contours differs somewhat from that in Fig. 4. For ground-based observations of Venus and Mercury, whose angular radii never exceed a few tens of arcseconds, the separation of conjugate points is achievable interferometrically, using two receiving antennas, as follows.

The echo waveform received at either antenna from one conjugate point is highly correlated with the echo waveform received at the other antenna from the same conjugate point. However, echo waveforms from the two conjugate points are largely uncorrelated with each other, no matter where they are received. Thus, echoes from two conjugate points can, in principle, be distinguished by cross-correlating echoes received at the two antennas with themselves and with each other and then performing algebraic manipulations on long time averages of the cross product and the two self-products.

The echo waveform from a single conjugate point experiences slightly different delays in reaching the two antennas, so there is a phase difference between the two received signals, and this phase difference depends only on the geometrical positions of the antennas and the target. This geometry changes as the earth rotates, but very slowly and in a predictable manner. The antennas are usually positioned so that contours of constant phase difference on the target disk are as orthogonal as possible to the constant-Doppler contours (which connect conjugate points). Phase difference hence becomes a measure of north–south position, and echoes from conjugate points can be distinguished on the basis of their phase relation.

The total number of fringes, or cycles of phase shift, spanned by the disk of a planet with diameter D and distance R from the radar is approximately $(D/R)/(\lambda/b_{PROJ})$, where b_{PROJ} is the projection of the interferometer baseline normal to the mean line of sight. For example, the Arecibo receiving interferometer links the main antenna to a 30.5-m antenna about 11 km further north. It places about 7 fringes on Venus, quite adequate for separation of the north–south ambiguity. The Goldstone 65-m dish has been linked to smaller antennas to perform three-element as well as two-element interferometry. Tristatic observations permit one to solve so precisely for the north–south location of a given conjugate region that one can obtain the region's elevation relative to the mean planetary radius. Altimetric information can also be extracted from bistatic observations using the time history of the phase information, but only if the variations in the projected baseline vector are very large.

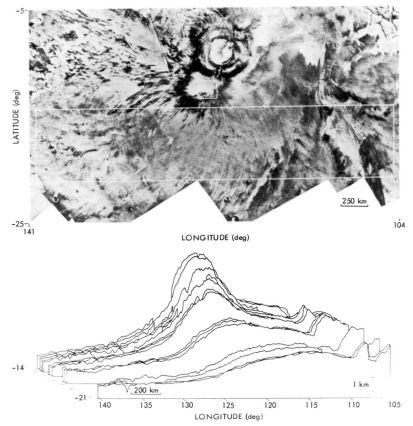

FIG. 14. Topographic contours for the southern flank (large rectangle) of the Martian shield volcano Arsia Mons, obtained from radar altimetry. (Courtesy L. Roth, G. S. Downe, R. S. Saunders, and G. Schubert.)

In constructing a radar map, we transform the unambiguous delay-Doppler distribution of echo power to planetocentric coordinates and fit a model to the data using a maximum-likelihood or weighted-least-squares estimator. The model contains parameters for quasi-specular and diffuse scattering as well as prior information about the target's dimensions and spin vector. For Venus, effects of the dense atmosphere on radar wave propagation must also be modeled. Residuals between the data and the best-fit model constitute a radar reflectivity map of the planet. Figure 15 shows a λ13-cm OC radar map of portions of Venus and Fig. 16 shows a λ3.8-cm SC radar map of a portion of the moon. Figure 17 shows a global map of Venus's λ17-cm SL radar reflectivity obtained by the Pioneer Venus orbiting spacecraft.

G. Physical Interpretation of Radar Reflectivities

Variations in radar reflectivities evident in delay-Doppler maps can be caused by many different physical phenomena, and their proper interpretation demands due attention to the radar wavelength, echo polarization, viewing geometry, prior knowledge about surface properties, and the nature of the target's mean scattering behavior. Similar considerations apply to inferences based on disk-integrated radar albedos.

In Fig. 15a, the alternating bright and dark bands probably result from modulation of the echo strength by preferential orientation of slopes toward and away from the radar. Since this area was nearly 70° from the subradar point and even a diffuse ($\cos^m \theta$) scattering law falls

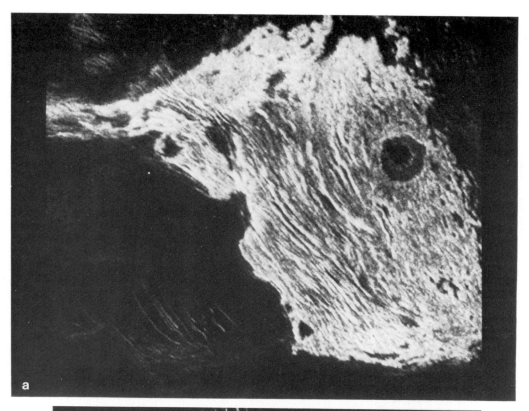

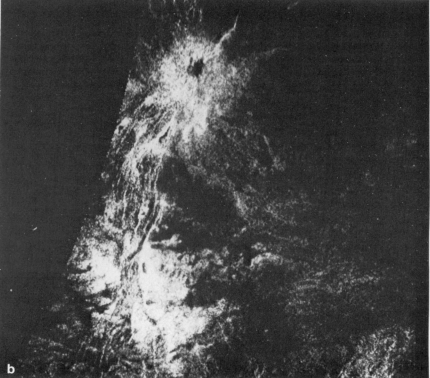

FIG. 15. Arecibo radar maps of portions of Venus. Maxwell Montes (a) is a mountain range produced by tectonic deformation. It is unknown whether the ~100-km-diameter circular feature, named Cleopatra, is of volcanic or impact origin. Beta Regio (b) contains a huge rift zone of linear faults connecting features that appear to be two giant shield volcanoes, Rhea Mons and Theia Mons. All these features are identified in the global map in Fig. 17. (Courtesy D. B. Campbell.)

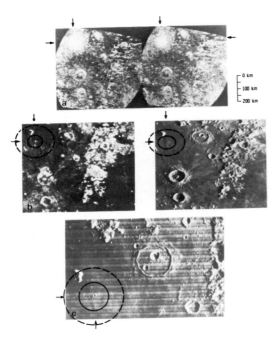

0 km
100 km
200 km

FIG. 16. Lunar crater Piton B (located by arrows) is surrounded by an ejecta blanket that is conspicuous in (a) the λ3.8-cm radar image (displayed in both continuous- and incremental-tone maps) but invisible in (b) earth-based and (c) Lunar Orbiter photographs. The sketched, 50- and 100-km-diameter circles are concentric to the crater and coplanar with the local mean surface. (Courtesy T. W. Thompson.)

off rapidly with angles of incidence that large. However, some of the contrast in Fig. 15a might arise from variations in decimeter-scale roughness.

In the SC lunar radar map in Fig. 16, the enhanced radar reflectivity around the crater Piton B is due to near-surface concentrations of wavelength-scale rocks ejected from the crater during its formation during a hypervelocity impact event. In other lunar radar maps there are reflectivity enhancements thought to be caused by surface chemistry (particularly by iron or titanium concentrations) rather than by structure.

Surface chemistry also appears to be the best available explanation for the approximately five-fold variation in the radar albedos ($\hat{\sigma}_{OC}$) of main-belt asteroids. These objects are thought to be blanketed with thick regoliths, and the variance in radar albedo suggests large variations in regolith porosity or metal abundance. The highest radar albedo estimated for an asteroid (16 Psyche) is consistent with porosities typical of the lunar regolith and a composition nearly en-

tirely metallic, suggesting that this 250-km-diameter object might be the largest piece of "refined" metal in the solar system.

In the Venus reflectivity map (Fig. 17), the brightest regions are more than five times brighter than the darkest regions and have normal-incidence Fresnel reflection coefficients equal to 0.3 or higher. The extremely bright areas might contain substantial amounts of basaltic rock enriched in such titanium- and iron-bearing minerals as ilmenite, magnetite, or pyrite. If this hypothesis is correct, those so-called high-dielectric minerals might be products of recent, if not currently active, volcanism. In any case, Venus's high disk-integrated reflectivity ($\hat{\sigma}_{OC} \sim 0.11$) suggests that Venus is dominated not by a global regolith but rather by extensive exposures of bedrock covered by variable, modest amounts of soil.

As is evident from Fig. 17, there is correlation between the elevation, rms slope, and radar reflectivity of Venus's surface. It has been suggested that some of the roughest, brightest, and most elevated terrain units (e.g., Rhea Mons and Theia Mons in Beta Regio and Maxwell Montes) are monumental volcanic constructs. Ground-based maps and recent maps obtained from Soviet Venera spacecraft show ancient impact craters as well as terrain that appears to be tectonically produced. The Magellan spacecraft, which was to be launched via the Space Shuttle as early as 1989 was expected to provide maps with a resolution (~200 m) much finer than in existing maps. Such maps would clarify the global geologic character of earth's intriguing sister planet.

H. JUPITER'S ICY GALILEAN SATELLITES

Among all the radar-detected planetary bodies in the solar system, Europa, Ganymede, and Callisto have the most bizarre radar properties. Their reflectivities are enormous compared with those of the moon and inner planets (Table II). Europa is the extreme example (Fig. 18), with a λ13-cm radar albedo (~1.0) that is indistinguishable from that of a metal sphere. Since the radar and optical albedos and estimates of fractional water-frost coverage increase by satellite—in the order Callisto, Ganymede, Europa—the presence of water ice presumably plays a critical role in determining the unusually high reflectivities. For any given porosity, ice is less radar-reflective and less absorbing than silicates.

In spite of the satellites' smooth appearances

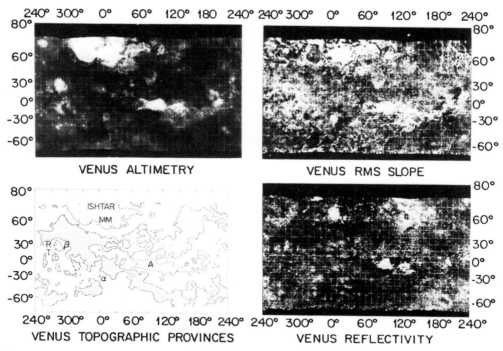

VENUS ALTIMETRY

VENUS RMS SLOPE

VENUS TOPOGRAPHIC PROVINCES

VENUS REFLECTIVITY

FIG. 17.   Venus altimetry, rms slope, and radar reflectivity maps derived from Pioneer Venus radar data. Brighter tones indicate higher, rougher, and more reflective areas, respectively. The map of topographic provinces shows (i) the distribution of highlands (vertical hatching), lowlands (shaded), and rolling plains (blank); and (ii) locations of Ishtar Terra, Aphrodite Terra (A), Alpha Regio (α), Beta Regio (α), Beta Regio (β), Rhea Mons (R), Theia Mons (T), and Maxwell Montes (MM). (See Fig. 15.) (Courtesy G. H. Pettengill.)

at the several-kilometer scales of Voyager high-resolution images, a high degree of near-surface structure at some scale(s) $\geq \lambda 13$ cm is suggested by (i) broad spectral shapes, indicative of a diffuse scattering process, and (ii) fairly large linear polarization ratios ($\mu_L \approx 0.5$).

The precise configurations of the satellite surfaces are constrained most severely by the man-

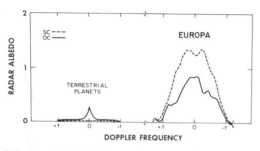

FIG. 18.   Typical $\lambda 13$-cm echo spectra for the terrestrial planets are compared with echo spectra for Jupiter's icy moon Europa. The abscissa has units of half the echo bandwidth.

ner in which they backscatter circularly polarized waves. Measured values of the $\lambda 13$-cm circular polarization ratio (μ_C) for Venus, the moon, and Mars average ~0.1, and one can easily image a surface so rough (e.g., Fig. 10) that incident waves would be completely unpolarized ($\mu_C \to 1$). However, μ_C actually exceeds unity for Europa, Ganymede, and Callisto. That is, the scattering largely preserves the transmitted handedness and the circular polarization is "inverted."

Weighted-mean $\lambda 13$-cm values of μ_C for Europa, Ganymede, and Callisto are within 0.2 of 1.6, 1.6, and 1.2, respectively, but the distributions of values of μ_C are wide, as indicated schematically in Fig. 19. Significant polarization and/or albedo features are present in the echo spectra, and in a few cases the feature's source can be identified tentatively in images acquired by Voyager spacecraft. Observations of Ganymede at $\lambda 3.5$ cm yield $\mu_C = 2.0 \pm 0.1$, indicating that the polarization inversion depends on wavelength.

If the radar echo arises from external reflec-

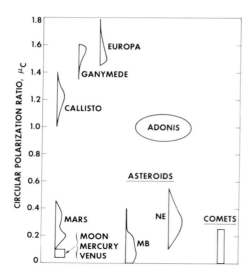

FIG. 19. Circular polarization ratios (μ_C) of radar-detected planetary targets. Approximate distributions of available estimates of μ_C are sketched for the icy Galilean satellites, Mars, mainbelt asteroids, and near-earth asteroids. The curve for near-earth asteroids excludes 2101 Adonis whose polarization ratio 1.0 ± 0.2 is close to values obtained for Callisto.

tions from the surface per se, the surface material must have a refractive index ~20% higher than that (1.8) of solid water ice and must be covered with nearly hemispherical craters. A different model postulates subsurface multiple scattering from randomly oriented planar interfaces between electrically dissimilar components (e.g., ice and void), with the regolith components' refractive indices in a ratio less than 1.4. In the hemispherical crater model, the polarization inversion arises from double-reflection backscattering from crater interiors. In the random facet model, the inversion results from many successive grazing reflections and particularly from total internal reflections in the denser medium.

Perhaps the most promising explanation for the polarization inversion and the huge reflectivities is a process involving, ironically, no reflections at all. Instead, the incident wave enters the surface, is refracted around volumes of material whose density (and hence refractive index) is somewhat higher than that of their surroundings, and reemerges from the surface with its incident polarization (SC) and power largely intact. If this refraction scattering theory is correct, the scale of the putative scattering centers

is probably between a centimeter and a meter. The variation in refractive index must be very smooth and continuous to avoid subsurface reflections, and the uppermost surface must be very tenuous, creating an impedance match between free space and the target. At present, we still lack precise information about the subsurface structure needed to produce refraction scattering and the particular geologic processes responsible for generating the requisite structures. However, one of the more plausible possibilities is that the unusual echoes arise because ejecta within the upper few decameters of the regoliths are thermally annealed. In this scenario, sharp dielectric boundaries between large solid fragments and fine-grained, porous debris have been changed into gentle gradations by exposure to temperature $\gtrsim 130$ K, so the large fragments become refraction scatterers.

As indicated in Fig. 19, the near-earth asteroid 2201 Adonis has polarization properties resembling those of Callisto. Since Adonis's orbit resembles those of some short-period comets, this object might be an extinct cometary nucleus, whose supply of near-surface volatiles (i.e., ices) is so exhausted that sunlight can no longer stimulate cometary (coma/tail) activity. Optical observations reveal no cometary activity for Adonis, and Adonis's μ_C is certainly not cometary. Of course, the same statements apply to Callisto, so it seems plausible that the surface of Adonis might be thermally altered and might resemble that of Callisto.

I. COMETS

Since a cometary coma is nearly transparent at radio wavelengths, radar is much more capable of unambiguous detection of a cometary nucleus than are optical and infrared methods, and radar observations of several comets (Table I) have provided useful constraints on nuclear dimensions. The radar signature of one particular comet (IRAS–Araki–Alcock, which came within 0.03 AU of earth in May 1983) revolutionizes our concepts of the physical nature of these intriguing objects. Echoes obtained at both Goldstone (Fig. 20) and Arecibo have a narrowband component from the nucleus as well as a much weaker broadband component from a cloud of particles ejected from the nucleus. Models of the echo suggest that most of the particles, which must be 1–5 cm in size, escaped

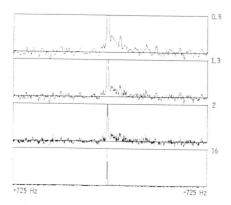

FIG. 20. Echo power spectrum of Comet IRAS–Araki–Alcock, shown here with four different vertical truncations and degrees of smoothing. Echo power is plotted against Doppler frequency. The narrow spike is echo from the comet's ~7-km nucleus. The broadband component, or skirt, is echo from a $\geq 10^3$-km debris swarm consisting of particles no smaller than a centimeter and, for the most part, not gravitationally bound to the nucleus. (Courtesy R. M. Goldstein, R. F. Jurgens, and Z. Sekanina.)

from the ~7-km-diameter nucleus to form a debris swarm whose size is $\geq 10^3$ km. Echo spectra for the nucleus reveal a nonspherical shape and show several features that are consistent with roughness on meter-to-kilometer scales. Thus, one envisions this object as extremely rough with an explosively active surface from which chunks of material are being blasted into space by subsurface sublimation of volatiles.

J. SATURN'S RINGS

Saturn's rings are the most distant radar-detected planetary entity and the only radar-detected ring system; neither distinction is likely to be relinquished in the foreseeable future. The rings are quite unlike other planetary targets in terms of both the experimental techniques employed and the physical considerations involved. For example, the relation between ring-plane location and delay-Doppler coordinates for a system of particles traveling in Keplerian orbits is different from the geometry portrayed in Fig. 4. The rings are grossly overspread (Table II), requiring the use of frequency-stepped waveforms in the more sophisticated radar studies.

Radar determinations of the rings' backscattering properties complement results of the Voy-ager spacecraft radio occultation experiment (which measured the rings' forward scattering efficiency at identical wavelengths) in constraining the size and spatial distributions of ring particles. The rings' circular polarization ratio is ~1.0 at λ3.5 cm at ~0.5 and λ13 cm, more or less independent of the inclination angle δ between the ring plane and the line of sight. Whereas multiple scattering between particles might cause some of the depolarization, the lack of strong dependence of μ_C on δ suggests that the particles are intrinsically rougher at the scale of the smaller wavelength. The rings' total-power (OC + SC) radar albedo shows only modest dependence on δ, a result that seems to favor many-particle-thick models of the rings over monolayer models. Delay-Doppler resolution of ring echoes indicates that the portions of the ring system that are brightest optically (the A and B rings) are also responsible for most of the radar echoes. The C ring has a very low radar reflectivity, presumably because of either a low particle density in that region or bulk compositions or particle sizes leading to inefficient particle scattering efficiencies.

IV. The Future of Planetary Radar Astronomy

During the remainder of this century new interplanetary spacecraft will carry radar experiments, and the sensitivity of ground-based telescopes will be improved, adding to the momentum of planetary radar astronomy as it pursues several exciting scientific directions.

For Mercury, ground-based delay-Doppler maps will provide our first looks at the hemisphere left unimaged by the Mariner 10 spacecraft a decade ago, and the dual-polarization radar techniques applied so successfully to the other terrestrial objects will characterize the planet's small-scale surface properties.

The next major step in spacecraft exploration of the inner solar system will be taken by the Magellan mission, which will obtain high-resolution images and altimetry for most of Venus. This mission will clarify the detailed nature and evolutionary history of the planet's surface as well as the principal geologic processes influencing the interior and the atmosphere.

Following Magellan, both Mars and the moon are to be studied intensely from spacecraft in polar orbits. Radar altimeters are key components of the instrument packages, which are designed to define the global elemental and mineralogical nature of the surface, the gravitational field, and the topography. These investigations are logical predecessors to Mars rover and sample-return missions and to a second generation of lunar landings by human beings.

The primary emphasis in ground-based radar astronomy will be on asteroids, comets, and the satellites of Mars, Jupiter, and Saturn. The goal of asteroid and comet science is to understand enormous populations; the strategy for reaching that goal is to use a few space missions to selected objects as fiducials to calibrate the interpretation of ground-based observations of many objects. In this context, radar observations will continue to refine our knowledge of the physical properties of individual objects and will play a critical role in the reconnaissance of the asteroid and comet populations. Radar astronomy will also play a role in selection of the target for a Near-Earth Asteroid Rendezvous Mission, analogous to its role in Viking Lander Site selection.

The Martian satellites Phobos and Deimos will be detectable with ground-based radar during several close approaches of Mars beginning in 1988. These objects, which were studied in great depth by the Viking spacecraft, may actually be captured asteroids, and measurement of their radar properties would help guide inferences about asteroid radar signatures. Phobos and Deimos are also very interesting in their own right and are potentially important as convenient, accessible outposts for the human exploration of Mars. One primary objective of radar observations would be to assess the bulk density and near-surface rock populations of the satellites' regoliths.

Ground-based observations of Jupiter's Galilean satellites will devote special attention to volcanically active Io, whose polarization properties have never been measured and whose radar albedo and angular scattering behavior are poorly known. For the icy moons, Europa, Ganymede, and Callisto, a major objective is to identify regions of polarization and albedo anomalies and to locate these regions in high-resolution (~20 m) visual images to be acquired by the Galileo Mission in the 1990s.

One of the most intriguing objects in the solar system is Saturn's largest satellite, Titan. Available constraints on Titan's surface temperature and pressure and on atmospheric chemistry suggest that the solid surface of this satellite might be composed largely of organic compounds and might be partially covered by ethane-rich oceans. However, the actual nature of Titan's surface, which is concealed by clouds, remains unknown. As with Venus, radar techniques provide our only means to discern the global character of the surface. Attempts to detect Titan from Goldstone and Arecibo will be carried out during the coming decade, and the results of these efforts will assist design of a radar instrument for a Titan spacecraft mission.

In summary, there are many exciting prospects for new observations, and the future of planetary radar astronomy is potentially very bright. However, a necessary condition for sustaining the vitality of ground-based efforts is major improvement in the sensitivity of the Arecibo and Goldstone telescopes. Activity involving spacecraft radars certainly should be vigorous and fruitful well into the 21st century.

BIBLIOGRAPHY

Campbell, D. B., Head, J. W., Harmon, J. K., and Hine, A. A. (1984). *Science* **226,** 167–169.

Eshleman, V. R. (1986). *Nature* **319,** 755–757.

Garvin, J. B., Head, J. W., Pettengill, G. H., and Zisk, S. H. (1985). *J. Geophys. Research* **90,** 6859–6871.

Goldstein, R. M., Jurgens, R. F., and Sekanina, Z. (1984). *Astronom. J.* **89,** 1745–1754.

Hagfors, T., Gold, T., and Ierkic, H. M. (1985). *Nature* **315,** 637–640.

Harmon, J. K., and Ostro, S. J. (1985). *Icarus* **62,** 110–128.

Jurgens, R. F. (1982). *Icarus* **49,** 97–108.

Ostro, S. J. (1985). *Publ. Astronom. Soc. Pacific* **97,** 887–884.

Ostro, S. J. (1982). "Satellites of Jupiter," D. Morrison, ed. University of Arizona Press, Tucson, pp. 213–236.

Ostro, S. J., Campbell, D. B., and Shapiro, I. I. (1985). *Science* **224,** 442–446.

Ostro, S. J., and Pettengill, G. H. (1984). In "Planetary Rings," A. Brahic, ed. Cepadues-Editions, Toulouse, France, pp. 49–55.

Pettengill, G. H., Eliason, E., Ford, P. G., Loriot, G. B., Masursky, H., and McGill, G. E. (1980). *J. Geophys. Research* **85,** 8261–8270.

Roth, L. E., Downs, G. S., Saunders, R. S., and Schubert, G. (1980). *Icarus* **42,** 287–316.

Shapiro, I. I., Campbell, D. B., and De Campli, W. M. (1979). *Astrophys. J.* **230,** L123–L126.

Simpson, R. A., and Tyler, G. L. (1982). *IEEE Trans. Antennas and Propagation AP-30,* 438–449.

Thompson, T. W., Zisk, S. H., Shorthill, R. W., and Cutts, J. A. (1981). *Icarus* **46,** 201–225.

PLANETARY SATELLITES, NATURAL

Bonnie J. Buratti   *JPL, California Institute of Technology*

GLOSSARY

Bond albedo: Fraction of the total incident radiation reflected by a planet or satellite.

Carbonaceous material: Carbon–silicate material rich in simple organic compounds. It exists on the surfaces of several satellites.

Differentiation: Melting and chemical fractionation of a planet or satellite into a core and mantle.

Geometric albedo: Ratio of the brightness at a phase angle of zero degrees (full illumination) compared with a diffuse, perfectly reflecting disk of the same size.

Greenhouse effect: Heating of the lower atmosphere of a planet or satellite by the transmission of visible radiation and subsequent trapping of reradiated infrared radiation.

Lagrange points: Five equilibrium points in the orbit of a satellite around its primary. Two of them (L4 and L5) are points of stability for a third body.

Magnetosphere: Region around a planet dominated by its magnetic field and associated charged particles.

Opposition effect: Surge in brightness as a satellite becomes fully illuminated to the observer.

Phase angle: Angle between the observer, the satellite, and the sun.

Phase integral: Integrated value of the function which describes the directional scattering properties of a surface.

Primary body: Celestial body (usually a planet) around which a satellite, or secondary, orbits.

Regolith: Surface layer of rocky debris created by meteorite impacts.

Roche's limit: Distance (equal to 2.44 times the radius of the primary) at which the tidal forces exerted by the primary on the satellite equal the internal gravitational forces of the satellite.

Synchronous rotation: Dynamical state caused by tidal interactions in which the satellite presents the same face towards the primary.

A natural planetary satellite is a celestial body in orbit around one of the nine principal planets of the solar system. The central body is known as the primary and the orbiting satellite its moon or secondary. Among the nine planets only Mercury and Venus have no known companions. There are 54 known natural planetary satellites in the solar system; there may exist many more undiscovered satellites, particularly small objects encircling the giant outer planets. The satellites range in size from planet-sized objects such as Ganymede and Titan to tiny, irregular bodies tens of kilometers in diameter (see Table I and Fig. 1).

I. Summary of Characteristics

A. DISCOVERY

The only natural planetary satellite known before the advent of the telescope was the Earth's moon. Phenomena such as the lunar phases and the ocean tides have been studied for centuries. When Galileo turned his telescope to Jupiter in 1610, he discovered the four large satellites in the Jovian system. His observations of their orbital motion around Jupiter in a manner analogous to the motion of the planets around the sun provided critical evidence for the acceptance of the heliocentric (sun-centered) model of the solar system. These four moons—Io, Europa, Ganymede, and Callisto—are sometimes called

TABLE I.   Summary of the Properties of the Natural Planetary Satellites

Satellite	Distance from primary (10³ km)	Revolution period (days) R = Retrograde	Orbital eccentricity	Orbital inclination (degrees)
Earth				
Moon	384.4	27.3	0.055	18 to 29
Mars				
M1 Phobos	9.38	0.32	0.018	1.0
M2 Diemos	23.50	1.26	0.002	2.8
Jupiter				
J14 Adrastea	128	0.30	0.0	0.0
J16 Metis	128	0.30	0.0	0.0
J5 Amalthea	181	0.49	0.003	0.4
J15 Thebe	221	0.68	0.0	0.0
J1 Io	422	1.77	0.004	0.0
J2 Europa	671	3.55	0.000	0.5
J3 Ganymede	1070	7.16	0.001	0.2
J4 Callisto	1880	16.69	0.010	0.2
J13 Leda	11110	240	0.416	26.7
J6 Himalia	11470	251	0.158	27.6
J10 Lysithea	11710	260	0.130	29.0
J7 Elara	11740	260	0.207	24.8
J12 Ananke	20700	617R	0.17	147
J11 Carme	22350	692R	0.21	164
J8 Pasiphae	23300	735R	0.38	145
J9 Sinope	23700	758R	0.28	153
Saturn				
S17 Atlas	138	0.60	0.002	0.3
S16 1980S27	139	0.61	0.004	0.0
S15 1980S26	142	0.63	0.004	0.1
S10 Janus	151	0.69	0.007	0.14
S11 Epimethus	151	0.69	0.009	0.34
S1 Mimas	186	0.94	0.020	1.5
S2 Enceladus	238	1.37	0.004	0.0
S3 Tethys	295	1.89	0.000	1.1
S14 Calypso	295	1.89	0.0	1?
S13 Telesto	295	1.89	0.0	1?
S4 Dione	377	2.74	0.002	0.0
S12 1980S6	377	2.74	0.005	0.2
S5 Rhea	527	4.52	0.001	0.4
S6 Titan	1220	15.94	0.029	0.3
S7 Hyperion	1480	21.28	0.104	0.4
S8 Iapetus	3560	79.33	0.028	14.7
S9 Phoebe	12950	550.4R	0.163	150
Uranus				
1986U7	49.7	0.33		
1986U8	53.2	0.37		
1986U9	59.2	0.43		
1986U3	61.8	0.46		
1986U6	62.7	0.47		
1986U2	64.6	0.49		
1986U1	66.1	0.51		
1986U4	69.9	0.56		
1986U5	75.3	0.62		
1985U1	86.0	0.76		
U5 Miranda	130	1.41	0.017	3.4
U1 Ariel	191	2.52	0.003	0.0
U2 Umbriel	266	4.14	0.003	0.0
U3 Titania	436	8.71	0.002	0.0
U4 Oberon	583	13.46	0.001	0.0
Neptune				
N1 Triton	355.5	5.89R	0.00	160
N2 Nereid	5567	359.9	0.749	27.7
Pluto				
P1 Charon	19.3	6.3R	0?	120

Radius (km)	Density (gm/cm³)	Visual geometric albedo	Discoverer	Year of discovery
1738	3.34	0.11		
14 × 10	1.9	0.05	Hall	1877
8 × 6	2.1	0.05	Hall	1877
20		<0.1	Jewitt *et al.*	1979
20		<0.1	Synott	1979/80
135 × 85 × 75		0.05	Barnard	1892
40		<0.1	Synott	1979/80
1815	3.55	0.6	Galileo	1610
1569	3.04	0.6	Galileo	1610
2631	1.93	0.4	Galileo	1610
2400	1.83	0.2	Galileo	1610
10			Kowal	1974
90		0.03	Perrine	1904/5
10			Nicholson	1938
40		0.03	Perrine	1904/5
5			Nicholson	1951
15			Nicholson	1938
20			Melotte	1908
15			Nicholson	1914
20 × ? × 10		0.4	*Voyager*	1980
70 × 50 × 37		0.6	*Voyager*	1980
55 × 45 × 33		0.6	*Voyager*	1980
110 × 95 × 80		0.6	Dollfus	1966
70 × 58 × 50		0.5	Fountain and Larson	1978
197	1.4	0.8	Herschel	1789
251	1.2	1.0	Herschel	1789
530	1.2	0.8	Cassini	1684
12 × 11 × 11		0.6	Space Telescope Tm.	1980
15 × 10 × 8		0.9	Smith *et al.*	1980
560	1.4	0.55	Cassini	1684
17 × 16 × 15		0.5	Laques and Lecacheux	1980
765	1.3	0.65	Cassini	1672
2575	1.88	0.2	Huygens	1655
205 × 130 × 110		0.3	Bond and Lassell	1848
730	1.2	0.4–0.08	Cassini	1671
110		0.06	Pickering	1898
~25			*Voyager 2*	1986
~25			*Voyager 2*	1986
~25			*Voyager 2*	1986
~30		~0.04	*Voyager 2*	1986
~30		~0.04	*Voyager 2*	1986
~40		~0.06	*Voyager 2*	1986
~40		~0.09	*Voyager 2*	1986
~60		~0.04	*Voyager 2*	1986
			Voyager 2	1986
85			*Voyager 2*	1985
242	1.2	0.22	Kuiper	1948
580	1.6	0.38	Lassell	1851
596	1.8	0.16	Lassell	1851
805	1.6	0.23	Herschel	1787
773	1.6	0.20	Herschel	1787
1750	2.8	0.4	Lassell	1846
300			Kuiper	1949
600	0.8		Christy	1978

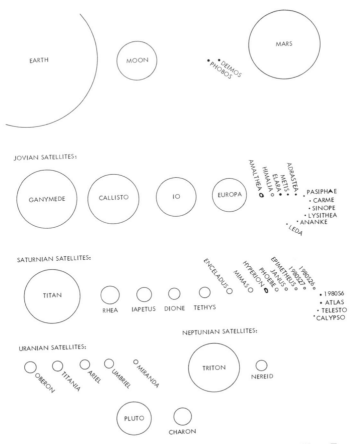

FIG. 1.   Relative shapes and sizes of the natural planetary satellites. For comparison, the relative sizes of smaller planets are shown. Jupiter and Saturn would span about twice the height of the figure and Uranus and Neptune would be about as wide as the figure.

the Galilean satellites. [*See* MOON (ASTRONOMY).]

In 1655 Christian Huygens discovered Titan, the giant satellite of Saturn. Later in the seventeenth century, Giovanni Cassini discovered the four next largest satellites of Saturn. It was not until over one hundred years later that the next satellite discoveries were made: the Uranian satellites Titania and Oberon and two smaller moons of Saturn. As telescopes acquired more resolving power in the nineteenth century, the family of satellites grew (see Table I). The smallest satellites of Jupiter and Saturn were discovered during flybys of the Pioneer and Voyager spacecraft (see Table II).

The natural planetary satellites are generally named after figures in classical Greek and Roman mythology who were associated with the namesakes of their primaries. They are also des-

ignated by the first letter of their primary and an Arabic numeral assigned in order of discovery: Io is J1, Europa, J2, etc. When satellites are first discovered but not yet confirmed or officially named, they are known by the year in which they were discovered, the initial of the primary, and a number assigned consecutively for all solar system discoveries, e.g., 1980J27.

When planetary scientists were able to observe and map geologic formations of the satellites from spacecraft images, they named many of the features after characters or locations from Western and Eastern mythologies.

B. PHYSICAL AND DYNAMIC PROPERTIES

The motion of a satellite around the center of mass of itself and its primary defines an ellipse

TABLE II.   Summary of Major Missions to the Planetary Satellites

Mission name	Object	Encounter dates	Type of mission
Luna 3 (USSR)	Moon	1959	Flyby (far side)
Ranger 7,8,9	Moon	1964–65	Crash landing; image return
Luna 9,13 (USSR)	Moon	1966	Soft landing
Luna 10,12,14 (USSR)	Moon	1966–68	Orbiter
Surveyor 1,3,5,6,7	Moon	1966–68	Soft landing
Lunar Orbiter 1–5	Moon	1966–68	Orbiter
Apollo 7–10	Moon	1968–69	Manned orbiter
Apollo 11,12,14–17	Moon	1969–75	Manned landing
Luna 16,20 (USSR)	Moon	1970–72	Sample return
Luna 17,21,24 (USSR)	Moon	1970–76	Rover
Mariner 9	Deimos	1971	Orbiter
	Phobos		
Pioneer 10	Jovian satellites	1979	Flyby
Pioneer 11	Jovian satellites	1979	Flyby
	Saturnian satellites	1979	Flyby
Viking 1,2	Phobos	1976	Orbiter
	Deimos		
Voyager 1	Jovian satellites	1979	Flyby
	Saturnian satellites	1980	Flyby
Voyager 2	Jovian satellites	1979	Flyby
	Saturnian satellites	1981	Flyby
	Uranian satellites	1986	Flyby

with the primary at one of the foci. The orbit is defined by three primary orbital elements (1) the semimajor axis, (2) the eccentricity, and (3) the angle made by the intersection of the plane of the orbit and the plane of the primary's spin equator (the angle of inclination). The orbits are said to be regular if they are in the same sense of direction (the prograde sense) as that determined by the rotation of the primary. The orbit of a satellite is irregular if its motion is in the opposite (or retrograde) sense of direction, or if it has a high angle of inclination. The majority of satellites move in regular, prograde orbits. Those satellites that do not move in regular, prograde orbits are believed to be captured objects (see Table I). [*See* CELESTIAL MECHANICS.]

Most of the planetary satellites present the same hemisphere toward their primaries, which is the result of tidal evolution. When two celestial bodies orbit each other, the gravitational force exerted on the near side is greater than that exerted on the far side. The result is an elongation of each body to form tidal bulges, which can consist of either solid, liquid, or gaseous (atmospheric) material. The primary will tug on the satellite's tidal bulge to lock its longest axis onto the primary-satellite line. The satellite, which is said to be in a state of synchronous rotation, keeps the same face toward the

primary. Since this despun state occurs rapidly (usually within a few million years), most natural satellites are in synchronous rotation.

The natural satellites are unique worlds, each representing a vast panorama of physical processes. The small satellites of Jupiter and Saturn are irregular chunks of ice and rock, perhaps captured asteroids, which have been subjected to intense meteoritic bombardment. Several of the satellites, including Phoebe (which is in orbit around Saturn) and the Martian moon Phobos, are covered with dark carbonaceous material believed to be representative of the primordial, unprocessed material from which the Solar System formed. The medium-sized satellites of Saturn and Uranus are large enough to have undergone internal melting and subsequent differentiation and resurfacing. The Saturnian satellite, Iapetus, presents a particular enigma: one hemisphere is ten times more reflective than the other. Three of the Galilean satellites show evidence of geologically active periods in their history; Io is presently undergoing intense volcanic activity. The Earth's Moon experienced a period of intense meteoritic bombardment and melting soon after its formation.

Saturn's largest satellite, Titan, has a predominantly nitrogen atmosphere thicker than that of the Earth. There is evidence that Triton, the large satellite of Neptune, also has an appreciable atmosphere. Io has a thin, possibly transient sulfur dioxide atmosphere that is thought to be related to outgassing from active volcanoes. None of the other satellites have detectable atmospheres. [*See* PLANETARY ATMOSPHERES.]

II. Formation of Satellites

A. THEORETICAL MODELS

Because the planets and their associated moons condensed from the same cloud of gas and dust at about the same time, the formation of the natural planetary satellites must be addressed within the context of the formation of the planets. The solar system formed 4.6 ± 0.1 billion years ago. This age is derived primarily from radiometric dating of meteorites, which are believed to consist of primordial, unaltered matter. In the radiometric dating technique, the fraction of a radioactive isotope (usually rubidium, argon, or uranium), which has decayed into its daughter isotope, is measured. Since the rate at which these isotopes decay has been measured in the laboratory, it is possible to infer the time elapsed since formation of the meteorites, and thus of the solar system. [*See* RADIOACTIVITY.]

The sun and planets formed from a disk-shaped rotating cloud of gas and dust known as the proto-solar nebula. When the temperature in the nebula cooled sufficiently, small grains began to condense. The difference in solidification temperatures of the constituents of the proto-solar nebula accounts for the major compositional differences of the satellites. Since there was a temperature gradient as a function of distance from the center of the nebula, only those materials with high melting temperatures (e.g., silicates, iron, aluminum, titanium, and calcium) solidified in the central (hotter) portion of the nebula. The Earth's Moon consists primarily of these materials. Beyond the orbit of Mars, carbon, in combination with silicates and organic molecules, condensed to form a class of asteroids known as carbonaceous chondrites. Similar carbonaceous material is found on the surfaces of Phobos, several of the Jovian and Saturnian satellites, and perhaps the Uranian satellites. Beyond the outer region of the asteroid belt, formation temperatures were sufficiently cold to allow water ice to condense and

remain stable. Thus, the Jovian satellites are primarily ice-silicate admixtures (except for Io, which has apparently outgassed all its water). On Saturn and Uranus, these materials are joined by methane and ammonia. For the satellites of Neptune and Pluto, formation temperatures were probably low enough for other volatiles, such as nitrogen and carbon monoxide, to exist in liquid and solid form. In general, the satellites, which formed in the inner regions of the solar system are denser than the outer planets' satellites, because they retained a lower fraction of volatile materials.

After small grains of material condensed from the proto-solar nebula, electrostatic forces caused them to stick together. Collisions between these larger aggregates caused meter-sized particles, or planetesimals, to be accreted. Finally, gravitational collapse occurred to form larger, kilometer-sized planetesimals. The largest of these bodies swept up much of the remaining material to create the protoplanets and their companion satellite systems. One important concept of planetary satellite formation is that a satellite cannot accrete within Roche's limit, the distance at which the tidal forces of the primary become greater than the internal cohesive forces of the satellite.

The formation of the regular satellite systems of Jupiter, Saturn, and Uranus is sometimes thought to be a smaller scaled version of the formation of the solar system. A density gradient as a function of distance from Jupiter does exist for the Galilean satellites (see Table I). This implies that more volatiles (primarily ice) are included in the bulk composition as the distance increases. However, this simple scenario cannot be applied to Saturn or Uranus because their regular satellites do not follow this pattern.

The retrograde satellites are probably captured asteroids or large planetesimals left over from the major episode of planet formation. Except for Titan and perhaps Triton, the satellites are too small to possess gravitational fields sufficiently strong to retain an appreciable atmosphere against thermal escape. [*See* PRIMITIVE SOLAR SYSTEM OBJECTS: COMETS AND ASTEROIDS.]

B. EVOLUTION

Soon after the satellites accreted, they began to heat up from the release of gravitational potential energy. An additional heat source was provided by the release of mechanical energy during the heavy bombardment of their surfaces by remaining debris. The satellites Phobos, Mimas, and Tethys all have impact craters caused by bodies that were nearly large enough to break them apart; probably such catastrophes did occur. The decay of radioactive elements found in silicate materials provided another major source of heat. The heat produced in the larger satellites was sufficient to cause melting and chemical fractionation; the dense material, such as silicates and iron, went to the center of the satellite to form a core, while ice and other volatiles remained in the crust.

Some satellites, such as the Earth's Moon, Ganymede, and several of the Saturnian satellites underwent periods of melting and active geology within a billion years of their formation and then became quiescent. Others, such as Io and possibly Enceladus and Europa, are currently geologically active. For nearly a billion years after their formation, the satellites all underwent intense bombardment and cratering. The bombardment tapered off to a slower rate and presently continues. By counting the number of craters on a satellite's surface and making certain assumptions about the flux of impacting material, geologists are able to estimate when a specific portion of a satellite's surface was formed. Continual bombardment of satellites causes the pulverization of the surface to form a covering of fine material known as a regolith.

III. Observations of Satellites

A. TELESCOPIC OBSERVATIONS

1. Spectroscopy

Before the development of interplanetary spacecraft, all observations from Earth of objects in the solar system were obtained by telescopes. One particularly useful tool of planetary astronomy is spectroscopy, or the acquisition of spectra from a celestial body.

Each component of the surface or atmosphere of a satellite has a characteristic pattern of absorption and emission bands. Comparison of the astronomical spectrum with laboratory spectra of materials which are possible components of the surface yields information on the composition of the satellite. For example, water ice has a series of absorption features between 1 and 4 microns. The detection of these bands on three of the Galilean satellites and several satellites of Saturn and Uranus demonstrated that water ice

is a major constituent of their surfaces. Other examples are the detections of SO_2 frost on the surface of Io, and methane in the atmosphere of Titan, and the possible detection of liquid nitrogen on Triton.

2. Photometry

Photometry of planetary satellites is the accurate measurement of radiation reflected to an observer from their surfaces or atmospheres. These measurements can be compared to light scattering models that are dependent on physical parameters, such as the porosity of the optically active upper surface layer, the albedo of the material, and the degree of topographic roughness. These models predict brightness variations as a function of solar phase angle (the angle between the observer, the sun, and the satellite). Like the Earth's Moon, the planetary satellites present changing phases to an observer on Earth. As the face of the satellite becomes fully illuminated to the observer, the integrated brightness exhibits a nonlinear surge in brightness that is believed to result from the disappearance of mutual shadowing among surface particles. The magnitude of this surge, known as the "opposition effect," is greater for a more porous surface.

One measure of how much radiation a satellite reflects is the geometric albedo, p, which is the disk-integrated brightness at "full moon" (or a phase angle of zero degrees) compared to a perfectly reflecting, diffuse disk of the same size. The phase integral, q, defines the angular distribution of radiation over the sky:

$$q = 2 \int_0^\pi \Phi(\alpha) \sin \alpha \, d\alpha$$

where $\Phi(\alpha)$ is the disk integrated brightness and α is the phase angle.

The Bond albedo, which is given by $A = p \times q$, is the ratio of the integrated flux reflected by the satellite to the integrated flux received. The geometric albedo and phase integral are wavelength dependent; whereas, a true (or bolometric) Bond albedo is integrated over all wavelengths.

Another ground based photometric measurement, which has yielded important information on the satellites surfaces, is the integrated brightness of a satellite as a function of orbital angle. For a satellite in synchronous rotation with its primary, the subobserver geographical longitude of the satellite is equal to the longitude of the satellite in its orbit. Observations showing significant albedo and color variegations for Io, Europa, Rhea, Dione, and especially Iapetus suggest that diverse geologic terrains coexist on these satellites. This view was confirmed by images obtained by the Voyager spacecraft.

3. Radiometry

Satellite radiometry is the measurement of radiation which is absorbed and re-emitted at thermal wavelengths. The distance of each satellite from the sun determines the mean temperature for the equilibrium condition that the absorbed radiation is equal to the emitted radiation:

$$\pi R^2 (F/r^2)(1 - A) = 4\pi R^2 \varepsilon \sigma T^4$$

or

$$T = \left(\frac{(1 - A)F}{4\sigma \varepsilon r^2} \right)^{1/4}$$

where R is the radius of the satellite, r is the sun-satellite distance, ε is the emissivity, σ is Stefan—Boltzmann's constant, A is the Bond albedo, and F is the incident solar flux (a slowly rotating body would radiate over $2\pi R^2$). Typical mean temperatures in degrees Kelvin for the satellites are: the Earth's Moon, 280; Io, 106; Titan, 97; the Uranian satellites, 60; and the Neptunian satellites, 45. For thermal equilibrium, measurements as a function of wavelength yield a blackbody curve characteristic of T: with the exception of Titan, the temperatures of the satellites closely follow the blackbody emission values. The discrepancy for Titan may be due to a weak greenhouse effect in the satellite's atmosphere.

Another possible use of radiometric techniques, when combined with photometric measurements of the reflected portion of the radiation, is the estimate of the diameter of a satellite. A more accurate method of measuring the diameter of a satellite from Earth involves measuring the light from a star as it occulted by the satellite. The time the starlight is dimmed is proportional to the satellite's diameter.

A third radiometric technique is the measurement of the thermal response of a satellite's surface as it is being eclipsed by its primary. The rapid loss of heat from a satellite's surface indicates a thermal conductivity consistent with a porous surface. Eclipse radiometry of Phobos, Callisto, and Ganymede suggests these objects all lose heat rapidly.

4. Polarimetry

Polarimetry is the measurement of the degree of polarization of radiation reflected from a satellite's surface. The polarization characteristics depend on the shape, size, and optical properties of the surface particles. Generally, the radiation is linearly polarized and is said to be negatively polarized if it lies in the scattering plane, and positively polarized if it is perpendicular to the scattering plane. Polarization measurements as a function of solar phase angle for atmosphereless bodies are negative at low phase angles; comparisons with laboratory measurements indicate this is characteristic of complex, porous surfaces consisting of multi-sized particles. In 1970, ground-based polarimetry of Titan that showed it lacked a region of negative polarization led to the correct conclusion that it has a thick atmosphere.

B. SPACECRAFT EXPLORATION

1. Imaging Observations

Interplanetary missions to the planets and their moons have enabled scientists to increase their understanding of the solar system more in the past 20 years than in the previous total years of scientific history. Analysis of data returned from spacecraft has led to the development of whole new fields of scientific endeavor, such as planetary geology. From the earliest successes of planetary imaging, which included the flight of a Soviet Luna spacecraft to the far side of the Earth's Moon to reveal a surface unlike that of the visible side, devoid of smooth lunar plains, and the crash landing of a United States Ranger spacecraft, which sent back pictures showing that the Earth's Moon was cratered down to meter scales, it was evident that interplanetary imaging experiments had immense capabilities. Table II summarizes the successful spacecraft missions to the planetary satellites.

The return of images from space is very similar to the transmission of television images. A camera records the level of intensity of radiation incident on its detector's surface. A series of scans is made across the detector to create a two-dimensional array of intensities. A computer onboard the spacecraft records these numbers and sends them by means of a radio transmitter to the Earth, where another computer reconstructs the image.

The first spacecraft to send pictures of a moon other then the Earth's was Mariner 9, which began orbiting Mars in 1971 and sent back images of Phobos and Deimos showing that these satellites are heavily cratered, irregular objects. Even more highly resolved images were re-

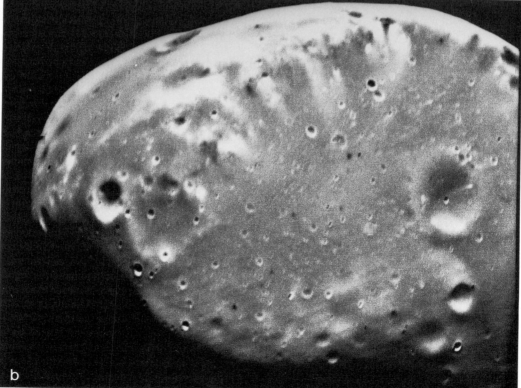

FIG. 2.   The two moons of Mars: (a) Phobos and (b) Deimos. Both pictures were obtained by the *Viking* spacecraft.

FIG. 3. A telescopic view of the moon, with the major features marked. (Photograph courtesy of Lick Observatory.)

turned by the Viking orbiters in 1976 (see Fig. 2). The Pioneer spacecraft, which were launched in 1972 and 1973 toward an encounter with Jupiter and Saturn, returned the first disk-resolved images of the Galilean satellites. By far the greatest scientific advancements were made by the Voyager spacecraft, which returned thousands of images of the Jovian and Saturnian satellites, the best of which are shown in Section IV. Color information for the objects was obtained by means of six broad-band filters attached to the camera. The return of large numbers of images with resolution down to a kilometer has enabled geologists to construct geologic maps, to make detailed crater counts, and to develop realistic scenarios for the structure and evolution of the satellites.

2. Other Experiments

Although images are the most spectacular data returned by spacecraft, a whole array of equally valuable experiments are included in each scientific mission. For example, a gamma ray spectrometer aboard the lunar orbiters was able to map the abundance of iron and titanium across the Moon's surface. Seismometers placed on the Moon recorded waves from small moonquakes. Measurements of heat flow from the interior of the Moon enabled scientists to understand something about its composition and present evolutionary state. On board the Voyager spacecraft were several experiments which are valuable for satellite investigations including an infrared spectrometer capable of mapping temperatures; an ultraviolet spectrometer; a photopolarimeter, which simultaneously measured the color, intensity, and polarization of light; and a radio science experiment that was able to measure the pressure of Titan's atmosphere by observing how radio waves passing through it were attenuated.

3. Future Missions

There are two currently scheduled spacecraft missions to the planetary satellites: the Voyager encounter to the satellites of Neptune (in 1989), and the Galileo mission to Jupiter. The Galileo spacecraft is due to be launched by the space shuttle in 1990 toward an encounter in 1996. It will consist of a probe to explore the Jovian atmosphere and an orbiter, which will make several close flybys of each Galilean satellite. The orbiter will contain both visual and infrared imaging devices, an ultraviolet spectrometer, and a photopolarimeter. The visual camera will be capable of obtaining images with 20 m resolution.

IV. Individual Satellites

A. THE EARTH'S MOON

1. Introduction

The Earth's Moon has played a key role in the lore and superstition of the world's peoples.

When Galileo first turned his telescope on the lunar disk in the early 1600s he clearly perceived the most obvious fact of lunar morphology: the demarcation of the Moon's surface into a dark smooth terrain and a brighter more mountainous terrain. This delineation is responsible for the appearance of the 'Man in the Moon'. The dark areas were called maria (Latin for seas; the singular is mare) because of their visual resemblance to oceans and the bright areas were called terrae (Latin for land; the singular is terra), or the lunar uplands. The lunar mountains were named after terrestrial mountain chains and the craters after famous astronomers. The appearance of the Moon through a large telescope is shown in Fig. 3.

The physical characteristics of the Moon are listed in Table 1. Since the Moon has been tidally despun, it is locked in synchronous rotation, and it was not until 1959 that the back side of the Moon was observed by a Soviet spacecraft. Because gravitational perturbations on the Moon cause it to wobble, or librate, in its orbit, about 60% of the surface of the Moon is visible from the Earth. The far hemisphere has fewer maria than the visible side.

According to theories of tidal evolution, the Moon is receding from the Earth and the Earth's spin rate is in turn slowing down. Thus in the past the lunar month and terrestrial day were shorter. Evidence for this is found in the fossilized shells of certain sea corals, which deposit calcareous material in a cycle following the lunar tides.

2. Origin

The three standard theories for the origin of the Moon are (1) Capture, in which the Moon forms elsewhere in the solar system and is gravitationally captured by the Earth; (2) Fission, which asserts that the Moon broke off from the Earth early in its formation; and (3) Coaccretion, which asserts that the Earth and Moon formed independently but nearly simultaneously near their present locations.

3. Early History of the Moon

Soon after the Moon accreted it heated up due to the reasons outlined in Section IIB. The result was the melting and eruption of basaltic lava onto the lunar surface between 3.8 to about 2.8 billion years ago to form the lunar maria. Such lava would be highly fluid under the weaker gravitational field of the Moon and could flow over vast distances.

Before Soviet and American spacecraft explored the Moon, there was considerable debate over whether the craters on the Moon were of impact or of volcanic origin. Morphological features, such as bright rays and ejecta blankets, around large craters show that they were formed by impacts.

The ringlike structures that delineate the maria are the outlines of impact basins, which filled in with lava. The maria are not as heavily cratered as the uplands because the lava flows which created them obliterated pre-existing craters.

4. Nature of the Lunar Surface

The physical properties of the Moon are known more accurately than those of any other satellite because of the extensive reconnaissance and study by spacecraft (see Table II). Rock samples have been returned by both American manned and Soviet unmanned missions.

The astronauts who walked on the Moon found that the upper few centimeters of the lunar surface was covered by fine dust and pulverized rock. This covering, or regolith, is the result of fragmentation of particles from constant bombardment of the Moon's surface during its history and may be structurally similar to the regoliths of other satellites.

Analysis of lunar samples revealed important chemical differences between the Earth and Moon. The lunar uplands consist of a low density, calcium-rich igneous rock known as anorthosite. The younger maria are composed of a dark basalt rich in the minerals olivine and pyroxene. Some of the lunar basalts, known as KREEP, were found to be anonymously rich in potassium (K), rare earth elements (REE), and phosphorous (P). Formations such as lava tubes and vents, which are similar to terrestrial volcanic features, are found in the maria.

The Moon has no water or atmosphere. Surface temperatures range from 100 to 380°K.

B. PHOBOS AND DEIMOS

Mars has two small satellites, Phobos and Deimos, which were discovered by the American astronomer Asaph Hall in 1877. In Jonathan Swift's moral satire *Gulliver's Travels* (published in 1726), a fanciful but coincidentally accurate prediction of the existence and orbital characteristics of two small Martian satellites was made. These two objects are barely visible in the scattered light from Mars in Earth-based telescopes. Most of what is known about Phobos and Deimos was obtained from the Mariner 9 and the Viking 1 and 2 missions to Mars (see Table II). Their physical and orbital properties are listed in Table I. Both satellites are shaped approximately like ellipsoids and are in synchronous rotation. Phobos, and possibly Deimos, has a regolith of dark material similar to that found on carbonaceous asteroids common in the outer asteroid belt. Thus the satellites may have been asteroids or asteroidal fragments, which were perturbed into a Mars crossing orbit and captured.

Both satellites are heavily cratered, which indicates that their surfaces are at least 3 billion years old (Fig. 2). However, Deimos appears to be covered with a fine, light-colored dust, which gives its surface a smoother appearance. The dust may exist because the surface is more easily pulverized by impacts, or simply because it is easier for similar material to escape from the gravitational field of Phobos, which is closer to Mars. The surface of Phobos is extensively scored by linear grooves that appear to radiate from the huge impact crater Stickney (named after the surname of Asaph Hall's wife, who collaborated with him). The grooves are probably fractures caused by the collision that produced Stickney. There is some evidence that tidal action is bringing Phobos, which is already inside Roche's limit, closer to Mars. The satellite will either disintegrate (perhaps to form a ring) or crash into Mars in about 100 million years. The suggestion that Phobos' orbit is decaying because it is a hollow extraterrestrial space station has no basis in fact.

C. THE GALILEAN SATELLITES OF JUPITER

1. Introduction and Historical Survey

When Galileo trained his telescope on Jupiter he was amazed to find four points of light which orbited the giant planet. These were the satellites Io, Europa, Ganymede, and Callisto, planet-sized worlds known collectively as the Galilean satellites. Analysis of telescopic observations over the next 350 years revealed certain basic features of their surfaces. There was spectroscopic evidence for water ice on the outer three objects. The unusually orange color of Io was hypothesized to be due to elemental sulfur. Orbital phase variations were significant, particularly in the cases of Io and Europa, which indicated the existence of markedly different terrains on their surfaces. Large opposition effects observed on Io and Callisto suggested their surfaces were porous, whereas the lack of an opposition effect on Europa suggested a smooth surface. The density of the satellites decreases as a function of distance from Jupiter (Table 1).

Theoretical calculations suggested the satellites had differentiated to form silicate cores and (in the case of the outer three) ice crusts. There is the possibility that the mantles of the outer three satellites are liquid water.

The Voyager missions to Jupiter in 1977 and 1979 (Table 2) revealed the Galilean satellites to be four unique geological worlds. Current knowledge of these objects is summarized in Fig. 4.

2. Io

About the size of the Moon, Io is the only body in the Solar System other than the Earth on which active volcanism has been observed. The Voyager spacecraft detected nine currently erupting plumes, scores of calderas (volcanic vents), and extensive lava flows consisting of nearly pure elemental sulfur (Fig. 5). As sulfur cools, it changes from dark brown, to red, to orange, and finally to yellow, which accounts for the range of colors on Io's surface. There are nearly black liquid sulfur lava lakes with floating chunks of solid sulfur. Sulfur dioxide is driven out of the volcanoes to condense or absorb onto the surface as white deposits. A thin, transient atmosphere with a pressure less than one millionth of the Earth's and consisting primarily of sulfur dioxide has been detected. There appears to be a total absence of water on the surface of Io, probably because it has all been degassed from the interior from extensive volcanism and escaped into space. The total lack of impact craters means the entire surface is young and geologically active.

The heat source for melting and subsequent volcanism is the dissipation of tidal energy from

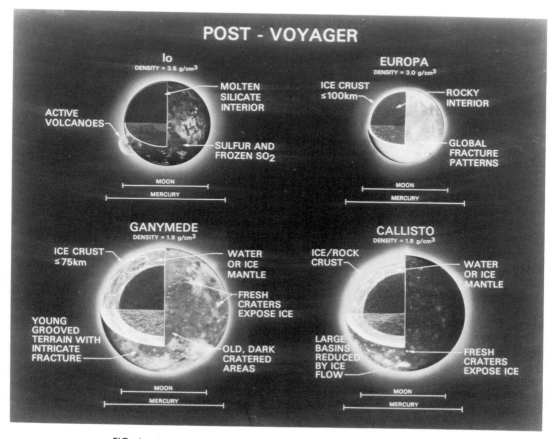

FIG. 4.   A summary of current knowledge of the Galilean satellites.

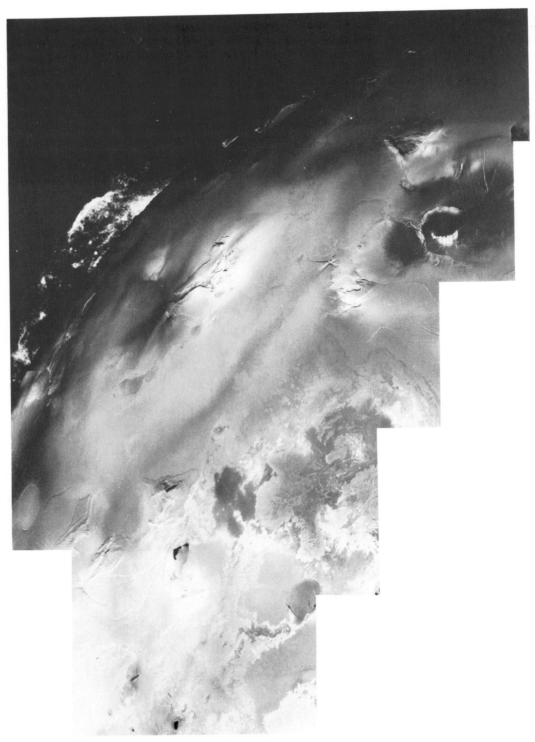

FIG. 5.   A highly processed *Voyager* image of the Galilean satellite Io, showing wispy structures in the volcanic plumes (upper left), volcanic vents, and sulfur lava flows.

Jupiter and the other Galilean satellites. As Io moves in its orbit, its distance from Jupiter changes as the other satellites exert different forces depending on their distance from Io. The varying tidal stresses cause Io to flex in and out. This mechanical energy is released as heat, which causes melting in Io's mantle.

A spectacular torus of ionized particles, primarily sulfur, oxygen, and sodium, co-rotates with Jupiter's strong magnetic field at Io's orbital position. High energy ions in the Jovian magnetosphere knock off and ionize surface particles, which are swept up and entrained in the field lines. An additional source of material for the torus is sulfur and sulfur dioxide from the plumes. Aurorae seen on Jupiter are caused by particles from the torus being conducted to the planet's polar regions. Because Io is a conducting body moving in the magnetic field of Jupiter it generates a flux tube of electric current between itself and the planet. Radio emissions from the Jovian atmosphere, which correlate with the orbital position of Io, appear to be triggered by the satellite.

3. Europa

When the Voyagers encountered the second Galilean satellite, Europa, they returned images of bright, icy plains crisscrossed by an extensive network of darker fractures (see Fig. 6). The existence of only a handful of impact craters suggested that geological processes were at work on the satellite until a few hundred million years ago or less. Europa is very smooth: the only evidence for topographical relief is the scalloped ridges with a height of a few hundred meters (see bottom of Fig. 6).

Part of Europa is covered by a darker mottled terrain. Dark features also include hundreds of brown spots of unknown origin, and larger areas, which appear to be the result of silicate laden water erupting onto the surface (bottom left of Fig. 6). The reddish hue of Europa is believed to be due to contamination by sulfur from Io.

The mechanism for the formation of cracks on Europa is probably some form of tidal interaction and subsequent heating, melting, and refreezing. Calculations show that Europa may still have a liquid mantle. Although some scientists have discussed the possibility of a primitive life form teeming in the mantle, there is no evidence that life does indeed exist there.

4. Ganymede

The icy moon Ganymede, which is the largest Galilean satellite, also shows evidence for geologic activity. A dark, heavily cratered terrain is transected by more recent, brighter grooved terrain (see Fig. 7). Although they show much diversity, the grooves are typically 10 km wide and one-third to one-half km high. They were implaced during several episodes between 3.5 and 4 billion years ago. Their formation may have occurred after a melting and refreezing of the core, which caused a slight crustal expansion and subsequent faulting and flooding by subsurface water.

The grooved terrain of Ganymede is brighter because the ice is not as contaminated with rocky material that accumulates over the eons from impacting bodies. The satellite is also covered with relatively fresh bright craters, some of which have extensive ray systems. In the cratered terrain there appear outlines of old, degraded craters, which geologists called palimpsests. The polar caps of Ganymede are brighter than the equatorial regions; this is probably due to the migration of water molecules released by evaporation and impact toward the colder high latitudes.

5. Callisto

Callisto is the only Galilean satellite that does not show evidence for extensive resurfacing at any point in its history. It is covered with a relatively uniform, dark terrain saturated with craters (Fig. 8). There is, however, an absence of craters larger than 150 km. Ice slumps and flows over periods of billions of years and is apparently not able to maintain the structure of a large crater as long as rocky material. One type of feature unique to Callisto is the remnant structures of numerous impacts. The most prominent of these, the Valhalla basin, is a bright spot encircled by as many as 13 fairly regular rings (as Figure 8).

6. The Small Satellites of Jupiter

Jupiter has eleven known small satellites, including three discovered by the Voyager mission. They are all probably irregular in shape (see Table I). Within the orbit of Io are at least three satellites: Amalthea, Adrastea and Metis. Amalthea is a dark, reddish heavily cratered object reflecting less than 5% of the radiation it receives; the red color is probably due to contamination by sulfur particles from Io. Little else

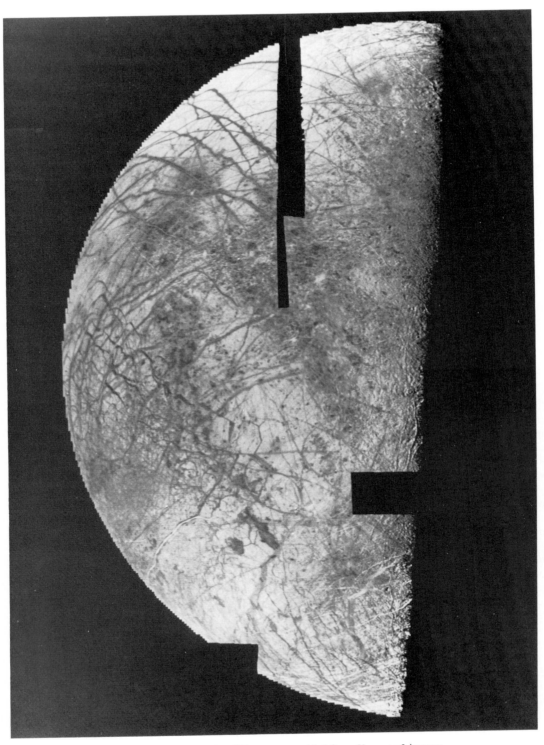

FIG. 6.   A photomosaic of Europa assembled from *Voyager 2* images.

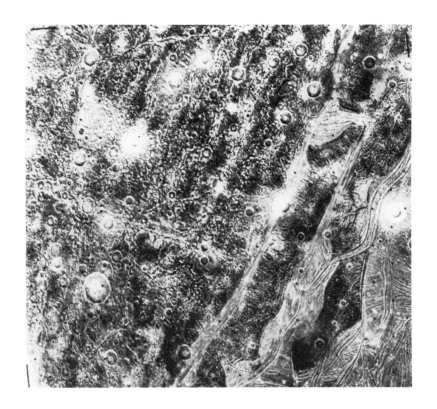

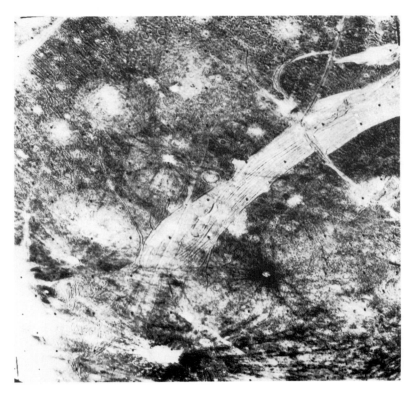

FIG. 7. *Voyager 2* images of Ganymede.

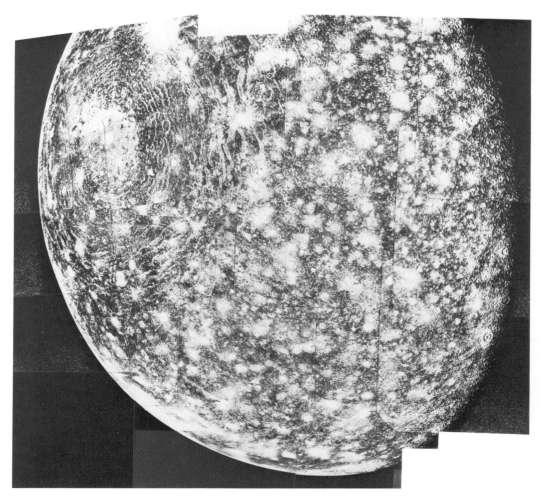

FIG. 8. A *Voyager 1* photomosaic of Callisto. The Valhalla impact basin, which is 600 km wide, dominates the surface.

is known about its composition except that the dark material may be carbonaceous.

Adrastea and Metis, both discovered by Voyager, are the closest known satellites to Jupiter and move in nearly identical orbits just outside the outer edge of the thin Jovian ring, for which they may be a source of particles. Between Amalthea and Io lies the orbit of Thebe, also discovered by Voyager. Little is known about the composition of these satellites, but they are most likely primarily rock-ice mixtures. The three inner satellites sweep out particles in the Jovian magnetosphere to form voids at their orbital positions.

Moving outward from Jupiter, we find a class of four satellites moving in highly inclined, orbits (Lysithea, Elara, Himalia, and Leda). They are dark, spectrally neutral objects, reflecting only 2 or 3% of incident radiation and may be similar to carbonaceous asteroids.

Another family of objects is the outermost four satellites, which also have highly inclined orbits, except they move in the retrograde direction around Jupiter. They are Sinope, Pasiphae, Carme, and Ananke, and they may be captured asteroids.

D. The Saturnian System

1. The Medium Sized Icy Satellites: Rhea, Dione, Tethys Mimas, Enceladus, and Iapetus

The six largest satellites of Saturn are smaller than the Galilean satellites but still sizable—as such they represent a unique class of icy satellite. Earth-based telescopic measurements showed the spectral signature of ice for Tethys, Rhea, and Iapetus; Mimas and Enceladus are close to Saturn and difficult to observe because of scattered light from the planet. The satellites' low densities and high albedos (Table I) imply that their bulk composition is largely water ice, possibly combined with ammonia. They have smaller amounts of rocky silicates than the Gali-

lean satellites. Resurfacing appears to have occurred on several of the satellites. Most of what is presently known of the Saturnian system was obtained from the Voyager flybys in 1980 and 1981.

The innermost medium-sized satellite Mimas is covered with craters, including one (named Arthur), which is as large as a third of the satellite's diameter (upper left of Fig. 9). The impacting body was probably nearly large enough to

FIG. 9.   The six medium-sized icy Saturnian satellites. From the upper left, in order of size: Mimas, Enceladus, Tethys, Dione, Rhea, and Iapetus.

break Mimas apart; such disruptions may have occurred to other objects. There is a suggestion of surficial grooves that may be features caused by the impact. The craters on Mimas tend to be high-rimmed, bowl shaped pits; apparently surface gravity is not sufficient to have caused slumping.

The next satellite outward from Saturn is Enceladus, an object that was known from telescopic measurements to reflect nearly 100% of the visible radiation incident on it (for comparison, the Moon reflects only about 11%). The

only likely composition consistent with this observation is almost pure water ice. When Voyager 2 arrived at Enceladus, it transmitted pictures to Earth which showed an object that had been subjected, in the recent geologic past, to extensive resurfacing; grooved formations similar to those on Ganymede were evident (see Fig. 10). The lack of impact craters on this terrain is consistent with an age less than a billion years. It is possible that some form of ice volcanism is presently active on Europa. The heating mechanism is believed to be tidal interactions, perhaps

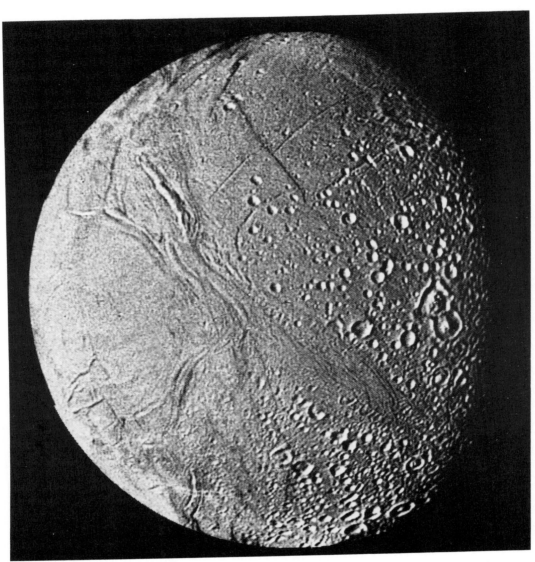

FIG. 10. A *Voyager 2* photomosaic of Enceladus. Both the heavily cratered terrain and the recently resurfaced areas are visible.

with Dione. About half of the surface observed by Voyager is extensively cratered and dates from nearly 4 billion years ago.

A final element to the enigma of Enceladus is the possibility that it is responsible for the formation of the E-ring of Saturn, a tenuous collection of icy particles that extends from inside the orbit of Enceladus to past the orbit of Dione. The position of maximum thickness of the ring coincides with the orbital position of Enceladus. If some form of volcanism is presently active on the surface, it could provide a source of particles for the ring. An alternative source mechanism is an impact and subsequent escape of particles from the surface.

Tethys is covered with impact craters, including Odysseus, the largest known impact structure in the solar system. The craters tend to be flatter then those on Mimas or the Moon, probably because of relaxation and flow over the eons under Tethys' stronger gravitational field. Evidence for resurfacing episodes is seen in regions that have fewer craters and higher albedos. In addition, there is a huge trench formation, the Ithaca Chasma, which may be a degraded form of the grooves found on Enceladus.

Dione, which is about the same size as Tethys, exhibits a wide diversity of surface morphology. Most of the surface is heavily cratered (Fig. 11), but gradations in crater density indicate that several periods of resurfacing events occurred during the first billion years of its existence. One side of the satellite is about 25% brighter than the other. Wispy streaks (see Figs. 9 and 11), which are about 50% brighter than the surrounding areas, are believed to be the result of internal activity and subsequent implacement of erupting material. Dione modulates the radio emission from Saturn, but the mechanism for this phenomenon is unknown.

Rhea appears to be superficially very similar to Dione (see Fig. 9). Bright wispy streaks cover one hemisphere. However, there is no evidence for any resurfacing events early in its history. There does seem to be a dichotomy between crater sizes—some regions lack large craters while other regions have a preponderance of such impacts. The larger craters may be due to a population of larger debris more prevalent during an earlier episode of collisions.

When Cassini discovered Iapetus in 1672, he noticed that at one point in its orbit around Saturn it was very bright; whereas, on the opposite side of the orbit it nearly altogether disappeared. He correctly deduced that one hemisphere is composed of highly reflective material, while the other side is much darker. Voyager images show that the bright side, which reflects nearly 50% of the incident radiation, is fairly typical of a heavily cratered icy satellite. The other side, which is centered on the direction of motion, is coated with a material with a reflectivity of about 3–4% (see Fig. 9).

Scientists still do not agree on whether the dark material originated from an exogenic source or was endogenically created. One scenario for the exogenic deposit of material entails dark particles being ejected from Phoebe and drifting inward to coat Iapetus. The major criticism of this model is that the dark material on Iapetus is redder than Phoebe, although the material could have undergone chemical changes after its expulsion from Phoebe to make it redder. One observation lending credence to an internal origin is the concentration of material on crater floors, which implies an infilling mechanism. In one model, methane erupts from the interior and is subsequently darkened by ultraviolet radiation.

Other aspects of Iapetus are unusual. It is the only large Saturnian satellite in a highly inclined orbit. It is less dense than objects of similar albedo; this implies a higher fraction of ice or possibly methane or ammonia in its interior.

2. Titan

Titan is a fascinating world that one member of the Voyager imaging team called 'a terrestrial planet in a deep freeze.' It has a thick atmosphere that includes a layer of photochemical haze (Fig. 12) and a surface possibly covered with lakes of methane or ethane. Methane was discovered by G. P. Kuiper in 1944: the Voyager experiments showed that the major atmospheric constituent is nitrogen, the major component of the Earth's atmosphere. Methane (which is easier to detect from Earth because of prominent spectroscopic lines) may comprise only a few percent or less. The atmospheric pressure of Titan is 1.5 times that of the Earth's; however, Titan's atmosphere extends much further from the surface (nearly 100 km) on account of the satellite's lower gravity. The atmosphere is thick enough to obscure the surface entirely.

Titan's density (Table I) implies a bulk composition of 45% ice and 55% silicates. It probably has a differentiated rocky core. Titan was able to retain an appreciable atmosphere, while the similarly sized Ganymede and Callisto were

FIG. 11.   The heavily cratered face of Dione is shown in this *Voyager 1* image. Bright wispy streaks are visible on the limb of the satellite.

not because more methane and ammonia condensed at Titan's lower formation temperature. The methane has remained; whereas, ammonia has been photochemically dissociated into molecular nitrogen and hydrogen, the latter being light enough to escape the gravitational field of Titan. The escaped hydrogen forms a tenuous torus at the orbital position of Titan. Although it has not been directly detected, argon may comprise a few percent of the atmosphere.

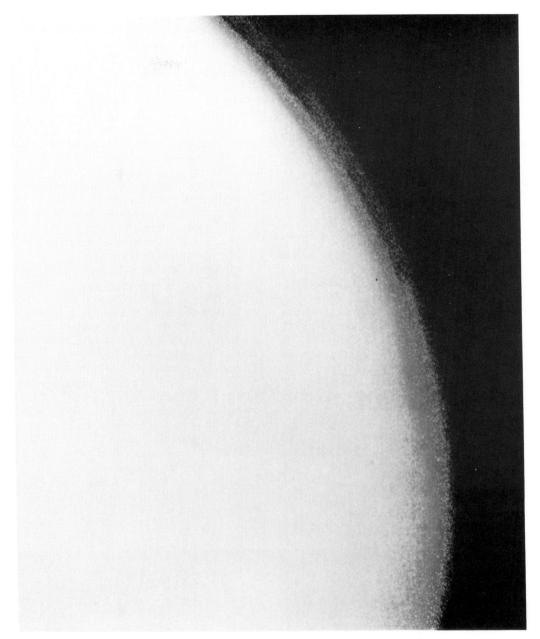

FIG. 12.   A *Voyager 1* image of Titan, showing the extended haze layer.

The infrared spectrometer on voyager detected nearly a dozen organic compounds such as acetylene (C_2H_2), ethane (C_2H_6), and hydrogen cyanide (HCN), which plays an important role in prebiological chemistry. These molecules, which are constantly being formed by the interaction of ultraviolet radiation with nitrogen and methane, constitute the haze layer of aerosol dust in the upper atmosphere. Much of this material, which gives Titan a reddish color, "rains" onto the surface and possibly forms lakes of ethane or methane. The role of methane on the surface and in the atmosphere of Titan may be similar to that of water on Earth; the triple point of methane (where it can coexist as a solid, liquid, or gas) is close to the surface temperature of Titan (93°K). In this scenario, methane in the atmosphere covers methane lakes and ice on the surface.

The northern polar area of Titan is darker than the southern cap. Ground-based observations detected long-term variations in the brightness of Titan. Both observations may be related to the existence of 30-year seasonal cycles on Titan. Even though only a small fraction of the incident solar radiation reaches the surface, Titan is probably subjected to a slight greenhouse effect.

3. The Small Satellites

The Saturnian system has a number of unique small satellites. Telescopic observations showed that the surface of Hyperion, which lies between the orbits of Iapetus and Titan, is covered with ice. Because Hyperion has a visual geometric albedo of 0.30, this ice must be mixed with a significant amount of darker, rocky material. It is darker than the medium-sized inner Saturnian satellites, presumably because resurfacing events have never covered it with fresh ice. Although Hyperion is only slightly smaller than Mimas, it has a highly irregular shape (see Table 1). This suggests, along with the satellite's battered appearance, that it has been subjected to intense bombardment and fragmentation. There is also evidence for Hyperion being in nonsynchronous rotation—perhaps a collision within the last few million years knocked it out of a tidally locked orbit.

Saturn's outermost satellite Phoebe, a dark object (Table I) with a surface similar to that of carbonaceous asteroids, moves in a highly inclined, retrograde orbit. Voyager images show definite variegations consisting of dark and bright (presumably icy) patches on the surface. Although it is smaller than Hyperion, Phoebe has a nearly spherical shape.

Three types of small satellites have been found only in the Saturnian system: the sheparding satellites, the co-orbitals, and the Lagrangians. All these objects are irregularly shaped (Fig. 13) and probably consist primarily of ice. The three shepherds, Atlas, 1980S26, and 1980S27, are believed to play a key role in defining the edges of Saturn's A and F rings. The orbit of Saturn's innermost satellite Atlas lies several hundred kilometers from the outer edge of the A-ring. The other two shepherds, which orbit on either side of the F-ring, not only constrain the width of this narrow ring, but may cause its kinky appearance.

The co-orbital satellites Janus and Epimetheus, which were discovered in 1966 and 1978, exist in an unusual dynamical situation. They move in almost identical orbits at about 2.5 Saturn radii. Every four years the inner satellite (which orbits slightly faster than the outer one) overtakes its companion. Instead of colliding, the satellites exchange orbits. The four-year cycle then begins over again. Perhaps these two satellites were once part of a larger body that disintegrated after a major collision.

The three remaining small satellites of Saturn orbit in the Lagrangian points of larger satellites: one is associated with Dione and two with

FIG. 13.   The small satellites of Saturn. They are, clockwise from far left; Atlas, 1980S26, Janus, Calypso, 1980S6, Telesto, Epimetheus, and 1980S27.

Tethys. The Lagrangian points are locations within an object's orbit in which a less massive body can move in an identical, stable orbit. they lie about 60 degrees in front of and in back of the larger body. Although no other known satellites in the solar system are Lagrangians, the Trojan asteroids orbit in two of the Lagrangian points of Jupiter.

E. THE SATELLITES OF URANUS: MIRANDA, ARIEL, UMBRIEL, TITANIA, AND OBERON

The rotational axis of Uranus is inclined 98 degrees to the plane of the solar system; observers on Earth thus see the planet and its system of satellites nearly pole-on. The orbits of Ariel, Umbriel, Titania, and Oberon are regular whereas Miranda's orbit is slightly inclined. Figure 14 is a telescopic image of the satellites. Theoretical models suggest the satellites are composed of water ice (possibly bound with carbon monoxide, nitrogen, and methane), and silicate rock. The higher density of Umbriel implies its bulk composition includes a larger fraction of rocky material. Melting and differentiation have occurred on some of the satellites. Theoretical calculations indicate that tidal interactions may provide an additional heat source in the case of Ariel.

Water ice has been detected spectroscopically on all five satellites. Their relatively dark albedos (Table I) are probably due to surficial contamination by carbonaceous material. Another darkening mechanism that may be important is bombardment of the surface by ultraviolet radiation. The four outer satellites all exhibit large opposition surges, which may indicate that the regoliths of these objects are composed of very porous material.

The Voyager 2 spacecraft encountered Uranus in January 1986 to provide observations indicating that at least some of the major satellites have undergone melting and resurfacing. One feature on Miranda consists of a series of ridges and valleys ranging from 0.5 to 5 km in height (Fig. 14b). Ariel, which is the geologically youngest of the five satellites, and Titania are covered with cratered terrain transected by grabens, which are fault-bounded valleys. Umbriel is heavily cratered and is the darkest of the major satellites, which indicates that its surface is the oldest. Oberon is similarly covered with craters, some of which have very dark deposits on their floors. The satellites are spectrally flat with

visual geometric albedos ranging from 0.2–0.4, which is consistent with a composition of water ice (or methane-water ice) mixed with a dark component such as graphite or carbonaceous chondritic material.

Voyager 2 also discovered 10 new small moons, including two which act as shepherds for the outer (epsilon) ring of Uranus (Table I). These satellites have visual geometric albedos of only 4–9%. They move in orbits that are fairly regularly spaced in radial distance from Uranus, and have low orbital inclinations and eccentricities.

F. THE SATELLITES OF NEPTUNE: TRITON AND NEREID

Neptune has an unusual family of satellites (Table I and Fig. 15). Triton, one of the largest satellites in the Solar System, moves in a highly inclined retrograde orbit. The orbit of Nereid, Neptune's other known satellite, is prograde and highly inclined. This situation implies an anomolous origin, perhaps involving capture of two remnant planetesimals by Neptune. The suggestion that Triton and Pluto were once both satellites of Neptune that experienced a near encounter, causing Triton to go into a retrograde orbit and the expulsion of Pluto from the system, is not plausible on dynamical grounds (see Section G). The detection of a third satellite in 1982 has not been confirmed.

The diameter and mass of Triton have not been well determined (Table I). The low formation temperature of the satellite constrains the composition to be primarily water ice, silicates, methane, and ammonia. The center of the satellite may consist of a differentiated rock-ice core, and a possibly liquid water-ammonia mantle. Methane ice has been detected on the surface and there is some spectral evidence for liquid nitrogen oceans or lakes. Because some of this liquid would evaporate, it can be deduced that Triton probably has a nitrogen atmosphere. If Triton was captured early in its history, it may show evidence of vast resurfacing events from a period of tidal heating.

Nereid is very faint and nothing about its surface composition has been directly observed.

G. THE PLUTO–CHARON SYSTEM

Soon after Pluto was discovered in 1930, scientists conjectured that the planet was an es-

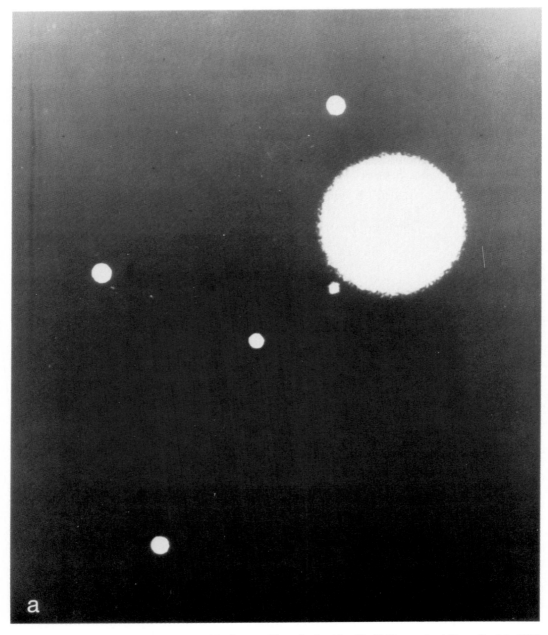

FIG. 14a.   Telescopic view of Uranus and its five satellites obtained by Ch. Veillet on the 154-cm Danish–ESO telescope. Outward from Uranus they are: Miranda, Ariel, Umbriel, Titania, and Oberon.

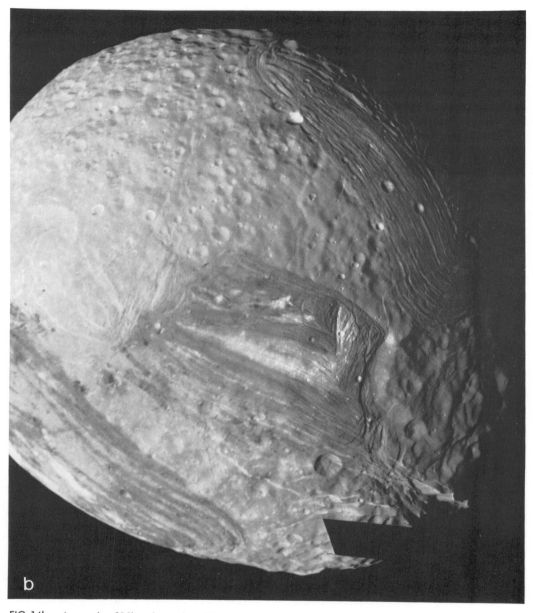

FIG. 14b.   A mosaic of Miranda produced from images taken by the *Voyager 2* spacecraft at 30–40 thousand km from the moon. Resolution is 560 to 740 m. Older, cratered terrain is transected by ridges and valleys indicating more recent geologic activity.

FIG. 15.   The two satellites of Neptune: Triton (near Neptune), and the faint Nereid. (Photograph courtesy of Lick Observatory.)

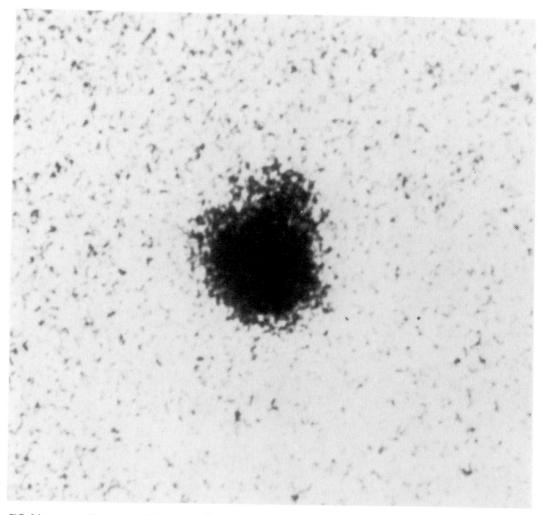

FIG. 16.   A negative image of the Pluto–Charon system obtained by J. W. Christy. Charon is the extended blob to the upper right of Pluto. (Photo courtesy of the U.S. Naval Observatory.)

caped satellite of Neptune. More recent calculations have shown that it is unlikely that Pluto could have acquired its present amount of angular momentum during an ejection event. Pluto is most likely a large planetesimal dating from the formation period of the solar system.

In 1978, Pluto was shown to have a moon of its own, which was named Charon (Fig. 16). Charon appears in specially processed high quality images as a fuzzy mass orbiting close to Pluto. It is more massive in comparison with its primary than any other satellite: its mass is between 5 and 10% that of Pluto. Nothing is known about the surface composition of Charon. Its low formation temperature implies it is probably a conglomerate of ices (including possibly solid nitrogen), and some small fraction of silicates. Between 1985 and 1991, Pluto and Charon will be undergoing a series of mutual eclipses, which will allow a more accurate determination of their radii, masses, and thus density. A more accurate knowledge of Charon's density can be used to constrain its bulk composition.

BIBLIOGRAPHY

Beaty, J. K., B. O'Leary, and A. Chaikin (eds.) (1981). *The New Solar System,* 2nd ed. Sky Publishing Corp., Cambridge Mass.

Burns, J. (ed.) (1977). *Planetary Satellites,* University of Arizona Press, Tucson.

Gehrels, T. (ed.) (1984). *Saturn*, University of Arizona Press, Tucson.

Hartmann, W. K. (1983). *Moons and Planets*, 2nd ed., Wadsworth, Belmont, California.

Morrison, D. (1980). *Voyage to Jupiter*, NASA SP–439. U.S. Government Printing Office, Washington, D.C.

Morrison, D. (1982). *Voyage to Saturn*, NASA SP–451. U.S. Government Printing Office, Washington, D.C.

Morrison, D. (ed.) (1982). *The Satellites of Jupiter*, University of Arizona Press, Tucson, 1982.

Soderblom, L. A. (1980). *The Galilean satellites of Jupiter*, Scientific American **242**, 88–100.

PLANETARY WAVES (METEOROLOGY)

Roland A. Madden   *National Center for Atmospheric Research**

GLOSSARY

Coriolis parameter: Twice the component of the earth's angular velocity about the local vertical, $2\Omega \sin \phi$, where Ω is the angular speed of the earth (7.292×10^{-5}/sec) and ϕ the latitude. It is equal to the component of the earth's vorticity about the local vertical, which is referred to as planetary vorticity in the text.

Del-operator (∇): Vector differential operator expressed in rectangular Cartesian coordinates as

$$\nabla = \mathbf{i}\,\frac{\partial}{\partial x} + \mathbf{j}\,\frac{\partial}{\partial y} + \mathbf{k}\,\frac{\partial}{\partial z}$$

where $\mathbf{i}$, $\mathbf{j}$, and $\mathbf{k}$ are unit vectors directed along the x, y, and z coordinates, respectively.

Divergence: Expansion or spreading out of a fluid. It is written as

$$\nabla \cdot \mathbf{V}$$

or

$$\frac{\partial u}{\partial x} + \frac{\partial v}{\partial y} + \frac{\partial w}{\partial z}$$

where ∇ is the del-operator, $\mathbf{V}$ the vector wind, and u, v, and w are wind components in the x, y, and z directions, respectively.

Equivalent depth: For the atmosphere, depth of a fluid of constant density whose free waves are similar to those of the atmosphere.

Exponential form: Alternate way to express waves, since $e^{i\theta} = \cos \theta + i \sin \theta$.

Inertial coordinate system: For purposes in meteorology, system with the origin on the axis of the earth and fixed with respect to the stars. When a coordinate system that is rotating with the earth is used, apparent forces arise in Newton's laws such as the Coriolis acceleration.

Laplace tidal equations: Equations, first developed by Laplace, that describe oscillations of an ocean of uniform depth covering a rotating planet.

Meridional direction: Northerly or southerly direction (e.g., along a longitude line).

Millibar (mb): Unit of pressure commonly used in meteorology. It is equal to 1 hectopascal or 100 Newtons/m^2.

Newton's second law: Time rate of change of momentum of a body is proportional to the resultant force acting on the body and in the direction of the resultant force.

Normal mode: Any one of a number of simple harmonic oscillations that can be superimposed or summed to give the oscillation of a linear system. Each normal mode has a characteristic structure and frequency and can exist independently of all other normal modes.

Pressure force: Force due to differences of pressure in a fluid. The pressure force per unit mass is given by $-(1/\rho)\,\nabla p$ where ρ is the density, ∇ the del-operator, and p the pressure. ∇p is the pressure gradient (meteorologists sometimes refer to $-\nabla p$ as the pressure gradient).

Relative vorticity: Vector measure of the local rotation in fluid flow, which is equal to the curl of the velocity vector

$$\mathbf{Q} = \nabla \times \mathbf{V}$$

where ∇ is the del-operator and $\mathbf{V}$ the vec-

* The National Center for Atmospheric Research is sponsored by the National Science Foundation.

tor wind. Of main concern is the component of **Q** along the local vertical, given in Cartesian coordinates by

$$\zeta = \frac{\partial v}{\partial x} - \frac{\partial u}{\partial y}$$

It is this vertical component that is referred to in the text. In the northern hemisphere relative vorticity is positive (negative) around low (high) pressure regions. The absolute vorticity is the relative vorticity plus the planetary vorticity given by the Coriolis parameter.

Shear: Variation of the wind in some direction (e.g., meridional shear is the shear in the north–south direction).

Stationary wave/force: Wave or force that is geographically fixed; that is, it does not move relative to a longitude or latitude line.

Streamline: Line whose tangent at any point in a fluid is parallel to the instantaneous velocity of the fluid at that point.

Vortex line: Curve that at each of its points is tangent to the instantaneous axis of rotation or vorticity vector of a fluid.

Vortex tube: Configuration marked out by all vortex lines that intersect a closed curve.

Zonal direction: Easterly or westerly direction (e.g., along a latitude line).

Planetary waves are, in the words of C.-G. Rossby, waves "... whose shape, wave length, and displacement are controlled by variations of the Coriolis parameter with latitude." They are large-scale waves such that the Coriolis parameter, or planetary vorticity, changes significantly across their dimensions. Unlike gravity waves, which exist with or without rotation, planetary waves can exist only on rotating planets. They are also called waves of the second class or second kind, terms coined at the end of the last century by researchers studying tides in the atmosphere and ocean. In contrast, waves of the first class are gravity waves. In the 1930s, Rossby, a meteorologist, studied waves on a uniform current in a flat, nondivergent fluid system, rotating with varying angular speed about the local vertical. Although these are considerable simplifications of an atmosphere on a rotating spherical planet, the waves of this model do capture the essence of planetary waves. They are called Rossby waves. As a result of his work, meteorologists and oceanographers often refer to planetary waves as Rossby waves. B.

Haurwitz studied both free and forced Rossby waves on a rotating plane and extended Rossby's work to include nondivergent waves on a rotating sphere. The latter are the limiting case of planetary waves in an infinitely deep fluid and are called Rossby–Haurwitz waves.

I. Theoretical Basis

A. BACKGROUND

Because of the complexities of atmospheric flow, accurate theoretical descriptions of planetary waves are difficult. Indeed, the most realistic treatments are relegated to computer calculations based on mathematical models of the atmosphere. The results of these calculations are often as complicated as the flow they represent. Nevertheless, numerical modeling of the atmosphere allows controlled experiments that provide important insights. At the other extreme are analytical approaches, which make considerable simplifying assumptions. Though relatively unrealistic, these simple models can isolate the essential dynamics. Some simple models are considered in detail here, and relevant results of more realistic models including the numerical calculations are only mentioned.

Equations of motion are the basis for a theoretical description of atmospheric flow. They represent an application of Newton's second law. Newton's laws apply to motion measured in a nonaccelerating or inertial coordinate system. Since we observe air motions relative to coordinate systems rotating with the earth, it is necessary to modify the equations of motion to take this into account. For meteorological considerations the most important effect of this rotation is equal to the product of the Coriolis parameter and the horizontal wind speed. This product is called the Coriolis acceleration. The Coriolis parameter itself is named after G. G. Coriolis, the French physicist who first analyzed it. The Coriolis parameter is equal to the component of the earth's vorticity about the local vertical. This planetary vorticity plays an essential role in atmospheric dynamics. Its value is shown as a function of latitude in Fig. 1.

B. CONSERVATION OF ABSOLUTE VORTICITY

A simple model that describes the fundamental dynamics of planetary waves is formulated in the so-called β-plane geometry illustrated in Fig.

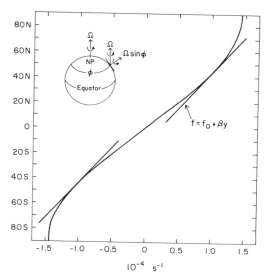

FIG. 1.   The Coriolis parameter or planetary vorticity ($f = 2\Omega \sin \phi$) as a function of latitude ϕ (curved line). Straight lines indicate the β-plane approximation to f assuming origins ($y = 0$) at 45°N or 45°S. The component of the earth's angular velocity about the local vertical is illustrated in the upper left.

1. The spherical earth is replaced by a plane with rectangular coordinates (x, y), where lines of constant y represent parallels or latitudes, and those of constant x are meridians or longitudes. The Coriolis parameter is assigned a value f and its derivative with respect to y a constant value β. Because Rossby first studied this model, β is called the Rossby parameter. The atmosphere is regarded as an incompressible, frictionless, and homogeneous fluid. Furthermore, orbits of fluid parcels are constrained to lie in horizontal planes. The appropriate equations of motion can then be written as

$$\frac{du}{dt} - fv = -\frac{1}{\rho}\frac{\partial p}{\partial x} \tag{1}$$

$$\frac{dv}{dt} + fu = -\frac{1}{\rho}\frac{\partial p}{\partial y} \tag{2}$$

Here t is time, u the velocity in the x-direction (positive when moving toward the east), and v the velocity in the y-direction (positive when moving toward the north). The pressure is p, and the density ρ is a constant due to the assumption of incompressibility and homogeneity. Terms $-fv$ and fu are components of the Coriolis acceleration, and du/dt and dv/dt are particle accelerations as measured in the rotating co-

ordinate system. Equations (1) and (2) are statements of Newton's second law of motion with accelerations per unit mass equal to applied forces per unit mass. Only one force, that of pressure which is always directed from high to low values, is considered. Meteorologists commonly replace the total derivatives representing particle accelerations by partial derivatives expressing velocity changes at a fixed point and so-called advective changes separately. Equations (1) and (2) become

$$\frac{\partial u}{\partial t} + u\frac{\partial u}{\partial x} + v\frac{\partial u}{\partial y} - fv = -\frac{1}{\rho}\frac{\partial p}{\partial x} \tag{3}$$

$$\frac{\partial v}{\partial t} + u\frac{\partial v}{\partial x} + v\frac{\partial v}{\partial y} + fu = -\frac{1}{\rho}\frac{\partial p}{\partial y} \tag{4}$$

The vertical component of the curl of the velocity vector is determined by differentiating Eq. (3) with respect to y and subtracting the result from Eq. (4) after it has been differentiated with respect to x. This eliminates the pressure dependence and gives

$$\frac{\partial \zeta}{\partial t} + u\frac{\partial}{\partial x}(\zeta + f) + v\frac{\partial}{\partial y}(\zeta + f)$$

$$+ (\zeta + f)\left(\frac{\partial u}{\partial x} + \frac{\partial v}{\partial y}\right) = 0 \tag{5}$$

where $\zeta = \partial v/\partial x - \partial u/\partial y$ and is the relative vorticity about the local vertical, f is the planetary vorticity, and $\zeta + f$ the absolute vorticity. By virtue of the assumed incompressibility and the constraint on particle velocities to lie in horizontal planes (vertical velocities equal zero), the horizontal divergence $\partial u/\partial x + \partial v/\partial y$ is everywhere zero, and Eq. (5) becomes

$$\frac{d}{dt}(\zeta + f) = 0 \tag{6}$$

This is the conservation of absolute vorticity principle. In the northern hemisphere an air parcel with no relative vorticity initially develops negative relative vorticity if it is displaced northward to latitudes where the planetary vorticity is larger (larger Coriolis parameter). This decrease of ζ can occur either through a change in horizontal shear of the wind or through an increase in clockwise turning of the streamlines. The conservation of absolute vorticity principle, derived for this simple model and in its more realistic forms, provides a basis for dynamic meteorology. Its implications for large-scale atmospheric flow are considered in the following mathematical discussions.

C. Some Fundamental Equations of Meteorology

Before proceeding to a wave equation based on the conservation of absolute vorticity principle, it is appropriate to expand the discussions of the equations of motion and introduce a few relations that are commonly used in meteorology. The geostrophic wind is derived from Eqs. (1) and (2) by assuming that accelerations are zero:

$$v_g = \frac{1}{\rho f} \frac{\partial p}{\partial x}$$

$$u_g = -\frac{1}{\rho f} \frac{\partial p}{\partial y}$$

The geostrophic wind is a wind for which the horizontal pressure force exactly balances the Coriolis acceleration. It is nearly equal to the actual wind associated with large-scale flow at middle and high latitudes. This is particularly true away from the surface where frictional forces due to irregular terrain are negligible.

In the northern hemisphere Coriolis accelerations deflect air particles to the right of their motion, so the geostrophic wind must be balanced by a pressure force directed to the left. That is the reason for the familiar rule of thumb that high pressure lies to the right of an observer looking downwind. (The opposite is true in the southern hemisphere.) As a consequence, lines of equal pressure drawn on a weather map indicate the approximate wind direction since geostrophic winds blow parallel to them. In addition, the pressure gradient or its spatial rate of change is proportional to the pressure force and the wind speed.

Above the surface, contours drawn on weather maps or upper air charts are not those of equal pressure at a constant height level, but rather they are contours of equal height on a constant pressure level. To write the geostrophic wind equations for a constant pressure surface it is necessary to introduce the hydrostatic equation:

$$\frac{\partial p}{\partial z} = -\rho g$$

Here z is the vertical coordinate and g the acceleration of gravity. The hydrostatic equation expresses a balance between vertically directed pressure and gravity forces (hydrostatic balance). It states that pressure falls off in the vertical direction in direct proportion to the air density.

An important implication of the hydrostatic equation can be understood by substituting from the perfect gas law for the density ρ. The perfect gas law is

$$p = RT\rho$$

where R is the gas constant for dry air and T the temperature. The hydrostatic equation becomes

$$\frac{1}{p} \frac{\partial p}{\partial z} = -\frac{g}{RT}$$

and it is seen that warming (cooling) a column of air increases (decreases) the distance between two constant pressure surfaces, or stretches (shrinks) a column of air. The hydrostatic equation also allows one to express the geostrophic wind equations given above as a function of the height Z of a constant pressure surface:

$$v_g = \frac{g}{f} \frac{\partial Z}{\partial x}$$

$$u_g = -\frac{g}{f} \frac{\partial Z}{\partial y}$$

Finally, the horizontal divergence is equal to the fractional change of the horizontal area occupied by the fluid per unit time. That is,

$$\frac{\partial u}{\partial x} + \frac{\partial v}{\partial y} = \frac{1}{A} \frac{dA}{dt}$$

where A is the area $dx\, dy$. For an incompressible fluid, conservation of mass requires that a volume element hA must remain constant, so that

$$h \frac{dA}{dt} + A \frac{dh}{dt} = 0$$

where h is the depth of the fluid element. From this, the continuity equation for an incompressible fluid can be written as

$$\frac{\partial u}{\partial x} + \frac{\partial v}{\partial y} = -\frac{1}{h} \frac{dh}{dt}$$

The continuity equation in this form allows, in an elementary way, isolation of the effect of mountain forcing on planetary waves, which is discussed in Section I,E, on forced waves.

D. Free Waves: The Homogeneous Wave Equation

To minimize mathematical difficulties, equations are linearized by assuming that the flow consists of a basic state that is steady, or unchanging with time, plus some small perturbations from that state. For example, the implications of Eq. (6) can be examined further by assuming that the wind along the x or zonal direction is a constant plus a small perturbation, and the wind along the y or meridional direction is simply a small perturbation from zero. That is,

$$u = \bar{u} + u' \tag{7a}$$

$$v = v' \tag{7b}$$

where $\bar{u}$ is assumed independent of t, x, and y. Equations (7) are substituted into Eq. (6), and u' and v' are further assumed to be independent of y. The result is a second-order, homogeneous, linear, partial differential equation of the form

$$\frac{\partial^2 v'}{\partial t\, \partial x} + \bar{u}\frac{\partial^2 v'}{\partial x^2} + \beta v' = 0 \tag{8}$$

Terms involving products of perturbation velocities have been dropped to arrive at Eq. (8), so it does not apply to large amplitude perturbations. The Rossby parameter β is equal to df/dy. Equation (8) states that local changes in relative vorticity ($\partial^2 v'/\partial t\, \partial x$) are determined by the zonal advection of relative vorticity ($\bar{u}\, \partial^2 v'/\partial x^2$) and the meridional advection of planetary vorticity ($\beta v'$). A solution to Eq. (8) takes the form of sinusoids independent of y, given by

$$v' = A e^{im(x-ct)} \tag{9}$$

where A is an arbitrary amplitude, m a wavenumber in x ($m = 2\pi/L$ where L is the wavelength), and c the wave speed in the x-direction relative to the earth (positive when moving toward the east). Substitution into Eq. (8) gives

$$c = \bar{u} - \frac{\beta}{m^2} \equiv \bar{u} - \frac{\beta L^2}{4\pi^2} \tag{10}$$

This is the Rossby wave equation. It states that, in the absence of any mean wind ($\bar{u} = 0$) Rossby waves move westward. Furthermore, the larger the scale of the wave (the smaller the wave number m) the greater is the size of the β term and the faster the westward propagation. In general, for u positive (typical eastward flow), the Rossby wave equation predicts that waves of very large scale would move westward ($\beta/m^2 > \bar{u}$), some of intermediate scale would be stationary ($\beta/m^2 = \bar{u}$), and waves of smaller scale would move eastward ($\beta/m^2 < \bar{u}$).

We can understand Rossby wave propagation in a qualitative way by considering air parcels that conserve absolute vorticity as in Eq. (6). The planetary vorticity f associated with a parcel moving north increases (Fig. 1), so the rela-

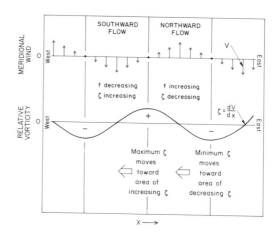

FIG. 2. Schematic showing the effect of northward- and southward-moving air parcels (top) on their accompanying relative vorticity (ζ) field (bottom). A consequence of the conservation of absolute vorticity principle is that ζ decreases for northward-moving parcels because their planetary vorticity f increases (see Fig. 1). The opposite is true for southward-moving parcels. Regions of decreasing (increasing) ζ lie west of minima (maxima) in ζ, so with no u wind the relative vorticity field moves to the west.

tive vorticity ζ must decrease. The opposite is true for southward-moving parcels. For a wave-like motion of northward- and southward-moving air as shown in Fig. 2, we find that northward-moving parcels and the resulting local decreases in relative vorticity are west of minima in relative vorticity. Similarly, southward-moving parcels and local increases in relative vorticity are west of maxima. With no $\bar{u}$ wind the relative vorticity field is displaced westward with time. If there is an eastward flow ($\bar{u} > 0$), then it advects the relative vorticity field to the east, and the wave displacement is determined by which effect is largest.

E. Forced Waves: The Inhomogeneous Wave Equation

An important consideration that is overlooked in Eq. (8) is the fact that there are geographically fixed sources and sinks of vorticity. The most important of these are due to mountains and heating and cooling patterns resulting largely from the distribution of sea and land. Equation (8) can be modified to include a general force acting to change the vorticity such that

$$\frac{\partial^2 v'}{\partial t\, \partial x} + \bar{u}\frac{\partial^2 v'}{\partial x^2} + \beta v' = K(x) \tag{11}$$

Now Eq. (11) is a second-order, inhomogeneous, linear, partial differential equation. $K(x)$ is a forcing function that is assumed, for simplicity, to depend only on x. A solution of Eq. (11) takes the form

$$v' = Ae^{im(x-ct)} + V(x) \qquad (12)$$

where the first term on the right-hand side is the general solution of the reduced, or homogeneous, equation [Eq. (8)], and $V(x)$ is a particular solution of the complete, or inhomogeneous, equation. Substitution of Eq. (12) into Eq. (11) gives a time-dependent part, already discussed in Section I,D, and a forced part that is independent of time, given by

$$\frac{d^2V}{dx^2} + \frac{\beta}{\bar{u}} V = \frac{K}{\bar{u}} \qquad (13)$$

Equation (13) is a second-order, inhomogeneous, linear, ordinary differential equation. Again, solutions of the reduced equation of Eq. (13) take the form $e^{imx} = \cos mx + i \sin mx$, with

$$m = \sqrt{\frac{\beta}{\bar{u}}} \qquad (14)$$

that is, the stationary wave ($c = 0$) of Eq. (10). The complete Eq. (13) can be solved by determining the Green's function through the method of variation of parameters. As a result,

$$V(x) = \frac{\sin(\sqrt{\beta/\bar{u}}x)}{\bar{u}\sqrt{\beta/\bar{u}}} \int_0^x K(x) \cos(\sqrt{\beta/\bar{u}}x)\, dx$$
$$- \frac{\cos(\sqrt{\beta/\bar{u}}x)}{\bar{u}\sqrt{\beta/\bar{u}}} \int_0^x K(x) \sin(\sqrt{\beta/\bar{u}}x)\, dx \qquad (15)$$

Here the boundary conditions are chosen such that $V = 0$ at the point $x = 0$ where the forcing $K(x)$ is assumed to start. With a knowledge of the forcing, the time-independent forced waves are described by Eq. (15).

To isolate the effect of mountain forcing it is necessary to relax the constraint of zero vertical velocity. This requires retaining the effect of horizontal divergence in Eq. (5). Equation (6) becomes

$$\frac{d}{dt}\left(\frac{\zeta + f}{h}\right) = 0 \qquad (16)$$

where h is the fluid depth. Equation (16) is a form of the conservation of potential vorticity principle. This, like Eq. (6), is a basic principle of dynamic meteorology. It is more realistic than Eq. (6) because it includes the effect of divergence. Beginning with Eq. (16) and following the steps outlined to go from Eq. (6) to Eq. (8) leads to

$$\frac{\partial^2 v'}{\partial t\, \partial x} + \bar{u}\frac{\partial v'}{\partial x} + v'\beta$$
$$= -\frac{(\partial v'/\partial x + f)}{h}\left(\bar{u}\frac{\partial h'}{\partial x} - v'\frac{\partial \bar{h}}{\partial y}\right) \qquad (17)$$

Here the depth of a constant density atmosphere is $h = \bar{h}(y) - h'(x)$, where $h'(x)$ is the height above a certain reference level (say sea level) of an idealized mountain. The mountain is assumed to extend infinitely in the y direction. The depth of the undisturbed atmosphere above the reference level is $\bar{h}(y)$, and it is a function of y to allow geostrophic $\bar{u}$. For a geostrophic wind of 10 m/sec, $\partial\bar{h}/\partial y$ is roughly 100 m per 1000 km at $\phi = 43°$, and v' is a small perturbation. The product $\bar{u}\,\partial h'/\partial x$ is normally much larger than that of $v'\,\partial\bar{h}/\partial y$, and typically $\partial v'/\partial x \ll f$, so the forcing function of the idealized mountain can be written as

$$K(x) = -\bar{u}\frac{f}{h}\frac{\partial h'}{\partial x} \qquad (18)$$

Equation (18) also assumes that the height of the mountain is much less than the depth of the undisturbed atmosphere ($h' \ll \bar{h}$).

In the case of mountain forcing it is easy to see that a vortex tube is shrunk moving up the mountain, and therefore its absolute vorticity or positive spin (northern hemisphere) must decrease; it stretches going down the mountain, and its absolute vorticity increases. This can be accomplished by changes in relative vorticity ζ or by movement to lower or higher latitudes changing the planetary vorticity f. The hydrostatic equation indicates that an analogous effect results from changes in heating. A vortex tube moving to a colder region cools and shrinks as the distance between two pressure surfaces decreases. Conversely, warming causes vortex tubes to stretch as the distance between pressure surfaces increases. Illustrations of mountain and heating forced vorticity change are shown in Fig. 3.

By way of example, consider a narrow continent 2000 km wide in the northern hemisphere that extends infinitely in the north–south direction. We specify a wind from the west, $\bar{u} = 10$ m/sec and $h = 8500$ m (the depth of a homogeneous atmosphere, $\rho = $ constant). Then Eq. (18) gives a negative vorticity forcing at $\phi = 43°$ of -3.5×10^{-11}/sec^2 for the first 1000 km, either through a cooling of the air column of -10 K/1000 km or through a rise in orography of 300 m/1000 km. For the second 1000 km it provides a force for positive vorticity in the opposite way.

The particular or forced solution to Eq. (11) is, from Eq. (15),

$$V(x) = \frac{K}{\beta}[1 - \cos(\sqrt{\beta/\bar{u}}x)]$$

for $0 \leq x \leq 1000$ km

$$V(x) = \frac{K}{\beta}[2\cos(\sqrt{\beta/\bar{u}}(x - 1000))$$

$$- 1 - \cos(\sqrt{\beta/\bar{u}}x)]$$

for 1000 km $\leq x \leq 2000$ km

$$V(x) = \frac{K}{\beta}[2\cos(\sqrt{\beta/\bar{u}}(x - 1000))$$

$$- \cos(\sqrt{\beta/\bar{u}}(x - 2000)) - \cos(\sqrt{\beta/\bar{u}}x)]$$

for $x \geq 2000$ km

with the forcing K equal to -3.5×10^{-11} and $\beta = 1.67 \times 10^{-11}$. The stationary wavelength for a Rossby wave at $\phi = 43°$ with $\bar{u} = 10$ m/sec is 4860 km and equal to $2\pi/\sqrt{\beta/\bar{u}}$. It should be noted here that including a y or meridional dependence in the wave solution given by Eq. (9) increases the stationary zonal wavelength. A stationary zonal wavelength equal to $2\pi/\sqrt{\beta/\bar{u}}$ is for waves of infinite meridional scale, and the smaller the meridional scale the more the stationary zonal wavelength exceeds this value.

Figure 4 shows the results of such forcing. V at $x = 0$ is zero but is directed southward in the region of negative vorticity forcing. An air parcel is displaced southward and to smaller values of planetary vorticity f as a result. The relative vorticity $\partial V/\partial x$ decreases too. The absolute vorticity $\partial V/\partial x + f$ continues to decrease for the first 1000 km, then it increases under the positive vorticity forcing for the next 1000 km. Thereafter it remains constant. Nevertheless, the forcing region has set up a downstream wave whose length is equal to that of the stationary wave. As air parcels move northward ($V > 0$) beyond the forcing region, their relative vorticity decreases in favor of an increasing f. When they cross the original latitude, their relative vorticity is zero, as it was before they encountered the forcing, but V is still directed northward. As a result, their relative vorticity becomes negative as their planetary vorticity increases until the parcels eventually turn southward. The oscillating motion continues indefinitely, exchanging vorticity between relative and planetary vorticity but keeping the absolute vorticity constant. The wavelength in the region of no forcing is equal to the stationary wavelength regardless of the size of the forcing; however, the nearer the size of the forcing region is to the stationary wavelength the larger is the amplitude.

F. Damped Waves: The Effect of Friction

Without a continuing source of energy, planetary waves once excited would not maintain a constant amplitude A but would rather decrease in amplitude with time. This is primarily a result of eddy friction which acts to weaken the flow and dissipate gradients. Besides eddy friction, infrared radiation, which is proportional to the fourth power of the temperature, acts to weaken temperature perturbations associated with the waves. The effect of such dissipating mechanisms can be illustrated by considering that they dissipate relative vorticity and that the rate of dissipation is linearly related to the relative vorticity itself. Equation (8) becomes

$$\frac{\partial^2 v'}{\partial t\, \partial x} + \bar{u}\frac{\partial^2 v'}{\partial x^2} + \beta v' = -\nu\frac{\partial v'}{\partial x} \qquad (19)$$

where ν is a proportionality constant called the coefficient of linear friction. Substituting from Eq. (9) into Eq. (19) results in a complex phase

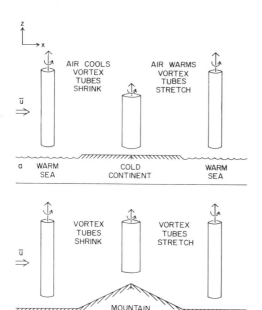

FIG. 3. Schematic of the effect of heating and mountains on vortex tubes. Shrinking causes a decrease in positive (northern hemisphere) vorticity while stretching causes an increase.

speed. The imaginary part represents an amplitude decay with time such that

$$v' = (Ae^{-\nu t})e^{im(x-ct)} \qquad (20)$$

These damped waves have a phase speed similar to the free waves, but their amplitudes decay exponentially in time. A parallel analysis of the steady state equation, Eq. (13), yields a complex wave number m in which the imaginary part represents a similar amplitude decay. In this case

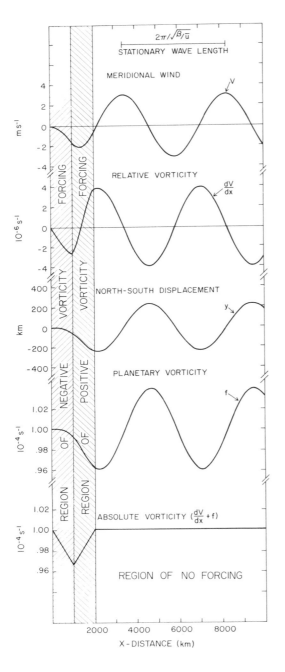

the amplitude decays exponentially with increasing distance from the stationary forcing. That is, the decay is a function of x rather than t. Friction increases the stationary wavelength slightly as well.

G. FREE WAVES ON A SPHERE

For very large-scale disturbances it is necessary to include the effect of the spherical shape of the earth. B. Haurwitz considered the conservation of vorticity in nondivergent flow on a sphere. He showed that in the absence of a mean current planetary waves would travel westward with a period τ, in days, of

$$\tau = \frac{n(n+1)}{2m} \qquad (21)$$

Here m is the zonal wave number and n an index that defines the meridional dependence of a given wave mode. In the β plane model the waves were unbounded in both the zonal and meridional directions. On the sphere they are bound to integer wave numbers in the zonal direction, and their pressure perturbations must have zero amplitude at the poles. Their zonal structures are sinusoids, and their meridional structures are described by sums of associated Legendre functions. That is,

$$p'_{m,n}(\phi) = \frac{n+1}{n}\sin(\phi)P_n^m - \frac{n-m+1}{n(n+1)}P_{n+1}^m$$

$$(22)$$

where $p'_{m,n}(\phi)$ is the meridional dependence of a pressure perturbation (or height perturbation of a constant pressure surface) of the m,nth mode, and P_n^m is an associated Legendre function of order m and degree n. The latitudinal structure

FIG. 4. Result of stationary forcing similar to that of Fig. 3. An air parcel moving eastward (toward larger x) is assumed to have no relative vorticity at $x = 0$. Forcing of negative vorticity equal to $-3.5 \times 10^{-11}/\text{sec}^2$ is imposed from $x = 0$ to $x = 1000$ km as might result from cooling or from rising up mountain range. Positive vorticity forcing of the same magnitude is imposed from $x = 1000$ km to x equals 2000 km. The stationary wavelength at $\phi \approx 43°$ with $\bar{u} = 10$ m/sec is shown at the top with the forced v-wind and relative vorticity just below. The north–south displacement of the parcel and the resulting planetary vorticity approximated by $f = f_{\phi=43} + \beta y$ are next. The lowest curve is the absolute vorticity, which is decreased in the first 1000 km and then increased to its original value in the second 1000 km. Waves produced downstream by the forcing have wavelengths equal to the stationary wavelength.

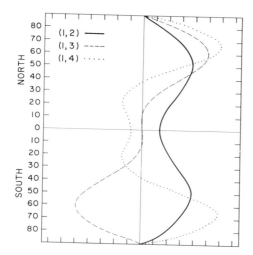

FIG. 5. Meridional structures of the pressure or height perturbation field for $m = 1$, $n = 2, 3, 4$ normal modes of the nondivergent vorticity equation on a sphere. Amplitudes are arbitrary. These structures are very nearly the same as those of the normal modes of Laplace's tidal equations (Hough modes) as well. (After R. Madden, *Rev. Geophys. Space Phys.* **17**, 1935–1949, 1979.)

of the $m = 1$, $n = 2, 3, 4$ modes determined from Eq. (22) are shown in Fig. 5. Waves with all m and n ($n \geq m$) values make up the normal modes of the system. These planetary waves are often called Rossby–Haurwitz waves.

The effect of divergence is important for large-scale waves, and Haurwitz pointed out that the nondivergent normal modes defined by Eqs. (21) and (22) are a special case of the divergent modes or waves of the second class studied by M. Margules in Germany and by S. S. Hough in England in the late nineteenth century. The waves of the first class are gravity waves, which exist with or without rotation. Waves of the second class result from the rotation of the planet, and they travel much more slowly than gravity waves. The waves of the second class are planetary waves and normal modes of Laplace's tidal equations. The mathematical expressions that define their horizontal structures are now called Hough functions. The waves are referred to as Hough modes and signified by H_n^m. They are sinusoids in the zonal direction and sums of associated Legendre polynomials in the meridional direction. For a constant density fluid they depend on its depth. In the case of the real atmosphere they depend on the equivalent depth, which itself is dependent on temperature and thought to range between 8 and 12 km. For the largest scale waves and this range of equivalent depths the coefficients of the associated Legen-

dre polynomials in a Hough function are small beyond the first two, so Eq. (22) and Fig. 5 describe the meridional structures of the corresponding Hough functions fairly well.

For equivalent depths of 8 to 12 km, propagation of the Hough normal modes is slower than that of the nondivergent Rossby–Haurwitz normal modes. Rossby–Haurwitz modes are identical in structure and propagation speed to Hough modes for infinite equivalent depths. Table I presents the propagation speeds and periods of some of the large-scale normal modes for nondivergent flow on a sphere (Rossby–Haurwitz modes) and for Laplace's tidal equations (Hough modes) with no mean wind [$\bar{u}(\phi) = 0$] and with a mean wind [$\bar{u}(\phi) \neq 0$]. Horizontal structures of the largest-scale normal modes of Laplace's tidal equations with realistic mean winds are similar to Hough functions. Important to note is that larger scales either in longitude (small m) or in latitude (small n) move westward faster. Furthermore, realistic winds lengthen the local period of all but the largest and fastest-moving modes. Relatively large-scale modes, as most of those listed in Table I, have discrete frequencies, but in the presence of winds with meridional shear smaller-scale modes do not. This feature makes it impossible to identify a stationary mode, as could be done for the Rossby wave equation [Eq. (10); see also Eq. (14)].

H. SUMMARY

The discussion has divided planetary waves into free and forced. However, even free waves need some forcing to excite them. They then propagate freely but would eventually decay as a result of dissipative processes without additional forcing. The largest scales move most rapidly westward. Free waves take the form of the normal modes of the atmosphere. The amplitude of an excited free wave depends on how closely the forcing projects onto a normal mode in space and time. That is, a forcing whose spatial distribution is similar to that of a given normal mode oscillating with a frequency near that of the normal mode would excite a large amplitude free wave. The more that the space and time variations of the forcing differ from a given mode the smaller is its excited amplitude.

Orography and irregular heating over the earth provide nearly stationary sources and sinks of vorticity, that force nearly stationary planetary waves. The effects of stationary heating evolve slowly from season to season, and

TABLE I.   Propagation Speeds and Periods of Some Large-Scale, Normal-Mode, Planetary Waves[a]

Source	\(n - m \)			
	1	2	3	4
m = 1				
Rossby–Haurwitz mode[b]	−120(3.0)	−60(6.0)	−36(10.0)	−24(15.0)
Hough mode[c]	−72(5.0)	−43(8.3)	−29(12.3)	−21(17.3)
Hough mode[d]	−74(4.9)	−36(9.9)	−20(18.4)	−13(28.1)
m = 2				
Rossby–Haurwitz mode[b]	−60(3.0)	−36(5.0)	−24(7.5)	−17(10.5)
Hough mode[c]	−48(3.7)	−30(5.9)	−21(8.5)	−16(11.5)
Hough mode[d]	−47(3.8)	−25(7.3)	−13(14.2)	−8(21.5)
m = 3				
Rossby–Haurwitz mode[b]	−36(3.3)	−24(5.0)	−17(7.0)	−13(9.3)
Hough mode[c]	−32(3.7)	−22(5.5)	−16(7.6)	−12(10.0)
Hough mode[d]	−28(4.3)	−16(7.4)	−9(13.7)	ND
m = 4				
Rossby–Haurwitz mode[b]	−24(3.8)	−17(5.2)	−13(7.0)	−10(9.0)
Hough mode[c]	−23(4.0)	−16(5.6)	−12(7.4)	−10(9.4)
Hough mode[d]	−17(5.2)	−11(8.2)	−7(13.6)	ND

[a] Propagation speeds are in degrees longitude per day (negative mean westward-propagating). Periods in days are in parenthesis.
[b] Rossby–Haurwitz modes, Eq. (22).
[c] Hough modes with 10-km equivalent depth and no mean wind (after M. S. Longuet-Higgins, *Phil. Trans. Roy. Soc. London, Ser. A,* **262,** 511–607, 1968).
[d] Hough modes with 10-km equivalent depth and December–January–February average 500-mb winds (after A. Kasahara, *J. Atmos. Sci.* **37,** 917–929, 1980). ND in the table means these modes do not have discrete periods.

even those of orography can change because they depend on the overall large-scale flow. An example of this dependence is included in the Rossby wave equation in the form of the zonal wind $\bar{u}$. It is natural that the stationary forced waves change slowly through the year, but the seasonal change in forcing is very slow relative to periods expected for most normal modes, so it is unlikely that it can effectively excite free waves. Any very slowly moving normal modes that might respond would be damped by dissipative processes in a small fraction of their expected period. Furthermore, it is unlikely that smaller scale, very slow modes can exist, because they would be seriously affected by the mean wind. However, there is an abundance of fast-varying phenomena in the atmosphere that could excite the large-scale fast-moving normal modes. Systems that we are accustomed to seeing on day-to-day weather maps introduce large-scale vertical motions and temperature gradients that can develop and dissipate quickly. The intensity of latent heat release is constantly changing from region to region, and the strength of the

mean flow impinging on mountain ranges varies continually as well. These can all serve as forcing functions for large-scale, fast-moving normal modes.

Theory suggests, then, that observations should reveal nearly stationary forced waves which bear a consistent relationship to mountains and heat sources, and occasional freely propagating, large-scale, normal modes which should exhibit certain structures and propagation speeds. These relationships, structures, and propagation speeds are well defined in the models so far discussed, but they are less so in the real atmosphere because of the complexity of the background wind and temperature fields in which they are embedded.

II. Observations

A. BACKGROUND

The analysis of the β-plane model leads to simple expectations about the structure of forced and free planetary waves in the atmosphere. Although the dynamical basis of these expectations is correct, comparisons with the atmosphere are not completely satisfactory. For example, the forcing of vorticity by irregular heating was illustrated through advective processes from warm to cold regions and vice versa. In reality, heating of air parcels also occurs through the release of latent heat and radiation processes, and the horizontal and vertical distribution of this heating is not known well enough to predict accurately its effects on the stationary waves. Furthermore, computer simulations based on relatively realistic mathematical models indicate that the effect of heating is important only in lower atmospheric levels. This means that above the lower portion of a vortex tube that cools and shrinks as it moves from warm sea to a cold continent (Fig. 3), there could actually be warming and stretching. These same simulations suggest it is the effect of mountains that dominates in upper levels of the atmosphere. In addition, the existence and structure of the free, traveling normal modes of the atmosphere cannot be predicted with certainty. However, computer simulations indicate that, at least in the troposphere (the lower 10–16 km of the atmosphere), the horizontal structures of the very largest-scale normal modes look like Hough functions and should be reasonably robust.

Accordingly, one might expect observations to reveal stationary planetary waves whose behavior is only qualitatively similar to that illustrated in Figs. 3 and 4 and large-scale traveling planetary waves similar to Hough waves. Meridional structures of some Hough modes are presented in Fig. 5, and likely propagation speeds and periods are contained in Table I under source d (Hough modes with realistic winds).

B. Observed Forced Waves

There is no certain way to isolate forced waves in the atmosphere, but by averaging weather maps over several days one can eliminate the migratory systems and get an approximate picture of the forced waves from the averaged flow. Figure 6 shows such averaged 500-mb height contours for January 15. There is a marked contrast between the two hemispheres. The northern hemisphere flow is more wavelike than that of the southern hemisphere. This presumably results from the fact that there are stronger vorticity sources in the north due to more irregular topography and heating there.

A crest in height contours appears over the west coast of North America and a trough near the east coast. This represents negative and positive relative vorticity, respectively, and is consistent with the theoretical expectation of forcing by the Rocky Mountain chain as well as that of a cold continent bounded by warm water. Downstream to the east there is another crest near the west coast of Europe and a minor

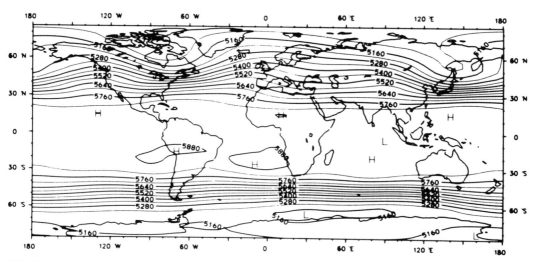

FIG. 6.   Average 500-mb height contours for January 15. Their values in meters are indicated on every other contour. Areas marked H and L are areas of relative maximum or minimum height. They correspond to high and low pressure regions at a constant height level of about 5500 m. Northern hemisphere analysis is based on 30 years of data while that in the southern hemisphere is based on eight.

trough over eastern Europe. The distance from the trough off the North American east coast to that over eastern Europe is about 110° of longitude at 45°N. That is approximately an 8000-km wavelength. The mean zonal wind at that latitude is 23 m/sec, and from Eq. (14) [or Eq. (10) with $c = 0$] the stationary wave number is $\sqrt{\beta/\bar{u}}$ and the wavelength $2\pi/\sqrt{\beta/\bar{u}}$, or 7500 km. Since stationary waves with finite meridional scales exceed this length, comparison with observations is surprisingly good. This is at least partly fortuitous, since there must be additional forcing between the trough off North America's east coast and that over central Europe in addition to the simplifications of the Rossby wave equation.

The mountains and highlands of the Mongolian Plateau and extreme eastern Siberia contribute to the crest–trough pattern extending from about 90°E to the east coast of Asia at midlatitudes. The relatively warm ocean water works in concert with topography to force positive vorticity and the coastal trough.

There is a weak trough immediately downstream from the Andes mountains of South America. It is considerably weaker than the one over eastern North America. This may reflect the fact that the region of forcing is relatively narrow and does not project well onto the likely preferred larger-scale stationary waves. For example, Eq. (10) or (14) indicates a stationary wavelength of about 8300 km for the observed $\bar{u} = 28$ m/sec at 45°S latitude. The width of South America is only about 800 km at that latitude. It should also be noted that in January in extratropical latitudes, the South American continent is slightly warmer than the surrounding ocean water, which would tend to force vorticity opposite to the forcing of the mountains.

C. Observed Free Waves

Because the amplitudes of the forced stationary waves are relatively large, it ordinarily requires some time and/or space filtering of data to isolate the free, traveling planetary waves. There often is ambiguity in identifying observed disturbances with a specific global-scale normal mode as well. However, one planetary wave that is ubiquitous is the so-called five-day wave. Figure 7 is an average picture of this wave. It appears as a single wave in longitude (zonal wave number one), and its meridional structure is similar to that of the H_2^1 mode shown in Fig. 5. It travels westward around the earth in five

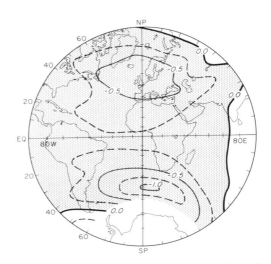

FIG. 7.   Average negative pressure perturbations of a westward-propagating wave. Positive perturbations are present on the other side of the earth. The pattern moves westward taking five days to move around the earth. The horizontal structure is similar to the H_2^1 normal mode ($m = 1$, $n = 2$ in Fig. 5). (After R. Madden and P. Julian, *J. Atmos. Sci.* **30**, 935–940, 1973.)

days, which is the period predicted for the H_2^1 mode.

Another regularly appearing disturbance that behaves as a free planetary wave is also a zonal wave number one in longitude; it takes one to three weeks to go around the earth. Its typical meridional structure, in the northern hemisphere from 850 mb (~1500 m) to 30 mb (~24,000 m), is presented in Fig. 8. It is not certain whether or not there is a symmetric or antisymmetric extension of this wave in the southern hemisphere. As a result it has not been unambiguously identified with a specific normal mode planetary wave. The analysis that produced Fig. 8 is restricted to one hemisphere, and so it may reflect both symmetric and antisymmetric modes. However, the average period of about 16 days (compare with Table I) and strong indications that the wave is out-of-phase between latitudes north and south of 45°, as in the H_4^1 mode of Fig. 5, suggest that this mode is dominant.

Important aspects brought out by Fig. 8 are the increase of amplitude with increasing height and the fact that the wave does not slope much with height. These are features predicted by theoretical treatments that consider the vertical structure of free planetary waves. The first of these means that they should have largest ampli-

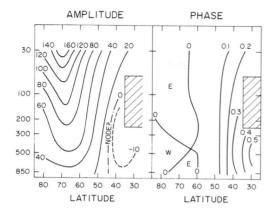

FIG. 8. Average amplitudes and phases in fractions of a cycle of an observed westward-propagating wave with an average period of 16 days. Phases are relative to the wave at 60°N and 500 mb. The vertical orientation of phase lines means the wave slopes vary little with height. The change from zero phase to 0.5 phase indicates that the wave is out-of-phase between latitudes north of 50° and those south of about 40°. Westward propagation was not evident at 30° latitude and 200 and 100 mb. The wave is zonal wave number one and below 300 mb is similar to the H_4^1 normal mode ($m = 1$, $n = 4$ in Fig. 5). (After R. Madden, *J. Atmos. Sci.* **40**, 1110–1125, 1983.)

tudes in the upper atmosphere. Not surprisingly then, the antisymmetric H_3^1 mode has been detected in the high atmosphere. Data from satellites provide coverage nearly from pole to pole. Figure 9 shows the antisymmetric nature of a

TABLE II. Observed Period in Days Determined by Projecting Actual Data onto the Structures of Normal-Mode Planetary Waves[a]

	$n - m$		
1	2	3	4
	$m = 1$		
4–6	7.5–20	12–30	NP[b]
	$m = 2$		
3.5–5	6–13	10–30	12–30
	$m = 3$		
4–5.5	6–14	NP	NP
	$m = 4$		
5–7.5	6–14	NP	NP

[a] After J. Alquist, *J. Atmos. Sci.* **39**, 193–202, 1982.
[b] NP in the table means these modes did not show regular westward propagation.

wave evident at 1 mb (~48 km) during April 1981. This particular wave moved westward with a period of 10 days in good agreement with the period predicted for the H_3^1 mode in the presence of realistic winds.

Finally, Table II represents a summary of some traveling planetary waves that have been detected. Periods associated with the various waves that appear in Table II were determined by first projecting observed height data onto normal modes of the form indicated in Fig. 5. Then the periods for each mode were estimated from

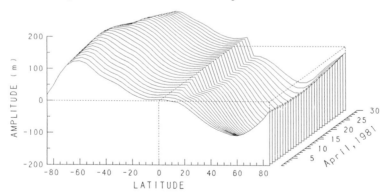

FIG. 9. Three-dimensional plot of the meridional structure of the height perturbation of a wave observed at 1 mb during April 1981. Data are derived from satellite measurements. Amplitudes are in meters. The fact that southern hemisphere (negative latitudes) amplitudes are positive and northern hemisphere amplitudes are negative reflects the fact that this wave is out-of-phase between the hemispheres. It takes the form of one wave around a longitude, and it propagates westward with a 10-day period. It is similar to the H_3^1 normal mode ($m = 1$, $n = 3$ in Fig. 5). (After T. Hirooka and I. Hirota, *J. Atmos. Sci.* **42**, 536–548, 1985.)

the data. They turn out to be reasonably consistent with predicted periods for the various modes. This method is in contrast to the one used to produce Figs. 7, 8, and 9 where the data, first filtered in time to isolate variations of a certain period, showed westward propagation and latitudinal structures consistent with the normal mode predicted to have that certain period.

In summary, two approaches have been used to isolate traveling planetary waves in observed data. Data filtered in time to examine a certain period reveal westward propagation and latitudinal structures that are predicted for that period, and data filtered in space by projecting onto the spatial structure of a certain normal mode reveal westward propagation and periods that are predicted for that certain mode. Only the very largest-scale traveling planetary waves have been identified in data.

III. Role of Planetary Waves in Weather and Climate

Migratory cyclones that are a major factor in our day-to-day weather are associated with smaller-scale waves than the forced, stationary and free, westward-propagating planetary waves. These smaller-scale waves typically form as a result of instabilities. That is, small perturbations amplify at the expense of the large-scale flow. Figure 10 is presented to illustrate the point. There is a crest–trough pattern over North America in approximately the same position of the forced planetary waves of Fig. 6. The pattern is distorted, however, by two smaller-scale waves which are evident over the west coast of North America and the southern tip of Greenland. They are shown to be linked with frontal zones at the surface and at 500 mb that separate cold air masses from warmer ones. These smaller waves and frontal systems move from west to east. On occasion they can amplify to an extent that they dominate the larger planetary waves; however, often their movement parallels the contours of the large-scale waves. Changes in the planetary waves can, as a result, have a profound effect on the day-to-day weather of an area. For example, during the northern winter of 1976–77, the stationary crest normally situated over the west coast of North America as shown in Fig. 6 was considerably stronger than normal. The trough that extends to the east of the Rocky Mountains was similarly anomalously strong. The result was drought in

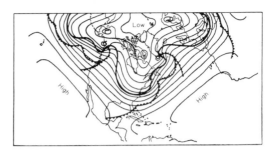

FIG. 10. Schematic 500-mb weather map. (After Palmén, in "Compendium of Meteorology," 1951.) Thin lines represent the height contours of the 500-mb pressure surface. The wind blows approximately parallel to these contours with high pressure to the right, or predominantly from west to east. There is a crest–trough pattern over North America similar to that expected for the forced, stationary planetary wave (compare with Fig. 6). Two smaller-scale waves are indicated by the troughs over the west coast of North America and the southern tip of Greenland. They are shown to be associated with frontal zones at 500 mb (smooth heavy line) and at the surface (irregular heavy line). These smaller scale waves and frontal systems typically move eastward, steered by the larger-scale upper-level flow.

California since normal rain-bearing storms from the Pacific were steered far to the north. Correspondingly, the eastern third of the United States experienced record low temperatures brought on by weather systems that originated in the Canadian arctic and followed the strong northerly flow of the deeper-than-normal trough southward.

In the northern hemisphere the forced stationary planetary waves transport heat from low and middle latitudes to high latitudes. This is accomplished because northward-moving air in the waves tends to be warmer than the zonal average while southward-moving air is colder. They similarly are able to transport zonal momentum, although the direction of this transport is dependent on latitude. These processes are the subject of considerable importance in better understanding wave–mean flow interactions.

The free planetary traveling waves transport little or no heat because their temperature and meridional wind perturbations are one quarter of a cycle out of phase. Similarly, they transport little or no zonal momentum. Interestingly, they may cause large fluctuations in transports as they move westward, constructively and destructively interfering with the forced stationary waves. There is speculation that this sort of interference may be partially responsible for the quasi-periodicities called index cycles that occur in the large-scale flow.

Another interesting aspect of planetary waves is their behavior on numerical weather predictions. Earliest numerical forecasts were based on Eq. (6). It was soon recognized that the largest-scale waves moved much more rapidly westward in the forecasts than in reality. Of course the westward propagation is a natural result of Rossby's wave equation [Eq. (10)]. Empirically determined terms were added to the forecast equation to keep the long waves stationary, and they were successful in improving short-range forecasts on a time scale of one to two days. A fundamental problem of forecast models based on Eq. (6) is that they included no mountains or heat sources to force stationary waves. As a result all the energy observed in the large-scale waves drifts westward as free planetary waves. Keeping all the energy of the large-scale waves stationary improved short-range forecasts, because the true large-scale stationary waves typically have more energy than free traveling waves that might be present, and because traveling waves do not move far in one or two days (the H_2^1 five-day wave excepted) in any case. [See WEATHER, PREDICTION, NUMERICAL.]

Efforts are under way to extend skillful numerical predictions to theoretical limits, thought to be about two weeks. The models that are used are much more sophisticated than Eq. (6). They include many vertical levels as well as mountains and various heat sources and sinks. It is generally recognized that if these efforts are to be successful, forecast models will need to simulate more faithfully forcing of vorticity by the mountains and irregular heating. In that way they will be able to predict more accurately the evolution of forced and free traveling planetary waves over longer forecast periods.

Planetary waves in the atmosphere have provided fertile ground for research by theoreticians and empiricists alike. Much is still to be learned. It is clear that as the understanding of these waves continues to improve so will the understanding and prediction of weather and climate.

BIBLIOGRAPHY

Gill, A. E. (1982). "Atmosphere–Ocean Dynamics." Academic Press, New York.

Holton, J. R. (1979). "An Introduction to Dynamic Meteorology," 2nd ed. Academic Press, New York.

Holton, J. R. (1975). "The Dynamic Meteorology of the Stratosphere and Mesosphere," Meteor. Monogr. No. 37. American Meteorological Society, Boston.

Hoskins, B. J., and Pearce, R. P., eds. (1983). "Large-Scale Dynamical Processes in the Atmosphere." Academic Press, New York.

Madden, R. A. (1979). *Rev. Geophys. Space Phys.* **17,** 1935–1949.

Platzman, G. W. (1968). *Quart. J. Roy. Meteor. Soc.* **94,** 225–248.

Salby, M. L. (1984). *Rev. Geophys. Space Phys.* **22,** 209–236.

PLANETS—*SEE* SOLAR SYSTEM, GENERAL

PLASMA CONFINEMENT

Allen H. Boozer *Princeton University*

GLOSSARY

Action-angle variables: Canonical coordinates θ, J, t, which are particularly useful in the study of Hamiltonian systems in which the motion remains in a finite region of space. In these coordinates, the Hamiltonian is a function of the action J alone with the angle θ obeying $d\theta/dt = \partial H/\partial J$. Action-angle coordinates do not always exist for the general one degree of freedom Hamiltonian $H(q, p, t)$, but if they do, the system is said to be integrable.

Ambipolar: Equal loss rate of ions and electrons. If there is more than one particle-loss process, the individual processes are generally nonambipolar. However, quasi-neutrality forces the sum of all the loss processes to be ambipolar.

Canonical coordinates: Any set of quantities q, p, and t that are related to each other by Hamilton's equations, $dp/dt = -\partial H/\partial q$ and $dq/dt = \partial H/p$. The function $H(q, p, t)$ is called the Hamiltonian. A canonical coordinate always has a so-called canonically conjugate quantity. The coordinates q and p are conjugate as are the canonical time t and the Hamiltonian H.

Canonical transformations: Let q, p, t and $\bar{q}$, $\bar{p}$, $\bar{t}$ be two sets of canonical coordinates with Hamiltonians H and $\bar{H}$ such that the barred quantities can be written as smooth functions of the unbarred quantities. That is, $\bar{q} = \bar{q}(q, p, t, H)$, etc. Then, the barred quantities are a canonical transformation of the unbarred quantities.

Degrees of freedom: Number of coordinatelike variables in the Hamiltonian. For example $H(q_1, q_2, p_1, p_2, t)$ has two degrees of freedom.

Generating function: Function $S(q, \bar{p}, \bar{t}, H)$, which produces a canonical transformation between q, p, t, H and $\bar{q}$, $\bar{p}$, $\bar{t}$, $\bar{H}$ through the relations $\bar{q} = \partial S/\partial \bar{p}$, $p = \partial S/\partial q$, $t = \partial S/\partial H$, $\bar{H} = \partial S/\partial \bar{t}$. An infinitesimal canonical transformation is produced by $s(q, \bar{p}, \bar{t}, H)$ with $S = q\bar{p} + \bar{t}H + \varepsilon s$ as ε goes to zero.

Hamiltonian mechanics: Motion of a particle of mass m and position q, which is subjected to a force $\partial\Phi(q, t)/dq$, is given by $md^2q/dt^2 = -\partial\Phi/\partial q$. This motion can also be described by the Hamiltonian $H(q, p, t) = p^2/2m + \Phi$ and Hamilton's equations, $dp/dt = -\partial H/\partial q$ and $dq/dt = \partial H/\partial p$. Hamiltonian mechanics is the study of the mathematical properties of Hamilton's equations with a general Hamiltonian $H(q, p, t)$ as well as the higher dimensional analogues. For example, with two degrees of freedom the Hamiltonian is $H(q_1, q_2, p_1, p_2, t)$ with $dp_1/dt = -\partial H/\partial q_1$, $dp_2/dt = -\partial H/\partial q_2$, $dq_1/dt = \partial H/\partial p_1$, and $dq_2/dt = \partial H/\partial p_2$.

KAM theorem: KAM stands for Kolmogorov, Arnold, and Moser, who showed that the topology of Hamiltonian trajectories is unchanged by sufficiently small nonresonant perturbations.

Poloidal: Short way around the torus, which is the θ direction of Fig. 4.

Toroidal: Long way around the torus, which is the ϕ direction of Fig. 4.

Virial theorem: Relation between the time-averaged kinetic and potential energy of bounded classical mechanics motion.

The confinement of a plasma in a laboratory requires the use of magnetic fields to balance the plasma pressure. In astrophysical plasmas, gravitational fields may balance the pressure forces, but the physics of gravitationally confined plasmas are generally studied under other topics such as stellar structure. Equilibrium, stability, and transport properties of plasmas confined by an embedded magnetic field are the major topics of research. The physics of plasma confinement is closely related to a number of other areas such as fluid mechanics, kinetic theory, and Hamiltonian mechanics.

I. Introduction

A plasma is a near ideal gas of electrons with an almost equal charge density of ions. Matter at temperatures sufficiently high compared to atomic ionization potentials, about $10^4\,°K$, forms a plasma over a broad range of density. [See PLASMA SCIENCE AND ENGINEERING.]

The confinement of a plasma by a magnetic field was originally studied to gain understanding of astrophysical phenomena, particularly phenomena of the solar corona. Although astrophysical plasmas remain an active area of study, plasma physics research, since the 1950s, has been dominated by the effort to achieve adequate plasma confinement for thermonuclear fusion. The thermonuclear fusion application imposes rather definite requirements on the plasma temperature, density, and energy confinement time as well as an upper limit on the magnetic field strength. These requirements have tended to define the plasma regimes of research interest. This is especially true since astrophysical plasmas are frequently in similar dimensionless parameter regimes. [See FUSION DEVICES; NUCLEAR FUSION POWER.]

The basic concept of magnetic confinement is simple. There is a force between an electric charge current, with density $\mathbf{j}$, and a magnetic field $\mathbf{B}$. This force is $\mathbf{j} \times \mathbf{B}$ per unit volume and can be used to balance the force due to the plasma pressure gradient, ∇p. The equation for plasma force balance is then $\nabla p = \mathbf{j} \times \mathbf{B}$. In a stationary plasma, Ampere's law relates the current density and the field by $\nabla \times \mathbf{B} = \mu_0\mathbf{j}$. These two equations, as well as the condition that the magnetic field be divergence free, define a large fraction of the physics of plasma confinement.

A simple result of plasma confinement theory is that a plasma cannot be confined purely by a self-produced magnetic field. Either there must be an external current-carrying circuit, which produces some part of the magnetic field, as in laboratory plasmas, or there must be an additional force, like gravitation in astrophysical plasmas. This result can be proved by assuming the contrary. Suppose there were a configuration in which both the plasma pressure and the magnetic field vanished outside of a finite region of space. Then, if the force balance equation is dotted with the position vector $\mathbf{x}$ and integrated by parts over all space, one finds

$$\int [3p + (1/2\mu_0)B^2]d^3x = 0$$

But, this is impossible because both the pressure and B^2 are greater than or equal to zero. This proof is based on a special case of the so-called virial theorem of classical mechanics.

Suppose a plasma is to be confined in some finite region of space by a magnetic field. There are a number of questions that should be answered. First, what properties must the magnetic field have in order to confine a plasma? Second, how high can the plasma pressure be made in comparison to field strength? The relevant dimensionless parameter is $\beta = (2\mu_0/B^2)p$, where β is a measure of the efficiency of utilization of the magnetic field for plasma confinement. Third, how long can the plasma be confined, or equivalently, what sources of energy and particles are required to maintain a given plasma configuration? It should be noted that in thermodynamic equilibrium the plasma must be Maxwellian, which implies, that in thermodynamic equilibrium, the current density is zero. Consequently, plasma confinement always implies entropy production, which is balanced either by the decay of the configuration or by externally supplied sources of energy, particles, and sometimes magnetic flux. Although these questions identify the types of issues which will be addressed, they represent too simple a view of plasma confinement to be answered definitively.

Important but subtle points constantly arise in the theory of plasma confinement. Consider a trivial consequence of the force balance equation, $\mathbf{B} \cdot \nabla p = 0$. That is, if there is a pressure gradient, then the magnetic field vector must lie in a surface of constant pressure. What are the implications? First, the constant pressure surfaces must be toroidal. It is a mathematical theorem that the only surface in three dimensions, which can have a finite vector field everywhere tangent to it, is a topological torus. Second, the integral curves of $\mathbf{B}$, which are the field line trajectories, satisfy the differential equation $d\mathbf{x}/d\tau$

= **B(x)**. The parameter τ is just a label for trajectory points with $d\tau = dl/B$ and dl the differential distance along a field line. The equation **B** · ∇p = 0 implies that a pressure gradient can only exist in regions of space in which the field line trajectories lie in nested surfaces forever. As will be shown in Section III,B, the differential equation for the integral curves of a divergence-free field is equivalent to a one degree of freedom, time-dependent, Hamiltonian mechanics problem. The condition that the field lines lie in nested surfaces is only satisfied by a magnetic field in special cases (corresponding to integrable Hamiltonians). An arbitrary magnetic field configuration will not confine a plasma. The presence of a symmetry can guarantee the existence of magnetic surfaces. However, the only symmetry, which is consistent with toroidal pressure surfaces, is toroidal which is the same as axial symmetry. In theoretical investigations, helical and cylindrical symmetry are also considered, but in both cases the plasma is not confined in a symmetry direction. Even if the externally imposed magnetic field, as required by the virial theorem, has axial symmetry, the natural, slowly decaying, plasma configuration may not be symmetric. See Section III,C.

The discussion has assumed that the plasma is confined for a long time compared to ion or electron collision frequencies. This implies that the distribution functions are nearly Maxwellian and are therefore described by a scalar pressure p. If the confinement time is only comparable to a collision frequency, the stress exerted by the plasma must be represented by a tensor rather than a scalar pressure. This case will be discussed in Section IV.

The conditions required for a magnetic field to confine a near-Maxwellian plasma are greater than just the existence of magnetic surfaces. For plasmas of thermonuclear fusion interest, the ions and electrons move a distance far longer than the size of the plasma between collisions. The particle trajectories, in the self-consistent magnetic and electric fields, must remain close to the constant pressure surfaces for near-Maxwellian distribution functions to be possible. Clearly, the condition that the particle trajectories which cross a pressure surface must remain close to it is not easily satisfied. This topic is discussed further in Section V.

The electric potential has a major role in determining the particle trajectories, but essentially no direct role in the overall plasma force balance. Why is this? The electric field **E** has a major effect on the particle trajectories when it has sufficient magnitude to balance the pressure force of the ions or the electrons. That is, $p/L \sim$ $en\mathrm{E}$ with L the gradient scale length, n the electron number density, and e the charge. The net charge density en_Δ is related to the electric field by Poisson's equation, $E \sim en_\Delta L/\varepsilon_0$. Together, these equations imply that $n_\Delta/n \sim (\lambda_\mathrm{d}/L)^2$, with the Debye length defined by $\lambda_\mathrm{d}^2 = \varepsilon_0 T/ne^2$. The plasma temperature $T = p/n$ is assumed to be in energy units. Typically, $\lambda_\mathrm{d}/L \lesssim 10^{-3}$, so the fractional charge imbalance is minute. This means an electric field exerts an almost equal and opposite force on the electrons and ions and consequently almost no force on the overall plasma. In plasma physics, the electric field is generally calculated using the quasi-neutrality constraint, $n_\Delta/n \to 0$, and is often called the ambipolar or self-consistent field. The large-scale (greater than λ_d) electric field normally has a magnitude that approximately balances the pressure of one of the species.

The self-consistent plasma electric field can have a beneficial effect on the particle trajectories and hence confinement. However, plasmas frequently establish a complicated, asymmetric electric potential even when the magnetic field is symmetric. A small asymmetric potential, $e\Phi/T$ ~ 0.01, can significantly alter the particle trajectories and degrade confinement. Such degradations in confinement are normally called anomalous transport. However, unless the plasma β is very small, such asymmetries in the electric field potential are associated with small-scale asymmetries in the magnetic field. Anomalous electron energy transport is frequently observed in laboratory plasmas.

II. Basic Plasma Physics

A. INTRODUCTION

This section covers the principles of basic plasma physics, which are required to understand the remainder of the article. The topics that will be considered are the motion of charged particles in given electric and magnetic fields, the effects of collisions between plasma particles and the plasma kinetic equation, and the two fluid equations.

B. PARTICLE DRIFT MOTION

In many plasmas of interest, the particles move a much greater distance than the size of the plasma between collisions. Plasma confinement, in these cases, is equivalent to the confinement of the particle trajectories in the self-consistent magnetic and electric fields. Unfortunately, particle trajectory calculations are generally quite difficult, even using numerical methods with given fields. Considerable insight and simplification results from the use of the asymptotic guiding center or drift equations for the particle motion. A vector form for these equations will be given in this section. The topic will be considered further in Section V,C after the structure of the magnetic field is discussed.

The equation of motion of a particle in a given magnetic and electric field appears simple enough

$$m(d\mathbf{V}/dt) = e\mathbf{V} \times \mathbf{B} + e\mathbf{E} \tag{1}$$

where m is the particle mass and e the charge. Indeed, if $\mathbf{B}$ and $\mathbf{E}$ are constant in space and time the motion is simple. The particles move in a circle of radius ρ with a frequency ω_c. Letting $\mathbf{V}_\perp$ be the part of the velocity perpendicular to $\mathbf{B}$, the radius, which is called the cyclotron or gyroradius, is $\rho = V_\perp/\omega_c$. The frequency, the cyclotron or gyrofrequency, is $\omega_c = eB/m$. The center of the circle, called the gyrocenter, moves with the velocity $\mathbf{V}_\parallel$ parallel to $\mathbf{B}$ and with a velocity $\mathbf{v}_E = (\mathbf{E} \times \mathbf{B})/B^2$ perpendicular to $\mathbf{B}$.

The particle trajectories are far more complex if the magnetic field depends on position $\mathbf{x}$. However, if the gyroradius ρ is small enough compared to the spatial variation, there is an approximate constant of the motion, the adiabatic invariant or magnetic moment

$$\mu \simeq (m/2)V_\perp^2/B \tag{2}$$

The existence of this invariant follows from a very general result of Hamiltonian mechanics. Suppose a Hamiltonian system has periodic motion in a canonical coordinate q. If the Hamiltonian is perturbed so that the parameters of the oscillatory motion change only slightly over a period, then $\oint p\, dq$ is an adiabatic invariant with p the canonically conjugate momentum. It is easy to see that the magnetic moment is just a constant times the standard Hamiltonian adiabatic invariant. The actual conservation properties of adiabatic invariants is a complex subject in Hamiltonian mechanics. Here it will suffice to say that if the fields are slowly varying analytic functions, then there is an invariant μ, accurately approximated by $mV_\perp^2/(2B)$, which is either conserved or its variation bounded except for exponentially small terms due to so-called Arnold diffusion. By a slowly varying field, we mean the gyroradius is small compared to the

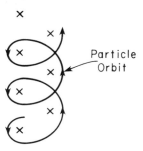

Field Strength Gradient Direction

Particle Orbit

FIG. 1. *Particle Orbits* The variation in the particle gyroradius with field strength causes the particle gyrocenter to drift, defining pathways known as particle drift orbits.

spatial scale along the magnetic field. A rapid spatial variation perpendicular to the magnetic field is irrelevant to the existence of an adiabatic invariant.

The constancy of the magnetic moment μ simplifies the evaluation of particle trajectories. Formally, this occurs through a reduction in the number of degrees of freedom of the Hamiltonian from three, for a general particle trajectory in three dimensions, to two. The canonical variables, which are associated with the two degree of freedom Hamiltonian, are a function of the three components of position as well as the average parallel velocity over a gyro-orbit. More intuitively, the spatial variation of the magnetic field causes a nonclosure of the gyro-orbits (see Fig. 1) and produces a slow drift, $\sim \rho V/L$, across the field lines. The lowest order expression for this drift velocity $\mathbf{v}$ across the field was given by Alfvén and is

$$\mathbf{v}_\perp = \frac{\mathbf{B}}{eB^2} \times (\mu\, \nabla B + mv_\parallel^2 \hat{b} \cdot \nabla \hat{b}) \tag{3}$$

with $B = |\mathbf{B}|$ and $\hat{b} = \mathbf{B}/B$. The Hamiltonian, or particle energy, is to lowest order

$$H = \tfrac{1}{2}mv_\parallel^2 + \mu B + e\Phi \tag{4}$$

as one would expect. However, the canonical coordinates have a nontrivial relation to the ordinary spatial coordinates: a canonical description will not be given until Section V,C. Provided that the drift velocity is derived from Hamiltonian equations, one would expect the qualitative features of the trajectories to be given correctly, even on a long-time scale, using the lowest order drifts, due to the so-called KAM theorem of Hamiltonian mechanics.

C. COLLISIONS AND THE KINETIC EQUATION

The motion of the ions and electrons, which form a plasma, is determined not only by the large-scale electric and magnetic fields but also by collisions. Collisions in a plasma have a different character than those in an ordinary gas. In an ordinary gas, a collision leads to a sudden, large change in the constants of the motion of a collisionless trajectory. In a plasma, the constants of the motion, like the energy and magnetic moment are constantly diffusing. The mathematical operator that represents the collisional effects will be discussed in this section as well as the kinetic equation that is used to find distribution of plasma particles in both position and velocity.

The standard collision operator for plasma problems is the Fokker–Planck operator, which is an integro-differential operator in velocity space on the electron and ion distribution functions. This operator conserves the energy and the momentum of the plasma as well as the number of ions and electrons. A simple model operator, the Lorentz operator, which unfortunately does not conserve momentum, helps clarify what is meant by a diffusive collision operator. Let $F(\mathbf{x}, \mathbf{V}, t)$ be the distribution function, that is, the phase–space density of particles with position $\mathbf{x}$ and velocity $\mathbf{V}$. The total number of particles is given by $\int f \, d^3x \, d^3V$. The Lorentz operator scatters the pitch λ of the particle velocity relative to some direction, usually taken to be the magnetic field direction. Let $V_{\parallel}$ be the component of the velocity along the field, then $\lambda = V_{\parallel}/V$. The Lorentz operator is

$$C_L(f) = \frac{\nu_\Omega}{2} \frac{\partial}{\partial \lambda} (1 - \lambda^2) \frac{\partial f}{\partial \lambda} \quad (5)$$

with ν_Ω the pitch angle scattering collision frequency. To understand the effect of this operator, suppose that a plasma were placed in a uniform magnetic field with an initial pitch distribution $f_0(\lambda) = 3(1 - \lambda^2)/2$. The equation that would give the time evolution of $f(\lambda, t)$, the kinetic equation, is $\partial f/\partial t = C_L(f)$. The solution is $f(\lambda, t) = 1 + [(\frac{1}{2})(1 - 3\lambda^2) \exp(-3\nu_\Omega t)]$. It should be noted that if $f(\lambda)$ is nonuniform in only a small region of width $\Delta\lambda$, f becomes uniform in a time $\sim(\Delta\lambda^2/\nu_\Omega)$, which is much smaller than a collision time $1/\nu_\Omega$. The analogous model collision operator for energy scattering is

$$C_E(f) = \frac{1}{V^2} \frac{\partial}{\partial V} \left[V^2 \nu_E \left(Vf + \frac{T}{m} \frac{\partial f}{\partial V} \right) \right] \quad (6)$$

with T the plasma temperature and m the mass of the particles being scattered.

The largest collision frequencies in a plasma are the electron pitch angle scattering on both the electrons and ions and the electron energy scattering on other electrons. The approximate value for these scattering rates for usual plasma parameters is

$$\nu_e \simeq \frac{5 \times 10^{-11}}{\text{sec}} \frac{n}{T_e^{3/2}} \quad (7)$$

with n the number of electrons per cubic meter and the electron temperature T_e measured in electron volts. The next largest collision frequencies are the ions pitch angle scattering and energy scattering on other ions. These collision frequencies ν are smaller than the electron collision frequency by roughly the square root of the electron to ion mass ratio. That is,

$$\nu_i \simeq \frac{10^{-12}}{\text{sec}} \frac{nZ^3}{T_i^{3/2}} \frac{1}{A^{1/2}} \quad (8)$$

with Z the charge of the ions, A the atomic number, and T_i the ion temperature. Equilibration of the ion and electron temperatures is very slow. This rate is smaller than the electron collision frequency by roughly the electron to ion mass ratio. Another peculiarity of plasma collisions is that if the velocity V of the particle being scattering is greater than the thermal velocity V_T of the species on which it is scattering, then the collision frequencies are smaller by approximately $(V_T/V)^3$.

The cause of plasma collisions is fluctuations from charge neutrality. The natural scale of these fluctuations is between the Debye length λ_d, which was discussed in Section I and the distance b at which the electrostatic energy equals the kinetic energy, $T \approx e^2/\varepsilon_0 b^2$. The density of electrons and ions in a Debye-length-scale region has a statistical fluctuation of roughly $\Delta n/n \sim (n\lambda_d^3)^{-1/2}$. The fluctuating electric field produced by the density fluctuation is about $\Delta E \sim e(\Delta n)\lambda_d/\varepsilon_0$. An individual particle interacts with this perturbation for either the time it takes to cross the fluctuation λ_d/V or the natural time for the fluctuation to change, which is the plasma frequency ω_p, the thermal velocity divided by the Debye length. The relative fluctuation in the particle velocity is thus $\Delta V/V \sim (\Delta n/n)(V_T/V)^2$ for a particle with velocity V greater than the thermal velocity. The scattering frequency ν is defined by the time it takes the small fluctuations to change the velocity of a particle by order unity. The contribution to the scattering by fluctuations of wave number k is proportional to $1/k$ for $1/\lambda_d < k < 1/b$, so an integration over all scales leads to a factor $\ln(\lambda_d/b)$. For V greater than V_T,

$$\nu \simeq \left(\frac{V}{\lambda_d}\right)\left(\frac{\Delta V}{V}\right)^2 \ln(\lambda_d/b) \simeq \frac{\omega_p}{n\lambda_d^3}\left(\frac{V_T}{V}\right)^3 \ln(\lambda_d/b) \tag{9}$$

The distribution of particles in a plasma as a function of position and velocity is determined both by trajectory and collisional effects. The kinetic equation for the distribution function is

$$\frac{\partial f}{\partial t} + V\cdot\nabla f + \frac{e}{m}(E + V\times B)\cdot\frac{\partial f}{\partial V} = C(f) \tag{10}$$

The right-hand side of the kinetic equation is the Fokker–Planck collision operator. It is easily shown that the characteristics of the operator on the left-hand side of the kinetic equation are the particle trajectories in the electric and magnetic fields E and B. If the drift velocity v is used to evaluate the particle trajectories instead of the exact velocity V, the kinetic equation is called the drift kinetic equation.

D. Two-Fluid Equations

Although the kinetic equations for the ions and electrons describe plasma confinement, their complexity obscures a number of basic properties of plasmas in which the gyroradii are small compared to the system size. These properties can be studied best by considering the first velocity moment of each kinetic equation with the additional assumption that the distribution function $f(x, V, t)$ is independent of the gyrophase. This moment gives the so-called two-fluid equations.

The first moment of the kinetic equation is evaluated by multiplying the equation by mV and integrating over velocity space. One finds for a species denoted by α,

$$\partial(m_\alpha n_\alpha u_\alpha)/\partial t + \nabla\cdot(n_\alpha m_\alpha u_\alpha) + \nabla\cdot P_\alpha$$
$$- e_\alpha n_\alpha(E + u_\alpha\times B) = -e_\alpha n_\alpha R_\alpha \tag{11}$$

with the following definitions

$$n_\alpha \equiv \int f_\alpha\, d^3V, \qquad u_\alpha = \frac{1}{n_\alpha}\int V f_\alpha\, d^3V$$

$$P_\alpha = \int m_\alpha(V - u)(V - u)f_\alpha\, d^3V \tag{12}$$

$$R_\alpha = -\frac{1}{e_\alpha n_\alpha}\int m_\alpha V C_\alpha(f)\, d^3V$$

The physical interpretation of the various definitions is obvious. The number density of the species is n_α, the mean velocity u_α, the pressure or stress tensor P_α, and the force between the species due to collisions $e_\alpha n_\alpha R_\alpha$.

The assumption that the distribution function is independent of gyrophase, which essentially follows from the smallness of the gyroradius compared to system size, restricts the form of the pressure tensor. The only nonzero components are

$$p_\parallel^{(\alpha)} = \int m_\alpha(V - u)_\parallel(V - u)_\parallel f_\alpha\, d^3V$$
$$p_\perp^{(\alpha)} = \frac{1}{2}\int m_\alpha(V - u)_\perp(V - u)_\perp f_\alpha\, d^3V \tag{13}$$

with the parallel and perpendicular signs meaning relative to the field B. If δ is the unit tensor and $\hat{b} = B/B$, then the pressure tensor is

$$P_\alpha = p_\parallel^{(\alpha)}\hat{b}\hat{b} + p_\perp^{(\alpha)}(\delta - \hat{b}\hat{b}) \tag{14}$$

It should be noted that the validity of the two-component form for the pressure tensor does not depend on the distribution function being near Maxwellian. Generally, in plasma problems the flow velocities u_α are much smaller than the thermal velocity $m_\alpha u_\alpha^2 \ll p^{(\alpha)}$. In this case, the inertial terms can be ignored giving

$$\nabla\cdot P_\alpha - e_\alpha n_\alpha(E + u_\alpha\times B) = -e_\alpha n_\alpha R_\alpha \tag{15}$$

for the ions and for the electrons. These are the two-fluid equations.

The equilibrium equation for a plasma is obtained by adding the electron and the ion fluid equations. The result is

$$\nabla\cdot p = j\times B \tag{16}$$

with j the current defined by

$$j = \sum_\alpha e_\alpha n_\alpha n_\alpha \tag{17}$$

In deriving the equilibrium equation, the quasineutrality condition

$$\sum_\alpha e_\alpha n_\alpha = 0 \tag{18}$$

was used as well as the condition that ion–electron collisions should exert equal and opposite forces on the two species.

In addition to the equilibrium equation, an equation called the generalized Ohm's law frequently arises in the discussion of plasma confinement. This equation is either the electron or the ion fluid equation. Usually one can assume that the electron fluid is isotropic, $p_\parallel^{(e)} = p_\perp^{(e)} =$

p_e, and that the electron temperature is constant along a field line. When these conditions hold, the electron fluid equation can be written as

$$\mathbf{E}_* + \mathbf{u} \times \mathbf{B} = \mathbf{R} \qquad (19)$$

with $\mathbf{E}_*$ defined by

$$\mathbf{E}_* = \mathbf{E} - \nabla \left[\frac{T_e}{|e|} \ln \left(\frac{n}{n_0} \right) \right] \qquad (20)$$

T_e and n_0 constants on each field line, and

$$\mathbf{u} = \mathbf{u}_e + \frac{\mathbf{B}}{B^2} \times \left\{ \frac{\nabla P_e}{|e|n} - \nabla \left[\frac{T_e}{|e|} \ln \left(\frac{n}{n_0} \right) \right] \right\} \qquad (21)$$

Actually, no distinction is usually made between $\mathbf{E}_*$ and $\mathbf{E}$, since both fields can be validly inserted into Faraday's law

$$\partial \mathbf{B}/\partial t = -\nabla \times \mathbf{E} \qquad (22)$$

In most plasma problems, Faraday's law is the only Maxwellian equation involving the electric field that is used. The quasi-neutrality assumption makes $\nabla \cdot \mathbf{E}$ irrelevant and the displacement current $\partial \mathbf{E}/\partial t$ can usually be neglected because of either quasi-neutrality or the slowness of field changes relative to the velocity of light.

The collisional force between the electrons and the ions, $e_\alpha n_\alpha \mathbf{R}_\alpha$, is clearly related to the relative motion of the ion and electron fluids, which is the current $\mathbf{j}$. It is therefore customary to write the dissipative term as

$$\mathbf{R} = \eta \mathbf{j} \qquad (23)$$

with η the resistivity. This equation is not precisely correct for two reasons. First, the resistivity is a diagonal tensor with distinct parallel and perpendicular components. Second, there are thermodynamic cross terms that contribute to the collisional force. For example, there is a force proportional to the flux of heat.

III. Toroidal Plasmas

A. INTRODUCTION

As noted in Section 1, the torus is the only plasma shape that can confine an isotropic plasma, $p_\parallel = p_\perp$. To simplify the discussion of toroidal plasmas, the pressure will be assumed isotropic. The effects of anisotropy are covered under mirror confinement.

Toroidal devices have always had a central position in the magnetic fusion program. The simplest system would have only a symmetric toroidal field, but as we will find, this field is not consistent with plasma equilibrium. A symmetric toroidal equilibrium, the tokamak (Fig. 2) can be produced by inducing a plasma current along an externally applied toroidal field. The toroidal current loop produces an outward force that is balanced by an externally applied vertical magnetic field. There are probably more tokamaks worldwide than all other plasma confinement devices combined. The two largest are the JET (Joint European Torus) device of the Euraton association at Culham, England, and the TFTR (Toroidal Fusion Test Reactor) device at Princeton in the United States. The basic parameters of these devices are listed in Table I. The central electron temperature in JET and in TFTR is ~3 keV and comes from the heating produced by the resistive or ohmic dissipation of the plasma current. The electron density is ~$10^{19}/\text{m}^3$. Large systems for additional heating are being brought into operation on these devices, which should push the plasma parameters close to those required for thermonuclear burning of the hydrogen isotopes deuterium and tritium (a density of

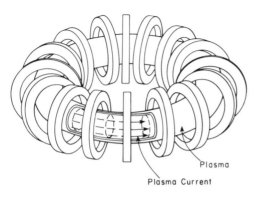

Plasma

Plasma Current

FIG. 2. The toroidal field coils and the plasma current, which produces the poloidal field, are illustrated by this tokamak. In addition, a vertical magnetic field is required to balance the outward force of the plasma current.

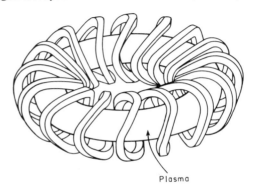

Plasma

FIG. 3. Field coils with a helical twist produce both a toroidal and a poloidal magnetic field, as shown by this stellarator.

TABLE I. Toroidal Devices

	JET	TFTR	Heliotron	Wendelstein VII-A
Major radius (m)	2.9	2.5	2.2	2.0
Average minor radius (m)	1.6	0.85	0.20	0.10
Field strength (T)	3.5	5.2	2.0	2.5
Plasma current (MA)	4.8	2.5	—	—
Pulse length (sec)	20	2	1	1
Energy confinement time (sec)	0.9	0.4	0.04	0.01

$\sim 10^{20}/m^3$, a temperature of about 10 keV, and an energy confinement time of ~ 2 sec). By adding a helical twist to the coils producing the magnetic field, the magnetic field lines can be given a twist, similar to that of a tokamak, but without the need for a net toroidal plasma current. Such configurations are called stellarators (Fig. 3). Existing stellarators are second only to tokamaks in their confinement properties. The best plasma conditions have been achieved on the Heliotron stellarator at Kyoto, Japan, and the Wendelstein VII-A at Garching, Germany. Table I lists the parameters of these stellarators. Both have large, nonohmic heating systems operating, electron temperatures above 1 keV, and densities of $\sim 10^{19}/cm^3$.

There are three topics which will be discussed in this section, the geometry of the magnetic field lines, the constraints of plasma equilibrium, and the energy principle. The geometry of the magnetic field lines is basic to toroidal confinement since equilibrium implies the pressure is constant along these lines. The field geometry is equivalent to that of a one degree of freedom, time-dependent, Hamiltonian system. In other words, the field line trajectories have identical mathematical properties to those of particle trajectories in one dimension with time-dependent forces. The Hamiltonian picture of the magnetic field separates the properties of the field, which can change arbitrarily rapidly to maintain force balance from those properties that are conserved except for slow resistive dissipation.

The equilibrium equation $\nabla p = \mathbf{j} \times \mathbf{B}$ and Ampere's law $\nabla \times \mathbf{B} = \mu_0 \mathbf{j}$ can be used to derive formulas that hold in equilibrium. These formulas for the current in the plasma have a singular mathematical form. The physical resolution of these singularities determines much of toroidal equilibrium, stability, and transport theory. For example, toroidal equilibria are frequently not uniquely defined by the boundary conditions. In the axisymmetric tokamak device, the plasma often has toroidal asymmetries, and these asym-

metries provide critical limitations on the obtainable plasma parameters. The energy principle will be considered as a powerful and traditional method for studying both the equilibrium and the stability of toroidal plasmas.

B. MAGNETIC GEOMETRY

The mathematical properties of the magnetic field line trajectories are determined by the condition that the magnetic field be divergence free. Three coordinates are, of course, required to describe the field. They are conventionally just the position vector $\mathbf{x}$. However, the magnetic field properties become more transparent if a more subtle coordinate system ψ, θ, ϕ is employed. The coordinates θ and ϕ can be any poloidal and toroidal angles (see Fig. 4) and $2\pi\psi$ is the amount of toroidal magnetic flux enclosed by a constant ψ surface. One can show that any divergence-free vector, like the magnetic field, can be written in the so-called canonical form,

$$\mathbf{B} = \nabla\psi \times \nabla\theta + \nabla\phi \times \nabla\chi \qquad (24)$$

The function $\chi(\psi, \theta, \phi)$ will be shown to the Hamiltonian of the field lines. Physically, $2\pi\chi$ is the poloidal magnetic flux outside a constant χ surface (Fig. 4).

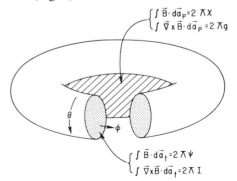

FIG. 4. The *Canonical coordinates* consist of the poloial angle θ, the toroidal angle ϕ, the poloidal magnetic flux $2\pi\chi$, and the toroidal flux $2\pi\psi$ are defined. The toroidal surface is either a constant ψ or a constant χ surface. The currents I and g are also defined.

To describe a field using the quantities ψ, θ, ϕ, and χ, it would appear that these quantities must be known as functions of the position $\mathbf{x}$. Actually, it is better to give the position as a function of the coordinates ψ, θ, ϕ. The equations $\mathbf{x}(\psi, \theta, \phi)$ are called the transformation equations. Mathematical identities, the dual relations, can then be used to express the gradients in terms of the derivatives of the transformation equations. For example,

$$\boldsymbol{\nabla}\psi = \frac{1}{J}\frac{\partial\mathbf{x}}{\partial\theta}\times\frac{\partial\mathbf{x}}{\partial\phi} \qquad (25)$$

with J the Jacobian, $J = (\partial\mathbf{x}/\partial\psi \times \partial\mathbf{x}/\partial\theta)\cdot\partial\mathbf{x}/\partial\phi$. One can also show that $1/J = \mathbf{B}\cdot\boldsymbol{\nabla}\phi$. We assume that $\mathbf{B}\cdot\boldsymbol{\nabla}\phi$ is finite for mathematical simplicity, but a general magnetic field can be treated in the Hamiltonian theory. One can specify a magnetic field by giving either $\mathbf{B}(\mathbf{x})$ or $\chi(\psi, \theta, \phi)$ and $\mathbf{x}(\psi, \theta, \phi)$.

The magnetic field lines are the trajectories $\mathbf{x}(\tau)$ given by the equation $d\mathbf{x}/d\tau = \mathbf{B}(\mathbf{x})$. The τ derivative of any function $\phi(\mathbf{x})$ is just $d\phi/d\tau = (d\mathbf{x}/d\tau)\cdot\boldsymbol{\nabla}\phi$. Therefore, the field line trajectories in the ψ, θ, ϕ coordinates are $d\psi/d\phi = (B\cdot\boldsymbol{\nabla}\psi)/(\mathbf{B}\cdot\boldsymbol{\nabla}\phi)$ and a similar expression for $d\theta/d\phi$. The canonical form for $\mathbf{B}$ [Eq. (24)] and some vector algebra, then given Hamilton's equations:

$$\frac{d\psi}{d\phi} = -\frac{\partial\chi}{\partial\theta}, \qquad \frac{d\theta}{d\phi} = \frac{\partial\chi}{\partial\psi} \qquad (26)$$

The poloidal flux function χ is the field line Hamiltonian with canonical coordinates ψ, θ, ϕ. The canonical position is θ, the canonical momentum ψ, and the canonical time is the toroidal angle ϕ. Since continuous transformations, such as $\mathbf{x}(\psi, \theta, \phi)$, preserve topological properties, questions, such as whether the field lines close on themselves, form surfaces, or fill a volume, are answered by the Hamiltonian $\chi(\psi, \theta, \phi)$ alone.

There are a number of Hamiltonian mechanics concepts that play a major role in the theory of toroidal plasmas. These include Poincaré plot, integrable fields, rotational transform, magnetic coordinates, magnetic islands, and stochastic regions. A Poincaré plot is illustrated in Fig. 5(a). Each time a field line passes around the torus a point is marked on a plane transverse to the torus. The Poincaré plot is, ideally, the plot of the points from an infinite number of traversals. If the points on the Poincaré plot of each field line life on a smooth curve, then the field is said to be integrable. If the field were associated with a plasma equilibrium, then these curves would

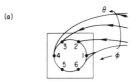

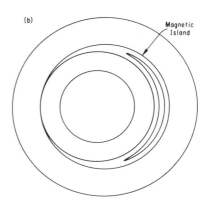

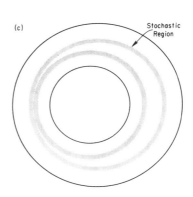

FIG. 5. The beginning of the construction of a Poincaré plot is illustrated in (a) using the first six field line intersections with the Poincaré plane. The Poincaré plot of a field that has a magnetic island is illustrated in (b), and a field with a stochastic region in (c).

just be the intersection of the constant pressure surfaces with the Poincaré plane. As the Poincaré plot of an integrable field is constructed, there is an average rate at which the poloidal angle θ advances for each toroidal traversal. This average advance is 2π times the rotational transform ι. When the field is integrable, one can always find canonical coordinates ψ, θ, ϕ such that χ is a function of ψ alone. These are called magnetic coordinates and are the action-angle variables of Hamiltonian mechanics. In these coordinates, the equation of a field line is ψ equal to a constant and $\theta - \iota(\psi)\phi$ equal to a constant with the rotational transform $\iota = d\chi/d\psi$.

A magnetic field $\mathbf{B}(\mathbf{x})$, which does not have special symmetries, is rarely exactly integrable (just as few Hamiltonians are). The Poincaré plot may show magnetic islands [Fig. 5(b)] or stochastic regions in which a single field line covers a finite volume [Fig. 5(c)]. Clearly, the regions of stochastic field line behavior must be limited in extent to obtain a finite pressure, plasma equilibrium.

When the magnetic field is time dependent, $\mathbf{B}(\mathbf{x}, t)$, the Hamiltonian $\chi(\psi, \theta, \phi, t)$ evolves slowly, on a resistive time scale, while the transformation equations $\mathbf{x}(\psi, \theta, \phi, t)$ evolve as rapidly as required to maintain force balance. These two time scales may differ by six orders of magnitude. The time development of a field is most easily examined in terms of the vector potential $\mathbf{A}$ with $\mathbf{B} = \nabla \times \mathbf{A}$. The canonical representation of the vector potential is $\mathbf{A} = \psi \nabla \theta - \chi \nabla \phi + \nabla G$ with G the gauge function (given any toroidal coordinate system, say ρ, θ, ϕ, it is essentially obvious that $\mathbf{A}$ can be written in the canonical representation with ψ, χ, and G single-valued functions, which proves the existence of the canonical form for $\mathbf{B}$). Coordinate transformation theory can be used to prove the identity

$$\frac{\partial \mathbf{A}(\mathbf{x}, t)}{\partial t} = -\left(\frac{\partial \chi}{\partial t}\right)_c \nabla \phi + \mathbf{v}_c \times \mathbf{B} + \nabla s \quad (27)$$

with the subscript c implying that the canonical coordinates ψ, θ, ϕ are held constant. The velocity of the canonical coordinates is $\mathbf{v}_c = (\partial \mathbf{x}/\partial t)_c$. The function $s(\psi, \theta, \phi)$ can be considered arbitrary due to the freedom of the gauge function G. Actually, it sometimes is useful to think of s as a function of four variables ψ, θ, ϕ, χ, defining the χ dependence by $\partial s/\partial \chi = -(\partial \phi/\partial t)_x$. If this is done, $s(\psi, \theta, \phi, \chi)$ is the generating function for infinitesimal canonical transformations in the extended phase space ψ, θ, ϕ, χ.

By far, the most important component of $\partial \mathbf{A}/\partial t$ is the component parallel to $\mathbf{B}$:

$$\mathbf{B} \cdot \frac{\partial \mathbf{A}}{\partial t} = -\left(\frac{\partial \chi}{\partial t}\right)_c \mathbf{B} \cdot \nabla \phi + \mathbf{b} \cdot \nabla s \quad (28)$$

This equation says that if a function s is found such that $\mathbf{B} \cdot \nabla s = \mathbf{B} \cdot \partial A/\partial t$, then the Hamiltonian does not depend on time. If such a function exists, then the topology of field does not change.

Faraday's law gives the electric field, $\mathbf{E} = -(\partial \mathbf{A}/\partial t + \nabla \Phi)$ with Φ the electric potential.

Ohm's law for a plasma [Eq. (19)], relates the electric field to the dissipation $\mathbf{R}$, which is usually taken to be $\mathbf{R} = \eta \mathbf{j}$ [Eq. (23)] with η the resistivity. That is, $\mathbf{E} + \mathbf{u} \times \mathbf{B} = \mathbf{R}$. Letting $\Phi_c = \Phi + s$, which is the electric potential transformed into the canonical coordinate space, one finds that

$$\mathbf{R} = (\partial \chi/\partial t)_c \nabla \phi + (\mathbf{u} - \mathbf{v}_c) \times \mathbf{B} + \nabla \Phi_c \quad (29)$$

This equation demonstrates the separation of time scales between the Hamiltonian χ and the transformation equations. The parallel component of $\mathbf{R}$, $\mathbf{R} \cdot \mathbf{B}$, determines the time evolution of χ. The perpendicular components of $\mathbf{R}$ only determine the perpendicular components of $(\mathbf{u} - \mathbf{v}_c)$, which is the velocity of the plasma relative to the canonical coordinates. The velocity of the canonical coordinates is not determined by Faraday's and Ohm's laws. This velocity can only be found by bringing in additional physics, namely force balance.

In the case of an integrable field, one can obtain an important equation for $\chi(\psi, t)$, which depends only on $\mathbf{B}$ being divergence free and on Faraday's law. This equation,

$$\left(\frac{\partial \chi}{\partial t}\right)_\psi = \frac{1}{(2\pi)^2} \frac{\partial}{\partial \psi} \int \mathbf{E} \cdot \mathbf{B} \, d^3x \quad (30)$$

follows obviously from Eq. (28) for $\mathbf{B} \cdot \partial \mathbf{A}/\partial t$ and the fact that the magnetic coordinate volume element is $d\psi \, d\theta \, d\phi/\mathbf{B} \cdot \nabla \phi$. The loop voltage in a plasma is defined by

$$V(\psi) = \frac{1}{2\pi} \frac{\partial}{\partial \psi} \int \mathbf{E} \cdot \mathbf{B} \, d^3x \quad (31)$$

so that $\partial \chi/\partial t = V/2\pi$ (note that V is an average of the parallel electric field in a surface of constant ψ). In a tokamak, a loop voltage is provided by changing the poloidal field in central hole of the torus. If the loop voltage is held constant for a long enough time, the magnetic field in the plasma can become time independent with $\chi(\psi, t) = \chi_0(\psi) + Vt$. Equation (30) implies that the loop voltage must be constant across the plasma in steady state. In other words, if the time scale for a change in the plasma is sufficiently short compared to the resistive time scale, then the Hamiltonian χ is conserved. But, if the time scale is sufficiently long, the loop voltage V becomes constant across the plasma with the value of V set by the change in the magnetic field outside of the plasma.

C. TOROIDAL EQUILIBRIUM

A magnetic field, which is consistent with equilibrium, $\nabla p = \mathbf{j} \times \mathbf{B}$, has important properties in addition to those of general magnetic fields. These properties can be expressed in terms of formulas for the plasma current **j**, which have a singular mathematical form. Much of toroidal equilibrium, stability and transport theory can be derived from these singularities.

While studying the current singularities, let us assume that the field is locally integrable and that **B** is given in magnetic coordinate form

$$\mathbf{B} = \nabla\psi \times \nabla\theta + \nabla\phi \times \nabla\chi(\psi) \quad (32)$$

In magnetic coordinates, the pressure is a function of ψ alone and $\mathbf{j} \cdot \nabla\psi = 0$. The condition $\mathbf{j} \cdot \nabla\psi = 0$ implies that the magnetic coordinates can be chosen so that in addition to the canonical form, **B** also has the representation

$$\mathbf{B} = g(\psi)\nabla\phi + I(\psi)\nabla\theta + \beta(\psi, \theta, \phi)\nabla\psi \quad (33)$$

The quantities g and I have a simple physical interpretation. The total poloidal current outside a ψ surface is $2\pi\mu_0 g$ and the total toroidal current inside is $2\pi\mu_0 I$ (see Fig. 4). The quantity β_* is closely related to the Pfirsch–Schlüter current, which is defined below. Given a **B(x)** that is consistent with equilibrium, the transformation **x**(ψ, θ, ϕ) and $\chi(\psi)$ can be found numerically.

The equations obeyed by an equilibrium current, $\nabla p = \mathbf{j} \times \mathbf{B}$ and $\nabla \cdot \mathbf{j} = 0$, are equivalent to $\mathbf{j}_\perp = (\mathbf{B} \times \nabla p)/B^2$ and

$$\mathbf{B} \cdot \nabla(\mathbf{j}_\parallel/B) = -\nabla \cdot \mathbf{j}_\perp \quad (34)$$

with the current written as $\mathbf{j} = (j_\parallel/B)\mathbf{B} + \mathbf{j}_\perp$. The interesting component of the current, $j_\parallel$, is the sum of two terms. The first, known as the Pfirsch–Schlüter current, is the special solution of the inhomogeneous differential equation for $j_\parallel$. The second term, known as the net current, is the solution of the homogeneous equation. In the special magnetic coordinate system of Eqs. (32) and (33), it is easily shown that the diamagnetic current $\mathbf{j}_\perp$ satisfies

$$\mathbf{j}_\perp = [g(\psi)\nabla\phi \times \nabla\psi - I(\psi)\nabla\psi \times \nabla\theta]$$
$$\times [1/B^2(\psi, \theta, \phi)] [dp(\psi)/d\psi]$$

It is useful to write $1/B^2$ in the Fourier decomposed form

$$\frac{1}{B^2(\psi, \theta, \phi)} = \frac{1}{B_0^2(\psi)} \left\{ 1 + \sum_{n,m}{}' \delta_{nm}(\psi) \right.$$
$$\left. \times \exp[i(n\phi - m\theta)] \right\} \quad (35)$$

with the prime implying that the $n = 0$, $m = 0$ term is omitted from the sum. The importance of magnetic coordinates [Eq. (32)] becomes apparent. The operation $\mathbf{B} \cdot \nabla$ operating on a Fourier

series in θ and ϕ gives a factor $(n - \iota m)$ in each term of the series. The Pfirsch–Schlüter current is found to be

$$\frac{j_{ps}}{B} = \frac{1}{B_0^2} \frac{dp}{d\psi} \sum{}' \frac{mg + nI}{n - \iota m} \delta_{nm} \exp[i(n\phi - m\theta)]$$
$$(36)$$

This equation has apparent singularities on every ψ surface on which ι is rational (the ratio of two integers). To avoid the singularities either $dp/d\psi$ must vanish or the resonant δ_{nm}'s must vanish. On a rational surface (ι rational) each magnetic field line closes on itself and therefore does not cover the surface. One can show that the vanishing of the resonant δ_{nm}'s on a rational surface is equivalent to $\oint dl/B$ being the same for every field line of the surface with dl the differential distance along a line. It is the form of the variation of $\oint dl/B$ that prevents a pure, symmetric, toroidal field from being consistent with equilibrium. However, before studying the Pfirsch–Schlüter current, let us consider the net current.

The net current j_n/B is the general solution to the differential equation $\mathbf{B} \cdot \nabla(j_n/B) = 0$. The general solution is

$$K = k(\psi) + \sum{}' k_{nm}\delta\left(\iota - \frac{n}{m}\right) \exp[i(n\phi - m\theta)]$$
$$(37)$$

with $K = \mu_0 j_n/B$, and $\delta(...)$ the Dirac delta function. To understand the role of the net current in toroidal equilibrium, it is best to consider an example, the pressureless plasma. In a pressureless plasma, force balance and Ampere's law can be combined as

$$\nabla \times \mathbf{B} = K\mathbf{B} \quad (38)$$

Let $\mathbf{B}_0$ be an integrable equilibrium field that has a smooth current distribution $K(\psi_0)$. Suppose a small change is made in the field surrounding the plasma so that the new equilibrium field is $\mathbf{B} = \mathbf{B}_0 + \mathbf{b}$ with the perturbation **b** having no special symmetries. The properties of the new equilibrium depend on the time scale. If the new equilibrium is observed on a time scale sufficiently short compared to the resistive time scale, then the field Hamiltonian χ and, therefore, the field topology are conserved. This conservation generally requires that delta function currents arise to prevent an island from opening at each rational surface. If the new equilibrium is observed after a resistive time interval, then the loop voltage V [Eq. (31)] will be a smooth function of ψ. However, with a smooth loop voltage, the rational flux surfaces break to form islands. The lack of smoothness in the flux surfaces, due to the islands, appears as a singularity in the perturbation analysis for the spatial distribution of the current. On the long time scale, the constancy of the loop voltage implies that the old and the new equilibrium should have the same current distribution K as a function of the toroidal flux, except within a few island widths of each magnetic island. Actually, for the perturbations of interest, the toroidal area integral $\int \mathbf{b} \cdot d\mathbf{a}_t$ inside an unperturbed flux surface vanishes as does the related poloidal area integral, so that the function $\chi(\psi)$ is conserved, except within a few island widths of the rational surfaces, through linear order in $\mathbf{b}$. (The actual change in $\chi(\psi)$ is proportional to $|\mathbf{b}|^2$ except near the islands where the change is proportional to $|\mathbf{b}|$.) This means that, the perturbation equations that describe the long and short time scales, relative to resistive time scales, are identical except in the method of handling the singularities that occur at the rational surfaces.

A perturbed pressureless equilibrium field $\mathbf{B}$ with toroidal flux function ψ is related to the unperturbed equilibrium field $\mathbf{B}_0$ with flux function ψ_0 by

$$\nabla \times \mathbf{b} = [K(\psi) - K(\psi_0)]\mathbf{B}_0 + K(\psi)\mathbf{b} \quad (39)$$

The perturbation $\mathbf{b} = \mathbf{B} - \mathbf{B}_0$ also satisfies $\nabla \cdot b = 0$. The current distribution K is assumed to be the same function of toroidal flux in the two equilibria. To lowest order $K(\psi) - K(\psi_0) = K'(\psi_0)\psi_1$ with $\psi_1 = \psi - \psi_0$. The function ψ_1 has a singular form due to the equation $\mathbf{B} \cdot \nabla\psi = 0$, which can also be written in lowest order as $\mathbf{B}_0 \cdot \nabla\psi_1 + \mathbf{b} \cdot \nabla\psi_0 = 0$. If the perturbation is Fourier decomposed as

$$\frac{\mathbf{b} \cdot \nabla\psi_0}{\mathbf{B} \cdot \nabla\phi} = \sum b_{nm} \cos(n\phi - m\theta) \quad (40)$$

then

$$\psi_1 = -\sum \frac{b_{nm}}{n - \iota m} \sin(n\phi - m\theta) \quad (41)$$

Of course, there is not a true singularity in ψ_1. The apparent singularity is resolved by the opening of a magnetic island around the rational surface. The island has a half-width in the unperturbed ι space of $\Delta_{nm} = |4m\iota' b_{nm}|^{1/2}$ with $\iota' = d\iota/d\psi$. The crudest resolution of the singularity is to replace $1/(n - \iota m)$ by $(n - \iota m)/[(n - \iota m)^2 + m^2\Delta_{nm}^2]$ to represent the finite island size. This will be useful during a later discussion.

To illustrate the essential features of the perturbed pressureless equilibrium, consider a cy-

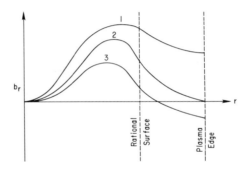

FIG. 6. The radial dependence of the magnetic perturbation $b_r = (f/r) \cos(n\phi - m\theta)$ is given for three different values of the force-free current gradient, dK/dr.

lindrical plasma that represents a toroidal system by being periodic along its axis with period $2\pi R$. For simplicity, assume that $R \gg r$ with r the cylindrical radial coordinate and that $B_\theta \sim (r/R)B_\phi$. In this case, the term $K(\psi)\mathbf{b}$ in Eq. (39) is negligible as is b_ϕ. The solution for $\mathbf{b}$ can be written in the form, $rb_r = f(r) \cos(n\phi - m\theta)$ and $mb_\theta = df/dr \sin(n\phi - m\theta)$. Using $\psi_0 \simeq \frac{1}{2}B_\phi r^2$, the differential equation for f is

$$\frac{1}{r}\frac{d}{dr}r\frac{df}{dr} - \frac{m^2}{r^2}f = -\frac{mR}{r}\frac{dK}{dr}$$

$$\times \frac{n - \iota m}{(n - \iota m)^2 + m^2\Delta_{nm}^2}f$$

$$(42)$$

The term on the right-hand side of the equation changes sign at $r = r_*$ where $\iota(r_*) = n/m$. In practice the current density is a maximum at the plasma center so dK/dr is negative as is $d\iota/dr$. For $r < r_*$, the right-hand side is, therefore, generally negative, which tends to make f oscillatory. For $r > r_*$, the right-hand side is positive, which gives f an exponential character. Figure 6 gives three solutions to the differential equation for f in the limit as Δ_{nm} goes to zero. The numbers, (1) to (3), refer to increasing values of dK/dr inside the singular surface at $r = r_*$. In case (1), a small perturbation at the boundary gives a small b_r at the rational surface and a small island, just as one would expect. However, in case (2), even a zero perturbation at the boundary gives a finite b_r at the singular surface, and case (3) is even more extreme. To interpret these solutions, consider a finite value of Δ_{nm}. Then, in case (3) one can find a value of Δ_{nm}, say Δ_0, such that b_r goes to zero at the plasma boundary. This means that the cylindrically symmetric equilibrium associated with solution (3) will, if infinitesimally perturbed, form an island of finite width, approximately Δ_0, on a resistive time scale. Such equilibria are called tearing mode unstable.

The term $m^2 f/r^2$ in Eq. (42) is stabilizing. Therefore, only low m numbers tend to be tearing mode unstable. In tokamaks, the $n = 1$, $m = 1$ mode prevents the transform in the center of tokamaks from being greater than ~ 1. The modes $n = 1$, $m = 2$ and $n = 2$, $m = 3$ are also important, and tokamak plasmas tend to lose equilibrium if the rotational transform near the edge of the plasma is much above a third, the so-called major disruption.

Equation (37) implies that a delta function current can be added at each rational surface. If this is done in the cylinder model, the delta function term in the toroidal current is

$$\mu_0 j_\phi = \frac{r_0}{m} \left[\frac{d\bar{b}_r}{dr} \right] \delta(r - r_*) \sin(n\phi - m\theta) \quad (43)$$

with [...] meaning the jump in the quantity at the surface $r = r_*$ and $\bar{b}_r = f/r$. The zero divergence condition on **b** implies $[\bar{b}_r] = 0$. By appropriately choosing $\Delta' \equiv [d\bar{b}_r/dr]/\bar{b}_r$ at $r = r_*$, one can make the solutions of Eq. (42) vanish at the plasma edge. In the section on the energy principle, we will find that $\Delta' > 0$ corresponds to energy release and therefore to a tearing instability.

The pressure gradient contribution to the parallel current, the Pfirsch–Schlüter current [Eq. (36)] has two related effects on toroidal equilibrium. First, in any toroidal plasma the magnetic field strength varies on a constant ψ surface. In a tokamak, for example, the field strength depends on the poloidal angle θ so that $b \approx b_0(\psi)(1 + \varepsilon \cos \theta)$ with $\varepsilon = r/R$, the local inverse aspect ratio (the minor radius divided by the major radius of a toroidal surface). In the expression for the Pfirsch–Schlüter current [Eq. (36)], the fourier coefficient δ_{01} multiplying $\cos \theta$ equals -2ε. By comparing the vertical field produced by this current with the tokamak poloidal field $B_\theta \approx (r/R)\iota B_\phi$, one finds that there is a large change in the shape of the flux surfaces when the volume-averaged plasma β ($\beta = 2\mu_0 p/B^2$) satisfies $\beta \gtrsim a\iota_a^2/R$ with a the plasma minor radius, ι_a the edge rotational transform, and R the major radius of the torus. For practical tokamak parameters, $\iota_a \approx \frac{1}{3}$ and $a/R \approx \frac{1}{3}$, there is a large change in the flux surface shape for $\beta \gtrsim 4\%$. The basic distortion of the surfaces is an outward shift of the center of the plasma, the magnetic axis, relative to the plasma edge.

The second effect of the Pfirsch–Schlüter cur-

rent is to cause the plasma to distort spontaneously just as the net current causes magnetic islands to open. The basic effect can be understood by generalizing the cylinder model used in the discussion of the pressureless plasma. The field strength in the unperturbed cylinder is $B_0(\psi_0)$. Assuming $2\mu_0 p/B_\theta^2 \ll 1$, the field strength changes little at a given point in space due to the perturbation. The perturbed field strength can be, therefore, approximated by $B^2(\psi, \theta, \phi) = B_0^2(\psi_0)$ or

$$\frac{1}{B^2} = \frac{1}{B_0^2(\psi_0)} \left(1 + 2\psi_1 \frac{1}{B_0} \frac{dB_0}{d\psi} \right) \quad (44)$$

The right-hand side of the differential equation for f [Eq. (42)] is just $m\mu_0 j_\phi$, so one obtains

$$\frac{1}{r} \frac{d}{dr} r \frac{df}{dr} - \frac{m^2}{r^2} f$$

$$= - \left[m \frac{R}{r} \frac{dK}{dr} \frac{n - \iota m}{(n - \iota m)^2 + m^2 \Delta_{nm}^2} \right.$$

$$\left. + \frac{2m^2 R^2}{(n - \iota m)^2 + m^2 \Delta_{nm}^2} \frac{dp}{d\psi} \frac{d \ln B_0}{d\psi} \right] f$$

$$(45)$$

The singularity of the Pfirsch–Schlüter current is regularized by introducing Δ_{nm} as before. The important feature of the Pfirsch–Schlüter term is the quadratic singularity $1/(n - \iota m)^2$ at each rational surface. Unlike the linear singularity in the force-free current term, the quadratic singularity can yield solutions on arbitrarily small spatial scales if $\Delta_{nm} = 0$. To demonstrate the point, let $r = r_* + x$ with $|x/r_*| \ll 1$. Then, the differential equation [Eq. (45)] can be reduced to $d^2 f/dx^2 = Df/x^2$ with $D = -2(dp/d\psi_0)(d \ln B_0/d\psi_0)R^2/(d\iota/dr)^2$. The solution to this equation is $f \propto x^\alpha$ with $2\alpha = 1 \pm (1 + 4D)^{1/2}$. The coefficient α becomes complex for $D < -\frac{1}{4}$. A complex α implies that a solution can be found that vanishes at any desired point. Clearly, for $D < -\frac{1}{4}$ distorted equilibria arise. This small spatial scale distortion of the plasma by the singularity of the Pfirsch–Schlüter current is a very general feature. The names used for these distortions are Suydam modes, Mercier modes, or ballooning modes depending on technical details.

In addition to driving a highly localized mode, the Pfirsch–Schlüter current singularity can also have a dominant effect on the low m modes driven by the net current term. When the island width Δ_{nm} is zero, stability is determined by the sign of D, regardless of how small D may be. This follows from one of the two solutions for f going as x^{-D} for $|D| \ll 1$ and $|x/r_*| \to 0$. Of course, the extreme sensitivity on D can be removed by allowing a finite island width Δ_{nm}. Unfortunately, the answer to the question of whether a plasma with axisymmetric boundary conditions will be in an asymmetric equilibrium may depend on the magnitude of the perturbations to which it has been subjected in the distant past.

D. ENERGY PRINCIPLE

A very powerful and traditional technique for studying toroidal equilibrium and stability is the energy principle. The energy W is the sum of the plasma and the field energy,

$$W = \int \left(\frac{p}{\gamma - 1} + \frac{B^2}{2\mu_0} \right) d^3x \qquad (46)$$

with $\gamma = \frac{5}{3}$, the adiabatic index of the plasma. The energy principle states that under certain constraints extrema of the energy are equilibria and that minima are stable equilibria. The most obvious constraint is that no energy enter or leave the integration volume. This constraint can be easily satisfied if the integration volume is bounded by a rigid boundary with perfect electrical conductivity that is tangent to the magnetic field. Such a boundary will be assumed for simplicity.

To evaluate changes in the energy, it is useful to describe the magnetic field using the transformation equations $\mathbf{x}(\psi, \theta, \phi, t)$ and the Hamiltonian $\chi(\psi, \theta, \phi, t)$ with the velocity of the canonical coordinates given by $\mathbf{v}_c = \partial \mathbf{x}/\partial t$. The pressure in the energy integral should be taken as a function of the electron number density n and the entropy per particle S. Changes in the pressure can then be shown to be

$$\left(\frac{\partial p}{\partial t} \right)_c = -\gamma p \boldsymbol{\nabla} \cdot \mathbf{v}_c + \gamma p \left(\frac{\partial \ln(nJ)}{\partial t} \right)_c$$

$$+ (\gamma - 1)p \left(\frac{\partial S}{\partial t} \right)_c \qquad (47)$$

by using the relations $\partial p/\partial n = \gamma p/n$, $\partial p/\partial S = (\gamma - 1)p$, and $(\partial J/\partial t)_c = J \boldsymbol{\nabla} \cdot \mathbf{v}_c$ with J the Jacobian of the canonical coordinates. The term $[\partial \ln(nJ)/\partial t]_c$ is the change in the number of particles in an infinitesimal canonical volume and would be zero if the particles were tied to the canonical coordinates. By differentiating W, expressing $\partial \mathbf{A}/\partial t$ in terms of $\mathbf{v}_c$ and $(\partial \chi/\partial t)_c$ using Eq. (27), and integrating by parts, one finds

$$\frac{\partial W}{\partial t} = \int \mathbf{v}_c \cdot (\boldsymbol{\nabla} p - \mathbf{j} \times \mathbf{B}) \, d^3x + \int \left[\left(\frac{\partial \sigma}{\partial t} \right)_c p \right.$$

$$\left. - \left(\frac{\partial \chi}{\partial t} \right)_c \mathbf{j} \cdot \boldsymbol{\nabla} \phi \right] d^3x \qquad (48)$$

with $\partial \sigma/\partial t = (\gamma/\gamma - 1)\partial \ln(nJ)/\partial t + \partial S/\partial t$.

The fact that equilibria are stationary points of the energy is demonstrated by the first integral in the expression for $\partial W/\partial t$. The constraint required, in addition to the boundary condition already discussed, is the vanishing of the second integral in the $\partial W/\partial t$ expression, or $(\partial \sigma/\partial t)_c = 0$ and $(\partial \chi/\partial t)_c = 0$. These constraints on σ and χ as well as the condition $(\partial S/\partial t)_c = 0$ are called the constraints of ideal MHD (magnetohydrodynamics). At equilibrium, a second variation of the energy, using the ideal MHD constraints, gives an equation that can be used to determine if an equilibrium is a minimum of the energy, which means ideal MHD stable, or just an extremum.

The second integral in the expression for $\partial W/\partial t$ gives conditions for stability in the presence of dissipation. Of the two terms in this integral, the term $(\partial \chi/\partial t)_c \, \mathbf{j} \cdot \boldsymbol{\nabla} \phi$ is the more important since the toroidal current density can become singular. For example, in the study of the net current in a perturbed cylinder, $j_\phi = Rj \cdot \boldsymbol{\nabla} \phi$ had a delta function singularity [Eq. (43)]. This illustrates an important point. Since the field topology is changing at the rational surface r_*, the radial field and χ must be related by $\partial b_r/\partial t = \hat{r} \cdot [\hat{z} \times \boldsymbol{\nabla} (\partial \chi/\partial t)]/R$ near $r = r_*$. Therefore, the energy driven by $(\partial \chi/\partial t)\mathbf{j} \cdot \boldsymbol{\nabla} \phi$ is

$$\frac{\partial W}{\partial t} = -\frac{\Delta' r_*}{m^2} V \frac{\partial}{\partial t} \left(\frac{\bar{b}_r^2}{2\mu_0} \right) \qquad (49)$$

with $V = 2\pi^2 r_*^2 R$ the volume of the rational surface and $\Delta' = [d\bar{b}/dr]/\bar{b}_r$ assumed independent of $\bar{b}_r$. If Δ' is positive, energy is released by the opening of a magnetic island and the equilibrium is tearing-mode unstable as noted earlier. This is known as Furth's energy criterion.

IV. Mirror Confined Plasma

A. INTRODUCTION

A stationary plasma, in which the gyroradius is small compared to the system size, does not in general have a scalar pressure but instead a pressure tensor $\mathbf{p} = p_\parallel \hat{b}\hat{b} + p_\perp(\delta - \hat{b}\hat{b})$. Pitch angle scattering tends to reduce the pressure anisotropy. One can show that $(p_\parallel - p_\perp)/(p_\parallel + p_\perp) \lesssim 1/(\tau_E \nu_a)^{1/2}$ with τ_E the energy confinement time and ν_a an energy-weighted-average collision frequency. Toroidal plasmas can have an anisotropic pressure, but the focus of this sec-

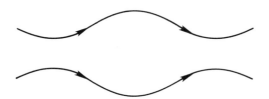

FIG. 7. The magnetic field lines in a simple mirror. The field strength is strongest where the field lines are closest together.

tion is on the confinement of anisotropic plasmas in nontoroidal, open-ended configurations.

There are two open confinement systems that have played a major role in the fusion program: the simple mirror (Fig. 7) and the tandem mirror (Fig. 8). A simple mirror confines the plasma ions by the variation in magnetic field strength coupled with the conservation of the magnetic moment μ. Remember that the energy of a small gyroradius particle is $H = mv_\parallel^2/2 + \mu B + e\Phi$ [Eq. (4)]. The electrons are generally isotropic due to their higher collision frequency and are confined by an ambipolar electric potential. There are two problems with the simple mirror. First, it is unstable to pressure driven instabilities, but this problem can be solved by a more complicated field configuration. Second, the confinement time of a simple mirror, $\sim 1/\nu_a$, is too short to confine an ignited fusion plasma.

The tandem mirror was invented to obtain longer confinement times in an open confinement device. The simplest version is given in Fig. 7. There is a long central cell that contains an almost isotropic plasma of density n_c. At each end are mirrors, which confine an anisotropic plasma of higher density n_m. The electrons are electrostatically confined throughout the device, so the electric potential and the density are related by $n \propto \exp(|e|\Phi/T_e)$. The central cell ions are also electrostatically confined along the field lines up to an energy to $T_e \ln(n_m/n_c)$. If this energy is greater than about five times the ion temperature, then adequate confinement is possible for thermonuclear ignition. Of course, the center cell must be very long compared to the end cells. However, by introducing extra complexity in the end cells, the required energy content of the ends cells can be greatly reduced. Tandem mirror experiments have been undertaken in the United States, the Soviet Union, and Japan, and the quality of the plasmas is rapidly improving.

B. MIRROR EQUILIBRIA

The equilibria of open confinement systems are evaluated using $\nabla \cdot \mathbf{p} = \mathbf{j} \times \mathbf{B}$. This equation can be written in a more useful form by evaluating the divergence of the pressure tensor to obtain

$$\nabla p_\parallel + \frac{p_\perp - p_\parallel}{B^2} \nabla \left(\frac{B^2}{2}\right) = \mathbf{J} \times \mathbf{B}$$

$$\mathbf{J} \equiv \frac{1}{\mu_0} \nabla \times (\sigma \mathbf{B}) \tag{50}$$

with $\sigma = 1 + \mu_0 (p_\perp - p_\parallel)/B^2$. This is the same equilibrium equation as that of a fluid with magnetic permeability μ_0/σ and pressure $p_\parallel$.

As in toroidal systems, magnetic coordinates greatly simplify the study of equilibria. However, the field lines never circle back on themselves inside an open confinement system; so one can always find magnetic coordinates, which are single valued in the plasma region, such that the Hamiltonian χ is zero. In other words, the field can always be written as $\mathbf{B} = \nabla \psi \times \nabla \theta$, which is well known as the Clebsch representation.

The simplest variables for the parallel pressure are $p_\parallel(\psi, \theta, B)$. In these variables

$$\frac{\partial p_\parallel}{\partial B} = -\frac{p_\perp - p_\parallel}{B}$$

and

$$\frac{\partial p_\parallel}{\partial \psi} \nabla \psi + \frac{\partial p_\parallel}{\partial \theta} \nabla \theta = \mathbf{J} \times \mathbf{B} \tag{51}$$

By properly choosing the flux function ψ and the angle θ, one can sometimes make $\partial p_\parallel/\partial \theta$ zero. This is possible if the drift orbits of particles that cross a constant ψ surface remain close to it. Although this is generally desirable, it is not required except when θ is a coordinate in a sym-

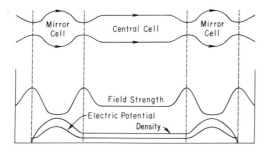

FIG. 8. The magnetic field lines in a tandem mirror are illustrated with the corresponding magnetic field strength, plasma density, and electric potential.

metry direction. If $\partial p_\parallel/\partial \theta$ is zero, which is the usual assumption, then the anisotropic equilibrium equations have many similarities to the isotropic equations. In particular, since $\mathbf{J} \cdot \nabla\psi = 0$, one can find a coordinate w that gives a simple covariant representation for $\mathbf{B}$, specifically $\sigma\mathbf{B} = \nabla w + \beta_*(\psi, \theta, w)\nabla\psi$. By arguments analogous to those for toroidal equilibria, one can show that the condition for equilibrium in an open system is that

$$\int \frac{\partial p_\parallel}{\partial \psi} \frac{dl}{B} \qquad (52)$$

be the same on every field line of a ψ surface.

C. Electric Potential

Although the electric potential in a plasma adjusts to ensure quasi-neutrality, $\Sigma\, e_\alpha n_\alpha = 0$, the actual evaluation of the potential is modified by the additional constraint of equilibrium. The simplest plasma consists of two species, ions and electrons, and equilibrium implies that both species are in equilibrium. If overall plasma equilibrium, $\nabla \cdot \mathbf{p} = \mathbf{j} \times \mathbf{B}$, is satisfied, then the additional equilibrium constraint can be taken as applying to the electrons. This equilibrium constraint is $\nabla p_e = -|e|n(\mathbf{E} + \mathbf{u} \times \mathbf{B})$ with $\nabla \cdot n\mathbf{u} = 0$. The small collisional force $en\mathbf{R} \approx en\eta\mathbf{j}$, between the electrons and the ions, has been ignored since it will be discussed under the topic of transport. The electron equilibrium constraint would appear to determine the velocity $\mathbf{u}$ and the potential Φ. Actually, these equations contain two arbitrary functions. In a torus, these arbitrary functions are functions of ψ and represent the arbitrary rates of toroidal and poloidal rotation. Mathematically, one arbitrary function is additive to $nu_\parallel/B$ and the other is additive to the potential. In open confinement systems, the net flow along the field is generally zero and this eliminates one arbitrary function. However, the arbitrariness of the potential remains. Let us consider the evaluation of the potential in some detail.

Ignoring transport effects, $\mathbf{E} = -\nabla\Phi$, and the parallel component of the electron force balance equation is $\mathbf{B} \cdot \nabla\Phi = (\mathbf{B} \cdot \nabla p_e)/|e|n$. In a toroidal plasma, this equation implies that the potential is a function of ψ alone, at least on the irrational surfaces. However, in open confinement systems the parallel component determines the functional dependence of Φ on only one of the three coordinates, namely the w dependence.

The constraint $\nabla \cdot n\mathbf{u} = 0$ is essentially automatic in isotropic toroidal plasmas, but it is an important constraint in open systems. General calculations are very complicated; so to simplify the mathematics, we assume that the electron temperature T_e is constant throughout the plasma. In the constant T_e case, the electron pressure balance can be rewritten as $\mathbf{u} \times \mathbf{B} = \nabla\Phi_*$ with $\Phi_* = \Phi - T_e \ln(n)$, so that $\mathbf{B} \cdot \nabla\Phi_* = 0$. We also assume that $\sigma\mathbf{B} = \nabla w + \beta_*\nabla\psi$ as well as $\mathbf{B} = \nabla\psi \times \nabla\theta$. The zero divergence condition, which can be written as $\sigma B^2 \partial(nu_\parallel/B)/\partial w = -\nabla \cdot n\mathbf{u}_\perp$, can be integrated over w to demonstrate that the potential Φ_* is an arbitrary function of $N \equiv \int(n/B)\, dl$. The function $\Phi_*(N)$ is set by transport effects. It should be noted that the constant N surfaces need not coincide with the surfaces given by the equilibrium constraint, Eq. (52). The dependence of the electric potential on θ is important since this dependence causes the particle drift orbits to cross the constant ψ surfaces and make the parallel pressure a function of θ as well as ψ.

V. Transport

A. Introduction

To maintain a steady-state plasma, sources of particles, energy, and sometimes magnetic flux are required. The evaluation of these requirements is the subject of transport. Particle and magnetic flux transport can be calculated using the force balance equations, which were used in the discussion of equilibrium, but with the inclusion of the small collisional force, $en\mathbf{R} \approx en\eta\mathbf{j}$, between the ions and electrons. If the particle mean-free path between collisions is longer than the size of the plasma, then small pressure anisotropy effects also make an important contribution to transport. The pressure anisotropy arises from the flows associated with currents in the plasma and is equivalent to a viscous drag on the parallel velocity component of each species. These are so-called parallel viscosity effects.

Unlike the particle and the magnetic flux, heat flux calculations require a different velocity moment of the kinetic equation, the $V^2\mathbf{V}$ moment. The equation for the heat flux $\mathbf{q}$, which is analogous to an isotropic two fluid equation, is

$$\nabla T - (2e/5p)\, \mathbf{q} \times \mathbf{B} = -K^{-1}\mathbf{q} \qquad (53)$$

The thermal conductivity $K \simeq (5p/2m)/\nu$ is analogous to $1/\eta$ with η the resistivity. The collision frequency ν in K has contributions from both like and unlike particle collisions. However, only unlike particle collisions contribute to the resistivity η due to the momentum conservation properties of the collision operator.

The fundamental transport coefficients K and η are evaluated by linearizing the kinetic equation in the deviation of the distribution function from a Maxwellian. The particle and heat flux calculations based on the force balance equation and the heat flux equation are in poor agreement with many experiments. This is particularly true for the electron heat flux perpendicular to the pressure surfaces. It is often two orders of magnitude larger than the predicted value. However, the transport of magnetic flux is generally close to the predicted value in plasmas that are stable to tearing modes. The explanation is apparently the existence of low-amplitude electric potentials and magnetic perturbations with a complicated spatial structure. The existence of these perturbations is closely related to the localized pressure driven perturbations discussed under toroidal equilibrium. However, additional effects are also important. For example, a complicated electric potential pattern can be established due to the different response of electrons and ions to a potential that varies across the field lines on a scale comparable to the ion gyroradius. These potential patterns tend to be very extended along the magnetic field lines to avoid electron shorting effects, and therefore tend to resonate with the magnetic field structure just as the pressure driven perturbations do. Both the lack of spatial symmetry and the complicated temporal behavior of these perturbations can cause the electron drift orbits to become stochastic, even in the absence of collisions, and account for the enhanced transport.

In the remainder of this section, the theory of transport will be discussed in the fluid limit, which means that the particle mean-free path between collisions is smaller than the plasma size, and in the opposite, low collisionality, limit. Although not all the predictions of the theory agree with experiments, the theory does clarify a number of transport features. The derived transport coefficients do give the correct rate of magnetic flux transport and generally the correct ion heat flux.

B. FLUID LIMIT

Particle and magnetic flux transport can be calculated in the fluid limit using the electron force balance equations called generalized Ohm's law, Eq. (19),

$$\mathbf{E} + \mathbf{u} \times \mathbf{B} = \eta \mathbf{j} \tag{54}$$

For simplicity in calculating the particle flux, consider a steady-state, toroidal plasma with no net current so that $\mathbf{E} = -\nabla\Phi$. If the generalized Ohm's law is dotted with $\mathbf{j}$, then, using $\nabla p = \mathbf{j} \times \mathbf{B}$ and $\nabla \cdot j = 0$, one finds that

$$\frac{dp}{d\psi} \int \mathbf{u} \cdot d\mathbf{a}_\psi = \frac{d}{d\psi} \int \eta j^2 \, d^3x \tag{55}$$

with $d\mathbf{a}_\psi = (\nabla\psi)J \, d\theta \, d\phi$ the flux surface area element and J the Jacobian. The square of the current j^2 is the sum of a so-called diamagnetic contribution $j_\perp^2$ and a Pfirsch–Schlüter contribution j_{ps}^2 with both terms proportional to $(dp/d\psi)^2$. Using the expressions for $\mathbf{j}$ derived in the section on toroidal equilibrium, the net particle flux $\Gamma \equiv \int n\mathbf{u} \cdot d\mathbf{a}_\psi$ can be written as $\Gamma = -(D_\perp + D_{ps}) \, dp/d\psi$ with

$$D_\perp = \eta \int \frac{\nabla\psi}{B^2} \cdot d\mathbf{a}_\psi$$

and

$$D_{ps} = 2\pi^2 \eta \, \frac{g + \iota I}{B_0^4} \sum{}' \left(\frac{mg + nI}{n - \iota m}\right)^2 \delta_{mn}^2 \tag{56}$$

The quadratic singularity in D_{ps} at the rational surfaces, $n\iota = m$, implies that unless the resonant δ_{nm}'s vanish, the transport will be sufficiently rapid to flatten the pressure gradient. It should be noted that if B^2 is an analytic function of ψ, θ, and ϕ, then the Fourier coefficients become exponentially small for large m and n. This means that the integral $\int d\psi/D_{ps}(\psi)$ generally exists and is finite, so that a finite pressure can be maintained by a finite flux Γ even if the pressure gradient vanishes at every rational surface. Actually, the Pfirsch–Schlüter transport is generally dominated by only one Fourier term, the $n = 0$, $m = 1$ component. Numerically, the particle confinement time τ_p in a torus with a small inverse aspect ratio, $\varepsilon = r/R$, is approximately

$$\tau_p \approx \frac{1}{6}\left(\frac{a}{\rho_e}\right)^2 \left[1 + \left(\frac{\delta}{2\varepsilon}\right)^2 \frac{1}{[\iota - (n/m)]^2}\right] \tau_e \tag{57}$$

The number of electron gyroradii across the plasma is a/ρ_e, τ_e is the electron collision time with ions, and δ is the dominant Fourier component with mode numbers n and m.

By magnetic flux transport is meant the motion of one flux relative to the other, $\partial\chi(\psi, t)/\partial t = V/2\pi$. The loop voltage V can be evaluated using Eq. (31) and the covariant expression for $\mathbf{B}$ [Eq. (33)],

$$V = \frac{2\pi\eta}{\mu_0}\left(g \frac{dI}{d\psi} - I \frac{dg}{d\psi}\right) \tag{58}$$

Heat flux calculations can be made in a manner analogous to the particle transport using Eq. (53) for $\mathbf{q}$. Assume the collision frequency is sufficiently small, so that $|\mathbf{q}|/K$ is much smaller in magnitude than the other terms in the equation. Under this assumption it is natural to first find the $K \to \infty$ heat flux, which satisfies the equation $\nabla T = (2e/5p)\mathbf{q}_\infty \times \mathbf{B}$. The heat flux must be constrained so that $\nabla \cdot \mathbf{q}_\infty = 0$, in order that the electron temperature be constant along the field lines as required for $|\mathbf{q}|/K \to 0$. Note, that the energy continuity equation is $\partial(3nT/2)\partial t = \nabla \cdot (\mathbf{q} + 5p\mathbf{u}/2)$. The solution $\mathbf{q}_\infty$ is clearly of the same form as the current $\mathbf{j}$ associated with equilibrium. In other words, there are apparent singularities in $\mathbf{B} \cdot \mathbf{q}_\infty$ at each rational surface. If Eq. (53) for $\mathbf{q}$ is dotted with $\mathbf{q}_\infty$, and $\mathbf{q}/K$ is approximated by $\mathbf{q}_\infty/K$, one finds

$$\frac{dT}{d\psi} \int \mathbf{q} \cdot d\mathbf{a}_\psi = \frac{d}{d\psi} \int \frac{1}{K} q_\infty^2 \, d^3x$$

The heat transport is predominately due to ions and is about $(m_i/m_e)^{1/2}$ larger that the particle transport. Numerically,

$$\tau_E \approx \frac{1}{6} \left(\frac{a}{\rho_i}\right)^2 \left[1 + \left(\frac{\delta}{2\varepsilon}\right)^2 \frac{1}{[\iota - (n/m)]^2}\right] \tau_i$$

with τ_i the ion collision time and ρ_i the ion gyroradius.

C. Low Collisionality Transport

If the mean-free path between collisions is longer than the plasma dimensions, the fluid limit calculation of dissipative effects is no longer valid. In a fluid limit calculation, drift effects are approximated by the perturbation that the drifts make to the Maxwellian enforced by collisions. If the mean-free path is long, as it would be in a fusion reactor, the distribution function must be a function of the constants of the motion of the drift trajectories. Nonetheless, the distribution function must be close to a Maxwellian if the confinement time is to be long when compared to this collision frequency. If τ_E is the local energy confinement time, then entropy production arguments imply that $|\langle \tilde{f}C(\tilde{f})\rangle| \approx 1/\tau_E$. The distribution function is $f = (1 + \tilde{f})f_M$, $C(\tilde{f})$ is the linearized collision operator, $\langle ... \rangle$ is a Maxwellian velocity space average, and f_M is a Maxwellian distribution. The closeness with which a Maxwellian can be approximated by the constants of the drift motion determines the long mean-free path limit, confinement time. In plasmas without symmetry directions, the drift trajectories are generally complicated with a range of characteristic time scales. Each of these characteristic time scales can give a different collisionality regime depending on whether collisions dominate on that scale.

The moment equations are, of course, valid in all collisionality regimes. The characteristic features of long mean-free path transport is determined by the so-called parallel viscosity, $\nabla \cdot \boldsymbol{\pi}$, with $\boldsymbol{\pi} = (p_\parallel - p_\perp)\hat{b}\hat{b}$, which is the pressure anisotropy. The important pressure anisotropy arises from the motion of the electrons and the ions through regions of varying field strength. The variation in field strength, through magnetic flux conservation, implies a compression and decompression of the flow. In a system with toroidal symmetry, like an ideal tokamak, the parallel viscosity cannot damp the angular momentum of the toroidal flow. The toroidal torque on a species is calculated by dotting $\partial \mathbf{x}/\partial \phi = J(\nabla\psi \times \nabla\theta)$ with the fluid equation and integrating over the volume between two differentially separated ψ surfaces. If this is done, one finds that the net particle flux is proportional to the toroidal torque on a species. Since only forces between the species can exert a toroidal torque in a symmetric tokamak, the ion and electron radial transport are always equal, independent of the size of the radial electric field. In an asymmetric plasma, however, the parallel viscosity damps both the poloidal and the toroidal rotation of the species, so that the radial diffusion rates of the ions and the electrons need not be equal from force balance considerations. Of course, quasi-neutrality implies the net outflow of ions and electrons must be equal. This equality is enforced by the electric potential $\Phi(\psi)$. If, for example, the natural rate of transport of one species is much larger than that of the other, the pressure gradient of the poorer confined species is balanced by the radial electric field, with $end \, \Phi/d\psi = -dp/d\psi$.

The parallel viscosity also causes a strong coupling between the net toroidal current and the pressure gradient. If the viscosity totally dominated the resistivity, then there would be no poloidal flows in an ideal tokamak. Force balance would then imply $j_\phi \approx -(dp/dr)/B_\theta$, and $\mathbf{E} + \mathbf{u} \times \mathbf{B} = 0$ would imply $u_r \approx -E_\phi/B_\theta$. So, the pressure gradient would give a net toroidal current, the bootstrap current j_b, and the electric field E_ϕ required to maintain the poloidal magnetic flux would give a net inward plasma flow, the Ware pinch u_w. Actually, the parallel viscosity is not totally dominant so that $j_b \approx -\varepsilon^{1/2}(dp/$

$dr)/B_\theta$ and $u_w \simeq -\varepsilon^{1/2}E_\phi/B_\theta$ with $\varepsilon = r/R$ the inverse aspect ratio.

Actual transport calculations are carried out by integrating the particle drift orbits. The issue is, how closely do the particle drift orbits follow the pressure surface? If every drift orbit that was started on a pressure surface stayed there, then the drift motion could cause no transport. It is, therefore, useful to study the drift trajectories in magnetic coordinates. Fortunately, the Hamiltonian drift equations are especially simple in these coordinates. As discussed earlier, the drift Hamiltonian is $H = mv_\parallel^2/2 + \mu B + e\Phi$ and has only two degrees of freedom. The natural canonical coordinates are the two angles θ and ϕ. The canonical momenta are the drift analogs of the exact canonical momenta of a particle in a magnetic field, the components of $\mathbf{P} = m\mathbf{V} + e\mathbf{A}$. To zeroth order in gyroradius the drift velocity is $\mathbf{v}_\parallel = (V_\parallel/B)\mathbf{B}$ while the canonical momentum conjugate to ϕ is $\mathbf{P} \cdot \partial\mathbf{x}/\partial\phi$. Therefore, the lowest order expression for the drift canonical momentum is $p_\phi = (v_\parallel/B)\mathbf{B} \cdot \partial\mathbf{x}/\partial\phi + e\mathbf{A} \cdot \partial\mathbf{x}/\partial\phi$. If the magnetic field is given in the forms $\mathbf{B} = g(\psi)\nabla\phi + I(\psi)\nabla\theta + \beta_*\nabla\psi$ and $\mathbf{B} = \nabla\psi \times \nabla\theta + \nabla\phi \times \nabla\chi$, then $p_\phi = (g/B)mv_\parallel - e\chi$ and $p_\theta = (I/B)mv_\parallel + e\psi$.

In a plasma with perfect magnetic flux surfaces, the properties of the particle drift orbits are determined by the field strength $B(\psi, \theta, \phi)$ alone. In a toroidally symmetric field $B(\psi, \theta)$ both p_ϕ and the energy H are conserved and one can easily find the drift orbit trajectories. Particles, which have a turning point $v_\parallel = 0$, are called trapped particles and make the largest excursions in ψ. These drift orbits have a characteristic banana shape in magnetic coordinates. The passing particles, the particles with no turn-

ing points, move rapidly along the field and therefore average out the radial drifts. If there are not symmetries, the trapped particles are very sensitive to the exact form of the field strength, [that is, B dependent on all three coordinates, $B(\psi, \theta, \phi)$]. Even a small breaking of the toroidal symmetry in a tokamak, by 1% or less, can greatly enhance the transport. On the other hand, the passing particles, which closely follow the field lines, are very sensitive to any stochasticity in the magnetic field, $\chi(\psi, \theta, \phi)$. The trapped particles are clearly little affected by magnetic field stochasticity due to their relative short excursions along the field lines between turning points.

BIBLIOGRAPHY

Baldwin, D. E. (1977). End-loss processes from mirror machines, *Rev. Mod. Phys.* **49**, 317–339.

Boozer, A. H. (1984). Three-dimensional stellarator equilibria by iteration, *Phys. Fluids* **27**, 2110–2114.

Freidberg, J. P. (1982). Ideal magnetohydrodynamic theory of magnetic fusion systems, *Rev. Mod. Phys.* **54**, 801–902,

Hugill, J. (1983). Transport in tokamaks a review of experiments, *Nucl. Fusion* **23**, 331–373.

Kovrizhnykh, L. M. (1984). Neoclassical theory of transport processes in toroidal magnetic confinement systems, with emphasis on non-axisymmetric configurations, *Nucl. Fusion* **24**, 851–936.

Liewer, P. C. (1985). Measurements of micorturbulence in tokamaks and comparisons with theories of turbulence and anomalous transport, *Nucl. Fusion* **25**, 543–621.

Miyamoto, K. (1978). Recent stellarator research, *Nucl. Fusion* **18**, 243–284.

Miyamoto, K. (1980). ''Plasma Physics for Nuclear Fusion,'' MIT Press, Cambridge, Massachusetts.

PLASMA DIAGNOSTIC TECHNIQUES

Charles B. Wharton *Cornell University*

GLOSSARY

Activation: Nuclear reaction in which a stable element is converted into a radioactive element by a collision.

Anomalous resistivity: Plasma state in which the conductivity is decreased by turbulence.

Charge exchange: Atomic reaction in which an electron from a cold atom transfers to an energetic ion, creating a fast atom and a cold ion.

Diagnostics: Technique of making multiple measurements in a mostly nonperturbing fashion.

Diamagnetic: Effect in a plasma in which particle motions decrease the applied magnetic field.

Doppler broadening: Widening of a spectral or resonance line caused by the relative velocities of emitting or scattering particles.

Faraday cup: Small cup used to collect charged particles, usually for current measurements.

Faraday shield: Conducting surface oriented such that electrostatic fields cannot penetrate but magnetic fields can.

Fresnel zone plate: Conducting screen having concentric ring slots spaced in a manner that the Fresnel zones at a particular distance are in phase. An artificial lens.

Inertial-confinement fusion: Method to achieve thermonuclear fusion in a time shorter than the disassembly time of the reacting region.

A small pellet is usually imploded by external forces to create sufficiently high temperatures and densities that ignition occurs.

Interferometer: Device arranged to compare the relative phases of two signals. The output may be electrical or electromagnetic.

Stark broadening: Widening of an emission line due to the presence of strong electric fields that distort the electron orbits.

Stripping cell: Device whose purpose is to remove one or more electrons from an energetic atom, generating an ion.

Radiometer: Instrument that measures the radiant energy from a source. Usually tunable in the frequency spectrum and spatially directive. Very sensitive.

Thomson scattering: Scattering of photons from a single electron or the uncorrelated scattering from an aggregate of electrons.

Plasma diagnostics is the science and art of making multiple, nonperturbing measurements in plasmas and inferring from them the physical and chemical properties of the plasma. The techniques employed involve a wide range of scientific disciplines and engineering technology.

I. Introduction

Plasma diagnostics, the science and art of obtaining nonperturbing measurements in plasma, has achieved considerable sophistication over the 30 or so years of its development. The basic physics of the measurements have remained essentially unchanged, but advances of technology over the years have allowed completely new areas to be addressed. Examples of these new developments of technology are lasers, electronic integrated circuits, millimeter wave and infrared techniques, optical fibers, computers, and high-speed electronic circuits in general. All of these advances have greatly facilitated data acquisition and analysis, especially in connection with large, complex plasma fusion research facilities, where the expense of large diagnostics systems can be justified.

But even a small plasma research group can bring to bear a powerful array of diagnostics instruments on its experiments, with a little ingenuity and at acceptable cost. The large data acquisition systems, which can consume much time and effort in their development, are not essential for modest intallations, and sufficient flexibility can be achieved by using simple but well-designed manual or microprocessor-controlled systems that are quite compact.

The advances made in solving diagnostics problems have developed into an important branch of applied physics and engineering. These advances have also had an influence on other fields of applied science and technology, providing incentive for new developments, and markets for new products.

A number of conferences and specialist schools are now devoted every year to the subject of diagnostics, and we can expect this trend to continue as new challenges and problems are faced. An example of such a challenge is the requirement for instrument "hardening" to withstand high radiation fluxes and other harsh environmental conditions associated with fusion reactor experiments. A number of interesting new features can thus be expected in the diagnostics picture within the next few years.

In this article, only the basic diagnostic techniques are described, leaving the applications to specific plasma installations for the individual investigator to consider. The broad categories of experiments are discussed, however, since some techniques are more applicable to certain plasma conditions than to others. For example, nuclear detection methods are of limited utility in gas discharge experiments, and conversely, intrusive probes are not generally useful in dense, fusion-condition plasmas. [See PLASMA CONFINEMENT; PLASMA SCIENCE AND ENGINEERING.]

II. Measurable Quantities

The tabulation given in this section lists several of the primary diagnostic methods that are commonly employed, grouped under categories of plasma parameters to be determined. Some of these methods are singled out for discussion in later sections. Very few of the measurements are simple, and some require careful, ingenious design to obtain usable data. Nevertheless, all of those listed have been employed on plasma experiments, and several are standard instruments by which the performance of major fusion experiments is judged.

In most experiments, diagnosticians measure each plasma quantity by more than one method for comparison. This approach is especially important when one method has high accuracy but no time resolution, and the other method is only semiquantitative but does have time resolution.

An example is the determination of electron temperature precisely by laser-light Thomson scattering (at one or two instants of time) or qualitatively but as a function of time by analysis of soft X-ray emission. Methods yielding data with good time resolution are identified with the abbreviation (tr).

Reference to the lists reveals that some methods appear under more than one plasma parameter heading, although one kind of data may be of more value than another. In such cases, the use giving the primary data is labeled (pri) and others are labeled (sec).

Key to Abbreviations in Tabulation

*	Discussed in later text
n_e	Electron density (N/m^3)
n_i	Ion density (N/m^3)
T_e	Electron kinetic temperature (eV)
T_i	Ion kinetic temperature (eV)
J	Current density (A/m^2)
B	Magnetic field (Tesla)
tr	Data usually have time resolution
spr	Data yield spatial resolution
pri	Primary application
sec	Secondary application

Tabulation of Measurable Quantities

1. Electron density

a. Microwave interferometer: 10^{16} ($n_e < 10^{21}$; pri, tr, spr, *

b. Microwave cavity perturbation: small column; $10^{14} < n_e < 10^{18}$; pri, tr

c. Langmuir probes: intrusive; $n_e = n_i$; $10^{14} < n_e < 10^{18}$; pri, tr, spr, *

d. Neutral beam attenuation or scattering: must know T_e; $10^{19} < n_i < 10^{21}$; pri, tr, spr, *

e. Optical interferometer: $10^{20} < n_e < 10^{25}$; pri, spr, *

f. Optical Balmer series limit: low T_e; $10^{18} < n_e < 10^{21}$; pri

g. Optical Thomson scattering: $10^{18} < n_e < 10^{25}$; pri, spr, *

h. Optical spectroscopic intensities; equilibrium; $n_e > 10^{19}$; sec, spr, tr,

i. RF conductivity probes: intrusive; quiescent, low-temperature plasma; $10^{14} < n_e < 10^{21}$; sec, spr, tr, *

j. Particle collector: beams or open-ended plasma; pri, spr, tr, *

2. Electron temperature

a. Optical Thomson scattering: $T_e > 5$ eV, $n_e > 10^{18}$; pri, spr, *

b. Langmuir probes: intrusive; $10^{14} < n_e < 10^{18}$; pri, spr, tr, *

c. Relative intensities of spectral lines: wall problems; $1 < T_e < 50$ eV; pri, tr

d. X-ray and VUV intensity; wall and line-radiation problems; $T_e > 100$ eV; pri, tr, *

e. Microwave radiation intensity: stable, high-collision plasma; $T_e > 0.1$ eV; pri, spr, tr, *

f. Far-infrared radiation intensity: cyclotron harmonics; $T_e > 10$ eV; pri, tr, *

3. Ion density

a. Langmuir probes (single and double): intrusive; $10^{14} < n_i < 10^{18}$; pri, spr, tr, *

b. Electron, ion, neutral-atom or neutron-beam probing: attenuation or scattering; $n_i > 10^{19}$; pri, spr, tr, *

c. Charge-exchange energetic neutral detector: $n_i > 10^{19}$; sec, spr, tr, *

d. Diamagnetic loop: requires knowledge of temperature and spatial distribution; sec, tr, *

e. Calorimetry: beams or open-ended plasma; must know velocity; sec, spr, tr,

4. Ion temperature or kinetic energy

a. Calorimetry: beams or open-ended plasma; total energy; must know density; pri, spr, tr, *

b. Doppler broadening of spectral lines: $T_i > 5$ eV; pri, spr, tr, *

c. Charge-exchange neutral-atom analyzer: single or multichannel; energy, momentum, and identity; $T_i > 80$ eV; pri, spr, tr, *

d. Diamagnetic loop: must know density and spatial distribution; pri, tr, *

e. Energy-momentum analyzer: escaping ions, beams; sec, spr, tr

f. Neutron flux: evidence of fusion reactions; sensitive to high-energy tail; pri, tr, *

g. Alpha particles: evidence of fusion reactions; sec

5. Inertial-confinement fusion

a. Time-of-flight: particle or shock-front velocity; pri, tr, *

b. Target activation: intense ion beams; $W_i > 300$ keV; pri, spr, *

c. Thompson parabola: intense ion beams; energy and identity; $W_i > 100$ keV; pri, spr, *

d. Film tracks: intense ion beams; energy and identity; pri, spr, *

e. X-ray and VUV imaging (pinhole or Fresnel plate): target temperature and dynamics; pri, spr, tr, *

f. Flash X-radiography: target dynamics; pri, spr

g. Rogovskii coil: beam current; pri, tr, *

h. Neutron analysis (pinhole, time-of-flight): pri, tr, *

6. Instabilities and turbulence

a. Microwave and IR scattering: scale-size, frequency, and magnitude of fluctuations; pri, spr, tr, *

b. Microwave radiation: nonthermal, anomalous effects; sec, tr, *

c. RF probes: intrusive, may induce oscillations; pri, spr, tr, *

d. Conductivity probes: intrusive; local resistivity; pri, spr, tr, *

e. Magnetic probes: resistive (anomalous) field penetration; pri, spr, tr, *

f. Anomalous Stark broadening: E-field due to turbulence; first-order or forbidden; pri, spr, tr

g. Particle beam scattering: deflection of halo due to fluctuations; pri, spr, tr

h. External $V–I$ measurements: anomalous resistivity; sec, tr

i. Charge-exchange neutral-atom energy-momentum: nonthermal distribution (e.g., high-energy tail); sec, spr, tr, *

7. Constituent identity; contaminants

a. Optical and atomic-resonance spectroscopy: incomplete ionization; pri, spr, tr

b. X-ray and VUV spectral emission: multiple ionization; pri, spr, tr, *

c. Ion or neutral-atom mass spectrometer: escaping particles; sec, spr, tr, *

III. Internal Probing

Physical probes are usually considered intrusive, at least in high-temperature plasmas. Nevertheless, their application is usually so direct and uncomplicated that their use is often justified.

A. LANGMUIR PROBES

One of the most venerable of plasma diagnostic techniques involves the current–voltage characteristics of metal probes immersed in the plasma. Both double and single probes are used, as indicated in the sketches in Fig. 1. The double probe has an advantage in that it is floating and cannot draw large currents to ground, making it less perturbing. The potential of the single probe is measured with respect to the chamber wall and does not require a floating metering circuit. When the probe voltage is made very negative, electrons are repelled, and only ions are collected. A sheath forms around the probe wire with a thickness proportional to the Debye length; thus the current collected depends more on the electron temperature T_e than on the ion temperature T_i. The ion current density at large negative voltage is directly proportional to the ion density:

$$J_i = 0.4 \, en_i \left(\frac{2k \, T_e}{n_i}\right)^{1/2} \qquad (1)$$

Thus the plasma density is calculated directly from the saturation ion current of Eq. (1), also indicated by the sketches on the right in Fig. 1.

When the probe voltage is made less negative, some electron current is drawn, increasing as the voltage goes through zero, until at large positive voltages a space-charge saturation occurs, as shown in Fig. 1. In the region between the two saturations, the current depends exponentially on the ratio of applied voltage to the plasma electron temperature. Thus the slope of a logarithmic plot of I versus V yields the electron temperature.

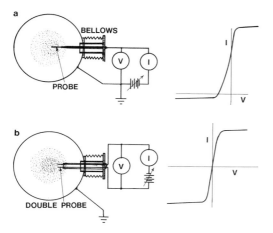

FIG. 1. Sketch of Langmuir probes and their $V\!-\!I$ characteristics: (a) is a single probe, with grounded metering circuit; (b) is a double probe with floating metering circuit.

B. MAGNETIC PROBES

A small coil inserted into a magnetized plasma can pick up changes in the magnetic field caused by internal currents or external influences. These coils are commonly used to measure the profiles of current channels, shocks, and MHD waves, also providing evidence of turbulence and fluctuations. The windings in such probes, which may be oriented in any direction, are usually mounted inside insulated tubes of ceramic or quartz and shielded by a Faraday shield to minimize electrostatic pickup. To obtain spatial resolution the probe position is controlled through a sliding vacuum seal or a small bellows (as shown in Fig. 1). Since the voltage induced in the coil is proportional to the rate of change of magnetic flux, the signal is integrated to give current information. Because of the small area and the necessity to integrate, such probes are rather insensitive, requiring large current densities to obtain readable signals.

C. RF FLUCTUATION PROBES

The same wire probes as used for Langmuir probes may be used to launch plasma waves or to pick up potential fluctuations. Several functions may be performed simultaneously by using filters to separate the signals. Measurement of the dispersion and damping of plasma waves can give information on plasma density and temperature and on the presence of electron streams or drifts. The spectrum of fluctuations gives information on instabilities and other nonthermal processes.

D. RF CONDUCTIVITY PROBES

A small coil driven by a high-frequency source induces currents in the surrounding plasma. These currents react back on the coil to change its inductance and losses in proportion to the reactive and real components of the plasma conductivity. A Faraday shield around the coil eliminates electrostatic coupling. Measurement of the change in coil Q and inductance give information on the plasma collision frequency (real or anomalous) and the plasma density.

As with other intrusive probes, RF probes cause a certain amount of perturbation.

E. FARADAY CUP PROBES

Miniature Faraday cups may be thought of as directional Langmuir probes, with the added feature that biased grids may be used to obtain particle energy distributions. Anisotropies and electron streaming are easily discerned, and the ratio of parallel to perpendicular energies can be measured. A thin foil may be placed across the entrance aperture to exclude low-energy particles. Cup diameters of a few millimeters may be inserted in some types of plasmas with only modest perturbation, but in other cases the probes are usable only outside the plasma.

IV. Beam Probes; Ion and Atomic Processes

A. Neutral Beam Attenuation

Beams of particles may be only mildly perturbing, depending on conditions. For example, observing the attenuation of a microampere energetic neutral helium atom beam due to impact ionization by plasma electrons yields information on the electron temperature and density with no noticeable effect on the plasma. Such a system is shown in Fig. 2. Substituting hydrogen for helium gives ion information since the major ionization is then due to ion charge exchange. Heavier particles, such as cesium or thallium, are also commonly used since they are subject to multiple ionizations, adding further information on plasma distribution conditions. In this case ion collector probes are located in appropriate places on the chamber wall to measure the number of ions formed instead of the number of neutrals lost.

B. Charge-Exchange Analysis

The equipment shown in Fig. 2 can also be used to measure the energy distribution of energertic neutral atoms arising from charge exchange. In this case the source is turned off, and the stripping cell gas pressure is optimized for the type and energy of neutal atoms to be detected.

Charge-exchange analysis is very complicated, relying on knowledge of several cross sections and atomic processes. The charge-exchange process itself occurs when an energetic ion exchanges an orbital electron with a neutral atom, producing an energetic neutral atom and a cold ion. Some neutral gas or an energetic neutral beam is required to be in the same region as the ions for the interaction to occur. The charge-exchange cross sections for most ion types have a maximum around 10 keV, and except for resonant interactions (e.g., $H + P^* \rightarrow H^* + P$), they fall off steeply at low energies and less steeply at higher energies. Usable data can be obtained down to 100 eV or so.

The stripping cell efficiency also falls off steeply at energies below a kilovolt imposing another low-energy restriction. The use of plasma or non-gas stripping cells may improve the overall performance, as well as reduce the gas backstreaming load into the plasma chamber.

The signal detected in a channel of the analyzer is proportional to the product of several terms: the ion charge-exchange cross section, the spatial distribution of both the ions and the neutral target atoms, the stripping cell efficiency, the efficiency of the ion detector, and the numerical aperture of the entrance hole. Very high sensitivity detectors are required; some experiments strive to be able to count individual ions, but most fusion plasma experiments yield pico- or nanoamperes of ion current at the detector.

Figure 3 shows a plot of the output signal amplitude versus energy for a typical fusion plasma. Note that the curvature indicates a decidedly non-Maxwellian ion energy distribution. The two solid curves are calculated responses for a plasma in the same device. Curve A is for a 500-eV Maxwellian in the center, with density and temperature variations radially. Curve B is for 150 eV. The experimental data have a tail with a slope exceeding 1000 eV, but the bulk of the distribution is at 200 to 400 eV.

Instead of relying on incidental neutral gas to provide target atoms, some experimenters inject an energetic neutral beam perpendicular to the viewing path of the analyzer. This method has several important advantages. First, the beam can be made up of atoms instead of molecules, providing a large interaction because of the resonant cross section. Second, the intersection of

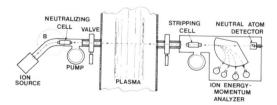

FIG. 2.   Neutral particle diagnostics system. At left is a source of energetic neutral atoms, with a detector at far right, used for beam attenuation measurements. At the right is a four-channel energy-momentum spectrometer for charge-exchange neutral atom analysis.

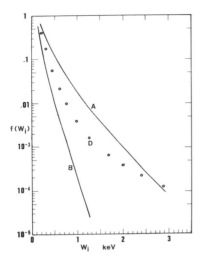

FIG. 3. Typical spectrum of ion energies in a plasma, obtained from charge-exchange analysis. Data points D would be characteristic of results from the system in Fig. 2 for a slightly nonthermal plasma. Curve A is for a 500-eV Maxwellian in the center, with density and temperature variations radially; curve B is for 150 eV.

the beam with the analyzer line of sight can be moved around the plasma volume, providing spatial resolution. Third, fully ionized plasmas can be diagnosed.

V. Electromagnetic Wave Probing

A. MICROWAVE DIAGNOSTICS

Microwave interferometry has become a standard method to determine the plasma density, the spatial distribution, and the collision frequency in laboratory plasmas. The quantity actually measured is the complex dielectric constant of the plasma, as a function of time and space.

The frequency employed depends upon the plasma density expected since waves cannot propagate at frequencies below cutoff. In an unmagnetized plasma or when the wave electric field is parallel to the confining magnetic field, the cutoff occurs at the plasma frequency

$$f_p = \frac{1}{2\pi}\left(\frac{ne^2}{m\varepsilon_0}\right)^{1/2} = 8.98\sqrt{n}\ \text{Hz} \qquad (2)$$

where n is the electron density per cubic meter, m the electron mass, e the electron charge, and ε_0 the permittivity, $10^{-9}/36\pi$. For example, X-band microwaves (10 GHz) are appropriate for

plasma densities of 10^{12} per cubic centimeter, and 3–4-mm wavelength microwaves (70–100 GHz) are appropriate for large fusion experiments or dense MHD chambers.

At frequencies above cutoff, the plasma complex dielectric constant depends on the electron density, the collision frequency, and the applied magnetic field in a complicated manner. Let us first introduce a simplified model to illustrate the principle of the method, and then discuss some of the complications. For a cold, uniform-density plasma, the complex dielectric constant $\check{\kappa}$ is

$$\check{\kappa} = \left(1 - \frac{\omega_p^2}{\omega^2 + \nu^2}\right) - j\left(\frac{\omega_p^2\,\nu/\omega}{\omega^2 + \nu^2}\right) \qquad (3)$$

where ω_p is the radian plasma frequency and ν the collision frequency. The complex refractive index is the square root of the dielectric constant, the real part contributing to phase shift and the imaginary part to wave attenuation. Figure 4 shows plots of the refractive index (real part) and attenuation index (imaginary part) as functions of density in a collisionless plasma (curves A). It is interesting to note that the refractive index is less than unity for a plasma, while for most dielectric media it is greater than one. There is no attenuation until cutoff is reached, at which density the refractive index goes to zero, and the wave is totally reflected. This behavior is similar to that in a lossless waveguide.

Curves B are for the case in which the plasma has a finite collision frequency. Attenuation is present at densities even below the critical density, and in addition, the refractive index no longer falls sharply to zero.

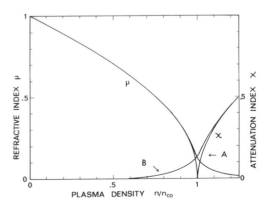

FIG. 4. Microwave refractive and attenuation indices as a function of plasma density normalized to the cutoff density. Curve A is for zero collision frequency, and curve B is for $\nu/\omega = 0.03$.

When there is a magnetic field present and the wave is propagating with its electric field perpendicular to the field lines, the propagation is described by the extraordinary mode. The dielectric constant then involves the electron cyclotron frequency as well as the plasma frequency:

$$\check{\kappa}_\perp = 1 - \frac{(\omega_p^2/\omega^2)(1 - j\nu/\omega)}{(1 - j\nu/\omega) - \omega_c^2/\omega^2} \quad (4)$$

where ω_c is the radian electron cyclotron frequency. The refractive index, the square root of $\check{\kappa}_\perp$, now goes through both a zero and a resonance as the density is increased. When the wave propagates along the magnetic field, it is called a cyclotron wave and can be represented by left-hand and right-hand circularly polarized waves, leading to Faraday rotation of the plane of polarization. Measurement of the angle of rotation of a cyclotron wave propagating along a magnetized plasma column provides another method of determining the electron density of the plasma. The rotation angle is

$$\psi = \frac{\pi d}{\lambda}$$
$$\left[\left(1 - \frac{\omega_p^2}{\omega^2 + \omega\,\omega_c}\right) - \left(1 - \frac{\omega_p^2}{\omega^2 + \omega\,\omega_c}\right)\right] \quad (5)$$

where d is the path length in plasma.

At large plasma density the right-hand mode becomes cut off, and only the left-hand mode remains, producing a whistler (named after a wave-type observed in the earth's magnetosphere).

A microwave interferometer, shown schematically in Fig. 5, can be used to measure both the phase shift and attenuation of a wave as it propagates through a plasma. The interferometer has two paths; the reference path provides a signal having a settable amplitude and phase, and the work path allows the wave to sample the plasma, where it suffers attenuation and phase shift in proportion to the integral of the damping and refractive indices over the plasma spatial distribution. Two detectors are shown, one sampling the transmitted amplitude (giving attenuation) and one detecting the phasor product of the signals from the two paths (giving phase shift). Microwave detectors usually have a square-law response, so the signal is proportional to power.

Figure 6 sketches the response that the circuit would have to a transient plasma, whose density rises through the value giving cutoff at the frequency in use and then decays back slowly toward zero. For the sketch shown, the plasma path would be five wavelengths long, giving five interference fringes in going from zero to cutoff density. The decrease in amplitude as cutoff is approached is actually caused by three effects: the plasma attenuation, the increase in reflection coefficient caused by the dielectric mismatch, and the divergent lens effect caused by the plasma cylindrical shape. At cutoff, the signal rests at one-forth maximum because of the square-law response of the silicon diode detector.

In large plasma installations, such as controlled fusion experiments, several propagation paths across chords of the plasma allow density profiles to be determined, generally employing data analysis involving Abel inversions.

B. FAR-INFRARED PROBING

Measurements based on the same principles as the microwave diagnostics above, but em-

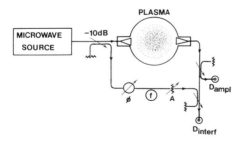

FIG. 5.   Schematic sketch of microwave interferometer. The reference path is split off with a 10-dB directional coupler and recombined with signals in the work path for detection in the interference detector. The amplitude of the signal after passing through the plasma is measured with the amplitude detector.

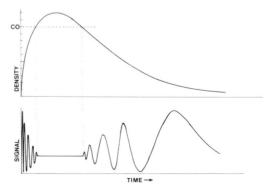

FIG. 6.   Typical signal response of the interferometer of Fig. 5 to a transient plasma density excursion. Square-law detector gives $\frac{1}{4}$ the maximum signal during plasma cutoff.

ploying quasi-optical techniques with molecular gas laser signal sources, are also routinely used in plasma laboratories. The appropriate conditions are dense plasmas or long path lengths in which sufficient phase shift occurs even though the waves are far from cutoff. Multiple paths are often used for such systems to obtain profile information. Detectors are usually solid-state devices cooled to liquid-nitrogen or even liquid-helium temperatures, although pyrolytic detectors or PIN diodes are used if the signal is large enough.

C. Optical Interferometry

A Mach–Zehnder interferometer, analogous to the microwave version in Fig. 5, or a Schlieren system can be used to show the density variations in very high density plasmas, such as theta pinches and shock tubes. The light source is usually a short-pulse ruby laser, sometimes operating in a multipulse mode, to take snapshots of the variations in refractive index caused by the plasma. Time resolution can be provided with a CW laser and a streak or framing camera.

D. Microwave Radiation from Plasma

A plasma emits radiations at various wavelengths, from the frequency of radio waves, through infrared, optical, and ultraviolet to the X-ray end of the spectrum. The radiant energy comes, in general, from the acceleration or from changes of quantum levels of electrons, but by a variety of mechanisms. Two broad categories of mechanisms can be distinguished: (1) the random (stochastic), incoherent emission of photons, leading to bremsstrahlung and blackbody radiation, and (2) coherent emission such as stimulated emission in a resonant system (lasers and masers) and nonthermal emissions due to plasma oscillations, instabilities, and other cooperative effects.

Incoherent emission in the microwave region of the spectrum has been studied for some time. It is this type of radiation that the radio astronomer usually views and that issues from gas-discharge microwave noise sources. The limiting emission power density within a frequency interval is the blackbody level. The radiant intensity (watts/unit solid angle $d\Omega$ per radian frequency interval $d\omega$) is given by the Planck function,

$$B_\omega = \frac{h\omega^3}{2\pi^2 c^2} \frac{1}{\exp(\hbar\omega/kT) - 1} \quad (6)$$

where h is Planck's constant and kT the plasma kinetic temperature. A microwave receiver generally has a directive antenna with only one polarization and a frequency low enough that the Rayleigh–Jeans approximation holds. If the antenna sees only the blackbody, and the depth of the radiating medium (plasma) is large with respect to the self-absorption length, the total power received in the detector is

$$P_\omega \, d\omega = kT \frac{d\omega}{2\pi} \quad (7)$$

The plasma may not be "black" at all frequencies but only at certain critical frequencies where it is optically thick and where the surface reflections are small. An example is the electron cyclotron frequency and its harmonics. Cyclotron frequency radiometers for high-magnetic field experiments (such as tokamaks) have been constructed for the frequency range 50–500 GHz using Michelson or Fabrey–Perot resonators and liquid-helium-cooled detectors, yielding valid electron temperature results. The emissivity was assessed, using Kirchhoff's law and the Einstein coefficients for each situation, to relate the radiation intensity to the plasma temperature.

Nonthermal radiation presents a problem in extracting temperature information from a microwave radiometer. Plasma instabilities or resonances between electron oscillations and physical structures produce coherent microwave emissions at superthermal intensities. For example, intense, high harmonics of the electron cyclotron frequency often are seen in plasmas having anisotropic velocity distributions.

E. Microwave Scattering

Figure 7 shows the geometry for a microwave scattering experiment, in which the receiving

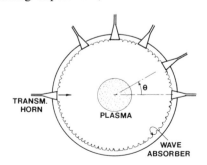

FIG. 7. Microwave horn antenna arrangement for scattering experiment. Absorber is essential to avoid stray reflections.

horn antenna directly opposite the transmitter ($\theta = 0$ degrees) can also be used for interferometry at the same time that scattering data are being taken.

The Thomson scattering from a plasma at microwave frequencies may be due to individual electrons or to bunches of electrons, grouped by waves, nonthermal fluctuations, instabilities, and so on. The method thus tends to tell more about plasma waves or the scale size and amplitude of turbulence than about electron temperature. A dimensionless parameter α is used to categorize scattering experiments:

$$\alpha = \frac{\lambda}{2\pi\lambda_D \sin \theta/2} = \frac{a}{2\pi\lambda_D} \quad (8)$$

where λ_D is the Debye length, θ the scattering angle, λ the wavelength, and a the scale size of the perturbation causing the scattering. When $\alpha \ll 1$ we get scattering from individual electrons (e.g., 90° laser scattering discussed below); when $\alpha \gg 1$ we get scattering from collective motions, the case in most microwave experiments.

Calculation of the scattered power involves taking the average of the 3-dimensional autocorrelation function of fluctuations, and the power spectrum is obtained from the Fourier transform of the autocorrelation function. The angle at which maximum signal is observed is usually calculated from the vector potential and occurs in the direction in which the phases of scattered wavelets are the most coherent. For a monochromatic disturbance the angle would be given by

$$\sin \frac{\theta}{2} = \frac{|\beta_w|}{2|\beta_i|} \quad (9)$$

where β_i is the wavenumber of the incident wave and β_w the wavenumber of the perturbation. For the monochromatic disturbance the frequency of the scattered wave f would be shifted as follows:

$$f_s = f_i \pm f_w$$
$$\beta_s = \beta_i \pm \beta_w \quad (10)$$

where the $+$ or $-$ sign depends on the direction of propagation of the disturbance.

The intensity of the scattered wave is usually very small, requiring a sensitive receiver and a transmitter having several watts of output. For example, turbulent density fluctuations of 1% of background density may require as much as a 110-dB signal difference between transmitter and receiver. It is best to obtain this difference

by having large transmitter power to reduce the problems with pickup of plasma radiation by a sensitive receiver.

F. Laser Thomson Scattering

The scattering of intense light beams from plasma electrons can yield the energy distribution and the density of the electrons, as well as indicate whether there are plasma instabilities present. The method is nonperturbing and has large frequency bandwidth, permitting rapid phenomena to be followed. Figure 8 shows schematically the equipment involved. A short-pulse laser having a power of 50 to 200 MW drives an amplifier, producing a light pulse of a few nanoseconds duration containing a few joules of energy. Because the Thomson cross section is very small, for modest plasma densities a small f-number system (large aperture) is required, followed by high quantum efficiency, low-noise detectors. In the system indicated, the exit slit of the monochromator has been replaced by a bundle of optical fibers leading to several photomultipliers, to provide an iterated profile of the scattered laser line.

The parameter α mentioned in the previous section is much less than unity for ruby laser light (694.3 nm), so the scattering is from noncorrelated electrons, and the scattered power spectrum mirrors the component of the electron velocity distribution along the direction of the light wave vector. The Doppler width of the line is thus directly proportional to the plasma electron temperature, and the amplitude is proportional to the electron density.

An extremely black beam dump is required to avoid stray light into the detectors, especially

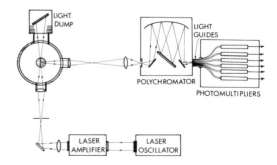

FIG. 8. Laser 90° Thomson scattering system. Six channels to sample the profile of the broadened line are shown, but any number that will fit in the exit slit housing is possible. Coated, dust-free lenses are required.

during calibration using Rayleigh scattering from a target gas. Since that scattering is not Doppler shifted, the detector light pipe position in the exit slit must be set right at the laser line, not detuned to either side as when looking at the broadened line.

VI. Magnetic Loops

The magnetic probes mentioned in Section III,B are inserted into the plasma and are thus a perturbation. To minimize the perturbation these probes are made small and are thus insensitive. Greater sensitivity, at the sacrifice of spatial resolution, can be obtained by using large-area loops that enclose the entire plasma. Two functions are observable with these loops, the total plasma current and the magnetic pressure.

A. Rogovskii Loop

This loop is used to measure the net transient current flowing in a plasma channel or in a conductor. The coiled coil is wrapped in a manner that couples to the azimuthal magnetic field, with a return conductor that does not couple, as sketched in Fig. 9. The voltage induced in the coil is proportional to number of turns × cross-sectional area of minor coil × rate of change of flux density.

The sketch shows the output connected to a differential amplifier, which is usually preferred to single-ended circuits to minimize stray pickup from ground loops. The signal passes through an integrator, yielding an output that is proportional to the net current encircled by the loop. Currents in the range 1 A to 100 kA are measureable by this method. At the largest currents a single turn of solid conductor is used instead of a coil of several turns.

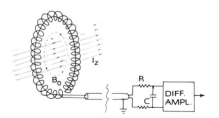

FIG. 9. Schematic sketch of a Rogovskii loop to measure current. Sensitivity depends on the diameter and number of turns of the small coil. A Faraday shield around the instrument may be required to reduce stray pickup.

A common fabrication method is to etch a spiral (by photoresist) in the outer conductor of semirigid coaxial cable, using the inner conductor for the current return. For greater sensitivity it is also possible to use a second, identical coil as the return. Because this application deals with time-varying currents it is important to shield the loop from transient electric fields, but it is equally important that the shielding be implemented in such a way that the induced EMF is able to penetrate into the coil. For slow or moderately fast phenomena, a thin foil of stainless steel or other highly resistive metal (having a skin time shorter than the characteristic times of the signal) generally provides the best shielding. For fast signals, a properly oriented Faraday shield is required.

Calibration is performed by passing a known oscillatory or pulsed current through the loop. The effects of axial asymmetry may be observed by placing the current-carrying calibrating conductor off-axis.

B. Diamagnetic Loop

Changes in plasma diamagnetism may be determined by measuring changes in the axial confining magnetic field B_0 as the perpendicular component of plasma pressure excludes magnetic field lines:

$$P_\perp + \frac{B_z^2}{2\mu_0} = \text{constant} \tag{11}$$

A simple loop encircling the plasma in a constant, uniform applied magnetic field generates a negative output voltage as the field perturbation B changes:

$$V_L = -\frac{d}{dt} \int \bar{B} \cdot dA \tag{12}$$

An integrator connected to the loop output yields a signal voltage that is proportional to the change in plasma energy density $\bar{W}_\perp$:

$$V_{\text{out}} = \frac{\mu_0}{RC} \frac{\bar{W}_\perp A}{B_0} \tag{13}$$

where RC is a large integration time constant, A the cross-sectional area of the plasma, and B_0 the steady background magnetic field. For a uniform plasma cross section A and length 1, the calibration factor for a single-turn loop is

$$\bar{W}_\perp = \frac{RCB_0 l}{\mu_0} V_{\text{out}} \text{ joules} \tag{14}$$

The loop sketched in Fig. 10 shows a series of small, counterwound coils outside the main coil

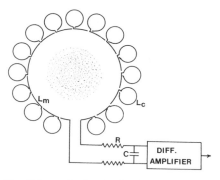

FIG. 10. Schematic sketch of a diamagnetic loop. The sum of areas of the compensation coils L_c is made equal to the area enclosed by the main loop L_w, to cancel pickup from the applied magnetic field while measuring the varying plasma component.

that do not couple to the plasma. They do couple to the background magnetic field, however, and if the sum of their areas is made equal to the area of the main coil, their induced voltage due to variations of the applied field just compensates the voltage induced by that field component into the main coil. The difference in induced voltages is that due to the plasma diamagnetism. This construction permits the loop to be used in systems that have time-varying applied magnetic fields.

VII. Nuclear Diagnostics

High-temperature plasmas emit radiations due to internal nuclear reactions or generate radiation and particles having sufficient energy to induce nuclear reactions externally. All of these cases provide diagnostic opportunities.

A. Soft X-Rays and XUV

Continuum radiation measurements in the soft X-ray energy range ($h\nu$ between 1 and 20 keV) provide a method to deduce the electron temperature; the slope of the continuum spectrum yields $(T_e)^{1/2}$ directly. The intensity of this bremsstrahlung unfortunately is very sensitive to the nuclear number Z of ions in the plasma, which must be accounted for to obtain valid conclusions.

Low-Z impurities, such as carbon and oxygen, have transitions below 1 keV, so a "dirty" discharge (vacuum walls covered with contaminants) does not generally produce distinct line radiation, and continuum slopes are reliably measurable. Often, however, after the walls have been extensively cleaned and low background pressure obtained, the continuum levels drop to such a low level that lines interfere with the slope analysis. But even if electron temperature data cannot be extracted, the lines permit identification of the impurities and may permit an assessment of the effective Z, a number that determines how much plasma energy is radiated at a particular electron temperature.

A somewhat different application of soft X-ray analysis is in the diagnostics of the target dynamics of inertial-confinement pellets. Here the application is imaging, either using the X-ray emission from the pellet itself, viewed by a diffraction grating or Fresnel zone plate, or using the highly divergent X-rays from a point source to produce magnified shadowgraphs. Special techniques are required to accommodate the picosecond time response and the submillimeter size of targets.

Instruments used to detect soft X-rays usually involve collimators, absorbers, and cooled detectors. Collimators may be for the purpose of localizing or imaging the source of radiation or to make multiple channels, each having its own absorber to provide energy analysis. Absorbers are usually aluminum or beryllium, and detectors may be lithium-drifted silicon crystals, cooled to liquid-nitrogen temperature to permit counting of single photons with pulse-height analysis. Some systems use PIN diodes or other types of signal-integrating detectors, and imaging systems use microchannel plates, image converters, or sensitive film plates.

B. Hard X-Rays

Pinhole photography at X-ray wavelengths has been a common method for observing interactions between high-energy electrons and objects for many years. In plasma confinement and particle beam experiments, the locations of electron leakage to walls and limiters can be determined by viewing a lead pinhole with a gated image converter, or if time resolution is not needed, simply with a film plate. Absorbers are used to block light and low-energy radiations and to provide an indication of the energies involved.

In pulsed, relativistic electron beam–plasma experiments a slightly different arrangement has been used in which a very small movable target of titanium foil is placed in the region to be studied. The X-ray emission from the target is viewed by an array of collimated photomultipliers, each having a different absorber to yield a time-resolved energy distribution. This method is appropriate for electron energies in the range 25 keV to 1 MeV.

C. NEUTRONS

High-temperature plasmas of deuterium, tritium, or helium-3 produce thermonuclear reactions, generating energetic neutrons and other reaction products. The thermonuclear reactions anticipated in a fusion experiment are D (d, n) T, D (d, p) ^{3}He, T (d, n) ^{4}He, ^{3}He (d, p) ^{4}He, where, to take an example, the notation D (d, n) T means that a deuteron striking a deuteron produces a neutron with a triton remaining.

Energy and momentum are conserved so the reaction products are all energetic. If thermal neutrons are to be detected, the 4–14-MeV neutrons must be moderated. This method introduces a delay, making it easier to distinguish neutron counts from gamma-ray signals or stray pickup.

True thermonuclear neutrons have an isotropic distribution since they are produced by a thermal distribution of ions. There are other sources of energetic neutrons (which the experimenter must beware not to confuse with the "real" ones) that are generated by runaway electrons, ions, or energetic neutral atom heating beams striking contaminated walls. Because the cross section for production is a steep function of energy, a slight nonthermal tail or a high-energy beam gone astray can predominate in the neutron generation.

The test is to see if the reaction products have an isotropic velocity distribution. Although in principle the production of neutrons can be a sensitive diagnostic technique for ion temperature, it must be verified that the source of the neutrons is from the bulk of a thermal distribution of ions. Thus neutron diagnostics cannot stand alone but must be coupled with charge-exchange and other ion diagnostics.

Fast neutron counting has some advantages over moderated counting, one being that the ion temperature can be read directly from the data. Special care must then be taken to avoid pickup of other prompt signals, by using detectors that have low sensitivity for them. An example is a ^{3}He ionization chamber connected to a pulse-height analyzer. Lead shielding absorbs X-rays, and ^{6}Li in a lithium carbonate collimator nonradiatively absorbs thermal neutrons.

Prompt neutron energies may also be analyzed in experiments producing large pulsed fluxes, such as particle-beam inertial confinement experiments, by collimation and time-of-flight. Sufficient resolution requires a rather long evacuated drift pipe with a fast detector at the end.

D. IONS

Particle-beam fusion is expected to use intense ion beams, probably in the low-mass range of carbon, lithium, and so on, at energies of 5 to 20 MeV. The diagnostics for these beams have special requirements because of the extremes in all parameters. Even the ion current is difficult to measure directly, for example, by using a single-turn Rogovskii loop, because the ion beam is nearly always partially neutralized by electrons that are picked up incidentally or by design. At currents up to a few kiloamperes, where space-charge effects are not severe, a foil-covered Faraday cup has been successful in stopping the slow electrons co-moving with the ions and collecting the total ion current. But at high currents such collectors fail.

Beam particle constituents may be identified with an instrument called a Thompson parabola, which also gives energy information. The beam passes through a crossed electric and magnetic field region, deflecting, for example, to the right due to the magnetic field and upward due to the electric field. The deflection path may be recorded on CR-39 plastic film or, for weak beams, on photographic film. This instrument also fails at current densities large enough that space charge diverges the beam.

Some indirect methods to measure intense ion beams have been developed that may be used for currents up into the megampere range. Nuclear activation is an example of a method sensitive to the ions in neutralized beams even when the net current is zero. Activation thresholds are very sharp, at distinct energies, with many of the reactions producing positron emitters that can be counted precisely in coincidence counters. An example is ^{7}Li (p, γ) ^{8}Be with a threshold at 0.42 MeV or ^{11}B (p, n) ^{11}C with a threshold at 3.2 MeV. The procedure is to bombard a sample in a vacuum lock with the beam and then quickly remove it to a counting room to record the positron decay process. This method works well up to current densities at which beam damage occurs on the sample. Material lost by the "blowoff" cannot be counted, and large errors can occur. A method of activation that avoids the blow-off problem is the coupled reaction ^{7}Li (p, γ) ^{8}Be and ^{63}Cu (γ, n) ^{62}Cu (β^+). The prompt 17-MeV gammas from the p–Li reaction escape the target before blowoff occurs, activating the copper target that is hidden from the beam. The copper is then counted.

The blowoff problem has been circumvented in another manner, that is, by beam attenuation through Rutherford scattering at an angle in a thin metal foil. Attenuation factors of 10^4 to 10^7 make it possible to use conventional methods for beam analysis.

Individual ion tracks in special film can be examined microscopically to determine energy. An example is CR-39 plastic, where proton tracks may be emphasized by etching, so that depth measurements are easier to make, yielding the proton energy.

BIBLIOGRAPHY

Akulina, D., Sindoni, E., Stott, P., and Wharton, C. B., eds. (1983). "Diagnostic for Fusion Reactor Conditions," 3-Volume Proceedings of International School of Plasma Physics, Varenna, Italy, 1982. Commision of European Communities Press, Brussels, Belgium.

Heald, M. A., and Wharton, C. B. (1965, 1978). "Plasma Diagnostics with Microwaves." Wiley, New York, 1965; 2nd ed., Krieger, Huntington, NY, 1978.

Huddlestone, R. H., and Leonard S. L. (1965). "Plasma Diagnostic Techniques." Academic, New York.

Podgornyi, I. M. (1971). "Topics in Plasma Diagnostics." Plenum, New York.

Sindoni, E., and Wharton, C. B., eds. (1979). "Diagnostics for Fusion Experiments," Proceedings of International School of Plasma Physics, Varenna, Italy, 1978. Pergamon Press, New York.

PLASMA SCIENCE AND ENGINEERING

J. L. Shohet   *University of Wisconsin*

GLOSSARY

Attachment: Phenomenon in which a neutral particle and an electron combine, producing an negatively charged ion.

De-excitation: Inverse of excitation. Radiation is usually emitted.

Diffusion: Phenomenon in which particles diffuse in position space, velocity space, and time.

Discharge machining: In this process, plasmas are used to provide a cutting surface between a thin wire and the work to be cut, usually by passing an arc between them through water.

Excitation: Process in which neutral particles or ions (that are not fully stripped) gain energy, which is evident by orbital electrons moving to higher energy states.

Fusion plasmas: Plasmas with relatively high temperatures and that are composed of light atoms, particularly hydrogen or its isotopes.

Industrial plasmas: Plasmas whose composition is generally of masses above hydrogen (they can have molecular weights of several thousand), and that are usually of two types: thermal (equilibrium) plasmas, in which the electron and ion temperatures are approximately equal, and nonequilibrium (glow-discharge) plasmas, which tend to have relatively high electron temperatures compared to the ion temperatures. Many chemical reactions can only take place in the plasma state.

Ionization: Inverse of recombination.

Ion milling: Beams of ions can be used to cut or "mill" narrow regions of materials to great accuracy.

Plasma: Collection of charged particles, usually of opposite sign, that tends to be electrically neutral. We often describe a plasma as the fourth state of matter. Adding energy to a solid melts it and it becomes a liquid; adding energy to a liquid boils it and it becomes a gas; adding energy to a gas ionizes it and it becomes a plasma.

Plasma-assisted chemical vapor deposition (CVD): Here, plasmas can be used to provide a mechanism to successfully deposit various chemicals on surfaces, either by treating the surface before deposition, or by providing a chemical pathway for successful deposition.

Plasma electronics: Includes applications in which the unique properties of plasmas are used directly in devices, such as arc melters, microwave sources, switchgear, plasma displays, welders, analytical instrumentation, arc lamps, and laser tubes.

Plasma etching: Process in which plasmas etch materials. That is, certain chemical reactions occur on surfaces, creating a volatile compound from the surface material and the plasma, which can then be pumped away.

Plasma polymerization: By ionizing a monomer gas, certain types of polymers can be made, which can be deposited as coatings on various materials. This process often occurs during etching, since by changing the ratio of the various plasma components, etching can be turned into polymerization.

Plasma processing: Encompasses applications in which plasmas or particle beams (charged or neutral) are used to alter an existing material, as in plasma etching, ion milling, ion implantation, or surface modification through plasma cleaning, hardening, or nitriding.

Plasma spray: Coating process that sputters heavy particles (clumps) from an arc system and then directs the spray of these particles to a surface for coating.

Plasma synthesis: Refers to applications in which plasmas are used to drive or assist chemical reactions to synthesize compounds, alloys, polymers, or other complex species starting from simpler starting materials.

Recombination: Phenomenon in which ions and electrons recombine to form neutral particles; radiation is sometimes emitted.

Sheath: Generally a region near a surface in which the plasma is not electrically neutral. Some sheaths can form in the main body of the plasma and are called double layers.

Sputter deposition: In this case, plasmas are used to sputter or knock off from a target electrode particles that are then deposited on a particular material.

Surface modification: Plasmas can be used to modify the properties of materials by interacting on the surface of those materials in several ways. For example, tool steel can be hardened considerably by subjecting the tools to a nitrogen plasma. Turbine blades can be plasma coated for improved mechanical and thermal properties.

Plasmas are composed of mixtures of electrons and positive and/or negatively charged ions as well as neutral particles. They are affected by electric and magnetic fields, which can be used to modify their properties. The temperature of such plasmas can be quite high. As a result, many interactions between particles are substantially different when they are in the plasma state. Thus, new materials can be manufactured that can have improved properties, new chemical compounds may be produced, and surfaces of existing materials can be altered. These aspects will have an increasingly significant role in the future of technology.

I. Introduction

Industrial use of plasma has applications that cover a broad range of activities and has a multibillion-dollar yearly impact in the economy. In order to understand how this occurs, we must first describe what a plasma is, how it behaves under the influence of electric and magnetic fields, and how it is characterized. Normally, we specify the following quantities when we describe a plasma: composition, electron and ion temperatures, and electron and ion densities. We divide plasmas into two general types as follows.

1. Industrial plasmas. The ions in these plasmas are generally composed of masses above hydrogen (the molecular weights can reach several thousand), and are usually of two types: thermal (equilibrium) plasmas, in which the electron and ion temperatures are approximately equal, and nonequilibrium (glow-discharge) plasmas, which tend to have relatively high electron temperatures compared to the ion temperatures.

2. Fusion plasmas. These plasmas have much higher temperatures than industrial plasmas and are composed of light atoms, particularly hydrogen or its isotopes, and are designed to produce energy by means of a thermonuclear reaction. [*See* NUCLEAR FUSION POWER.]

The graph shown in Fig. 1 indicates some general groupings of industrial plasmas according to their application and how they compare with fusion plasmas. One axis of the graph is proportional to temperature and the other to density.

At present, industrial plasma applications are largely empirical in nature. Further progress will require a much more thorough understanding of the plasma behavior as well as of the interaction of the plasma with solid materials. Design tools, transportable diagnostics, and models are needed.

We characterize industrial applications in three broad and somewhat overlapping areas. They are:

1. *Plasma processing,* which encompasses applications in which plasmas or particle beams (charged or neutral) are used to alter an existing material, as in plasma etching, ion milling, ion implantation, or surface modification through plasma cleaning, hardening, or nitriding.

2. *Plasma synthesis,* which refers to applications in which plasmas are used to drive or assist chemical reactions to synthesize compounds, alloys, polymers, or other complex species starting from simpler starting materials. This could also include the inverse processes of plasma decomposition. *Many chemicals and/or chemical*

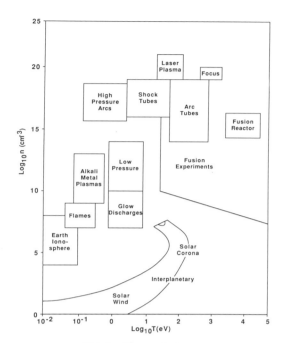

FIG. 1.   Types of plasmas.

reactions can only take place in the plasma state.

3. *Plasma electronics,* which includes applications in which the unique properties of plasmas are used directly in devices, such as arc melters, microwave sources, switchgear, plasma displays, welders, analytical instrumentation, arc lamps, or laser tubes.

There are four common fundamental requirements needed for progress in industrial applications of plasmas. They are:

1. Theory, modeling, and systems concepts.
2. Plasma chemistry and interactions.
3. Plasma diagnostics and characterization.
4. Plasma generation and confinement.

It is the purpose of this article to provide a general introduction to plasma science and engineering in order to show both the applications and how advances in this field can be made. The following short description of industrial applications of plasma processing and technology shows how widespread the impact of plasma technology is.

Plasma polymerization. By ionizing a monomer gas, certain types of polymers can be made, which can be deposited as coatings on various materials. There is an important application of this work in the biotechnology field, since biocompatible polymers can be used to coat various implant materials that would otherwise be rejected by the body. Various pharmaceuticals and other "exotic" chemicals can only be made with this process, which is often a result of the combination of ion and free radical generation by the plasma.

Plasma-assisted CVD (chemical vapor deposition). Here, plasmas can be used to provide a mechanism to successfully deposit various chemicals on surfaces, either by treating the surface before deposition or by providing a chemical pathway for successful deposition.

Sputter deposition. In this case, plasmas are used to sputter from a target electrode particles that are then deposited on a particular material.

Plasma etching. The major application of this work is to the semiconductor industry. As the spacing between lines in integrated circuits shrinks to 1 μm and below, conventional "wet" etching using chemicals begins to fail. This is because such processing acts in a spherical direction and undercuts the walls between the etch regions. Appropriately designed plasma etching (dry etching), perhaps combined with electric and magnetic fields or ion beams, offers a dramatic improvement in the etch process, and it is believed that the future of the entire semiconductor fabrication industry will rest with plasma processing.

Ion milling. Beams of ions can be used to cut or "mill" narrow regions of materials to great accuracy.

Surface modification. Plasmas can be used to modify the properties of materials by interacting on the surface of those materials in several ways. For example, tool steel can be hardened considerably by subjecting the tools to a nitrogen plasma. Turbine blades can be plasma coated for improved mechanical and thermal properties.

Welding. The use of plasmas in welding, especially in arc welding, has been known for some time. However, many problems continue to exist with welding, and much of it is due to the lack of understanding of the plasma composition, the plasma temperature and density, and the electric field and current distribution in the welding arc.

Discharge machining. In this process, plasmas are used to provide a cutting surface between a thin wire and the work to be cut, usually by passing an arc between them through water. This process was used to cut a magnet coil that was required to be in a particular twisted shape.

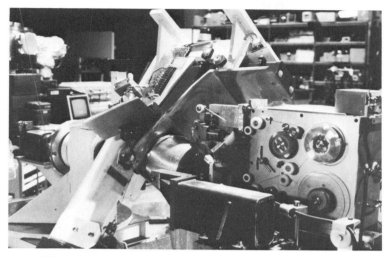

FIG. 2.   Electrical discharge machining (EDM) of a magnet coil.

The tooling and resulting magnet coils are shown in Figs. 2 and 3.

Plasma displays. Many new uses for such displays are being found. A portable computer whose screen is a full plasma display has recently been introduced.

Arc devices. A major component of American industry has involved the use of arc technology in the electrical power system field. Switchgear today is still being designed empirically without the understanding of plasma-surface interactions. The U.S. switchgear industry is suffering greatly from foreign competition as a result. Arc illumination is a major application of this technology as well, with activity above $20 billion per year.

Arc melting. Arc furnaces have been in use for many years. Their applications to the refining and extraction of ores have been many. Yet there can be major improvements made if the proper understanding of the interaction of plasmas with the ores and metals are understood. For example, in the melting of iron ore, 20 lb of graphite electrode are used up per ton of ore. If

FIG. 3.   Three-dimensional twisted magnet coils, made by electrical discharge machining. There is not a single bend in the coils, since they are made by "slicing" a solid block of aluminum in this shape.

just 1 lb graphite were saved, $20 million per year savings would result.

Plasma spray. This is a coating process that sputters heavy particles (clumps) from the cathode of an arc system and then directs the spray of these particles to a surface for coating. It has applications where spray coatings are required.

Much needs to be done to formulate appropriate understanding of the plasma and plasma–surface interactions. Measurements have often found that intuitive understandings of plasma behavior have not been borne out when the actual measurements are made. However, if the measurement system itself results in a perturbation of the plasma, the results may be unclear, and thus noninvasive diagnostics need to be developed.

II. Basic Plasma Properties

A. Density, Temperature, Composition

The mixture of ions, electrons, and neutral particles making up a plasma must be describable in terms of quantities that can be used to describe them so that their properties can be analyzed. There are several of these quantities that provide a useful comparison. They are:

Density, described by n_e, n_{ix}, and n_x, which refer to the electron density, the ion density of species x, and the neutral density of species x, respectively. It is important to note that most plasmas contain ions of several different species (positively and/or negatively charged), and the number density (usually expressed in units of particles per cubic centimeter) is a very important quantity. The density is usually a function of both position and time. It should be measured experimentally and is then often modeled theoretically, depending on the nature of the various processes that act to change them. In particular, we refer to the following processes:

Attachment, in which a neutral particle and an electron combine producing a negatively charged ion.

Diffusion, in which particles diffuse in position space or velocity space. Thermal diffusion is related to particle diffusion, but refers to energy, not particle transport.

Recombination, in which ions and electrons recombine to form neutral particles; radiation is sometimes emitted.

Ionization, the inverse of recombination.

Excitation, in which neutral particles or ions (that are not fully stripped) gain energy, which is evident by orbital electrons moving to higher energy states.

De-excitation, the inverse of excitation. Often, radiation is emitted.

There are many ways in which these processes can occur, such as ionization by electron impact, chemical ionization, or radiation absorption.

It is often surmised that a plasma is electrically neutral, but such a condition usually does not occur when a plasma is in contact with a surface. Under these circumstances, a "sheath" is developed in which either electrons or ions are the dominant species. Usually, this results in

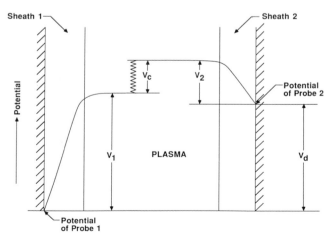

FIG. 4. Electrostatic potential between the plasma and an electrode in the sheath region.

a net electric field in the sheath. Sheaths have considerably different properties than do the neutral plasma, and care must be taken to understand them. Figure 4 shows the electrostatic potential between a plasma and a conducting material in the sheath region. The conductivity of a plasma may actually be quite high, often greater than metals.

Another important quantity that is needed to characterize a plasma is the temperature of the individual components, i.e., T_e, T_{ix}, and T_x. Temperature is also a quantity that is not usually constant in space or time.

The composition of the plasma is of paramount importance. Here, one needs to know the mass numbers of all of the ions and neutral particles in the plasma. In many cases, it is desirable to know this as a function of both time and position, and the nature of the diagnostic devices needed to determine this ranges from very simple to very sophisticated, indeed. [*See* PLASMA DIAGNOSTIC TECHNIQUES.]

B. Plasma Production

Plasmas may be generated by passing an electric current through a gas. Normally, gases are electrical insulators, but there are always a few charge carriers present that can be accelerated by the electric field and can then collide with neutral particles, producing an avalanche breakdown, thus making the plasma. The electric field needed for breakdown can be made with a potential set up between a pair of electrodes, with an "electrodeless" rf (radio-frequency) induction coil, with shock waves, with lasers, or with charged or neutral particle beams. The latter processes can also produce gaseous plasmas if they impinge on a solid target. In addition, heating various materials (usually alkali metals) in ovens or furnaces will cause not only evaporation of neutral particles but also ionization, and plasmas may be made in this way. Many chemical processes can also cause ionization. [*See* DIELECTRIC GASES.]

III. Plasma Physics

A. Plasma Dynamics

The dynamics of the motion of the charged particles in a plasma are governed by the fundamental equation of motion:

$$\mathbf{F} = q(\mathbf{E} + \mathbf{v} \times \mathbf{B}) \qquad (1)$$

where $\mathbf{F}$ is the force on the charged particle, $\mathbf{E}$ is the electric field, $\mathbf{B}$ is the magnetic field, and $\mathbf{v}$ is the charged particle's velocity. If one knew all of the values of the electric and magnetic fields, including those produced by all of the particles (self-consistent fields) at the location of each particle, and if collisions between particles could be neglected, then the trajectories of all the particles could be obtained simultaneously with a computer. However, as can be seen from Fig. 1, such a computer cannot exist, since literally billions of equations would need to be solved simultaneously. In addition, collisions between particles, including charged particle–neutral particle collisions, must occur, so such a method cannot be totally practicable. However, much understanding can be made with numerical simulation of "clumps" or "clouds" of plasmas using modern supercomputers, where upward of 1 million simultaneous equations can be solved over a reasonable time period.

Thus, we need to develop a means to consider collisional interactions between particles making up the plasmas. Such interactions are classified into two types—elastic and inelastic collisions. In elastic collisions, kinetic energy, linear momentum, and angular momentum of the two colliding particles are conserved. In inelastic collisions, some of the energy and momentum is changed into or from internal vibration energy, chemical energy (such as chemical bonds), or to conduct the processes of ionization, excitation, recombination or de-excitation and potentially generate electromagnetic radiation during the process as well.

The most common representation for such interactions is called the collision cross section. We develop this formulation as follows. Let us assume that a beam of particles of density n particles/cm^3 travelling with a velocity v and of cross section A passes through the plasmas a distance dx. Let N be the number of plasma particles/cubic centimeter. The number of beam particles colliding with plasma particles per unit time may then be written as

$$\frac{dn}{dt} = -[(N\sigma A\, dx)/(A\, dx)]nv \qquad (2)$$

The term $N\sigma A\, dx$ is the probability of collision in the volume $A\, dx$, and nv is the particle current density of incoming beam particles; σ is the collision cross section for this particular process. We may also write $v\, dt = dx$ and $N\sigma = p_0 p_c$, and we can rewrite Eq. (2) to be

$$n = n_0 \exp(-p_0 P_{cx}) \qquad (3)$$

It is also convenient to write

$$n = n_0 \exp(-p_0 P_c v t) = n_0 \exp(-\nu_c t) \quad (4)$$

introducing an average collision frequency $\nu_c = p_0 P_c v$. In the above equations, P_c is the *probability of collision* for a particular process and p_0 is the "reduced" pressure $= 273 p/T$, which expresses a concentration $N/V = 3.54 \times 10^{16} p_0$ molecules/cm^3. The term $p_0 P_c$ has units of $1/$ length, or $1/(p_0 P_c) = \lambda$, the mean free path.

Each process has its own cross section, which in many cases can only be determined experimentally. [*See* COLLISION CROSS SECTIONS (ATOMIC PHYSICS).]

1. Particle Diffusion

a. Free Diffusion. To examine this condition, we first consider particles as *free* to move in a plasma of similar particles. For simplicity, we assume particle flow in one dimension along which a density gradient dn/dx has been established. By using Newton's second law, we can write, for the change in momentum due to *elastic* collisions between the particles, each of mass m

$$\frac{d(nmv_x)}{dt} = -mv_x n\nu_m \quad (5)$$

The velocity v_x is acquired by the particle due to random collisions with the other particles and is therefore independent of the coordinates, as long as the particles are all assumed to be in energy equilibrium. Thus, we may write

$$\frac{d(nv_x)}{dt} = v_x \frac{d(nv_x)}{dx} = v_x^2 \frac{dn}{dx} = -nv_x\nu_m \quad (6)$$

When we average over v_x and define the *particle current* to be $\Gamma_x = nv_x$ and make the approximation that $v_x^2 = v_T^2/3$, where v_T is the average thermal speed, we obtain

$$\Gamma_x = -(v_T^2/3\nu_m)\frac{dn}{dx} = -D \; dn/dx \quad (7)$$

which is the free diffusion equation, and D is the free diffusion coefficient,

$$D = (v_T^2/3\nu_m) = (v\lambda/3) \quad (8)$$

The value of D is not normally equal for electrons and ions, since, if they have the same temperature, the average thermal speed v_T is not equal.

Written in three dimensions, if the plasma is isotropic, we obtain $\Gamma = -D \, \nabla n$, where ∇ is the gradient operator. In this case Γ is a vector. If there are spatial variations of the diffusion coeffi-

cient, we should write:

$$\Gamma = -\nabla(Dn) \quad (9)$$

The diffusion coefficient for a given species of particles could have components that are due to several processes, each of which give spatial diffusion. For example, electrons are simultaneously colliding with other electrons, ions, and neutral particles. Each process contributes a "mean free path" and a collision frequency, which must be added together, reciprocally in the case of the former and directly in the case of the latter, to obtain the combined mean free path and free diffusion coefficient for all of the combined processes.

If we consider that the diffusion has not reached a steady state, then we can calculate the time variation in concentration in any element of space, if there is no source or sink of particles, from the equation

$$\frac{\partial n}{\partial t} + \nabla \cdot \Gamma = 0 \quad (10)$$

which leads directly to the time-dependent diffusion equation

$$\frac{\partial n}{\partial t} = D \, \nabla^2 n \quad (11)$$

This equation is separable and can often be solved as a boundary value problem. Inclusion of other effects that generate or destroy particles in an element of space requires a modification of Eq. (10) to

$$\frac{\partial n}{\partial t} = D \, \nabla^2 n + \alpha n^2 - \beta n^2 \quad (12)$$

where α and β are the source and sink terms, respectively. These can be ionization (source), recombination (sink), etc.

b. Mobility and Time-Varying Fields. In the presence of time-varying electric fields and collisions, we may approximate Eq. (1) as

$$m(dv/dt) + (mv_m)v = qE_0 \exp(j\omega t) \quad (13)$$

where q is the charge of the particle, and is negative for electrons and either positive or negative for ions, depending on their nature. The steady-state velocity is then

$$v = \left(\frac{q/m}{j\omega + \nu_m}\right) E \quad (14)$$

We may take the quotient of v/E, which is defined as the *mobility*,

$$\mu = v/E = \left(\frac{q/m}{j\omega + \nu_m}\right) \quad (15)$$

If we are interested in the case of a dc electric field where $\omega = 0$, or a case at high pressure where the collision frequency is much larger than the ac frequency of the electric field, then the mobility reduces to

$$\mu = q/(m\nu_m) \qquad (16)$$

In many cases, the mobilities of the different species of particles in a plasma must be measured. Neutral particles, being uncharged, do not have a mobility.

The ratio of the diffusion coefficient to the mobility is an important quantity because it is a measure of the average particle energy. In a dc electric field, the ratio is

$$\frac{D}{\mu} = \frac{v_T^2 m}{3q} = \frac{2}{3q}\left(\tfrac{1}{2}mv_T^2\right) = \frac{2}{3q} u_{\text{avg}} \qquad (17)$$

The quantity u_{avg} is the average kinetic energy of the particles, and the numerical constant depends upon how the averaging is carried out. The value $\tfrac{2}{3}$ is correct for a Maxwellian distribution of velocities.

If we write the particle flow equation for the particles discussed previously and add the contribution through the mobility from the electric field, we obtain

$$\Gamma = -D\,\nabla n - n\mu E \qquad (18)$$

Assuming that the electric field is in the z direction and that a steady-state condition applies $(d/dt = 0)$, we may apply Eq. (16) to the continuity equation to obtain

$$\nabla^2 n = -\frac{\mu}{D}\,\mathbf{E}\cdot\nabla n \qquad (19)$$

c. Ambipolar Diffusion. So far, we have assumed that the ions and electrons have diffused freely with no interaction between them. When the density is high, this is not the case. That is, the free diffusion coefficients, not normally being equal, will result in the enhanced transport of one species of particle. Charge separation will occur, and, as a result, an ambipolar electric field is established that will affect the motion of the particles directly through Eq. (17). We may write, for the positively charged particles,

$$\Gamma_+ = -D_+\,\nabla n_+ + \mu_+ \mathbf{E}_s n_+ = n_+ \mathbf{V}_+ \qquad (20)$$

where $\mathbf{E}_s$ is the space-charge field. In terms of the velocity,

$$\mathbf{V}_+ = -\frac{D_+}{n_+}\,\nabla n_+ + \mu_+ \mathbf{E}_s \qquad (21)$$

By writing a similar expression for the negatively charged particles,

$$\mathbf{V}_- = -\frac{D_-}{n_-}\,\nabla n_- - \mu_- \mathbf{E}_s \qquad (22)$$

If we eliminate $\mathbf{E}_s$ between these equations and set $n_+ = n_- = n$ as well as the gradients and velocities equal, we obtain the result

$$\mathbf{V} = -\left(\frac{D_+\mu_- + D_-\mu_+}{\mu_+ + \mu_-}\right)\nabla n \qquad (23)$$

The quantity in the parentheses is a diffusion coefficient for the two signs of particles interacting on each other *so that they both diffuse together*. This quantity is called the *ambipolar diffusion coefficient* and is defined as

$$D_{\text{amb}} = \frac{D_+\mu_- + D_-\mu_+}{\mu_+ + \mu_-} \qquad (24)$$

The ambipolar time-dependent diffusion equation (for both species of particles) is thus

$$\frac{\partial n}{\partial t} = D_{\text{amb}}\,\nabla^2 n \qquad (25)$$

2. Heat Transport

The above discussion centered on particle transport. In addition to the transfer of particles by diffusion, *energy* can also be transported. We write the energy flux, in a way similar to the particle flux, as

$$Q = n(\tfrac{1}{2}mv^2)v \qquad (26)$$

Let E denote the thermal energy of a particle. Then $\bar{E}$, the mean thermal energy of particles at a given point, is a function of the local temperature T at that point, and, as in the particle diffusion case, we shall assume only a one-dimensional variation for this development. The *specific heat* of the plasma is given by the relation

$$c_v = \frac{d}{dT}(\bar{E}/m) \qquad (27)$$

The net *particle* flux passing a given point, for particles passing the origin of the x direction leaving the region between x and $x + dx$ without having made a collision, is

$$d\Gamma_{0x} = +\frac{v}{6\lambda}\,n(x)e^{-(x/\lambda)}\,dx \qquad \text{if } x > 0 \quad (28)$$

Each of the particles passing the origin carries with it a thermal energy equal, on the average, to the value of $\bar{E}$ at the regon between x and $x + dx$. We may write a similar expression for the particle flux from the points where $x < 0$.

We previously expanded the density $n(x)$ in a Taylor series about the origin. We shall do a similar thing to obtain the energy flux, and expand the local energy in terms of the energy at the origin. In order to do this, we need the average velocity in a given coordinate direction, which is in the, say, positive or negative direction of that coordinate. Note that the average velocity otherwise would be zero. The number density of those particles is, on the average, half the actual density. Assuming a Maxwellian distribution of velocities, we may write the total average velocity to be

$$n\bar{V}_+/2 = \int Vf\, d^3V$$
$$= (2kT/\pi m)^{1/2} = \tfrac{1}{2}V_T \quad (29)$$

where V_T is the thermal velocity $(3kT/m)^{1/2}$. Thus, the total thermal energy that the particles from x and $x + dx$ carry across the origin is

$$nV_T\bar{E}/4 = \tfrac{1}{4}nV_T(\bar{E} + \lambda u\ \partial\bar{E}/\partial x) \quad (30)$$

and for particles produced in the region to the left of the origin, the energy flux is

$$nV_T\bar{E}/4 = \tfrac{1}{4}nV_T(\bar{E} - \lambda u\ \partial\bar{E}/\partial x) \quad (31)$$

In these last two equations, the energy is now written in terms of the energy at the origin. In addition, λ is the mean free path, and u is a numerical constant roughly of order unity. We now obtain the net rate of flow of energy from the negative to the positive side as

$$\tfrac{1}{4}nV_T(\bar{E} - \lambda u\ \partial\bar{E}/\partial x) - \tfrac{1}{4}nV_T(\bar{E} + \lambda u\ \partial\bar{E}/\partial x)$$
$$= -\tfrac{1}{2}nV_T\lambda u\ \partial E/\partial x \quad (32)$$

We can now obtain this result in terms of the specific heat and the temperature to be

$$Q = -\tfrac{1}{2}mnV_T\lambda u c_v\ \partial T/\partial x \quad (33)$$

Thus, heat flow is proportional to the temperature gradient, rather than the density gradient. The thermal conductivity χ is then

$$\chi = \tfrac{1}{2}mnV_T\lambda u c_v \quad (34)$$

so the heat flux is then:

$$\mathbf{Q} = -\chi\ \boldsymbol{\nabla}T \quad (35)$$

3. Effects of AC Electric and Magnetic Fields

In order to determine the response of the plasma to an electromagnetic wave, we must first examine its response to ac electromagnetic fields. We assume that the plasma is "cold" in that the motion of the individual charged particles can be considered to be entirely due to the electric fields they experience, in this case, externally imposed fields only. We do this by reexamining the equation of motion of a charged particle:

$$\mathbf{F} = q\mathbf{E} = m\frac{d\mathbf{v}}{dt} \quad (36)$$

If we assume that the electric field is driven by an ac source, we may adopt a "phasor" notation for the ac part of the signal. That is, we shall assume all quantities vary as

$$\exp(-j\omega t) \quad (37)$$

as far as their time variation, where ω is the driving frequency. Thus, the electric field, for example, is:

$$\mathbf{E} = \mathbf{E}_0 \exp(-j\omega t) \quad (38)$$

We only need to solve for the spatial part of the variation. Thus, Eq. (38) becomes

$$qE_0 = -j\omega v_0 \quad (39)$$

We use Eq. (38) to find the ratio of velocity to electric field, that is, the mobility. It is simply

$$\mu = |\mathbf{v}_0/\mathbf{E}_0| = \frac{-q}{j\omega m} \quad (40)$$

The *conductivity* of the plasma is determined by noting that the electric current density may be written as

$$\mathbf{J} = nq\mathbf{v} = nq\mu\mathbf{E} = -\frac{nq^2\mathbf{E}}{j\omega m} = \sigma\mathbf{E} \quad (41)$$

where σ is the conductivity of the plasma in the cold plasma approximation. If a plane monochromatic wave propagates through a plasma, we find that the wave will cut off (no longer propagate) where the density of the plasma reaches a certain value. To determine this, we examine Maxwell's equation:

$$\boldsymbol{\nabla} \times \mathbf{H} = \mathbf{J} + \frac{\partial\mathbf{D}}{\partial t} \quad (42)$$

If we continue to use the exponential time notation, then we may write Eq. (42) as

$$\boldsymbol{\nabla} \times \mathbf{H}_0 = \sigma\mathbf{E}_0 - j\omega\varepsilon_0\mathbf{E}_0 \quad (43)$$

or

$$\boldsymbol{\nabla} \times \mathbf{H}_0 = -j\omega\varepsilon_0\left(1 - \frac{\sigma}{j\omega\varepsilon_0}\right)\mathbf{E}_0 \quad (44)$$

Using Eq. (41) we finally obtain

$$\boldsymbol{\nabla} \times \mathbf{H}_0 = -j\omega\varepsilon_0\left(1 - \frac{n_e q_e^2}{m_e\varepsilon_0\omega^2}\right)\mathbf{E}_0 \quad (45)$$

The effective permittivity ε of the plasma is then

$$\varepsilon = \varepsilon_0 \left(1 - \frac{n_e q_e^2}{m_e \varepsilon_0 \omega^2}\right) = \varepsilon_0[1 - (\omega_{pe}^2/\omega^2)] \quad (46)$$

Equation (46) shows that whenever the electron plasma frequency, $\omega_{pe} = [n_e q_e^2/(m_e \varepsilon_0)]^{1/2}$, is greater than the driving frequency ω, the effective permittivity becomes negative and the wave no longer propagates. Note that it is always *less* than the permittivity of free space as well. If $\varepsilon/\varepsilon_0 < 0$, then we do not have normal propagation of waves, but evanescence.

B. Types of Plasmas

In examining Fig. 1, we can now consider the properties of those plasmas that are of direct interest to modern industrial problems. We concentrate on two types: glow (nonequilibrium) and arc (thermal) discharges. In general, if one considers a set of electrodes across which a dc potential is applied, one may classify the glow and arc discharges roughly according to the graph shown in Fig. 5.

The vertical axis is the voltage across the discharge, and the horizontal axis is the current. Note that as the current increases, the discharge goes from non-self-sustaining, to a glow discharge, to an abnormal glow, to an arc discharge. The voltage across the electrodes drops as the abnormal glow region is entered, rises, and drops again as the arc region is entered.

1. Glow Discharges

Figure 6 displays a typical picture of a glow discharge plasma between planar electrodes. The appearance of this discharge is complicated. It is maintained by electrons produced at the cathode by positive-ion bombardment. In the

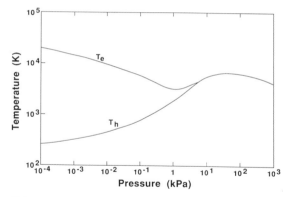

FIG. 6.   Transition between thermal and nonequilibrium plasmas.

Aston dark space there is an accumulation of these electrons, which gain energy through the Crookes dark space. The cathode glow results from the decay of excitation energy of the positive ions on neutralization. When the electrons gain sufficient energy in the Crookes dark space (also called the cathode fall, cathode dark space, or Hittdorf dark space) to produce inelastic collisions, the excitation of the gas produces the negative glow. The end of the negative glow corresponds to the range of electrons with sufficient energy to produce excitation, and in the Faraday dark space the electrons once more gain energy as they move to the anode.

The positive column is the ionized region that extends from the Faraday dark space almost to the anode. It is not an essential part of the discharge, and for very short discharge tubes it is absent. In long tubes it serves as a conducting path to connect the Faraday dark space with the anode. This portion of the discharge is nearly electrically neutral, and the main electron and ion loss occurs by ambipolar diffusion. In the last few mean free paths, the electrons may gain energy high enough to excite more freely as the positive ions are forced away from the anode, producing the anode glow.

It is important to note in Fig. 6 the curves for electric field strength and net space charge. In general, most plasma processing tends to be done with the object to be processed placed on the cathode, other electrodes, or a "target" electrode. As a result, knowledge of the electric field and potential distributions near these electrodes is very important for an understanding of the nature and energies of the particles as they traverse the sheath region around the cathode.

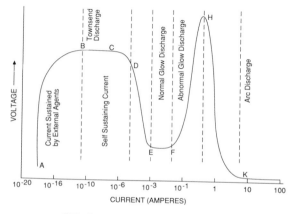

FIG. 5.   Glow and arc discharges.

2. Arc Discharges

A true arc discharge is characterized by a low cathode fall of the order of the ionization potential of the neutral atoms in the plasma. In glow discharges, the cathode fall tends to be much higher, perhaps of the order of 200 V. From Fig. 5, one can see that the current voltage characteristic is falling and the current density of the arc is very high. From this information it is clear that electron emission from the cathode must be governed by some mechanism other than positive-ion bombardment of the cathode. Usually the transition from glow to arc is a rapid and discontinuous process.

Arc discharges may be classified by the emission process that occurs at the cathode. Four types of arcs may be defined.

1. The thermionic arc in which the cathode is heated by the discharge and the arc *is* self-maintaining.
2. The thermionic arc in which the cathode is heated by an external source and the arc *is not* self-maintaining.
3. Field-emission arcs in which the electron current at the cathode is due to a very high electric field at the cathode surface.
4. Metal arcs.

In addition, arcs are frequently classified as high- or low-pressure arcs if the gas pressure is roughly above or below 1 atm. A high-pressure arc is characterized by a small, intensely brilliant core surrounded by a cooler region of flaming gases, sometimes called the aureole. If the arc occurs between highly refractory electrodes, such as carbon or tungsten, both the anode and the cathode are incandescent.

At low pressures, the appearance of the column depends upon the shape of the discharge tube. A constricted tube gives rise to a highly luminous column even at low pressures. The most important difference between low- and high-pressure arcs is in the temperature of the positive column. The high-pressure column is at a very high temperature (5000 K or higher). The ions, electrons, and neutral atoms of the high pressure positive column are in *local thermal equilibrium,* and therefore such plasmas are often called *thermal plasmas.*

The neutral gas temperature of the low-pressure arc is never more than a few hundred kelvins, whereas the electron temperature may be of the order of 40–50,000 K. (1 eV = 11,600 K.)

The difference between glow and arc plasma is shown in Fig. 6.

The current density at the cathode of an arc is very much greater than that of the glow discharge (Fig. 7). In many cases, the cathode current density is practically *independent* of the arc current. As a result, the plasma tends to concentrate in a small area near the cathode and produces *cathode spots.*

Thus low-pressure arc and glow discharges are sometimes called nonequilibrium (nonthermal) discharges.

3. RF Discharges

In most cases, rf and/or microwave radiation can be used to break down and maintain a discharge. The advantage of this process is that it is not necessary to have electrodes in contact with the plasma.

An important use of rf discharges occurs when we might wish to coat an electrode with an electrically insulating material (dielectric), by means of a plasma sputtering or chemical vapor deposition process. Normally, the cathode of the discharge would be the electrode that would receive the positive ions for the coating process. However, if a dc glow discharge is used to produce the plasma, the dielectric that is depos-

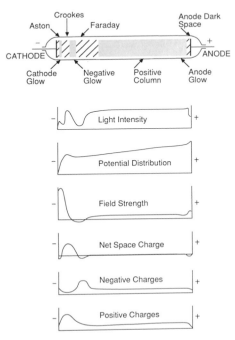

FIG. 7.   The glow discharge.

ited on the cathode will charge up, and the fluxes of both ions and electrons to the surface will become equal, regardless of the potential applied to the electrode itself. The result is that electrons and ions will combine at the dielectric, and no current will be drawn through the electrode to sustain the discharge and it will go out.

An ac discharge, if its frequency were high enough, could be used to maintain the discharge, because as the potential on the electrode backing the dielectric is reversed the charge on the dielectric will leak off. Then when the cycle is reversed again, the deposition process will reoccur. Typically, rf frequencies greater than about 1 MHz are required to maintain a discharge. Below this frequency, the discharge will be extinguished before the potential is reversed.

4. Breakdown

As the electric field in a discharge tube increases from zero, a small dark current (Townsend discharge) is drawn, but at some point there is a sudden transition to one of the several forms of self-sustaining discharge. Figure 8 shows the critical breakdown voltage as a function of the pressure–distance product, where the distance is measured between the electrodes.

From Fig. 8, it is seen that in all cases shown a minimum value of the voltage required for breakdown appears at a critical value of the pressure–distance product. At low pressures, below the critical value of the pressure–distance product (pd), the discharge will take place in the *longer* of two possible paths. This means that bringing electrodes closer together at the lower pressures will provide better insulation.

However, in bringing metal parts closer together in order to provide a higher breakdown voltage, care must be taken in the disposition of solid insulating material because the electrostatic field on the surface of the electrode will tend to increase. Above a certain value, cold or "field" emission of electrons will take place, and a vacuum arc (metal arc) will begin.

At high pressures, the minimum breakdown distance is so short that it can be virtually impossible to make any measurements. The minimum breakdown voltage is nearly equal to the normal cathode fall potential in the glow discharge.

C. Magnetic Fields

If a dc magnetic field is superimposed on the plasma, the basic equation of motion [Eq. (1)] now includes the magnetic field vector **B**, which will greatly affect the motion of the particles. First, the equation of motion will now be different depending on whether the component equation will be parallel to or perpendicular to the magnetic field. In the direction parallel to the field, the motion is nearly as though there were no magnetic field. This is not totally the case, since if the energy and magnetic moment of the charged particles are conserved, the orbit in the spatially varying magnetic field will result in a coupling between the parallel and perpendicular components of motion.

The basic effect in a uniform field is that the particles tend to follow the magnetic field lines and orbit around them. The result is a collimation along the field lines, which has important consequences. The plasma conductivity becomes a tensor, and hence the plasma in a dc magnetic field is anisotropic.

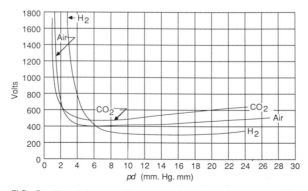

FIG. 8. Breakdown voltage as a function of pressure–distance product.

D. Plasma Potential

In a nonthermal glow-discharge plasma, we can make several conclusions about the temperatures and densities of the three components, electrons, ions, and neutrals. The electron and ion densities will tend to be equal. Usually, glow-discharge plasmas are not highly ionized, so the neutral density tends to be much larger than the plasma (ion plus electron) density. The thermal speed of the electrons tends to be much greater than the ions or the neutrals.

Suppose we suspend a small, electrically isolated dielectric material into the plasma. Initially it will be struck by electrons and ions with current densities

$$J_e = q_e n_e v_e \qquad J_i = q_i n_i v_i \qquad (47)$$

However, if $v_i \ll v_e$, then the dielectric tends to build up a negative charge and its potential becomes negative with respect to the plasma. The resulting electric field surrounding this material will have a marked effect on the motions of the particles near the material. Since the potential is negative, electrons will tend to be repelled from and ions attracted to the dielectric.

The dielectric tends to keep charging up negatively until the electron flux is reduced and the ion flux increased sufficiently so that the electron and ion fluxes balance. We call the potential of the material at this point V_p or the plasma potential.

E. Debye Length

The Debye length is a measure of how far into the plasma the potential of an electrode or probe is observed. It can be expressed as

$$\lambda_{de} = (\varepsilon_0 k T_e / n_e q_e^2)^{1/2} \qquad (48)$$

for the electrons, where ε_0 is the permittivity of free space, k is Boltzmann's constant, and q_e is the charge of the electron. A similar expression for the Debye length of the ions can be written with the appropriate ion quantities. Normally, electrons tend to congregate around a positive potential, so the electron Debye length will be important under these conditions, and vice versa for the ions. Often $\lambda_{di} = \lambda_{de}$.

IV. Plasma Diagnostics

In order to understand what is happening in a plasma, it is necessary to "diagnose" it. We may break down the various diagnostic measurement techniques into two parts: invasive

TABLE I.   Diagnostic Techniques

Invasive	Noninvasive
Langmuir probes	Radiation spectroscopy
Magnetic probes	Optical
Current probes	Microwave
Beam probes	X Rays
Radiation probes	Far infrared
Optical, micro-	Ultraviolet
wave, etc.	Particle Collectors
	Energy analysis
	Mass analysis

and noninvasive techniques. Table I lists some of these.

A noninvasive technique only "listens" or collects what comes out of the plasma. In this case, either radiation or particles are expelled, and we design instrumentation that can detect and analyze them. Invasive diagnostics require either the insertion of a probe or the injection of a particle or radiation beam into the plasma to have the diagnostic work. All other things being equal, noninvasive diagnostics are the most desirable, since they will perturb the plasma the least. However, some of the beam and radiation probes usually perturb the plasma so slightly that for many practical purposes they can be considered noninvasive diagnostics.

A. Probes

Figure 9 shows a sketch of a Langmuir probe and its associated circuit. The battery supplies a

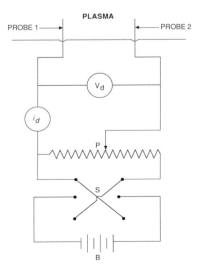

FIG. 9.   A Langmuir probe circuit.

potential to the probe, and the return circuit to the battery must come from the plasma. The method by which this happens is not always obvious. Such probes are usually very simple devices, consisting only of an insulated wire. The problems with such probes are that sheaths can form around the wire and the resulting measurements will be significantly changed by their presence. A "typical" plot of the probe current versus probe voltage characteristic is shown in Fig. 10.

The characteristic is taken either continuously, by varying the battery voltage, or point-by-point in a pulsed discharge, if the probe bias is changed from pulse to pulse. In addition, the probe voltage can be rapidly swept to obtain a characteristic dynamically.

At the point V_s, the probe is at the same potential as the plasma. The electric field across the sheath is zero at this point, and charged particles travel to the probe based entirely on their own thermal velocities. Since electrons move much faster than ions, because of their small mass, if the temperatures of the electrons and ions are about the same then what is collected is primarily electron current.

If the probe voltage is made positive with respect to the plasma, electrons are accelerated toward the probe and the ions are repelled. Thus, the small ion current present at potential V_s now vanishes, and an electron-rich sheath builds up until the total net negative charge in the sheath is equal to the positive charge on the probe. The thickness of the sheath is of the order of the Debye length, and outside of it there is very little electric field, so that the bulk of the

plasma is undisturbed. The electron current is the current that enters the sheath through random thermal motions, and since the area of the sheath is relatively constant as the probe voltage is increased, we have the fairly flat portion A of the characteristic. This is called the electron saturation current region.

If the probe is now made negative relative to V_s, we begin to repel electrons and accelerate ions. The electron current falls as V_p decreases in region B, which we call the transition region. If the electron velocity distribution were Maxwellian, the shape of the curve here after the contribution from ions is subtracted would be exponential.

At large negative values of probe potential, almost all of the electrons are repelled. We then have an *ion sheath* and *ion saturation current*, as shown in region C. This is similar to region A, but there are two differences between the ion and electron saturation currents caused by the mass difference. The first is that often the ion and electron temperatures are not equal, and as a result the Debye lengths are unequal, and the sheath widths are considerably different. The second problem is that if a magnetic field is present, the motion of the electrons is affected much more than the motion of the ions.

The shape of part B of the characteristic is related to the distribution of electron energies and can be used to determine T_e if the distribution is Maxwellian. The magnitude of the electron saturation current is proportional to $n(kT_e/m_e)^{1/2}$, from which n can be obtained (if T_e is previously found). The magnitude of the ion saturation current depends on n and kT_e, but only slightly on kT_i if $T_i \ll T_e$, so ion temperature is not easily measured with probes.

The space potential is found by locating the "knee" or junction between regions A and B of the curve, or by locating the point V_p (the zero current floating potential), and calculating V_s.

Experimental complications for the successful use of probes are many. A partial list of them is as follows:

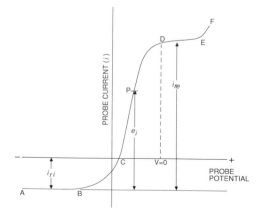

FIG. 10. Current–voltage characteristic of a Langmuir probe.

Surface Layers	
Perturbation of the Plasma	
Change of Probe Area	Effect of the Probe
Reflections	Shield
Macroscopic Gradients	Oscillations
Metastable Atoms	Photoemission
Secondary Emission	Negative Ions
and Arcing	Ion Trapping

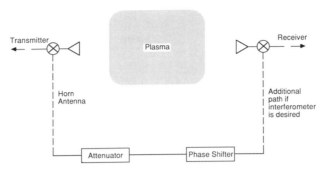

FIG. 11.   A microwave interferometer.

B. Radiation Probes

An example of a radiation probe of a plasma is a microwave interferometer. Figure 11 shows a drawing of such an interferometer.

The plasma introduces a phase shift of the microwave signal that passes through it. By comparison with the reference arm of the interferometer, it yields a signal propositional to the phase difference, which can then be calibrated to yield density directly.

The reason for the phase shift is easily seen if we adopt an exponential notation for the time and spatial variation of a plane monochromatic wave. That is, a wave will propagate as

$$\exp j(kz - \omega t) \qquad (49)$$

where k is the wave number. The wave equation

$$\frac{\partial^2 E}{\partial z^2} = \frac{1}{c^2} \frac{\partial^2 E}{\partial t^2} \qquad (50)$$

then becomes

$$-k^2 E = -(\omega^2/c^2)E \qquad (51)$$

and thus the relationship between k and ω can be determined to be

$$k = \omega/c = \omega(\varepsilon \, \mu)^{1/2} \qquad (52)$$

or

$$k = \omega(\varepsilon_0\mu_0)^{1/2} [1 - (\omega_p/\omega)^2]^{1/2} \qquad (53)$$

The phase shift is the difference in the product of the wave number k in the presence of the plasma, as defined above, times a fixed path length L with the free-space wave number $k_0 = \omega(\varepsilon_0\mu_0)^{1/2}$ times that same path length. That is,

$$\Delta\phi = (k - k_0)L \qquad (54)$$

which is directly related to density. The interferometer will detect density as it increases until cutoff is obtained, whereupon no further information is available. The higher the density required to be measured, the higher the frequency of the interferometer.

There are other restrictions with such a system. For one thing, we are assuming plane monochromatic waves. If the plasme has a dimension that is comparable to the wavelength, the plane-monochromatic wave approximation may not be valid. The plasma is also assumed to be cold. Hot plasmas can significantly affect the results. Finally, the interferometer measures *integrated* line density along its path, so the extent of spatial resolution is severely limited. Nevertheless, a microwave interferometer is one of the more important diagnostics.

C. Radiation Spectroscopy

The ultimate object in using spectroscopy in a plasma is the interpretation of all spectroscopic observations in terms of a fully consistent theoretical plasma model. Because spectra arise from atomic processes, the model must emphasize the particulate nature of the plasma, rather than the fluid aspects that characterize many of the properties of the plasmas.

Development of an appropriate model must take account of numerous atomic collision processes, so that it depends on atomic physics for its basic data. The step from atomic physics to a theoretical model for the plasma consists of discovering ways of taking into account the numerous possible atomic processes to give a composite picture of the spectrum. In many ways, the problems of laboratory spectroscopy are similar to those encountered in interpreting astrophysical spectra. A major difference might be in the possibility that laboratory plasmas have properties that change rapidly with time, so that time-dependent, rather than steady-state, solutions are required.

It is convenient to divide the treatment into

two parts: (1) considerations of the intensities of lines and continuum, and (2) the *shape* of the lines. We begin with the first part.

1. Plasma Models

The spectroscopic radiation that we shall be concerned with is emitted when an electron makes a transition in the field of an atom or ion. The observed intensity of the radiation thus emitted depends on three processes:

1. The probability of there being an electron in the upper level of the transition.
2. The atomic probability of the transition in question.
3. The probability of the photons thus produced escaping from the volume of the plasma without being reabsorbed.

Considerable simplification is achieved if the effect of the interaction of radiation (process 3) with the plasma is considered separately. There are, in fact, physically realizable circumstances where this effect may be neglected (optically thin plasmas).

2. The LTE Model

In the local thermodynamic equilibrium (LTE) model, it is assumed that the distribution of electrons is determined exclusively by particle collision processes and that the latter take place sufficiently rapidly so that the distribution responds *instantaneously* to any change in the plasma conditions. In such circumstances, each process is accompanied by its inverse, and these pairs of processes occur at equal rates by the principle of detailed balance. Thus, the distribution of energy levels of the electrons is the same as it would be in a system in complete thermodynamic equilibrium. The population of energy levels is therefore determined by the law of equipartition of energy and does not require knowledge of atomic cross sections for its calculation.

Thus, although the plasma temperature and density may vary in space and time, the distribution of population densities at any instant and point in space depends entirely on *local* values of temperature, density, and chemical composition of the plasma. The uncertainties in predictions of spectral line intensities from an LTE-model plasma depend on the uncertainties in the values of the plasma parameters and of the atomic transition probabilities.

If the free electrons are distributed among the energy levels available to them, then, according

to statistical mechanics, their velocities have a Maxwellian distribution. The number of electrons of mass m and with velocities between v and $v + dv$ is

$$dn_e = n_e 4\pi(m/2\pi kT_e)^{3/2} \exp(-mv^2/2kT_e)v^2\, dv \tag{55}$$

where n_e is the total density of free electrons and T_e is the electron temperature. For the **bound levels,** the distributions of electrons are given by the Boltzmann and Saha equations, which are, respectively,

$$\frac{n(p)}{n(q)} = \frac{\omega(p)}{\omega(q)} \exp[\chi(p, q)/kT_e] \tag{56}$$

$$\frac{n(z + 1)n_e}{n(z, g)} = \frac{\omega(z + 1, g)}{\omega(z, g)} 2(2\pi kT_e/h^2)^{3/2}$$
$$\times \exp[\chi(z, g)/kT_e] \tag{57}$$

where $n(p)$, $n(z + 1, g)$, and $n(z, g)$ are the population densities of various levels designated by their quantum numbers p, q, and g (the last for the bound level) and ionic charge $z + 1$ and z. The term $\omega(z, p)$ is the statistical weight of the designated level, $\chi(p, q)$ is the energy difference between levels p and q, and $\chi(z, g)$ is the ionization potential of the ion of charge z in its ground level g. Equations (55)–(57) describe the state of the electrons in a LTE model plasma.

If the plasma is optically thin, then the intensity $I(p, q)$ of the spectral line emerging from a transition between bound levels p and q is given by

$$I(p, q) = \frac{1}{4\pi} \int n(p)A(p, q)h\nu(p, q)\, ds \tag{58}$$

where $A(p, q)$ is the atomic transition probability and $h\nu(p, q)$ is the photon energy. The integration is made over that depth of the plasma that is viewed by the detector, and the intensity of radiation $I(p, q)$ is measured in units of power per unit area per unit solid angle.

D. Particle Analysis

Determination of the composition of plasma and neutral particles is important in plasma science and engineering, since the composition determines the nature of the physical and chemical processes that can occur. In addition, analysis of the products of the reaction is important for a determination of the effectiveness of a particular process.

Several different methods are currently in use for this purpose. We discuss one particular

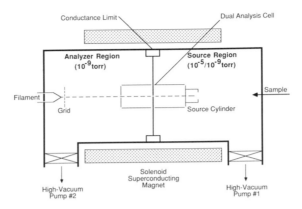

FIG. 12.   Fourier transform mass spectrometer.

method, which is a procedure for analyzing the masses of different ions based on the fact that the cyclotron frequency in a dc magnetic field is proportional to the charge-to-mass ratio of that ion. In this case, the ions can be either single ionized atoms or complex molecules.

Figure 12 shows the diagram of the analysis device, called a Fourier transform mass spectrometer (FTMS). A sample of material is placed at the right-hand side of the cell and an electron beam or a laser is directed at the material. Ions produced by this reaction can be electrostatically confined between the "trap" plates shown in the figure if a small voltage is applied (approximately 1–2 V).

The ions orbit at their appropriate cyclotron frequencies. An rf excitation pulse that is "swept" over a large range of possible cyclotron frequencies is applied between the "excitation" plates in the figure. This will excite the ions so that their orbits increase in radius and *ions of equal charge-to-mass ratio will move coherently,* so that a "tube" of ions moves around the cell. This tube induces an alternating charge pulse between the "detection" plates in the figure, and the output signal from these plates is Fourier analyzed and the resulting spectrum is displayed on an oscilloscope.

These devices are capable of extremely high resolution, which means that particles of very high mass numbers that are similar (such as mass numbers 1000 and 1001) can be displayed and analyzed. Other mass spectrometers, using different principles, can be used in other applications.

V. Plasma–Surface Interactions

Surfaces in contact with plasmas are bombarded by electrons, ions, neutral particles, and photons. Electron and ion bombardment is particularly important, especially because the particles are so energetic compared with neutral particles.

Several effects occur at once during this process: nondissociative chemisorption, physical adsorption, surface diffusion, dissociative chemisorption, and formation of product molecules. Figure 13 shows a schematic representation of these processes.

A. Plasma-Assisted Chemical Vapor Deposition (PACVD)

In this case, we desire to deposit a thin film of some material on the surface of a material, which we call a substrate. Previously, vaporizing the material to be deposited and allowing the vapor to come in contact with the substrate under vacuum resulted in the deposition of a material film on the substrate. However, in order for good deposition to occur, it was often necessary to heat the substrate to elevated temperatures.

PACVD allows lower substrate temperatures. This is particularly important in electronics applications where coatings are deposited onto device structures. In this process, a plasma is placed in contact with the substrate, and either the plasma particles themselves or neutral particles may form the coating. In general, the method of deposition is far from being well understood.

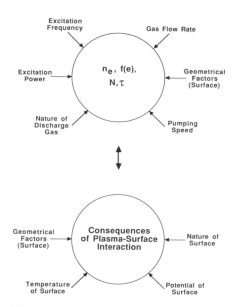

FIG. 13.   Processes occurring in plasma etch.

B. ETCHING

1. Physical Aspects

Figure 14 shows cross sections of films etched with liquid or plasma etchants. The isotropic profile represents no *overetch* and can be generated with either wet or dry (plasma) etching techniques. The anisotropic profile requires plasma etching.

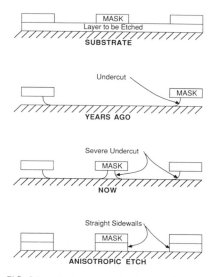

FIG. 14.   Variation of etching conditions.

Until recently, liquid (wet) etching techniques have been the main method of pattern delineation. This is because of two reasons: (1) the technology involved in liquid etching is well established, and (2) the selectivity (ratio of the etch rate of the film being etched to the etch rate of the underlying film or substrate) is generally infinite with typical liquid etchant systems.

Unfortunately, wet etching presents several problems for micrometer and submicrometer geometries. Because of the acid environments of most etchant solutions, photoresists can lose adhesion, thereby altering patterning dimensions and preventing linewidth control. In addition, as etching proceeds downward, it proceeds laterally at essentially an equal rate. This undercuts the mask and generates an isotropic etch profile, as shown in Fig. 14. Because the film thickness and etch rate are often nonuniform, a certain degree of *overetching* is required. If the film thickness is small, relative to the minimum pattern dimension, undercutting is inconsequential. However, if the film thickness is comparable to the lateral film dimension, as is the case for current and future devices, undercutting can be intolerable.

In addition, as device geometries decrease, spacings between stripes of resist also decrease. With micrometer and submicrometer patterns, the surface tension of etch solutions can cause the liquid to *bridge* the space between two resist stripes, and etching of the underlying film is eliminated.

Plasma etching has demonstrated viable solutions to essentially all the problems encountered with liquid etching. Adhesion does not seem to be critical with dry etching techniques. Undercutting appears to be controllable by varying the plasma composition, the gas pressure, and the electrode potentials.

Two additional considerations are in favor of dry etching. First, wet etching requires the use of relatively large volumes of dangerous acids and solvents, which must be handled and ultimately recycled or disposed of. Dry etching uses relatively small amounts of chemicals, although many of the gases used in these processes are also toxic.

In making plasmas for dry etching, a fill gas is broken down by means of application of an external electric field. As the electric field increases, free electrons, whose velocities increase by the action of the field, gain energy. However, since they lose this energy by collisional processes, an increase in pressure, which decreases the mean free path, then decreases the electron energy. What is important therefore, is a measurement of the velocity of an electron or ion as a function of the ratio of E/p (electric field divided by pressure). Figure 15 is a graph of the drift velocity (in an electric field) as a function of this ratio E/p.

The glow-discharge plasmas currently used for microelectronic applications can be characterized by the following parameters:

$$\text{Pressure} = 50 \text{ mtorr to 5 torr}$$
$$n_e = 10^9 - 10^{12} \text{ cm}^{-3}$$
$$T_e = 1 - 10 \text{ eV}$$

Usually the electron temperature is greater than the ion temperature by a factor of about 10. The plasmas are usually very weakly ionized, well below 1%.

These characteristics give the plasma special properties. The electron temperatures are high enough so that chemical bonds can be broken by electron-neutral collisions. As a result, highly reactive chemical species can be produced for etching or deposition. In addition, the *surface chemistry* occurring in glow discharges is generally modified by the impingement of ions and electrons (and photons) onto the film being etched. The combination of both of these processes results in etch rates and etch profiles unattainable with either process individually.

The energy of ions and electrons striking surfaces in a glow discharge is determined by the potentials established within the reaction chamber. Etching and deposition are generally carried out in a plasma produced by rf, which is often capacitively coupled to the plasma. The important potentials are the *plasma potential*, the *floating potential*, and the potential of the *externally biased (powered) electrode*.

Usually, under these circumstances, the electrode surfaces are at a negative potential with respect to the plasma. The result of this is that positive ions bombard the surfaces. The energy of the bombarding ions is established by the difference in potential between the plasma and the surface which the ion strikes. Because these potentials may range from a few volts to a few thousand volts, surface bonds can be broken, and, in certain instances, sputtering of film or electrode material can occur.

In addition, exposure of materials to energetic radiation can result in *radiation damage*. Positive ions can cause *implantation* or displacement damage, while electrons, X rays, or ultraviolet photons can result in ionization. Defects thus created can serve as trapping sites for electrons or holes, resulting in an alteration of the electrical properties of the materials. Such alterations may be very beneficial and can improve the surface properties of many materials considerably.

Figure 16 shows a capacitatively coupled configuration used for etching. Consider an rf field established between the two plates. On the first half cycle of the field, one electrode is negative and attracts positive ions; the other is positive and attracts electrons. At the frequencies used (50 kHz to 40 MHz), and because the mobility of electrons is greater than that of the ions, the electron current is much larger than the ion current. This causes a depletion of electrons in the plasma and thus a positive plasma potential. On

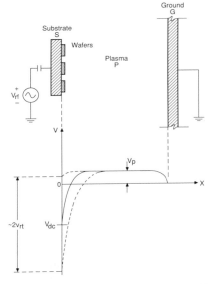

FIG. 16.   Capacitively coupled plasma etching reactor.

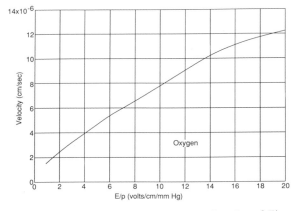

FIG. 15.   Electron drift velocity as a function of E/p.

the second half cycle, a large flux of electrons flows to the electrode that received the small flux of ions. Further, since plasma etching systems generally have a dielectric coating on the electrodes, and/or have a series (blocking) capacitor between the power supply and the electrode, no net charge can be passed. Therefore, on each subsequent half cycle, negative charge continues to build on the electrodes (and on other surfaces in contact with the plasma), so that electrons are repelled and positive ions are attracted to the surface.

When a sufficiently large negative bias is achieved on the electrodes, such that the fluxes of electrons and positive ions striking these surfaces are equal, then the transient situation ceases. At this point, the time-averaged (positive) plasma and (negative) electrode potentials are established.

The plasma potential is essentially uniform throughout the observed glow volume in an rf discharge. Between the glow and the electrode is a narrow region wherein a change from the plasma potential to the electrode (or surface) potential occurs. This is the sheath or dark space, and ions that reach the edge of the glow region are accelerated across the potential drop and strike the electrode or substrate surface.

Because of the series capacitor and/or the dielectric coating of the electrodes, the negative potentials established on the two electrodes in the system may not be the same. For example, the ratio of the voltages on the electrodes has been shown to be inversely proportional to the fourth power of the ratio of the relative electrode areas:

$$V_1/V_2 = (A_2/A_1)^4 \qquad (59)$$

If V_1 is the voltage on the powdered electrode and V_2 is the voltage on the grounded electrode, the voltage ratio is the inverse ratio of the electrode ares to the fourth power. However, for typical etch systems, the exponent is generally less than 4. Although the actual electrodes in a plasma reactor often have the same area, in Eq. (59) A_2 is the *grounded* electrode area, that is, the area of *all grounded* surfaces in contact with the plasma. Because of this, the average potential distribution is similar to that shown in Eq. (59).

In this case, the energy of the ions striking the powered electrode will be higher than that of ions reaching the grounded electrode.

Other plasma parameters can also affect the electrical characteristics. For example, rf power levels and frequency can radically change

things. As the frequency is raised, there will be a point at which the ions can no longer follow the alternating voltage, so that the ion cannot traverse the sheath in one half cycle. Above this frequency, ions experience an accelerating field that is an average over a number of half cycles. Such an average motion is described with the oscillation center approximation. The drift for such a motion is given by

$$\mathbf{r}_0 = -\nabla\Phi \qquad (60)$$

where Φ is the *ponderomotive* potential,

$$\Phi = \frac{q^2}{m^2\omega^2} \{E_0^2/2\} \qquad (61)$$

Thus, the net drift, which is *independent* of the sign of the charge, is toward lower average electric fields. The oscillation center drifts in the high-frequency field as if subjected to this potential.

At lower frequencies, the ions are accelerated by instantaneous fields and can attain the maximum energy corresponding to the maximum instantaneous field across the sheath. As a result, for a constant sheath potential, ion bombardment energies are higher at lower frequencies.

2. Chemical Aspects

Figure 13 showed the primary processes occurring during a plasma etch. There are six required steps, and if any one of them does not occur, the entire processing stops. They are (1) generation of reactive species, (2) diffusion to surface, (3) adsorption, (4) reaction, both chemical and physical (such as sputtering), (5) desorption, and (6) diffusion into bulk gas.

The reactive species must be generated by electron–molecule collisions. This is a vital step, because many of the gases used to etch thin-film materials do not react spontaneously with the film. For instance, carbon tetrafluoride, CF_4, does not etch silicon. However, when CF_4 is dissociated via electron collisions to form fluorine atoms, etching of silicon occurs rapidly.

The etchant species diffuse to the surface of the material and adsorb onto a surface site. It has been suggested that free radicals have fairly large *sticking coefficients* compared with relatively inert molecules such as CF_4, so adsorption occurs easily. In addition, it is generally assumed that a free radical will *chemisorb* and react with a solid surface. *Surface diffusion* of the adsorbed species or the produce molecule can also occur.

Product desorption is a crucial step in the etch process. A free radical can react rapidly with a solid surface, but *unless the product species has a reasonable vapor pressure so that desorption occurs, no etching takes place.* For instance, when an aluminum surface is exposed to fluorine atoms, the atoms adsorb and react to form AlF_3. However, the vapor pressure of AlF_3 is approximately 21 torr at 1240°C, and thus etching is precluded at room temperatures.

The chemical reactions taking place in glow discharges are often extraordinarily complex. However, two general types of chemical processes can be categorized. They are (1) homogeneous gas-phase collisions and (2) heterogeneous surface interactions. In order to completely understand and characterize plasma etching, one must understand the fundamental principles of both processes.

Figure 17 shows how two etch processes may result in a synergism in which the resulting etch rate is greater than the sum of the two. In this case a X_eF_2 plasma and an argon ion beam are used together.

a. Homogeneous   Gas-Phase   Collisions. These represent the manner in which reactive free radicals, metastable species, and/or ions are generated. As shown in Table II, electron impact can result in a number of different reactions.

Due to the electronegative character of many of the etch gases currently used (O_2, CF_4, CHF_3, CCl_4, BCl_3, etc.), electron attachment often takes place, thereby generating negative as well as positive ions in the plasma. *Although these negative ions affect the plasma energetics, they probably have little if any effect on surface reactions, because they are repelled by the negative electrode potential.*

C. PLASMA POLYMERIZATION

In the plasma etching process, a competing process that can dominate over etching can occur, which is called *polymerization*. A polymer is defined as a high-molecular-weight compound made up from a small repeating organic unit called a monomer. The magnitude of the molecular weight ranges from 1000 to several million atomic mass units (amu) and, depending on conditions, the reaction product could have a statistical distribution of molecular weights.

In order for monomers to form polymers one of two reactions must occur: condensation and addition. The condensation reaction usually results in the loss of a small portion of the original starting molecules. Addition polymers are those that result from the reaction of an unsaturated monomer with an initiator that begins a chain reaction at an actiated site to start the growing polymer chain.

As the ratio of fluorine to carbon is increased, polymerization ceases and etching begins at a critical value, which depends on the potential applied to the surface.

Two general schemes have been proposed for organization of chemical and physical information on plasma etching and polymerization. Both have dealt primarily with carbon-containing gases, but with slight modifications can be easily applied to other gases. Figure 18 is a schematic of the influence of the fluorine-to-carbon ratio and electrode bias on etching and polymerization.

This model does not consider the specific chemistry occurring in a glow discharge, but rather views the plasma as a ratio of fluorine to carbon species which can react with a silicon

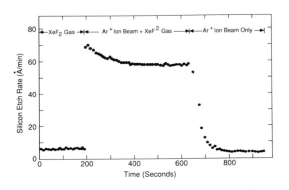

FIG. 17.   Etching synergism of two processes.

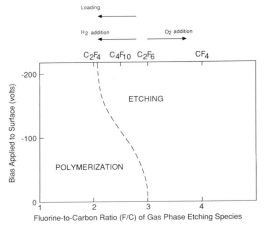

FIG. 18.   Etching rate as a function of electrode bias.

TABLE II. Results of Electron Impact

Excitation (rotational, vibrational, electronic)	$e + A_2 \rightarrow A_2 + e$
Ionization	$e + A_2 \rightarrow A_2^+ + 2e$
Dissociative ionization	$e + A_2 \rightarrow A^+ + A + 2e$
Dissociation	$e + A_2 \rightarrow 2A + e$ or $A^- + A^+ + e$
Dissociative attachment	$e + A_2 \rightarrow A^- + A$

surface. The generation or elimination of these active species by various processes or gas additions then modifies the initial fluorine-to-carbon ratio of the inlet gas.

The F/C ratio model accounts for the fact that in carbon-containing gases, *etching and polymerization* occur simultaneously. The process that dominates depends on etch-gas stoichiometry, reactive-gas additions, amount of material to be etched, and electrode potential, and on how these factors affect the F/C ratio. For instance, the F/C ratio as seen in Fig. 18 determines whether etching or polymerization is favored. If the primary etchant species for silicon (F atoms) is consumed either by a loading effect or by reaction with hydrogen to form HF, the F/C ratio decreases, thereby enhancing polymerization. Such effects are caused primarily by enhanced energies of the ions striking these surfaces.

In the *etchant-unsaturate* model described by Eqs. (62)–(65), specific chemical species derived

Saturated   Unsaturated

$$e + \text{Halocarbon} \rightarrow \text{saturated radicals} \\ + \text{unsaturated radicals} + \text{atoms} \tag{62}$$

$$\left.\begin{array}{l}\text{Reactive atoms} \\ \text{Reactive molecules}\end{array}\right\} + \text{unsaturates} \rightarrow \text{saturates} \tag{63}$$

$$\text{Atoms} + \text{surfaces} \rightarrow \left[\begin{array}{l}\text{chemisorbed layer} \\ \text{volatile products}\end{array}\right. \tag{64}$$

$$\text{Unsaturates} + \text{surfaces} \rightarrow \text{films} \tag{65}$$

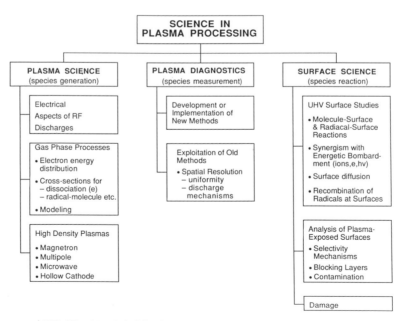

FIG. 19. Needed skills for plasma processing of semiconductors.

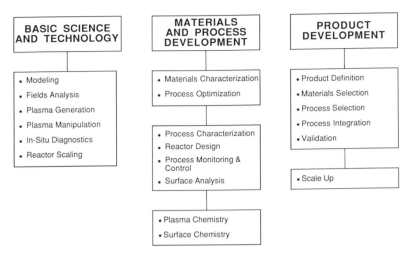

FIG. 20.   Needed skills for plasma processing in the pharmaceutical industry.

from electron collisions with etchant gases are considered.

Application of this model to a CF_4 plasma results in the chemical scheme described by

$$2e + 2CF_4 \rightarrow CF_3 + CF_2$$
$$+ 3F + 2e \quad (66)$$

$$F + CF_2 \rightarrow CF_3 \quad (67)$$

$$4F + Si \rightarrow SiF_4 \quad (68)$$

$$nCF_2 + \text{surface} \rightarrow (CF_2)_n \quad (69)$$

Depending on the particular precursors generated in the gas phase, etching, recombination, or film formation (i.e., polymerization) can occur. Also, gas-phase oxidant additives (O_2, F_2, Cl_2, etc.) can dissociate and react with unsaturate species. As an example, O_2 can undergo the following reactions in a CF_4 plasma:

$$e + O_2 \rightarrow 2O + e \quad (70)$$

$$O + CF_2 \rightarrow COF_{2-x} + xF \quad (71)$$

Mass spectrometer studies of oxidant additions in fluorocarbon and chlorocarbon gases have demonstrated that the relative reactivity of atoms with unsaturate species in a glow discharge follows the sequence F = O > Cl > Br. Of course, the most reactive species present will preferentially undergo saturation reactions that reduce polymer formation and that may increase halogen atom concentration. Ultimately, determination of the relative reactivity of the plasma species allows prediction of the primary atomic ethants in a plasma of specific composition.

VI. Conclusion

There are many areas of importance for successful progress in these fields. Two views from industry (from the semiconductor and pharmaceutical industries) have provided the charts of Figs. 19 and 20, which show the various problems needed to successfully advance the field.

BIBLIOGRAPHY

Blaustein, B. C., ed. (1969). "Chemical Reactions in Electrical Discharges." American Chemical Society, New York, Advances in Chemistry Series no. 80.

Brown, S. C. (1959). "Basic Data of Plasma Physics." Wiley, New York.

Bunshah, R. F., ed. (1982). "Deposition Techniques for Films and Coatings." Noyes, Park Ridge, N.J.

Chen, F. F. (1974). "Introduction to Plasma Physics." Plenum, New York.

Coburn, J. W. (1980). *Plasma Chem. Plasma Process.* **2**, 2.

Holloban, J. R., and Bell, A. T., eds. (1974). "Techniques and Applications of Plasma Chemistry." Wiley, New York.

Knights, J. C., ed. (1984). "The Physics of VLSI." American Institute of Physics, New York.

Mucha, J. A., and Hess, D. W. (1983). "History of Dry Etching." American Chemical Society, New York.

Mucha, J. A., and Hess, D. W. (1983). "Plasma Etching." American Chemical Society, New York.

National Materials Advisory Board. (1985). "Plasma Processing," Report of the National Materials Advisory Board. National Academy Press, Washington, D.C.

Nowogrozki, M. (1984). "Advanced III–V Semiconductor Materials Technology Assessment." Noyes, Park Ridge, N.J.

Shohet, J. L. (1971). "The Plasma State." Academic Press, New York.

Venugopalan, M., ed. (1971). "Reactions under Plasma Conditions," Wiley, New York.

PLASTICITY*

Jules L. Routbort   *Argonne National Laboratory*

GLOSSARY

Climb: Motion of a dislocation perpendicular to slip plane.

Creep: Plastic deformation under conditions of constant stress or load.

Dislocation: Line defect that exists between the slipped and unslipped regions of a crystal.

Engineering strain: Change in length divided by original length.

Engineering stress: Applied load divided by original cross-sectional area.

Fatigue: Plastic deformation under conditions of cyclic stress or strain.

Fatigue limit: Stress level below which no fatigue damage occurs; also called endurance limit.

Glide: Motion of a dislocation parallel to slip plane.

Resolved stress: Component of true stress resolved on the slip plane and in the slip direction.

Steady-state creep: Creep during which the dislocation structure is constant; also called secondary creep.

Strain rate: Rate of strain change per unit time.

True strain: ln(instantaneous length/original length).

True stress: Applied load divided by instantaneous area.

Ultimate tensile stress: Maximum engineering stress before failure.

Yield stress: Stress at which macroscopic plasticity begins.

Plasticity is the capacity of a solid to deform permanently on loading. In contrast to elastic deformation, which involves the stretching of atomic bonds and which is reversible (i.e., on unloading, the body returns to its original shape), plastic deformation is a nonreversible process in which the atomic positions are altered. This occurs by the application of forces, greater than a critical value, that can shear atomic planes such that the atoms move from their initial positions to equivalent adjacent lattice positions. On unloading, the material retains its shape change with a small correction for the elastic component.

I. Introduction

The irreversible process of plastic deformation occurs whenever a shear stress exceeds a critical value and causes permanent changes in atomic positions. Plastic deformation will be assumed to occur at constant volume, as is generally the case for metals. However, this may not be true for ceramics, which generally contain as-fabricated porosity and which can densify under the application of a stress at high temperatures. This article is restricted to a discussion of plasticity in crystalline solids, but it should be mentioned that polymers and glasses can be plastically deformed in a time-dependent process called viscous deformation, which involves the permanent rearrangement of atomic and molecular bonds. Viscous flow can occur at room temperature in some polymers but requires very high temperatures in metals and ceramics. [*See* METALLURGY, MECHANICAL.]

* Work supported by the U.S. Department of Energy, Basic Energy Sciences, Materials Science, under Contract W-31-109-ENG-38.

Plasticity is a very complex property that depends on many variables. Plastic deformation is very sensitive to discrete defects, both microscopic and atomic, in a material. Because the atomic positions change during plastic deformation, the stress required to maintain the flow of material is quite sensitive to crystal structure. Slipping of one layer of atoms over another to accommodate plasticity is a localized phenomenon and is, therefore, sensitive to such crystal lattice defects as dislocations, vacant atomic sites, solute atoms, second-phase particles, and grain boundaries. Furthermore, the strength of the atomic bonds depend on temperature. Thus, plasticity is temperature dependent and, by a similar argument, dependent on the rate of loading. Plasticity is also a strong function of the composition of the material. Because the structure of the material changes continuously during deformation, the plastic deformation depends on the loading history.

The objective of all studies of plastic deformation of materials is to gain sufficient insight into the processes controlling the deformation that a functional relationship among all of the pertinent variables can be established. The mechanical equation of state thus derived can then be used by the engineers responsible for the fabrication or the application of a material to predict the plastic response of the material to a load. This predictive capability becomes increasingly important in a high-technology society where machinery operates at elevated temperatures, which increase the probability of plastic flow even in normally brittle materials such as ceramics. Problems in plasticity typically arise in materials processing. The most common metal-processing techniques (i.e., drawing, rolling, extrusion, explosive forming, stamping, and forging) involve some form of permanent deformation. Furthermore, in many applications of both metals and ceramics, plasticity is important. For example, lead rain gutters used in England in the nineteenth century have undergone permanent shape changes under their own weight; the plasticity of a fissioning ceramic nuclear fuel is one of the properties that control the lifetime of a fuel element; and the large plasticity of aluminum allows for rapid formation of the seamless beer can. [See METAL FORMING.]

The following discussion is divided into several sections. The first section deals with the concept of dislocations, which control plasticity. The next section deals with some basic phenomena in the deformation of crystalline solids, and subsequent sections deal with the effects of the manner in which stress is applied. For example, a stress (generally restricted experimentally to simple tension or compression, but in practice a complex combination) may be imposed under conditions of constant strain rate (Section III), constant load (creep, Section IV), or cyclic load (fatigue, Section V).

II. Dislocations

A. HISTORY

It is well established that plastic deformation of a crystalline solid is achieved by the motion of two-dimensional crystal lattice defects called dislocations. Their existence in a wide variety of materials has been directly observed by several techniques. Furthermore, the fact that plastic deformation is accommodated by the creation and movement of dislocations has been positively established for all crystalline solids.

The first suggestions of the existence of dislocations were provided by observations in the nineteenth century that the deformation of metals was preceded by the formation of slip bands; that is, one portion of the specimen sheared with respect to the other under the application of a stress that had a shear component. The shear stress τ is the applied force per area acting to shear the sample. This is to be contrasted with a hydrostatic stress, which acts normal to the shear planes. The unit of stress is the pascal (Pa) (1 Pa = 1 newton per meter^2), which is related to the English unit of stress (pounds per inch^2, psi) as follows: 1 psi = 6894 Pa = 6.894×10^{-3} MPa, where M = 10^6.

The discovery that X rays could be diffracted by metals established the crystalline nature of metals and led theoreticians, particularly J. Frenkel in 1926, to calculate the theoretical shear strength of metals on the basis of a model that allowed one plane of atoms to be sheared relative to the other. However, the experimental value of shear strength was at least a factor of 10 lower than theoretically predicted. This discrepancy led Orowan, Polanyi, and Taylor in 1934 to propose the existence of a dislocation that would allow the displacement to occur over part of the atomic plane, so that less force was required for deformation. Indeed, dislocations were later observed, and their behavior has been studied by such techniques as etch pitting, X-ray transmission and diffraction, field ion microscopy, and transmission electron microscopy. The last technique, which was developed in the early 1950s, has proved invaluable in establishing the details of material structure required to understand plasticity. Typical examples of dislocations as revealed by the transmission electron microscope are shown in Fig. 1.

B. GEOMETRY

The atoms in crystalline solids are arranged in a periodic structure. The three most common crystal structures are face-centered cubic (fcc), body-centered cubic (bcc), and hexagonal close-packed, although in practice metals and ceramics have other, more complex structures as well. The primary step in deformation is the translation or slip of one part of the crystal over the other. Thus, one set of crystallographically equivalent planes of atoms, designated the slip plane, slips over another plane in one of several possible crystallographically equivalent slip directions. The slip system that is activated will depend on the crystal structure, type of bonding, and temperature. Materials with similar structures tend to slip on the same planes. [*See* BONDING AND STRUCTURE IN SOLIDS; CRYSTALLOGRAPHY.]

A two-dimensional representation of an edge dislocation, the type postulated by Orowan, Polanyi, and Taylor, is shown in Fig. 2. Dislocations are created by inserting an extra half-plane of atoms. The dislocation in Fig. 2 is denoted by the symbol ⊥, and the intersection of the extra half-plane with the plane of the paper is shown by a row of shaded atoms. The extra half-plane is above the slip plane. A dislocation of opposite sign is denoted by the symbol ⊤ and has the extra half-plane below the slip plane. The Burgers vector **b** of the dislocation is an important parameter and is defined by drawing a clockwise circuit around the dislocation, consisting of an equal number of steps in each of the four directions. The direction and distance from the first to the last atom in such a circuit give the Burgers vector for the dislocation. For example, if the circuit is started at atom 1 in Fig. 2 and proceeds through five atoms to the right, five down, five left, and five up, it will end at atom 2, and the Burgers vector is seen to be parallel to the slip direction. Also, it is perpendicular to the dislocation line, which is the intersection of the extra half-plane with the slip plane. An edge dislocation always has **b** perpendicular to the dislocation line. The other type of dislocation, a screw, first described by Burgers in 1939, has its slip direction parallel to the dislocation line. Dislocations in all materials are combinations of screw and edge dislocation components.

Dislocations are formed as a result of internal stress caused by either impurities or thermal gradients or by impingement of different parts of the growing interface (i.e., a change in lattice structure) or by the formation and subsequent collapse of vacancy platelets. The grown-in dislocation density of a well-annealed metal single crystal is small, but finite. Engineering materials that have been formed by plastic deformation contain high dislocation densities as a result of the deformation. Dislocations can be nucleated by localized stress concentrations in the material (heterogeneous nucleation) or in a defect-free region by a uniform high stress (homogeneous nucleation). Grain boundaries are also sources of dislocations.

The fact that a crystal containing dislocations can deform more easily than one that has no dislocations can now be explained. When a perfect crystal is subjected to a shearing force, each atomic bond is stretched like a spring. When each atom has been translated to a point midway between its original equilibrium position and a new position, the atomic bonds will break and new ones will form because of the interatomic bonding forces. Therefore, without dislocations, plastic deformation would require the cooperative movement of all atoms above the slip system. This necessitates a very high force.

However, if a crystal contains a dislocation, a much smaller force is required to shear the crystal by one atomic distance. The movement of the dislocation requires only a small shear force because only the atoms immediately adjacent to the dislocation are displaced large distances. For example, if the atom immediately to the right of the dislocation moves one position to the left, the dislocation effectively moves to the right. Continued motion results in a step on the crystal surface and the removal of the dislocation. The result is the same as that produced by shearing a dislocation-free crystal, but a much lower force is needed. The motion of a dislocation is analogous to the movement of a large rug by the introduction of a wrinkle across the rug and the propagation of the wrinkle rather than by the application of a massive force to the whole rug.

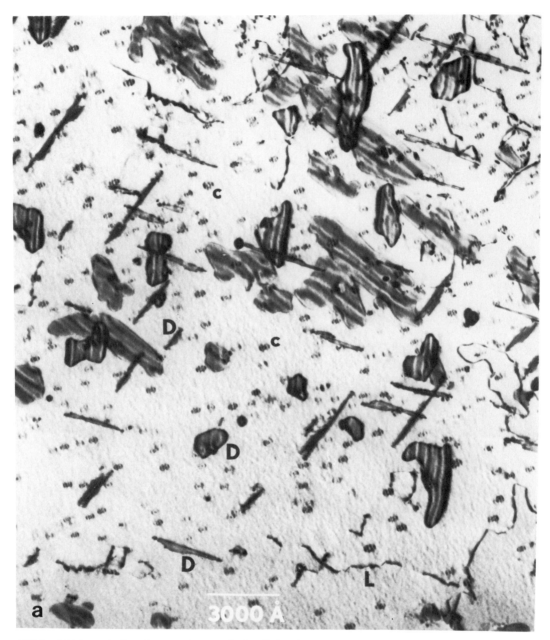

FIG. 1.   (a) Transmission electron micrograph of radiation damage structure produced in a stainless steel alloy after bombardment with 1-MeV electrons in a high-voltage electron microscope. The structure consists of dislocation loops (D), line dislocations (L), and cavities (c). The average cavity diameter is $\sim 10^{-8}$ m, or 100 Å. (Courtesy of P. Okamoto, Argonne National Laboratory, Argonne, Illinois.) (b) Screw dislocations in a low-angle twist boundary in NiO. (Courtesy of N. L. Peterson, Argonne National Laboratory, Argonne, Illinois.)

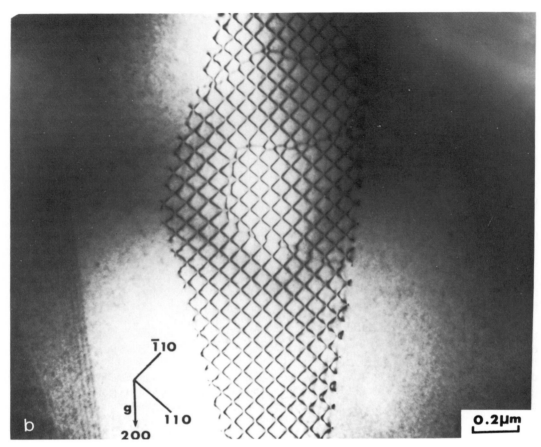

110

g   110

200

b

0.2μm

FIG. 1.   (*Continued*)

Single crystals deform by slip in close-packed directions on planes that are close-packed. Table I lists both the Burgers vector (the slip direction) and the primary operative slip plane at room temperature for several common crystal structures. It should be mentioned that Table I presents the most general cases and that there are exceptions. For example, PbTe, an NaCl-type crystal, slips on a {100} plane rather than a {110} because of charge effects that are present in ionic solids. The slip systems change with temperature, because increased thermal energy makes other systems energetically favorable. The shear stress τ is that portion of the applied stress σ that acts on the slip plane.

Single crystals deform anisotropically, and the translations leave linear traces, called slip lines, on the surfaces. However, normally produced metals and ceramics are polycrystalline; they contain a large number of individual grains, which vary in size from 10^{-2} to 10^{-6} m. The presence of grain boundaries allows a body to deform isotropically under a stress. The grain size of a material is one of the microstructural features that the manufacturer can control in order to alter the plastic response of the solid.

TABLE I.   Slip Directions and Planes for Some Common Crystal Structures

Crystal structure	Slip direction[a]	Primary slip plane[a]
Face-centered cubic (e.g., Al, Cu)	$\langle 110 \rangle$	$\{111\}$
Body-centered cubic (e.g., Nb, V)	$\langle 111 \rangle$	$\{110\}$
Hexagonal close-packed (i.e., Zn, Cd)	$\langle 11\bar{2}0 \rangle$	$\{0001\}$
Diamond cubic (i.e., Si, Ge)	$\langle 110 \rangle$	$\{111\}$
Rocksalt (i.e., NaCl, MgO)	$\langle 110 \rangle$	$\{110\}$

[a] See crystallography for an explanation of notation.

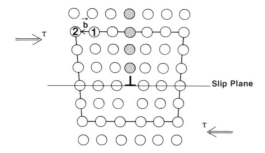

FIG. 2.   Edge dislocation in a simple cubic crystal.

C. Basic Theory of Plasticity

Several basic properties of dislocations must be explored before a comprehensive theory of plasticity can be developed. These can be divided into static and kinetic properties. The motion of a dislocation is a kinetic property, and the static properties are those related to the forces, stresses, and strains on dislocations.

It is obvious from Fig. 2 that an edge dislocation can move most easily by breaking the bond of an atom adjacent to it. The dislocation will then move one lattice spacing in the opposite direction along the slip plane, also called the glide plane. This conservative motion is called glide, and the glide of many dislocations results in slip, the basic step in plasticity. However, an edge dislocation can also move perpendicular to the glide plane by a process called climb. If a row of atoms above the dislocation normal to the plane of the paper is removed, the dislocation will move up one lattice spacing. This nonconservative motion is clearly more difficult than glide because it requires the transport of atoms. Atomic diffusion becomes likely at high temperatures only, and therefore climb is important at high temperatures only. In general, the transport of a single atom rather than a row of atoms occurs. This results in the formation of steps, called jogs, along dislocations.

The glide velocity of a dislocation depends on the applied shear stress, type and purity of crystal, temperature, and type of dislocation. In a series of brilliant experiments, Johnson and Gilman directly measured dislocation velocities as a function of stress and temperature in LiF. In these experiments, a crystal containing fresh dislocations introduced by indenting the surface with a diamond was subjected to a stress pulse for a given time at a fixed temperature. The velocity of the dislocation was calculated by dividing the distance the dislocation had moved by the duration of the stress pulse. The distance moved was determined by etching the sample before and after the application of the stress pulse. (The intersection of a dislocation and a free surface, when attacked by an aggressive solution, results in the formation of an etch pit, which clearly identifies the location of the dislocation.) The dislocation velocity v is given by the empirical relationship

$$v = A\tau^m \tag{1}$$

where A is a material constant and m varies from 1 to ~50, depending on material, purity, and temperature.

The velocity versus stress curves have several interesting features: Below a certain stress, called a critical stress, dislocations do not move; and dislocations do not move faster than the velocity of sound in a material. Dislocations move faster at low temperatures because resistance to dislocation motion due to lattice vibrations (phonons) decreases as temperature decreases. Thus, plasticity, which is a macroscopic property that requires dislocation motion and therefore atomic motion, is sensitive to microscopic properties such as atomic bonding and impurities, as well as to macroscopic properties such as microstructure and crystal structure.

Linear plastic strain is defined as a permanent change in length per initial length and is therefore dimensionless. That is, under a shear stress, deformation characterized by a shear strain γ occurs. The determination of the relationship between stress and plastic strain is the goal of most of the research efforts in plasticity. Also of importance is the plastic strain rate: the time rate of change of the plastic strain, denoted by $\dot{\gamma}$. The strain produced by a group of dislocations is given by

$$\gamma = b\rho_m \bar{X} \tag{2}$$

where b is the magnitude of the Burgers vectors, ρ_m the mobile dislocation density (number per unit area), and $\bar{X}$ the average distance moved by a dislocation. Only the dislocations that are free to move contribute to the plastic strain. The plastic strain rate $\dot{\gamma}$ is given by

$$\dot{\gamma} = b\rho_m v \tag{3}$$

where v is the average dislocation velocity and is given by Eq. (1).

The atoms in a crystal containing dislocations are displaced from their perfect lattice sites, and the resulting distortion produces a stress field around the dislocation. For example, the region above the slip plane in Fig. 2 contains an extra half-plane and is in compression while the region below is in tension. If the solid is assumed to be an ideal isotropic elastic body and the distortions around the core of a dislocation are ignored, these stresses can be calculated. Once the stresses are known, the energy of a dislocation, the forces between dislocations, and the forces between dislocations and point defects can be calculated. However, since a crystal is always anisotropic and core effects can be large, the simple calculations are only qualitatively correct. The stress field of both an edge and a screw dislocation at a position r from the dislocation is proportional to the shear modulus G times the Burgers vector divided by the distance r. In the case of a screw dislocation, the stress consists of a pure shear, whereas for an edge dislocation the stress field has a dilational as well as a shear component. The strain energy of a dislocation (energy per unit length of dislocation) is proportional to Gb^2. Thus, a dislocation can be viewed as a stretched string with a line tension equal to Gb^2.

The stress fields around dislocations cause dislocations to interact with one another and with solute atoms. For example, the force F_x between two parallel edge dislocations of opposite signs with parallel Burgers vectors is shown in Fig. 3. The figure indicates that they attract one another (negative F_x) for $x > y$ and repel one another for $x < y$. The configuration is stable

($F_x = 0$) only for $x = y$. Hence, unlike edge dislocations tend to form dipole pairs, whereas like edge dislocations align above one another. It is not surprising that, because of the interaction forces, plasticity tends to induce very definite dislocation substructures. The study of these structures by the use of transmission electron microscopy forms an important part of the study of plasticity.

III. Microscopic Basis of Plasticity in Crystalline Solids

A. YIELD AND STRESS–STRAIN BEHAVIOR

In a single crystal, the applied stress σ and the resulting plastic strain ε are related to the shear stress τ and the shear strain γ, respectively, by a proportionality constant called the Schmidt factor. This geometric factor resolves σ into the slip plane in the slip direction. However, most engineering materials are aggregates of single crystals. These aggregates, of course, can be randomly or preferentially oriented and can vary in size. For polycrystals, the average stress can be related to the shear stress by the Taylor factor.

One of the most fundamental experiments in plasticity involves the measurement of the plastic strain and stress as a specimen is deformed in tension at a constant rate and at a fixed temperature. Such experiments can be routinely performed in the universal testing machine shown in Fig. 4. The specimen to be tested is gripped between an upper and a lower fixed crosshead while the upper crosshead is moved away at a fixed rate by means of two drive screws. The resultant stress is measured by a load cell in the upper crosshead. Strain gauges or extensiometers can be used to measure the strain.

A schematic result typical of an fcc metal single crystal at room temperature is shown in Fig. 5, where τ is plotted as a function of γ. The crystal begins to exhibit macroscopic plasticity when the stress exceeds a critical resolved shear stress τ_c. There is some subjectivity about the determination of τ_c: It is exactly the stress required to move dislocations and is a fundamental materials parameter and corresponds to a yield stress in polycrystals; it is quite sensitive to the test temperature and imposed strain rate.

The τ–γ curve in Fig. 5 is divided into three stages after τ_c is reached. The plastic strain in stage I results from the dislocations moving on

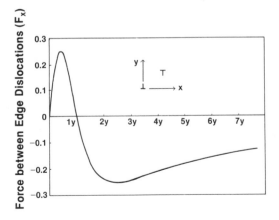

FIG. 3.   Force between parallel edge dislocations with parallel Burgers vector for dislocations of opposite sign. Force is in units of $Gb^2/2\pi(1 - \nu)y$, where ν is Poisson's ratio. [After Cottrell, A. H. (1953). "Dislocations and Plastic Flow in Crystals," Oxford University Press, New York.]

FIG. 4. Model 1125 universal testing machine. (Courtesy of Instron Corp., Canton, Massachusetts.)

the primary slip system (Table I), which is the slip system with the highest resolved stress. Stage I is characterized by a low rate of work hardening, which is the slope of the τ versus γ curve. The units of the work-hardening rate θ are the same as for stress. The extent of stage I, as well as the extent of the other stages, depends on the crystallographic orientation of the sample in relation to the tensile axis, on the crystal structure and purity, and on temperature. Stage I is followed by stage II, in which θ increases. The strain in stage II results from the motion of dislocations that are formed on the secondary slip systems when the applied stress (aided by the internal stress of the primary dislocations and the crystal orientation) is such that the criti-

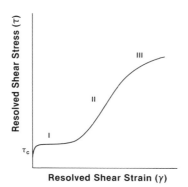

FIG. 5. Typical tensile stress–strain curve for an fcc single crystal measured at room temperature.

cal stress to activate dislocations on the secondary systems is reached. Finally, stage II ceases and stage III, a region of dynamic recovery, begins. This region is characterized by a continually decreasing θ and is the result of cross slip (the process by which screw dislocations switch from one crystallographic plane to another) of screw dislocations at room temperature or of climb at high temperatures. The extent of this region depends strongly on temperature and on a fundamental materials parameter called the stacking-fault energy (SFE; to be discussed next).

A dislocation can be split into components called partials. When two such partial dislocations are created by the dissociation of a dislocation, they cannot stay next to each other because of the repulsive force acting between them. As a consequence, the partial dislocations separate and the region between the partials has a different atomic stacking sequence than the bulk of the crystal. This is called a stacking fault, and energy is required to form it. The magnitude of the SFE determines the separation between partial dislocations, and since cross-slip requires no dissociation into partials, the SFE also determines the ease of cross-slip and therefore the extent of stage III. The stress at which stage III starts decreases exponentially with increasing temperature and with increasing SFE. Stacking-fault energies can vary within one crystal structure (e.g., SFE $\approx$ 140 mJ m^{-2} for aluminum and SFE $\approx$ 20 mJ m^{-2} for gold, both fcc metals) and with crystal structure. Some metals and ceramics have such a high SFE that separation into partials is not generally observed.

Figure 5 was drawn for a typical fcc metal. Similar results have been obtained for other crystal structures and for ceramic single crystals. Figure 6 shows the results obtained at room temperature for an MgO single crystal. MgO has a considerably higher melting temperature T_M than most metals ($T_M = 3070$ K for MgO as compared with 993 and 1808 K for aluminum and iron, respectively). MgO is also a ceramic and cannot be deformed to as large a strain as a ductile metal without fracture. Hence, stage III in MgO appears only at higher temperatures.

A schematic load–elongation curve for a typical polycrystalline metal pulled in tension at a constant rate is shown in Fig. 7a. Three features are important: (1) The yield stress is generally not sharply defined, (2) a maximum load L_M is observed, which corresponds to the transition

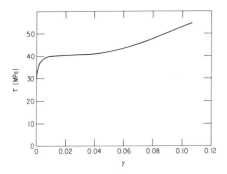

FIG. 6. Resolved shear stress–strain data for ⟨100⟩ MgO deformed at 25°C at a strain rate of 4×10^{-5} sec^{-1}. (From J. Routbort, Argonne National Laboratory, Argonne, Illinois.)

TABLE II. Room-Temperature Values of Strength Parameters for Some Common Polycrystalline Metals

Material	σ_y (MPa)	σ_{UTS} (MPa)	Elongation to fracture (%)
Aluminum—6061 TS	276	310	17
Steel—stainless AISI 304	241	586	55
Steel—cold drawn AISI C1020	352	420	15
Steel—cold drawn AISI C1040	490	586	12
Steel—hot rolled AISI C1095	455	827	10
Titanium	483	552	20
Titanium–9% Manganese	965	1034	13

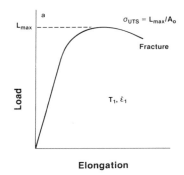

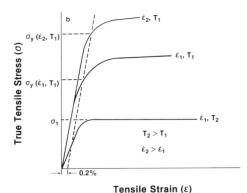

FIG. 7. (a) Typical load–elongation curve for a polycrystalline metal. (b) Results of the load–elongation data of (a) converted to true tensile stress and tensile strain. Also shown is the effect of increasing temperature from T_1 to T_2 and strain rate from $\dot{\varepsilon}_1$ to $\dot{\varepsilon}_2$. The schematic data are for a test performed at elevated temperatures.

between uniform plastic flow and neck formation, and (3) the sample eventually fractures. The ultimate tensile stress σ_{UTS} is the maximum load per original area and is a measure of the strength of the material.

The data of Fig. 7a are replotted to account for the changes in length and area that occur during testing and are shown in Fig. 7b as a plot of true tensile stress versus tensile strain.[1] The yield stress σ_y is generally defined as the intersection of the σ–ε curve with a line drawn parallel to the elastic line, but offset by 0.2% strain. Some typical values of σ_y, σ_{UTS}, and the elongation to fracture for metals are shown in Table II. The strength parameters have a wide range of values. They depend on such materials parameters as crystal structure, grain size, composition, and dislocation arrangement and on the test parameters of temperature and strain rate.

The elongation to fracture is the amount of plastic tensile strain that a material can withstand before fracture occurs. Usually, as the strength increases, the elongation-to-fracture decreases. This is readily seen for the case of the addition of 9% manganese to titanium (Table II), which increases σ_y and σ_{UTS} by 100% but at the

[1] True stress and strain are distinguished from engineering stress and strain as follows. The engineering stress is the load per unit original cross-sectional area, whereas the true stress is the load per instantaneous area. The engineering strain is $(l - l_0)/l_0$ and the true strain is $\ln(l/l_0)$, where l_0 is the original length and l is the instantaneous length.

expense of a 35% decrease in the elongation to fracture. A material that has a large elongation to fracture is called ductile, and one that exhibits little or no plasticity before fracture is called brittle. At room temperature, aluminum is very ductile, and polycrystalline ceramics are very brittle. The ductility of a material increases with increasing temperature. The fracture mode of a material refers to whether the fracture occurs by a ductile or by a brittle mechanism or, in general, a combination.

Dislocation motion is a statistical process. If the energy barriers that a dislocation must overcome in order to move are small enough to be aided by thermal energy, the stress required to cause deformation will be lower at finite temperatures than at absolute zero. Under such conditions, an increase in temperature or a decrease in strain rate $\dot{\varepsilon}$ will reduce the stress required to produce plastic flow. To a good approximation, however, for T less than $\sim\frac{1}{3}$ the melting point (in kelvin), deformation of an fcc metal is nearly independent of $\dot{\varepsilon}$.

Another type of stress–strain curve is shown schematically in Fig. 8. The material exhibits both an upper yield point σ_u and a lower yield stress σ_L. This behavior is an example of a discontinuous yield, which occurs in certain metals, notably low-carbon steels. In materials that exhibit such unstable yielding, plastic flow occurs in only part of the specimen. This Lüders band, in which the dislocation density has rapidly multiplied, then propagates throughout the crystal at $\sigma = \sigma_L$. Upper-yield points have also been observed in ceramic single crystals, such as Al_2O_3, CoO, MnO, NaCl, and LiF. The mechanism responsible for this behavior has been attributed to the interaction between impurity atmospheres and dislocations or to a rapid increase in dislocation density and the commensurate decrease in velocity of dislocations necessary to keep the strain rate constant according to Eq. (3). Another discontinuous yield phenomenon is the Portevin–Le Chatelier effect. Materials that deform in this mode undergo serrated yielding over a certain temperature and strain-rate range (see Fig. 9). The effect is caused by an impurity–dislocation interaction.

B. Strengthening Mechanisms

One of the important goals of material scientists is to increase the strength of materials without sacrificing their ductility or fabricability and without increasing their weight or cost. These goals have led to the development of hundreds of new metals, alloys, ceramics, and composites in the past several decades. Without these developments, our advanced technological society would not be possible. Several common methods employed to strengthen materials will now be discussed; several have been alluded to in previous sections.

Even in pure single crystals, dislocations encounter a finite resistance to motion. This force is known as the Peierls force or Peierls stress and is a result of the periodicity of the crystal structure: All positions of the dislocation are not equivalent. The energy differences in the various dislocation positions result in a periodic force that acts on the dislocation. In the absence of thermal activation, the Peierls stress is the minimum stress required to cause plastic deformation. The magnitude of the Peierls stress de-

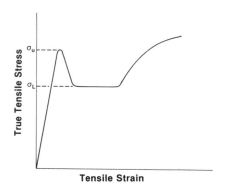

FIG. 8. Systematic stress–strain curve for a metal that exhibits an upper yield point.

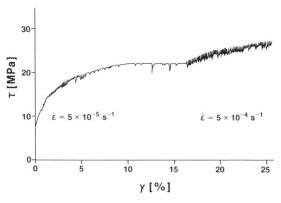

FIG. 9. Multiple yielding (Portevin–Le Chatelier effect) in MnO measured at 900°C. (Courtesy of K. C. Goretta, Argonne National Laboratory, Argonne, Illinois.)

pends on crystal structure, atomic bonding, slip plane, and Burgers vector, and is usually greater for ceramics than metals and is larger in bcc metals than in fcc metals. The Peierls stress is one of the prime factors that determine on what planes slip occurs. In general, a dislocation can readily overcome the Peierls barrier at elevated temperatures.

Strengthening methods rely either on the use of physical obstacles to dislocation motion or on the creation of internal stress fields. Both methods increase the applied stress necessary to cause plastic flow and hence strengthen the material. Alloying a material, which is called solution hardening, creates internal stress fields by the replacement of solvent atoms by foreign atoms or by the addition of atoms that occupy interstitial sites. In the first case, the internal stress is a result of the differences in atomic radii between the host and the substitutional atom. In the second case, the stress is the result of the distortions caused by the displacements of the atoms surrounding the interstitial atom. In both cases, the applied stress must overcome the internal stress and hence the flow stress is increased.

An example of solution hardening caused by a substitutional alloy is the case of brass, a zinc–copper alloy. Zinc forms a complete substitutional solid solution with copper up to ~30% by weight. The tensile stress of copper can be increased ~50% by the addition of zinc. Nitrogen, on the other hand, because of its small size, forms an interstitial alloy with iron as, does a small (<0.02%) amount of carbon in iron. In both cases, there is a significant strengthening effect.

The second major type of strengthening is called precipitation hardening. In any two-phase material where the second phase exists as a dispersion of fine particles in the major-phase matrix, dislocation motion will be hindered. The actual mechanism of hardening depends on the nature of the precipitate. If the second-phase particle is nonshearable and brittle, the dislocations cannot pass through the obstacles. However, they overcome the barrier by the Orowan process, whereby they encounter the obstacle and bow out. Eventually, a dislocation surrounding the particle and a free dislocation, which moves until additional particles are encountered, are created. If the particles are shearable but have a crystallographic structure incompatible with the matrix, a dislocation will be pinned until an interface dislocation is cre-

ated whose Burgers vector accounts for the difference between the Burgers vector in the matrix and the Burgers vector in the particle. In both these cases the precipitates act as dislocation barriers and increase the flow stress.

Particles can also have associated stress fields or, in the case of a modulus difference, act to distort the applied stress. Generally it is found that the flow stress increases with the square root of the concentration of the solute or precipitate, in accordance with theory. It should be mentioned that irradiation with neutrons or other charged particles can harden a metal but cause loss of ductility.

Polycrystalline material can be hardened by decreasing the grain size d in accord with the Hall–Petch relation,

$$\sigma_y = \sigma_0 + kd^{-1/2} \qquad (4)$$

where σ_0 and k are materials constants. Grain boundaries, which are regions of large atomic misfit, act as dislocation barriers. The piling up of dislocations at a grain boundary produces a stress that can nucleate dislocations in neighboring grains. This mechanism is used to explain the Hall–Petch relation.

A grain cannot deform independently; it must accommodate the shape changes of its neighbors or else fracture will result. If, for example, a two-phase material consists of large grains of a deformable phase and smaller grains of a nondeformable phase, the material can have a very high strength. Indeed, many steels have been designed with a microstructure to exploit this strengthening effect. A construction-grade AISI C1020 (0.2% carbon) steel consists of two constituents: a bcc phase (ferrite), which is ductile, and a lamellar structure, Fe + Fe_3C (called pearlite), which because of the cementite (Fe_3C) is brittle.

C. Work Hardening

The concept of work hardening has been introduced in conjunction with Fig. 5. After initial yielding (see Figs. 5–7), the stress required for continued flow increases. That is, more stress is needed to produce an additional increment of strain. It is easy to bend a small metal rod once, but each successive bend requires more force until the metal rod fractures. Work hardening dictates that forming operations that produce large shape changes be conducted at elevated temperatures so that the work-hardened dislocation structures can relax.

Work hardening is the result of dislocation interactions. As deformation proceeds, the dislocation density increases, and the moving dislocations can interact, so that their distribution changes as well. Dislocations can multiply by a variety of mechanisms. Therefore, the dislocation structure continually changes. Dislocation–dislocation interactions fall into three general types: junction reactions, jog formation, and stress–field interactions.

Plastic deformation of a polycrystal requires slip on multiple systems. Dislocations on one slip system therefore interact with dislocations on another. When this occurs there is an interaction that may be attractive or repulsive, depending on whether the dislocations combine over the intersection length. The combination will occur whenever the energy of the combined dislocation is less than the sum of the energy of the separate dislocations. Nevertheless, in both cases, the junction reaction produces a retarding force on the moving dislocations, causing work hardening.

Jog formation occurs when dislocations intersect. If the cutting dislocation is a screw, the resultant jog will not be able to glide. In this case, motion can occur only by climb, and the energy required to produce vacancies or interstitials to allow climb is very high. The presence of a relatively immobile jog will produce a drag on the dislocation and will cause work hardening.

A common cause of work hardening is stress–field interactions. Dislocations of opposite sign attract and, if during deformation they approach one another, the applied stress may not be sufficiently large to pull them apart. This is called a dislocation dipole, and since the forces produced by the applied stress on the two dislocations of the dipole balance, the applied stress produces no net force to move the dipole. The dipole will trap dislocations as they move in its vicinity, and therefore the dislocations in the group tend to become tangled and the stability and trapping efficiency of the dipole increase. Eventually, for large strains, these groups form small cells, which are very effective barriers to dislocation motion.

Work hardening is very sensitive to temperature. As the temperature is increased, thermal activation aids the applied stress in overcoming the hardening caused by dislocation–dislocation interactions. At sufficiently high temperature, the dislocation structure may recover simultaneously with deformation, and hence a steady-state is reached at which hardening and recovery balance and the work-hardening rate becomes zero.

IV. Creep Deformation

A. GENERAL

For $T \gtrsim T_M/3$, where the temperature is in absolute degrees, the deformation of a metal or ceramic becomes time dependent. In particular, the flow stress becomes more sensitive to strain rate for elevated temperatures. Creep is the time-dependent deformation of a solid under a constant stress. Creep and the fracture that results from the accumulated creep deformation are important factors in designs of all metal or ceramic parts that operate under a nonhydrostatic stress at elevated temperatures. Creep will become increasingly important as operating temperatures of machinery are raised in order to improve efficiency and conserve energy.

In the laboratory, creep deformation is usually measured with an apparatus that allows the sample to be subjected to a constant stress at a fixed temperature. The resulting elongation is measured as a function of time. A typical apparatus is shown in Fig. 10. The continually decreasing cross-sectional area of the sample is compensated for by the cam arrangement, which decreases the load at the same rate, thus keeping the stress constant until sample necking occurs. The strain can be measured in a number of ways, most commonly by the use of an extensiometer. In practice, many creep tests are performed under a constant load. A schematic of a test performed at elevated temperature at a constant stress on a metal or a ceramic is shown in Fig. 11.

After the initial elastic deformation, the creep curve usually consists of three distinct regions: a primary, a secondary, and a tertiary regime. The creep rate is defined as the instantaneous strain rate or the slope of the curve divided by the original length (engineering) or the actual length (true) at any given time. Primary (also called transient) creep is characterized by a decreasing creep rate. Secondary creep begins where the primary region ends and is characterized by a constant creep rate. This is also referred to as the steady-state creep region. Finally, the creep rate begins to increase with increasing time (and strain) until failure occurs. The extent of each region depends primarily on the temperature, stress, and previous deformation history. Some materials exhibit a different

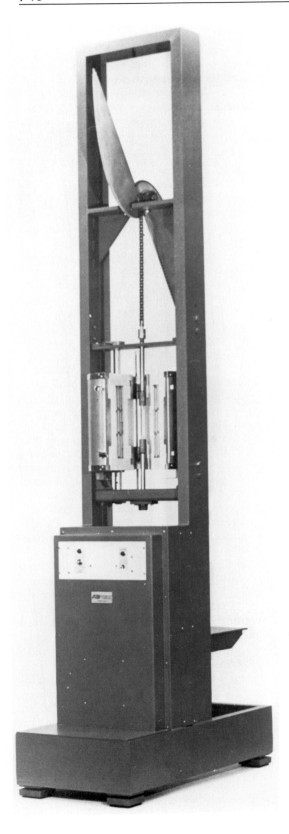

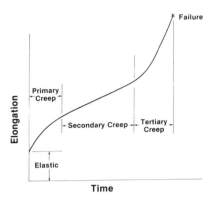

FIG. 11. Schematic representation of an elongation versus time curve, showing the typical creep of a metal or ceramic at elevated temperature.

elongation versus time behavior called sigmoidal creep, which does not have a steady state: that is, there is a direct transition from a decreasing creep rate with strain to an increasing creep rate.

B. Creep Mechanisms

The transient part of the creep curve results from the establishment of dislocation substructure. The creep strain at elevated temperatures is frequently related to the time of loading t by

$$\varepsilon = \beta t^{1/3} \tag{5}$$

where β is a function of stress, temperature, and history.

Transient creep can be qualitatively understood in terms of dislocation theory. Initially, the dislocation structure has been set by specimen history. At high temperatures, many of the dislocations aided by thermal activation can move in response to the applied stress. Hence, the dislocation structure readjusts itself to a configuration of lower energy over a period of time. The first dislocations to move are those that have the shortest distance to move. These dislocations have the highest velocity and thus produce the highest creep rate. Since the process of rearrangement is essentially statistical in nature, the strain rate will continue to decrease until a steady-state structure is established, at which point primary creep becomes steady-state

FIG. 10. Constant-stress creep testing machine. (Courtesy of Applied Test Systems, Butler, Pennsylvania.)

creep. Although in general primary creep accounts for only a small fraction of the total strain, it cannot be neglected, especially under conditions in which the temperature or stress may be changing. In this case, every change of conditions produces another transient, and the accumulated strain may be very large. This is illustrated by a nuclear fuel element in the core of a reactor that is periodically shut down or whose power level is frequently adjusted.

During steady-state creep, the dislocation structure is constant. This means that the work-hardening rate, which is governed by such processes as dislocation–dislocation interactions, dislocation–obstacle interactions, and the increase in dislocation density due to dislocation generation, is balanced by recovery processes such as annihilation of dislocations, overcoming of obstacles, and annealing of dislocation structures. Most of these processes require diffusional transport and hence occur at high temperatures only. Under these conditions a stress–strain curve of the type shown in Fig. 7b for the case of $\dot{\varepsilon} = \dot{\varepsilon}_1$ and $T = T_2$ (lower curve) is observed: For a sample subjected to a constant stress σ_1, the steady-state strain rate is $\dot{\varepsilon}_1$. That is, at steady state, a creep curve and a stress–strain curve yield equivalent information.

A convenient representation of steady-state creep is a deformation mechanism map. These are available for a wide variety of materials (see "Deformation Mechanism Maps—The Plastic-

ity of Metals and Ceramics," by H. J. Frost and M. F. Ashby, Pergamon Press, 1982). A deformation map for pure polycrystalline nickel of 1-mm grain size is shown in Fig. 12. The ordinate is shown as the shear stress normalized by the temperature-dependent shear modulus, a dimensionless quantity. The abscissa is the homologous temperature (also dimensionless), which is the temperature divided by the melting temperature, both in absolute degrees. The light lines are lines of constant strain rate, and the dark lines delineate the regions of operation of the various deformation mechanisms: plasticity, power-law creep (high and low temperature), and diffusional creep (lattice and grain-boundary diffusion). The transitions between the various mechanisms are not clearly defined, and near the boundaries more than one mechanism operates. Such maps are usually constructed from theoretical models through the use of experimental parameters and are verified (over a limited range) by creep tests. If, for example, a nickel part is to operate at $T = 0.6T_M$ and $\sigma_S/G = 10^{-3}$, the steady-state creep rate would be $\sim 10^{-2} \ \mathrm{sec}^{-1}$, that is, 1% sec^{-1}, a very high rate. If the stress were reduced by a factor of 10 to $\sigma_S/G = 10^{-4}$, the resulting strain rate would be $\sim 10^5$ lower, or $\sim 10^{-7} \ \mathrm{sec}^{-1}$; that is, to reach 1% strain would take 28 hr. Notice that in both cases the sample deforms by power-law creep, although the mechanisms are different. For constant T, the sample is in the low-temperature region for

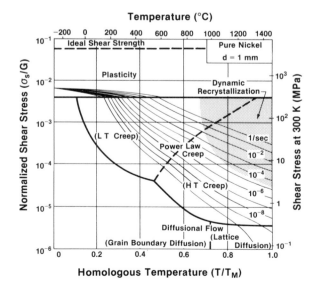

FIG. 12. Deformation map drawn for pure nickel with a grain size of 1 mm. (Courtesy of M. F. Ashby, University of Cambridge, Cambridge, England.)

the first case, but it is in the high-temperature region in the second. If the temperature were changed to $T/T_M = 0.4$, the creep rate would be reduced to $\dot{\varepsilon} \sim 10^{-6}$ sec^{-1}.

The high-stress region of the deformation map, labeled plasticity, has been previously discussed under yield and stress–strain behavior since it does not represent creep. In this region, plastic flow is controlled by the conservative motion of dislocations (glide) and is limited by the interaction of the mobile dislocations with other dislocations, precipitates, solutes, and grain boundaries, or by the Peierls stress or even by interactions with the lattice vibrations.

At higher temperatures ($T > 0.3T_M$), the strain is caused by glide, but the process is limited by the climb velocity. This climb-controlled glide region requires mass transport, which can readily occur at elevated temperatures. This mechanism of creep is called power-law creep. The steady-state creep rate is given by

$$\dot{\varepsilon} = A_1 D_{\text{eff}}(Gb/kT)(\sigma/G)^n \qquad (6)$$

where A_1 is a material constant, G the shear modulus, k Boltzmann's constant, and n the stress exponent. The value of the stress exponent depends on how the dislocation density ρ depends on stress (normally, $\rho \propto \sigma^2$) and on the details of the climb process. Theoretical values of n range from 3 to 6 for single crystals. D_{eff} is the effective diffusion coefficient. At high temperatures ($T > 0.6T_M$), diffusion through the lattice is fast enough to allow for climb, and $D_{\text{eff}} \approx D_v$, where D_v is the volume diffusion coefficient. This process is high-temperature (HT) power-law creep. At lower temperatures, bulk diffusion is too slow, but mass transport can take place along dislocation cores. Then $D_{\text{eff}} \approx D_c$, where D_c is the core-diffusion coefficient. This represents the low-temperature (LT) power-law creep region. Since a diffusion process is thermally activated, the steady-state creep rate can be written

$$\dot{\varepsilon} = A_2(Gb/kT)(\sigma/G)^n \exp(-Q_c/kT) \qquad (7)$$

where A_2 is a materials constant and Q_c the activation energy for creep, which is related to bulk or core diffusion. For usual values of n and Q_c, small changes in T or σ result in large changes in $\dot{\varepsilon}$.

At very high temperatures and low stresses, dislocation processes are slow. However, diffusion becomes so rapid that direct mass transport by diffusion becomes the rate-controlling process. In this diffusional flow region, atoms flow from grains under compression to grains under tension, to cause a net elongation of the sample in the tensile direction. When diffusion occurs through the grains by lattice diffusion, the process is called Nabarro–Herring creep. The process is called Coble creep if mass transport is along the grains. In the diffusional flow region, the steady-state creep rate is given by

$$\dot{\varepsilon} = \frac{13.3\sigma\Omega}{kTd^2} D_v \left(1 + \frac{\alpha\pi\delta}{d}\frac{D_B}{D_v}\right) \qquad (8)$$

where Ω is the atomic volume, D_B the grain-boundary diffusivity, α a constant, and δ the width of the grain boundary. In this creep region, the steady-state strain rate is proportional to stress (a phenomenon called viscous flow) and depends on grain size and temperature. Notice that fine-grained materials creep much more quickly than large-grained materials in the diffusional flow region. For example, if the grain size of the nickel shown in Fig. 12 were decreased to 0.1 mm, the creep rate would increase by 100 to 1000 depending on temperature.

Creep deformation can also occur by other mechanisms that are important but that are not addressed by the deformation maps—for example, twinning, dynamic recrystallization, and grain-boundary sliding. These processes, which usually do not contribute much to the creep strain, are discussed in the more specialized references in the Bibliography.

Many of the metallurgical techniques for improving strength can be used to increase creep resistance. For example, the addition of precipitates or other barriers to dislocation motion will decrease $\dot{\varepsilon}$ for a given T and σ. In addition, decreasing the diffusion coefficients will reduce $\dot{\varepsilon}$. This may be done, for example, by changing the anion-to-cation ratio (stoichiometry) in a binary metal oxide. Grain size and microstructure are also important parameters in the tailoring of creep-resistant materials.

The basic ideas presented above have been verified experimentally in a wide variety of metals and ceramics. Ceramics, however, behave differently from metals in one important respect. Diffusion in metals occurs by the motion of metal vacancies or interstitials. Ceramics contain both anions and cations, and diffusion of both species must occur to conserve charge. Therefore, diffusional creep or power-law creep will be rate-limited by the slowest-moving species, and the activation energy may reflect minority defect diffusivities.

Ceramics also have, in general, much smaller grain sizes than metals and usually have considerably higher activation energies for diffusion. As-fabricated structural ceramics contain porosity and can simultaneously creep and densify. Because ceramics are made from naturally occurring minerals, they are usually less pure than metals. The impurities frequently form deleterious phases on the grain boundaries, which change creep properties. Ceramics usually have a high resistance to plastic deformation and creep up to very high temperatures.

C. CREEP RUPTURE

Fracture in creep is usually termed rupture. This occurs at the end of the tertiary creep region and is preceded by an increasing strain rate at a constant temperature and stress. Although many designs can accommodate small plastic strains, none can tolerate fracture. It is unlikely that tertiary creep is due solely to specimen necking since many materials, ceramics in particular, fail in creep at strains too small to produce necking.

For most materials the beginning of tertiary creep coincides with the formation of internal flaws or pores. "Creep cavitation" can occur in metals and ceramics. The exact mechanisms are not certain, but in general, sliding grains develop high tensile stresses at the boundaries. These stresses can nucleate cracks and pores, which because of their small initial size have no effect on the earlier stages of creep. The pores can grow by volume or boundary diffusion of vacancies and can link up to form interconnected porosity along a boundary, which leads to final fracture.

For design purposes, the data are usually plotted as applied stress versus the logarithm of the time to fracture. Such a plot yields a reasonably straight line. The time to failure decreases with both increasing stress and temperature. Fracture is a complex process, but in general there are a number of material modifications that can be made to increase the time to failure. First among these would be to prolong the steady-state creep region by increasing the creep resistance through the use of metallurgical techniques discussed in the previous section.

V. Fatigue

Fatigue, or fracture under the repeated cyclic application of stress or strain, also involves plasticity. It may occur after only a few cycles of large plastic strain amplitude, in which case it is called low-cycle fatigue. On the other hand, parts may fail after thousands of cycles even if the strain amplitude is always below the elastic limit. This high-cycle fatigue was discovered in the mid–nineteenth century in Europe when failures of railway axles became widespread. Early research on fatigue concentrated on generating data in order to predict empirical fatigue lifetimes. More recently, however, research has concentrated on the microscopic mechanisms. It is recognized that fatigue consists of two stages: crack initiation and crack propagation. The crack initiation process is governed by dislocation movement and therefore is a plastic-deformation process. Crack propagation is a fracture-mechanics problem and as such falls outside the scope of this article. However, brief mention of this topic was made at the end of the creep-rupture section.

A close-up view of a typical apparatus used to study fatigue is shown in Fig. 13. The ram of this machine is operated by a servo-motor-controlled hydraulic system with a very fast and accurate response. The machine is capable of cycling in either stress or strain control and can be programmed to perform complex loading cycles. Tests are frequently performed at a fully reversed stress or strain amplitude, where the levels of stress or strain are equal during the tension and compression halves of the cycle. The load and extension are measured and controlled by a load cell and extensometer, respectively. Schematic results for two typical metals

FIG. 13.   Close-up view of a specimen (within the coil) undergoing fatigue at elevated temperature in a typical servo-hydraulic apparatus. The coil is attached to a radio-frequency generator, and the sample is heated. (Courtesy of MTS, Minneapolis, Minnesota.)

are shown in Fig. 14. Such plots are called *S–N* curves. In some metals such as steels, the fatigue process seems to stop at amplitudes about equal to half of the tensile strength. Therefore, they exhibit a fatigue or endurance limit. Failure will not occur for stresses below this limit. Other metals, such as aluminum, have no such limit. When no fatigue limit exists, the fatigue strength is sometimes defined as the stress that gives a lifetime of 10^8 cycles. Because of the experimental difficulties involved in testing brittle solids in tension, few fatigue data for ceramics exist. What follows, then, is based on experiences with the fatigue of metals.

Under conditions of constant-strain-amplitude low-cycle fatigue, irreversible plastic deformation will occur in each cycle. Initially, the sample will work-harden rapidly as a dislocation network is established. In a fatigue test, the increasing work hardening is reflected in that each cycle requires more stress to produce the same level of plastic strain. As damage accumulates, the work-hardening rate decreases to zero, implying that a pseudo-steady-state has been achieved. Indeed, transmission electron microscopy reveals that dislocation cells have been formed. However, as additional strain is imposed, cracks that will eventually cause failure are initiated. It is believed that the initiation of cracks results from the stress concentrations caused by grain boundaries, second-phase particles, or surface flaws. The cracks are formed locally, although most of the structure has reached a steady state.

Under high-cycle fatigue conditions, the stress amplitude is usually held constant at a level less than the yield stress so that there should be no plastic strain and therefore no fatigue. However, it has been stated that yield phenomena occur over a range of stresses (see Fig. 7b). Microstrain will occur for applied stresses below the nominal yield stress, and although the microstrain may be small compared with the elastic strains, a large number of cycles will produce measurable damage and can result in fatigue failure. The microstrain is not homogeneous and is concentrated in a few relatively isolated slip bands that are formed during the first few thousand cycles. These are called persistent slip bands, and their plastic strain amplitude is quite large compared with the strain amplitude averaged over the entire sample. The damage mechanism seems to be related to the formation of intrusions and extrusions within the band. The bands are relatively free of dislocations, but are divided up by dislocation walls that cross the band thickness.

Once formed, the cracks can propagate to cause ultimate failure, much as in the manner described in the creep-rupture section. The propagation of cracks is a fracture mechanism and as such is outside the scope of this article. It should be mentioned, however, that plastic flow in metals and ceramics occurs at the tip of a crack because of the high stress concentrations.

BIBLIOGRAPHY

Cottrell, A. H. (1953). "Dislocations and Plastic Flow in Crystals." Oxford University Press, New York). The classical text on this subject.

Frost, H. J., and Ashby, M. F. (1982). "Deformation-Mechanism Maps: The Plasticity and Creep of Metals and Ceramics." Pergamon Press, New York. Contains a large number of deformation maps and an excellent summary of how they are produced.

Hirth, J. P., and Lothe, J. (1968). "Theory of Dislocations." McGraw-Hill, New York. A complete mathematical theory of dislocations for material scientists.

Hull, D., and Bacon, D. J. (1984). "Introduction to Dislocations," 3rd ed. Pergamon Press, Oxford. A thorough introduction to dislocations.

McClintock, F. A., and Argon, A. S. (1966). "Mechanical Behavior of Materials." Addison-Wesley, Reading, Mass. A comprehensive text written for engineers.

Rogers, H. C. (1977). In "Encyclopedia of Science and Technology," Vol. 10. McGraw-Hill, New York, pp. 463–471.

Suh, N. P., and Turner, A. P. L. (1975). "Elements of the Mechanical Behavior of Solids." Scripta, Washington, D.C. An excellent introduction written for engineers.

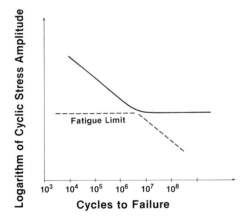

FIG. 14. Schematic *S–N* curves for a metal that shows a fatigue limit and one that does not.

PLASTICIZERS

N. W. Touchette and J. Kern Sears   *Monsanto Company*

GLOSSARY

Compatibility: Ability of two or more substances to mix together without objectionable separation.

Dry blend: Dry free-flowing mixture of resin with plasticizers and other additives, prepared by blending the components at temperatures below the fusion temperature.

Fusion: Condition attained in the course of heating a mixture of resin and plasticizer when all the resin particles have dissolved in the plasticizer, so that upon cooling a homogeneous solid results.

Gel: Condition attained in the course of heating a mixture of resin and plasticizer when the plasticizer has been absorbed by the resin to an extent resulting in a dry but weak and crumbly mass.

Low-temperature flexibility: Denotes a temperature at which the composition attains a specified apparent modulus of elasticity as measured by a torsional test.

Plastisol: Suspension of a finely divided polyvinyl chloride in a liquid plasticizer that has little or no tendency to dissolve the resin at normal temperatures but becomes a solvent for the resin when heated.

Polyvinyl chloride: Resin made by polymerization of vinyl chloride. The pure polymer is hard, brittle, and difficult to process but can be made flexible by addition of plasticizers.

Solvation: Process of swelling, gelling, or fusion of a resin by a plasticizer as a result of mutual attraction.

Plasticizers are modifiers used with polymeric substances to alter processing or end-property characteristics by improving workability, flexibility, or extensibility of the system. Plasticizers can be either external or internal in classification. External plasticization based on organic esters is the more widely used of the two and is the subject of this article. Examples of plasticization are evident from earliest times where clay was modified with water to make pottery to the present use of water in gelatin for foodstuff applications. Water in each of these cases acts as a temporary or fugitive plasticizer.

I. Historical

The beginning of modern plasticizer usage occurred in 1856 when the Hyatt brothers used camphor to plasticize nitrocellulose. Camphor proved to be fugitive due to volatility, leaving finished articles stiff and brittle after short time periods. The need to overcome the aging problem of camphor plus requirements of other synthetic resins have led to development of synthetic organic ester plasticizers with a variety of performance and permanence properties.

Of the many resins, polyvinyl chloride (PVC) is the main polymer utilizing plasticizers. The response of PVC to plasticization allows development of a continuum from rigid to flexible systems with highly desirable properties. The patent in 1933 on plasticization of PVC by Semon of B. F. Goodrich was coincident with the patent by Kyrides of Monsanto on di-2-ethylhexyl phthalate (DOP) as a plasticizer for several resins. DOP was the dominant PVC plasticizer for years.

To be effective, plasticizer and resin must be mixed into a coherent, homogeneous material that shows no phase separation on aging. Blending of plasticizer and resin can be accomplished by heat, with or without simultaneous mixing or pressure, or by dissolution in a common solvent followed by removal of the solvent.

II. Classification of Plasticizers

Plasticization of a resin or plastic can be accomplished by external or internal means. The examples given above are external, which refers to the addition to a resin of a plasticizer that is usually a liquid ranging in molecular weight from under 300 to 5000. External plasticizers maintain their molecular identity and are bonded to the resin by secondary valence forces such as hydrogen bonding and van der Waals forces. Variation in concentration or type of plasticizer with a given resin has a direct influence on the final properties of the system. This gives the formulator the ability to vary the final composition over a wide range based on a limited number of raw materials.

Internal plasticization is accomplished by addition of a comonomer during the polymerization step in resin manufacture or by grafting a side chain into the resin backbone. The resulting polymer is less ordered and has fewer points of polymer–polymer interaction. In general both internal and external plasticization produce the desired end result, but since each change in plasticization level in internal plasticization requires a new base polymer, growth of internal plasticization has not occurred to the same degree as external plasticization.

External plasticizers are classified in several ways: chemical family, primary/secondary, and by end-use property. The classification may change for a product depending upon the resin system under consideration.

Based on chemical structure, plasticizers can be organic esters, hydrocarbons, and resinous compounds. Organic esters may be further subdivided into monomeric and polymeric where monomeric denotes either mono-, di-, or tri-functional esters with molecular weight generally less than 1000. Polymeric on the other hand characterizes polyesters based on dibasic acids and glycols with molecular weights ranging from about 800 to 5000. Hydrocarbons are generally low polarity oils or waxes that are used at low concentrations as extenders or in elastomers as processing oils. Chlorinated paraffin is an example of a hydrocarbon type that is used as an extender in PVC compositions. Resinous compounds are typically solid polymeric substances such as ethylene vinylacetate copolymer, ethyl-

enevinyl acetate terpolymer, nitrile rubber, or chlorinated polyethylene.

Plasticizers can also be ranked on their compatibility with a resin. Those that exhibit a high degree of compatibility are classed as primary while those of lower compatibility, which when used alone lead to exudation of droplets or liquid surface film upon aging, are secondary or extender plasticizers. Primary plasticizers generally have sufficient compatibility to be used as the sole plasticizer in a system, while a secondary is used along with a primary. The classification may change dependent upon the resin system, so a plasticizer that is primary in one resin may be a secondary in another system.

Plasticizers may also be distinguished by the desirable properties imparted to the compounded resin system. Thus, some plasticizers are designated as low temperature plasticizers, meaning they produce better than normal flexibility at subambient temperatures. Other end-use classifications include fast fusion, general purpose, flame retardant, oil resistant, mar resistant, nontoxic, and stain resistant.

III. Theory of Plasticization

Four general theories are used to explain the effects plasticizers have on resins. The lubricity theory proposes that the rigidity or the resistance of a resin to deformation results from intermolecular friction. The plasticizer acts to facilitate movement of resin macromolecules over one another. It gives internal lubricity through very weak bonding between plasticizer and resin molecules. [See RUBBERLIKE ELASTICITY.]

The gel theory was developed for amorphous polymers and proposes an internal three-dimensional honeycomb or gel structure that gives a polymer its resistance to deformation. This gel is formed by loose attachments occurring at intervals along the polymer chains. The dimensions of the cells determine the movement of the polymer chains. The plasticizer breaks these points of attachments and masks the centers of force by solvation of the polymer chain at these points. The resulting system appears to have fewer points of attachment between polymer chains.

The mechanistic theory assumes that a dynamic equilibrium exists between solvation and desolvation of the polymer chains by plasticizer. Plasticizers of different classes are attracted to the polymer by forces of different magnitudes, but this attraction or bonding is not permanent.

There is a continuous exchange wherein a plasticizer molecule becomes attached to a given active group on the chain only to be dislodged and replaced by another.

The free-volume theory can be used to explain both internal and external plasticization. This theory assumes that each polymer molecule has free space associated with it. This free space or free volume increases with increased molecular motion. Small molecules have a high proportion of free volume compared to high-molecular polymers. Introduction of a plasticizer molecule into a polymer increases the fractional free volume in the blend compared to the polymer alone. This greater free volume permits easier internal motion of the polymer chains. The fact that free volume increases with molecular motion is useful in explaining the internal plasticization achieved by side-chain addition where each side chain acts as a small molecule, and free volume of the system is increased. The free volume is fixed in a particular location in the polymer molecule. External plasticizers added to a polymer affect a larger free volume at any location for the amount added. The size and shape of the molecule and the nature of its atoms and grouping of those atoms (i.e., polarity, H-bonding ability, and density) determine how it functions as a plasticizer.

IV. Process of Plasticization

Four distinct steps have been proposed to explain the mechanism of addition of a small plasticizer molecule to a much larger polymer molecule. The four steps of plasticization might not occur for every plasticizer–polymer combination. Furthermore, they may appear at times to overlap.

Initially there is rapid, irreversible plasticizer movement into the porous areas of the resin. Then comes an absorption step during which the total volume of resin–plasticizer may decrease although the resin particles themselves swell slowly on the outside, setting up strong internal strains. Rate studies during early particle swelling indicate low to moderate activation energies (5–50 kcal) and half-lives directly proportional to vapor pressure of the plasticizer. This indicates a simple diffusion process.

There follows the third step with higher activation energies (68–111 kcal), during which more severe changes take place inside the particle, with little or no overall volume change. This appears to be a diffusion process with half-lives

much longer than the first but again proportional to the vapor pressure of the plasticizer. From the order of magnitude of activation energies, it appears that polymer molecules are being disentangled and separated. In the upper end of the activation energy range the energies involved are sufficient to break primary carbon–carbon bonds. In fact, bond rupture has been observed in some high-molecular-weight polymers on swelling, even without mechanical mixing.

At the start of this third step, most or all of the plasticizer has been incorporated in the resin, and while slight volume changes have taken place, plasticization is not complete. Plasticizer is present as clusters of molecules in between clusters or bundles of polymer segments (e.g., crystallites). As energy is applied to the system during this third step, a marked change occurs. The volume remains the same, but the dielectric constant rises significantly, showing that the polymer molecules are no longer rigidly held together but their dipoles are free to move; the polymer is now in its rubbery state rather than its glassy state. The plasticizer molecules have now penetrated the clusters or bundles of polymer molecules, and solvation is essentially complete. The energy barrier that must be passed to permit dipole orientation is reported to be higher than the energy barrier for movement of polymer segments to accommodate the diffusing plasticizer.

The fourth step occurs when the plasticization process has been obtained by heating the mixture until fluid. Some toughening structure must be formed upon cooling. This may result from van der Waal's forces or hydrogen bonding of polymer segments either directly or through the plasticizer, or from entanglement of polymer chains or from crystallization of the polymer segments.

These four steps of plasticization may not be identified for each polymer/plasticizer combination. But in some systems the steps are very distinct, as demonstrated in the combination of PVC with an ester type plasticizer. Plasticization can proceed at room temperature through the first two steps but is prohibited thermodynamically from proceeding further until a threshold temperature is reached. Although production of plasticized PVC with good physical properties requires these four steps, good use is made of intermediate steps during processing. These can occur in dry blending or plastisol operations. [*See* POLYMER PROCESSING.]

The key parameter of a plasticizer/resin system for useful performance is compatibility. Compatibility for our purposes is defined as the ability of two or more substances to mix together to form a homogeneous composition with useful properties. The first criteria for compatibility is miscibility of the system. According to thermodynamic theory, two substances are miscible in an ideal system when the free energy of mixing is negative as determined by

$$\Delta G = \Delta H - T \Delta S \qquad (1)$$

where ΔG is the free energy of mixing, ΔH the heat of mixing, ΔS the entropy of mixing, and T the temperature. In a polymer solution due to the high molecular weight of the polymer, the deviations from an ideal molecular weight are significant. This has led to the development of the Hildebrand solubility parameter δ and the Flory–Huggins interaction parameter χ. The Hildebrand theory was designed for solutions in which solvent and solute molecules are approximately the same size. This results in ideal entropy, and solubility is controlled by ΔH. The solubility parameter affords one means to approximate ΔH. The Flory–Huggins interaction parameter χ is a means of correcting for nonideality of both heat and entropy of mixing. The exact value of χ depends upon the concentration at which it is determined.

V. Raw Materials and Chemical Reactions

Plasticizer raw materials are either petrochemical based or naturally occurring chemicals with the greatest part being petrochemical-derived. The principal technology in plasticizer manufacture is esterification. However, most manufacturers are basic in alcohol, acid, or both.

Alcohols used in plasticizer manufacture were initially available only from natural sources. Due to the high demand and uncertainty of supply, synthetic routes to these alcohols from olefins were developed based on oxo, aldol condensation, Ziegler, or dimerization technology. [See ORGANIC CHEMICALS, INDUSTRIAL PRODUCTION.]

1. The oxo process is the commercial application of a chemical reaction called oxonation or, more accurately, hydroformylation. An olefin is reacted with hydrogen and carbon monoxide to produce an aldehyde containing one more carbon atom than the olefin

$$2\ RCH{=}CH_2 + 2\ CO + 2\ H_2 \rightarrow$$
$$RCH_2CH_2CHO + RCH(CHO)CH_3 \qquad (2)$$

The mixture of aldehydes is hydrogenated to a mixture of the corresponding alcohols.

2. Aldol condensation of *n*-butyraldehyde is the basic reaction used to produce 2-ethylhexyl alcohol. The *n*-butyraldehyde starting material is obtained from propylene by the oxo process.

3. The Ziegler process is used to produce straight-chain, primary, even-numbered carbon alcohols from ethylene. Triethyl-aluminum is produced from aluminum, hydrogen, and ethylene. Further reaction with ethylene lengthens the alkyl chains in two-carbon increments. Oxidation and hydrolysis yields the primary alcohols.

4. The dimerization process reacts *n*-butylenes to form octene, which is subsequently converted into nonyl alcohol via an oxo process.

The principal acid moiety used in the esterification process for plasticizers is phthalic anhydride. This is petrochemical-based and obtained by oxidation of *o*-xylene or naphthalene. Most processes are based on *o*-xylene.

The derivation of acids and alcohols for phthalate plasticizers from petroleum-refining operations is shown in Fig. 1. This flow chart shows the overall relationship of phthalate plasticizer raw materials rather than a comprehensive step-by-step reaction sequence and does not show the wide range of other chemicals available.

Other dibasic acids used in plasticizer production include adipic, glutaric, and azelaic acids in the aliphatic series. Trimellitic anhydride represents a trifunctional acid molecule.

Phosphate plasticizers are reaction products of phosphorus oxychloride with the appropriate alcohols, phenol, or alkylated phenols.

Other reactions involve epoxization of naturally occurring oils or fatty acid derivatives to introduce the oxirane oxygen

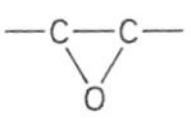

$$-C-C-$$
$$\backslash\!O\!/$$

into the molecule.

For most plasticizers, manufacture involves reaction of an acid anhydride or acid with an alcohol or mixed alcohols:

$$RCOOH + R'OH \rightarrow RCOOR' + H_2O \qquad (3)$$
$$R{-}(COOH)_2 + 2\ R'OH \rightarrow$$
$$R{-}(COOR')_2 + 2\ H_2O \qquad (4)$$

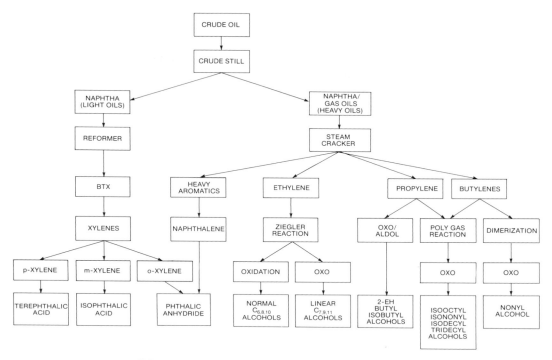

FIG. 1.   Flow chart of phthalate plasticizer raw materials.

Specifically for phthalates, the equation

$$\text{phthalic anhydride} + 2ROH \longrightarrow \text{diester} + H_2O \quad (5)$$

is in reality a two-step process with formation of a half-ester when the ring is opened with one mole of alcohol and subsequent reaction with another mole of alcohol to form the diester. If mixed alcohols are used, the resulting product is a statistical distribution of symmetrical and unsymmetrical esters. This combination results in products with unique properties.

VI. PVC Plasticization

PVC, by nature of its toughening supermolecular structure, responds better to plasticization than other amorphous polymers such as polystyrene, poly(vinyl butyral), poly(vinyl acetate), and acrylics. Crystalline polymers such as nylon and poly(vinylidene fluoride) benefit from small additions of plasticizers to aid processing or to affect the small amorphous areas so as to improve flexibility, toughness, adhesion, and other properties.

As a consequence of the good response of PVC to plasticization, usage in PVC represents the dominant end use for plasticizers. This response of PVC is unique in its acceptance of large amounts of plasticizers with gradual change in physical properties from a rigid solid to a soft gel or viscous liquid. Plasticization allows production of various materials, from rigid siding to flexible wire insulation or from clear blister packages to clear flexible food wrap, all based on PVC resin. The level and type of plasticizer has a profound influence on the final performance of the plasticized system.

Plasticizers are incorporated into PVC in three major ways: hot compounding, plastisol, and solvent methods. In the first, the resin/plasticizer mix progresses quickly to a final solvated state by rapid input of heat energy and vigorous mixing. The resin/plasticizer mix in an incompletely solvated state called dry blend, in which the plasticizer that is adsorbed on the surface of the resin particle can be isolated to allow versatility in processing. In the plastisol technique the resin/plasticizer mix is converted to the fully solvated state by heat; again, the isolation of an incompletely solvated state called gel allows versatility in processing. In the solvent method, dissolution of both resin and plasticizer in a common solvent permits solvation of the resin without heating. Processing of the solution and subsequent removal of the solvent yields the completely plasticized system.

In either hot compounding or plastisol techniques, a processing temperature for the PVC/plasticizer mix of 150–170°C must be achieved to obtain good solvation, as evidenced by fully developed mechanical properties. The process of solvation of resin by plasticizer occurs to a limited degree even at room temperature for some combinations of resin and plasticizer. Evidence of low-temperature solvation is seen best in plastisols by an increase in viscosity and ultimate formation of a hard gel. Incompletely solvated systems exhibit low elongation and tensile strength though having a good overall appearance. Proper processing temperatures to achieve maximum performance is a necessity in PVC technology.

A. PERFORMANCE CHARACTERISTICS

1. Compatibility

Beyond the theoretical consideration of mutual compatibility of plasticizers and resins there is the practical aspect of compatibility, where two or more substances (i.e., plasticizer and resin) must mix with each other to form a homogeneous composition with useful properties. The term useful properties encompasses not only initial compatibility but long-term compatibility when the plastic mix is subjected to potentially hostile environments of stress, solvent attack, or oxidative conditions.

Stress imposed by bending or pressure upon a compatible plasticizer/PVC system may cause plasticizer exudation upon the surface. This change in compatibility is attributed to compression of the gel structure of PVC and to formation of more polymer–polymer bonds which exclude plasticizer. Due to the dynamic state of plasticizer/PVC, the process is reversible in that plasticizer can be reabsorbed upon relief of stress. Under many conditions the plasticizer redistributes itself during bending stress to remain compatible.

Some plasticizer/PVC systems, when exposed to solvent environments (including water), develop plasticizer exudation at the surface. This incompatibility results from solvent penetration of the plasticizer/PVC system with formation of a solvent/plasticizer blend that is less compatible with the PVC. Removal of the solvent allows a return to a compatible state but at a lower plasticizer level than earlier.

Plasticizers that are susceptible to oxidation may change compatibility with PVC upon exposure to air. Exudation of the oxidized plasticizer plus unmodified plasticizer is evident as a tacky or oily film on the surface.

2. Fusion and Processability

Plasticizer choice and concentration influences fusion of the plasticizer/PVC system. Ease of fusion is technically and economically important in determining processing conditions best suited for each fabrication technique.

Fusion is a time/temperature process. As a consequence, much attention is given to plasticizers that reduce either time or temperature of the fusion process. In general, aromaticity in the plasticizer molecule improves fusion over aliphatic molecules (i.e., the 2-ethylhexyl derivative of phthalic acid is faster fusing than the same derivative of adipic acid). Introduction of a second ring in the phthalic molecule to give an alkyl benzyl phthalate further improves fusion. In the phthalate family, the alcohol choice influences fusion such that increasing molecular weight alcohols tend to retard fusion. A listing of plasticizers in order of decreasing ease of fusion, with exceptions, is as follows: phosphates, phthalates, isophthalates, terephthalates, adipates, azelates, and sebacates.

The ease of function achieves importance in all aspects of PVC processing. However, in some operations the effect may be unrecognized due to the high temperature/shear of the process. In dry blending, cycle-time is controlled by the fusion characteristics of the plasticizer. In a similar fashion those operations using the gel stage of plastisol processing are regulated by fusion characteristics of the plasticizer. Storage stability of plastisols, as measured by viscosity increase, is a function of plasticizer fusion properties.

The fusion effect is additive, so a blend of rapid- and slow-fusion plasticizers yields fusion characteristics intermediate between the two.

3. Efficiency

Plasticizer addition above a plasticization threshold makes the resin softer and more flexible, reduces modulus and tensile strength, and increases elongation. The degree of change in these properties is a function of both plasticizer type and concentration. The response curve as a function of plasticizer concentration can be near linear, sigmoidal, or more complex.

At levels below the plasticization threshold, antiplasticization occurs where the resin becomes harder and more brittle with a reduction in elongation and increase in tensile properties. The response curves peak at the plasticization threshold, which is dependent upon plasticizer type and temperature. Generally, this threshold is in the range 10–20 phr. Antiplasticization in PVC is unwanted.

Another measure of efficiency is improvement in subambient temperature performance. Plasticizer addition lowers the glass transition temperature T_g of the polymer system. Lowering of T_g is almost a linear function of plasticizer concentration. Plasticizer type influences the slope of the stiffness or modulus versus concentration curve. Linear molecules produce a less steep curve than do bulky molecules, and thus adipates and linear alcohol phthalates are used where good low-temperature performance is needed.

Plasticizer efficiency refers to its ability to develop the desired effect in a resin. The more plasticizer required to produce the effect the less efficient is the plasticizer. Many attempts have been made to compare the efficiency of one plasticizer with that of another by a one-number efficiency factor. Results are highly dependent upon the test method and the end-use property under consideration. As indicated earlier, the response curves for various physical properties vary according to plasticizer type and concentration. In spite of these difficulties, meaningful efficiency numbers for specific end-use properties can be developed. High efficiency of a plasticizer may not always be desirable. This is the case when the plasticizer is less costly than the resin on a weight or volume basis, and the properties at higher plasticizer level continue to be acceptable.

4. Permanence

Because the plasticizer is not held permanently to the resin by chemical bonds, it is subject to loss. The more important ways in which the plasticizer is lost are volatility, extraction, migration, chemical attack, and biochemical attack.

a. Volatility. Plasticizer loss causes plasticized PVC to shrink and become stiff. With low molecular weight plasticizers such as dibutyl phthalate, this can occur quickly at room temperature. As the molecular weight of plasticizers is increased, permanence at room or elevated temperature improves. Volatile loss from plasticized PVC is subject to two major controls governing transfer of matter. There is resistance to the movement of plasticizer from inside the mass to the surface and a resistance to loss from the surface. In some cases rate of diffusion may control volatile loss, while in others ease of vaporization may control. For most commercial plasticizers, vaporization controls volatility, and thus volatility can be correlated with vapor pressure of the plasticizer.

b. Extraction. Plasticizers may be extracted from PVC by various liquids. As with volatility, extraction may be theoretically controlled by rate of loss from the surface or by rate of diffusion inside the PVC, but true extraction processes become much more complex due to the extractant.

Liquid in contact with plasticized PVC may penetrate the matrix itself or may remain essentially outside the PVC, depending upon its molecular size and its thermodynamic compatibility with PVC. Customarily, in plasticized PVC both of the above processes are termed extraction, although only the first is true extraction while the second is migration.

An extractant medium that is a good solvent for plasticizer removes any diffused plasticizer from the surface before it can diffuse back into the PVC as dynamic equilibrium would require. The extraction process is diffusion-controlled under these conditions.

An extractant that is a good solvent for both plasticizer and PVC tends to swell the resin and at the same time truly extract plasticizer. Extraction rates depend upon diffusion of the solvent into the resin as well as diffusion of plasticizer/solvent out of the resin.

Plasticized PVC in end-use applications can be subjected to many extractants (i.e., water, soapy water, kerosene, gasoline, oils, and even dry-cleaning solvents). Plasticizer performance varies with the structure of the plasticizer. Generally, the effect of plasticizer concentration upon extraction is a typical S-shaped curve. At low plasticizer concentrations, the PVC system is hard, and the little plasticizer present is held tightly. As plasticizer concentrations increase, with resultant softer PVC and more open gel structure, the plasticizer is more mobile, and greater quantities are extracted. However, some traces are never fully extracted. The concentration at which the vertical segment of the extraction curve appears is determined by the plasticizer type.

c. Migration. Plasticizers may migrate from plasticized PVC by contact to other polymeric substances if the resistance at the interface is low and if the plasticizer is compatible with the second polymer. Mobility, ease of diffusion, efficiency of the plasticizer, and temperature help determine the magnitude of the effects. The primary concern is for the change occurring in the material to which plasticizers migrate, but the service life or permanence of the PVC article is also controlled by migration. Serious migration may result in substantial loss of plasticizer with resultant stiffening of the PVC. An example of plasticizer migration is softening of nitrocellulose or acrylic lacquer following prolonged contact with plasticized PVC. In plasticized PVC film used for tape application, plasticizer migration to the adhesive layer can cause softening and loss of adhesion.

d. Chemical Attack. Unplasticized PVC is characterized by excellent chemical or corrosion resistance. Plasticizer type and concentration largely govern chemical resistance of plasticized PVC. Water permeability of the PVC system is a factor, with high water permeability favoring greater attack by chemical agents. By class of plasticizers, phthalates offer best resistance to chemical attack, followed by phosphates, adipates, and polymerics.

e. Biochemical Attack. Unplasticized PVC and most synthetic polymers are inert to biological systems. Plasticizers may impart to the inert PVC their own biological properties. With the large number of destructive microbiological agents and the adaptability of each, it is highly probable that there are no plasticizers that are completely free from fungal or bacterial attack. Some are readily utilized by microorganisms as sources of carbon; others are sufficiently resistant that they are useful in warm, moist, or wet environments. As the fungus grows, secreting digestive enzymes onto the plasticized films, degradation precedes the advancing mycelial growth. Destruction of the plasticizer results in tack and exudation, embrittlement, weight loss, and discoloration.

Fungus growth in plasticized PVC may be luxuriant at times. At other times it is detected more by the pigments or dyes that are produced by the organism, which may dissolve into the plasticizer, spread, and even migrate to other materials. Proper choice of plasticizer and, if need be, the use of a fungicide in the formulation can control fungal attack. Alkyl phthalates and phosphates are among the more resistant commercial plasticizers. Epoxy plasticizers based on natural products are especially susceptible to fungus growth. This is not the case with epoxy resin materials which can replace epoxy plasticizers in stabilizing concentrations.

Bacterial attack is similar to fungal attack but requires more moisture and is geographically more widely distributed, since destructive bacteria may live in anaerobic environments, deep in the soil or in deoxygenated water. Phthalates are useful here.

B. TEST METHODS

Plasticizer testing falls into two categories: per se and in-compound. Per se testing involves use of classical chemical and physical methods to determine purity, specific gravity, color, water content, boiling point, vapor pressure, acidity, refractive index, viscosity, surface tension, flash and fire points, and solubility. As an example, ASTM D-1045 details a number of these tests for plasticizers while ASTM D-1249 sets specifications for DOP.

In-compound testing is concerned with performance in PVC, and many times the methods are product end-performance tests. For an initial screening of the value of a plasticizer in PVC, a limited number of tests suffice, as shown in Table I. From this list, the benchmark tests are

penetration hardness (shore A Hardness),
low-temperature flexibility (T_f),
volatility (weight or plasticizer lost), and
bending stress compatibility (loop test).

For intensive testing in addition to benchmark tests, end-use applications have their own important tests. For example, in the electrical insulation field, retention of tensile properties after oven aging and electrical resistivity tests are critical. In the flooring area, important tests are stain resistance and surface integrity as measured by gloss. Upholstery must offer migration- and oil-resistance and low-temperature flexibility. Food film requires low-temperature flexibility and low toxicity. Interior automotive systems must withstand volatilization to prevent fogging of interior window glass and offer good low-temperature flexibility and long-term heat stability.

TABLE I. Test Methods for Plasticized PVC

Property	Measurement	ASTM method
Compatibility	Bending stress	D-3291
	Humidity	D-2383
Processability	Dry blending	D-2396
	Fusion	D-2538
	Viscosity	D-1823,
		D-1824
Efficiency	Penetration hardness	D-676
	Low-temperature modulus	D-1043
	Tensile properties	D-638
Permanence	Volatility	D-1203
	Water extraction	D-1239
	Kerosene extraction	D-1239

The test methods employed generally have been developed to measure a critical performance parameter of the end use. In some cases they achieve a high degree of acceptance such as ASTM standardization, while many remain only an industry test used for a limited segment of the market.

C. MAJOR FAMILY PERFORMANCE

The important plasticizer families for PVC applications in decreasing order of production volumes and the major performance benefits of each are given in Table II.

Configuration of the alcohol moiety influences plasticizer performance in PVC, and as was shown in Fig. 1, plasticizer alcohols are derived by several processes that give rise to structural

TABLE II. Plasticizer Families and Benefits in PVC

Family	Benefits
Phthalates	General purpose
	Good balance of properties
	Cost efficient
Phosphates	Flame retardant
	Fast fusion
	Efficient
	Solvent resistant
Adipates	Low-temperature
Polymerics	Low volatility
	Solvent resistant
	Migration resistant
Trimellitates	Low volatility

differences. With 2-ethylhexyl alcohol (2EH) as the benchmark, the structure of other plasticizer alcohols is given in Table III. In the case of 2EH and ''nearly linear'' alcohols, branching is limited to the 2-position, whereas in both ''branched'' and ''lightly branched'' alcohols, branching can occur at random and may be as high as three branches per molecule. Many of the critical properties of plasticizer performance depend upon the length of the unbranched chain in the alcohol. For example, totally linear or nearly linear products offer better low-temperature flexibility, lower volatile loss, better efficiency, and lower plastisol viscosity. Conversely, branched-chain products offer superior electrical resistivity and better solvent resistance. These are generalizations since in actual practice plasticizers of the same chain length are not readily available.

1. Phthalates

Phthalates have been the dominant plasticizer for PVC, offering an attractive performance/price balance for many applications. Phthalates may be subdivided according to the alcohol moiety into dialkyl, alkylaryl, and diaryl phthalates. Of these three classes, dialkyl phthalates have dominated the PVC market. With the exception of one product, all commercial phthalates are *o*-phthalic based. The exception is di-2-ethylhexyl terephthalate (DOTP) which is based on *p*-phthalic. A listing of commercial dialkyl phthalates is given in Table IV. Commercial alkylaryl phthalates are limited to three benzyl phthalate derivatives; butyl, octyl, and branched twelve-carbon alcohols. Diphenyl phthalate represents a diaryl that finds little use in PVC.

Per se properties and performance in PVC for the phthalates of importance in PVC are given in Tables V and VI, respectively. The data in Table VI show the relationship of plasticizer molecular weight to volatility and solvent loss. Efficiency and low temperature are influenced by molecular weight and structure of the alcohol. Both BBP and DHP find utility in PVC due to their efficiency and fast-fusion characteristics, which offset their high volatile loss.

DOP has properties that made it the product of choice for many vinyl applications where reasonable service life was expected. As specifications and requirements became more severe, DINP, DNP, and DIDP offered improved volatility over DOP. As both volatility and low-temperature performance improvement was

TABLE III. Structural Configuration of Plasticizer Alcohols

Name	Process	Branching Position	Branching Type
2EH	Aldol	2	100% ethyl
Branched	Polygas	Varied	40–45% monomethyl 60–55% dimethyl and trimethyl
Lightly branched	Dimerization	Varied	60–65% monomethyl 40–35% dimethyl and trimethyl
Linear	Ziegler	—	—
Nearly linear	Ziegler	2	30% monomethyl

needed, DHNUP and DNODP were used. Applications requiring superior performance on low volatile loss were filled by DTDP and DUP.

2. Phosphates

In addition to imparting a degree of flame retardance to PVC, phosphates are also efficient plasticizers in softening PVC. In general, triaryl (TAP) and alkyldiaryl phosphates (AAP) are used in PVC, with limited use of the trialkyl derivative, tri-2-ethylhexyl phosphate (TOF). Performance results in PVC for the more important TAP and AAP products are given in Table VII. DOP is included as a reference point.

Because of its high chlorine content, PVC burns with difficulty. Plasticizers contribute to burning by increasing fuel value of the system. Compared with other plasticizers, both triaryl and alkyldiaryl phosphates inhibit burning of plasticized PVC. Phosphates can be blended with other types of plasticizers to obtain an acceptable balance of flame resistance and physical properties.

The method of inhibition differs for the two classes of phosphates. The mechanism of polymer burning involves a step in which a heat source pyrolyzes the solid plastic to produce gases that fuel the fire. Fire retardants work in both the condensed and the vapor phase to inter-

TABLE IV. Commercial Dialkyl Phthalates

Name	Abbreviation	Carbon atoms in alcohol
o-Phthalates		
Dimethyl	DMP	1
Diethyl	DEP	2
Dibutyl	DBP	4
Dihexyl	DHP	6
Di-2-ethylhexyl	DOP	8
Di heptyl, nonyl, undecyl	DHNUP	7,9,11
Di normal octyl, decyl	DNODP	6,8,10
Di isononyl	DINP	9
Di nonyl	DNP	9
Diisodecyl	DIDP	10
Di nonyl, decyl, undecyl	DNDUP	9,10,11
Diundecyl	DUP	11
Ditridecyl	DTDP	13
p-Phthalates		
Di-2-ethylhexyl terephthalate	DOTP	8

TABLE V. Commercial Phthalates

Name[a]	Molecular weight	Vapor pressure (mm Hg @ 200°C)	Specific gravity	Refractive index	Viscosity (cS @ 25°C)
DHP	335	3	1.008	1.491	30
DOP	391	1.3	0.982	1.485	58
DHNUP	414	0.6	0.971	1.482	41
DNODP	NA	NA	0.970	1.481	34
DINP	418	0.5	0.969	1.485	72
DNP	418	0.38	0.972	1.486	80
DIDP	447	0.35	0.965	1.483	88
DNDUP	450	0.34	0.958	1.482	48
DUP	475	0.2	0.954	1.481	54
DTDP	530	0.08	0.951	1.484	215
DOTP	391	1.2	0.981	1.487	63
BBP[b]	312	1.9	1.119	1.538	42

[a] See Table IV for identification.
[b] Butyl benzyl phthalate.

rupt melting of the polymer and burning of the gases. TAPs, because of their stability, function well in the vapor phase. AAPs appear to decompose in the flame front to form poly(phosphoric acid) which remains in the condensed phase to form a char that reduces flammability and smoke evolution.

Phosphates have an advantage over antimony oxide in flame-retarding PVC compositions. Phosphates produce clear compositions whereas antimony oxide, due to its tinctorial character, yields translucent or opaque compositions.

Performance of flame-retarded PVC in an ac-tual fire situation is subject to many variables. Thus, many laboratory tests have been devised in an attempt to predict and measure both flame spread and smoke. In general, four tests are used: oxygen index (ASTM D-2863), NBS smoke chamber (E-662), vertical burn (D-568), and flame spread (modified D-3806). Table VIII shows performance of PVC plasticized with phosphates by means of these tests, where high values are desired on oxygen index while low numbers are desirable on the other tests. These results demonstrate the need for a balance of properties when PVC is compounded since out-

TABLE VI. Properties of PVC Plasticized with 40% Phthalates

Plasticizer[a]	Shore hardness	Volatility (one day) % weight loss	T_f(°C)	Extraction % weight loss
DHP	69	5.2	−41	3.4
DOP	73	1.8	−39	17.6
DHNUP	74	0.72	−48	>28
DNODP	73	0.72	−49	>28
DINP	76	0.64	−37	26.8
DNP	74	0.60	−42	>28
DIDP	79	0.48	−38	>28
DNDUP	76	0.44	−51	>28
DUP	80	0.40	−51	>28
DTDP	82	0.36	−35	>28
DOTP	76	0.72	−41	>28
BBP[b]	71	3.4	−24	1.4

[a] See Table IV for identification.
[b] Butyl benzyl phthalate.

TABLE VII.   Properties of PVC Plasticized (40% Plasticizer)

Plasticizer[a]	Shore hardness	Volatility (one day) % weight loss	$T_f(°C)$	Extraction % weight loss
DOP	73	1.8	−39	17.6
2-EHDPP	72	3.0	−39	3.1
IDDPP	72	1.2	−35	3.0
PTBPDPP	80	1.2	−8	1.0
IPPDPP	78	1.3	−15	1.0
DOA	71	5.2	−64	>28
DHNA	71	3.9	−68	>28
DINA	76	2.0	−62	>28
DOZ	72	2.3	−67	>28
DOS	69	1.7	−69	>28
Adipic polyester Drapex® 334F (MW 2000)	74	0.2	−22	0.3
TOTM	72	0.2	−33	>28
THNTM	76	0.2	−39	>28

®Trademark of Witco Co.
[a] 2-EHDPP = 2-ethylhexyl diphenyl phosphate
   IDDPP   = isodecyl diphenyl phosphate
   PTBPDPP = *p,t*-butylphenyl diphenyl phosphate
   IPPDPP   = isopropylphenyl diphenyl phosphate
   DOA   = di-2-ethylhexyl adipate
   DNHA   = di-heptyl, nonyl adipate
   DINA   = di-isononyl adipate
   DOZ   = di-2-ethylhexyl azelate
   DOS   = di-2-ethylhexyl sebacate
   TOTM   = tri-2-ethylhexyl trimellitate
   THNTM   = tri-(heptyl, nonyl) trimellitate

standing performance in all properties is not obtainable.

3. Adipates, Azelates, and Sebacates

These plasticizers, which are based on linear dicarboxylic acids with 6, 9, and 10 carbon chains, respectively, are used principally because of their good low-temperature performance.

The advantage of the linear structure of these products over the corresponding cyclic phthalate is shown in Table VII in a comparison of low-temperature flexibility (LTF) of DOA, DOZ and DOS with DOP. The same trends seen in phthalates as the alcohol type is changed is seen in the three adipates: DOA, DHNA, and DINA. Volatility is principally a function of molecular weight, while alcohol structure influences LTF and efficiency. None of these esters offers good

TABLE VIII.   Comparison of Phosphate Performance in Flame Tests

Plasticizer (33.3%)[a]	Oxygen index	Vertical burn afterflame (sec)	Smoke chamber flaming mode	Tunnel flame spread (in.)
2EHDPP	28.1	0.5	186	3.2
IDDPP	27.8	0.4	198	3.5
PTBPDPP	31.2	13.3	247	2.2
IPPDPP	31.8	8.8	347	3.7

[a] See Table VII for identification.

solvent resistance, and they are less compatible in PVC than phthalates.

4. Polymerics

Polymeric plasticizers are a class offering good characteristics of volatility and of solvent and migration resistance to plasticized PVC. Polymerics usually are high molecular weight, linear, saturated, terminated polyesters of adipic, azelaic, or glutaric acids and propylene or butylene glycols. In most cases, termination is done by monocarboxylic fatty acids or monohydric alcohols. Molecular weight ranges from about 800 to over 5000.

Polymerics have the following structures:

$$T_1-[G-A]_n-G-T_1 \qquad (6)$$

$$T_2-[A-G]_n-A-T_2 \qquad (7)$$

where A is a dibasic acid, G a glycol, T_1 a fatty acid, T_2 a monohydric alcohol, and n the number of repeating units.

Choice of acid, glycol, and terminators is very important since the total architecture of the molecule is critical to achieve good compatibility with PVC because of the high molecular weight of the polyester.

Due to poor efficiency and high cost relative to phthalates, polymerics are used frequently in combination with phthalates to improve migration resistance and low oil extraction characteristics over the phthalate system. Upholstery and aircraft seating subject to extraction by body or hair oil are examples of useful applications. Refrigerator gasketing and oil-resistant electrical wire use high levels of polymerics frequently as the sole plasticizer.

Screening-type performance for a typical polymeric is given in Table VII to illustrate the value of polymerics compared with DOP. Polymerics that differ in structure or molecular weight give slightly varied performance.

In the class of polymeric plasticizers are included a group of resinous high molecular weight products such as nitrile butadiene rubbers, ethylene-vinyl acetate copolymers, ethylene-vinyl acetate carbon monoxide terpolymers, and chlorinated polyethylene. These all function as inefficient plasticizers for PVC and suffer from difficulty in processing but do offer a degree of permanence not obtainable by other means.

5. Trimellitates

Trimellitates such as tri-(2-ethylhexyl) trimellitate (TOTM) and nearly linear heptyl, nonyl

mixed ester trimellitate (THNTM) were developed to provide extremely low volatility while maintaining a good balance of properties similar to phthalates. Their low volatility, good oxidation resistance, and relatively low polarity make them especially useful in electrical wire and cable insulation and jacketing. Structurally they resemble a phthalate with an added functional group, as shown in structure (8) for TOTM.

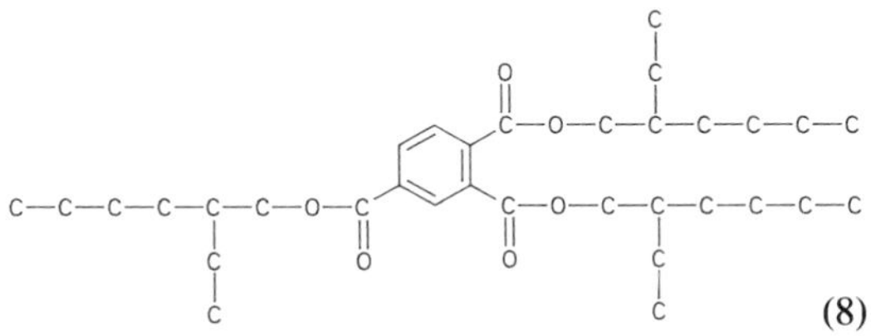

$$(8)$$

Performance of TOTM and THNTM is shown in Table VII.

The impact of incremental additions of TOTM to a high molecular weight phthalate (DUP) in improving performance in an electrical insulation typically used in residential applications is shown in Table IX. TOTM is beneficial in reten-

TABLE IX. Performance of TOTM/DUP Blends in Wire Insulation

TOTM, weight %	8.5	14.15	19.8
DUP, weight %	19.8	14.15	8.5
Low temp flex, °C	−30	−26	−22
Volume resistivity ($\times 10^{13}$ cm)			
@ 23°C (dry)	69	157	407
@ 23°C (wet)	18	27	45
Insulation resistance in water (megohms/1000 feet #14 wire)			
6 hours @ 15.6°C	1418	2613	3387
1 week @ 75°C	1.90	2.01	2.31
Mechanical properties			
Elongation, %	359	333	320
Modulus @ 100%, psi	1904	1983	2049
Tensile strength, psi	2927	2923	3003
Retention of mechanical properties, % (oven aged, 7 d @ 136°C)			
Elongation	55	69	77
Modulus	159	144	135
Tensile	104	103	100
Formulation PVC	56.78 Wt. %		
Plasticizer	28.3		
Basic lead siliate sulfate	1.7		
Lead chlorosilicate	3.4		
Bisphenol A	0.14		
Calcium carbonate	3.4		
Calcined clay	4.5		
Antimony oxide	1.7		
Stearic acid	0.08		

uenesulfonamide as a flow aid to avoid degradation and improve processing. For nylon coatings or very flexible filaments, larger amounts of plasticizer are required.

Nylon is plasticized by small amounts of water, even water absorbed from air, which results primarily in increased ease of segmental motion. Since nylon is predominantly crystalline, plasticization produces large changes in a small amount of polymer. The overall effect is small but significant in some properties. Nylons must be very dry when processed to avoid serious degradation. Dry nylon is very brittle and may be post-treated with water to return nylon to a moisture content in equilibrium with atmospheric humidity. Sulfonamide plasticizers have much the same effect as water but are more easily controlled and are more permanent. Sulfonamides tend to exclude additional water pickup to prevent subsequent plasticization.

G. POLYURETHANE FOAMS

Plasticizers such as phthalates improve elasticity, compression, and recovery properties of flexible polyurethane foam. In rigid polyurethane foams, small additions of 2–5 phr of BBP produce foams of equal density but with higher compressive strength, finer cell structure, and tougher cell wall than the unplasticized polymer. The resulting foam is stronger, insulates better, and is resistant to environmental degradation.

H. STYRENE–BUTADIENE

The copolymers of styrene and butadiene fulfill important surface coating requirements. They have a good balance of properties since polystyrene alone would be brittle and polybutadiene would be soft and tacky. Plasticization occurs internally in the polystyrene backbone by incorporation of butadiene, but additional flexibility may be had by use of external plasticizers. Almost all of the PVC plasticizers are compatible with these copolymers.

I. MISCELLANEOUS VINYL RESINS

Polyvinyl acetate may be internally plasticized by copolymerization with flexibilizing monomers such as fumarates, maleates, and acrylates. These copolymers may be further plasticized with DBP, BBP, benzoates, triphenyl phosphate, sulfonamides, and some polyesters for specific adhesive or coating uses.

Polyvinyl alcohol is plasticized very significantly by water vapor from air. Plasticizers deliberately added are glycerol, sorbitol and other polyols, and low levels of alkyl diaryl phosphates.

Polyvinyl butyral is compatible with many plasticizers of the type used in PVC to improve performance in coatings and paper-saturating uses.

Polyvinylidene chloride and its copolymers are blended with moderate amounts of plasticizers that contribute high but controlled blocking action in temporary food wraps. The copolymers provide excellent barriers against moisture. Glycolates, citrates, and phosphates are useful.

J. LINEAR POLYESTERS

Polyethylene terephthalate (PET) and related polyesters are crystalline polymers and are difficult to plasticize. PET is plasticized by water in a manner similar to that of nylon but to a lesser degree. Little or no external plasticizers are used for PET in high-strength film and textile fiber applications.

Injection molding of PET containing 35 weight % reinforcing fiberglass is made possible by use of plasticizers and nucleating agents. The plasticizer at about a 5% level provides the free volume in the PET molecules for essentially complete crystallization during the short mold-dwell time at acceptably low mold temperatures. Plasticizers of choice include neopentyl-glycol dibenzoate, N-ethyl-o,p-toluenesulfonamide and higher molecular weight sulfonamides, trimellitates, and diphenyl sulfone.

K. ELASTOMERS

Plasticizers for elastomers generally are of two types: ester and petroleum oils. Broadly speaking ester plasticizers contribute processing advantages to polar elastomers while petroleum oils are useful with nonpolar elastomers. Petroleum oils, usually called extender oils, are further classed as either paraffinic, naphthenic, or aromatic. Depending upon the rubber type, the degree of unsaturation in the petroleum oil must be considered so as not to interfere with cure rates and subsequently affect the final vulcanizate properties.

Petroleum oils are the larger volume plasticizer usage in elastomers, finding utility in natural rubber, SBR, EPR, and EPDM and neoprene. NBR represents the elastomer using the bulk of ester plasticizers in elastomers although other types frequently use esters for specialized applications. Most PVC plasticizers such as phthalates, adipates, glutarates, sebacates, phosphates, polymerics, trimellitates, and epoxy compounds are used. In those applications requiring high-temperature exposure as well as good flexibility at low temperature, high molecular weight adipates and sebacates offer good performance. As in PVC formulating, solvent resistance is imparted by polymerics.

VIII. Heath and Safety Factors

Most plasticizers are characterized by low skin sensitization and eye irritation under normal use or handling conditions. DOP and diheptyl, nonyl, and undecyl phthalate are degraded quickly in the river die-away and the semicontinuous activated sludge tests.

In spite of many years of use without any evidence of ill health effects of exposed employees, the safety of phthalates and adipates is under question due to animal study results released in 1980 by the National Toxicology Program (NTP). In particular, DOP, DOA, and BBP were the subject of NTP rodent-feeding studies. The validity of these studies, due to considerations of the spontaneous growth of tumors in the species of mice used and lack of carcinogenic response in rats, is questionable. Additionally, the result is further clouded by interspecies metabolism differences between rodents (i.e., rats and mice) and primates (i.e., monkeys and man). NTP and the Chemical Manufacturers Association have programs to retest in rats and expand the test program to include monkeys.

Material Safety Data Sheets available from plasticizer manufacturers should be consulted for up-to-date information on safe handling and toxicity of plasticizers.

BIBLIOGRAPHY

"Modern Plastics Encyclopedia," Vol. 62 (No. 10a). McGraw-Hill, New York, October 1985.

Nass, L. I., ed. (1977). "Encyclopedia of PVC," 3 vols. Marcel Dekker, New York.

Sears, J. K. and Darby, J. R. (1982). "The Technology of Plasticizers." Wiley, New York.

Sears, J. K., Touchette, N. W., and Darby, J. R. (1985). "Applied Polymer Science" (R. W. Tess and G. W. Poehlein, eds.), 2nd ed., pp. 611–614. American Chemical Society, Washington, D.C.

PNEUMATIC TRANSPORT

George E. Klinzing   *University of Pittsburgh*

GLOSSARY

Choking: Unstable regime experienced in vertical transport that occurs when the transport gas velocity is insufficient to overcome the gravitation forces of the particles.

Dense phase: Pneumatic transport flow conditions having high solids loading with considerable solid–solid and solid–wall interactions. Generally, the voidage values are lower than 0.95 for dense phase flow.

Dilute phase: Conditions of flow where the solids concentration is low giving a voidage close to 1.0. The main energy loss in such a system is still dominated by the transport gas velocity. Oftentimes the loading ratios are in the range of 1.0.

Particle velocity: Steady-state velocity achieved by the particles when transported by the gas. The particle density and diameter affects the final particle velocity.

Phase diagram: Variety of diagrams that present the various flow conditions possible in pneumatic transport systems. Operability and unstable regimes are generally designated in the phase diagrams. One possible diagram relates the pressure drop to the transport gas velocity at constant solids flow rates.

Saltation: Condition in a horizontal transfer line that permits the solids to deposit on the bottom of the pipe.

Slip velocity: Difference between the transport gas velocity and the particle velocity.

Solid loading: Ratio of solids flow rate to transport gas flow rates. This parameter is some-

times used to give a rough estimate of the flow conditions.

Solid pressure drop: Pressure drop in a pneumatic conveying process that can be attributed to the solids interactions with themselves and the wall.

Transport gas velocity: Gas velocity used to convey the solid material through the pipeline.

Voidage: Void volume occupied by the gas in a gas–solid system. The voidage determines the regime of flow of the pneumatic transport system.

In many processes the movement of solids between units is essential. Pneumatic transport of these solids by a gas stream in an enclosed conduit is an efficient, economical method to deliver these solids to the right place at the right time. Pneumatic transport involves three basic steps: (1) the delivery of the particles to be conveyed for contacting with the transport gas, (2) the conveying of the solids in the pipeline, and (3) the collection of the particles at the end of the transport section. Fine particles of less than 1 μm to large particles in the range of 0.1 m can be conveyed pneumatically and these transported distances can vary from a few meters to a few kilometers.

I. Introduction

Pneumatic transport can be traced back to ancient time; however, the first large-scale use of this gas–solid system has been the transport of materials for cement over a distance of 2 miles in a 12-in. line for the construction of Grand Coolie Dam. Since that time, a multitude of materials have been moved pneumatically. Generally, the pipelines vary in size from 1 to 14 in. with conveyed distances of 10 to 2000 ft being common place. Diameter size of particles transported range from micrometers to a few inches.

Some advantages of pneumatic conveying are its complete enclosure of the material transported, adaptation to a wide variety of materials, lower initial cost, ease in automation, less maintenance, and savings on bulk shipments.

On the negative side, however, power costs have been cited as a detriment to the use of pneumatic conveying. These costs are usually associated with the air moving equipment. New advances in this area have reduced these costs significantly. Besides the high power cost, in order to operate a pneumatic conveying system higher technology is required.

In order to carry out pneumatic transport a gas supply is necessary as well as a solid feeder and collection devices at the end of transfer. Gas supply can be provided by a fan, a blower, or a compressor. The solids feeders are many and varied in nature of operation. Some of these are the rotary feeder, the screw feeder, the pressure tank, and venturi feeder, and the fluidized bed feeder.

Pneumatic transport can be classified as either vacuum delivery or pressure delivery, or by the phase of transport, such as dilute or dense. A vacuum system picks up solids at several points and delivers them to one site. For pressure conveying pickup is made at a single point and delivered to multiple points. Combinations of the two systems are also possible, giving multiple pickups and multiple deliveries. For dilute and dense designations these correspond roughly to the amounts of solids delivered per volume of gas. No precise definition exists for these phase designators.

Transport lines can be arranged in many geometric configurations. Horizontal, vertical, and inclined lines are the usual designations.

II. Basic Analytic Formulation

In order to analyze pneumatic transport in an analytic manner the forces acting on the system must be described in detail. For an unsteady operation these forces are balanced by the acceleration terms of the system. The basic forces acting on the gas–solid flow system are the drag forces, the pressure force, gravity forces, frictional forces, and electrostatic forces, if present. One can write this force balance as

$$\Delta m \, dU/dt = F_{\text{drag}} - \text{pressure} \cdot \text{area} - F_{\text{gravity}}$$
$$- F_{\text{frictional}} - F_{\text{electrostatics}} \qquad (1)$$

where Δm is differential mass, U velocity, and t time. This expression applies to both the gas phase and solid phase, with a different formulation for each phase.

For the solid phase the expression is written as

$$(1 - \varepsilon)\rho_p \frac{dU_p}{dt} = \frac{3}{4} C_{\text{DM}} \rho_f \frac{(U_f - U_p)^2}{(\rho_p - \rho_f)D_p} (1 - \varepsilon)\rho_p$$
$$- \frac{dP}{dx}(1 - \varepsilon) - g(1 - \varepsilon)\rho_p$$
$$- 2f_s \frac{U_p^2(1 - \varepsilon)\rho_p}{D} \qquad (2)$$

and for the fluid phase as

$$\varepsilon\rho_f \frac{dU_f}{dt} = -\frac{3}{4} C_{\text{DM}} \frac{\rho_f(U_f - U_p)^2}{(\rho_p - \rho_f)D_p} (1 - \varepsilon)\rho_p$$
$$- \frac{dP}{dx}\varepsilon - g\varepsilon\rho_f - \frac{2f_g U_f^2 \varepsilon\rho_f}{D} \qquad (3)$$

In these equations ε is voidage, ρ_p density of the solid, U_p particle velocity, C_{DM} drag coefficient of a multiple particle system, ρ_f density of the fluid, U_f fluid velocity, D_p diameter of the particle, D diameter of the pipe, P pressure, X axial distance, g gravitational constant, f_s solid frictional term, and f_g gas frictional term. Note that the drag term in each phase is the same value but of opposite sign. For steady state the left-hand side of the equations are equal to zero. Combining the two expressions the drag terms cancel each other producing

$$-\frac{dP}{dx} = g(1 - \varepsilon)\rho_p + g\varepsilon\rho_f + \frac{2f_s U_p^2(1 - \varepsilon)\rho_p}{D}$$
$$+ \frac{2f_g U_f^2 \varepsilon\rho_f}{D} \qquad (4)$$

This is the functional equation for the design of pneumatic transport systems. The expression can be written in a shorthand notation as

$$\Delta P_{\text{total}} = \Delta P_{\text{gravity}} + \Delta P_{\text{frictional}} \qquad (5)$$

Some specific observations are essential in order to evaluate Eqs. (2)–(5). In most cases the gas gravitational contribution is small in comparison to the solid gravitational force. Thus this term can be often neglected. The drag coefficient in the above expressions are based on a multiple-particle system and can be expressed by

$$C_{\text{DM}} = \varepsilon^{-4.7} C_D \qquad (6)$$

The drag coefficient of the single particle system, C_D, can be in a variety of regimes depending on the slip-velocity-based Reynolds number

$$Re_p = D_p[(U_f - U_p)/\mu_f]\rho_f \qquad (7)$$

where μ_f is fluid viscosity. The voidage term, ε, arises also in these basic equations. The relationship to other system parameters is given as

$$\varepsilon = 1 - (W_s/A\rho_p U_p) \qquad (8)$$

where W_s is solid mass flow rate.

For the frictional representations the gas and solid friction factor must be defined (i.e., f_g and f_s). For the gas friction factor one can use a variety of expressions; one such convenient form is that given by Koo

$$f_g = 0.0014 + 0.125\, Re^{-0.32} \qquad (9)$$

where the Reynolds number is that based on the tube dimension. In order to assess the solid friction factor one finds themselves in an evolving field. The detail gas–solid flow structure should be known in order to use the appropriate solid frictional term. The regions of dilute and dense surface has significant effects in determining the solid frictional representation. A few expressions for this f_s term is seen in Table I. In regard to the solid frictional term one will note that the particle velocity squared is the dynamic term for this expression. Recent considerations have suggested that slip velocity $(U_f - U_p)^2$ is the appropriate term to more accurately describe the frictional action of the system.

One difficulty in applying this theory is determining the particle velocity, since rarely do systems have the facility to measure this value. Some empirical expressions can be used to address this issue. Table II has a listing of particle velocity expressions that can be used in the design of these systems.

The condition of unsteady state was first addressed in the basic dynamic equations govern-

ing gas–solid transport. This condition of unsteadiness arises more often than one would suspect in the design of a transfer line. The initial injection of particles into the system is, of course, an unsteady process for accelerating the particles to a steady-flow condition. Every bend or change in flow direction of a system causes accelerations and decelerations. For long pipelines a pressure drop in the system causes the gas velocity to increase and thus continuous changing or unsteadiness of these variables. Equation (2) can be considered for the unsteady particle behavior. One can substitute for $d(t)$ the expression (dL/U_p) and integrate Eq. (2) to obtain

$$L = \int_{U_{p_1}}^{U_{p_2}} U_p\, dU_p \left/ \left[\frac{3}{4} C_{DM} \frac{\rho_f(U_f - U_p)^2}{(\rho_p - \rho_f)D_p} \right.\right.$$
$$\left.\left. -\frac{dP}{dx}\frac{1}{\rho_p} - g - \frac{2f_s U_p^2}{D} \right] \qquad (10)\right.$$

Equation (10) can give the acceleration length necessary to recover the steadiness of the particulate system. The pressure loss over this transient length L_a can be given as

$$\Delta P_{accel} = \int_0^{L_a} \rho_p(1 - \varepsilon)g\, dL + \int_0^{L_a} \frac{2f_g\rho_f U_f^2\, dL}{D}$$
$$+ \int_0^{L_a} \frac{2f_s\rho_p(1 - \varepsilon)U_p^2}{D}\, dL$$
$$+ \rho_p(1 - \varepsilon)U_p^2 \quad \text{at } L_a \qquad (11)$$

Some concern exists as to the appropriateness of the frictional representation for the unsteady regimes. This, however, has not been fully addressed to date.

Bends were mentioned as regions of decceleration and acceleration in the pipeline. The previous equations could be applied to these regions. The effect of solid particles on bend losses have been studied by few investigators. One such correlation has been presented by P. Schuchart. The pressure loss due to the solids in the bend is given as ratio to that for a straight pipe section.

$$\left(\frac{\Delta P_{bend}}{\Delta P_{straighten}}\right)_{solids} = 210 \left(\frac{2R_B}{D}\right)^{-1.15} \qquad (12)$$

This expression is for a variety of bend configurations where R_B is the bend radius. In order to calculate the total pressure loss in a bend the gas contribution must be added to the solid contribution. Ito has developed reliable pressure loss expression for single-phase pressure losses around

TABLE I.   Frictional Expression for Pneumatic Transport

Dilute phase (Konno–Saito)	$f_s = 0.0285\,\sqrt{gD}/U_p$
Horizontal	$f_s = 0.00315\,\dfrac{(1 - \varepsilon)}{\varepsilon^3}\left[\dfrac{(1 - \varepsilon)U_f}{U_f - U_p}\right]^{-0.979}$
Vertical (Yang)	$f_s = 0.0293\,\dfrac{(1 - \varepsilon)}{\varepsilon^3}\left[\dfrac{(1 - \varepsilon)U_f}{\sqrt{gD}}\right]^{-1.15}$
Dense phase (Klinzing–Mathur)	$f_s = \dfrac{5.55\, Dt^{1.07}}{U_g^{0.643}D_p^{0.259}\rho_p^{0.909}}$

TABLE II.   Particle Velocities

$D_p < 40 \ \mu m$	$U_p = U_g - U_t$
Hinkle modified IGT	$U_p = U_g(1 - 0.68 \ D_p^{0.92} \rho_p^{0.5} \rho_f^{-0.2} D_t^{-0.54})$
Yang	$U_p = U_f - U_t \ \sqrt{\left(1 + \dfrac{2 f_s U_p^2}{gD}\right)} \ \varepsilon^{-4.7}$

bends. This equation for the single-phase loss is

$$\Delta P_{\text{single fluid phase}}$$

$$= \frac{0.029 + 0.304[\text{Re}(r_0/R_B)^2]^{-0.25}}{(R_B/r_0)^{1/2}} \frac{L\rho_f \bar{U}^2}{2D} \quad (13)$$

where r_0 is radius of the pipe.

III. Phase Classifications

In the Section I the phase terms "dilute" and "dense" were used as flow classifications. More detailed phase diagram behavior is necessary in order to understand pneumatic transport. Before addressing the phase diagrams the character of the solids and their classification is beneficial to the practitioner of pneumatic transport. For this purpose the Geldart diagram generally used in fluidization will be presented. Figure 1 shows the four regions of particle classification according to their size, density, and ability to be fluidized. Group A is termed aeratable powder, which is easily fluidized, group B is a sandlike material, which also can be fluidized but having a larger particle size than group A. Group C powder is a fine cohesive powder that has considerable interparticle forces present and needs extraordinary means to cause fluidizations. Group D consists of larger sized particles that form a spoutable classification that has not been in one-to-one correspondence with pneumatic transport. Lohrmann has attempted to investi-

gate a wide variety of materials and associate their group properties to blow-vessel behavior, while Canning and Thompson considered the classification in relation to dense-phase flow. Figure 2 is a diagram of the materials studied by Lohrmann and their relationship to the Geldart diagram. Group A are the best candidates for dense-phase conveying. Group B powder were found to be suitable for pulsed phase conveying, while not suitable for dense-phase conveying. Group D powders with a low density form stable plugs, while denser materials and large particles cause vibrations. These powders can be conveyed in a dense phase provided that there is a narrow size distribution. Group C powder can be troublesome in pneumatic conveying, just as they are in fluidization; generally dense phase is not recommended for this powder.

Zenz has proposed a convenient phase diagram behavior with plots of pressure losses versus the gas velocity at a diagram of this type. The minimum point in the pressure loss is noteworthy. Ideally this would be a convenient place to design for a system. To the left of the minimum is a sharply increasing segment of the dia-

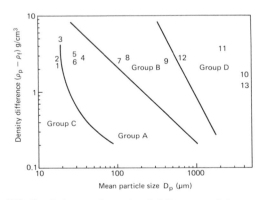

FIG. 2.   Lohrmann's study of different particles on a Geldart diagram. Code: (1) pulverized fly ash, (2) ground calcine mineral, (3) manganous oxide, (4) unslaked lime, (5) cement, (6) ferrosilicon baghouse dust, (7) sodium sulfate, (8) electric arc furnace dust, (9) foundry sand, (10) maize grits, (11) magnesite, (12) milled asbestos waste, and (13) ice.

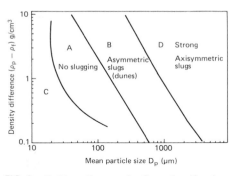

FIG. 1.   Geldart diagram for flow classification.

gram. This region is unstable and often times designers want to stay far from this region and suggest operations in the region far to the right of the minimum point. The crossover between dense and dilute phase operations is sometimes designated by the minimum point. The unstable region to the left of the minimum point can represent the region of choking and slugging operation. If the energy is sufficient a more continuous densely packed flow can take over. If the energy available is not sufficient the flow will plug and stop.

In the field of phase diagrams Matsen has suggested another avenue of pursuit. This diagram has an analogy in thermodynamics in the vapor–liquid equilibrium behavior. Figure 3 shows one of these diagrams with the dilute–dense phases noted. The transition between the curves is termed the choking locus. This diagram is more specific in its designation of the terms dilute and dense phase. It is essential to have a knowledge of the voidage term in order to construct these phase diagrams. [*See* ENGINEERING THERMODYNAMICS.]

Tuba, Mathur, and Klinzing have used the basic premise of Matsen and generalized and reduced the phase diagram in terms of a flux format. The reduced fluxes are given as

$$j_g^* = j_g \sqrt{\frac{\rho_f}{gD(\rho_p - \rho_f)}}$$

$$j_p^* = j_p \sqrt{\frac{\rho_p}{gD(\rho_p - \rho_f)}}$$

where j_g is volumetric gas flow per area and j_p volumetric solid flow per area. Upon analysis of the phase relationship and similarity with thermodynamics an equation of state similar to van der Waal's equation has been suggested. Thus

$$j_p^* = \frac{R'}{(1 - B\phi)} j_g^* - A\phi^2 \qquad (14)$$

where $\phi = (1 - \varepsilon)$. Figure 4 shows a plot of a pneumatic system that has been analyzed in this manner. The three parameters A, B, and R are unique to the particular powder being conveyed. These parameters may be a technique for characterizing pneumatic transport materials. Thus these diagrams are useful in noting the phase relationships and closeness to choking phenomena. In order to construct such diagrams the voidage term must be known. This involves generally determination of the particle velocity and then calculation of the voidage as

$$\varepsilon = 1 - W_s/\rho_p U_p A \qquad (15)$$

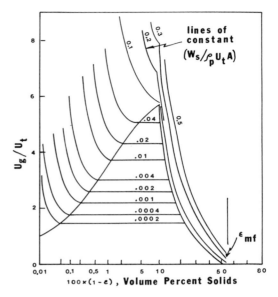

FIG. 3.   Matsen-type phase diagram.

where A is cross-sectional area of the pipe. Work in the modeling area has suggested that a combination of the Matsen-type diagram with the Zenz diagram is able to show the phase behavior and pressure loss at the same time. Figure 5 shows the initial attempt at this synthesis.

IV. Flow Measurements

One of the most challenging tasks in pneumatic transport is the ability to be able to measure the flow of solids at any particular time.

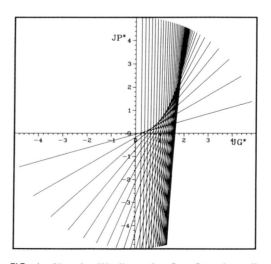

FIG. 4.   Van der Waal's analog flux–flux phase diagram: Parameters A, B, and C are 108.49, 1.148, and 14.456, respectively. The void increment is 0.02.

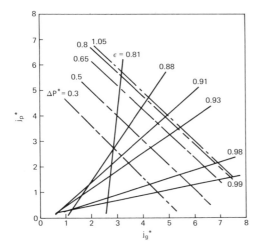

FIG. 5. Phase diagram with pressure drop representation.

These relatively instant measurements will permit the use of automatic control in order to regulate solid–gas flow in much the same manner as we control single-phase, gas- and liquid-flow systems.

The simplest forms of flow meters follow the same instruments used in normal flow metering, namely the orifice and venturi meters. A combination of the two in series has been utilized to measure solid flow. The orifice responding to the gas flow and the venturi responding to the solids. Work on a single venturi meter as a solid-flow measuring device has been renitiated by Crowe and Davies. These meters have been tested only on systems of low values of solid loading. Erosion by the solid particles presents itself as a concern in the orifice and venturi meters.

The detection means defines the basis of several other flow measuring devices. Light, sonic, electrical charge, and dielectric variations are the basis of a number of flow meters.

The principle of Doppler shift due to the solids present in a gas flow has been applied both on a sonic and light basis. These techniques are able to give the particle velocities very accurately. For this Doppler shift principle only dilute phase systems are applicable.

A microphone's ability to pick up the vibrations generated by solids flowing in a pipe has been used by Argonne National Laboratories to determine the flow of solids in a line.

A light absorption meter using a halogen lamp source and a coupled photodiode op/amp was used by Klinzing to follow solid loading in a coal-gas transport system. The extinction of light through the solids flow follows Beer's logarithmic absorption law.

Soo and his colleagues have proposed a number of solid-flow meters based on the electrostatic and glow discharge principles. There are a variety of geometric configurations for these meters. Those inserted into the flow stream will also suffer from erosion limitations. King also has explored the analysis of electrostatics in gas–solid flows. He also used the generated charges in a cross-correlation manner to determine the particle velocities. More recently Smith and Klinzing have used modern electronic and computer technology to automate the cross-correlation technique.

The dielectric property can be conveniently measured by the Auburn international meter. This unit has the same diameter as the transport pipe, and as such, suffers little erosion as well as not affecting the flow pattern of the solids. Mathur and Klinzing have used this as a direct-flow meter and a cross-correlation technique for the particle velocity.

A novel use of physics in a solid-flow meter has been utilized by Micromotion base on a coriolis force generated in flow around a 180° bend. Mathur has tested this unit on coal-gas flow and found it to be adequate for solids flow at moderate and high loadings. Erosion in such a unit has not been addressed but could be sizable in use with abrasive solids.

New advances in the area of solid meters are most welcome and needed. We seem to be limited only by our imagination in this field.

V. Electrostatic Effects

Should one be transferring solid material pneumatically under conditions at which the relative humidity of the transport gas is high and the conveying lines are metals that are well grounded, the electrostatic effects in pneumatic transport may never occur. In ungrounded systems with low relative humidities for the transport gas, electrostatic forces can be generated in a pneumatic transport system that can affect the overall energy requirement of the system. As the particle size decreases the size of the charging increases. Above about 400 μm, little electrostatic charging is seen. For electrostatics to be generated, rapid contact and breakage of contact between dissimilar materials is essential in triboelectrification. Charging on particles can increase until the breakdown voltage of air is exceeded. If a discharge occurs and the solid concentration and oxygen content of the gas are sufficient, a powder or dust explosion can occur. The discharge potential for air is 3000 kV/m.

The charge per unit mass on particles can be given by

$$Q/m_p < 9 \times 10^6 \, \varepsilon_0/r_p\rho_p \qquad (16)$$

where ε_0 is permittivity of free space and r_p radius of the particle. Once particles charge they can retain this charge for a relatively long period of time, depending on the resistivity of the powder. The time constant for decay of charge is given as

$$T = \bar{\varepsilon}\varepsilon_0 R \qquad (17)$$

In pneumatic transport Masuda et al. have developed a charge expression to account for contact times of the particles with the containing wall and showed that

$$Q = C \, \Delta V[1 - \exp(\Delta t/\tau)] \qquad (18)$$

where C is capacitance, ΔV potential difference, τ relaxation time, and Δt contact time. Charge transferred on contact of particles with a wall are most conveniently measured by current. Knowing the rate of collisions of the particle and the wall one can write

$$i = QN_I \qquad (19)$$

where i is current and N_I collision rate. The collision rate is proportional to the solid-flow rate.

Soo was first to realize the interrelationship between particle mechanics, heat transfer, and electrostatic charge transfer. Masuda and co-workers expanded this approach, considering both elastic and inelastic collisions between particles and particles with the confining wall. The main purpose of these exploits was to determine the contact area at collision. For the current generated on an insulated pipe wall of length L the contact area is given as

$$I = -W_s\frac{\Delta Q}{m_p} = -\frac{6W_s\varepsilon_0 V_c A \, \Delta t}{\pi D_p^3 \rho_p Z_0 \tau}\frac{\Delta n}{\Delta x}L \qquad (20)$$

where ΔQ is charge increment on particle by impaction, V_c contact potential, A contact area, Z_0 gap distance, $\Delta n/\Delta x$ change in number of collision with length, and L length of pipe.

Masuda experimentally tested Eq. (20) by measuring the current generated on an isolated section of metal pipe. Figure 6 shows the agreement he achieved in his analysis. The current shows a quadratic behavior with mass flow rate. It is noteworthy that the sign of the current changes at a certain solid-flow rate.

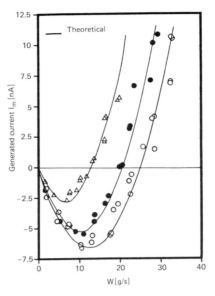

FIG. 6.   Current from pipe bend to ground; ($\triangle$) 10 m/sec, ($\bullet$) 20 m/sec, and ($\bigcirc$) 30 m/sec.

In a comprehensive study of current generated by electrostatics in pneumatic conveying Masuda et al. found the current to be proportional to the diameter of the tube and inversely proportional to the particle diameter, while the velocity had an almost square dependency. A triboelectric series was suggested on the materials they studied as (listed in decreasing order)

flour
morundum
glass
PVC
steel
quartz sand
teflon

Smeltzer, Weaver, and Klinzing have proposed a film-penetration model borrowed from heat and mass transfer to explain charge transfer between particles and the wall. The basic expression for the current is given as

$$i = h_e A \, \Delta V$$

The term h_e is called an electrostatic transfer coefficient and can be developed for various contact time distributions. For the gamma distribution one obtains

$$h_e = \frac{\sigma\beta^{1/2}\Gamma(\alpha + 1/2)}{(\sigma m_p/c\rho_p)^{1/2}\pi^{1/2}\Gamma(\alpha + 1)} \qquad (21)$$

where σ is conductivity, α and β coefficients for the gamma distribution, c capacitance, and m_p mass of particle.

An additional pressure drop loss can be added to the normal pressure loss when the electrostatic effects are present. Basically this can be expressed as

$$\Delta P_{electrostatic} = E_x(Q/m_p)(1 - \varepsilon) \qquad (22)$$

where E_x is the potential gradient.

The potential gradient is difficult to assess in most situations; however, the charge on the particle can be found from capture in a Faraday cage.

The particle velocity is also affected by the static charges on the particles. Using the force balance at steady state, the particle velocity can be related to the electric field and charging as

$$U_p = U_f - U_t \left[\frac{2f_s U_p^2}{gD} + \frac{E_x Q}{gm_p}\right]^{1/2} \varepsilon^{2.35} \qquad (23)$$

where U_t is terminal velocity of the particle. Ally has developed the pressure loss due to electrostatics from the use of a force balance and the application of Gauss's law. His development gives

$$\Delta P_{electro} = \frac{45}{16} \frac{(1 - \varepsilon)^2 Q^2 D}{\pi^2 \varepsilon_0 D_p^6} \qquad (24)$$

In addition he found that the electrostatic pressure drop falls off in a logarithmic fashion with the moisture content of the system.

When one begins to review the way in which charging occurs and recalls the earlier concepts of electron transfer between different materials, one sees that charging is unique to almost each situation considered. The charge is related to the chemistry of the particle, the wall confining the flow (the work function), the diameter of the particle, the length of the transport line, the moisture content of the system, and the number of particles flowing in the system. Thus each system has a unique work function, depending on whether the system is a glass particle—plastic tube, metal particle, plastic tube, etc. The exact expressions for the dependence of each parameter on the charge are not entirely known. From recent data some estimates can be made for a number of the factors. The precise expression for the charge dependence on the particle size is not known but it appears that the smaller particles generate a larger charge in a flowing system. This result is shown from the number density considerations of the flow for two sizes of particles at the same loading conditions. In addition, one needs to consider that the maximum charge on a particle is proportional to the diameter squared of the particle. Thus one has, for example,

$$\frac{Q_{75\,\mu m}}{Q_{150\,\mu m}} \alpha \left(\frac{D_{p\,150\,\mu m}}{D_{p\,75\,\mu m}}\right)^3_{\text{number density}}$$

$$\left(\frac{D_{p\,75\,\mu m}}{D_{p\,150\,\mu m}}\right)^2_{\text{maximum charge}} \alpha \left(\frac{D_{p\,150\,\mu m}}{D_{p\,75\,\mu m}}\right)$$

$$(25)$$

In analysis of data for the development of charge, it appears that the charge achieves an equilibrium value in the first few feet of travel length. Experimental evaluation of the effect of moisture on the generated charge indicates a rather profound dependency. It appears that this effect above all seems to dominate the particle charging. Exploring this further from the pressure drop generated due to the electrostatics effects one also finds a similar effect of the moisture on the pressure drop. It has been noted that when the moisture ratio of kilograms water to kilograms solids exceeds 0.1, no electrostatic contributions are seen.

Each individual system will have a unique behavior of charging and resultant pressure drop increases. This effect, of course, has its basics in the work function of the system.

When one thinks of electrostatics, one often has an adverse connotation. Discharges into powders have been cited as the cause of many violent explosions. With the use of more electronic components in a variety of technologies electrostatic shielding and grounding is imperative. The use of electrostatics as a means to measure flow rates has been cited before. New developments in this area are imminent.

BIBLIOGRAPHY

Geldart, D. (1973). *Powder Technol.* **7**, 285.

Ito, H. (1959). *Trans. ASME: J. Basic Eng. Ser. D* **81**, 123.

Ito, H. (1960). *Trans. ASME: J. Basic Eng. Ser. D* **82**, 131.

Klinzing, G. E. (1981). "Gas–Solid Transport," McGraw-Hill, New York.

Masuda, H., Komatsu, T., Mitsui, N., and Iinoya, K. (1976/1977). *J. Electrost.* **2**, 341.

Mathur, M. P., Klinzing, G. E., and Tuba, S. T. (1982). *Pneumatech* **I**, Stratford, England (May 1982).

Matsen, J. M. (1981). *Proc. AIChE Meet., New Orleans* (preprint).

Schuchart, P. (1968). *Chem. Ing. Tech.* **40**, (21/22), 1060.

Soo, S. L. (1967). "Fluid Dynamics of Multiphase Systems," Blaudell, Waltham, Massachusetts.

Yang, W. C. (1978). *AIChE J.* **24**, 548.

POLLUTION, ENVIRONMENTAL

M. E. Baur   *University of California, Los Angeles*

GLOSSARY

Biological oxygen demand (BOD): Quantity of oxygen used by microorganisms in a water sample during a standard time interval, usually 5 days. Also called biochemical oxygen demand.

Curie: Unit of intensity of radioactivity. A curie of material is the amount that undergoes 3.70×10^{10} nuclear disintegrations per second.

Free radical: Molecular fragment having one or more unpaired electrons. Free radicals are usually highly reactive.

Half-life: Time required for half of a set of unstable nuclei to undergo radioactive decay.

LD_{50} (50% lethal dose): Amount of a toxic material fatal to 50% of the animals in a test sample; usually expressed in milligrams of toxic material per kilogram body weight of the test animal.

Rem (roentgen equivalent man): Quantity of radiation that produces the same biological damage in humans as one rep of gamma radiation; a rep represents a radiation energy input of 97 ergs per gram of body tissue.

Residence time: Average time a given chemical species spends in a reservoir.

Environmental pollution deals with the modification of the natural chemical environment of the earth by human activity; its study encompasses the sources, distribution, and effects of pollutants, that is, chemical substances introduced into the terrestrial environment in loca-tions and/or at levels not observed in the absence of human intervention. Environmental pollution as a discipline originates in the recognition that the terrestrial soil/hydrosphere/atmosphere/biosphere system is an interlinked set of reservoirs with complex connective pathways, so that modification of chemical condition in one locale can propagate in unintended ways into others, producing unexpected effects.

I. Introduction

Manipulation of the chemical environment by humans may be said to have begun with the domestication of fire. Use of toxic heavy metals, most notably lead and mercury, is believed to have caused deleterious effects on human health in some ancient societies, and deterioration of air quality in medieval urban settings due to use of heavily polluting fuels is documented. Profound large-scale modification of the chemistry of the earth's soils, waters, and atmosphere began with the industrial revolution of the eighteenth century and has accelerated with the development of chemical and nuclear technology, until now no portion of the earth's surface is entirely free of the chemical signs of human activity.

The first task of the specialist in environment pollution studies is identification of the chemical substances that cause deleterious effects. This is followed by an effort to identify the pathways by which the substances have reached their locus of action, to find their sources, and finally to eliminate or minimize those sources. The tools for identification and measurement of the concentration of such a substance are those of analytical chemistry, mainly various forms of spectroscopy, gas and liquid chromatography, and neutron activation analysis. These subjects will not be considered here.

The major known sources of pollutants are

summarized in Section II. This is followed by a survey of the major reservoirs into which the environment is conventionally divided—the atmosphere, the hydrosphere, and the biosphere—with a description of the chemical perturbations of importance in each. Issues connected with toxic waste disposal in dump sites are touched on in the last section. No specific consideration is given to substances—cosmetics, pharmaceuticals, and food additives—that directly affect the chemical/biochemical state of the human population and appear so far to have had limited impact on or distribution through the biosphere and environment outside this population. A complete and final catalog of pollutant species cannot be given; new substances are constantly entering the chemical cycle, and negative effects often become apparent only after a considerable time. The scale of the problem may be appreciated from Fig. 1, where the number of chemical substances for which reasonably complete environmental analysis is available is seen to be a small portion of the whole. There are no foolproof procedural algorithms in the field; environmental scientists must employ all their skills to foresee the consequences of even seemingly benign or neutral alterations in the chemical environment, and each pollutant substance has its own particular chemical features.

II. Sources of Chemical Pollutants

A. PROCESSES RELATED TO ENERGY GENERATION

1. Combustion

Combustion (the rapid oxidation of fuels) has been employed for warmth and food and materials processing since prehistoric times, but it assumed particular importance with the development of the heat engine in the eighteenth century. A heat engine is a cyclic device by means of which a quantity of heat generated in a hot reservoir is extracted and partially converted to useful work (a remnant of this heat is necessarily lost to a cold reservoir). The efficiency of a heat engine increases with the temperature of the hot reservoir, other things being equal, and it is thus advantageous to operate at as high a temperature as possible.

The chemical process of combustion may be represented as

$$CH_xO_y + \frac{x - 2y + 4}{4} O_2 = CO_2 + \frac{x}{2} H_2O$$

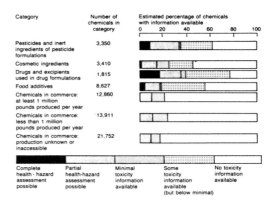

FIG. 1. Availability of health-effects information for categories of chemicals. [From Conservation Foundation. (1984). "State of the Environment, an Assessment at Mid-Decade." Washington, D.C. Reproduced by permission of the National Academy of Sciences.]

where CH_xO_y is the general stoichiometric formula for a fuel. For high-grade coal, x and $y \cong 0$; for petroleum hydrocarbons, $x \cong 2$ and $y \cong 0$; for cellulose (wood), $x = 2$, $y = 1$; for methyl alcohol, $x = 4$, $y = 1$; for natural gas (methane), $x = 4$, $y = 0$. In this simplified scheme, combustion appears innocuous, as both CO_2 and H_2O are normal components of the atmosphere. However, the amount of CO_2 that has been generated, principally in the twentieth century, is believed to be sufficient to perturb the earth's heat balance (Section III,C,1). Also, engine combustion is not complete, and varying amounts of carbon monoxide (CO) and residual hydrocarbon (HC) are released in the exhaust gases. The CO is directly toxic, as it bonds to hemoglobin, the oxygen-transporting pigment in blood, displacing oxygen and resulting in asphyxiation. The volatile organic compounds (VOC) in the HC mix enter the gas phase directly, while less volatile compounds may attach to particles (soot, dust, ash, aerosols) and remain in the atmosphere for a considerable time. Inhalation of the HC species can be directly harmful; further, many of them contribute to the complex chemistry of smog (Section III,C,1).

Two other characteristics of combustion are of importance for environmental pollution:

1. At the elevated temperature encountered in engines, the major constituents of air, nitrogen (N_2) and oxygen (O_2), partially react to form a series of compounds, notably nitrogen oxide (NO). These compounds, collectively denoted

NO_x, are important in the development of photochemical smog and acid rain, as well as being undesirable components of the atmosphere in their own right.

2.   Natural fossil fuels contain impurities and trace constituents that enter the atmosphere on combustion and can cause deleterious effects. The most important is sulfur, which occurs in coals and petroleum hydrocarbons at levels up to several percent by weight; combustion yields the gaseous species SO_2 and other sulfur oxides (collectively denoted SO_x). Atmospheric processes convert SO_2 to SO_3 and, by addition of water, to sulfuric acid (H_2SO_4), a principal constituent of acid rain, snow, and fog (Section IV,B). In addition to natural impurities, fuel additives whose function is to increase the efficiency of engine operation may be present. The most notable such additive is lead, in the form of the compound lead tetraethyl, $Pb(C_2H_5)_4$, added to some gasoline to increase octane (antiknock) rating. Lead from ethyl gasoline is believed to have been a major contributor to the lead burden in the bodies of urban residents in developed countries.

The internal combustion engine (gasoline and diesel engines) has assumed an important role in modern technology because of its relative simplicity and high efficiency. Efforts to reduce the levels of atmospheric contaminants produced in its operation encounter the obstacle that increased efficiency and reduction of all pollutant species tend to be mutually inconsistent. A high operating temperature favors efficiency but also increases NO_x production; manipulation of the air–fuel mixture ratio to reduce NO_x tends to augment HC and CO levels, as shown in Fig. 2. Nevertheless, some improvement in the amounts of these substances released to the environment by vehicles has been achieved by improved engine design, modification of fuels, and the introduction of catalyst devices in the exhaust pathway. Figure 3 shows trends in the release of various primary atmospheric pollutants for the United States between 1970 and 1981. Vehicle engines are the principal contributors to CO and NO_x, while stationary sources (power plants and factories) are the largest sources of SO_x and particulates.

2. Nuclear Reactions

The development of techniques for the controlled release of nuclear energy in reactors after World War II was regarded as an important ad-

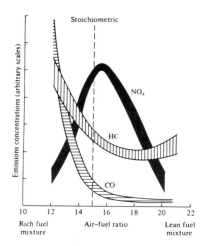

FIG. 2.   Effect of air–fuel ratio on exhaust emissions. [From Hodges, L. (1977). "Environmental Pollution," 2nd ed. Holt, Rinehart & Winston, New York. Reproduced by permission of source, Battelle Memorial Laboratories, Columbus, Ohio.]

vance in the direction of nonpolluting energy production. Unfortunately, new environmental problems have arisen from the operation of such reactors; these include the release of radioactive materials in leak episodes or occasional catastrophic accidents, and the storage of waste materials.

The only nuclear process that has been found practicable for energy production is the fission of a radioactive species, either naturally occurring (^{235}U) or artificially generated (^{239}Pu), to produce neutrons whose capture induces further energy-releasing radioactive decay in a chain re-

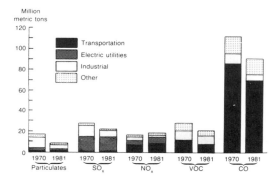

FIG. 3.   Air emissions by source and type of pollutant, 1970 and 1981. [From Conservation Foundation. (1984) "State of the Environment, an Assessment at Mid-Decade." Washington, D.C. Reproduced by permission of the Conservation Foundation.]

action, for example,

$$^{235}U + {}^1n \text{ (neutron)} = \text{two daughter nuclei}$$
$$+ 3{}^1n + \text{energy}$$

The fuel is contained in rods jacketed with a cladding (e.g., steel or zirconium). For capture to occur efficiently, the neutrons must be slowed (thermalized) by a moderator substance (water, heavy water, graphite). Control rods of a neutron absorber (cadmium, boron) remove enough of the thermal neutrons to hold the reactor in a steady state. The heat generated is removed from the reactor core by a coolant, usually pressurized water in U.S. reactors. The heated water is used in a secondary loop to operate an engine.

Fission of a ^{235}U nucleus releases about 200 MeV (million electron volts) or 3.2×10^{-4} ergs of energy, of which about 30% can be captured as useful work. A reactor consumes about 3 moles (700 g) of ^{235}U fuel per megawatt of power generating capacity per year. About 6 moles/MW-year of radioactive daughter species are thereby generated; these are short-lived and undergo further decay with emission of neutrons, α particles (^4He nuclei), β particles (electrons), and γ radiation (high-energy electromagnetic radiation) to produce longer-lived further product nuclei. The irradiation of the moderator, control rod, coolant, and structural material in the reactor also generates radioactive nuclides; reactor operation thus results in a substantial multiplication of the amount of radioactive material existing on the earth.

Short-lived daughter nuclides represent an environmental hazard only where accidental discharge of material from a rector occurs. Of particular concern in such cases are ^{131}I (half-life 8.07 days) and ^{136}Cs (half-life 13 days), both β emitters that efficiently enter the food chain and are thereby transmitted to animals and humans. Medium- and long-lived species represent a hazard in leakage episodes as well, but are mainly of concern because they must be kept in secure storage for long periods of time. Reactors typically generate about 1700 MCi (megacuries) of such species per 1000 MW per year, mostly in the form of ^{90}Sr (670 MCi/1000 MW-year, half-life 40.4 years), ^{137}Cs (930 MCi/MW-year, half-life 43.2 years), ^{85}Kr (95 MCi/1000 MW-year, half-life 15.2 years), and ^3H or tritium (6 MCi/1000 MW-year, half-life 17.7 years). It is estimated that the total long-term steady-state radiation load from reactor operation, representing a balance between generation and decay, will be 69 $\times$ 10^3 MCi per 1000 MW of reactor power generated. At the beginning of 1986 there were approximately 350 major commercial power reactors in operation worldwide (about one-third in the United States) with a total capacity of about 150,000 MW. These represent a long-term steady-state load of about 10^7 MCi. For comparison, the natural radiation loading of soils and seawater, primarily due to ^{40}K and ^{87}Rb, is approximately 5×10^5 MCi worldwide. A roughly similar amount of radiation results from impingement of cosmic rays on the earth. Thus dispersal over the earth of the radioactive waste generated in reactors would increase this radiation burden by about 10-fold. It is evident that the proper storage and containment of radioactive waste is an extremely important problem (Section VI).

Reactors are subject to unintended releases of radioactivity. If coolant flow through the core fails (a loss-of-coolant accident or LOCA), core temperatures may rise sufficiently to cause fuel and control rods to melt, thus destroying the means of controlling heat generation. Water in contact with the core under such circumstances may react with molten metal, forming an explosive mix of hydrogen and oxygen. Essentially this occurred at the Three Mile Island reactor near Harrisburg, Pennsylvania, on March 29, 1979. While an explosion was averted, the reactor core was partially exposed and emitted radioactive gases and water to the environment. A total of about 0.1 MCi of radioactive gas was emitted, and radiation dose rates in excess of 1000 mrem/hr occurred locally in gas plumes. Some of the population in the vicinity probably were exposed to doses of 100 mrem/hr or more (compared with an average yearly normal exposure of 200 mrem) and a subsequent excess infant mortality of about 600 individuals has been inferred from public health statistics for Pennsylvania and New York. A much more serious incident occurred on April 26, 1986, at one of four graphite-moderated reactors at Chernobyl near Kiev in the USSR. An explosion, presumably of hydrogen generated in a thermal excursion, ignited a fire. As no containment dome surrounded the reactor, massive releases of radioactivity occurred. No exact figures are yet available, but total emission probably approached 100 MCi. As of 1 month after the incident, 24 individuals had died of radiation exposure incurred while attempting to bring the reactor under control.

Releases of radiation such as those of Three

Mile Island and Chernobyl will no doubt continue to occur and cause damage to local human populations and ecological networks, but a nuclear war would produce a far more serious perturbation of the world environment. Explosion of a 1-megaton nuclear bomb can produce 10^3 to 10^5 MCi of radiation, depending on conditions; 10,000 such weapons could generate 10^7 to 10^9 MCi, much of it released to the atmosphere. Evenly distributed over the earth, this would multiply the background radiation level by a factor of as much as 10^3, increasing the average exposure of human populations from the present figure of 100–200 mrem/year to perhaps 100 rem/year for several years; many survivors would be exposed to much higher totals. Such levels are sufficient to produce radiation illness in some individuals and significant gene damage, increase in incidence of cancer, and susceptibility to debilitating infection in all. Similar effects on plants and animals would result in widespread disruption of ecological networks.

B. PROCESSES RELATED TO MINERAL RECOVERY AND PROCESSING

Fossil fuels are produced by the slow chemical modification or maturation of deposits of organic matter at moderate temperatures (100–200°C) in suitable mineral substrates (usually shales or limestones). The nature of the fuel formed depends on the ratio of carbon, hydrogen, and oxygen in the starting material; organic debris with a high H/C and low O/C ratio, as provided by bacteria and algae, tends to produce petroleum, while that with a low H/C and high O/C ratio from woody plants leads to peat or coals of various rank. Methane (CH_4) gas is formed to some extent in conjunction with both petroleum and coals, but especially as an end product of prolonged high-temperature maturation. Since all living cells contain nitrogen, sulfur, and phosphorus, these elements occur in combination in varying proportions in fossil fuel deposits. A typical approximate stoichiometric formula for a bituminous coal, for instance, is $C_{100}H_{85}S_2N_{1.5}O_{9.5}$.

Coal recovery involves the use of large amounts of water for cooling, washing, and lubrication, followed by discharge as effluent. During recovery and processing a portion of the sulfur content of coal is oxidized to SO_x and dissolves in this effluent, producing sulfurous acid (H_2SO_3) and sulfuric acid (H_2SO_4). Release of such contaminated water to the environment causes large-scale acidification of lakes, streams, and ground water. The problem is particularly severe with open-pit mining or strip-mining unless steps are taken to sequester that part of the sulfur (usually about half) in the form of iron pyrite, FeS_2, which is easily oxidized to SO_x in air.

Production and refining of petroleum hydrocarbons release large quantities of pollutant chemicals into the environment. These include saltwater brines, which accompany crude oil in deposits and are pumped out with the latter in wells, various sulfur compounds, and halogenated and nitrogenated hydrocarbons. Refining operations and evaporation of petroleum in transfer and storage release about 7×10^{10} kg of volatile hydrocarbons into the atmosphere per year globally, about 2×10^9 kg in the United States. Transport of petroleum by ship is accompanied by leakage of hydrocarbons into the marine environment; total oil spill amounts are estimated at 5×10^9 kg per year worldwide. The effects of hydrocarbon contamination of water on the biota are poorly understood, although prompt kills of fish and birds from massive discharges are frequent. Bacterial degradation of petroleum in natural waters is most efficient with straight-chain hydrocarbons and aromatics; thus crude oil spills are probably a smaller long-term perturbation than are spills of refined petroleum enriched in branched compounds.

Ore mining presents many of the same possibilities for chemical modification of the environment as does fossil fuel recovery. Sulfur-containing minerals frequently accompany ores and, unless properly recovered and treated, promote the formation of acid waters and contamination of the atmosphere with SO_2 and H_2S. Acidic sludges and slurries are produced in conjunction with the recovery and processing of iron, copper, zinc, and lead; cyanide salts are used in the recovery of gold and silver and are frequently discharged into the environment.

C. INDUSTRIAL PROCESSES

Industrial operations are an important source of pollutants in both the atmosphere and the hydrosphere. The principal contributor to atmospheric perturbation is combustion, stationary plant combustion being a major source of atmospheric SO_x (Fig. 3). Virtually all industrial operations have an impact on the composition of natural waters; it is not possible here to do more than give an overview. The totality of industry in the United States is estimated to generate 5×10^{10} cubic meters (5×10^{13} liters or 1.4×10^{13} gal) of wastewater annually, containing 8×10^9 kg of solids. The chemical industry is an important source of organic chemicals; these enter the environment mainly by discharge of spent solvents and solvent evaporation. The latter contributes about 10^{10} kg of material worldwide per year. A particularly troublesome group of such organics is the polychlorinated biphenyls (PCBs), used as solvents, plasticizers, and insulating fluids because of their exceptional chemical and thermal stability. However, they thus have long persistence in natural waters. They are believed to cause an increased incidence of birth defects and possibly cancer and to adversely affect resistance to disease in animals and humans. Their manufacture in the United States was discontinued in 1977. Other solvents of importance in industry that are known health hazards include benzene (C_6H_6) a skin irritant, carcinogen, and general systemic poison, and tetrachloroethylene or TCE (C_2Cl_4), a mutagen and carcinogen. Mineral processing and plating industries generate wastes containing cadmium (high human toxicity due to its tendency to replace zinc in enzymes), copper (slight toxicity to humans and animals, toxic to plants), cyanides (acute toxins), mercury (acute and chronic toxicity to humans), silver (slight toxicity to humans), and zinc (plant toxin at high levels).

Paper and pulp processing operations have severe impacts on the aqueous environment. The manufacture of paper requires that the cellulose fraction in wood be separated from branched aromatic compounds (collectively termed *lignin*), which stabilize the wood structure. Treatment with solutions of bisulfite ion (HSO_3^-) or a strongly alkaline solution of NaOH and Na$_2$S is used for this purpose. The resulting effluent is a concentrated dark mix of organic waste, sulfur compounds, and alkali that changes the pH, oxidation–reduction balance, and oxygen content of the waters into which it is released. Textile processing plants have similar, if less pronounced, effects.

Especially important in its impact on human health has been the widespread use of asbestos in construction, insulation, and transport (in brake linings). Asbestos is a generic term for several fibrous minerals. The most common is chrysotile, a serpentine silicate with approximate composition $Mg_3Si_2O_5(OH)_4$. A second form is amphibole asbestos; like chrysotile, it is a hydrated chain structure silicate, but it has

more varied composition, its principal chemical difference from the latter being that it contains a substantial amount of aluminum ion. The utility of asbestos lies in its resistance to heat and fire and the ease with which it can be fabricated into thin flexible sheets. Mining and processing of the mineral generate airborne microfibers, which are easily inhaled by those working with or using asbestos products. A typical such fiber (length 11 μm) is shown in Fig. 4. These small fibers evade the filtering apparatus of the upper respiratory tract and pass into the alveoli of the lung, where they lodge permanently. Either by chemical action or direct mechanical abrasion (the precise causative mechanism is not established) they cause breakdown of lung tissue, leading to impairment of respiratory function (asbestosis), lung cancer, or in some cases mesothelioma (a cancer of the lung lining that is only known to occur in individuals exposed to asbestos). Although these effects were known or strongly suspected by the 1920s, production and use of asbestos continued to be actively promoted until about 1975. Its use is now being reduced in the United States.

D. AGRICULTURE

Effects on the chemical environment associated with food production can be broadly divided into two areas: those due to the use of biological promotors (fertilizers) and those due to biological inhibitors (biocides for the removal of unwanted plants, animals, or microorganisms).

The limiting factor in growth of crop plants is usually a low level of one or more soil ingredients—nitrogen, phosphorus, or an essential trace mineral—or of availability of water. In-

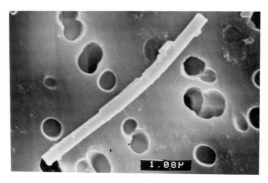

FIG. 4. Asbestos microfiber, length about 11 μm. Electron micrograph, ×9300.

creased yield is achieved by augmenting the limiting ingredient. Common synthetic fertilizers used to increase availability of nitrogen and/or phosphorus in soils include ammonium nitrate (NH_4NO_3), urea [(NH_2)$_2CO$], sodium nitrate ($NaNO_3$), and ammonium phosphate ($NH_4H_2PO_4$) and polyphosphates. Since all these materials are highly water-soluble, a portion is leached from the soils to which they are applied and enters the hydrosphere, where it tends to promote changes in the chemical composition and ecology of lakes, streams, and rivers. Irrigation water frequently contains high quantities of dissolved salts, which remain in the soil and cause loss of fertility; in some cases trace minerals in irrigation water build up to toxic levels. This has occurred with selenium in some areas of the Central Valley of California.

A wide array of biocides is used in agriculture. General systemic poisons such as arsenicals have been largely supplanted by compounds in the chlorinated hydrocarbon, carbamate, or organophosphorus groups for use against insects, soil parasites, and weeds. Structural diagrams of representative members of each group are shown in Fig. 5. The insecticides pass through the insect cuticle (epidermis) and exert their toxic function primarily though disruption of nerve function; many organophosphorus compounds are related to nerve gases developed for military use.

Movement of biocides from the point of application through water runoff into the broader environment is a potential source of ecological disruption. Most of the organophosphorus compounds undergo rapid chemical breakdown, but the halogenated hydrocarbons tend to be more stable. Table I shows some of the more common biocides, their LD_{50} for rats in milligrams of compound per kilogram of body weight, and their persistence in river water in terms of percent remaining in the water 1 week and 8 weeks after an initial concentration of 1×10^{-5} g/liter was monitored.

E. DOMESTIC OPERATIONS

Certain very stable species that penetrate the environment and produce significant chemical modifications there are widely applied for domestic purposes in developed countries. Synthetic detergents have largely replaced soaps for cleaning and washing. A detergent consists of a surfactant, a species that lowers the surface tension of water, facilitating the emulsification (dispersal in very small droplets) of dirt and stains, plus builders, compounds that complex divalent ions (which make water "hard"). Early surfactants were chemically very stable and tended to persist and enter natural water supplies, causing foaming and other deleterious effects. "Biodegradable" (less stable) surfactants are now employed. Many builders, notably polyphosphates, act as nutrients for plants and microorganisms in natural waters that they eventually reach, thus perturbing the ecological balance.

Chlorofluorocarbons (CFCs) or *Freons,* such as Freon-12 (CCl_2F_2) and Freon-11 (CCl_3F), have been widely used as the heat transfer fluids

TABLE I.   Properties of Some Common Biocides

	LD_{50} (mg/kg)	Persistence in river water (%)	
		1 week	8 weeks
Chlorinated hydrocarbons			
Aldrin	54–56	100	20
DDT	420–800	100	100
Dieldrin	50–55	100	100
Endrin	5–43	100	100
Heptachlor	90	100	0
Carbamates			
Sevin (carbaryl)	540	90	0
Organophosphorus			
Malathion	480–1500	100	0
Parathion	4–30	100	0

ORGANOCHLORINE INSECTICIDES

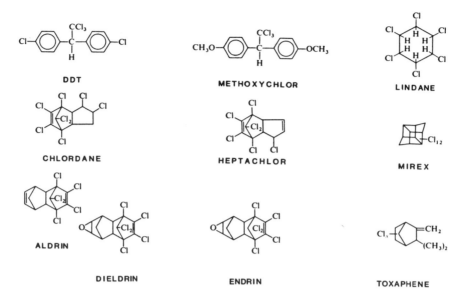

ORGANOPHOSPHORUS INSECTICIDES

CARBAMATE INSECTICIDE

HERBICIDES

FIG. 5. Chemical structures of representative biocides. [Adapted with permission from Manahan, S. E. (1984). "Environmental Chemistry," 4th ed. Willard Grant Press, Boston.]

in refrigerators and as a propellant medium in spray cans because of their low cost, absence of toxicity, and exceptional stability. These compounds permeate the atmosphere and gradually migrate to the stratosphere, where they seem to be perturbing the ozone layer, which shields the surface of the earth from dangerous ultraviolet radiation (Section III,C,2).

III. The Atmosphere

A. NATURAL STRUCTURE AND CHEMICAL COMPOSITION

Table II gives the standard sea-level composition of dry air in terms of mole fraction (i.e., mixing ratio) and actual total number of moles in the atmosphere as a whole for each species, assuming vertical chemical uniformity. In addition to the components shown in Table II, natural air contains water vapor in variable amounts, ranging in mole fraction from near 0 to 0.03.

This composition represents the culmination of an evolutionary process unlike that of any other known planet. The oxygen component in the atmosphere is believed to derive almost entirely from the photosynthetic action of green plants and was absent before this form of photosynthesis arose about 2.5×10^9 years ago. The concentration of CO_2, on the other hand, is low compared with, say, that of the planet Venus; this is because most of the terrestrial CO_2 inventory is in solution in seawater (3×10^{18} moles) or incorporated in carbonate minerals (6×10^{21} moles).

The composition of the atmosphere has probably varied over more recent geologic time as well, and it is not clear precisely what mechanisms act to control component concentrations. It is therefore difficult to know the degree of atmospheric buffering against chemical perturbation by human activity. A major such perturbation has been the addition of 1.7×10^{16} moles of CO_2 from combustion of fossil fuels since about 1860. This has caused the atmospheric concentration to rise from 290 to 340 ppm (1980), which corresponds to a net increase of only 0.9×10^{16} moles in the atmosphere. The remaining half of the CO_2 from combustion must have moved to the seawater reservoir, consistent with the fact that atmosphere–hydrosphere CO_2 transfer is efficient, the residence time of CO_2 in the atmosphere being only on the order of 7 years. Whether this partitioning will continue in the long term is unknown.

Concern has been expressed about depletion of the earth's oxygen supply if deforestation on land and kills of marine algae by oil spills reduce the photosynthetic production of O_2. In the short term, this is not a significant problem. Fossil fuel combustion requires about 4×10^{14} moles of O_2 per year, and the amount of O_2 consumed by the respiration of animals and by inorganic reactions with minerals is not more than a few percent of this; so even if all produciton of O_2 were to cease, the inventory in the atmosphere would suffice to sustain human activity for some 10,000 years before a 10% reduction in oxygen concentration occurred.

The pressure of the atmosphere decreases uniformly with altitude, but the temperature exhibits vertical structure. It decreases through the lowest portion, or troposphere, to an inversion point, the tropopause, whose height varies from about 10 km at the poles to 15 km at the equator. Above the tropopause, temperature increases with altitude through the stratosphere;

TABLE II.   Sea-Level Composition of Dry Air

Chemical species	Molecular weight	Mole fraction	Total atmospheric amount (moles)
Nitrogen (N_2)	28.01	78.0840	1.42×10^{20}
Oxygen (O_2)	32.00	20.9460	3.81×10^{19}
Argon (Ar)	39.94	0.9340	1.70×10^{18}
Carbon dioxide (CO_2)	44.01	3.4×10^{-4} (340 ppm)	6.1×10^{16}
Neon (Ne)	20.12	1.8×10^{-5} (18 ppm)	3.3×10^{15}
Helium (He)	4.01	5.2×10^{-6} (5.2 ppm)	9.5×10^{14}
Methane (CH_4)	16.04	1.5×10^{-6} (1.5 ppm)	2.7×10^{14}
Krypton (Kr)	83.80	1.1×10^{-6} (1.1 ppm)	2.0×10^{14}
Nitrous oxide (N_2O)	44.01	5×10^{-7} (0.5 ppm)	9×10^{13}
Hydrogen (H_2)	2.02	5×10^{-7} (0.5 ppm)	9×10^{13}
Xenon (Xe)	131.30	8×10^{-8} (0.08 ppm)	1.5×10^{13}

this heating is mainly due to absorption of solar ultraviolet energy by the dissociation of ozone:

$$O_3 \overset{\underset{\text{solar UV}}{\text{radiation}}}{=} O_2 + O$$

A second inversion occurs at a height of about 90 km, at the mesopause, between the mesosphere and thermosphere; the heating in the latter is mainly due to absorption of far-UV solar radiation by dissociation of N_2 and O_2. These temperature inversions separate the atmosphere into distinct reservoirs, since they act as barriers to convective mixing; material passes between troposphere and stratosphere mainly by the relatively slow process of diffusion. Local temperature inversions also occur in the lower troposphere to form regional reservoirs in which contaminant chemicals can build up to high concentrations; a well-known local inversion phenomenon is that which occurs at elevations ranging from 300 m to 2 km over the city of Los Angeles.

B. LOWER-ATMOSPHERE EFFECTS

The introduction of species from fossil fuel combustion (Section II,A,1) into the lower atmosphere leads to an extensive series of chemical reactions, most of which are of free-radical type. They can be divided into three categories: daylight (D) reactions, requiring solar near-UV radiation; night (N) reactions, which are important in the absence of sunlight; and 24-hr (DN) reactions.

The fundamental light-requiring process is

$$NO_2 + h\nu \text{ (below 398 nm)} = NO + O \quad \text{(D)}$$

The O atom then forms ozone:

$$O + O_2 = O_3 \quad \text{(DN)}$$

and ozone reacts with NO to form NO_2:

$$NO + O_3 = NO_2 + O_2 \quad \text{(DN)}$$

Most emission of NO_x by combustion is in the form of NO, and direct air oxidation of NO to NO_2 is too slow to be of much significance; at typical concentrations the formation of O_3 as an intermediate permits this oxidation on a time scale of an hour or so in a typical urban environment. A further D process is

$$O_3 + h\nu \text{ (below 308 nm)} = O^* + O_2 \quad \text{(D)}$$

where O^* denotes an excited state (the ^1D state) of the oxygen atom, followed by

$$O^* + H_2O = 2OH\cdot \quad \text{(DN)}$$

whereas at night ozone reacts further with NO_2:

$$O_3 + NO_2 = NO_3 + O_2 \quad \text{(N)}$$

A further source of the hydroxyl radical OH· is

$$2NO_2 + H_2O = HNO_2 + HNO_3 \quad \text{(DN)}$$

$$HNO_2 + h\nu \text{ (295–410 nm)} = OH\cdot + NO$$

The nitrate radical formed at night leads to production of nitric acid, HNO_3:

$$NO_3 + NO_2 = N_2O_5 \quad \text{(N)}$$

$$N_2O_5 + H_2O = 2HNO_3 \quad \text{(N)}$$

Nitric acid is also produced from OH·:

$$OH\cdot + NO_2 = HNO_3 \quad \text{(DN)}$$

Thus solar irradiation of a mix of air and NO leads to a buildup of the oxidant O_3, the reactive free radical OH·, and the strongly acidic HNO_3; after sunset, OH· production ceases but that of NO_3 begins, leading through N_2O_5 to HNO_3. The latter acidifies water droplets, causing the formation of rain, snow, mist, or fogs with low (i.e., acid) pH. Ozone and OH· undergo further reactions in the presence of hydrocarbons RH:

$$OH\cdot + RH = R\cdot + H_2O$$

$$R\cdot + RCHO = RH + RCO\cdot$$

$$RCO\cdot + O_2 = RC(O)OO\cdot$$
$$\text{(peroxyacyl radical)}$$

$$RC(O)OO\cdot + NO_2 = RC(O)OONO_2$$
$$\text{(peroxyacyl nitrate or PAN)}$$

The PAN compounds formed in the last step are potent eye irritants; it is the mix of such substances with O_3 and the brown NO_2 that constitutes "smog." Carcinogenic nitrosamines can be formed from secondary amines with HNO_2:

$$R_2NH + HNO_2 = R_2NNO \text{ (nitrosamine)} + H_2O$$

Carcinogenic condensed-ring hydrocarbons such as benzo[a]pyrene

are emitted in the exhaust mix of unburned HC; they undergo further reactions with O_3 or N_2O_5

to yield mutagenic epoxides or nitro compounds.

Particulates, especially the soot emitted in incomplete combustion or by diesel engines, play a role in the pollution chemistry of the lower atmosphere. They absorb carcinogens of low volatility such as the condensed aromatics, thereby retaining them in the atmospheric reservoir available for inhalation, and also catalyze the further reaction of these species with oxidants. Conversion of SO_2 to SO_3, leading to sulfuric acid formation, is slow in the gas phase but is probably accelerated when the SO_2 is adsorbed on soot grains. Thus particulates contribute to acid precipitation.

C. Upper-Atmosphere Effects

1. Ozone Layer

Ozone, objectionable when present in the lower atmosphere, is a vital ingredient of the stratosphere; by absorbing light at wavelengths below about 330 nm it protects the biota from exposure to this damaging radiation and maintains the tropopause temperature inversion, which is important in shaping terrestrial meteorological patterns. It is produced in the upper atmosphere by

$$O_2 + h\nu \text{ (below 242 nm)} = O + O$$

$$O_2 + O + M \text{ (third body)} = O_3 + M$$

Maximum O_3 levels of as much as 10 ppm are attained around 25 km altitude. However, these high concentrations result from kinetic factors, the formation reactions being rapid compared with natural removal processes, either absorption of ultraviolet light or reaction with O or OH·:

$$O_3 + O = O_2 + O_2$$

$$O_3 + OH\cdot = O_2 + HOO\cdot$$

Introduction of additional material capable of reacting with O_3 will therefore diminish its steady-state concentration.

NO_x introduced into the stratosphere by aircraft and missiles or (in the event of nuclear war) by large explosions may have such an effect:

$$N_2O + h\nu = N_2 + O \qquad O + O_3 = O_2 + O_2$$

$$NO + O_3 = NO_2 + O_2$$

Of more concern is the effect of CFCs (Section II,E). These compounds are so stable that a significant fraction of them, when released to the lower atmosphere, survives to migrate by convection and diffusion to the stratosphere, where they are readily dissociated by near-UV light to form free radicals:

$$CCl_2F_2 \text{ (Freon-12)} + h\nu = CClF_2\cdot + Cl\cdot$$

$$Cl\cdot + O_3 = ClO + O_2$$

The original analysis of these processes (1974) suggested that ozone levels in the stratosphere might drop by as much as 15% as a result of the CFCs then present. The ensuing controversy led to a number of other estimates, both lower and higher. Definitive evidence for ozone depletion has recently been obtained; a drop of about 40% in O_3 concentration was detected over Antarctica in October 1985, and although there was a recovery of concentration later, it appears certain that large ozone depletions will occur in future, at least transiently and locally.

2. Greenhouse Effect

The temperature of the earth's surface represents a balance of incoming solar energy, largely in the visible and ultraviolet regions of the spectrum, against outgoing radiation from the atmosphere, hydrosphere, and solid surface of the earth, largely in the infrared. This balance is sketched in Fig. 6. The temperature that a blackbody must have to radiate away exactly the amount of energy that the earth receives from the sun is only 255 K ($-18°C$); the actual average temperature of the earth's surface (15°C) is higher because infrared radiation is partially trapped in the atmosphere and reradiated back to the ground. That is, the atmosphere is relatively transparent to short-wavelength light and relatively opaque to infrared light; this is termed the *greenhouse effect*. The bulk of the effect is due to water vapor; but absorption by CO_2 between 12 and 16.3 μm contributes about a quarter of the 33 K excess greenhouse temperature. The increase in atmospheric CO_2 levels (Section II,A) of about 17% since 1860 should enhance the greenhouse effect and lead to a warming of the earth by 2–3°C on average; however, because of the natural variability in terrestrial temperatures and the presence of other factors that can influence the radiation balance, this increase has not yet been conclusively demonstrated. It is also possible that other trace atmospheric species that are increasing due to human activity, notably methane (CH_4) and the same CFCs implicated in stratospheric ozone depletion, will add to the greenhouse effect.

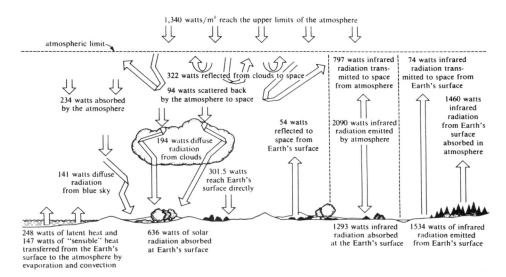

FIG. 6. Terrestrial radiation energy balance. Overall solar energy flux is 1340 W/m². [From Manahan, S. E. (1984). "Environmental Chemistry," 4th ed. Willard Grant Press, Boston. Reproduced by permission.]

A profound perturbation of the earth's heat balance would be a probable consequence of nuclear war. The explosions in such a war might inject enough soot into the stratosphere to greatly reduce the amount of solar radiation reaching the surface for many weeks or months. The resulting temperature drop has been termed "nuclear winter." A similar cooling may have occurred at the end of the Cretaceous period, 65 million years ago, as a result of dust produced by the collision of a massive meteorite with the earth and may have contributed to the wave of extinctions in the biota at that time.

IV. The Hydrosphere

A. NATURAL STRUCTURE AND CHEMISTRY

Whereas the atmosphere is uniform laterally and varies in structure only with altitude, the hydrosphere consists of many distinct reservoirs of differing size. These are the seas and oceans, with a limited range of variation in composition and a definite vertical structure; lakes, with a wide range of chemical variation and also vertical stratification; rivers, with highly variable composition but little structure; precipitation (rain, snow, fog, mist), whose only important compositional variable is pH (degree of acidity); and ground water, about which little of general character can be said.

The compositions of seawater and average river water with respect to major solution species are given in Table III with concentrations in molarity (M). The composition of seawater seems to have changed little since mid-Precambrian time, about 2×10^9 years ago, although the pH may have varied slightly. The principal factor regulating the concentrations of the major ionic species is probably chemical exchange with mantle magma at midocean crustal spreading centers, and there is no evidence for any human-related effect on bulk oceanic composition.

Oceans are divided vertically into

1. the photic zone, the top few meters into which light penetrates and photosynthesis is possible;
2. the remainder of the surface or mixed layer, extending to about 100 m depth, in which temperature, composition, and nutrient levels are somewhat variable and convective mixing occurs;
3. the thermocline layer, from 100 to about 1500 m, in which the temperature decreases sharply; and
4. the deep ocean, below 1500 m, with essentially uniform temperature near 3°C.

The thermocline acts as a barrier separating reservoirs of the surface and deep ocean zones.

TABLE III. Compositions of Seawater and Average River Water

Species	Concentration (M)	
	Seawater (surface)	Average river
Na^+	0.47	2.7×10^{-4}
K^+	0.010	5.9×10^{-5}
Ca^{2+}	0.010	3.8×10^{-4}
Mg^{2+}	0.054	3.4×10^{-4}
Cl^-	0.56	2.2×10^{-4}
SO_4^{2-}	0.028	1.2×10^{-4}
HCO_3^-	2.4×10^{-3}	1×10^{-3}
CO_3^{2-}	2.7×10^{-5}	1×10^{-4}
Total nitrogen	2.0×10^{-5}	
Total phosphorus	1.5×10^{-6}	
pH	8.15	7.5–8.5

Chemical perturbations associated with human activity mainly reach only the surface zone.

Lakes also exhibit thermal stratification. In the summer months a warm upper layer called the epilimnion is separated by a thermocline from a cool bottom zone or hypolimnion. Autumn cooling of the epilimnion may make the surface water denser than the bottom water, provoking a turnover and bringing chemical species from bottom sediments to the top. Such recirculation can have severe biological consequences; rapid turnover of a lake in East Africa brought so much water supersaturated with CO_2 to the surface that a CO_2 cloud formed and suffocated a number of area residents.

B. Perturbation of pH

Human activity has not so far affected global oceanic pH, although discharges of acidic wastes into shallow near-shore waters can cause destructive local pH fluctuations. On the other hand, discharge of combustion-related SO_x into the atmosphere (Section II,A,1) and production of HNO_3 (nitric acid) from NO_x in engine exhaust (Section III,B) have acted to markedly shift the pH of precipitation and lake and river water in large areas of North America and Europe. Natural rainwater has a pH of about 5.7 because some atmospheric CO_2 dissolves in it to form the weak acid H_2CO_3. Atmospheric SO_3 and HNO_3 also readily dissolve in rainwater, the SO_3 forming sulfuric acid, H_2SO_4. Both sulfuric acid and nitric acid are strong acids and their impact on rainwater pH is far greater for a given molar quantity dissolved than is that of CO_2. In consequence, the average pH of rainfall in the

northeastern United States has decreased to about 4, and values as low as 2 have been observed in such "acid rain." Lakes, rivers, and ground water in the affected areas are becoming more acid, resulting in decreases in fish populations and damage to forests. Alleviation is difficult, since the excess acidity may be carried long distances by air mass movement and it is not generally possible to pinpoint the source industries or plants.

C. Perturbation of Redox Balance

The chemical state of a water body is sensitive to its redox (oxidation–reduction) level, which in turn depends on the amount of dissolved oxygen (DO) the water contains—9 mg O_2/liter for water saturated with air (25°C). If nutrient levels and oxygen-consuming biological activity are low in a body of water undergoing efficient exchange with the atmosphere (as in a rapidly flowing stream), DO levels remain near the maximum and the body remains aerobic (top of Fig. 7); but if excessive nutrient material requiring oxygen for its utilization is introduced, the DO may be exhausted and the body converted to the anaerobic cycle (bottom of Fig. 7), characterized by microbial fermentation and sulfur metabolism. Addition of nutrients by the runoff of fertilizer from agricultural land (Section II,D) or of industrial and domestic waste in sewage from urban areas thus works to enhance the biological oxygen demand (BOD) in adjacent water bodies and cause them to become entirely or partially anaerobic. Addition of such nutrients to lakes accelerates their eutrophication, that is, their conversion to anaerobic bogs. Anaerobic conditions are frequently encountered under natural

Aerobic cycle

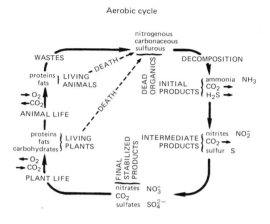

Anaerobic cycle

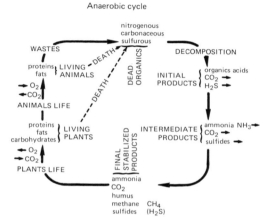

FIG. 7. Aerobic and anaerobic nutrient cycles. [From Vesilind, P. A. (1975). "Environmental Pollution and Control." Ann Arbor Science, Ann Arbor, Michigan. Reproduced by permission.]

conditions as well, in bodies of water with limited circulation such as the Black Sea.

D. Perturbation by Toxic Materials

The overtly toxic materials present in the hydrosphere include biocides, industrial solvents, by-products of the petroleum industry, and trace metals. Materials in the first three classes, most notably chlorinated hydrocarbons such as DDT and the PCBs, though undergoing slow microbial degradation, are present in sufficient concentration in many aquatic environments to severely affect the biota. In favorable cases trace metals may be removed by deposition in the sediments. However, they can remain in the hydro-

sphere because of the formation of complexes with other pollutant species or with naturally produced chemicals. An example is the mobilization of mercury (Hg) in ocean waters. Mercury is a component of numerous products employed in industry, especially fungicides; it had been assumed that dumping of such substances in ocean water would lead to sequestering of mercury as the insoluble salt HgS in anaerobic bottom sediments. Instead, some of the mercury reacts with methylcobalamin, a compound produced by anaerobic microorganisms abundant in bottom sediments, to form the soluble $HgCH_3^+$ (methylmercury) ion, which may reach high concentrations in the tissues of fish and other aquatic organisms and expose human populations to mercury poisoning.

V. The Biosphere

The biosphere is both a chemical reservoir in its own right and the fundamental entity of concern in judging the importance of environmental chemical perturbations. Many chemical substances entering the global environment have accumulated in high concentration in living tissue and produced severe effects on animal or plant populations. Examples are the buildup of mercury in certain marine fish and of various trace metals in shellfish. Perhaps most dramatic is the tendency of chlorinated hydrocarbon pesticides to accumulate in body lipids of animals that feed on insects or plants to which these substances have been applied. The process of concentration continues with consumer animals higher in the food chain. In the case of DDT, a typical situation is to have DDT concentrations of 0.001 ppm in water and bottom sediments, 0.05 ppm in water plants and plankton, 0.5 ppm in fish and aquatic invertebrates, and 10–100 ppm in aquatic birds such as ospreys or gulls. In many avian species the resulting body load of pesticide interferes with reproduction by inhibiting calcium metabolism and interfering with eggshell formation.

Massive deaths in the biota do occur as a consequence of perturbations in the chemical environment, but it is often difficult to identify the specific causative agent. In the mid-1980s *Waldsterbe* (forest death) has affected much of Europe, killing large numbers of forest trees, especially in Germany. Many of the perturbing factors discussed elsewhere in this article have been invoked to account for this phenomenon, but a definitive explanation is not yet available.

VI. Toxic Wastes

It will be appreciated from earlier parts of this article that a large portion of the chemical perturbation of the environment is due to the discharge of waste materials: combustion exhaust, effluents from mining, spent solvents, and by-products of chemical manufacture. In some cases these wastes have been found to have unexpectedly high toxicity, an example being 2,3,7,8-tetrachloro-p-dioxin (TCDD or "dioxin"), a by-product of the manufacture of the pesticide intermediate hexachlorophene and the herbicide 2,4-D. TCDD is an acute direct toxin and mutagen involved in several mass poisoning episodes.

Uncontrolled discharge of such materials to the environment is unacceptable, and accordingly their elimination or sequestering is receiving much current attention. Elimination by selective chemical transformation is possible in the case of molecular compounds, but this is usually very expensive; combustion is undesirable as some incompletely burned material will be released to the atmosphere and hydrosphere. The alternative is storage in a dump site until biological degradation can occur. Exposure of wastes in the dump site to the atmosphere to permit fast, aerobic microbial activity may entail transfer of toxic volatiles to the atmosphere and cannot usually be permitted. Therefore, mainly slow anaerobic activity takes place, sometimes generating further toxic material in unpredictable fashion. Thus it is required that the toxic dump be well sealed, that is, be a reservoir with a very long residence time for all contents (very slow transfer to the environment). Most dumps have not satisfied this criterion and have allowed extensive transfer of toxics to the ground water, which is a ubiquitous component of soils. Once in the ground water or aquifer, waste chemicals are disseminated throughout the surrounding area, contaminating well water, evaporating into the atmosphere, and negatively affecting the health of humans and the biota generally.

Nuclear waste presents even more serious problems. No method of elimination is possible; radioactive material must be sequestered until it decays, which for some common by-products of nuclear reactor operation will take thousands of years. Continual exposure to radiation promotes deterioration of materials, so nuclear wastes must frequently be transferred to new containers. Burial of such wastes is of dubious feasibility, since the likelihood that they can be kept from reaching an aquifer for a sufficiently long period is small. Nevertheless, there may be geological formations that are sufficiently impervious to water penetration that they can serve as a disposal matrix for radioactive wastes. This remains to be established.

BIBLIOGRAPHY

Bond, R. G., and Straub, C. P. (eds.). (1972–1978). "Handbook of Environmental Control," Vols. I–V, CRC Press, Cleveland, Ohio.

Chiras, D. D. (1985). "Environmental Science," Benjamin/Cummings, New York.

Hunt, V. D. (1982). "Handbook of Energy Technology," Van Nostrand–Reinhold Co., Princeton, New Jersey.

Manahan, S. E. (1984). "Environmental Chemistry," 4th ed. Willard Grant Press, Boston.

Pitts, B. J. F., and Pitts, J. N., Jr. (1986). "Atmospheric Chemistry," Wiley/Interscience, New York.

Sax, N. I., and Lewis, R. J., Sr. (1986). "Rapid Guide to Hazardous Chemicals in the Workplace," Van Nostrand–Reinhold, Princeton, New Jersey.

Thibodeaux, L. J. (1979). "Environmental Movement of Chemicals in Air, Water, and Soil." Wiley, New York.